T0213841

Lecture Notes in Computer Science 9373

Commenced Publication in 1973
Founding and Former Series Editors:
Gerhard Goos, Juris Hartmanis, and Jan van Leeuwen

More information about this series at http://www.springer.com/series/7407

Lecture Notes in Computer Science 9373

Commenced Publication in 1973
Founding and Former Series Editors:
Gerhard Goos, Juris Hartmanis, and Jan van Leeuwen

More information about this series at http://www.springer.com/series/7407

Marijn Janssen · Matti Mäntymäki
Jan Hidders · Bram Klievink
Winfried Lamersdorf · Bastiaan van Loenen
Anneke Zuiderwijk (Eds.)

Open and Big Data Management and Innovation

14th IFIP WG 6.11 Conference on
e-Business, e-Services, and e-Society, I3E 2015
Delft, The Netherlands, October 13–15, 2015
Proceedings

 Springer

Editors

Marijn Janssen
Faculty of Technology, Policy
    and Management
Delft University of Technology
Delft
The Netherlands

Matti Mäntymäki
Turku School of Economics
University of Turku
Turku
Finland

Jan Hidders
Faculty of Electrical Engineering,
    Mathematics and Computer Science
Delft University of Technology
Delft
The Netherlands

Bram Klievink
Faculty of Technology, Policy
    and Management
Delft University of Technology
Delft
The Netherlands

Winfried Lamersdorf
Department of Informatics
Hamburg University
Hamburg
Germany

Bastiaan van Loenen
The Faculty of Architecture and the Built
    Environment
Delft University of Technology
Delft
The Netherlands

Anneke Zuiderwijk
Faculty of Technology, Policy
    and Management
Delft University of Technology
Delft
The Netherlands

ISSN 0302-9743   ISSN 1611-3349 (electronic)
Lecture Notes in Computer Science
ISBN 978-3-319-25012-0   ISBN 978-3-319-25013-7 (eBook)
DOI 10.1007/978-3-319-25013-7

Library of Congress Control Number: 2015950044

LNCS Sublibrary: SL1 – Theoretical Computer Science and General Issues

Springer Cham Heidelberg New York Dordrecht London

Printed on acid-free paper

Springer International Publishing AG Switzerland is part of Springer Science+Business Media
(www.springer.com)

Preface

Since its inception in 1998, the I3E Conference has brought together researchers and practitioners from all over the world. The I3E conference main area is in the field of e-business, e-services, and e-society, which are abbreviated as the three Es. The I3E conference series is truly multidisciplinary covering areas ranging from computer science to information systems and data science. The 2015 I3E conference was the 14th consecutive I3E conference. This year's conference received submissions from Asia, Europe, and South America.

The mission of IFIP Working Group 6.11 is to organize and promote exchange of information and co-operation related to all aspects of electronic business, electronic services, and electronic society (the three Es). Working Group 6.11 has members with diverse backgrounds including information systems, computer science, and business studies. The I3E conference series the is flagship event of the Working Group.

The theme of the 2014 I3E conference was "Open and Big Data Management and Innovation" Information economy and openness have been the dominating terms over the last couple of years. Data have become widely available and many innovations are based on the utilization of data. The Internet of Things (IoT) enables the availability of large volumes of data, while people create data using social media. Today we have a worldwide exchange of data and this influences our society, resulting in new business opportunities and new services. This has already resulted in the rise of "data science," which aims at better understanding how to use all these data in our information society and how it can co-evolve with its information and communication technology (ICT). This needs to make the theory of complex systems applicable to the information society.

Furthermore, there is a lack of tools and instruments to deal with the vast amount of data. Big data will have a major influence on the operating of businesses and the functioning of society. This conference contributes to advancements in areas such as data analytics, mining, visualization, sensor networks, information retrieval, and information extraction; research relating to data extraction and analytics, statistical inference, data quality, and issues such as bias, missing data, endogeneity, user interface, and visualization are encouraged. Papers are in the field of technology, business, society, or combinations thereof.

The host of the IFIP I3E conference was the Faculty of Technology, Policy and Management (TPM) at Delft University of Technology, The Netherlands. TU Delft is the largest and most comprehensive university of engineering sciences in The Netherlands. The I3E 2015 conference was organized in cooperation between three faculties of Delft University of Technology. The faculties of Technology, Policy and Management (TPM), The Faculty of Architecture and the Built Environment, and the Faculty of Electrical engineering, Mathematics and Computer Science (EEMS). Big and open data (BOLD) is an important topic at Delft University of Technology and there are two main initiatives within Delft University of Technology in this area. Delft

Data Sciences (DSS) is a framework initiative for research activities in data science (http://www.delftdatascience.tudelft.nl/). DDS is characterized by the focus on the engineering aspects of data science, in line with TU Delft's key strengths. Furthermore there is the Knowledge Center Geoinformation Governance, which conducts research on institutional, legal, and organizational aspects of sharing and (re)using geographical information (http://www.bk.tudelft.nl/en/about-faculty/departments/otb-research-for-the-built-environment/knowledge-centre-geoinformation-governance/).

Making a successful conference requires resources and commitment. We would like to thank the authors for their submissions. We also wish to thank the reviewers for ensuring the academic standard of the conference. We wish to extend our thanks to Jo-Ann Karna, Laura Bruns, Diones Supriana, and Martijn Milikan. Finally, we would like to thank everyone involved in organizing the conference.

Enjoy reading!

August 2015

Marijn Janssen
Matti Mäntymäki
Jan Hidders
Bram Klievink
Winfried Lamersdorf
Bastiaan van Loenen
Anneke Zuiderwijk

Organization

Conference Chairs

Marijn Janssen	Delft University of Technology, The Netherlands
Jan Hidders	Delft University of Technology, The Netherlands
Bastiaan van Loenen	Delft University of Technology, The Netherlands
Matti Mäntymäki	University of Turku, Finland

Steering Committee

Winfried Lamersdorf	University of Hamburg, Germany
Wojciech Cellary	Poznan University of Economics, Poland
Matti Mäntymäki	University of Turku, Finland

Witness Workshop Chairs

Bram Klievink	Delft University of Technology, The Netherlands
Stefan Henningsson	Copenhagen Business School, Denmark
Yao-Hua Tan	Delft University of Technology, The Netherlands

Local Support

Ricardo Matheus	Delft University of Technology, The Netherlands
Dhata Praditya	Delft University of Technology, The Netherlands
Diones Supriana	Delft University of Technology, The Netherlands
Sélinde van Engelenburg	Delft University of Technology, The Netherlands
Agung Wahyudi	Delft University of Technology, The Netherlands

Organization Chair

Anneke Zuiderwijk	Delft University of Technology, The Netherlands

I3E 2014 Keynote Speakers

Sunil Choenni	Ministry of Security and Justice, The Netherlands
Yogesh Dwivedi	Swansea University, UK
Liesbet van Zoonen	Erasmus University Rotterdam, The Netherlands

I3E 2014 Program Committee

Esma Aimeur	University of Montreal, Canada
Americo Amorim	Daccord SA
Hernán Astudillo	Universidad Tecnica Federico Santa María, Chile
Evandro Baccarin	University of Londrina, Brazil
Aaron W. Baur	ESCP Europe Wirtschaftshochschule Berlin, Germany
Khalid Benali	LORIA - Université de Lorraine, France
Salima Benbernou	Université Paris Descartes, France
Djamal Benslimane	Lyon University, France
Markus Bick	ESCP Europe Wirtschaftshochschule Berlin, Germany
Peter de Bruyn	University of Antwerp, Belgium
Wojciech Cellary	Poznan University of Economics, Poland
Antonio Cerone	IMT Institute for Advanced Studies Lucca, Italy
Narciso Cerpa	University of Talca, Chile
François Charoy	Université de Lorraine - LORIA – Inria, France
Caspar Chorus	Delft University of Technology, The Netherlands
Joep Crompvoets	University of Leuven, Belgium
Tomi Dahlberg	University of Turku/Turku School of Economics, Finland
Bruno Defude	Télécom sudparis, France
Georgios Doukidis	Greece
Christos Douligeris	University of Pireaus, Greece
Enrico Ferro	Istituto Superiore Mario Boella, Italy
Mila Gasco	ESADE, Spain
Yiwei Gong	Wuhan University, China
Adam Grzech	Wroclaw University of Technology, Poland
Mohand-Said Hacid	Université Claude Bernard Lyon 1 - UCBL
Teresa M. Harrison	State University of New York at Albany, USA
Jan Hidders	Delft University of Technology, The Netherlands
Geert-Jan Houben	Delft University of Technology, The Netherlands
Alexandru Iosup	Delft University of Technology, The Netherlands
Marijn Janssen	Delft University of Technology, The Netherlands
Jonna Järveläinen	Turku School of Economics, Finland
Atsushi Kanai	Hosei University, Japan
Ranjan Kini	Indiana University Northwest, USA
Bram Klievink	Delft University of Technology, The Netherlands
Irene Krebs	University of Technology Cottbus, Germany
Winfried Lamersdorf	University of Hamburg Germany
Hongxiu Li	Turku School of Economics, Finland
Yong Liu	Aalto University, Finland
Euripidis Loukis	University of the Aegean, Greece
Luis Felipe Luna-Reyes	University at Albany, USA
José Machado	Centro ALGORITMI, University of Minho, Portugal
Matti Mäntymäki	Turku School of Economics, Finland
Tiziana Margaria	University of Limerick and Lero, Republic of Ireland

Contents

Keynote Panel

Driving Innovation Using Big Open Linked Data (BOLD) Panel 3
   Yogesh K. Dwivedi, Vishanth Weerakkody, Marijn Janssen,
   Jeremy Millard, Jan Hidders, Dhoya Snijders, Nripendra P. Rana,
   and Emma L. Slade

Adoption

Adoption of Mobile Banking in Jordan: Exploring Demographic
Differences on Customers' Perceptions 13
   Ali Abdallah Alalwan, Nripendra P. Rana, Yogesh K. Dwivedi,
   Banita Lal, and Michael D. Williams

Young-Elderly Travellers as Potential Users and Actual Users of Internet
with Mobile Devices During Trips 24
   Niklas Eriksson and Susanna Fabricius

An Empirical Study on the Adoption of Online Household e-waste
Collection Services in China. 36
   Shang Gao, Jinjing Shi, Hong Guo, Jiawei Kuang, and Yibing Xu

Perceptions of Teachers and Guardians on the Electronic Record
in the School-Family Communication 48
   António Abreu, Álvaro Rocha, and Manuel Pérez Cota

Cryptocurrencies as a Disruption? Empirical Findings on User Adoption
and Future Potential of Bitcoin and Co 63
   Aaron W. Baur, Julian Bühler, Markus Bick, and Charlotte S. Bonorden

A Systematic Review of Impediments Blocking Internet of Things
Adoption by Governments 81
   Paul Brous and Marijn Janssen

Understanding the Adoption of Mobile Internet in the Saudi Arabian
Context: Results from a Descriptive Analysis 95
   Abdullah Baabdullah, Yogesh Dwivedi, Michael Williams,
   and Prabhat Kumar

Conceptualising and Exploring User Activities in Social Media 107
   Marcel Rosenberger, Tobias Lehmkuhl, and Reinhard Jung

Understanding the Determinants of Privacy-ABC Technologies Adoption
by Service Providers   119
    Ahmad Sabouri

Consumers' Perceptions of Social Commerce Adoption in Saudi Arabia   133
    Salma S. Abed, Yogesh K. Dwivedi, and Michael D. Williams

Big and Open Data

Linking Operational Business Intelligence with Value-Based Business
Requirements   147
    Tom Hänel and Carsten Felden

Operationalizing Data Governance via Multi-level Metadata Management . . .   160
    Stefhan van Helvoirt and Hans Weigand

A MapReduce Based Distributed Framework for Similarity Search
in Healthcare Big Data Environment   173
    *Hiren K.D. Sarma, Yogesh K. Dwivedi, Nripendra P. Rana,
    and Emma L. Slade*

Big Data, Big Opportunities: Revenue Sources of Social Media Services
Besides Advertising.   183
    Julian Bühler, Aaron W. Baur, Markus Bick, and Jimin Shi

Big Data Analytics as a Service for Business Intelligence   200
    Zhaohao Sun, Huasheng Zou, and Kenneth Strang

Linked Relations Architecture for Production and Consumption of Linksets
in Open Government Data   212
    Petar Milić, Nataša Veljković, and Leonid Stoimenov

Budgetary Data (in an Open Format) Benefits, Advantages, Obstacles
and Inhibitory Factors in the View of the Intermediaries of this System:
A Study in Latin American Countries   223
    Gisele da Silva Craveiro and Cláudio Sonáglio Albano

Transparency Dimensions of Big and Open Linked Data: Transparency
as Being Synonymous with Accountability and Openness   236
    Ricardo Matheus and Marijn Janssen

Open Data Landscape: A Global Perspective and a Focus on China   247
    *Charles Shen, Zainab Riaz, Madhuri S. Palle, Qiurui Jin,
    and Feniosky Peña-Mora*

Open Data Platforms and Their Usability: Proposing a Framework for
Evaluating Citizen Intentions   261
    Kawaljeet Kapoor, Vishanth Weerakkody, and Uthayasankar Sivarajah

E-Business, E-Services and E-Society

Enabling Flexible IT Services by Crowdsourcing: A Method for Estimating
Crowdsourcing Participants 275
  Yiwei Gong

Mining Learning Processes from FLOSS Mailing Archives 287
  Patrick Mukala, Antonio Cerone, and Franco Turini

Private-Collective Innovation and Open Source Software: Longitudinal
Insights from Linux Kernel Development......................... 299
  Dirk Homscheid, Jérôme Kunegis, and Mario Schaarschmidt

Towards a Set of Capabilities for Orchestrating IT-Outsourcing
in the Retained Organizations................................ 314
  Bas Kleinveld and Marijn Janssen

Why Do Small and Medium-Size Freemium Game Developers Use Game
Analytics? 326
  Antti Koskenvoima and Matti Mäntymäki

Dynamic IT Values and Relationships: A Sociomaterial Perspective 338
  Leon Dohmen

Designing Viable Multi-sided Data Platforms: The Case of Context-Aware
Mobile Travel Applications 354
  Mark de Reuver, Timber Haaker, Fatemeh Nikayin, and Ruud Kosman

The Conceptual Confusion Around "e-service": Practitioners' Conceptions... 366
  *Eva Söderström, Jesper Holgersson, Beatrice Alenljung, Hannes Göbel,
  and Carina Hallqvist*

Social Customer Relationship Management: An Architectural Exploration
of the Components 372
  Marcel Rosenberger

Removing the Blinkers: What a Process View Learns About G2G
Information Systems in Flanders (Part 2) 386
  Lies Van Cauter, Monique Snoeck, and Joep Crompvoets

A Visual Uptake on the Digital Divide 398
  Farooq Mubarak and Reima Suomi

Sentiment Analysis of Products' Reviews Containing English and Hindi
Texts.. 416
  Jyoti Prakash Singh, Nripendra P. Rana, and Wassan Alkhowaiter

Adaptive Normative Modelling: A Case Study in the Public-Transport
Domain   423
    Rob Christiaanse, Paul Griffioen, and Joris Hulstijn

Business Process as a Service (BPaaS): Model Based Business and IT
Cloud Alignment as a Cloud Offering   435
    Robert Woitsch and Wilfrid Utz

Witness Workshop

IT-Enabled Resilient, Seamless and Secure Global Supply Chains:
Introduction, Overview and Research Topics   443
    Bram Klievink and Gerwin Zomer

Determining the Effects of Data Governance on the Performance and
Compliance of Enterprises in the Logistics and Retail Sector   454
    Nick Martijn, Joris Hulstijn, Mark de Bruijne, and Yao-Hua Tan

Data Quality Assurance in International Supply Chains: An Application
of the Value Cycle Approach   467
    Yuxin Wang, Joris Hulstijn, and Yao-Hua Tan

Towards a Federated Infrastructure for the Global Data Pipeline   479
    Wout Hofman

Key Design Properties for Shipping Information Pipeline   491
    Jensen Thomas and Yao-Hua Tan

Enhancing Awareness on the Benefits of Supply Chain Visibility Through
Serious Gaming   503
    *Tijmen Joppe Muller, Rainer Müller, Katja Zedel, Gerwin Zomer,
    and Marcus Engler*

Author Index   513

Keynote Panel

Driving Innovation Using Big Open Linked Data (BOLD) Panel

Yogesh K. Dwivedi[1(✉)], Vishanth Weerakkody[2], Marijn Janssen[3], Jeremy Millard[2,4], Jan Hidders[5], Dhoya Snijders[6], Nripendra P. Rana[1], and Emma L. Slade[1]

[1] School of Management, Swansea University, Swansea, UK
{y.k.dwivedi,N.P.Rana,e.l.slade}@swansea.ac.uk
[2] Business School, Brunel University London, Uxbridge, UK
Vishanth.Weerakkody@brunel.ac.uk, jeremy.millard@3mg.org,
Jeremy.Millard@brunel.ac.uk
[3] Section of Information & Communication Technology, Faculty of Technology,
Policy, and Management, Delft University of Technology, Delft, The Netherlands
M.F.W.H.A.Janssen@tudelft.nl
[4] Danish Technological Institute, Aarhus, Denmark
[5] Web Information Systems Group, The Software and Computer Technology Department,
Delft University of Technology, Delft, The Netherlands
A.J.H.Hidders@tudelft.nl
[6] Stichting Toekomstbeeld der Techniek (STT),
The Dutch Study Center for Technology Trends, Den Haag, The Netherlands
snijders@stt.nl

Abstract. Governments have always retained public service data internally in their own systems with only limited information provided to the public and other stakeholders such as the business, charitable and NGO communities. However, the rapid advancement of ICTs coupled with electronic publishing via the Internet in the last decade in particular has enabled governments to exploit the potential of wider distribution and use of such data previously held in internal systems. The panellists will discuss how Big, Open and Linked Data (BOLD) can be utilized to drive innovation and what obstacles and challenges may be encountered. Empowering citizens, potential mis-use in identity theft, policy manipulation or market distortion, and the need to combine open data with closed sources will be discussed.

Keywords: Big data · Open data · BOLD · Panel

1   Introduction

Traditionally, governments have retained public service data internally in their own systems and only a limited amount of information was provided to the public and other stakeholders such as the business, charitable and NGO communities. However, the rapid advancement of ICTs coupled with electronic publishing via the Internet in the last

© IFIP International Federation for Information Processing 2015
M. Janssen et al. (Eds.): I3E 2015, LNCS 9373, pp. 3–9, 2015.
DOI: 10.1007/978-3-319-25013-7_1

decade in particular has enabled governments to exploit the potential of wider distribution and use of such data previously held in internal systems. Recognising these developments, a European Union directive to encourage greater realisation of the economic value of public data through its reuse entered into force in 2003, thus, paving the way for governments to open up to the public previously closed data. This sought to both encourage provision and regulate licensing and charging for information by Member States. In 2004, the OECD recommended that all publicly funded research data should be made available.

The dual motivations of transparency of government and the economic potential of the reuse of data have since motivated politicians to promote the adoption of "open data" as a practice for their governments. In the US, the Presidential Open Government Directive in December 2009 required the use of open formats by federal US agencies, and in May 2010 the UK Prime Minister set out plans for opening up government data through the data.gov.uk website. Subsequently, the European Commission published a Communication on Open Data in 2011, and in the same year the USA, UK and initially six other countries were signatories to the Open Government Declaration.

Although selected public sector data has always been available to public bodies and other organisations, albeit with costs and restrictions, progressive moves by governments in countries such as the UK and US to improve availability and ease of reuse (through machine-readability and technical standards) has removed many such barriers. One of the unforeseen effects of the open data movement has been to make more data easily accessible to other actors in the policy space, including researchers, think-tanks and, most significantly, other parts of the public sector and governmental systems, including local governments. However, the availability of open data relating to various public services has led to more transparency around these services and allowed the general public to hold government departments accountable. This has encouraged stakeholders (including citizens, businesses, charities and NGOs) to take an active interest in the way services are currently delivered and stimulated thinking around how to improve services. Yet the actual use of open data is cumbersome and each stakeholder has to do it themselves as there are no proven solutions facilitating the process of using open data for services transformation or co-creation. Moreover, although the open data movement has gathered much momentum in Europe, the majority of open data repositories (e.g. data.gov.uk, etc.) are too generic and not of direct interest to individual citizens.

Although big or large volumes of raw open data when published in an electronic format is machine readable and can be shared online and re-used, on its own open data offers limited potential for decision making. However, when open data is linked (Big, Open and Linked Data – BOLD) with extra information that provides more context, this offers greater opportunities for stakeholders to exploit the data for innovative purposes, for example through collaboration and co-creation. BOLD could also increase the reach of statistical and operational information, and deeper analysis of outcomes and impacts. Indeed, BOLD offers an opportunity to discover new ways of assessing policy and service outcomes, healthcare and wellbeing, measuring development and making appropriate interventions through innovative solutions where needed.

Although the use of BOLD to date remains limited and at an early stage there are many examples that show its potential. When, in 2011, the UK's mapping authority released all its geodata rather than selling it, the organization completely transformed its business model from being simply a purveyor of raw data for others to use to becoming its customers' partner in collecting, analysing, tailoring, visualizing and applying linked data in meeting their needs. This new value adding and problem solving role boosted and diversified the authority's income. In another example, Washington DC was one of the first cities to make BOLD available, running hackathons and competitions as part of its 'Apps for Democracy' initiative, which eventually saved the City authorities $2.3 million at a cost of $50,000.

In responding to the theme of The 14th IFIP Conference on e-Business, e-Services and e-Society (I3E 2015), the panellists will discuss how Big, Open and Linked Data (BOLD) can be utilised to drive innovation and what obstacles and challenges may be encountered regarding this. The discussion will be based on empirical evidence, using insights from cases; there will be a focus on analytical models, building on conceptual and theoretical arguments, and our aim is to offer directions for future research.

2   Programme

Professor Yogesh Dwivedi will open the panel by presenting an overview of the topic and subsequently will moderate the panel discussion.

Professor Vishanth Weerakkody will evaluate the role of stakeholders in the BOLD arena and discuss some of the challenges facing public agencies and potential users in exploiting public open data. The discussion will particularly centre on the Hype Vs Reality in the BOLD debate and use several European Commission funded R&D projects to discuss the present technical and user problems faced by local governments who are engaged in promoting BOLD for improved decision making. The discussion will also examine how universities, professional institutes and organisations are preparing their graduates and professionals to deal with the BOLD challenge. In the public service space, who are the main users of big and open data? Are citizens interested or prepared to exploit open data? Are conventional statistical tools equipped to deal with BOLD? Which tools are emerging as the front runners for exploiting BOLD? These are some of the other questions that will be debated.

Professor Marijn Janssen will argue that BOLD can be used to empower citizens and to create an ecosystem in which governments, businesses and citizens can strengthen each other. Yet driving innovations from data is a complex process in which both the available data and the users' demands need to be taken into account. The name "data-driven" already implies that data instead of customers' needs are leading. Furthermore all stakeholders have different interests and concerns. Policy-makers want to use the results for their policy-making, businesses want to create new services to enrich their existing products and services, whereas citizens want to know what is going on and have free access to services without having to up their privacy. For this purpose platforms creating a community, providing an overview of the data and collaborative development environment to develop new ideas, are needed. However, the opening and use of data might make apparent the low data quality of sources and the focus might be on developing tools and platforms without having the user in mind.

Jeremy Millard's discussion will focus on two main but related issues. First, one of the wicked challenges of BOLD is its potential mis-use, either through negligence, corruption or criminality resulting, for example, in identity theft, policy manipulation or market distortion. Although this is a battle that can never be won, what are the technical, legal, regulatory and ethical challenges we face and how might they be addressed, also taking account of both the Snowden and Assange cases? Second, BOLD becomes even more interesting and disruptive when public data is mixed with data from other sources, including, for example, the private data of businesses or citizens. The smart disclosure approach starts from the premise that people, when given access to data and useful decision tools built for example by governments, can use both their own personal data disclosed by them together with other appropriate data. This can be used to make decisions about their own lives, such as healthcare choices, as well as to self-regulate and be able to hold governments and other actors to account, as well as to cooperate and engage with them.

Jan Hidders will discuss the technological data management challenges of BOLD. In order for BOLD to become effective and usable there must be tools that allow ordinary users, people interested in using the data, to select, transform, combine and visualise the data in an ad-hoc fashion. In some sense this has always been the goal of data management tools, but with the advent of BOLD new challenges have been added such as the scale of the data (in number of data sources, sizes of schemas as well as sizes of actual data content), the possible lack or presence of structure and semantics, and the large heterogeneity of the data sources. This puts all kinds of new demands on the existing data management tools that are currently not being met. These demands are not only in terms of scalability and ability to deal with the semantics of the data, but also in terms of interfaces and languages that allow users to deal with such data in a more intuitive way than now is possible. A key question is what underlying data model these tools should be based upon? Should it be XML, JSON or perhaps RDF which comes with semantic abilities? Or should we perhaps go back to classical models such as the relational model? These and other issues will be discussed in this section.

Dhoya Snijders will focus on the ways in which BOLD is altering how citizens and government relate to each other. Already, BOLD pleads for new types of civil servant with different skills, different educations and, some argue, a different epistemological standpoint. The speed with which data can be created and with which correlations within the data can be made is impacting government's deep-rooted quest for causality. Similarly, BOLD is creating new types of citizen, who are data-focused and data-driven. Citizens are increasingly connecting, measuring, analysing and testing public data themselves. The relations between government and citizens are hereby mediated by BOLD. And, as BOLD is itself becoming intelligent, data is not only opened up, linked, and analysed by human actors - machine-learning is picking up speed and quality. Both citizens and governments will increasingly have to deal with non-human actors in the form of intelligent data-driven systems. To do so we need to develop what sociologists have dubbed double contingency in which we as humans need to understand how intelligent machines will understand us and vice versa.

Nripendra Rana and Emma Slade will assist in organisation and coordination of the panel and note taking of the panel content.

3 Panellist Bios

Yogesh K Dwivedi is a Professor of Digital and Social Media and Head of Management and Systems Section (MaSS) in the School of Management at Swansea University, Wales, UK. His research interests are in the area of Information Systems (IS) including the adoption and diffusion of emerging ICTs (e.g. broadband, RFID, e-commerce, e-government, m-commerce, m-payments, m-government) and digital and social media marketing. His work on these topics has been published (more than 100 articles) in a range of leading academic journals including: CACM, EJIS, IJPR, GIQ, JORS, ISF, ISJ and IJICBM. He has co-edited more than ten books on technology adoption, e-government and IS theory and had them published by international publishers such as Springer, Routledge, and Emerald. He acted as co-editor of fourteen special issues; has organised tracks, mini-tracks and panels in leading conferences; and served as programme co-chair of IFIP WG 8.6 Conference at the prestigious IIM Bangalore, India in 2013. He is Associate Editor of GIQ, EJM and EJIS, Assistant Editor of JEIM and TGPPP, Senior Editor of Journal of Electronic Commerce Research and member of the editorial board/review board of several journals. He is a life member of the IFIP WG8.6 and 8.5.

Vishanth Weerakkody is Professor of Digital Governance in the Business School of Brunel University in London, United Kingdom. His research experience is focused in the area of public sector service transformation through technology. Professor Weerakkody has published more than 100 peer reviewed articles and has guest edited several special editions of journals and books on this theme. He is co-founder of the e-Government track at the Americas Conference on Information Systems (AMCIS) and Transforming Government Workshop at Brunel University, London. Vishanth has many years of experience in R & D projects in the area of digital governance and is currently an investigator in several European Commission and Internationally funded projects on the use of ICTs in the public sector (e.g. LiveCity, OASIS, DAREED, PolicyCompass, UBiPol, CEES, EGovPoliNet, I-MEET, SI-DRIVE). He is the Editor-in-Chief of the International Journal of Electronic Government Research and one of the two international board members of the Digital Government Society of North America. He is a fellow of the UK Higher Education Academy and combines more than 25 years of experience in industry, teaching and research leadership.

Prof. Dr. Marijn Janssen is full Professor in ICT & Governance and head of the Information and Communication Technology section of the Technology, Policy and Management Faculty of Delft University of Technology. His research interests are in the field of orchestration, (shared) services, intermediaries, open and big data and infrastructures within constellations of public and private organizations. He was involved in EU funded projects in the past (a.o. EGovRTD2020, eGovPoliNet and Engage), is Co-Editor-in-Chief of Government Information Quarterly, Associate Editor of the International Journal of Electronic Business Research (IJEBR), Electronic Journal of eGovernment (EJEG), International Journal of E-Government Research (IJEGR), is conference chair of IFIP EGOV2015 and IFIP I3E2015 conference (about big and open data innovation) and is chairing mini-tracks at the DG.o, ICEGOV, HICCS and AMCIS conferences. He was ranked as one of the leading e-government researchers in a survey in 2009 and 2014 and has published over 320 refereed publications. More information: www.tbm.tudelft.nl/marijnj.

Jeremy Millard is a Senior Research Fellow at Brunel University (London) and Senior Policy Advisor, Danish Technological Institute (Denmark). He has forty years' global experience working with governments, development agencies, and private and civil sectors in all parts of the world. In the last twenty years he has focused on how new technical and organisational innovations transform government and the public sector. Work with the European Commission includes research and studies on eGovernment, administrative burden reduction, and on developing business models for ICT and ageing. He also recently led an impact assessment of the European eGovernment Action Plan, led the large-scale Europe-wide survey and analysis of eGovernment eParticipation, and developed the 2020 Vision Study on Future Directions of Public Service Delivery. Jeremy has also worked since 2008 as an expert for the UN on their successive global eGovernment development surveys, and provided inputs to both the World Bank and the OECD on eGovernment developments, for example through a survey on back-office developments in support of user-centred eGovernment strategies, as well as ICT-enabled public sector innovation. He works extensively outside Europe in these and related areas, including in the Gulf, the Western Balkans, Georgia, Japan, and India.

Dr Jan Hidders is Assistant Professor in the Web Information Systems group of the Software and Computer Technology department of Delft University of Technology. His research interests are in data integration, data indexing, data linking and large scale data processing, particularly in the domain of graph processing. He was involved in several EU projects such as GRAPPLE and ImREAL where he worked on data integration in the e-learning domain. He has published over 60 refereed publications in conferences such as ICDE, CCGRID, ICDT, CIKM, IUI, FoIKS, ICWE and ISWC, and in journals such as Information Systems, Fundamenta Informaticae, Theory of Computing Systems, BPM Journal and Journal of Computer and System Sciences. He has edited special issues in Fundamenta Informaticae and the Journal of Computer and System Sciences. He has co-organised several workshops such as SWEET (Scalable Workflow Enactment Engines and Technology) at the SIGMOD conference and BeyondMR (Beyond MapReduce) at the EDBT conference. He is currently one of the co-organizers of the 2015 EDBT Summer School on Graph Data Management.

Dr Dhoya Snijders works for STT, the Dutch Study Center for Technology Trends and is currently carrying out a technology foresight study on big data. He holds an MA in Philosophy from the University of Amsterdam and an MA and PhD in Organizational Sciences and Public Administration from the VU University. His research mainly focuses on classifications and their consequences. He worked as a consultant for some years in the field of e-Government within the Dutch Ministries of Healthcare and Justice. There he carried out national and international projects on ICT governance, eHealth, Open Data, and Big Data. During this time he published on the implementation of ICT in a context of multilevel governance.

Dr Nripendra P Rana is a Lecturer at the School of Management at Swansea University in the UK. He holds a BSc in Mathematics (Hons), an MCA, an MTech, and an MPhil degree from Indian universities. He also holds an MBA with distinction and a PhD from Swansea University. His current research interest is in the area of technology and e-Government adoption and diffusion. He has published his work in some refereed journals including ISF, ISM, ESJ, IJBIS, IJICBM, IJEGR, and TGPPPP. He has varied work experience of teaching in the area of computer science at undergraduate and postgraduate levels. He also possesses a good experience of software development and leading successful software projects.

Emma Slade is a Research Officer in the School of Management at Swansea University. She holds a BSc (Hons), MSc with distinction, and PhD in Business Management from Swansea University. Emma's research interests include consumer and merchant mobile payment adoption, e-government adoption, and consumer forgiveness. Her research has been published in Psychology & Marketing, Journal of Strategic Marketing, and Journal of Computer Information Systems. She also presented at the 2014 Academy of Marketing Science Conference and the 2013 UK Academy for Information Systems Conference. Emma has been invited to review papers for Information Systems Frontiers, Internet Research, and a number of conferences.

Adoption

Adoption of Mobile Banking in Jordan: Exploring Demographic Differences on Customers' Perceptions

Ali Abdallah Alalwan[1], Nripendra P. Rana[2],
Yogesh K. Dwivedi[2(✉)], Banita Lal[3], and Michael D. Williams[2]

[1] Amman College of Banking and Financial Sciences, Al-Balqa' Applied University,
P.O. Box 1705 Amman Jordan
alwan.a.a.ali@gmail.com
[2] School of Management, Swansea University, Swansea, SA2 8PP, UK
{nrananp,ykdwivedi}@gmail.com, m.d.williams@swansea.ac.uk
[3] Nottingham Business School, Nottingham Trent University, Nottingham, NG1 4BU, UK
banita.lal@ntu.ac.uk

Abstract. This study aims to explore whether Jordanian customers' perceptions on intention and adoption of Mobile banking (MB) services varies according to their demographic characteristics. As per the prior literature, five demographic factors, namely age, gender, income, education and customer's experience with computer and Internet have been considered in the current study. The required data were collected from the field survey questionnaires administered to a convenience sample of Jordanian banking customers. The major statistical results (mean and standard deviation) demonstrate that the customers' perceptions on intention and adoption of MB are likely to vary due to customers' demographic differences. According to the current study's findings, it was also noticed that despite the fact that the most of the sample respondents express a high intention to adopt MB, the adoption rate for the majority of MB services was low.

Keywords: Mobile banking · Customer · Jordan · Demographic differences · Adoption · Behavioral intention

1 Introduction

Mobile banking (MB) is identified as the "use of mobile terminals such as cell phones and personal digital assistants to access banking networks via the wireless application protocol (WAP)" [44, p. 760]. Such innovative banking channel has been progressively implemented over the banking context worldwide to launch customers a wide range of the higher quality banking services (i.e. balance enquiries, fund transfers, payment of bills) without any time or place restrictions [42].

In Jordan, banks seem to be more motivated to adopt MB as an essential banking channel to increase the geographical coverage, customer's satisfaction and loyalty as well as to minimise the operational and labour costs related to traditional branches [4, 26, 32, 33, 42]. Further, the mobile and telecommunication area is growing phenomenally; where there are four mobile services providers along with up to 8.984 million of

© IFIP International Federation for Information Processing 2015
M. Janssen et al. (Eds.): I3E 2015, LNCS 9373, pp. 13–23, 2015.
DOI: 10.1007/978-3-319-25013-7_2

mobile subscriptions [37]. Practically, out of 26 banks working in Jordan, 15 banks have Launched MB services [26].

However, banks worldwide express their concerns regarding the lower adoption rates of MB services by customers. By the same token, Jordanian banking customers seem to be not fully motivated to adopt MB [5]. According to recent study by Alafeef et al. (2011), 6 % of the total Jordanian banking customers have actually used MB services. Alafeef et al. [3] also mentioned that 31 % of banking customers are not aware of the existence of MB services introduced by their banks.

Accordingly, it could be concluded that the main challenge pertaining to successful implementation of MB is conceiving bank clients to move from using human encounter to fully adopt MB [44]. Thus, there is always a necessity to identify the main factors that could hinder or enhance customers' intention and adoption of MB. However, MB related issues have been rarely examined in Jordan. Furthermore, banks need to have more information regarding customers' perception on intention to use and adoption of MB that could differ as per the demographic differences. This, in turn, will help these banks to conduct a useful market segmentation strategy that could accelerate the customers' intention and adoption of such technology [14, 16, 31, 34]. Accordingly, this study aims to conduct an empirical examination to discover if there are considerable variations in the Jordanian customer's perception on the intention and adoption of MB.

2 Literature Review

Theoretically, many studies have endeavored to interpret the most important factors that could hinder or foster customer intention and acceptance of Mobile banking. For instance, Püschel et al. [30] claimed that Brazilian customers' attitudes towards Mobile banking were significantly affected by relative advantage, followed by compatibility; ultimately enriching the customers' intention to adopt Mobile banking. Brown et al. [7] also found that banking customers are more enthused to adopt Mobile banking by relative advantage, trialability, and consumer banking needs. Perceived benefits and governmental regulations have been confirmed as the key positive enablers of customer attitudes towards Mobile banking in Indonesia [29]. In line with this, customers were found to be more motivated to use Mobile banking if they recognised Mobile banking as being useful in their daily life, compatible with their habits and other technologies, and less expensive [18, 42]. By the same token, customer intention to adopt Mobile banking was significantly determined by the role of perceived usefulness, monetary cost, self-efficacy and perceived ease of use [23]. Later, Zhou [43] empirically supported the considerable role of a bank's reputation, information quality, self-efficacy, service quality, and system quality in shaping the customers' initial trust in Mobile banking. More recently, Hanafizadeh et al. [18] all supported the crucial role of perceived usefulness and ease of use in motivating customers to adopt Mobile banking.

With regard to the role of demographic factors, Laukkanen and Pasanen [22] indicated that customers' adoption of Mobile banking was exclusively predicted by the customers' age and gender, but was not predicted by education level, career, family size and income. Riquelme and Rios [35] revealed that the males' intention to adopt Mobile banking was strongly affected by the role of perceived usefulness, while females paid particular attention to aspects related to ease of use. Chiu et al. [9] indicated that there were no statistical differences in the role of these factors on customers' intention. Laukkanen [21] also discussed that regardless of age differences, customers seem to be more reluctant to adopt Mobile banking due to the negative role of value and usage barriers, while mature customers were observed to be more concerned about barriers pertaining to risk, image, and traditional barriers.

Nevertheless, there is a dearth of literature addressing customer intention and usage of Mobile banking by Jordanian banking customers [5, 19]. Both Khraim et al. [19] and Awwad and Ghadi [5] have found that Mobile banking characteristics - trialability, complexity, compatibility, relative advantages, and risk - are the key predictors of Jordanian customer intention and adoption of Mobile banking. Yet, there is still a need to clarify and empirically examine the important role of the customers' characteristics (i.e. age, gender, income, education, and technology) in shaping the Jordanian customers' perception on intention to use MB and adoption of such emerging systems. Thus, in order to fill this gap, *current study intends to empirically test and explore whether Jordanian customers' perceptions on intention and use of MB could vary according to their demographic characteristics: age, gender, income, education, customers' experience with computer and Internet.*

3   Theoretical Basis

The impact of demographic factors has received a great deal of attention from information systems studies, which assert that variations in customers' reactions to and perceptions of technology could be attributed to the variation in the customers' demographic characteristics [11, 28]. Therefore, this study examines how the demographic factors (such as age, gender, income, education, and customers' experience with computer and Internet) could reflect differences in the Jordanian customers' perception on intention to use and actual adoption of Mobile banking services. These five factors have been widely studied and examined over the online banking area [12, 24, 25]. Further justification and discussion regarding each of these factors are provided in the following subsections.

3.1   Age

Theoretically, age has been debated either as an independent variable or as a factor that could lead a variation on the persons' perception toward a certain kind of individual or collective behaviour and actions [17]. Likewise, prior studies over the information system area have paid a particular interest for the important role of age in shaping the individual perception on the aspects related to technology (i.e. usefulness, ease of use, behavioural intention and actual usage of technology [39–41]. According to Venkatesh et al. [38], actual adopters of computer were found largely to be within the age range of 15 to 35 years. In the self-service technology (SST) context, Dabholkar et al. [11] observed that customers are more likely to be varying on their intention towards and usage of SST according to their age. Therefore, *it could be expected that the Jordanian banking customers' perceptions on intention and adoption of MB would differ according to their age groups.*

3.2 Gender

According to Morgan [27], gender could be debated either as a descriptive variable or as an explanatory variable. Over the information system (IS) area, a number of authors [39–41] have examined the role of gender in moderating or directly predicting the individual' perception, intention and behaviour toward technology. In keeping with Venkatesh et al. [38], the usage of computer is more likely to be in the higher level among males than females. Differently, some IS researchers [12, 38, 40, 41] have indicated that men are more likely to accept a new technology based on the benefits and advantages perceived; while women usually pay more attention to aspects related to complexity, facilitated resources, and assurance. Therefore, *it could be expected that the Jordanian banking customers' perceptions on intention and adoption of MB would differ according to their gender.*

3.3 Education

In the line with Burgess's [8] proposition, individuals who enjoy an adequate education level are more likely to have a positive perception and ability to conduct a set of complicated actions. Therefore, it has been largely claimed that the well-educated people are more likely to have a positive reaction and perception toward new innovations, thereby; they are more likely to adopt a new technology in comparison with those who are at the less educational level [8, 36]. This thought has been also approved by number of IS studies [2, 28, 38]. For example, Al-Somali et al. [2] empirically proved that Saudi banking customers, who have an adequate level of education, are more likely to have positive attitudes toward Internet banking. Therefore, *it could be expected that the Jordanian banking customers' perceptions on intention and adoption of MB would differ according to their education level.*

3.4 Income

Instead of the employee context where the cost could be restricted in terms of time and effort, customers are more likely to be sensitive to the financial issues that could form their perception toward using the technology [24]. Generally speaking, customers with higher income are more likely to be able carry the financial cost associated with using a new technology [36]. Indeed, using MB services could require customers to pay a cost for using such services in addition to the other cost of having a smart phone, Internet access, and using specified applications [35]. From this perspective, banking customers' perception toward such novel technology could be different according to the customers' income level. This proposition was confirmed by Meuter et al. [25] who empirically approved income level as a key determinant of both customer readiness and customer experiment of different kinds of self-service technologies. Al-Ashban and Burney [1] also empirically approved income level as a considerable positive predictor of the customers' acceptance of Tele-banking in Saudi Arabia. Likewise, according to Kolodinsky et al. [20], the adoption rate of Internet banking channels was observed to be in the higher level among customers who have a higher income level instead of lower income customers. Therefore, *it could be expected that the Jordanian banking customers' perceptions on intention and adoption of MB would differ according to their income level.*

3.5   Experience

Given the particular nature of MB as a self-service banking channel requiring customers to produce the financial services without any assistance from banking staff, the adequate levels of experience and skills with technology could be the important prerequisites to successfully apply this technology (Meuter et al., 2005). Hence, customer experience has been identified by a number of studies as a crucial determinant of customer percep-tion and behaviour towards SST [24, 25]. With reference to Meuter et al. [25], prior experience strongly influences the customer's decision to try a self-service technology, either directly or indirectly, through the mediating impact of customer readiness. Further, customer experience was found to be one of the most influential factors predicting customer propensity toward mobile ticketing in a transportation context [24]. Based on empirical results established by Curran et al. [10] and Chiu et al. [9] customer familiarity in dealing with self-service technology has a significant and positive impact on customers' intention and orientation toward this technology. Therefore, *it could be expected that the Jordanian banking customers' perceptions on the intention and adop-tion of MB would differ according to customer's experience with computer and Internet.*

4   Research Methodology

Either in information system area or the MB context, it has been highly noticed that the field survey is one of the most prevalent and commonly adopted methods for testing an individual's intention and behaviour towards such an emerging system [13, 15]. Further, this study was conducted with the aim to test and explain the Jordanian banking customers' perception on intention and adoption of Mobile banking. Therefore, the field survey was found to be the most suitable and cost-effective research method allowing access to a large number of Jordanian banking customers in different places within a reasonable time [6]. This, in turn, led to observe that the self-administered questionnaire was a suitable data collection method to obtain the required data from Jordanian banking customers.

As mentioned before, the self-administered questionnaire was selected to derive responses from Jordanian banking customers regarding their perception of the aspects related to behavioural intention and use of Mobile banking. The seven-point Likert scale was used to measure the behavioural intention items with anchors ranging from '1 - strongly agree' to '7 - strongly disagree'. A set of six common financial services was adopted to measure the adoption of Mobile banking. These services have been widely adopted by relevant studies that have examined customers' use or adoption of Internet banking, Mobile banking, and Telebanking [23, 30, 44]. The seven-point time scale was adopted to measure the use behaviour toward these services with anchors including: 'never', 'once a year', 'several times a year', 'once a month', 'several times a month', 'several times a week', 'several times a day' [41]. Furthermore, six close-ended questions were used for demographic variables such as age, gender, income, education level, Internet experience, and computer experience.

5   Results

5.1   Descriptive Analysis of Usage Behavior

We provide a statistical description regarding the usage patterns of the six Mobile banking services. Balance enquiries and downloading bank statements seem to be the most frequently used Mobile banking services that are applied by the respondents. Indeed, of the 343 valid responses, 105 (30.6 %) used Mobile banking several times per month to look at their bank balance or to download the balance on their bank statement. Yet, balance enquiries and downloading of bank statements via mobile banking have never been used by 85 (24.8 %) of the respondents. In summary, the average mean usage of balance enquiries and downloading bank statements performed by using Mobile banking was 3.65 and the standard deviation was 1.83.

Paying bills was the next widely used Mobile banking service as 110 respondents (i.e. 32.1 %) have used Mobile banking to pay bills once a month while, 130 (37.9 %) valid responses mentioned that they have never used Mobile banking to pay bills. In addition to these, the usage mean of paying bills performed by Mobile banking was 2.77. The third Mobile banking service used by respondents was funds transfer. Even though 145 (42.2 %) of the respondents mentioned that they have never transferred funds through Mobile banking, 118 (34.4 %) [60 + 58] of the respondents indicated that they had utilised Mobile banking for funds transfer about once a month or several times a year. The usage mean of this service was 2.41 and its standard deviation was 1.52. A total of 79 [44 + 35] (23 %) respondents used Mobile banking to request a chequebook or bank certificates several times per year or once a year. Yet, many more respondents (i.e. 179, 52.1 %) have never used the Mobile-banking channel to receive the same services. The mean of using Mobile banking to obtain these services was 2.21 and the standard deviation was 1.52.

Of the few who have used mobile banking for payment of instalments of loans and mortgages, 51 (14.8 %) of the respondents have used Mobile banking for these services once per month. However, the vast majority of respondents (i.e. 224, 65.3 %) reported that they have never applied for these services via Mobile banking. The mean of using mobile banking to conduct these services was too low (about 1.88) with the standard deviation of 1.35.

Moreover, 237 respondents (69 %) have never used these services. Only 94 [33 + 33 + 28] (27.4 %) used Mobile banking to perform these services once a year, several times per year, and once a month respectively. Moreover, these services had the lowest usage mean (i.e. 1.68) among the Mobile banking services and its standard deviation was 1.18.

5.2 Respondent's Demographic Characteristics and Customers' Perception Relating to Intention to Use and Adoption of Mobile Banking

Both mean and standard deviation (SD) are tested in the current study to see how the customers' perception regarding the issues related to behavioral intention and adoption of MB could be vary according to respondent's demographic characteristics. In the term of age difference, all age categories express a high intention to use MB. Yet, according to the average mean accounted for age group 25–30 and 31–40, younger generation seems slightly more interested in using MB in future. The actual adoption mean extracted for age categories indicted the lower adoption rates of MB services by Jordanian customers. However, the adoption rates of MB were noticed in their highest level for age group 25–30 followed by age group of 31–40.

As for gender difference, both male and female seem to be more willing to adopt MB due to the highest mean score accounted in this regard, yet; the mean accounted for male was higher than for female. On the other hand, the usage mean accounted for both male and female was low with value of 2.27 for male and another lower value for female (i.e. 2.012).

According to main statistical outcomes regarding the refection of educational level, as expected, the highest rate of behavioural intention was noticed regarding those respondents who have a PhD degree with mean value of 6.33 followed by those who have masters (5.39) and Bachelors (5.37) degrees. By the same token, the largest mean of the adoption behaviour was accounted for respondents who have a PhD with mean value of 2.28. Lower adoption means were also recorded for the all education groups.

In the term of computer experience, the largest mean value of behavioral intention (i.e. 5.62) was accounted for those respondents who have three years of experience or above with computer while the value of 4.25, which is the lowest value was accounted for those who have less than three years of experience with computer. Similarly, the highest mean value of adoption (i.e. 2.26) was noticed in the case of respondents who have experience of three years and above.

As for Internet experience, the largest mean (i.e. 5.47) of behavioural intention was accounted by those respondents who have experience with Internet for three years and above. On the other hand, the lowest mean (i.e. 4.25) of behavioural intention was in the case of respondents with Internet experience less than one year. By the same token, the largest mean (i.e. 2.22) of adoption of MB services was in the case of Internet experience of three years or above whereas the lowest one (i.e. 1.92) in this regard was in the case of experience group of 1–2 years.

6   Discussion

Generally, the main results extracted in the current study were found to be in line with what has been discussed and approved by prior literature regarding the role of demographic factors on the perception toward technology. For instance, the age categories of 25–30 and 31–40 express a higher willingness toward using MB in comparison with older customers (i.e. 50–60 and those above 61). Even though the adoption rates of MB services are too low over all age groups, younger customers were observed as highly involved in using MB in comparison to older customers. Such issues could possibly return to the fact that older customers are more likely to not have the important skills and experience that enable them to properly use MB and most of them they do not have smart phones needed to use MB services. Younger customers seem to have more capability and confidence to interact with such sophisticated technology like Mobile banking. This could be regarded to the fact that younger customers have more interaction with technology and Mobile innovation and having an adequate level of technological savvy, awareness, skills, and knowledge, [41, 42].

As for gender, the findings indicated that there are slight differences between males and females in their perception toward intention and adoption. Nevertheless, it could be difficult in the current stage to argue that males are more interested and hence can heavily adopt MB services than females. This could be attributed to the fact that the influence of gender differences are more likely to vanish over the highly evolved communities where both males and females have equal opportunities to be educated, to work, and to get interacted with technologies. In parallel with these results, there are several studies [11, 12] that have empirically disapproved a variation between males and females in their intention and reaction toward technology.

In line with what has been expected regarding the role of education, highly educated respondents seem to be more motivated to use MB and more involved as well in adopting such technology. As discussed earlier in this study, people with a good level of education are more likely to have the sufficient knowledge and skills that could help them to cope with new technology rather than less educated people. Theoretically, this proposition has been highly supported by different studies conducted in the same area of interest [2, 28, 38].

Respondents were also observed to be different in their intention and adoption of MB according to the variation in their income level. While the higher income respondents seem to be more interested and active users of MB services, lower income respondents are less interested and motivated in this regard. Over the marketing literature, it has been highly argued that the higher income customers are less sensitive for the cost issues [25, 36]. Those customers are more able to carry the financial cost associated with buying the important facilities and resources required to use Mobile banking [20].

As it is expected, customer perceptions toward both behavioural intention and adoption are more likely to be different according to the customers' experience with technology and Internet. Indeed, the results regarding the willingness to adopt MB and the actual adoption of MB were able to reach the highest level among respondents who have a good level of technology experience for more than three years. As discussed in Sect. 3.5, the MB channel as self-service technology requires customers to independently conduct all the process to produce and transport the MB services. Therefore, adequate level of experience with technology is a very important aspect to let the customers be more confident in their ability to deal with such complicated systems. These results are consistent with other IS studies, which assured the important role of customers' experience in formulating the customers' perception and reaction to cope with different kinds of systems [10].

This study comprises an important contribution by exploring the main demographic features of adopters and potential adopters of Mobile banking in Jordan as more emerging system calling for further explanation and examination. Primarily, a theoretical contribution was captured in the current study by synthesising the relevant literature of IS area and Mobile banking as well. Further, such theoretical propositions were empirically tested via collecting sufficient amount of data from the Jordanian banking customers. Practically, the results of the current study alert the Jordanian banks about the current state of the Jordanian customers' intention and adoption of MB services. Therefore, by conducting an empirical study to discover the demographic features of adopters and potential adopters of MB, this study was hoping to provide the Jordanian banks with relevant guidelines that would facilitate an effective implementation and acceptance of MB in proportion to the customer's category and their demographic characteristics. The main results of mean and standard deviation of behavioural intention obviously suggest that the majority of the respondents seem to be more motivated to adopt MB. In addition to this, most of them enjoy an adequate level of education and experience with the Internet and the computer, thus; moving them as actual users of MB will not be expensive and difficult. Therefore, allowing customers to try using these applications through experimental accounts rather than using their own accounts could create a positive experience and let customers actually experience how much they will benefit by using these valuable, useful and easier applications [13].

7   Conclusion

The fundamental intention of the current research was to discover how the Jordanian banking customers' perceptions on aspects related to their intention and adoption of MB could be differing because of their demographic features. Five common demographic variables, namely age, gender, income, education, and technology experience were identified and tested in the current study. The empirical findings clearly suggest that Jordanian banking customers are more likely to be different in their perceptions toward intention and adoption of MB according to their demographic differences.

7.1   Limitations and Future Research Directions

One of the main limitations of the current study is that the data was derived using a convenience sample of banking customers from two cities in Jordan: Amman and Al-Balqa, raising a concern regarding the applicability of the current study results for other banking customers in different regions. Accordingly, it would be more useful for future studies to capture the required data from a large sample size covering the most parts in Jordan. This study only focuses on the customers' demographic features while it does not pay attention to the psychological and behavioral factors (such as habit, innovativeness, customer readiness, and self-efficacy). Therefore, examining such factors along with demographic features could provide a rich understanding of the customer's reaction and perception toward Mobile banking. Future studies should look at the impact of cultural aspects on the Jordanian customers' perceptions toward MB especially when such aspects have not been examined in the area of Mobile banking.

References

1. Al-Ashban, A.A., Burney, M.A.: Customer adoption of Telebanking technology: the case of Saudi Arabia. Int. J. Bank Mark. **13**(5), 191–201 (2001)
2. Al-Somali, S., Gholami, R., Clegg, B.: An investigation into the acceptance of online banking in Saudi Arabia. J. Bus. Res. **29**(2), 130–141 (2009)
3. Alafeef, M., Singh, D., Ahmad, K.: The influence of demographic factors and user interface on mobile banking adoption: a review. J. Appl. Sci. **12**(20), 2082–2095 (2012)
4. Alalwan, A.A., Dwivedi, Y.K., Rana, N.P., Lal, B., Williams, M.D.: Consumer adoption of Internet banking in Jordan: examining the role of hedonic motivation, habit, self-efficacy and trust. J. Financ. Serv. Mark. **20**(2), 145–157 (2015)
5. Awwad, M.S., Ghadi, M.Y.: Investigating of factors influencing the intention to adopt mobile banking services in Jordan. Dirasat Adm. Sci. **37**(2), 545–556 (2010)
6. Bhattacherjee, A.: Social Science Research: Principles, Methods, and Practices, 2nd edn. Florida, USA (2012)
7. Brown, I., Cajee, Z., Davies, D., Stroebel, S.: Cell phone banking: predictors of adoption in South Africa—an exploratory study. Int. J. Inf. Manag. **23**(5), 381–394 (2003)
8. Burgess, R.: Key Variables in Social Investigation. Routledge, London (1986)
9. Chiu, Y., Fang, S., Tseng, C.: Early versus potential adopters: exploring the antecedents of use intention in the context of retail service innovations. Int. J. Retail Distrib. Manag. **38**(6), 443–459 (2010)
10. Curran, J.M., Meuter, M.L., Surprenant, C.F.: Intentions to use self-service technologies: a confluence of multiple attitudes. J. Serv. Res. **5**(3), 209–224 (2003)
11. Dabholkar, P.A., Bobbitt, M.L., Lee, E.J.: Understanding consumer motivation and behavior related to self-scanning in retailing: implications for strategy and research on technology based self-service. Int. J. Serv. Ind. Manag. **14**(1), 59–95 (2003)
12. Dean, D.H.: Shopper age and the use of self-service technologies. Managing Serv. Qual. **18**(3), 225–238 (2008)
13. Dwivedi, Y.K., Irani, Z.: Understanding the adopters and non-adopters of broadband. Commun. ACM **52**(1), 122–125 (2009)
14. Dwivedi, Y.K., Lal, B.: Socio-economic determinants of broadband adoption. Ind. Manag. Data Syst. **107**(5), 654–671 (2007)
15. Dwivedi, Y.K., Choudrie, J., Brinkman, W.P.: Development of a survey instrument to examine consumer adoption of broadband. Ind. Manag. Data Syst. **106**(5), 700–718 (2006)
16. Dwivedi, Y.K., Wastell, D., Laumer, S., Henriksen, H.Z., Myers, M.D., Bunker, D., Elbanna, A., Ravishankar, M.N., Srivastava, S.C.: Research on information systems failures and successes: status update and future directions. Inf. Syst. Front. **17**(1), 143–157 (2015)
17. Finch, J.: Age. In: Burgess, R. (ed.) Key Variables in Social Investigation. Routledge, London (1986)
18. Hanafizadeh, P., Behboudi, M., Koshksaray, A.A., Tabar, M.J.S.: Mobile-banking adoption by Iranian bank clients. Telematics Inform. **31**(1), 62–78 (2014)
19. Khraim, H.S., Shoubaki, Y.E., Khraim, A.S.: Factors affecting Jordanian consumers' adoption of mobile banking services. Int. J. Bus. Soc. Sci. **2**(20), 96–105 (2011)
20. Kolodinsky, J.M., Hogarth, J.M., Hilgert, M.A.: The adoption of electronic banking technologies by US consumers. Int. J. Bank Mark. **22**(4), 238–259 (2004)
21. Laukkanen, T.: Internet vs Mobile banking: comparing customer value perceptions. Bus. Process Manag. J. **13**(6), 788–797 (2007)
22. Laukkanen, T., Pasanen, M.: Mobile banking innovators and early adopters: How they differ from other online users? J. Financ. Serv. Mark. **13**(2), 86–94 (2008)

23. Luarn, P., Lin, H.H.: Toward an understanding of the behavioral intention to use mobile banking. Comput. Hum. Behav. **21**(6), 873–891 (2005)

24. Mallat, N., Rossi, M., Tuunainen, V.: An empirical investigation of mobile ticketing service adoption in public transportation. Pers. Ubiquit. Comput. **12**(1), 57–65 (2008)

25. Meuter, M.L., Bitner, M.J., Ostrom, A.L., Brown, S.W.: Choosing among alternative service delivery modes: an investigation of customer trial of self-service technologies. J. Mark. **69**(2), 61–83 (2005)

26. Migdadi, Y.K.A.: The developing economies' banks branches operational strategy in the era of e-banking: the case of Jordan. J. Emerg. Technol. Web Intell. **4**(2), 189–197 (2012)

27. Morgan, D.H.J.: Gender. In: Burgess, R. (ed.) Key Variables in Social Investigation. Routledge, London (1986)

28. Proença, J., Rodrigues, M.A.: A comparison of users and non-users of banking self-service technology in Portugal. Manag. Serv. Qual. **21**(2), 192–210 (2011)

29. Purwanegara, M., Apriningsih, A., Andika, F.: Snapshot on Indonesia regulations in mobile internet banking users' attitudes. Procedia Soc. Behav. Sci. **115**, 147–155 (2014)

30. Püschel, J., Mazzon, J.A., Hernandez, J.M.C.: Mobile banking: Proposition of an integrated adoption intention framework. Int. J. Bank Mark. **28**(5), 389–409 (2010)

31. Rana, N.P., Dwivedi, Y.K., Williams, M.D.: A meta-analysis of existing research on citizen adoption of e-government. Inf. Syst. Front. **17**(3), 547–563 (2015)

32. Rana, N.P., Dwivedi, Y.K., Williams, M.D., Lal, B.: Examining the success of the online public grievance redressal systems: an extension of the IS success model. Inf. Syst. Manag. **32**(1), 39–59 (2015)

33. Rana, N.P., Dwivedi, Y.K., Williams, M.D., Weerakkody, V.: Investigating success of an e-government initiative: validation of an integrated IS success model. Inf. Syst. Front. **17**(1), 127–142 (2015)

34. Rana, N.P., Dwivedi, Y.K.: Citizen's adoption of an e-government system: validating extended social cognitive theory (SCT). Gov. Inf. Quart. **32**(2), 172–181 (2015)

35. Riquelme, H.E., Rios, R.E.: The moderating effect of gender in the adoption of mobile banking. Int. J. Bank Mark. **28**(5), 328–341 (2010)

36. Rogers, E.M.: Diffusion of Innovations. The Free Press, New York (1995)

37. The Jordan Times.: Mobile phone penetration projected to reach 200 % (2013). http://jordantimes.com/mobile-phone-penetration-projected-to-reach-200 on 2 June 2015

38. Venkatesh, A., Shih, E., Stolzoff, N.: A longitudinal analysis of computing in the home. In: Sloane, A., van Rijn, F. (eds.) Home Informatics and Telematics, pp. 205–215. Springer, US (2000)

39. Venkatesh, V., Morris, M.G.: Why don't men ever stop to ask for directions? Gender, social influence, and their role in technology acceptance and usage behaviour. MIS Q. **24**(1), 115–140 (2000)

40. Venkatesh, V., Morris, M., Davis, G., Davis, F.: User acceptance of information technology: toward a unified view. MIS Q. **27**(3), 425–478 (2003)

41. Venkatesh, V., Thong, J.Y.L., Xu, X.: Consumer acceptance and use of information technology: extending the unified theory of acceptance and use of technology. MIS Q. **36**(1), 157–178 (2012)

42. Wessels, L., Drennan, J.: An investigation of consumer acceptance of M-banking. Int. J. Bank Mark. **28**(7), 547–568 (2010)

43. Zhou, T.: Understanding users' initial trust in Mobile banking: an elaboration likelihood perspective. Comput. Hum. Behav. **28**(4), 1518–1525 (2012)

44. Zhou, T., Lu, Y., Wang, B.: Integrating TTF and UTAUT to explain mobile banking user adoption. Comput. Hum.Behav. **26**(4), 760–767 (2010)

Young-Elderly Travellers as Potential Users and Actual Users of Internet with Mobile Devices During Trips

Niklas Eriksson[✉] and Susanna Fabricius

Department of Business Management and Analytics,
Arcada University of Applied Sciences, Jan-Magnus Janssonin aukio 1,
00560 Helsinki, Finland
{niklas.eriksson,susanna.fabricius}@arcada.fi

Abstract. The population is rapidly ageing in countries such as Finland. However, little research has been conducted to better understand older travellers' use of Internet and mobile devices. This qualitative study aims at exploring young-elderly (aged 60–75) travellers as potential users and actual users of the Internet with mobile device during trips. The results identify a range of possible drivers and barriers for the use of Internet with mobile devices and their impact on the travel experience during trips. The study also suggests that there is a substantial number of young-elderly travellers that are quite advanced in their mobile usage behaviour.

Keywords: Older travellers · Travel experience · Mobile devices · Technology adoption and use · User behaviour · Digital services

1 Introduction

Many travellers are nowadays acting as their own travel agents and they build their own travel packages and trip itineraries [21]. Mobile technology such as tablet devices and smart phones is taking the digital development even further. A large scale survey by Hjalager and Jensen [10] confirms that many travellers want to be online before the trip, during the trip and after the trip. However, individuals adopt technology based innovations (services) very differently [20]. One consumer segment, which is becoming very relevant, is the so called young-elderly (aged 60–75) segment [1, 4]. Population ageing is faced by most developed countries. For example in Finland the proportion of persons aged 65 or over in the population is estimated to rise from the present 18 % to 26 % by 2030 and to 28 % by 2060 [25]. Even globally the population aged 60 or over is the fastest growing [29]. However, only little research has been conducted on the influence of Internet in the older tourist market [17, 30]. On the other hand an ever growing number of older people in countries such as Finland are Internet users. This appears from Statistics Finland's survey on use of information and communications technology. In 2013, the upper age limit of the survey sample was raised from 74 to 89 years [26]. The number of mobile devices has also quickly increased in the oldest age groups. In fact, the

© IFIP International Federation for Information Processing 2015
M. Janssen et al. (Eds.): I3E 2015, LNCS 9373, pp. 24–35, 2015.
DOI: 10.1007/978-3-319-25013-7_3

adoption rates of smart phones for people aged 60–75 have passed 50 % in Finland [4]. Once older adults join the digital world it tends to become an integral part of their lives [23].

Based on the fact that individuals adopt and use online services differently, the lack of research of older tourists' online behaviour and the increased use of Internet and mobile devices in this segment, it seems necessary and timely to investigate young-elderly travellers' online behaviour. This study will focus on exploring young-elderly travellers as potential users and actual users of Internet with mobile devices during trips.

2 Literature

2.1 Online Activities and Older Travellers

Before the trip travellers are generally focused on assessing the destination and planning transportation and accommodation, whereas in the during trip settings the search strategies of tourists are primarily focused on planning the venue on-site, such as activities to undertake [22]. Seniors research the Internet for travel information in different ways, e.g. through Google searches and by visiting travel sites [27]. However, according to the same authors seniors rarely use social networks such as Facebook and Twitter for travel planning, but many of them have entered an 'e-buyer' era, where they actually buy travel products online. Generally speaking social media is growing in importance as an influence on the online travel information search process [34]. For example travel sites such as Tripadvisor provide global platforms for rating travel services and sharing experiences. Tourists often want to recall memories and share them with others with e.g. photos and stories both during the trip and after the trip [28]. Nowadays numerous travellers come pre-loaded with apps and content for mobile use. Mobile services have the potential to support tourists in different stages of the trip and mobile devices are used for many online travel activities, e.g. search, book and reflect [7]. Many travellers have developed new routines in during trip settings due to the use of a smart phone. These new routines include for example finding information about the travel, book tickets and taking photos and sharing with others immediately [33]. In fact using a personal mobile device for online purposes during trips is seen by many travellers as important [10]. Minazzi and Mauri [14] point out that the use of mobile devices and applications affect the travel experience in different stages of the traveller life cycle. On the other hand Pesonen et al. [17] found that Finnish seniors seldom use their mobile devices to connect to the Internet during trips. Others have also emphasized that we should be careful to overestimate the extent of travellers' use of mobile devices [11]. However, neither should we underestimate the number of technology savvy older travellers. Research conducted by Reisenwitz et al. [19] shows that seniors are online more hours and more frequently and feel more comfortable online. Niemelä-Nyrhinen [15] concluded that elderly (baby boomers) in Finland have, generally speaking, a low level of technology anxiety.

2.2 Drivers and Barriers of Technology Use and Impact on Travel Experience

Several models have been developed to better understand individuals' adoption and use of information technology. One of the most widely used models is the technology acceptance model (TAM) by Davis [6] which is based on the theory of reason action (TRA) by Fishbein et al. [8] with routes in psychology theories and its extension theory of planned behavior (TPB) by Ajzen [2]. Other often used models in technology adoption research are the diffusion of innovations theories (DOI) by Rogers [20], the unified theory for the acceptance and use of technology (UTAUT) by Venkatech et al. [31] and the unified theory for consumers' acceptance and use of information technology (UTAUT2) by Venkatech et al. [32]. These models have been applied to explain the acceptance of different types of technologies by older adults [5]. Mallenius et al. [12] found through the lens of UTAUT, that Expected benefit, Perceived security, Usability, Anxiety, Training, Guidance, Price barriers and Social influence are relevant when investigating elderly individuals' mobile device and service use in Finland.

It should be noted that when investigating adoption and use of technology it is very important to distinguish between *potential users* and *actual users*. Gerpott [9] found that innovation-based attributes explained mobile Internet acceptance better for actual users than for potential users. Recent studies have also emphasized the need for research on the impact of technology on the tourist experience among those who actually uses their mobile devices [16]. The same authors concluded that emotional responses, missed opportunities, monetary burden and behavioural consequences may lead to negative travel experience effects. In fact, Wang et al. [33] emphasized that not only drivers of smart phone use in travel should be examined in order to understand its impact on travel experience, but also barriers (resistance) to use. Experiences of technology while travelling may also induce anxieties and tensions, due to technology addiction [18].

3   Methodology

We conducted a semi-structured qualitative study with 14 pensioners within the age group 60–75 to better understand young-elderly travelers as potential users and as actual users of the Internet with mobile devices during trips. The sample was drawn from a Swedish speaking pensioners' association in Helsinki Finland. In order to gain as broad as possible view of the target segment both low-proficiency users of the Internet and high-proficiency users of the Internet were selected in collaboration with the association. Also low frequency travelers and high frequency travelers were selected. The selection of the sample can hence be described as purposive [13]. All informants were strangers to the two interviewers and the interviews lasted for about an hour on average. The interviews were conducted in spring 2015 and according to ethical standards and confidentiality by providing advance information to the informant regarding the purpose of the study, as well as about who will have access to the data and confidentiality guidelines of the project.

An instrument was developed to guide the interviews. The questions analyzed in this study are listed below:

- Describe how you mostly search for information during your trips
- Do you use the Internet with your smart phone and/or tablet device during your trips?
- What do you do with them? Which device do you prefer?
- Is there a difference in how you use the device abroad compared to your use in Finland?
- Why do you use the Internet with your smart phone and/or tablet device during your trips?
- Why do you not use the Internet with your smart phone and/or tablet device during your trips?
- Do you see that that the use of the Internet with a smart phone and/or tablet device during your trips impacts your travel experience?
- How do you share your travel experiences during your trips?

The questions are flexible in character which allowed us to account for individual differences and take the advantage of the iterative nature of interviewing [13]. The interviews were conducted in Swedish and voice recorded. Direct citations have been translated to English by the authors.

The sample may be biased towards people with a higher educational degree and a larger household income than the average Finnish 60–75 aged individuals (cf. [17]). On the other, hand our purpose was to select as diverse informants as possible based on travel frequency and online proficiency and not based on educational and economic situation. From Table 1 we can see that the informants have a wide distribution in perceived online proficiency, annual travel frequency and numbers of trips during the past three years. Furthermore we are not aiming at generalizing results but to explore possible drivers and barriers for the potential use and actual use of the Internet with mobile devices during trips.

4   Results

According to the theory discussed and based on the results we have divided the sample into *potential users* and *actual users* of the Internet with mobile devices in during trip settings. Five informants (F1, F2, M7, F11 and M13) belong to potential users and nine (F3, F4, M5, M6, M8, F9, F10, M12 and F14) belong to actual users. With the theoretical discussion on technology use and impact on travel experience in mind we analysed the results of the two groups.

4.1   Potential Users

The results are sub-divided into themes that represent the drivers and barriers of potential users of the Internet with mobile devices during trips.

Expected Added-Value. F2, M7 and F11 said that they see themselves using a mobile device, either smart phone or tablet device, during the trip in the near future. They all three see potential added value by using a mobile device during a trip, e.g. fast and convenient access to information, locating themselves on a map. In fact two of them (F11, M7) had already purchased a smart phone, but had not taken it into use. F11 said that the smart phone was purchased specifically for the needs of an upcoming trip

Table 1. Interview informants

Nr	Gender	Age	Retired	Education	Annual household income	Family	Travel Frequency	Trips past 3 years, Domestic	Trips past 3 years, Abroad	Perceived online proficiency	Internet importance during trip	Main purpose to travel	Main travel style	Preferred destination	Main travel partner	Barriers to travel
F1	Female	75	2003	Vocational	12.000	Living alone	Once a year	8	3	Poor	Not at all	Holiday, visit friends, children / grandchildren	Self-organized	New and Known	Spouse	None
F2	Female	69	2008	Vocational	25.000	Living alone	2-7 times a year	15	10	Moderate	Slightly	Senior dance trips	Packaged trips	New	Senior dancers	None
F3	Female	71	2010	University	90.000	Married	2-7 times a year	10	12	Excellent	Very	Holiday, visit friends, children / grandchildren	Self-organized	New and Known	Spouse	None
F4	Female	62	-	Vocational		Widow	2-7 times a year	1	2	Good	Medium	Holiday, Senior dance trips	Packaged trips	New and Known	Friends, Children / grandchildren	None
M5	Male	62	-	University	90.000	Married	2-7 times a year	4	11	Moderate	Very	Holiday, Work	Self-organized	New	Friend	Economy
M6	Male	73	2005	Vocational	100.000	Married	2-7 times a year	10	10	Moderate	Very	Holiday, visit friends, children / grandchildren	Self-organized	Known	Spouse	None
M7	Male	74	2002	Vocational	50.000	Married	>= once/ month	4	3	Good	Not at all	Senior trips	Packaged trips	Known	Pensioners	None
M8	Male	60	2014	University	35.000	Living alone	2-7 times a year	5	10	Moderate	Medium	Holiday and visit friends	Self-organized	New	Alone	None
F9	Female	60	2014	University	110.000	Married	2-7 times a year	4	7	Moderate	Extremely	Holiday and visit friends	Self-organized	New and Known	Friend and spouse	Economy
F10	Female	72	2003	University	80.000	Married	2-7 times a year	6	3	Mediocre	Very	Holiday and visit friends	Self-organized	New	Friend	Family
F11	Female	69	2006	University	55.000	Married	>= once/ month	25	1	Mediocre	Not at all	Summer house	Self-organized	Known	Alone	Family
M12	Male	69	2013	Vocational	70.000	In relationship	2-7 times a year	10	7	Excellent	Extremely	Holiday and visit friends	Self-organized	New	Spouse	None
M13	Male	72	2005	Vocational	60.000	Married	Once a year	2	4	Excellent	Very	Holiday	Packaged trips	New	Friends, Spouse, Children / grandchildren	Health
F14	Female	71	2008	University	45.000	Widow	2-7 times a year	0	15	Good	Medium	Holiday, visit friends, children / grandchildren	Self-organized	New and Known	Alone, Friends, Children / Grandchildren	Economy

"I need to be able to locate myself on a map during my trip to Spain and therefore I have purchased a smart phone, but I have not taken it into use yet." (F11)

F1 on the other hand found it more challenging to perceive the travel experience enhancement of using smart phones, rather she finds them making people anti-social and distracted from the world around them.

"I cannot really see the purpose of people sitting in trains, busses etc. with their heads down .... Phones should be used only as phones ..." (F1)

M13 said he owns an old smart phone that he does not use, because it is slow and he does not really need it.

Travel Style and Partner. F1 and F2 see that their travel style influences their behaviours and that they form different roles with their travel partners during the trip. In fact, because their travel partner uses a mobile device they do not need to use one themselves.

"I mostly travel together with my sister... My sister is searching for information with a mobile device during trips so I don't really need to." (F2)
"When we were in Stockholm my friend used her tablet device to find information on where to go and what to do... I do the talking with locals and she looks for information." (F1)

M7, F1 and F2 mostly go on packaged trips (see Table 1). M7 sees that because he mostly takes part in organised senior-trips he does not really need a mobile device to arrange things during the trip. F2 on the other hand sees herself using a personal mobile device when she goes on trips without her sister.

"I probably need to start using one, as I cannot go on every trip with my sister." (F2)

Knowledge and Guidance. Four of the informants (F1, F2, M7 and F11) expressed that technology in general can be struggling and frustrating to use. They admit that their personal technology skills are limited but that it can be hard for them to admit to others that they have problems using technology appliances. M7 said that he has a smart phone waiting for him, but he has not taken it into use. The reason is that he needs help in learning how to use it.

"I have been struggling to make the effort to get it up and running. I should ask help from my daughter so that I will be able to use it." (M7)

Anxiety. Anxiety towards technology does not seem to be a hindrance for F2, F11 and M13 to become actual mobile device users during trips, as they are looking for and booking travel on the Internet in the pre-trip stage. F11 is mostly purchasing routine trips on the Internet (train, ferry and bus tickets) and F2 and M13 have purchased flights, accommodation and travel packages online. M7 is using the Internet for finding travel information but he has not tried to purchase travel online. M7 feels more comfortable using traditional travel agents to make trip arrangements and prefers to go on packaged senior-trips, rather than to organize them himself online.

"I call the local travel agent to make trip arrangements... I can trust them as I have used them a lot before." (M7)

F1 concluded that she is trying to stay away from technology appliances and even her children have told her to stay away from technology. She even considered her-self lucky to be out of working life due to so much now being computer-based and she could not cope with that. Her only point of contact with the Internet is if someone wants her to check some information on a web-site.

4.2   Actual Users

The results are sub-divided into themes that represent actual drivers and barriers of using the Internet with mobile devices during trips and their impact on travel experience.

Added-Value. The informants use Internet with mobile devices during trips to find local sights, check reviews, weather and opening hours, find the shortest routes to places etc. Map services are the most widely used mobile services. The informants find the mobile device convenient and time-saving as they have instant access to information in any situation. F14 said that she is very attached to her iPhone and F9 that it feels like she has inside information about local places as she can check other travelers' experiences online with her mobile device. This kind of information was not available to her before or it was not easily accessible. However, it can also have a negative effect on the travel experience as it may passivate travelers from asking locals etc., making travelers actually missing out on experiences.

> "…on the other hand mobile devices may passivate us to ask locals, as we start to believe that we already know everything about the place." (F9)

In fact, M8 sometimes intentionally leaves his mobile device out and asks e.g. the hotel reception for local tips as they may have some really valuable knowledge to offer. According to M8 visiting the online top rated attractions may not give him the authentic travel experience he is looking for. Also digital map services may be found too efficient.

> "Nowadays we do not get lost and stumble into interesting things like we used to do." (F3)

The informants also kill or fill time in transportation by using their mobile devices. Only one informant (M12) mentioned that he uploads pictures with his mobile device in social media during trips. All other informants prefer to share their experiences after the trip. However, social media is not extensively used for sharing travel experiences after the trip either, primarily due to unwillingness to share private information.

Mobile devices also give the informants a feeling of safety, that the traveler can be reached and that they can reach travel partners and people at home (e.g. SMS, e-mail, WhatsApp, Facebook). On the other hand the awareness of things at home may also increase tensions during the trip (see technology tensions below).

Travel Style. All nine informants reported that they mainly organize their trips themselves, rather than taking part in packaged trips (see Table 1). They find the mobile device is a key tool for their style of traveling and experiencing things. The following two narratives represent their responses well.

> "I want to organize things myself and then the mobile device comes in handy." (M8)

"I can imagine that if I would take part in an organized trip I would not really need my mobile device as everything then is organized." (F9)

Some of them even feel reluctant to take part in organized senior-trips. It may even be hard for them to see or they don't want to see themselves as seniors by definition.

"That sounds a little bit boring... I think that is for people who cannot travel in another way....it is good that they are arranged but that is not for me." (M8)

Usability. All nine informants found mobile devices to be mostly easy to use and bring along while travelling. Most of them bring both a tablet device and a smartphone with them on the trip. However, there are different situations for using these two devices to enhance the travel experience. The tablet device is mostly used at the hotel (F3, F10, M5). For example M12 uses the tablet device at the hotel e.g. in the morning to plan what to see, while he carries his smart phone while wandering around during the day. The smartphone is smaller and therefore easier to carry around.

"I don't think I need to bring my tablet device on my next trip as we are going to backpack... the smartphone is easier to carry along." (F10)

On the other hand M5 and F3 find the tablet device easier to use due to its screen size. Aging may cause changes in visual capacity and other restrictions for self-arrangements.

"The tablet device is more convenient and sharper, I can see better with it." (M5)

One other hand F14 concluded that she does not need a tablet device as on her iPhone she can e.g. re-size the text.

Many stated that mobile devices are an important information and communication channel during trips nowadays. In fact, mobile devices have to some extent replaced e.g. traditional brochures, paper maps and sending postcards. As M6 stated

"They are so versatile, one can do anything with them; take pictures, search for information, communicate..." (M6)

On the other hand F9 finds it important to have a backup plan e.g. paper map just in case something goes wrong with the mobile device. She feels that she cannot totally rely on her mobile devices. Many of M6s friends find it very nice if he sends them a traditional postcard as no one else is doing that nowadays.

Monetary Burden. Informants said that roaming charges abroad are a problem and depending on the destination they worry about them. For example M12 and F14 have partly guarded themselves from roaming charges in the Nordic and Baltic countries by purchasing a subscription that allows for data transfer at the same price as in Finland. The following three narratives represent well the informants' worries about the potential monetary burden.

"I have to turn off some updates on my phone while travelling abroad." (F10)

"It is better to be careful with the use abroad so that I don't have a big bill waiting for me at home." (F4)

"When I went to the US I had to set the device in flight mode due to terrible roaming-costs." (F3)

Tensions. Both M8 and M12 say that it is very important that the hotel has a proper Wifi, it can even be decisive of whether they will stay at the hotel or not. M12 feels that he will complain if the Wifi is not working properly or if it is over-charged.

"… I remember once in Sweden they over-charged for the Wifi, then I posted a complaint about that." (M12)

M12 brings his mobile devices everywhere, except perhaps to the beach. M12 also admits that he is addicted to his devices.

"I have to admit that I'm addicted …. I feel half naked if I forget my phone." (M12)

Some of them also admit that it can be frustrating to be in network-dead zones, being unable to connect to the Internet. Informants also mentioned that mobile devices make them not really getting away from home (they read what is going in the news at home, read their e-mails, Facebook updates etc.).

"One should turn off the phone…there is not really a need to be online all the time…" (F10)

5   Discussion and Conclusions

First we studied the potential users (five informants) of the Internet with mobile devices during trips. We found a range of possible drivers and barriers, but we sub-divided the results according to four thematic factors influencing the potential use; expected added-value, travel style and partner, knowledge and guidance and anxiety. Generally speaking there was a consensus in this group that there is added-value to use the Internet with mobile devices during a trip. Neither did technology anxiety seem to be a great hurdle, except for one informant. In fact, usage barriers seemed to be more related to their style of traveling, their personal knowledge of using technology and availability of support by e.g. a younger family member. This is in line with previous research that some guidance may be needed to push the adoption and use of mobile device and services by elderly [12]. It may, however, be hard for this group to admit that they need help with technology appliances, which may constitute a greater barrier than being anxious about learning new technologies.

Second we studied the actual users (nine informants) of the Internet with mobile devices during trips. Also in this group we identified a range of possible drivers and barriers, but we sub-divided the results according to five thematic factors influencing the actual use and the travel experience; added-value, travel style, usability, monetary burden and tensions. The findings indicate there is a consensus in the group that there definitely is added-value in using the Internet and mobile devices during trips and it affects the travel experience positively. However, sharing their travel experiences with their mobile devices in social media is not widely used. They also see that they may passivate as travelers and that they may miss out on authentic travel experiences due to their use of mobile devices. Usability is generally not a problem, rather they reflect about which mobile device(s) to use and bring along. Two of them preferred a tablet device (larger screen) due to restrictions in visual capacity. They all found that a mobile device is handy for their most common style of traveling, self-organized trips. The usage barriers and the negative effects on travel experience are to be found in technology tensions that are caused by e.g. roaming costs in international travel contexts, poor Wifi and dead zones. However, these usage barriers or travel experience barriers are also typical for younger travelers [7, 16]. One of the informants even admitted that he is addicted to his mobile devices.

To sum it up, in this study the conventional description of seniors as technology anxious and highly reluctant to use technology [15, 17] was not widely recognized. Rather this study suggests that there is a substantial number of young-elderly travelers' who are quite advanced in using the Internet with mobile devices during trips. Hence, the study indicates that Internet is nowadays an integral part of many young-elderly travelers' travel experience. However, we should not generalize the results to a total population due to the explorative and qualitative research approach and the purposive sample selection. Quantitative research could empirically verify the results on a larger scale and test dependency between the variables suggested in this study. The analysis of the interview results could also be extended and interpreted to a theoretical model. Limiting the sample to 60–75 year old seniors, here referred to as young-elderly, obviously decreases the importance of health and physical capacity issues to use technology. A 75 + sample may indeed give us very different results [23]. On the other hand, according to Mallenius et al. [12], age is not really the key, rather it is the functional capacity that matters when determining how mobile devices and services are perceived. When studying the behavior of elderly consumers, not only the chronological age should be discussed, but also the cognitive age (how old one "feels" to be) [3, 24]. In this study a substantial number of the informants could not see themselves as seniors by definition.

References

1. Allmér, H., Råberg, M.: Young-elderly and digital use. In: IRIS36 - 36th Information Systems Research Seminar in Scandinavia: "Digital Living", Oslo, Norway (2013)
2. Ajzen, I.: The theory of planned behavior. J. Organ. Behav. Hum. Decis. Process. **50**, 179–211 (1991)
3. Barak, B., Leon, G.S.: Cognitive age: a non chronological age variable. NA – Adv. Consum. Res. **08**, 602–606 (1981)
4. Carlsson, C., Walden, P., Vogel, D., Merne, M.: Young elderly – progressive market for digital service? In: Panel Discussion in the 27th Bled eConference, Bled, Slovenia (2014)
5. Chen, K., Chan, A.H.S.: A review of technology acceptance by older adults. Gerontechnology. **10**(1), 1–12 (2011)
6. Davis, F.D.: Perceived usefulness, perceived ease of use and user acceptance of information technology. J. MIS Q. **13**(3), 319–340 (1989)
7. Eriksson, N.: User categories of mobile travel services. J. Hospitality. Tour. Technol. **5**(1), 17–30 (2014)
8. Fishbein, M., Ajzen, I.: Belief, Attitude, Intention and Behavior: An Introduction to Theory and Research Reading. Addison-Wesley, Massachusetts (1975)

9. Gerpott, T.J.: Attribute perceptions as factors explaining mobile internet acceptance of cellular customers in Germany – an empirical study comparing actual and potential adopters with distinct categories of access appliances. Expert Syst. Appl. **38**(3), 2148–2162 (2011)
10. Hjalager, A.-M., Jensen, J.M.: A typology of travellers based on their propensity to go online before, during and after the trip. In: Fuchs, M., Ricci, F., Cantoni, L. (eds.) Information and Communication Technologies in Tourism 2012, Helsingborg, Sweden, pp. 96–107. Springer, Wien (2012)
11. Douglas, A., Lubbe, B.: Mobile devices as a tourism distribution channel. In: Tussyadiah, I., Xiang, Z. (eds.) Information and Communication Technologies in Tourism 2014, Dublin, Irland, pp. 855–867. Springer, Wien (2014)
12. Mallenius S., Rossi M., Tuunainen V.K.: Factors affecting the adoption and use of mobile devices and services by elderly people -results from a pilot study. In: 6th Annual Global Mobility Roundtable, Los Angeles (2007)
13. McGehee, N.G.: Interview techniques. In: Dwyer, L., Gill, A., Seetaram, N. (eds.) Handbook of Research Methods in Tourism. Edward Elgar, Cheltenham (2012)
14. Minazzi, R., Mauri, A.G.: Mobile technologies effects on travel behaviours and experiences: a preliminary analysis. In: Tussyadiah, I., Inversini, A. (eds.) Information and Communication Technologies in Tourism 2015, pp. 507–522. Springer, Lugano (2015)
15. Niemelä-Nyrhinen, J.: Baby boom consumers and technology: shooting down stereotypes. J. Consum. Mark. **24**(5), 305–312 (2007)
16. Neuhofer, B., Buhalis, B., Ladkin, A.: Technology as a catalyst of change: enablers and barriers of the tourist experience and their consequences. In: Tussyadiah, I., Inversini, A. (eds.) Information and Communication Technologies in Tourism 2015, Lugano, Switzerland, pp. 789–802. Springer, Switzerland (2015)
17. Pesonen, J., Komppula, R., Riihinen, A.: Senior travellers as users of online travel. In: Tussyadiah, I., Inversini, A. (eds.) Information and Communication Technologies in Tourism 2015, Lugano, Switzerland, pp. 831–846. Springer, Switzerland (2015)
18. Paris, C.M., Berger, E.A., Rubin, S., Casson, M.: Disconnected and unplugged: experiences of technology induced anxieties and tensions while traveling. In: Tussyadiah, I., Inversini, A. (eds.) Information and Communication Technologies in Tourism 2015, Lugano, Switzerland, pp. 803–816. Springer, Switzerland (2015)
19. Reisenwitz, T., Iyer, R., Kuhlmeier, D., Eastman, J.: The elderly´s internet usage: an updated look. J. Consum. Mark. **24**(7), 406–418 (2007)
20. Rogers, E.M.: Diffusion of Innovations, 5th edn. Free Press, New York (2003)
21. Sigala, M.: Measuring customer value in online collaborative trip planning processes. Mark. Intell. Planning **28**(4), 418–443 (2010)
22. Sirakaya, E., Woodside, A.G.: Building and testing theories of decision making by travellers. Tour. Manag. **26**(6), 815–832 (2005)
23. Smith, A.: Older Adults and Technology Use. PewResearchCentre (2014). http://www.pewinternet.org/2014/04/03/older-adults-and-technology-use/. Accessed 15 April 2015
24. Sudbury, L., Simcock, P.: A multivariate segmentation model of senior consumers. J. Consum Mark. **26**(4), 251–262 (2009)
25. Statistics Finland: Population projection. http://tilastokeskus.fi/til/vaenn/2012/vaenn_2012_2012-09-28_tie_001_en.html. Accessed 5 March 2015
26. Statistics Finland: Over one-quarter of persons aged 75 to 89 use the Internet. http://www.stat.fi/til/sutivi/2013/sutivi_2013_2013-11-07_tie_001_en.html. Accessed 15 January 2015

27. Thébault, M., Picard, P., Ouedraogo, A.: Seniors and tourism: an international exploratory study on the use of the internet for researching recreational information use. Int. Bus. Res. **6**(3), 22 (2013)
28. Tussyadiah, I.P., Fesenmaier, D.R.: Mediating tourists experiences-access to places via shared videos. Ann. Tour. Res. **36**(1), 24–40 (2009)
29. United Nations: World Population ageing (2013). http://esa.un.org/unpd/wpp/Documentation/pdf/WPP2012_HIGHLIGHTS.pdf. Accessed 5 March 2014
30. Vigolo, V., Confente, I.: Older tourists: an exploratory study on online behavior. In: Tussyadiah, I., Xiang, Z. (eds.) Information and Communication Technologies in Tourism 2014, Dublin, Ireland, pp. 439–449. Springer, Wien (2014)
31. Venkatesh, V., Morris, M.G., Davis, G.B., Davis, F.D.: User acceptance of information technology: toward a unified view. J. MIS Q. **27**(3), 425–478 (2003)
32. Venkatesh, V., Tong, J.Y.L., Xu, X.: Consumer acceptance and use of information technology: extending the unified theory of acceptance and use of technology. J. MIS Q. **36**(1), 157–178 (2012)
33. Wang, D., Xiang, Z., Fesenmaier, D.R.: Smartphone use in everyday life and travel. J. Travel Res. 1–12 (2014)
34. Xiang, Z., Gretzel, U: Role of social media in online travel information search. Tour. Manag. **31**(2), 179–188 (2010)

An Empirical Study on the Adoption of Online Household e-waste Collection Services in China

Shang Gao[1(✉)], Jinjing Shi[2], Hong Guo[3], Jiawei Kuang[2], and Yibing Xu[2]

[1] Department of Computer and Information Science,
Norwegian University of Science and Technology, Trondheim, Norway
shanggao@idi.ntnu.no
[2] School of Business Administration,
Zhongnan University of Economics and Law, Wuhan, China
{ariel1236,xyb_mx2013}@163.com,
harveykuang.yali@gmail.com
[3] School of Business Administration, Anhui University, Hefei, China
homekuo@gmail.com

Abstract. Online household e-waste collection services are emerging as new solutions to disposing household e-waste in China. This study aims to investigate the adoption of online household e-waste collection services in China. Based on the previous technology diffusion theories (e.g., TAM, UTAUT), a research model with six research hypotheses was proposed in this research. The research model was empirically tested with a sample of 203 users of online household e-waste collection services in China. The results indicated that five of the six research hypotheses were significantly supported. And the most significant determinant for the behavioral intention to use online household e-waste service was effort expectancy. However, facilitating condition did not have significant impact on users' behavior of using online household e-waste collection services.

Keywords: Adoption · UTAUT · Online household · e-waste collection services

1 Introduction

Today, electrical and electronic products become increasingly important in peoples' daily lives. However, it also produces a tremendous amount of e-waste in the mean time. To build a smart and sustainable city, the disposal of waste is a crucial aspect that should be taken into consideration. For example, E-waste has become a serious problem in China in terms of both quantity and toxicity, exacerbated by the development and advancement of electronic industry. A large amount of e-waste was generated in China due to the fast consumption rates of electrical and electronic products. E-waste was largely collected by the informal sector. According to the previous study (e.g., [22]), it was found that the majority of electrical and electronic

© IFIP International Federation for Information Processing 2015
M. Janssen et al. (Eds.): I3E 2015, LNCS 9373, pp. 36–47, 2015.
DOI: 10.1007/978-3-319-25013-7_4

products were often sold to peddlers. However, most peddlers tended to focus on waste fractions of the collected e-waste with which they can make a profit instead of focusing on raised environmental problems.

With the rapid economic growth in China, consumers' demand for electrical and electronic products is booming simultaneously. We have witnessed exponential increase of the quantity of output of four major household appliances (e.g., TV) in the past two decades. For instance, the output of TVs has increased to 133.82 million units in 2014, from 33.70 million units in 1998. When these household appliances eventually reach the ends of their life circles, they will become obsolete and pose a potential threat to both natural environment and human health if they were not properly disposed. The recycling chain for e-waste is classified into three main subsequent steps: collection, sorting/dismantling and pre-processing (including sorting, dismantling and mechanical treatment) and end-processing [20]. Being the first step of the whole recycling chain, collection serves as a crucial process of e-waste recycling. Currently, there exist six major options for household e-waste disposal: second use, storage, discarding, disassembly, formal collection and informal collection [6]. Online household e-waste collecting and recycling platform is aiming to become a convenient way of collecting e-waste and to offer higher recycling price than official price notwithstanding. However, it seems that most Chinese electrical and electronic products consumers are not familiar with this platform.

Online household e-waste collection revolutionizes the traditional ways of collecting and disposing e-waste. It integrates information flows, logistics and capital flows, constructing a closer connection between e-waste sellers and recyclers, which facilitates the e-waste collecting process and ultimately improves the effectiveness and efficiency of the whole recycling chain for e-waste.

The objective of this research is to investigate consumers' behavior intention to use online household e-waste collection services in China. Based on previous technology diffusion theories (e.g., TAM [7], UTAUT [26]), a research model with six research hypotheses was proposed. And the research model was empirically tested using data collected from a survey of users of an online household e-waste collection service called Taolv365 in China.

The remainder of this paper is organized as follows. The related literature review is reviewed in Sect. 2. The research model and hypotheses are presented in Sect. 3. In Sect. 4, an empirical study is carried out to examine the research model. This is followed by a discussion of the findings and limitation of the study in Sect. 5. Section 6 concludes this research work and points out directions for future research.

2 Literature Review

2.1 Technology Diffusion Theories

An important and long-standing research question in information systems research is how to accurately explain user adoption of information systems [8]. Several models have been developed to test the users' attitude and intention to adopt new technologies or information systems. These models include the Technology Acceptance Model

(TAM) [7], Theory of Planned Behavior (TPB) [1], Innovation Diffusion Theory (IDT) [19], Unified Theory of Acceptance and Use of Technology (UTAUT) [26], and Mobile Services Acceptance Model (MSAM) [11, 12]. UTAUT was developed through a review and consolidation of the constructs of eight models (e.g., TAM, TRA [9], TPB, IDT) that earlier research had employed to explain IS usage behavior. Moreover, variables that influence users' behavioral intention employed in TAM3 [23] were classified into the following four types: individual differences, system characteristics, social influence and facilitating conditions [24].

Due to its unified perspective, although UTAUT has not been as widely used as TAM, partly resulting from its complexity and intricacy to test its applicability, it has gradually drawn researchers' attentions and has been recently applied to exploring user acceptance of mobile technologies and mobile devices (e.g., [4, 14, 17]). We built our research model based on UTAUT to assess the adoption of online household e-waste collection services in China.

2.2   Research on e-waste Collection Services in China

As the largest exporter of electrical and electronic products and importer of waste electrical and electronic products around the world, China plays a key role in the global life cycle of electronics [28]. Being concerned about the dispose of e-waste, many studies have been conducted during the last few decades.

As shown in Table 1, previous studies have provided us with different aspects to understand the current situations of e-waste collection and recycling system in China, including the legislative policy, e-waste collection channels, household recycling behaviors, etc. However, they were mainly focusing on the traditional methods of e-waste dispose while neglecting online e-waste collection services.

This research aims to complement and extend existing studies on the adoption of e-waste collection services by focusing on the emerging channel of disposing e-waste: online household e-waste collection services. The online household e-waste collection services can be seen as information systems. Therefore, we would like to use the existing technology diffusion theories to examine the adoption of online household e-waste collection services in China.

3   Research Model and Hypotheses

A research model that identifies important factors that impact users' intention to use online household e-waste collection services was developed in this research. The proposed research model (see Fig. 1) is a simplified version of UTAUT. We have developed the following six research hypotheses (labeled in Fig. 1) based on the research model.

Performance Expectancy and Effort Expectancy. Performance expectancy is defined as the degree to which an individual believes that using the system will help him or her to attain gains in job performance [26]. It reflects user perception of performance improvement by using online household e-waste collection services such as convenience, fast response, and service effectiveness. The performance expectancy

Table 1. Literature review on e-waste collection system in China

Literature	Research purpose	Findings
Streicher-Porte, Geering, 2009 [22]	Analyze household data and compares literature sources about how Chinese households dispose of obsolete electrical and electronic equipment.	The findings indicated that the informal collection sector of obsolete EEE in China mainly in terms of door-to-door collection by peddlers played a major role in the current management of this waste stream.
Yu, Williams, Ju, & Shao, 2010 [28]	Review the existing framework for e-waste management in China including regulatory policies and pilot projects.	Two alternate policies were proposed: shared responsibility with deposit to incentivize consumer participation and integrating informal collection/reuse with formal dismantling/recycling.
Chi, Streicher-Porte, Wang, & Reuter, 2011 [5]	Gather information on informal e-waste management, take a look at its particular manifestations in China	The findings revealed the actual situation of e-waste management in China. Moreover, it indicated that the improvement of informal recycling sector lied on dedicated efforts from economic, technical and social aspects.
Chi, Wang, & Reuter, 2014 [6]	Investigate the collection channels of e-waste and household recycling behaviors in Taizhou city of China.	The authors suggested that rather than directly competing with the informal collection sector, a better solution was to harness its strengths and incorporate it into a more accountable and regulated e-waste collection system.

construct is the strongest predictor of behavioral intention and remains significant at all settings [26]. When users feel that online household e-waste collection services are useful and can bring them convenience to dispose their e-waste, their intention to use to the services are likely to be high.

Effort expectancy is defined as the degree of ease associated with the use of the system [26]. It reflects user perception of how difficult it is to use online household e-waste collection services. According to earlier research (e.g., [26]), effort expectancy positively affects performance expectancy. When users feel that online household e-waste collection services are easy to use and do not require much effort, they will have a high expectation toward acquiring the expected performance. Otherwise, their performance expectancy is likely to be low. Thus, we formulate the following research hypotheses:

H1: Effort expectancy has a positive influence on performance expectancy.

H2: Performance expectancy has a positive influence on users' behavioral intention of using online household e-waste collection services.

H3: Effort expectancy has a positive influence on users' behavioral intention of using online household e-waste collection services.

Social Influence. Social influence is similar to subjective norm of TRA and reflects the effect of environmental factors such as the opinions of a user's friends, relatives, and superiors on user behavior [15]. In addition, recommendation as well as word-of-mouth effect will also affect consumers' behavioral intention [10]. Thus, we formulate the following research hypothesis:

H4: Social influence has a positive influence on users' behavioral intention of using online household e-waste collection services.

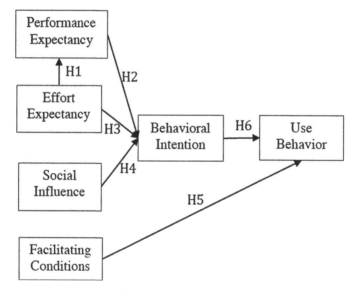

Fig. 1. Research model

Facilitating Conditions. Facilitating conditions are similar to perceived behavioral control of TPB and reflect the effect of a certain user's knowledge, ability, and resources [26]. Online household e-waste service is a rather new service for e-waste recycling, but similar to other e-commerce services. In order to use this service, users need some basic infrastructures like PC, smart phones, Internet, etc. In addition, users need to bear usage costs such as data service and logistics fees when using online household e-waste collection services. If users do not have these necessary devices as well as financial resources and operational skills, they are not likely to adopt or use online household e-waste collection services. Thus, we formulate the following research hypothesis:

H5: Facilitating conditions have a positive influence on users' behavior of using online household e-waste collection services.

Behavior Intention. Consistent with the underlying theory for all of the intention models [21], we expect that behavioral intention will have a significant positive influence on usage behavior [26]. If the intention of a certain user to use online household e-waste collection services is high, he or she will be more likely to use this service. Otherwise, the user is unlikely to adopt this service. Thus, we formulate the following research hypothesis:

H6: Behavioral intention has a positive influence on use behavior.

4 An Empirical Study with the Research Model

In this empirical test, our research model was examined through the use online household e-waste collection service called Taolv365.

4.1 The Online Household e-waste Collection Service: Taolv365

Taolv365 aimed to build online household recycling system with Chinese characteristics. It is one of the first pioneers using the Internet as the basis for the reverse supply service platform. In addition, it has largest amount of registered users in China. Therefore, this website is a good case for instrument testing and identifying relevant implications for practitioners.

Based on the national environmental standards, Taolv365 gradually establishes the system of online household e-waste collection services to manage the chaotic e-waste collection market. Taolv365 also aimed to deal with some existing recycling problems (e.g., the lack of channels). Their ambition is to create an intelligent and green recycling system to collect e-waste in a more convenient and environmental-friendly way.

Taolv365 provides the following three services to the users:

- Information Service: Taolv365 can provide information to the clients in the following channels: call center, online messaging service, and the third platform collecting channels which can offer 7days/24 h help to users. Anyone who has e-waste can choose one of the channels for help.
- Price Inquiry Service: E-waste is classified on Taolv365. Users can easily get the price of their e-waste according to the standard classification system. It also provides SMS platform for immediate feedback, professional customer service so that users can timely grasp of recycling market and trends. Users can also donate their devices or electronics to charities via Taolv365.

- The Transaction Channel Service: The website integrates recycling industry buyers through formal channels. The buyers bid online, commit the transaction volume and shipments, and classify the standards carefully so as to achieve the optimal price. As long as submitting the order, the service can get the real-time tracking of the logistics, trading and payment situations.

4.2   Instrument Development

The validated instrument measures from previous research were used as the foundation to create the instrument for this study. Previous studies were reviewed to ensure that a comprehensive list of measures were included. In order to ensure that the instrument better fit this empirical study, some minor words changes were made to ensure easy interpretation and comprehension of the questions. All the items were adopted from prior studies [26] and modified to fit the domain of online household e-waste collection services. As a result, 17 measurement items[1] were included in the questionnaires. In addition, a seven-point Likert scale, with 1 being the negative end of the scale (strongly disagree) and 7 being the positive end of the scale (strongly agree), was used to examine participants' responses to all items in the survey.

4.3   Samples

The data for this study was collected through paper-based questionnaires from 20th April to 30th April 2015 in the biggest city in the central China. People were asked to participate in the survey voluntarily. Firstly, we explained who we were, what they were supposed to do during our survey, and the purpose of the survey. The participants were also informed that the results would be reported only in aggregate and their anonymity would be assured. After participants experienced the online household e-waste collection service provided by Taolv365, they were asked to complete the questionnaires and submitted them to us. 210 completed questionnaires were collected, among which 203 of them were valid questionnaires (i.e., valid respondent rate 96.7 %). Among the participants, 86 of the participants were male, and 117 were female. In terms of age, 120 participants were 25 years old and under 25 years old, while 83 participants were over 25 years old.

4.4   Descriptive Results

Some key findings from the descriptive results are summarized in this section. The means for each of the items in the survey are presented in Table 2. The mean value for the measurement value PE1 is 5.30. 145 of the 203 respondents (71.43 %) agreed that the service of online household e-waste collection was useful. In the same time, 144 respondents felt that online household e-waste collection service improved both their efficiency and convenience of handling e-waste.

---

[1] The survey items are available at this link: http://www.idi.ntnu.no/~shanggao/ewaste.html.

According to the results, most respondents thought that it would be easy for them to use online household e-waste collection services. 132 respondents indicated that it is easy for them to learn how to use online household e-waste collection services. Moreover, gender differences had an obvious effect on peoples' reaction towards social influence. Men were more likely to be affected by someone who were important to or had influence on them than women. Similarly, effort expectancy has a more significant likelihood to result in peoples' behavioral intention among men than women.

Item FC1 has the highest mean value in the construct facilitating conditions. Most respondents indicated that they had the necessary resources to use online household e-waste collection services. Furthermore, many of them thought that they had the necessary knowledge to use online household e-waste collection services.

Furthermore, many respondents gave us a neutral response when we were asking the question on their frequency to use online household e-waste collection services to deal with their household appliances or electronic devices. This means that there is still a huge business market for online household e-waste collection services in China.

Table 2. Means, factor loadings, composite reliability, and AVE for each item

Construct	Item	Mean	Factor loading	Composite reliability	AVE	Cronbach's alpha
Performance expectancy	PE1	5.30	0.850	0.927	0.810	0.882
	PE2	5.16	0.925			
	PE3	5.26	0.923			
Effort expectancy	EE1	4.68	0.932	0.950	0.827	0.930
	EE2	4.58	0.919			
	EE3	4.98	0.898			
	EE4	4.65	0.888			
Social influence	SI1	4.51	0.965	0.968	0.937	0.933
	SI2	4.47	0.917			
Facilitating conditions	FC1	5.17	0.867	0.913	0.778	0.867
	FC2	5.15	0.877			
	FC3	4.70	0.901			
Behavioral intention	BI1	4.88	0.939	0.946	0.854	
	BI2	4.86	0.918			0.914
	BI3	4.72	0.914			
Use behavior	UB1	4.02	0.974	0.973	0.947	0.944
	UB2	4.11	0.973			

4.5   Measurement Model

The quality of the measurement model is determined by (1). Content validity, (2). Construct reliability and (3). Discriminant validity [2]. To ensure the content validity of our constructs, a pretest with 6 researchers in the field of information systems was carried out in March 2015. And we found that the questionnaire was well understood by all the researchers.

Table 3. Discriminant validity

Variables	BI	EE	FC	PE	SI	UB
BI	0.924					
EE	0.755	0.909				
FC	0.825	0.803	0.882			
PE	0.730	0.732	0.713	0.900		
SI	0.760	0.714	0.667	0.649	0.968	
UB	0.670	0.593	0.538	0.521	0.693	0.973

Note: Diagonals represent the average variance extracted, while the other matrix entries represent the squared correlations.

To further test the reliability and validity of each construct in the research model, the Internal Consistency of Reliability (ICR) of each construct was tested with Cronbach's Alpha coefficient. As a result, the Cronbach's Alpha values range from 0.867 to 0.944. A score of 0.7 is marked as an acceptable reliability coefficient for Cronbach's Alpha [18]. All the constructs in the research model were above 0.70. Consequently, the scales were deemed acceptable to continue.

Convergent validity was assessed through composite reliability (CR) and the average variance extracted (AVE). Bagozzi and Yi [3] proposed the following three measurement criteria: factor loadings for all items should exceed 0.5, the CR should exceed 0.7, and the AVE of each construct should exceed 0.5. As shown in Table 3, all constructs were in acceptable ranges.

4.6   Structural Model and Hypotheses Testing

The structural model was tested using SmartPLS. Table 4 presents the path coefficients, which are standardized regression coefficients. Five (H1, H2, H3, H4, H6) of the six research hypotheses were significantly supported. According to the results, performance expectancy, effort expectancy and social influence were found to have a statistically significant effect on users' behavior intention to use online household e-waste collection services, while facilitating condition did not have significant impact on users' behavior of using online household e-waste collection services.

Table 4. Test of hypotheses based on path coefficient

Hypothesis	Path coefficient	Hypothesis result
H1.Effort expectancy to performance expectancy	0.732***	Supported
H2.Performance expectancy to behavior intention	0.279**	Supported
H3.Effort expectancy to behavior intention	0.281**	Supported
H4.Social influence to behavior intention	0.379**	Supported
H5.Facilitating conditions to behavior intention	–0.044	Not supported
H6.Behavior intention to use behavior	0.706***	Supported

$*p < 0.05$; $**p < 0.01$; $*** p < 0.001$

The R^2 (R square) denotes to coefficient of determination. It provides a measure of how well future outcomes are likely to be predicted by the model, the amount of variability of a given construct. In our analysis, the R^2 coefficient of determination is a statistical measure of how well the regression coefficients approximate the real data point. According to the result, 70.4 % of the variance of behavior intention can be explained by the research model.

5 Discussion

In this research, we studied the adoption of online household e-waste collection services in China. From the theoretical perspective, this research contributed to the literature on the adoption of new technologies (i.e., online e-waste collection services) in China by building upon previous technology diffusion theories. To our best knowledge, most previous studies were mainly focusing on the traditional channels of household collection as well as the conflicts between formal and informal e-waste recycling systems (e.g., [6, 22]). This research is one of the first studies to examine users' adoption of online household e-waste collection services in China. From a practical perspective, it offered some insights to the use and adoption of online household e-waste collection services in China.

Since effort expectancy has an indirect effect on users' behavioral intention through performance expectancy, the total effect of effort expectancy on behavioral intention is 0.281 + 0.732 * 0.279 = 0.485. Therefore, the most important determinant for the behavioral intention to use online household e-waste service was effort expectancy, while social influence and performance expectancy were ranked as second and third respectively. As to the significant impact of social influence on users' behavior intention, one possible explanation is that Chinese consumers are considered to be more concerned about their social images. Moreover, the effect of word-of-mouth is an important influencer in China [27]. In addition, effort expectancy has an significant positive influence on performance expectancy, which is consistent with previous research findings [26]. However, facilitating conditions is not proved to be the determinant for the use behavior. This can be partially attributed to the fact that most participants in this study were between 15 to 25 years old. They were quite skillful at using online services and unlikely to ask for help from their fellows. This had been further confirmed in the follow-up interview session. Some participants also indicated that online household e-waste collection services were easy to use so that they did not need help from others. Thus, facilitating conditions did not have a significant positive influence on use behavior. This is also consistent with the previous findings (e.g., [13, 25]).

When promoting the online household e-waste collection services, practitioners could make advantage of social influence, such as celebrity worship and the effect of word-of-mouth, especially in countries like China. Moreover, most participants believed that online household e-waste collection services improve the efficiency to deal with household e-waste. It was also interesting to note that people with different gender and age groups responded differently to social influence. The results indicated that men were much more likely to be affected by someone who were important or have influence on them than women.

However, we were also aware of some limitations. Firstly, we only tested the research model and research hypotheses with samples from one province in China. This sample might not be fully representative of the entire population in China. Secondly, all the data were collected using self-reported scales in the research. This may lead to some caution because common method variance may account for some of the results that has been cited as one of the stronger criticisms of tests of theories with TAM and TAM-extended research [16]. However, our data analysis with convergent and discriminant validity does not support the presence of a strong common methods bias. Last but not least, the services provided by Taolv365 may not represent all the online household e-waste collection services.

6 Conclusion and Future Research

This research was designed to explore users' adoption of online household e-waste collection services in China. A research model with six research hypotheses was proposed. The results indicated that five of the six research hypotheses were significantly supported. And the most significant determinant for the behavioral intention to use online household e-waste service was effort expectancy. However, facilitating condition did not have a significant positive impact on users' behavior of using online household e-waste collection services.

Continuing with this stream of research, we plan to examine some additional constructs' (e.g., trust) influence on the adoption of online household e-waste collection services. Future research is also needed to empirically verify the research model with larger samples across China.

References

1. Ajzen, I.: The theory of planned behavior. Organ. Behav. Hum. Decis. Process. **50**(2), 179–211 (1991)
2. Bagozzi, R.P.: The role of measurement in theory construction and hypothesis testing: toward a holistic model. In: Ferrell, O.C., Brown, S.W., Lamb, C.W. (eds.) Conceptual and Theoretical Developments in Marketing, pp. 15–32 (1979)
3. Bagozzi, R.P., Yi, Y.: Specification, evaluation, and interpretation of structural equation models. J. Acad. Mark. Sci. **40**(1), 8–34 (2012)
4. Carlsson, C., Carlsson, J., Hyvonen, K., et al.: Adoption of Mobile Devices/Services - Searching for Answers with the UTAUT Proceedings of the 39th Annual Hawaii International Conference on System Sciences - Volume 06. IEEE Computer Society (2006)
5. Chi, X., Streicher-Porte, M., Wang, M.Y., et al.: Informal electronic waste recycling: a sector review with special focus on China. Waste Manag. **31**(4), 731–742 (2011)
6. Chi, X., Wang, M.Y., Reuter, M.A.: E-waste collection channels and household recycling behaviors in Taizhou of China. J. Clean. Prod. **80**, 87–95 (2014)
7. Davis, F.D.: Perceived usefulness, perceived ease of use and user acceptance of information technology. MIS Q. **13**(3), 319–340 (1989)
8. DeLone, W., McLean, E.: Information systems success: the quest for the dependent variable. Inf. Syst. Res. **3**(1), 60–95 (1992)

9. Fishbein, M., Ajzen, I.: Belief, Attitude, Intention and Behavior: An Introduction to Theory and Research. Addison-Wesley, Reading (1975)
10. Fitzsimons, G.J., Lehmann, D.R.: Reactance to recommendations: When unsolicited advice yields contrary responses. Mark. Sci. **23**(1), 82–94 (2004)
11. Gao, S., Krogstie, J., Gransæther, P.A.: Mobile Services Acceptance Model. In: The Proceedings of International Conference on Convergence and Hybrid Information Technology. IEEE Computer Society (2008)
12. Gao, S., Krogstie, J., Siau, K.: Adoption of mobile information services: an empirical study. Mob. Inf. Syst. **10**(2), 147–171 (2014)
13. Gao, S., Krogstie, J., Yang, Y.: Differences in the adoption of smartphones between middle aged adults and older adults in China. In: Zhou, J., Salvendy, G. (eds.) ITAP 2015. LNCS, vol. 9193, pp. 451–462. Springer, Heidelberg (2015)
14. Gao, S., Yang, Y., Krogstie, J.: The adoption of smartphones among older adults in china. In: Liu, K., Nakata, K., Li, W., Galarreta, D. (eds.) ICISO 2015. IFIP AICT, vol. 449, pp. 112–122. Springer, Heidelberg (2015)
15. López-Nicolás, C., Molina-Castillo, F.J., Bouwman, H.: An assessment of advanced mobile services acceptance: contributions from TAM and diffusion theory models. Inf. Manag. **45**(6), 359–364 (2008)
16. Malhotra, N.K., Kim, S.S., Patil, A.: Common method variance in IS research: a comparison of alternative approaches and a reanalysis of past research. Manage. Sci. **52**(12), 1865–1883 (2006)
17. Park, J., Yang, S., Lehto, X.: Adoption of mobile technologies for Chinese consumers. J. Electron. Commer. Res. **8**(3), 196–206 (2007)
18. Robinson, J.P., Shaver, P.R., Wrightsman, L.S.: Criteria for scale selections and evaluation. Academic Press, San Diego (1991)
19. Rogers, E.M.: The diffusion of innovations. Free Press, New York (1995)
20. Schluep, M., Hagelueken, C., Kuehr, R., et al.: Sustainable Innovation and Technology Transfer Industrial Sector Studies: Recycling–from E-waste to Resources. United Nations Environment Programme and United Nations University, Bonn (2009)
21. Sheppard, B.H., Hartwick, J., Warshaw, P.R.: The theory of reasoned action: a meta-analysis of past research with recommendations for modifications and future research. J. Consum. Res. **15**(3), 325–343 (1988)
22. Streicher-Porte, M., Geering, A.-C.: Opportunities and threats of current e-waste collection system in China: a case study from Taizhou with a focus on refrigerators, washing machines, and televisions. Environ. Eng. Sci. **27**(1), 29–36 (2010)
23. Venkatesh, V., Bala, H.: TAM 3: advancing the technology acceptance model with a focus on interventions. Manuscript in Preparation (2013). http://www.vvenkatesh.com/IT/organizations/Theoretical_Models.asp
24. Venkatesh, V., Bala, H.: Technology acceptance model 3 and a research agenda on interventions. Decis. Sci. **39**(2), 273–315 (2008)
25. Venkatesh, V., Davis, F.D.: A theoretical extension of the technology acceptance model: four longitudinal field studies. Manag. Sci. **46**(2), 186–204 (2000)
26. Venkatesh, V., Morris, M.G., Davis, G.B., et al.: User acceptance of information technology: toward a unified view. MIS Q. **27**(3), 425–478 (2003)
27. Wiedemann, D.G., Haunstetter, T., Pousttchi, K.: Analyzing the basic elements of mobile viral marketing-an empirical study. In: 7th International Conference on Mobile Business, ICMB 2008, pp. 75–85. IEEE (2008)
28. Yu, J., Williams, E., et al.: Managing e-waste in China: policies, pilot projects and alternative approaches. Resour. Conserv. Recycl. **54**(11), 991–999 (2010)

Perceptions of Teachers and Guardians on the Electronic Record in the School-Family Communication

António Abreu[1], Álvaro Rocha[2(✉)], and Manuel Pérez Cota[3]

[1] ISCAP-IPP, Rua Jaime Lopes Amorim,
s/n 4465-004 S. Mamede de Infesta,
Matosinhos, Portugal
aabreu@iscap.ipp.pt
[2] DEI, University of Coimbra,
Pólo II – Pinhal de Marrocos,
3030-290 Coimbra, Portugal
amrocha@dei.uc.pt
[3] EEI, University of Vigo, Campus Torrecedeira,
Rúa Torrecedeira 86, 36280 Vigo, Spain
mpcota@uvigo.es

Abstract. This paper presents the first stage of an investigation work whose purpose is to introduce a new form of communication between the school and the family, through an electronic record (ER). We tried to identify, with a questionnaire, the perceptions of Teachers and Guardians towards the adoption of an ER. Both groups underlined the importance of communication between the school and the family, and expressed interest in the ER, as a communicational tool.

Keywords: School · Family · School-family relationship · Communication · Electronic record

1 Introduction

The present study, focused on the communication, management and dissemination of information in digital environments, is of great significance and interest. We intent to give our contribution to the communication between the school and the family, with the development of an Electronic Record (ER).

Indeed, the technological revolution imposed changes both in society [1, 11, 12, 16, 17, 28] and in the forms of communication, which exceed barriers of space and time [6, 7, 22].

Schools should use electronic resources to develop new forms of communication between the school and the family [3, 4, 17, 23–25, 28], hence de ER [12–15, 18, 29].

Currently, this communication is established via the traditional student record (TR), in paper format, which is not the most direct contact vehicle, as it requires the mediation of the student and, moreover, limits the communication content [18].

© IFIP International Federation for Information Processing 2015
M. Janssen et al. (Eds.): I3E 2015, LNCS 9373, pp. 48–62, 2015.
DOI: 10.1007/978-3-319-25013-7_5

In order to render this communication easier [17], the need make the TR a more dynamic and attractive tool, both to families and the school, emerged, and this may be achieved with the ER.

We intend to develop a study that assumes as a starting point three investigation questions:

- Is the school-family communication fundamental for the teaching-learning process?
- Is the TR a fundamental tool for school-family communication?
- Does the ER improve and increase the potential of school-family communication?

This study comprised three stages. The first stage, the focus of this paper, involved a feasibility study as to the adoption of the ER in school-family relationships, based on the perceptions of Guardians and Teachers.

2 Investigation Methodology

In this stage of the study, we adopted the survey investigation method, to the extent that this constitutes a viable and desirable method within a school environment and generates accurate and reliable measures which allow for a statistical analysis [6]. The adopted tool was the questionnaire, divided in two parts: the first focused on the characterization of the sample and the latter being the questionnaire itself, which, in its turn, included Part B, pertaining to the "Student Record", and Part C, concerning the "Inclusion of items in an Electronic Report". We adopted a *Likert* type ordinal scale with five answer alternatives (from "1" to "5") that ranged from "Completely disagree" to "Completely agree".

For this investigation work we developed the following hypothesis:

H1:   The educational community values communication between the school and the family;
H2:   The TR is a fundamental tool in school-family communication;
H3:   The ER improves school-family communication.

We conducted a pre-test in a group of thirty Teachers and Guardians, in order to clarify and validate the questionnaire. As the questionnaire did not offer doubts, we moved to the questionnaire distribution stage, in paper format and on-line, amongst the participating schools and school groupings, considering the answering availability on the part of the respondents and according to the preferences of School Principals.

The survey was carried in the form of a questionnaire in schools and cooperative/ private and public schools located in the North, Centre and South of the country, between 17 November 2013 and 8 March 2014. In an informal meeting the School Principals were informed as to the purpose of the study and the entire questionnaire administration process, and they in turn, together with the responsible parties of the school groupings, had to explain Teachers, class directors, elements from parent associations and Guardians the purpose of the investigation work and how they could participate.

The obtained data was treated with the SPSS computer program, version 20. Besides the descriptive data analysis, we carried out a correlational analysis, crossing some of the variables comprised by the questionnaire, in order to find the answers to the three laid out investigation hypothesis [19]. We also used absolute (standard deviation) and relative (coefficient of variation) measures of dispersion, useful when describing the observed variation in the values of a set and informing us as to the homogeneity. The smaller (<1) the standard deviation and (<=20 %) the coefficient of variation, the smaller the dispersion and, consequently, the more homogenous the answers.

2.1   Characterization of the Sample

The sample included 1002 Guardians of students attending basic and secondary education and 300 Teachers exercising their professional activity in cooperative/private and public educational institutions.

2.2   Factor Analysis

We carried out a data analysis to the questions comprising Part B and Part C of the Guardians and Teachers questionnaires, adopting data treatment techniques that allowed us to agglomerate the initial information, with a view to render the analysis easier [20].

We selected the principal components method, in order to isolate the original correlating variables, constituting factors [21].

The factor analysis method analyses a set of variables with the purpose of verifying the possibility of grouping answers that are similarly interpreted by the elements of the sample, in order to determine their position within the set of variables [9, 26, 27]. This allows us to turn the variables of a scale into a smaller number of factors: the principal components. To define the number of components to be retained, we carried out a preliminary analysis, applying the factor analysis and interpreting the obtained results [8].

In part B of the Teachers and Guardians questionnaires, concerning the "Student Record", we drew our factors from the 15 variables presented in Table 1.

We began our analysis with the de *Kaiser Meyer Olkin (KMO)* and the *Bartlett* tests, which determine the quality of the factor analysis to the sample data [2, 10].

After analysing the test results, the *KMO* revealed a value of 0.887 and the *Bartlett* a significance value below 5 %, the factor analysis to component B of the Guardians and Teachers questionnaires was considered feasible [11].

Using the principal components method, we proceeded with the extraction of factors from the 15 variables, adopting the *Kaiser* criterion (eigenvalues above one), complying with the requirement that factors should explain at least approximately 60 % of the total variation observed in the original variables [10]. We retained four factors, which explain 66.7 % of the total variation observed in the original 15 variables.

Through the *Varimax* method we were able to obtain the extreme values of the coefficients relating each variable to the retained factors, in order to associate each variable with a single factor [5]: Factor 1 – ER; Factor 2 – Impact of the school-family relationship in the teaching learning process; Factor 3 – TR; Factor 4 – School-family relationship difficulties.

Table 1. Part B of the Teachers and Guardians questionnaires - "Student Report".

Part B of the Teachers and Guardians questionnaires – "Student Record".	
#	Item
1	The relationship between the family and the school is fundamental for the teaching-learning process
2	The relationship between the school and the family is not always easy and stands out as a complex relationship
3	Whenever the parents are involved the development of the students is potentially improved
4	The participation of parents in the school benefits the Teachers and contributes to a more constructive work on their part
5	Parents assume more favourable attitudes towards Teachers when they cooperate constructively with each other
6	Parents can easily head to the school during the receiving hours appointed by Class Directors and in meeting days
7	The Traditional Student Record constitutes an important tool in family-school communications
8	Teachers regularly use the Traditional Record in their communications with Guardians
9	It is important to find a tool that will allow for the intensification of and an easier communication between the School and the Guardian
10	The adoption of an Electronic Report may contribute to an improved communication between the school and the Guardian
11	An Electronic Record may constitute an important work tool for Class Directors, Teachers, Parents and School Principals
12	An Electronic Report may constitute a communication resource between every Teacher and Guardians
13	An Electronic Report may be used by School Principals to contact Teachers
14	An Electronic Report may be used by School Principals to contact Guardians
15	An Electronic Record offers more advantages than the Traditional Record where the communication between the multiple school community actors is concerned

Table 2 presents the factorial matrix after the *Varimax* rotation, where the saturation between each factor and the principal components can be observed. The saturations of the variables in each factor are always higher than the required minimum of 40 %. The value presented in Table 2 – 0,661, is owed to the fact that the respective item is in a reverse scale. In schools, at the beginning of the school year, the Class Director defines, according to his school hours, a day and a time to receive Guardians. These receiving hours usually coincide with the opening hours of the educational establishment. However, for most of the Guardians this schedule overlaps their office hours. To make this situation worse, many Guardians work outside their areas of residence, where educational establishments are usually located, which prevents them from attending the pre-scheduled meetings. Being a negative value, the inclusion of item 6 in factor 4 is carried out in a reverse scale.

In part C of the questionnaires ("Inclusion of items in an Electronic Report"), Teachers and Guardians were inquired about 7 items, as seen in Table 3.

The *Kaiser Meyer Olkin (KMO)* and *Bartlett* tests revealed a *KMO* value of 0.881 and a Bartlett significance value below 5 %, which allowed us to carry out the factor analysis, as the results were considered significant [2, 10].

After confirming the possibility of carrying out a factor analysis, and adopting the method of the principal components, we obtained our factors from the 7 variables, employing the *Kaiser* criterion (eigenvalues above one) and complying with the criterion that factors should explain at least approximately 60 % of the total variation observed in the original variables [10]. We retained two factors, which explain 72.6 % of the total variation observed in the 7 original variables [27].

Table 2. Rotation method: Varimax with Kaiser normalization. Rotation converged in 5 iterations. N = 1302. KMO = 0.887.

Items	Factor			
	1	2	3	4
1. The relationship between the family and the school is fundamental for the teaching-learning process	0.050	**0.702**	0.017	–0.057
2. The relationship between the school and the family is not always easy and stands out as a complex relationship	0.065	0.067	0.092	**0.830**
3. Whenever the parents are involved the development of the students is potentially improved	0.124	**0.741**	0.084	0.125
4. The participation of parents in the school benefits the Teachers and contributes to a more constructive work on their part	0.166	**0.724**	0.104	–0.023
5. Parents assume more favourable attitudes towards Teachers when they cooperate constructively with each other	0.101	**0.749**	0.117	–0.052
6. Parents can easily head to the school during the receiving hours appointed by Class Directors and in meeting days	–0.095	0.155	0.295	**–0.661**
7. The Traditional Student Record constitutes an important tool in family-school communications	0.024	0.220	**0.808**	–0.074
8. Teachers regularly use the Traditional Record in their communications with Guardians	0.052	0.042	**0.873**	–0.018
9. It is important to find a tool that will allow for the intensification of and an easier communication between the School and the Guardian	**0.474**	0.286	0.100	0.341
10. The adoption of an Electronic Report may contribute to an improved communication between the school and the Guardian	**0.886**	0.086	–0.019	0.109

(Continued)

Table 2. (*Continued*)

Items	Factor			
	1	2	3	4
11. An Electronic Record may constitute an important work tool for Class Directors, Teachers, Parents and School Principals	**0.899**	0.137	–0.011	0.074
12. An Electronic Report may constitute a communication resource between every Teacher and Guardians	**0.904**	0.118	–0.021	0.074
13. An Electronic Report may be used by School Principals to contact Teachers	**0.698**	0.146	0.147	–0.014
14. An Electronic Report may be used by School Principals to contact Guardians	**0.896**	0.080	0.045	0.017
15. An Electronic Record offers more advantages than the Traditional Record where the communication between the multiple school community actors is concerned	**0.880**	0.038	–0.033	0.052
% Explained variance	31.8	15.7	10.5	8.6
Total explained variance	66.7			

Table 3. Part C of the Teachers and Guardians questionnaires – "Inclusion of items in an Electronic Report"

Part C of the Teachers and Guardians questionnaires - "Inclusion of items in an Electronic Report"	
Number	Item
1	Absence justifications by the Guardians
2	Absence management by the Class Director
3	Mid-term evaluation record
4	Evaluation criteria and tools for each subject
5	Student self-assessment forms
6	Communication management (Messages, *SMS* and *Emails*) between Class Directors, Teachers and Guardians
7	Videoconference between Class Directors and Guardians

Through the *Varimax* method we were able to obtain the extreme value of the coefficients relating each variable to the retained factors, connecting each variable to a single factor [5]: Factor 1 – evaluation and absence records, Factor 2 – student self-assessment and communication.

Table 4 shows the factorial matrix after the *Varimax* rotation, where the saturations between each factor as well as the principal components can be observed. The saturation of the variables in each factor stand above the required minimum of 40 %.

Table 4. Rotation Method - Varimax with Kaiser normalization. Rotation converged in 5 iterations. N = 1302. KMO = 0,887.

Items	Factor	
	1	2
1. Absence justifications by the Guardians	**0.861**	0.141
2. Absence management by the Class Director	**0.868**	0.248
3. Mid-term evaluation record	**0.767**	**0.405**
4. Evaluation criteria and tools for each subject	**0.588**	**0.571**
5. Student self-assessment forms	**0.439**	**0.667**
6. Communication management (Messages, *SMS* and *Emails*) between Class Directors, Teachers and Guardians	**0.458**	**0.668**
7. Videoconference between Class Directors and Guardians	0.054	**0.884**
% Explained variance	40.5	32.1
Total explained variance	72.6	

In Part C of the questionnaires – "Inclusion of Items in an Electronic Report", the inclusion of two factors was substantiated. However, several items converged in more than one, and this justifies a subsequent analysis of the items individually (Table 4).

2.3   Internal Consistency Analysis of the Adopted Scales

An internal consistency analysis, which allows us to study the properties of the measure scales and the questions they comprise, was carried out in Part B of the questionnaires.

The adopted scale was an ordinal *Likert* type scale, with five answer alternatives (from "1" to "5") ranging from "Completely disagree" to "Completely agree". It included 15 items, organized into 4 factors (Table 5).

In the first dimension, Factor 1 – ER, *Cronbach's Alpha* value is over 0.80, and the unidimensional data can be considered adequate (Table 6).

In Factor 2 – Impact of the school-family relationship in the teacher-learning process, *Cronbach's Alpha* is over 0.70, and the unidimensional data can be considered acceptable (Table 6). In Factor 3 – TR, *Cronbach's Alpha* value comes close to 0.70, and the unidimensional data can be considered acceptable (Table 6). Finally, in Factor 4 – School-family relationship difficulties, *Cronbach's Alpha* is below 0.70, and the unidimensional data cannot be considered acceptable (Table 6).

The scale used in Part B – "Student Report", is suitable to measure the factors that were constructed for this sample, with the exception of factor 4; despite this limitation the scale was nevertheless used.

Additionally, we asked the Teachers and Guardians to indicate the advantages and disadvantages of using an ER. Suggested advantages by Teachers and Guardians concerning the use of an ER amounted to 84.7 % and 84.3 %, respectively. As to disadvantages, we registered 24.3 % for Teachers and 20.2 % for Guardians. The most frequently referred advantages in the adoption of an ER, by Teachers and Guardians, were: "Faster, more efficient and safer communication"; "Saves paper"; "Decreases repetitive tasks"; "Customized access to information, at any given time and place";

Table 5. Organization of factors and respective items.

Factors	Items
Factor 1 – ER	9. It is important to find a tool that will allow for the intensification of and an easier communication between the School and the Guardian
	10. The adoption of an Electronic Report may contribute to an improved communication between the school and the Guardian
	11. An Electronic Record may constitute an important work tool for Class Directors, Teachers, Parents and School Principals
	12. An Electronic Report may constitute a communication resource between every Teacher and Guardians
	13. An Electronic Report may be used by School Principals to contact Teachers
	14. An Electronic Report may be used by School Principals to contact Guardians
	15. An Electronic Record offers more advantages than the Traditional Record where the communication between the multiple school community actors is concerned
Factor 2 – Impact of the school-family relationship in the teaching-learning process	1. The relationship between the family and the school is fundamental for the teaching-learning process
	3. Whenever the parents are involved the development of the students is potentially improved
	4. The participation of parents in the school benefits the Teachers and contributes to a more constructive work on their part
	5. Parents assume more favourable attitudes towards Teachers when they cooperate constructively with each other
Factor 3 – TR	7. The Traditional Student Record constitutes an important tool in family-school communications
	8. Teachers regularly use the Traditional Record in their communications with Guardians
Factor 4 – School-family relationship difficulties	2. The relationship between the school and the family is not always easy and stands out as a complex relationship
	** 6. Parents can easily head to the school during the receiving hours appointed by Class Directors and in meeting days

Table 6. Cronbach's Alpha results for each factor

Factor 1		Factor 2	
Cronbach's Alpha	Nr. of Items	Cronbach's Alpha	Nr. of Items
0.922	7	**0.730**	4
Factor 3		**Factor 3**	
Cronbach's Alpha	Nr. of Items	Cronbach's Alpha	Nr. of Items
0.661	4	**0.661**	4

"Daily follow-up" and "Virtual Presence". In their answers given to the question involving the disadvantages of adopting an ER the respondents were objective, indicating the inexistence of disadvantages, the possible lack of internet access by some and, mainly, the lack of computer skills.

Part C of the Guardians and Teachers questionnaire focused on the inclusion of items in an ER. We observed that 95 % of the Guardians and 94 % of the teachers believe that all the items of the TR should be included in an ER.

We asked the Guardians and the teachers to indicate their level of agreement as to the inclusion of the following items in a prospective ER, as shown in Table 7.

From the obtained answers, all items reveal a high average agreement, higher for statements 3, 1 and 2 followed by 4, 6, 5 and 7.

From the answers given by the Teachers in our study sample, the items that reveal a higher average agreement involve the statements 2, 3, 4, 6 and 1 followed by 5, and lower for 7. We also asked our study participants to identify some of the items that, in their view, should be included in an ER. Faced with multiple responses, we opted for the inclusion of the most significant, namely the record of absences, mid-term evaluations, record of school entries and exits and communication between Guardians.most welcome.

3   Presentation and Discussion of the Results Concerning the Investigation Questions

The developed study was based on a set of questions included both in the questionnaire given to Guardians and in the questionnaire given to Teachers, which were subject to analysis.

The first investigation questions, "School-family communication is fundamental for the teaching-learning process", relates to Factor 2 – Impact of the school-family relationship in the teaching-learning process, and comprises the following statements: 1, 3, 4 and 5 (Table 5).

The majority of Guardians (96.7 %) believes that the relationship between the school and the family is fundamental for the teaching-learning process. This is also the perception expressed by the majority of Teachers (99.3 %).

Table 7. Inclusion of items in a prospective ER

Items
1. Absence justifications by the Guardians
2. Absence management by the Class Director
3. Mid-term evaluation record
4. Evaluation criteria and tools for each subject
4. Student self-assessment forms
5. Communication management (Messages, SMS and Emails)
6. Videoconference

The observed average values reveal the presented variations, with all items showing a high level of agreement, both for Guardians and Teachers. Factor 2 – Impact of the school-family relationship in the teaching-learning process, thus reveals an average value of 4.49 for Guardians and of 4.55 for Teachers, which come significantly close to the maximum possible value. In conclusion, we can say that both for Guardians and Teachers, the communication between the school and the family is fundamental for the teaching-learning process.

The analysis of the second investigation question "The TR is a fundamental tool in school-family communication" relates to Factor 3 – TR, and comprises statements: 7 and 8 (Table 5). Teachers regularly use the student TR when communicating with Guardians. Indeed, both Guardians (88,8 %) and Teachers (87,9 %) perceive the student TR as an important school-family communication tool.

The observed average values reveal the presented variations, with all items showing a high level of agreement, both for Guardians and Teachers. Factor 3 – TR, thus reveals an average value of 4.20 for Guardians and of 4.13 for Teachers, which come significantly close to the maximum possible value. In conclusion, we can say that both for Guardians and Teachers, the TR constitutes a fundamental school-family communication tool.

The analysis of the investigation question "The ER improves school-family communication", relates to the analysis of Factor 1 – ER, and comprises statements: 9, 10, 11, 12, 13, 14 and 15 (Table 5).

In our sample, Guardians (83.9 %) and Teachers (92.3 %) are unanimous as to the relevancy of finding a tool that allows for the intensification of and an easier communication between the school and the Guardian. Guardians (78 %) and Teachers (87.3 %) go so far as to admit that an ER may become a communication and work resource, between the school and the family. The majority of Guardians (73 %) and Teachers (82 %) believe that the ER may bring more benefits when compared with the TR.

The observed average values reveal the presented variations, with all items showing a high level of agreement, both of Guardians and Teachers. Factor 1 – ER, thus reveals an average value of 4.11 for Guardians and of 4.33 for Teachers, which come significantly close to the maximum possible value.

In conclusion, we can say that both Guardians and Teachers believe that the ER promotes and improves the communication between the school and the family.

The herein presented measures of dispersion reveal relatively low values for the standard deviation, with values below one or coefficient of variation percentages of twenty or less (Tables 8, 9 and 10). Therefore, everything suggests a good dispersion of the obtained results, pointing to a greater homogenization of answers and, consequently, an average value that is more representative of reality.

Table 8. Measures of Dispersion of Factor 2 – Influence of the School/Family relationship in the teaching-learning process and its items.

	Guardians				Teachers			
	N	Average	Standard deviation	Coefficient of variation	N	Average	Standard deviation	Coefficient of variation
Factor 2 – Influence of the school-family relationship in the teaching-learning process	**1002**	**4,49**	**0,54**	**12 %**	**300**	**4,55**	**0,47**	**10 %**
1. The relationship between the school and the family is fundamental for the teaching-learning process.	1002	4,79	0,53	11 %	300	4,82	0,44	9 %
3. Whenever the parents are involved, the development conditions of the students are potentially improved	1001	4,37	0,79	18 %	299	4,47	0,74	17 %
4. The participation of the parents in the school benefits the Teachers, and contributes to a more constructive work on their part	1001	4,34	0,79	18 %	299	4,31	0,77	18 %
5. Parents assume more favourable attitudes towards Teachers when they cooperate constructively with each other	1001	4,47	0,74	16 %	299	4,60	0,61	13 %

Table 9. Measures of Dispersion of Factor 3 – TR.

	Guardian				Teacher			
	N	Average	Standard deviation	Coefficient of variation	N	Average	Standard deviation	Coefficient of variation
Factor 3 – Traditional Record	**1002**	**4,20**	**0,83**	**20 %**	**300**	**4,13**	**0,73**	**18 %**
7. The Traditional Student Record constitutes an important tool in family-school communications	1002	4,49	0,82	18 %	299	4,33	0,79	18 %
8. Teachers regularly use the Traditional Record in their communications with Guardians	1002	3,91	1,08	28 %	300	3,94	0,91	23 %

Table 10. Measures of Dispersion of Factor 1 – ER.

	Guardian				Teacher			
	N	Average	Standard deviation	Coefficient of variation	N	Average	Standard deviation	Coefficient of variation
Factor 1 – Electronic Report	**1002**	**4,11**	**0,85**	**21 %**	**300**	**4,33**	**0,71**	**16 %**
9. It is important to find a tool that will allow for the intensification of and an easier communication between the school and the Guardian	1002	4,32	0,91	21 %	299	4,57	0,73	16 %
10. The adoption of an Electronic Report may contribute to an improved communication between the school and the Guardian	1002	4,03	1,09	27 %	300	4,35	0,81	19 %
11. An Electronic Record may constitute an important work tool for Class Directors, Teachers, Parents and School Principals	1002	4,12	1,02	25 %	298	4,43	0,81	18 %

(*Continued*)

Table 10. (*Continued*)

	Guardian				Teacher			
	N	Average	Standard deviation	Coefficient of variation	N	Average	Standard deviation	Coefficient of variation
12. An Electronic Report may constitute a communication resource between every Teacher and Guardians	1001	4,13	1,01	24 %	300	4,37	0,82	19 %
13. An Electronic Report may be used by School Principals to contact Teachers	1002	4,05	1,01	25 %	300	4,04	1,14	28 %
14. An Electronic Report may be used by School Principals to contact Guardians	1002	4,13	1,00	24 %	300	4,28	0,90	21 %

4   Conclusion

The collaboration between the school and the family, two fundamental institutions for the education of the child, is essential to improve the teaching-learning process, and promotes the desired educational success. Education is a task that must be shared by Teachers, Guardians and community institutions [19]. Because we believe that, in the teaching-learning process as a whole, communication and, particularly, the means with which we communicate, are essential to educational success, we directed our study to understand the opinion of Guardians and Teachers as to the feasibility of adopting an ER in the school-family relationship.

As a starting point for the development of this study we posed three hypothesis: H1: The educational community values communication between the school and the family; H2: The TR is a fundamental tool in school-family communication; H3: The ER improves school-family communication.

We concluded, based on the collected answers, that Teachers and Guardians perceive school-family communication as being fundamental in the teaching-learning process, to the extent that the factor concerning the impact of the school-family relationship in the teaching-learning process obtained an average of 4.49 amongst Guardians and 4.55 amongst Teachers, which comes significantly close to the maximum value possible. For Teachers and Guardians, the TR constitutes a fundamental tool in school-family communications. Currently, the TR is, for Guardians, the only vehicle of communication with Teachers, as they cannot head to school during opening hours. From the point of view of the Guardians and Teachers who participated in the study, the ER can promote and improve school-family communications. Indeed, the analysis carried out to the answers collected from our study participants led us to conclude that the Factor 1 – ER, reveals an average value of 4.11 amongst Guardians and of 4.33

amongst Teachers, which come significantly close to the maximum value possible. The majority of Teachers and Guardians asserts that the ER should include the items that are already available in the TR. They also stress the importance of including the following items: record of absences, mid-term evaluations, record of entries and exits from school, as well as communication amongst Guardians.

Considering the obtained results, we observed that the three hypothesis were reinforced, as the majority of the respondents (Guardians and Teachers), believe that the communication between the school and the family is fundamental for the teaching-learning process; moreover, the TR emerges as fundamental tool in school-family communications and the ER promotes and improves this school-family communication.

Therefore, the ER may assume, in a ground breaking, dynamic and assertive way, a facilitating role in school-family communications, allowing for the development of partnerships.

In this sense, the development of a web application, the ER, is justified, to enable the consultation of the entire body of information concerning the student and promote an easier, quicker and safer communication between the school and the family.

References

1. Abrantes, B.: Conceção e Desenvolvimento de um Ambiente de Aprendizagem Pessoal Baseado em Ferramentas Web 2.0. Universidade de Aveiro (2009)
2. Bartlett, M.S.: The effect of standardization on a chi square approximation in fator analysis. Biometrika **38**, 337–344 (1951)
3. Berto, R.M.V.S., Nakano, D.N.: A Produção Científica nos Anais do Encontro Nacional de Engenharia de Produção: Um Levantamento de Métodos e Tipos de Pesquisa. Revista Produção **9**(2), 65–76 (2000)
4. Boonen, A.: Pourquoi utiliser les technologies de l'information et de la communicationdans le domaine de l'éducation? In: Scheffknecht, J.J. (ed.) Les technologies de l'informationà l'école: raisons et stratégies pour un investissement. Strasbourg, Conseil de l'Europe (2000)
5. Costello, A., Osborne, J.: Best practices in exploratory fator analysis: four recommendations for getting the most from your analysis. Pratical Assess. Res. Eval. **10**(1), 1–9 (2005)
6. Groves, R.M., Brick, M.J., Couper, M., Kalsbeek, W., Harris-Kojetin, B., Kreuter, F., Tourangeau, R.: Survey Methodology. Wiley, Hoboken (2004)
7. Guimarães, R.C., Sarsfield Cabral, J.A.: Estatística, 2ª Edição. Verlag Dashöfer (2010)
8. Hair, J.F., Anderson, R.E., Tatham, R.L., Black, W.C.: Análise multivariada de dados, 5th edn. Bookman, Porto Alegre (2005)
9. Jiménez, E.G., Flores, J., et al.: Cuadernos de estadística: análisis factorial. La Muralla, Salamanca (2000)
10. Kaiser, H.F.: An index of fatorial simplicity. Psychometrika **39**(1), 31–36 (1974)
11. Leong, F.T.L., Austin, J.T.: The Psychology Research Handbook, 2nd edn, p. 516. Sage Publications, Thousand Oaks (2006)
12. Lima, L., Sá, V.: A participação dos pais na governação democrática das escolas. In: Lima, J. (Org.). Pais e professores, um desafio à cooperação. Edições Asa, Porto (2002)
13. Lourenço, L.P.R.: Envolvimento dos Encarregados de Educação na Escola: Conceções e Práticas. Universidade de Lisboa (2008)

14. Marques, R.: Colaboração Escola-Famílias: um conceito para melhorar a Educação. Ler Educação, 8 (1992)
15. Marques, R.: A escola e os pais: como colaborar?. Texto Editora, Lisboa (1993)
16. Mattelart, A.: The Information Society. Sage Publications, London (2003)
17. McLean, N.: Technology can bridge the gap between parents and schools. The Independent, 5 November (2009)
18. Montadon, C., Perrenoud, P.: Entre pais e professores, um diálogo impossível? Para uma análise sociológica das interações entre a família e a escola. Celta, Oeiras (2001)
19. Paro, V.: Qualidade do Ensino: A Contribuição dos Pais. Xamã, São Paulo (2003)
20. Pereira, A.: Análise de dados para Ciências Sociais e Psicologia. Edições Sílabo, Lisboa (2006)
21. Pereira, C.B.: O marketing do lugarzinho: uma aplicação exploratória da técnica de índice de preços hedônicos a jovens consumidores de restaurantes na cidade de São Paulo. Tese de Doutoramento em Administração – Faculdade de Economia, Administração e Contabilidade, Universidade de São Paulo. São Paulo, 165 (2004)
22. Salvador, P., Rocha, Á.: An assessment of content quality in websites of basic and secondary portuguese schools. In: Rocha, Á., et al. (eds.) New Perspectives in Information Systems and Technologies. Advances in Intelligente Systems and Computing 275, vol. I. Springer International Publishing, Switzerland (2014)
23. Sarmento, T.: (RE)pensar a interação escola família. Revista Portuguesa de Educação 18, 53–75 (2005)
24. Sarmento, T., Marques, J.: A Escola e os Pais, Coleção Infans. Centro de Estudos da Criança, Braga (2002)
25. Silva, P.: Escola-família: Tensões e potencialidades de uma relação. In: J.Á. de Lima (Org.) Pais e Professores, Um Desafio à Cooperação. Porto, Edições ASA (2002)
26. Stevens, J.: Applied Multivariate Statistics for the Social Sciences, 3rd edn. Lawrence Erlbaum, Mahway (1996)
27. Tabachnik, B.G., Fidell, L.S.: Using Multivariate Statistics, 5th edn. Pearson Education, Boston (2006)
28. Villas-Boas, M.: A parceria entre a escola, a família e a comunidade: reuniões de pais. Edição do Departamento de Avaliação Prospetiva e Planeamento do Ministério da Educação, Lisboa (2000)
29. Villas-Boas, M.: Escola e família: uma relação produtiva de aprendizagem em sociedades multiculturais. Escola Superior João de Deus, Lisboa (2001)

Cryptocurrencies as a Disruption? Empirical Findings on User Adoption and Future Potential of Bitcoin and Co

Aaron W. Baur[✉], Julian Bühler, Markus Bick,
and Charlotte S. Bonorden

ESCP Europe Business School Berlin, Berlin, Germany
{abaur,jbuehler,mbick}@escpeurope.eu,
charlotte.bonorden@edu.escpeurope.eu

Abstract. In this paper, we examine cryptocurrencies as a potentially disruptive sort of payment method. Due to its relative importance, we focus in particular on Bitcoin. Through an inductive, exploratory interview approach with 13 individuals in three distinct groups, the determinants usability, usefulness, and subjective norm that could make Bitcoin a game-changer are explored. The results reveal that most stakeholders consider perceived ease of use still rather low, with perceived usefulness varying according to the user group. The notion of Bitcoin as having much future potential as a payment method is confirmed across all interviewees. Interestingly, the underlying concept of a blockchain is also seen as a potential revolutionary way to create a more just society based on open platforms and open data. However, the reasons of why Bitcoin is actually a disruption to existing solutions varies widely.

Keywords: Electronic payment · Bitcoin · Cryptocurrency · Digital currency · e-commerce · Technology adoption · TAM · Blockchain

1 Introduction

The Internet and the massive growth of e-commerce have bred various new online payment and money transaction methods in the last several years. This is being spurred by a new hype in 'fin-tech' startups in the global technology hubs. Even though famous quotes like "banking is necessary, banks are not" made by former Wells Fargo CEO Richard Kovacevich [1] are already quite dated, technology finally seems ready to enable real innovations in this sector. A new generation of founders is motivated to revolutionize how the financial industry works, what society thinks about the concept of money, and how the future of monetary transactions will look.

As the number of alternative payment types has grown, traditional means of online payment, like credit cards, are on the downgrade as they are fighting with competition that offers higher user-friendliness, more security, and lower costs [2, 3]. As Koley [4] reports, even though the virtual duopoly of Visa and MasterCard is still experiencing healthy growth rates (9.5 % and 9.6 %, respectively) and remains the preferred method for (offline) point-of-sale (POS) transactions, firms like JCB and UnionPay outperform

© IFIP International Federation for Information Processing 2015
M. Janssen et al. (Eds.): I3E 2015, LNCS 9373, pp. 63–80, 2015.
DOI: 10.1007/978-3-319-25013-7_6

Rank	Name	Market Cap
1	Bitcoin	$ 3,372,991,146
2	Ripple	$ 213,123,917
3	Litecoin	$ 56,741,480
4	Dash	$ 15,396,138
5	Stellar	$ 13,645,326
6	Dogecoin	$ 11,934,891
7	BitShares	$ 9,554,346
8	Nxt	$ 9,217,893
9	BanxShares	$ 7,508,688
10	Peercoin	$ 5,772,497

Fig. 1. Top ten cryptocurrencies and market cap in USD, as of May 15, 2015 (Source: http://coinmarketcap.com/currencies/views/all/)

them by far (growing 20.7 % and 44.8 %, respectively). This provides hints of altered consumer behavior including a perceived viable alternative to credit cards and a gradual change in the payment market [5]. PayPal, which was originally meant for facilitating eBay transactions, is now one of the major players in online payment [6]. Additionally, new firms that are home to completely different industries than banking, like Google and Apple, offer solutions that have already gained a remarkable level of user acceptance and dissemination among society and businesses alike [7, 8]. However, all of these solutions remain tied in one way or another to the user's regular bank account and a traditional notion of the concept money.

Another, potentially far more disruptive innovation in this field applies to crypto-currencies [9]. Advocates of these new means of payment claim various advantages, like a fully decentralized, peer-to-peer transaction system, elimination of chargeback risks, lower associated transaction costs, increased level of security, greater ease of use, and full support for mobile devices [10, 11]. Of the several hundred different crypto-currencies,[1] Bitcoin is the most well-known, most discussed, and most widely traded one with a market capitalization of more than three bn USD (see Fig. 1 for the top ten cryptocurrencies as of May 15, 2015). Hence, in this paper we focus on Bitcoin and use 'Bitcoin' and 'cryptocurrency' somewhat interchangeably.

The body of scientific research in the field of cryptocurrencies is still very manageable (see Related Work section below). One stream that especially lacks work is the question of user adoption and the relevant drivers of it (e.g., [12, 13]). Academic research has mostly neglected the user perspective, which connects the technological infrastructure with the established economies [12]. In particular, the decisive impact on society as a whole that users expect of these blockchain technologies has not been gathered. To the best of our knowledge, there has not been any empirical research applying detailed interviews to gain 'rich knowledge' about these kinds of questions. With this paper, we contribute to this under-researched turf by means of an exploratory, qualitative approach using 13 semi-structured interviews.

---

[1] The website https://coinmarketcap.com lists 560 as of May 15, 2015.

In particular, the study aims to examine the following research questions:

1. What are the perceived advantages and disadvantages of Bitcoin as compared to other forms of (electronic) payment?
2. What are the drivers and barriers Bitcoin users, merchants, and experts see in the adoption of Bitcoin?
3. How do users, merchants, and experts evaluate the future potential of Bitcoin as a serious means of currency, unit of account, asset, and disruption for society?

The remainder of this paper is structured as follows: After giving an overview of the applicable literature on cryptocurrencies and present payment options, the applied research methodology is presented. In the following part, the interview findings are discussed. The paper closes with a conclusion section and also discusses contributions, limitations, and future avenues to advance this research area.

The aim is to explore Bitcoin's main advantages and disadvantages compared to current payment types, to identify the drivers and barriers of Bitcoin's adoption, and to evaluate its future potential as seen by the three distinct groups of (pure) users, merchants, and experts.

2 Related Work

In order to find empirical works on Bitcoin adoption and usage, we conducted a thorough literature search following the frameworks of vom Brocke et al., Levy and Ellis, and Webster and Watson [14–16]. Search terms were 'Bitcoin', 'crypto AND currency[ies]', and 'cryptocurrency[-ies]'. As suggested by Chen et al. [17], we searched the databases *Web of Science, EBSCO Business Source Complete, IEEE Xplore, ScienceDirect*, and the *ACM Digital Library*. To be as exhaustive as possible, the important IS conferences ICIS, ECIS, HICSS, PACIS, and AMCIS were also explored with the identical keywords. As the 'mother paper' of Bitcoin by Satoshi Nakamoto (most likely a group pseudonym) was just released in 2008 [18], we did not limit the search timeframe. In fact, it was not until 2011 that research papers about cryptocurrencies were published in journals and from conferences. From 2013, larger peer-reviewed journals have been picking up the topic and have started to accept papers more frequently.

The results, combined with the work of Scott [19] and Böhme et al. [20], revealed four main streams of research in regard to Bitcoin, addressing three different layers:

- *Technical:* This stream includes technical details about the 'back-end' of cryptocurrency, like cryptography, mathematical models, and system design. It consists of two sub-streams: On the one hand, conceptual and prototyping work in regard to the *protocol* layer (e.g., [21–23]), and on the other, research concerning the *network* layer, i.e., mainly security-related works made up of experimental and quantitative research designs (e.g., [24–26]).

The other three streams address the *ecosystem* layer:

- *Economic:* This body of research looks at cryptocurrency from an economist's perspective, and mainly includes traditional economic models, portfolio theory, incentive structures, and the like (e.g., [20, 27–31]).
- *Regulatory:* Here, researchers discuss legal, fiscal, tax, and regulatory issues of Bitcoin as a new form of currency (e.g., [10, 32–34]).
- *Social Science:* Along this road, this research is about sociology, trust, anthropology, ethics, and politics in regard to the new phenomenon of cryptocurrencies (e.g., [35–38]).

Surprisingly, the *Social Science* branch is the least developed, but probably the most relevant for Information Systems (IS) research. Hence, this paper focuses on this human-centered aspect of cryptocurrency.

2.1 Traditional Digital Payment Solutions

For a long time, credit cards have been the dominant means of payment on the Internet. Credit cards like Visa, MasterCard, American Express, and Discover still denote the highest volumes of all forms of payments. For instance, daily transaction volumes of Visa amount to almost 18 bn USD (Fig. 2). Due to certain drawbacks of credit cards to merchants and consumers—e.g., the possibility of chargebacks, high fees, and a significant fraud risk—a wide array of alternative means of online payment has developed. Important players include China Union Pay, PayPal, and increasingly Bitcoin, albeit with a huge gap when compared to their old-school credit card competitors.

New players from the 'fin-tech' startup realm enter the market and push the financial market to become more digital and more efficient, encouraging higher transaction volumes and enhancing security. Payment service providers advertise better security standards, a better ease of use, and lower fees. However, none of the current alternatives could so far solve the three inherent major flaws completely.

- Payments through alternative providers still run through customers' credit/debit cards or bank accounts, so transaction fees cannot be cut entirely, and the savings are sometimes even marginal. Furthermore, cross-border transactions still involve significant fees, regardless of whether the money is transferred directly via credit card or via a service like PayPal or Skrill.

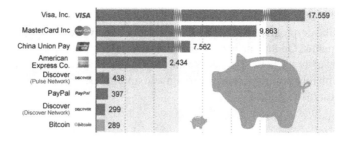

Fig. 2. Daily transaction volume in mn USD, as of the end of 2013 (Source: http://statista.com)

- The risk of fraud is still present, as every time a customer purchases goods online, his or her personal data like name, address, and financial information are transferred over the Internet. This information therefore remains prone to identity theft, no matter whether it is stored only at one single point (PayPal) or at several.
- Usability or ease of use is still a major weakness, especially concerning older people who would like to purchase goods online.

These flaws and other ideas have led to a completely new approach to e-payment: cryptocurrencies.

2.2 Overview of Cryptocurrency

From a technical perspective, cryptocurrency can be defined as a "type of digital currency which relies on cryptography, usually alongside a proof-of-work scheme, in order to create and manage the currency. A decentralized network of peer-to-peer computer nodes working in sync creates and verifies transactions of transfer of said currency within the network" [39]. A more practical definition is given by Ametrano [40], stating that cryptocurrency "can be transferred instantly and securely between any two parties, using the Internet infrastructure and cryptographic security with no need for a trusted third party. Its value is not backed by any single government or organization." Furthermore, Wiatr [41] gives the definition that "a cryptocurrency is a modern digital medium of exchange. It is a new decentralized, limited and peer-to-peer payment system. Most cryptocurrencies are created to introduce new units of currency, whose total amount is limited. All cryptocurrencies use cryptography to control the creation and transfer of money." Summarizing the above, we basically follow Spenkelink [42] and find four main characteristics of cryptocurrencies. First, cryptocurrency works through a decentralized network, meaning that it is free of any external regulations. Second, it has a strong P2P-approach. This guarantees the direct transaction of money between two parties, regardless of whether it is B2B, B2C, or C2C. Third, it uses the public Internet, which provides speed, efficiency, and resilience. Finally, cryptocurrency uses public-key cryptography to make money transactions as secure as possible.

After the introduction of Bitcoin as the first digital currency in 2009, the number of cryptocurrencies began to rise, and it quickly became a global phenomenon [43, 44]. Besides Bitcoin, there are hundreds of alternatives, each with different strengths. For example, *Litecoin* uses the *scrypt* encryption and claims to transact four times faster than Bitcoin, making it a viable alternative for high-speed applications like financial trading [20]. The so-called *Altcoins* can be regarded as 'forks' of either Bitcoin or Litecoin, since they are based on the same type of algorithms. Examples include *Feathercoin*, *ChinaCoin*, and *Dogecoin* [39].

Even though in theory Bitcoins can be mined, i.e., created by everyone, the system behind Bitcoin ensures certain characteristics that are important conditions for valid (fiat) currencies. First, scarcity: Bitcoins are limited to 21 million, keeping inflation low [45]. Second, security: Bitcoin follows the highest security standards using public and private keys [46]. After a person registers as a Bitcoin provider, he or she gets a digital wallet identification number (ID) that will be shown in public, once the person transfers Bitcoin. All past transactions are recorded in the blockchain. Third, simplicity: Transferring Bitcoins is easy, requiring a Bitcoin application or wallet on a mobile device and the required amount of Bitcoins, normally received through a Bitcoin exchange or an ATM.[2] The transfer then only takes minutes, regardless of geographical distance.

2.3 Value-Add and Risks of Cryptocurrencies

Most authors support the argument that Bitcoin has a significant future and will become a larger threat for credit card companies [2, 47, 48], but for now the potential is estimated to be limited to only specific fields [13, 49]. Grinberg [11] underlines the fact that Bitcoin is likely to be the new, revolutionary payment method for micropayments, because those are hardly profitable when credit card fees are applied to such small notional amounts. Böhme et al. [20] argue that it is becoming less certain whether the design of Bitcoin meets the requirements to replace credit card payments for the everyday consumer, particularly with regard to security standards. Furthermore, they illustrate the fact that there are numerous competing virtual currencies, for instance Litecoin that accelerates payment confirmations, or 'NXT' that reduces computational requirements. To become an essential part of the customer's daily life, Bitcoin first needs to achieve the overall reliance in value offer and user adoption.

Spenkelink [42] follows a broader perspective by concluding that the future of cryptocurrencies is very unclear in general and one can only speculate about its development. He further claims that many different usage scenarios and different stakeholder needs prevail, which makes the estimation of Bitcoin's mass adoption even harder. Descôteaux [50] holds the opinion that for the future development of Bitcoin, an appropriate legal and regulatory framework is extremely important in order to overcome the existing risks and threats slowing down the adoption of cryptocurrency. The literature also discusses the drawbacks and disadvantages of Bitcoin, such as the possibility of money laundering [51–54], trade in various illegal goods [55, 56], potential financing of terrorism [51, 57], Bitcoin loss or theft [21, 58], and tax evasion or enormous volatility [27, 43, 59]. These aspects were also mentioned by our interview partners.

3 Research Methodology

As described above, only limited research on Bitcoin's advantages, disadvantages, disruptive potentials, and adoption drivers exists. Therefore, we approached the research by following an entirely open, inductive, exploratory research design [60, 61]. Accordingly, an interview guideline was designed based on the literature review, but also including very open questions to motivate interviewees to come up with their own ideas, thoughts, anecdotes, and reasoning. To get insights into how different groups see

---

[2] The website http://coinatmradar.com/ lists 389 ATMs globally as of May 15, 2015.

Table 1. Sample description

Individual	Group	Occupation	Bitcoin Experience	Means of Interview
C1	Consumer	Student	none	Face-to-Face
C2		Entrepreneur	medium	Face-to-Face
C3		Entrepreneur	extensive	Skype
C4		Scientist	none	Face-to-Face
C5		Architect	none	Face-to-Face
M1	Merchant (E-Commerce)	Innovation Manager, Retail Industry	extensive	Skype
M2		CEO, Wholesale Industry	none	Face-to-Face
M3		Head of Digital, Media Industry	little	Face-to-Face
M4		Market Researcher, Retail Industry	none	Skype
B1	Bitcoin Exchange	CEO	extensive	Skype
B2		CTO	extensive	Skype
B3		Senior Marketer	medium	Face-to-Face
B4		Accountant	medium	Face-to-Face

and experience Bitcoin, the guideline was used to conduct semi-structured interviews in November and December 2014. To gain maximum diversity and coverage, 13 interviews with individuals from three distinct groups were used as a sample [62, 63]. These are end-consumers, e-commerce merchants, and employees of Bitcoin exchanges with differing levels of experience with Bitcoin (see Table 1 for the sample description). The selection of the interview partners was based on a purposive sampling strategy [64]. This ensured the inclusion of a wide array of aspects of the evidence gathered from informants [65, 66]. The interviews were recorded and subsequently transcribed [66]. They lasted between 34 and 56 min.

Interview transcripts were analyzed using the open-coding technique [67]. Relevant passages were identified and axial coding was subsequently used to refine the categories. The analysis was diligently carried out to fulfill the common requirements for qualitative research (e.g., [68–70]). The interpretation process was rule-driven and documented ("procedural validity"; [70]) In addition, we triangulated the data [70] by interviewing people with different perspectives on and knowledge about the phenomenon (i.e., three distinct user groups). Therefore, the results can be assumed to be reliable [71].

4 Analysis and Results

While analyzing the interview transcripts in order to find answers to the research questions, we decided to apply the established constructs *perceived ease of use* and *perceived usefulness* of the Technology Acceptance Model TAM [72, 73] and *subjective norm* of its successor, TAM2 [74, 75], as they work well as ordering categories.

Another one, *future potential*, emerged out of the data. Table 2 summarizes the findings, which are organized around the four research constructs and the case codes. We have included original quotes, which is the 'beauty of qualitative research' [76] and more often than not provide the deepest insight into the matter. In the following paragraphs, we discuss the condensed quintessence of our findings.

4.1 Perceived Ease of Use

Consumers found mobile wallets and Bitcoin in general difficult to use, therefore requiring some training, although one interviewee regarded Bitcoin as easy and self-explanatory. As buying, selling, or sending Bitcoin is very similar to using a Google Wallet, it is possible that more consumers will come across technical difficulties when using Bitcoin for the first time. They agreed that especially the younger generation

Table 2. Findings and case coding

Research Constructs	Codes	Opinions of Individuals C = Consumer M = Merchant B = Bitcoin Exchange	Exemplifying Quotes
Perceived Ease of Use	Mobile Wallets	useful, but cumbersome (C1/2/3); needs more development and refinement (C2); Bitcoin best use case for wallets (B3)	**C2:** "Google Wallet was really complicated and not easy to use" **C5:** "I have not yet dared to install some kind of Wallet"
	Usage / Implementation Process	easy and self-explanatory (C2); potentially easy (C1/C4/M2); easy and fast, hard at POS (M1/3); easy, but not easy enough for mainstream (B1/2/3)	**C1:** "sounds pretty easy" **B1:** "the general mainstream user has no clue about Bitcoin"
	Technical Understanding (Explanation)	easy handling (C2); merchants also need training (B2); easy, but concept and idea not yet understood by society (B1/4)	**M1:** "not a technical, but a real societal challenge"
	Capability (Education/Age)	some learning necessary (C3); only accepted by generation Y/Z (B3/4)	**C3:** "you have to learn how to use it" **C1:** "certainly, the under 40 crowd is far more likely to embrace it, if it is convenient in the end"
	Perceived Risks	volatility (M1/2); advances in technology and usability of other mobile payments (B1); security and storage (B2); use as speculative trading device (B1/2/3)	**B3:** "the volatility of Bitcoin and the frictions of moving money in and out of Bitcoin are a limiting factor in its adoption" **B2:** "Bitcoin also comes with the burden of keeping it secure and accessible at the same time. No really good solution exists yet that solves both problems satisfactorily"

(Continued)

Table 2. (*Continued*)

	Convenience	cuts out annoyance of searching for cash at POS (C1); Bitcoin more convenient than rest (C2/3); with QR code very convenient (B1/3)	**M1:** "convenient mobile payments arose at the expense of security standards" **B1:** "paying with Bitcoin is both convenient and secure. We just need to scan a QR code in these days and our payments are transferred successfully. Anywhere, anytime" **B2:** "just scan the QR code, done!"
Perceived Usefulness	Anonymity	big problem with CCs, Bitcoin may be solution (C1/2/3); anonymity not an issue for consumers, only important for special target industries (M1/2/3, B1/2/3/4)	**C2:** "people are more and more paranoid about their privacy" **M1:** "maybe relevant in gambling and sex business" **M2:** "higher anonymity will not increase payment share of Bitcoin"
	Security / Fraud	involved in CC fraud (C1), high perceived level of security with Bitcoin (C1); 100% security not possible (all); fraud of CCs not main argument (M1/2/3); security a very strong Bitcoin bonus (B1/2/3)	**C1:** "American CCs are not as secure as European ones that have chips in them, we have worse technology than Mongolia" **M1:** "CC fraud covered by insurance" **M2:** "Bitcoin accounts can be hacked as well and used for transactions, hence fraud risk still exists" **B2:** "blockchain technology serves as a trust anchor for both parts of a trade"
	Transaction Fees	alternatives to CC highly needed (C2); lower fees most persuasive argument for Bitcoin (C1), wiring fees exorbitant (C1); payout in local currency will have high fees (M1); low fees essential (M1/2/3); esp. relevant to merchants (B1/2/3)	**C2:** "airlines ripping people off when paying with CC, very frustrating" **M1/2/3:** "most promising argument for Bitcoin" **M3:** "interesting for micropayments" **B1:** "transaction costs have 'skyrocketed' in these days and hence leave merchants from certain industries at a competitive disadvantage"
	Instant Transactions	more relevant to merchants (C1/2/3); instant transaction very helpful (M3/4); time and no chargebacks strong arguments pro Bitcoin (B1/2/3)	**M2:** "receiving payout faster from customers to us has its benefits and we would prefer it" **M3:** "time is money, so great" **B1:** "for merchants, Bitcoin displays a direct money inflow of their revenues, whereas with CC payment they need to wait for around a week to receive their payments"
	Global Reach & Cross-Border	great not to search for ATM abroad (C1/3); Bitcoin attractive for remittance (C3); international speed and simplicity very persuasive (B1/2/3)	**B1:** "SEPA has failed, a technology based system like Bitcoin needed to increase speed and decrease fees at the same time" **B2:** "transfer money everywhere in the world at velocity of sound. Payment providers like Western Union have made that possible too, however at high costs"

(*Continued*)

Table 2. (*Continued*)

	Trading / Investment	Bitcoin as a trading currency (C1/2); if not mainstream payment, then alternative investment tool (B2/4)	**B2:** "a secure and efficient, trustful decentralized monetary transfer system with an inherent currency unit that is of non-governmental and non-corporate origin, perfect for investment"
	Emerging Economies	high attractiveness for developing world (all); challenges very different to our market (C2/3); no political restrictions, embargos, capital controls (B2/3)	**B2:** "highly necessary for countries with a high percentage of underbanked people. Why shouldn't a web shop be able to sell products wherever it can ship it? Most major package delivery services have a wider reach than PayPal or MasterCard"
Subjective Norm	Innovation / Technology	Bitcoin on the forefront (C2); makes mobile wallets attractive (C1/3); Fin-Tech innovations very important (C1/2/3, B1/2/3); innovation reason nr 1 for Bitcoin (M1); innovation crucial for us (M1/2/3)	**M1:** "innovation is all we have, Bitcoin helps!" **M3:** "in my industry (press and media), innovation is key. We can no longer survive if we don't go digital. We will definitely continue focusing on becoming more innovative" **B2:** "the financial industry has not come up with a good candidate for an online currency / payment system, so Bitcoin is it"
	Competitive Advantage	competitive pressure not that large yet (M2/3); merchants will feel the pressure soon with higher penetration (B1/2/3)	**M3:** "accepting Bitcoin is a competitive advantage right now, for how long remains to be seen" **M2:** "we are looking into Bitcoin but will not expect to adopt it for another 1-2 years. So far most competitors don't accept Bitcoin as it is still a product for early adopters / innovators and misses scale and acceptance in the market"
	Peer Influence	friend trusted more than journalist (C1/5); p2p character of Bitcoin fosters peer influence (B2)	**C1:** "if someone says that Bitcoin is a very practical, efficient and secure way of doing things, I would check it out" **B1:** "word of mouth is needed in order to spread the word and explaining the beauty behind Bitcoin"
	Latest Trends	trends not of relevance (C1/2); people more conservative with latest trends in financial issues (B2); direct, peer-to-peer transfers may be considered cool (B1)	**M1:** "exciting thing, especially for Techies, but from a bottom line perspective, not relevant yet" **B2:** "the traditional financial system has not been very keen on adopting new technologies and when it comes to money, people tend to be more conservative than with other things"
	Lifestyle / Marketing	Bitcoin great marketing asset (M1/3/4)	**M3:** "if a whole industry does it, it makes sense to adopt it. As for now, we us it, but Bitcoin is not a big thing yet" **B1:** "Bitcoin could become a lifestyle product, people will find it increasingly cool to use"
	Alternative: PayPal	no viable alternative (C1/2); good alternative (all others)	**C1:** "takes time to register" **C2:** "another account to keep track of"
	Alternative:	favorite means of payment	**C1/2:** "typing in the details again and again is

Table 2. (*Continued*)

Future Potential	Credit Card (CC)	(C1/2); insecure (B1/2/3); OK (M1/2/3)	annoying" **M1:** "used by majority of costumers"
	Alternative: Others	bank transfer / wire (all); Master-pass / Sofortueberweisung (all but C1/2/3)	**B3:** "other alternatives will of course remain, but become less and less important"
	Current Subjective Market Penetration	not clear who accepts Bitcoins currently (C1/2/3); still a niche form of payment, low demand (M1/3); not really accepted (M2); customer demand drives penetration (B1/3/4)	**C2:** "recognizing more 'Pay with Bitcoin'-buttons on official websites people would boost credibility" **B3:** "current lack of easy-to-use and secure solutions for storing and using Bitcoin without third-party services do not make Bitcoin's advantages apparent to new users"
	Subjective Future Market Potential	payout in local currency will boost acceptance (M1/2); more and more solutions pave way for great potential (B2)	**C2:** "will be the future", "easiest, most secure and fastest way of payment" **M1:** "will become a big thing in the future" **M2/3:** "will be big, but never replace CCs" **B3:** "Bitcoin is a very promising candidate for the currency of the internet"
	Time Until Widespread Adoption	quick adoption (C3); very difficult to guess (all but C3); probably very different speed in different regions of world (M1/2, B1/2/3)	**M1:** "in IT stuff, I stopped giving prognoses" **B1:** "hard to tell, maybe 3 to 5 years, also depending on regulation/taxation/legislation"

should manage to handle Bitcoin and related software/apps, however. These findings are similar to those of Spenkelink [42]. Risks were not mentioned, but instead consumers focused on convenience, especially at offline stores (POS).

One merchant complained about the implementation process at POS and stressed that it's not a technical, but rather a societal challenge to switch to Bitcoin. Implementation in online shops was seen as fast and easy, and similar to other payment options. Most saw volatility as the main threat and agreed that current mobile payment solutions are convenient, but insecure; maybe Bitcoin can be a remedy here.

The representatives of Bitcoin exchanges saw Bitcoin as the best use case for mobile wallets. Even though they judged usage and implementation as easy, they still found it not easy enough. In fact, the whole cryptographic system behind Bitcoin was deemed just too complex to be understood by users and merchants, and it discouraged them from giving Bitcoin a try. However, their own companies and the majority of other exchanges already focus on emphasizing the easy set-up, integration, and usage of Bitcoin. This has been confirmed by a check on the landing pages of Bitpay, Cubits, Coinbase, and Safello.[3] Here, tutorials, videos, and other training and marketing activities help users get started.

---

[3] https://bitpay.com; https://cubits.com; https://www.coinbase.com; https://safello.com.

4.2 Perceived Usefulness

For consumers, anonymity and security rank very high. They feel threatened by credit card fraud and the old technology used, and consumers hope Bitcoin can be a remedy here. Due to high transaction fees of credit cards and especially money wiring (fees here seen as exorbitant), the lower costs are seen as the major argument for using Bitcoin. The instantaneous character of transactions was seen most relevant to merchants. The fact of doing away with ATM searches abroad once Bitcoin is also widely accepted at POS, and the possibility of doing remittances as well as using Bitcoin as an investment vehicle were all seen as very positive. Finally, Bitcoin was found especially suitable for emerging economies, as they face very different challenges (e.g., lack of availability of traditional banking services).

Merchants had a quite different view on anonymity and fraud: Anonymity was only crucial for special target groups, e.g., in online gambling or the sex business, not for mainstream costumers, thus stating the opposite of Pagliery [57]. The security of Bitcoin was also not necessarily seen as higher than that of credit cards, as Bitcoin accounts can also be hacked, which has happened before. Low transaction fees were the single most important argument for merchants, also in regard to micropayments. Having the payout of orders immediately on their accounts was also seen as very advantageous, as they are well aware that time is money.

Bitcoin exchange staff is more convinced about Bitcoin's usefulness regarding the elimination of fraud. All experts believe that security will become an increasing problem in the near future, which is why more secure payment solutions are necessary, and this is a big benefit of Bitcoin. In terms of anonymity, the necessity of finding ways to get rid of skyrocketing transaction fees, and the beneficial effect of instant transactions, the experts strongly agreed with the merchants. In terms of global and cross-border reach, the managers quoted pure speed and simplicity as persuasive arguments in favor of Bitcoin. Whereas one interviewee judged SEPA as failed, another mentioned Western Union as working, but at very high costs. Bitcoin as investment was regarded as ideal, as there is no government or corporate involvement. This is in strong opposition to the findings of Baek and Elbeck [27] and Garcia et al. [77], who see Bitcoin as far too speculative to be used by a non-professional investor. Lastly, emerging economies were mentioned as a suitable target market, since online shops often times can ship to remote locations, but cannot receive funds from there due to capital market restraints.

4.3 Subjective Norm

In terms of innovation and technology, consumers rated Bitcoin to be very innovative, and it helps making mobile wallets attractive. In terms of peer influence, the interviewees so far unfamiliar with the (Bitcoin) process validated that they would give it a try upon recommendation of a friend. Two consumers denied that they give any relevance to current trends or lifestyle issues, i.e., Bitcoin plays no role here.

Merchants focused on the huge importance of innovation for their business, making this category the most decisive factor to offer Bitcoin payments. However, competitive pressure itself is currently still manageable, with one merchant confirming that they will look into it, but will probably not introduce it for another 1–2 years. One merchant claims Bitcoin to be a competitive advantage. In terms of trends, one merchant mentioned that it is a cool thing, especially for early-adopters/techies, but has no effect on revenue or profit, i.e., the bottom line. This statement basically confirms the research of Kostakis and Giotitsas [78]. The merchant also claims it to be somewhat a marketing tool, a thing to arouse interest and stand out from the crowd.

As the last group, Bitcoin exchange executives also stress fin-tech innovations as the key driver for competitiveness,[4] making Bitcoin an important part of it. From their point of view, merchants should be preparing themselves, as competitive pressures to offer it as a payment solution will rise soon, triggered through increased word-of-mouth between users. Whether Bitcoin can be considered a 'cool trend', there are two different opinions: One manager confirms this, insisting that Bitcoin could become a lifestyle product, but another reminds us that when it comes to money issues, people are by far more conservative when compared to other areas of life.

4.4 Future Potential

The customers' overall opinion about Bitcoin is very positive; they perceive it as very promising. Nevertheless, credit cards, PayPal, and some other payment services are still and will for some time in the future be the most popular online payment method. Customers are aware of the benefits offered by Bitcoin and alternative means of payments in general, but experience a certain level of effort, through a cumbersome registration and administration effort. Issues with credit cards, e.g., manually typing in the card number and security code, seem to be tolerated until better solutions are developed and more transparency is created as to which shops actually accept Bitcoin.

Merchants strongly support cryptocurrency. Regardless of all arguments and restrictions discussed above, from a business perspective, Bitcoin seems to provoke large interest, and is seen as the future (therefore in line with Van Alstyne [33]). Hence, already two of the four interviewees accept Bitcoin. Nevertheless, the current customer demand is very low, so Bitcoin is still a pure niche product and will take time to develop.

As expected, the managers of Bitcoin exchanges have a very optimistic and positive view of Bitcoin's potential. However, the lack of secure storage and usage solutions hinders market penetration. A widespread initiative to push Bitcoin would be welcomed, and the type and extent of regulation will be decisive for its market penetration.

The high importance of educating people and businesses about the actual purpose and functioning of Bitcoin can be regarded as the unifying factor of all groups (in line with Papilloud and Haesler [79]). All interviewees also agreed that the actual timing of widespread adoption cannot be judged at all.

Interestingly, interviewees of all groups also mentioned the underlying blockchain as an idea and technology that could have a far-reaching, disruptive potential outside of

---

[4] European Commission: http://europa.eu/rapid/press-release_MEMO-13-719_en.htm.

the payment area. Applications as diverse as document version control, proof of whether a person's vote has been counted or clear identification for e-government and open data solutions were addressed.

5 Conclusion

5.1 Contributions

With this paper, we have contributed to answering several research questions relating to Bitcoin's relative advantages and disadvantages, the drivers and barriers of adoption, and the perceived future potentials from user and professional perspectives.

First, perceived ease of use among stakeholders is still considered rather low. Issues of usability of mobile wallets, implementation, storage, and transfer as well as offering user training are not solved satisfactorily yet. Risks like volatility, security, and accessibility remain. However, people who are using it confirm its high level of convenience. In total, some homework needs to be done here.

Second, perceived usefulness is confirmed for the main part, albeit not yet for the majority of online shoppers. Low transaction fees are central and pivotal to all interviewed individuals. A smaller influence is given to a somewhat higher anonymity and lower risk of fraud as well as faster claim of payments for merchants. Potential international reach and acceptance were also convincing. Bitcoin as a new investment class is controversial, but the high added value for less-developed countries is not.

Third, subjective norm is somewhat split. To use Bitcoin in innovation and in creating competitive advantage is for all groups of high importance. Peer-influence, the fact of being trendy or not, as well as lifestyle issues have less influence on the subjective norm of using Bitcoin.

Fourth, future potential was questioned by comparing it with alternatives. All of them are deemed important, but with a diminishing degree. Current market penetration is judged as very low, recognizing Bitcoin as a niche phenomenon. Subjective future potential is quite the opposite, with all groups assigning a boost in penetration and eventually becoming 'a big thing'. The actual timing of this boost was, however, not answered by anyone.

Fifth, this piece of research has found some evidence that Bitcoin, due to its completely new way of working without a central institution overseeing it, is about to become a serious new player in the online payment market. However, society and businesses are still far from embracing Bitcoin in their daily lives, making educating people about Bitcoin's advantages and use necessary. But Bitcoin's conditions seem good to become fully accepted and trusted by first addressing certain industries and later convincing the masses. One step in that direction may be seen in the latest tech-giant Microsoft accepting Bitcoin for Apps, Windows licenses, Windows Phone, and Xbox in December 2014.[5] Hence, policy-makers and merchants should be prepared.

---

[5] http://blogs.microsoft.com/firehose/2014/12/11/now-you-can-exchange-bitcoins-to-buy-apps-games-and-more-for-windows-windows-phone-and-xbox.

Finally, the interviews brought additional areas of use of the blockchain scheme to the surface. The possibility to issue unique identifiers to things and people alike opens up immense opportunities in the future.

From a theoretical point of view, we help to expand the body of knowledge about users' acceptance, views, opinions, and feelings about cryptocurrencies—a very new social phenomenon. Additionally, online shops, policy-makers, and regulators should be alarmed that Bitcoin is indeed not just a crazy idea from some IT nerds, but a possible game changer for the future. They should take adequate measures to handle it.

5.2 Limitations and Future Research

Our paper suffers from some limitations. First, only 13 individuals were interviewed, which makes generalizations difficult, as is the case with most inductive, qualitative research. Based on these interview findings, as a follow-up, hypotheses could be formulated that are then tested in a large-scale quantitative survey in a deductive research setting. These survey results would then make generalizations possible. Second, there may be a bias that almost one-third of the interviewees work for Bitcoin exchanges and hence have an apparent interest in seeing cryptocurrencies as too positive. However, very new and potentially disruptive innovations demand to include the voice of professionals, who can judge the whole scope of the research object [65].

Overall, the paper lays a first foundation to spark future research avenues. Comparing developments in different parts of the world could debunk differences due to risk-aversion (e.g., very high in Germany), methods of payment (e.g., widespread credit card use in the US), or technical availability (e.g., lack of traditional banking infrastructure in developing countries). Lastly, including the opinions of people from regulatory and administrative authorities could help in getting the complete picture of cryptocurrencies. And lastly, focusing more on the non-payment aspects of the blockchain technology to revolutionize society in the future may yield great additions to the body of knowledge in the social sciences.

These are just a few ideas contributing to the massive research effort that is still needed to completely understand the emergent and possibly disruptive nature of cryptocurrencies, and to see whether the blockchain can also contribute to a more just society, based on open platforms and open data with access for everyone.

References

1. Foster, G., Gupta, M., Palmer, R.: Business-to-Business Electronic Commerce. Cases in Strategic-Systems Auditing (1999)
2. Bourgeois, R.: Impact of PayPal, Google, Amazon & Emerging Payment Providers on Visa, MasterCard & Payment Industry. The Long-View: 2010 Edition – U.S. Perspectives. Bernstein Global Wealth Management, pp. 195–206 (2010)
3. Olsen, Ø.: Annual Report on Payment Systems 2012 (2013)
4. Koley, T.: End of duopoly in credit card payment scheme industry. J. Econ. Finance 4, 67–76 (2014)

5. Gonggrijp, S., Geerling, M., Mallekoote, P.: Successful introduction of new payment methods through 'co-opetition. J. Payments Strategy Syst. **7**, 136–149 (2013)
6. Conrad, L.: En garde! banks and paypal will clash. U.S. Banker **117**, 102 (2007)
7. Pogue, D.: The future is plastic. Sci. Am. **312**, 35 (2015)
8. Dempsey, P.: Is it time to take a reality check on the mobile payments market. Eng. Technol. **9**, 20 (2014)
9. Cusumano, M.A.: The Bitcoin ecosystem. Commun. ACM **57**, 22–24 (2014)
10. Brito, J., Castillo, A.: Bitcoin: a primer for policymakers. Policy **29**, 3–12 (2014)
11. Grinberg, R.: Bitcoin: an innovative alternative digital currency. Hastings Sci. Technol. Law J. **4**, 160–208 (2011)
12. Glaser, F., Zimmermann, K., Haferkorn, M., Weber, M.C., Siering, M.: Bitcoin - asset or currency? revealing users' hidden intention. In: European Conference on Information Systems, pp. 1–14 (2014)
13. Bohr, J., Bashir, M.: Who uses Bitcoin? an exploration of the Bitcoin community. In: Twelfth Annual International Conference on Privacy, Security and Trust (PST), pp. 94–101 (2014)
14. vom Brocke, J., Simons, A., Niehaves, B., Riemer, K., Plattfaut, R., Cleven, A.: Reconstructing the giant: on the importance of rigour in documenting the literature search process. In: European Conference on Information Systems, pp. 2206–2217 (2009)
15. Levy, J., Ellis, T.J.: A systems approach to conduct an effective literature review in support of information systems research. Informing Sci. J. **9**, 181–212 (2006)
16. Webster, J., Watson, R.T.: Analyzing the past to prepare for the future: Writing a literature review. MIS Q. **26**, 13–23 (2002)
17. Chen, H., Chiang, R.H.L., Storey, V.C.: Business intelligence and analytics: from big data to big impact. MIS Q. **36**, 1165–1188 (2012)
18. Nakamoto, S.: Bitcoin: A Peer-to-Peer Electronic Cash System (2008)
19. Scott, B.: Peer-to-Peer Review: The State of Academic Bitcoin Research (2014). http:// suitpossum.blogspot.de/2014/12/academic-bitcoin-research.html
20. Böhme, R., Christin, N., Edelman, B., Moore, T.: Bitcoin: economics, technology, and governance. J. Econ. Perspect. **29**, 213–238 (2015)
21. Verbücheln, S.: How Perfect Offline Wallets Can Still Leak Bitcoin Private Keys (2015)
22. Andrychowicz, M., Dziembowski, S., Malinowski, D., Mazurek, L.: Secure multiparty computations on Bitcoin. In: 2014 IEEE Symposium on Security and Privacy (SP), pp. 443–458 (2014)
23. Babaioff, M., Dobzinski, S., Oren, S., Zohar, A.: On Bitcoin and red balloons. In: Faltings, B., Leyton-Brown, K., Ipeirotis, P. (eds.) EC 2012 Proceedings of the 13th ACM Conference on Electronic Commerce, pp. 56–73. ACM, New York (2012)
24. Feld, S., Schönfeld, M., Werner, M.: Traversing Bitcoin's P2P network: insights into the structure of a decentralised currency. Int. J. Comput. Sci. Eng. (forthcoming)
25. Biryukov, A., Khovratovich, D., Pustogarov, I.: Deanonymisation of clients in Bitcoin P2P network. CoRR abs/1405.7418 (2014)
26. Decker, C., Wattenhofer, R.: Information propagation in the Bitcoin network. In: 2013 IEEE Thirteenth International Conference on Peer-to-Peer Computing (P2P), pp. 1–10 (2013)
27. Baek, C., Elbeck, M.A.: Bitcoins as an investment or speculative vehicle? a first look. Appl. Econ. Lett. **22**, 30–34 (2015)
28. Wu, C.Y., Pandey, V.K.: The value of Bitcoin in enhancing the efficiency of an investor's portfolio. J. Financ. Planning **27**, 44–52 (2014)
29. Ali, R., Barrdear, J., Clews, R., Southgate, J.: Innovations in payment technologies and the emergence of digital currencies. Bank Engl. Q. Bull. **54**, 262–275 (2014)

30. Kristoufek, L.: BitCoin meets google trends and wikipedia: quantifying the relationship between phenomena of the internet era. Sci. Rep. **3**, 1–7 (2013)
31. Kondor, D., Csabai, I., Szüle, J., Pósfai, M., Vattay, G.: Inferring the interplay between network structure and market effects in Bitcoin. New J. Phys. **16**, 1–10 (2014)
32. Raiborn, C., Sivitanides, M.: Accounting issues related to Bitcoins. J. Corp. Acc. Finance **26**, 25–34 (2015)
33. van Alstyne, M.: Why Bitcoin has value. Commun. ACM **57**, 30–32 (2014)
34. Holmquist, E.: Bitcoin and the coming revolution in financial transactions. RMA J. **97**, 23–29 (2014)
35. Angel, J.J., McCabe, D.: The ethics of payments: paper, plastic, or Bitcoin? J. Bus. Ethics, 1–9 (2014)
36. Zarifis, A., Efthymiou, L., Cheng, X., Demetriou, S.: Consumer trust in digital currency enabled transactions. In: Abramowicz, W., Kokkinaki, A. (eds.) BIS 2014 Workshops. LNBIP, vol. 183, pp. 241–254. Springer, Heidelberg (2014)
37. Lin, P., Chung, P.-C., Fang, Y.: P2P-iSN: a peer-to-peer architecture for heterogeneous social networks. IEEE Netw. **28**, 56–64 (2014)
38. Maurer, B., Nelms, T.C., Swartz, L.: "When perhaps the real problem is money itself!": the practical materiality of Bitcoin. Soc. Semiot. **23**, 261–277 (2013)
39. Ahamad, S.S., Nair, M., Varghese, B.: A survey on crypto currencies. In: 4th International Conference on Advances in Computer Science, AETACS, pp. 42–48 (2013)
40. Ametrano, F.M.: Hayek Money: The Cryptocurrency Price Stability Solution. SSRN J., 1–54 (2014)
41. Wiatr, M.: Bitcoin as a Modern Financial Instrument (2014)
42. Spenkelink, H.: The adoption process of cryptocurrencies. Identifying factors that influence the adoption of cryptocurrencies from a multiple stakeholder perspective (2014)
43. Vigna, P., Casey, M.J.: The Age of Cryptocurrency. How Bitcoin and Digital Money are Challenging the Global Economic Order. St. Martin's Press, New York (2015)
44. Dinu, A.: The Scarcity of Money: The Case of Cryptocurrencies. Master' Thesis (2014)
45. Papp, J.: A medium of exchange for an internet age: how to regulate Bitcoin for the growth of e-commerce. Pittsburgh J. Technol. Law Policy **15**, 33–56 (2014)
46. Bayern, S.: Dynamic common law and technological change: the classification of Bitcoin. Wash. Lee Law Rev. Online **71**, 22–34 (2014)
47. Citi GPS: Disruptive Innovations II – Ten More Things to Stop and Think About (2014)
48. Andersson, G., Wegdell, A.: Prospects of Bitcoin. An evaluation of its future (2014)
49. Antonopoulos, A.M.: Mastering Bitcoin: Unlocking Digital Cryptocurrencies. O'Reilly & Associates, Sebastopol (2015)
50. Descôteaux, D.: Bitcoin: More Than a Currency, Potential for Innovation. (2014)
51. Dostov, V., Shust, P.: Cryptocurrencies: an unconventional challenge to the AML/CFT regulators? J. Fin. Crime **21**, 249–263 (2014)
52. Evans-Pughe, C., Novikov, A., Vitaliev, V.: To bit or not to bit? Eng. Technol. **9**, 82–85 (2014)
53. Stokes, R.: Virtual money laundering: the case of Bitcoin and the Linden dollar. Inf. Commun. Technol. Law **21**, 221–236 (2012)
54. Bryans, D.: Bitcoin and money laundering: mining for an effective solution. Indiana Law J. **89**, 441–472 (2014)
55. Brezo, F., Bringas, P.G.: Issues and risks associated with cryptocurrencies such as Bitcoin. In: Berntzen, L., Dini, P. (eds.) SOTICS 2012. The Second International Conference on Social Eco-informatics, pp. 20–26. IARIA, USA (2012)
56. Trautman, L.J.: Virtual currencies: Bitcoin & what now after liberty reserve and silk road? Richmond J. Law Technol. **20** (2014)

57. Pagliery, J.: Bitcoin and the Future of Money. Triumph Books, Chicago (2014)
58. Krugman, P.R.: Bitcoin is evil. New York Times 2012 (2013)
59. Cheung, A., Roca, E., Su, J.-J.: Crypto-currency bubbles: an application of the Phillips–Shi–Yu (2013) methodology on Mt. Gox bitcoin prices. Appl. Econ. **47**, 2348–2358 (2015)
60. Ghauri, P.N., Grønhaug, K.: Research Methods in Business Studies. A Practical Guide. Financial Times Prentice Hall, Harlow (2005)
61. Creswell, J.W., Plano Clark, V.L.: Designing and Conducting Mixed Methods Research. Sage, Thousand Oaks (2007)
62. Denzin, N.K., Lincoln, Y.S.: The Sage Handbook of Qualitative Research. Sage, Thousand Oaks (2011)
63. Hair, J.F., Wolfinbarger, M., Money, A.H., Samouel, P., Page, M.J.: Essentials of Business Research Methods. M.E. Sharpe, Armonk (2011)
64. Miles, M.B., Huberman, A.M.: Qualitative Data Analysis. Sage, Thousand Oaks (1994)
65. Eisenhardt, K.M.: Building theories from case study research. Acad. Manag. Rev. **14**, 532–550 (1989)
66. Kvale, S., Brinkmann, S.: InterViews. Learning the Craft of Qualitative Research Interviewing. Sage, Thousand Oaks (2015)
67. Strauss, A., Corbin, J.: Grounded Theory: Grundlagen Qualitativer Sozialforschung (Basics of Qualitative Research). Beltz, Weinheim (1996)
68. Miles, M.B., Huberman, A.: Michael: drawing valid meaning from qualitative data: toward a shared craft. Educ. Researcher **13**, 20–30 (1984)
69. Hesse-Biber, S.N., Leavy, P.: The Practice of Qualitative Research. Sage, Thousand Oaks (2006)
70. Flick, U.: Managing Quality in Qualitative Research. Sage, London (2008)
71. Shenton, A.K.: Strategies for ensuring trustworthiness in qualitative research projects. Educ. Inf. **22**, 63–75 (2004)
72. Davis, F.D.: Perceived usefulness, perceived ease of use, and user acceptance of information technology. MIS Q. **13**, 319–340 (1989)
73. Davis, F.D., Bagozzi, R.P., Warshaw, P.R.: User acceptance of computer technology: a comparison of two theoretical models. Manag. Sci. **35**, 982–1003 (1989)
74. Venkatesh, V.: Determinants of perceived ease of use: integrating control, intrinsic motivation, and emotion into the technology acceptance model. Inf. Syst. Res. **11**, 342 (2000)
75. Venkatesh, V., Davis, F.D.: A theoretical extension of the technology acceptance model: four longitudinal field studies. Manag. Sci. **46**, 186 (2000)
76. Silverman, D.: Doing Qualitative Research. Sage, London (2013)
77. Garcia, D., Tessone, C.J., Mavrodiev, P., Perony, N.: The digital traces of bubbles: feedback cycles between socio-economic signals in the Bitcoin economy. J. R. Soc. Interface **11**, 1–28 (2014)
78. Kostakis, V., Giotitsas, C.: The (a)political economy of Bitcoin. TripleC **12**, 431–440 (2014)
79. Papilloud, C., Haesler, A.: Distinktion: the veil of economy: electronic money and the pyramidal structure of socieries. Scand. J. Soc. Theory **15**, 54–68 (2014)

A Systematic Review of Impediments Blocking Internet of Things Adoption by Governments

Paul Brous[1,2(✉)] and Marijn Janssen[1]

[1] Delft University of Technology, Delft, The Netherlands
{P.A.Brous,M.F.W.H.A.Janssen}@tudelft.nl
[2] Rijkswaterstaat, Delft, The Netherlands

Abstract. The Internet of Things (IoT) has high promises and might provide many benefits, yet has been given scant attention in e-government literature. Within the IoT, physical objects, "things", are networked and connected to the Internet. These "things" are able to identify themselves to and communicate with other devices or "things". There are many impediments blocking the adoption of IoT, and there is limited insight in these barriers. In this paper, impediments for the adoption of IoT are investigated by conducting a literature review and carrying out two case studies. The impediments found in literature were confirmed and extended using the case studies. Results show that impediments are interrelated and occur on the strategic, tactical and operational level. For adoption the impediments needs to be addressed in concert. Research on e-governance can benefit from understanding these interrelated impediments.

Keywords: Internet of things · IoT · Adoption · Open data · e-governance · e-government · Smart cities · Impediments · Barriers · Challenges

1 Introduction

The term, the Internet of Things (IoT) refers to the increasing network of physical objects that feature an IP address for internet connectivity, and the communication that occurs between these objects and other Internet-enabled devices and systems [1–3]. The IoT makes it possible to access remote sensor data and to monitor and control the physical world from a distance, allowing many physical objects to act in unison, though means of ambient intelligence [1]. These devices and the communication between these devices can benefit e-government by providing enough quality data to generate the information required to make the right decisions at the right time [3], but in order to achieve this, a variety of impediments need to be overcome. For example, Inductive loops embedded in the road surface are a key technology for traffic detection. An inductive loop is a simple and reliable way to detect the movement of vehicles over a road surface and is extensively used in traffic responsive traffic signal systems to collect traffic data to optimise signal timings accordingly [2]. Such loops provide data on traffic density, flows and speeds for trend analysis as well as providing a key input to real-time traffic models which predict queues or delays. However, installing these loops is costly and the flow of traffic can be obstructed during installation. Another example is that of the application of sensors for the inspection and testing of levees

© IFIP International Federation for Information Processing 2015
M. Janssen et al. (Eds.): I3E 2015, LNCS 9373, pp. 81–94, 2015.
DOI: 10.1007/978-3-319-25013-7_7

(smart levees) in levee management [4]. The sensors embedded in the levees supply a wide range of data. This data is centrally stored and used for the real time visualization of the measurements in a dashboard displaying the sensor results. The data is then directly interpreted for detection and warning systems. These sensors are increasingly being used for the management and monitoring of water barriers, but the technology and the models required to fully analyse the data are still in their infancy and managers are unable to fully trust the system.

The benefits of IoT for governments are known [3] and often emphasized in work. However, less attention has been given to the impediments, or barriers, of IoT, especially with regards to the management and maintenance of large physical infrastructure, have till now not been investigated systematically. Several researchers mention the need for further research in this area [6]. This research explores systematically the potential impediments of the IoT by investigating real world case studies and reviewing state of the art literature.

The methodology used in this research is described in section two. A first overview of IoT impediments will be presented in section three on the basis of state of the art literature. Explorative case studies at the Directorate General of Public Works and Water Management of the Netherlands will be presented in Sect. 4. The Directorate General of Public Works and Water Management of the Netherlands is commonly known within The Netherlands as "Rijkswaterstaat", often abbreviated to "RWS", and is referred to as such within this research. RWS is part of the Dutch Ministry of Infrastructure and the Environment and is responsible for the design, construction, management and maintenance of the main infrastructure facilities in the Netherlands. The results of the literature review and the case studies, and the potential impediments of IoT adoption in e-governance will be discussed in section five. The results show that IoT has a variety of potential impediments at the strategic, tactical and operational levels. Finally conclusions will be drawn in section six.

2   Research Method

We followed two main research steps to determine the potential impediments of IoT for e-governance. First the common impediments of IoT were identified from a rigorous review of literature. The keywords: "Internet of Things" (or "IoT"), "impediments" (or "barriers"), and "e-governance", returned zero hits within the databases Scopus, Web of Science, IEEE explore, and JSTOR. When we replaced the keyword "e-governance" with "governance", we retrieved the same result. The query [all abstract: ("impediments" OR "barriers") AND "internet of things" AND "e-governance"] searching between 2000 and 2015 returned fifty-five hits in Google Scholar. Removing the word "governance" totally from the search string returned more results (67, 0, 13, 6, and 3170 hits respectively).

We found that a great deal of these articles mentioned IoT as being a potential facilitator for achieving the goal of a Smart City based on IoT technology, and some touched on the impediments, barriers or challenges of the implementation of IoT, but most articles relied on anecdotal evidence. Very few articles found were of a general nature. We then filtered these results and performed a forward and backward search and selected relevant articles based on the criteria that they specifically referred to potential impediments, barriers or challenges with regards to the use or implementation of IoT within potential e-governance applications. The results of the literature lead to a framework within which we developed the case studies.

We used explorative case studies to extend and refine the list from literature the potential impediments of IoT within e-governance applications as the second main research method. Two cases were studied within the context of RWS, which gave the researchers access to subject matter experts and internal documentation for all the cases. This helped ensure the construct validity of the case studies [8]. The cases were selected based on their use of IoT for e-governance purposes – the unit of analysis being programmes within RWS which use and develop IoT for e-governance purposes. The Netherlands is an e-participation leader according to the United Nations e-government survey [9] (2014). This contributes to the validity of the cases as being good representations of e-governance. The cases under study were selected from different domains within RWS in order to ensure diversity and external validity through replication logic [8], in which each case serves as a distinct experiment that stands on its own as an analytic unit. The domains selected were road management and water management respectively.

We studied two separate cases to refine and extend the list of benefits from literature. In the Netherlands there are many similarities, but also, subtle differences in how processes are managed between the "wet" or, water management domain and the "dry", or road management domain. For example, when dealing with objects in the water domain, it is not always possible to be highly accurate with regards to location, as objects placed in water are less static and are often more difficult to physically get to than objects on the ground. We felt it necessary to select cases from both these domains in order to gain a more rounded perspective of the implementation of IoT within e-government in the Netherlands. The cases selected were: 1. Sensor information gathered for the purpose of road management; 2. Sensor information gathered for the purpose of water management. The first case deals with sensor information gathered by RWS with regards to traffic and road management. The second case study deals with sensor information gathered by RWS with regards to water management. The case studies were explorative in method and descriptive in nature. Unstructured interviews were held with managers, subject matter experts, and consultants within RWS. Internal documentation was also studied. Finally, the results of the cases were shared with and verified by subject matter experts within RWS. The pattern-matching technique [11] was used to analyse the case study evidence. Such logic compares an empirically based pattern (findings from the case studies) with a predicted pattern suggested by the literature review, strengthening the internal validity of the research [8]. The technique was applied in the following way. First the common impediments of IoT found in literature were listed. These common impediments were then compared with the evidence of the impediments of IoT from the case study analysis. There were several iterations throughout the research as each case introduced new potential impediments. The potential impediments of the IoT are expressed in italics within this paper.

3 Literature Background

Public and private organizations are increasingly turning to the IoT as new sources of data. However, there are several technological and regulatory challenges that need to be addressed. Scarfo (2014) believes that the most important of them are related to data ownership, security, privacy and sharing of information [12]. It is clear that the implementation of IoT for e-governance faces a variety of impediments. We list the possible impediments of IoT according to strategic/political, tactical and operational divisions. This is a popular division [13, 14], which is suitable for e-governance research.

3.1 Strategic/Political

Skarmeta et al. (2014) consider security and privacy to be the main obstacles for a full acceptance of IoT [15]. IoT devices generate a huge amount of data. The sensitivity levels of the information, is a crucial aspect to be considered by the access control mechanism. Disclosure of user data could reveal sensitive information such as personal habits or personal financial information. The unauthorized access to this information can severely impact user privacy [12, 15–20].

Data produced by IoT devices can be combined, processed and analysed, creating additional insights, so it is important to allow access to data generated by other IoT devices, whilst preventing the unauthorized access and misuse of this information [15]. However, as the IoT becomes more widespread, new security issues become evident [21]. Ortiz et al. (2013) believe that whilst these technologies have been widely investigated for traditional technologies such as relational databases, so far there are no convincing solutions for providing fine grained access control. This hinders the uptake of IoT in e-governance applications dealing with sensitive data [12, 15–22]. In this way, IoT requires novel approaches to ensure the safe and ethical use of the generated data [23], requiring a strong data governance [12, 18, 19, 24, 25]. A weak form of data governance can impede the safe and ethical use of data generated by IoT devices.

A lack of, or poorly coordinated, policies and regulations regarding IoT can also greatly impede the implementation and application of IoT. According to Misuraca (2009), IoT brings with it a wealth of new business opportunities. There is enormous scope for developing applications and selling new services [26]. Governments need to develop policy and regulations and position themselves carefully within this arena [19, 22, 25]. In this regard, public organisations should consider carefully the role they play in the enabling IoT development in the private sector. Market forces of supply and demand can play substantial roles in the success or failure of IoT [17, 26–28]. For example, according to Qiao et al. (2012) the IoT industry, in the short term, will demonstrate an inevitable outbreak growth at the growth stage of the Industry Lifecyle Theory [29]. The internal mechanism of explosive growth is that the whole networking industry chain achieves linkage development between supply and demand [27], but there is a danger that IoT may miss this linkage development due the chain of IoT industry being blocked by a tactical barriers such as a lack of technology break-throughs, standards bottlenecks and cost barriers [27].

3.2   Tactical

Although reduction in overall costs is an often cited benefit of IoT for e-governance [3], many researchers also cite high development and implementation costs as an important impediment to the implementation and application of IoT for e-governance [17, 19, 22, 27, 30]. According to Yazici (2014), high maintenance costs are often rated as the largest impediments to IoT implementation. A fully functional IoT system based on RFID technology can be substantial. By way of example, Yazici (2014) quotes Wal-Mart's vendors as having spent US$1 to US$3 million on a RFID implementation.

The Internet of things is more than one device, application or network. In order to ensure sustainable connectivity, all interfaces and communication protocols require unified industry standards [17]. However, Fan et al. (2014) believes that the large number of standards-setting organizations has led to a situation in which the top standard has not yet been set. Vendors are free to choose which standard they find best fits their production line, leading to a wide variety of available types which impedes interoperability and integration of data [12, 17, 22, 25, 28, 31]. According to Zeng et al. (2011), there are two methods to integrate things to the Internet: direct integration and indirect integration.

Home appliances are usually directly integrated whilst RFIDs are indirectly integrated through a RFID reader with an embedded server. It is not uncommon for a system to utilise both methods. But IoT requires that a large number of devices be integrated with the existing Internet. These devices can be diverse in terms of data communication methods and capabilities, computational and storage power, energy availability, adaptability, mobility, etc. The heterogeneity at the device level is, in this way, a serious impediment to IoT adoption [16]. This is especially complex as consumers of data are also heterogeneous [16]. Their needs vary in terms of capabilities and data quality. Furthermore, different applications might implement disparate data processing or filtering [16]. Zeng et al. (2011) believe that it is these heterogeneity traits of the overall system that make the design of a unifying framework and the communication protocols a very challenging task, especially with devices with different levels of capabilities. This issue is exacerbated in a large distributed environment.

According to Zeng et al. (2011), Universal Plug and Play (UPnP) is currently the most popular solution for personal network implementation [16]. However, there is no authentication protocol proposed for UPnP. All devices are allowed to configure the other devices on the personal network, without any user control. This can result in a critical security issue when the smart things become available on the Internet. The attention given to security by a number of authors [12, 15–18, 22] suggests that a lack of security standards is becoming a serious impediment to IoT implementation. Whilst there are many standard technologies and protocols to address many security threats, the severe constraints on the IoT devices and networks prevent a straightforward implementation of these solutions [15]. Furthermore, IoT devices generally have to work in harsh, uncontrolled environments, where they may be prone to attacks, misuse or malicious intentions [15]. If a mission critical system is hacked or becomes unavailable, this can lead to a breakdown of trust in the system [15, 17].

According to Kranenburg et al. (2014), the success of user-centric services based on IoT technology depends primarily on people participating and sharing the information flows [20]. Willingness on the part of people to participate in these systems is therefore required [16, 17, 19, 21, 24, 28, 30, 32]. Kranenburg et al. (2014) believe that this willingness is predominantly dependent on the perception of people: the perceived trust and confidence in IoT and the perceived value that the IoT generates for them. The greater the trust of users in the IoT, the greater their confidence in the system and the more willing they will be to participate [20]. A lack of trust in the system can be a strong impediment to the effectiveness of IoT.

3.3   Operational

Operational barriers include human capital issues such as difficulty in employing qualified personnel, lack of specialists, and personnel skill shortage to operate new applications [19, 22, 32], as well as insufficient IoT oriented training and educational activities [22]. Harris et al. (2015) also identify personnel reluctance to change or to learn new technology as a barrier. A lack of understanding about how IoT works, the possible benefits, and how to make the business case for IoT implementation were also found to be barriers by a number of researchers [19, 32–34]. Reyes et al. (2012) also includes calculating the return on investment and the payback period in this category [34].

Operational barriers also include technical issues such as limitations in information technology (IT) infrastructural capabilities [12, 16–20, 28, 35]. According to Scarfo (2014), the main technological challenges include architecture, energy efficiency, security, protocols and quality of service [12]. An important enabler for the IoT is to permit others to access and use the things that have been published publicly on the internet. It should be possible for users to make use of things that others have shared and to make use of things in their own applications, perhaps in ways unanticipated by the owner of the thing [31]. This requirement means we need a sophisticated set of mechanisms to publish and share things and ways to find and access those things [31]. A lack of these mechanisms as well as the level of knowledge required to implement, manage and maintain the available toolsets can form an important barrier to implementing IoT for e-governance purposes.

Data management issues are also of concern. Public organisations are often faced with a complex legacy of data and applications when implementing IoT solutions [24]. Many public organisations may have several generations of systems running in parallel, and much of the data fed into the system has been done manually, with associated risks in terms of data quality [24, 25, 31].

In short, IoT faces a variety of barriers related to the proper use (privacy and security for example) and proper management of the data collected by the vast number of interconnected things. *Strategic/political barriers are: data privacy issues, data security issues, weak or uncoordinated data policies, weak or uncoordinated data governance, and conflicting market forces. Tactical barriers include: costs, interoperability and integration issues, acceptance of IoT, and trust related issues. Operational issues are: a lack of sufficient knowledge regarding IoT, IT infrastructural limitations, and data management issues.*

4   Case Studies

The goal of the case study research was to refine and extend the list of impediments from literature and to understand the real life impediments of IoT in the most complete way possible. The case study research therefore involved the use of multiple data collection methods. The cases were selected from the primary processes of RWS. Generally IoT is implemented in RWS with the specific intention of ensuring the good working of the primary processes to achieve the primary objectives. In RWS there is a subtle divide in how processes are managed between the water management domain and the road management domain. In order to gain a rounded perspective of the benefits of IoT within RWS, it was believed necessary to select cases from both these domains.

4.1   Case Study 1: Road Management Data Collection at RWS

RWS builds, manages and maintains the Dutch national highways. Correct data is required to do this effectively. Over the years, RWS has developed several methods for obtaining the necessary data from the highways it manages, collecting, processing and making the data available to traffic and road management teams. Measurements are generally made by placing sensors in the road in many different locations. These sensors produce large amounts of data which is mainly used in mid-term planning, long term projections, air quality predictions and noise calculations which have an impact on health and safety measures as well as the environmental impact, and improving service efficiency with regards to road works management.

RWS has created a national network of monitoring points, the "Weigh in Motion" (WIM) network. At present, RWS estimates that at least 15 percent of freight traffic on the Dutch national road network is overloaded. Overloading of heavy vehicles causes road pavement structural distress and a reduced service lifetime [36]. Effectively reducing overloading reduces the damage to the road infrastructure, lengthening the road's lifetime and reduces the frequency of maintenance. The WIM network, consisting of measuring stations in the road on which the axle loads of heavy traffic is weighed, is used to support the enforcement of overloading by helping the enforcement agency to select overloaded trucks for weighing in a static location. The WIM system is one of the most advanced measuring systems in the world. Between 2010 and 2013, RWS built a nationwide WIM network with a total of 18 measuring points. The network consists of 6 newly remodelled measuring stations and 12 new measuring stations. The network provides access to the actual load of the main road, about peak times when it comes to overcharging and it provides RWS with the ability to collect information concerning the compliance behaviour of individual carriers. This forms the basis for business inspections and legal follow-up programs.

RWS faces and has faced a variety of impediments and challenges during the implementation and maintenance of the WIM network. There are different perceptions of the level of ambition pursued by the WIM maintenance process. For example, According to RWS officials, RWS has not yet implemented a structured learning cycle with regards to data quality – "the quality of the data has not been quantified, and solving data quality issues is incident driven". In this regard, learning takes place in practice and is not formally addressed. Although there have been no direct accusations made between departments, there is also little inter-departmental trust exhibited. According to RWS officials, "Implementation of new technology takes too long, and the implementation process is difficult to follow". There appears to be insufficient knowledge and expertise within the CID to independently manage the WIM systems [38]. The CID reports only on the technical availability of the systems and no information can be provided regarding the performance of the WIM network. RWS is unable to guarantee the reliability of the data due to a lack of a framework of standards. Requirements that exist for managing WIM are not included in the project tender. In 2011, the management of the database with the WIM data was transferred to the Inspectorate General for the Environment and Transport (IET). However the related expertise was not successfully shared. At the present moment, RWS has no access to the data held in the IET databases. The technical requirements were incomplete and some still need to be developed. IET systems are not yet ready to automatically manage the data. There are several legacy issues as technical management is only focused on the availability of the current IT systems. At the time of writing there were technical problems with the license plate recognition system. Governance and mandate appears unclear – it is unclear how the process is coordinated, and the IT supply organisation, the Central Information Department (CID), is unaware that they are also responsible for the sharing of data within the WIM systems. There is no single substantive authority that brings the parties from the entire supply chain together (there is no single authority that assumes responsibility for the entire chain). There is no well-designed change process; changes in the maintenance process are difficult to implement. The representative of the CID in the steering committee has no mandate.

The impediments for the adoption of IoT by e-governance identified in this case are: *1. Strategic/political barriers: data privacy issues, weak or uncoordinated data policies, weak or uncoordinated data governance, and conflicting market forces. Tactical barriers include: interoperability and integration issues, acceptance of IoT, and trust related issues. Operational issues are: a lack of sufficient knowledge regarding IoT, IT infrastructural limitations, and data management issues.*

4.2 Case Study 2: Water Management Data Collection at RWS

Information regarding water quantity and water quality is essential for the primary processes of Rijkswaterstaat: ensuring that flooding does not occur, sufficient clean water and smooth and safe traffic on the water. RWS also collects data on biology and chemistry, measuring nutrients and (micro) pollutants in surface water, suspended matter, sediment and aquatic animals.

The National Water Measurement Network, at RWS known as "Landelijk Meetnet Water" (LMW), is a facility that is responsible for the acquisition, storage and distribution of data for water resources. LMW has more than 400 data collection points using a nationwide system of sensors. The data is then processed and stored in the data centre and is made available to a variety of systems and users. The LMW was created from the merger of three previous existing monitoring networks: the Water Monitoring

Network, which monitored inland waterways such as canals and rivers; the Monitoring Network North, which monitored North Sea oil platforms and channels; and the Zeeland Tidal Waters Monitoring Network which monitored the Zeeland delta waterways. Four main types of measurement activities can be identified: water quantity, water quality, meteorological data and control information on infrastructure. The LMW measures a wide variety of hydrological data such as water levels, flow rates, wave heights and directions, flow velocity and direction, and water temperature. The LMW also measures meteorological data such as wind speed and direction, air temperature and humidity and air pressure amongst others. This meteorological data is collected in close collaboration with the Dutch Royal Meteorological Institute. The LMW provides a complete technical infrastructure for gathering and distribution of data and delivers the data to various stakeholders within and outside RWS.

According to RWS officials, there is often not clear who is ultimately responsible for the entire information chain. It has been discovered that the different departments work with different targets [39]. For example, the goals of the project organisation are to get production legitimacy of payment and to execute system contract management whilst the goals of the asset manager are to prevent flooding, to show demonstrable compliance with the statutory requirements and to reduce probability of failure. This results in different levels of the organization addressing different points of discussion. RWS officials report that a major impediment is "the lack of agreement and decision-making regarding vital strategic choices such as the form of contract management, the tender strategy, and outsourcing of personnel". The experience is that there is a failure to align the implementation with business targets as management teams tend to focus purely on the supply chain with a lack of awareness for possible risks or opportunities that occur outside of the supply chain. Whilst agreements occur between management teams, collaboration does not always occur in operations. There appear to be significant impediments regarding the coordination of activities and projects. RWS officials also quoted a lack of trust in the private sector to be able to manage and maintain the systems adequately. The perspective of RWS is that the private sector is not adequately developed regarding the necessary technical knowledge required by such an intricate system. RWS officials have stated that one of the reasons for this perspective is the intricacy of the RWS technical architecture itself. The experience is that the current architecture and legacy data make future integration of data very difficult.

The impediments for the adoption of IoT by e-governance identified in this case are: *1. Strategic/political barriers: data security issues, weak or uncoordinated data policies, weak or uncoordinated data governance, and conflicting market forces. Tactical barriers: interoperability and integration issues, acceptance of IoT, and trust related issues. Operational issues: a lack of sufficient knowledge regarding IoT.*

5   Discussion

The objective of this research was to identify potential impediments of the IoT for e-governance purposes. The IoT is important because a physical (or sensor) object that is able to communicate digitally is able to relate not only to a single entity, but also becomes connected to surrounding objects and data infrastructures. This allows for a situation in which many physical objects are able to act in unison, by means of ambient intelligence [1]. These devices and the communication between these devices can benefit e-government by providing enough quality data to generate the information required by government and citizens to make the right decisions at the right time.

We used two main research methods: (1) a literature review, (2) analysis of two IoT case studies. The literature review provided us with an overview of the existing body of knowledge, allowing us to analyse where gaps in knowledge or focus occur. It also provided definitions for the key concepts and helped develop a broader knowledge base in the research area. Case study research is a widely used qualitative research method in information systems research, and is well suited to understanding the interactions between information technology-related innovations and organizational contexts [40]. Following the advice of Yin (2003), the protocol used in the case study included a variety of data collection instruments. In order to counter the possible influences of bias, multiple research instruments were employed to ensure construct validity through triangulation [8].

The results of the literature review and the case studies demonstrate that the IoT is faced with a variety of impediments with regards to adoption b which correlates with impediments identified in the literature review. Table 1 below lists the main impediments of IoT, differentiating between strategic, tactical and operational benefits.

Formulating strategy requires defining goals and initiatives based on available resources and an assessment of the internal and external environments in which the organization competes [41]. Strategic impediments can therefore exert an important influence on an organization's likelihood of success. IoT is capable of providing a continuous stream of "trusted" data which managers can use to make informed, decisions, but the adoption of IoT for this purpose needs to be carefully coordinated by strong data policies and strong governance of the data with a purposeful awareness of opposing market forces and the capability of the private sector to provide critical services. Public sector organisations also need to address data privacy and data security issues for IoT adoption to be successful. These issues are interrelated as legal frameworks and strong policies provide guidelines within which organisations can face the pitfalls placed by security and privacy issues.

The main impediments appear to manifest during implementation, once organisations decide to operationalize the business plan. At that stage it becomes clear that although the technology is ready for widespread implementation, IoT remains an innovation which needs to not only integrate with current legacy systems but for which standards, policy, and legal frameworks still need to be developed with regards to social, technical and ethical issues.

Achieving a strategic plan or objective requires the administrative process of selecting among appropriate ways and means. Tactical planning is short range planning that emphasizes the current operations of various parts of the organization [42]. The case studies show that a good deal of attention should be paid to coordinating "soft", organisational issues such as trust and acceptance of IoT solutions as well as to the coordination of harder, technical, issues which require standardization in order to ensure interoperability and integration of data and systems. Significantly, costs were

Table 1. Impediments of IoT for e-governance in relation to the case studies.

		Impediments	Literature	Case 1	Case 2
Strategic	Social Responsibility	Data privacy issues	✓	✓	
		Data security issues	✓		✓
		Lack of legal framework	✓	✓	✓
	Productivity	Weak or uncoordinated data policies	✓	✓	✓
		Weak or uncoordinated data governance	✓	✓	✓
	Market standing	Conflicting market forces	✓	✓	✓
Tactical	Profitability	Costs	✓		
	Physical resources	Interoperability and integration issues	✓	✓	✓
		Lack of a framework of standards	✓	✓	✓
	Worker attitude	Acceptance of IoT	✓	✓	✓
		Trust related issues	✓	✓	✓
Operational	People	Lack of sufficient capabilities/knowledge	✓	✓	✓
	Technology	IT infrastructural limitations (issues with legacy systems)	✓	✓	
	Processes	Data management issues (data quality issues)	✓	✓	

only mentioned in the case studies as an impediment with regards to a negative business case. Since the primary processes of RWS are directly connected with health, safety and security of Dutch citizens, cost was disregarded as secondary to achieving the primary objective. It is possible that cost may be a more significant impediment in countries with less accessibility to the necessary funding than RWS.

A primary use of IT in government is to improve the efficiency of government operations [43]. As with many other organisations RWS uses IoT as a tool in industrial automation, in which simple manual tasks such as opening and closing bridges are automated. This reduces very low-level coordination work that was previously executed by humans, but complete automation or outsourcing of work can lead to a lack of sufficient knowledge within the organisation regarding the technique and the management of the data. This situation can develop into a significant impediment to the maintenance of IoT in e-governance applications. Technology continues to advance, but whilst many RWS officials were confident that technology was generally not a serious impediment, it is important to ensure that chose technique is compatible with the IoT architecture.

6 Conclusion

This paper represents one of the first papers on IoT for e-government. There has been limited research in the field of e-government regarding IoT, and there is much potential as expressed by the potential benefits, but the adoption of IoT within e-governance applications requires careful preparation and coordination. The IoT makes it possible to access remote sensor data and to monitor and control the physical world from a distance, and combining and analysing captured data allows governments to develop and improve services which cannot be provided by isolated systems, but this can only be achieved by addressing the potential impediments of to IoT adoption at all levels.

This research provides a systematic insight into the potential impediments of the IoT for e-government purposes by means of case study analysis and a review of literature. The research shows that impediments range from the political to the operational level. Specifically impediments for e-government can be attributed to data privacy issues, data security issues, weak or uncoordinated data policies, weak or uncoordinated data governance, and conflicting market forces, costs, interoperability and integration issues, acceptance of IoT, and trust related issues, a lack of sufficient knowledge regarding IoT, IT infrastructural limitations, and data management issues.

Many of the issues are interrelated; interoperability and integration issues have a direct impact on costs and on trust in the systems, and many issues can be resolved with sufficient knowledge and capabilities within the organisation. But the issues do need to be resolved in concert. It is important that governments address dominant impediments, such as privacy and security issues, within public policy and legal frameworks to assist public organisations with implementation of IoT. Similarly, technical and knowledge issues are very much interrelated with a lack of standards and impediments regarding interoperability and integration of data.

Acknowledgements. We acknowledge and thank the people of the Rijkswaterstaat who gave of their time and expertise during the case study research. This research is funded by Rijkswaterstaat.

References

1. Ramos, C., Augusto, J.C., Shapiro, D.: Ambient intelligence-the next step for artificial intelligence. IEEE Intell. Syst. **23**, 15–18 (2008)
2. Hounsell, N.B., Shrestha, B.P., Piao, J., McDonald, M.: Review of urban traffic management and the impacts of new vehicle technologies. IET Intell. Transp. Syst. **3**, 419–428 (2009)
3. Brous, P., Janssen, M.: Advancing E-Government Using the Internet of Things: A Systematic Review of Benefits. Presented at the IFIP Egov (2015)
4. Stichting Ijkdijk: livedijk-utrecht. http://www.ijkdijk.nl/nl/livedijken/livedijk-utrecht
5. RTV Utrecht: Proef met dijksensoren in Woerden en Nieuwegein. http://www.rtvutrecht.nl/nieuws/859256/proef-met-dijksensoren-in-woerden-en-nieuwegein.html
6. Marche, S., McNiven, J.D.: E-Government and E-Governance: the future isn't what it used to be. Can. J. Adm. Sci. Rev. Can. Sci. Adm. **20**, 74–86 (2003)

7. Haller, S., Karnouskos, S., Schroth, C.: The internet of things in an enterprise context. In: Domingue, J., Fensel, D., Traverso, P. (eds.) FIS 2008. LNCS, vol. 5468, pp. 14–28. Springer, Heidelberg (2009)
8. Yin, R.K.: Case Study Research: Design and Methods. SAGE Publications, Chicago (2003)
9. United Nations, Department of Economic and Social Affairs: United Nations e-government survey 2014: e -government for the future we want. United Nations, New York (2014)
10. Eisenhardt, K.M.: Building Theories from Case Study Research. Acad. Manage. Rev. **14**, 532–550 (1989)
11. Trochim, W.M.: Outcome pattern matching and program theory. Eval. Program Plann. **12**, 355–366 (1989)
12. Scarfo, A.: Internet of things, the smart X enabler. In: 2014 International Conference on Intelligent Networking and Collaborative Systems (INCoS), pp. 569–574 (2014)
13. Ivanov, D.: An adaptive framework for aligning (re)planning decisions on supply chain strategy, design, tactics, and operations. Int. J. Prod. Res. **48**, 3999–4017 (2010)
14. Ackoff, R.L.: Towards a system of systems concepts. Manag. Sci. **17**, 661–671 (1971)
15. Skarmeta, A.F., Hernandez-Ramos, J.L., Moreno, M.V.: A decentralized approach for security and privacy challenges in the internet of things. Presented at the 2014 IEEE World Forum on Internet of Things, WF-IoT 2014 (2014)
16. Zeng, D., Guo, S., Cheng, Z.: The web of things: a survey. J. Commun. **6**, 424–438 (2011)
17. Fan, P.F., Wang, L.L., Zhang, S.Y., Lin, T.T.: The research on the internet of things industry chain for barriers and solutions (2014)
18. Hummen, R., Henze, M., Catrein, D., Wehrle, K.: A cloud design for user-controlled storage and processing of sensor data. Presented at the CloudCom 2012 - Proceedings: 2012 4th IEEE International Conference on Cloud Computing Technology and Science (2012)
19. Yazici, H.J.: An exploratory analysis of hospital perspectives on real time information requirements and perceived benefits of RFID technology for future adoption. Int. J. Inf. Manag. **34**, 603–621 (2014)
20. van Kranenburg, R., Stembert, N., Moreno, M., Skarmeta, A.F., López, C., Elicegui, I., Sánchez, L.: Co-creation as the Key to a public, thriving, inclusive and meaningful EU IoT. In: Hervás, R., Lee, S., Nugent, C., Bravo, J. (eds.) UCAmI 2014. LNCS, vol. 8867, pp. 396–403. Springer, Heidelberg (2014)
21. Ortiz, P., Lazaro, O., Uriarte, M., Carnerero, M.: Enhanced multi-domain access control for secure mobile collaboration through linked data cloud in manufacturing. In: IEEE 14th International Symposium and Workshops on a World of Wireless, Mobile and Multimedia Networks (WoWMoM), 2013, pp. 1–9 (2013)
22. Harris, I., Wang, Y., Wang, H.: ICT in multimodal transport and technological trends: Unleashing potential for the future. Int. J. Prod. Econ. **159**, 88–103 (2015)
23. Roman, R., Najera, P., Lopez, J.: Securing the internet of things. Computer **44**, 51–58 (2011)
24. Gilman, H., Nordtvedt, J.-E.: Intelligent energy: the past, the present, and the future. SPE Econ. Manag. **6**, 185–190 (2014)
25. Stephan, E.G., Elsethagen, T.O., Wynne, A.S., Sivaraman, C., Macduff, M.C., Berg, L.K., Shaw, W.J.: A linked fusion of things, services, and data to support a collaborative data management facility. In: 2013 9th International Conference Conference on Collaborative Computing: Networking, Applications and Worksharing (Collaboratecom), pp. 579–584 (2013)
26. Misuraca, G.: Futuring e-Government: governance and policy implications for designing an ICT-enabled knowledge society. In: Proceedings of the 3rd International Conference on Theory and Practice of Electronic Governance, pp. 83–90. ACM, New York (2009)
27. Qiao, H., Wang, G.: An analysis of the evolution in Internet of Things industry based on industry life cycle theory (2012)

28. Wiechert, T.J.P., Thiesse, F., Michahelles, F., Schmitt, P., Fleisch, E.: Connecting mobile phones to the internet of things: a discussion of compatibility issues between EPC and NFC. Presented at the Association for Information Systems - 13th Americas Conference on Information Systems, AMCIS 2007: Reaching New Heights (2007)
29. Audretsch, D.B., Feldman, M.P.: Innovative clusters and the industry life cycle. Rev. Ind. Organ. **11**, 253–273 (1996)
30. Nam, T., Pardo, T.A.: The changing face of a city government: a case study of Philly311. Gov. Inf. Q. **31**(Supplement 1), S1–S9 (2014)
31. Blackstock, M., Lea, R.: IoT mashups with the WoTKit. In: 2012 3rd International Conference on the Internet of Things (IOT), pp. 159–166 (2012)
32. Speed, C., Shingleton, D.: An Internet of cars: Connecting the flow of things to people, artefacts, environments and businesses. Presented at the Sense Transport 2012 - Proceedings of the 6th ACM Workshop on Next Generation Mobile Computing for Dynamic Personalised Travel Planning (2012)
33. Pedro, M.: Reyes, patrick jaska: is RFID right for your organization or application? Manag. Res. News. **30**, 570–580 (2007)
34. Reyes, P.M., Li, S., Visich, J.K.: Accessing antecedents and outcomes of RFID implementation in health care. Int. J. Prod. Econ. **136**, 137–150 (2012)
35. Prasad, K.H., Faruquie, T.A., Joshi, S., Chaturvedi, S., Subramaniam, L.V., Mohania, M.: Data cleansing techniques for large enterprise datasets. Presented at the SRII Global Conference, San Jose, USA, April 29 (2011)
36. Mulyun, A., Parikesit, D., Antameng, M., Rahim, R.: Analysis of loss cost of road pavement distress due to overloading freight transportation. J. East. Asia Soc. Transp. Stud. **8**, 1020–1035 (2010)
37. Bagui, S., Das, A., Bapanapalli, C.: Controlling vehicle overloading in BOT projects. Procedia Soc. Behav. Sci. **104**, 962–971 (2013)
38. Evert, H., van der Valk, S.W.: A3 Weigh in Motion (2013)
39. Hopman, F., Meijer, E., Cyt, A., Piarelal, S., Hoogervorst, O.: RWS LMW HHS Adviesrapport. docx
40. Janssen, M.: Designing electronic intermediaries: An agent-based approach for designing interorganizational coordination mechanisms. Doctoral Dissertation, Delft University of Technology, Delft, The Netherlands (2001)
41. Nag, R., Hambrick, D.C., Chen, M.-J.: What is strategic management, really? Inductive derivation of a consensus definition of the field. Strateg. Manag. J. **28**, 935–955 (2007)
42. Tactical planning vs strategic planning. https://managementinnovations.wordpress.com/2008/12/10/tactical-planning-vs-strategic-planning/
43. Castro, D.: Digital Quality of Life: Government. Available SSRN 1285002 (2008)

Understanding the Adoption of Mobile Internet in the Saudi Arabian Context: Results from a Descriptive Analysis

Abdullah Baabdullah[1], Yogesh Dwivedi[1(✉)], Michael Williams[1], and Prabhat Kumar[2]

[1] Management and Systems Section (MaSS), School of Management, Swansea University, Swansea SA2 8PP, UK
{685177,y.k.dwivedi,m.d.williams}@swansea.ac.uk
[2] Department of Computer Science and Engineering, National Institute of Technology, Patna, India
prabhat@nitp.ac.in

Abstract. Utilising Mobile Internet (M-Internet) services would increase socio-economic benefits. Hence, it is necessary to consider the factors that may increase the adoption of M-Internet services within the context of Saudi Arabia. This research aims to examine potential users' intentions towards different variables that may be significant for supporting higher usage of M-Internet services in the domain of the Kingdom of Saudi Arabia. This study embraces the following variables: perceived risk, innovativeness; performance expectancy, effort expectancy, social influence, facilitating conditions, perceived value, hedonic motivation and behavioural intention. Data was collected by means of a questionnaire on a convenience sample that consisted of 600 subjects with a response rate of 69.5 %. The findings gathered from a descriptive analysis suggested that the related variables are perceived as significant by participants and they have strong behavioural intention to use M-Internet services.

Keywords: Saudi Arabia · M-Internet · UTAUT2 · Perceived risk · Innovativeness

1 Introduction

The importance of studying M-Internet within the context of Saudi Arabia is derived from the statistics that the Kingdom has experienced considerable growth of marketplaces in Information and Communication Technology (ICT) in the Middle East; predominantly in the retail and financial services' sectors as well as domestic broadband penetration (STC Group, 2011). For example, between 2005 and 2011, domestic broadband penetration had risen from zero to 44 % and M-Internet penetration had increased from 60 % to 191 % over the same period [26]. The statistics of using various types of M-Internet services in Saudi Arabia refers to the growth in the adoption of this new technology in areas such as M-Advertising. For example, in 2010, it reached a compound annual growth of 31 % according to Analysys Mason [6]. Also, the number of M-Internet users in Saudi Arabia had reached 11.5 million at the end of 2011

© IFIP International Federation for Information Processing 2015
M. Janssen et al. (Eds.): I3E 2015, LNCS 9373, pp. 95–106, 2015.
DOI: 10.1007/978-3-319-25013-7_8

representing 40.5 % of the population [16]. However, the adoption of other types of M-Internet in Saudi Arabia is still in its early stage. For example, only 39 % of M-Internet users in Saudi Arabia used this service for commercial purposes in 2011 [6]. Indeed, enhancing the level of behavioural intention toward using M-Internet in Saudi Arabia is a challenging task especially in the business arena due to the strong presence of cultural and technological parameters [6]. Tackling the factors that might predict behaviour intention towards using M-Internet among potential users will have a positive impact on raising the rate of using M-Internet across the Kingdom. This research will consider performance expectancy, effort expectancy, social influence, facilitating conditions, perceived value, hedonic motivation, perceived risk and innovativeness as probable predictors of behavioural intention amongst potential adopters of M-Internet in the domain of Saudi Arabia. By doing this research, this study will pinpoint the factors that can predict behaviour intention of M-Internet and thereby enhancing the use of M-Internet in the Saudi Arabian context. This paper is structured as follows: section two mentions an overview of M-Internet literature; then the researchers will give an overview of the theoretical basis in the third section. Next, methodology, results and discussion are outlined in the fourth, fifth and sixth sections respectively. In section seven, this study gives a synopsis.

2 Literature Review

Defining M-Internet is quite important before having any theoretical or empirical discussion as there are a lot of mobile strands that have different characteristics [10]. M-Internet can be defined as the "access to the Internet with devices that offer wireless connectivity" [18]. M-Internet relies on the three interconnected elements: the mobile device, a mobile network, and mobile content [18]. Consequently, the M-Internet makes a world where everyone can be connected regardless of time and space [19]. These distinctive traits, and their link to M-Internet, have generated an extraordinary level of M-Internet adoption worldwide. Indeed, there are plenty of studies that have focused on studying M-Internet (e.g. [7, 19, 21]. Within the context of Saudi Arabia, Alwahaishi and Snášel [4, 5] provided their new framework to pinpoint variables that might affect the acceptance of M-Internet and use in Saudi Arabia. Nonetheless, M-Internet reluctance can be attributed to different factors such as the lack of widespread use of mobile phones [32] and inaccessibility of the technological support [12]. However, within the context of Saudi Arabia and from 2004, use of the M-Internet has grown due to the development in wireless technology which led to a number of advantages such as discarding the need for landlines; minimising repetitive software installations; and solving the problem of temporary loss of high-speed Internet in some areas through using mobile devices [6]. Nevertheless, this advancement of M-Internet in Saudi Arabia has recorded a lower level compared with neighbouring countries such as Kuwait, UAE and Qatar. As such, a study was held in 2010 about the broadband subscriptions by using M-Internet, and found that mobile subscriptions in Qatar, UAE, Kuwait, Saudi Arabia were 72 %, 68 %, 65 % and 42 % respectively [9]. The adoption of M-Internet in Saudi Arabia still lags behind neighbouring states for many reasons such as network problems; that is to say, wireless access is limited in remote areas in

Saudi Arabia. Furthermore, even if there was access, frequent network problems have been recorded. According to a study conducted by Analysys Mason [9] in Saudi Arabia, it showed that there is a need to "harmonise the mobile broadband spectrum" in the "700/800 and 2.6 GHz bands for the use by mobile operators." As M-Broadband refers to Internet access (wireless communications) that has taken the place of a mobile phone, this shortage impacts negatively on the mobile broadband services and eventually on M-Internet services [9]. Although the adoption of M-Internet services has experienced a permanent growth in the Saudi context, the magnitude of this growth is still modest when compared with other states. According to the STC Group [26] regarding individual/household purchasing on the Internet in 2009 by either using the M-Internet or a fixed Internet, Saudi Arabia was way behind the UK and Norway with 2 %, 66 %, and 70 % respectively; Saudi Arabia are also behind Italy (12 %) and Greece (10 %). Interestingly, this modest adoption of M-Internet amongst Saudi citizens was in contrast with the huge increase in the budget of the Saudi telecom sector. As such, in 2009, the per capita telecom investment in Saudi telecom exceeded US $400 m compared with US$150 m in the UK. This information is important to notice that although the telecom revolution industry which is the basis for any activity through M-Internet, the actual use of M-Internet services in Saudi Arabia was way behind the UK. These evidences suggest that M-Internet services have not been heavily used yet, and Saudi users have not yet perceived these services. So, this study will examine the factors that may affect the adoption of this technology by the potential users through examining the factors that might influence behaviour intention of the potential users towards using M-Internet in the Saudi context. The outcomes of this current paper will be as follows: this research will consider UTAUT2 factors as well as perceived risk (PR) and innovativeness (INN) to pinpoint accurately the variables that change the behavioural intention of Saudi potential users in order to use the M-Internet.

3 Theoretical Basis

The Unified Theory of Acceptance and Use of Technology (UTAUT) proposes to predict adopters' intentions to adopt IT. It embraces four variables: performance expectancy (PE), effort expectancy (EE), social influence (SI), and facilitating conditions (FC). In this theory, PE, EE, and SI are direct determinants of behavioural intention (BI) and actual usage, and FC is a direct determinant of actual usage. Venkatesh et al. [30] confirmed that UTAUT predicted 70 % and 50 % of the variances in BI and usage respectively. Furthermore, a considerable number of writers have applied UTAUT in order to predict BI and usage in the field of M-Internet (e.g. [4, 5, 21, 32]. The UTAUT has been internationally tested, modified and developed over the years in order to explain users' behaviour towards the adoption of the M-Internet, e.g. Venkatesh et al. [29] adopted UTAUT and UTAUT2. Venkatesh et al. [29] developed UTAUT2 for investigating users' adoption of the M-Internet by extending UTAUT to UTAUT2 to bridge the void that occurred in the users' adoption domain. UTAUT2 adds new constructs: i.e. hedonic motivation (HM); price value (PV); and habit (HT) in addition to the UTAUT variables including PE, EE, SI, FC, BI and usage. The addition of these new variants gives more capability of estimation of the variables' intention to use compared with the UTAUT model. Moreover, in UTAUT2, association between FC and BI has been applied in order to offer a method to test the significance of FC in effecting BI of potential adopters in this study rather than testing the influence of FC over usage as applied in the UTAUT model. Thus, this paper holds UTAUT2 variables for examining the use of the M-Internet by potential adopters in the Saudi Arabian domain. UTAUT2 assembled factors of other models in order to offer a higher level of predictability about the behaviour of potential adopters when using the M-Internet. In detail, PE and EE in UTAUT2 replaced perceived usefulness and perceived ease of use

Table 1. Description of factors

Factor examined	Definition	Example citations that have tested this construct for examining the adoption of mobile applications
Performance Expectancy (PE)	"The degree to which using a technology will provide benefits to consumers in performing certain activities" [28, p. 159]	[29, 30]
Effort Expectancy (EE)	"The degree of ease associated with consumers' use of technology" [28, p. 159]	[3, 28]
Facilitating Conditions (FC)	"Refer to consumers' perceptions of the resources and support available to perform a behaviour" [28, p. 159]	[29, 34]
Social Influence (SI)	"The extent to which consumers perceive that important others (e.g. family and friends) believe they should use a particular technology" [28, p. 159]	[20, 29]
Hedonic Motivation (HM)	"The fun or pleasure derived from using technology and it has been shown to play an important role in determining technology acceptance and use" [9, p. 402]	[27, 29]
Price Value (PV)	"Consumers' cognitive trade-off between the perceived benefits of the applications and the monetary cost for using them" [28, p. 161]	[22, 29]
Innovativeness (INN)	"The willingness of an individual to try out any new information technology" [34, p. 229]	[2, 22]
Perceived Risk (PR)	"The expectation of losses associated with purchase and acts as an inhibitor to purchase behaviour" [17, p. 454]	[24, 34]
Behaviour Intention (BI)	"The intention of the user to use the technology" [33, p. 160]	[29, 33]

Table 2. Scale items of the selected factors

Construct	Item	Source
Performance Expectancy (PE)	PE1. M-Internet will be useful in my daily life	[29]
	PE2. Using M-Internet will help me accomplish things more quickly	
	PE3. Using M-Internet will increase my productivity	
	PE4. Using M-Internet will increase my chances of achieving things that are important to me	
Effort Expectancy (EE)	EE1. Learning how to use M-Internet will be easy for me	
	EE2. My interaction with M-Internet will be clear and understandable	
	EE3. M-Internet will be easy to use	
	EE4. It will be easy for me to become skilful at using M-Internet	
Social Influence (SI)	SI1. People who are important to me think that I should use M-Internet	
	SI2. People who influence my behaviour think that I should use M-Internet	
	SI3. People, whose opinions that I value, prefer that I use M-Internet	
Facilitating Conditions (FC)	FC1. I have the resources necessary to use M-Internet	
	FC2. I have the knowledge necessary to use M-Internet	
	FC3. M-Internet is compatible with other technologies I use	
	FC4. I can get help from others when I have difficulties using M-Internet	
Hedonic Motivation (HM)	HM1. Using M-Internet will be fun	
	HM2. Using M-Internet will be enjoyable	
	HM3. Using M-Internet will be very entertaining	
Price Value (PV)	PV1. M-Internet is reasonably priced	
	PV2. M-Internet is good value for the money	
	PV3. At the current price, M-Internet provides good value	
Perceived Risk (PR)	PR1. Providing M-Internet service with my personal information would involve many unexpected problems	[34]
	PR2. It would be risky to disclose my personal information through this service provider	
	PR3. There would be high potential for loss in disclosing my personal information to this service provider	
	PR4. Using M-Internet services subjects your checking account to financial risk	[17]
	PR5. I think using M-Internet puts my privacy at risk	[33]

(Continued)

Table 2. (*Continued*)

Construct	Item	Source
Innovativeness (INN)	INN1. If I heard about M-Internet technology, I would look for ways to experiment with it	[22]
	INN2. Among my peers, I am usually the first to explore new technologies	
	INN3. I like to experiment with new technologies: i.e. M-Internet	
	INN4. In general, I am not hesitant to try out new information technologies	[1]
	INN5. Compared to my friends, I seek out a lot of information about M-Internet services	[2]
	INN6. I would try the new M-Internet service even if, in my circle of friends, nobody has trialled it before	
Behavioural Intention (BI)	BI1. I will use M-Internet in the future	[29]
	BI2. I will always try to use M-Internet in my daily life	
	BI3. I will plan to use M-Internet frequently	

in TAM respectively. FC in UTAUT2 interchanged with perceived behavioural control (PBC) in TPB [11]. Similarly, HM in UTAUT2 replaced the perceived enjoyment variable. Likewise, social impact in UTAUT2 is similar to subjective norm and family influence [29]. Finally, PV stands for price and cost [23]. Venkatesh et al. [29] suggested examining the appropriateness of the UTAUT2 in various contexts. This research adopts the UTAUT2 model factors in addition to two new factors: i.e. perceived risk (PR) and innovativeness (INN) to critically evaluate the impact of these factors over BI of the potential adopters [25]. After reviewing the literature [14, 31], this study will adopt a conceptual model that will consist of candidate variables that might highly affect the adoption of potential users of the M-Internet in the Saudi Arabian context [8]. In order to check these variables, this research followed a quantitative study [15]. Unlike UTAUT2, this study does not consider habit as an independent variable as well as usage as a dependent variable. The reason for not considering habit and usage lies in the fact that this study focuses on the potential user rather than the actual usage of the M-Internet. Hence, it discards actual usage. Furthermore, according to Venkatesh et al. [29], habit has a direct effect on usage; consequently, there is no need for considering the habit variable when not studying usage. The mentioned constructs are viewed in Table 1 as follows:

4 Methodology

According to Table 2, this research tests nine variables: i.e. INN, PR, PE, EE, SI, FC, HM, PV and BI. It refers to the writers to whom this study relies on when considering 35 items. The level of responses was measured by implementing the seven-point Likert scale. Regarding demographic variables, this research embraces gender, age, education, occupation and monthly income. The researchers employed convenience sampling in three cities: i.e. Jeddah, Riyadh and Dammam.

5   Results

5.1   Response Rate

Table 3 outlines the distributed sample for M-Internet technology, returned responses, incomplete and problematic responses, and finally valid responses.

Table 3.   Response Rate

	M-Internet	%
Sample	600	100
Returned responses	436	72.67
Incomplete and problematic responses	19	3.17
Valid responses	417	69.5

5.2   Respondents' Profile and Characteristics

Table 4 below shows thedemographic characteristics of M-Internet respondents:

5.3   Respondents' Profile and Characteristics

According to Table 5, the average mean score and standard deviation were calculated for studied constructs as well as for items of these constructs.

6   Discussion

The descriptive statistics gave mean and standard deviation for each related variable and its items. Generally speaking, reading the average mean and average standard deviation for each variable can be classified into one of these four categories: high mean and high standard deviation, low mean and low standard deviation, high mean and low standard deviation, and low mean and high standard deviation [13]. The related variables, i.e. PE, EE, SI, FC, HM, PV, INN and BI, had a high average mean and a low standard deviation within the accepted limits (±1SD). The high mean, which is combined with the low standard deviation, indicated that the respondents' answers tended to be 'strongly agree' with a high certainty. In contrast, the low mean, which is combined with the low standard deviation, indicated that the respondents' answers tended to be 'strongly disagree' with a high certainty; i.e. perceived risk (PR) as participants seem to be less concerned regarding risks related to M-Internet services. Consequently, it is recommended that future studies should re-examine the significance

Table 4. Respondents' profile and characteristics

Demographic profile	Number of respondents (N = 417)	Percentage (%)
Gender		
Male	247	59.2
Female	170	40.8
Total	417	100
Age		
> = 18–20	55	13.2
21–29	242	58
30–39	68	16.3
40–49	33	7.9
50 and above	19	4.6
Total	417	100
Education		
Less than High School	14	3.4
High School	77	18.5
Diploma	74	17.7
Bachelor	191	45.8
Postgraduate	61	14.6
Total	417	100
Occupation		
Student	53	12.7
Government employee	218	52.3
Private sector employee	97	23.3
Self employed	49	11.7
Total	417	100
Monthly income (Saudi Riyals)		
1000–4000	36	8.6
4001–8000	91	21.8
8001–14000	183	43.9
14001–20000	73	17.5
More than 20000	34	8.2
Total	417	100

level of each one of these variables to obtain a better prediction over the adoption of the M-Internet in the Saudi Arabian context.

6.1   Contribution

Very few studies have been conducted in Saudi Arabia regarding M-Internet. This research has investigated statistically the variables that affect the BI towards adopting the M-Internet amongst potential adopters; thus, this study can suggest the best way to increase the BI of the potential users which is an essential step towards actually

Table 5. Descriptive analysis of measurement items

Construct	Item	Mean	Standard Deviation
Performance expectancy	PE1	6.66	.566
	PE2	6.67	.563
	PE3	6.64	.590
	PE4	6.67	.573
	Average	6.66	.573
Effort expectancy	EE1	6.56	.641
	EE2	6.61	.591
	EE3	6.56	.641
	EE4	6.57	.624
	Average	6.60	.583
Social influence	SI1	6.47	.686
	SI2	6.47	.672
	SI3	6.46	.696
	Average	6.46	.684
Facilitating conditions	FC1	6.45	.671
	FC2	6.46	.646
	FC3	6.45	.678
	FC4	6.20	.885
	Average	6.39	.72
Hedonic motivation	HM1	6.81	.403
	HM2	6.81	.408
	HM3	6.82	.401
	Average	6.81	.404
Price value	PV1	6.01	.732
	PV2	5.97	.751
	PV3	5.96	.770
	Average	5.98	.751
Innovativeness	INN1	6.38	.721
	INN2	5.50	1.312
	INN3	6.42	.743
	INN4	6.52	.676
	INN5	6.19	.953
	INN6	6.46	.710
	Average	6.24	.852
Perceived risk	PR1	1.98	.554
	PR2	1.94	.547
	PR3	1.98	.539
	PR4	2.04	.589
	PR5	2.11	.673
	Average	2.01	.580

(*Continued*)

Table 5. (*Continued*)

Construct	Item	Mean	Standard Deviation
Behavioural intention	BI1	6.70	.517
	BI2	6.71	.514
	BI3	6.70	.518
	Average	6.70	.516

adopting the M-Internet. This research included PR and INN to establish a suggested gap in UTAUT2 that disregarded the empirical findings about the significance of these two variables. Indeed, this is compatible with Venkatesh et al. [29] who suggest investigating UTAUT2 applicability on the M-Internet in different countries.

7   Conclusion

This study attempts to pinpoint the various factors that are considered to be effective on the behavioural intention of potential users when it comes to using the M-Internet in Saudi Arabia. Nine variables were selected, i.e. PE, EE, SI, FC, HM, PV, PR, INN and BI. Through running a descriptive analysis for every variable after giving the questionnaire to the participants, the findings of the analysis indicated that the related factors are effective when it comes to the behavioural intention of the participants.

7.1   Limitations and Future Research Directions

First of all, it will not be possible to generalise the findings to be representative of the overall population in Saudi Arabia because of the employment of a descriptive analysis rather than an inferential one which depends on using a Structural Equation Modelling (SEM). As this study considers investigating the behaviour intention instead of the actual use of the M-Internet, it is not going to provide a complete view about the actual usage of this service in Saudi Arabia. Moreover, conducting a longitudinal research is going to provide a better understanding regarding M-Internet in addition to clarifying the effects of the variables under study whose effects are stable and lasting over time. Consequently, any future study has to take into account these issues.

References

1. Agarwal, R., Prasad, J.: A conceptual and operational definition of personal innovativeness in the domain of information technology. Inf. Syst. Res. **9**(2), 204–215 (1998)
2. Aldás-Manzano, J., Ruiz-Mafé, C., Sanz-Blas, S.: Exploring individual personality factors as drivers of M-shopping acceptance. Ind. Manage. Data Syst. **109**(6), 739–757 (2009)
3. Alkhunaizan, A., Love, S.: What drives mobile commerce? An empirical evaluation of the revised UTAUT model. Int. J. Manage. Mark. Acad. **2**(1), 82–99 (2012)

4. Alwahaishi, S., Snášel, V.: Consumers' acceptance and use of information and communications technology: A UTAUT and flow-based theoretical model. J. Technol. Manage. Innov. **8**(2), 61–73 (2013)
5. Alwahaishi, S., Snášel, V.: Factors influencing the consumers' adoption of mobile Internet. In: The Third International Conference on Digital Information and Communication Technology and its Applications DICTAP 2013, pp. 31–39. The Society of Digital Information and Wireless Communication (2013b)
6. Analysys Mason.: Final Report to the GSMA: the Socio-Economic Benefit of Allocating Harmonised Mobile Broadband Spectrum in the Kingdom of Saudi Arabia (2012). http://www.gsma.com/spectrum/wp-content/uploads/2012/05/GSMA_report_on_KSA_DD2_6GHz_2012-04-30.pdf. Accessed: 01 February 2015
7. Baabdullah, A.M., Dwivedi, Y.K., Williams, M.D.: IS/IT adoption research in the saudi arabian context: analysing past and outlining the future research directions. In: European, Mediterranean, and Middle Eastern Conference on Information System (EMCIS 2013), Windsor, United Kingdom, 17–18 October 2013 (2013)
8. Baabdullah, A. M., Dwivedi, Y. K., Williams, M. D.: Adopting an extended UTAUT2 to predict consumer adoption of M-technologies in Saudi Arabia. In: UK Academy for Information Systems Conference Proceedings 2014. Paper 5 (2014)
9. Brown, S.A., Venkatesh, V.: A model of adoption of technology in the household: A baseline model test and extension incorporating household life cycle. Manage. Inf. Syst. Q. **29**(3), 4 (2005)
10. Chae, M., Kim, J.: What's so different about the mobile Internet? Commun. ACM **46**(12), 240–247 (2003)
11. Chen, S.C., Li, S.H.: Consumer adoption of e-service: Integrating technology readiness with the theory of planned behavior. African J. Bus. Manage. **4**(16), 3556–3563 (2010)
12. Cheong, J.H., Park, M.C.: Mobile Internet acceptance in Korea. Internet Res. **15**(2), 125–140 (2005)
13. Dancey, C.P., Reidy, J.: Statistics Without Maths for Psychology. Pearson Education, Harlow (2007)
14. Dwivedi, Y.K., Wastell, D., Laumer, S., Henriksen, H.Z., Myers, M.D., Bunker, D., Elbanna, A., Ravishankar, M.N., Srivastava, S.C.: Research on information systems failures and successes: status update and future directions. Inf. Syst. Frontiers **17**(1), 143–157 (2015)
15. Dwivedi, Y.K., Choudrie, J., Brinkman, W.P.: Development of a survey instrument to examine consumer adoption of broadband. Ind. Manage. Data Syst. **106**(5), 700–718 (2006)
16. Ethos Interactive.: Internet and mobile users in Saudi Arabia KSA (2012). http://blog.ethosinteract.com/2012/02/06/Internet-user-in-saudi-arabia-ethos-interactive. Accessed: 03 February 2015
17. Featherman, M.S., Pavlou, P.A.: Predicting e-services adoption: a perceived risk facets perspective. Int. J. Hum Comput Stud. **59**(4), 451–474 (2003)
18. Fogelgren-Pedersen, A.: The mobile Internet: The pioneering users' adoption decisions. In: Proceedings of the 38th Annual Hawaii International Conference on System Sciences, HICSS 2005, p. 84. IEEE (2005)
19. Hsu, H.H., Lu, H.P., Hsu, C.L.: Adoption of the mobile Internet: An empirical study of multimedia message service (MMS). Omega **35**(6), 715–726 (2007)
20. Huang, J., Liu, D.: Factors influencing continuance of mobile virtual community: Empirical evidence from China and Korea. In: 2011 8th International Conference on Service Systems and Service Management (ICSSSM), pp. 1–6. IEEE (2011)
21. Jiang, P.: Adopting mobile Internet: A demographic and usage perspective. Int. J. Electron. Bus. **6**(3), 232–260 (2008)

22. Karaiskos, D.C., Kourouthanassis, P., Lantzouni, P., Giaglis, G.M., Georgiadis, C.K.: Understanding the adoption of mobile data services: Differences among mobile portal and mobile Internet users. In: 8th International Conference on Mobile Business, ICMB 2009, pp. 12–17. IEEE (2009)

23. Mallat, N., Rossi, M., Tuunainen, V.K., Oorni, A.: The impact of use situation and mobility on the acceptance of mobile ticketing services. In: Proceedings of the 39th Annual Hawaii International Conference on System Sciences, HICSS 2006, vol. 2, p. 42. IEEE (2006)

24. Shin, D.H.: The evaluation of user experience of the virtual world in relation to extrinsic and intrinsic motivation. Int. J. Hum.-Comput. Interact. **25**(6), 530–553 (2009)

25. Slade, E., Dwivedi, Y.K., Piercy, N.L., Williams, M.D.: Modeling consumers' adoption intentions of remote mobile payments in the UK: Extending UTAUT with innovativeness, risk and trust. Psychol. Mark. **32**(8), 860–873 (2015)

26. STC Group.: ICT in Saudi Arabia: A socio-economic impact review (2011). http://www.enlightenmenteconomics.com/Reports/assets/ICTinSaudi%20Arabia.pdf. Accessed: 05 February 2015

27. Thong, J.Y., Hong, S.J., Tam, K.Y.: The effects of post-adoption beliefs on the expectation-confirmation model for information technology continuance. Int. J. Hum Comput Stud. **64**(9), 799–810 (2006)

28. Thong, J.Y., Venkatesh, V., Xu, X., Hong, S.J., Tam, K.Y.: Consumer acceptance of personal information and communication technology services. IEEE Trans. Eng. Manage. **58**(4), 613–625 (2011)

29. Venkatesh, V., Thong, J., Xu, X.: Consumer acceptance and use of information technology: Extending the unified theory of acceptance and use of technology. MIS Q. **36**(1), 157–178 (2012)

30. Venkatesh, V., Morris, M.G., Davis, G.B., Davis, F.D.: User acceptance of information technology: Toward a unified view. MIS Q. **27**, 425–478 (2003)

31. Williams, M.D., Rana, N.P., Dwivedi, Y.K.: The unified theory of acceptance and use of technology (UTAUT): A literature review. J. Enterp. Inf. Manage. **28**(3), 443–488 (2015)

32. Wiratmadja, I.I., Govindaraju, R., Athari, N.: The development of mobile Internet technology acceptance model. In: 2012 IEEE International Conference on Management of Innovation and Technology (ICMIT), pp. 384–388. IEEE (2012)

33. Wu, J.H., Wang, S.C.: What drives mobile commerce? An empirical evaluation of the revised technology acceptance model. Inf. Manag. **42**(5), 719–729 (2005)

34. Zhou, T.: Examining location-based services usage from the perspectives of Unified Theory of Acceptance and Use of Technology and privacy risk. J. Electron. Commer. Res. **13**(2), 135–144 (2012)

Conceptualising and Exploring
User Activities in Social Media

Marcel Rosenberger[1](✉), Tobias Lehmkuhl[2], and Reinhard Jung[1]

[1] Institute of Information Management, University of St. Gallen,
St. Gallen, Switzerland
marcel.rosenberger@student.unisg.ch,
reinhard.jung@unisg.ch
[2] St. Gallen, Switzerland
tobias.lehmkuhl@swisscom.com

Abstract. A growing number of companies are recognising the benefits of using social media in customer relationship management. At the same time, the consumers' expectations are rising: short response times, individual communication, real interaction with humans, and participation. It is a challenge to observe the many different user activities on many different social media sites. The aim is to reduce the complexity of integrating multiple social media sites with enterprise systems. Therefore, a conceptualisation of user activities in social media is presented. A user activity is a cross-over of an action invoked on an object and a user who acts in a certain context. The 40 user activity types are compared with actual features of ten social media sites. We find out that a substantial share of them can be integrated technically using the social media site's Application Programming Interfaces (APIs).

Keywords: Social media · User activities · Conceptualisation · Integration

1 Introduction

Social media have become a noticeable part of society. This development attracts attention of companies that aim to take advantage of the opportunities, such as improvement of the reputation and marketing efficiency, support cost reduction, and product innovation from co-creation [1–5]. At the same time, consumers benefit from participating companies, e.g. through relationship advantages, interaction and exchange, and influence on business processes. Examples are discounts, special promotions, and the acceleration of the fulfilment of support requests. The many active users perform various actions in social media and create a lot of data therewith. This information overload is a challenge for companies, because the increasing number of content, user profiles, and connections cannot be timely assessed manually [1].

Information systems (IS) are needed to manage the social media initiatives, providing functions to publish, observe and analyse social media data and integrate it with company data [6]. A preliminary step is to identify business-relevant user activities and to process the related data. Examples of user activities are joining groups, placing like- or dislike flags, adding others to the friend list, reading specific texts, watching videos, and changing

© IFIP International Federation for Information Processing 2015
M. Janssen et al. (Eds.): I3E 2015, LNCS 9373, pp. 107–118, 2015.
DOI: 10.1007/978-3-319-25013-7_9

profile information. These user activities comprise business opportunities in the form of leads (i.e. potential customers), enriched customer profile information, and a better understanding of interests and markets. Fliess & Nesper [4] state that "activities of customers can be considered as an economic resource". Similarly, Holts [7] highlights that user activities in social media create value and stimulate companies' revenues.

The more social media sites are considered by the company, the higher is the media penetration. Consequently, there is a need to integrate multiple social media sites mutually. However, the social media sites are diverse and facilitate different user activities. Posts, tweets, pins, profiles, groups, and pages, which are posted, tweeted, pinned, modified, added, or viewed, are only a small proportion. There are no common social media data structures, on which the integrations could be built.

Research on user activities in social media is contemporary and there are a number of existing conceptualisations [17–19, 22]. These are valuable to understand the user's motivation of being active and show some features of social media. However, the proposals are inappropriate to guide the implementation of integration software between multiple social media and an enterprise system. For this purpose, the existing conceptualisations are too abstract or they are exemplary and not exhaustive. Especially the related data of the user activities and the technical accessibility had not been researched yet. Our proposed conceptualisation and exploration of user activities closes this gap.

The topic is motivated from practitioners. The research is part of a joint social Customer Relationship Management (CRM) program with scientists and practitioners from companies of the insurance industry. The status quo of the companies shows that social media tools are isolated and not technically integrated into existing CRM systems. Social media monitoring tools are used to capture developments and to extract aggregated metrics, such as number of posts, likes, and age distribution on single sites. Relevant posts can be identified automatically based on tags, keywords, and rules. However, software solutions that recognise business-relevant user activities in multiple social media and invoke adequate business processes automatically are not yet implemented. Profiting business areas are customer service, sales, marketing, human resources, and research and development [3]. A conceptualisation of user activities is useful for designing general integration solutions. The audience are practitioners and researchers of social media and information systems.

Chapter 2 gives the conceptual background. Chapter 3 connects to existing knowledge and related work. The research methodology is presented in Chap. 4. Then, the user activity types in social media are described (Chap. 5) and compared with actual features of ten social media sites (Chap. 6). The final chapter states implications, limitations, and guides further research.

2 Related Work

Web 2.0 is an economic, social, and technology concept of the Internet, which enables users to create content and build a network with other users [8]. The results from user participation, e.g. posts, friend lists, and profiles, are accessible by other parties of the community. As stated in the definition by Kaplan & Haenlein [9], "social media is a group of Internet-based applications that build on the ideological and technological foundations of web 2.0". We use the term to refer to the sites/platforms that are built on the web 2.0 concept (e.g. Facebook, Google +, and Twitter). The terms "social media site" or "social media sites" are only used when an emphasis on singular or plural is necessary. A basic principle of social media is to connect to others and share information [10]. Social media and enterprise systems are heterogeneous systems, which can be connected through system integration. According to Hasselbring [11], heterogeneity leads to complexity, which is an issue for the integration task.

Küpper et al. [12] show results from a market study of 40 vendor solutions for social media tools. The findings indicate that most tools provide features to capture and analyse aggregated social media data. The capturing and analysis of individual data (i.e. single posts, user profiles, etc.) as well as the integration into enterprise systems is sparse. Similarly, other authors state that the integration of social media with enterprise systems is still insufficient [13, 14]. In particular, Trainor et al. [15] identify a lack of interaction between CRM systems and social media technology. For example, customer data and user data in social media are not interrelated and business-processes are not triggered from incidents in social media automatically.

Atig et al. [16] conceive user activity as the time when the user is active in social media. The authors classify users based on activity profiles and thereby do not differentiate between what the users are actually doing when they are active. Heinonen [17] conceptualises consumers' social media activities based on two dimensions: consumer motivation and consumer input. The consumers' motivation to use social media falls into one of three categories: information processing, entertainment activities, and social connection. The consumer input has three main types, which are consumption, participation, and contribution. The author's framework allows classifying users' activities. For example, "creating and managing a social network" is motivated from the need for social connection and requires creating a profile and linking to friends (productive consumer input). The proposed framework is abstract and does not allow deriving the related data of the activities. Pankong et al.'s [18] ontology for social activities is more concrete. In principle, the ontology is an entity-relationship-model, which shows entities (e.g. users, posts, likes, and topics) and its relationships (e.g. "is a", "has a", and "related to"). Some entities, however, are ambiguous (e.g. reply, retweet, and comment). Besides, the viewing of content is not included in the ontology. The model facilitates a snapshot-view of the social media graph. The circumstances in which the users create the content is not incorporated. This is justifiable considering that the authors focus on existing explicit and implicit relationships of users, similarly to Yang et al. [19]. However, the location and time of an activity are also expedient to determine the business-relevance [20]. Hotho & Chin [21] analyse the circumstances of user activities. Available sensors of a smartphone are used to conceive the current situation of the user (e.g. installed applications, busy status, missed calls count, position from Graphical Position System (GPS) sensor, remaining battery power, and ringtone volume). Richthammer et al. [22] identify 11 online social network (OSN) activities. Examples are "User posts Item/Comment", "User sends messages to Contact/Page", "User is linked to Item/Comment", and "Contact/Page views User's Profile". However, these are only "fundamental user activities on OSNs" and are not complete. For example, the sharing, deletion and modification of content is not considered.

3 Methods

The literature review follows vom Brocke et al.'s [23] methodology, which comprises three process steps, being (1) definition of review scope, (2) conceptualisation of topic, and (3) literature search. The authors highlight that not only results should be presented, but, to allow replicability, also details about the approach. The scope (1) of the literature review is characterised by six aspects borrowing from Cooper [24] (Table 1).

Table 1. Taxonomy of the conducted literature review (borrowing from [24])

Characteristic	Categories			
(a) focus	research outcomes	research methods	theories	applications
(b) goal	integration	criticism	central issues	
(c) organisation	historical	conceptual	methodological	
(d) perspective	neutral representation		espousal of position	
(e) audience	specialised scholars	general scholars	practitioners	general public
(f) coverage	exhaustive	exhaustive and selective	representative	central/pivotal

The focus (a) is on existing research results concerning user activities in social media. The goal (b) is to connect to existing knowledge on a conceptual level (c). The perspective (d) can be characterised as neutral representation, because the position is unbiased. Practitioners and researchers of social media are the target audience (e). The results are representative (f) for the IS community, because prominent data sources have been queried.

The conceptualisation of the topic (2) includes a "working definition of [the] key variable(s)" [25]. A keyword search (3) in the databases of AISel, EBSCO, Emerald, IEEE, JSTOR, ProQuest, and Web of Science in the title (TI), topic (TO), abstract (AB), keyword (KW), and full text (TX) fields was applied using the search string: *"social media" AND ("user actions" OR "user activities")*. The initial list of publications has been filtered by reading the titles and abstracts. Relevant papers were analysed based on the full texts. Table 2 shows the numerical results of the keyword searches.

The development of the user activity types comprised a study of features of large, popular social media. The sample of sites for analysis has been selected on the following criteria: (1) large number of active users per month (>100 m.); (2) English localisation of the platform; (3) availability of a public API; and (4) permission for commercial use. The initial list of contemplable sites has been compiled of studies and rankings of social media [26, 27]. The listed sites have been evaluated against the aforementioned criteria, based on information from press releases, technical notes, terms of use, and responses from enquiries to the providers. Possible user activities

Table 2. Numerical results of the keyword searches

Data source	Search fields	Publications	
		Total	Relevant
AISeL	TI, AB	11	2
EBSCO	TI, AB, KW, TX	22	5
Emerald	TI, AB, KW	30	1
IEEE	TI, AB, KW	4	1
JSTOR	TI, AB, KW, TX	9	1
ProQuest	TI, AB, KW	4	1
Web of Science	TI, TO	4	1
Total[a]			**10**

[a] The total number is not equal to the column sum, because duplicates have been counted only once.

have been gathered by analysis of the features and functions. They have been grouped according to the philosophical idea of family resemblance and following an abstraction-based modelling approach [28, 29].

4 User Activity Types in Social Media

The user activity types shown in Fig. 1. represent the actions that users perform in social media. Activities take place in a context, in which the user is situated, defined by time, location, social media site, device, and application. The combination of an object type and an action is termed a user activity type in social media. The complex graph structure of social media is broken down into an activity log, which contains entries of the form: user u invoked action a on object o (on site s with device d in application p from location l at time t).

Five actions can be applied to eight object types. All user-generated content results from the Create-, Update-, or Share-action. The content is displayed on the screens of the users' devices via the View-action. The Delete-action removes content. The variety of features across different social media that facilitate the creation, modification and viewing of content is reducible to 8×5 user activity types.

4.1 Social Media Object Types

The idea of family resemblance is adduced to group similar objects. The most prototypical objects constitute an object type. An object type subsumes all objects, which have most functions and structure in common with that object type, and have least commonalities with other object types [28].

Kietzmann et al. [30] present a framework of functional building blocks of social media, which are identity, conversations, sharing, presence, relationships, reputation, and groups. The seven blocks are facets of user experience in social media and give an orientation to gather object types and functions. Table 3 identifies social media object types by analysing its structure and functions [31].

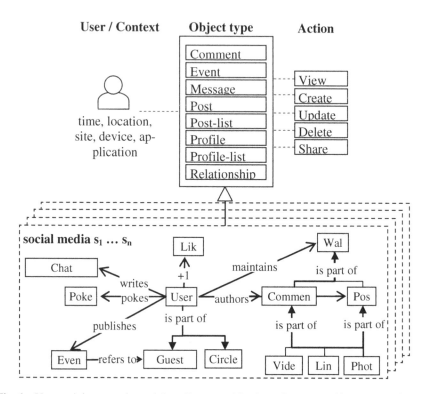

Fig. 1. User activity types in social media: a combination of context, object type and action

An object type is enabling a function, if it supports the intention behind the concept of the building block. It is partly enabling, if the intended user experience of the functional building block is a side-effect. An object type is not contributing to a functional building block, if it does not enable the function.

The eight object types may have variants that share similar concepts, but have a different terminology or are used to distinguish the same concept that is used in different contexts on the same site. An example is a Comment, which contains multimedia content and has a parent, which may be a Post, Post-list, Comment or another object type. Some sites use a Comment object type to represent answers, like Tumblr, or a job application, like Xing.

The same object type is also used in different variants on the same social media site, such as on Facebook, where both comments and reviews exist. A Comment primarily contributes to the functional building blocks Conversations and Sharing. An Event object type defines a happening, which has relatedness to time. It can be a birthday party, a music festival, a meeting, and so on. Events facilitate to meet people (Presence), build communities (Groups), and relate to each other (Relationships). A Message is multimedia content that is addressed to a specified set of receivers. Posts subsume a main content entry found on all social media sites. They may be termed tweet, job, or pin, and engage interaction by allowing adding Comments and Relationships.

Table 3. Technical identity of social media objects

Object types	Structure / data	Functions						
		Identity	Conversations	Sharing	Presence	Relationships	Reputation	Groups
Comment	- contains content (e.g. text, video, audio and image) - refers to another object	O	●	●	◐	O	◐	O
Event	- contains descriptive information about a happening - has a relatedness to time	O	O	O	●	●	◐	●
Message	- contains content (e.g. text, video, audio and image) - has a sender and recipient (list)	O	●	●	◐	O	O	O
Post	- contains content (e.g. text, video, audio and image)	O	●	●	◐	O	◐	O
Post-list	- is a collection of posts	O	●	●	◐	◐	O	◐
Profile	- contains descriptive information about an actor	●	O	◐	●	◐	●	O
Profile-list	- is a collection of profiles	◐	◐	O	◐	●	●	●
Relationship	- connects two objects	O	◐	O	◐	●	●	◐

Legend: Object type is ... O not enabling the function, ◐ partly enabling the function, ● enabling the function

A Post-list is a collection of Posts. A Profile is a representation of an entity of the real life, such as a person, company, or a community. A Profile-list is a collection of Profiles with possible variants, such as circle, contact list, and friend list. A Relationship connects two other objects. An example is a bookmark, which can be described as a Relationship between a Profile and a Post. The poke feature in Facebook can be treated as a Relationship between two Profiles.

4.2 Actions on Social Media Objects

Table 4 lists actions, which can be invoked on social media objects referring to Hypertext Transfer Protocol (HTTP) methods. HTTP is the underlying, technological protocol of social media sites [32]. Five basic actions on social media objects can be

Table 4. Actions on social media objects

Action	Description	HTTP methods
View	View is triggered when content of an object is loaded and displayed on the user's screen (e.g.: a video is played)	GET
Create	The Create-action occurs when something new is added opposed to the Update-action when a change to an existing object is done by a user	POST
Update	The Update-action results in a modified, existing social media object	PUT/MOVE
Delete	When an object is removed on social media an event with action Delete is raised	DELETE
Share	The Share-action occurs when existing content, usually originated from another user, is put into a different context or is exposed to additional users on the same platform. It is a copy of an already existing entity	COPY

identified. Sharing is something particularly found in social media [30]. The citing of a text phrase or the re-tweet of content on Twitter is an example of the Share-action.

5   Empirical Exploration

Table 5 shows the results of the empirical exploration of the user activities in ten social media.

The APIs of large social media define access options to functions and data using web-services. They include formats and provide methods to publish posts, resolve connections between users, and retrieve comments, for example. Dark underlined numbers signify that the user activity type can be monitored in the specified social media using the provided API. Dark numbers that are not underlined mean that the type exists on the site, but the APIs of the site do not provide access to monitor it. For example, in Facebook a user can view a post. However, this activity cannot be monitored using the public API of Facebook in the recent version of the Graph API V2.1 [33]. On the other hand, it cannot be ruled out that access is included in upcoming versions. Furthermore, using the APIs is not the only access approach to social media data. Instead of using the API, the View-action of own and shared posts can be recognised by linking a Facebook post with external content from a corporate website, where the company can evaluate page requests (by observing the HTTP/1.1 GET-method).

Six user activity types are theoretical constructs, which do not occur in the analysed social media. These are Message/Update, Message/Delete, Message/Share, Profile-list/Share, Relationship/Update, and Relationship/Share. Firstly, a Message is private, because it cannot be shared. Secondly, a Message, once sent, cannot be fetched back, removed, or edited. A Profile-list cannot be shared by others. Access privileges of the Profile-lists are maintained by the owners only. A Relationship does either exist or does not exist. It cannot be modified; but it can be deleted.

Table 5. Empirical exploration of user activities in social media

Action / Object type	View	Create	Update	Delete	Share
Comment Answer, Recommendation, Job Application, Review	1 2 3 4 5 6 7 8 9 10	1 2 3 4 5 6 7 8 9 10	1 2 3 4 5 6 7 8 9 10	1 2 3 4 5 6 7 8 9 10	1 2 3 4 5 6 7 8 9 10
Event Meeting, Happening	1 2 3 4 5 6 7 8 9 10	1 2 3 4 5 6 7 8 9 10	1 2 3 4 5 6 7 8 9 10	1 2 3 4 5 6 7 8 9 10	1 2 3 4 5 6 7 8 9 10
Message Chat, Fanpost, Gift	1 2 3 4 5 6 7 8 9 10	1 2 3 4 5 6 7 8 9 10	1 2 3 4 5 6 7 8 9 10	1 2 3 4 5 6 7 8 9 10	1 2 3 4 5 6 7 8 9 10
Post Job, Life event, Pin, Project, Question, Status, Tweet	1 2 3 4 5 6 7 8 9 10	1 2 3 4 5 6 7 8 9 10	1 2 3 4 5 6 7 8 9 10	1 2 3 4 5 6 7 8 9 10	1 2 3 4 5 6 7 8 9 10
Post-list Blog, Board, Page, Photo album, Wall	1 2 3 4 5 6 7 8 9 10	1 2 3 4 5 6 7 8 9 10	1 2 3 4 5 6 7 8 9 10	1 2 3 4 5 6 7 8 9 10	1 2 3 4 5 6 7 8 9 10
Profile Community, Company, User	1 2 3 4 5 6 7 8 9 10	1 2 3 4 5 6 7 8 9 10	1 2 3 4 5 6 7 8 9 10	1 2 3 4 5 6 7 8 9 10	1 2 3 4 5 6 7 8 9 10
Profile-list Circle, Contact list, Friend list, Group, Guest list, Partner list	1 2 3 4 5 6 7 8 9 10	1 2 3 4 5 6 7 8 9 10	1 2 3 4 5 6 7 8 9 10	1 2 3 4 5 6 7 8 9 10	1 2 3 4 5 6 7 8 9 10
Relationship Bookmark, Favourite, Follow, Invitation, Join, Like, Poke, Rating	1 2 3 4 5 6 7 8 9 10	1 2 3 4 5 6 7 8 9 10	1 2 3 4 5 6 7 8 9 10	1 2 3 4 5 6 7 8 9 10	1 2 3 4 5 6 7 8 9 10

Legend: 1-Facebook, 2-Flickr, 3-Google+, 4-LinkedIn, 5-Meetup, 6-Pinterest, 7-Tumblr, 8-Twitter, 9-Xing, 10-YouTube

Event type is … ○ not existent on the social media site, ● existent on the social media site, ● existent on the social media site, and can be accessed using the API

The majority (70 %) of user activity types that exist on a social media site, can also be monitored using the API and thus can be integrated with enterprise systems using a public, recommended access approach. The View-actions are usually not provided; only Google + has custom activities, which can be triggered by developers in case an entity is read. On most social media Post/Create is observable by subscription to Post-lists. The Update-, and Delete-actions could be identified by periodic polling, whereby known objects are checked regularly to notice if they are still existent or modified. Comment, Post, Profile, and Relationship exist in all analysed social media. Thus, these are essential object types. Facebook (83 %), Google + (80 %), and Xing (80 %) feature the most complete set of user activity types. The APIs of Google +, LinkedIn, and YouTube provide the most complete set of access options, covering 95 %, 80 %, and 79 % of applicable user activities of each site.

6 Discussion and Conclusions

The user activity types define user activities in social media. They specify what users do in social media when they create or consume content. Hence, the user activity types advance from existing definitions of user activities that conceive user activity as the time when the user is active in social media [16]. A user activity type is a crossover of a social media object type and an action and takes place in a specific user's context. The object types reveal the underlying structure and data, which large social media sites share. The actions are operations that users perform with an object type. The user-context describes the situation in which the user resides while invoking an action on an object.

The results are useful to design and develop integration software that facilitates to process user activities of multiple different social media sites. Middleware-based solutions require similar structuring of information. The presented user activity types support that purpose, because they allow to consolidate the different user activities of different, large social media sites. There are technical restrictions limiting the feasibility to capture "everything", because some user activity types cannot be captured using the APIs. Moreover, as also highlighted by other authors, users' permission and privacy need to be considered [34]. It must be a major concern of all business-oriented social media initiatives, because of the risk to destroy relationships to customers in case of an accident. An example is unintended data exposure to unauthorised parties. As a result, not every user activity that can be monitored technically should also be tracked.

The user activity types originate from the abstraction of individual features collected from a study of ten social media sites. They have an empirical basis and rely on publicly available data. The issue, caused by the underlying induction of the abstraction, is that the user activity types are only certainly valid for the analysed social media, and are not necessarily generalisable to all available sites.

Further research is encouraged to concretise the user activity types in terms of a canonical data schema, which defines data types and attributes. Based on the detailed level, (business-specific) rules can be proposed for filtering user activities. Monitoring of user activities in social media leads to a reactive system [35]. A fully integrated IS, however, should comprise functions to interact, requiring both directions of a communication. This is not contrary to the research results, but is a possible extension.

References

1. Smith, T.: Conference notes – the social media revolution. Int. J. Mark. Res. **51**, 559 (2009)
2. Baird, C.H., Parasnis, G.: From social media to social customer relationship management. Strateg. Leadersh. **39**, 30–37 (2011)
3. Cappuccio, S., Kulkarni, S., Sohail, M., Haider, M., Wang, X.: Social CRM for SMEs: current tools and strategy. In: Khachidze, V., Wang, T., Siddiqui, S., Liu, V., Cappuccio, S., Lim, A. (eds.) iCETS 2012. CCIS, vol. 332, pp. 422–435. Springer, Heidelberg (2012)
4. Fliess, S., Nesper, J.: Understanding patterns of customer engagement – How companies can gain a surplus from a social phenomenon. J. Mark. Dev. Compet. **6**, 81–93 (2012)
5. Jahn, B., Kunz, W.: How to transform consumers into fans of your brand. J. Serv. Manag. **23**, 344–361 (2012)
6. Acker, O., Gröne, F., Akkad, F., Pötscher, F., Yazbek, R.: Social CRM: How companies can link into the social web of consumers. J. Direct Data Digit. Mark. Pract. **13**, 3–10 (2011)
7. Holts, K.: Towards a Taxonomy of Virtual Work. In: 31st International Labour Conference 2013, vol. 7, pp. 31–50. Rutgers Univ. New Brunswick, New York (2013)
8. Musser, J., O'Reilly, T.: Web 2.0 Principles and Best Practices. O'Reilly Media, Sebastopol (2006)
9. Kaplan, A.M., Haenlein, M.: Users of the world, unite! The challenges and opportunities of Social Media. Bus. Horiz. **53**, 59–68 (2010)
10. Ang, L.: Is SCRM really a good social media strategy? J. Database Mark. Cust. Strateg. Manag. **18**, 149–153 (2011)
11. Hasselbring, W.: Information system integration. Commun. ACM **43**, 32–38 (2000)
12. Küpper, T., Lehmkuhl, T., Jung, R., Wieneke, A.: Features for social CRM technology – An organizational perspective. In: AMCIS 2014 Proceedings, pp. 1–10 (2014)
13. Sarner, A., Thompson, E., Sussin, J., Drakos, N., Maoz, M., Davies, J., Mann, J.: Magic Quadrant for Social CRM, pp. 1–20 (2012)
14. Reinhold, O., Alt, R.: How companies are implementing social customer relationship management: insights from two case studies. In: Proceedings of the 26th Bled eConference, pp. 206–221 (2013)
15. Trainor, K.J., Andzulis, J., Andzulis, J.M., Rapp, A., Agnihotri, R.: Social media technology usage and customer relationship performance: a capabilities-based examination of social CRM. J. Bus. Res. **67**, 1201 (2013)
16. Atig, M.F., Cassel, S., Kaati, L., Shrestha, A.: Activity profiles in online social media. In: IEEE/ACM International Conference on Advances in Social Networks Analysis and Mining (ASONAM 2014), pp. 850–855 (2014)
17. Heinonen, K.: Consumer activity in social media: managerial approaches to consumers' social media behavior. J. Consum. Behav. **10**, 356–364 (2011)
18. Pankong, N., Prakancharoen, S., Buranarach, M.: A combined semantic social network analysis framework to integrate social media data. In: Proceedings of 2012 4th Internatioanl Conference Knowledge Smart Technology KST 2012, pp. 37–42 (2012)
19. Yang, C.C., Tang, X., Dai, Q., Yang, H., Jiang, L.: Identifying implicit and explicit relationships through user activities in social media. Int. J. Electron. Commer. **18**, 73–96 (2013)
20. Yu, Y., Tang, S., Zimmermann, R., Aizawa, K.: Empirical observation of user activities. In: Proceedings of the First International Workshop on Internet-Scale Multimedia Management - WISMM 2014, pp. 31–34 (2014)

21. Woerndl, W., Manhardt, A., Schulze, F., Prinz, V.: Logging user activities and sensor data on mobile devices. In: Atzmueller, M., Hotho, A., Strohmaier, M., Chin, A. (eds.) MUSE/MSM 2010. LNCS, vol. 6904, pp. 1–19. Springer, Heidelberg (2011)

22. Richthammer, C., Netter, M., Riesner, M., Sänger, J., Pernul, G.: Taxonomy of social network data types. EURASIP J. Inf. Secur. **2014**, 11 (2014)

23. Vom Brocke, J., Simons, A., Niehaves, B., Riemer, K., Plattfaut, R., Cleven, A., Von Brocke, J., Reimer, K.: Reconstructing the giant: on the importance of rigour in documenting the literature search process. In: 17th European Conference on Information Systems (2009)

24. Cooper, H.M.: Organizing knowledge syntheses: a taxonomy of literature reviews. Knowl. Soc. **1**, 104–126 (1988)

25. Webster, J., Watson, R.T.: Analyzing the past to prepare for the future: writing a literature review. MIS Q. **26**, xiii–xxiii (2002)

26. Singh, N., Lehnert, K., Bostick, K.: Global social media usage: insights into reaching consumers worldwide. Thunderbird Int. Bus. Rev. **54**, 683–700 (2012)

27. Statista: Leading social networks worldwide as of June 2014. ranked by number of active users. http://www.statista.com/statistics/272014/global-social-networks-ranked-by-num-ber-of-users/

28. Rosch, E., Mervis, C.B.: Family resemblances: studies in the internal structure of categories. Cogn. Psychol. **7**, 573–605 (1975)

29. Bussler, C.: Modeling Methodology. B2B Integration: Concepts and Architecture. Springer, Heidelberg (2003)

30. Kietzmann, J.H., Hermkens, K., McCarthy, I.P., Silvestre, B.S.: Social media? Get serious! understanding the functional building blocks of social media. Bus. Horiz. **54**, 241–251 (2011)

31. Faulkner, P., Runde, J.: Technological objects, social positions, and the transformational model of social activity. MIS Q. **37**, 803–818 (2013)

32. Gourley, D., Totty, B., Sayer, M., Aggarwal, A., Reddy, S.: HTTP: The Definitive Guide. O'Reilly Media, Inc, California (2002)

33. Facebook: Graph API Reference. https://developers.facebook.com/docs/graph-api/reference/v2.1

34. Woodcock, N., Broomfield, N., Downer, G., McKee, S.: The evolving data architecture of social customer relationship management. J. Direct Data Digit. Mark. Pract. **12**, 249–266 (2011)

35. Sarkar, A., Waxman, R., Cohoon, J.P.: High-level system modeling. In: Bergé, J.-M., Levia, O., Rouillard, J. (eds.) High-Level System Modeling: Specification Languages, pp. 1–34. Springer, US (1995)

Understanding the Determinants of Privacy-ABC Technologies Adoption by Service Providers

Ahmad Sabouri[✉]

Deutsche Telekom Chair of Mobile Business and Multilateral Security,
Goethe University Frankfurt, Theodor-W.-Adorno-Platz 4, 60323 Frankfurt, Germany
ahmad.sabouri@m-chair.de

Abstract. As using online services penetrates deeper in our everyday life, lots of trust-sensitive transactions are carried out electronically. In this regard, a big challenge is to deal with proper user authentication and access control without threatening the users' privacy. However, commonly used strong authentication schemes fail to address important privacy requirements. In this paper, we focus on an emerging type of digital certificates, known as Privacy-preserving Attribute-based Credentials (Privacy-ABCs), which allow privacy and security go hand-in-hand. So far, there has been no systematic study on the potential factors that have influence on the adoption of Privacy-ABCs by service providers. Thus, we developed a conceptual model of the relevant factors based on well-established theories and our practical experience with trialing Privacy-ABCs, and evaluated the model through expert surveys.

Keywords: Privacy-preserving attribute-based credentials · Anonymous credentials · Technology adoption · Expert surveys

1 Introduction

Nowadays, usernames and passwords are the most commonly used authentication schemes. However, the hassle of managing different usernames and passwords grows as the number of electronic services increases. This, on the one hand, raises security risks because many users tend to reuse the same password for different services. On the other hand, it introduces privacy threats for cross-linking activities of the users in different domains as it is highly probable to be able to correlate different identifiers of the same person [35] because they typically prefer to choose the same or similar usernames for their various accounts.

An alternative solution to improve the security problem is to employ strong authentication techniques such as digital certificates. Nonetheless, the most commonly used strong authentication techniques do not follow the Privacy-by-Design [6] principle of *Data Minimization*. For instance, the use of X509 certificates causes "Over-Identification", as it mandates the users to reveal all the attested

© IFIP International Federation for Information Processing 2015
M. Janssen et al. (Eds.): I3E 2015, LNCS 9373, pp. 119–132, 2015.
DOI: 10.1007/978-3-319-25013-7_10

attributes in the certificate so that the validity of the digital signature is preserved, even if only a subset of attributes is required for the authentication purpose. Using online federated authentication and authorization techniques such as OpenID, SAML, Facebook Connect, and OAuth could support the minimal disclosure principal and allow the users to provide the service providers with only the requested information rather than the whole user's profile stored at the Identity Service Provider (IdSP). However, all these protocols suffer from a so-called "Calling Home" problem, meaning that for every authentication transaction the user is required to contact the IdSP (e.g., Facebook, Gmail, OpenID Provider). This introduces privacy risks to both users and service providers.

The focus of this paper is on a promising type of digital certificates called Privacy-preserving Attribute-based Credentials (Privacy-ABCs) that provide a strong basis for secure yet privacy-enhanced access control systems. Privacy-ABCs offer a solution to cope with *Minimal Disclosure* of attributes as well as supporting *Partial Identities*. Privacy-ABC users can obtain credentials from their IdSPs and when authenticating to different service providers, they can produce *unlinkable Privacy-ABC tokens* containing only the required subset of information available in the credentials without involving the IdSP or any third party in the process. Therefore, they can help overcome the "Over-Identification" and "Calling Home" problems. The prominent instantiations of such Privacy-ABC technologies are Microsoft U-Prove[1] and IBM Idemix[2]. Both of these technologies are studied in depth by the EU-funded project ABC4Trust, where a common architecture for Privacy-ABCs was designed, implemented and verified in two real-life trials [27].

Privacy-ABCs are emerging technologies that are not yet properly adopted. There have been a handful of proposals on how to realize a Privacy-ABC system in the literature [4,5]. However, the diversity of their features and implementations hindered their practical use. As Privacy-ABCs are in the pre-adoption phase, our rigorous literature review on drivers and inhibitors of Privacy-ABCs using well-known databases such as JStore, MISQ, AISnet, ACM and IEEE ended up in a limited set. Borking investigated the adoption of Privacy Enhancing Technologies (PETs) in general [3]. Nevertheless, PETs can be very different in their characteristics and their adoption schemes. For instance, Tor[3] is also an example of PETs that can be employed directly by the end users, while in order to have Privacy-ABC technologies operational, at least three entities have to adopt or accept the technology: (1) Credential issuers, which are typically organizations such as governments, banks, and telco operators who have authentic source of data about the users, (2) Service providers, which perform access control to their resources relying on the credential attested by the issuers, (3) Users, who consider using such kind of credentials. Therefore, Privacy-ABCs have special characteristics and effects that make them deserve a separate study. Therefore, this paper focuses on the adoption factors influencing **service providers** and

---

[1] http://microsoft.com/uprove.
[2] http://idemix.wordpress.com/.
[3] https://www.torproject.org/.

launches the first systematic work based on well-established theories to investigate the (future) adoption of Privacy-ABCs. Understanding the determinants of adoption by service providers is very important in the sense that identifying these factors can facilitate building guidelines for the supporting bodies to pave the road for the further adoption of Privacy-ABCs. Therefore, we developed a conceptual model based on the existing innovation adoption theories and evaluated the factors through expert surveys.

The rest of this paper is organized as follows. In Sect. 2, we introduce the features and concepts of Privacy-ABCs in more details and also deliver an overview of the theories in the literature explaining innovation adoption. Later, we present our conceptual model of the determinants in Sect. 3. Then, in Sect. 4, we present our empirical evaluation of the factors, and later in Sect. 5 discuss our findings and their implications. In the end, we conclude the paper in Sect. 6.

2   Theoretical Background

2.1   How Privacy-ABCs Work

A *Credential* is defined to be "a certified container of attributes issued by a credential Issuer to a User" [1]. An *Issuer* vouches for the correctness of the attribute values for a *User* when issuing a credential for her. In an example scenario, Alice as a *User*, contacts the Bundesdruckerei (the German authority responsible for issuing electronic IDs) and after a proper proof of her identity (e.g. showing her old paper-based ID), she receives a digital identity credential containing her first name, surname and birth-date. In the next step, she can seek to access an online Discussion Forum. The service provider provides Alice with the access policy that requires her to deliver an authentic proof of her first name. Using Privacy-ABCs features, Alice has the possibility to derive a minimal authentication token from her identity credential that contains only the first name. As a result, her privacy is preserved by not disclosing unnecessary information (i.e. surname and birth-date). Note that the commonly used digital certificates do not offer such capability as any change in those certificates invalidates the issuers' signature. Another example where Alice could use her Privacy-ABC might be with an online movie rental website, which requires age verification. Alice is able to provide such a proof without actually disclosing her exact birth-date. The proof is done based on complex cryptographic concepts that can show her birth-date attribute in her credential is before a certain date.

2.2   Innovation Adoption

A prominent approach to investigate adoption of new technologies is covered by the Diffusion of Innovation theory (DOI), presented by Rogers [26]. DOI theory sees innovations as being communicated through certain channels over time and within a particular social system. The approach focuses on the way in which a new technological invention migrates from creation to use. Rogers identified five important attributes of innovations that might influence the decision for their adoption or rejection. The five characteristics of innovations are relative advantage, compatibility, complexity, trialability, and observability, which are valid for both individual and organizational adoption of technology.

Technology-Organization-Environment (TOE) framework was presented by Tornatzky [33] to study the adoption of technological innovations. The framework considers a threefold context for adoption and implementation of technological innovations: technological context, organizational context, and environmental context. The technological context relates to the technologies relevant to the firm such as the current internal practices and equipment, as well as the set of relevant technologies external to the firm. The organizational context describes the characteristics of an organization including firm size, degree of centralization, formalization, complexity of its managerial structure, the quality of its human resources, and the amount of slack resources available internally. Comparing to Rogers' model, TOE includes a new and important component, environmental context. The environment context is the arena in which a firm conducts its business such government and the competitors.

Iacovou et al. [15] presented a model to investigate the interorganizational systems (IOSs) characteristics that influence firms to adopt IT innovations in the context of Electronic Data Interchange (EDI). In this model, Perceived Benefits is a different factor from the TOE framework, whereas organizational readiness is a combination of the technology and organization context of the TOE framework. Nevertheless, Iacovou et al. included and highlighted external pressure as an important factor.

We have identified the Institutional Theory also to be relevant for our research. Institutional factors including schemas, rules, norms, and routines are crucial in shaping organizational structure and organizational decisions [28]. According to the institutional theory, organizational decisions are not driven purely by rational goals of efficiency, but also by social and cultural factors and concerns for legitimacy. It is posited by DiMaggio and Powell [9] that Coercive isomorphism, known as the pressures from other organizations, Mimetic isomorphism, known as the imitation of structures adopted by others in response to pressures, and Normative isomorphism, known as conformity to normative standards established by external institutions, potentially have influence on the behaviour of an organization.

3 Conceptual Model for Adoption of Privacy-ABCs

Based on the theories explained in the previous section, we constructed a combined conceptual model of the relevant factors that are potentially applicable to Privacy-ABCs adoption. We also propose some factors that are new and specific to the domain of Privacy and characteristics of Privacy-ABCs. The conceptual model presented in Fig. 1 incorporates thirteen factors categorized in five groups.

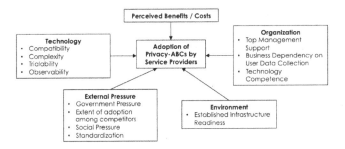

Fig. 1. Conceptual model for factors influencing adoption of Privacy-ABCs by service providers

Technology

Privacy-ABCs are a kind of new technologies for privacy-respecting access control and their characteristics may have a strong influence on the decision of the potential adopters.

Compatibility: Refers to the degree to which an innovation is perceived as consistent with the existing values, needs, and past experiences of the potential adopters [26]. Various published studies examined the role of compatibility, and considered it to be an essential determinant of IT innovation adoption [23,31,32,36] and several of them (e.g. [32,36]) found it as a significant driver. Regarding Privacy-ABCs, it is probable that a higher compatibility of their specifications with the existing Identity and Access Management (IAM) protocols and standards that are commonly used increases the likelihood of a positive decision to adopt them, as any change in the IAM processes may affect a wide range of the subsystems.

Complexity: Refers the degree to which an innovation is perceived as relatively difficult to understand and use [26]. There have been several works considering the role of complexity in innovation adoption [7,16,32]. Privacy-ABCs are based on difficult cryptographic concepts, which are not easy for people beyond the inventors to understand. On the one hand, following the claim by Borking [3], understanding of Privacy-ABCs for the purpose of adoption requires technical and legal knowledge. On the other hand, Wästlund et al. [34] claim that the users have difficulties in using these novel technologies. Therefore, we consider Complexity to be a relevant factor in adoption of Privacy-ABCs.

Trialability: Refers to the degree to which an innovation may be experimented with and tested on a limited basis [26]. In another word, it concerns how easy it would be for a potential adopter to test (or partially test) the features that the new technology provides. There is always a level of uncertainty for the adopters when they decide to invest on a new innovation and reducing this uncertainty by allowing them to try out the innovation would probably influence their decision. Trialability has been identified to be important in a number studies concerning adoption of a new technology [16,18,24]. Borking [3] also highlights the role of Trialability for the adoption of PETs in general. Therefore we envision that Privacy-ABCs adoption can be influenced by their Trialability.

Observability: Refers to the extent to which the innovations are visible for the outside world [26]. In this regard, Moore and Benbasat [20] consider demonstrability as one type of observability. Unlike many other innovations that have visible results and can be well demonstrated, Privacy-ABCs are very challenging to present. They are not like standalone products and are always integrated into another service in order to perform access control. Therefore, demonstrators have difficulties showing all the added values of Privacy-ABCs in demos.

Perceived Benefits/Costs

It has been indicated in both DOI and Iacovou models that the likelihood of the allocation of the managerial, financial, and technological resources necessary to use that innovation are increased when there is better managerial understanding of the relative advantage of an innovation increases. Employing Privacy-ABCs comes with some direct and indirect costs such as for implementation, education, and change of processes. It can be challenging to find business cases that are enabled by Privacy-ABCs directly, nevertheless, compliance with data protection regulation, less investment in personal data storage and protection, and reduced risk of privacy breaches are the perceptions that can influence Privacy-ABCs adoption.

Organization

Beside the characteristics of an innovation itself, several organizational characteristics of the potential adopters have an influence on their decision to adopt or reject an innovation.

Top Management Support: It has been shown that technology innovation adoption can be influenced by top management support and their attitudes towards change [7,22,32]. The lack of top management support was identified as a key inhibitor in B2B deployment of e-commerce by [30]. Borking [3] also mentioned that top management's attitude towards changes caused by PETs can influence the adoption of PETs. Consequently, it may be the case that top management support increases the likelihood of adoption of Privacy-ABCs.

Business Dependency on User Data Collection: The role of industry sector in which a firm operates has been investigated and identified to have an influence on adoption of IT technologies in [11,19,25]. Indeed, Privacy-ABCs have not been invented only for the use of businesses and enterprises and can be adopted by any other organization. Nevertheless, we experienced in the context of the ABC4Trust EU research project that it is usually more challenging to convince businesses to integrated Privacy-ABCs into their services compared to other organizations such as non-profit ones. That increased the curiosity to have a special look at the case of commercial adopters that might result in a useful impression of the influencing factors. Essentially, the *Business Model* of a company defines its roadmap. Osterwalder [21] defines the business model to be a conceptual link forming a triangle between strategy, business organization and ICT. Among the elements of a business model, employing Privacy-ABCs can influence the following:

- Product: Adopting Privacy-ABCs allows the customers to reveal less personal information. Therefore, if the business model value proposition is shaped around the users' data (such as Social Networks), the company will be probably more reluctant to employ such kind of technologies. As a result, we expect that dependency of the business to the collected users data plays an important role in the decision for adopting Privacy-ABCs.
- Customer Interface: It is a common practice to conduct targeted marketing and advertisement based on the extra information collected from the users, such as demographic data. Using Privacy-ABCs heavily influence this part as they prevent the service providers from such kind of data collection.

Technology Competence: Technological resources have been consistently identified as an important factor for successful information systems adoption [8,32]. A higher perceived technical competence was also identified by [17] as a key factor in adoption of electronic data interchange. The work by [36] also demonstrated that technology competence significantly drives e-business usage. So, the role of technology competence has been proven in the literature in adoption of many IT innovations. We consider technical competence to be relevant for adoption of Privacy-ABCs as we also experienced in the context of ABC4Trust pilots that typical developers had difficulties to integrate Privacy-ABCs into some services on their own and constant support of technology providers was needed, while developers with scientific background and technical understanding of the technology went through the integration process smoothly. Hence, lack of technical competency can hinder Privacy-ABCs adoption.

External Pressure

As we mentioned earlier, various sources of external pressure may influence the adoption of new innovations. Here we briefly introduce the ones that are relevant for Privacy-ABCs.

Regulatory Pressure: A regulatory body may be the source of coercive pressures [29]. In this regard, there has been movements in the regulatory sectors in Europe introducing more restriction on users' data collection and processing (Art. 6 and 7 of Directive 95/46/EC) as well as secure storage of the collected data (Art. 16 and 17 of Directive 95/46/EC). Consequently, it can be foreseen that organizations will soon feel pressure to start reconsidering their data collection schemes and look for secure solutions that reduces their liability for protecting the users data.

Social Pressure: There have been major incidences recently which we expect them to have an influence on adoption of privacy enhancing technologies in general. The most well-known incidence was brought up by Edward J. Snowden[4], which indeed highly stimulated the public opinion on the need for a raise of privacy in online environments. So, we expect that social pressure on the service providers will increase and therefore urges them towards employing mechanisms that reduce personal data collection in their processes.

Extent of Adoption among Competitors: The existence of mimetic pressures toward the adoption of innovations by organizations is confirmed in [10,13]. Knowing a competitor has adopted an innovation and it has been a success, the firm tends to adopt the same innovation [14]. The work by [29] confirmed the strong role of this factor in adoption of E-Procurement System. It could happen that offering more privacy becomes an advertising parameter especially in countries with more privacy protection culture. Therefore adoption of Privacy-ABCs by the competitors of a firm can motivate the decision makers to follow the same approach not to lose on the trust reputation.

Standardization: It is very typical for industries to employ procedures, processes or protocols that are standardized in order to ensure interoperability and sustainability of their products and services. In this regard, Standardization can become a source of normative isomorphism. There have been standardization projects that are very relevant to Privacy-ABCs and the ABC4Trust architecture [12]. For instance, ISO/IEC 24760 focuses on a framework for identity management and is conducted in 3 parts covering *Terminology and concepts*, *Reference architecture and requirements*, and *Practice*. Such standards have a good potential to influence the future adoption of Privacy-ABCs.

Environment

Here with environment we refer to the external conditions that do not introduce any pressure but can facilitate or hinder adoption of an innovation. For instance, it is more likely to succeed in implementing the idea of a remote movie rental company in a country that has cheaper, faster and more reliable postal services around.

Established Infrastructure Readiness: Electronic IDs have been implemented in various countries around the world, and therefore use of digital certificates for authentication and access control have been leveraged for service providers. Privacy-ABCs have been demonstrated their capabilities to be integrated with the existing eID infrastructure [2]. Furthermore, the European Commission also considered investing on the research for integration of Privacy-ABCs into future electronic IDs[5]. Consequently, having the global infrastructure ready to support Privacy-ABCs, the integration of these technologies into authentication and access control of service providers will be facilitated.

---

[4] http://en.wikipedia.org/wiki/Edward_Snowden.
[5] http://www.futureid.eu/.

4   Empirical Evaluation

4.1   Methodology

As Privacy-ABCs are not yet adopted, it is not possible to survey the service providers (adopters/non-adopters) in order to discover the drivers and the inhibitors. Thus, we decided to follow a forecast approach and collect the opinion of the experts from the relevant fields on the importance and influence level of the potential factors we introduced in our conceptual model. We designed a questionnaire containing quantitative and used a 5-point Liker scale from "not important at all" or "not at all influential" to "extremely important" or "extremely influential". Moreover, based on our experience of the ABC4Trust pilots, we made some of the factors more granular and presented them in two questions. That includes the Cost factor, which we presented as *Cost of Integration* and *Cost of Education*, Complexity factor, divided into *Complexity for Developers* and *Complexity for Users*, and the Government Pressure, presented as *Regulations for Data Collection* and *Regulations for Securing the Collected Personal Data*.

We refined the questionnaire in an iterative process performed in four steps with the help of two groups, one *with* dominant knowledge of Privacy-ABCs, and one *without* dominant knowledge of Privacy-ABCs. In the first step, a person with dominant knowledge in the field reviewed the questionnaire to check the technical correctness and readability of the questions. After rounds of discussions a version was ready to be reviewed by the people without dominant knowledge to validate the readability of the questions. After receiving their feedbacks, the questions were modified to improve the readability. In the next step, again a person with dominant knowledge reviewed the changes and the proposed updates. The next version was then distributed to the people without dominant knowledge and as we did not receive further clarification requests, the questionnaire was finalized. In this questionnaire, the respondents were asked to evaluate their level of expertise using the five-level Dreyfus model of skill acquisition. They were also requested to select their domains of expertise from relevant list including "Privacy and Identity Management", "Data Protection", "Policy Maker", and "Software and Services", or specify it if it was not on the list (multiple selection was allowed).

The survey was performed during the ABC4Trust summit event, on 20th of January 2015 in Brussels. The event was one of the best opportunities to get into contact with the experts of the relevant domains as it was broadly advertised via various important channels such as the one from the European Commission. Furthermore, having prestigious guest speakers also increased the chance of attracting stack-holders to the event. During this event, we gave a full day tutorial of Privacy-ABCs to the participants, covering various aspects such as limitation of current Identity Management Systems, how Privacy-ABCs work in theory, their implementation on computers, smartcards and mobile phones, as well as four real-time demos of some scenarios where Privacy-ABCs could improve users' privacy. These demos addressed a wide range of scenarios, namely "online university course evaluation system", "school community interaction platform", "online movie streaming"[6], and "hotel booking"[7]. It is important to note that most of the tutorials and presentations were performed by the partners who were not involved in this study so we avoided unintentional biasing of the audience.

4.2   Results

At the end of the day, the participants were asked to answer the provided questionnaire. From over 80 participants, 20 completed the questionnaire, of which we excluded 3 as the respondents evaluated themselves below "Proficient" (below 4 out of 5). From the remaining respondents (the experts), 10 chose "Privacy and Identity Management", 3 chose "Data Protection", 3 chose "Policy Maker", and 5 chose "Software and Services" as their fields of expertise (multiple selection was allowed).

Table 1 summarizes the influence/importance level of the factors from the experts' perspective along with their ranking. The items ranked from 1 to 7 have a mean value over 3.0, meaning that the experts considered them on average "very" or "extremely" important or influential. The results show that in experts' opinion, "Technical Competency" of the adopters has the least effect on their decision among the others. Nevertheless, all the factors received a mean score over the average (2.0).

5   Implications and Discussion

From a practical point of view, the results give directions to supporting communities showing them where to put their future efforts. To foster adoption of Privacy-ABCs, priority shall be given to the items ranked from 1 to 7 (*mean* > 3.0). In this regard, the opinion of the experts can be reflected in two dimensions:

First, the Privacy-ABCs technology developers shall enrich the implementation of Privacy-ABCs in terms of

- Usability and Risk Communication: Privacy-ABCs such as different anonymity levels or applying predicates over attributes did not exist in the previous generation of access control mechanisms. Thus user interfaces shall be enhanced to appropriately communicate such features. In addition to that, Privacy-ABCs are user-centric approaches and their implementation essentially requires a piece of software to run on behalf of the users. This urges the users to install a client agent to represent them in the protocol steps, which consequently reduces the mobility of the users as they need to have this software on every device they use. In this regard, new deployment schemes reducing the need for client side installation can support reducing the complexity for the users.
- Agile Trial Platforms: Having online services that allow the interested parties to rapidly and with minimal effort integrate Privacy-ABCs for trial purposes can significantly improve their trialability.

---

[6] https://idemixdemo.mybluemix.net/.
[7] https://abc4trust.eu/demo/hotelbooking.

Table 1. Ranking of the relevant factors for adoption of Privacy-ABCs by service providers based on the experts' opinions

Rank	Factor	Mean	Var.
1	Business model dependency to data collection	3,71	0,22
2	Complexity for users	3,53	0,39
3	Observability	3,29	0,60
3	Top management support	3,29	1,47
5	Trialability	3,24	0,32
6	Cost of integration	3,19	0,83
7	Complexity for developers	3,12	0,74
7	Regulations for data collection	3,12	0,61
9	Regulations for securing the collected personal data	2,94	0,68
10	Established infrastructure readiness	2,88	0,99
11	Social pressure	2,71	1,10
12	Compatibility with existing IdM infrastructure	2,65	1,49
13	Competition among service providers	2,59	0,76
14	Cost of education	2,53	0,89
14	Standardization	2,53	1,26
16	Technical competency	2,35	1,12

- Designing Comprehensive Demos: In our questionnaire we asked the experts to select the most informative demos they saw during the day. The school community interaction platform received the most points (9 votes). The experts mainly mentioned they liked the fact that it was a complete set of scenarios and there were very many roles and a richer set of credentials. This allowed to show similarities and differences in the policies and implementations. However, the university course evaluation demo received the second highest point (6 votes) and the given reason was that the smaller scope made the scenario basic, very clear and easy to understand the benefits.
- Plug-and-Play Libraries: providing robust, rich and plug-and-play libraries along with appropriate documentation can notably facilitate the integration process for the software developers and consequently decrease the integration costs.

The second dimension relates to the dissemination strategies. More effort shall be put to target high-ranked managers and provide them with supporting materials that raise their understanding and awareness of Privacy-ABCs such as what these technologies can offer, how Privacy-ABCs can influence their processes, and what is needed for them to employ Privacy-ABCs. Moreover, the data protection bodies and the policy makers shall try to disseminate the capabilities of Privacy-ABCs to the regulatory authorities so that they become aware of the technical means to enforce minimal data collection regulations.

From a theoretical perspective, our results contribute to the existing theories by delivering a reduced conceptual model (Fig. 2) as a result of the expert surveys. Compared to the literature, our conceptual model introduces a new potential factor for adoption of technologies that limit service providers' access to users' data. The low variance clearly confirms that most of the experts had similar opinion on the role of "Business Model Dependency to Data Collection" and considered it as a key factor for shaping the desire of the service providers to adopt Privacy-ABCs.

The conceptual model also triggers theoretical research to boost the identified factors. More specifically, we see open questions on the methods to efficiently, transparently and explicitly communicate identity and attribute disclosure risks to the users via corresponding user interfaces of Privacy-ABCs.

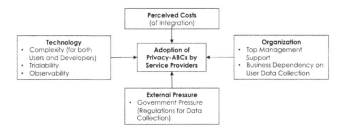

Fig. 2. Reduced conceptual model for factors influencing adoption of Privacy-ABCs by service providers

6 Conclusion

Privacy-ABC technologies are promising mechanisms that allow privacy and security go hand-in-hand. They provide various privacy features such as minimal attribute disclosure as well as unlinkable partial identities. Privacy-ABCs have passed the trial phase and proved their applicability and it is now important to understand how we can push these technologies forward. In this work, we investigated the potential factors that influence adoption of Privacy-ABCs by service provider and empirically evaluated the developed conceptual model through expert surveys. We collected the opinion of the experts during an especial international event where they received a full-day tutorial of various aspects of Privacy-ABCs.

The statistics of the collected opinions show that *Business Model Dependency to Data Collection, Complexity for User, Top Management Support, Observability, Trialability, Cost of Integration, Regulations for Data Collection*, and *Complexity for Developers* are the most important or influential factors impacting the decision of the service providers to employ Privacy-ABCs. These findings put lights on the directions towards which the supporting community should move and imply, despite the common beliefs of recognizing Privacy-ABCs as a redeemer to fight Social Networks, Privacy-ABCs may have higher chance to succeed in their adoption if they first target the service providers in the markets that are not based on users' data.

References

1. Sabouri, A. (ed.): Architecture for attribute-based credential technologies - final version. Deliverable D2.2, The ABC4Trust EU Project (2014). https://abc4trust.eu/download/Deliverable_D2.2.pdf. Accessed 08 November 2014
2. Bjones, Ronny, Krontiris, Ioannis, Paillier, Pascal, Rannenberg, Kai: Integrating Anonymous Credentials with eIDs for Privacy-Respecting Online Authentication. In: Preneel, Bart, Ikonomou, Demosthenes (eds.) APF 2012. LNCS, vol. 8319, pp. 111–124. Springer, Heidelberg (2014)
3. Borking, J.J.: Why adopting privacy enhancing technologies (pets) takes so much time. In: Gutwirth, S., Poullet, Y., de Hert, P., Leenes, R. (eds.) Computers, Privacy and Data Protection: An Element of Choice, pp. 309–341. Springer, Netherlands (2011)
4. Brands, S.: Untraceable off-line cash in wallet with observers. In: Helleseth, T. (ed.) Advances in Cryptology - CRYPTO 1993, vol. 765. Springer, Heidelberg (1994)
5. Camenisch, Jan L., Lysyanskaya, Anna: Signature Schemes and Anonymous Credentials from Bilinear Maps. In: Franklin, Matt (ed.) CRYPTO 2004. LNCS, vol. 3152, pp. 56–72. Springer, Heidelberg (2004)
6. Cavoukian, A., et al.: Privacy by Design: The 7 Foundational Principles. Information and Privacy Commissioner of Ontario, Canada (2009)
7. Chong, A.Y.L., Ooi, K.B., Lin, B., Raman, M.: Factors affecting the adoption level of C-Commerce: an empirical study. J. Comput. Inf. Syst. **50**(2), 13 (2009)
8. Crook, C.W., Kumar, R.L.: Electronic data interchange: a multi-industry investigation using grounded theory. Inf. Manag. **34**(2), 75–89 (1998)
9. DiMaggio, P.J., Powell, W.W.: The iron cage revisited: institutional isomorphism and collective rationality in organizational fields. Am. Sociol. Rev. **48**(2), 147–160 (1983)
10. Fligstein, N.: The spread of the multidivisional form among large firms, 1919–1979. Adv. Strat. Manag. **17**, 55–78 (1985)
11. Goode, S., Stevens, K.: An analysis of the business characteristics of adopters and non-adopters of world wide web technology. Inf. Technol. Manag. **1**(1–2), 129–154 (2000)
12. Hansen, M., Obersteller, H., Rannenberg, K., Veseli, F.: Establishment and prospects of Privacy-ABCs. In: Rannenberg, K., Camenisch, J., Sabouri, A. (eds.) Attribute-based Credentials for Trust, pp. 345–360. Springer International Publishing, Switzerland (2015). http://dx.doi.org/10.1007/978-3-319-14439-9_11
13. Haunschild, P.R., Miner, A.S.: Modes of interorganizational imitation: the effects of outcome salience and uncertainty. Adm. Sci. Q. **42**(3), 472–500 (1997)
14. Haveman, H.A.: Follow the leader: mimetic isomorphism and entry into new markets. Adm. Sci. Q. **38**(4), 593–627 (1993)
15. Iacovou, C.L., Benbasat, I., Dexter, A.S.: Electronic data interchange and small organizations: adoption and impact of technology. MIS Q. **19**(4), 465–485 (1995)
16. Kendall, J.D., Tung, L.L., Chua, K.H., Ng, C.H.D., Tan, S.M.: Receptivity of singapore's SMEs to electronic commerce adoption. J. Strat. Inf. Syst. **10**(3), 223–242 (2001)
17. Kuan, K.K., Chau, P.Y.: A perception-based model for edi adoption in small businesses using a technology-organization-environment framework. Inf. Manag. **38**(8), 507–521 (2001)
18. Martins, C.B., Steil, A.V., Todesco, J.L.: Factors influencing the adoption of the internet as a teaching tool at foreign language schools. Comput. Edu. **42**(4), 353–374 (2004)

19. Miller, N.J., McLeod, H., Ob, K.Y.: Managing family businesses in small communities. J. Small Bus. Manag. **39**(1), 73–87 (2001)
20. Moore, G.C., Benbasat, I.: Development of an instrument to measure the perceptions of adopting an information technology innovation. Inf. Syst. Res. **2**(3), 192–222 (1991)
21. Osterwalder, A., et al.: The business model ontology: a proposition in a design science approach (2004)
22. Premkumar, G., Ramamurthy, K., Crum, M.: Determinants of EDI adoption in the transportation industry. Eur. J. Inf. Syst. **6**(2), 107–121 (1997)
23. Premkumar, G., Roberts, M.: Adoption of new information technologies in rural small businesses. Omega **27**(4), 467–484 (1999)
24. Ramdani, B., Kawalek, P.: SME adoption of enterprise systems in the Northwest of England. In: McMaster, T., Wastell, D., Ferneley, E., DeGross, J. (eds.) Organizational Dynamics of Technology-Based Innovation: Diversifying the Research Agenda, IFIP International Federation for Information Processing, vol. 235, pp. 409–429. Springer, USA (2007)
25. Raymond, L.: Determinants of web site implementation in small businesses. Internet Res. **11**(5), 411–424 (2001)
26. Rogers, E.: Diffusion of innovations. Free Press, New York (2003)
27. Sabouri, A., Krontiris, I., Rannenberg, K.: Attribute-Based Credentials for Trust (ABC4Trust). In: Fischer-Hübner, S., Katsikas, S., Quirchmayr, G. (eds.) TrustBus 2012. LNCS, vol. 7449, pp. 218–219. Springer, Heidelberg (2012)
28. Scott, W.R.: Institutions and Organizations (Foundations for Organizational Science). SAGE Publications Inc., New York (2000)
29. Soares-Aguiar, A.: Palma-dos Reis, A.: Why do firms adopt E-Procurement systems? using logistic regression to empirically test a conceptual model. IEEE Trans. Eng. Manag. **55**(1), 120–133 (2008)
30. Teo, T.S., Ranganathan, C., Dhaliwal, J.: Key dimensions of inhibitors for the deployment of web-based business-to-business electronic commerce. IEEE Trans. Eng. Manag. **53**(3), 395–411 (2006)
31. Teo, T.S., Tan, M., Buk, W.K.: A contingency model of internet adoption in Singapore. Int. J. Electron. Commer. **2**(2), 95–118 (1997)
32. Thong, J.Y.: An integrated model of information systems adoption in small businesses. J. Manag. Inf. Syst. **15**(4), 187–214 (1999)
33. Tornatzky, L.G., Fleischer, M., Chakrabarti, A.K.: Processes of technological innovation (1990)
34. Wästlund, E., Angulo, J., Fischer-Hübner, S.: Evoking Comprehensive Mental Models of Anonymous Credentials. In: Camenisch, J., Kesdogan, D. (eds.) iNetSec 2011. LNCS, vol. 7039, pp. 1–14. Springer, Heidelberg (2012)
35. Zafarani, R., Liu, H.: Connecting users across social media sites: a behavioral-modeling approach. In: The 19th ACM SIGKDD International Conference on Knowledge Discovery and Data Mining, KDD 2013, Chicago, IL, USA, 11–14 August 2013, pp. 41–49 (2013)
36. Zhu, K., Dong, S., Xu, S.X., Kraemer, K.L.: Innovation diffusion in global contexts: determinants of post-adoption digital transformation of european companies. Eur. J. Inf. Syst. **15**(6), 601–616 (2006)

Consumers' Perceptions of Social Commerce Adoption in Saudi Arabia

Salma S. Abed, Yogesh K. Dwivedi$^{(\boxtimes)}$, and Michael D. Williams

Management and Systems Section (MASS),
School of Management, Swansea University, Swansea, UK
{717185, y.k.dwivedi, m.d.williams}@swansea.ac.uk

Abstract. This study aims to examine the factors that affect consumer adoption of social commerce technologies in the context of Saudi Arabia. The factors descriptively explored in this research include: performance expectancy, effort expectancy, social influence, hedonic motivation, habit, trust, consumer innovativeness, information quality and behavioural intention. The survey data utilised in this research was collected through a self-administered questionnaire within a convenience sample. The results obtained through a descriptive analysis confirmed that Saudi consumers perceive the abovementioned factors as important and they have strong behavioural intention to use social commerce technologies.

Keywords: Social media · Social commerce · Customer · UTAUT2 · Adoption · Saudi Arabia

1 Introduction

Social media plays an important part in the countries' economic development [23] as it offers novel ways for both organisations and customers to link with each other. Businesses began to embrace social media websites as a technique to improve communication, information sharing, and collaboration by applying many innovative and vital business practices [34]. Social media motivated companies to work faster by creating and operating more interdependencies in global markets [1]. Therefore, the development of social media has enhanced a new e-commerce model named social commerce. The term of social commerce is defined as a concept of Internet-based social media that allows people to contribute actively in the selling and marketing of diverse products and services in online marketplaces [20]. This dynamic process assists consumers to get better information about different products and services delivered by companies [15].

Saudi Arabia has witnessed the biggest growth of diffusion of social media platforms, which is the strongest supporting factor to e-commerce adoption within the country [9]. Many large businesses, as well as small companies and new ventures, have set-up their organisations and group profiles on Facebook, LinkedIn and other similar websites. In fact, Saudi Arabia ranks the second in the Arab countries, after Egypt, for registering 5,240,720 Facebook users [5]. In addition, Saudi Arabia represents the largest percentage of Twitter users with around 830,300 users or 38 % of total Arab

© IFIP International Federation for Information Processing 2015
M. Janssen et al. (Eds.): I3E 2015, LNCS 9373, pp. 133–143, 2015.
DOI: 10.1007/978-3-319-25013-7_11

users [5]. Moreover, statistics have revealed that Saudi Arabia has contributed 90 million video views per day. This is the maximum number of YouTube viewings worldwide per Internet user [9]. These amazing evidences have created a new background for business owners, managers, and marketers to reach their potential consumers. However, even though consumers in other countries such as Hong Kong, China, South Korea, and Thailand apply online shopping activities actively, mostly Saudi consumers use online social media simply to help them make buying decisions [3]. The remaining sections of the paper include the relevant literature review and theoretical basis in the third section. These are followed by the methodology, results, and discussion outlined in the fourth, fifth and sixth sections, respectively. Finally, the research contribution, conclusion and limitations are delivered in section seven eight, and nine.

2 Literature Review

Limited empirical studies have been conducted within this context of social commerce as it is still new. [14, 15] extended the Technology Acceptance Model (TAM) to measure social commerce adoption by consumers. The researcher examined some components of social commerce that affected consumers' intention to buy. The model tested ratings and reviews, referrals and recommendations, and forums and communities aiming to help to introduce new business plans for e-vendors. The study also indicated that trust is a continuing issue in e-commerce and can be examined in social commerce constructs. The researcher collected survey data and applied structural equation modelling (SEM) for analysis. The results pointed out that perceived usefulness and forums and communities have a positive impact on trust. In addition, the findings showed that trust has a significant effect on intention to buy [14, 15].

In the context of Saudi Arabia, Only two studies have been found that focused on the use of online social media to help overcome e-commerce adoption barriers. [4] presented research that is designed to demonstrate a conceptual framework by extending it from TAM. The framework aims to examine social media effects and perceived risk as the moderating effects between purchasing intention and an actual online purchase in Saudi Arabia. Furthermore, [17] evaluated the effectiveness of online social networking by entrepreneurs in the Arabian Gulf including Bahrain, Kuwait, United Arab Emirates, Saudi Arabia, Qatar, and Oman. The research used a qualitative approach by interviewing a sample size of 50 business entrepreneurs in the Arabian Gulf who used online social networks as a method of promoting their products. The study found that online social networks are a cheap and easy method of advertising and would give all entrepreneurs a better chance of reaching their target market as well as succeeding in their ventures. In addition, entrepreneurs can now target their markets using online social networks. Moreover, social networking websites allow businesses to introduce their products to different market segments, with a low chance of failure and low expense. Finally, the researchers added that the rise in web-based social interaction can change the way businesses operate in the future. Finally, empirical studies conducted in the context of social commerce are very limited, which this study aims to address. This study proposes a new model that can be developed and extended by applying more recent and comprehensive technology adoption theories as well as adding other more appropriate constructs especially in the consumer context.

3 Theoretical Basis

In order to explain social commerce adoption effectively from the customers' perspective, the conceptual model should provide a clear image of social commerce features. From the analysis of the common theories in the field of technology acceptance, [31, 33] developed the Unified Theory of Acceptance and Use of Technology (UTAUT) by merging eight IT acceptance models. The UTAUT has four main constructs including performance expectancy, effort expectancy, social influence, and facilitating conditions that influence behavioural intention to use a technology and usage behaviours. UTAUT was capable to explain around 70 per cent of the variance of behavioural intention and around 50 per cent of the variance of use behaviour [31, 33]. Recently, [32] proposed an additional three constructs to the original UTAUT model including hedonic motivation, price value, and habit. [32] claimed that the suggested additions to UTAUT2 present major changes in the variance described in behavioural intention and use behaviour especially within the consumers' context.

Considering the limits of the earlier investigated constructs for social commerce adoption [14, 15], UTAUT2 is more suitable in the context of this study. This is because UTAUT2 was developed on UTAUT, which has been credited as the most parsimonious and comprehensive predictive model [31, 32]. Furthermore, UTAUT2 is proposed mainly for explaining technology acceptance from the customers' contexts other than the organisational use [32]. Besides, the UTAUT2 has investigated factors influencing users' acceptance of iPad phones [16], mobile payment uses [29], and mobile learning acceptance [19] that shares similar technological characteristics with social commerce. As a result, the UTAUT2 has been selected as the theoretical foundation of the proposed conceptual model in order to understand the antecedents of customers and their preference to buy from social media websites, in a new cultural context (Saudi Arabia). This shadows [32] the suggestion that future research should apply UTAUT2 in diverse countries. This also follows the call by [32] for future research to investigate the UTAUT2 on different technologies. Furthermore, to be consistent with the recommendations of [32], other external factors (trust, consumer innovativeness, and information quality) will be measured along with UTAUT2 constructs (performance expectancy, effort expectancy, social influence, hedonic motivation, habit) in the same conceptual model. These other constructs have been developed according to the literature review.

Furthermore, discard the two independent variables facilitating condition and price value and the dependent variable use behaviour which are considered a part from UTAUT2 model's constructs. Discarding both independent variables is a logical step as social commerce use does not need any technical infrastructure support as in system adoption; apart from the Internet cost, there is no monetary cost for using social media because most of the time, wireless connections are available free of charge. Thus, discarding the dependent variable use behaviour is because the literature indicated that social commerce technologies are new in the Arabian Gulf generally and in Saudi Arabia specifically and they are still not fully used [2]. The suggested constructs are demonstrated below in Table 1:

Table 1. Description of factors

Constructs examined	Definition
Performance expectancy	"The degree to which an individual believes that applying the technology will help him or her to gain in job performance" [31].
Effort expectancy	"Extent of ease connected with the use of system" [31].
Social influences	"The extent to which an individual perceives that important others believe he or she should apply the new system" [31].
Hedonic motivation	"The feeling of cheerfulness, joy, and enjoyment, which is stimulated by applying technology" [32].
Habit	"The extent to which, people tend to perform behaviours automatically because of learning" [32].
Trust	"Individual willingness to depend based on the beliefs in ability, benevolence, and integrity" [12]
Consumers' innovativeness	"The degree to which the individual is willing to adopt innovations such as goods and services or new ideas without communicating with others' previous purchasing experience" [24]
Information quality	"The consumers' general perception of the accuracy and completeness of website information as it relates to products and transactions" [21].
Behavioural intention	"The extent to which an individual intends to adopt the technology in the future" [31].

4 Methodology

This study utilised a total of thirty-six scale items that were derived from the literature of technology adoption in order to measure the selected variables; i.e. Trust, consumers innovativeness and information quality in addition to other variables in the UTAUT2 model. To do that, it is vital to make items that will help in measuring the characters of the related variants. Several items were used in order to measure the variables of the UTAUT2 which were adapted from [32]. The additional variables have used items that were adopted by a number of writers. Therefore, trust was selected from [12, 18, 21, 26]. In addition, this paper selected the consumer innovativeness items from [27]. Furthermore, information quality items have been adopted from [21]. The degree of responses was estimated using the seven-point scale ranging from strongly agree to strongly disagree. In regards to the language of the data collection tool, the questionnaire was translated into Arabic to overcome the cultural and linguistic differences [6]. Then, a pilot study was conducted using 20 questionnaires that were distributed to Saudi social media users who were asked to give their feedback in case faced any difficulties in answering the questionnaire [10]. Accordingly, the questionnaire's items were rechecked in terms of clarity, language simplicity, and length.

When it came to the sampling, this study implemented a convenience sampling as the researcher does not have a list of social commerce potential users. Additionally, convenience sampling is cost-effective [10, 11]. Furthermore, the results of a convenience sample can be generalised more appropriately since it allowed for the presence of a variety of profiles and characters of potential users [11]. The population that was going to be sampled were all from the regions of Saudi Arabia including big cities and small towns. This has been achieved by distributing the survey questionnaires with both hard copies and online as web links. In the hard copy survey, most of the respondents of the questionnaires were students, as the questionnaires were distributed to distance-learning students in the Management School at King Abdul-Aziz University in Jeddah. The students were present for two weeks at the university campus for the final exams. The researcher took the chance to distribute the survey to students. Distance-learning students enrolled in the programme are from different regions of the Kingdom including large cities and small towns; there were different age groups; and from both genders. Therefore, they were representative of the diverse population of Saudi Arabia. In the soft copy of the survey, the researcher used the online survey software Qualtrics for distributing the web-based survey. The web link was sent to different Saudi Arabian e-commerce groups on social media such as social networking sites including Facebook and LinkedIn; as well as the micro blogging service such as Twitter. Due to space constraints scale items/measurements cannot be provided in the paper but will be available upon the request.

5   Results

5.1   The Response Rates

As mentioned earlier, the survey questionnaires were circulated in both hard copies and online as web links. In the hard copy survey, participation was completely voluntary. Respondents were requested to complete a questionnaire based on their perception and/or acceptance of social commerce. A total of 700 survey questionnaires were distributed and the returned completed surveys were 417 with a 59 per cent response rate. In the soft copy of the survey, the researcher used the online survey software Qualtrics for distributing the web-based survey. One of the features the online survey software Qualtrics provided was that it showed how many respondents started the survey, but did not complete it. The total number of consumers who participated in the survey was 225. The total number of consumers who completed the survey was 120, with a response rate of 53 per cent. As a result, a total of 537 survey questionnaires were collected from both the paper-based survey and web-based survey. The questionnaires were carefully checked before entering the data using SPSS 22.0. Out of the 537 questionnaires collected, only 507 were used; 27 were considered unusable and discarded due to the huge amount of missing data resulting from missing pages or incomplete sections. In addition, three questionnaires were also discarded due to them having the same answer to all questions. These responses were considered as invalid, and they were removed in the data-editing process stage [28].

5.2   Respondents' Profile and Characteristics

When it comes to respondents' profile and characteristics, this research adopts the following demographic information: gender, age, and education (see Table 2). The demographic details of the main survey sample show that the majority of the respondents were female, forming 65.1 % of the whole sample, while males are represented by only 34.9 %. In regard to the respondents' age, the descriptive statistics demonstrate that the largest age population was within 21–29 years old with 62.7 %, followed by the age group of > = 18–20 with 17.9 %. The rest of the percentages were divided among the age group of 30–39 (13.4 %) and 5.5 % for those who were between 40–49, whereas the smallest percentage was 0.4 % as only two respondents were at the age of 50 and above. Regarding the educational level, the majority of respondents hold a Bachelor's degree, representing 45.0 % of the total sample. The second largest group were high school holders (38.9 %) followed by 10.5 % as postgraduates and 5.5 % as diploma holders. A very small percentage of respondents held less than high school qualification with (0.2 %). Table 2 shows the demographic details of the respondents in the main survey sample.

5.3   Descriptive Analysis and Normality Assumption

According to Table 3 below, the descriptive statistics show that there are three items devoted to measure consumers' perceptions on performance expectancy (PE). PE2 achieved the highest mean score of 5.61 (±1.302). In contrast, the lowest mean was 5.36 (±1.436) as a value recorded for PE3. There are four items identified on effort expectancy (EE). As seen, the largest mean scores were 5.56 (±.1.430) for EE3 and the lowest mean is for EE2, 5.24 (±.1.445). Social influence (SI) was measured by three

Table 2. Respondents' profile and characteristics

Variable	Group	Frequency	Percent
Gender	Male	177	34.9
	Female	330	65.1
	Total	**507**	**100.0**
Age	> = 18–20	91	17.9
	21–29	318	62.7
	30–39	68	13.4
	40–49	28	5.5
	50 and above	2	0.4
	Total	**507**	**100.0**
Education	Less than high school	1	0.2
	high school	197	38.9
	Diploma	28	5.5
	Bachelor's degree	228	45.0
	Postgraduate	53	10.5
	Total	**507**	**100.0**

Table 3. Descriptive and normality tests

Constructs	Descriptive statistics							
	Items	N	Mean	Std. deviation	Skewness		Kurtosis	
					Statistic	Std. error	Statistic	Std. error
Performance expectancy (PE)	PE1	507	5.56	1.333	−.861	.108	.583	.217
	PE2	507	5.61	1.302	−1.126	.108	1.389	.217
	PE3	507	5.36	1.436	−.707	.108	−.032	.217
Effort expectancy (EE)	EE1	507	5.46	1.440	−.891	.108	.468	.217
	EE2	507	5.24	1.445	−.769	.108	.210	.217
	EE3	507	5.56	1.430	−1.137	.108	1.091	.217
	EE4	507	5.33	1.539	−.936	.108	.303	.217
Social influence (SI)	SI1	507	4.47	1.678	−.428	.108	−.417	.217
	SI2	507	4.79	1.600	−.443	.108	−.437	.217
	SI3	507	5.09	1.487	−.652	.108	−.048	.217
Hedonic motivation (HM)	HM1	507	5.55	1.505	−1.121	.108	.887	.217
	HM2	507	5.63	1.382	−1.109	.108	1.113	.217
	HM3	507	5.56	1.422	−.857	.108	.144	.217
Habit (HT)	HT1	507	4.65	1.720	−.377	.108	−.615	.217
	HT2	507	4.15	1.892	−.191	.108	−.956	.217
	HT3	507	4.79	1.536	−.341	.108	−.727	.217
	HT4	507	4.71	1.776	−.514	.108	−.598	.217
Trust (TR)	TR1	507	4.33	1.826	−.293	.108	−.855	.217
	TR2	507	4.45	1.652	−.287	.108	−.577	.217
	TR3	507	4.37	1.737	−.288	.108	−.766	.217
	TR4	507	4.38	1.612	−.260	.108	−.590	.217
	TR5	507	4.91	1.536	−.550	.108	−.162	.217
Consumer innovativeness (CI)	CI1	507	5.16	1.542	−.672	.108	−.038	.217
	CI2	507	5.39	1.484	−.920	.108	.532	.217
	CI3	507	5.48	1.521	−1.010	.108	.501	.217
	CI4	507	4.94	1.688	−.608	.108	−.366	.217
	CI5	507	4.71	1.584	−.384	.108	−.409	.217
	CI6	507	4.66	1.679	−.377	.108	−.561	.217
Information quality (IQ)	IQ1	507	4.92	1.496	-.324	.108	−.446	.217
	IQ2	507	5.41	1.327	−.596	.108	−.058	.217
	IQ3	507	4.72	1.568	−.435	.108	−.266	.217
	IQ4	507	4.91	1.461	−.415	.108	−.215	.217
	IQ5	507	4.81	1.435	−.309	.108	−.367	.217
Behavioural intention (BI)	BI1	507	5.28	1.479	−.641	.108	−.060	.217
	BI2	507	5.05	1.534	−.551	.108	−.227	.217
	BI3	507	5.12	1.552	−.577	.108	−.315	.217
Valid N (listwise) 507								

items. The highest mean scores were 5.09 (±1.487) for SI3 and 4.79 (±.1.600) for SI2 followed by 4.47 (±1.678) for SI1 as the lowest mean. Moreover, there are three items allocated to measuring consumers' perceptions on hedonic motivation (HM). The highest mean value is 5.63 (±1.382) recorded for HM2, while the lowest mean value is 5.55 (±1.505) recorded for HM1. Habit (HT) was measured by four items with 4.79 as the highest score recorded for HT3 and the lowest mean value is 4.15 for HT2. Table 3 also shows that there are five items identified to measure consumers' perceptions on trust (TR). TR5 had the largest mean value of 4.91 (±1.536) compared to TR1 that had the lowest mean score of 4.33 (±1.826). In addition, six items were identified to measure consumers' perceptions on consumer innovativeness (CI). CI3 recorded the highest value with 5.48 (±1.521). In contrast, the lower score recorded was for CI6, 4.66 (±1.679). Finally, three items were adopted to measure the behavioural intention (BI) construct. The lowest mean was for BI2 with a value of 5.05 (±1.534) while the highest mean score was 5.28 (±1.479) for BI1.

Screening the data for assessing the variables normality is a crucial step in the analysis [13, 22, 30]. Normality means the shape of normal distribution of metric variable and its correspondence [13]. Normality of a single variable can be measured statistically or graphically [7, 25, 30]. The failure to achieve normality can result from invalid statistical tests. This study has tested skewness and kurtosis at the item level. Skewness refers to the symmetry of distribution; the test indicates if the distribution is shifted or unbalanced to one side [30]. There are two types of skewness: positive skewness, when the distribution is shifted to the left; and negative skewness when it is shifted to the right [13]. Kurtosis refers to the peakness of distribution [30]. Peaked distributions are termed leptokurtic, whereas, flatter distributions are termed platy-kurtic. The values of skewness and kurtosis are zeroes when variables have normal distributions. Consequently, positive or negative values indicate a deviation from normality. The range of values for suitable deviations is affected by sample size. In small samples less than 30, slight deviations can be serious, whereas with large sample sizes with more than 200 it can be ignorable [13]. On the other hand, the most generally acceptable critical value for kurtosis and skewness distribution is ± 2.58 [13]. Table 3 indicates that skewness and kurtosis variables fall within the acceptable range.

6   Discussion

The presented descriptive results in this study help to visualise what the data revealed. After overviewing the literature review of the studies that have used IS theories to examine social commerce adoption, only one study has used the TAM model [14, 15]. As a result, this study adopted the UTAUT2 variables as a more recent and compre-hensive technology adoption theory. Other more suitable constructs were added especially in consumer context (trust, consumer innovativeness, and information quality). Furthermore, this study has identified items to test the proposed variables, which has been examined in previous research in the literature. By collecting empirical data from 700 paper-based survey participants and 225 web-based survey participants, the study identified 507 valid participants response. The findings provided a summary regarding the response rate, respondents' profile and characteristics, and a descriptive analysis and normality tests.

Regarding the descriptive analysis of the measurement items, the standard deviation is used to quantify the amount of variation or dispersion of the examined data values. When the standard deviation is close to 0, it indicates that the data points lean very close to the mean of the examined data values, but a high standard deviation shows that the data is spread over a wider range of values [8]. In other words, the low value of standard deviation reflects that there is a high certainty that most of the participants have similar views towards the variable. In this study, the average mean and standard deviation of all examined variables were in their recommended level. In addition to the normality test, Table 3 indicates that skewness and kurtosis variables fall within the acceptable range. As a result, it seems that the items of PE, EE, SI, HM, HT, TR, CI, IQ and BI were able to capture a high average mean with a suitable normality test results. Accordingly, it is worth stating that the majority of the survey questionnaire's respondents positively perceive the aspects associated to these constructs. Therefore, future research should take this into consideration so that the significance level of the dependent constructs over the behaviour intention to use social commerce can be examined; this will certainly guide the Saudi organisations to give more consideration towards the most significant factors that affects consumers to use social commerce technologies.

6.1   Research Contribution

The current study makes a significant contribution by proposing the UTAUT2 model for examining the adoption of social commerce technologies, which is a novel modern technology. Furthermore, the study also expanded the applicability of UTAUT2 by focusing on a new cultural context (that is: Saudi Arabia). Finally, this study is able to extend the theoretical horizon of UTAUT2 by including other external factors from the technology adoption literature.

7   Conclusion

This study aims to identify the important factors that influence the adoption of social commerce by Saudi customers. UTAUT2 has been identified as a suitable theoretical foundation for proposing a conceptual model. The study has added other significant and frequently used factors (trust, consumer innovativeness, and information quality) along with UTAUT2 constructs to formulate the model. In order to achieve the study's objectives, a quantitative field survey was conducted to obtain data from a convenience sample of Saudi customers; the data collection used a self-administered questionnaire. Finally, the researcher did a descriptive analysis for each one of the investigated variables. The findings indicated that these factors play a significant role in the behavioural intention for the participants.

7.1   Limitations and Future Research Directions

This study aimed at investigating behaviour intention to use social commerce technologies in the context of Saudi Arabia. First, conducting a descriptive analysis instead of inferential analysis will not allow the extension of the findings to the whole of the Saudi Arabian population. Consequently, this study will guide to assume the hypotheses in regards to the relations between factors, as well as using the structural equation modelling (SEM) to test the measurement model, structural model, and model fitness. Second, to consider investigating the behaviour intention rather than the actual use of social commerce will not give an overall view about using this technology in Saudi Arabia. Therefore, these issues should be taken into consideration in future research. This may assist organisations in Saudi Arabia to select the best strategy for encouraging consumers to use social commerce technologies, which will benefit their businesses. Finally, a comparative research should be conducted between developing and developed countries; also, the cultural context should be taken into consideration in the comparative research.

References

1. Abed, S., Dwivedi, Y.K., Williams, M.D.: Social media as a bridge to E-Commerce adoption in SMEs: a systematic literature review. Forthcoming Mark. Rev. **15**(1), 39 (2015)
2. Abed, S., Dwivedi, Y.K., Williams, M.D.: SMEs' adoption of E-Commerce using social media in a Saudi Arabian context: a systematic literature review. Int. J. Bus. Inf. Syst. **19**(2), 159–179 (2015)
3. ALMowalad, A., Putit, L.: Factors influencing saudi women consumer behavior in online purchase. J. Emerg. Econ. Islamic Res. (JEEIR) **1**(2), 1–13 (2013)
4. Al-Mowalad, A., Putit, L.: The extension of TAM: the effects of social media and perceived risk in online purchase. In: 2012 International Conference on Innovation Management and Technology Research (ICIMTR), pp. 188–192. IEEE, May 2012
5. Arab ICT Use and Social Networks Adoption Report.: Madar Research & Development, KACST (2012). http://www.kacst.edu.sa/en/about/publications/Other%20Publications/Arab%20ICT%20Use%20Report%202012.pdf. Accessed 05 January 2015
6. Brislin, R.: Comparative research methodology: cross-cultural studies. Int. J. Psychol. **11**(3), 215–229 (1976)
7. Coakes, S.J., Steed, L., Ong, C.: SPSS Version 16.0 for Windows. Wiley, Australia (2009)
8. Dancey, C.P., Reidy, J.: Statistics without Maths for Psychology. Pearson Education, Harlow (2007)
9. De Kerros Boudkov Orloff, A.: Ecommerce in Saudi Arabia: driving the evolution adaptation and growth of ecommerce in the retail industry. Sacha Orloff Consulting Group, SOCG, 17 June 2012. Accessed http://www.scribd.com/doc/136654512/E-Commerce-in-Saudi-Arabia-Driving-the-Evolution-Adaptation-and-Growth-of-Ecommerce-in-the-Retail-Industry-SOCG-2012June17
10. Dwivedi, Y.K., Choudrie, J., Brinkman, W.P.: Development of a survey instrument to examine consumer adoption of broadband. Indus. Manag. Data Syst. **106**(5), 700–718 (2006)
11. Franzosi, R.: From Words to Numbers: Narrative, Data, and Social Science, vol. 22. Cambridge University Press, Cambridge (2004)

12. Gefen, D., Karahanna, E., Straub, D.W.: Trust and TAM in online shopping: an integrated model. MIS Q. **27**(1), 51–90 (2003)
13. Hair, J.F., Black, W.C., Babin, B.J., Anderson, R.E.: Multivariate Data Analysis, 7th edn. Pearson, Upper Saddle River (2010)
14. Hajli, M.: Social commerce adoption model. In: UK Academy for Information Systems Conference Proceedings 2012, Paper 16 (2012). http://aisel.aisnet.org/ukais2012/16
15. Hajli, M.N.: Social commerce for innovation. Int. J. Innov. Manag. **18**(4), 1–24 (2014)
16. Huang, C.Y., Kao, Y.S., Wu, M.J., Tzeng, G.H.: Deriving factors influencing the acceptance of pad phones by using the DNP based UTAUT2 framework. In: Technology Management in the IT-Driven Services (PICMET), 2013 Proceedings of PICMET 2013, pp. 880–887. IEEE (2013)
17. Indrupati, J., Henari, T.: Entrepreneurial success using online social networking: evaluation. Edu. Bus. Soc.: Contemp. Middle East. Issues **5**(1), 47–62 (2012)
18. Jarvenpaa, S.L., Tractinsky, N., Saarinen, L.: Consumer trust in an internet store: a cross-cultural validation. J. Comput.-Mediated Commun. **5**(2), 0 (1999)
19. Kang, M., Liew, B.Y.T., Lim, H., Jang, J., Lee, S.: Investigating the determinants of mobile learning acceptance in Korea using UTAUT2G. In: Chen, G., Kumar, V., Kinshuk, Huang, R., Kong, S.C. (eds.) Emerging Issues in Smart Learning, pp. 209–216. Springer, Heidelberg (2015)
20. Kim, S., Park, H.: Effects of various characteristics of social commerce (S-Commerce) on consumers' trust and trust performance. Int. J. Inf. Manag. **33**(2), 318–332 (2013)
21. Kim, D.J., Ferrin, D.L., Rao, H.R.: A trust-based consumer decision-making model in electronic commerce: the role of trust, perceived risk, and their antecedents. Decis. Support Syst. **44**(2), 544–564 (2008)
22. Kline, R.B.: Principles and Practice of Structural Equation Modelling, 2nd edn. The Guilford Press, New York (2004)
23. Lea, B.R., Yu, W.B., Maguluru, N., Nichols, M.: Enhancing business networks using social network-based virtual communities. Indus. Manag. Data Syst. **106**(1), 121–138 (2006)
24. Midgley, D.F., Dowling, G.R.: Innovativeness: the concept and its measurement. J. Consum. Res. **4**(4), 229–242 (1978)
25. Pallant, J.: SPSS Survival Manual. McGraw-Hill, New York (2010)
26. Pavlou, P.A.: Consumer acceptance of electronic commerce: integrating trust and risk with the technology acceptance model. Int. J. Electron. Commer. **7**(3), 101–134 (2003)
27. Roehrich, G.: Consumer innovativeness: concepts and measurements. J. Bus. Res. **57**(6), 671–677 (2004)
28. Sekaran, U.: Research Methods for Business: A Skill Building Approach, 4th edn. Wiley, New York (2003)
29. Slade, E.L., Williams, M.D., Dwivedi, Y.K.: Devising a research model to examine adoption of mobile payments: an extension of UTAUT2. Mark. Rev. **14**(3), 310–335 (2014)
30. Tabachnick, B.G., Fidell, L.S.: Using Multivariate Statistics. Pearson, Boston (2006)
31. Venkatesh, V., Morris, M., Davis, G., Davis, F.: User acceptance of information technology: toward a unified view. MIS Q. **27**(3), 425–478 (2003)
32. Venkatesh, V., Thong, J., Xu, X.: Consumer acceptance and use of information technology: extending the unified theory of acceptance and use of technology. MIS Q. **36**(1), 157–178 (2012)
33. Williams, M.D., Rana, N.P., Dwivedi, Y.K.: The unified theory of acceptance and use of technology (UTAUT): a literature review. J. Enterp. Inf. Manag. **28**(3), 443–488 (2015)
34. Yates, D., Paquette, S.: Emergency knowledge management and social media technologies: a case study of the 2010 Haitian earthquake. Int. J. Inf. Manag. **31**(1), 6–13 (2011)

Big and Open Data

Linking Operational Business Intelligence with Value-Based Business Requirements

Tom Hänel$^{(\boxtimes)}$ and Carsten Felden

Department of Management and Information Systems,
TU Bergakademie Freiberg, Freiberg, Germany
{tom.haenel,carsten.felden}@bwl.tu-freiberg.de

Abstract. Operational business intelligence (OpBI) integrates data of business processes to analyse their performance in relation to organizational goals. The consequent decision-making concerns a timely recognition and execution of actions to maintain performant business processes. OpBI systems can be designed according to a firm-specific definition of requirements guided by considerations from business model, business process and information system perspective. However, there is no approach to link the design of OpBI jointly with characteristics of business models and business processes, yet. The paper uses therefore an action research method and proposes a business approach that combines e³value with the work system framework to set up conceptual application designs for an OpBI-reliant decision support. We report on results of a long-term research project to demonstrate the development and application of our approach in four different business scenarios. The findings include implications towards a business-oriented application design of OpBI systems.

Keywords: Operational Business Intelligence · e³value · work system · ADAPT

1 Introduction

Organizations measure business processes using performance indicators in terms of time, quality, or cost [1]. The maintenance of performant business processes has to be closely linked to business strategy so that process improvements are valuable and lead to competitive advantages [2]. Management activities of process performance are thereby associated with IT to collect and analyse data about business processes [3]. Such IT capabilities need to be correspondent and compatible to business strategy, too, to avoid missing of expected performance results [2]. One possible concept to analyse business processes is OpBI dealing with an integration of daily business data [4]. This supports business operation's managers in gaining relevant knowledge to evaluate business process performances [5]. Management actions taken in consequence of an OpBI-reliant decision-making have to bring benefits to the manner of an organization creating value in its business environment. The paper's goal is therefore to investigate a linkage of OpBI with firm-specific business requirements.

The current discussion about OpBI provides no conceptual insights to consider business requirements for designing analytical systems in a particular case. For

© IFIP International Federation for Information Processing 2015
M. Janssen et al. (Eds.): I3E 2015, LNCS 9373, pp. 147–159, 2015.
DOI: 10.1007/978-3-319-25013-7_12

instance, the analysis requirements of insurance companies differ from issues of automotive suppliers from business perspective, although the technical system components can be quite similar. It is not obvious for application developers, how an OpBI system needs to be logically designed in order to maintain and improve performant business processes from a perspective of business operation's managers. A specification of OpBI systems can benefit from a value-based requirements engineering so that business value models initialize requirements for business processes and IT systems [6]. We investigate such a value-based requirements definition for OpBI systems and propose an approach to link the logical design of analytical databases with firm-specific business value models. The paper contributes with a development and application of our approach to the scientific discussion using participatory action research in context of four different organizations. This offers collaborative insights for research and practice to the discourse about business approaches so that operational management actions are beneficial for performant business processes.

Chapter 2 refines the problem of research and analyses related areas. The research method is presented in Chap. 3. Chapter 4 introduces our approach and Chap. 5 reports on its application during an action research project in four different business scenarios. Finally, a conclusion summarizes findings and further research activities.

2 Status quo

OpBI is understood as a decision support concept for business operation's managers to analyse business processes in favour of continuous improvements of process design and execution [4]. OpBI supports an identification of control actions based on timely relations between process performance and the status of goal achievement. [5]

2.1 Problem Refinement

OpBI integrates data emerging in or flowing into IT systems during operational task fulfilment [4, 5]. From a technical viewpoint, OpBI systems can be equipped with IT providing business operation's managers access to manifold sources of information and analytical options in combination with high performance data processing. The discussion about Hadoop [7], cloud computing [8], combinations of transactional and analytical databases [9], or data virtualization techniques [10] points to a variety of technical options. However, these advancements will only lead to a successful decision support, if the performance analysis and action taking using an OpBI system is consistent to business goals and value creation processes. This requires a conceptual modelling of analytical requirements for OpBI systems in compliance with operational concerns of an organization. We conducted a literature review using the databases of Business Source Complete, IEEE, AIS, ACM, Emerald, and Science Direct to examine scientific publications according to MIS rankings [11]. The reviewed publications do not discuss a conceptual modelling of analytical requirements for a successful application of OpBI. A lack of discussion about conceptual modelling of operational information is evident, yet.

2.2 Related Research Areas

OpBI addresses performance management (PM), BI, and business process management (BPM) [12]. PM structures business strategies and translates them into goals and ratios [13]. Process PM (PPM) monitors business processes using performance indicators [14]. The PPM concept is not limited to a specific IS support, but BPM or BI systems are discussed therein currently [1]. Monitoring business processes has a technical background coming from the BPM perspective [15]. BPM systems log transactions and events for execution tracking and process modelling [16]. The analysis of log data is limited, yet [17]. This extends especially in contexts of sophisticated processes with distributed tasks [18]. Due to an early stage of PM in the area of BPM, an integration of BI and BPM is taken into consideration [3]. From a BI perspective, the analysis of process data has a different focus. Business Process Intelligence supports the design and redesign of processes of an organization [19]. This affects a small range of users making strategic or tactical decisions. In contrast, process-centric BI concerns an integration of BI applications into process executions [20]. This affects the process performance due to accelerations and improvements of a process execution. BI provides analytical information to fulfil process related tasks. This differs from our OpBI understanding by using BI techniques for an analysis and control of business processes. Process-centric BI does not address a consideration of analytical information for an immediate measurement of process performance, an investigation of deviations, or a derivation of control actions.

3 Research Method

We apply an action research method, because this has been used successfully to model business requirements and to align them with IT characteristics [6]. This is similar to our area of discourse by a conceptual modelling of OpBI systems. We extend the methodological knowledge and refer it to a participatory form of action research [21] - researchers and practitioners participate in a research process collectively. The collaboration allows a combination of modelling knowledge with practical experiences about analysing and controlling business processes. Action research supports a solution of immediate performance problems and a consolidation of conceptual knowledge on designing OpBI systems. Participatory action research has been successfully applied, too, in order to ensure that IT implementations result in business benefits [22]. Our intention is quite similar as we want to link the conceptual design of OpBI systems with value-based business requirements. Therefore, we deduce a practicability of participatory action research to deliver a business contribution in consequence of an OpBI-reliant decision-making. In a three-year research period, we performed an iterative and collaborative research process together with four organizations. Assumptions on designing OpBI systems were refined in cycles of diagnosis, action, evaluation, and reflective learning [21]. An approach to link value-based business models and OpBI systems emerged in consequence of our experiences. The approach builds upon the findings and multi-perspective view on requirements engineering of Gordijn and Akkermans [6].

4 Linking Value-Based Business Models and OpBI Systems

Our approach consists of different activities resulting in e^3value models [6], a classification of business process requirements according to the work system framework [23] and ADAPT models [24]. Figure 1 classifies the elements of the approach into the perspectives of a value-based requirements engineering [6].

An e^3value model describes an exchange of value objects between business actors in a commercial network. Such a network consists for instance of an organization anywhere in a value chain with its potential customers and suppliers. Business actors with an equal value proposition can be grouped to market segments. The value objects to be exchanged in a commercial network are trading items (products, services) in consideration of economic equivalents (money). Value activities model specific performance areas, in which an organization creates or adds value to yield profits. To dig deeper in the particular mechanisms of value activities, we bridge to a consideration of the business process perspective using the work system framework. Both approaches consider an internal and external view on organizations. Table 1 demonstrates the coincidence of e^3value and the work systems framework. A work system considers participants carrying out business processes by use of information and technology.

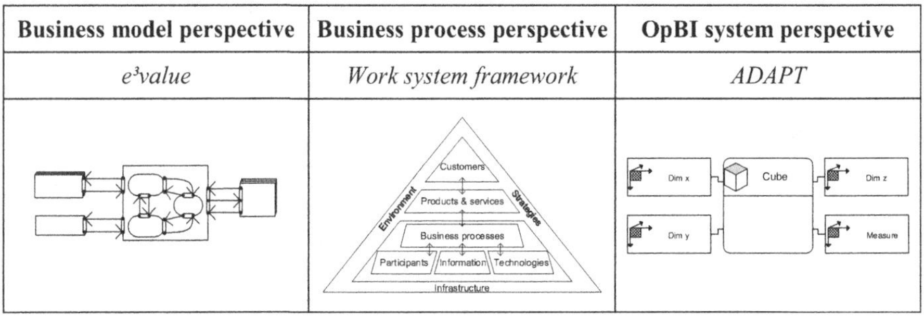

Business model perspective	Business process perspective	OpBI system perspective
e^3value	*Work system framework*	*ADAPT*

Fig. 1. Elements of our approach

Table 1. Mapping of e^3value and work system framework

View on organizations	Elements of e^3value	Elements of work system framework
External	The whole e^3value model	Strategies
	Customers, external stakeholders, partners, or suppliers modelled as market segments or actors	Environment, customers
	Value exchanges, especially value objects	Products & services
Internal	Concerning organization performing specific value activities	Infrastructure
	Value activities representing areas of performance	Business processes
		Participants
		Information
		Technologies

These core elements of work system performance characterize together with general infrastructure components an insider's view on an organization's business value model. The performance output are products or services, which are the objects of value exchange with customers and the value chain environment. Strategic considerations influence the insider's and outsider's view regarding to work system performance. We use the elements of the work system framework and e^3value to deduce requirements for an analysis and control of value activities from an IT system perspective. Therefore, we use the ADAPT notation to develop logical data models as measurement and structuring instrument for value activity information in operational decision contexts. The work system and e^3value elements are assigned to dimensions and measures of an ADAPT model. The dimensions span a cube consisting of a set of measures having a clear reference to the value objects of the business model. The relationships of dimensions and measures follow the criteria of creating and exchanging values.

5   Action Research Results

We present the results of an action research project that was carried out from August 2012 to February 2014 in Germany in order to develop and apply our linking approach. Four organizations participated in three subsequent cycles of action research. The considered organizations were a machine tool manufacturer, a service provider for IT and communication (ICT) products, a hydraulics engineering company, and an insurance agency. The first cycle refers to activities of interaction, application, and reflection from a business models perspective and results in e^3value models. The outcome of the second research cycle is represented by a work system classification. The third cycle of action research lead to ADAPT models for an OpBI database design. Illustrations of e^3value and ADAPT models are presented only in context of the machine tool manufacturer due to the limited space of the paper.

5.1   Research Cycle 1: Creation of Value-Based Business Models

Machine Tool Manufacturer. The organization modernizes gear hobbing machines. Equipment upgrades happen according to individual customer orders with negotiated budgets, period and quality requirements. The value activities (cf. Fig. 2) include a deployment of new components, such as control units or milling heads. Once the transfer of a customer's machine happens, a dismantling in machine components takes place. Specific and standard parts are cleaned and listed. The employees record geometrical data and take pictures in case of incomplete drawings. Decisions about a rework or a remanufacturing depend on the machine state. Finally, the execution of the re-assembling happens. Disturbance variables are the individuality and the unpredictability of the machines and their states. Different projects and suppliers must be coordinated in consideration of compliance in time and cost conditions.

ICT Service Provider. Logistical services are performed to distribute ICT products from different brands through different channels. The product procurement involves manufacturers or network operators. Devices are customized according to specified

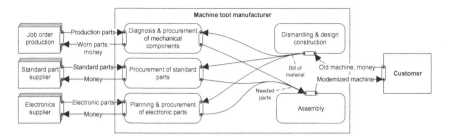

Fig. 2. e³value model of a machine tool manufacturer

requirements, e.g. exchanges of electronic parts. A recovery resets returned devices to factory settings and performs functional checks. If, necessary, a partner company repairs defect devices. Final products are completed and packaged for shipment. Disturbance variables are fluctuating throughput quantities, a changing staff, heterogeneous products and fast price slumps. Especially velocity and cost efficiency are important control aspects.

Hydraulics Engineering Company. The organization produces hydraulic or pneumatic cylinders and job orders. The customer segment includes industrial trucks, rail vehicles, agricultural machinery, printing presses, or injection moulding machines. The manufacturing concerns activities of metal processing like milling, drilling, turning, welding, or laundry. The assembly of finished products includes functional tests, colouring, and shipment. A quality control records complaints during production and decides for rework, sorting, claim, or scrap. Disturbance variables are order withdrawals, missing materials, troubles of external manufacturers, unavailable labours, or malfunctions of e.g. automatic welders or CNC machining centres. Such disturbances lead to delays of planning cycles. Considering monthly value creation targets should overcome the uncertainties. This means that alternative outputs have to compensate adverse circumstances, if, for example, an order is cancelled.

Insurance Agency. Insurance products are distributed on behalf of an insurance group. The strategic goals of the insurance group concern high premium customer portfolios, optimized trading results, and excellent business processes. The insurance agency has to fulfil the goals by efficient service actions. A planning and scheduling of sales conversations concludes insurance contracts for different products with commercial or individual customers. The agency coordinates, supervises and settles customer claims. Disturbance variables are manifold. Expiring insurance contracts or premature dismissals reduce the number of customers. Failing approaches to agree conversation dates or cancellations counteract attempts to sustain or increase sales revenues. Delays or contradicting information impair the handling of claims due to a missing communication between different contact points, which record claims or requests.

5.2   Research Cycle 2: Classification of Business Process Characteristics

Table 2 classifies the studied organizations into the work system framework. The processes need to be dynamic with a certain variability. The business processes are deterministic and repeatable, while the performance results differ for changing situations. The tasks depend on knowledge and experience of the employees executing, guiding, and instructing operational activities. The information refers to reference inputs, control indicators, resources, products, or stakeholders.

Information technologies mentioned in Table 2 refer to ERP, product data management, warehouse management, or collaborative portal solutions. Important is the availability of data collection techniques. The infrastructure includes a low to medium specialized technical equipment. Human resources are specialists and executive staff

Table 2. Classification of case studies into work system framework

	Insurance agency	Hydraulics engineer	Machine tool manufacturer	ICT service provider
Processes and activities	Consulting, claim settlement, sales conversations	Manufacturing, quality control, assembly	Dismantling, cleaning, rework, assembly	Customization, recovery, shipment
Participants	Senior manager, back office, sales representatives, call centre agents	Engineers, assemblers, operators, supervisors	Project teams with assemblers, engineers, project leader	Shop floor and temporary staff, supervisors and unit manager
Information	Customer records, availability and history, cross selling ratio, claims, expense ratios, premium targets and incomes, contracts	Time data, design drawings, bill of materials, defect reports, article data, consumption rates, target/actual quantities, expense ratios, added value	Time data, design drawings, geometrical data, bill of materials, orders, delivery dates, quality indications, budged limits	Time data, expense ratios, target quantities, delivery dates, article master data, consumption rates, actual quantities, defective products
Technologies	Platform to prepare and manage proposals, policies issues, portfolios and accountings	ERP, Product data management, Machine data acquisition, Time keeping	ERP, Product data management, Time keeping, Project management system,	ERP, Warehouse management system, Machine data acquisition, Time keeping

(Continued)

Table 2. (*Continued*)

	Insurance agency	Hydraulics engineer	Machine tool manufacturer	ICT service provider
Infrastructure	Office equipment with interfaces to the insurance group, four employees	Office and production equipment, 100 employees, staff involvement	Office and production equipment, 70 employees, project hierarchies	Office and logistics equipment, 1,500 employees, flat hierarchies
Strategies	Increase of shareholder values, high premium customers	High quality, flexibility and velocity, reliability to customers	Specialization, focus on customer, undercutting of original prices	Diversification of sales, service and repair, high quality at low costs
Environment	Insurance group, financial markets, changing commercial and legal conditions, regional sales area	Supplier relations, high competitive pressure, growing international market	Supplier relations, high competitive pressure, deadline and cost pressure, international market	Supplier and partner relations, international market, varying order situations, fast slumps
Customers	Individual and business clients	Machine building companies	Metal processing companies	Retailers and resellers
Products and services	Insurance products, financial services	Hydraulic cylinder, job orders	Gear hobbing machines	ICT products

organized in problem-oriented communication hierarchies. Customer relations are business-to-business and business-to-customer. The organizations offer specialized products or services in different price segments with a medium to high complexity. They have heterogeneous configurations and consist of sophisticated features. The environment is characterized by competitive pressure and changing conditions in regional and international distribution areas. External factors are the behaviour of suppliers, partners, or associated companies. Strategies of the studied organizations include specialization, diversification, quality excellence, flexibility, velocity, and customer orientation.

5.3   Research Cycle 3: Logical Application Design of OpBI Systems

Machine Tool Manufacturer. The OpBI system supports the budgeting and scheduling of modernization projects. Data gathering happens manually due to the heterogeneity of working activities. A tracking system records the corresponding working times. The database design (cf. Fig. 3) points out expenses for performing the value

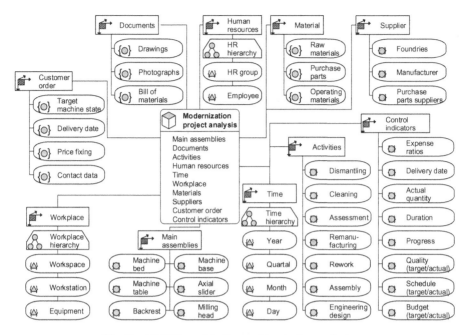

Fig. 3. ADAPT model of a machine tool manufacturer

activities on different levels of detail. Planning and management of project workflows happen simultaneously. Current states of a machine, incurred costs, spent working times, and delivery progress of needed assemblies are demonstrated.

A comparison of actual performances to target indicators enables staff to intervene in case of deviations. The procurement combines supplier information with required rework orders and the quality of finished parts. This rating of providers eases a selection for similar constructed parts. The restored machine features of the individual overhaul projects are comparable so that cost estimations become more confident.

ICT Service Provider. The OpBI system evaluates cost transparency and efficiency to react fast and flexibly on changing order quantities. The affected IT systems are an ERP and a warehouse management system. The data collection occurs with scanners, light barriers, and a machine time tracking. The designed OpBI database provides a basis to derive management actions for an adjustment of order cycles according to product groups. The consequence is a coordination of logistical cost and product-specific price slumps. The data model facilitates a combination of production batches with similar or equal features to improve processing times. Faced by staff changes, performance targets are determined according to human resource groups. These targets depend on product groups and periods of employment. The calculation of product-specific delivery times leads to higher planning reliability as consequence of specifiable agreements for repair services in context of an outsourced repair service.

Hydraulics Engineering Company. The OpBI system determines a value added of manufacturing activities and cost ratios of quality issues. The underlying IT systems refer to an ERP system with integrated data acquisition. Terminals collect production data using card readers and barcode scanners. A quality assurance tool collects internal quality complaints. The logical designed database supports an incremental accretion measurement of components and products during manufacturing and assembly. This ensures a constant review of value creation targets. Differences will lead to immediate decisions. A consideration of expenses to create specific features improves the employment of resources, materials, and technologies. Constructors get information to determine prices for new products or add-ons during the design phase based on needed product features. The quality assurance derives actions by costs-by-cause principles using the different process perspectives. The logical model enables a calculation of expenses for rework, sorting, or scrap for internal quality complaints.

Insurance Agency. The OpBI system combines information of more than 1,800 customers with allocated service tasks. A platform for proposal preparation, policy issues, portfolio management, and accounting supports semi-standardized information records. Sales representatives or office employees enter this information manually. The OpBI's data model considers reasons for unsuccessful approaches to agree conversations. For example, holidays or shift work lead often to calls at inconvenient customer situations. The scheduling is managed according to reachability of customers, now, and appointments are located in nearby sales regions to reduce travel cost. The data model supports a customer-specific control of claim handling to achieve a well-founded settlement. This depends on extent of loss or damage, underlying insurance contracts, and customer behaviour. The agency monitors deadlines for claim review to accelerate handling times. It is measurable whether a customer has already reported claim information and how far the reports coincide. A comparison of monthly premiums with a number of contracts per customer leads to a prioritization of claims or a consideration of goodwill. This is beneficial to decide about win-back actions in notice management, too.

5.4   Lessons Learned

The action research cycles demonstrate methodological and organizational issues to design OpBI database systems based on value-based business requirements. This delivers insights on measuring and evaluating the performance of business processes in four business scenarios. The conjoint reflection of business models, business processes, and IT systems has proven to be advantageous. Valuable results were achieved in all four organizational settings despite of different situational characteristics. Figure 4 repeats the relation between the perspectives of our approach.

The joined elements of e^3value and work systems are linking an organization's strategy with the maintenance of performant business processes. The association to OpBI is represented at the bottom of Fig. 4. ADAPT models are instruments to collect, elaborate, and analyse data about business processes and build the basis to configure management actions. An important aspect learned from our research is the context-sensitive enrichment of these common descriptive perspectives (cf. Table 3).

The linkage of OpBI with firm-specific business requirements is irreducible complex by observational research methods, because it is necessary to involve situated and practical knowledge resulting from collaboration activities between researchers and

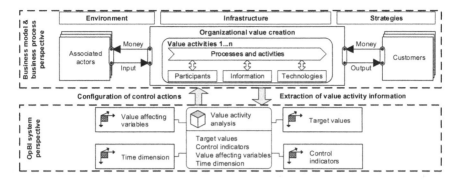

Fig. 4. Relation of business model, business process, and OpBI system perspective

Table 3. Firm-specific and common aspects of our results

Firm-specific aspects	Common aspects
• Business contexts and strategies	• Methodological building blocks
• Business process descriptions	• Action research cycles
• Performance management situations	• Collaboration of research and practice
• OpBI database designs	• Business process orientation
• Management control actions	• Use of operational IT systems

practitioners. Participatory action research enables such a reference to practical contexts. Implications concerning a performance management of business processes depend thereby always on specific organizations. However, our conceptual findings allow a broader consensus on modelling OpBI systems, although they are not object of a rigorous generalization. Especially the work system framework helped us to learn about common aspects like operational IT systems and repeatable business processes.

6   Conclusion

OpBI will support the management of performant business processes, if the analytical concerns are in concurrence to the business requirements of an organization. The paper's contribution enhances a discussion about conceptual aspects of linking OpBI systems design with value-based business requirements. We developed and applied a management approach in coherent action research cycles to provide a conceptual basis for designing OpBI systems from a business perspective.

The paper's arguments shift the discussion about an operational decision-making from technical aspects to a consideration of business strategies. Such a view on information systems is in line with contributions about the impact of IT on business process performance [2]. The novel conceptual approach of value modelling, work system analysis, and analytical design is relevant for application developers and business operation's managers. This supports a definition and evaluation of requirements for an operational decision-making in an organization's business context. The gained conceptual and practical experience from our action research project refers to four different business scenarios. The collaboration of researchers and practitioners has produced a valid conceptual approach and meaningful outcomes in practical contexts. One learning effect is that a consideration of such collaborative efforts leads to firm-specific implications and to reproducible conceptual insights.

This paper builds its evidence on action research, so that its findings and implications have a qualitative nature. The investigated organizations represent typical scenarios of manufacturing and service provision. This indicates a certain resilience of the action research method and is intercessional for a confident replication logic in additional business scenarios. Upcoming research activities should therefore further consolidate conceptual considerations about the integration of analytical concerns and business value perspectives. This allows taking charge of changing analytical technologies and digital opportunities based on a given business logic or value constellation.

References

1. Blasini, J.: Critical success factors of process performance management systems: results of an empirical research. In: 21st European Conference on Information Systems, Utrecht, vol. 158, pp. 1–12 (2013)
2. Trkman, P.: The critical success factors of business process management. Int. J. Inf. Manage. **30**, 125–134 (2010)
3. Vukšić, V.B., Bach, M.P., Popovič, A.: Supporting performance management with business process management and business intelligence: a case analysis of integration and orchestration. Int. J. Inf. Manage. **33**, 613–619 (2013)
4. Davis, J., Imhoff, C., White, C.: Operational Business Intelligence: The State of the Art. Beye NETWORK Research, Boulder (2009)
5. Hänel, T., Felden, C.: towards a stability of process oriented decision support concepts using the example of operational business intelligence. In: Pre-ICIS BI Congress 3: Driving Innovation Through Big Data Analytics, Orlando (2010)
6. Gordijn, J., Akkermans, H.: Value based requirements engineering: exploring innovative e-commerce ideas. Requirements Eng. J. **8**, 114–134 (2002)
7. McAfee, A., Brynjolfsson, E.: Big Data: The Management Revolution. Harvard Bus. Rev. **90**, 60–66 (2012)
8. Juan-Verdejo, A., Baars, H.: Decision support for partially moving applications to the cloud: the example of business intelligence. In: International Workshop on Hot Topics in Cloud Services, New York, pp. 35–42 (2012)
9. Plattner, H.: A common database approach for OLTP and OLAP using an in-memory column database. In: ACM SIGMOD International Conference on Management of data, Providence, pp. 1–2 (2009)
10. Van der Lans, R.: Data Virtualization for Business Intelligence Systems: Revolutionizing Data Integration for Data Warehouses. Morgan Kaufmann, Waltham (2012)
11. MIS Journal Rankings. http://aisnet.org/?JournalRankings
12. Cunningham, D.: Aligning business intelligence with business processes. What Works **20**, 50–51 (2005)

13. Otley, D.: Performance management: a framework for management control systems research. Manage. Acc. Res. **10**, 363–382 (1999)
14. Kueng, P., Krahn, A.: Building a process performance measurement system: some early experiences. J. Sci. Ind. Res. **58**, 149–159 (1999)
15. Janiesch, C., Matzner, M., Müller, O.: Beyond process monitoring: a proof-of-concept of event-driven business activity management. Bus. Process Manage. J. **18**, 625–643 (2012)
16. van der Aalst, W.M.P.: Process Mining: Discovery. Conformance and Enhancement of Business Processes. Springer Publishing Company, Berlin (2011)
17. Kang, B., Kim, D., Kang, S.H.: Periodic performance prediction for realtime business process monitoring. Ind. Manage. Data Syst. **112**, 4–23 (2012)
18. Cheung, M., Hidders, J.: Round-trip iterative business process modelling between BPA and BPMS tools. Bus. Process Manage. J. **17**, 461–494 (2011)
19. Felden, C., Chamoni, P., Linden, M.: From process execution towards a business process intelligence. In: Abramowicz, W., Tolksdorf, R. (eds.) BIS 2010. LNBIP, vol. 47, pp. 195–206. Springer, Heidelberg (2010)
20. Bucher, T., Gericke, A., Sigg, S.: Process-centric business intelligence. Bus. Process Manage. J. **15**, 408–429 (2009)
21. Baskerville, R.L.: Investigating information systems with action research. Commun. AIS **2**, 1–32 (1999)
22. Breu, K., Peppard, J.: The participatory paradigm for applied information systems research. In: 9th European Conference on Information Systems, Bled, pp. 243–252 (2001)
23. Alter, S.: Work system theory: overview of core concepts, extensions, and challenges for the future. J. Assoc. Inf. Syst. **14**, 72–121 (2013)
24. Getting Started with ADAPT. http://www.symcorp.com/downloads/ADAPT_white_paper.pdf

Operationalizing Data Governance via Multi-level Metadata Management

Stefhan van Helvoirt and Hans Weigand[✉]

Tilburg School of Economics and Management, Tilburg, The Netherlands
svhelvoirt@gmail.com, h.weigand@tilburguniversity.edu

Abstract. Today's rapidly changing and highly regulated business environments demand that organizations are agile in their decision making and data handling. At the same time, transparency in the decision making processes and in how they are adjusted is of critical importance as well. Our research focusses on obtaining transparency by not only documenting but also enforcing data governance policies and their resultant business and data rules by using a multi-level metadata approach. The multi-level approach makes a separation between different concerns: policy formulation, rule specification and enforcement. This separation does not only give more agility but also allows many different implementation architectures. The main types are described and evaluated.

Keywords: Data warehouses · Data governance · Metadata · Business rule enforcement

1 Introduction

The amount of data that is available in the digital universe is growing at an exponential rate and will only continue to grow with the rise of new technologies such as the Internet of Things. Nowadays data is more important than ever before due to the speed of business change. This is emphasized with the rise and use of Master Data Management (MDM) systems in the last decade. MDM adds a new dimension to the data that focusses on establishing integration and interoperability of heterogeneous databases and applications in a business oriented manner [1, 2].

Recent studies have shown that organizations that are capable of effectively utilizing and analyzing their data outperform their competitors [5]. In order to actively use the data that is available both within and outside the organization, the organization must find a way to actively and sufficiently tag the data with metadata [9]. Especially in a new digital world in which organizations are rapidly integrating data from various heterogeneous sources. This need is emphasized with the rise of new data warehouse platforms such as IBM's Data Reservoir. Having a proper data governance program in place is crucial for effectively managing the data that resides in such aggregated environment [3].

What is important in today's highly regulated business environments is not only effective data governance but also that the governance is transparent and auditable. For instance, exporting and importing shippers need to comply with tax regulations and

© IFIP International Federation for Information Processing 2015
M. Janssen et al. (Eds.): I3E 2015, LNCS 9373, pp. 160–172, 2015.
DOI: 10.1007/978-3-319-25013-7_13

customs security controls. It is very hard for companies to prove compliance if the data infrastructure is not well-controlled in a transparent way. Sometimes data governance is mandatory by law as with BASEL BCBS 239, effective from 1/1/2016.

Our research goal is obtaining adaptability and transparency by not only documenting but also enforcing data governance policies and their resultant business and data rules. In this paper, we introduce a multi-level framework and use it to evaluate the current capabilities of IBM's InfoSphere package as used in its Data Reservoir solution, while also providing incentives to further extent the governance capabilities. Additional layers of logic are added to reify governance policies in data movement, applications and databases. A preamble on Data Governance and metadata is provided in Sect. 2, to lay the foundation for our multi-level metadata framework discussed in Sect. 3. Section 4 continues with an overview of various implementation styles to establish a Data Governance environment and Sect. 5 evaluates IBM InfoSphere offerings to establishing operationalized Data Governance.

2 Background

2.1 Data Governance versus Data Management

According to Khatri and Brown [6], based on Weill and Ross, "governance refers to what decisions must be made to ensure effective management and use of IT (decision domains) and who makes the decisions (locus of accountability for decision making). Management involves making and implementing decisions." Data management activities focus on the development and execution of architectures, policies, practices and procedures to enhance and manage the information lifecycle within a specific application and mostly during data entry/creation. Data Governance on the other hand also includes aggregated and integrated data that is made available as a data asset within the organization.

The Data Governance domain consists of three focus areas; people, processes and technology. Many publications on data governance focus primarily on the people and processes aspects of implementing a data governance program. We can use IBM's holistic approach to Big Data governance as an example, which consists out of the following six sets: define business problem, obtain executive sponsorship, align teams, understand data risk and value, implement analytical/operational projects and measure results. Its focus has been primarily on the first four steps. Our focus is mainly on the "implement analytical/operational projects" from a technical perspective as this appears to be research gap. However, all steps in the holistic approach are needed in order to have an efficient and reliable data governance program. Capturing and enforcing business rules, without the proper knowledge of the available data, its value and the interdependencies between data is undoable and undesirable. We do not underestimate the political change that is needed to transform the organization into an information-driven environment. An overall transition needs to be made from thinking and developing individual applications to a unified acceptance and usage of data as the foundation of information and knowledge [8].

2.2 Data Quality and Trust

Governance is more than achieving compliance [7]. Achieving data governance has to do with adopting practices and principles that increase data quality and trust. Having established data quality and trust, the organization can start using their data in a reliable and controlled manner and evaluate its data usage and governing capabilities by implementing appropriate metrics. A valuable data quality standard currently in development is ISO 8000. It focusses on data characteristics and exchange in terms of vocabulary, syntax, semantics, encoding, provenance, accuracy and completeness.

2.3 Metadata as Indispensable Enabler

The importance and utilization of metadata has been increasing rapidly over the last decade as metadata is making its transition from a technical aspect to a business necessity. Metadata is needed for establishing data quality and turning data into understandable information that can be consumed by both business/IT users and software for automation.

Looking at publications on metadata form the past fifteen to twenty years shows that there are various types and classifications of metadata, each with its own specific purpose and granularity. This paper uses the metadata framework as presented by Ron Klein (KPMG) at the 2014 ECCMA conference [7]. This metadata framework consists of three vertical levels and three horizontal levels. The vertical levels are *business*, *technical* and *operational*, while the horizontal levels consist of the categories *descriptive*, *administrative* and *lineage*. Business metadata includes business terms, data owners, stewards, and governance policies and business rules governing the data. Technical metadata is used for tool integration to manage, transform and maintain the data. Examples of technical metadata are database system names, table and column names, code values and derivation rules. Operational metadata contains run-time information e.g. last load, usage statistics and log reports. In short, business metadata has a value and meaning for business oriented users, technical metadata is used primarily by Extract, Transform, Load (ETL) developers while operational metadata is used to provide insights in data usage and rule validation.

3   Multi-level Metadata Management

In the following, we will focus on the policies *behind* the meta-data as such, for instance, policies and access control, policies on quality requirements, or policies on the use of semantic standards, rather than the meta-data tags themselves. Our goal is transforming descriptive data governance policies into implementable rules by using a multi-level metadata approach. The multi-level metadata model consists of four levels as depicted in Fig. 1. Distinguishing these levels leads to maximal adaptability and transparency (cf. [4]).

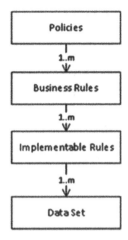

Fig. 1. Multi-level Metadata Management framework

Policies are abstract formulations of business goals, desirable behavior, guidelines and generally accepted practices. Business rules are a formalization of a (partial) aspect of a policy, stating the trigger and follow-up action in a structural natural language consisting of business terminology. Implementable rules are executable objects that contain the logic needed to enforce the business rule. How a business rule is implemented and enforced is entirely dependent on enforcement strategy of that specific rule. As mentioned by Weigand et al. [14], "although business rules are more formal than policies, they are still at the level of business requirements (...), rather than execution. They model "what" is required, rather than "how" it should be implemented" [14]. The same paper emphasizes the need to clarify the enforcement strategy and provides four types of enforcement: preventive, punitive, corrective and adhortative. Here adhortative means that the responsible user is requested to solve the violation when it is detected; the system does not prevent or correct it itself. Punitive means that a sanction is given on violation of the rule whereas corrective means that the system automatically corrects the violation and moves forward to a consistent state, typically by means of compensation. When the enforcement is separated from the business rule specification as such, this allows for great flexibility: the company can switch rather easily from a more loose adhortative approach to a strict preventive approach (or vice versa) depending on the desired compliance levels and operational costs.

Where the enforcement of a business rule takes place is highly dependent on the chosen enforcement strategy and goal of the business rule. For example, a policy stating that all telephone numbers should be formatted according to the applicable standard of the country that it applies to, will have a business rule declaring the use of a given standard for telephone numbers within a given region. This rule could be enforced at the point of data creation using a preventive strategy, or when data is analyzed, transformed and moved to a different location using a corrective strategy.

Business rules are defined as condition action (CA) rules and require a structural transformation to become executable condition action (ECA) rules.

4 Implementation Archetypes

4.1 Business Rule Extraction

For the enforcement of business rules, we start from the generally accepted approach expressed, among others, by Pierre Bonnet in his book on Enterprise Data Governance [2] in which the business knowledge is extracted from the software (hard-coding) and is presented to the business users in an environment that they can (partially) control. "Maintaining knowledge, in particular within complex and evolving organizations that characterize modern companies, cannot survive the trap set out by fixed and stratified hard-coded software, nor informal (textual) documentation, rarely up to date and non-executable". According to Bonnet, a software package must first be able to interact with an MDM system, before it can demonstrate its ability to enforce the relating business rules (BRMS) that will eventually affect the processes (BPM). "First the data, then the rules and finally the processes" [2]. This way, rules are defined per data domain and not based on the software package that uses the data.

4.2 Enforcement Architectures

Isolating data governance rules from the code is one thing, but still leaves many choices on how to enforce the rules. Based on our analysis, we distinguish between a *decentralized*, *centralized* and *leveled* implementation archetype. The archetype that is most applicable to a given situation depends on the available resources and business requirements [15]. For example, an analytical driven environment will have a specific way of enforcing policies as data is collected from various sources and ingested into one or multiple repositories designed and optimized for specific analytical computations (e.g. IBMs Data Reservoir). The enforcement of policies in such an environment could largely occur at the processing of data movement. On the other hand, enforcing governance policies on the actual applications/databases that create/store the data would require a different approach to integrating and enforcing policies. The difference in these three implementation styles as described in this section, is purely in the area of policy *enforcement*. Our base assumption is that policy and asset descriptions are high level and should not be restricted or influenced by the underlying technology and infrastructure. Furthermore, capturing, defining and maintaining the definition and description of policies and assets at domain or organizational level allows for greater consistency, transparency and manageability. However, this integration also has its costs and concerns. One of the concerns is that responsibility for some resource, including data, should not be taken away from the agents owning the data.

We start off by illustrating and defining the decentralized implementation style, displayed in Fig. 2. In this example we have four data storages, each containing the (business) definitions of applicable policies (no pattern) and the resulting implementation code (striped pattern). The decentralized implementation is very common in situations where data governance maturity is low. This implementation style has some benefits and limitations as illustrated in the Table 1.

Fig. 2. Decentralized implementation

Table 1. Benefits and limitations of decentralized implementation

Benefits	Limitations
Enforcement of policies as close to the source of data as possible	Siloed knowledge resulting in a lack of reusability and increased risk of inconsistency among separated data storages
Less dependencies and decreased systematic risk	Monitoring compliance and conducting audits is costly and time consuming

Although there are some benefits to mention for the decentralized implementation style, these do not outweigh the limitations. Especially in today's rapidly growing digital ecosystem, in which data is being created by an increased amount of utilities both within and outside the organization. A leveled or centralized implementation style would deliver a more feasible and desirable approach to enforcing data governance, however this requires the presence of a central governance catalog like system for centrally storing and defining data governance policies and assets (cf. [11]).

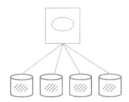

Fig. 3. Leveled implementation

The leveled (Fig. 3) and centralized (Fig. 4) implementation styles make use of a central governance catalog. The leveled implementation style uses a centralized governance catalog repository which is capable of storing the data asset definitions and data governance policies. These asset descriptions describe both the business characteristics of a dataset (business definition using business terminology, owner, steward etc.). Business assets and policies are linked to denote which policies should apply to a specific asset. The implementation and enforcement of these policies is conducted at the System of Record/Reference (SoR). This approach allows for the creation and maintenance of both asset descriptions and policies at a central level, allowing for greater transparency and consistency. At the same time, the implementation can make use of the tools most efficient for the particular SoR. However, there are also some drawbacks to this implementation style as shown in Table 2. To address the consistency problem, one could imagine an automated update system that pushes any changes in the policy definitions forward to the SoRs. However, when the diverse SoRs use different local enforcement tools, such an update may also require as many compilations as there are different SoRs.

Table 2. Benefits and limitations of leveled implementation

Benefits	Limitations
Consistency in asset and policy definitions	Gap between the definition of a policy and the actual implementation which could result in misinterpretation and incorrect enforcement
Increased transparency in the available data and the rules that shape the data and its use throughout the data lifecycle	Lack of consistency in the enforcement of policies due to high diversity of SoR sources

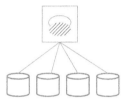

Fig. 4. Centralized implementation

Lastly, we have the centralized implementation style as displayed in Fig. 4. A centralized implementation requires the presence of a governance catalog repository with enhanced and additional capabilities. We can distinguish at least three variants of centralized systems capable of establishing and enforcing governance policies on data assets. These variants differ in how users get access to the data: distributed or intermediated. The first being an environment in which end users direct their requests to the various source applications or databases that in turn call a *service* running on the governance catalog system to evaluate, conduct and if needed enforce a policy. In the second environment, the governance catalog functions as an *intermediary* that ingest data from various sources and uses ETL practices to conduct and enforce governance policies before providing the data to the requesting end user. Lastly is a *hybrid* environment of having both the capabilities of data movement/integration (second environment) and service-like enforcement (first environment).

The hybrid environment contains the highest level of complexity to implement as it requires functionality for two entirely different environments. The first service-based environment excels in a landscape in which policies are defined for enforcing rules designed for the data creation phase (rather than retrieval phase), while the second "intermediary" environment excels in a more analytical landscape in which data needs to be collected, aggregated and delivered to an end user or analytical application (e.g. SPSS). In the first environment we described, the governance catalogs functions primarily as a rule engine. In the second environment the governance catalog functions like a true catalog that controls the data flows from source systems to end users based on the governance policies that are defined. Both environments are needed to establish a holistic data governance solution. Table 3 summarizes the benefits and limitations of a centralized implementation strategy.

Table 3. - Benefits and limitations of centralized implementation

Benefits	Limitations
Optimal consistency in asset and policy definitions and enforcement	May lead to higher network load, possibly lower enforcement efficiency, Single Point of Failure
Allows full integration of policies	Requires high level of integration (organizational and technical)

Technological advances and the use of Enterprise Application Integration (EAI) and Service Oriented Architecture (SOA) to develop new applications and services help in establishing this service-oriented environment for discovering and analyzing of data. EAI allows for the extraction of business policies and rules from the applications, creating increased flexibility and agility. SOA is a framework to "address the requirements of loosely coupled standards-based and protocol-independent distributed computing, mapping enterprise information systems appropriately to the overall business process flow" [10]. Technological advances include, amongst others, new ways of processing data (e.g. NoSQL, in-memory, Hadoop), a decrease in storage costs and increase in memory and computing power to perform the needed operations and a move to semantic systems. SOA can be enhanced with semantic technologies, for instance, to improve service identification [13].

4.3   Catalogue Virtualization

Although a centralized governance catalogue has important management advantages, the drawback is that business users – in particular, the managers responsible for the data –are set on a distance. This can be remedied by virtualizing the catalogue. This means that the various data policies are stored in a distributed way, under the control of the business user. These business users are at various levels: company-wide standards are maintained at corporate level, other policies at division of department level. In the simplest form of virtualization, these distributed data policies are just synchronized regularly with the central governance catalog. Alternatively, there is only a virtual central catalog, the combination of all distributed policies. In both cases, we assume that policy owners receive feedback (dashboard) on the actual policy compliance.

A critical issue in such a solution is the consistency of the policies. Policies may be conflicting. For instance, a corporate policy may be that all management reports are readable for the internal audit group, whereas a manager may want to restrict access to members of his own department only. One business user may want to express weights in kg and another one in pounds. In the context of this paper, we just mention a few alternative solutions which roughly correspond to the general rule enforcement strategies that we mentioned in Sect. 3. One is to accept inconsistencies as a fact of life and include meta-rules for solving them. A meta-rule can be based on the company hierarchy where corporate policies overrule local ones. This corresponds to a corrective approach because it effectively makes changes in the policies – not in their formulation, but in their application. Alternatively, we can take an adhortative approach that accepts inconsistencies but stimulates policy owners to avoid them at specification time. Closely related, a lazy evaluation (corresponding to a detective approach) can be used that detects conflicts when they actually occur and reports them back to the policy owners. This can be a pragmatic approach in situations of relative low governance where the probabilities of actual conflicts are low. Finally, the most rigid approach is to prevent any inconsistency by using a consistency checker before any policy is deployed. This is a challenge in a distributed environment, although in principle, such a checker is not different from the checkers in a centralized catalog. Last but not least, it is not necessary to choose only one approach. For instance, the company may use a preventive approach for all data standard policies and a detective approach for data access policies.

Once a virtual solution is in place, a next step can be to relax the centralization of the governance catalog. In large companies, a completely centralized approach is not realistic. Some distribution in "regions" or "zones" is unavoidable. In such a situation, a business user may be connected to one region, but also with more regions. Locally, he can manage his policies for both. Data traffic between regions is based on agreements that appear as policies in each of the regions involved.

The virtual solution described in this section can be combined smoothly with a strict distinction between "policy" and "rule" level, as sketched in Sect. 3. This means that the business users publish policies in a user-friendly policy language that is translated to formal business rules on the central catalog (physical or virtual).

5   State-of-the-Art Governance Solutions: IBM InfoSphere

In this section, we analyze in depth one commercially available solution in data governance, IBM InfoSphere. Since this is considered state-of-the-art technology, it can be seen as representative. Our goal for this study was not to compare it with other products, but to see to what extent a multi-level governance model is or can be implemented with this solution. Our analysis is based on the system documentation, expert interviews, and user experience.

5.1   Description

Our evaluation of the IBM InfoSphere suites capability to define and enforce governance policies focuses on IBM InfoSphere Information Governance Catalog (formerly known as InfoSphere Business Information Exchange) and IBM InfoSphere DataStage. Additional tools such as IBM InfoSphere Information Analyzer, IBM InfoSphere Optim and IBM InfoSphere Guardium are used to illustrate specific enforcement examples. Information Governance Catalog is designed to contain both the business glossary (terminology) as well as a list of all available information assets (e.g. dataset, table, policies, and rules) and a variety of additional metadata to describe and define the asset. An information asset is defined as "a body of information, defined and managed as a single unit, so that it can be understood, shared, protected and exploited effectively. Information assets have recognizable and manageable value, risk, content and lifecycles" [12]. Information Governance Catalog allows for the creation of a hierarchical structure to define the relations between policies and rules. These rules can be assigned to a business term. The business term defines and references to the actual source of the authoritative data.

Enforcing data governance policies focusses primarily on achieving a compliance layer. The compliance layer consists of four areas; Policy Administration, Policy Implementation, Policy Enforcement and Policy Monitoring. A *policy* is a (natural language) description of business intent for a class of assets to adhere to a

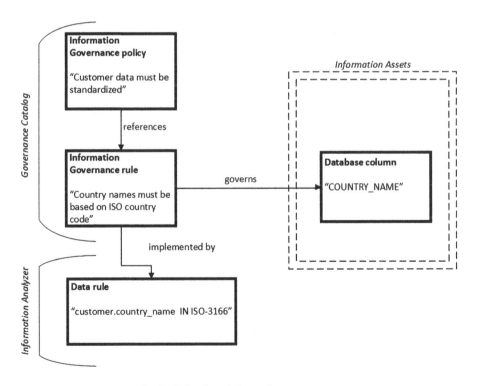

Fig. 5. Infosphere information governance

certain behavior. A *rule* defines how a policy will be implemented, it contains the policy response. Additionally to policies and rules are the *control* and *enforcement points*. A control point is a collection point for evidence that a policy is being complied with. An enforcement point is where a policy implementation (rule implementation or process) is executing. Enforcement points can be seen as hard enforcement measures that guarantee preventive compliance. Control points provide the soft enforcement which is used in the remediation that occurs after the fact. Soft enforcements are used in policies defined at a more abstract level that cannot be hard enforced.

Figure 5 illustrates how information governance policies, governance rules, data rules and the information assets are related. There is some correspondence to the multi-level meta-data framework described above. One difference is that the information assets are related to the information governance rules (business rules), rather than implementable data rules, but if there is 1–1 relationship between data rule and information governance rule, the two representations are equivalent. However, there is no formal representation for the information governance policies and rules, and hence it is not possible to check the consistency of the rules or adapt them automatically.

5.2   Evaluation

Does the IBM InfoSphere suite support multi-level metadata to enable governance, and to what extent? To answer this question, we have looked at the capabilities to validate and enforce rules in the process of moving data (ETL) by using primarily IBM Info-Sphere DataStage. Within DataStage we have the capability to create jobs for performing various ETL activities. These jobs can be assigned to a rule in the Governance Catalog as the implementable artifact. Executing a DataStage job results in the creation of operational metadata which is used to establish lineage and provide metadata to the governance dashboard. The operational metadata is mapped to IBM's private proprietary metadata model called XMeta. Besides generating data lineage the capabilities include measuring data quality and values using IBM InfoSphere Information Analyzer. Having insights into the quality and usage of data creates a tremendous increase in transparency for both business and technical users.

The Information Governance Catalog should be used as the central storage point for all the metadata that is needed for providing sufficient insights in definitions, usage, accountability and compliance. However, the current capabilities of enforcing rules and measuring their results requires a lot of technical expertise (ETL development etc.), which is undesirable in an environment that should be business driven. A more formal (semantic) approach to defining the rules should empower the business users with more capabilities to governing "their" data. Policy and rule administration are currently defined in free-text format, which could result in misinterpretations during the implementation and enforcement phase, and creates an opaque environment. A BRMS that utilizes the capabilities of defining rules in a natural structured language reduces opaque and misinterpretation, resulting in a more transparent environment. Answering the research question: IBM InfoSphere suite supports multi-level metadata to enable governance, but there is still a lot of room to enhance these capabilities to increase transparency and formalization. At the moment, it supports typically a leveled approach, not full centralization.

6 Conclusion

In order to operationalize data governance, the implementing organization needs to have the resources and capabilities in place to define and enforce data governance policies and rules. Using a multi-level metadata framework we created an insightful segregation between defining the policy and rule specifications and the resultant implementation of rules and jobs. With this segregation in place, and the capability of empowering qualified business users to define governance specifications, allows for better adaptability and transparency of data governance. As this paper presented, there are various ways of implementing a multi-level metadata framework. Having the capability to enforce policies and rules both in a centralized and decentralized manner, allows for the most flexibility. However, specific software might need to be purchased to establish an environment for this in the form of a governance catalog. IBM InfoSphere suite provides most of the capabilities needed to start operationalizing a multi-level data governance program, but formalization of policies and rules is needed in order to get to a higher level, in particular, to one supporting self-adaptation.

References

1. Baca, M., Gill, T., Gilliland, A.J., Whalen, M., Woodley, M.S.: Introduction to Metadata, Revised edn. Getty Publications, Los Angeles (2008)
2. Bonnet, P.: Enterprise Data Governance: Reference and Master Data Management Semantic Modeling. Wiley, New York (2013)
3. Cheong, L., Chang, V.: The need for data governance: a case study. In: Toowoomba: 18th Australasian Conference on Information System (2007)
4. Gong, Y., Janssen, M.: From policy implementation to business process management: principles for creating flexibility and agility. Gov. Inf. Q. **29**, S61–S71 (2012)
5. IBM Center for Applied Insights. Outperforming in a data-rich hyper-connected world. New Orchard Road: IBM Corporation (2012)
6. Khatri, V., Brown, C.V.: Designing data governance. Commun. ACM **53**(1), 148–152 (2010)
7. Klein, R.: Metadata is 'not' a technical term anymore: frame to work. In: 2014 International Data Quality Summit, ECCMA 2014: KPMG, pp. 1–18 (2014)
8. Marco, D.: Practical steps for overcoming political challenges in data governance. In: IDQSummit 2014 ECCMA 2014: EWSolutions, pp. 1–37 (2014)
9. NISO. Understanding Metadata. National Information Standards Organization (2004)
10. Papazoglou, M., van den Heuvel, W.-J.: Service oriented architectures: approaches, technologies and research issues. VLDB J. **16**(3), 389–415 (2007)
11. Singh, G., Bharathi, S., Chervenak, A., Deelman, E., Kesselman, C., Manohar, M., .Pearlman, L.: A metadata catalog service for data intensive applications. In: Supercomputing ACM/ IEEE Conference (2003)
12. The National Archives. The Role of the Information Asset Owner: a Practical Guide. National Archives (2010)

13. Vitvar, T., Peristeras, V., Tarabanis, K.: Semantic Technologies for E-Government: an Overview. Springer, Berlin (2010)
14. Weigand, H., van den Heuvel, W., Hiel, M.: Business policy compliance in service-oriented systems. Inf. Syst. **36**(4), 791–807 (2011)
15. Wende, K., Otto, B.: A Contingency Approach to Data Governance. In: MIT Information Quality: ICIQ (2007)

A MapReduce Based Distributed Framework
for Similarity Search in Healthcare Big Data Environment

Hiren K.D. Sarma[1], Yogesh K. Dwivedi[2(✉)], Nripendra P. Rana[2], and Emma L. Slade[2]

[1] Department of Information Technology, Sikkim Manipal Institute of Technology,
Rangpo Sikkim, India
hirenkdsarma@gmail.com
[2] School of Management, Swansea University, Swansea, SA2 8PP, UK
{y.k.dwivedi,n.p.rana}@swansea.ac.uk,
emmaslade@hotmail.co.uk

Abstract. Similarity search in the big data environment is a challenging task. Patient Similarity search (PaSi) is an important issue in healthcare network and data. The results of PaSi search may be highly useful for drawing different conclusions and decisions to improve healthcare systems. Such findings can also be useful for choosing the treatment paths for new patients. In this paper, we propose a MapReduce based framework as a solution to the PaSi problem in the context of a healthcare network imagined to be implemented considering the healthcare centers of India. It is assumed that such a healthcare network will be implemented in future over the Government of India cloud known as GI cloud or 'MeghRaj'. The paper also discusses the associated implementation challenges of the proposed framework and the query handling approach for the proposed framework to solve the PaSi problem is stated. Finally, the paper outlines the future scope of the work.

Keywords: Big data · MapReduce · Similarity search · Patient similarity (PaSi) · Cloud · Framework

1 Introduction

In today's world, the volume of digital data generated by different information and communication technology (ICT) related applications, and other applications in the domain of meteorology, scientific instruments, healthcare or medical networks, etc., is enormous [1]. Such huge volume of data, which is generated largely over networks, is not practical to handle through classical database management approaches [3]. For these systems, capturing, storing, processing and retrieval of appropriate data in a timely manner are some extremely important issues. Centralized solutions to these problems are not suitable and distributed solutions have their own problems [2]. Some problems of distributed processing include network bottlenecks, requirements of global information locally, extra communication overheads, etc. Such applications have given birth to the concept of big data. Big data has attracted the attention of researchers and data scientists in recent times. Novel solution approaches are required to handle big data related issues [1].

© IFIP International Federation for Information Processing 2015
M. Janssen et al. (Eds.): I3E 2015, LNCS 9373, pp. 173–182, 2015.
DOI: 10.1007/978-3-319-25013-7_14

Healthcare network is one example, which generates a huge volume of data every day. One of the processing issues connected to healthcare data is to find out Patient Similarity (PaSi). It is defined as the rate of similarity between two or more patients in terms of their symptoms, treatment procedures, personal information, etc. [1]. A typical PaSi solution will find out those patients who have the greatest amount of information in common. Then the treatment paths followed for such patients can be adapted for new patients. These data related to patients are stored in different databases of patient information systems maintained across healthcare networks. There could be several issues in solving PaSi. The data format used for different patients could be different. This is due to the lack of predefined record structure applicable to all patients. The volume of such data to be processed will be colossal. Moreover, some data related to patients can be uncertain and the data will be generated at a very high rate across the healthcare network. As a result, we need big data solutions to address the PaSi problem of healthcare networks. Hence, there is a need to think of some distributed and scalable solution approaches in order to address this problem. MapReduce is a tool that can be used to develop distributed and scalable solutions against big data problems [2, 3]. MapReduce has also been used to solve some healthcare problems [1].

'MeghRaj' is a cloud computing environment developed by the Government of India [4]. There is scope to implement a healthcare network connecting different health centers or hospitals spread across the country over this cloud. If such a system is implemented, it is going to generate big data. Thus, we need different big data solutions to handle different issues related to data processing and storage of these data.

In this paper, we consider a healthcare network that can be implemented over the 'MeghRaj' cloud and address the issue of finding PaSi. We propose a framework, which is based on MapReduce, to address the PaSi problem. We assume that the patient information will be stored in an unstructured manner. Even in the same machine or data source, two different patients' information can be differently structured. This framework is a proposal and its performance evaluation through simulation is undertaken by us.

The rest of the paper is organized as follows. Section 2 presents the background on big data followed by Sect. 3 in which related works are mentioned and the problem undertaken here is stated formally. In Sect. 4 the proposed framework is discussed and Sect. 5 describes the implementation challenges present in the proposed framework. Finally, Sect. 6 concludes the paper with an outline of the future scope of this work.

2   Background

MapReduce is a programming model and an associated implementation for processing and generating large datasets [3]. It is possible to handle big data through MapReduce, programmers find the system easy to use, and this parallel data processing tool has been made popular by Google. It is a scalable and fault-tolerant data processing tool that makes it possible to process a massive volume of data in parallel with the association of many low end computing systems [2]. Users specify the computation task at hand in terms of a map and a reduce function, and then the underlying runtime system processes the given task by distributing the computation tasks across large scale clusters of computation nodes. This tool can handle machine (i.e. computation node) failures and can also make efficient use of network and disks by appropriate scheduling mechanisms. A decomposable algorithm, partitionable data, and sufficient small data partitions are required for effective use of MapReduce [6]. There are some enhancements to MapReduce. For example, in the work [5], classic MapReduce was optimized to decrease the data transformation load. A shared area for information was considered in this approach. Such an approach is suitable for solving problems like k-nn and top k queries. In [8], a method was developed to handle workloads in hierarchical MapReduce architecture. Haloop proposed in [7] is another type of MapReduce structure suitable for handling iterative problems. iMapreduce proposed in [10] also supports iterative processes. The work presented in [12] is aimed at reducing the amount of data transferred in the MapReduce network. Here, MPI (Message Passing Interface) was used for message passing in a MapReduce structure. The work presented in [11], replaces Hadoop File System (HDFS) with a concurrency optimized data storage layer. This layer is based on the BlobSeer data management service. It is essential to estimate the input/output (I/O) behavior of MapReduce applications and the work presented in [9] is a model that can be used to estimate I/O behavior of MapReduce applications.

In this section, we also establish the relationship of healthcare data with big data. If we look at the networked environment considering the hospitals across a country like India, then we visualize that patients' data will be generated at an exponential rate. These data will have different formats and standards. In healthcare networks, various data related to patients' health, diseases and recovery processes could be made available. Such data will be of great help for the treatment of other patients. Of course, for this to happen, there is a need of processing these data from different perspectives.

Big data is characterized by four 'Vs' namely volume, variety, velocity, and veracity [1]. Data generated through healthcare networks exhibit all the above four characteristics. In such systems huge volume of data is generated in various formats with a high velocity. Moreover, for many patients we get uncertain data in the data generated by healthcare networks and this fact leads to veracity of healthcare data. Thus all four Vs of big data are present in healthcare data. Therefore, big data solutions are required to solve different data processing problems of healthcare data.

Looking at the high volume of data in healthcare networks, big data solutions are necessary for data analysis [1]. According to [13], processing costs can be reduced by using big data analytics in healthcare. In the work presented in [14], problem like selection of appropriate treatment paths is addressed and solution for improvement of healthcare systems has been proposed. A scalable knowledge discovery platform for healthcare big data is proposed in [15].

3   Related Work and Problem Statement

Finding Patient Similarity (PaSi) is the major task considered in this paper. We consider a very specific healthcare network system, which is yet to be built but must be a reality in the near future. The system under consideration is the healthcare network of India. Different healthcare units i.e., hospitals spread across the country will be connected in a hierarchical manner. One has to find out PaSi of two or more patients in terms of their symptoms, treatments, personal information, etc. The objective in PaSi is to identify those patients who have the greatest amount of information in common. Using the result of PaSi, new patients can be treated by following treatment processes adapted for those previous patients.

PaSi solutions can be found out by either of the two approaches as mentioned in [1]. First is the use of machine learning and data mining algorithms and the second being information retrieval by simple search or by entity-relationship graphs. A brief survey related to these two techniques can be found in [1].

In [1], a MapReduce based method for finding PaSi solution is proposed. The method is scalable and distributed - named as ScaDiPasi - takes small execution time and is implementable over big data related to healthcare networks. The experimental results reported in the paper show that the ScaDiPasi would be able to produce PaSi solutions over big data of healthcare networks.

3.1 Cloud of Government of India

With an aim of exploiting the benefits of cloud computing, the Government of India has initiated a very ambitious GI cloud project. This cloud has been named as 'MeghRaj' [4]. As mentioned in [4], the objectives of GI cloud are: optimum utilization of infra-structure; speeding up the development and deployment of e-governance applications; easy replication of successful applications across different states of the country to avoid duplication of effort and cost in development of similar applications; and, making the certified applications following common standards available in one place.

The GI cloud consists of multiple national and state clouds. A detailed discussion on the architecture of GI cloud and projects to be implemented under GI cloud can be found in [4].

This GI cloud infrastructure can definitely be used for implementing a healthcare network across the country. Such a healthcare network will generate healthcare big data. Any stakeholder of the healthcare network can share the benefits of processing such big data with different objectives within less time. Therefore, the advantages and benefits of such a system can be significant.

3.2 Problem Statement

Finding Patient Similarity (PaSi) is an important problem. As already discussed, there are different approaches for finding PaSi solutions. Interestingly, there exists no single solution applicable to all kinds of problems. Although there are several PaSi solutions already proposed for different situations, we believe, those will not be directly applicable to the big data to be generated by the cloud based healthcare network system of India that is being considered in the current research. Therefore, we need a novel solution to solve the PaSi problem of the healthcare network over GI cloud. Hence, the problem statement of this paper is:

To design a big data based framework for addressing the Patient Similarity (PaSi) problem considering the future healthcare network that can be built over the GI cloud.

4 Proposed Framework

In this section, we provide a framework for similarity search related to the healthcare big data environment. The proposed framework is based on MapReduce [3]. We consider the cloud environment 'MeghRaj' [4]. It is assumed that a healthcare network considering all health centers of India spread across the country will be implemented over 'MeghRaj'. The structure of the healthcare network will be hierarchical as shown in Fig. 1. Different layers in this hierarchy are the hospitals at different levels such as (i) at panchayat level, (ii) at block level, (iii) at sub-division level, (iv) at district level, (v) at state level, (vi) at region level, and (vii) at national level. Although the type of hospitals can be categorized as elementary health center, public health center, medical college and hospital, private hospital, etc., at this stage we do not discriminate the hospitals based on type. We assume that all hospitals are similar with respect to generation of patients' data. We mainly need to work on the patient data irrespective of the facilities and infrastructure of the hospitals, which are generating such data.

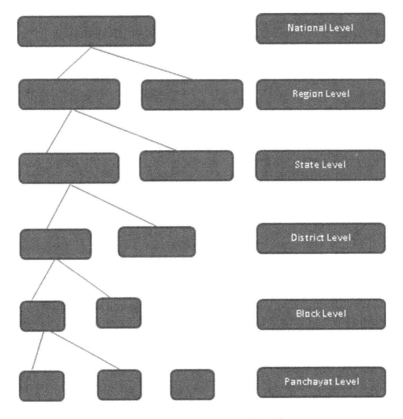

Fig. 1. Hierarchical organization of health centers

The data generated by such a healthcare network system will be stored in the cloud 'MeghRaj'. It is assumed that the healthcare network will be implemented over the Internet. As an end result, a legitimate user of the healthcare network should be able to throw a query related to similarity search problem to the healthcare network and should receive the result back from the network as quickly as possible. This scheme is shown in Fig. 2.

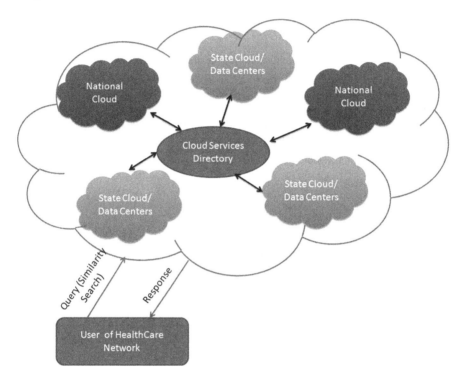

Fig. 2. Query sending to healthcare network to be implemented over MeghRaj cloud

4.1   How MapReduce Will Work

In order to optimize the solution process of similarity search problem, we propose a load-balancing module, which keeps track of distributed load among the computing nodes of the network. Moreover, this module tries to balance the computing load among the nodes.

As we assume that the data sources can be of heterogeneous nature, we propose to have a data format middleware, which brings different data formats to a homogenous common format before any kind of processing task takes place.

Bottlenecks due to maximum message exchanges in distributed processing are an issue, which is also unavoidable. We propose a bottleneck assessment and control module that takes care of the possibilities of bottleneck occurrence. In the presence of a bottleneck in certain nodes, this module will divert the necessary workload to some other nodes so that the bottleneck can be controlled temporarily.

Data aggregation module will aggregate different data against different queries into some aggregated state. In an aggregated state the volume of data will be reduced significantly and intermediate code to represent data will be generated. This aggregated data in encoded form with reduced volume will be moving across the network reducing amount of data traffic in the network.

All these four modules mentioned above i.e., load balancing module, data format middleware, bottleneck assessment and control module, and data aggregation module, will be working outside the MapReduce environment. The output of these modules will be integrated with MapReduce for optimal performance of the entire system. This framework is depicted in Fig. 3.

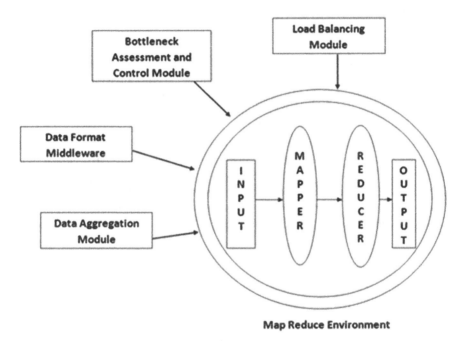

Fig. 3. Proposed MapReduce based framework for similarity search in healthcare big data

5 Implementation Challenges

The major challenges in total implementation of the proposed framework are as follows:

Challenge Set 1. Appropriate technique for load estimation and appropriate algorithm for load balancing are to be designed and associated theoretical complexity analysis is to be carried out. Proper task scheduling algorithm for load balancing is to be designed and analyzed.

Challenge Set 2. Data format middleware is to be designed, which will be highly specific to the structures of different databases present in the healthcare network.

Challenge Set 3. Proper algorithm for bottleneck assessment is to be designed. Moreover, bottleneck has to be controlled and this may lead to migration of processes or computation tasks from one node to another lightly loaded node. Thus there is a necessity to design proper process migration algorithm.

Challenge Set 4. Data aggregation algorithm considering the patient databases is to be designed and analyzed for its performance. Proper encoding mechanism has to be designed for it.

5.1 Solution for Similarity Search

The Patient Similarity (PaSi) search is the similarity search problem considered here. A user throws a query, and then this query is translated into an appropriate uniform format through an intermediate query building process. It works on the MapReduce framework. We propose to have three phases of query processing to solve the PaSi problem. In each phase, Mapper and Reducer functions are to be implemented along with a Ranker function. The Ranker function at each phase evaluates the similarity level of the output of each phase with the input query. This is proposed to be a dynamic decision regarding forwarding of the output of the previous phase to the next phase for further processing in search of more similar results. A threshold level of similarity can be set as per the wish of the user with respect to his/her query. The second phase will continue to be executed until the similarity equates or exceeds the threshold level set by the user. This evaluation is carried out by the Ranker function to be implemented in each phase and a decision is made dynamically regarding the forwarding of the processing task to the next phase. This scheme is shown diagrammatically in Fig. 4.

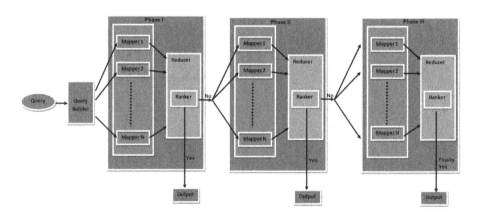

Fig. 4. Proposed query processing model based on MapReduce

6 Conclusion

In this work we address the problem of similarity search in a healthcare network using 'big data'. We focus on the Patient Similarity search popularly known as PaSi, and propose a framework for addressing the PaSi problem in healthcare data. Implementation issues are discussed thoroughly and a MapReduce based model of query handling is also proposed. The proposed framework and the query handling model are designed considering the Government of India cloud also known as 'MeghRaj'. It is assumed that a healthcare network will be implemented considering all the hospitals spread across India and will be deployed over 'MeghRaj' cloud. As for future scope of this work, it is noteworthy that various algorithms can be designed to address the implementation challenges outlined in Sect. 5. Moreover, the query handling model can be implemented over MapReduce framework considering some suitable patients database and various performance parameters, like execution time and accuracy, can be measured and analyzed.

References

1. Barkhordari, M., Niamanesh, M.: ScaDiPaSi: an effective scalable and distributable MapReduce-based method to find patient similarity on huge healthcare networks. Big Data Res. **2**(1), 19–27 (2015)
2. Lee, K.H., Lee, Y.J., Choi, H., Chung, Y.D., Moon, B.: Parallel data processing with MapReduce: a survey. AcM sIGMoD Rec. **40**(4), 11–20 (2012)
3. Dean, J., Ghemawat, S.: MapReduce: simplified data processing on large clusters. Commun. ACM **51**(1), 107–113 (2008)
4. GI Cloud Initiative (2015). http://deity.gov.in/content/gi-cloud-initiative-meghraj
5. Ding, L., Xin, J., Wang, G., Huang, S.: ComMapReduce: an improvement of MapReduce with lightweight communication mechanisms. In: Lee, S.-G., Peng, Z., Zhou, X., Moon, Y.-S., Unland, R., Yoo, J. (eds.) DASFAA 2012, Part II. LNCS, vol. 7239, pp. 150–168. Springer, Heidelberg (2012)
6. Highland, F., Stephenson, J.: Fitting the problem to the paradigm: algorithm characteristics required for effective use of MapReduce. Procedia Comput. Sci. **12**, 212–217 (2012)
7. Bu, Y., Howe, B., Balazinska, M., Ernst, M.D.: HaLoop: efficient iterative data processing on large clusters. Proc. VLDB Endowment **3**(1–2), 285–296 (2010)
8. Martha, V.S., Zhao, W., Xu, X.: h-MapReduce: a framework for workload balancing in MapReduce. In: 27th International Conference on IEEE Advanced Information Networking and Applications (AINA), pp. 637–644 (2013)
9. Groot, S.: Modeling I/O interference in data intensive Map-Reduce applications. In: 12th International Symposium on IEEE/IPSJ Applications and the Internet (SAINT), pp. 206–209 (2012)
10. Zhang, Y., Gao, Q., Gao, L., Wang, C.: Imapreduce: a distributed computing framework for iterative computation. J. Grid Comput. **10**(1), 47–68 (2012)
11. Nicolae, B., Moise, D., Antoniu, G., Bougé, L., Dorier, M.: BlobSeer: bringing high throughput under heavy concurrency to hadoop Map-Reduce applications. In: 2010 IEEE International Symposium on Parallel & Distributed Processing (IPDPS), pp. 1–11 (2010)
12. Mohamed, H., Marchand-Maillet, S.: MRO-MPI: MapReduce overlapping using MPI and an optimized data exchange policy. Parallel Comput. **39**(12), 851–866 (2013)

13. Srinivasan, U., Arunasalam, B.: Leveraging big data analytics to reduce healthcare costs. IT Prof. **15**(6), 21–28 (2013)
14. Jee, K., Kim, G.H.: Potentiality of big data in the medical sector: focus on how to reshape the healthcare system. Healthc. Inf. Res. **19**(2), 79–85 (2013)
15. Metaxas, O., Dimitropoulos, H., Ioannidis, Y.: AITION: a scalable KDD platform for Big Data Healthcare. In: 2014 International Conference on IEEE-EMBS Biomedical and Health Informatics (BHI), pp. 601–604 (2014)

Big Data, Big Opportunities: Revenue Sources of Social Media Services Besides Advertising

Julian Bühler[✉], Aaron W. Baur, Markus Bick, and Jimin Shi

ESCP Europe Business School Berlin, Berlin, Germany
{jbuehler,abaur,mbick}@escpeurope.eu,
jimin.shi@edu.escpeurope.eu

Abstract. Facebook, Twitter, Instagram, and other players in the social media world have been on the rise during the last couple of years. In contrast to their popularity, their underlying business models are vague and often only linked to advertising. In this explorative study we identify new revenue sources for social media service providers besides advertising. Based on three use cases with Facebook, Tencent, and LinkedIn, we identify three possibly fruitful ways to extend existing social media business models. Subsequently, a survey with 301 respondents changes perspectives on the user's willingness to pay in order to identify usage-related differences evoked by cultural and external circumstances. Four derived hypotheses lead the way to avenues of further research especially in terms of Big Data analytics with new e-commerce trends like Facebook's Buy Button.

Keywords: Social media · Advertising · Business models · Big data · Case study

1 Introduction

On 30 January 2015, one of the world's leading social media services Facebook [1] updated its terms and data policy significantly by introducing new advertisement rules [2]. Users had to accept these changes in order to be able to still use the service. The modifications helped Facebook be more efficient in online behavioral advertising by collecting user information from various websites automatically, with the help of cookies. This behavior suggests that advertising is still one of the most important revenue pillars in the social media world. For instance, Facebook reported that income from advertising represented 88 % of its total revenue in Q2/2013 [3] and increased to 93 % in Q4/2014, compared to "payments and other fees revenue," comprising only 7 % [4]. However, tracking user reactions to this personalized advertising automatically leads to a massive overload of information which require modern analytic processes like certain Big Data algorithms. Still it remains unclear how and if Facebook interprets the tracked data, as not necessarily storage or analysis, but effective and efficient transformation of Big Data into reliable information causes problems [5].

In contrast to this development, Clemons [6] already forecasted six years ago that advertising would fail to play a role as the leading revenue source for Internet-based companies. The author suggests that consumers tend to ignore advertising including

© IFIP International Federation for Information Processing 2015
M. Janssen et al. (Eds.): I3E 2015, LNCS 9373, pp. 183–199, 2015.
DOI: 10.1007/978-3-319-25013-7_15

online variations for three main reason: no trust in advertising, no willingness to see advertising, and no need for advertising at all to become informed [6]. The last argument is especially linked to the Internet as it has become easier for customers to collect all necessary information to decide on product purchases via search engines and product comparison websites. Therefore, it might be doubtful whether the focus on Big Data algorithms in advertising contexts is worth deeper research.

Since Clemons' analysis in 2009, the rise of social media has led to an unforeseen dominance of services including Facebook, Twitter, YouTube, and Instagram amongst others on the Internet. Thus, in this study we aim to identify existing alternative business models for revenue generation besides advertisement, as especially smaller or more specialized social media services struggle to capitalize on this single business model in the same way. The starting point for us to answer this question is a valuable categorization by Zambonini [7] that sheds light on other potential business models. However, social media was not considered explicitly despite the dominant position it has, nor did other promising studies recently (e.g., [8]). Until now, information systems (IS) or marketing research has not yet focused on operationalization of alternative revenue sources that trigger users' perceptions appropriately. In this empirical study we analyze different existing revenues, seeking to answer our main research question: *Which revenue sources besides advertising should social media service providers utilize?* Directly linked to this question are the underlying techniques in terms of suitable Big Data analytics routines [9].

The structure of this study is as follows. We first briefly discuss the term *social media* and its cognates as well as advertising in the context of this study (Sect. 2). Then we explain our research method, which contains three case studies of Facebook, Tencent, and LinkedIn and a survey with 301 participants (Sect. 3). Afterwards, we present three alternative revenue sources besides advertising based on the case study results and analyze them from the perspective of social media users (Sect. 4). We conclude this explorative study by deriving research hypotheses for accurate theory development in this field of study and with respect to Big Data analytics.

2   Conceptual Background

2.1   Definition of Business Model and Social Media Services

Before analyzing existing business models besides advertising in the social media environment, it is necessary to briefly clarify our understanding of these terms. We follow the definition introduced by Osterwalder and Pigneur [39, p. 14] who define a business model as 'the rationale of how an organization creates, delivers, and captures value". In addition and according to Kaplan and Haenlein [10, p. 61], 'social media services' can be understood as "a group of Internet-based applications that build on the ideological and technological foundations of Web 2.0, and that allow the creation and exchange of User Generated Content." The term 'social media services' is also linked to Web 2.0 by other authors and frequently used similarly to the expression 'social network' or 'social network site' (SNS) (e.g., [11–13]). We follow the first definition from Kaplan and Haenlein as it addresses the most relevant attributes of social media, but we consistently use the term social media services instead of social network (sites) in this paper. In the context of Big Data infrastructure, social media services are an appropriate object of study as they can serve as a certain 'source type'.

2.2 Advertising as Financial Source for Social Media Services

Of 16 business models analyzed by Zambonini [7], advertising is the most common financial source for all web-based services and applications and can be classified as third-party supported revenue, which also includes social media services. Facebook as one of the largest social media services "had third quarter advertising revenue of $1.8 billion" [14, p. 1], being the largest share of overall revenues of $2.02 billion. Traditional online advertising is randomly displayed on the screen of a user, unrelated to the person's demographic characteristics, cultural background, or user preferences. While this type of advertising was used by many providers in the first years of their existence, only a few – including one of the largest microblogging services from China, Sina Weibo – are still using this advertising strategy.

A more common variation nowadays is user-related advertising. It is a kind of targeted advertising that exploits user data to personalize the ads shown with the goal of increasing the click through rate and the conversion rate. It entails elements of users' data (e.g., language and location) [15], browser history (by means of cookies saved in browsers), and social media activities. All of these social media related activities are tracked, recorded, and registered into a social graph, which provides advertisers with relevant data. However, a problematic "industry-driven obsession with the 'social graph'" [16] can occur if the underlying Big Data analysis procedures are too technical and lead to invalid interpretations of activities. Such activities may comprise the action of adding contacts, Facebook Likes on a page or post, as well as comments on other users' timelines and actions that involve online applications such as games, music, or news. Facebook's latest terms of January 2015 stretched the area of influence significantly, allowing the company also to retrieve remote information from other websites or digital services. Mobile advertising which is not explicitly addressed in this study adds the component of accessibility and local optimization to this concept. This enables location-based service functionality and, thus, timed place-sensitive advertisements such as notifications of sales and special events [17].

3 Research Method

3.1 Research Design: Case Study Analysis

We chose a case study approach because it is suitable for our explorative setting. According to Yin et al., "A case study is an empirical inquiry that investigates a contemporary phenomenon in depth and within its real-life context, especially when the boundaries between phenomenon and context are not clearly evident" [18, p. 13]. Following the meta-analysis on case study research by Dubé and Paré [19] for methodological procedure, we postulated a clear research question in the introduction. Based on our general social media service definition, we decided to apply a *multiple case design* to accurately address both the diverse functionality of services and cultural diversity of users. All main elements, which are described in this and the next subchapter, are summarized in the complete visualization of our research model and approach (Table 1).

Table 1. Research model and approach

Research Design		
Clear Research Question [20–23]	*Multiple Case Design [18, 20, 24]*	*Unit of Analysis [18]*
"Which revenue sources besides advertising should social media service providers utilize?"	▪ Facebook ▪ Tencent ▪ LinkedIn	▪ Type A: entertainment-oriented social media services ▪ Type B: business-oriented social media service
Data Collection		
Multiple Data Collection [18, 25]		
▪ Website analysis of three social media services (incl. terms of use, sitemaps, FAQs, and user accounts) ▪ Questionnaire addressing social media users and their willingness to pay for features, functions, or services		
Data Analysis		
Logical Chain of Evidence [18, 20]	*Cross-Case Patterns [21, 23, 26]*	
▪ Identification of relevant social media services (Step 1) ▪ Identification of existing revenue sources (premium features, services, etc.) besides advertising (Step 2) ▪ Analysis of users' willingness to pay for these offers (Step 3)	▪ Similarities and differences of willingness to pay for offers for both social media service types A and B ▪ Empirical analyses as the basis	

We used the Alexa website ranking [1] to identify the most globally used social media services and to distinguish between two types of social media services for our cases: entertainment-oriented social media services and business-oriented social media services. For us, entertainment-oriented social media services focus on general social networking tasks like picture sharing, pin boards, and gaming and emphasize predominantly leisure activities. On the other hand, we consider a business-oriented social media service as a service designed to share people's skills and professional interests, and to promote oneself. Together, these two variations form our *unit of analysis* [19].

Following the Alexa ranking, we focused on the two leading entertainment services and the largest business service. First in the ranking is Facebook, which is not only the leading entertainment-oriented social media service according to our categorization, but also the most viewed website overall, second only to Google. In addition, we selected Tencent as a second entertainment service because it is present in a large, but partly restricted, market—China—and can reveal unknown phenomena. This cultural and legal distinction is in line with Yin's definition of adequate case selection [18]. Tencent as a portal is also present with two services in the Top 20, Qq.com and Weibo.com. Regarding the business-oriented social media services, we chose LinkedIn, which is the unchallenged worldwide leader and reaches the 13th rank in the Alexa ranking of all websites (not only social media services) [1].

3.2 Multiple Data Collection: Website Analysis and User Survey

According to Dubé and Paré's [19, p. 615] analysis of Sawyer [25] and Yin [18], "A major strength of case study data collection is the opportunity to use many different sources of evidence to provide a richer picture of the events". Hence, a *multiple data collection* approach was used as we first analyzed the three selected social media service websites with regard to offered revenue sources besides advertising. In addition, we developed a questionnaire to identify users' perceptions and adoption of these business models especially with respect to cultural and legal differences.

The general data collection phase was thus split into two parts (A & B) as well, starting with an exhaustive website analysis including terms of use and sitemaps (A). We additionally created user accounts on all three services to gain an overview of all offered functions and services that could potentially serve as alternative revenue sources for the service providers. After the selection of Facebook, Tencent, and LinkedIn (Step 1), this identification of revenue streams ties in afterwards as a second step within a *logical chain of evidence*. We then distributed a questionnaire to a non-specific target group of social media users in a second part of the data collection phase (B) to identify their willingness to pay for additional services and functions (Step 3).

The questionnaire itself is separated into three main sections and follows a funnel approach design [27], starting with questions on the demographic and socioeconomic situation of the participants. In the second section, we address the general usage of social media services and ask participants about their level of activity on associated social media websites. For all social media services they use with an individual account, conditional questions are designed to gather information on their willingness to pay for certain extra features, which we identified during the website analysis. The third section focuses on identifying social media services in professional or business surroundings in particular. We aimed to gain insights on user behavior here because we especially expected services such as LinkedIn to earn money with alternative business models besides advertising. We ask participants to give reasons why they use these services and which types of premium functions, upgrades, or other functionalities are subject to costs. Both for entertainment- and business-oriented services, participants are also requested to state prices they expected to pay for these extra features. This allows us to identify *cross-case patterns* between the three case studies. The final part of our questionnaire addresses how users intend to use these services in the future.

As this is an explorative approach, we have the goal of making inferences on a broad and random sample of the population and do not define a concrete target group for this study. Thus, possible exclusions of participants due to potential group bias issues can be avoided *a priori* [27, 28]. The only characteristic participants have to fulfill is that they are Internet users and/or have Internet access regularly. This single criterion is helpful to address participants who are already online and probably active on or familiar with social media services. In line with the population we strive for, i.e., Internet users in general, we designed our questionnaire to be digital, not paper-based. Even though a few users might not be able to access the survey without hindrance due to technical problems or lack of support [27, 28], it is an appropriate distribution means in our context. Therefore, the link to the corresponding survey was spread through various channels. This includes several mailing lists, social media services, and instant messengers as well as offline notices with short links and QR codes to our survey at many physical locations like universities and stations, being accessed by heterogeneous visitors.

4 Analysis and Results

4.1 Case Studies: Identification of Alternative Financing Strategies

Case Study 1: Virtual Goods on Tencent. In our first case study, we focused on one of the largest social media service providers worldwide, the Chinese company *Tencent*. It offers several services via its platform qq.com (Alexa rank 10) like China largest social networking site similar to Facebook called *Q Zone*, a microblogging service similar to Twitter called *Weibo*, and other services like instant messaging (*WeChat*) or games (*QQ Game*). We decided to analyze Tencent besides its high ranking as it represents a large target group of Asian social media users as well as a broad variety of services and functions.

We analyzed the website including subpages and logged in as regular users to receive insights from their perspective. Besides advertising, we initially learned that Tencent has created thousands of different virtual goods for their numerous services. These virtual goods are offered to users for money and can be seen as "intangible objects purchased in order to be used in online communities such as SNS sites or online games. They also comprise virtual money (or virtual currency) which is used to purchase these intangible or physical goods" [29]. Ho and Wu [30, p. 208] investigated user's intention to purchase virtual goods especially in online games and according to the authors they became "a major source of income" for service providers. As especially the "social dimension [...] is likely to influence the purchasing intentions of individuals" [31, p. 790], virtual goods can play an important role not only for online games but social media service providers in general.

From what we found, the virtual goods offered by Tencent can be divided into two categories by their functions. The site offers goods with customization purposes, such as wallpapers and decorations for personal space. Virtual goods of this category are mostly offered on the social networking service Q Zone, but can also be found on WeChat in terms of emoticon stickers for chatting or outfits and cosmetics for customizing one's avatar. The second category consists of virtual goods offered for gaming purposes, such as tools and credits which users can utilize in certain games. According to Guo and Barnes [32], this is both an emerging trend for South Korea and China as Asia's biggest gaming markets as well as the USA (currently the largest e-gaming market). Unlike other social media services, e.g., Facebook, Tencent does not open itself as a platform for third party developers to offer their own games but soleily develops its own game.

After analyzing the offered types of virtual goods, it is important to clarify how Tencent is actually earning money with them. The company uses its own virtual money called Q coin which users can buy to purchase the virtual goods. The value of Q coin is connected to the Chinese currency Renminbi (CNY) and 10 Q Coins are worth 10 CNY (or ≈ 1.6 USD). This virtual money can be bought via different channels, such as QQ's official online payment site, mobile credits, post offices, convenient shops, or even newspaper stands around China. Purchased this way, Q coin serves like a prepaid card that users need to purchase in advance to deposit the equivalent money into their account. A second option is offered by Tenpay, which is a third party payment platform launched by Tencent connected with users' debit/credit card account or other online payment solutions such as PayPal. Once bought, users can share the same account information to purchase virtual goods with Q Coins among all the social media services of Tencent.

From this case study analysis we summarize that Tencent follows a strategy which allows its users to have free registration in order to maximize users' acquisition. Afterwards Tencent starts to push its charged virtual goods to the users which upgrade their accounts or improve their social media experience. This paid upgrade of user accounts to a premium one, which entails the possibility to exploit several different features precluded to the free accounts, is the main characteristic of the popular business model denominated "freemium" [33]. The term "freemium" describes a business model which combines "'free' and 'premium' consumption in association with a product or service" [34, p. 1]. Figure 1 shows that unlike most other social media service providers, the biggest proportion of Tencent's revenues is not advertising but virtual goods. Sold with the help of the freemium concept, they had a share between 70 % (minimum, Q4/2013) and of 81 % (maximum, Q3/2014) over the total revenues in the last two years [35]. These results suggest that Tencent was successful in establishing a financing method for its social media service besides advertising.

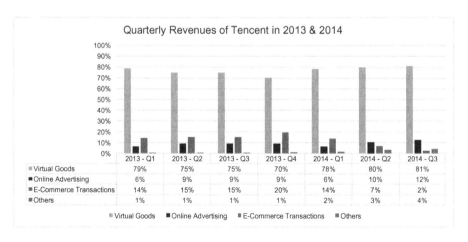

Fig. 1. Comparison of revenue sources for Tencent in 2013 & 2014 (Source: [35])

Case Study 2: e-commerce and Social Commerce Services on Facebook and Tencent. Social media services can also provide users with e-commerce functions within their environment, for example online trading and payment services. Functionalities including transactions between businesses and customers (B2C, as well as B2B and C2C) are often offered on the website of the social media service or as a third party payment platform on other websites. We analyze these activities which are sometimes referred to as social commerce in our second case study.

In the context of e-commerce functionality, Facebook seems to follow a strategy of trying out several approaches very fast according to our analyses. During the last couple of years, Facebook first had started with a service called Credits. Facebook Credits was a payment service for its platform of games and apps, developed and offered by third party developers. Users could purchase Facebook Credits and then convert them automatically into the in-game or in-app items they need. Facebook served as an online payment service and all purchases of virtual goods for applications and games on Facebook's platform were processed through Facebook Credits, the only payment option [36]. This service started in 2011 and developers earned 70 % of the value while Facebook retained 30 %. It was later replaced in 2013 by a service called Local Currency Payments, a service that converts the value of a virtual good into local currencies of users. The most common payment channels are credit card/debit cards, PayPal and mobile technologies (e.g., Google Wallet), but alternative local payment methods exist, varying from country to country [37]. For example, Moneybookers is available in the U.S. and Giropay in Germany. The service fee charged from game and app developers remained unchanged. The latest changes include Facebook Gifts, an online trading platform offered by Facebook in September 2012, where users could buy physical 3rd party company gifts or vouchers for their friends. This service was closed in 2014 and at the moment, users can only buy vouchers in the form of game cards in some countries. After various tests, Facebook will most likely introduce a new e-commerce feature called Buy Button in 2015, which allows third party companies to publish posts directly via the Facebook profile. The buying process for users is supposed to be easier as purchases can be made without leaving Facebook's website.

Besides the virtual goods business described in the first case study, our website analyses reveal that Tencent also play an important role in China's e-commerce sector. There are two individual services, PaiPai.com and Tenpay, which both are connected with Tencent's social media services like Q Zone. According to Tencent, PaiPai.com is a B2C & C2C online trading platform launched in 2006 where users can find products from different categories, for example, clothing, electronic or education products [38]. They can log-in with their account from Tencent and every purchase can be shared among the various services. The business model of PaiPai.com is not based on account fees, which do not exist for sellers, but on added values. Sellers can buy trust certificates and receive in return e.g. extra space for larger product pictures in high resolution. The second e-commerce service offered by Tencent is Tenpay, which can not only be used to buy Q Coins, but also for regular bank transactions like online payments, money transfers, or others. Unlike Facebook's payment methods, Tenpay is a third party payment method available not only for Tencent's social media services, but also for the platforms of other companies.

Results of this case study with both companies reveal that major social media services are currently active in the e-commerce sector and even though Facebook changed its strategy over the years, e-commerce seems to be a promising alternative financing strategy to advertising.

Case Study 3: Account Upgrade Services on LinkedIn. In our third case study, we analyze LinkedIn[1] as the world's largest representative of other business-orientated social media services like Xing or Viadeo. LinkedIn offers a previously described freemium business model including four types of certain premium and account upgrade services. The first one is named *job seeker* and is designed for individual users looking for a new job opportunity. The second option is labeled as *business plus*, where experienced LinkedIn users can use extra functionality, such as additional search features. *Sales plus* is supposed to be used by salespeople looking for new leads or business partners and *Recruiter Lite* is dedicated to company representatives seeking for new employees via the LinkedIn platform. In addition to these four premium subscriptions, LinkedIn offers two additional premium services called *Talent Solutions* and *Marketing Solutions*. Talent Solutions is designed to support larger companies during their recruiting process while Marketing Solutions allows customers to present their content in a prominent position both on LinkedIn's website and the mobile app. Once users subscribe to one of the upgrade options, they get access to features included in the premium package.

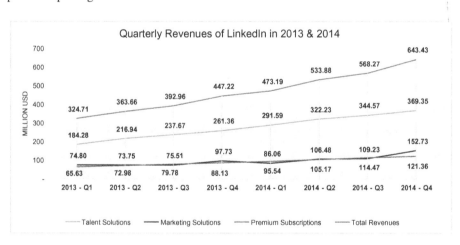

Fig. 2. Comparison of different revenue sources for LinkedIn in 2013 & 2014 (Source: LinkedIn revenue reports, http://investors.linkedin.com/results.cfm)

We consolidated LinkedIn's information of the quarterly revenue reports[2] and Fig. 2 illustrates the revenue sources for 2013 & 2014. They indicate a successful adop-

---

[1] Service description of LinkedIn: https://www.linkedin.com/about-us.

[2] LinkedIn: Quarterly Earnings 2013/2014: http://investors.linkedin.com/results.cfm? Quarter=&Year=2013 and http://investors.linkedin.com/results.cfm?Quarter=&Year=2014.

tion of the freemium concept with steadily growing revenue over the years. Anticipating similar results for 2015 following the recent trend, we consider this business model being a potential alternative financing strategy to advertising for social media service providers as well. The business model *advertising* was previously categorized as third-party supported. In contrast, the analyzed business model account *upgrade services* comes within limits of subscription, pay for additional premium content or advanced features. Both *virtual goods* (here: virtual products) and *e-commerce* belong to payments according to [7]. All three alternative models identified in the case studies build the underlying structure for our second empirical part of this paper, a survey which includes participants in the core target markets of social media services, China, the USA, and Europe.

4.2 Survey Results

Demographic Structure. After a period of six weeks we received 333 responses to our online survey in total. Out of these participants, 31 did not fully complete the survey and were excluded from our sample and further analyses. One additional participant was removed from our sample due to insufficient data reliability, which results in a final sample size of n = 301 respondents. The demographic distribution of our final sample is illustrated in Table 2, revealing non-representativeness but wide range regarding the age and varying nationalities of the participants, as strived for:

Table 2. Demographic distribution of survey respondents

Demographic category	Distribution
Age	33.7 years (sd: 11.1)
Gender	61.1 % female, 38.9 % male
Nationality	20.27 % from North America (incl. 14.95 % from the US, 4.64 % from Canada)
	31.23 % from Asia (incl. 26.25 % from China)
	45.18 % from Europe (incl. 13.62 % from Germany, 6.98 % from Italy, 6.31 % from France)
	03.32 % from other continents

Another important characteristic of the sample of respondents is their status of employment. The majority of participants stated they are employed for wages (44.19 %), followed by 21.93 % who are students and 12.62 % who are interns. The remaining participants are self-employed (8.31 %), currently looking for a job (5.32 %), retired from work (3.65 %), or in other employment (3.99 %). This sample structure is relevant especially with respect to the sections of the question-naire dealing with willingness to pay. More than half of the participants—at least employees for wages and self-employed people—can be considered to earn money with their job and, therefore, have the purchasing power to pay for virtual goods or premium functionality.

Entertainment-Oriented Social Media Usage. The popularity of social media usage among our survey participants is clearly visible according to the percentage of people with a subscription and account for at least one of the entertainment-oriented services. We provided predefined answers for seven important social media services, but respondents could name additional services in a text box. Results show that 92.69 % (n = 279) of the respondents have subscribed to a social media service and an additional 2.33 % is planning to do so. Only less than five percent (4.98 %) of the sample is not interested in using social media services at all, right now, or in the future. Overall, results indicate a high level of activity as 265 of the 279 social media users in our questionnaire use the services at least once a week, 98.61 % of them even once or several times a day. These results were measured on a 7-point Likert scale ranging from –3 ("I never use social media service") to +3 ("I use social media service a couple of times a day").

Going deeper into the analysis, the distribution of subscriptions to different social media services throughout our sample shows the clear supremacy of Facebook, followed by YouTube and Twitter. These results are in line with common rankings based on page impressions, e.g., the Alexa ranking, and confirm the dominant position occupied by this services globally. A more detailed look results from splitting the sample group into geographical clusters by continent membership. While Facebook reaches a saturation of more than 73 % in all continents including Asia, even though access is technically restricted in China, other services vary significantly. As anticipated, all three Chinese-based social media services—Renren, Sina Weibo, and Tencent—are used by nearly 50 % or more users in Asia but are virtually not recognized by those coming from North America or Europe. Figure 3 gives an overview of the saturation of main social media services by continent.

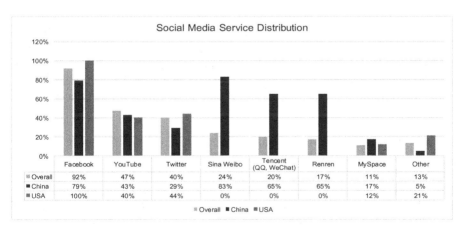

Fig. 3. Social media service distribution in China, the US, and worldwide (Source: own survey results)

Although entertainment-oriented social media services are being used quite heavily, our results indicate that purchases are relatively rare in this field. Only 31 respondents (10.80 %) bought additional functions or products. Age differences are not noticeable, but regarding the continent-based distribution, results clearly show that Chinese survey participants are the majority who purchased virtual products including virtual money on social media services. In our survey setting, participants were provided with certain offers like virtual goods we identified within the case studies, but they could also name other services or virtual goods. Respondents from Asia paid predominantly for premium accounts (e.g., Sina Weibo VIP accounts; Tencent QQ diamond account), virtual products (e.g., Tencent QQ avatar outfits, pets, and new themes; Tencent WeChat emoticon stickers), and virtual money (e.g., Tencent Q Coins). On the other hand, only survey participants from the US purchased Facebook gift cards, while Europeans had no particular key feature. On average, these 31 respondents spent $21.84 for the additional functions and products (Fig. 4). These results clearly indicate a cultural difference in terms of usage interests between Asian (in particular Chinese) and European or North American social media users.

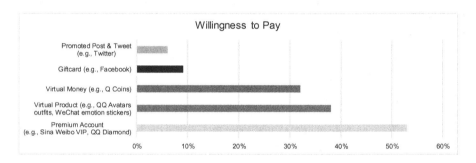

Fig. 4. Willingness to pay for selected functions and products (Source: own survey results)

The last part of this second section of our questionnaire addressed the willingness to pay money for additional functions and products. Overall, social media services in this field have low potential here according to our results because only 3.97 % of existing and prospective users stated their interest in spending money for these features on entertainment-oriented social media services. Of all offers, virtual products in the Asian market have the highest potential compared to others.

Business-Oriented Social Media Usage. The final part of our questionnaire addressed business-oriented social media services. We provided the respondents with ten predefined answer options according to the ten most used services in Alexa ranking (Absolventa, Biznik, Cofoundr, Ecademy, E.Factor, LinkedIn, Ryze, Tianji, Xing, and Ziggs) and included an additional option to name "other service." The overall percentage of respondents using these services is significantly lower compared to entertainment-oriented services, but still 64.78 % (n = 195) subscribe to one and 14.29 % (n = 43) plan to subscribe in the future. LinkedIn is the unchallenged leading service in this field with 168 users (55.81 %) within our total sample and 24 (7.97 %) of them from China. This is interesting as LinkedIn released its beta version of a simplified Chinese site only in

February 2014 and a large new target group of Chinese-speaking members potentially joined the service since then. Other services than LinkedIn, in contrast to entertainment-oriented social media services, only play a role in individual regions and countries according to our results. For example, XING, a business network with similar functionality as LinkedIn, is used by 61 participants (20.27 %) even though it is only available in the German-speaking market and Spain. Viadeo was named by only three participants.

The next block of survey questions aimed at identifying reasons why users tend to use business-oriented social media services. Among the many possible reasons for subscription, three seem to be the most popular among our survey participants. Throughout all continents and countries, "networking with professionals" (72.82 %) is extremely important to our survey participants (Fig. 5). Business-oriented services in the social media world have given a completely new playground to the professional networking experience, so that many of our survey participants claim to use their accounts to find new professional contacts and share views, ideas, and know-how, alongside a new name in their contact lists. A second reason for 56.92 % in our sample is related to the recruiting purpose of such services, which is to "find a job." A detailed look at this option reveals that together with students (18.97 %) and interns (16.41 %), the survey participants who use a social media service like LinkedIn to find a job are mostly employed for wages (50.29 %). These results suggest that LinkedIn and others are nowadays an important tool used by employed people to steer their careers and to find new job opportunities. Additionally, visibility within professional communities ("Become visible active in professional communities," 51.28 %) as the third important option supports this line of argumentation.

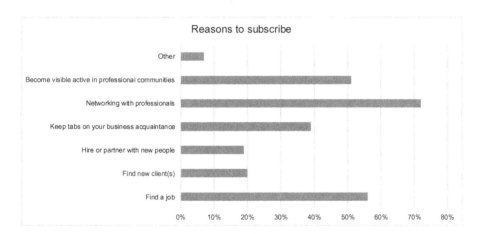

Fig. 5. Reasons to subscribe to business-oriented social media services (Source: own survey results)

These findings directly lead to the last part of this survey, which addresses the willingness to pay for premium functions in business-oriented social media services. Overall, 12.82 % (n = 25) of the 195 users of these services stated that they already paid for premium features, and many of them confirmed in an open question design that gaining access to more features and functions in order to get in contact with new business partners was the main reason. Especially LinkedIn's direct mail function called InMail, which comes along with all payment models—analyzed in the third case study—seems to be a key function users are willing to pay for. Additionally, extended search functions including new algorithms (offered by, e.g., LinkedIn and XING) are important for both job seeking users and recruiters from companies. According to our results, the majority of the current users of such premium features (88.00 %) is satisfied with them and will keep the subscription.

Our survey participants could state in a text box the amount of money they are potentially willing to pay per month in total for business-oriented social media services. The answer range is quite broad, from a minimum of $1.81 to a maximum of $113.00. A distinct pattern with regard to willingness to pay does not exist according to these findings, since the results show a pretty high standard deviation ($21.04). But the average value of $17.00 for all respondents, including potentially new members of these services, is comparable to the more frequently used entertainment-oriented social media services.

5   Conclusion

5.1   Findings and Hypotheses

With explorative case studies, we first identified three alternative business models, i.e., *virtual goods*, *e-commerce*, and *account upgrade services*. This extends the more general research by Zambonini [7] and Vukanovic [8]. We then analyzed the current user behavior on prominent entertainment- and business-oriented social media services. A clear focus was set in this second empirical part on questions directly related to willingness to pay for these extra functions or products. Results indicate that alternatives exist to challenge the cash cow advertising. From our survey and our second case study we confirmed that virtual goods are particularly used for customization and gaming purposes, e.g., on Facebook or Tencent's Q Zone. On the other hand, account upgrades are bought by users of both types of social media services analyzed in this study, but predominantly on business-oriented ones like LinkedIn. Therefore we derive the following hypotheses, which can be tested in confirmative follow-up studies:

H1: *Users of entertainment-oriented social media services tend to buy virtual goods while those of business-oriented social media services spend money on account-upgrade offers.*

We could also identify differences in the use behavior influenced by what we think can be seen as three moderators: geographical location (H2), age (H2, H4), and availability of a service in a certain language (H3). These moderators change the relationship postulated in H1:

H2: *Asian users of entertainment-oriented social media services, especially younger ones, tend to buy more virtual goods and less account upgrades compared to Europeans and North Americans.*

H3: *Availability in a native language moderates the effect of users striving for business-oriented social media services for all users. The more languages offered by a business-oriented service provider, the more users will join the service.*

H4: *Age moderates the effect of striving for social media presence for all users. The older a user grows, the more he or she tends to be present in business-oriented social media services.*

Business-oriented social media services are less present in Asia compared to Europe and North America, but this market penetration seems to have changed rapidly since the market leader LinkedIn made its services available in both the Japanese and Chinese languages. We think that this decision will boost memberships accompanied by a high potential of selling account upgrade services. With regard to our initially postulated research question, providers of entertainment-oriented social media services should concentrate on virtual goods in the gaming sector, especially with particular focus on Asian users. According to our results, they are open minded to alternative services and can set a trend, which could then be followed by European and North American users. In contrast, business-oriented service providers should utilize premium functions and account features as part of the well-established freemium concept.

5.2   Limitations and Further Research

In general, we could prove that users are willing to pay for additional features within social media services. However, this environment is absolutely fast moving with new ideas, functions, and implementations aspiring daily, like Facebook being on the brink of introducing its Buy Button after trying out several precursors. We had to set limits on this initial explorative approach, and we think that in addition to leading players like Facebook, Tencent, and LinkedIn, more and new emerging social media services like Instagram, which struggles to make money at the moment, should be examined as well.

From the empirical data perspective, we gained a suitable sample size of 301 respondents for this explorative setting. The hypotheses deduced are based on data from individuals from all around the leading social media markets in North America, Asia, and Europe. This already covers a decent percentage of general cultural clusters. In upcoming studies, however, we suggest researchers narrow the scope of their empirical design to one of the core markets in order to shed light on specific cultural characteristics. We think further research with individual analyses, especially on users' willingness to spend money for extra services in these markets, could provide new insights.

The initial part of this research paper described Facebook's radical move of introducing new advertisement rules at the beginning of this year. Rather than being a gift for its users, this leading company in the social media environment consequently took the next step on the way to optimized user-related advertising. Right now advertising is by far the most profitable financing strategy of social media services. However, it is not the only effective one, as some players like Tencent demonstrate. Advertising is not the whole story when it comes to revenues, and new ways like virtual goods or freemium concepts are on the horizon. In the end, convenient e-commerce functionalities offered directly within a social media service—like Facebook's Buy Button—could be a condign revenue competitor to advertising. But regardless of the revenue source used, tracking the activities will be of utmost importance. Social media service providers should have this in mind, especially as "managing context in light of Big Data will be an ongoing challenge." [16, p. 671].

References

1. Alexa Internet Inc.: The top 500 sites on the web. http://www.alexa.com/topsites
2. Facebook: updating our terms and policies: helping you understand how facebook works and how to control your information. https://www.facebook.com/about/terms-updates
3. Melanson, D.: Facebook reports $1.81 billion in revenue for Q2 2013, 1.15 billion monthly active users. http://www.engadget.com/2013/07/24/facebook-q2-2013-earnings/
4. Statista: Facebook's global revenue as of 1st quarter 2015, by segment (in million U.S. dollars). http://www.statista.com/statistics/277963/facebooks-quarterly-global-revenue-by-segment/
5. Moorthy, J., Lahiri, R., Biswas, N., Sanyal, D., Ranjan, J., Nanath, K., Ghosh, P.: Big data: prospects and challenges. J. Decis. Makers **40**, 74–96 (2015)
6. Clemons, E.K.: Business models for monetizing internet applications and web sites: experience, theory, and predictions. J. Manag. Inf. Syst. **26**, 15–41 (2009)
7. Zambonini, D.: Monetizing your web app: business model options. http://www.boxuk.com/blog/monetizing-your-web-app-business-models/
8. Vukanovic, Z.: New media business models in social and web media. J. Media Bus. Stud. **8**, 51–67 (2011)
9. Goes, P.B.: Big data and IS research. MIS Q. **38**, iii–viii (2014)
10. Kaplan, A.M., Haenlein, M.: Users of the world, unite! The challenges and opportunities of social media. Bus. Horiz. **53**, 59–68 (2010)
11. Chiang, I., Huang, C., Huang, C.: Characterizing web users' degree of web 2.0-ness. J. Am. Soc. Inform. Sci. Technol. **60**, 1349–1357 (2009)
12. boyd, d., Ellison, N.B.: Social network sites: definition, history, and scholarship. J. Comput.-Mediated Commun. **13**, 210–230 (2007)
13. Maier, C., Laumer, S., Weitzel, T.: Although I am stressed, I still use IT! Theorizing the decisive impact of strain and addiction of social network site users in post-acceptance theory. In: Proceedings of 34th International Conference on Information Systems, Milan, Italy (2013)
14. Wilhelm, A.: Facebook's desktop ad revenues fell $26 m, in Q3 as its mobile ad revenue surged $226 m. http://techcrunch.com/2013/10/30/facebooks-desktop-ad-revenues-fell-26m-in-q3-as-its-mobile-ad-revenue-surged-226m/
15. Gatto, K.: How facebook sells your personal information. http://news.discovery.com/tech/gear-and-gadgets/how-facebook-sells-your-personal-information-1301241.htm
16. boyd, d., Crawford, K.: Critical questions for big data. Inf. Commun. Soc. **15**, 662–679 (2012)
17. Johnston, K.: What are the advantages of mobile advertising? http://smallbusiness.chron.com/advantages-mobile-advertising-55314.html
18. Yin, R.K.: Case Study Research. Design and Methods. Sage Publications, Thousand Oaks (1994)
19. Dubé, L., Paré, G.: Rigor in information systems positivist case research: current practices, trends, and recommendations. MIS Q. **27**, 597–635 (2003)
20. Benbasat, I., Goldstein, D.K., Mead, M.: The case research strategy in studies of information systems. MIS Q. **11**, 369–386 (1987)

21. Eisenhardt, K.M.: Building theories from case study research. Acad. Manag. Rev. **14**, 532–550 (1989)
22. Mays, N., Pope, C.: Qualitative research: rigour and qualitative research. BMJ **311**, 109–112 (1995)
23. Miles, M.B., Huberman, A.M.: Qualitative Data Analysis. An Expanded Sourcebook. Sage Publications, Thousand Oaks (1994)
24. Lee, A.S.: A scientific methodology for MIS case studies. MIS Q. **13**, 33–50 (1989)
25. Sawyer, S.: Analysis by long walk: some approaches to the synthesis of multiple sources of evidence. In: Trauth, E.M. (ed.) Qualitative Research in IS. Issues and Trends. Idea Group Publishing, Hershey (2001)
26. Nisbett, R., Ross, L.: Human Inference. Strategies and Shortcomings of Social Judgment. N.J. Prentice-Hall XVI, Englewood Cliffs (1980)
27. Hair, J.F.: Multivariate Data Analysis. Prentice Hall, Upper Saddle River (2010)
28. Stopher, P.R.: Collecting, Managing, and Assessing Data using Sample Surveys. Cambridge University Press, Cambridge (2012)
29. Javelin strategy and research: virtual currency and social network payments – the new gold rush: how emerging virtual transactions will alter the payments landscape forever (2011). https://www.javelinstrategy.com/brochure/212
30. Ho, C.-H., Wu, T.-Y.: Factors affecting intent to purchase virtual goods in online games. Int. J. Electron. Bus. Manag. **10**, 204–212 (2012)
31. Animesh, A., Alain, P., Sung-Byung, Y., Wonseok, O.: An odyssey into virtual worlds: exploring the impacts of technological and spatial environments on intention to purchase virtual products. MIS Q. **35**, 789–810 (2011)
32. Guo, Y., Barnes, S.: Virtual item purchase behavior in virtual worlds: an exploratory investigation. Electron. Commer. Res. **9**, 77–96 (2009)
33. Anderson, C.: Free: the past and future of a radical price. Hyperion, New York (2009)
34. Niculescu, M.F., Wu, D.J.: When should software firms commercialize new products via freemium business models? (2011) http://rady.ucsd.edu/faculty/seminars/papers/Niculescu.pdf
35. Tencent: tencent announces 2014 third quarter and interim results. http://www.tencent.com/en-us/at/pr/2014.shtml
36. Eldon, E.: Facebook sets July, 1, 2011 deadline to make credits sole canvas game payment option. http://www.adweek.com/socialtimes/facebook-sets-July-1-2011-deadline-to-make-credits-sole-canvas-game-payment-option/255581?red=if
37. Facebook: facebook payments. https://developers.facebook.com/docs/payments/overview
38. Tencent: tencent - products and services – ecommerce. http://www.tencent.com/en-us/ps/ecommerce.shtml
39. Osterwalder, A., Pigneur, Y.: Business Model Generation: A Handbook for Visionaries, Game Changers, and Challengers. Wiley, New Jersey (2013)

Big Data Analytics as a Service for Business Intelligence

Zhaohao Sun[1][✉], Huasheng Zou[2], and Kenneth Strang[3]

[1] Department of Business Studies, PNG University of Technology, Lae 411, Papua New Guinea
zsun@dbs.unitech.ac.pg, zhaohao.sun@gmail.com
[2] College of e-Commerce, Ningbo Dahongying University, Ningbo 315175, Zhejiang, China
zoufan99@163.com
[3] School of Business and Economics, State University of New York, Plattsburgh at Queensbury,
NY 12804, USA
kenneth.strang@plattsburgh.edu

Abstract. This paper proposes an ontology of big data analytics and examines how to enhance business intelligence through big data analytics as a service by presenting a big data analytics services-oriented architecture (BASOA), and applying BASOA to business intelligence, where our surveyed data analysis showed that the proposed BASOA is viable for developing business intelligence and enterprise information systems. This paper also discusses the interrelationship between business intelligence and big data analytics. The proposed approach in this paper might facilitate the research and development of business analytics, big data analytics, and business intelligence as well as intelligent agents.

Keywords: Big data analytics · e-commerce · Business intelligence · Intelligent agents

1   Introduction

Big data and big data analytics has become one of the important research frontiers [1]. Big data and its emerging technologies including big data analytics have been not only making big changes in the way the e-commerce and e-services operate but also making traditional data analytics and business analytics bring new big opportunities for academia and enterprises [2]. Big data analytics is an emerging big data technology, and has become a mainstream market adopted broadly across industries, organizations, and geographic regions and among individuals to facilitate data-driven decision making for business and individual's hedonism [3, 4].

Business intelligence (BI) has received widespread attention in academia, e-commerce, and business over the past two decades [5]. BI has become not only an important technology for improving business performance of enterprises but also an impetus for developing e-commerce and e-services [15]. However, BI is facing new challenges and opportunities because of dramatic development of big data and big data technologies [6, 7]; that is, how to use big data analytics to enhance BI becomes a big issue for business, e-commerce, e-services, and information systems.

© IFIP International Federation for Information Processing 2015
M. Janssen et al. (Eds.): I3E 2015, LNCS 9373, pp. 200–211, 2015.
DOI: 10.1007/978-3-319-25013-7_16

Big data analytics and BI are the top priorities of chief information officers (CIOs) and comprise a $12.2 billion market [8]. According to a study of Gartner, worldwide BI and analytics software, consisting of BI platforms, analytic applications and advanced analytics, totalled $14.4 billion in 2013, an 8 % increase from 2012 revenue [9]. This fact attracts unprecedented interest and adoption of big data analytics. According to the annual survey results of 850 CEO and other C-level executives of global organisations, McKinsey [2] concludes that 45 % of executives put "big data and advanced analytics" as the first three strategic priorities in both strategy and spending in three years' time and more than one thirds of executives will now spend or in three years' time in this area. IDC (International Data Corporation) predicts that the business analytics software market will grow at a 9.7 % compound annual growth rate over the next five years from 2012 to 2017 [4].

The above brief discussion and literature review implies that there is a close relationship between big data analytics and BI. However, the following two important issues have not been drawn significant attention in the scholarly peer-reviewed literature:

- What is the relationship between big data analytics and BI?
- How can big data analytics enhance BI?

This paper will address these two issues through extending our early research on analytics service oriented architecture [2]. To address the first issue, we propose an ontology of big data analytics in Sect. 2 through overviewing our early work on data analytics and big data analytics. To address the second issue, we examine big data analytics as a technology for supporting BI through examining the relationship between big data analytics and BI in Sect. 3. We then present a big data analytics service oriented architecture (BASOA), in which we also explore how to apply big data analytics as a service to enhance BI, where we show that the proposed BASOA is viable for developing BI based on our surveyed data analysis.

The remainder of this paper is organized as follows. Section 2 looks at the fundamentals of big data analytics by proposing an ontology of big data analytics and discussing the relationships of big data analytics and data analytics. Section 3 discusses BI and its relationships with big data analytics. Section 4 presents BASOA. Section 5 applies proposed BASOA to BI. The final sections discuss the related work and end this paper with some concluding remarks and future work.

2   Fundamentals of Big Data Analytics

This section proposes an ontology of big data analytics and looks at the interrelationship between big data analytics and data analytics. To begin with, this section first examines the fundamentals of big data analytics.

Big data analytics is an integrated form of data analytics and web analytics for big data [2]. According to [7, 10], big data analytics can be defined as the process of collecting, organizing and analyzing big data to discover patterns, knowledge, and intelligence as well as other information within the big data. Similarly, big data analytics can be defined as techniques used to analyze and acquire knowledge and intelligence from big data [7]. Big data analytics is an emerging science and technology involving the multidisciplinary state-of-art information and communication technology (ICT), mathematics, operations research (OR), machine learning (ML), and decision sciences for big data [1, 2]. The main components of big data analytics include big data descriptive analytics, big data predictive analytics and big data prescriptive analytics [11].

- Big data descriptive analytics is descriptive analytics for big data [12], and is used to discover and explain the characteristics of entities and relationships among entities within the existing big data [13, p. 611]. It addresses the problems such as what happened, and when, as well as what is happening. For example, web analytics for pay-per-click or email marketing data belongs to big data descriptive analytics [14].
- Big data predicative analytics is predicative analytics for big data, which focuses on forecasting trends by addressing the problems such as what will happen, what is likely to happen and why it will happen [12, 15]. Big data predicative analytics is used to create models to predict future outcomes or events based on the existing big data [13, p. 611]. For example, big data predicative analytics can be used to predict where might be the next attack target of terrorists.
- Big data prescriptive analytics is prescriptive analytics for big data, which addresses the problems such as what we should do, why we should do it and what should happen with the best outcome under uncertainty [11, p. 5]. For example, big data prescriptive analytics can be used to provide an optimal marketing strategy for an e-commerce company.

An ontology is a formal naming and definition of a number of concepts and their interrelationships that really or fundamentally exist for a particular domain of discourse [16]. Then, an ontology of big data analytics is an investigation into a number of concepts and their interrelationships that fundamentally exist for big data analytics. Based on the above discussion, we propose an ontology of big data analytics, as illustrated in Fig. 1. In this ontology, big data analytics is at the top while big data and data analytics are at the bottom. Big data descriptive analytics, big data predictive analytics, and big data prescriptive analytics are at the middle level as the core parts of any big data analytics.

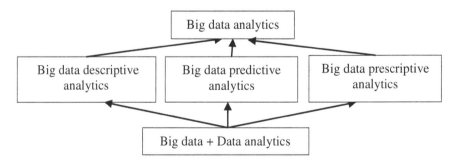

Fig. 1. An ontology of big data analytics

It should be noted that the above-proposed ontology of big data analytics is still simple. We will extend it by adding another level between the second level and the third level in Fig. 1. This added level will elaborate big data descriptive, predictive and prescriptive analytics taking into account the corresponding real-world examples, methods and techniques.

In Fig. 1, data analytics refers to as a method or technique that uses data, information, and knowledge to learn, describe and predict something [15, p. 341]. In brief, data analytics can be then considered as data-driven discoveries of knowledge, intelligence and communications [12]. More generally, data analytics is a science and technology about examining, summarizing, and drawing conclusions from data to learn, describe and predict something [2].

The fundamentals of big data analytics consists of mathematics, statistics, engineering, human interface, computer science and information technology [1, 2]. The techniques for big data analytics encompass a wide range of mathematical, statistical, and modeling techniques [13, p. 590]. Big data analytics always involves historical or current data and visualization [17]. This requires big data analytics to use data mining (DM) to discover knowledge from a data warehouse (DW) or a big dataset in order to support decision making, in particular in the text of big business and management [15, p. 344]. DM employs advanced statistical tools to analyze the big data available through DWs and other sources to identify possible relationships, patterns and anomalies and discover information or knowledge for rational decision making [13, p. 590]. DW extracts or obtains its data from operational databases as well as from external open sources, providing a more comprehensive data pool including historical or current data [13, p. 590]. Big data analytics also uses statistical modelling (SM) to learn something that can support decision making [2]. Visualization techniques as an important part of big data analytics make knowledge patterns and information for decision making in a form of figure or table or multimedia. In summary, big data analytics can facilitate business decision making and realization of business objectives through analyzing current problems and future trends, creating predictive models to forecast future threats and opportunities, and analyzing/optimizing business processes based on involved historical or current data to enhance organizational performance using the mentioned techniques [12]. Therefore, big data analytics can be represented below.

$$
\text{Big data analytics} = \text{Big data} + \text{data analytics} + \text{DW} + \text{DM} + \text{SM}
$$
$$
+ \text{ML} + \text{Visualization} + \text{optimization} \tag{1}
$$

Where + can be explained as "and". This representation reveals the fundamental relationship between big data, data analytics and big data analytics, that is, big data analytics is based on big data and data analytics, as illustrated in Fig. 1. It also shows that computer science and information technology play a dominant role in the development of big data analytics through providing sophisticated techniques and tools of DM, DW, ML and visualization [2]. SM and optimization still plays a fundamental role in the development of big data analytics, in particular in big data prescriptive analytics [11].

It should be noted that the above equation is a concise representation for the technological components of big data analytics whereas the proposed ontology of big data analytics in this Section is to look at what big data analytics constitutes at a relatively high level. We will consider the big data descriptive, predictive and prescriptive analytics as one dimension, and the technological components of big data analytics as another dimension. Then we will provide this 2-dimension analysis as a future research work.

3   Business Intelligence and Big Data Analytics

This section will examine business intelligence (BI) and its relationships with big data analytics.

BI has drawn increasing attention in academia and business over the past two decades, although the term was already coined in 1958 by an IBM scientist [18]. There are many different definitions on BI. For example,

- BI is defined as providing decision makers with valuable information and knowledge by leveraging a variety of sources of data as well as structured and unstructured information [19].
- BI refers to as a collection of information systems (IS) and technologies that support managerial decision makers of operational control by providing information on internal and external operations [15].
- BI is a framework consisting of a set of concepts, theories, and methods to improve business decision making by using fact-based support systems [5].

The first definition of BI emphasizes information and knowledge for decision makers. The second definition stresses "a collection of ISs and technologies" while specifies the decision makers to "managerial decision makers of operational control", and information to "information on internal and external operations". The last definition emphasizes "a set of concepts, theories, and methods to improve business decision making". Based on the above analysis, BI can be defined as a set of theories, methodologies, architectures, systems and technologies that support business decision making with valuable data, information and knowledge. This definition reflects the evolution of BI and its technologies from decision support systems (DSS) and its relations with data warehouses, executive information systems [8].

The principal tools for BI include software for database query and reporting (e.g. SAP ERP, Oracle ERP, etc.), tools for multidimensional data analysis (e.g. OLAP), and data mining (e.g. predictive analysis, text mining, web mining) [20]. Data warehousing is also considered as a foundation of BI [5].

Based on the previous subsection's discussion, big data analytics can be considered a part of BI [5], because it "supports business decision making with valuable data, information and knowledge" [2]. Both BI and big data analytics are common in emphasizing either valuable data or information or knowledge. BI involves interactive visualization for data exploration and discovery, for them Tableau, QlikView and Tibco's Spotfire are BI tools for interactive visualization for data exploration and discovery [21]. These BI tools are also considered as the tools of big data analytics. This implies that BI and big data analytics share some common tools to support business decision making.

Currently, BI is based on four cutting-age technology pillars of cloud, mobile, big data and social technologies [17, 22], each of these pillars corresponds to a special kind of web services, that is, cloud services, mobile services, big data services and social networking services; all these constitute modern web services [17]. Each of these services has been supported by analytics services and technologies [2]. They are effectively supported also by big data analytics as a service and technology, as shown in Fig. 2.

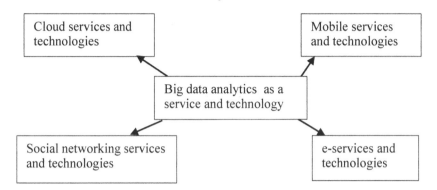

Fig. 2. Interrelationship between big data analytics and web services [2]

It should be noted that for the state-of-art web services, Sun and Yearwood [17] explores that web services mainly consist of mobile services, analytics services, cloud services, social networking services, and service as a web service. In reality, each of them involves sophisticated ICT technologies. Then, technologies are added to mobile services, analytics services, cloud services and social networking services in [2]. Here we emphasize big data analytics as a service and technology at the center to support cloud services and technologies, social networking services and technologies, mobile services and technologies, e-services and technologies to reflect the big data and analytics as an emerging new service and technology. The readers can easily find practical examples to reflect this trend. We do not go into it anymore because of limitation of space.

Based on IDC's prediction for the IT market in 2014 [22], spending on big data will explode and grow by 30 %, to $14 + billion, in which, big data analytics services will experience an explosive growth. The spending on big data analytics services will exceed $4.5 billion, growing 21 %. The number of providers of big data analytics services will triple in three years. This means that big data analytics as a service and technology has become an important emerging market, together with cloud services, mobile services and social networking services [2]. All these four services and the technologies shape the most important markets for e-commerce and e-services [17].

Furthermore, BI is a more general concept for improving business performance and business decision making. Big data analytics is a pivotal part for developing BI, at least from a technological viewpoint and data viewpoint. From a technological viewpoint, big data analytics is data-driven and business oriented technique and facilitates business decision making and then improves BI [2]. From a data viewpoint, big data analytics relies on data analytics and big data which have become a strategic natural resource for every organization, in particular for multinational organizations as well as for e-commerce and e-services. Discovering secrets from databases, data warehouses, data marts and the Web has become the central topics for business operations, marketing and BI. This is just the task of big data analytics.

4   BASOA: Big Data Analytics Services Oriented Architecture

This section proposes a big data analytics service oriented architecture (BASOA) and then examines each of the main players in the BASOA. Different from the traditional SOA [23], the proposed BASOA specifies general services to big data analytics services, as showing in Fig. 3. We use BA in this architecture, BASOA, to represent big data analytics, which implies that big analytics (BA) can represent big data analytics briefly. This is reasonable because we have big data and big analytics, both are originally from data and analytics respectively.

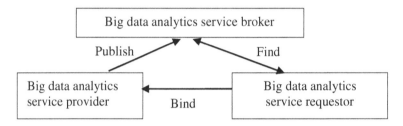

Fig. 3. BASOA: a big data analytics SOA

In this BASOA, big data analytics service provider, big data analytics service requestor, big data analytics service broker are three main players. In what follows, we will look at each of these in some detail, taking into account BI.

Big data analytics service requestors include organizations, governments and all level business decision makers such as CEO, CIO and CFO as well as managers. Big data analytics service requestors also include business information systems and e-commerce systems. Big data analytics service requestors require big data analytics services including information analytics services, knowledge analytics services, business analytics services with visualization techniques to provide knowledge patterns and information for decision making in a form of figure or table or report [24]. More generally, big data analytics service requestors include people who like to make decisions or acquire information based on analytical reports provided by big data analytics service provider [2]. Therefore, a person with smartphone receiving analytics services is also a big data analytics service requestor [12].

Big data analytics service brokers are all the entities that facilitate the development of big data analytics services, which include popular presses, traditional media and social media, consulting companies, scholars and university students, and so on [2]. All these use a variety of methods and techniques to improve the better understanding of big data analytics services in general and data analytics, business analytics, web analytics, and their services in particular [2]; all these have been offered to university students as a course material or content in business and computing areas to some extent in recent years. McKinsey Consulting (http://www.mckinsey.com/), Boston Consulting Group

(BCG), and IDC as big data analytics service brokers have played an important role in pushing big data analytics in businesses and enterprises, just as they promote "big data". Gartner and Forrester are also famous big data analytics service brokers in the world [12].

Big data analytics service providers include analytics developers, analytics vendors, analytics systems or software and other intermediaries that can provide analytics services. Recently, web analytics service (WAS) providers are important big data analytics service providers. A WAS provider, for example, Adobe Marketing Cloud (http://www.adobe.com/au/solutions/digital-marketing.html), aggregates and analyses blog data about the online behaviors of users who visited the client's website, then they evaluate a variety of analytical reports concerning the client's customer online behaviors that the client wishes to understand. This can then facilitate their strategic business decision making [25]. Application service providers (ASPs) can also provide web analytics in a hosted ASP model with quicker implementation and lower administrative costs [25]. Analytics developers provide analytic tools with extensive data extraction, analytics and reporting functionality such as Piwik, CrawlTrack [26]. Google is not only a search engine provider, but also a WAS vendor, because Google provides Google Analytics (http://www.google.com/analytics/), a big data analytics, with good tracking tools. In fact, most hosting websites, like Baidu, also provide these similar big data analytics services. A mobile phone company can provide big data analytics services to the customers with smartphone [12]. For example, Mobile App Analytics (http://www.google.com/analytics/mobile/), a part of Google Analytics, is also a mobile big data analytics services provider that helps the smartphone customers to discover new and relevant users through traffic sources reports. Mobile App Analytics plays a role of integration and gets engaged through event tracking and flow visualization, and sets and tracks the goal conversions one wants most: purchases, clicks, or simply time spent on the app. More generally, many information systems have contained an analytics app as a system component to generate table, diagram or report. All these kinds of information systems can be considered as big data analytics service providers. The big data analytics services providers on the Web include Amazon, Google and Microsoft [2].

5   Applying BASOA to BI

This section looks at how to apply the proposed BASOA to enhance BI in some detail.

Analytics as a service (AaaS) is a relatively new concept that has emerged as a rapidly growing business sector of web analytics industry, which provides efficient web log analytic services for firm-level customers [17]. BAaaS (Big data analytics as a service), as discussed in the BASOA above, means that an individual or organization or information system uses a wide range of analytic tools or apps wherever they may be located [12]. BAaaS has the ability to turn a general analytic platform into a shared utility for an enterprise with visualized analytic services [12]. An analytics service can be available on the Web or used by smartphone. Therefore, big data analytics services include e-analytics services or web analytics services (WAS) [2]. BAaaS is gaining popularity rapidly in business, e-commerce, e-service, and management in recent years. For example, BAaaS model has been adopted by many famous web companies such as

Amazon, Microsoft, and eBay [12]. The key reason behind it is that the traditional hub-and-spoke architectures cannot satisfy the demands driven by increasingly complex business analytics [12]. BAaaS promises to provide decision makers with visualizing much needed big data. Cloud analytics is an emerging alternative solution for big data analytics [2].

As previously defined, BI is "a set of theories, methodologies, architectures, systems and technologies that support business decision making with valuable data, information and knowledge". BASOA is an architecture for supporting business decision making with big data analytics services. The theory of big data analytics providers, brokers and requestors of the BASOA can facilitate the understanding and development of BI and business decision making. For example, from a deep analysis of the BASOA, an enterprise and its CEO can know who are the best big data analytics providers and brokers in order to improve his business, market performance, and competition.

We surveyed 71 information technology managers at the Association for Education in Journalism and Mass Communication (AEJMC) in Montreal during August 6–9, 2014 [2], to collect data concerning the enterprise-level acceptability of the BASOA concept. These results indicate some preliminary support for the BASOA concept of having service brokers work with service requesters and providers similar to the way private mortgage and loans work in the USA. Based on this preliminary enterprise acceptability of this BASOA model, we propose that more research be done to investigate how it could be used.

6   Related Work and Discussion

We have mentioned a number of scholarly researches on data analytics, big data analytics, and BI. In what follows, we will focus on related work and discussion on ontology of big data analytics, and the work of SAP as well as incorporation of big data analytics into BI.

Why does big data analytics really matter for modern business organizations? There are many different answers to this question from different researchers. For example, Davis considers that the current big data analytics has embodied the state of art current development of modern computing [27], which has been reflected in Sect. 2. Gandomi and Harder [7] discuss how big data analytics has captured the attention of business and government leaders through decomposing big data analytics into text analytics, audio analytics, video analytics, social media analytics, and predictive analytics. This implies that data has been classified into text data, audio data, video data, and social media data in [7].

Big data analytics and BI have drawn an increasing attention in the computing, business, and e-commerce community. For example, Lim et al. [5] examine business intelligence and analytics by focusing on its research directions. They consider business intelligence and analytics (BIA) as a current form replacing the traditional BI, whereas we still consider BI and big data analytics are two different concepts, although they have close relationships and share some commons. Fan et al. [6] provide a marketing mix framework for big data management through identifying the big data sources, methods, and applications for each of the marketing mix, consisting of people, product, place, price and promotion. However, what is the relationship between marketing intelligence and BI in terms of big data analytics should have been mentioned in their work [6].

Ontology has been important in computer science and artificial intelligence [16]. A basic search in Google scholar (i.e. article title and key words) reveals that there are few publications entitled "ontology of big data analytics". We then explored it and put it as a part of this research through updating our early work on data analytics, business analytics and big data analytics [2]. We explore the interrelationship among big data analytics, big data descriptive analytics, big data predictive analytics, and big data prescriptive analytics using the proposed ontology. The results reported in this paper on ontology of big data analytics and big data analytics equation are an extension and development of our early work [2] by adding optimization and ML to the equation. This is only a beginning for providing a relatively comprehensive ontology of big data analytics. In this direction, we will investigate more academic reviewed sources as a future work to develop an ontology of big data analytics with three levels for each related analytics: big data, methods and applications based on the method of Fan et al. [6]. Such an investigation would become an important guide for the research and development of big data analytics and BI.

SAP, one of the leading vendors of ERP [28], has introduced its enterprise service-oriented architecture [20, p. 383]. SAP's architecture specifies general services to enterprise services whereas our BASOA model specifies general services to big data analytics services. Big data analytics services should be a part of state-of-the-art e-commerce services [17], and then the proposed BASOA can be considered as a concrete application for the enterprise service-oriented architecture of SAP. However, SAP's enterprise systems focus on key applications in finance, logistics, procurement and human resources management as an ERP system. We conceive that our BASOA will be incorporated into the next generation enterprise systems integrating SCM, CRM, and KM systems, and e-commerce systems. This is also the motivation of our proposed BASOA.

7   Conclusion

This paper proposed an ontology of big data analytics, and looked at the relationship between big data analytics and BI. This paper also presented a big data analytics service oriented architecture (BASOA) and discussed how to use BASOA to enhance BI. The preliminary analysis on the collected data shows that this proposed BASOA is viable for facilitating the development of BI. The proposed approach in this paper might facilitate research and development of big data analytics, business analytics, BI, e-commerce, and e-services.

In the future work, besides mentioned in the previous sections, we will analyse the foregoing collected data vigorously and explore enterprise and e-commerce acceptability of BASOA for BI. We will also explore big data analytics and its applications in e-commerce and cloud services, and realize BASOA using intelligent agents technology [29], where we will also look at some implementation related issues such as how to collect, store, and process big data – by whom, for what, access rights, and many more.

References

1. Chen, C.P., Zhang, C.-Y.: Data-intensive applications, challenges, techniques and technologies: a survey on big data. Inf. Sci. **275**, 314–347 (2014)
2. Sun, Z., Strang, K., Yearwood, J.: Analytics service oriented architecture for enterprise information systems. CONFENIS 2014, Hanoi, 4–6 December 2014. In: Proceedings of iiWAS2014, pp. 506-518. ACM Press (2014)
3. Sun, Z., Firmin, S., Yearwood, J.: Integrating online social networking with e-commerce based on CBR. In: Proceedings of the 23rd ACIS 2012, Geelong, 3–5 December 2012
4. Vesset, D., McDonough, B., Schubmehl, D., Wardley, M.: Worldwide Business Analytics Software 2013–2017 Forecast and 2012 Vendor Shares (Doc # 241689), 6 2013. [Online]. http://www.idc.com/getdoc.jsp?containerId=241689. Accessed 28 June 2014
5. Lim, E., Chen, H., Chen, G.: Business intelligence and analytics: research directions. ACM Trans. Manage. Inf. Syst., vol. 3, no. 4, Article 17, pp. 1–10 (2013)
6. Fan, S., Lau, R.Y., Zhao, J.L.: Demystifying big data analytics for business intelligence through the lens of marketing mix. Big Data Res. **2**, 28–32 (2015)
7. Gandomi, A., Haider, M.: Beyond the hype: big data concepts, methods, and analytics. Int. J. Inf. Manag. **35**, 137–144 (2015)
8. Holsapplea, C., Lee-Postb, A., Pakath, R.: A unified foundation for business analytics. Decis. Support Syst. **64**, 130–141 (2014)
9. van der Meulen, R., Rivera, J.: Gartner Says Worldwide Business Intelligence and Analytics Software Market Grew 8 Percent in 2013, 29 4 2014. http://www.gartner.com/newsroom/id/2723717. Accessed 28 June 2014
10. Beal, V.: Big data analytics 2014. http://www.webopedia.com/TERM/B/big_data_analytics.html. Accessed 20 June 2014
11. Minelli, M., Chambers, M., Dhiraj, A.: Big Data, Big Analytics: Emerging Business Intelligence and Analytic Trends for Today's Businesses, Chinese Edition 2014 edn. Wiley & Sons, Hoboken (2013)
12. Delena, D., Demirkanb, H.: Data, information and analytics as services. Decis. Support Syst. **55**(1), 359–363 (2013)
13. Coronel, C., Morris, S., Rob, P.: Database Systems: Design, Implementation, and Management, 11th edn. Course Technology, Cengage Learning, Boston (2015)
14. Cramer, C.: How Descriptive Analytics Are Changing Marketing, 19 May 2014. http://www.miprofs.com/wp/descriptive-analytics-changing-marketing/. Accessed 6 July 2015
15. Turban, E., Volonino, L.: Information Technology for Management: Improving Strategic and Operational Performance, 8th edn. Wiley, Danvers (2011)
16. Gruber, T.: Toward principles for the design of ontologies used for knowledge sharing. Int. J. Hum.-Comput. Stud. **43**(5–6), 907–928 (1995)
17. Sun, Z., Yearwood, J.: A theoretical foundation of demand-driven web services. In: Demand-Driven Web Services: Theory, Technologies, and Applications, IGI-Global 2014, pp. 1–25 (2014)
18. Luhn, H.P.: A Business Intelligence System. IBM J. Res. Dev. **2**(4), 314–319 (1958)
19. Sabherwal, R., Becerra-Fernandez, I.: Business Intelligence: Practices, Technologies, and Management. John Wiley & Sons Inc, Hoboken (2011)
20. Laudon, K.G., Laudon, K.C.: Management Information Systems: Managing the Digital Firm, 12th edn. Pearson, Harlow (2012)
21. Brust, A.: Gartner releases 2013 BI Magic Quadrant 2013. http://www.zdnet.com/gartner-releases-2013-bi-magic-quadrant-7000011264/. Accessed 14 February 2014

22. IDC: IDC Predictions 2014: Battles for Dominance — and Survival — on the 3rd Platform, December 2013. http://www.idc.com/getdoc.jsp?containerId=244606. Accessed 13 February 2014
23. Papazoglou, M.P.: Web Services: Principles and Technology. Pearson Prentice Hall, Harlow (2008)
24. Kauffman, R.J., Srivastava, J., Vayghan, J.: Business and data analytics: new innovations for the management of e-commerce. Electron. Commer. Res. Appl. **11**, 85–88 (2012)
25. Park, J., Kim, J., Koh, J.: Determinants of continuous usage intention in web analytics services. Electron. Commer. Res. Appl. **9**(1), 61–72 (2010)
26. Laudon, K., Laudon, J.: Management Information Systems-Managing the Dgital Firm. Person, Boston (2012)
27. Davis, C.K.: Viewpoint beyond data and analytics- why business analytics and big data really matter for modern business organizations. CACM **57**(8), 39–41 (2014)
28. Elragal, A.: ERP and Big Data: The Inept Couple. Procedia Technol. **16**, 242–249 (2014)
29. Sun, Z., Finnie, G.: Intelligent Techniques in E-Commerce: A Case-Based Reasoning Perspective, p. 2010. Springer, Heidelberg (2004)

Linked Relations Architecture for Production and Consumption of Linksets in Open Government Data

Petar Milić[(✉)], Nataša Veljković, and Leonid Stoimenov

Faculty of Electronic Engineering, University of Niš,
Aleksandra Medvedeva 14, Niš, Serbia
milicpetar86@gmail.com,
{natasa.veljkovic,leonid.stoimenov}@elfak.ni.ac.rs

Abstract. Linking open data in government domain, can lead to creation of new services and information as well as discovery of new ways to perform queries and get results in accessible, machine processable and structured manner. To reach the full potential of open government data more relations between data should be discovered. The interconnection of open government data and semantic description of their relations can bring new aspect of producing and consuming the data. In this paper we investigate issues for producing and utilizing open government data with special focus on dataset relations. We have proposed the Linked Relations (LIRE) architecture for relations creation between datasets and a basic RDF model of relation between two datasets. The architecture contains different modules that perform analysis of datasets attributes and suggest the type of relation between the datasets. It can be utilized by open data portals for creating relations between datasets belonging to different public agencies and government sectors. An idea presented in this paper is made available as CKAN plugin.

Keywords: Open government data · Linked data · Linkset · Dataset relations · CKAN open data portal

1 Introduction

Publication of open government data (OGD) leads to more openness, transparency and efficiency of public administration. It also brings benefits for citizens by influencing development of government services for society, hence producing better public service outcomes. Open data philosophy suggests that data should be published in open formats and in ways that make them accessible, readily, reusable and available to the public, business and government sector [1]. Following this approach, data can be easily consumed by both web developers and common users. Having in mind that most of the published data comes in original (raw) format, beneficiary contribution is not negligible, we can even say that it is immense. Every reuse of data adds new value and creates new knowledge enabling data lifecycle to expand and evolve.

Web of Data represents decentralized and heterogeneous sources of information interlinked through typed links. This is achieved by publishing structured data in

© IFIP International Federation for Information Processing 2015
M. Janssen et al. (Eds.): I3E 2015, LNCS 9373, pp. 212–222, 2015.
DOI: 10.1007/978-3-319-25013-7_17

Resource Description Format (RDF) using URIs. RDF is the format on which is based Semantic Web, as its use of URIs allows data to be identified by reference and linked with other relevant data by subject, predicate or object. Making OGD available in the Web of data, makes them publicly available, and practically expand Web of data space, allowing their discovery and usage. Kalampokis et al. [2] claim that real value of OGD is revealed with linking data which provides unexpected and unexplored insights into different domains and problem areas. Linked Government Data (LGD) are actually OGD that runs on Semantic Web's kerosene – metadata. Metadata provide documentation, context and necessary background information.

According to Sheridan and Tennison [3] the Semantic Web standards are mature and powerful, but there is still a lack of practical approaches and patterns for publishing OGD. Nevertheless, over the last few years there is increasing number of governments that are publishing their OGD data as linked data. The adoption of the LGD has led to the extension of the government open data space, connecting data from diverse domains such as economy, finance, medicine, statistics and others to enable new types of applications. With LGD paradigm users can browse one data source and then navigate along links to related data sources. Most promising implementation of LGD is based on Semantic Web philosophy and technologies. Tim Berners-Lee [4] noted this and he gave instructions how to link data and to include government data into the Web of Data.

By following the Open Government movement [5] in the world, many governments have published their open data through the open data platforms [6]. One of the most utilized open-source solutions for publishing open datasets is the CKAN's (Comprehensive Knowledge Archive Network) open data platform [7]. This platform enables both back-end and front-end interface, used respectively for publishing/modifying and searching/reviewing open datasets. What this platform doesn't offer is the possibility to link datasets between each-other, creation of meaningful relations between them and publishing this data as linked data. This tackled our minds and we wanted to generalize this problem and create a common architecture that could be applied to other open data platforms as well and to suggest possible relations between datasets and enable their linking.

In this paper we explore area of dataset relations for producing and utilizing linked government data. We describe architecture that determines type of relation between datasets, in order to take advantage of the relations for dataset linking. The architecture contributes to defining a model of semantic representation of dataset relations and their automatic production. This leads towards production of quality LGD in line with their interlinking and integration. The remaining of the paper is organized as follows. In Sect. 2 we review related work. In Sect. 3 we propose an architecture for modelling and linking dataset relations. Section 4 presents the visual tool for creating relations between datasets, developed on the basis of architecture described in Sect. 3 that shows benefits of its use. Finally, in Sect. 5 conclusions are drawn along with the future work.

2 Related Work

A group of authors [8] proposed architecture for integrating Public Sector Information (PSI) catalogs via the activities and components essential for discovery, allowing the presentation of catalogs in standardized form, facilitating search and retrieval across resources. This architecture requires downloading and transforming catalogs with retrievable records into a common schema language format along with addressing semantic heterogeneity with schema matching and statistical analysis of ontology structures. Pioneers in LGD, UK and USA have shown that creating high quality Linked Data from raw data files requires considerable investment into reverse-engineering, documenting data elements, data cleanup, schema mapping and instance matching [9].

If we want that government datasets become linked government datasets, semantics must be added to them. Appropriate rules tell us how to describe and how to establish links between them. For linking datasets, Alexander et al. [10] propose Vocabulary of Interlinked Datasets (VOID), a vocabulary that allows to formally describe linked RDF datasets. It defines classic LOD and 3rd-party case. In the LOD case the linkset is a subset of one of the two involved datasets, while in 3rd-party case a third dataset is involved that actually contains the linkset.

Interoperability between government datasets and bringing them closer to the Web of Data is also discussed in [11]. The authors designed DCAT vocabulary, based on exploration of seven existing open data portals to allow expression of datasets in the RDF data model. They conduct feasibility study to prove their claim that different catalogues can be rendered in proposed vocabulary.

Many authors in literature deals with proposing systems for production and consumption of LGD. Ding et al. [12] suggest the use of LGD ecosystem for LGD data production and consumption as a Linked Data – based system where users manage and consume open government data in connection with online tools, services and societies. It supports large-scale LGD production, promote LGD consumption and grow of the LGD community. TWC LGD ecosystem from [12] is based on converting raw OGD datasets into linked data and their integration with other resources. Kalampokis et al. [13] give classification scheme for OGD, where they showed technological and organizational approach for provision of linked data based on relevant literature. The proposed architecture links decentralized data with maintaining a list of available resources in the area and assigning a URI to each of them. This solution is intended to use in single government cases to link data in different datasets belonging to different public agencies and government sectors.

Schmachtenberg et al. [14] present an overview of the linkage relationships between datasets in the form of an updated LOD cloud diagram based on data that can actually be retrieved by a Linked Data crawler. They consider that two datasets can be linked if there exists at least one RDF link between resources belonging to the datasets. Using metadata with appropriate metadata architecture can bring benefits for LGD publication and use, along with improving the ability for finding and interpreting of LGD data, creating order within datasets, comparing, correct interpretation, accessibility, visualization and other benefits, as discussed in [15].

Janssen, Estevez and Janowski state in [16] that successfully linking of datasets requires understanding the data's context sensitive meaning. They claim that collected data from organizations which do not always anticipate its full potential use, might not sufficiently align with other datasets or possible relationships among datasets are unknown.

To the best of our knowledge, there are not work that deals with automatic production of linked datasets. Some of them [8, 9, 12] requires intervention of users which is a time-consuming process and does not specifies rules for their interlinking. Kalampokis [13] and Schmachtenberg [14] discuss on linking datasets based on their semantic description without going deeper in the OGD datasets. This work does not tackle informations hidden in published OGD datasets, which by our opinion can help in linking OGD datasets. Zuiderwijk et al. [15] claims that metadata can give potential in realization of all benefits from linked data that exists in literature. Similarly, Janssen et al. [16] points to the fact that the relations between datasets are unknown and that they can contribute to better linking of datasets.

Based on the approaches for linking OGD datasets mentioned in previous paragraphs, we got the idea to explore relations between datasets to check whether they can be used to produce linked datasets. According to that, we have developed an architecture for visual creation of relations between datasets, and their automatic linking. The so called LIRE (Linked Relations) architecture was created at first to enrich CKAN open data portal with a tool that enables automatic creation and deletion of linksets. In the following section we will explain more thoroughly the proposed architecture.

3   Linked Relations Architecture – LIRE

LIRE system architecture, outlined in Fig. 1, consists of different interconnected components, each with specially assigned tasks. Implemented functionalities are available through a single workbench. LIRE enables users to find, manage, integrate, publish and reuse relations between datasets. It promotes production and consumption of linksets, semantic data that describe relations between datasets.

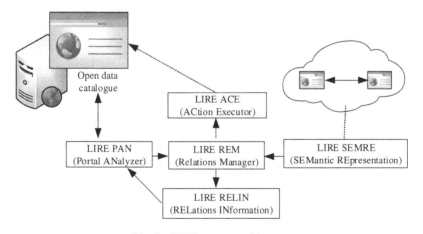

Fig. 1. LIRE system architecture

LIRE consists of the following modules:

- Portal Analyzer – PAN. This module is the entry point of LIRE. PAN enables filtering of datasets on different parameters (by tag, organization, group, all datasets or random datasets). It prepares datasets for processing for REM module and gives necessary information about datasets to RELIN module for their visual display.
- Relations Manager – REM. Creating, managing and determining type of relations between datasets based on their metadata are carried out in REM. This module implements a model on which is based determination of the type of relation, described in detail in the next section. REM examines datasets metadata to determine similarity between them for possible relation creation. Module does not limit the user in choosing the type of relations, so that the user does not have to apply the type of relation that is suggested by REM, but can apply selected, if it differs from the proposed one. It calls ACE module to execute pending actions to the OGD portal and RELIN module to refresh display after editing.
- Action Executor ACE – This module executes actions created by REM to store results of editing in the OGD platform. Supported actions are "CREATE" and "DELETE".
- Relations Information – RELIN is module used for visualization of relations. It enables short preview of information of datasets in REM module to enable user to decide whether to relate datasets or not, and user interface for managing datasets relations. Also incorporates jQuery and CSS libraries for visualizing datasets relations and their graphical management. Every dataset is represented with graphical element that contains information on dataset's description, tags, formats and existing relations.
- Semantic Representation – SEMRE module creates semantic representation of any existing relations. This semantics is created based on the model of RDF graph, described in detail in Sect. 3.2. Implemented RDF graph model is based on voID (vocabulary of Interlinked Datasets) vocabulary, because of its simplicity for describing linked datasets.

3.1 Creating Relations with LIRE

Relations Manager deals with managing of relations between datasets. It examines data that describe datasets to determine type of relation. After examination, determined type is suggested to user who can accept suggested solution or choose another one. Using developed relation suggestion models, described in Table 1, that are based on presence/absence of selected datasets metadata, REM module of LIRE architecture can determine whether datasets can have one of the following relations: parent_of, child_of, links_from, links_to. The child_of relation model consists of thirteen conditions, listed as C1-C13, where each condition examines certain dataset property or combination of properties on a true/false basis. If all conditions are met then relation between two datasets is of type child_of. The conditions for child_of relation can be used also for determining whether the relation between two dataset is of type parent_of, but with following modifications: conditions C4, C5, C7 and C8 should be less than, while C10-C12 should be greater than.

Table 1. Models for relations child_of and links_from

CHILD_OF		LINKS_FROM	
C1. Number of same/similar tags between two datasets	>0	C1. Number of same/similar tags between two datasets	>0
C2. Do they belong to the same organization	true	C2. Whether they are open	true
C3. Do they belong to the same group	true	C3. Whether the number of the same/similar resource formats of the first dataset is greater than the number of the same/similar resource formats in the second dataset	>
C4. Whether the number of the same/similar tags of the first dataset is greater than the number of the same/similar tags in the second dataset organization	>	C4. Whether the five star index of the both datasets is higher than 3	>3
C5. Whether the number of the same/similar tags of the first dataset is greater than the number of the same/similar tags in the second dataset group	>	C5. Whether they have at least one linked format in its resources	true
C6. Are they linked via links in extra field	true	C6. Whether they have at least one machine processable format	true
C7. Whether the number of the same/similar resource formats of the first dataset is greater than the number of the same/similar resource formats in the second dataset	>	C7. Whether the first dataset was created before the second	<
C8. Whether the first dataset was created after the second	>	C8. Whether the descriptions of two datasets are similar	>n
C9. Whether the descriptions of two datasets are similar	>n		
C10. Whether the number of total views of the first dataset is less than the number of total views of the second dataset	<		
C11. Whether the number of recent views of the first dataset is less than the number of recent views of the second dataset	<		
C12. Whether the five star index of the first dataset is less than the five star index of the second dataset	<		
C13. Whether they are open	true		

For links_from relation there are eight conditions, listed as C1-C8. If all conditions are met then relation between two datasets is of type links_from. These conditions can be used for the determining whether the relation between two datasets is of type links_to with following modifications: condition C3 should be less than and C7 greater than.

3.2 Creating Semantics of Relations with LIRE

Modelling relations between OGD datasets using linked data principles and techniques can add more semantics to government data, enabling thus easier search and retrieval of information by using semantic tools. Adding semantics to OGD is achieved through RDF description of datasets with help of Dublin Core and DCAT vocabularies [17, 18]. Dublin Core expresses metadata that describe dataset in RDF for direct machine processing through most well-known and basic terms, while DCAT facilitates interoperability and increases discoverability for easy consume of LGD.

LIRE architecture has SEMRE component which carries out modelling of relations with use of voID (vocabulary of Interlinked Datasets) vocabulary, because voID is one of the most widespread vocabularies for LGD and has a feature called linkset. A linkset is a collection of RDF links where an RDF triple has subject and object described in different datasets [19]. Vocabulary voID is convenient for use in our case because it is naturally intended for describing linked datasets. Knowing linkset structure, we can define a basic RDF model of relation between two OGD datasets implemented in SEMRE (Fig. 2).

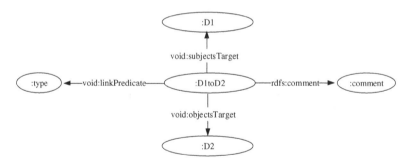

Fig. 2. RDF graph of relation between two datasets implemented in LIRE SEMRE

Implemented RDF graph model can be described using two semantic web data formats Turtle and RDF + XML. In Turtle, it would be:

```
@prefix void: <http://rdfs.org/ns/void#> .
@prefix rdfs: <http://www.w3.org/2000/01/rdf-schema#> .
:D1toD2 a void:Linkset;
void:subjectsTarget :D1;
void:objectstarget :D2;
void:linkPredicate :type;
```

```
rdfs:comment :comment;
```

Represented with RDF+XML, it would look like:

```
<rdf:RDF>
<void:linkSet>
<void:subjectsTarget>D1</void:subjectsTarget>
<void:objectsTarget>D2</void:objectsTarget>
<void:linkPredicate>type</void:linkPredicate>
<rdfs:comment>comment</rdfs:comment>
</void:linkSet>
</rdf:RDF>
```

In Turtle code D1toD2 represents the name of linkset, i.e. the name of the relation, and it is identified by void:linkset statement. It is also a part of URI of appropriate RDF/XML syntax. Terms D1 and D2 are dataset's names represented by void:subjectsTarget and void:objectsTarget statements respectively. In both data formats type is the type of relation (parent_of, child_of, etc.) represented by void:linkPredicate. Relation description is contained in comment element identified by rdfs:comment.

The appropriate mapping between dataset relation elements and voID vocabulary is illustrated on Fig. 3.

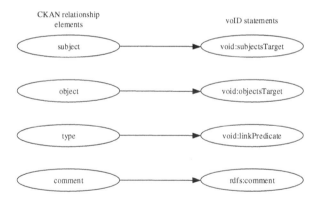

Fig. 3. Mapping between CKAN relations properties and voID statements conducted in SEMRE

SEMRE uses voID description of dataset relations with aim to offer users more data related with it and to enable easy access, search and retrieval of information. In this manner SEMRE enables access to LGD from semantic web applications. It also offers a mechanism to implement a semantic description of dataset relations into the open data catalog.

The RELIN component of architecture enables user interface for creating related datasets. Every dataset is represented with graphical element that contains information on dataset's description, tags, formats and existing relations.

4   LIRE Architecture Deployment for CKAN Data Catalogue

To demonstrate the value of LIRE architecture, we have deployed the architecture as plugin for CKAN open data portal. The plugin is in beta phase now, but it will be soon available from the CKAN's online plugin repository. To present the use case of using LIRE as plugin we have installed CKAN platform on local computer, and uploaded to it few datasets from datahub.io. To see existing relations between datasets, user filters the display by using one of the following parameters: datasets per tag, per group, per organization or random number of datasets. If he skips filtering, all datasets from portal and their relations will be loaded. RELIN component gives output depicted on Fig. 4. All datasets matching user filters and their related datasets are presented in the page. With the given datasets, user can: create, update and delete relations between the datasets and to select datasets for which want to obtain a semantic description of the relations in selected format.

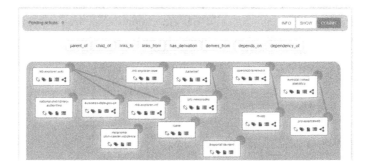

Fig. 4.   Visual representation of dataset relations

User can perform following operations: remove or create relation between datasets. Every user action is saved in the list of actions pending to be executed in CKAN. After user finishes changes on the datasets, he needs to commit actions to CKAN. Removing existing relation is enabled through the user interface by clicking on the relation and choosing option delete from the context menu. When a user chooses to create relation between two datasets, for example *rkb-explorer-ieee* and *rkb-explorer-wiki* with child_of relation, firstly he needs to choose relation type from the application menu. Direction of the connection determines subject and object of the dataset relation, starting dataset is subject and ending dataset is object. Application examines meta-data of these datasets based on model described in Table 1 and returns results to the user. In these concrete example, user pick was child_of, and this type is not matched with the one proposed by the model links_from (Fig. 5). User can choose to proceed with his action and relate datasets in child_of relation or take the suggestion from LIRE and go for links_from relation.

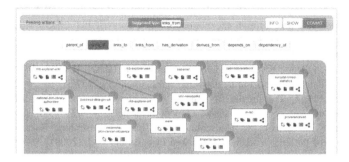

Fig. 5. Relating datasets

5 Conclusion

In this paper we presented an architecture for relating datasets and modelling or managing their relations with linked data principles and techniques along with model for their semantic representation. It reduces effort needed to preview datasets in order to relate them based on characteristics and data that describe them. As we have shown in Sect. 2, so far there is no research in the area of datasets relations and their modelling by using linked data. Dataset relations offer local level of relating, but if we describe them by linked data, dataset can be enriched with new data and information and with added semantics. For that purpose an RDF graph model for describing relations between two datasets is defined using voID vocabulary. Mapping between CKAN relations properties and voID statements shows that there are simple way for producing linked datasets. The modular nature of the proposed architecture makes it applicable to other portals except CKAN, but it requires additional time to review and analyze the data that describe the datasets, which are an essential element of our architecture. Future work includes investigation of the possibility for the development of the model and tool for semantic management of datasets and their relations and platform for accessing linked datasets based on defined RDF model. Also, incorporating proposed linked datasets model into the CKAN will be of great help to the users and developers in creating semantic applications.

References

1. Ayers, D.: Evolving the link. IEEE Internet Comput. **11**(3), 95–96 (2007)
2. Kalampokis, E., Tambouris, E., Tarabanis, K.: Linked open government data analytics. In: Wimmer, M.A., Janssen, M., Scholl, H.J. (eds.) EGOV 2013. LNCS, vol. 8074, pp. 99–110. Springer, Heidelberg (2013)
3. Sheridan, J., Tennison, J.: Linking UK government data. In Proceedings of the WWW 2010 Workshop on Linked Data on the Web (2010)
4. Berners-Lee, T.: Linked data. http://www.w3.org/DesignIssues/LinkedData.html
5. Veljković, N., Bogdanović-Dinić, S., Stoimenov, L.: Benchmarking open government: an open data perspective. Gov. Inf. Q. **31**(2), 278–290 (2014). Elsevier

6. Veljković, N., Bogdanović-Dinić, S., Stoimenov, L.: Exploring collaboration between public administrations through the notion of open data (ICIST 2015), Kopaonik, 8–11 March, vol. 1, pp. 122–127 (2015)
7. CKAN Open data portal. http://ckan.org/
8. Shadbolt, N., O'Hara, K., Berners-Lee, T., Gibbins, N., Glaser, H., Hall, W.: Linked open government data: lessons from data.gov.uk. IEEE Intell. Syst. **27**(3), 16–24 (2012)
9. Government, H.M.: Putting the frontline first: smarter government. https://www.gov.uk/government/uploads/system/uploads/attachment_data/file/228889/7753.pdf
10. Alexander, K., Cyganiak, R., Hausenblas, M., Zhao, J.: Describing linked datasets. In: Proceedings of the 2nd Workshop on Linked Data on the Web (2009)
11. Maali, F., Cyganiak, R., Peristeras, V.: Enabling interoperability of government data catalogues. In: Wimmer, M.A., Chappelet, J.-L., Janssen, M., Scholl, H.J. (eds.) EGOV 2010. LNCS, vol. 6228, pp. 339–350. Springer, Heidelberg (2010)
12. Ding, L., Lebo, T., Erickson, S.J., DiFranzo, D., Williams, T.G., Li, X., Michaelis, J., Graves, A., Zheng, G.J., Shangguan, Z., Flores, J., McGuinness, L.D., Hendler, A.J.: TWC LOGD: a portal for linked open government data ecosystems. Web Semant. Sci. Serv. Agents World Wide Web **9**(3), 325–333 (2010)
13. Kalampokis, E., Tambouris, E., Tarabanis, K.: A classification scheme for open government data: towards linking decentralised data. Int. J. Web Eng. Technol. **6**(3), 266–285 (2011)
14. Schmachtenberg, M., Bizer, C., Paulheim, H.: Adoption of the linked data best practices in different topical domains. In: Mika, P., et al. (eds.) ISWC 2014, Part I. LNCS, vol. 8796, pp. 245–260. Springer, Heidelberg (2014)
15. Zuiderwijk, A., Jeffery, K., Janssen, M.F.: The potential of metadata for linked open data and its value for users and publishers. JeDEM-e-J. e-Democracy Open Gov. **4**(2), 222–244 (2012)
16. Janssen, M., Estevez, E., Janowski, T.: Interoperability in big, open, and linked data-organizational maturity, capabilities, and data portfolios. Computer **47**(10), 44–49 (2014)
17. Dublin Core Metadata Initiative: DCMI Metadata Terms. http://dublincore.org/documents/dcmi-terms/
18. W3C: Data Catalog Vocabulary (DCAT). http://www.w3.org/TR/vocab-dcat/
19. Alexander, K., Cyganiak, R., Hausenblas, M., Zhao, J.: Linkset. http://www.w3.org/TR/void/#linkset

Budgetary Data (in an Open Format) Benefits, Advantages, Obstacles and Inhibitory Factors in the View of the Intermediaries of this System: A Study in Latin American Countries

Gisele da Silva Craveiro[1(⊠)] and Cláudio Sonáglio Albano[2]

[1] Universidade de São Paulo (USP), São Paulo, Brazil
giselesc@usp.br
[2] Universidade Federal do Pampa (UNIPAMPA), Bagé, Brazil
claudio.albano@unipampa.edu.br

Abstract. Governments are under pressure to meet new social demands and seek new forms of management. As well as this, the fact that they make considerable use of information technology, has led to the growth of their databases, and made governments and the respective government organizations, fertile ground for open data initiatives. Having access to budgetary data and being able to make use of it (with regard to revenue and expenditure), has traditionally always aroused great interest in society. The purpose of this study is to determine the potential benefits and possible obstacles that can affect the intermediaries who take action on the basis of budgetary data. A number of intermediaries (members of society and government policymakers) from four Latin-American countries were interviewed. The results show that the structural barriers of governments have an adverse effect on their activities because they impair the quality of the information that is made available. Nonetheless, the benefits that allow a greater degree of transparency as well as the ability to reveal more knowledge of the inside operations of governments, encourage the intermediaries to take part in this "ecosystem".

Keywords: Benefits · Obstacles · Open government data · Intermediaries

1 Introduction

Public organizations (in this study regarded as governments) are being confronted with new demands on the part of society. These demands are linked to the need for greater transparency in the management of public funds, and a greater control over the services provided, as well as a greater degree of accountability on the part of their managers. In this climate, information and communication technology (ICT) is beginning to play a key role.

The Internet can be cited as one of the most important resources for this technology. The set of factors outlined here – a greater demand by society for public organizations, a widespread use of information technology and the growing use and importance of the Internet – have made feasible the rise of a platform called open government. In the

© IFIP International Federation for Information Processing 2015
M. Janssen et al. (Eds.): I3E 2015, LNCS 9373, pp. 223–235, 2015.
DOI: 10.1007/978-3-319-25013-7_18

view of [6], under the aegis of open government, public organizations are seeking to meet the requirements of society by offering information as a means of finding a solution to some of the problems.

Among an array of opportunities created by open government initiatives, there is the prospect of being able to use open data (OD) which according to [3] means making information available on the Internet in a way that can allow it to be recycled for third parties. According to [12], open government data (OGD) involve the publication and dissemination of information of the public sector (i.e. governments) in the Web, which is shared in a rational and comprehensible format in a way that allows its reuse in digital applications or in other words, makes it legible for machines.

Following the rise of concepts (or technological platforms) of open data and open government, new players are emerging in the scene who may require new ways of using and appropriating the information made available, as well as reshaping traditional requirements. Thus the OGD projects must form a network between the government and society. It is only in this way that the assurances about the prospects of using open data effectively can be fulfilled, which means it is necessary to create and maintain an environment where the different players can interact [11]. Thus, governments and society must create an environment which benefits both and allows these sectors to grow and become more involved. Several authors claim that there are potential benefits that can be derived from open government data: economic growth and an improvement in the provision of public services among other factors.

From what has been outlined so far, undertaking this work is justified in so far as it seeks to achieve the following general objective: to find out the advantages, benefits, obstacles and inhibitory facts in the view of the OG intermediaries, who are involved in the ecosystem of budgetary data in Latin American countries. In attaining this goal, it is hoped that an answer can be found to the following research question: by having a knowledge of the expectations of the OGD intermediaries, can governments be granted better conditions to carry out projects of open government data in the budgetary "ecosystem"?

By undertaking this study it is hoped that a contribution can be made to the background of open government data by obtaining information about how the inter-mediaries take action, what benefits are sought that can overcome obstacles and other factors that pervade the activities carried out in the budgetary "ecosystem" of open data. Another key issue is the context of the work since it focuses on representative countries from the Latin American continent, namely Argentina., Brazil, Mexico and Uruguay.

The exposed here, we consider important to clarify that intermediaries terms and ecosystem in this work. Intermediate are players (individually or representative of governments and civil society organizations), which operate with public data, released in an open format. Ecosystem as a group of organizations and individuals (government and society), as well as structures (example - public authorities and their agencies, non-profit society organizations), technological tools (e.g. - software), legal compo-nents (example - laws) and values of society that interact in a particular government sector and society by making use of public data released in an open format.

2 Theoretical Framework

The theories and models which act as a frame of reference, are employed for this study. This entails defining concepts about open data, its eco-system and the performance of intermediaries. The framework is outlined in the following sections together with its advantages, benefits, obstacles and inhibitory factors in an OGD ecosystem which supports the data collection data and analysis.

2.1 Open Data, Its Ecosystem and Intermediaries

The advent of the platform (or concept) of open data originates in a historical process, where public organizations were always great users of information technology. This fact allied to new requirements by society for greater efficiency in public bodies and the growing use of the Internet in modern society, have given rise to a new concept called open data. The focal point of this study is open government data (OGD) for the [13], the publication and dissemination of information in the public sector (governments) in the Web, which can be shared in a rational and comprehensible format in a way that can allow it to be recycled in digital applications or in other words, made legible for machines.

Open government and open data are grounded on three pillars: transparency, participation and collaboration. Transparency imposes the responsibility to inform citizens about what exactly the government is doing and what activities it intends to carry out. Participation allows citizen to offer their ideas and skills and thus assists the government to devise more effective and far-reaching policies and also provide more information to society. Collaboration lays stress on the effectiveness of government and encourages closer cooperation between society and the different levels of government.

Until the time when all the initiatives and/or opportunities that the government had for making information available to society (more recently through the use of Internet technology), these questions were traditionally addressed with a view to ensuring transparency, greater control and a better exercise of citizenship. However, with the advent of open data, new opportunities have arisen for society to go beyond these traditional requirements in areas such as the following: developing new goods and services to bring about financial opportunities (through these goods and services), making improvements in social welfare and assisting the government (by providing feedback) through an interaction made possible by these new goods and services.

Authors such as [6, 11] stress that it cannot be expected that simply by opening up new data, governments will begin to produce goods and services and create economic opportunities. For this to take place an attempt must be made to study, understand and suggest improvements or corrections to the proceedings of governments and society in general, with a view to fostering new activities, facing challenges and reducing (or ideally overcoming) unforeseen problems.

Researchers into the environment of information have used the metaphor of the eco-system to stand for the wide and intricate web of relationships between the suppliers, users, data, material infrastructure and institutions. In the opinion of [5], an eco-system is "a system of people, activities, values and technologies in a particular environment". [5] think that the owners of an eco-system are located in three entities: (a) the policies and practices of governments; (b) the web users, companies and citizens of society; (c) the innovators. These can interact in several ways and influence the way the eco-system evolves: the policies and practices of government can interact with the web users, civil society and companies.

An ecosystem cannot be established in a satisfactory way without the presence of intermediaries. These might be government policymakers or originate in the society where there is a capacity and desire to act in the eco-system of open government data by having access to data and being able to handle and disseminate it. In other words, these data can be made useful for third parties for society through its most varied kinds of representation or organization.

Studies such as those of [8, 9], have underlined the importance of the intermediaries in this ecosystem. Setting out from these studies, these authors recognize that it is important for a wide range of players to take part because several technical activities are needed to handle the available information and make it accessible to the public. They state that a good deal is required to handle the available information and that this entails a close collaboration between the participants of this movement.

For the purposes of this study the authors define the term "intermediaries" as referring to "all the players (in an individual way or representatives of governments and social organizations), who are involved with public data that are released in an open format. They may or may not make use of technological, legal or structural artifacts in their activities. In making use of open data, the intermediaries aggregate value to the data to ensure that they can be understood more easily (and hence have a greater value) for third parties after their intervention". The intermediaries can and must supply goods and/or services with the public database for the government and society/third parties.

2.2 Possible Benefits and Advantages

It can be argued that there are several advantages for society and government in using government data in an open format. As well as having the chance to exercise a greater degree of citizenship, people are given new economic opportunities by having access to public information. The study by [10] estimates that when it is spread to cover areas such as education, transport, health, finance, fuel, and electricity, open data has the potential to add three trillion dollars to the global economy every year.

Governments also benefit from the use of open data. [4, 6], cite the following as possible advantages to governments: greater internal efficiency - by having greater/ better access to information, governments can make use of these data to undertake their activities and thus increase their efficiency by using information that is difficult to obtain (even by the governments themselves or other governments); and greater efficiency owing to the feedback provided by society that uses open data can and must "act as an informer" of the government about the possible approach that should be adopted for a better/greater use of public resources. [7] classify these possible benefits and/or advantages in three groups which are as follows: political and social; economic and technical; and operational. These benefits and/or advantages were used for an analysis of the main questions.

2.3 Possible Obstacles and Inhibitory Factors

Some problems can arise in OGDs such as obstacles or possible inhibitors. Governments make great use of ICT; currently, they deploy different technological platforms in which the databases are stored. These databases should be the source of the data from open data projects. Governments have to carry out technical tasks that can be distinguished from the implementation of projects and involve the "conversion" of the data from these bases to formats that are suitable for publication in accordance with the rules of open data.

Another area that serves as a possible obstacle or inhibitory factor concerns the "understanding" of data from society (which is addressed in the issue – the quality and usability of available content). How can one ensure that the data are reliable and that the content is up-to-date? Since the open data must be published in different formats from those stored in the database, the task of "conversion" is a necessary investment when forming the teams [5].

The structure of government bodies can also have an adverse effect on these initiatives since even when there are legal guarantees for the publication of the data, the governments need the operational conditions to carry out this task. It is recognized that there is a need for training and to adapt the culture of the public servers to this end as well as a suitable functional and technological framework for a correct understanding of the system and to meet the legal, technological and structural requirements. [7, 13] classify the obstacles and possible inhibitory factors into six large groups which are as follows institutional, complexity of tasks, use and participation, legislation, quality of information and techniques. These obstacles and inhibitory factors were drawn on for the data analysis.

3 Methodology

In the light of its objectives, this can be characterized as an exploratory study, as defined by [2], who believes that exploratory studies are suited to broadening the researcher's knowledge of a relatively unknown phenomenon and thus providing a greater understanding of it. Another feature of this kind of work is that the researcher does not expect to obtain definitive answers to the problem addressed.

In forming the sample, the authors of the study made use of their findings from a previous work on the "ecosystem" of open government data. Nineteen interviews were conducted, some in person and others via Skype; the interviews were recorded and subsequently transcribed. Four interviews were conducted in Argentina, six in Brazil, five in Mexico and four in Uruguay; all the interviews were carried out between September and December 2014.

The study was based on analysis assumptions with regard to content and a previous categorization of the aims of the research questions according to [1]. This author thinks that conducting an analysis of content is an investigative technique which is employed with a view to obtaining an objective and systematic description of the issue that is being communicated. One of the ways of putting into effect an analysis of content is to categorize the texts that emerge from the data collection. This categorization can be carried out in an a priori way and is grounded on a theoretical basis.

The categories must be valid and there must be a consistent classification of each element. A valid kind of categorization must be significant with regard to the content of the materials which are being analyzed and conform to the goals set out for the study. A qualitative approach was adopted for handling the data analysis which sought to determine what citations or references to previously established categories occurred according to [2]. In Table 1, there is a chart showing the categories that were drawn on for the analysis

Table 1. Categories for analisys.

Category	Characteristics of each category
Profile of the respondent	Terms referring to the profile of the respondent
Action taken in the course of the study	Terms used in making citations to the activities that were carried out
	What activities were carried out? However they carried out? Why were they carried out? What were the reasons for the course of action taken?
Benefits and/or advantages	Particular issues: political, social, economic and operational or technical
Obstacles and/or inhibitory factor	Particular issues: institutional, complexity of tasks, use and participation, legislation, quality of information and techniques
Results attained	Terms that can define the main results that were attained

In Table 2 there is a description of the instrument that was used, together with the purpose of each question and its relation to the categories defined for the analysis. The first part of the interview consisted of questions to determine the profile of the interviewee. The other questions were aimed at finding out about the activities that were carried out, the benefits and/or advantages, obstacles or inhibitory factors regarding the OGD ecosystem, especially with regard to budgetary data.

Table 2. Questions of instrument.

Questions	Purpose of the questions	Category analysis
Identification of the respondent	To find out the name and organization of the respondent	Profile of respondent
How would you define your use of open data and how much time have you spent on open data?	To form a profile of the respondents through their performance	Performance
What are the factors that have led you to play your role (i.e. from the action you have taken) in using open data?	To find out what factors have driven the respondents to take action in this area	Benefits and/or advantages

(Continued)

Table 2. (*Continued*)

Questions	Purpose of the questions	Category analysis
What are the main expected benefits (on the basis of your activities) of using open data (for governments and society in general)?	To obtain a better understanding of the driving-force behind the intermediaries in the eco-system	Benefits and/or advantages
What are the main difficulties encountered when using open data?	To find outwhat factors serve as obstacles or inhibitory factors in the eco-system and may, in a general way, impair the performance of the intermediaries	Obstacles and/or inhibitory factors
Is the information available suitable and adequate for carrying out your activities?	To find out the obstacles or inhibitory factors with regard to making information available	Obstacles and/or inhibitory factors
How would you define access to tools and/or resources for taking action to ensure budgetary transparency?	To determine the obstacles or inhibitory factors with regard to resources or tools	Obstacles and/or inhibitory factors
What results stand out from your performance?	This question seeks to find out what results/experiences of the intermediary need highlighting	Results attained a

4 Results and Analysis

In this section there is an examination of the results and respective analysis which follows the order of the questions in the instrument. In this way, the profiles of the respondents are shown together with the way they behaved during the project. Following this, the benefits and advantages are outlined together with obstacles and inhibitory factors.

4.1 The Profile of the Respondents and their Way of Behaving

The respondents basically had professional backgrounds and came from the following areas: public policymaking, economics, law, journalism and information technology Attention should be drawn to the presence of journalist in this group. With regard to the question of their performance, there were actors (intermediaries) of governments, who stressed that their main aim was to support the task of publishing data. On the other hand, there were key players who sought to publish the academic subject of the research they were carrying out.

The intermediaries in the area of information technology stated that they acted in this area to develop tools that could make it easier to access and handle the data. Several actors (intermediaries) stated that they were involved in the observation of public policies and in the context of the interviews, the issues they focused on were financial matters or the public budget. Another issue that was often cited to justify the work of the intermediaries, were questions related to transparency.

4.2   Interviews: Benefits and Advantages

Table 3 shows the benefits and/or advantages that were revealed by the interviews. In all the tables there is a number beside the factor that is cited, which corresponds to the number of citations given for it. The number of citations of each category is the sum total of the citations of its respective advantages and/or benefits. The results are arranged to include all four countries and each interviewee is able to cite more than one benefit or advantage.

This way of displaying the results can be explained by the fact that the main purpose of the study is to find out the advantages, benefits, obstacles and inhibitory factors in the view of the OGD interviewees who are involved in the ecosystem of budgetary data.

Table 3. Benefits and/or advantages

Category	Benefits and/or advantages cited – number of citations
Political and social (53)	Greater transparency(11); increased participation and self-accountability of citizens (web users) (11); creation of trust in the government (5); public participation in producing data (6); equality of access to the data (1); new government services for the citizens (1); improvement in the formulation of policies (7); creation of new kinds of knowledge in the public sector (5) and new social services (innovations) (6)
Financial (11)	Assists in improving procedures, goods and/or services (1); development of new products or services (3) and the use of collective knowledge (7)
Operational and technical (14)	Improvement in policymaking (10); improvement in decision-making which allows comparisons to be made (1); makes it easier to access and find data (2) and the creation of new databases by combining data (1)

It is clear from the results that the greatest benefits and/or results are linked to political and/or social factors where there are good prospects of bringing about greater transparency. This is one of the three pillars of open government and its citation was either made in a direct form or by mentioning other factors such as: the prospect of encouraging greater participation among the citizens or providing the public with greater access to the data.

Financial gains were also mentioned in this category and there were clear signs of another pillar of open data being found in this category- collaboration. This was clear when the interviewees stated that they were able to derive advantages or benefits from the use of collective knowledge through the use of an intermediary.

The factor highlighted above converge when obtaining operational and technical benefits or advantages in so far as there is a recognition that policymaking can be improved and that to achieve this a greater participation and collaboration between society and the government is needed through the main players (intermediaries) of these sectors.

With regard to political and/or social issues, it should be noted that some benefits or advantages were cited that could be obtained by governments such as: the creation of trust in the government and the discovery of new knowledge in the public sector. In addition, society can profit from open data in the following areas: new government services and social services for the general public. This confirms the underlying assumption of certain authors that everybody can derive benefits or advantages from the use of public data in an open format.

4.3   Possible Obstacles and/or Inhibitory Factors

Table 4 shows the obstacles and/or inhibitory factors that emerged from the interviews. In all the tables, there is a number beside the factor being cited which corresponds to the number of its citations. The number of citations of each category is the total sum of the mentions of the respective obstacles and/or inhibitory factors. The results have been arranged to include all four countries and each interviewee can cite more than one benefit or advantage.

Several authors stated that they had a lot of difficulties in publishing public data in an open format which originated from the inner structures of the government. This is

Table 4. Obstacles and/or inhibitory factors

Category	Obstacles and/or inhibitory factors – number of citations
Institutional (25)	Lack of uniformity in the policy for publishing data (15); priority given to the interests of organizations to the detriment of the interests of private citizens (5); failure to define the procedures adopted for working and/or interacting with the web users (3) the existence of doubtful standards for working with the web user (2)
Complexity of the task (38)	Duplication of data, the data made available in different ways either before or after the processing, resulting in uncertainty about the source (9); even when the data can be found, the users may not be aware of their potential uses (6); the data formats and datasets are much too complex to be handled and used easily (17); a lack of support for the tools or helpdesk (5) and the focus is on making use of individual datasets whereas the real value may come from a combination of several datasets (1)
Use and participation (40)	A lack of incentives for the user (9); public organizations fail to react to the input of the user (2); the costs are unexpectedly high (1); lack of time to use open data (1); a lack of knowledge about how to use or make sense of the data (15); lack of the necessary capacity to make use of the information (9); no statistical knowledge or understanding of the the potential value or limitations of statistics (3)

(*Continued*)

Table 4. (*Continued*)

Category	Obstacles and/or inhibitory factors – number of citations
Legislation (3)	Written permission required to have access to the data or to reproduce them (2) and the question of the renewal of contracts/agreements (1).
Quality of the information (32)	Lack of information (10); lack of precise information (8); information that is incomplete, only a part of the total picture shown or only conveyed at particular intervals (6); the loss of essential information (1) and the fact that similar data stored in different systems produce different results (7)
Techniques(29)	The data should be in a well defined format which is easily accessible: whereas the data formats are arbitrary, the data format must be strictly defined (1); absence of standards (11); absence of a support to make the data available (1); lack of goals and standards (1); no standard software for the processing of open data (5) and fragmentation of software and applications (10)

confirmed by the interviewees when the following factors had several citations: a lack of uniformity in the policy for publishing data and the fact that priority was given to the interests of organizations to the detriment of those of private citizens.

Other obstacles that were often cited by various authors were linked to the quality of the published information in various ways. This is also confirmed by the interviewees where there were a large number of citations: duplication of data; data being made available in different ways; the data formats and the fact that the datasets are much too complex to handle and use easily; lack of precise information, absence of standards; and the fragmentation of the software and applications. At the same time, attention should be drawn to the low number of citations regarding obstacles or inhibitory factors related to legislation.

There are many references to factors regarding the interests of society and the following had a lot of citations: a lack of incentives for the user; a lack of knowledge about how to use or make sense of the data and a lack of the skills needed to use the information. This is supported by the mention made of the factors related to the ability of the government to work together with society and the difficulties experienced by those wishing to obtain support or clarification of the content made available.

With regard to other factors, the driving-force behind the desire of society to participate in the OGD initiatives, as well as its qualifications for using these data, is to some extent related the factor of participation. These difficulties might be aggravated by a lack of understanding of the content made available and by the skills (some of which are technical) that are needed to have access to the information and be able to handle it.

4.4   Results Attained

Some questions about the instrument seek to find out what results can be attained. The purpose of these questions was to establish the results that were obtained and allow a parallel to be made between the benefits, advantages, obstacles or inhibitory factors and these results and learning experiences.

The main results that were mentioned, corroborate the fact that this issue is still in its early stages. This is because several of the interviewees stated that the principal result was the publication of documents with a view to publishing material about the question which included documents such as books, lists of statistics, and catalogues. Other results are as follows: it has had a small impact but is gradually being formed; there is more discussion of the issue and the community which is being formed around the subject is growing although only gradually; the fact that open data is an important question but still does not have great popular appeal; the fact that the issue is still in its early stages in the countries that are the focal point of this study.

In the previous subject, various advantages and benefits were cited by the interviewees as being susceptible to inclusion in this ecosystem and the results confirm some of these advantages such as: improving policymaking; a rise in the number of accusations of misdemeanors, which might lead to a better use of public resources; a heightened awareness in society of public expenditure and a greater trust in the government.

Some results confirmed that there were obstacles and difficulties, especially with regard to the limited interest and capacity of society to have access to the data or be able to use it. The interviewees cited the following results in support of this fact: a greater concern and awareness of the importance of allowing society to have access to data and the respective use of this data; in some areas there was a greater interest in the use of the data/information made available in this format as, for example, public transport; finally the formation of alliances between organizations of society increases the interest in the subject.

The results that were cited strengthen the need for the formation of networks between governments and society and allows the intermediaries to play a stronger role in this ecosystem. This is because mention was made of the effort and work necessary to obtain some results which must be carried out by the government and key players in society. As a result, certain activities were carried out such as public events and courses aimed at bringing about a greater degree of integration.

5 Final Considerations

Before giving our final thoughts about this work, an attempt will be made to outline some of its limitations and make suggestions for further studies. One drawback of the work which should be mentioned, is the fact that the sample was formed in an intentional or rather, non-probabilistic way which might have caused a degree of bias in the answers. In an attempt to overcome this problem of bias, the authors decided to interview key players (or intermediaries) from different social origins and professional backgrounds, although all of them were linked to the ecosystem of budgeting.

The authors are aware that in future studies it would be valuable to carry out research into other ecosystems of open government data such as in education, and health among other areas, as a means of determining what results are different and/or similar. Forming a sample with intermediaries from other ecosystems will allow greater assistance to be given to public managers in future planning of schemes for the publication of data in an open format.

Undertaking this study has made it possible to confirm some of the advantages and/or benefits that are proclaimed by several authors, especially with regard to the open data which has the potential to exercise transparency and allow society to have more control over the acts of government and thus establish a historic trend. At the same time, it also confirms that few opportunities have been found in the economic field because not many interviewees referred to benefits and/or advantages in this area.

The results make clear that society still has little interest in the issue and the interviewees made several references to this fact. Several corroborative reasons were mentioned such as the difficulties of understanding the context of the data and the need for a higher level of education and culture to include the social players in this eco-system. It should be underlined that there was also mention of the need for the inter-mediaries to have knowledge and technical skills (related to the information technology software sector), and a list was made of the technical difficulties of having access to the data or being able to handle it.

It is also evident from the results that the presence and involvement of governments is important since they can derive benefits and advantages from publishing the data in an open format. Governments should also give priority to mitigating factors that might harm the publication of data in a way and thus circumvent obstacles and technical or structural difficulties with regard to the publication of information. Government actions can also help to overcome the limited interest of society in available information, as well as to encourage activities that can lead to an understanding of the material that is published.

It is believed that this study has attained its intended objective of identifying the advantages, benefits, obstacles and inhibitory factors, from the standpoint of the OGDs who are involved in the ecosystem of budgetary data in Latin America. At the same time, the results have made it possible to find answers to the research question since identifying the factors listed above (as well as the results and lessons learned by the key players in the ecosystem of opexn government data) allow the policymakers to profit from the improved conditions for carrying out projects of open government data.

One of the contributions that can be made by the managers is as follows: it is extremely important to create the conditions that can enable organizations like those studies in this work, to be able to carry out their activities and in particular, to have an environment (or an ecosystem) that allows organizations to be combined so that part-nerships can be formed in a way that is complementary to their individual capacities. These partnerships can also be formed to obtain benefits or advantages and as a way of overcoming possible obstacles and inhibitory factors through activities that encourage innovation and collaboration, which are the two basic principles of open data.

Acknowledgements. The authors acknowledge the support of the Iniciativa Latino Americana de Datos Abiertos (ILDA, the Latin American Open Data Initiative) and Avina Foundation that made possible this research study.

References

1. Bardin, L.: Content Analysis. 9th edn. 70 editor, Lisboa (2009)
2. Creswell, J.W.: Research project. Qualitative, quantitative and mixed method, 3rd edn. Artmed editor, Porto Alegre (2009)

3. Eaves, D.: The three laws of open government data (2009). http://eaves.ca/2009/09/30/three-law-of-open-government-data/
4. Espinoza, J.F., Recinos, I.P., Morales, M.P.: Datos Abiertos: oportunidades y desafíos paraCentroamérica con base en unacadena de valor. Open Data For Development in Latin America and the Caribbean (OD4D), Montevideo, Uruguay (2013)
5. Harrison, T.M., Pardo, T.A., Cook, M.: Creating open government ecosystems: a research and development agenda. Future Internet **4**, 900–928 (2012). http://www.mdpi.com/journal/futureinternet
6. Helbig, N., Creswell, A.M., Burke, B.G., Pardo, T.A., Luna-Reyes, L.F.: Modeling the informational relationships between government and Society. In: Open Government Consultative Workshop – CTG, Albany, NY, US (2012)
7. Janssen, M., Charalabidis, Y., Zuiderwijk, A.: Benefits, adoption barriers and myths of open data and open government. Inf. Syst. Manag. **29**, 258–268 (2012)
8. Kuk, G., Davies, T.: The roles in assembling open data complementarities. In: Thirty Second International Conference on Information Systems, Shanghai, vol. 32 (2011). http://soton.academia.edu/TimDavies/Papers/1216268/The_Roles_of_Agency_and_Artifacts_in_Assembling_Open_Data_Complementarities
9. Mayer-Schoenberger, V., Zappia, Z.: Participation and power: intermediaries of open data. In: 1st Berlin Symposium on Internet and Society, Proceedings of 1st Berlin Symposium. Berlin: Alexander von Humboldt Institutfuer Internet und Gesellschaft (2011). http://berlinsymposium.org/sites/berlinsymposium.org/files/participation_and_power.pdf
10. McKinsey, C.: GI. Open data: Unlocking innovation and performance with liquid information (2013). http://www.mckinsey.com/Insights/MGI/Research/Technology_and_Innovation
11. Prince, A., Jolías, L., Brys, C.: Análisis de la cadena de valor del ecosistema de DatosAbiertos de la Ciudad de Buenos Aires, Montevideo, Uruguay (2013). http://www.princeconsulting.biz/pdf/7.pdf
12. W3C. Improving access to government with the best web use (2009). http://www.w3c.br/divulgacao/pdf/gov-web.pdf
13. Zuiderwijk, A., Janssen, M., Choenni, S., Meijer, R., Alibaks, R.S.: Socio-technical Impediments of open data. Electron. J. e Gov. **10**(2), 156–172 (2012). Academic Publishing International Ltd. Zuiderwijk. http://www.ejeg.com

Transparency Dimensions of Big and Open Linked Data

Transparency as Being Synonymous with Accountability and Openness

Ricardo Matheus[(⊠)] and Marijn Janssen[(⊠)]

Delft University of Technology, Delft, The Netherlands
{R. Matheus, M. F. W. H. A. Janssen}@tudelft.nl

Abstract. Although one of the main reasons to open data by governments is to create a transparent government, many initiatives fail to deliver transparency. Making data available does not automatically yield up transparency. Furthermore transparency is an ill-defined concept and understood in different ways. Added on this puzzle, Big Data Analytics are becoming reality in governments and on society due the quantity of open data available and the evolution of techniques and instruments used to analyze data. This paper develops a Big and Open Linked Data (BOLD) Framework identifying categories, dimensions and sub-dimensions that influence transparency. Our framework conceptualizes transparency as a process of data disclosure and usage. Transparency is based on the two major synonymous concept used on literature, namely accountability and openness. Accountability means revealing important details for transparency to control governments financially and operationally, whereas openness reveals details of what, how and why politics took the decision, without revealing important parts of the political game inside government such in military and nuclear area.

Keywords: Big and Open Linked Data (BOLD) · Transparency · Evaluation · Accountability · Openness

1 Introduction

One of the main purposes of opening data by governments is to create transparency [1, 2]. Transparency is often viewed as a condition *sine qua non* for Democracy [3–5]. Since the mid-nineties politicians and governments started to create portals to publish government information including budgeting and statistical data [4]. Some governments created Application Interface Programming (APIs) to enable streaming data. Recently, world watched a boom of Big Data Analytics usage of techniques for some reasons [6] e.g. the sufficient possible quantity of data to be collected by millions of users' devices (smartphones, social networks on Internet), processed at enterprises or opened by government [7], further the fast evolution of hardware to collect, storage, treat and analyze all the data in a small period of time with high velocity of processing [8]. For this we name on this paper Big and Open Linked Data (BOLD).

© IFIP International Federation for Information Processing 2015
M. Janssen et al. (Eds.): I3E 2015, LNCS 9373, pp. 236–246, 2015.
DOI: 10.1007/978-3-319-25013-7_19

All these effort are aimed to create transparency, although the actual contribution to creation of transparency can be challenged [9, 10]. The merely opening of data without providing any user-interface might result in an inability to use, whereas the inclusion of a predefined user-interface and visualization might only result in a biased picture. The more data is opened the higher the information overload. Disclosing open data does not result in transparency, only the actual use of the open data can result in transparency.

A lack of transparency is usually caused by information asymmetry between the government (agent) and the citizens (principal) [11]. Information asymmetry refers to the situation in which one party has more information than the other [12]. In our situation this occurs as government has more information than its constituents. Information asymmetry prevents that all the stakeholders have the same insight on the governmental issues, being able to control the public policies and participate on the public management. The opening of data should overcome the information asymmetry to some extent and provide better insight into the inner part of the government.

Although transparency is an intuitive appealing concept, there is no uniform view on what constitutes transparency. Some authors even define transparency as magical concept [13, 14], whereas others use the term as synonymous with accountability [15–17] and/or openness [9, 18]. In this research we contribute to this discourse by investigating the aspects of transparency and dimensions for opening data to enable transparency. This paper is structured as follows. First we present the research approach that uses literature review and interviews as the primary research instrumented. This is followed by the literature background, which results in the conceptualizations of transparency. Next a Big and Open Linked Data Framework is presented to identify dimensions and subdimensions that influences transparency. This results in two dimensions and theirs sub-dimensions: data dosclosure and data usage. Finally conclusions are drawn.

2 Research Approach

Our aim is to get insights into the factors influencing and conditions for creating transparency. First a literature review of transparency concept was included by searching on the terms "transparency + government" "Transparency + accountability + government" and "Transparency + openness + government". Papers published in the e-government top journals were reviewed including Government Information Quarterly (GIQ), International Journal of E-Government Research (IJEGR), Transformational Government, People, Processes and Policy (TGPPP), E-Government, an International Journal (EGIJ), International Journal of Public Administration in the Digital Age (IJPADA) and Information Polity (IP). The review was limited to the first 50 papers as presented by Google Scholar and Scopus. The main reason for this limitation is that after the fifth page, the number of publications started to repeat itself and the remaining works was not considered as influential due to the limited number of citations. Another limitation was that some books were not accessible online. Those publications were excluded. A total of 200 publications were found of which 85 publications were considered to be relevant for our research after scanning the papers but only 54 papers were selected as source and reference.

Results having only the keywords inside the list of references were excluded. The literature review revealed two mainstreams: transparency as being synonymous with accountability [15–17] and as synonymous with openness [9, 18]. Next the content of these papers was analyzed for factors affecting and conditions necessary for data disclosure and data usage. The literature review results in transparencyfFramework considering dimensions that influences data disclosure and data usage to promote transparency, identified as a majority flow mechanism to achieve transparency.

3 Views on Transparency

Our literature review revealed that he concept of transparency is complex due to its ambiguity and various usage in the literature and practice [3]. Transparency has been used as a magic concept by governments to improve efficiency [13] or as synonymous with accountability [15–17] and synonymous with openness [9, 18]. In practice the concept of transparency is often misunderstand and more talked about than practiced [19].

In our view transparency is aimed at overcoming the information asymmetry between the government and the public. Transparency refers to the ability for the public to understand the various aspects of government. It is about the ability to see the inner working of the government. This means that who and how decisions are made and what evidence is used are transparent. For this purpose data about the functioning of the government should be released. From this puzzle, two main categories on transparency were identified at literature view: Transparency as synonymous with accountability and transparency as synonymous with openness.

3.1 Transparency as Synonymous with Accountability

Accountability normally involves a relationship between two or more parties, where one party holds the responsibility of performance given certain objectives pre-stablished or planned, taking in consideration public principles such as effectiveness and efficiency use of resources to realize the purposed objective. Accountability implies answerability for one's actions or inactions and the responsibility for their consequences [20]. Accountability means also taken responsibility for decisions. Elections are a case of accountability in governments, where people can judge the past actions of politics after managing the state during some period of time [21]. Accountability concerns the comparison of objectives with the realized performance and deviations [16].

Some actions, which are often considered as part of transparency, are in fact actions necessary for accountability and keeping politicians and public officials accountable. Many open data are just a publicity of data or the published data cannot be used for accountability, as its characteristics are not suitable for this. For example too low quality or only providing insight into one aspect. Transparency as accountability is also used to identify when it is possible to enable anti-corruption in government [22, 23]. For the practice perspective, the Transparency International [24] has lead its objective of transparency toward the anti-corruption goal, identifying who has not a suitable work within public management. In their index the term transparency is used to advocate its objective, the anti-corruption. Nevertheless, in our view transparency is the way to enable anti-corruption tools. In conclusion, information through transparency of governments is the raw material to enable accountability or anti-corruption tools. Accountability uses data/information [25] and do something with the information that publicity created [16]. Yet accountability does not need complete transparency. Some activities might be hidden, as they are no needed for being accountable. Only surrogates are published which are necessary to keep one accountable. No knowledge of the inner working of the public system needs to be published. Those parts remain hidden and are not transparent. From this we conclude that although accountability and transparency are overlapping they are also distinct.

3.2 Transparency as Synonymous with Openness

After being sworn in the United States of America, the President Barack Obama created the Memorandum of Open Government with the aim of "creating an unprecedented level of openness in Government" [26]. One of the underlying goals was to create an open and transparent government. Openness does not represent automatically the result of increasing transparency in governments, however, it influenced the creation of open government data portals and legislations as a freedom of information act. For example a box can be open, but still it might not be transparent and you cannot see inside it. On the other hand openness might be necessary condition for transparency. If the system is closed there cannot be any transparency. From this we conclude that although openness and transparency are overlapping but distinct concepts.

Scientific literature points out that openness is also close to open government initiatives [27, 28]. From this, scientific literature and practical people are facing an operational and theoretical definition toward openness as Open Government concept. For governments it is not an initial stage of the word and the concept, however, for military/civil usage of nuclear weapon/energy [29, 30] and the finances [28], openness is not also on initial phase. Both of them can be inspired on how processes can be at same time transparent, open and not show evidence of core business or private information, what basically is citizens' need to do accountability and politics for governance without revealing important parts of the political game inside government.

4 Towards a Transparency Framework

4.1 Basic Framework

Basically the BOLD has three-steps: Collecting data, from data internal databases, spreadsheets, document files, sensors spread over a city or a social network on Internet. Secondly, storage of data, that requires advanced and unique data storage, management [31] and thirdly analysis and visualization technologies [6]. In the opening of data there are two important stakeholders the publishers and users of open data. The publishers and users often are unaware of each other needs and encounter different challenges and barriers [32]. The data publishers' main activity is the disclosure of data, which is necessary before data can be used. Yet only releasing data does not result in any transparency. Only the actual use of data results in transparency. Both steps are influenced by a large amount of factors. In conclusions, transparency can only be created when both data disclosure and usage happens. In Fig. 1 the basic transparency framework in the form of these two essential activities are presented. Hereafter we delve into the details of factors and conditions impacting data disclosure and usage.

Fig. 1. - Basic transparency framework

4.2   BOLD Transparency Framework

From the basic framework, was possible to identify that each dimension could be deep described with sub-dimensions. The Fig. 2 reveals it and each dimensions identified was deeply described on the Sect. 5.3

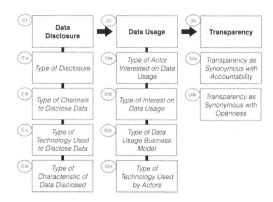

Fig. 2. The BOLD Transparency Framework

4.3   Data Disclosure Category

Being transparent requires the disclosure of data. The disclosure of data is a condition and a first principle for creating transparency. Yet simple making data online is not sufficient. Meta-data about the information quality, the way information is disclosed influences the actual transparency. Four dimensions were identified as follow and summarized at Fig. 2.

A. Type of Data Disclosure. The literature review [33] identified that disclosure of data and information in government occurs when there is Proactive dissemination by the government, Release of requested materials by the government, Public meetings and Leaks from whistleblowers. The disclosure of data prompts two "types of data disclosure", a first sub-dimension of dimension "data": (i) formal and (ii) informal. Furthermore it was identified the existence of different channels to disclose, sometimes, the same kind of data, or eventually, on different formats and conditions.

B. Type of Channels to Data Disclosure. On the dimension "types of channels to disclose data" the following types of channels were identified: (i) transparency portals [34], (ii) freedom of information access (FOIA) using all kind of channels [35], (iii) open government data portals [36], (iv) governmental portals, (v) dashboard of services advertising, (vi) outdoors with accountancy expenditures data in public works, (vii) public financial statements (newspaper, (viii) paper based at blackboards on City Hall, (ix) Internet based on portals), (x) call center to provide information and (xi) call center to provide access and demand to public services [37]. The channels afore-mentioned can use different kinds of technologies and some of them were explicated such as public financial statements, with newspaper, paper based at blackboards on City Hall and Internet based via portals.

C. Type of Technology Used to Disclose Data. The third dimension, "Type of technology used to disclose data", is part of the dimension "data" disclosure and a list of technology identified on literature is presented by: (i) politics discourses and civil servant responses, (ii) printed based (paper, newspapers, outdoors of public works and services, signs, etc.), (iii) electronic formats of data (static web portal, downloadable files, etc.) [38] and (iv) real-time electronic accessable data (direct databases access and APIs) [39].

D. Type of Characteristic of Data Disclosed. This implies on the fourth dimension "type of characteristic of disclosed data", that comprehends factors and conditions of data, such as the "quality of data", taking in consideration the type of disclosure, type of channels and technology used and characteristics identified at literature. It is important to highlight the types identified were not deeply discussed at this point and will be approached on next publication, deepening the dimensions and sub-dimensions found here and presented below: (i) data accuracy [40], (ii) data timely [41], (iii) data acessibility [42], (iv) data completeness [43], (v) data security [44], (vi) data trustiness, (vii) data free [26], (viii) data documentation, (ix) data permanently and history, (x) data primarily [41], (xi) data metadata and interlinked [45], (xii) data non-proprietary and non-discrimnatory [41, 46], (xiv) data license-free [26, 41], (xv) data machine processability [41, 46], xvi) portal simple language [47], (xvii) open data policy and license [26, 41].

4.4   Data Usage Category

Only publishing data does not create transparency. The second step of the framework flow of transparency is the Data Usage, in which the public usage data to address solutions to solve theirs interests through the best technology they consider. Taking this principle, the dimension "Data Usage" has four sub-dimensions, factors and conditions as follow and summarized at Fig. 2.

A. Type of Actor Interested on Data Usage. Anyone or any organization need to use data for some specific interest, running a determined business model and with a chosen technology to collect, treat and analyze the data. The scientific and practical literature [48, 49] identified five actors that have been using transparent data, whatever the types already described at Data Usage category and its dimensions: (i) Academics, (ii) Enterprises, (iii) Governments, (iv) Journalists and (v) Organized Civil Society.

B. Type of Interest on Data Usage. From the actors that use disclosed data, it is necessary to comprehend their interest to use the data. The scientific and practical literature [37, 49, 50] identified four types of interest on data usage dimension: (i) Service Delivery, (ii) Accountability, (iii) Advocacy and (iv) Participation.

C. Type of Data Usage Business Model. To sustain the data usage is necessary a business model. The scientific and practical literature [37, 49, 50] identified six types of data usage business model dimensions: (i) Big Data Analysis in governments, (ii) Governmental Portals and procedures for participation, (iii) Governmental portals for social control (accountability), (iv) Organized Civil Society portals and procedures for participation, (v) Organized Civil Society Data Visualization portals to intermediate relationship between governments and civil society and (vi) Private applications to improve service delivery ran by advertising or supported by civil society organizations.

D. Type of Technology Used by Actors. Taking in consideration that actor has an interest and a business model to use the data, the last type of dimensions is technology. The scientific and practical literature [7, 51] identified six types of technology that can be used: (i) Computer programming languages, (ii) Data Visualization, (iii) Geography Coding and Mapping, (iv) Networking Analysis, (v) Business Intelligence and (vi) Data Mining.

4.5   A Framework with the Dimension of Transparency

From the BOLD Transparency Framework, is possible to addres the Table 1 - BOLD Framework Summary Dimensions and Sub-Dimensions.

Table 1. - BOLD framework summary dimensions, sub-dimensions and types of specifications and characteristics

# ID	Name of category and dimensions	Sub-dimensions identified
D1	Data disclosure	–
D1a	*Type of disclosure*	*Formal* (Proactive dissemination by the government, Release of requested materials by the government, Public meetings) and *Informal* (Leaks from whistleblowers)
D1b	*Type of channels to disclosure data*	Transparency portals freedom of information access (FOIA) using all kind of channels open government

(Continued)

Table 1. (*Continued*)

# ID	Name of category and dimensions	Sub-dimensions identified
		data portals, governmental portals, dashboard of services advertising, outdoors with accountancy expenditures data in public works, public financial statements (newspaper, paper based at blackboards on City Hall, Internet based on portals), call center to provide information and call center to provide access and demand to public services
D1c	*Type of technology used to disclose data*	politics discourses and civil servant responses, printed based (paper, newspapers, outdoors of public works and services, signs, etc.), electronic formats of data (static web portal, downloadable files, etc.) and real-time electronic accessible data (direct databases access and APIs)
D1d	*Type of characteristics of data disclosed*	data accuracy, data timely, data acessibility, data completeness, data security, data trustiness, data free, data documentation, data permanently and history, data primarily, data metadata and interlinked, data non-proprietary and non-discriminatory, data license-free, data machine processability, portal simple language, open data policy and license
D2	Data usage	–
D2a	*Type of actor interested on data usage*	Academics, Enterprises, Governments, Journalists and Organized Civil Society
D2b	*Type of interest on data usage*	Service Delivery, Accountability, Advocacy and Participation
D2c	*Type of data usage business model*	Big Data Analysis in governments, Governmental Portals and procedures for participation, Governmental portals for social control (accountability), Organized Civil Society portals and procedures for participation, Organized Civil Society Data Visualization portals to intermediate relationship between governments and civil society and Private applications to improve service delivery ran by advertising or supported by civil society organizations
D2d	*Type of technology used by actors*	Computer programming languages, Data Visualization, Geography Coding and Mapping, Networking Analysis, Business Intelligence and Data Mining

5 Conclusions

Transparency is a multi-facetted concept and stakeholders give different meanings to the concept. An important contribution is the identification of two transparency concepts that are often used in the literature. One concept is accountability and the other openness. Both are overlapping with transparency, but are distinct concepts. We define transparency as the level of insight into functioning of the government. For this purpose data should be disclosed.

Although open data disclosure is a condition for transparency only the actual use can result in transparency. Hence a framework consisting of open data disclosure and use resulting in transparency was proposed where dimensions and sub-dimensions were revealed.

The identified dimensions help to understand what influences the synonymous types of transparency concept. We recommend combining information quality literature with the framework presented in this paper. Furthermore we suggest to filter the dimensions identified in this paper and determine its magnitude of influence on transparency. The next research paper will provide a case study based on the BOLD Transparency Framework identified to find refine the dimensions and sub-dimensions.

References

1. Bertot, J.C., Jaeger, P.T., Grimes, J.M.: Using ICTs to create a culture of transparency: E-government and social media as openness and anti-corruption tools for societies. Gov. Inf. Q. **27**, 264–271 (2010)
2. Dawes, S.S., Helbig, N.: Information strategies for open government: challenges and prospects for deriving public value from government transparency. In: Wimmer, M.A., Chappelet, J.-L., Janssen, M., Scholl, H.J. (eds.) EGOV 2010. LNCS, vol. 6228, pp. 50–60. Springer, Heidelberg (2010)
3. Bersch, K., Michener, G.: Conceptualizing the quality of transparency. In: 1st Global Conference on Transparency Rutgers University, Newark (2011)
4. CETIC: Survey on the use of information and communication technologies in Brazil : ICT households and enterprises 2013. Brazilian Internet Steering Committee - Regional Center for Studies on the Development of the Information Society (Cetic.br), Sao Paulo (2014)
5. Hamilton, A., Madison, J., Jay, J., Pole, J.R.: The Federalist. Hackett Publishing, Indianapolis (2005)
6. Russom, P.: Big data analytics. TDWI Best Practices Report, Fourth Quarter (2011)
7. Chen, Y.-C., Hsieh, T.-C.: Big data for digital government: opportunities, challenges, and strategies. Int. J. Public Adm. Digital Age (IJPADA) **1**, 1–14 (2014)
8. Salvador, E., Sinnott, R.: A cloud-based exploration of open data: promoting transparency and accountability of the federal government of Australia. In: 1st Symposium on Information Management and Big Data, p. 22 (2014)
9. Bannister, F., Connolly, R.: The trouble with transparency: a critical review of openness in e-Government. Policy Internet **3**, 1–30 (2011)
10. Janssen, M., Charalabidis, Y., Zuiderwijk, A.: Benefits, adoption barriers and myths of open data and open government. Inf. Syst. Manage. **29**, 258–268 (2012)
11. Eisenhardt, K.M.: Agency theory: an assessment and review. Acad. Manage. Rev. **14**, 57–74 (1989)
12. Jensen, M.C., Meckling, W.H.: Theory of the firm: managerial behavior, agency costs, and ownership structure. In: Brunner, K. (ed.) Economics Social Institutions, vol. 1, pp. 163–231. Springer, The Netherlands (1979)
13. Ward, S.J.: The magical concept of transparency. In: Ethics for Digital Journalists: Emerging Best Practices, p. 45 (2014)

14. Baume, S., Papadopoulos, Y.: Bentham revisited: transparency as a "magic" concept, its justifications and its skeptics. In: Transatlantic Conference on Transparency Research, University of Utrecht, 7 June 2012 (2012)
15. Armstrong, E.: Integrity, transparency and accountability in public administration: recent trends, regional and international developments and emerging issues. United Nations, Department of Economic and Social Affairs, pp. 1–10 (2005)
16. Naurin, D.: Transparency, publicity, accountability-the missing links. Swiss Polit. Sci. Rev. **12**, 90 (2006)
17. Fox, J.: The uncertain relationship between transparency and accountability. Dev. Pract. **17**, 663–671 (2007)
18. Bugaric, B.: Openness and transparency in public administration: challenges for public law. Wis. Int'l LJ **22**, 483 (2004)
19. Hood, C., Heald, D.: Transparency: The Key To Better Governance? Oxford University Press, Oxford (2006)
20. Roberts, N.C.: Keeping public officials accountable through dialogue: resolving the accountability paradox. Public Adm. Rev. **62**, 658–669 (2002)
21. Przeworski, A., Stokes, S.C., Manin, B.: Democracy, Accountability, and Representation. Cambridge University Press, Cambridge (1999)
22. Ball, C.: What is transparency? Public Integrity **11**, 293–308 (2009)
23. Florini, A., Birdsall, N., Flynn, S., Haufler, V., Lipton, D., Morrow, D., Sharma, S.: Does the invisible hand need a transparent glove? the politics of transparency. In: World Banks Annual Conference on Development Economics, pp. 163–184 (2000)
24. Index, C.P.: Transparency International (2007)
25. Michener, G., Bersch, K.: Conceptualizing the quality of transparency. In: paper apresentado na 1ª Conferência Global sobre Transparência, ocorrida na Rutgers University, Newark, em maio de (2011)
26. Obama, B.: Transparency and open government. Memorandum for the heads of executive departments and agencies (2009)
27. Coglianese, C.: The transparency president? the Obama administration and open government. Governance **22**, 529–544 (2009)
28. Birchall, C.: Introduction to 'secrecy and transparency': the politics of opacity and openness. Theor. Cult. Soc. **28**, 7–25 (2012)
29. Gonçalves, M.E.: Transparency, openness and participation in science policy processes. In: Pereira, A.G., Vaz, S.G., Tognetti, S. (eds.) Interfaces Between Science and Society, pp. 180–188. Greenleaf Publishers, Sheffield (2006)
30. Goncalves, O.D.: Openness and transparency, stakeholder involvement. In: Proceedings of an International Conference on Effective Nuclear Regulatory Systems: Further Enhancing the Global Nuclear Safety and Security Regime (2010)
31. Chen, H., Chiang, R.H., Storey, V.C.: Business intelligence and analytics: from big data to big impact. MIS Q. **36**, 1165–1188 (2012)
32. Zuiderwijk, A., Janssen, M.: Barriers and development directions for the publication and usage of open data: a socio-technical view. In: Gascó-Hernández, M. (ed.) Open Government, vol. 4, pp. 115–135. Springer, New York (2014)
33. Piotrowski, S.J., Van Ryzin, G.G.: Citizen attitudes toward transparency in local government. Am. Rev. Public Adm. **37**, 306–323 (2007)
34. Matheus, R., Ribeiro, M.M., Vaz, J.C., de Souza, C.A.: Using internet to promote the transparency and fight corruption: Latin American transparency portals. In: Proceedings of the 4th International Conference on Theory and Practice of Electronic Governance, pp. 391–392. ACM (2010)

35. Jaeger, P.T., Bertot, J.C.: Transparency and technological change: ensuring equal and sustained public access to government information. Gov. Inf. Q. **27**, 371–376 (2010)
36. Zuiderwijk, A., Janssen, M.: Open data policies, their implementation and impact: a framework for comparison. Gov. Inf. Q. **31**, 17–29 (2014)
37. Matheus, R., Vaz, J.C., Ribeiro, M.M.: Open government data and the data usage for improvement of public services in the Rio de Janeiro City. In: Proceedings of the 8th International Conference on Theory and Practice of Electronic Governance, pp. 338–341. ACM (2014)
38. Zuiderwijk, A., Janssen, M., Choenni, S., Meijer, R., Alibaks, R.S.: Socio-technical impediments of open data. Electron. J. e-Gov. **10**, 156–172 (2012)
39. Wong, A., Liu, V., Caelli, W., Sahama, T.: An architecture for trustworthy open data services. In: Damsgaard Jensen, C., Marsh, S., Dimitrakos, T., Murayama, Y. (eds.) IFIPTM 2015. IFIP AICT, vol. 454, pp. 149–162. Springer, Heidelberg (2015)
40. Loiacono, E.T., Watson, R.T., Goodhue, D.L.: WebQual: a measure of website quality. Mark. Theor. Appl. **13**, 432–438 (2002)
41. Obama, B.: Making Open and Machine Readable the New Default for Government Information. Executive Order, Washington DC (2013)
42. Perdue, R.R.: Internet site evaluations: the influence of behavioral experience, existing images, and selected website characteristics. J. Travel Tourism Mark. **11**, 21–38 (2001)
43. Dawes, S.S.: Stewardship and usefulness: policy principles for information-based transparency. Gov. Inf. Q. **27**, 377–383 (2010)
44. Pfleeger, C.P., Pfleeger, S.L.: Security in Computing. Prentice Hall Professional Technical Reference, Upper Saddle River (2002)
45. Berners-Lee, T., Chen, Y., Chilton, L., Connolly, D., Dhanaraj, R., Hollenbach, J., Lerer, A., Sheets, D.: Tabulator: exploring and analyzing linked data on the semantic web. In: Proceedings of the 3rd International Semantic Web User Interaction Workshop, Athens, Georgia (2006)
46. Strong, D.M., Lee, Y.W., Wang, R.Y.: Data quality in context. Commun. ACM **40**, 103–110 (1997)
47. Mendonça, D., Jefferson, T., Harrald, J.: Collaborative adhocracies and mix-and-match technologies in emergency management. Commun. ACM **50**, 44–49 (2007)
48. Zuiderwijk, A., Janssen, M., Poulis, K., van de Kaa, G.: Open data for competitive advantage: insights from open data use by companies. Proceedings of the 16th Annual International Conference on Digital Government Research, pp. 79–88. ACM, Phoenix (2015)
49. Zuiderwijk-van Eijk, A., Janssen, M.: Participation and data quality in open data use: open data infrastructures evaluated. In: Proceedings of the 15th European Conference on e-Government, pp. 18–19, June 2015, Authors version. ACPI, Portsmouth (2015)
50. Matheus, R., Ribeiro, M.M., Vaz, J.C.: New perspectives for electronic government in Brazil: the adoption of open government data in national and subnational governments of Brazil. In: Proceedings of the 6th International Conference on Theory and Practice of Electronic Governance, pp. 22–29. ACM (2012)
51. Gray, J., Chambers, L., Bounegru, L.: The Data Journalism Handbook. O'Reilly Media Inc., Sebastopol (2012)

Open Data Landscape: A Global Perspective and a Focus on China

Charles Shen[(⊠)], Zainab Riaz, Madhuri S. Palle, Qiurui Jin,
and Feniosky Peña-Mora

Advanced ConsTruction and InfOrmation techNology (ACTION) Laboratory,
Civil Engineering and Engineering Mechanics,
Columbia University, New York, USA
charles@cs.columbia.edu

Abstract. Governments are producing significant public data that, if made open, is expected to create enormous social and commercial value as well as improve the civil governance. Unleashing the true power of open public data requires a much better understanding of its ecosystem than is known currently. This paper surveys the global open data landscape by taking into account the Open Data Barometer (ODB) ranking system and its three sub-indexes - readiness, implementation and impact. These indexes are compared and analyzed on the basis of income levels of the ODB ranked countries. Finally, using air quality open data, data availability in developing countries like China is compared with countries of better practices such as UK and US. The comparison helps in understanding the current situation and barriers in opening data in China.

Keywords: Open data · Data availability · Air quality data

1 Introduction

Open data is data that can be used, re-used and distributed amongst the people without any legal, technological or social restriction [1]. It is, therefore, becoming a philosophy where data is accessible to the public for free. This approach induces a sense of accountability and transparency by building a bridge between the people and the government/organizations. Furthermore, it seeks to move beyond transparency, towards a problem solving platform in which open data can become a stepping stone to: drive more effective decision-making and efficient service delivery; spur economic activity; and empower citizens to take an active role in improving their own communities [2].

Researchers have looked at open data from many different perspectives. Some researchers focus on the relevant initiative in individual countries [3, 4] or cities [5]. Other researchers have revealed the political, social and economic impact of open data [6, 7]. Another interesting topic on open data concerns its business aspect. Hartmann et al. [8] looked at the types of business models amongst companies relying on data as their key business resource, and discussed capturing value through data driven business models. Magalhaes and Manley [9] examined 500 US firms that use Open Government

© IFIP International Federation for Information Processing 2015
M. Janssen et al. (Eds.): I3E 2015, LNCS 9373, pp. 247–260, 2015.
DOI: 10.1007/978-3-319-25013-7_20

Data (OGD) and classified them into three categories of business models: enablers, facilitators and integrators. Success stories of many practitioners of open data companies and governments have also been presented [2, 10]. In addition, integrating OGD into the Web of Linked Data has also been investigated [11] where Linked Data describes a method of publishing structured data such that it becomes more useful through semantic queries.

2 Research Motivation and Approach

The motivation behind this particular research is to tackle urban challenges from an open data perspective. According to UN [12], the percentage of population residing in urban areas is expected to increase from 30 % in 1950 to 66 % in 2050. Among the four groups of countries with different income levels [13], countries that are experiencing the fastest pace of urbanization since 1950 are upper-middle-income countries such as Brazil, China and Mexico. Recently, the Chinese government has released a plan for integration of the Beijing-Tianjin-Hebei (Jing-Jin-Ji) regions, which together is the largest mega-region covering 216,000 sq. km and affecting more than 100 million people [14]. The unprecedented urbanization poses ever-increasing sustainable development challenges to cities and the newly urbanized population.

Our research on Open Data for Sustainable Urbanization (ODSU) aims to tackle the urban challenges of these developing countries with a focus on China. In particular, we want to understand whether and how the open data ecosystem can play a role towards a more sustainable and efficient urbanization process. This research is planned in three phases. In the first phase the global open data landscape is surveyed using secondary data. On global scale, countries are categorized based on four different income levels; at the county scale we focused particularly on China, US and UK. In the ongoing second phase of our research, we are collecting and analyzing an extensive list of open data urban applications around world's major cities, identifying the best practices and evaluating their impact on urbanization efficiency. In the third phase of this research, we plan to propose and implement appropriate open data use cases in China's Jing-Jin-Ji area as a pilot study.

This paper, however, only looks into the first phase of the planned research where it attempts to survey the global landscape of open data by taking into account the Open Data Barometer (ODB) ranking index and its associated sub-indexes namely, readiness, implementation and impact [15]. An analysis is performed in order to understand the relationship between these indexes and the income levels of ODB ranked countries. Finally, to realize the current open data situation in China, a comparison is provided with the trendsetters in the ODB ranking index i.e. UK and US.

2.1   Choice of ODB

A number of open data benchmarks have been developed such as World Bank's Open Data Readiness Assessment (ODRA) [16], World Wide Web Foundation's Open Data Barometer (ODB) [15], Open Knowledge Foundation's Open Data Index (ODI) [17] and Capgemini Consulting's Open Data Economy (ODE) [18], to name few global and widely used benchmarks. However, each of these benchmarks serves a different purpose and focus. Susha et al. [19] suggest that ODB provides a more comprehensive perspective since it not only includes measures at various stages like readiness, implementation, and impact but also highlight the importance of involvement of major stakeholders and challenges throughout the open data process.

According to the authors [19], the ODB offers an insightful analysis of the entire chain (readiness, implementation, and impacts) and is a goal-oriented measure that can be used to realize how to modify implementation so as to accomplish a particular impact (economic, social, or political). However, the authors also suggest that most open data benchmarks (except for ODRA of the World Bank) produce results that are generic and ambiguous and the ranks of countries should not be expected to convey a strictly numeric position of a country but rather an approximation of reality. They specifically consider ODB more argumentative when it comes to open data diffusion.

For this particular research ODB is selected since it offers a snapshot of open data diffusion worldwide. Moreover, the research objective is to see the role of open data when dealing with unprecedented urbanization challenges particularly in developing countries like China. ODB, in addition to readiness, also offers perspective on the implementation and impact stages, which are considered useful when it comes to understanding the urban applications in developed countries and to develop guidelines for countries like China.

Furthermore, UK and US have been identified for comparison with China in this research since China is an example of the upper-middle-income countries with the fastest urbanization rate while US and UK are two high-income countries with high urbanization rate. Also, these two countries have been consistently highlighted as trendsetters by all major open data indexes. UK is ranked highest by the ODB [15], the ODI [17], the PSI Scoreboard [20] and identified as one of the trendsetters by Open Data Economy [18]. The US, in addition to being ranked 2nd by ODB and 8th by ODI, ranks highest in terms of data availability and data portal usability [18].

3 Open Data Overview

The Open Data Barometer (ODB) ranking system is a part of the World Wide Web Foundation's work on the common assessment methods for open data [15]. The weightage of each sub-index is given in Table 1. Using the ODB scores heat maps[1] are developed and presented (Figs. 1, 2 and 3) where the country colour depicts whether it's sub-index is high, moderate or low. The heat map was made so as to divide the total number of countries equally into 3 different layers so as to allow proper comparison. The lightest layer represents those countries in the bottom 1/3rd of the index ranking, the moderate layer represents those in the middle 1/3rd, and the darkest layer represents those countries in the top 1/3rd of the index ranking.

---

[1] Tool acknowledgement: www.knoema.com.

Table 1. ODB sub-indexes and their weightage (Source: ODB second edition 2014)

ODB Sub-Index	Sub-Index components and weightage
Readiness (25 %) The readiness index measures how the country in consideration is using existing open data and the extent to which open data is available.	Government (33.3 %)
	Entrepreneurs and Business (33.3 %)
	Citizens and Civil Society (33.3 %)
Implementation (50 %) The implementation sub-index is dependent on: Innovation, Social Policy and Accountability	Accountability Dataset Cluster (33.3 %)
	Innovation Dataset Cluster (33.3 %)
	Social Policy Dataset Cluster (33.3 %)
Impact (25 %) The impact sub-index is dependent on three categories: political, social and economic	Political (33.3 %)
	Economic (33.3 %)
	Social (33.3 %)

As seen on the heat map (Fig. 1) North America, a large part of Europe, Australia, Japan and South Korea have strong readiness sub-indexes. This means that these countries have strong government open data initiatives along with entrepreneur, business and citizen participation. By plotting the scored implementation sub-index of all 86 countries on a heat map, it can be observed that a few more countries fall under the high range on the heat map for the implementation index including Russia, Chile and Brazil (Fig. 2). These countries have high implementation index even though they do not have a high readiness sub-index, which leads us to question the dependency between these two sub-indexes. Finally for the impact sub index (Fig. 3), it can be seen that the only countries that seem to have a high impact sub-index are those in Europe, US, Canada and New Zealand

Fig. 1. Readiness sub-index heat map

Fig. 2. Implementation sub-index heat map

A general realization here is that many countries have a lower impact sub-index compared to their other two sub-indexes, questioning the impact that open data has on their political, economic and social standings of these countries. Moreover, another observation from the heat maps is that sub-indexes need not show complete dependency on one another. Following are few examples that reinforce this interpretation:

Fig. 3. Impact sub-index heat map

- Brazil has a moderate readiness sub-index, a high implementation sub-index and a low impact sub-index;
- Australia has a great readiness and implementation indexes but not too strong impact index doesn't;
- China and India along with some more other Asian countries have decent readiness and implementation indexes, but low impact indexes;
- Russia, Ecuador and Chile have moderate readiness sub-index but strong implementation index.

4   Open Data Overview Based on Income

The 86 countries listed in the ODB ranking system are divided into four levels of income categories as per World Bank [13] - low, lower-middle, upper-middle and high. Heat Maps for ODB score of all countries in all four categories are developed and presented (Figs. 4, 5, 6 and 7).

The low-income group consists of 16 countries out of the 86 ODB ranked countries. The average ODB score of this group is 11.69. As can be seen from the heat map (Fig. 4) mostly the countries have a low ODB score in this category with a few exceptions in Africa and Indian sub-continent. The lower-middle income group (Fig. 5) consists of 14 countries with an average ODB score of 17.66. As seen from the heat maps (Figs. 4, 5) the low and lower-middle income-ODB rank categories constitute mostly of countries from Asia and Africa.

The Upper-Middle income layer comprises of 21 countries with an average ODB score of 28.57. From the heat map (Fig. 5), it can be seen that the countries of this group that are part of Asia and Africa generally have a lower ODB score than those that fall under South America and North America.

The high-income group consists of 35 countries with an average ODB score of 57.14. An interesting observation is that all countries in the EU are not dark blue on the map. It can be seen that even though EU in general has a high open data standing [15], these practices are not standardized across the region. Furthermore, high ODB scores of countries such as the US and UK result in an overall data skew of this income class.

Figure 8 compares the lowest, average and highest ODB scores for a particular income region. The graph shows that the most significant rise is from the upper middle layer to the high layer - the low, average and high bars increase by around 50 %. Another interesting trend is observed when low to lower-middle and lower-middle to upper-middle layers are compared. Here the average and high values seem to increase by around 30 % which is also a significant figure. Therefore, on a general note, it can be

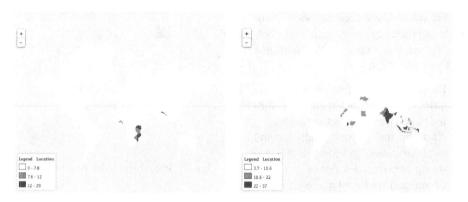

Fig. 4. Heat map for ODB scores of low income countries

Fig. 5. Heat map for ODB scores lower-middle income countries

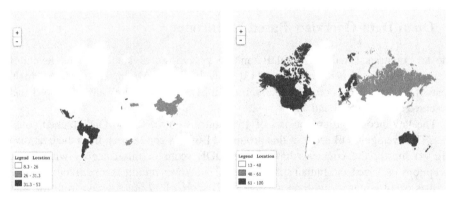

Fig. 6. Heat map for ODB score of upper-middle income countries

Fig. 7. Heat map for ODB score of high income countries

concluded that as the income increases, it is likely that ODB scores of the countries in that category also increase. This graph consists of the overall ODB score and makes an interesting case to look at the three sub-indexes separately to understand the sharp rise from the upper-middle to high income class.

Next, the three separate sub-indexes are analyzed for individual trends. The first sub-index, the readiness index (Fig. 9), has a uniform increase across all the income classes for the low, average and high values. As can be seen from the graph in Fig. 9, most changes are in the range of 33–47 %. Also, the average rise in all income levels is considerably uniform as compared to the overall ODB index analysis. Therefore, it can be concluded that the sharp rise from the upper-middle to the high class of the overall ODB scores is not dependent on the readiness sub-index.

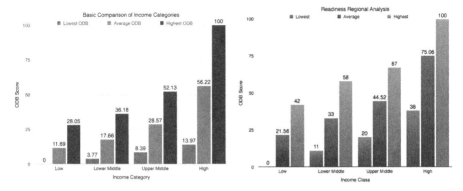

Fig. 8. Basic comparison of income categories

Fig. 9. Readiness regional analysis

Now, observing the graph for the implementation sub-index (Fig. 10) shows that the lowest, average, and highest values across all the income categories are increasing, however, not uniformly. In fact, the rise from the upper-middle class to the high class appears to be very similar to that of the increase seen in the overall ODB analysis earlier. Since the implementation sub-index weighs 50 % of the entire ODB index, it can be concluded that this jump plays an important role in the upsurge portrayed in the overall ODB analysis. Moreover, it can also be observed from Fig. 10 that the rise in the lower category of income is not following the same trend as the overall ODB and the readiness sub-index pattern (both have a constant increase in all levels). In fact, the percentage increase of the implementation index actually decreases for the lower-middle category. Also, from the graph it can be noticed that some countries, despite their reasonably high income, have not implemented open data as efficiently as one would expect them to do so. In this income group, 9 countries have a sub-index of less than 40, which is 25 % of the total number of countries in the class.

The impact sub-index (Fig. 11), on the other hand is very different from the other two sub-indexes. Every income class has at least one country with a zero sub-index. This fact, along with the minimal averages, leads one to believe that this is by far the weakest index. Although countries are showing to have reasonable readiness and implementation sub-indexes, their impact sub-index is below par, proving that more open data initiatives are needed in the economic, social and political sectors. The only similarity with the other two sub-indexes is the surge from the upper-middle income class to the high income class. Although the impact sub-index weighs only 25 % of the total ODB score of a country, this rise in the impact index plays a considerable role in the jump seen in the overall ODB analysis because the percentage increase is extremely large. Therefore, it can be concluded that the implementation sub-index and the impact sub-index are responsible for the increase in the overall ODB scores from the upper-middle income class to the high income class.

In order to understand the overall landscape of the ODB ranked countries along with the representation of the three sub-indexes a line graph (Fig. 12) is developed. Here, the overall ODB country rank, 86 in total, is plotted on the x-axis whereas the

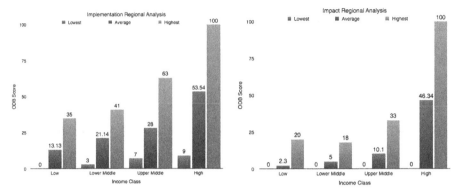

Fig. 10. Implementation regional analysis   **Fig. 11.** Impact regional analysis

y-axis highlights the ODB sub-index score. Hence, one point on the x-axis represents the country's ODB rank and the three corresponding coloured points on the y-axis are translated into the respective sub-index scores. The graph shows that countries have higher ODB ranks due to better performing sub-indexes. It can be seen from the line graph that there are certain anomalies. For example, Chile has an overall ODB rank of 15 with the readiness sub-index of 69, the implementation sub-index of 73 and the impact sub-index of 8. Here the implementation sub-index is higher than the readiness sub-index. In fact, the impact sub-index is very low for the country as it is a part of the high income layer and has a reasonably high ODB rank as well. The open data hasn't had a noticeable impact on government efficiency, social policies and the economy mostly due to the lack of government initiatives and low entrepreneurial activity in the country.

Another similar example is that of Brazil with an overall of ODB rank of 21 with a moderate readiness sub-index, high implementation sub-index but a very low impact sub-index when compared with other counties in similar range of ODB rank. It is observed that even though the openness in Brazil's 2013 ODB results is pretty large for categories such as census, government spending, international trade etc. none of these categories actually adhere to the full open data standards hence giving it an overall moderate readiness factor. Also, it has been observed in case of Brazil that although there are a number of open data policies in place by government, the policies do not really pay much attention to the actual user perspective or overcoming the impediments of the use of open data [21]. This means that in order to improve the impact sub-index, the policies must be refined in a way that only fully open data is released benefiting the civil society at large.

We also looked at the percentage difference from lower income tier to higher income tier countries in terms of their income vs. ODB values. The World Bank classification of the four income country categories is based on GNI per capita [13]. Therefore, in Table 2 we listed these percentage differences of the average values of GNI, GNI per-capita and ODB. It can be seen that in the bottom tier jump (low to lower middle) and middle tier jump (lower middle to upper middle), the ODB value increase falls significantly behind the GNI increase. For example, at the low to lower middle

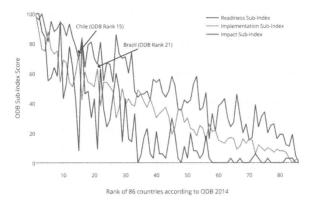

Fig. 12. ODB sub-index line graph

jump, the average ODB value percentage increase is only half of the average GNI per capita increase, and nearly 1/3 of the average GNI increase. At the top tier, however, the numbers are much more consistent. In the jump from upper middle to high class, the average ODB value percentage increase is nearly the same as that of the average GNI, and is over 60 % of the average GNI per capita increase. These numbers probably suggest that higher income countries generally put up more open data efforts, resulting in their open data status more likely matching their income level status. A more solid conclusion would require future work that looks into more details of the breadth and depth of various income and ODB parameters.

Table 2. National income and ODB values - percentage differences among income groups (11 countries without updated GNI values are not considered)

ODB sub-index	Avg. GNI (US$)	Avg. GNI per Capita (US$)	Avg. ODB values
Upper middle to high	45.17 %	78.85 %	48.18 %
Lower middle to upper middle	60.17 %	74.29 %	38.19 %
Low to lower middle	90.9 %	66.32 %	33.8 %

5   UK, US and China Comparison

After reviewing the global open data landscape, this paper compares highly ranked ODB countries like UK and USA with China in more depth. China has been taken in this analysis as it is an epitome of the developing world. Taking into consideration it's size, population and gross domestic product, it should indeed be releasing vast amounts of data, contributing to the society and making use of areas such as machine learning and business intelligence. However, government initiatives for open data pose a major challenge to this contribution. This section analyses the current open data situation of the country and the barriers that it needs to overcome.

5.1 ODB Sub-indexes and Their Relationship

As observed previously, there is a substantial jump in the average ODB scores of countries from the upper middle income category to the countries of the high income. In the previous section, it has been concluded that this was mainly due to the implementation and impact sub-indexes. In this section, a comparison is performed between the overall ODB scores, readiness sub-index, implementation sub-index and impact sub-index of UK, US and China (Fig. 13). The US and UK can be considered examples of the high income layer, and China is an example of the upper-middle income layer. UK and US are ranked first and second respectively in the ODB rankings of 2014 [15]. The reason as to why UK and US rank so high is because of legislations in their respective countries [22]. In recent few years,

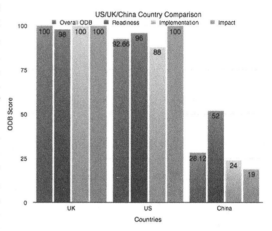

Fig. 13. UK, US, China Country comparison

these governments have launched a number of initiatives that basically target health, energy, climate, education, finance, public safety and development sectors thereby, improving open data initiatives [22].

China, on the other hand ranks 46th in the ODB ranking. Although its readiness sub-index is just twice as low as the other two countries, it's lagging behind in the implementation and impact sub-indexes, which is 75 % of the overall ODB weightage. Therefore, it can be observed that China needs to work on factors such as making datasets fully open data compliant in the fields of innovation, social policy and accountability, implementing strong open data legislation as well as maximizing impact in the fields of political, social, and economic importance.

5.2 National Data Portal

Both US [23] and UK [24] have created national data portals where data.gov and data. gov.uk have released 14,008 and 22,385 datasets respectively. The most common machine-readable formats for US and UK are XML and CSV, while popular non-machine readable formats for the two are HTML and PDF respectively. In addition, US also offers a significant number of datasets in zip format as compared to UK.

Unlike US and UK, China does not have a national data portal yet. However, certain open data is available through different agencies. One such example is availability of open data through National Bureau of Statistics of China (NBSC), which offers both a Chinese and an English version. The Chinese version is organized into monthly, quarterly, annual, regional, international and census data. The English version only consists of four categories i.e. monthly, quarterly, annual and regional data. All data from the English version is in machine-readable format. However, the same doesn't apply for the Chinese version, as it is only available in HTML format.

5.3   Dataset Example – Air Quality Index

In this section, we look at an example dataset common in all three countries, the air quality metrics, which measures the air pollutants level in daily air quality. The data is from the US Environment Protection Agency [25], U.K. Department for Environment Food and Rural Affairs [26], and the Chinese Ministry of Environmental Protection [27]. In U.S. and U.K. the metrics is reported by Air Quality Index (AQI), while in China, prior to 2012 the metrics was reported by Air Pollution Index (API). China switched to AQI in 2012 [27]. From the datasets it is clear that for UK and US AQI data is consistently available all regions and major cities since year 2000.

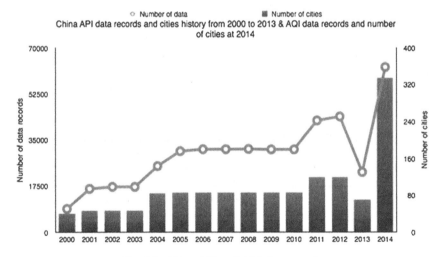

Fig. 14. China API and AQI data records

China's air quality metrics, on the other hand, has been reported with API since 2000. Figure 14 shows that in 2013 the number of reporting cities as well as the number of API datasets reduced dramatically due to the transition process from API to AQI. However, there is an obvious improvement afterwards as the number of cities with available AQI data increased to 335 in 2014 from 120 in 2012 with API data.

According to the ODB measurement methodology and weightage mechanism [15], we have calculated the ODB score for AQI data for the three countries. Our findings were, on a full score of 100, UK obtains 95 with the only limitation of no linked data URLs. U.S. and China receive scores of 80 and 50 respectively; the main reasons that pull China's score down are related to machine-readability and ability to download data. Since the Chinese dataset is already provided in HTML, technically it is not difficult to incorporate both of these aspects. Another reason for reduced score is an explicit link for open-license, which should be even easier to address.

6   Conclusion

We presented a heat-map analysis of the global open data landscape based on the ODB ranking system. We looked at the overall ODB indexes and its associated readiness, implementation and impact sub-indexes, for countries of the low, lower-middle, upper-middle and high income-levels, respectively. Our results show that in many countries the three sub-indexes do not exhibit dependency on each other. The impact sub-index is found to be most often the weakest part of the three, and in quite a few cases extremely low compared to the readiness and implementation indexes. This observation shows that, on one hand, governments around the world are establishing more and more open data initiatives and citizens are engaging in an increasing number of open data activities; on the other hand, tangible political, social and economic benefits from open data remain to be seen. This may be because we are still in the early stage of the entire open data life cycle where harvests are yet to be reaped. However, the extremely large gaps between the impact and other two sub-indexes in some countries may warrant a thoughtful review of the existing open data initiatives to more effectively align the investments with the expected results of open data.

In addition to the global open data perspective, we also provide a comparison of China with leading open data advocates, UK and US. We found that although China is lagging behind the two other countries in the three sub-indexes, gap in the implementation and impact sub-index is much larger than the gap in readiness sub-index. This shows the Chinese government has made good progress in facilitating open data initiatives from the policy and regulatory front, but more needs to be done especially on how to put those policies into execution, which is crucial for a positive impact of open data. Follow-up example of air quality data further confirmed China's clear progress in making data ready, but not yet providing data optimal for implementation. In summary, we believe that a number of natural steps can be taken to boost China's open data status to its next level, e.g., establishing a national and regional data portals will facilitate interested parties to find the right data; making available more machine-readable data will dramatically improve its usability and value.

References

1. Open Knowledge. What is Open? (2015). https://okfn.org/opendata/. Accessed 12 Feb 2015
2. Goldstein, B., Dyson, L.: Beyond Transparency: Open Data and the Future of Civic Innovation, 1st edn. Code for America Press, San Francisco (2013)

3. Huijboom, N., Van den Broek, T.: Open data: an international comparison of strategies. Eur. J. ePractice 12, March/April 2011
4. Heimstädt, M., Saunderson, F., Heath, T.: Conceptualizing Open Data ecosystems: A timeline analysis of Open Data development in the UK, Discussion Paper, School of Business & Economics: Management, No. 2014/12
5. Kassen, M.: A promising phenomenon of open data. Gov. Inf. Q. **30**(4), 508–513 (2013)
6. Janssen, M., Charalabidis, Y., Zuiderwijk, A.: Benefits: adoption barriers and myths of open data and open government. Inf. Syst. Manage. **29**(4), 258–268 (2012)
7. Zuiderwijk, A., Janssen, M.: Open data policies, their implementation and impact: a framework for comparison. Gov. Inf. Q. **31**(1), 17–29 (2014)
8. Hartmann, P.M., Zaki, M., Feldmann, N., Neely, A.: Big Data for Big Business? Working Paper, Cambridge Service Alliance, University of Cambridge (2015). http://www.cambridgeservicealliance.org/. Accessed Feb 2015
9. Magalhaes, G., Manley, L.: Business models for open government data. In: Proceedings of the 8th Annual International Conference on Theory and Practices of Electronic Governance (ICEGOV), Guimarães, Portugal, Oct 2014
10. WWW Foundation (2014). http://opendataresearch.org/sites/default/files/posts/Common%20Assessment%20Workshop%20Report.pdf. Accessed 20 Feb 2015
11. Shadbolt, N., O'Hara, K., Berners-Lee, T., Gibbins, N., Glaser, H., Hall, W., Schraefel, M.C.: Linked open government data: lessons from Data.gov.uk. IEEE Intell. Syst. **27**(3), 16–24 (2012)
12. World Urbanization Prospects - The 2014 Revision (2014). http://esa.un.org/unpd/wup/Highlights/WUP2014-Highlights.pdf. Accessed 5 Mar 2015
13. The World Bank. Data (2015). http://data.worldbank.org/about/country-and-lending-groups. Accessed 21 Mar 2015
14. President Xi's Requirements on Jing-jin-ji Coordinated Development Plan (2015). http://news.xinhuanet.com/politics/2014-02/27/c_119538131.htm. Accessed 19 Jun 2015
15. Open Data Barometer, 2nd edn. (2015). http://www.opendatabarometer.org/report/about/method.html. Accessed 5 Mar 2015
16. The World Bank – Open Government Data Working Group (2013). http://data.worldbank.org/sites/default/files/1/od_readiness_-_revised_v2.pdf. Accessed 20 Feb 2015
17. Open Knowledge Foundation – Open Data Index (2014). https://index.okfn.org/country. Accessed 19 April 2015
18. Capgemini Consulting – The Open Data Economy: Unlocking Economic Value by Opening Government and Public Data (2013). http://ebooks.capgemini-consulting.com/The-Open-Data-Economy/#/4/. Accessed 19 April 2015
19. Susha, I., Zuiderwijk, A., Janssen, M., Gronlund, A.: Benchmarks for evaluating the progress of open data adoption: usage, limitations, and lessons learned. Soc. Sci. Comput. Rev. (2014). doi:10.1177/0894439314560852
20. ePSI Platform – The PSI Scoreboard (2015). http://www.epsiplatform.eu/content/european-psi-scoreboard. Accessed 19 April 2015
21. Matheus, R., Ribeiro, M.M.: Open data in legislative: The case of Sao Paulo (2014). http://www.opendataresearch.org/content/2014/665/open-data-legislature-case-s%C3%A3o-paulo-city-council. Accessed 25 Mar 2015
22. The White House. Open Data Initiative (2013). https://www.whitehouse.gov/blog/2014/05/09/continued-progress-and-plans-open-government-data. Accessed 06 Apr 2015
23. Data.Gov (2015). http://www.data.gov/. Accessed 2 Feb 2015
24. Data.Gov.UK Opening Up Government (2015). http://data.gov.uk/. Accessed 2 Feb 2015

25. EPA. Air Quality Report (2014). http://www.epa.gov/airquality/airdata/ad_rep_aqi.html. Accessed 20 Apr 2015
26. Department for Environment Food & Rural Affairs (2015). http://uk-air.defra.gov.uk/air-pollution/daqi. Accessed 19 Apr 2015
27. Ministry of Environmental Protection of the People's Republic of China (2015). http://datacenter.mep.gov.cn/. Accessed 12 Apr 2015

Open Data Platforms and Their Usability: Proposing a Framework for Evaluating Citizen Intentions

Kawaljeet Kapoor[✉], Vishanth Weerakkody,
and Uthayasankar Sivarajah[✉]

Brunel Business School, College of Business,
Arts and Social Sciences, Brunel University London, London, UK
{Kawaljeet.Kappor,Vishanth.Weerakkody,
Sankar.Sivarajah}@brunel.ac.uk

Abstract. Governments across the world are releasing public data in an effort to increase transparency of how public services are managed whilst also enticing citizens to participate in the policy decision-making processes. The channel for making open data available to citizens in the UK is the data.gov.uk platform, which brings together data relating to various public services in one searchable website. The data.gov.uk platform currently offers access to 25,500 datasets that are organized across key public service themes including health, transport, education, environment, and public spending in towns and cities. While the website reports 5,438,159 site visits as of June 2015, the average time spent on the site has been recorded at just 02:12 min per visitor. This raises questions regarding the actual use and usability of open data platforms and the extent to which they fulfill the stated outcomes of open data. In this paper, the authors examine usability issues surrounding open data platforms and propose a framework that can be used to evaluate their usability.

Keywords: Open data · Citizen · Usability · Evaluation · Public services

1 Introduction

The push for making public services data available to the community started around mid 2000s with the European Union directive encouraging greater realisation of the economic value of public data through its reuse in 2003. This directive, combined with the advancement of Internet and associated ICT tools facilitating data analytics [29], has paved way for governments to open up data to the community. Conventionally, government departments retained public service data within their systems, with limited information being released to citizens and other stakeholders (businesses, charitable organizations, and NGO communities). However, since the last decade, the spread of digital governance and associated norms such as responsiveness, accessibility and efficiency of public services, transparency and accountability, have motivated governments to exploit the potential of wider distribution and use of such data [28]. One of the first countries to mandate the use of open data was the United States. The Presidential Open Government Directive in December 2009 required the use of open

© IFIP International Federation for Information Processing 2015
M. Janssen et al. (Eds.): I3E 2015, LNCS 9373, pp. 261–271, 2015.
DOI: 10.1007/978-3-319-25013-7_21

formats by all federal US agencies. Six months later, the UK followed with their own plans for open public data, with the Prime Minister announcing the setting up of the data.gov.uk website. Subsequently, the European Commission published a Communication on Open Data in 2011, and in the same year, USA, UK, and initially six other countries were signatories to the Open Government Declaration.

The UK is considered as one of the leading countries in Europe for open data. The data.gov.uk is one of the most comprehensive open data repositories making available non-personal UK government data about public services, ranging from health, social services, education, transport to crime and other geo-environmental data. It was launched in closed-beta in September 2009 and publicly launched in January 2010 (data.gov.uk/blog/the-new-datagovuk). When the data.gov.uk website was officially launched in January 2010, ordnance survey data which provides information on geographical locations was one of the key datasets that was opened up as part of the project [3]. Subsequently, in June 2010, the Treasury released the Combined Online Information System (COINS), which operates as the UK Government's central accounting system detailing the spending of all government departments and their major spending programmes [38]. As of June 2015, the data.gov.uk website contains over 25500 datasets. The data can be used by individuals, businesses and other stakeholders under the conditions that the copyright and the source of the data is acknowledged by including an attribution statement specified by data.gov.uk.

One of the motivations of the open data movement has been to make more data easily accessible to diverse stakeholders with a view of enticing them to participate and contribute to the public policy-making space. For example, it is anticipated that researchers, think-tanks, entrepreneurs, businesses leaders, representatives of public services, NGOs, charities, community groups and citizens, at large, will use the open data to contribute to the policy decision making process, particularly across local governments. Indeed, this has encouraged these stakeholders to take an active interest in the way services are currently being delivered, and has stimulated thinking around how to improve services. Although countries such as the UK and US have taken proactive steps to improve the availability and ease of use (through machine-readability and technical standards), there still remain several barriers to accessibility and usability of open data. Moreover, the actual use of open data is cumbersome and stakeholders have to do it themself. In order to fully exploit the potential of open data, users will usually require a certain level of applied skills. The fact that there is no existing easy-to-use, proven solution which can help citizens exploit the open data for decision making regarding their own lives, or contribute to the wider public policy making debate, does not help. Such issues are further compounded by the generic nature of open data repositories such as www.data.gov.uk and www.epsiplatform.eu, and thus their relevance and direct interest to citizens, in particular.

The motivation for this paper lies in the reasoning that although the availability of open data offers many opportunities for citizens, no research exists that questions the usability of open data platforms, particularly from a citizen's perspective. In this paper, we set out to examine and discuss some of the salient factors that influence the usability of open data by citizens and propose a conceptual framework to encapsulate these factors. In order to do this, we first review the benefits and challenges of open data followed by examining the role of open data platforms, and motivations for using such platforms. Thereafter, we identify potential measures for evaluating the usability of open data platforms and propose a framework to capture these. We conclude the paper by offering a discussion to synthesise the main arguments presented in the paper, identifying the main limitation, and pointing at future research directions.

2 Open Data: An Overview

The goal of Open Data initiatives has been to open all non-personal and non-commercial data, especially data collected and processed by government organizations [1]. It can be seen as a movement very similar to the Open Source or Open Access phenomenon. In the course of this trend, public sector organisations have started making governmental data available on web portals, as web services so that the public have access to these data at a single point of access to official datasets. The increase in availability of open data initiatives has been seen as mainly due to the growing pressure imposed by governments on all kinds of public organisations to release their raw data [18]. The key motivators to encourage public organisations for publishing data revolves around government's perception that the open access to publicly-funded data provides: (a) greater economic returns from public investment [11], (b) provides policy-makers with data needed to address complex problems [7], (c) generate wealth through the downstream use of outputs [18], and (d) help involve citizens in analysing large quantities of datasets [30]. In general, the overarching arguments for stimulating open data are highlighted as the increase in political economic growth and the contribution to public values (i.e. transparency and accountability).

2.1 Benefits and Challenges in Using and Accessing of Open Data

Many scholars believe that Open data can be a valuable resource of information if published in a useful manner (e.g. [1, 2]). Some of the key benefits and challenges identified in the extant literature (e.g. [10, 11, 19, 32] are synthesised and presented in Tables 1 and 2, respectively. These tables do not provide an extensive list of benefits and challenges of open data use, but highlight the prominent opportunities and concerns discussed in the literature.

Although open data can potentially provide many benefits, its use also comes with a number of challenges. Some of the key challenges identified in the extant literature are presented in the following table.

2.2 Use of Open Data Platforms

The main purpose of open data platforms has been to promote access to government data and encourage development of creative tools and applications to engage and serve the wider community [22]. In doing so, enabling civic engagement by providing opportunity for citizens, public sector organisations, businesses and independent developers to use systematically-updated stream of open data is being encouraged.

Table 1. Benefits of open data

Benefits	Description	References
Increased transparency and accountability	Making government data transparent should increase public trust in government and civil servants and also allow citizens to hold the government officials accountable	[11, 19, 32]
Economic growth	Opening government data is believed to bring a range of economic benefits such as encouraging the marketplace to develop products and services, which increase productivity, offer employment, and bring revenue back to government in the form of taxation revenue	[1, 19]
Societal benefits	Potential to allow citizens to interact with government in a more informed and interactive manner	[32]
Cost reduction and efficiencies	Sourcing data is often significantly costly in both time and money for organisations. Opening government data can significantly reduce the costs associated with acquiring data	[33]
Improved data quality	Leads to improved data quality via crowdsourcing of corrections or by filling gaps in data	[11]
Simulating innovation	Opening government data encourages developers and the general public to explore and play with new data that might lead to development of innovative solutions	[19, 32]

The governments perceive that making this data available on the web would lead to more transparency, participation, and innovation throughout society [10, 19]. Often open data platforms publish datasets covering a wide range of domains, from environmental data over employment statistics to the budgets of municipalities. Publishers of these datasets can be individual government agencies or providers of larger repositories that collect public datasets and make them available in a centralized and possibly standardized way. Governments and publishers of open data expect the users to exploit these data in many ways as possible for the benefit of the society [12]. For example, general public (non-technical users) may use it simply to analyse trends over time from one policy area, or to compare how different parts of government go about their work. On the other hand, technical users such as software developers are encouraged to create useful applications out of the raw data files, which can then be used by everyone benefitting the wider society.

In terms of the process to find the available open data, end-users of these platforms (i.e. citizens, businesses) who wish to access and use Open Data need to first identify relevant datasets manually or by visiting a central repository/platform (e.g. data.gov. uk). In the case of finding datasets manually, this includes finding organizations or agencies that publish open datasets on platforms that provide a central and responsive entry point where users can search for data. If a single dataset can be found, that

Table 2. Challenges of open data

Challenges	Description	References
Increasing public interest	Challenge of raising the capacity and awareness of civil servants, citizens and the private sector on their rights to access and re-use public data initiatives	[32, 40]
Cost of opening up data	Time and resource costs are seen as obstacles for government departments in opening their data, especially as they were often experienced as upfront costs	[11, 32]
Data ownership risks and legality concerns	In the past, if councils or government departments contracted a third party to gather data for them, or purchased data directly from the third party, they often licensed the data and did not own the intellectual property rights, and thus could not directly release it under open copyright	[18, 41]
Uncertainty about data stream continuity	If a user is not positive that a data stream will be maintained in the future, this creates uncertainty around any project using that data stream. This reduces the chances that an organisation or individual will be willing to invest the time and resources into a product or application that uses this data	[18, 32]
Data quality concerns	Government departments may be reluctant to release data that they see as low quality. Some agencies are worried about the potential liabilities of releasing their data concerning information accuracy, up-to-dateness etc.	[10, 39]
Privacy violation	Data that includes private or potentially sensitive information on citizens; there can be concerns over whether and how the data can be anonymised, what can be released, to whom and under what copyright	[9, 11]

contains all the relevant data, the user can directly extract the required information. However, it is rather unlikely to find all relevant data in a single file. The way people access and use Open Data is greatly influenced by the way the data is published [1]. Many government agencies or organizations collect large amounts of data. In its original, raw form, this data is often not very useful for end users. Therefore, many datasets are cleaned and customized before being published. While some publishers prefer the data to be in a human-readable format, others prefer a machine-readable format. Apart from accessing data from these platforms, users (e.g. organisations) are also encouraged to submit useful data that can be published to the general public. Government open data initiatives are also encouraging users in a number of ways to be involved as part of these projects dependent on their background or interest. For example, one of the challenges is making existing data come to life, and users are encouraged to combine and reorganise existing data to offer new insights resulting in useful visualisations of these data [12].

3 Developing Measures for Evaluating the Acceptance of Open Data Platforms

Websites such as data.gov.uk make it easy for citizens to access governmental data and other offered services whilst increasing citizens' potential of contributing to democratic processes [14]. According to Wangpipatwong et al. [37], citizen use of such websites substantially reduces the management and operational costs for the government. This study aims to empirically investigate the use of the aforementioned open data website from a citizen's perspective. A suitable mix of measures will be borrowed from the available innovation adoption models to evaluate the citizens' continued use intentions of such websites. This will be undertaken by gathering the opinions of those who already have the experience of using data.gov.uk along a set of measures identified from the literature.

Available literature shows that very few studies have attempted to empirically evaluate the performance of open data websites. There are, however, evidences of other studies using different measures of innovation adoption to investigate the performance of different websites. For instance, Wangpipatwong et al. [37] use the Technology Acceptance Model (TAM) alongside self-efficacy as an added measure, to evaluate the use of an e-government website. Fang and Holsapple [15] focus on the navigation structure of a website and their impact on the usability of that website by using factors defining its usability. Wang and Senecal [36] used ease of use, speed, and interactivity to measure the usability of a website and its subsequent impact on user attitudes and intentions.

The literature is rich with theoretical models, mostly developed from the psychology and sociology theories, which assist in analysing the acceptance of a service or a product [34, 35]. Some of the most used models come from the following theories: Diffusion of Innovations theory (DOI) by Rogers [25], Theory of Reasoned action (TRA) by Fishbein and Ajzen [16], Theory of Planned Behaviour (TPB) by Ajzen [4] and Ajzen and Fishbein [5], Technology Acceptance Model (TAM) by Davis [13], Decomposed Theory of Planned Behavior, Extended Technology Acceptance Model, and Unified Theory of Acceptance and Use of Technology by Venkatesh et al. [35]. The DOI theory is regarded as a principal theoretical perspective on technology adoption offering a conceptual framework for discussing adoption at a global level. Rogers [27] has synthesized sixty years of innovation-adoption research in developing this theory. His DOI model has been well received in the world of innovative solutions, and it is one of the most used theories in the field of innovation diffusion [20, 31].

Rogers [27] identified the following five attributes as the perceived attributes of innovations within DOI – relative advantage, compatibility, complexity, trialability, and observability. It can be easily observed from the attributes used in the aforementioned models that the TPB model is an extension of the TRA model, and the decomposed TPB model shares similarities with TAM. Fishbein and Ajzen [16] incorporated attitudes, subjective norms and behavioural intention in their TRA model.

TAM is also regarded as an adaptation of the TRA model and the TAM model also shares two attributes with the DOI model (relative advantage/perceived usefulness and complexity). Davis [13] identified perceived usefulness and ease of use alongside the effects of attitude on intention in their TAM model, as the factors influencing the acceptance of a technology. Giving due consideration to all of these innovation adoption models, the following attributes were shortlisted depending upon their relevance to the case of open data website being covered within this study (Fig. 1): perceived usefulness, compatibility, ease of use, result demonstrability, trust, risk, social approval, visibility, and behavioural intentions.

Perceived usefulness, also referred to as the relative advantage, will help assess if the information available on the website is relatively better across multiple aspects in comparison to the same data that a citizen can access via other physical offices and platforms. In measuring the advantages of a new service, users tend to evaluate the pluses and minuses of using that service. This characteristic is known to determine the ultimate rate of most innovation adoptions in the long run [24]. In terms of compatibility, the website will be assessed for the type of information it offers to the citizens with respect to the type of information the citizens are interested in, or are expecting to, access using such open data platforms. Rogers [27] describes compatibility to be the degree with which the introduced innovation manifests itself as being consistent with users' past experiences, present values, and their future needs.

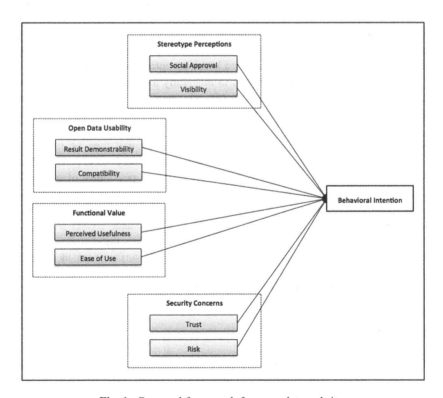

Fig. 1. Proposed framework for open data websites

Users' knowhow of a service tends to dictate their perception of the level of effort involved in using that service. The less complex a service is to use/operate, the more easily it is accepted. The ease of using the data.gov.uk website will be evaluated from a perspective of optimizing user experience; the design of the interface, time required to look up for the desired information, understandability of different features/tabs offered within the website, and any other navigation complexities will be explored using this attribute (page layout, scrolling and paging, text appearances, links, search and so on). Result demonstrability will be measured along users' satisfaction of the quality of information that they can retrieve using the open data website. The trust factor will be used to measure the extent to which the users trust the validity of the information that they are accessing using such open data websites. A user's uncertainty about the quality of information being gathered can potentially lead to anxiety, which can come in the way of their adoption decision. The expected social or economic loss caused from using a new system constitutes perceived risk [21, 26]. In this context, the risk factor will be used to measure users' apprehensions of inputting personal information onto such websites, and also their confidence in using the information available on such websites.

The members of a social system generally tend to display a sense of belonging by being a part of the activities that are regarded as a norm within their social system [23]. Social interaction and information exchange can play critical roles in promoting the use of a new system, in effect, motivating individuals to accept that system [6]. Social approval is a construct that will help measure approval from a user's social circle (friends, families, and peers) regarding the use of open data platforms and their authenticity. Visibility of a system encourages peer discussions of that system, which collectively contributes towards achieving a better acceptance rate for that system [27]. Thus, the visibility construct will be used in this study to help identify the awareness that the citizens have about the existence of such open data platforms that they can utilize to their benefit on a daily basis.

The effects/influences of perceived usefulness, compatibility, ease of use, result demonstrability, trust, risk, social approval, and visibility will then be individually measured across users' behavioural intentions. Behavioural intention is perceived as an instinctive probability that a user relates directly with the possibility of a particular behavioural option being chosen [8]. Some models of innovation adoption and diffusion (TRA and TPB) acknowledge this attribute to be the best immediate predictor of the acceptance of a given service/product [23]. These eight characteristics are expected to positively and significantly impact users' intentions towards the usage of open data platforms. The positive or negative correlations that will surface post the empirical evaluations will then be logically reasoned and analysed for their significance in influencing users' intentions towards using the data.gov.uk website.

4 Discussion and Conclusions

This paper focuses specific interest on open data platforms to establish an understanding of its usability from a citizen perspective. In the UK, data.gov.uk is the functional open data website available to be accessed by the public. The UK government has introduced an action plan for a smarter and more efficient government to open the government and promote transparency in empowering citizens and their roles in the civic society; their focus is on redefining the relationship between the frontline and the centre to better manage finances via efficient delivery [33].

In terms of theoretical implications, this study broadly touches upon the streams of open government, ICT literature, and digital governance. The framework presented in this paper has been developed from well-established and most used theories in innovation adoption. This framework can be applied across different open data domains to assess the impact of transparent governance on empowering and encouraging citizen engagement in open government data initiatives. Other researchers can use this framework to build upon, as required, to run in-depth analyses of specific aspects (such as trust in available information, level of risk involved in utilizing that information, and so on) of open data and its usability across different contexts.

In terms of practical implications, a significant milestone of this research is the development of a framework that would enable a reliable assessment of the open data platforms. A 2015 report from a four million project funded by the European commission revealed that one of the issues of such open data platforms is that most data owners in the public sector hold a poor understanding of the relative benefits of different data formats [17]. Most owners tend to adopt a path of least resistance and publish the data in its original format, ignoring the potential of making the data available in more reader-friendly capable formats. Nevertheless, such open data holds numerous practical implications for international data standard forums [17]. The framework will assist policymakers, publishers of open data, IT practitioners (application developers), and other proactive citizens in analyzing the usability of open data platforms. This framework will help evaluate the perceived usefulness of readily available open data, whilst measuring its compatibility with user needs. The stakeholders will also be able to assess the quality of information available in these websites across the risk and trust perceptions of the involved users. Other issues such as user friendliness and usefulness will also be measurable across the aspects of ease of use and result demonstrability. Overall, the constructs put together in the framework will help practitioners to summarize the effectiveness of the specific open data platforms being evaluated, to be eventually bettered for future use by the citizens.

This study suffers from the limitation of no empirical evidence supporting the validity of the proposed framework. This is an ongoing research, and having established a framework for evaluating open data, our future research will be focused on empirically assessing the validity of this framework in the context of a UK open data website: data.gov.uk. Exclusive scenarios will be defined prior to the empirical evaluations; for instance, the users will be directed to explore specific categories (housing, environment, taxes, health, and so on) in the targeted website to assess the retrieved results against the framework proposed within this study. This will help analyze and identify problem areas, if any, to be marked for improvement by the publishers of such open data.

References

1. Borzacchiello, M.T., Craglia, M.: The impact on innovation of open access to spatial environmental information: a research strategy. Int. J. Technol. Manage. **60**(1–2), 114–129 (2012). doi:10.1504/ijtm.2012.049109
2. Braunschweig, K., Eberius, J., Thiele, M., Lehner, W.: The State of Open Data Limits of Current Open Data Platforms (2012)
3. BBC News. Ordnance Survey offers free data access. http://news.bbc.co.uk/1/hi/8597779. stm (2009). Accessed 10 Jun 2015
4. Ajzen, I.: From intentions to actions: a theory of planned behaviour. In: Kuhl, J., Beckman, J. (eds.) Action-Control: from Cognition to Behavior, pp. 11–39. Springer, Heidelberg (1985)
5. Ajzen, I., Fishbein, M.: Understanding Attitudes and Predicting Social Behavior. Prentice-Hall, Englewood Cliffs (1980)
6. Bandura, A.: Social Foundations of Thoughts and Action: a Social Cognitive Theory. Prentice-Hall, Englewood Cliffs (1986)
7. Arzberger, P., Schroeder, P., Beaulieu, A., Bowker, G., Casey, K., Laaksonen, L., Wouters, P.: An international framework to promote access to data. Science **303**, 1777–1778 (2004)
8. Chiu, R.K.: Ethical judgment and whistleblowing intention: examining the moderating role of locus of control. J. Bus. Ethics **43**(1–2), 65–74 (2003)
9. Choenni, S., van Dijk, J., Leeuw, F.: Preserving privacy whilst integrating data: applied to criminal justice. Inf. Polity **15**, 125–138 (2010)
10. Conradie, P., Choenni, S.: On the barriers for local government releasing open data. Gov. Inf. Q. **31**, S10–S17 (2014)
11. Cranefield, J., Robertson, O., Oliver, G.: Value in the mash: exploring the benefits, barriers and enablers of open data apps. In: Proceedings of the European Conference on Information Systems (ECIS) 2014, Tel Aviv, Israel, 9–11 June 2014. ISBN 978-0-9915567-0-0
12. Data.gov.uk.: About data.gov.uk. http://data.gov.uk/about (2015)
13. Davis, F.D.: Perceived usefulness, perceived ease of use, and user acceptance of information technology. MIS Q. **13**, 319–340 (1989)
14. Fang, Z.: e-government in digital era: concept, practice, and development. Int. J. Comput. Internet Manage. **10**(2), 1–22 (2002)
15. Fang, X., Holsapple, C.W.: An empirical study of web site navigation structures' impacts on web site usability. Decis. Support Syst. **43**(2), 476–491 (2007)
16. Fishbein, M., Ajzen, I.: Belief, attitude, intention, and behavior: an introduction to theory and research. Addison-Wesley, Reading (1975)
17. Glidden, J.: The Citadel Reveals Open Data Findings. https://opensource.com/government/15/6/citadel-open-government-data-results (2015)
18. Janssen, K.: Open government data and the right to information: opportunities and obstacles. J. Community Inform. **8**(2) (2012). http://www.ci-journal.net/index.php/ciej/article/view/952/954
19. Janssen, M., Charalabidis, Y., Zuiderwijk, A.: Benefits, adoption barriers and myths of open data and open government. Inf. Syst. Manage. **29**(4), 258–268 (2012)
20. Kapoor, K., Dwivedi, Y.K., Williams, M.D.: Role of innovation attributes in explaining the adoption intention for the interbank mobile payment service in an indian context. In: Dwivedi, Y.K., Henriksen, H.Z., Wastell, D., De', R. (eds.) TDIT 2013. IFIP AICT, vol. 402, pp. 203–220. Springer, Heidelberg (2013)
21. Labay, D.G., Kinnear, T.C.: Exploring the consumer decision process in the adoption of solar energy systems. J. Consum. Res. **8**(3), 271–278 (1981)

22. Martín, A.S., de Rosario, A.H., Pérez, C.C.: Open government data: a european perspective. In: Information and Communication Technologies in Public Administration: Innovations from Developed Countries, vol. 195 (2015)
23. Ozaki, R.: Adopting sustainable innovation: what makes consumers sign up to green electricity? Bus. Strategy Environ. **20**(1), 1–17 (2011)
24. Pannell, D.J., Marshall, G.R., Barr, N., Curtis, A., Vanclay, F., Wilkinson, R.: Understanding and promoting adoption of conservation practices by rural landholders. Aust. J. Exp. Agric. **46**(11), 1407–1424 (2006)
25. Rogers, E.M.: Diffusion of Innovations. The Free Press, Glencoe (1962)
26. Rogers, E.M., Shoemaker, F.F.: Communication of Innovations. Free Press, New York (1971)
27. Rogers, E.M.: Diffusion of Innovations, 5th edn. Free Press, New York (2003)
28. Sivarajah, U., Irani, Z., Weerakkody, V.: Evaluating the use and impact of Web 2.0 technologies in local government. Gov. Inf. Q. (2015). ISSN 0740-624X, http://dx.doi.org/10.1016/j.giq.2015.06.004
29. Sivarajah, U., Irani, Z., Jones, S.: Application of Web 2.0 technologies in e-government: a United Kingdom case study. In: 47th Hawaii International Conference on System Sciences (HICSS), pp. 2221–2230. IEEE (2014)
30. Surowiecki, J.: The Wisdom of Crowds: Why the Many are Smarter than the Few and How Collective Wisdom Shapes Business Economies, Societies and Nations. Doubleday, New York (2004)
31. Tornatzky, L.G., Klein, K.J.: Innovation characteristics and innovation adoption-implementation: a meta-analysis of findings. IEEE Trans. Eng. Manage. **29**(1), 28–43 (1982)
32. Ubaldi, B.: Open Government Data: Towards Empirical Analysis of Open Government Data Initiatives, OECD Working Papers on Public Governance, No. 22, OECD Publishing, Paris (2013). doi:http://dx.doi.org/10.1787/5k46bj4f03s7-en
33. Veljković, N., Bogdanović-Dinić, S., Stoimenov, L.: Benchmarking open government: an open data perspective. Gov. Inf. Q. **31**(2), 278–290 (2014)
34. Venkatesh, V., Morris, M.G., Davis, G.B., Davis, F.D.: User acceptance of information technology: toward a unified view. MIS Q. **27**, 425–478 (2003)
35. Venkatesh, V., Thong, J., Xu, X.: Consumer acceptance and use of information technology: extending the unified theory of acceptance and use of technology. MIS Q. **36**, 157–178 (2012)
36. Wang, J., Senecal, S.: Measuring perceived website usability. J. Internet Commerce **6**(4), 97–112 (2007)
37. Wangpipatwong, S., Chutimaskul, W., Papasratorn, B.: Understanding citizen's continuance intention to use e-government website: a composite view of technology acceptance model and computer self-efficacy. Electron. J. e-Gov. **6**(1), 55–64 (2008)
38. Wilcox, J.: Government drops first set of COINS. PublicTechnology.net (2010). http://legacy.publictechnology.net/sector/central-gov/govt-drops-first-set-coins
39. Zhang, J., Dawes, S.S., Sarkis, J.: Exploring stakeholders' expectations of the benefits and barriers of e-government knowledge sharing. J. Enterp. Inf. Manage. **18**, 548–567 (2005)
40. Zuiderwijk, A., Janssen, M.: Open data policies, their implementation and impact: a framework for comparison. Gov. Inf. Q. **31**(1), 17–29 (2014)
41. Zuiderwijk, A., Janssen, M., Choenni, S., Meijer, R., Alibaks, R.S.: Socio-technical impediments of open data. Electron. J. e-Gov. **10**(2), 156–172 (2012)

E-Business, E-Services and E-Society

Enabling Flexible IT Services by Crowdsourcing: A Method for Estimating Crowdsourcing Participants

Yiwei Gong[(⊠)]

School of Information Management,
Wuhan University, Wuhan, Hubei 430072, People's Republic of China
yiweigong@whu.edu.cn

Abstract. Crowdsourcing has become an increasingly attractive practice for companies to execute business processes in open contexts with on-demand workforce and higher level of flexibility. One of the challenges is the identification of the best-fit crowdsourcing participant from a group of online candidates. This paper presents a method of AHP-TOPSIS based on Grey Relation Analysis for estimating participants of a crowdsourcing task based on their online profiles and proposals. This method is tested by an experiment on a dataset of 348 completed IT service crowdsourcing tasks. An analysis on the matching between the test result and the actual selection result reveals the accuracy and efficiency of this method. Companies can use this method to facilitate the quality control at the beginning of crowdsourcing and keeps the selection of participants easy. This paper contributes to the design of a software agent for crowdsourcing platforms to automatically rank the participants of a task.

Keywords: Crowdsourcing · Flexibility · AHP · TOPSIS · Grey relation analysis

1 Introduction

The concept of crowdsourcing is first coined in 2006 and simply means outsourcing certain tasks and problem formulations to an undefined (and generally large) network of people in the form of open calls [1]. With today's development of Internet and mobile technologies, and the explosion of social media, companies are able to have a better engagement of distributed crowds of individuals for their innovation and problem-solving needs [2]. As a result, an increasing number of companies, ranging from small startups to those listed in Fortune 500, are trying to make use of crowdsourcing to access knowledge and skills that previously unavailable to them and to solve parts of business processes formerly executed in-house [3]. In this way, companies can have a more flexible workforce and higher knowledge absorptive capacity and business processes can be adapted on-demand, which results in higher level of process flexibility [4].

Crowdsourcing can be considered as an online and distributed problem-solving model [5] and suggests that engaging crowds can help companies develop solutions to a variety of business challenges. As the business challenges and tasks vary, so do the

© IFIP International Federation for Information Processing 2015
M. Janssen et al. (Eds.): I3E 2015, LNCS 9373, pp. 275–286, 2015.
DOI: 10.1007/978-3-319-25013-7_22

knowledge and skills that crowdsourcing participants have. Unlike simple and low-priced tasks that commonly require general skills, IT service tasks are knowledge intensive and require crowdsourcing participants with special skills and knowledge. This makes the identification of suitable participants a challenge. Matching skills to tasks often relays on sophisticated online crowdsourcing platforms to manage distributed workers and support task providers [2]. However, prior research on crowdsourcing focuses on its business models [6, 7] or brand-related and marketing alike activities [8, 9], taken the online crowdsourcing platforms and their functionalities as given [10]. The state of the art in crowdsourcing practice still lacks approaches for automated estimating participants considering their skills and knowledge [11].

In this paper, we propose a method of AHP-TOPSIS based on Grey Relation Analysis to help companies estimate and identify the best-fit participant for their IT service crowdsourcing tasks. The method is tested on a dataset of 348 completed crowdsourcing tasks in the IT service domain. A post-hoc analysis on the matching between the estimation result and the actual decision made by task providers reveals the accuracy and efficiency of this method.

2   Research Context

2.1   Forms of Crowdsourcing

The way of using crowdsourcing to abstain flexible workforce for business process execution is similar to cloud computing where computing capacity is provided on demand [11, 12]. A typical form of crowdsourcing is publishing the request for proposals through an online marketplace platform with the details of the needed service and its expected duration and (a range of) cost. Then potential participants bid on the task by submitting their proposals. Although many proposals would be received for a task, only the best-fit candidate will be selected to carry out the task. At the end, company can decide to accept and pay for the work, or refuse it if it does not fulfil the expectation. This marketplace form of crowdsourcing enables companies to access the vast potential of workforce with various backgrounds, while it has more flexibility and less risk than having a fixed outsourcing contract [13].

There are also other forms of crowdsourcing such as knowledge contributions (e.g. Wikipedia), rating (i.e. participants 'vote' on a given topic) and micro-task (e.g. Galaxy Zoo) where participants complete the task voluntarily. In addition, contest-based crowdsourcing is used for obtaining innovative ideas or solutions, in which all the participants make their effort and results are determined on a comparative basis and probably only the top contributor(s) would be reworded. Those forms of crowdsourcing are out of the scope of this paper, as they are less effective in providing business process flexibility.

2.2   Related Work

Estimation issue in crowdsourcing has been observed by researchers, and there are some studies for automatically estimating different submissions to find out those with sufficient quality for a task. For example, Tarasov et al. [14] proposed a dynamic estimation of worker reliability in rating-based crowdsourcing. This approach is for detecting noisy and incompetent workers by estimating their submissions, instead of estimating the workers before the task was taken into execution.

Mechanisms for estimating crowdsourcing participants for business process execution can be found in the research of BPEL4People in social networks [15], in which a ranking method based on Hyperlink-Induced Topic Search (HITS) algorithm is provided to estimate the expertise of works in a social network. In this method, a certain skill, its expected level and the importance of a task are used as input, and the ranking result presents a list of all the suitable crowdsourcing works in a social network. The underlying concept of BPEL4People is that the flexibility of traditional SOA-based business process systems can be enhanced by enabling human-based services with very the same API used by software-based Web services. In this way, tasks would be able to match to suitable workers that are registered and active on the crowdsourcing social platform [11].

However, not all crowdsourcing platforms take the form of social networks where crowdsourcing participants work with each other in a joint task and some of them would take the role of supervisor or coordinator. Instead, many crowdsourcing platforms have a form of marketplaces, where a task is completed by only one participant exclusively. In this case, many candidates will compete for the same task by submitting their proposals and the task provider has to choose the best-fit one from them. The more candidates a task has, the more difficult for the task provider to manually estimate and identify the best-fit participant, because of information overload. It is therefore important to have an estimation method to rank all the candidates for the task provider to choose. And a reliable estimation should be a necessary functionality of online marketplace crowdsourcing platforms to support task providers for quality control and solving the problem of information overload.

2.3   AHP, TOPSIS and GRA

Selecting the best-fit crowdsourcing participant from a group of candidates is a typical Multiple Attribute Decision Making (MADM) problem [16] in which decision-making is for selecting the most appropriate one from many feasible solutions. Analytical Hierarchical Process (AHP) [17] is one of the most outstanding MADM methods, which first estimates the relationship among criteria weight and then the total value of each choice based on the obtained weight [18]. The Technique for Order of Preference by Similarity to Ideal Solution (TOPSIS) is another outstanding MADM method, which is based on the concept that the best choice should have the shortest Euclidean distances from the positive ideal and the farthest from the negative ideal [16]. AHP and TOPSIS can be used in combination [19] where AHP is used to calculate the weights of the parameters and these weights are later used in TOPSIS.

The drawback of TOPSIS method is its linear variation of each alternatives, which cannot provide an accurate ranking between two alternatives that have the same distances to the ideal. This problem can be solved by Grey Relational Analysis (GRA) which is an effective method to solve decision making problems by generalizing estimates under limited samples and uncertain conditions [20]. GRA is a kind of flexible measurement of curve similarity. By using GRA, the nonlinear relationship between the sequences of each alternative can be well reflected, which can compensate the inaccuracy problem of TOPSIS method.

AHP-TOPSIS based on GRA has been proved to be useful in solving MADM problems [21]. In this study, it is employed for estimating crowdsourcing participants for given tasks.

3 Estimation Method

In this section a method of AHP-TOPSIS based on GRA is proposed for estimating crowdsourcing participants. This method has the following three phases.

3.1 Phase 1: Identifying Estimation Parameters

The estimation parameters used by an algorithm-based method should be quantitative, otherwise they cannot be calculated. In addition, the data of parameters should be easy to access, otherwise the desired automation in the ranking of participants cannot be achieved. In this study, there are two underlying assumptions. The first one is that task providers and candidate participants do not know each other in actual life. This means that a task provider makes its selection decision only based on the related candidate participants' information that is available online. The second one is that task providers will insist on looking for the best-fit participant rather than shifting to other strategy like choosing the first acceptable candidate. This means that all related candidate participants should be involved in the consideration during the decision-making. In a typical marketplace crowdsourcing model, participants' online information comes from either their online profiles or the proposals that they submitted to the task. Both these two sources of information are involved in the formulation of the estimation parameters in this study. Afterwards, the parameters that cannot be quantitated has to be ignored, and the parameters that reflect the same property are merged. At the end, parameters are categorized into benefit parameters (the larger the value is, the better the solution is) and cost parameters (the smaller the value is, the better the solution is).

3.2 Phase 2: Using AHP to Calculate the Weight of Parameters

In AHP, the multi-attribute weight measurement is calculated via pair-wise comparison of the relative importance of two factors. Assuming that there are N number of decision parameters, denoted as $(P_1, P_2, \ldots, P_n)$, its judgment matrix would be $A = [a_n]$, in which a_n represents the relative importance of P_1 and P_2. Using the row vector average normalization proposed by Satty [17], the weight of P_i is calculated as:

$$W_i = \frac{\left(\prod\limits_{j=1}^{n} a_{ij}\right)^{\frac{1}{n}}}{\sum\limits_{i=1}^{n}\left(\prod\limits_{j=1}^{n} a_{ij}\right)^{\frac{1}{n}}} i,j = 1,2,\ldots,n.$$

3.3   Phase 3: Using GRA-Based TOPSIS to Estimate Participants

In this phase the algorithm has the following ten steps.

1. Normalizing of Initial decision matrix $X = (x_{ij})_{m \times n}$, get the normalization matrix $Z = (z_{ij})_{m \times n}$. $(i = 1,2,\ldots,m; j = 1,2,\ldots n$

   For benefit parameters:

$$Z_{ij} = \frac{x_{ij} - \min\limits_{i} x_{ij}}{\max\limits_{i} x_{ij} - \min\limits_{i} x_{ij}} \tag{1}$$

   For cost parameters:

$$Z_{ij} = \frac{\max\limits_{i} x_{ij} - x_{ij}}{\max\limits_{i} x_{ij} - \min\limits_{i} x_{ij}} \tag{2}$$

2. Calculating   the   weighted   decision   matrix $S = (s_{ij})_{m \times n}$, $(i = 1,2,\ldots,m; j = 1,2,\ldots n)$.

$$S_{ij} = w_{ij} z_{ij}$$

3. Calculating the positive ideal solution S^+ and negative ideal solution S^-

$$S^+ = \left(s_1^+, s_2^+, \ldots, s_n^+\right); S^- = \left(s_1^-, s_2^-, \ldots, s_n^-\right)$$

   Where $s_j^+ = \max\limits_{i} s_{ij} = w_j; s_j^- = \min\limits_{i} s_{ij} = 0, i = 1,2,\ldots,m; j = 1,2,\ldots,n.$

4. Calculating the Euclidean distance between each solution and positive/negative ideal solution d_i^+, d_i^-

$$d_i^+ = \sqrt{\sum_{j=1}^n \left(s_{ij} - s_j^+\right)^2}, d_i^- = \sqrt{\sum_{j=1}^n \left(s_{ij} - s_j^-\right)^2}$$

where $i = 1, 2, \ldots, m; j = 1, 2, \ldots, n.$

5. Calculating the grey relation coefficient matrix of each solution and positive/negative ideal solution L^+, L^-:

$$L^+ = \left(l_{ij}^+\right)_{m \times n}, L^- = \left(l_{ij}^-\right)_{m \times n}$$

Where $l_{ij}^+ = \dfrac{\min\limits_i \min\limits_j \left|s_j^+ - s_{ij}\right| + \theta \max\limits_i \max\limits_j \left|s_j^+ - s_{ij}\right|}{\left|s_j^+ - s_{ij}\right| + \theta \max\limits_i \max\limits_j \left|s_j^+ - s_{ij}\right|}$

$$l_{ij}^- = \dfrac{\min\limits_i \min\limits_j \left|s_j^- - s_{ij}\right| + \theta \max\limits_i \max\limits_j \left|s_j^- - s_{ij}\right|}{\left|s_j^- - s_{ij}\right| + \theta \max\limits_i \max\limits_j \left|s_j^- - s_{ij}\right|}$$

Where $\theta \in (0, 1)$, is distinguishing coefficient, Here the value of θ is set to be 0.5.

Simplifying the formulas : $L_{ij}^+ = \dfrac{\theta}{\left|z_{ij} - 1\right| + \theta}; L_{ij}^- = \dfrac{\theta}{\left|z_{ij} + 1\right| + \theta}$

6. Calculating the grey relation grade of each solution and positive/negative ideal solution l_i^+, l_i^-:

$$l_i^+ = \frac{1}{n}\sum_{j=1}^n l_{ij}^+; l_i^- = \frac{1}{n}\sum_{j=1}^n l_{ij}^-$$

7. Applying nondimensionalization to d_i^+, d_i^-, l_i^+ and l_i^-

$$D_i^+ = \frac{d_i^+}{\max\limits_i d_i^+}, D_i^- = \frac{d_i^-}{\max\limits_i d_i^-}$$

$$L_i^+ = \frac{l_i^+}{\max\limits_i l_i^+}, L_i^- = \frac{l_i^-}{\max\limits_i l_i^-}$$

Where $i = 1, 2, \ldots, m$

8. Calculating the relative closeness degree:

$$P_i^+ = \frac{D_i^+}{D_i^+ + D_i^-}, U_i^+ = \frac{L_i^+}{L_i^+ + L_i^-}$$

9. Combining P_i^+ and U_i^+: $Q_i^+ = v_1 P_i^+ + v_2 U_i^+$

Where v_1 and v_2 reflect the degree of preference of decision makers, $v_1 + v_2 = 1$, and $v_1 = v_2 = 0.5$

10. Sorting solutions by the value of Q_i^+. The better Q_i^+ is, the better the solution is, and vice versa. $\max(Q_i^+)$ is the final decision.

4 Experiment

In order to evaluate the proposed method of AHP-TOPSIS based on GRA on its accuracy and efficiency in estimating crowdsourcing participants, an experiment was carried out on data from a popular Chinese crowdsourcing marketplace platform, Epweike (http://www.epweike.com/). This experiment used the proposed method to estimate crowdsourcing participants of certain tasks, and then the estimation result was compared with the actual selection decision made by the task provider. This comparison allows for an analysis on the accuracy of the method and the impact of the number of candidates on the actual decision-making which reflects the efficiency of this method.

4.1 Dataset

In this experiment, a dataset of 348 valid and completed tasks between 2010 and 2015 for IT services crowdsourcing is used. The content of those task includes software/mobile application development, website construction, database and system design, server maintenance, etc. In those tasks the number of openly visible proposals is more than 3. Tasks that have less than 4 proposals are ignored, because the GRA-based TOPSIS algorithm does not return a meaningful ranking result when the number of candidates is less than 4.

4.2 Approach

Phase 1: Identifying Estimation Parameters. The following 10 parameters of crowdsourcing participants are identified for the estimation, and the data of these parameters can be accessed from the website openly (Table 1).

These parameters describe the information of either the participant itself or the proposal it provided. Among those parameters, P1, P2, P3, P4, P5, P9 and P10 are benefit parameters; while P6, P7 and P8 are cost parameters.

Table 1. The description of each parameter

Parameters	Description
P1	The total volume of the participant in the history
P2	The praise rate accumulates from the assessments made by task providers for each task completed by the participant
P3	The number of biddings that the participant participates
P4	The degree of matching between the required skills for the task and the skills that the participant has
P5	The website's evaluation of the participant on its intelligence, authenticity, trusted transactions and public praise
P6	The website's overall evaluation of the participant
P7	The price proposed by the participant
P8	The task duration proposed by the participant
P9	The number of visitors of participant's homepage
P10	VIP level

Phase 2: Using AHP to Calculate the Weight of Parameters. By AHP, the weight of the 10 parameters is calculated:

$$w_i = (0.140911227, 0.194917168, 0.051377508, 0.041047892, 0.038491227,$$
$$0.030836462, 0.255857835, 0.209338244, 0.016223402, 0.020999035).$$

Phase 3: Using GRA-based TOPSIS to Estimate Participants. For space reason, only the calculation of Task1 is given as an example. This task has 23 candidate participants and 18 openly available proposals. The other 5 proposals are closed and only visible to the task provider, and therefore those 5 participants are not taken into account in the experiment. The best one is identified from the rest 18 candidates by using the GRA-based TOPSIS method. Through the estimation, the final value of Q_i^+ is presented in the following Table 2.

By descending the order of Q_i^+, the sort of participants of Task 1 can be get.

According to the result presented in Table 3, the best candidate for Task 1 is Participant 20. The values of each of its parameters are:

$$P_{20} = (17000, 100\ \%, 4, 4, 40, 5, 800, 5, 12624, 1)$$

But the subjective choice made by the task provider is Participant 1, and the values of its parameters are:

$$P_1 = (9200, 100\ \%, 4, 3, 39, 6, 500, 2, 2588, 1)$$

Table 2. Value of Q_i^+ for each participant of Task 1

Participants	Q_i^+		Participants	Q_i^+
1	0.481154071		10	0.459536143
2	0.486380377		11	0.481035623
3	0.425769924		12	0.503450868
4	0.512654617		13	0.499094489
5	0.488922997		14	0.619961894
6	0.509868043		15	0.505601337
7	0.479417731		16	0.434555178
8	0.51068885		17	0.603533483
9	0.441772416		18	0.479186991

Table 3. The ranking of participants of Task 1

Order	Participants		Order	Participants
1	20		10	2
2	23		11	1
3	4		12	15
4	9		13	24
5	7		14	8
6	21		15	14
7	16		16	12
8	17		17	22
9	6		18	3

In this example, the estimation result is not matching with the actual selection.

In this experiment, Phase 3 was repeated on all the 348 tasks. Then the result of the estimation was compared with the actual selection made by the task provider to find out whether they are matchable. The latter is openly available on the crowdsourcing website.

4.3   Result Analysis

In order to evaluate the accuracy of the proposed method, the result of estimation is compared with the actual selection decision made by task providers. A metric of matching rate is used to indicate the percentages of matchable tasks. In this experiment, the overall matching rate is 88.22 % (307 out of 348 tasks). The matching rates under different number of participants are presented in the following table.

The matching rates presented in Table 4 indicate that the proposed method has a high accuracy when the number of participants is between 4 and 10. There are 248 tasks (71.26 % of the total 348 tasks) fall into this range and the matching rates are above 90 %. Specially, in the simplest situation where only 4 participants competing a task, the estimation result is 100 % matched with the manual decision-making. This means the proposed method can achieve an estimation result that is very similar with the manual decision-making, when the manual decision-making is simple and the information overload problem does not appear.

Table 4. Matching rates under different number of participants

Number of Participants	Number of Tasks	Number of Matched tasks	Matching Rate	Number of Participants	Number of Tasks	Number of Matched tasks	Matching Rate
4	65	65	100%	31	11	8	72.73%
5	55	54	98.18%	32	11	8	72.73%
6	24	23	95.83%	37	8	5	62.5%
7	18	17	94.44%	38	3	3	100%
8	61	57	93.44%	39	5	3	60%
9	15	14	93.33%	40	4	2	50%
10	10	9	90%	42	2	1	50%
11	5	5	100%	44	2	1	50%
12	7	6	85.71%	47	2	1	50%
15	4	4	100%	52	3	1	33.33%
17	5	4	80%	57	3	1	33.33%
22	5	4	80%	60	1	0	0%
23	4	3	75%	62	3	1	33.33%
29	8	6	75%	63	4	1	25%

Technically speaking, the accuracy of this method will not be influenced by the number of participants. However, the matching rates generally declines along with the increase of participants. A possible explanation is that when the number of participants increase, it is more difficult to manually identify the best-fit participant. The task provider might have to rely on its own experience or even instinct in the decision-making for participant selection rather than an objective comparison between candidates. When the number of participants is more than 50, manually identifying the best-fit participant becomes very difficult and the actual selection result deviates very much from the estimation result, which results in matching rates below 33.33 %. The following curve demonstrate the decline of the matching rate. Under the assumption that the accuracy of manual decision-making is mainly and negatively influenced by information overload, the proposed method could solve this problem and improve the decision-making for crowdsourcing tasks with a large number of participants. To solidly prove this statement, investigation on the factors that impact manual decision-making for selecting crowdsourcing participants is desired in the future (Fig. 1).

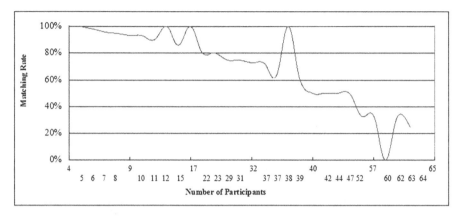

Fig. 1. The matching rates decline along with the increase of participants

5   Conclusions and Future Work

Companies want to adopt crowdsourcing to obtain on-demand IT services and enable flexible business processes. The challenge to overcome the information overload problem in estimating and identifying the best-fit crowdsourcing participant out of a large group of candidates. This paper presents a method of AHP-TOPSIS based on GRA for estimating crowdsourcing participants. The proposed method has been tested on a dataset of 348 valid and completed IT service crowdsourcing tasks from a Chinese crowdsourcing online marketplace platform. For the tasks with a small number of candidates, a matching between the estimation result and the actual selection made by task providers proofs the accuracy of the proposed method. Although the matching rate generally declines along with the increase of the number of candidates, this reflects the information overload problem in manual decision-making. The proposed method could solve this problem by providing an accurate and objective estimation which is uninfluenced by the number of participants. Employing this method allows task providers to quickly and easily identify the best-fit crowdsourcing participant. Furthermore, this method facilitates the design of a software agent for crowdsourcing platforms to rank the participants of a task automatically.

In the future work, it would be interesting to contact the task providers who had to make decision with a large number of candidates, especially those chose a participant other than the one recommended by the proposed method. Then invite them to interviews or surveys with questions such as how long it took them to make a decision on selecting the best-fit participant, what are the factors impacting their decision, and whether they think this method would help them in overcoming the information overload problem.

References

1. Howe, J.: The rise of crowdsourcing. Wired Mag. **14**(6), 1–4 (2006). Condé Nast
2. Lakhani, K.R.: Using the crowd as an innovation partner. Harvard Bus. Rev. **91**, 60–69 (2013)
3. Afuah, A., Tucci, C.L.: Crowdsourcing as a solution to distant search. Acad. Manage. Rev. **37**, 355–375 (2012)
4. Gong, Y., Janssen, M.: From policy implementation to business process management: principles for creating flexibility and agility. Gov. Inf. Q. **29**, 61–71 (2012)
5. Brabham, D.C.: Crowdsourcing as a model for problem solving. Convergence: Int. J. Res. New Media Technol. **14**, 75–90 (2008)
6. Prpic, J., Shukla, P.P., Kietzmann, J.H., McCarthy, I.P.: How to work a crowd: developing crowd capital through crowdsourcing. Bus. Horiz. **58**, 77–85 (2015)
7. Simula, H., Ahola, T.: A network perspective on idea and innovation crowdsourcing in industrial firms. Ind. Mark. Manage. **43**, 400–408 (2014)
8. Djelassi, S., Decoopman, I.: Customers' participation in product development through crowdsourcing: issues and implications. Ind. Mark. Manage. **42**, 683–692 (2013)
9. Gatautis, R., Vitkauskaite, E.: Crowdsourcing application in marketing activities. Procedia – Soc. Behav. Sci. **110**, 1243–1250 (2013)
10. Majchrzak, A., Malhotra, A.: Towards an information systems perspective and research agenda on crowdsourcing for innovation. J. Strateg. Inf. Syst. **22**, 257–268 (2013)
11. Satzger, B., Psaier, H., Schall, D., Dustdar, S.: Auction-based crowdsourcing supporting skill management. Inf. Syst. **38**, 547–560 (2013)
12. Nevo, D., Kotlarsky, J.: Primary vendor capabilities in a mediated outsourcing model: can IT service providers leverage crowdsourcing? Decis. Support Syst. **65**, 17–27 (2014)
13. Gefen, D., Gefen, G., Carmel, E.: How project description length and expected duration affect bidding and project success in crowdsourcing software development. J. Syst. Softw. (20015)
14. Tarasov, A., Delany, S.J., Namee, B.M.: Dynamic estimation of worker reliability in crowdsourcing for regression tasks: making it work. Expert Syst. Appl. **41**, 6190–6210 (2014)
15. Schall, D., Satzger, B., Psaier, H.: Crowdsourcing tasks to social networks in BPEL4People. World Wide Web **17**, 1–32 (2014)
16. Hwang, C.-L., Yoon, K.: Multiple Attribute Decision Making: Methods and Applications a State-of-the-Art Survey. Springer, Berlin (1985)
17. Saaty, T.L.: The Analytic Hierarchy Process: Planning, Priority Setting. Resource Allocation, Mcgraw-Hill (1980)
18. Ngai, E.W.T., Chan, E.W.C.: Evaluation of knowledge management tools using AHP. Expert Syst. Appl. **29**, 889–899 (2005)
19. Patil, S.K., Kant, R.: A fuzzy AHP-TOPSIS framework for ranking the solutions of knowledge management adoption in supply chain to overcome its barriers. Expert Syst. Appl. **41**, 679–693 (2014)
20. Deng, J.: Introduction to grey system. J. Grey Syst. **1**, 1–24 (1989)
21. Chen, M.-F.: Combining grey relation and TOPSIS concepts for selecting an expatriate host country. Math. Comput. Model. **40**, 1473–1490 (2004)

Mining Learning Processes from FLOSS Mailing Archives

Patrick Mukala[1(✉)], Antonio Cerone[1,2], and Franco Turini[1]

[1] Department of Computer Science, University of Pisa, Pisa, Italy
{mukala, cerone, turini}@di.unipi.it
[2] IMT Institute for Advanced Studies, Lucca, Italy
antonio.cerone@imtlucca.it

Abstract. Evidence suggests that Free/Libre Open Source Software (FLOSS) environments provide unlimited learning opportunities. Community members engage in a number of activities both during their interaction with their peers and while making use of these environments. As FLOSS repositories store data about participants' interaction and activities, we analyze participants' interaction and knowledge exchange in emails to trace learning activities that occur in distinct phases of the learning process. We make use of semantic search in SQL to retrieve data and build corresponding event logs which are then fed to a process mining tool in order to produce visual workflow nets. We view these nets as representative of the traces of learning activities in FLOSS as well as their relevant flow of occurrence. Additional statistical details are provided to contextualize and describe these models.

Keywords: FLOSS learning processes · Learning activities in open source · Mining software repositories · Process mining · Semantic search

1 Introduction

Currently a number of studies provide evidence that suggests the existence of learning opportunities in FLOSS environments [1, 10, 12–17, 22]. As part of this substantiation, FLOSS communities have been established as environments where successful collaborative and participatory learning between participants occurs [14, 16, 17].

Moreover, the levels of interest as well as the aura created around the occurrence of learning within FLOSS have attracted practitioners in tertiary education to consider incorporating participation in FLOSS projects as a requirement for some Software Engineering courses [12, 14, 24]. A number of pilot studies have been conducted in order to evaluate the effectiveness of such an approach in traditional settings of learning [10–13, 20]. To aid in this endeavor, in our previous study, we put an emphasis on how learning occurs in terms of phases [2, 19]. To this end, it has been proposed that a typical learning process in FLOSS occurs in three main phases: Initiation, Progression and Maturation. In each phase, a number of activities are executed though interactions between Novices and Experts. A Novice is considered as any participant in quest of knowledge while the knowledge provider is referred to as the Expert. Figure 1 depicts the categorization of the learning phases with the Initiation Phase synonymously

© IFIP International Federation for Information Processing 2015
M. Janssen et al. (Eds.): I3E 2015, LNCS 9373, pp. 287–298, 2015.
DOI: 10.1007/978-3-319-25013-7_23

corresponding to understanding on the x axis as a learning stage, while Progression and Maturation correspond to practicing and developing respectively. The gray area in Fig. 1 represents the progression with regards to users as they progressively perform the types of activities on the y axis.

In this paper, we present an approach for mining these learning phases from FLOSS data. For illustrative purposes, we detail our approach and present the results for the understanding (Initiation) phase, which is at the bottom of the scale in Fig. 1. In this phase, FLOSS participants get involved in the projects by reviewing and communicating with the purpose of understanding contents without producing any tangible contributions. Initiation is a critical stage as the participant accesses project repositories and exchanges emails and posts messages seeking information and posting any requests. Figure 1 also shows how, in the practicing and developing phases, the participants' activities gradually move from simply using to posting and making significant contributions through commits [2].

FLOSS repositories, such as CVS, Bug reports, mailing archives, Internet relay chats etc., contain all traces of participants' activities as they work in these environments. Singh et al. [6] argue that the FLOSS environment typically includes discussion forums or mailing lists to which users can post questions and get help from developers or other users. These forums are unrestricted and act as a learning environment for novices and experts alike. While many studies have provided invaluable insights in this direction, their results are mostly based on surveys and observation reports [7, 8, 21–23, 24]. Our paper proposes to contribute in this context by studying learning activities from FLOSS repositories using process mining. In particular, the paper focuses on tracing and visualizing the learning activities as well as their flow of occurrence collected in mailing archives of a FLOSS platform called OpenStack [25].

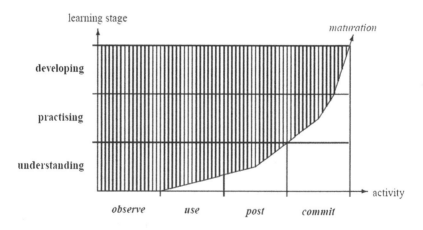

Fig. 1. Learning stages and participants' learning progression in OSS communities [2]

Our major contribution is the approach used in analyzing the data, the application of process mining and mostly the discovered empirical evidence of learning activities' traces. The rest of the paper is structured as follows. Section 2 provides preliminary details on mining the data and constructing the log and succinctly describes the Initiation Phase of the learning process. In Sect. 3 we discuss the data collection and analysis and then present the empirical results. Section 4 concludes the paper.

2   Preliminaries: Mining Data and Catalog of Key Phrases

In order to identify activities and construct the event logs needed for our analysis, we undertake a number of tasks. The first task is analyzing the contents of emails. Text mining appears to be the most direct solution for this task as we need to analyze the contents of a post/email and deduct a corresponding activity. Current text mining tools such as Carrot2, GATE, OpenLP, RapidMiner and KH Coder appear not to be appropriate for the kind of analysis we want to conduct. Tracing learning activities requires semantic interpretation of email contents and this could not be achieved by using any of these tools.

Therefore, we considered making use of Semantic search with MS SQL as a fit alternative. Semantic search improves search by understanding the contextual meaning of the terms and tries to provide the most accurate answer for a given text document. However, this also requires the use of key phrases to steer the search [18]. Our choice of key phrases is based on a number of studies conducted in FLOSS with regards to the kinds of questions and answers that are asked in FLOSS communication environments [3–5, 27]. We start from this categorization, following questions and responses categories; then we deduct a number of key phrases. We try as much as we can to include all the identified key phrases and expressions within the context of identifying learning activities and establishing the learning process across its three phases, although in this paper we only present the details of the first phase.

We make use of previous findings [3, 4], a formal model of learning activities in FLOSS communities [19], as well as lexical semantics to draw a catalog of key phrases with respect to our endeavor. Lexical semantics builds from synonyms of terms and their homonyms to derive the meaning of words in specific contexts. Hence, making use of semantic search is paramount and promises to capture the meaning of message contents as much as possible in identifying activities. Figure 2 presents a catalog that contains the key phrases that semantically identify activities as categorized according to the participants' roles in the Initiation Phase of the learning process.

Principal activities gravitate around observing and making contacts in the Initiation Phase of the learning process [19]. Ideally, this step constitutes an opportunity for the Novice to ask questions and get some help depending on the requests while the Expert intervenes at this point to respond to such requests.

On the one hand, a Novice seeking help can execute a number of activities. These include *FormulateQuestion*, *IdentifyExpert*, *PostQuestion*, *CommentPost* or *PostMessage*, *ContactExpert* and *SendDetailedRequest*. On the other hand, the main activities as undertaken by the Expert during the same period of time include *ReadMessages* on the mailing lists/Chat messages, *ReadPost* from forums, *ReadSourceCode,* as any participant commits code to the project, or *CommentPost*, *ContactNovice* and *CommentPost*.

STATES	GLOBAL KEYWORDS	PARTICIPANTS	ACTIVITIES	KEYPHRASE/CONDITIONAL ACTIVITY
Observation	'problem', 'help', 'error'	NOVICE	FormulateQuestion	If PostQuestion = true
			IdentifyExpert	"How did you do this", "I saw your code", "I need your help", "this does not work for me", "I", "is this possible to do this", "can this be done", "very helpful", "very well"
			PostMessage	If IdentifyExpert = true
			PostQuestion	"How can I do ?", "How to?", "don't understand how", "could help?", "what is wrong?", "my code is not running", "code not executing", "question", "How to", "what is wrong", "where can I", "Any ideas how to solve this problem?", "I have tried doing", "search for this", "but have had little luck", "any help?", "any suggestions?", "everything I could", "new to the"
			CommentPost	"does not work", "not executing", "this does not work for me", "do not know what is wrong with my code", "here is my code", "in short my problem", "step by step", "details provided", "works as follows", "I want it to", "expect it to", "my question is like this", "what I mean"
		EXPERT	ReadMessages	If CommentPost = true
			ReadPost	If CommentPost = true
			ReadSourceCode	"syntax error", "maybe you should...", "it seems to work for me", "do not know what is wrong with the code" or "not sure it can") or "running your code"
			CommentPost	"system details needed", "more details needed", "more problems details needed", "more details needed of what is on the screen", "Did it work before?", "provide exact step by step details"
ContactEstablishement	"can I get your help", "can you help", "send question", "contact details", "send email", "send file", "more details"	NOVICE	ContactExpert	If SendDetailedRequest = true
			SendDetailedRequest	"actually the code is like this...", "I tried this", "I don't know how", "I don't understand how?", "can you help", "your help", "you explain", "as you asked", "so my question is", "I wanted to know", "what I meant is", "my screenshot looks", "I get this error", "how do I fix this"
		EXPERT	ContactNovice	If CommentPost/SendFeedback = true
			CommentPost/Send Feedback	"does not work", "not executing", "maybe you should...", "it seems to work for me", "this does not look right", "you need to delete this", "the syntax is not correct", "send me your code", "what is your problem", "this works for me", "Did it work before", "I think it should work"

Fig. 2. Catalog of key phrases for initiation phase

In order to conduct our analysis, we need to identify the most appropriate repository in this regard. The main criteria in making such a decision lies on the existence of some form of communication exchange between FLOSS members on any candidate repository. Mailing Archives contain email messages between FLOSS members about discussions on topics relevant to the community. Some of these topics involve general questions or specific requests about files, pieces of code or even the use of new plug-ins etc. Hence, these Mailing Archives provide adequate details to track activities and explain their flow of occurrence in the Initiation Phase. Moreover, it is worth noting that the same approach can be applied to mine the remaining phases on other repositories such as source code or commits.

3 Data Collection and Analysis

The FLOSS platform used in our analysis is OpenStack [25]. According to Wikipedia, "OpenStack is a free and open-source software cloud computing software platform. Users primarily deploy it as an infrastructure as a service (IaaS) solution. The technology consists of a series of interrelated projects that control pools of processing, storage, and networking resources throughout a data center—which users manage through a web-based dashboard, command-line tools, or a RESTful API that is released under the terms of the Apache License" [25].

We considered this platform mainly due to the availability of data about email archives and also because it is still an active platform. This database is made up of 7 tables that store data pertaining to compressed files (source_code file, bugs), the mailing lists as per group discussions and topic of interests, the number of messages exchanged as well as details of the individuals involved in these exchanges as shown in Table 1.

This repository contains exactly 54762 emails exchanged between 3117 people who are registered on 15 distinct mailing lists. These emails were sent during a period of time spanning from 2010 to 2014. The length of the messages considered is of typical email length specifically with an average of 3261 characters, the longest email was of 65535 characters and the shortest message yields a single character length.

In order to analyze this data set, we make use of process mining techniques. The key in Process Mining is to identify events. An event is a tuple made up essentially of case ID, performer, activity and any relevant attributes we need for our analysis. In our case, we include the phase of the learning process, date as well as the role (Novice, Expert). Other key components include the catalog of key phrases as shown in Fig. 2 and the data set. Based on all these elements, we generated our event log, which is the set of all identified events.

An event E is a sextuple (t, a, p, d, s, r) such that: t is the case in the event and can be either a topic on emails or an issue number on code and bug reports; a is the activity; p is the participant; d is the relevant date of occurrence; s is the state of the learning process; and r is the participant's role in the process.

Moreover, we refer to the catalog introduced earlier to retrieve the mappings between key phrases, activities, states and participants. Let c_1, c_2 and c_3 be catalogs respectively for Initiation, Progression and Maturation. We distinguish between key

Table 1. Details of mailing archives elements from OpenStack

Tables	Records
dbo.compressed_files	401
dbo.mailing_lists	15
dbo.mailing_lists_people	4434
dbo.messages	54762
dbo.messages_people	54762
dbo.people	3117
dbo.people_upeople	3117

phrases for activities and states. We refer to key phrases for states as *gl_key* (global keys) while the key phrases that help distinguish activities are referred to as *lc_key* (local keys). We define catalogs as sextuples $(C, c_i, gl\_key, state, lc\_key, activity, role)$ such that: C is the set of all our catalogs, $c_i \in C$ is a single catalog, *gl_key* is the key phrase for the identification of a state, *state* is the state as it appears in the catalog, *lc_key* is the key phrase used to identify an activity, *activity* is the corresponding activity in the catalog, and *role* is the role as it appears in the catalog. Using such information, we generate the event log to be analysed through process mining.

3.1 Process Mining Mailing Archives

In order to process mine these records, we choose Disco (Discover Your Processes) [26], an appropriate tool for analyzing the identified events and providing efficient visualizations to demonstrate the workflow of occurrence of activities in these processes. Disco is a toolkit for process mining that enables the user to provide a preprocessed log specifying case, activities, originator and any other needed attributes. The tool performs automatic process discovery from the log and outputs process models (maps) as well as relevant statistical data.

In essence, Disco applies process mining techniques in order to construct process models based on available logging data that is organized into an event log. This logging data is all the details about transactions that can be found in log file or transaction databases. Therefore, an event log can take a tabular structure containing all recorded events that relate to executed business activities [26].

Making use of Disco, we produced the Process Models representing the occurrence of learning activities as documented by their corresponding email messages.

For simplicity, we choose to represent the models through a graphical workflow as well as the statistical information as provided by Disco. Disco offers the possibility for a process model to be represented with frequency metrics that explain the flow of occurrence of events.

The main objective of the frequency metrics is the depiction of how often certain parts of the processes have been executed. We can distinguish three levels of frequency: absolute frequency, case frequency and maximum repetitions. We consider these metrics to model learning activities executed by both the Novices and Expert.

Additional details regarding the numerical measures such as events over time, active cases during this given period of time, case variants, the number of events per case as well as case duration could be plotted as needed. However, for simplicity and effectiveness, we represent only major statistical details that are most representative of the presence, impact and occurrence of learning activities in FLOSS over the chosen period of time.

3.2   Empirical Results

Before we unpack details about process models for both the Novice and Experts, we give some crucial details about the overall Initiation Phase. It should be noted that in Figs. 3 and 4 the numbers, the thickness of the arcs or edges, and the coloring in the model illustrate how frequent each activity or path has been performed. For the purpose of this paper, we retained the topic of emails, the message itself, the people involved in exchanging these emails, the resulting activities and classification of where such activities fall in our defined learning curve to build events and produce the event log used for model extraction.

The analysis of the Initiation Phase of the learning process is carried out on data that refer to the period between the 11th of November 2010 and the 6th of May 2014. During this time, we note that a total of 123401 events were generated. An event represents a tuple made up of the case (in this context, the discussion topic), the email senders as well as the relevant learning activities. With about 565 cases, a total of 14 activities are executed with an average time per case of 69.9 days while the median duration is of 57.8 days.

We can also point out that participants in quest for knowledge claim the majority of activities with a total of 122838 amounting to 99.54 % of all executed activities at this point in contrast with Experts who intervene at a lower rate of 0.36 % with 440 activities, slightly ahead of people doing something other than exchanging knowledge with 123 activities.

The process model depicted in Fig. 3 represents a workflow for all the activities performed by the Novice during the first phase of the learning process. On average, how often an activity has been executed in this process by the Novice as well as how often an activity links to another (path) can be noted through the numbers, the thickness of the arcs or edges, and the coloring in the model. We note that the Novice in OpenStack has engaged in a number of learning activities throughout this period of time. Figure 3 demonstrates that in 51 cases the process would start from formulating a question, posting the question, commenting on post (and this has occurred about 27 times), posting a message, which indicates that an expert has been identified, contacting that expert and sending a detailed request to the expert through commenting on a post. The numerical argument between the transitions from one activity to another indicates how many times on average this has happened.

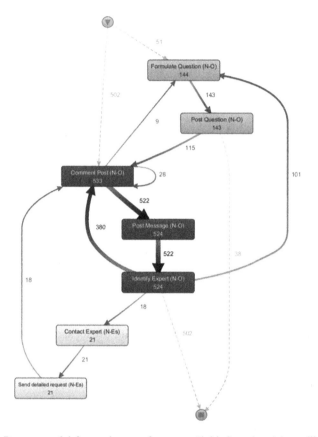

Fig. 3. Process model for novice–per frequency [initiation phase] in mailing lists

Moreover, Fig. 3 also shows that in 502 cases, a Novice starts by commenting on a post first, then formulates a question and follows the process as explained above. Sometimes (101 times in the depicted process map), after identifying an expert, a Novice could go back to formulating another question (or follow-up questions) or even go back to just commenting on the post as part of the interaction.

The process model depicted in Fig. 4 represents a workflow for all the activities performed by the Expert during the first phase of the learning process. One should note that 6 main activities are undertaken by the Expert. In some cases, the Expert would contact the Novice, by commenting on a post or giving feedback regarding a request from the Novice, then read messages and posts, commenting on these posts as well as reading source code, especially if the Novice's requests have to do with source code. The assumption here is that the Expert is referred to as such because of the nature of the reaction activity. Every time an Expert comments, it is in response to a Novice request or to request further details on an already posted question.

In some instances, the Expert would go back to reading messages after commenting on a post or reading source code, or sometimes contacting the Novice again after commenting on a post involved in the exchange. In some instances, on average in 9

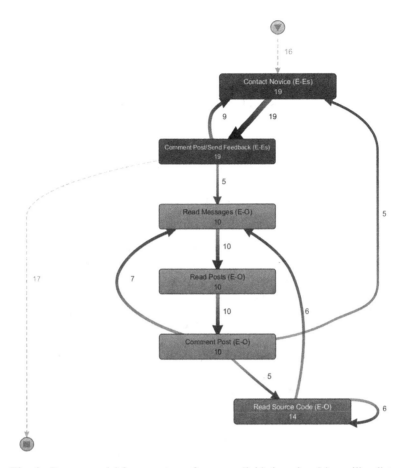

Fig. 4. Process model for expert–per frequency [initiation phase] in mailing lists

cases, the Expert goes back to contacting the Novice after providing feedback. This is where further details about the request or clarification might be requested.

4 Conclusion

FLOSS environments appear to provide learning opportunities for participants. While this aspect has been previously investigated [7, 8], we believe that there is a need for empirical support for such findings. An important remark [9] emphasizes that in such previous work, content data had been collected using surveys and questionnaires or through reports from observers who have been part of the community for a defined period of time. Research suggests that a growing number of participants are highly motivated to engage in these platforms through discussions and email exchange. These interactions produce massive volumes of data that contain evidence of learning. Since these learning activities are not directly observable in the repositories, we made use of semantic search in MS SQL to identify activities from message texts and constructed the event log that we then analyzed with the help of the Disco process mining tool.

Using a combination of text mining techniques (semantic search), key phrases and a set of rules, we believe that our approach has provided insights on how to find traces of interaction and learning activities in FLOSS email messages. We can thus say that it is feasible to trace these activities and that process mining can play a catalyst role in identifying these activities in FLOSS. Our work aimed to give evidence about the existence of learning processes in FLOSS environments through the empirical analysis of OpenStack Mailing Archives.

Our results demonstrate how these learning processes are extracted and how each activity fits within the global picture. For a Novice, we can note that the process in some cases spans from formulating a question, posting the question, commenting on post, posting a message, which indicates that an expert has been identified, contacting that expert and sending a detailed request to the expert through commenting on a post. In most cases, a Novice starts by commenting on a post first, and then formulates a question before following the process as described above. Six main activities are performed by the Expert starting from contacting the Novice, through commenting on a post or giving feedback regarding a request from the Novice. The Expert will then read messages and posts, commenting on these posts as well as reading source code, especially if the Novice's requests have to do with source code. In some instances, the Expert would go back to reading messages after commenting on a post or reading source code, or sometimes contacting the Novice again after commenting on a post involved in the exchange. In some other instances the Expert goes back to contacting the Novice after providing feedback. This is where further details about the request or clarification might be requested.

Finally, more experiments using this approach can be conducted in order to trace activities in the next two phases. These phases include the Progression and Maturation phases. Figure 1 indicates how FLOSS contributors start by just observing, in the Initiation Phase, and gradually evolve to more tangible activities such as posting and committing software artifacts and source code [2]. The types of activities in the Progression phase include *Revert*, *Post* and *Apply*. After the Expert makes contact, the Novice provides additional details about the previous request through activity *Revert*. The Expert gets further involved by providing guidance and the required help to the Novice through *Post,* while the Novice implements the new acquired knowledge through *Apply*. During the Maturation phase, activities include *Analyze, Commit, Develop, Review* and *Revert*. In this phase, the Novice's acquired skills are expressed through the execution of advanced activities such as producing new code, reviewing new commits and providing assistance during discussions, thus gradually performing a transition from Novice to Expert [19]. The Expert performs the same activities with an emphasis on transferring knowledge and providing help when needed rather than applying new skills.

References

1. Sowe, S.K., Stamelos, I.: Reflection on knowledge sharing in F/OSS projects. In: Russo, B., Damiani, E., Hissam, S., Lundell, B., Succi, G. (eds.) Open Source Development, Communities and Quality. IFIP, vol. 275, pp. 351–358. Springer, Heidelberg (2008)

2. Cerone, A.: Learning and activity patterns in OSS communities and their impact on software quality. In: Proceedings of the 5th International Workshop on Foundations and Techniques for Open Source Software Certification, (OpenCert 2011). Electronic Communications of the EASST, vol. 48. The European Association of Software Science and Technology (2011)
3. Lakhani, K.R., Von Hippel, E.: How open source software works: "free" user-to-user assistance. Res. Policy 32(6), 923–943 (2003)
4. Steehouder, M.F.: Beyond technical documentation: users helping each other. In: Proceedings of Professional Communication Conference, IPCC 2002, pp. 489–499. IEEE International (2002)
5. Singh, V., Twidale, M.B., Nichols, D.M.: Users of open source software-how do they get help? In: hicss, pp. 1–10). IEEE, December1899
6. Singh, V., Twidale, M.B., Rathi, D.: Open source technical support: a look at peer help-giving. In Proceedings of the 39th Annual Hawaii International Conference on System Sciences, HICSS 2006, vol. 6, pp. 118c–118c. IEEE, January 2006
7. Glott, R., Meiszner, A., Sowe, S.K., Conolly, T., Healy, A., Ghosh, R., Karoulis, A., Magalhães, H., Stamelos, I., Weller, M.J., West, D.: FLOSSCom-Using the Principles of Informal Learning Environments of FLOSS Communities to Improve ICT Supported Formal Education
8. Glott, R., Meiszner, A., Sowe, S.K.: FLOSSCom phase 1 report: analysis of the informal learning environment of FLOSS communities. FLOSSCom Project (2007)
9. Cerone, A., Fong, S., Shaikh, S.A.: Analysis of collaboration effectiveness and individuals' contribution in FLOSS communities. In: Proceedings of the 5th International Workshop on Foundations and Techniques for Open Source Software Certification (OpenCert 2011). Electronic Communications of the EASST, vol. 48. The European Association of Software Science and Technology (2011)
10. Papadopoulos, P.M., Stamelos, I.G., Meiszner, A.: Enhancing software engineering education through open source projects: four years of students' perspectives. Educ. Inf. Technol. 18(2), 381–397 (2013)
11. LeBlanc, R.J., Sobel, A., Diaz-Herrera, J.L., Hilburn, T.B.: Software Engineering 2004: Curriculum Guidelines for Undergraduate Degree Programs in Software Engineering. IEEE Computer Society Press, Washington, D.C. (2006)
12. Sowe, S.K., Stamelos, I.G.: Involving software engineering students in open source software projects: experiences from a pilot study. J. Inf. Syst. Educ. 18(4), 425 (2007)
13. Cerone, A.K., Sowe, S.K.: Using free/libre open source software projects as e-learning tools. In: Proceedings of the 4th International Workshop on Foundations and Techniques for Open Source Software Certification (OpenCert 2010). Electronic Communications of the EASST, vol. 33. The European Association of Software Science and Technology (2010)
14. Fernandes, S., Cerone, A., Barbosa, L.S.: Analysis of FLOSS communities as learning contexts. In: Counsell, S., Núñez, M. (eds.) SEFM 2013. LNCS, vol. 8368, pp. 405–416. Springer, Heidelberg (2014)
15. Fernandes, S., Cerone, A., Barbosa, L.S.: FLOSS Communities as Learning Networks. Int. J. Inf. Educ. Technol. 3(2), 278–281 (2013)
16. Fernandes, S., Cerone, A., Barbosa, L.S., Papadopoulos, P.M.: FLOSS in technology-enhanced learning. In: Cerone, A., Persico, D., Fernandes, S., Garcia-Perez, A., Katsaros, P., Ahmed Shaikh, S., Stamelos, I. (eds.) SEFM 2012 Satellite Events. LNCS, vol. 7991, pp. 121–132. Springer, Heidelberg (2014)
17. Meiszner, A., Glott, R., Sowe, S.K.: Free/libre open source software (FLOSS) communities as an example of successful open participatory learning ecosystems. UPGRADE Eur. J. Inform. Prof. 9(3), 62–68 (2008)

18. Larson, B.: Delivering Business Intelligence with Microsoft SQL Server 2012. McGraw-Hill Osborne Media, New York (2012)
19. Mukala, P., Cerone, A., Turini, F.: An abstract state machine (ASM) representation of learning process in FLOSS communities. In: Canal, C., Idani, A. (eds.) SEFM 2014 Workshops. LNCS, vol. 8938, pp. 227–242. Springer, Heidelberg (2015)
20. Meiszner, A., Mostaka, K., Syamelos, I.: A hybrid approach to computer science education– a case study: software engineering at Aristotle University. In: CSEDU 2009 - Proceedings of the First International Conference on Computer Supported Education, Lisboa, Portugal (2009)
21. Dillon, T., Bacon, S.: The potential of open source approaches for education. FutureLab opening education reports (2006). http://www.futurelab.org.uk/resources/publicationsreports-articles/opening-education-reports/Opening-Education-Report200
22. Sowe, S.K., Stamelos, I., Deligiannis, I.: A framework for teaching software testing using F/OSS methodology. In: Damiani, E., Fitzgerald, B., Scacchi, W., Scotto, M., Succi, G. (eds.) Open Source Systems. IFIP, vol. 203, pp. 261–266. Springer, Heidelberg (2006)
23. Sowe, S.K., Stamelos, I., Angelis, L.: An empirical approach to evaluate students participation in free/open source software projects. In: IADIS International Conference on Cognition and Exploratory Learning in Digital Age CELDA 2006, pp. 304–308 (2006)
24. Meiszner, A., Stamelos, I., Sowe, S.K.: 1st international workshop on: 'designing for participatory learning' building from open source success to develop free ways to share and learn. In: Boldyreff, C., Crowston, K., Lundell, B., Wasserman, A.I. (eds.) OSS 2009. IFIP AICT, vol. 299, pp. 355–356. Springer, Heidelberg (2009)
25. OpenStack: In Wikipedia, the free encyclopedia, 3 November 2014. http://en.wikipedia.org/w/index.php?title=OpenStack&oldid=632224644. Accessed 10 November 2014
26. Günther, C.W., Rozinat, A.: Disco: discover your processes. In: Proceedings of the Demonstation Track of BPM 2012. CEUR Workshop Proceedings, vol. 940, September 2012. CEUR-WS.org
27. Elliott, M.S., Scacchi, W.: Free software development: cooperation and conflict in. Free/open Source Softw. Dev. 152 (2005)

Private-Collective Innovation and Open Source Software: Longitudinal Insights from Linux Kernel Development

Dirk Homscheid[1,2(✉)], Jérôme Kunegis[2], and Mario Schaarschmidt[1,2]

[1] Institute for Management, University of Koblenz-Landau, Koblenz, Germany
[2] Institute for Web Science and Technologies,
University of Koblenz-Landau, Koblenz, Germany
{dhomscheid,kunegis,
mario.schaarschmidt}@uni-koblenz.de

Abstract. While in early years, software technology companies such as IBM and Novell invested time and resources in open source software (OSS) development, today even user firms (e.g., Samsung) invest in OSS development. Thus, today's professional OSS projects receive contributions from hobbyists, universities, research centers, as well as software vendors and user firms. Theorists have referred to this kind of combined public and private investments in innovation creation as private-collective innovation. In particular, the private-collective innovation model seeks to explain why firms privately invest resources to create artifacts that share the characteristics of non-rivalry and non-excludability. The aim of this research is to investigate how different contributor groups associated with public and increasing private interests interact in an OSS development project. The results of the study show that the balance between private and collective contributors in the Linux kernel development seems to be changing to an open source project that is mostly developed jointly by private companies.

Keywords: Open source · Open source community · Private-Collective innovation · Linux kernel

1 Introduction

Open source software (OSS) has changed how researchers and practitioners look at software development and related business models [1–3]. OSS differs from traditional in-house software development, as its outcome, the software, is freely accessible by anyone. As a result, major companies such as Facebook, Google and Twitter use open source technologies. In 2012 one million open source projects were catalogued and this figure was projected to double within two years [4]. In addition, according to IDC research, the OSS market should be worth approx. $8 billion in 2013 [5].

Due to this success, firms have also started to actively engage in OSS development [6, 7]. While in early years, software technology companies such as IBM and Novell invested time and resources in OSS development, today even user firms (e.g., Samsung)

© IFIP International Federation for Information Processing 2015
M. Janssen et al. (Eds.): I3E 2015, LNCS 9373, pp. 299–313, 2015.
DOI: 10.1007/978-3-319-25013-7_24

invest in OSS development [8]. Thus, today's successful OSS projects receive contributions from hobbyists, universities, research centers, as well as from software vendors and user firms [9, 10]. Theorists have referred to this kind of combined public and private investments in innovation creation as private-collective innovation [11]. This concept asserts that a private investment model – where firms create and commercialize ideas themselves – and a collective invention model, where multiple economic actors create public goods innovations, may coexist under certain circumstances [12]. In particular, the private-collective innovation model seeks to explain why firms privately invest resources to create artifacts that share the characteristics of non-rivalry and non-excludability [13].

The private-collective model also implicitly assumes that private and public investments in innovations are approximately equal. However, successful OSS projects receive more than 75 % of their code from contributors[1] who are paid by a company [8] and the majority of code is written between 9 am and 5 pm – again indicating that contributions are predominantly provided by firms [14]. These figures contrast with the picture of private-collective innovation as an invention mode where public and private interests manifest equally. The aim of this research therefore is to investigate how different contributor groups associated with public and increasing private interests interact in an OSS development project.

In order to study the interplay of both interest groups we not only need to consider demographic characteristics of the community but also the structural patterns of interactions in it. To achieve this goal, we analyze developers active in the Linux kernel (LK) development community from a social network point of view, as the interaction between the members of a software development community reflects the structure of their collaboration. In particular, we investigate degree distributions and the Gini coefficient in the contributor network with respect to the private and collective contributor groups. Network centrality measures are important indicators of influence in OSS development and are known to deviate according to having firm sponsorship or not [15].

We start with detailing what volunteer and firm-sponsored (i.e., employed) developers motivate to participate in OSS development. Then, we discuss the private-collective innovation model in more detail. Based on a dataset of mailing list communication of LK developers from 1996 to 2014 we calculate network measures for each type of developer (e.g., firm-sponsored, hobbyist, university-affiliated, etc.) and compare them for each year. We discuss implications for research and provide further avenues for research concerning private-collective innovation.

2   Theoretical Background

2.1   Open Source Software Contributors

OSS Communities. An OSS project relies on contributors who make up the core element of an OSS community. OSS is commonly understood as a type of software that

---

[1] In this study the terms *contributor, developer, actor, participant* and *programmer* are used synonymously to denote people who are active in OSS projects and in Linux kernel development.

can be used, changed, and shared by any person. The software itself is in most cases developed by a heterogeneous group of people and distributed under specific licenses, which guarantee the above-mentioned characteristics of OSS [16].

In general, a community arises when different people come together and share a common interest [17]. Thus, von Hippel and von Krogh [11] conceptualize OSS development communities as *"Internet-based communities of software developers who voluntarily collaborate in order to develop software that they or their organizations need"* [11, p. 209]. Besides the fact that OSS communities consist of hobbyists, who voluntarily provide their resources to the community, the definition also involves another important contributor group – organizations. Organizations differ from hobbyists in terms of their motivation to engage and are represented in the community by their employed developers. In turn, employed developers might be considered as proxies for firm interests in the community.

Motivation of Voluntary OSS Developers. The pertinent literature specifies intrinsic and extrinsic motivation as major drivers for hobbyists to engage in OSS projects (e.g., [18–21]). Intrinsic motivation is the execution of an activity due to the accompanying enthusiasm and not for the achievement of specific results [22]. A behavior is extrinsically motivated when an activity is performed for reward, recognition or because of an instruction from someone or an obligation [22]. Although researchers agree on different forms of intrinsic and extrinsic motivation, there is often disagreement about their relevance. The most relevant forms of both motivation types in the context of OSS developers are described briefly in the following.

In connection with OSS developers, researchers investigated a plethora of intrinsic motivators. Among these, *joy-based intrinsic motivation* is the strongest and most prevalent driver of OSS contributors [19]. Joy-based motivation is closely linked to the creativity of a person. Frequently, contributors to OSS projects have a strong interest in software development and related challenges [18].

Another fundamental aspect of intrinsic motivation is *altruism*, which is the desire to help others and to improve their welfare. In OSS communities, developers code programs, report bugs, etc., at their own expense, which includes the invested time and opportunity costs. They participate in the OSS community, without taking advantages of its outcome [18, 21, 23].

In addition, the *OSS ideology* plays a crucial role for many contributors and involves

- joint collaborative values, such as helping, sharing and collaboration,
- individual values, such as learning, technical knowledge and reputation,
- OSS process beliefs, such as code quality and bug fixing and
- beliefs regarding the importance of freedom in OSS, such as an open source code and its free availability and use for everyone [24].

Besides these distinguishing aspects of participants' intrinsic motivation to engage in OSS projects, researchers have found that extrinsic stimuli can also have an impact on the activities of actors in communities (e.g., [18–21]).

An extrinsic stimulus is given through a *personal need* of a developer. Many OSS projects are launched because the initiators needed software with specific functions that are not available to date, and they have the willingness and knowledge to develop these [21].

OSS communities offer the possibility for developers to *improve their programming skills* and their knowledge through participation in a project. Programmers are free to choose in which tasks they participate according to their interests and abilities. As a result, the self-learning participants are experiencing a continuous learning curve and build a repertoire of experiences, ways and means to solve specific software development tasks [21].

In addition, "signaling incentives" as described by Lerner and Tirole [25] can also be a reason for people to participate in OSS communities. The incentives cover, inter alia, the *recognition by other members* of the community and the *improvement of the professional status*.

Motivation of Firms Involved in OSS Development. In addition to hobbyists, companies are also active in OSS communities. While voluntary OSS contributors are driven by intrinsic and extrinsic values, economic and technological aspects motivate firms to participate in OSS projects. In recent times, companies open outwards to organize their innovation activities more effectively and efficiently. A means to complement their own resource base are innovation communities. In the case of software companies, OSS communities form a resource pool these firms can benefit from – depending on the strategy they pursue [6, 26]. Literature investigating motivational aspects of companies active in OSS projects reveals that economic theory is not sufficient to explain the relation between firms and their OSS engagement. Andersen-Gott et al. [27] have reviewed this issue and identified the following three categories of motivational factors that are relevant for companies active in OSS communities.

1. *Innovative Capabilities.* If the involvement of a company in an OSS project is aligned with the business model it maintains, the interaction with the community can lead to better or new products which imply a competitive advantage. The inclusion of external contributors increases the firm's innovative capacity.
2. *Complementary Services.* The dominant way for firms to appropriate from OSS is by providing complementary services to customers (e.g., training, technical support, consultancy and certifications [28]) aligned with their business strategy [29]. Firms pursuing this concept deploy own employees that also contribute to the open source project and community work. Thus, the company (1) acquires external knowledge through their own employees active in OSS development and (2) has access to complementary resources in the community, which are difficult to replicate internally [30].
3. *Cost Reduction.* Companies can publish the source code of their proprietary software under an OS license, try to attract external developers and build a community around the software. In this case, the company will get, for example, ideas for new features, bug reports, documentation, and extensions of the software from external contributors without having to pay for it [25, 31]. Further, in the long run, the code is maintained by the community, such that the firm has lower costs than its competitors with proprietary software [32]. However, it should be noted that establishing an ecosystem and an active community around released source code is no easy task as rivals could pursue similar strategies [26, 33].

2.2   Private-Collective Innovation

In organization science, two different modes of innovation are dominant, namely the private investment and the collective action model.

The *private investment model* is associated with a rather closed innovation behavior. Innovators tend to protect their internally developed proprietary knowledge as this is the source of their profits and competitive advantage [11]. Here innovation is clearly seen as a closed process driven by private investments in order to lead to private returns for the innovator [34].

The *collective action model* of innovation is connected to the provision of a public good. Innovators collaborate in order to develop a public good under conditions of market failure. The produced good is characterized by non-excludability and non-rivalry [35]. This model requires that innovators supply their collected knowledge about a project to a common knowledge base and thus make it a public good. This innovation method can unfortunately be exploited by free riders, who wait until other contributors have done the work and use the outcome for free [11, 35].

OSS communities are an example for a mixture of both mentioned innovation models. OSS contributors freely reveal their privately developed source code as a public good. The developers do not make commercial use of their property rights, although the source code is created as a result of private investments. This innovation behavior is termed private-collective innovation [11]. To get a deeper understanding of how OSS communities combine the best of both models, OSS innovation is first considered from the private investment and second from the collective action point of view.

From the *private investment model perspective*, OSS deviates in two major aspects from the conventional private investment model. First, software contributors are the actual innovators in OSS rather than commercial software developers, because they create software that is needed either by themselves or by the community. Second, OSS developers freely reveal the source code, which they have developed by private means; this manner stands in contrast to the classical innovation behavior. Due to the lack of a commercial market for the sale or licensing of OSS, it is made openly available as a public good [11]. Rewards for the developers are provided in forms other than money or commercialization of property rights. Contributors gain private profits such as reputation, experience or reciprocity [25, 36].

From the *collective action model view*, the community produces a public good with its attributes of non-excludability and non-rivalry [35]. Taking the above given description of the collective action model into account, the non-excludability would bring a dilemma with it because free riders benefit from the software but do not contribute to the good compared to the developers. This circumstance is not a problem, as in line with the OS ideology people voluntarily participate in OSS development and share the results without costs [11]. Moreover, contributors obtain benefits, for example problem solving expertise, learning and enjoyment, from the participation on developing a public good, which the free rider cannot get [25, 37]. The benefits in form of selective incentives are connected to the development process of the good and thus only accessible for the participants. Therefore, OSS contributions cannot be seen as pure public goods as these have significant private elements that evolved out of the ideology, which support the community [11].

In sum, the private-collective model of innovation combines the advantages of both private investment and collective action model. Table 1 compares the most important aspects of the three innovation models from an economic perspective and in relation to OSS development.

Table 1. Comparison of different aspects for the private, collective and private-collective innovation model (Source: adapted from Demil and Lecocq [38] and Schaarschmidt et al. [39])

	Private	Collective	Private-Collective
License	Proprietary	Open	Open
Copyright ownership	Company	Collective	Collective
Number of participating companies	One	None	One or more
Revenue stream	Direct	None	Indirect
Control intensity	High	Low	Low
Knowledge sharing intensity	Low	High	High

3 Method

3.1 Research Context

To find a relevant OSS project for our research, we have taken different aspects into account (e.g., size of the project, activity and continuity, company involvement, availability of a large set of data). Finally, we chose the LK project as our research context, which has served as an example for OSS in many previous studies (e.g., [40]).

The LK project was initialized by Linus Torvalds in 1991 and has been one of the most active OSS projects since its beginning. There are software releases every three month on average, which are possible because of the fast-moving development process and the broad foundation of contributors, ranging from hobbyists to companies. Thus, it involves more people than any other OSS project. The kernel itself makes up the core component of any Linux system and is used in operating systems for mobile devices right up to operating systems for supercomputers. Typically, a new release of the kernel comprises over 10,000 patches contributed by over 1,100 developers representing over 225 companies and is published under the GNU General Public Licence v2 [8].

Besides the fact that the LK is one of the largest cooperative software projects ever started, it has also an economic relevance, as many companies have business models that rely on the LK or on software working on top of the LK, respectively. Many of these companies do actively participate in the improvement of the kernel and thereby take effect on the orientation of the development. Very active companies in the kernel development, among others, are RedHat, Intel, IBM, Samsung, Google and Oracle [8].

3.2 Data Collection and Coding of Contributor Categories

To obtain the data needed for our research, we crawled the LK mailing list web archive[2]. We use mailing list data as it is suitable to calculate network positions that

---

[2] http://marc.info/?l=linux-kernel.

represent developers' influence [15]. The "linux-kernel"[3] mailing list has the purpose of discussing LK development topics as well as of reporting bugs. The observation period of the LK community ranges from 1996 (beginning of the web archive) to 2014.

We identified actors that occur multiple times on the list, for example with different email addresses, but identical sender names. We have mapped these to one person object related to the email address s/he has used when sending a message to the list for the first time.

The identified people interacting in the LK mailing list act partly on behalf of companies. To get a deeper understanding if the actors in the mailing list are affiliated with a firm we used the domain name of the email addresses to assign people to a contributor category. Developers sending messages from a domain indicating that the person is employed by the corresponding company are classified as employed contributors, whereas people using email addresses from public email providers such as yahoo.com were classified as hobbyists. Likewise, we identified developers with email addresses indicating universities and research institutions. Assigning LK actors to a contributor category was done in a semi-manual and semi-automated process in order to obtain a high accuracy of the attributions. Detailed information about the different contributor categories is provided in Table 2.

Table 2. Contributor categories

Contributor category	Description	Examples
Companies	People with email addresses from companies	intel.com, redhat.com
Hobbyists	People with private email addresses	gmail.com, yahoo.com
Universities	People from universities	columbia.edu, duke.edu
Research institutions	People from research institutions and public authorities (e.g., IEEE, government, military)	ieee.org, nasa.gov, army.mil

The cleaned dataset comprises 1,941,119 communication replies for the total time period with overall 86,509 contributors involved. The overall distribution of the contributor groups is made of 37.96 % of company developers, hobbyists represent 51.22 %, universities account for 9.65 % and research institutions make up 1.17 %. Descriptive information about the dataset is given in Figs. 1 and 2. Figure 1 shows the quantity of identified contributors per contributor group and year from 1996 to 2014. Figure 2 states the amount of messages sent per contributor group and year for the investigated period.

3.3 Social Network Analysis

A social network represents persons connected by edges. Social networks can represent friendship relationships, communication, interaction contacts or other types of social

---

[3] http://vger.kernel.org/vger-lists.html#linux-kernel.

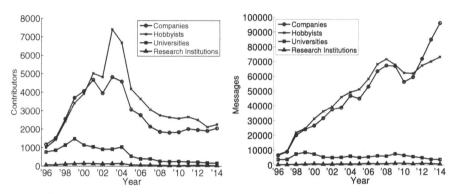

Fig. 1. Contributors per group and year

Fig. 2. Amount of messages sent per group and year

relationships. Social network datasets are widely used, not only in the area of social network analysis, but also in the areas of data mining, sociology, politics, economics and other fields [41].

In order to study the interactions of developers in the LK community, we perform an analysis of the LK mailing list's communication network. Communication within the Linux developer community can be modelled as a directed network, in which nodes are developers and directed edges (i.e., arcs) are a reply of one developer to another. In our dataset, we ignore all messages that are not replies to other developers. A relationship between the sender of a starter message and others does only emerge when one or more people reply to the starter message. We perform a structural analysis of this network to study the interplay of developers interacting.

The directed network of replies we consider is annotated with two additional metadata:

- For replies, the posting timestamp is known. This allows us to make a longitudinal analysis of the considered network statistics.
- For developers, we know their company, university or other affiliation, if any, allowing us to identify four categories of developers, as described in Sect. 3.2.

We perform social network analysis with Matlab and the KONECT Toolbox [42].

The contribution of one user in a directed social network can be used to measure both the activity and the importance of that user in the community. We achieve this by considering the following network-based measures, each of which is defined for individual nodes:

- The in-degree of a node equals the total number of replies received by a developer. The in-degree can thus be interpreted as a measure of importance of a developer.
- The out-degree of a node equals the total number of replies written by a developer, and can thus be interpreted as a measure of the activity of a developer.

- As a network-wide measure, we additionally define the Gini coefficient of the in-degree distribution [43], which denotes the inequality of the in-degrees. It is zero when all developers have equal in-degrees and one when a single developer received all replies. It can thus be interpreted as a measure of diversity of the community [44].

4   Results

4.1   Comparison of In-Degree and Out-Degree

In a first analysis, we compare the in-degree and the out-degree of all developers, i.e., the number of replies given vs. the number of replies received. Figure 3 shows the results of this analysis. We can observe that both measures are highly correlated – developers who receive many replies also write many replies. Thus, for the LK community the activity and the importance of developers correlate highly.

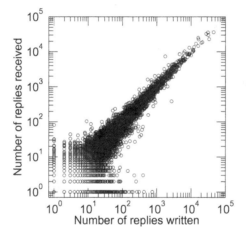

Fig. 3. Comparison of in-degree and out-degree

4.2   Comparison of Degree Per Group

In this analysis, we want to find out whether the developer-based measures of activity and importance vary from one group to another. We compute the distributions of out-degree and in-degree, for each group for the whole dataset aggregated over all years. The results are shown in Figs. 4 and 5.

The plots show that:

- The highest activity as measured by the out-degree is achieved by company developers, then hobbyists, and the lowest activity is given by developers from research institutions and universities.

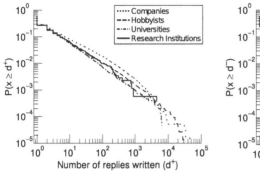

Fig. 4. Out-degree distribution

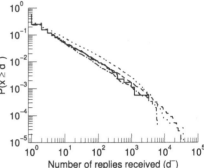

Fig. 5. In-degree distribution

- The measure of importance, the in-degree, correlates and shows the same pattern as for the activity: company developers have the most importance, then hobbyists, and finally developers from research institutions and universities.

These results are consistent with the observation that the measures of activity and importance correlate.

To verify the statistical significance of our results, we perform pairwise Mann–Whitney U tests, testing whether values of each statistic for one type of developer are statistically different from the values for another group. The group differences are statistically significant ($p < 0.05$; company developers vs. hobbyists: $p < 0.10$ for the out-degree), except for developers from companies vs. developers from research institutions and hobbyists vs. developers from research institutions for the in-degree and out-degree; developers from universities vs. developers from research institutions for the out-degree.

4.3 Longitudinal Analysis

In order to study the change of the community over time, we compute three group-wide measures of activity and importance for each individual year in the range 1996 to 2014.

- The average value of the out-degree and the in-degree of all developers in each group, restricted to all replies given and received, respectively during a given year.
- The Gini coefficient of the in-degree distribution of all developers of a given group, restricted to all replies received during a given year.

The results of the analysis are shown in Figs. 6, 7 and 8.

The average out-degree and in-degree (Figs. 6 and 7) show a consistent result with the degree distribution shown in Figs. 4 and 5. The average out-degree standing for activity of the developers of the different groups increases for the developers from companies as well as hobbyists over time and does not chance significantly for the developers of the other contributor categories. The measure of importance, the in-degree, shows a similar behavior. The values for the developers from companies and hobbyists increase and do not change significantly for the other types of developers.

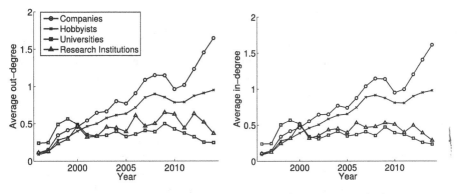

Fig. 6. Average out-degree   Fig. 7. Average in-degree

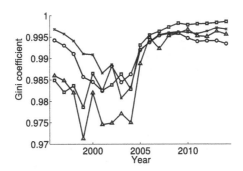

Fig. 8. Gini coefficient

The network wide measure Gini coefficient (Fig. 8) decreases slightly for developers from companies in the last years. The high fluctuation of the Gini coefficient for universities and research institutions is related to the small group sizes.

Across all developer groups and times, the Gini coefficient is very near to one (>98 %, up to fluctuations). This value is higher than in the large majority of social networks [43], indicating that the importance in the community is concentrated in a very small number of actors when compared to other typical social networks.

5 Discussion and Conclusion

5.1 Discussion

This study provides results of an activity analysis of different contributor groups in the LK development from 1996 to 2014. The aim of this study was to investigate how different contributor groups associated with public and increasing private interests interact in an OSS development project. To achieve this goal, we analyze developers active in the LK development community from a social network perspective, as the interaction between members of a software development community reflects the structure of their collaboration.

The first result of our analysis shows that the out-degree, as a measure for activity, and the in-degree, as a measure of importance, correlate highly both in general and for each contributor group individually. Thus, developers who write many replies to the mailing list also receive many replies. This phenomenon in the LK mailing list differs from forum communication, where a variety of user roles with different communication patterns can be identified [45]. Both the highest activity and importance, is achieved by company developers followed by hobbyists. The lowest activity and importance can be attributed to developers from research institutions and universities. Connecting these results to the amount of contributors per contributor category (Fig. 1) it can be seen that although the amount of developers from companies is less than the amount of hobbyists the impact made by employed developers on the LK community is larger. The mentioned impact is expressed by the amount of messages sent per contributor group and year (Fig. 2) as well as stated by the measures of activity (Fig. 6) and importance (Fig. 7), clearly seen especially from 2010 on.

The second result of our analysis shows that the Gini coefficient, as a network wide measure, decreases in recent years for developers from companies and remains constant for the other groups. Although the decrease is small, this can be seen as a tendency that the importance in the community for the group of developers from companies is distributed to more actors. When considering the early years of the LK development (from 1996 until 2000) it can be seen that university members were the most active and important force in the project (Figs. 6 and 7). From 2001 the activity and importance of hobbyists and companies increases alike. Although the amount of company contributors remains relatively stable for the period starting from 2010 until 2014 (Fig. 1) the activity (Figs. 2 and 6) and importance (Fig. 7) in that time increases sharply.

5.2   Conclusion

The aforementioned observations help to answer the research question of how different contributor groups associated with public and increasing private interests interact in an OSS development project, here the LK development. In the beginning, the LK project was driven by intrinsic motivated enthusiasts who are hobbyists and university members (e.g., students) and are associated with pure collective interests. As software and services that were built on top of the LK got more and more influence, the participation of firms, for example, motivated by offering complementary services (see Sect. 2.1), increased. As a consequence the private interests in this project increased, too, especially from 2010 on with the diffusion of mobile devices powered by Android [46], which uses the LK as foundation of the operating system.

Summarizing the results of this study, it can be concluded that the balance between private and collective contributors in the LK development seems to be changing to an open source project that is mostly developed jointly by private companies. These firms need the kernel for their products and services. Thus, the LK project is no longer just an open source alternative for hobbyist to develop open source software for the reason not be locked in by proprietary software. The advantages for the participating companies outweigh the drawbacks in terms of collective copyright ownership and less control in the project (see also Sect. 2.2, Table 1) as the LK community can be utilized to complement their own resource base for innovations.

5.3 Implications for Research

Our findings can be classified into the context of private-collective innovation. The LK development project is an outstanding OSS project. The engagement of companies has increased in the last years, as more and more firms have business models that rely on the kernel. Although companies cannot dictate what the community should do, they can in a way influence the trajectory of the project by assigning employed developers to the project [10, 47]. As the results of our study for the LK project show, employed developers can take key positions in a community due to the intensity of the commitment, expressed by activity and importance, for the community. With this in mind, future research should more thoroughly discuss the nature and structure of firm presence in OSS development. The majority of early research on OSS somehow neglected firm presence and centered on developer motivation while later research discussed emerging OSS business models (e.g., [1]). However, our study calls for more longitudinal studies on firm presence in OSS as (1) firm engagement varies over time and (2) former hobbyists might transform into employed developers (the latter was no focus of this study).

5.4 Limitations and Suggestions for Future Research

Our research has some limitations that have to be considered when utilizing our study's outcomes. The categorization of LK mailing list actors by the hostname of their email addresses into four contributor groups is an approximate but sufficient classification. It is known that there are developers that do not use their company email address while contributing and actors may do personal work out of the office [8]. We mapped developers that occur more than once on the LK mailing list to one person object related to the email address s/he has used when sending a message to the list for the first time. Multiple occurrences happen if the message sender uses different email addresses over time, but the same sender name (e.g., because of company changes). We have not considered these dynamics in our analysis. Furthermore, it has to be considered that the LK project is a unique OSS project, so that our conclusions cannot directly be transferred to other OSS projects where firm-sponsored developers are involved. Further research can consider to investigate the different types and content of the interaction as well as the aforementioned dynamics of developers or compare the multi-vendor project LK to single-vendor OSS projects.

References

1. Bonaccorsi, A., Giannangeli, S., Rossi, C.: Entry strategies under competing standards: hybrid business models in the open source software industry. Manage. Sci. **52**, 1085–1098 (2006)
2. Capra, E., Francalanci, C., Merlo, F., Rossi-Lamastra, C.: Firms' involvement in open source projects: a trade-off between software structural quality and popularity. J. Syst. Softw. **84**, 144–161 (2011)

3. Riehle, D.: The single-vendor commercial open source business model. Inf. Syst. e-Business Manage. **10**, 5–17 (2012)
4. Guignard, C.: Ten facts and figures about open source software. http://blog.nxcgroup.com/2013/ten-facts-and-figures-about-open-source-software/
5. IDC: Worldwide Open Source Software 2009–2013 Forecast (2009)
6. Grand, S., Von Krogh, G., Leonard, D., Swap, W.: Resource allocation beyond firm boundaries: a multi-level model for open source innovation. Long Range Plan. **37**, 591–610 (2004)
7. Schaarschmidt, M., Von Kortzfleisch, H.: Divide et Impera! The role of firms in large open source software consortia. In: Proceedings of the 15th Americas Conference on Information Systems (AMCIS), San Francisco, USA (2009)
8. Corbet, J., Kroah-Hartman, G., McPherson, A.: Linux Kernel Development. How Fast It is Going, Who is Doing It, What They are Doing, and Who is Sponsoring It. Linux Foundation, San Francisco (2013)
9. Teigland, R., Di Gangi, P.M., Flåten, B.-T., Giovacchini, E., Pastorino, N.: Balancing on a tightrope: managing the boundaries of a firm-sponsored OSS community and its impact on innovation and absorptive capacity. Inf. Organ. **24**, 25–47 (2014)
10. Schaarschmidt, M., Von Kortzfleisch, H.: Firms' resource deployment and project leadership in open source software development. Int. J. Innov. Technol. Manage. **12**, 1–19 (2015)
11. Von Hippel, E., Von Krogh, G.: Open source software and the "private-collective" innovation model: issues for organization science. Organ. Sci. **14**, 209–223 (2003)
12. Stürmer, M., Späth, S., Von Krogh, G.: Extending private-collective innovation: a case study. R&D Manage. **39**, 170–191 (2009)
13. Alexy, O., Reitzig, M.: Private-collective innovation, competition, and firms' counterintuitive appropriation strategies. Res. Policy **42**, 895–913 (2013)
14. Riehle, D., Riemer, P., Kolassa, C., Schmidt, M.: Paid vs. volunteer work in open source. In: Proceedings of the 47th Hawaii International Conference on System Science (HICSS), Hilton Waikoloa, Big Island (2014)
15. Dahlander, L., Wallin, M.W.: A man on the inside: unlocking communities as complementary assets. Res. Policy **35**, 1243–1259 (2006)
16. Open Source Initiative: The open source definition. http://opensource.org/osd
17. Preece, J.: Online Communities. Designing Usability and Supporting Sociability. Wiley, New York (2000)
18. Hars, A., Ou, S.: Working for free? Motivations for participating in open-source projects. Int. J. Electron. Comm. **6**, 25–39 (2002)
19. Lakhani, K.R., Wolf, R.: Why hackers do what they do: understanding motivation and effort in free/open source software projects. In: Feller, J., Fitzgerald, B., Hissam, S.A., Lakhani, K.R. (eds.) Perspectives on Free and Open Source Software, pp. 3–22. MIT Press, Cambridge (2005)
20. Roberts, J.A., Hann, I.-H., Slaughter, S.A.: Understanding the motivations, participation, and performance of open source software developers: a longitudinal study of the apache projects. Manage. Sci. **52**, 984–999 (2006)
21. Wu, C.-G., Gerlach, J.H., Young, C.E.: An empirical analysis of open source software developers? Motivations and continuance intentions. Inf. Manage. **44**, 253–262 (2007)
22. Ryan, R.M., Deci, E.L.: Intrinsic and extrinsic motivations: classic definitions and new directions. Contemp. Educ. Psychol. **25**, 54–67 (2000)
23. Baytiyeh, H., Pfaffman, J.: Open source software: a community of altruists. Comput. Hum. Behav. **26**, 1345–1354 (2010)
24. Stewart, K., Gosain, S.: The impact of ideology on effectiveness in open source software development teams. MIS Q. **30**, 291–314 (2006)

25. Lerner, J., Tirole, J.: Some simple economics of open source. J. Ind. Econ. **50**, 197–234 (2002)
26. Dahlander, L., Magnusson, M.: How do firms make use of open source communities? Long Range Plan. **41**, 629–649 (2008)
27. Andersen-Gott, M., Ghinea, G., Bygstad, B.: Why do commercial companies contribute to open source software? Int. J. Inf. Manage. **32**, 106–117 (2012)
28. Fitzgerald, B.: The transformation of open source software. MIS Q. **30**, 587–598 (2006)
29. Dahlander, L.: Appropriation and appropriability in open source software. Int. J. Innov. Manage. **9**, 259–285 (2005)
30. Riehle, D.: The economic motivation of open source software: stakeholder perspectives. Computer **40**, 25–32 (2007)
31. Henkel, J.: Open source software form commercial firms: tools, complements, and collective invention. Zeitschrift für Betriebswirtschaft **4**, 1–23 (2004)
32. Hawkins, R.E.: The economics of open source software for a competitive firm. NETNOMICS Econ. Res. Electron. Netw. **6**, 103–117 (2004)
33. Ågerfalk, P.J., Fitzgerald, B.: Outsourcing to an unknown workforce: exploring opensourcing as a global sourcing strategy. MIS Q. **32**, 385–409 (2008)
34. Liebeskind, J.P.: Knowledge, strategy, and the theory of the firm. Strat. Manage. J. **17**, 93–107 (1996)
35. Olsen, M.: The Logic of Collective Action. Harvard University Press, Cambridge (1967)
36. Von Krogh, G.: The communal resource and information systems. J. Strat. Inf. Syst. **11**, 85–107 (2002)
37. Raymond, E.S.: The Cathedral and the Bazaar: Musings on Linux and Open Source by an Accidental Revolutionary. O'Reilly Media Inc., Sebastopol (2001)
38. Demil, B., Lecocq, X.: Neither market nor hierarchy nor network: the emergence of bazaar governance. Organ. Stud. **27**, 1447–1466 (2006)
39. Schaarschmidt, M., Walsh, G., MacCormack, A., Von Kortzfleisch, H.: A problem-solving perspective on governance and product design in open source software projects: conceptual issues and exploratory evidence. In: Proceedings of the International Conference on Information Systems (ICIS), Research-in-Progress, Milan, Italy (2013)
40. Lee, G.K., Cole, R.E.: From a firm-based to a community-based model of knowledge creation: the case of the Linux kernel development. Organ. Sci. **14**, 633–649 (2003)
41. Wasserman, S., Faust, K.: Social Network Analysis. Methods and Applications. Cambridge University Press, Cambridge (1994)
42. Kunegis, J.: KONECT - the Koblenz network collection. In: Proceedings of the 22nd International Conference on World Wide Web Companion, Rio de Janeiro, Brazil (2013)
43. Kunegis, J., Preusse, J.: Fairness on the web: alternatives to the power law. In: Proceedings of the Web Science Conference 2012, Evanston, IL, USA (2012)
44. Kunegis, J., Sizov, S., Schwagereit, F., Fay, D.: Diversity dynamics in online networks. In: Proceedings of the 23rd Conference on Hypertext and Social Media, Milwaukee, WI, USA (2012)
45. Chan, J., Hayes, C., Daly, E.: Decomposing discussion forums using common user roles. In: Proceedings of WebSci 2010: Extending the Frontiers of Society On-Line, Raleigh, North Carolina (2010)
46. Google: Android @ i/o: the playground is open. Official Google blog. http://googleblog. blogspot.de/2012/06/android-io-playground-is-open.html
47. Schaarschmidt, M., Walsh, G., Von Kortzfleisch, H.: How do firms influence open source software communities? A framework and empirical analysis of different governance modes. Inf. Organ. **25**, 99–114 (2015)

Towards a Set of Capabilities
for Orchestrating IT-Outsourcing
in the Retained Organizations

Bas Kleinveld[1,2](✉) and Marijn Janssen[2]

[1] The Orchestrate Group B.V., Lichttoren 32, 5611 BJ Eindhoven
The Netherlands
bas.kleinveld@orchestrate-group.com
[2] Faculty of Technology, Policy and Management,
Delft University of Technology, Jaffalaan 5, 2628 BX Delft, The Netherlands
M.F.W.H.A.Janssen@tudelft.nl

Abstract. Managing outsourcing processes requires a retained organization to orchestrate an organization's IT functions into a concerted whole. Nowadays multiple outsourcing vendors need to be managed. The purpose of this research is to identify capabilities affecting orchestration of outsourced IT functions. An existing framework for outsourcing capabilities is used as a starting point. Due to the scarcity of research a qualitative research approach based on two case studies is taken. The findings shows that the core IS capabilities as found in the literature are found to be too abstract, ambiguous and that several essential capabilities are missed. A framework containing a refined and extended set of capabilities is derived. Four new capabilities were found including: demand, financial, (service) delivery and service portfolio management.

Keywords: Capabilities · IT outsourcing · Multi-vendor outsourcing · Capability · Orchestration · Retained organization

1 Introduction

Many organizations have outsourced (part of) their IT functions to one or more outsourcing vendors. The IT outsourcing and services offering have changed over time. Due to technology developments such as cloud and market developments companies do not rely on a single vendor any more. Many organizations struggle with their sourcing strategies; retained capabilities and the way function to orchestrate the sourcing providers should be embedded in the organization. In practice, existing capabilities fall short and retained organizations should consider improving their capability in orchestrating sourced IT functions to achieve their objectives. This paper investigates capabilities needed by retained IT organizations for orchestrating sourced IT functions.

When outsourcing IT to other parties the IT still need to be managed by the outsourcing party. For this purpose the client needs a *retained organization*, which orchestrates the various outsourcing vendors. Troost (2009) defined the retained organization as the mediator between the client organization and the internal and

© IFIP International Federation for Information Processing 2015
M. Janssen et al. (Eds.): I3E 2015, LNCS 9373, pp. 314–325, 2015.
DOI: 10.1007/978-3-319-25013-7_25

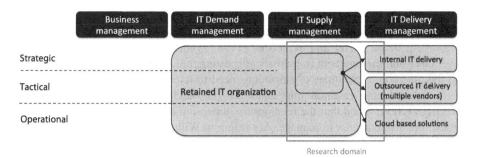

Fig. 1. Research domain

external service suppliers when outsourcing activities. The Fig. 1 shows the retained organization, which is matching the demands from the business (IT demand management) and with the delivery of IT services by either internal or external delivery partners (IT supply management).

The retained organization needs capabilities for managing the interdependencies with the multiple outsourcing vendors. In this paper we focus on strategic and tactical capabilities. The main reason for focusing on strategic and tactical capabilities is that operational capabilities are associated with delivery functions performed mostly by suppliers or operational delivery teams, whereas the strategic and tactical aspects of the capabilities are relatively unknown. Lots of non-academic literature on management and governance frameworks is available, but limited scientific literature was identified investigating capabilities needed for orchestrating IT functions. The *goal* of this research is to identify capabilities affecting the orchestration of sourced IT functions.

This paper is structured in sections as follows. First we present the reviewed literature on the capabilities of retained organizations and orchestrating the sourced IT-function. Next the research approach is presented followed by the description of case studies. The case studies include the analysis of the orchestration function, as well as the identified capabilities. These capabilities are elaborated on and further discussed. Finally, conclusions are drawn.

2   Literature Background

A study performed by Gartner (2003), shows that enterprise organizations have some key challenges across IT within their organizations. In planning IT outsourcing, organizations seem to have difficulties in deciding which specific roles and capabilities have to be retained in the organization in order to mitigate risk, manage the IT supplier and manage the business demand appropriately Joha (2003). Organizations have a set of capabilities to remain competitive and conduct their activities. Willcocks et al. (2006) defined a capability as a distinctive set of human resource-based skills, attitudes, motivations and behaviors that contribute to achieving specific activities and influencing business performance. In 1998, Feeny and Willcocks developed an IT governance and management framework out of two streams of research. The first stream concerned characteristics of high performers in the Information Technology (IT) function; the second examined the retained capabilities needed to run effective IT outsourcing deals. Feeny and Willcocks (1998) synthesized the findings into an IT governance and management framework suggesting four tasks of the future IT function and the nine core capabilities organizations need to retain in-house.

In detailed case research into major IT outsourcing arrangements Lacity and Willcocks (2000) found that the relationship dimension between client and suppliers a critical, but also complex to manage. According to Willcocks et al. (2006) governance has a central task to devise organizational arrangements (structures, processes and staffing) to successfully manage the interdependencies, and ensure that the IT function delivers value for money. The model from Willcocks et al. (2006) presents a number of serious human resource challenges. It requires high performers in each role. Furthermore, in contrast to the more traditional skills found in IT functions there needs to be a much greater emphasis on business skills and orientation in all but the two very technical capabilities, i.e. architecture planning and making technology work. There is a significantly increased requirement for 'soft' interpersonal skills across all roles, all roles demand high performers, and each role requires a specific set of people behaviors, characteristics and skills (Willcocks et al. 2006). Managing the outsourcing processes required a retained organization to orchestrate the IT functions in a concerted whole. Empirical research remains scarce and there is little known about the underlying theories and management principles. One exception is Plugge and Janssen (2014) who investigated multi-vendor governance by focusing on how resources are coordinated and which resources the organizations are dependent on.

When orchestrating a portfolio of capabilities some capabilities will be executed adequately, others poorly; but a few must be superior to the competition if the business is to sustain a market position that is valuable and difficult to match (Day 1994). Managers must become thought leaders and strategic architects in determining the company's core functions, building on an in-depth understanding of the potential competitiveness and value contributions of external providers (Spiller et al. 2014).

In multivendor outsourcing, vendors experience a strong degree of distrust between each other, due to lack of unclear strategic roles and responsibilities and lack of structure (e.g. meetings, forecasts). In particular, the vendors' inability or unwillingness to cooperate in delivering IT services to the client affected their relationship negatively. The various roles, activities and responsibilities of each party are often not clear (Plugge and Bouwman 2013). Given the strategic nature of their IT-outsourcing goals, it is important for outsourcing-driven and outsourcing-centric customers to make relational investments in their partnerships with vendors (Mehta and Mehta 2010). The coordination of IT activities that reflects the relationship between the client and the vendors, however, reveals that no particular coordination mechanism is used. Based on analyses by Plugge and Janssen (2014) research demonstrates that clients and vendors need to develop and implement clear governance structures and mechanisms to coordinate the delivery of IT services successfully. The executive management of both client and vendors has to implement these governance structures to be able to proactively monitor changes in a multivendor landscape.

Governing multivendor outsourcing arrangement is a continuous process that requires regular management attention (Plugge and Janssen 2014). Managing complex vendor relationships and contracts requires levels of project management skills typically expected of category managers. This creates significant opportunities to lead increasingly complex sourcing relationships and manage these external relations to maximize value. According to Coltman and Devinney (2013) the need to match supply with demand has gained a prominent position in the service operations and the wider operations management literature. However, the role that managers play in the alignment process has been rarely investigated. Several companies have recognized the need to improve the end-to-end orchestration of their value chains (Spiller et al. 2014).

The findings of a study by Plugge et al. (2013) demonstrates that outsourcing capabilities and organizational dimensions are perceived to be critical factors in achieving quality performance, and that a fit between them is paramount. The outcome of this study demonstrates that the outsourcing experts perceived that the client's need was less important in influencing organizational structure (Plugge et al. 2013). The results suggest that monitoring and assessing changing client circumstances regularly is a prerequisite for providers to be an agile organization. Consequently, they must be willing and able to adapt their outsourcing capabilities and organizational structure to achieve high quality performance and thus to remain competitive.

3   Research Approach

The goal of this research is to identify capabilities affecting the orchestration of sourced IT functions. This research focuses on situations in which services are continuously delivered. The main reason for this focus is that orchestrating continuous service delivery is complex in nature, since it is related to multiple internal and external suppliers working together. Case study research is employed to gain insight into the capabilities desired and the actual capabilities an organization has to orchestrate. The case studies were investigated by reading reports and conducting interviews. In total two case studies were investigated meeting the following requirements. The companies should be an international company with more than 20,000 employees. The organization should have a mix of internal delivery teams and external multiple vendors and the IT budget should exceed 100 million euro's.

Semi-structured interviews have been conducted on the required core capabilities in orchestrating sourced IT functions will be held. The interviews were conducted between January and May 2015, with 6 senior managers and took 60 to 90 min and are important since interviews enable the collection of in-depth information from senior managers. The interview scheme is included as appendix to this paper. Interviews were conducted with the service delivery manager, IT director Applications, EMEA IT manager, program director, director of sourcing, and IT director Platforms.

4   Case Studies

The case studies help us to gain results gain in-depth insight in the nature of orchestration function in IT organizations and the capabilities needed and actual capabilities for orchestration.

4.1   Case 1

The company is one of Europe's largest electricity companies with approximately 31,000 employees. Operations are conducted in different European countries with revenues exceeding 15 billion euro in 2013. The business activities are divided into several Business Units.

The IT organization is divided in a demand and supply management structure. How the demand and supply interact with one another is a central factor for the success of the IT strategy and orchestration function. The IT organization is based on a federated organizational model, where all corporate resources are allocated to IT. The company has over 2,500 applications, of which around 1,200 applications are supported (and financed) by IT. IT supply and delivery management is divided in 3 main IT delivery pillars: IT infrastructure, and 2 application solutions groups.

4.2   Case 2

This company is a leading global manufacturing company based in The Netherlands. The company supplies industries and consumers worldwide with innovative products. Its portfolio includes well-known international brands. The company is consistently ranked as one of the leaders in the area of sustainability. The company has global operations with more than 50,000 employees. The revenue in 2013 exceeded 15 billion euro. The company is organized in three main business units.

The IT organization is divided in a demand and supply management structure. Demand and Supply Management manages the interface between business and IT, maintaining the balance of demand and supply of IT services. IT supply is divided in 3 main delivery pillars: Infrastructure services, and two application services groups.

4.3   Capabilities in the Orchestration Function

When analyzing the orchestration function of both cases, the companies struggle with their sourcing strategies, related capabilities and the way orchestration function should be embedded in the organization. In interviews and documentation, core capabilities to perform the orchestration function were identified. In both cases the core capabilities are primarily performed by external resources. The main reason is the lack of required skills within the organization, and lack of available resources to perform the activities. As a consequence orchestration function in organizations did not have a high maturity level. Further research in this domain is recommended (Table 1).

Table 1. Overview of capabilities

Type of capabilities	Case: Company X	Case: Company Y
IT-Architecture	• IT architecture consists of enterprise architecture, solution architecture and domain knowledge/ architecture • Guarding existing templates/systems while keeping-up the pace with new technologies	• IT architecture is considered enterprise architecture, solution architecture, and to a limited extend domain and technical architecture • Although architecture is considered a core capability, it can be sourced from externals in certain situations
Contract management and sourcing	• Managing contracts between key suppliers and the business • Not implemented at Company X, but is considered key	• Managing (outsourced) contracts and vendors • Contract management is performed by Sourcing (purchasing), not by IT
Supplier and vendor management	• Managing supplier performance • Managing the vendors and the related IT services • Supplier management is not implemented during times of this research	• Managing suppliers on the operational performance, KPI management and operational budget Contract management is performed by IT
Team and relationship management	• Managing relationships, stakeholder and people is a capability often underestimated, it remains a people business • Right team balance between performance, capabilities, internal and external resources	• Close relationships with the business on their needs and requirements will improve the orchestration function related to sourced IT functions
Content knowledge related to business domains and (ERP) systems	• Content knowledge is understanding of the business processes and map to functionality and IT solutions • Performed by many externals	• Content knowledge is understanding of the business processes and map to functionality of the applications in use (strategic applications) • Performed by many externals, although there is a clear strategy to replace externals on this capability
Demand management	• Stakeholder management • Understanding of business needs • Managing the business demand for IT	• Managing the demand of IT needs from the business • Mapping the needs to right IT groups

(Continued)

Table 1. (*Continued*)

Type of capabilities	Case: Company X	Case: Company Y
Service delivery management	• Managing the IT solutions operational delivery (on both internal and external side) on a day-to-day basis	• Managing the operational delivery on a day-to-day basis including quality assurance
Financial management	• Mapping the costs to activities, matching this to the actual and forecasted costs made by suppliers	• Charge out model for projects and support services • Manage supplier statement of works and service level agreements to invoicing and financial bookings • Although this is considered a core capability, this activity is not optimal performed
Service portfolio management	• Not indicated as capability by interviewees	• Defining the services to be delivered to the business • IT departments will act more as service providers to their internal customers • Ability to understand what to deliver to internal customers

4.4   The Orchestration Function Analyzed

During interviews, the interviewees were asked which capabilities influenced their orchestration function in IT. These capabilities were collected and analyzed using a data matrix. Table 2 visualizes the capabilities provided by the interviewees from the case studies, including additional capabilities. Some capabilities in the case studies have been confirmed during the interviews, although there is still a gap between case study capabilities and the capabilities identified in literature.

In our research we found 4 capabilities in addition to the capabilities suggested by Feeny and Willcocks (1998) and Willcocks et al. (2006). Demand management, financial management, (service) delivery management and service portfolio management were found as capabilities, which were mentioned by the interviewees as shown at the bottom of the table. Several capabilities have been integrated to a single capability based on the interview outcomes. Making technology work and architecture planning are IT-architecture, while informed buying, contract facilitation, contract monitoring and vendor development are considered supplier management, contract management and sourcing. Relationship management has been extended to team and relationship management. The capabilities from the case studies and indicated by interviewees are described hereafter.

4.5   IT-Architecture

IT-Architecture was mentioned by almost all interviewees consists of enterprise architecture, solution architecture, and to a limited extend domain and technical architecture. Interviewees indicate that although architecture is considered a core capability, it can be sourced from externals in certain situations. An interviewee quotes the importance of architecture as: "There are always excuses for not using the defined architectural standards. Finding reasons for exceptions creates complexity and issues in organizing IT services. The challenge is that IT delivery functions need to keep up the pace with new technologies while keeping alive the legacy technologies and systems".

4.6   Supplier Management, Contract Management and Sourcing

Supplier management, contract management and sourcing are recognized as a core capability by all interviewees to perform the orchestration function of IT. Although this is recognized, several interviewees see developments in this domain. It is recognized as core capability while interviewees indicate that supplier and contract management is a domain for improvement.

4.7   Team and Relationship Management

Interviewees recognize the need for relationship management and good team management. Managing relationships, stakeholder and people is a capability often underestimated, as all performed activities is still about people. Close relationships with the business on their needs and requirements will improve the orchestration function related to sourced IT functions (interviewee). "At the moment we have a balance and right mix of capabilities. Externals contribute to the right mix of this balance in team management".

4.8   Content/Technical Knowledge Related to Business Domains/Functions

Many interviewees indicate that content and technical knowledge related to business domains and functions is a key capability in IT departments, and a key capability in orchestrating sourced IT functions. Content knowledge relates to understanding of the customer's business and translate this to IT solutions, but also related to the understanding of processes and company dynamics. In addition resources with technical knowledge are needed to be able to judge supplier and internal delivery performance, related to their objectives. Content and technical knowledge related to business functions and domains can relate back to architecture. Some capabilities that have been indicated by interviewees do not match the capabilities from literature and the case studies. Service integration management was a capability from the case studies that was not mentioned by interviewees.

Table 2. Identified capabilities

Identified Capabilities		TOTAL	case 1			case 2		
			Customer organization P1	Customer organization P2	Customer organization P3	Customer organization P4	Customer organization P5	Customer organization P6
Capability from literature (Feeny & Willcocks, 1998)								
1. Architecture planning	IT-Architecture (4.5)	5	x		x	x	x	x
2. Making technology work	IT-Architecture (4.5)	5	x		x	x	x	x
3. Informed buying	Contract management and sourcing (4.6)	6	x	x	x	x	x	x
4. Contract facilitation	Contract management and sourcing (4.6)	6	x	x	x	x	x	x
5. Contract monitoring	Contract management and sourcing (4.6)	6	x	x	x	x	x	x
6. Vendor development	Supplier & vendor management (4.6)	4		x	x	x	x	
7. Relationship building	Team & relationship management (4.7)	4	x	x	x	x		
8. Business system thinking	Content & technical knowledge related to business functions (4.8)	5		x	x	x	x	x
Capabilities not found in literature								
Demand management (4.9)		3	x			x		x
Financial management (4.10)		2	x		x			
(Service) delivery management (4.11)		3	x	x	x			
Service portfolio management (4.12)		1	x					

4.9   Demand Management

Demand management is not a capability from literature, but indicated to be important from the cases. Managing the business demand has been indicated as a capability that influences the orchestration function by half of the interviewees. Customer intimacy and business information management are given as areas that are important for demand management.

4.10   Financial Management

Almost all interviewees have indicated that cost optimization if an important driver in their objectives and goals of IT departments. In the findings that influence capabilities, interviewees mentioned cost optimization as important issue. Although this is an important issue, financial management of their IT functions is only recognized by a number of interviewees, Customer organizations should consider financial management as capability in orchestrating sourced IT functions related to their financial objectives.

4.11   (Service) delivery management

The delivery of IT services to the business is considered a capability to organizations, as interviewees indicate that it will become even more important. "When outsourcing more services, the capability of delivery management changes. We've experiences this and now adapt the philosophy: eyes on, hands off", according to an interviewee. Mechanisms that can improve delivery functions mentioned across several interviewees is to introduce shared KPI's across multiple IT towers in delivery and more output measurement rather then activity management. Focus on end result rather then activity.

4.12   Service Portfolio Management

One firm believes that service portfolio management will become a more important capability now and in the future, as IT departments will act more as service providers to their internal customers. Interviewee quotes: "This is a core capability organizations should have to be able to understand what they deliver to their internal customers". Cloud solutions drives the business to buy services directly from the market, while the value for IT departments will decrease if they don't deliver services to their internal customers, rather then resources.

5   Discussion and Conclusions

In the two case studies the organizational members are aware of the importance of orchestrating sourced IT functions and the need to build the right capabilities. Although they are aware of which capabilities are needed, the existing capabilities fall short. In particular in the domains as contract, vendor and supplier management. Retained organizations should consider financial management as capability in orchestrating sourced IT functions related to achieving their financial objectives. Many interviewees indicated that content and technical knowledge related to business functions is a key capability for their IT department, and a key capability in orchestrating sourced IT functions. Two interviewees indicated that although architecture is considered a core capability, it can be sourced from externals in certain situations.

Due to lack of required skills, resources and capabilities, the orchestration func-tion in organizations does not have the desired maturity level. When analyzing the orchestration function of the cases, the companies struggle with developing their or-chestration capabilities and the way orchestration function should be organized in their organizations. In the interviews, core capabilities to perform the orchestration function are clearly identified, but these capabilities are often performed by external people or by persons not equipped to perform the role. The main reason for this is the lack of people having the required skills within the organization, and lack of people to perform the activities. The orchestration function is identified as an abstract capability con-sisting of a subset of capabilities that IT organizations should retain after outsourcing services to external providers.

When comparing to the findings of the case studies with the literature a clear gap can be seen. Feeny and Willcocks synthesized the findings into an IT governance and management framework suggesting four tasks of the future IT function and the nine core capabilities organizations need to retain in-house. The model from Feeny and Willcocks (1998) presents a number of serious human resource challenges. However literature limited attention is paid to invest in the relationship, while the cases show that relationship and team management are an essential capability for orchestration. Fur-thermore, in contrast to the more traditional skills found in IT functions there needs to be a much greater emphasis on business skills and orientation in all but the two very technical roles. Although the four tasks and nine capabilities (Feeny and Willcocks, 1998) show the need of these, the cases show that the capabilities identi-fied need to be practically applicable, and some of the capabilities are formulated at a too high level of abstraction for practical applicability in organizations. In our re-search we found 4 capabilities to the capabilities of Feeny and Willcocks (1998), which are demand management, financial management, (service) delivery manage-ment and service portfolio management. In our research we extended the capability relationship man-agement with team and relationship management.

References

Coltman, T., Devinney, T.M.: Modeling the operational capabilities for customized and commoditized services. J. Oper. Manag. **31**, 555–566 (2013)

Day, G.S.: The capabilities of market-driven organizations. J. Mark. **58**(10), 49–63 (1994)

Feeny, D.F., Willcocks, L.P.: Core IS capabilities for exploiting information technology. Sloan Manag. Rev. **39**(3), 9–21 (1998)

Gartner Executive Report Series, IT Spending: How Do You Stack Up (2003). Retrieved from the Gartner Inc. http://www.gartner.com

Gartner predicts limited IT outsourcing growth and increased volatility (2013). http://www.cio.com/article/737472/Gartner_Predicts_Limited_IT_Outsourcing_Growth_and_Increased_Volatility

Joha, A.: The retained organization after IT outsourcing, the design of its organizational structure. Master Thesis, Delft University of Technology, Delft, The Netherlands (2003)

Lacity, M., Willcocks, L.: IT outsourcing relationships: a stakeholder perspective. In: Zmud, R. (ed.) Framing the Domains of IT Management Research. Glimpsing the Future Through the Past. Pinnaflex Educational Resources, Cincinnati (2000)

Mehta, N., Mehta, A.: It takes two to tango: how relational investments improve IT outsourcing partnerships. Commun. ACM **53-2**, 160–164 (2010)

Plugge, A.G., Bouwman, H.: Fit between sourcing capabilities and organizational structure on IT outsourcing performance. Prod. Plan. Control **24-4**(5), 375–387 (2013)

Plugge, A.G., Janssen, M.: Governance of multivendor outsourcing arrangements: a coordination and resource dependency view. Doctoral dissertation, Delft University of Technology, Delft, The Netherlands (2014)

Plugge, A.G., Bouwman, W.A.G.A., Molina-Castillo, F.J.: Outsourcing capabilities organizational structure and performance quality monitoring: towards a fit model. Inf. Manag. **50**(6), 275–284 (2013)

Spiller, P., Reinecke, N., Ungerman, D., Teixeira, H.: New role for the CPO: orchestrating the end-to-end value chain. Supply Chain Manag. Rev. **18**, 27–33 (2014)

Strategic Road Map for Outsourcing Competencies. Gartner research G00250619 (2013). Retrieved from the Gartner Inc. www.gartner.com

The Reality of IS Lite. Gartner research G00117022 (2003). Retrieved from the Gartner Inc. http://www.gartner.com

Troost, M.: Sourcing the retained organization in IT outsourcing. Master Thesis, Delft University of Technology, Delft, the Netherlands (2009)

Willcocks, L., Fenny, D., Olsen, N.: IS Capabilities: Feeny-Willcocks IT governance and management framework revisited. Eur. Manag. J. **24**(1), 28–37 (2006)

Why Do Small and Medium-Size Freemium Game Developers Use Game Analytics?

Antti Koskenvoima and Matti Mäntymäki[✉]

Turku School of Economics, Rehtorinpellonkatu 3, 20500 Turku, Finland
{antti.koskenvoima,matti.mantymaki}@utu.fi

Abstract. The increased use of the freemium business model and the introduction of new tools have made analytics pervasive in the video game industry. The research on game analytics is scant and descriptive. Thus, reasons for employing game analytics are not well understood. In this study, we analyze data collected with a set of in-depth interviews from small and medium-sized freemium game developers. The results show that game analytics is used to (1) assist design, (2) to reduce the risks associated with launching new games, and (3) to communicate with investors and publishers. The study advances the research on the business value of game analytics.

Keywords: Analytics · Freemium · Games

1 Introduction

Use of big data and analytics has become pervasive in the video game industry [1]. The adoption of analytics has been driven by the fast development of cost-effective solutions that enable basic analytics even for start-up sized game developers. The second driver of game analytics is the diffusion of the freemium business model that has created a need to accurately measure, predict, and intervene player behavior[1]. For example, Supercell, the company behind the top-grossing freemium games Clash of Clans and Boom Beach, generated revenue of 1.8 billion USD from in-game purchasing in 2014[2].

The term freemium is a combination of words "free" and "premium" [2] that describes a business model in which a service or a product is offered a for free but a premium is charged for advanced features, functionality or related products and services. Game analytics refers to applying analytics and big data in the gaming context [3]. Game analytics can be used to improve the players' gaming experience, or to maximize in-game purchases by tweaking various aspects of the game [4, 5]. For example, the time taken for a specific task, the cost of a specific item or the power of a specific weapon [6, 7] can be optimized with game analytics.

---

[1] http://www.theguardian.com/technology/2014/dec/09/clash-of-clans-billion-dollar-mobile-games.

[2] http://www.theguardian.com/technology/appsblog/2013/jul/23/clash-of-clans-supercell-free-mium.

© IFIP International Federation for Information Processing 2015
M. Janssen et al. (Eds.): I3E 2015, LNCS 9373, pp. 326–337, 2015.
DOI: 10.1007/978-3-319-25013-7_26

Game analytics is a new field with limited research coverage [8]. Prior research has focused on describing methods of data gathering and analyzing [3] as well as the role of analytics in game development [9, 10]. The limited amount of prior literature can be partly explained with the fact that game developers consider analytics confidential and are thus often reluctant to share information on their analytics processes [11].

As a result, the business aspects of game analytics are not well understood. Consequently, this study seeks to investigate *why and how do small and medium-sized freemium game developers use game analytics?* To answer the research question we have conducted a set of in-depth interviews among freemium game developers. Our results show that game analytics is used to (1) assist design, (2) to reduce the risks associated with launching new games, and (3) to communicate with investors and publishers. The study advances the research on the business value of game analytics.

2   Literature Review

The Freemium Model. The freemium business model is extensively employed with digital products or services such as software, games, and web services. As a result, companies employing the freemium model strive to monetize their user base by striving to convert the users of the free version into paying customers [12–14].

Freemium games typically employ micro-transactions and advertising as the main monetization strategies [15]. Micro-transactions refer to buying virtual items or services that can be used and have value only inside a specific gaming environment [cf. 13]. Second, the players can be provided with the opportunity to buy "time", either to bypass waiting enforced by the game mechanics or to skip repetitive "grinding" phases [15], or to unlock additional levels or areas of the game. Third, virtual items and benefits compared to the non-paying users can be bundled into a premium user account or a subscription [12]. Fourth, in-game advertising and potentially offering an ad-free upgraded version against a fee can be used to monetize the players. Fifth and finally, game operators can sell information about the players to third parties, typically marketers [16].

Freemium games are typically designed to engage the player immediately, because there is no initial cost causing a lock-in effect [17]. In addition, the game mechanics used in freemium games motivate the players to make in-game purchases [18]. As a result, sustained play and customer lifetime value are critical for the economic success of freemium games.

Game Analytics. Game analytics is a subset of analytics applied to the game development [19]. Analytics in turn refers to using business intelligence in the process of discovering and communication patterns from the data and using the recognized patterns in solving business problems [20].

The need for game analytics has increased as games have become more sophisticated and complex [21], and the rise of mobile gaming and the freemium business model has also had its effect. Gross et al. [22] maintain that studying online games is important from a managerial viewpoint "to understand ways that interactions while gaming can be improved, in order to make better games."

Today's video games, particularly online games and social media games, can collect data about almost all players' in-game activities [4]. Game developers can use this data to obtain information about e.g. potential sources of player frustration [23]. For example, Replica Island, a game for Android devices, employed a player tracking system to identify instances where players were facing difficulties (e.g. player deaths). The whole metrics system was implemented by a one-man team at virtually no monetary cost, and has proven to be extremely valuable in identifying problems with the game design [23]. However, according to Drachen et al. [24], game analytics is far from the traditional value chain of the video game development and hence not very highly prioritized.

Prior game analytics research has employed the purchase funnel concept to illustrate the challenges the freemium model poses [3, 15, 25]. The AIDA-model (awareness, interest, desire, and action) from consumer behavior literature describes the process of new product adoption [26]. In application marketplaces where people typically download freemium games the customer can go through the process from awareness to action in seconds. However, out of the people who have downloaded a freemium game, only a fraction, e.g. 5 per cent will pay anything [15]. As a result, successful employment of the freemium model requires sustained player engagement and efficiently managing a large pool of non-paying players towards conversion during the course of the play. A freemium game can be a hit without ever being profitable, and thus monetization is essential [27].

3   Empirical Research

Data Collection. The empirical data was collected with five semi-structured interviews from experienced professionals from companies developing freemium games. The companies they worked in were based Finland. Obtaining the empirical data was challenging since finding the number of seasoned professionals with in-depth knowledge on freemium games as business in general and game analytics in particular is small. At the time of the interview, each of the five interviewees was in the process of developing a freemium game and using software to collect and analyze gameplay data. In addition, topics related to game analytics were frequently considered company confidential and hence many prospective informants declined the interview.

The theme of the interview was presented to the interviewees in advance. A semi-structured approach was selected due to the relative novelty of the research area. According to May [28] the semi-structured approach allows the interviews to be flexible and interactive.

The interviews were done face-to-face in sessions from fifty minutes to one hour twenty minutes. All interviews were recorded and transcribed by the first author. In addition, notes were taken during the interviews. Appendix 1 summarizes the data collection.

Analysis. The analysis process started already during the interviews as notes about potentially insightful themes were taken. In the first round of analysis, the interviews were transcribed and coded based on the research question [28]. Through iterations of the data, a constant-comparative method was applied to identify, elaborate, and clarify categories [29]. Emergent categories were examined within and across interviews to determine salience and recurrence. Interconnections and discrepancies between the interviews and previous literature and between different interviews were also coded. The recurrence of certain key themes and limited number of new ones emerging over coding process indicated that the data exhibited saturation.

Results. We identified two main themes that describe the different roles of game analytics for small and medium-sized freemium game developers. These are analytics as a communication tool and analytics as a decision support tool. Based on these two themes we derived an emergent theme, analytics as necessity.

Analytics as Commnunication Tool

> "[Analytics] are a kind of tool for studios to justify their decisions, for example why a certain game is not ready for launch yet, because we need to improve this metric. The investors are more willing to give extra time, when they can see that in the long run the game will make more money if improved." (P4)

As the quotation above illustrates, investors and publishers are very keen on analytics. The interviewees clearly stated that the key performance indicators have to be reported frequently to investors and publishers, and lack of improvement in these metrics will lead to questions. On the other hand, analytics help game designers make their case when they feel the launch schedule needs to be postponed as the game is promising but behind schedule.

> "If I were a game publisher, I would ask teams to soft launch and provide retention and ARPDAU [average revenue per daily average users] numbers before I would invest anything." (P3)

Freemium games are typically continuously improved through scheduled updates and developers want to see the impact of changes as soon as possible. Sizeable and potentially risky adjustments are tested out with a small subgroup of players so that regular service would not be disturbed. Similarly, before the game is launched worldwide, it can be released in smaller market (e.g. Canada or Finland). The purpose of these soft launches is to collect data to ensure that the game will be profitable.

> "We used to have a thing where once a week every studio [under that publisher] would report to the headquarters in California in an hour long conference call and give a pre-formatted presentation in which the key metrics were analyzed, future plans to improve them laid out etc." (P4)

The direct business benefits of game analytics include informed financial decisions, such as rationalizing marketing spending and budgeting for launch. The interviewees were very conscious that the video game industry has had several high-profile costly flops, i.e. projects that went overtime and over budget and made considerable losses. The interviews also indicated some of which could potentially have been avoided with the help of analytics. In addition, the interviewees recognized pressure from many stakeholders, competitors, players, and publishers on the gaming companies to adopt game analytics.

> "Retention is the most important metric in the game industry" (P2)

The interviewees considered retention rate as the most important individual metric of business success in freemium game development. Consequently, all interviewees maintained that they are keen on tracking and monitoring retention figures. Retention rate refers to how players keep playing a game for subsequent sessions. A simple example of measuring retention is tracking the number of game overs per player. Typically retention was measured over a time period (e.g. 7 days, a month). The interviews also revealed that retention is a key metric for monetization as well as the basis of funnel analysis. In addition, retention is used to improve first impressions and the tutorials.

Interestingly, yet there was a consensus regarding importance and benefits from game analytics, the interviews also indicated that increasing volumes of data make it more challenging to extract relevant information from it. The interviewees wanted to keep the analytics process as simple as possible and focus on the key performance indicators (KPIs) and their trends over time and different versions.

"The most common wisdom in this free-to-play model is that if you don't have retention you are never going to make money. Retention stems exactly from that the game itself has some interesting aspects and is in some way fun."(P4)

In addition to retention, metrics that measure monetization and additional retention metrics, e.g. average revenue per user (ARPU), conversion rates, tutorial funnels and day 1, day 7, day 14 and day 30 retention rates, were mentioned as examples of metrics that most game developers monitor, or at least should monitor.

Additionally, customer lifetime value is used to measure to what extent the costs related to customer acquisition are covered. Customer lifetime value is an aggregate of other metrics, namely cost-per-install and ARPU, and is used to guide marketing spending. The following quotation depicts the role of customer lifetime value:

"It is mostly based on things like is it profitable to invest in marketing the game. LTV (Lifetime Value) will tell you that." (P1)

Analytics as a Decision Support Tool

"You cannot make a good game with just analytics, it's very challenging – [everything new] comes from the creative side" (P2)

The informants, particularly the ones who had been actively involved in actual game design, stressed quite strongly that analytics does not, and it also should not, drive game design. The interviewees maintained that analytics can support decision-making by e.g. occasionally disqualifying the intuition of the designers, but that analytics does not offer ready solutions. Further underlining that the design philosophy is not data-driven but data-supported, the interviewees stressed that game developers can and will outsource data collection, crunching and storage, but never game design. The analytics and metrics do not make the games, but they can help in making them better. The following quotation describes the role of analytics in relation to design:

"[Regarding Game Analytics] maybe the larger benefit, at least in our case, is that instead of driving development, they are used to spot errors in the code and clear design mistakes. - - It's more about monitoring – first you design, then you code and then you monitor how well did it go." (P5)

We observed an interesting controversy regarding the interviewees' views on the role of game analytics. While all believed in the value of analytics and some emphasized tracking as much as possible to ensure having sufficient amount of data whenever needed, others preferred a more strict selection of metrics to avoid information overload and collecting what they referred as "vanity metrics". These interviewees also more deliberately emphasized the importance of evaluating the benefits against the respective costs:

A recurring theme in the interviews was that the informants considered players unpredictable and that they often behave differently than the designers expected, often seemingly irrationally. Analytics can help designers to understand players' perspective of the gaming experience. For example, one interviewee described a mobile game that had a design mechanic punishing players for erroneous behavior by reducing his/her points. An alternative solution was to end the game immediately after an error was made and force the players to start over. Somewhat counter-intuitively the latter version, which the designers felt was more "hardcore-oriented", was more appealing to players and led to higher retention levels. As a result, game analytics proved that, against game conventions and designers' expectations, the harder and more punishing version was more popular.

The analysis tools and methods in use were largely uniform among informants. Maybe for this reason they did not see analytics as a real source of competitive advantage. As all our informants worked for small or medium-sized companies with very limited human and financial resources to be spend on analytics, they had simply adopted 'off-the-shelf' analytics tools. Since most small and medium-sized game companies use a similar set of tools, the informants considered the expertise and proficiency in using tools and the ability to draw the right creative conclusions the best way to derive value from game analytics. The informants also stated that the use of analytics differs considerably between smaller and larger companies in the freemium games market.

There was a strong consensus among the informants that data allows real and accurate insights about player behavior. All of the interviewees had experience from primarily quantitative analysis and metrics. The informants' view was that game companies seldom utilize qualitative data, as it was considered more taxing, less effective, and harder to implement with third-party solutions. The informants also stated that interviewing players about their gameplay habits can lead to misleading results since people can seldom tell what they really want and would use.

> "We don't do analytics because they are cheaper, but because they are better. – – I never trust people who say "If you would develop this, I would use it all the time". Only when I can measure that they really use it, I will believe it." (P3)

All informants stated that the analytics data is actively shared and communicated within the development team, but the interviews also indicated that game companies seldom share their sales numbers or metric data in public. Companies can however compare the KPIs of new launches with their prior games. This way game analytics also help guide the portfolio management of game publishers.

The data-supported design process is iterative. First, a visible problem is noticed in high-level metrics. Then the designers seek for possible causes by drilling down into the data. Specific changes are made to the game, and then the effects are measured. There is a constant loop for validating design decisions.

> "[We tend to find a] – –high-level issue and then try to find one specific user experience issue that you think you change, and then you iterate the process." (P4)

A typical way to tackle the issues that are found during the development process is to create two or more different versions of the game. These versions are then randomly distributed to players and their respective performance is measured. This is referred to as A/B testing. It cannot be used extensively for every decision since developing each alternative consumes resources and the inferior versions are a wasted effort. Acquiring relatively reliable results from the A/B tests also takes considerable time.

> "When something new is added to the game, it is done in two different ways and half of the players get version A and the other half gets version B and then the metrics are compared. There may be a hypothesis behind the test, but it is more about trying to find what works and then developing the better version further, leave it as is or abandon both." (P1)

Funnel analysis was another common analysis method used by the informants. Tutorials are a typical example of a funnel. Throughout the tutorial phase gradually fewer and fewer players reach each subsequent step. They have not necessarily paid anything for the game yet, so it is essential that they do not churn out this early. Measuring the return rate or retention of players in subsequent steps (or levels) in the game was the most common way of analyzing the quality of said funnels. When the designers note a notch in the retention, they know that there is a problem in that part of the funnel.

Analytics as a Necessity. The findings strongly suggest that game analytics plays an important role in communicating with investors and publishers as well as in supporting game development and manager's decision-making. Hence, analytics was seen as a necessity in freemium game development. Every informant pointed out several benefits of game analytics. In fact, operating without utilizing analytics was considered "flying blind" and analytics was seen as a means to reduce the risk of failures. Interestingly, the interviewees also repeatedly stated that there had been no need to state an explicit business case to justify investments in game analytics – the benefits were that clear.

> "It is clear that – especially in this free-to-play model – you cannot operate under the mentality that you just launch a game and hope for the best." (P1)

However, utilization of game analytics is characterized by the lack of resources. This explains why game analytics was seen as something that is important but not as a differentiating factor or a potential source of competitive advantage. Table 1 summarizes the main findings.

Table 1. Summary of findings

Analytics as a communication tool	Investors and publishers follow certain key metrics The metrics provide a common ground for discussion Retention is the most important metric to track Average revenue per user (ARPU) and Lifetime Value (LTV) are also actively followed
Analytics as a decision support tool	Help to reduce the risk of total failure Role is to support design decisions, not drive them Analytics affect the whole development process, there is a constant loop of changes, assessment and improvement Developers use game analytics to e.g. identify problems in the game design and/or bugs in the code that lead to player churn
Analytics as a necessity	Analytics viewed necessary but not as a source for competitive advantage Reliance on third-party game analytics tools and software. Lack of resources (time, skills, and money) restricts the use of more advanced analytics The emphasis is on quantitative data

4   Discussion

Theoretical Implications. The findings indicate that game analytics play a pivotal role in freemium game development. The interviewees emphasized a combination of designer creativity and analytics is required to obtain the best results and repeatable results. Our results imply that retention is the most important metric to track. Game analytics was not viewed as source of competitive advantage, at least for smaller companies relying on standard third-party analytics tools, but considered more as a risk reduction tool. The possible competitive advantage of utilizing game analytics is derived from the experience and creativity of the analysts and designers.

Prior research has presented the vast possibilities and introduced sophisticated solutions for game analytics [1, 3]. However, our results imply that the analytics processes used by small and medium-sized freemium developers are rather simple. There are two opposing philosophies in data gathering. One stresses that data cannot be tracked retroactively, and tracking as much as possible ensures you have the data when you need it. The other warns about collecting "vanity metrics" that do not contribute to game design and only confuse the analysis process [see e.g. 30].

Very large volumes of data call for increasingly sophisticated and powerful analysis tools, which in turn increase the costs. Prior research has promoted utilizing a combination of data sources (e.g. analytics, interviews, biometric measuring) to improve results [10]. However, our results show that small and medium-sized freemium developers primarily focus on measuring retention and a few other key metrics. Furthermore, the data analysis methods were simple and many of the professionals cited that they do not have advanced database management skills.

Altogether, these observations align with prior research arguing that alongside being actionable, the results from analytics should be human-readable and easily interpretable [as in e.g. the model presented in 31].

Retention rates were further utilized in evaluating which alternative was better in A/B tests and improving funnels, especially tutorials. The methods used resembled the split testing used by Andersen et al. [32] to recognize the most engaging design choice from three alternatives. The small and medium-sized freemium developers also showed interest in adapting more complex analytical procedures, but did not have either the skills or resources to implement them. The interviewees did not share much about the gathering of qualitative data, which further emphasizes that the focus is on directly actionable, quantitative metrics.

Prior literature on game analytics has recognized the managerial importance of actionable results [33]. Our results add on this body of research by emphasizing the value of the predictability and decision support in analytics. In addition, our results demonstrate that for freemium game developers, game analytics provided important tools for investor communications. Third, game analytics can help in optimizing marketing spending by customer lifetime value calculations or campaigning for more time from the publisher citing promising retention trends.

Managerial Implications. Compared to game industry's prior reluctance to adopt analytics [7], our results indicate a change in mindset. Our set of interviews among game industry professionals indicated that the benefits from game analytics are so self-evident that they are not even always explicitly stated. Since basic analytics are today available for even startup companies, the initial costs of implementing game analytics are relatively low. For example, applying basic telemetry data analysis is one the most cost-efficient way to do user research, and even one-man teams can afford it [34].

Our results imply that game analytics is viewed as a risk reduction tool. Using analytics during the development stage can decrease the risk of total failure in the launch stage. In addition, even the basic level of analytics with the standard key performance indicators can assist companies in e.g. terminating projects that are unlikely to generate sufficient revenue to become profitable.

Certain metrics have become close to industry standards. Key performance indicators that measure monetization and retention, such as ARPU, conversion rates, tutorial funnels and day 1, day 7, day 14 and day 30 retention and customer lifetime value were mentioned as examples of the key metrics. These metrics are also used to evaluate and predict the financial success of the game.

Many of the interviewees actually stated that investors require key ratios such as retention rate and ARPU when they discuss potential investments with the teams. Similarly, publishers expect results and want hard numbers to confirm the feasibility of the game concept. For example, certain leading publishers also already demand weekly reports that cite key metrics from their developers.

Limitations and Future Research. Like any other empirical research, the present study is subject to a number of limitations. First, the empirical data was collected solely from small and medium sized companies developing freemium games that were based in Finland. For example, large freemium developers such as Supercell are well known for their advanced use of game analytics. Thus, we suggest future research with a broader empirical coverage.

Second, the freemium model as well as analytics is employed in other fields such as online music as well as more traditional fields such as insurance. Future research could examine how analytics is utilized in other sectors to find best practices that could be applicable across industries.

Appendix 1: Table of Interview Subjects

	Position	Description	In development	Company size
P1	Product Lead	Experience from multiple mobile game companies. Has been utilizing game analytics since 2009.	Freemium mobile game	SME 20-49 employees
P2	Chief Product Officer	Has led his own game studio for many years. First touched simple forms of analytics in 2006, has been responsible for analytics at many companies.	Freemium mobile game	SME 5-9 employees
P3	Chief Executive Officer	Started a gaming company after an engineering career. Begun with analytics in 2011 and has utilized them in two successful game projects.	Freemium mobile game	SME 5-9 employees
P4	Product Manager	Begun as a community manager, and utilized analytics in that role. Has now shifted focus to e.g. balancing in-game economy.	Freemium mobile game	SME 20-49 employees
P5	Chief Executive Officer	Started a small game development company with friends. Few launched games; the new project is the first time analytics are used systematically.	Freemium PC game	SME 5-9 employees

References

1. Bauckhage, C., Kersting, K., Sifa, R., et al.: How players lose interest in playing a game: an empirical study based on distributions of total playing times, pp. 139–146 (2012)
2. Wilson, F.: The Freemium business model. A VC Blog, vol. 23 (March 2006)
3. El-Nasr, M.S., Drachen, A., Canossa, A.: Game Analytics: Maximizing the Value of Player Data. Springer Science & Business Media, Berlin (2013)
4. Drachen, A., Thurau, C., Togelius, J., et al.: Game data mining. In: El-Nasr, M.S., Drachen, A., Canossa, A. (eds.) Game Analytics, pp. 205–253. Springer, Berlin (2013)
5. McAllister, G., Mirza-Babaei, P., Avent, J.: Improving gameplay with game metrics and player metrics. In: El-Nasr, M.S., Drachen, A., Canossa, A. (eds.) Game Analytics, pp. 621–638. Springer, Berlin (2013)
6. Zoeller, G.: Game development telemetry in production. In: El-Nasr, M.S., Drachen, A., Canossa, A. (eds.) Game Analytics, pp. 111–135. Springer, Berlin (2013)
7. Mellon, L.: Applying Metrics Driven Development to MMO Costs and Risks. Versant Corporation, Redhood City (2009)
8. Drachen, A., Canossa, A.: Towards gameplay analysis via gameplay metrics, pp. 202–209 (2009)
9. Canossa, A., El-Nasr, M.S., Drachen, A.: Benefits of game analytics: stakeholders, contexts and domains. In: El-Nasr, M.S., Drachen, A., Canossa, A. (eds.) game analytics, pp. 41–52. Springer, Berlin (2013)
10. Gómez-Maureira, M.A., Westerlaken, M., Janssen, D.P., et al.: Improving level design through game user research: a comparison of methodologies. Entertainment Comput. 5, 463–473 (2014)
11. Wallner, G., Kriglstein, S., Gnadlinger, F., et al.: Game user telemetry in practice: a case study, vol. 45 (2014)
12. Mäntymäki, M., Salo, J.: Why do teens spend real money in virtual worlds? A consumption values and developmental psychology perspective on virtual consumption. Int. J. Inf. Manag. 35, 124–134 (2015)
13. Mäntymäki, M., Salo, J.: Purchasing behavior in social virtual worlds: an examination of Habbo Hotel. Int. J. Inf. Manag. 33, 282–290 (2013)
14. Wagner, T.M., Benlian, A., Hess, T.: The advertising effect of free–do free basic versions promote premium versions within the freemium business model of music services?, pp. 2928–2937 (2013)
15. Fields, T., Cotton, B.: Mobile & Social Game Design: Monetization Methods and Mechanics. Morgan Kaufmann, Los Altos (2012)
16. Clemons, E.K.: Business models for monetizing internet applications and web sites: experience, theory, and predictions. J. Manag. Inf. Syst. 26, 15–41 (2009)
17. Zauberman, G.: The Intertemporal dynamics of consumer lock-in. J. Consum. Res. 30, 405–419 (2003)
18. Hamari, J., Lehdonvirta, V.: Game design as marketing: how game mechanics create demand for virtual goods. Int. J. Bus. Sci. Appl. Manag. 5, 14–29 (2010)
19. Drachen, A., El-Nasr, M.S., Canossa, A.: Game analytics–the basics. In: El-Nasr, M.S., Drachen, A., Canossa, A. (eds.) Game analytics, pp. 13–40. Springer, Berlin (2013)
20. Davenport, T.H., Harris, J.G.: Competing on Analytics: The New Science of Winning. Harvard Business Press, Cambridge (2007)
21. Hullett, K., Nagappan, N., Schuh, E., et al.: Data analytics for game development: NIER track, pp. 940–943 (2011)

22. Gross, S., Hakken, D., True, N.: Getting real about games: using ethnography to give direction to big data (2013)
23. Pruett, C.: Hot failure: tuning gameplay with simple player metrics. Game Dev. Mag. **19** (2010)
24. Drachen, A., Canossa, A., Sørensen, J.R.M.: Gameplay metrics in game user research: examples from the trenches. In: El-Nasr, M.S., Drachen, A., Canossa, A. (eds.) Game Analytics, pp. 285–319. Springer, Berlin (2013)
25. Moreira, Á.V., Vicente Filho, V., Ramalho, G.L.: Understanding mobile game success: a study of features related to acquisition, retention and monetization. SBC **5**, 2–12 (2014)
26. Webster, F.E.: New product adoption in industrial markets: a framework for analysis. J. Mark. **33**, 35–39 (1969)
27. Davidovici-Nora, M.: Paid and free digital business models innovations in the video game industry. Commun. Strateg. **94**, 83–102 (2014)
28. May, T.: Social Research: Issues, Methods and Process. McGraw-Hill Education, UK (2011)
29. Corbin, J., Strauss, A.: Strategies for Qualitative Data Analysis: Basics of Qualitative Research. Techniques and Procedures for Developing Grounded Theory. SAGE Publications, Thousand Oaks (2008)
30. Canossa, A.: Meaning in gameplay: filtering variables, defining metrics, extracting features and creating models for gameplay analysis. In: El-Nasr, M.S., Drachen, A., Canossa, A. (eds.) Game Analytics, pp. 255–283. Springer, Berlin (2013)
31. Xie, H., Kudenko, D., Devlin, S., Cowling, P.: Predicting player disengagement in online games. In: Cazenave, T., Winands, M.H., Björnsson, Y. (eds.) CGW 2014. CCIS, vol. 504, pp. 133–149. Springer, Heidelberg (2014)
32. Andersen, E., Liu, Y., Snider, R., et al.: On the harmfulness of secondary game objectives, pp. 30–37 (2011)
33. Canossa, A., Cheong, Y.: Between intention and improvisation: limits of gameplay metrics analysis and phenomenological debugging. In: DiGRA Think, Design, Play (2011)
34. Canossa, A.: Interview with Nicklas "Nifflas" Nygren. In: El-Nasr, M.S., Drachen, A., Canossa, A. (eds.) Game Analytics, pp. 471–473. Springer, Berlin (2013)

Dynamic IT Values and Relationships: A Sociomaterial Perspective

Leon Dohmen[(⊠)]

CGI Nederland B.V., Eindhoven, Netherlands
Leon.dohmen@cgi.com

Abstract. Management scholars are criticized for ignorance and the wrong approach when studying the impact of technology in organizational life. Impact of technology in this paper is interpreted as IT values created or achieved from equivalent and contingent interaction between human (people) and non-human agents (technology, organization). Researchers and theorists propose to include a sociomaterial perspective and to develop general and broader, empirical based patterns across different contexts. Based on a literature review containing publications of theoretical considerations and empirical research this paper introduces a first general and sociomaterial based overview and taxonomy of IT values and their relations. IT values have a techno-economic or socio-techno orientation, are dynamically entangled and competitive, and complementary or overlapping. IT values are related to time, sponsor and, hierarchy. The identified IT values are ordered into a framework which has to be treated as a starting point to discuss further the definition, dynamics and relations of IT values from a sociomaterial perspective.

Keywords: Impact of technology · IT values · Sociomateriality · Relationship · Entanglement · Emergence · Techno-economic · Socio-techno

1 Introduction

Management scholars are criticized in two ways when researching the *impact of technology* in organizational life. Either they ignore it[1] or they prefer a linear approach by separating technology, organization and people [1]. Sociomaterial theory is proposed as alternative. "*Sociomateriality* stands out as a symbol for the interest in the social and the technical, and in particular, the subtleties of their contingent intertwining" [2]. One of the key concepts of sociomateriality is based on Actor Network Theory. Human and non-human agents are inseparable connected maintaining equivalent relationships and "enact continuously relational effects" [2]. But also sociomateriality is subject of criticism. Sociomaterial oriented theorists and researchers argue to unlock broader, general patterns across different contexts [2] and to acquire more empirical evidence.

---

[1] Orlikowski refers to a study of Zammuto et al. including a survey of four journals: Academy of Management Journal, Academy of Management Review, Administrative Science Quarterly and, Organization Science. Only 2.8 % of the research articles in these journals focused on technology and organizations.

© IFIP International Federation for Information Processing 2015
M. Janssen et al. (Eds.): I3E 2015, LNCS 9373, pp. 338–353, 2015.
DOI: 10.1007/978-3-319-25013-7_27

This paper interprets impact of technology as *IT values* achieved from the equivalent and contingent interaction between technology, organization and people. To meet the objection of linear research, IT values will be analyzed through a sociomaterial lens. First IT values will be identified, collected and ordered based on literature review and discussions with subject matter experts. Secondly this paper introduces the relational dimensions between collected IT values.

2   Dynamic IT Values through a Sociomaterial Lens

Assumed is that IT values are depending from the sociomaterial context and are emergent as attitude, learning processes and skills play a crucial role [3]. Here we accept that values are in a "state of becoming" rather than a status quo [2] assuming a fully relational ontology, where IT values exist in relation to each other.

To acquire insights in the dynamics of IT values when applying IT-facilities this study, identifying and collecting IT values, proposes a new sociomaterial based IT values framework and taxonomy. The question we focus on for this inventory is: *Which IT values can be identified and how are they related?*

2.1   Research Method

To search for broader and general sociomaterial patterns zooming out technique is used for a literature review containing theoretical expositions and/or empirical research. A zooming approach provides ideas about how to extend qualitative research methods for investigations of sociomateriality [2]. Exploration and selection of IT values and composition of the framework and taxonomy happened from March 2011 until April 2015. The study started with a consultation of subject matter experts with an academic and/or business background. Following their literature suggestions revealed a heterogeneous representation of the concept of IT value.[2] Due to the zooming out approach publications from outside the IT domain like product design were included. Books and articles showed different approaches and definitions to describe and explain IT value (related) concepts. These concepts showed for example quantitative and qualitative research approaches and/or objective and subjective (perceived) definitions of IT value. Another difference was that some articles focus more on the process of value creation and conditions rather than on IT value. Because the contribution of this paper is to compose a framework and taxonomy for IT values we focused especially on studies and considerations with primary attention for explaining and defining IT value. After composing the first draft of the framework further consultation of subject matter experts took place which led to new literature suggestions. Additional literature review was done using keywords (see footnote 2) searching on the Internet for relevant articles,

---

[2] Recommended literature, which some of is included in the list of references, led to an extensive list of concepts related to or representing IT value: business/IT alignment, business value, company value, contribution of IT, customer satisfaction, customer value, information system success, IT absorption, IT adoption, IT value perception, net benefits, technology acceptance, user acceptance, use of IT, user value and, value of IT.

papers and books dealing with the subject of IT value. New IT values discovered became part of the collection. To construct and order collected IT values, terms and definitions of the public values discussion [4] are adopted for the IT values framework. Public values are values in governance and public service [4]. The public values discussion involves the relational and entangled dimension between values which corresponds with important principles of sociomateriality. Strong point of this long term and evolving zooming out approach is the broad coverage of the concept of IT value. A weak point is due to the extensive range of related key words (see footnote 2), important relevant literature can be missed. Despite this weakness, because of the need to unlock broader general patterns as contribution to support sociomaterial theory it is chosen for this approach.

The framework presented in this paper consists of eight nodal values, sixteen neighbour values and sixty four co-values (Table 1).[3] Neighbour values are the bridge between nodal values, which are referred to as starting point and co-values are determined as promoter or contributor [4]. Co-values can be positive or negative. The interpretation of an IT value and position in the framework is partly derived from the categories of user value from Boztepe [5].[4] Because of the readability of the table references are omitted here. These are explicitly mentioned in the accompanying text.

We prosecute our discourse about IT values with a brief description and definition of chosen values when appropriate supported by illustrative examples from practice.

2.2   Utility Value

Utility value is the consequence of using a product and encompasses neighbour values *convenience, safety, quality (and performance), economy* [5] and, *service* [6]. IT values to a large extend are connected to the material aspects of the product and have a techno-economic orientation.

A utility is acquired to fulfil (convenience) needs – including *accessibility, appropriateness* and (physical) *compatibility* – of the user and *avoid unpleasantness* [5]. Technological *availability* [7] should also be considered as an important co-value for convenience. The human role is shaping as well as being shaped by *time* [8]. Orlikowski and Yates [8] define time as "people produce and reproduce what can be seen to be temporal structures to guide, orient and coordinate their ongoing activities."

Security, health and *comfort* are co-values of safe usage [7]. Paro, the robot seal used in healthcare for people suffering from dementia, is an interesting reference in practice for these co-values [9]. Paro reduces stress and leads to a positive mood. Serving road safety, *compliance* is meeting governmental laws and regulations [10].

Durability and *reliability* of materials used are important co-values of quality and performance [5]. *Efficiency* is also performance related [5]. Investments in basic

---

[3] Selected values follow the definition of the referred to literature in the reference list, in some cases supported by an example or key word between brackets.

[4] Boztepe originally distinguishes four categories which are positioned as nodal value in this paper: utility, social significance, emotional and spiritual. These categories are used as starting point for the framework and further supplemented with other relevant nodal values.

Table 1. General sociomaterial IT values framework

Nodal IT values	Neighbour IT values	IT co-values
Utility	Convenience	– Accessibility – Appropriateness – (Physical) compatibility – Availability – Time management – (Avoidance of) sensory unpleasantness
	Safety	– Security – Health (e.g. reducing stress) – Comfort – Compliance
	Quality and performance	– Durability – Reliability – Fit for purpose (usefulness) – Agility (flexibility) – Speed – Effectiveness – Efficiency (ease of use)
	Economy	– Use economy – Purchase economy (price value) – Objective financial indicators (e.g. net margin, profitability, operational expenses, etcetera) – Share value
	Service	– Assurance – Responsiveness – Empathy – Relationship
Social	Social prestige	– Influence – Power – Impression management (face saving acts) – Respect
	Identity	– Role fulfilling – Group belongingness
	Ethics	– Right conduct – Moral principles – Honesty
Emotional	Pleasure	– Fun – Enjoyment – Beauty – (Job)satisfaction – Attachment – Affection (love) – Detachment – Addiction – Nomophobia

(Continued)

Table 1. (*Continued*)

Nodal IT values	Neighbour IT values	IT co-values
		– Panic
		– Anger
	Sentimentality	– Memorability
		– Nostalgia
Cognitive	Stimulation	– Excitement
		– Curiosity
		– Self-actualization
	Growth	– Independent thought and action (independence)
		– Creating new innovative things
		– Diffusion (of gained knowledge)
Universal	Welfare	– Social innovation
		– Tolerance
	Protection	– Sustainability
		– Care for people
Traditional	Loyalty	– Commitment (deep attachment)
		– Respect
Spiritual	–	– Good luck
		– Superstition
Singularity	Super-humanity	– Super-intelligence
		– Immortality
		– Personalized food

infrastructures have a different purpose compared to investments in innovative applications [11]. So the utility should fit the *purpose* of use. Business operation changes permanently due to increased competition, new rules in law, etcetera. To adapt smoothly to these changes *agility* [12] or *flexibility* is another important aspect of quality and performance [13]. Increasing *speed* to access knowledge and service delivery is experienced as an important gain when applying social tools [14]. Al-Maskari and Sanderson [15] refer to a general term like *effectiveness* to express utility value. From an expectancy perspective Venkatesh et al. [16] refer to ease of use (*effort*) and *usefulness* (performance). Usefulness is interpreted as similar meaning as fit for purpose.

Economy is a next neighbour value of utility value used in as well as the business context [9, 17, 18] as consumer context [5, 16]. Where Boztepe [5] refers to *purchase economy* Venkatesh et al. [16] introduce the term *price value* as an indicator of technology use in a consumer context. Sneller [9, 18] and Kersten [17] emphasize the importance of *use* economy when discussing utility value. Kersten [17] urges to replace legacy systems ('old' IT) by modern technologies ('new' IT). Economy value includes the life cycle of a technology and is measured with *objective financial indicators* like *net margin* and *profitability*. Implementing an ERP-system increases *share value* [18].

Besides applying a utility or product, service is determined as an important neighbour value of utility value. DeLone and McLean [6] adapted their previous IS

Success Model. Service quality contains similar co-values as identified for product quality and performance. Additional important co-values are *responsiveness, assurance* and *empathy. Relationship* between actors when providing services to a product also impacts user value perceptions [19].

2.3 Social Value

Social value involves socially oriented benefits. From a sociomaterial perspective here the material is used to derive or gain social advantage. This includes *social prestige* and construction and maintenance of one's *identity* [5]. *Ethics*, another identified neighbour value, is of increasing importance [20].

"Social significance (prestige) value refers to the socially oriented benefits attained through ownership and experience" [5]. Product benefit examples lead to social associations (*impression management, face saving acts*) between family and other social groups with increase in *respect, influence* and *power* as consequence [21]. "Possession of a trendy object is often seen as sufficient to communicate a certain image of self (identity)" [5]. Social significance or influence becomes meaning in relations with others which concerns *group belongingness* and *role fulfilling* [5, 21]. Companies can build up a company image (identity) by chasing IT fashions [22].

Ethics refers to a set of (local) principles of *right conduct* or a theory or system of *moral* values. IT solutions have a big impact on the work of others. Engineers of IT solutions should embed an ethical (value) dimension (*honesty*) in the requirements of the to build solution [20].

2.4 Emotional Value

Emotional value is about aroused feelings of affective states like *pleasure* and is triggered by co-values like *fun* and sensory *enjoyment* [16, 21]. Also hedonistic values like *beauty* initiate pleasurable experiences and belong to this neighbour value pleasure. *Memorability* can arouse a *sentimental* feeling which is also associated with emotional value [5].

Attachment is a positive (pleasurable) emotional state in the relation between user and product [23]. Opposite to it *detachment* is a negative emotion which indicates the lack of linkage between an individual and a product [23]. Socio-technical studies see *job satisfaction* and productivity as important outcomes manipulated by social and technical factors [7]. Automation leads to controlling and deskilling while empowering and upskilling are a result of informate. Both – automate and informate – are different purposes when applying IT leading to different values [7]. Robot seal Paro is an interesting example how people suffering from dementia can get *attached* to it and develop *affection* for Paro [9]. On the other hand attachment can evolve into habit for example in mobile phone usage [16] and *addiction*. Venkatesh, Thong and Xu [16] introduce habit for technology use in a consumer context to extend UTAUT.[5] Habit

---

[5] UTAUT is Unified Theory of Acceptance and Use of Technology.

here is defined as prior behaviour or automatic behaviour. *Nomophobia* [24] is detected as a new 'illness'. When people have lost or forgotten their mobile phone emotional feelings like *panic* become part of them.

Using a screensaver or background picture on a device with a family photo or *memorable* event – *nostalgia* – is an experience of emotional feelings. Sentimentality is here appointed as a separate neighbour value [5].

Emotional value benefits arise from affective experiences related to aesthetic, giving meaning and provoke feelings as *love* (affection) and *anger*. Emotional co-values are person related, subjective and intrinsic [5]. These contributors are mainly assigned to the socio side of sociomateriality.

2.5 Cognitive Value

"IT has given a boost to knowledge related activities which are a continuation of the written word and printed book. This has been sometimes referred to as the information revolution" [25]. Heng [25] provides a classification scheme for IT-applications and addresses the *cognitive* value of IT. This nodal value is lacking in the overview of Boztepe [5]. The value of IT recognized as a source to contribute to knowledge creation and distribution is linked to the network era [26], social technologies and the networked organization [27]. *Stimulation* and *growth* are neighbour values of cognitive value. Cognitive value is also designated as epistemic value [21].

"IT's contribution (stimulation) to the knowledge enterprise enable employees to create, store and disseminate knowledge on a scale hitherto unknown" [25]. *Curiosity, excitement* and *self-actualization* are important elements to acquire (new) knowledge and support the *creation* and *growth* of *new (innovative) things* [21]. Value sensitive design here is linked *to independent thought and action* decoupled from group values [21]. The more rapidly (individual) innovative IT capabilities can be deployed, the more rapidly (business) value will grow [28]. While legacy research highlights random adoption in a social network Baldwin and Curley [28] claim that *diffusion* of innovations can be actively directed and accelerated, especially for IT systems.

Cognitive value – diffusion and access to information and knowledge – is a value which evolved from efficiency and information value [26]. Cognitive value is primarily intrinsic and subjective but can be made explicit and objective. Applying IT in education becomes more and more popular and impacts the learning process [29]. Savas [23] appoints *independence* as a positive emotion possibly leading to attachment to a product. This link to emotional value shows that values are not perceived isolated but are dynamic and (closely) interwoven [5].

2.6 Universal Value

To *care for people* in emerging countries platforms on the Internet like Get It Done [30] provide opportunities to fund projects or create your own projects. These crowd funding initiatives supported by IT are occurrences associated with *universal* values like *welfare* and *protection*. Welfare and protection can also be associated with public values [4]. However, the discussion about public values goes far beyond IT values only. The application of IT by national and local administrations leads besides IT value also to transformations in relationships between governmental institutions reciprocally and to transformations in relationships between governmental institutions and their citizens [31]. The latter is another example of dynamic emergence [2] of IT values.

Social innovation [32] and *tolerance* reflect welfare. People are able to connect with social tools [27] to whoever they like to. The application of IT is also seen as a social innovation issue in Belgium to support the increasing aging of the baby boom generation [33].

Sustainability shows care for planet and nature. A great example of protection is the 'volmeld'-system of the city Groningen in the Netherlands [9]. Underground waste containers are equipped with a system which transfers twice a day a message of the degree of filling to the central computers. If the percentage reaches seventy percent the container is automatically included in the route of the garbage trucks. Besides time savings (efficiency) which is usually a business objective the environmental pollution is decreased due to the reduction of $co2$ emission by seventeen percent [9]. U-city concerns the environmentally friendly and sustainable smart (or knowledge) city which makes the ubiquitous computing available amongst the urban elements such as people, building, infrastructure and open space [34]. This is an ongoing example of the evolving entanglement of technology, things and people like described and explained in sociomateriality theory and the dynamic values it creates [2].

Universal value is associated with care for people and planet [21]. From a socio-materiality perspective this nodal value is very much linked to the socio part of materiality. However, these values can easily be mixed with organizational value objectives like improving efficiency.

2.7   Traditional Value

Respect and *commitment* are related to acceptance of customs and ideas that *traditional* culture and religion impose on themselves [21]. Commitment (deep attachment) is related to co-value attachment which is associated with nodal value emotional.

Besides involving content also the support in users' tasks in maintaining ideas and customs is an example of traditional value. Commitment can evolve to *loyalty* to a product and recommendation. Commitment can also lead to repeat purchase of a product [35] or increased (intention of) use [16]. "Positive experience with use will lead to greater user satisfaction in a causal sense. Similarly, increased user satisfaction will lead to increased intention of use and thus use" [6].

Within traditional value the socio and material are very close related. Irritation and frustration – as experiences of negative value perceptions – can be linked directly to the (material) product when it does not work properly or as expected.

2.8   Spiritual Value

"*Spiritual* value refers to *good luck* and *sacredness* (*superstition*) enabled by a product" [5]. According to Boztepe [5] examples show that communication technologies are increasingly becoming enablers of spiritual experiences too. For instance, several websites have been set up that serve Muslims who live away from their home countries, allowing them to pay online and make sacrifices on their behalf.

2.9 Super-Humanity Value

Technological *singularity* [36] which can be achieved via biomedical science and nanotechnology will create *super-intelligence* and *super-humanity* [37]. Although mainly envisioned and to date hardly proven, the concept seems to be able to abolish biological limitations and create immortality [38]. Vinge [37] described several appearances of singularity and super-human being based on artificial intelligence, intelligence amplification and, biomedical, Internet and, digital Gaia[6] scenarios. For this paper, singularity is appointed as a provisional end point in the evolution of IT (see Fig. 1).

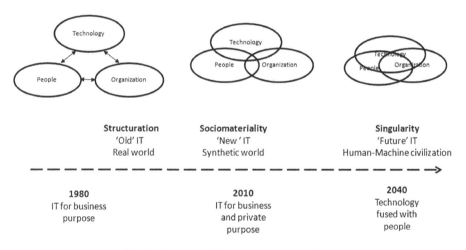

Fig. 1. Emergent IT values and time relation

A further application of IT that should not be ignored is 3D food printing [39]. Data-driven recipes (Jeffrey Lipton, Cornell Creative Machines Lab) provide possibilities for older people with chewing and swallowing problems to create a meal with a modified food structure[7] and nutritional hardness (Pieter Debrauwer, TNO). 3D (food) printing adds a new dimension to the application of IT. While old applications and IT systems transfer data into reports and documents – focusing mainly on data, transaction and transfer – 3D printing transforms data into physical products. Hereby 3D food printing creates also new sensory experiences. The printed food can be tasted and smelled. Other values mentioned during the 3D Food Printing Conferences [39] were

---

[6] The Digital Gaia Scenario: The network of embedded microprocessors becomes sufficiently effective to be considered a superhuman being.

[7] In the food industry this is also referred to as food texture.

contribution to flexibility, sustainability and comfort. Instead of feeding elderly people with pureed food which sometimes leads to social isolation while people feel ashamed, feeding people with *personalized food* that looks regular but which structure is adapted increases group belongingness (nodal value social). To date it can be determined that IT values are able to trigger all five human sensory experiences: feel; see; hear; smell, and; taste.

With this involvement of biological (food) aspects it could be considered to include the biological aspect in sociomaterial theory. Also the nodal value universal and neighbour value protection is linked to biological value when involving a co-value like sustainability.

2.10   The Sum of IT Values: Net Benefits

With respect to the ongoing theoretical and philosophical discussion of foundations for sociomateriality [40] our study to explore, define and categorize IT values contains a broad orientation and generalization leading to an overview and ordering of IT values from a sociomaterial perspective. Herewith we respond to the call and need for sociomaterial generalization across different contexts [2].

Same as for public values [4] IT values seem to be in competition and are not strictly separated but overlapping each other. From utility value – being the material representative – a number of socio related IT values are triggered, sometimes as an undesired and negative side-effect like addiction or when improving efficiency consequence can be a decrease in working pleasure – informate versus automate [7]. This means that IT values do not exist in isolation, are dynamically entangled, maintain *relationships* and are the sum up of positive and negative values in a certain context. To express the sum of positive and negative IT values we propose to connect the term *net benefits* [6, 41] to sociomaterial theory when discussing and researching IT values. Net benefits have to be defined depending on the context, are probably the most accurate descriptor of value and, should address the level of analysis [6].

The term cyborg[8] [2] is suggested as a useful metaphor that supports exploration and explanation of sociomaterial reality in future organizational life. But we also observe a further entanglement or even fusion of the synthetic[9] and real world [1]. IT values are in an ongoing state of becoming subjected to dynamics of emergence [2].

3   IT Values and Relationships

As explained before values do not appear in isolation but are closely interwoven [5] and subject to dynamics of emergence [2, 3]. In our interpretation of the relational ontology *relations* between values can be competitive, complementary or overlapping

---

[8] A hybrid of machine and organism. See also: http://www.zdnet.be/nieuws/161067/belgische-cyborg-opent-deuren-met-ingeplante-chips/.

[9] Orlikowski [1] here refers to the world known as MKP20 within Sun Microsystems. Sun Microsystems is acquired by Oracle in 2010. See also: http://virtualworldsforum.com/.

[4]. The terms nodal, neighbour and co-values [4] suggest that IT values are related and entangled in a value network. At least three types of relationship can be distinguished in our relational ontology: time, stakeholder and hierarchy. We zoom in further to these relation types to discover, describe and explain the relational tension between IT values.[10]

3.1   IT Values and Time Relation

The value of IT shifts over time from focus on efficiency to focus on effectiveness and flexibility and customer satisfaction [26, 42]. Explaining the difficulties and complexity of IT an article in the Dutch professional magazine CFO refers to 'old' and 'new' IT [17]. Examples mentioned for old IT are enterprise resource planning and mainframes. Social media is referred to as a new IT trend. Pictured by Fig. 1, let us put the emergence of IT and IT values in a *time* perspective. We include the theoretical origin and expected continuation of sociomateriality.

The start of this timeframe is around 1970 denoted as the fifth technological revolution [43] and a (provisional) destination is called singularity whose estimated range is around 2040 [37, 38]. In the beginning of the technological revolution IT was applied mainly inside organizations only for business purposes. IT from that era like enterprise resource planning applications, mainframe systems and terminals is typed as old IT. Around 1990 the structuration theory – theory of structure and agents – supports social research focusing on the impact of IT on business processes and work executed by people within organizations [7]. Much has changed in the decades after 1980.

Sociomateriality saw light as successor of structuration theory because technology, organization, and people became more entangled and inseparable [44]. IT became widespread into as well as organizational as private lives. Besides physical presence IT offers the possibility for virtual presence which introduced the phenomena of synthetic world [1]. In synthetic worlds, deployed within organizations, people can collaborate and communicate real time [1]. IT created in this era is referred to as new IT in this paper. Appearances of new IT are social media and mobile devices like smart phone and tablet computer. This is also the era the paper positions phenomena like digital business, services and society: IT for business and private purpose.

But the end is not reached yet. Researchers and scientists predict that people and technology will fuse which will create new and unprecedented opportunities which are referred to as 'future' IT. IT converged with other technologies will overcome biological and cognitive limitations and interaction between humans and non-humans will change dramatically [36–38].

IT was once a mean to improve operational efficiency, to date IT has changed in a multipurpose and multiform vehicle covering a broad spectrum of socio and material related values. The entanglement and dynamic emergence of IT and IT values clarify the objection to linear and quantitative research [1]. Referring to the general IT values

---

[10] This paper contains a first orientation, interpretation and description of the sociomaterial relation between IT values. Further theoretical and empirical research is needed to determine this sociomaterial relational concept in more detail.

framework in Table 1 we can conclude that the framework contains a mix of old (e.g. efficiency), new (e.g. social innovation) and future (e.g. super-humanity) IT values.

3.2 IT Values and Stakeholder Relation

Besides time related IT values are also assigned to different *stakeholders* [45]. Reviewing and re-specifying the DeLone and McLean model of IS success the term net benefits [41] distinguishes different types of stakeholders: societal, organization (management and groups) and individual. IT values listed in Table 1 depend from the stakeholder point of view and can vary by situation. Stakeholder relation is an important source for competition between values. When striving for organizational value as efficiency this can lead to deskilling [7]. Involving IT outsourcing causes multiple customer and supplier stakeholder relations.

Chau et al. [45] compared IT value studies conducted in Asia and Europe. They note that in IT value research most studies involve the organizational level. Fewer studies focus on the individual level. Chau et al. [45] distinguish objective measures – e.g. accounting and financial indicators, costs, return on investment and, firm value – besides perceptual measures – e.g. increased decision quality, better alignment with business strategy, etcetera. They observe a general shift from using objective measures to perceptual measures to study IT value.

A special entry to societal value is the debate about public values [4]. This includes at one hand the broader discussion about the common good referring to contributors to value like public interest and social cohesion. On the other hand national and local governments apply technology serving public values. Technology shapes the intra-organizational aspects of public administration institutions and the relationship between public administration institutions and citizens leading to IT induced public sector transformation [31]. Due to increase of digital fraud and other crime (cyber-crime) security seems to become a value of increasing societal importance.

3.3 IT Values and Hierarchy Relation

The third relational dimension is *hierarchy*. Hierarchy is related to the relative primacy or importance of a value [4]. Relative primacy depends on the context. From a public value point of view liberty may be more important than efficiency. Within an organization, especially when acting in private competition, efficiency may be of more importance instead of employee's job satisfaction. Hierarchy of values and their relations are considered to be inseparable. Hierarchy between values is designated as prime values and instrumental values. Prime values are seen as temporary[11] conditions whereas instrumental values are consequences [4]. The hierarchical relation between IT values listed in Table 1 in this paper can be different per situation.[12]

---

[11] We add here the word temporary before condition because this reflects the sociomaterial principle that IT values are in an ongoing state of becoming subjected to dynamics of emergence.

[12] This relational view and entanglement of IT values – based on sociomaterial principles – is important and meaningful to understand the concept of IT values in this paper. However, more investigation is needed to deepen out this relationship and to create a stronger theoretical fundament.

Referring once again to Paro – the robot seal used in healthcare to accompany patients suffering from dementia – how should value be expressed? Paro is reducing stress and has a positive impact on the mood of patients. Should we primarily look at the aspects of value or should we first look at the cost of development and maintenance of Paro [9]? In the Paro example the techno-economic view [43] and socio-techno perspective [44] become unified.

4   Summary and Looking Forward

Primary contribution of this paper is to unlock general and broader patterns regarding the impact of technology which is interpreted as IT value. Herewith this paper encompasses a response to the call and need to fundament sociomaterial theory [2] because the concept of sociomateriality is extremely theoretical and philosophical [40]. A second purpose for contribution is to close a bit of the gap in management research ignoring the impact of technology in organizational life [1]. The study is limited to IT value as a result of the interaction between technology, organization and people. The process of value creation and conditions is excluded.

The answer to the research question *"Which IT values can be identified and how are they related?"* delivers a taxonomy and framework of IT values containing eight nodal, sixteen neighbour and sixty four co-values. The sociomaterial relationship between IT values is dimensioned to time, stakeholder and hierarchy. Grounding theories of sociomateriality like Actor Network Theory are part of ongoing discussions [2]. Due to dynamics of emergence and the continuous state of becoming the concept of IT values is a spectacular and interesting sociomaterial subject for social research. IT values, their entanglement and relations are far more than a linear relationship between two or a (limited) number of variables and their causality and therefore a difficult to capture phenomenon.

IT values are the result of interaction between technology, organization and people [1]. The result of this extensive study is a generalized IT value framework based on sociomaterial theory and guiding principles. This overview should be seen as a starting point for further discussing IT values and their relationships. The principles below are accompanying the general sociomaterial IT value framework (see Table 1):

1. The study to explore, collect, define and categorize IT values to a general framework is the answer to the call to search for generalizations and broader patterns to support sociomaterial theory.
2. The constructed framework should be seen rather as a starting point for further discussion and research than as an end point.
3. IT values have a socio-techno or techno-economic orientation.
4. IT values are measured subjectively (perceived) or objectively.
5. IT values are classified as nodal, neighbour and co-values.
6. Due to dynamics of emergence IT values should be understood as a state of becoming (temporary condition) ontology instead of a solid state.
7. IT values exist not isolated but are entangled and maintain dynamic emergent, competitive, complementary and overlapping relationships.

8. IT values are time, stakeholder and hierarchy related.
9. IT values can be either positive or negative.
10. Net benefits are the sum of positive and negative IT values.
11. It should be considered to include biological aspects into sociomaterial theory.

In a next paper the theoretical foundations of sociomaterial IT values and their relations are applied in a digital business situation. Subjects like new IT, mobile IT, user behaviour, use patterns, security and IT values in a digital culture will be deepened. This next paper will contribute to sociomaterial theory by associating empirical evidence to this paper dealing with IT values and their dynamics and relationships.[13]

Acknowledgements. The author thanks Bert Kersten, professor at Nyenrode Business University and Marijn Janssen, professor at Technical University Delft for their sublime and extended guidance and support during this study. Without this great help this paper could not have been written.

References

1. Orlikowski, W.J.: The sociomateriality of organisational life: considering technology in management research. Camb. J. Econ. **34**(1), 125–141 (2009)
2. Cecez-Kecmanovic, D., Galliers, R.D., Henfridsson, O., Newell, S., Vidgen, R.: The sociomateriality of information systems: current status, future directions. MIS Q. **38**(3), 809–830 (2014)
3. Deursen van, A.J., Dijk van, J.A.: CTRL ALT DELETE: Productiviteitsverlies door ICT-problemen en ontoereikende digitale vaardigheden op het werk. Enschede: Universiteit Twente/Center for e-Government studies (2012)
4. Jørgensen, T.B., Bozeman, B.: Public values: an inventory. Adm. Soc. **39**(3), 354–381 (2007)
5. Boztepe, S.: User value: competing theories and models. Int. J. Des. **1**(2), 55–63 (2007)
6. DeLone, W.H., McLean, E.R.: Information systems success revisited. In: Proceedings of the 35th Hawaii International Conference on System Sciences 2002, Waikoloa, Hawaii, pp. 2966–2976. IEEE (2002)
7. Orlikowski, W.J.: The duality of technology: rethinking the concept of technology in organizations. Organ. Sci. **3**(3), 398–427 (1992)
8. Orlikowski, W.J., Yates, J.: It's about time: temporal structuring in organizations. Organ. Sci. **13**(6), 684–700 (2002)
9. Sneller, L.: Over de waarde van IT. Nyenrode Business Universiteit, Breukelen (2012)
10. Janssen, M.: Technologische krachten in het bestuur. Technical University Delft, Delft (2013)
11. Goldstein, P., Katz, R.N., Olson, M.: Understanding the value of IT. EDUCAUSE Q. **26**(3), 14–18 (2003)

---

[13] Personal reflections, findings and insights collected during the study to IT values and their relations are captured and expressed in a number of (smaller) articles and blogs on media like Slideshare, ManagementSite and Blogit: http://www.slideshare.net/ldohmen; https://www.managementsite.nl/auteurs/leon-dohmen; http://www.blogit.nl/author/leon-dohmen.

12. Verniers, H., Teunissen, W.: IT Service Portfoliomanagement: Maximaliseer de Waarde van IT. Van Haren Publishing, Zaltbommel (2011)
13. Symons, C.: Measuring the Business Value of IT. Forrester Research, Cambridge (2006)
14. Bughin, J., Chui, M., Manyika, J.: Clouds, big data, and smart assets: ten tech-enabled business trends to watch. McKinsey Q. **56**, 75–86 (2010)
15. Al-Maskari, A., Sanderson, M.: A review of factors influencing user satisfaction in information retrieval. J. Am. Soc. Inf. Sci. Technol. **61**(5), 859–868 (2010)
16. Venkatesh, V., Thong, J.Y., Xu, X.: Consumer acceptance and use of information technology: extending the unified theory of acceptance and use of technology. MIS Q. **36**(1), 157–178 (2012)
17. Kersten, B.: De werkelijke bijdrage van IT aan de business, pp. 20–23. CFO (2011)
18. Sneller, L.: Does ERP Add Company Value. HAVEKA, Alblasserdam (2010)
19. Sun, Y., Fang, Y., Lim, K.H., Straub, D.: User satisfaction with information technology service delivery: a social capital perspective. Inf. Syst. Res. **23**(4), 1195–1211 (2012)
20. van den Hoven, J.: Over duimschroeven en kant-en-klaarmaaltijden. de IT-auditor **3**, 5–7 (2011)
21. Kujala, S., Väänänen-Vainio-Mattila, K.: Value of information systems and products: understanding the user's perspective and values. J. Inf. Technol. Theor. Appl. **9**(4), Article 4 (2009)
22. Wang, P.: Chasing the hottest IT: effects of information technology fashion on organizations. MIS Q. **34**(1), 63–85 (2010)
23. Savas, Ö.: A perspective on the person-product relationship: attachment en detachment. In: McDonagh, D., Hekkert, P., van Erp, J., Gyi, D. (eds.) Design and Emotion, pp. 317–321. Taylor & Francis, London (2004)
24. Szpakow, A., Stryzhak, A., Prokopowicz, W.: Evaluation of threat of mobile phone - addition among Belarusian university students. Prog. Health Sci. **1**(2), 96–101 (2011)
25. Heng, M.: Three Dimensions of Information Technology Applications: A Historical Perspective. Vrije Universiteit, Amsterdam (1993)
26. van der Zee, H.: Business Transformation and IT. Dutch University Press, Tilburg (2001)
27. Bughin, J., Byers, A.H., Chui, M.: How social technologies are extending the organization. McKinsey Q. **20**, 1–10 (2011)
28. Baldwin, E., Curley, M.: Managing IT Innovation for Business Value. Intel Press, Hillsboro (2007)
29. Puentedura, R.R.: SAMR and TPCK: a hands-on approach to classroom practice. Ruben R. Puentedura's weblog (2014). http://www.hippasus.com/rrpweblog/archives/2014/12/11/SAMRandTPCK_HandsOnApproachClassroomPractice.pdf. Accessed 3 January 2015
30. Verboom, H.: Get It Done. Retrieved on July 1, 2012, Get It Done: http://www.getitdone.org/ (2009)
31. van Veenstra, A.F.: IT-induced Public Sector Transformation. BOXpress, 's-Hertogenbosch (2012)
32. van den Broek, C., Dohmen, L., van der Hooft, B.: Changing IT in Six. Royal Van Gorcum, Assen (2010)
33. Mousaid, S., Leys, M.: ICT in Wel en Wee. Vrije Universiteit Brussel, Brussel (2012)
34. Lee, S.H., Han, J.H., Leem, Y.T., Yigitcanlar, T.: Towards ubiquitous city: concept, planning, and experiences in the Republic of Korea. In: Yigitcanlar, T., Velibeyoglu, K., Baum, S. (eds.) Knowledge-based Urban Development: Planning and Applications in the Information Era, pp. 148–170. Information Science Reference, Hershey (2008)
35. Lam, S.Y., Shankar, V., Erramilli, M.K., Murthy, B.: Customer value, satisfaction, loyalty, and switching costs: an illustration from a business-to-business service context. J. Acad. Mark. Sci. **32**(3), 293–311 (2004)

36. Vinge, V.: The Coming Technological Singularity. Feedbooks, Paris (1993)
37. Vinge, V.: Signs of the singularity: hints of the singularity's approach can be found in the arguments of its critics. IEEE Spectrum's SPECIAL REPORT: THE SINGULARITY (2008)
38. Kurzweil, R.: Reinventing humanity: the future of machine-human intelligence. Futurist **40**, 39–46 (2006)
39. 3D Food Printing Conference: The future of 3D food printing for professionals and consumers. In: 3D Food Printing Conference (2015). http://3dfoodprintingconference.com/. Accessed 20 April 2015
40. Leonardi, P.M.: Theoretical foundations for the study of sociomateriality. Inf. Organ. **23**(2), 59–76 (2013)
41. Seddon, P.B.: A Respecification and extension of the DeLone and McLean model of IS success. Inf. Syst. Res. **8**(3), 240–253 (1997)
42. Nelson, M.R.: Assessing and communicating the value of IT. Res. Bull. EDUCAUSE Cent. Appl. Res. **2005**(16) (2005)
43. Perez, C.: Technological Revolutions and Techno-Economic Paradigms. The other canon foundation, Norway and Tallinn University of Technology (2009)
44. Orlikowski, W.J., Scott, S.V.: Sociomateriality: Challenging the Separation of Technology, Work and Organization. London School of Economics and Political Science, London (2008)
45. Chau, P.Y., Kuan, K.K., Liang, T.-P.: Research on IT value: what we have done in Asia and Europe. Eur. J. Inf. Syst. **16**, 196–201 (2007)

Designing Viable Multi-sided Data Platforms: The Case of Context-Aware Mobile Travel Applications

Mark de Reuver[1(✉)], Timber Haaker[1,2], Fatemeh Nikayin[1], and Ruud Kosman[2]

[1] Delft University of Technology, Engineering Systems and Services, Delft, The Netherlands
{g.a.dereuver,t.haaker,f.a.nikayin}@tudelft.nl
[2] InnoValor B.V., Enschede, The Netherlands
{timber.haaker,ruud.kosman}@innovalor.nl

Abstract. Advances in data semantification and natural language querying are enabling new generations of context-aware mobile applications. Such applications would rely on platforms that integrate heterogeneous sets of user data from a range of applications and systems. Designing these platforms is challenging as they should serve multiple user groups at the same time. In this paper, we analyze who should subsidize multi-sided data platforms that enable mobile context-aware travel applications. After analyzing the different user groups and revenue models, we assess end-user acceptance of these revenue models through a survey among 197 potential users. Results show that users willing to share data with app developers are more inclined to use data-driven mobile travel apps but are less inclined to pay for them. This paradoxical result explains why premium-pricing as well as data-monetization strategies can both be viable.

1 Introduction

While app developers have long struggled to seize the opportunity [1], they now commonly use context information to enrich their mobile applications. While GPS is the most prominent examples, app developers also rely on aggregated past behavior, inferred preferences and past transactions. However, if app developers share such context information among each other, even smarter context-aware applications can be envisioned. New technologies for linking and semantifying large sets of data as well as natural language querying approaches are driving these opportunities. One promising application area is that of travel applications: whereas most existing traveling apps focus either on booking, transport advice or local events, sharing user context data between such applications would enable far more integrated and coherent travel advice.

Sharing context information generated by a mobile application with other app developers risks violating user privacy as well as harming competitive position of the individual app developer. These issues could be circumvented by instantiating shared and trusted data platforms that aggregate user data. However, designing such platforms is challenging since they should satisfy multiple user groups: end-users, app developers

© IFIP International Federation for Information Processing 2015
M. Janssen et al. (Eds.): I3E 2015, LNCS 9373, pp. 354–365, 2015.
DOI: 10.1007/978-3-319-25013-7_28

and potentially even others. A core issue is then who should subsidize the platform, i.e. what is the source of revenues going to be [2, 3, 4]. User willingness to pay for more advanced context-aware travel advice is often problematic. Monetizing user data by selling it to third parties can be a risky approach as well.

The objective of this paper is *to analyze who should subsidize multi-sided data platforms that enable mobile context-aware travel applications*. We do so by analyzing the case of 3cixty: a data platform for mobile context-aware travel applications, developed within the context of 3cixty. We outline different revenue models and subsequently analyze user acceptance of such revenue models by analyzing the results of a survey among 197 early adopters in the Netherlands.

Section 2 develops a theoretical background based on multi-sided platform literature. Section 3 provides a background on the 3cixty case and technical platform. Section 4 sketches potential value networks and revenue models, which are subsequently assessed through user research in Sect. 5. Section 6 concludes the paper, draws limitations and implications for future research.

2 Background on Multi-sided Platforms

In general, platforms can be seen "as building blocks (they can be product, technologies or services) that act as a foundation upon which an array of firms can develop complementary products, technologies or services" [5]. A platform serving multiple user groups, such as buyers and sellers, is typically denoted as a multi-sided platform [6]. The objective of a multi-sided platform is to facilitate the transactions between different user groups, such as consumers and app developers. Multi-sided platforms can be analyzed using the framework of two-sided markets [3]. The focal artifact in this paper can be seen as a multi-sided platform as it connects multiple groups of actors (i.e., end-users and app developers) while providing generic functionality on which services can be developed (i.e., shared databases and querying tools).

Such shared platforms can generate considerable network externalities, which implies that the value of the platform depends on the number of users [7, 8]. Network externalities imply that the value of a system depends on the installed base of users [9, 10]. Increasing adoption levels can thus lead to a positive feedback cycle that further increases the usefulness of the technology [11].

Network externalities are direct if the value of the product increase by others buying, connecting, or using the same platform or services provided via the platform. The utility of the platform increases with the number of other using it. Examples of direct network effects are social media, which become more valuable if more end-users join the platform. But also because platforms have interchangeable components, users can share the benefits of the same technical advance. Backward compatibility, interoperability and interface standards are therefore crucial. Typically, direct network effect refers to positive effect between users in a group; however, when more consumers for a platform reduce the value for similar consumers, the platform entails a negative direct network effect.

Externalities are indirect when the value of the platforms depends on the number of users in a different user group. For instance, video game consoles become more valuable for consumers if there are more developers creating games for that console. Indirect network effects may also be negative, for instance more advertisers on a search engine platform decrease its value for searchers of independent advice. These indirect effects are typical for multi sided markets, where service providers, developers and consumers meet. As service providers as well as developers make the platform's various components better, the platform gets more attractive over time. As consumers use the platform more, they make the market larger. Once the adoption of a product or technology has started, these network externalities provide benefits to both new and existing users such as reduced price, lower uncertainty about future versions of platforms as well as complementary services, communities of users, higher quality products, new market opportunities.

Network externalities are important since they call into question which user group should subsidize the platform. If one user group of the platform is considered to be of more value than the other, it may be that they are subsidized by lowering prices or offer access for a fee [3]. In practice, different subsidization models are being used. The concept of marquee user specifically refers to those user groups that have such great value for the other user groups that their adoption of the platform should be subsidized [12].

3 The 3cixty Case

The objective of the project is to provide a data platform for apps and services for city visitors (i.e., tourists as well as citizens). The platform will retrieve data from websites (e.g. hotels, restaurants, events, sights), other platforms (e.g. social media), smartphones (e.g. apps for exploring a city) and/or sensors. The platform offers various data services to app developers, for instance a cross domain querying language, a crowdsourcing mechanism, a mobility profiling service, query augmentation and social media mining.

Apart from these enabling services, the platform provides a clean data repository about a city that can empower specific applications. The data is "semantified" (i.e. semantics of the data has been made explicit) and "reconciled" (i.e. heterogeneous data coming from various sources has already been integrated). App developers can access this clean data repository via a single application programming interface (API) which is web developer friendly (i.e. apps developer do not need to interact with many specific data sources APIs). App developers can access this clean data repository via queries which provide an additional level of expressivity and enable to provide answers to queries that no single data source can do. The platform can answer questions like "Give me the list of hotels reachable within 10 min by metro of the venue of the Franz Ferdinand concert". The specific features of the platform are listed in Table 1.

Table 1. 3cixty platform features

	Frequency of city trips
High-level querying language with a cross-domain knowledge base	Execute mixed-/cross-domain (tourism, mobility, etc.) queries in your app that combine diverse types of information, including the types of interlinked data provided by the other 3cixty services
Query augmentation service	Automatically personalize queries with restrictions based on data provided by other services (e.g., reviews by friends of users)
Generic crowdsourcing mechanism	Efficiently extend the 3cixty core APIs to enable users to contribute information about aspects of the city. Access such information provided by others and through on-demand crowdsourcing, ask people in real-time to contribute data to help others
A "parallel exploration" graphical user interface	Enable users to construct and save trees of interrelated queries that enable them to explore several aspects of the city simultaneously
A "wish list" service	Allow users to indicate where they may want to go and (optionally) when and how Store this information in the cloud so that it can be accessed on appropriate occasions, even from other devices
Social media mining	Enable users to see 'nearby buzz', i.e. what people are talking about on social media in the neighborhood or in the city
Mobility profiling service	Track users' movements within the city, including their use of modes of transportation Give users access to information from their mobility profiles - and, with permission, to those of relevant other persons

Using enabling services on the platform and the clean data repository, app developers can develop applications such as city trip and accommodation planning, traffic update, cultural and entertainment updates and information about the city.

4   Revenue Models

To roll out and commercialize the platform, different roles are required. The generic value network shows the business roles (blocks) and the value exchanged between them in the form of service and revenue flows (arrows). Typically different roles are required in the research phase and roll out phases of the services, see Fig. 1.

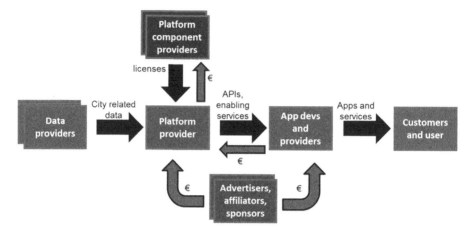

Fig. 1. Basic value network

Note that the 'user' is the actual person that uses the app while the 'customer' is the party actually paying for the app. For example a 'mobility solution provider' could act as service provider and commission an app developer to develop a 3cixty enabled mobility app. The app could be part of a mobility solution that is offered to companies ('customers'). The company's employees ('users') would use the app to manage their mobility. Of course the roles of service provider and customer need not be separated; some party may simply order an app with an app developer and offer it to users directly via an app store. Or an app developer may even directly develop and offer an app to users; in that case app developer, app owner/service provider and customer are the same actor. Further detailing of business roles is possible as well, as for instance the roles of app owners and service providers can be separated (accommodating for white label business models).

The production of the app is a joint effort of the parties on the left hand side of the picture with the platform provider (system integrator) playing a pivotal role. Several parties provide information to the platform which can, amongst other things, be stored, combined, enriched and aggregated in the platform. Possibly, the services could be subsidized or even paid for by finance providers like insures, (local) governments or other providers. As an extra revenue stream, advertisements could play a role in the value network as well. Advertisement agencies could offer profile based advertisements to their customers and the advertisement provider is in practice often a value chain in itself.

For defining the revenue model, two main issues have to be dealt with. First, it should be decided whether the platform is offered in a profit or non-profit approach. Second, it should be decided whether the platform is open or closed to third party app developers. The motivation for this distinction is that open and closed platforms as well as privately and publicly owned platforms occur in practice and provide relevant but distinct options for the 3cixty platform. In the closed model the platform is used internally as a service platform and in the open model the platform is offered 'as-a-service' to 3rd parties like app developers. The different scenarios are depicted in

Fig. 2 together with some actual examples of platforms and/or apps that follow such business scenario. One may also consider a situation in which there is no platform provider at all, but where (some of) the developed 3cixty software is made available as open source. In that case any interested organisations could use the software.

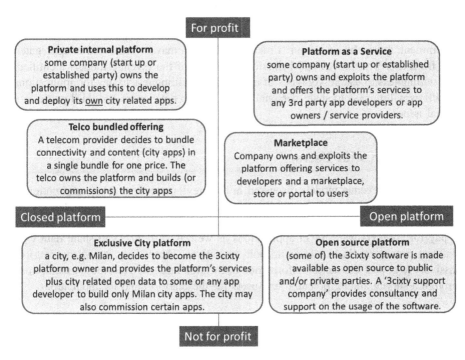

Fig. 2. Exploitation scenarios

Several revenue models can be considered to commercialize the platforms. The app market is dominated by the freemium model as over 90 % of all apps are for free, so user willingness to pay is often low. One approach to monetizing a service platform is that the platform is owned by some app developer and used as an internal platform to create and deploy its own apps. In this scenario the roles of platform provider and app developer are played by the same actor. The platform is closed in business sense, i.e. it can only be used by the app developer owning the platform. Other app developers or service providers cannot make use of the platform to build apps. Many providers use service platforms to efficiently develop and deploy their services and apps. A different approach is that the roles of platform provider and app developer are separated. In this case the services of the 3cixty platform are made available to app developers/service providers to support the creation and deployment of apps. Third parties can use part or all of the 3cixty functionality and data to develop new apps or enrich existing apps. Within these two scenarios still many business models are possible with different actors taking up the role of platform provider (private vs public party) and supported by different revenue models, i.e. bundling, pay-per-use, affiliation, advertisement, subsidy and sponsoring, open source etc. However, the idea behind the development of the 3cixty platform is geared towards an open model.

Given these features many revenue models could be applicable. Affiliation models where commission fees from companies that appear in the traveling app and that realize transactions, e.g. bookings, via the app. Advertising models, in which contextual advertising based on user profiles (e.g. preferences following from searches or a wish list) and levels of context awareness (location, time). Or data driven model: local government, tourist board or tourist trade organisation may be interested in aggregated data from user preferences and behaviour. For any of the revenue models identified, user willingness to use 3cixty types of applications, willingness to pay for applications as well as willingness to share personal data with app developers and third parties are crucial issues.

5 Validation with Users

For any of the revenue models discussed in the previous section, a key assumption is willingness of end-users to use mobile travel apps that are based on integrated sets of context data. To further validate the revenue models, key concerns are user willingness to pay for context-aware travel applications as well as willingness to share data with platform providers. Such data could both be actively shared (e.g. crowdsourcing) or passively (e.g. past transaction information or location data). We test these assumptions through a survey among potential end-users for the platform and its applications.

5.1 Method

The intended population comprises young people who have a smartphone and who have a habit of traveling to cities. A convenience sample of Dutch students is used. Invitations to participate in the survey were posted on the Facebook page of various students of a bachelor course in December 2014. To obtain a homogenous group, we only include students possessing a smartphone in our sample. 197 valid responses were received, most of them being bachelor students. Age varies between 17 and 29, with average 20.6 years old. Gender was balanced with 53 % male. Respondents make 2.3 city trips per year, with standard deviation 2.1 and a maximum of 12.

Constructs were operationalized into self-developed, reflective scales, see Table 2. Although respondents were not exposed to the exact 3cixty service mockups, they were asked to reflect on 3cixty type of applications that provide comprehensive travel information and booking possibilities. All items were measured on a 7-point Likert scale. Confirmatory factor analysis was carried out using WarpPLS, see Table 1. Convergent validity is acceptable with average variance extracted exceeding .5 benchmark, and standardized factor loadings exceeding .6. Construct reliability is acceptable.

Table 2. Measures.

Construct	Measure	Std factor loading	AVE	Construct reliability
Frequency of city trips	I like to go to large cities on vacation	.83	.69	.82
	I like to go on vacation within Europe	.83		
Willingness to receive local travel experiences	I find earlier travel experiences of others at the same location relevant for my travel	.83	.68	.75
	When on vacation, I am open to suggestions about local activities	.83		
Willingness to actively share travel experiences	I like to explore the area when on vacation	.76	.58	.81
	When on vacation, I like to share my travel experiences	.76		
Willingness to passively share travel data	Please indicate to what extent you are willing to share with a mobile application: Location (e.g. for restaurants/sights nearby, routes)	.67	.50	.74
	Social media (Facebook, Twitter for a.o. opinions on sites, events etc.)	.80		
	Transaction history (for a.o. offers based on transactions, preferences etc.)	.64		
Intention to use	Assume a mobile app would be offered which combines existing travel apps (e.g. finding hotels, restaurants, public transport, events). Would you use such an application?			
Willingness to pay	How much would you be willing to pay for such an integrative travel app?			

Discriminant validity is acceptable as correlations between latent variables do not exceed the square root of average variance extracted, see Table 3.

5.2   Results

A structural model is assessed using WarpPLS. The advantage of using WarPLS is that it allows not only interval but also categorical and nominal variables, such as gender. The model shows acceptable overall fit according to the fit indices generally

Table 3. Correlations among latent variables (diagonals show square root of average variance extracted).

	Frequency of city trips	Willingness to receive local travel experiences	Willingness to actively share travel experiences	Willingness to passively share travel data
Frequency of city trips	.83			
Willingness to receive local travel experiences	.30	.83		
Willingness to actively share travel experiences	.34	.35	.76	
Willingness to passively share travel data	.07	.05	.06	.71

recommended with WarpPLS (Tenenhaus GoF = .359, Sympson's paradox ratio = .846), and low multicollinearity (Average block VIF = 1.215), see Fig. 3.

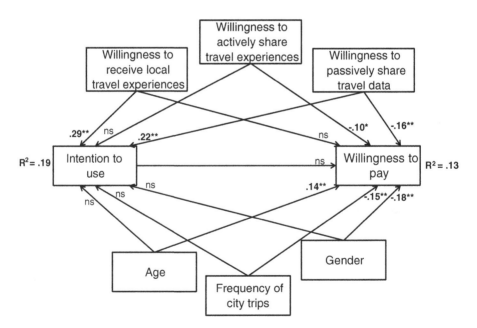

Fig. 3. Structural model (standardized regression weights on the arrows) * p < .05; ** p < .01

Explained variance of both intention to use and willingness to pay is low. We find that intention to use depends on the degree to which respondents plan their trips (in advance or ad hoc) and the degree to which they are inclined to share data via social media. On the other hand, willingness to pay depends on travel frequency, gender and age, while willingness to share data has a negative impact. Intention to use does not significantly affect willingness to pay.

The tendency to share data via social media also affects the intention to use. This implies that users of a travel application will be more likely to share data, which opens opportunities to utilize such data for monetization.

Willingness to pay, on the other hand, depends on how frequent a person travels, as well as gender and age. Usage intention does not contribute to willingness to pay, which suggests that even if people are willing to use the application, they are not willing to pay for it.

Interestingly, intention to share has a negative impact on willingness to pay. Apparently, those that are less inclined to share data on social media are more willing to pay for a travel app.

Another striking finding is that intention to use the travel app has no significant impact on willingness to pay for the application. Overall, the findings illustrate a dilemma for this type of social media traveling apps. Factors that drive intention to use completely differ from factors that drive intention to pay. This explains why alternative revenue models need to be considered in which app providers do not depend on willingness to pay. Apparently, one should either focus on those willing to pay for a travel app, but are not willing to actively share their data via social media. And when one focuses on people that are willing to share their data, willingness to pay will generally be lower. As such, the findings suggest a dilemma between two revenue models: a free app with monetization of shared data or a paid app without sharing of data with third parties.

6 Conclusions

Designing data platforms for context-aware mobile services entails several issues that are typical for multi-sided platforms. For the case being studied, we identified at least four different user groups that may or may not subsidize the platform: end-users, advertisers, app developers and government organizations. Who should subsidize the platform strongly depends on the exploitation scenario, i.e. whether a profit or non-profit model is chosen and whether the platform should be open or closed.

To evaluate the revenue models developed in this paper, we studied the end-user (i.e. consumer) perspective. We found support for a typical paradox observed in this domain: the reasons why consumers are willing to use applications are completely different from the reasons why they would be willing to pay. In fact, the more a user is willing to share data and travel experiences, the less that user is willing to pay for a context-aware travel app. As such, providers of data platforms can target one segment of end-users with a premium-priced app or another segment with a for-free app in which data is being shared for commercial purposes with third parties.

The findings are in line with earlier research on privacy concerns in information systems which suggests that people make trade-offs between utility, price and privacy harm of an application [13, 14]. How such compensation of privacy concerns with discounts or added value from apps works out in a big data era where consequences of disclosing personal data cannot always be foreseen warrants further research.

The paper contributes to understanding of the complexity in designing multi-sided data platforms. Platforms that link data providers, end-users, application developers and advertisers create highly complex sets of network externalities between and within the different user groups. In the case studied in this paper, it became clear that without prior demarcations on profit-orientation and openness, a multifold of revenue models can be considered in parallel. Even after our user research, different revenue models are still relevant.

We used a convenience sample gathered through links on social media. As our research model included willingness to share travel experiences including via social media, there may be systematic bias in the sample. However, considering the high adoption rate of social media in the population of young technical students, we assume this will not be a major threat.

The present paper focused on end-users for the validation of the revenue models. Our next step will be studying the app developers' perspective. Earlier workshop and informal interviews with app developers suggest they may be willing to pay for data platforms if they truly add value for their business. Whether this is the case depends strongly on technological issues such as how to structure the data, how to form the APIs and which of the technical features of the data platform become available first.

Acknowledgements. We are grateful for comments and suggestions from Anthony Jameson. The paper has been developed within the EIT Digital innovation activity 3cixty (14523). We appreciate the contributions by several BSc-students from the course TB232A at Delft University of Technology in collecting data for the user study reported in this paper.

References

1. De Reuver, M., Haaker, T.: Designing viable business models for context-aware mobile services. Telematics Inform. **26**, 240–248 (2009)
2. Armstrong, M.: Competition in two-sided markets. RAND J. Econ. **37**, 668–691 (2006)
3. Rochet, J.-C., Tirole, J.: Platform competition in two-sided markets. J. Eur. Econ. Assoc. **1**, 990–1029 (2003)
4. Rochet, J.-C., Tirole, J.: Two-sided markets: a progress report. RAND J. Econ. **37**, 645–667 (2006)
5. Gawer, A.: Platform dynamics and strategies: from products to services. Platforms, Markets and Innovation, pp. 45–76. Edward Elgar Publishing, Cheltenham (2009)
6. Evans, D.S.: Some empirical aspects of multi-sided platform industries. Rev. Netw. Econ. **2**, 19 (2003)
7. Church, J., Gandal, N.: Platform competition in telecommunications (2004)
8. Shapiro, C., Varian, H.R.: The art of standards wars. Managing in the modular age **41**, 247–272 (1999)

9. Katz, M.L., Shapiro, C.: Network externalities, companition and compatibility. Am. Econ. Rev. **75**, 424–440 (1985)
10. Shapiro, C., Varian, H.R.: Information rules: a strategic guide to the network economy. Harvard Bus. Sch. Press, Boston (1998)
11. Arthur, W.B.: Competing technologies, increasing returns, and lock-in by historical events. Econ. J. **89**, 116–131 (1989)
12. Eisenmann, T., Parker, G., Van Alstyne, M.W.: Strategies for two-sided markets. Harvard Bus. Rev. **84**, 92 (2006)
13. Hann, I.-H., Hui, K.-L., Lee, S.-Y.T., Png, I.P.: Overcoming online information privacy concerns: an information-processing theory approach. J. Manag. Inf. Syst. **24**, 13–42 (2007)
14. Laudon, K.C.: Markets and privacy. Commun. ACM **39**, 92–104 (1996)

The Conceptual Confusion Around "e-service": Practitioners' Conceptions

Eva Söderström[1(✉)], Jesper Holgersson[1], Beatrice Alenljung[1],
Hannes Göbel[2], and Carina Hallqvist[2]

[1] School of Informatics, University of Skövde, Skövde, Sweden
{eva.soderstrom,jesper.holgersson,beatrice.alenljung}@his.se
[2] Section for Information Technology, University College of Borås, Borås, Sweden
{hannes.gobel,carina.hallqvist}@hb.se

Abstract. The e-service concept has been a central concern in many research and practitioner areas in recent years. There are expectations of citizens, customers, commercial companies and public organizations of what e-services are, their functionality and benefits. However, there is conceptual confusion that may hamper collaboration and research viability. This paper explores the conceptual vagueness and presents an empirical investigation of how the e-service concept is treated in practice, along with its kindred concept "IT service". Results show that public and commercial organizations approach e-services differently, that translation problems can cause lack of comparability in research results, and that additional concepts may be introduced instead of e-service.

1 Introduction

The need for providing services using information and communication technology (ICT) has multiplied concurrently with the growth and increased importance of ICT in society, for public administrations as well as business organizations. Nowadays, customers and citizens expect services to be electronically available, and the e-service concept has been a central concern in several research areas in recent years, e.g. e-business [1], IT development and maintenance [2], and e-government [3]. It has been described by practitioners as well as researchers, and quite a few researchers have tried to explain what "e-service" is [4, 5]. However, a universal definition is lacking, and "e-service" hence suffers from conceptual vagueness. The purpose of this paper is to investigate the conceptual vagueness and its consequences from a practitioner perspective, and to revitalize the conceptual discussion about e-services. In particular, the discussion will be conducted in relation to its kindred concept IT service.

2 Framing the Concepts e-service and IT Service

A "service" is traditionally seen as a set of activities provided by a provider to a consumer in order to generate value for both parties [6]. However, "service" is associated with a wide variety of meanings, not the least depending on the current context, and is thus

© IFIP International Federation for Information Processing 2015
M. Janssen et al. (Eds.): I3E 2015, LNCS 9373, pp. 366–371, 2015.
DOI: 10.1007/978-3-319-25013-7_29

burdened with a clutter of meanings. There is no commonly agreed definition of *"e-service"* [4], but in a broad sense, e-service is seen as service delivered via electronic networks [4, 7]. Most research also agrees that e-services are based on interactivity and "driven by the customer and integrated with related organizational customer support processes and technologies" [8, p. 186]. It is a consumer who initiates interaction by requesting a service from an e-service provider. Researchers define e-services differently. For example, Javalgi et al. [7] says that e-services are interactive services delivered via the Internet whereas Rowley [4, p. 341] defines e-service as "deeds, efforts or performances whose delivery is mediated by information technology". Many researchers, however, take the concept for granted and do not define it at all. Instead, e-service is treated as something that is commonly known [e.g. 9]. Traditionally, IT has had a supporting role for businesses. By combining new technologies with the "new" service dominant logic paradigm [e.g. 10] new opportunities for service innovation emerge [e.g. 11]. For IT services, the field of IT Service Management (ITSM) is a key point of origin. ITSM is a widespread area where private and public sectors both have to manage and maintain IT-systems and processes as services. Within this field, Information Technology Infrastructure Library (ITIL) is one of the recognized large and extensive frameworks [12]. ITIL views an IT service as a service offered by an IT service provider. In contrast to the ITIL view of IT services, Jia and Reich [13] claim that the IT service concept traditionally has been described as a human mediated service delivered by IT personnel to business clients. This insinuates that an IT service is only related to the support provided to a user by a helpdesk function. It is also a more narrow view of the IT service concept than the one suggested by ITIL, thus emphasizing the conceptual confusion in the area.

3   Research Design

The study has a qualitative research approach in which the conceptual views and interpretations of e-service and IT service in different organizations have been investigated. Data was collected from 7 municipalities, 5 small and medium-sized IT enterprises (SMEs), and 1 regional alliance of municipalities. The interviewees were chosen based their potential to provide rich information concerning the concepts in focus.

3.1   Interviews

Open-ended interviews were conducted in which a semi-structured interview guide was used [14]. This ensured a solid basic part of the interviews, and gave flexibility to add questions when needed. The questions covered: (a) if some of the concepts e-service or IT service are used in the organization, (b) if other related similar concepts are used, (c) the interviewees perception of the concepts, and (d) if there are organization-collective definitions of the concepts. The interviews have been performed by various combinations of researchers, thus allowing for investigator triangulation [14]. Most interviews were conducted at the participant's workplace, some through the phone or email due to geographical distance. Each lasted for about 15–20 min, were recorded and subsequently transcribed.

3.2   Data Analysis

The qualitative data analysis was conducted in three steps, with an emphasis on researcher triangulation [14]. (1) Each researcher separately walked through the transcripts for their own perception of the material without being influenced by the others. (2) The researchers agreed to review the material from these dimensions: (a) similarities and differences within public organizations; (b) Similarities and differences within commercial organizations; and (c) similarities and differences between public and commercial organizations. (3) The researchers conducted a joint analysis using a whiteboard and color coding. Each respondent was given an identifier (letter + number): C for companies, M for municipalities, and LGF for the regional alliance.

4   Empirical Conceptual Elaboration

4.1   Similarities and Differences Within Public Organizations

In public organizations, the e-service concept is widely used and mostly referring to the same thing: services that previously were handled manually are now also offered via the Internet. The following quotation is an example of this view:

> "...E-service for me is something that is targeting citizens digitally."(ME)

The focus is on citizens, but also companies. Public organizations often speak of citizens as external end users of the e-services, as illustrated by this quotation:

> "An e-service is [...] a self-service that I can use to keep in contact with the municipality or a public authority, [...] and that I can do it anytime. If the e-service is really good, I think it should have connection straight into the business systems so that it results in more efficiency" (LGFA).

Accordingly, e-services need to provide value, for citizens and/or commercial organizations, for the municipality, or for both. This view is in contrast to a more general perspective in the e-government research community emphasizing that e-service mostly is provided by public administrations as a means to enhance internal efficiency. However, some municipalities claim that "real" e-services must provide value for both citizens and municipalities:

> "I do not think it is a real e-service when it is only the citizen that benefits, while the internal handling is the same as before. You spend the same amount of time." (MC)

Mutual value is illustrated by the Swedish Association for Local Authorities and Regions, who say there is evidence that new and efficient e-services have contributed to reduce administrative costs for commercial organizations with 7 billion SEK, and that the e-services have reduced wrongful payments to citizens with 150 million SEK per year [15]. Some municipalities view digitized forms as e-services, while others view e-services as being those who cover an entire chain from citizens into the organization's ICT systems:

> "We are talking smart e-services [...] that get into the various organizational systems." (MD)

For some municipalities, mutual value is key while others are satisfied with increased value for only citizens. In contrast to "e-service", the majority of the municipalities do not use IT service at all, they simply state that they are not familiar with or are not using that as a concept in their organizations. Those who do use it or relate to it in some way view "IT service" as the internal IT department and helpdesk service:

"IT support is what we use, you say computer support or IT support but this is more practical. You want help with something concerning IT." (MF)

4.2   Similarities and Differences Within Commercial Organizations

In commercial organizations, an e-service can be defined in many different ways, but primarily connected to the Internet and to end-users and what they can do online:

"A traditional service that is accessible via a network-based interface, typically implemented using web technology. Preferably services provided by public authorities." (CG)

The focus of companies seems to be on IT, and on service offerings using IT. In commercial organizations, "e-service" is not as prominent as "IT service". A common view of the IT-concept is in line with definitions provided by existing frameworks:

"Yes, the idea is that we must create value by managing the results that the client wants without the need to take ownership of specific "risks". IT Services is really this concept but applied to people and technology in an IT organization." (CE)

Hence, the definition of IT service is wide and does not focus only on end-users. One reason may be that existing frameworks such as ITIL are commonly used in commercial organizations, who therefore inherit the definitions used in the frameworks. It should be noted that not all commercial organizations use the term IT service, but rather have a plethora of service types that they discuss:

"I actually think we mostly use the service concept [...] that is because we know what area we operate within and what area we focus on [...] Well we know we work with IT services so perhaps that is why we do not define it so explicitly." (CB)

Some companies differentiate between IT service and e-service in a different way, but referring to IT services as something internal and e-service as being external. Others, however, view e-services as being for organizational development instead. One very common view of e-services in commercial organizations is that the concept is associated with public authorities rather than commercial organizations:

"We don't use those concepts [e-service and IT service] in our organization, they are more used within the public sector." (CF)

4.3   Similarities and Differences Between Public and Commercial Organizations

When merging material for the two organizational types, several similarities and differences can be identified. The ITIL framework, for example, colors the commercial organizations' view of the service concept, which differs somewhat from how public authorities define it. One key aspect we identified is that of translation ambiguity of the concept "service" to other languages. Our research was conducted in a Swedish setting, and the Swedish language can translate "service" in two ways: One is focused on what is performed rather than on the technology mediating the service, while the other is technology-focused in terms of the technology used being the center of attention, such as in the ITIL definition of IT service. Commercial organizations are to some extent aware of the dual meaning of the "service" concept, while municipalities mainly refer to the service concept in relation to "support". This is natural since public organizations always have been focused on servicing their citizens and commercial organizations. The commercial organizations base and develop services focused on IT, involving IT technology, as well as processes and people that use the technology.

5   Concluding Analysis

Public and commercial organizations both differentiate between internal and external e-services. For example, CA expressed that e-services are services to end customers, which indicates that there are other services that are internal. People attach different meanings to concepts, and a common definition is often lacking:

> *"We do not have a common definition [of the e-service concept] and we suffer from that."* (CA).

A consequence of different meanings is that respondents may answer questions originating from one meaning, while researchers collecting data, or the collaborating organization had a different meaning in mind. The risk is that the interpretation does not represent the actual views of the other, which can make e.g. research results flawed and difficult to compare. Our study shows that there is a conceptual confusion based on both language and interpretation, and that definitions and scope vary within and between public and commercial organizations. Failing to ensure that collaborating partners, customers and providers, etc. mean the same thing can thus result in great problems. A common ground needs to be documented in any collaboration, in particular if translation is an issue. Our findings showed translation problems between Swedish and English, but this problem may hold true for other languages as well. Whether or not this is the case can only be established when a common ground is in place. Another dimension of the e-service conceptual discussion is what counts as a "real" e-service and what does not. Opinions vary, and even if research has discussed this issue to some extent, there is a difference with how it is discussed in public and commercial organizations. Future research should adopt a practitioner's perspective and conduct studies focused on empirical application of these levels. The purpose of this paper was to draw attention to the problem of conceptual vagueness and its consequences, and to revitalize the conceptual discussion about e-services. Our findings are a start of such a discussion, and future research needs to deepen and expand e-service research concerning its vagueness and confusion.

References

1. Janita, M.S., Miranda, F.J.: The antecedents of client loyalty in business-to-business (B2B) electronic marketplaces. Ind. Mark. Manag. **42**, 814–823 (2013)
2. Lu, J., et al.: Recommender system application developments: a survey. Decis. Support Syst. **74**, 12–32 (2015)
3. Fakhoury, R., Aubert, B.: Citizenship, trust, and behavioural intentions to use public e-services: the case of Lebanon. Int. J. Inf. Manag. **35**, 346–351 (2015)
4. Rowley, J.: An analysis of the e-service literature: towards a research agenda. Internet Res. Electron. Netw. Appl. Policy **16**(6), 879–897 (2006)
5. Rust, R.T., Lemon, K.N.: E-service and the consumer. Int. J. Electron. Commer. **5**(3), 85–101 (2001)
6. Grönroos, C.: Service logic revisited: who creates value? And who co-creates? Eur. Bus. Rev. **20**, 298–314 (2008)
7. Javalgi, R., Martin, C., Todd, P.: The export of e-services in the age of technology transformation: challenges and implications for international service providers. J. Serv. Mark. **18**(7), 560–573 (2004)
8. de Ruyter, K., Wetzels, M., Kleijnen, M.: Customer adoption of e-service: an experimental study. Int. J. Serv. Ind. Manag. **12**(2), 184–207 (2001)
9. van Velsen, L., et al.: Requirements engineering for e-Government services: a citizen-centric approach and case study. Gov. Inf. Q. **26**(3), 477–486 (2009)
10. Vargo, S.L., Lusch, R.F.: Evolving to a new dominant logic for marketing. J. Mark. **68**, 1–17 (2004)
11. Barret, M., Davidson, E., Prabhu, J., Vargo, S.L.: Service innovation in the digital age: key contribution and future directions. MIS Q. **39**(1), 135–154 (2015)
12. Cannon, D.: ITIL Service Strategy. The Stationary Office, Norwich (2011)
13. Jia, R., Reich, B.H.: IT service climate, antecedents and IT service quality outcomes: some initial evidence. J. Strateg. Inf. Syst. **22**(1), 51–69 (2013)
14. Patton, M.Q.: Qualitative Evaluation and Research Methods, 2nd edn, p. 532. Sage, Newbury Park (1990)
15. SKL, Strategy for the e-society (in Swedish), S.A.o.L.A.a. Regions, Editor (2011)

Social Customer Relationship Management: An Architectural Exploration of the Components

Marcel Rosenberger[✉]

Institute of Information Management, University of St. Gallen,
St. Gallen, Switzerland
`marcel.rosenberger@student.unisg.ch`

Abstract. In the recent years, social media have rapidly gained an increasing popularity. Companies have recognised this development and anticipate advantages from using social media for commercial purposes. Social customer relationship management (CRM) professionalises the use of social media and aims at integrating customers into operational procedures. This induces changes of existing structures, e.g. culture and organisation, business processes, information systems (IS), data structures, and technology. The intended transformation from CRM to social CRM is a complex task, because different aspects are affected, which also are mutually dependent. A prerequisite for the successful implementation of social CRM is understanding these aspects and its dependencies. Separation of concerns is a useful means of addressing complexity. The conglomeration of different issues is dissolved by conceptualising components and its relationships. This paper separates the concerns of social CRM using architectural perspectives and aims at building a better understanding. The research method is a literature review in which artefacts are gathered and assigned to five layers, which are business, process, integration, software, and technology. The conclusion states that social CRM is an emergent research field and comprises a call for more artefacts that concretise abstracted components of the business-layer.

Keywords: Social CRM · Design science · Enterprise architecture · Artefacts · Literature review

1 Introduction

Social media have gained interest and popularity in the past years. They are applications that build on web 2.0, which is a concept that encourages connecting, participation, and collaboration of users and sharing of content over the Internet [1]. The high number of social media users attracts attention of companies, which aim to profit from the potentialities [2, 3]. At a first glance, with only little effort companies can use social media to publish advertisements that reach many people, which improves the marketing efficiency. This view, however, is short-sighted. A closer study reveals more opportunities, which are enabled by the integration of social

© IFIP International Federation for Information Processing 2015
M. Janssen et al. (Eds.): I3E 2015, LNCS 9373, pp. 372–385, 2015.
DOI: 10.1007/978-3-319-25013-7_30

media and the consumers into operational procedures. Examples of the potentialities are support-cost reduction, product innovation, and improvement of the reputation [4–6]. Social customer relationship management (CRM) is a philosophy and business strategy that professionalises the relationship to customers using social media and aims at realising the opportunities [7].

The transformation of an organisation from CRM to social CRM is a complex task, because many different aspects are affected, which also are mutually dependent. For example, Askool and Nakata [8] highlight that a strategy must be developed to govern social CRM initiatives. The management's task is to provide for a supportive company culture and implement organisational changes [9]. Existing information systems (IS) need adjustments to enable and enhance business processes. Finally, Social CRM requires integrations on functional and technological level [10–12]. Without understanding social CRM and its components, implementation projects are likely to fail. However, a holistic view of the components is still missing. The existing literature either focuses on single aspects or provides an abstracted overview of multiple aspects without giving details [13, 14]. This is justifiable in consideration of the complexity of social CRM. Still, a complete picture is desirable. Separation of concerns is a useful means of addressing complexity. The conglomeration of different issues is dissolved by conceptualising the components and its relationships. Artefacts, which are the results of design science, document components and its relationships of a domain of interest. They contribute to actual design-oriented business problems and support the implementation of technology-based solutions [15, 16]. This paper aims to answer the following research question (RQ).

RQ: What are the components of social CRM from an architectural perspective?

The intention is to build a better understanding of social CRM from an enterprise architect's view. Instead of proposing another abstracted framework or deep diving into parts of the complex, the present paper reuses existing research results and integrates the findings into a holistic view. The concerns of social CRM are separated using connected architectural perspectives. This allows investigating social CRM focused on specific aspects and in its entirety. The research method is a literature review, which allows determining the current state of research. The artefacts of the discovered publications are assigned to five layers, which represent the architectural perspectives business, process, integration, software, and technology. The layers are adopted from Enterprise Architecture (EA), which is a holistic framework that helps representing an enterprise's artefacts. Each layer is a view from the perspective of a specific concern. All artefacts of all layers represent the entire body of knowledge within the research scope. The target groups of this paper are IS architects and researchers of social CRM.

The paper is structured in five sections. Section 2 gives the background of the underlying concepts, which are social CRM, artefacts, and EA. The research method is described in Sect. 3. Then, the findings are presented considering each architectural layer. Finally, the paper concludes with a discussion and interpretation and guides further research.

2   Conceptual Foundation

Figure 1 provides an overview of the architectural exploration of the artefacts of social CRM. Each EA-layer may contain artefacts, and a single artefact may also address concerns of multiple layers mutually. Artefacts of social CRM and artefacts of CRM are included to broaden the perspective.

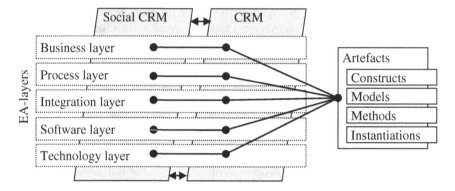

Fig. 1. Overview of the architectural exploration

2.1   Social CRM

Web 2.0 is a concept of the internet, which enables users to create content collaboratively and build a network with other users. The characteristics are user participation, openness, and network effects [1]. Openness means that results from user participation (e.g. posts, comments, and profiles) are accessible by other parties of the community. The concept is the foundation for social media, which are applications of the web 2.0 concept.

A basic feature of social media is to connect to other users to share information with them [17]. This principle is beneficial for the CRM of the company, which has the objective to establish and maintain profitable relationships to key customers and customer segments. CRM is a strategic approach that "involves identifying appropriate business and customer strategies, the acquisition and diffusion of customer knowledge, deciding appropriate segment granularity, managing the co-creation of customer value, developing integrated channel strategies and the intelligent use of data and technology solutions to create superior customer experience" [18].

Companies participate in the social network of users connecting to its target group. This facilitates the opportunity to gain business-relevant insights from the accessible data of the communication between the users. These insights help to intensify the relationship and to align the business with consumer needs. The integration of CRM with social media leads to the term "social CRM", which is a philosophy and business strategy [7]. Customers are engaged to participate in business processes with the result of a value-added for both: the company and the customer.

2.2   Artefacts

Design science research is a paradigm that aims at solving real-world problems by designing general solutions. It is a fundamental IS discipline, which develops artefacts that improve the capabilities of organisations [15]. Generality means that an artefact solves a class of problems instead of an individual problem of a single organisation. March and Smith [19] identify four artefact-types, which are constructs, models, methods, and instantiations.

Constructs are the basic language of concepts needed to describe phenomena. Models build on constructs and relate them with each other. Methods describe activities to meet specified targets. These forgoing artefacts can be instantiated in specific implementations representing the fourth artefact-type. The two main evaluation criteria are that artefacts are innovative and valuable [20]. Artefacts are ideal candidates to answer the research question, because they make components and its relationships explicit.

2.3   Enterprise Architecture

EA is a holistic framework, which provides views of an organisation's system from the perspective of specific concerns [21]. According to ISO/IEC/IEEE 42010:2011(E) [22] architecture is defined as the "fundamental concepts or properties of a system in its environment embodied in its elements, relationships, and in the principles of its design and evolution". The elements of a system can be related to five EA-layers, which are business-layer, process-layer, integration-layer, software-layer, and technology-layer [23].

The business-layer represents the strategy and subsumes organisational goals and success factors, products/services, targeted market segments, core competencies and strategic projects. The process-layer contains models to represent organisational units, business locations, business roles, business functions, metrics and service flows, for example. Applications and enterprise services are associated with the integration-layer. The software-layer contains software-components and data resources. Hardware units and network notes operate on the level of IT infrastructure (technology-layer). Relationships exist between components associated to the same layer and across layers. In the context of social CRM, sales and support are connected processes, for example, which both are associated to the process-layer. Social media are applications, which ultimately run on hardware. Thus, a connection between components of the integration-layer and components of the technology-layer exists. Aier et al. [21] identify the dissolution of information silos as an exemplary means of use of EA, which is the intention of use in this paper.

3   Method

The method for finding the existing artefacts is a literature review. Vom Brocke et al. [24] propose guidelines of a rigour process of literature reviews. They state that not only results should be presented, but, to allow replicability, also the approach. Table 1 characterises the conducted literature review following the taxonomy proposed by Cooper [25]. The focus (1) is on existing constructs, models, methods, and instantiations that support the design, implementation and governance of social CRM. The goal (2) is to connect to existing knowledge to solve the research problem on a conceptual level (3).

The perspective (4) can be characterised as neutral representation, because the position is unbiased. Practitioners and researchers of social CRM are the target audience (5). The results are representative (6) for the IS community because prominent data sources have been queried.

Table 1. Taxonomy of the conducted literature review (borrowing from [25]).

Characteristic	Categories			
(1) focus	research outcomes	research methods	theories	applications
(2) goal	integration	criticism	central issues	
(3) organisation	historical	conceptual	methodological	
(4) perspective	neutral representation		espousal of position	
(5) audience	specialised scholars	general scholars	practitioners	general public
(6) coverage	exhaustive	exhaustive and selective	representative	central/pivotal

A keyword search in the databases of AISeL, EBSCO, Emerald, IEEE, JSTOR, ProQuest, and Web of Science in title (TI), topic (TO), abstract (AB), keyword (KW) and full text (TX) fields was applied. The first search-string was built to find specific design science results containing the term "social CRM" in particular. The total number of hits without duplicates was low (24). As a consequence, the search-string has been broadened. The second search-string includes design science results that consider CRM and also social media or web 2.0. This ensures the inclusion of research results that are applicable to the research scope whereas the term "social CRM" is not used. Only reviewed publications have been considered to ensure the level of quality. Duplicate publications of the two searches have been removed. The relevance of the distinct papers has been determined by reading the full texts. For example, publications that defined the term "CRM" as "component reference model" or "core reaction model" have been treated as not relevant. Only original publications written in English have been incorporated. The artefact-type of the found artefacts has been determined and the publication has been assigned to an architectural-layer. In cases where no unequivocal assignment could be made, multiple assignments of the same publication to all fitting layers have been made.

4   Findings

Table 2 shows the numerical results of the literature review. The two search-strings, which ultimately lead to the relevant publications, are the following.

Search-string (1):   *"social crm" AND "design science"*

Search-string (2):   *(crm OR "customer relationship management")*
                         AND ("web 2.0" OR "social media") AND "design
                         science"

The keyword "design science" proved to be eligible, because it allows an efficient and effective search. Prior searches with the keywords "architecture" or "integration" did not lead to noteworthy results. Applying the keyword "model" shows results, but this term is more often used in the context of quantitative research and signifies statistical models and does not lead to the sought architecture elements. Both terms "elements" and "components" are too broad and do not reduce the results sufficiently.

Table 2. Numerical results of the literature review

Data source	Search fields	Search string		Publications	
		(1)	(2)	Total[a]	Relevant
AISeL	TI, AB, KW, TX	21	38	53	12
EBSCO	TI, AB, KW, TX	2	53	53	5
Emerald	TI, AB, KW	--	1	1	--
IEEE	TI, AB, KW	1	4	5	2
JSTOR	TI, AB, KW, TX	--	5	5	--
ProQuest	TI, AB, KW	3	46	46	1
Web of Science	TI, TO	--	--	--	--
Total[a]		**24**	**137**	**151**	**21**

[a]The total numbers are not equal to the column and row sums respectively, because duplicates have been counted only once.

Applying the search-string (1) to the data source AISeL displays 21 results. This indicates two different things. Firstly, social CRM is a present research field of the IS community. Secondly, design science is a common research paradigm of this research field. Apart from that, Emerald, JSTOR, and Web of Science display no results for search-string (1). A possible reason is that Emerald and JSTOR include mainly journal publications and no conference papers. The publication-period of conference proceedings is usually shorter. Hence, "social CRM" is a novel term that is not yet established in journals. The fact that JSTOR has results for "design science", but no results for "social CRM", supports that argument. Emerald does not feature to search the full texts of publications and the occurrence of "social CRM" and "design science" in the metadata is non-existent. This also applies to Web of Science, where only title (TI) and topic (TO) fields are searchable. In total, search-string (1) leads to 24 unique publications.

Applying search-string (2) displays results in all data sources, except Web of Science due to the limitation to search in the abstracts or the full texts of the publications. The high total number of unique results (137) indicates that CRM is better established in research than social CRM. More precisely, the term "social CRM" is not as widely used as the combination of CRM and web 2.0 or social media. This is not surprising, because all three terms CRM, social media, and web 2.0 are the foundation and a prerequisite to define social CRM. EBSCO, ProQuest, and AISeL are the data sources with the highest totals for search-string (2).

In summary, 21 relevant publications describe artefacts that represent components of social CRM. Figure 2 shows the yearly distribution of the publications. The findings indicate that social CRM is a contemporary research field. All artefacts have been published in the past six years. The year 2012 is remarkable, because the number of publications increased threefold compared to the previous year. Since then, the number of publications that include artefacts relevant to social CRM amounts five to six per year in the queried data sources.

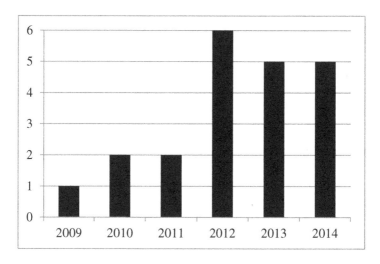

Fig. 2. Distribution of the relevant publications per year

Table 3 lists the publications, the artefact-types, the artefact-names, and the addressed EA-layers. A model is the most commonly occurring artefact-type among the explored publications. Eleven publications present models, five describe instantiations, three develop a method, and two publications propose constructs. Models, methods, and instantiations are built on constructs [19]. Consequently, every publication that describes an artefact contains or relies on constructs. In cases where models, methods, and instantiations implicitly rely on artefacts of other types, only the targeted artefact of the publication is incorporated. As a result, the number of counted constructs can be less than the number of the other artefact-types.

The business-layer includes the most artefacts. Ten artefacts address aspects of the business-layer, nine thematise processual issues, nine target applications of social CRM (integration-layer), seven outline software issues, and no artefact depicts components of the IT infrastructure (technology-layer). The artefacts of ten publications address concerns of multiple EA-layers. Examples are framework-models, which provide different views of components and its relationships [13, 26, 27]. Five artefacts address aspects of the integration-layer and software-layer in conjunction. The combined consideration of the business-layer and process-layer occurs four times and is second most therewith. This indicates a close architectural relationship between these layers.

A model of the business-layer is most frequent counting seven occurrences. Four publications develop models that highlight processual issues and four other publications describe instantiations that relate to the integration-layer implementing applications.

Table 3. Explored publications, artefact-types, artefact-names, and addressed EA-layers

Publication	EA-layer					Artefact-type				Artefact
	Business	Process	Integration	Software	Technology	Constructs	Model	Method	Instantiation	
[28]	○	○	●	●	○	○	●	○	○	*Framework of an ontology-based social media analysis*
[29]	○	●	○	○	○	○	○	○	●	*Text mining application for exploring the voice of the customer*
[30]	○	○	○	●	○	○	●	○	○	*Data model of data objects of social networks*
[31]	○	○	●	○	○	○	●	○	○	*Framework for gathering business intelligence from blogs*
[32]	○	●	●	●	○	○	○	○	●	*Social CRM tool framework*
[33]	●	○	○	○	○	○	○	●	○	*Method for developing a taxonomy of social media*
[14]	○	○	●	●	○	○	○	○	●	*User interface design for Twitter*
[34]	●	●	○	○	○	○	○	●	○	*Social media innovation method*
[35]	○	○	●	○	○	○	○	○	●	*Social data analytics tool (SODATO)*
[26]	●	●	○	○	○	○	●	○	○	*Social CRM framework*
[36]	●	○	○	○	○	●	○	○	○	*Web 2.0 factors and value drivers*
[37]	●	●	●	○	○	○	●	○	○	*Maturity model for the adoption of social media*
[38]	●	●	○	○	○	○	●	○	○	*System dynamics model and word-of-mouth effects*
[39]	○	●	○	○	○	●	○	○	○	*Ontology for IS sentiment analysis*
[13]	●	○	○	○	○	○	●	○	○	*Enterprise 2.0 management framework*
[40]	○	●	●	●	○	○	●	○	○	*Blueprint of an analytical social CRM system*
[41]	○	○	○	●	○	○	●	○	○	*Social network data model*
[42]	●	○	○	○	○	○	●	○	○	*Customer satisfaction theoretical framework*
[43]	○	○	●	●	○	○	○	○	●	*Multimedia platform providing social e-services*
[27]	●	○	○	○	○	○	●	○	○	*Social media strategy framework*
[44]	●	○	●	○	○	○	○	●	○	*Social app prototypes*

Legend: ○ not a focus of the publication; ● focus of the publication

4.1 Business-Layer

The artefacts of the business-layer address abstracted concepts, strategic aspects, organisational goals or success factors relevant to social CRM. Emamjome et al. [33] propose a method for developing a taxonomy of social media in an organisational context. The findings help to create an understanding of the concepts and support building a foundation for further research. Patten and Keane [13] conceive social CRM as a combination of three dimensions of a company-wide concept. Their Enterprise 2.0 Management Framework relates the dimensions (1) technology, tools and capabilities, (2) user-generated content, and (3) employee and customer applications, which are governed by culture and policies.

The maturity model of organisational adoption of social media allows positioning and comparing of the companies' achievements along the dimensions strategy, processes, IS, culture, and governance [37]. The artefact is useful to determine and adjust the approach to social media from a strategic perspective. Factors and value drivers of web 2.0 support the creation of a social CRM strategy, because they show causes and effects in a business context. Lehmkuhl and Jung [36] identify social networking, interaction orientation, user-added value, and customisation/personalisation as factors, which have a varying relevance for a specific company depending on the business model. A commerce-oriented online marketplace, for example, should encourage the customer-company and customer-customer communication (interaction orientation). A content-oriented online newspaper can profit from gathering and exchanging information and opinions of its readers by enabling social networking. According to Werder et al. [27], the social media strategy should include three components, which are scope, capabilities and governance. The scope is defined by actors, platforms and interaction. Social media objectives and activities are conceived as capabilities. The governance-component addresses value, resources and risks. Customer experience and customer satisfaction are further strategic focus areas that need attention and allocation of funds [42].

Yoon et al. [44] develop a conceptualisation of social commerce identifying the components user contribution, participation, collaboration and technological features. By reference to the customer life cycle model they argue that web 2.0 supports business goals. Interesting to note, social commerce addresses similar aspects of what Greenberg [7] terms social CRM. A clear differentiation between both terms is missing. In summary, the artefacts of the business-layer help to understand basic concepts, identify strategic aspects for planning and organising social CRM, include recommendations for governance, and "[stimulate] thinking about the impact of social media beyond the marketing function" [27].

4.2 Process-Layer

In order to implement social CRM in organisations the management needs to introduce, adjust, and evaluate business processes. A framework model helps to scope the tasks and structure work packages. The core processes for planning social CRM are readiness assessment, strategy development, value creation, multichannel management, information management and performance measurement [26]. The implementation-activities are governed by project management and change management and employee engagement. Helms et al. [34] focus on user participation in the innovation process developing a social media method for matching innovation tasks with social media characteristics. The characteristics are organised in three dimensions, which are audience, content and time. Botzenhardt et al. [29] focus on supporting the product development process. Their instantiation is a text mining software that analyses the unstructured content of customers' posts in social media. Maier and Reinwald [38] support the decision-making in the complaint management process. The authors propose a system dynamics model and incorporate the influence of social media on word-of-mouth effects and the customers' repurchase behaviour. Online social networks act as an accelerator and can have both: positive and negative effects.

The artefacts of the process-layer represent only some processes of social CRM. Other relevant processes are not covered, e.g. customer support, lead management, upselling/cross-selling, and market research.

4.3   Integration-Layer

The artefacts of the integration-layer propose models of applications and describe real-life instantiations of social CRM applications. The framework of an ontology-based social media analysis is a social CRM system model [28]. The central component is the Text Mining Framework, which has social media interfaces to access data of social media and database interfaces to enrich the data of enterprise systems (e.g. CRM system). The Ontology Engineering component extracts domain concepts of the company (i.e. ontology) from the data of its enterprise systems. The extracted company information and information about products is forwarded to the text mining application, which uses the ontology for filtering relevant and irrelevant social media data. The insights from social media data can be used to enhance products or plan marketing campaigns, for example. Chau and Xu [31] propose a framework for collecting and analysing business intelligence in blogs. This model is more concrete and regards blogs, which are a type of social media. The approach is the same: content analysis techniques are used to gather insights from the user-created content. Deng et al. [32] apply network analysis technologies to enhance marketing and sales processes and implement an application. The publication contains details of the data resources and software components and thus is additionally assigned to the software-layer. The Social Data Analytics Tool is a software-instantiation that fetches the social graph and the social text from social media [35]. The social graph represents actors and actions. Sentiments, keywords, and topics are extracted from the social text. Spagnoletti and Resca [43] implement a multimedia online platform and highlight that an online-community is a valuable tool for CRM.

Most artefacts of the integration-layer concentrate on gathering insights from social media data using analytical technologies. The represented integrations between enterprise systems and social media have in common that they follow an extract-transform-load (ETL) approach. This leads to a unidirectional connection from social media to the enterprise systems. However, a communication needs a two-way integration.

4.4 Software-Layer

The artefacts of the software-layer model components of social CRM applications. Examples of components are the user interface and the data model [14]. Rosemann et al. [41] include social media data in a Business Intelligence (BI) system that is capable to report on characteristics, needs, wishes and demands of customers. These insights are extracted by analytical operations on a combined set of the data of the CRM system, the data warehouse and social media. Key activities and components of the analytical social CRM system can be organised by the groups social web, multi-channel-management, analytical CRM, data, and operational CRM [40]. Analytical operations on the data are analysis, reporting, monitoring, and generating. The operational activities are planning, executing, and controlling CRM processes.

The identified artefacts are valuable to describe the structure and function of social CRM applications. The models and instantiations either focus on single components of a social CRM application in detail or give an overview of multiple components and its relationships.

4.5 Technology-Layer

No artefact represents technological aspects of IT infrastructure, such as hardware units, network nodes, and physical servers. It is undisputed that technology is an important component of a social CRM system [7]. However, the artefacts discovered in the literature review do not provide details about necessary or recommended IT infrastructure components.

5 Conclusion

Social CRM is an emergent research field with an increasing number of publications that present artefacts. However, the term is not commonly established. Besides "social CRM", researchers use the terms "Enterprise 2.0", "CRM 2.0", and "social commerce" without clear differentiation [13, 44, 45]. The findings help to establish a better understanding of social CRM, because the explored artefacts reveal the components and its relationships. Social CRM, in entirety, comprises aspects of the business-, process-, integration-, software-, and technology-layers. No single artefact, however, covers all components of all layers mutually. This is explainable by the complexity of social CRM and, in consequence, the need to examine the philosophy and business strategy on different levels of abstraction and by separation of concerns. No discovered artefact represents components of the technology-layer. A possible reason is that the setup of hardware units, network nodes and servers are specific to an organisation and depend on company-size and individual organisational requirements. However, research results should also be applicable to other (similar) situations. Artefacts provide a general solution to a problem in a specified context [15, 19]. Hence, it is not surprising that the technology-layer is under-represented in research. On the contrary, artefacts of the technology-layer are a possible research output. Examples are design principles of successful infrastructure-setup to accommodate the high load of the analytical social media data processing.

The findings have managerial impacts. Not all artefacts, however, are relevant to all stakeholders of social CRM. For example, the management of an organisation might want to adduce artefacts of the business-layer to refine the social media strategy, while developers receive conceptual guidance from the models and instantiations of the integration-layer and software-layer in particular.

A limitation of this research is that the quality of the discovered artefacts has not been evaluated. Furthermore, due to the research question, only design science results, i.e. artefacts, are included in the literature review.

Further research may focus on components of a single layer, multiple layers, or the connection of the layers. A higher layer gives orientation and determines aspects, which need further concretisation on a lower layer. For example, culture and governance are mentioned as important issues of the business-layer [13, 37]. However, no artefact on the process-layer continues these aspects. It is not sufficient to identify *what* the important components of social CRM are, but also *how* the components function. In summary, the findings lead to a call for more artefacts that concretise the components of the business-layer. Especially constructs, which define a common terminology and methods that guide the successful implementation of social CRM components are sparse.

References

1. Musser, J., O'Reilly, T.: Web 2.0 principles and best practices (2006)
2. Smith, T.: Conference notes – the social media revolution. Int. J. Mark. Res. **51**, 559 (2009)
3. Baird, C.H., Parasnis, G.: From social media to social customer relationship management. Strateg. Leadersh. **39**, 30–37 (2011)
4. Cappuccio, S., Kulkarni, S., Sohail, M., Haider, M., Wang, X.: Social CRM for SMEs: current tools and strategy. In: Khachidze, V., Wang, T., Siddiqui, S., Liu, V., Cappuccio, S., Lim, A. (eds.) iCETS 2012. CCIS, vol. 332, pp. 422–435. Springer, Heidelberg (2012)
5. Fliess, S., Nesper, J.: Understanding patterns of customer engagement – how companies can gain a surplus from a social phenomenon. J. Mark. Dev. Competitiveness **6**, 81–93 (2012)
6. Jahn, B., Kunz, W.: How to transform consumers into fans of your brand. J. Serv. Manag. **23**, 344–361 (2012)
7. Greenberg, P.: CRM at the Speed of Light: Social CRM 2.0 Strategies, Tools, and Techniques for Engaging your Customers. McGraw-Hill Osborne Media, New York (2010)
8. Askool, S., Nakata, K.: A conceptual model for acceptance of social CRM systems based on a scoping study. Ai Soc. **26**, 205–220 (2010)
9. Kuikka, M., Äkkinen, M.: Determining the challenges of organizational social media adoption and use. In: ECIS 2011 Proceedings, Paper 248 (2011)
10. Acker, O., Gröne, F., Akkad, F., Pötscher, F., Yazbek, R.: Social CRM: how companies can link into the social web of consumers. J. Direct Data Digit. Mark. Pract. **13**, 3–10 (2011)
11. Reinhold, O., Alt, R.: How Companies are implementing social customer relationship management: insights from two case studies. In: Proceedings of the 26th Bled eConference, pp. 206–221 (2013)
12. Trainor, K.J., Andzulis, J.M., Rapp, A., Agnihotri, R.: Social media technology usage and customer relationship performance: a capabilities-based examination of social CRM. J. Bus. Res. **67**, 1201–1208 (2013)
13. Patten, K., Keane, L.: Enterprise 2.0 management and social issues. In: AMCIS 2010 Proceedings, Paper 395 (2010)

14. Gruzd, A.: Emotions in the twitterverse and implications for user interface design. AIS Trans. Hum. Comput. Interact. **5**, 42–56 (2013)
15. Hevner, A.R., March, S.T., Park, J., Ram, S.: Design science in information systems research. MIS Q. **28**, 75–105 (2004)
16. Müller, B., Olbrich, S.: The artifact's theory – a grounded theory perspective on design science research, pp. 1176–1186 (2011)
17. Ang, L.: Is SCRM really a good social media strategy? J. Database Mark. Cust. Strateg. Manag. **18**, 149–153 (2011)
18. Frow, P.E., Payne, A.F.: Customer relationship management: a strategic perspective. J. Bus. Mark. Manag. **3**, 7–27 (2009)
19. March, S.T., Smith, G.F.: Design and natural science research on information technology. Decis. Support Syst. **15**, 251–266 (1995)
20. Peffers, K., Tuunanen, T., Rothenberger, M.A., Chatterjee, S.: A design science research methodology for information systems research. J. Manag. Inf. Syst. **24**, 45–77 (2007)
21. Aier, S., Gleichauf, B., Winter, R.: Understanding enterprise architecture management design – an empirical analysis, pp. 645–654 (2011)
22. ISO/IEC/IEEE: Systems and software engineering - Architecture description, Reference number ISO/IEC/IEEE 42010:2011(E). IEEE Comput. Soc. (2011)
23. Winter, R., Fischer, R.: Essential layers, artifacts, and dependencies of enterprise architecture. In: Proceedings of 2006 10th IEEE Int. Enterprise Distributed Object Computing Conference Work, EDOCW2006, pp. 1–12 (2006)
24. Vom Brocke, J., Simons, A., Niehaves, B., Riemer, K., Plattfaut, R., Cleven, A., Brocke, J. Von, Reimer, K.: Reconstructing the giant: on the importance of rigour in documenting the literature search process. In: 17th European Conference on Information Systems (2009)
25. Cooper, H.M.: Organizing knowledge syntheses: a taxonomy of literature reviews. Knowl. Soc. **1**, 104–126 (1988)
26. Lehmkuhl, T.: Towards social CRM - a model for deploying web 2.0 in customer relationship management (2014)
27. Werder, K., Helms, R., Slinger, J.: Social media for success: a strategic framework (2014)
28. Alt R., Wittwer M.: Towards an ontology-based approach for social media analysis. In: Proceedings 22nd European Conference on Information Systems. Tel Aviv, pp. 1–10 (2014)
29. Botzenhardt, A., Witt, A., Maedche, A.: A text mining application for exploring the voice of the customer (2011)
30. Braun, R., Esswein, W.: Corporate risks in social networks–towards a risk management framework. In: 18th American Conference on Information Systems, pp. 1–12 (2012)
31. Chau, M., Xu, J.: Business intelligence in blogs: understanding consumer interactions and communities. MIS Q. Manag. Inf. Syst. **36**, 1189–1216 (2012)
32. Deng, X.L., Zhang, L., Wang, B., Wu, B.: Implementation and research of social CRM tool in mobile BOSS based on complex network. In: 2009 1st International Conference on Information Science and Engineering, ICISE 2009. pp. 870–873 (2009)
33. Emamjome, F., Gable, G.G., Bandara, W., Rabaa'i, A.: Understanding the value of social media in organisations: a taxonomic approach. In: PACIS 2014 Proceedings, Paper 59 (2014)
34. Helms, R.W., Booij, E., Spruit, M.: Reaching out: involving users in innovation tasks through social media. In: ECIS 2012 Proceedings, Paper 193 (2012)
35. Hussain, A., Vatrapu, R.: Social data analytics tool (SODATO). In: Tremblay, M.C., VanderMeer, D., Rothenberger, M., Gupta, A., Yoon, V. (eds.) DESRIST 2014. LNCS, vol. 8463, pp. 368–372. Springer, Heidelberg (2014)

36. Lehmkuhl, T., Jung, R.: Value creation potential of web 2.0 for Sme — insights and lessons learnt from a European producer of consumer electronics. Int. J. Coop. Inf. Syst. **22**, 1340003 (2013)

37. Lehmkuhl, T., Baumöl, U., Jung, R.: Towards a maturity model for the adoption of Social Media as a means of organizational innovation. In: Proceedings of the Annual Hawaii International Conference on System Sciences, pp. 3067–3076 (2013)

38. Maier, M., Reinwald, D.: A system dynamics approach to value-based complaint management including repurchase behavior and word of mouth. In: ECIS 2010 Proceedings, Paper 55 (2010)

39. Park, E., Storey, V., Givens, S.: An ontology artifact for information systems sentiment analysis (2013)

40. Reinhold, O., Alt, R.: Analytical social CRM: concept and tool support. In: Proceedings 24th Bled eConference, pp. 226–241 (2011)

41. Rosemann, M., Eggert, M., Voigt, M., Beverungen, D.: Leveraging social network data for analytics CRM strategies - the introduction of social BI. In: ECIS 2012 Proceedings (2012)

42. Seng, W.M.: E-government evaluation: the use of design science approach to build and evaluate customer satisfaction theoretical framework. In: 2012 Seventh International Conference on Digital Information Management (ICDIM), pp. 266–273 (2012)

43. Spagnoletti, P., Resca, A.: A design theory for IT supporting online communities. In: 2012 45th Hawaii International Conference on System Sciences, pp. 4082–4091 (2012)

44. Yoon, S.Y., Studies, P., Technologies, E.: Empirical investigation of Web 2.0 technologies for social commerce and implementation of social app prototypes (2013)

45. Trainor, K.J.: Relating social media technologies to performance: a capabilities-based perspective. J. Pers. Sell. Sales Manag. **32**, 317–331 (2012)

Removing the Blinkers: What a Process View Learns About G2G Information Systems in Flanders (Part 2)

Lies Van Cauter[1]([⌗]), Monique Snoeck[2], and Joep Crompvoets[1]

[1] KU Leuven, Public Governance Insitute, Leuven, Belgium
{Lies.VanCauter,Joep.Crompvoets}@soc.kuleuven.be
[2] KU Leuven, Research Centre for Management Informatics, Leuven, Belgium
Monique.Snoeck@kuleuven.be

Abstract. Information sharing across the public sector is a precondition for innovation. The reality today is that data are scattered throughout administrative services. Creating government-to-government (G2G) information systems (IS) has the potential to sustain fluent data flows. Despite this potential, G2G IS projects fail to deliver the expected benefits. Factor research partially explains why so many G2G information systems fail. In this paper we take a broader perspective by applying process research to study six recurrent problems of Flemish G2G IS in their dynamic context. We test whether Sauer's needs and support-power analysis can provide additional management insights concerning G2G IS projects. Our results, based on interviews and focus groups, show that seemingly controllable problems have much deeper roots that require managers' action.

Keywords: IS failure · G2G · Process management

1 Introduction

In November 2014 the Organisation for Economic Co-operation and Development (OECD) called for action to enable public sector innovation. A core precondition is free flowing data, since shared information provides a basis for simplification, accountability or collaboration and allows organisations to learn collectively [17].

The potential added-value of free flowing data is high, but unfortunately, the reality today is that data are scattered throughout administrative services. The Weberian bureaucracy, characterized by its strict task allocation and hierarchy, has led to fragmentation of policy and service delivery. This problem is pervasive: the need for information sharing exists both across different levels of governments (vertical dimension) as among different governmental agencies (horizontal dimension).

ICT is perceived as an important driver of change because the creation of digital government-to-government (G2G) information systems (IS) has the potential to sustain free data flows [2]. Unfortunately G2G IS projects continue to fail to deliver expected benefits [4, 24]. This problem has heavily been researched during the last decades, without resulting in a great improvement of failure rates. One cause might be that for a long time a rational and technical view on failure dominated [19]. Positivistic researchers

© IFIP International Federation for Information Processing 2015
M. Janssen et al. (Eds.): I3E 2015, LNCS 9373, pp. 386–397, 2015.
DOI: 10.1007/978-3-319-25013-7_31

and project managers believed that problems can be eliminated if failure factors are listed and if management can detect and eradicate these linear failure factors. Yet eliminating failure factors does not warrant success. Rational factor research ignores the context of an IS as well as the dynamic non-linear interactions of (non-)technical factors such as legislation, politics, economic or cultural factors [24]. Rational project managers tend to follow fixed goals and try to minimise the risk of random context events, as such they only see a part of the IS failure puzzle [12].

In 2014 Dwivedi et al. [8] called for research that includes a larger part of the failure puzzle by incorporating local contingencies and the dynamic environment of ISs. A growing research stream that connects to this call is the 'process perspective' (e.g. Lee and Liebenau, Markus and Robey, Sauer [8]). Process managers look at the interaction between an IS, its stakeholders and context factors. An IS project does not exist in a vacuum [12, 25], compatibility with a given environment is a key precondition for innovation [5]. Stakeholders of G2G ISs interact dynamically and may have diverse interests which can e.g. result in sabotage of project goals. A major cause of failure is the inability to deal with these [2]. The ability to adapt to environmental developments and changing stakeholders' needs, determines governmental innovation [2, 12]. Sauer believes that managers who are confronted with troubled ISs, could start with an analysis of their situation by conducting a needs assessment: What problems need to be solved and what stakeholder support would meet these. This should be followed by a support power analysis: Who has the power to provide the required support? By conducting such analysis, managers will be more aware about context and dynamic stakeholder interactions that influence their ISs (i.e. a process perspective) [22]. We elaborate further on Sauer's work, as the main research question is:

Can the needs and support-power analysis of Sauer provide additional insights for G2G IS management in Flanders?

In a previous article [23] we conducted a needs and support-power analysis, but limited ourselves to the study of recurrent technological and political problems of G2G IS projects in Flanders. The analysis showed that factor research can appoint recurrent problems but that Sauer's process perspective provides extra insights concerning the context of these problems and support (difficulties) of relevant stakeholders. This article investigates whether conducting the same analysis for recurrent economic and juridical problems of G2G IS projects in Flanders, can add insights to the prior analysis results. The specific research question of the paper is: *Does the analysis of the economic and juridical recurrent problems, via the needs and support-power analysis of Sauer, provide additional insights for G2G IS management in Flanders?*

The structure of this paper is as follows. The theoretical framework is sketched in Sect. 2. Methodology is described in Sect. 3. The actual analysis is presented in Sect. 4, it is followed by a discussion of the results in Sect. 5. We conclude in Sect. 6.

2   Theoretical Framework

Sauer sees ISs as the product of a process which is open to flaws. This process consists of an initiation, development, implementation and operational phase, it may be problematic.

An innovation process can be split up in a need and a support management process which may be influenced by contextual uncertainties. In order to continue the innovation process, a project organisation requires enough support: support is searched during the support management process. If there is too little support, the endurance of the innovation process and the whole IS is threatened [22].

Sauer modelled a triangle of dependences: ISs exist to serve stakeholder's interests. They require a variety of support if they are to function at all. The IS's project organisation has a special role in innovating the system. Support for carrying out this role will only be given if supporters' interests are served. Managers could use a needs analysis to define the problems with which the project organisation will be confronted during the innovation process. This analysis further pictures the context influences and available problem solving mechanisms. A support-power analysis helps additionally to determine who has the potential power to provide the required support.

2.1 Needs Analysis

The project organisation's needs analysis will consist of two parts: (a) an analysis of problems and (b) an analysis of the required support to solve these problems:

a. Analysis of problems is twofold: (1) map the problems to be solved, (2) do a context scanning. Context helps to define problems but constraints originating in this context may make the innovation process problematic. The context is analysed along six dimensions: 1. human factors, 2. history, 3. technological process, 4. structure, 5. politics and 6. environment. Environment is subdivided in: 6.1. customers, 6.2. suppliers, 6.3. competitors, 6.4. technology, 6.5. regulators, 6.6. interests and 6.7. culture [22].

b. The analysis of support looks at available problem-solving mechanisms G2G ISs are often ineffective in transferring data between organisations. These systems are confronted with complex combinations of problems. Managers should understand problems in depth, in order to deal with these in an effective manner. In the project organisation the idea champion takes up this vital task. The potential of an innovation is also dependent on the context in which information sharing takes place [22].

2.2 Support Power Analysis

The support power analysis investigates who is able to provide support identified in the needs analysis and what other relations may affect stakeholders [22]. It can be applied at any stage of a G2G IS project. Sauer advices to conduct this analysis often. An idea champion should acknowledge his dependence on others. Information sharing requires both thinking about the own organisation and of external actors [17]. The latter may support/obstruct a project while trying to protect their core values and may react in an unpredictable way to interventions. Costs- benefits are not evenly divided, some stakeholders win, some lose. Idea champions should be sensitive for chances of random decision making, managing these becomes in itself dynamic, management strategies are dependent on the situation at hand and stakeholders' reactions [7, 12].

3 Methodology

3.1 Data Collection

In 2012 an exploratory research on trends and challenges of G2G IS projects was conducted. 20 experts of all Belgian governmental levels (i.e. local, provincial, regional and federal) were interviewed on this matter. In 2014 32 idea champions of G2G ISs in Flanders (Belgium) were brought together in five focus groups to discuss IS challenges, trends and the management thereof. We detect an overlap between the findings of both studies: managers in G2G IS projects face recurrent problems, which we structured via Sauer's 'needs and support-power analysis framework'. Technical/political problems were discussed in a previous paper [23]. This paper analyses recurrent economic/juridical problems and focuses primarily on vertical ISs between the Flemish regional government and local governments. However, since for some IS projects several Flemish organisations were involved, this adds a horizontal dimension. The next paragraphs describe data collection techniques in more detail.

Interviews 2012. Interviewing is a common data collection technique in IS research. In 2012, 20 experts on G2G IS projects were interviewed face-to-face on trends and challenges of G2G IS projects. Interviews were carried out over a three-month period. The interviewees represented local or provincial stakeholder groups, managed a successful G2G IS project or tried to monitor several G2G ISs. They worked for 14 different organisations at all governmental levels, this was a deliberate choice: by collecting different points of view, the risk of attribution bias was reduced. The interviewer asked open-ended questions to probe interviewees when interesting topics surfaced [20]. Both (non-)verbal language was captured. Each interview lasted between one and two hours and was transcribed with permission. The policy documents and legislation interviewees referred to, were studied as well. All interviewees received an end report. After 20 interviews, a point of saturation was reached.

Focus Groups 2014. A focus group (FG) is a group of individuals assembled by researchers to discuss and comment on a certain topic. It allows to obtain a variety of perspectives from a single data-gathering session [20]. Mainly the last decade FGs are gaining visibility and acceptance in IS research [3]. In 2012 we created an inventory of existing G2G ISs in Flanders. Based on this inventory, 40 IS idea champions were invited to participate in FG discussions. Five refused cooperation, five others did not show up. We slightly over-recruited the number of idea champions and reached as such the optimal number of 6 or 7 participants per session. A pretested questioning route was used to guide the conversations. The moderator briefed the participants, tried to create an informal sphere and ensured that everyone could have a say. She encouraged dialoguing via follow-up questions and by showing a stimulating body language. A senior researcher took up the role of assistant moderator. She observed body language, took notes and summarised the viewpoints. After 5 sessions saturation was achieved [3]. For a more detailed description of the data collection see [23].

3.2 Data Analysis Method

The interview and FG questions are not based on a specific theoretical model so that the data could speak for itself (an inductive approach). The five stages model of Krueger [14] was used to interpret the data. (1) Familiarisation: The researcher gets familiar with the major themes by reading the transcripts. (2) Themes: She develops categories within the major themes based on a questioning route. (3) Indexing: Data within and between cases are compared. (4) Charting: Data are reduced, important quotes rearranged under new codes. (5) Mapping and interpretation: Links between quotes are interpreted to make sense of the data as a whole. By studying the data, a series of G2G IS problems surfaced. We grouped these in 4 main categories: (A) technological, (B) political, (C) economic and (D) juridical recurrent problems. (A) Recurrent technological problems have to do with the business case, IT infrastructure, developers, planning and security. (B) Political recurrent problems involve top management support, user involvement and the skills/position of the idea champion. (C) Economic problems are about the need for (in)tangible resources. (D) Finally too much/less change in legislation, involuntary use and privacy form recurrent juridical problems. 287 pages of transcripts were coded in the qualitative data analysis programme NVivo. Data were analysed in two stages. (1) The problems detected from the interview data of 2012 were compared to the FG results of 2014. (2) Recurrent problems were compared to Sauer's framework by applying a needs and support analysis.

4 Analysis

The data analysis of interviews and FGs reveals an overlap in problems concerning political, technological, economic and juridical issues. Considering that these problems reoccur in both studies, we assume that they are rather structural and widespread for Flemish G2G IS projects. Factor research lists these problems too but ignores their interaction and context. It misses as such a part of the IS failure puzzle. In contrast, we take a process view by conducting a needs and support-power analysis on the recurrent problems. This enables us to research whether additional insights can be found via the process perspective. We found that the need and support analysis of political and technological recurrent problems indeed provides additional insights [23]. In the next paragraphs we will analyse recurrent economic and juridical issues and research whether these too provide additional insights. Due to space limits only the six most prominent economic and juridical recurrent problems are presented. Per problem three main things are described: (1) the problem to be solved (= problem description), (2) which context elements influence the problem (= context), and (3) which mechanisms can solve these problems and whose support is relevant therefor (= support). Every problem is influenced by several context elements. These elements are numbered and the applicable context category is mentioned between brackets. These numbers are referred to in the description of the support in order to motivate which support element relates to which element of the context.

4.1   Economic Agreements

To set up and run a G2G IS project, stakeholders must agree to provide tangible resources such as money/personnel and intangible resources (i.e. data, information) [4].

Problem 1: Money. *Problem Description.* A sufficient amount of money should be spent on a G2G IS for development, maintenance and adaptions: Who will finance what and when?

Context. Several context factors influence this need. (1) A macro factor is the economic crisis, due to budget cuts the willingness to do something for another government dropped (environment) and (2) funding to stimulate information sharing is under pressure (environment). (3) The budgetary capacity of governments differs widely but is relevant for obtaining adequate hardware, software and IT knowledge (structure). (4) Funds might stimulate municipalities with a small capacity, but in the past, Flemish funds were sometimes unilaterally abandoned, creating local distrust (history). (5) The configuration of Flemish departments stimulates silo creation: every department has its own budget. Information sharing challenges the classic revenue model (structure). (6) Flemish politicians see it as a means for cost reduction but *"a G2G IS is a current account. If there is a change in the IS, stakeholders have to invest money to adapt to these changes."* A lack of invested resources leads to suboptimal solutions. Ministerial priorities can influence the annual budget of an IS too (politics).

Support. (1–2) Funds to stimulate use are not desirable in times of budget cuts but may convince stakeholders to support an IS that has to outgrow technical problems. G2G ISs require immediate investment costs, benefits are only obtained over time. Funds are useful to bridge the period when other benefits cannot yet be reaped. (3) Cities have more means and a stronger bargaining power than small sized municipalities. (4) Support is given more easily when increased performance is expected. (5–6) If a ministerial cabinet supports an IS, it is easier to ask for more resources/cooperation of other departments. Respondents advise to prevent regular IS changes and to explain why these changes are needed.

Problem 2: Personnel. *Problem Description.* Every participating organisation in a G2G IS project should invest a sufficient amount of personnel time at all stages of the project, in order to tackle interoperability or IT problems and in order to enable data input and analysis.

Context. (1) In a G2G IS project the number of data inputters differs widely per organisation. 75 % of the Flemish municipalities has less than 20.000 citizens and takes up many tasks with few people. Cooperation in G2G ISs is cost demanding for small municipalities, resulting in higher investments than gains(structure). *"They chose their own priorities: not all data requests from the Flemish government will be answered."* (2) Governments often lack IT skilled personnel (structure). (3) E-government is demanding as it often requires a duplication of services (e.g. due to the digital divide or during the transition to an IS) (human factors and technological process). (4) Every governmental organisation works and evolves on its own speed (history).

Support. (1) Due to their size, cities have more negotiation power than smaller municipalities. Their ISs may be more advanced than the Flemish ones. Cities claim that if the Flemish government wants their support, it should recognise their expertise instead of imposing ISs. The respondents think that users with a small capacity can be motivated to support an IS by the availability of a help desk. (2–4) Difference in capacity/speed might be tackled by the creation of several entrance levels.

Problem 3: Intangible Resources. *Problem Description.* G2G ISs need to be fed with data from different parties.

Context. (1) Stakeholders in G2G context often face a lack of shared goals (environment). (2) The Flemish government aims to collect policy info for the whole of Flanders. The required data are available at local level but municipalities do not intend to spend their limited resources on making their data available to the Flemish government when their benefit is uncertain (environment). (3) This problem is exacerbated by previous experiences that it is hard to obtain data in return from the Flemish level (environment). (4) The Flemish government recently agreed to standardize on 'open data', but not all departments welcome this strategy. Some ISs are financed by 'pay per data use', but who will pay for 'open' ISs is not yet clear (environment).

Support. (2–3) The respondents notice that if the Flemish government wants local governments to support an IS, it should see them as data sharing partners and not as data subordinates. (4) Respondents think that stakeholder support for opening ISs is only realistic if politicians provide an alternative finance model for 'pay per use'.

4.2   Legal Agreements

Rules and legislation are mostly created to ensure quality, equity or responsible resource use. But these can restrict innovation if they cannot be easily adapted to specific needs or a dynamic environment [17]. In our research we found 3 recurrent legal problems that concern change in legislation, voluntariness of use and privacy issues.

Problem 4: Change in Legislation and Regulations. *Problem Description.* Legal or jurisdictional aspects may hinder the progress of e-government. Legislation and regulations need to be altered or sometimes developed.

Context. (1) The legal status of a digital G2G IS might not be recognised, even if it is, compared to its paper counterpart, the authentic source (structure). (2) Digitisation often comes after legislation. IS stakeholders may find their activities prohibited by formal rules. The option to simplify legislation is regularly ignored as idea champions lack time or the juridical capacity to perform such an exercise. As a consequence digitisation does not simplify the multiple adapted/expanded rules but builds yet another level of legislation (structure and environment). (3) Politicians change legislation/regulations regularly, which brings along an adaption cost for ISs (environment).

Support. (1–2) The respondents are less likely to support complex ISs. As such they believe that reengineering legislation is necessary. *"IT'ers tend to see legislation as hard to programme, holy and untouchable. They should realise it can be adapted."*

Problem 5: Voluntariness of Use. *Problem Description.* An IS needs to be used to be successful. Voluntariness of use differs per case, rights and obligations may be laid down in rules and regulations.

Context. (1) The Flemish government often legally obliges IS use for local governments. Non-use is not always reprimanded (regulators). The "bell tower principle" states that if the Flemish government asks for municipal efforts, it should provide a financial reward, this is often ignored in practice (history). (2) Another option is using the carrot by funding data exchange, yet this created perverse effects in the past (e.g. minimal data import to get the money) (environment). (3) A third option is voluntary participation in an IS based on a win-win (environment). (4) The need of local data for the Flemish government often originates from European requests (regulators).

Support. (1) Due to a lack of resources many local governments will not provide support to low priority G2G ISs. Use of legislation alone is an indolent solution for getting support, it may result in ISs plagued with poor data quality. As local governments become more emancipated, they tend to refuse support even if it is legally obliged. (2) Support can drop severely when funding stops. (3) Municipalities are tired of double data requests, they want to support data reuse. The respondents also advice to ask potential users who refuse to cooperate why this is the case (Fig. 1).

CONTEXT	
HUMAN FACTORS	TECHNICAL PROCESS
- Digital divide: duplication of services induces duplication of data input efforts	- Duplication of services induces duplication of development efforts
HISTORY	POLITICS
- Unilaterally adapted funds, low local trust - Organisations work at their own speed - Ignorance 'bell tower principle'	- Politicians underestimate (recurrent) costs - Ministerial priorities influence finances - Changes legislation ⇔ adaption cost ISs
ENVIRONMENT	STRUCTURE
- Eco. crisis, dropped willingness to do sth. for others - Culture: hard to locally obtain Flemish data - Interests: distrust, lack of shared goals, desire for benefits in return for data ⇔ obliged use - Interests: win/ funds as stimulus ⇔ perverse effects - Interests: Vagueness privacy as an excuse for not sharing data, open data ⇔ pay for use - Regulators: no reengineering, data sharing may be prohibited or obliged. Problem legal status digital ISs - Regulators: European data requests	- Differing budgetary & IT capacity - Budgetary configuration stimulates silos - Low (IT) personnel capacity 75% municipalities, own local priorities - Time intensive to change legislation, lack of juridical personnel capacity - Time consuming to get permission of (conflicting) privacy commissions

Fig. 1. Overview of legal and economic context constraints of Flemish G2G IS projects

Problem 6: Privacy. *Problem Description.* Governments that share information have to respect the fundamental right of privacy.

Context. (1) In order to share information, G2G IS projects often need a permission of a privacy commission, but obtaining permissions is time consuming. The more because the Belgian federal structure may make several privacy commissions competent according to the applicable policy level or policy domain. Respondents complain that the advice of different commissions may conflict (structure). (2) Data may not be shared or reused due to data protection regulations. It is still rather unclear if reused data may be enriched. The respondents see privacy as a vague issue (environment).

Support. (1–2) An IS often needs the support of a kind of privacy commission(s), which can be time consuming. The respondents ask for more clarity on privacy issues. They propose to only let the most relevant commission decide, when several privacy commissions are authorised. The respondents believe that on the other hand, privacy protection may also be a misused argument to refuse support to provide data.

5   Discussion

3 economic and 3 legal recurrent problems for Flemish G2G IS projects were uncovered by the analysis. Economic agreements have to be made to prevent problems with (in)tangible resources such as (1) money, (2) personnel, (3) data/information. Agreements on legal/juridical issues are crucial to prevent problems with: (4) too much/few changes in legislation/regulation, (5) voluntariness of use and (6) privacy. Several researchers confirm the importance of these problems: In terms of the economic perspective, previous research shows that it is hard to decide who will bear the costs of G2G IS projects [13], and that a lack of capacity is indeed a major barrier [9, 16, 18]. Financial support may promote ISs but obtaining funds can be tough [13]. Intangible resources are valuable, stakeholders want a benefit in return [1]. Concerning the juridical perspective, other scholars also found that G2G ISs may mismatch formal rules [11, 15] and require a time-intensive creation/adaption of legislation/regulations [11, 19]. They confirm that rules may be used as an excuse to block projects [17]. While ISs based on voluntary collaboration have more chance to succeed. Finally, the lack of clarity in privacy policies is a barrier for G2G ISs [9, 25].

This article investigated recurrent economic and juridical problems of G2G ISs in Flanders. It is an addition to a previous article about recurrent technological and political problems. Taking a look at the whole picture is interesting. All four categories of problems are based on the study of interviews in 2012 and focus groups in 2014. In total we found 14 recurrent problems. Each problem on itself is not new, factor research has listed these as well [21]. Yet, we go further by describing influencing context and support issues of these problems. By conducting a needs and support-power analysis it became clear that seemingly controllable problems have much deeper roots. Even more, context and support elements of different problems are interrelated.

In the previous paper elements pointed to: (1) a tendency of Flemish idea champions to merely focus on Flemish interests, (2) a lack of coordination of IS initiatives and (3) a political disinterest in ISs. These discourage local stakeholders to support Flemish G2G ISs. This paper studied if a needs and support-power analysis of economic and juridical recurrent problems can provide additional insights: First, the previous findings are confirmed, the 4 categories of problems seem to have similar roots: (1) The ignorance of the bell tower principle, a unilateral adaption of funds and a difficulty to locally obtain Flemish data, point to the tendency to merely focus on Flemish interests. (2) A budgetary configuration that stimulates siloisation and a lack of (G2G) shared goals and benefits, point to a lack of integration. (3) The unrealistic expectation of quick savings without much expenditures points to a political disinterest and lack of knowledge of ISs. Second, the analysis of economic and juridical factors adds two root elements. Local stakeholders seem also discouraged to support Flemish G2G IS projects because of (4) major differences in organisational capacity (e.g. differences in budget, IT, personnel capacity or e-government speed) and (5) juridical complications such as complex piled legislation, the need for duplication and the insecure legal status of ISs. Flemish idea champions may become discouraged by conflicting privacy commissions and by the prohibition of data sharing. Knowing why local stakeholder support may be discouraged is a start, but it is essential to know which action Flemish idea champions can take to tackle recurrent problems at their roots:

Action 1: Move from Hierarchical Project Management to Network Process Management. The main focus on the own Flemish interests and a lack of coordination be-tween different departments point to a 'hierarchical-project' way of managing. Project managers focus on clear goals and a predefined output while minimizing the risk of random events and ignoring the process [7]. This approach is not compatible with the environment of G2G IS project in Flanders. A 'network-process' management approach would allow to actively involve local stakeholders, as such Flemish idea champions could get an image of their core values. They may also get more conscious of the world around them and might better coordinate actions which allows better conflict anticipation. Local and Flemish stakeholders continually interact and adapt to their environment. These dynamics make G2G IS projects difficult to manage. Managers must also realise that the involvement of many people risks scope creep [12].

Action 2: Go for the Win-Win: Manage Needs, Capacities and Speeds. Often legislation mandates without providing resources, which negatively influences mutual trust [11]. We notice a trend of municipalities to refuse cooperation even if it is legally obliged. They argue they are busy with their own business and have other needs [13]. Flemish idea champions are dependent from other parties. Our research results show that local and Flemish stakeholders expect some benefits in exchange of their data (e.g. funds, increased performance, less duplication or data reuse). If funds or a legal obligation are the only drivers to provide data, perverse effects (e.g. poor data quality) might surface. A win-win between partners is definitely more stimulating [15]. Bigdeli et al [4] state that the resources of local governments are more limited than those of central/regional governments. Yet even within one group of stakeholders, in this case municipalities, there is a large capacity difference: the smallest Flemish municipality counts 80 inhabitants, the largest 480.000. Populous local governments, with many resources, are more likely to adopt e-government than smaller less resourceful municipalities [25]. In general, it is hard for Flemish idea champions to deal with the variety of needs, capacities and speeds [10]. This might be overcome by involving small, medium and large local governments and providing several IS entrance levels.

Action 3: Make Politicians More Aware of their Juridical Deeds and the Interaction with ISs, Dare to Redesign Legislation. Bekkers believes that the design of an IS should come before the creation of new legislation [1]. Flemish politicians are continually generating new legislation, which influences organisational flexibility [4]. If there is a change in legislation, business processes and their supportive systems have to be adapted [19]. G2G IS projects are burdened by a web of detailed and even conflicting rules/ regulations [1]. The short term orientation of Flemish politicians and the bureaucratic nature of public organisations (e.g. focus on predictability, legal security and equality) frustrates innovation [10]. Flemish idea champions should try to make politicians aware of their ignorance of ISs and the related juridical complications.

6 Conclusion and Future Research

This paper studied the roots of six recurrent economic and juridical problems of Flemish G2G IS. We aimed to extend the body of knowledge by investigating how local contingencies and support-power relations affect the likelihood of failure of Flemish G2G IS projects. The research findings show that a needs and support power analysis (i.e. a process perspective) provides additional insights for G2G IS management in Flanders. The analysis takes a broader look on IS failure than the classic factor perspective by incorporating the dynamic IS context and its stakeholders.

A previous study of recurrent technological and political problems of Flemish G2G IS, showed that apparently controllable problems have deeper roots which discourage local stakeholders to support ISs (i.e. the focus on Flemish interests by idea champions, political disinterest in technology and a lack of coordination). In this study it appears that economic and juridical problems have the same roots. Yet these are also rooted in major differences in organisational capacities and juridical complications.

Insight in the deeper roots of recurrent IS problems adds, compared to factor research, a new piece to the complex IS failure puzzle. Given this knowledge, which action could managers take? Managers should consider that the 'network process' management approach might be more suitable than an 'hierarchical project' management approach concerning Flemish G2G IS projects, they should have an eye for needs/capacity/speeds and could make politicians more aware of juridical consequences of their deeds, and in particular for the interplay of juridical aspects with ISs.

Our research findings are limited to Flemish G2G IS projects. Future research could study the roots of recurrent problems of G2G ISs in other regions or countries.

References

1. Bekkers, V.: The three faces of e-government: innovation, interaction and governance. In: Anttiroiko, A-V., et al. (eds). Innovations in Public Governance. Series Innovation in the Public Sector, pp. 194–216. IOS Press, Amsterdam (2011)

2. Bekkers, V., van Duivenboden, H., Thaens, M.: ICT and Public Innovation. Assessing the ICT-Driven Modernization of Public Administration. IOS Press, Amsterdam (2006)
3. Belanger, F., Tech, V.: Theorizing in Information Systems Research: using focus groups. Australas. J. Inf. Syst. **17**(2), 109–135 (2012)
4. Bigdeli, Z., Kamel, M., deCesare, S.: Inter-organisational electronic information sharing in local G2G settings: a socio-technical issue. In: European Conference on ISs, paper 79 (2011)
5. Dawes, S.: E-government and innovation: a fresh look at experience. In: IFIP EGOV (2014)
6. Dawes, S., Cresswel, A.M., Pardo, T.A.: From "need to know" to "need to share": tangled problems, information boundaries, and the building of public sector knowledge networks. Public Adm. Rev. **69**(3), 392–402 (2009)
7. De Bruyn, H., ten Heuvelhof, E., in 't Veld, R.: Process Management, Why Project Management Fails in Complex Decision Making. Springer, Heidelberg (2010)
8. Dwivedi, Y.K., Wastell, D., Laumer, S., Zinner, H., et al.: Research on IS failures and success: status update and future directions. Inf. Syst. Front. **17**, 143–157 (2014)
9. Ebrahim, Z., Irani, Z.: E-government adoption: architecture and barriers. Bus. Process Manage. J. **11**(5), 589–611 (2005)
10. Edmiston, K.L.: State and local e-government. Prospects and challenges. Am. Rev. Public Adm. **33**(1), 20–45 (2003)
11. Fountain, J.E.: Challenges to organizational change: multi-level integrated information structures (MIIS). In: National Center for Digital Government, Paper 15 (2006)
12. Janssen, M., van der Voort, H., van Veenstra, A.F.: Failure of large transformation projects from the viewpoint of complex adaptive systems: management principles for dealing with project dynamics. Inf. Syst. Frontiers **17**(1), 15–29 (2014)
13. Jing, F., Pengzhu, Z.A.: Field study of G2G information sharing in Chinese context based on the layered behavioral model. In: HICCS 42, pp. 1–13 (2009)
14. Krueger, R.: Focus Groups: A Practical Guide for Applied Research. Sage Publications, Thousand Oaks (1994)
15. Lam, W.: Barriers to e-government integration. J. Enterp. Inf. Manage. **18**(5), 511–530 (2005)
16. Moon, M.J.: The evolution of e-government among municipalities; rhetoric or reality? Public Adm. Rev. **62**(4), 424–433 (2002)
17. OECD: Innovating the public sector: from ideas to impact, Paris, 1–22 (2014)
18. Pardo, T.A., Nam, T., Burke, G.B.: E-government interoperability: interaction of policy, management and technology. Soc. Sci. Comput. Rev. **30**, 1–17 (2011)
19. Pardo, T.A., Scholl, H.J.: Walking atop the cliffs: avoiding failure and reducing risk in large scale e-government projects. In: HICCS 35 (2002)
20. Peters, R.M., Janssen, M., van Engers, T.M.: Measuring e-Gov impact existing practices and shortcomings. In: 6th International Conference of ECommerce, pp. 480–489 (2004)
21. Petter, S., DeLone, W., McLean, E.R.: Information Systems Success: The Quest for the Independent Variable. Journal of MIS **29**(4), 7–61 (2013)
22. Sauer, C.: Why IS Fail: A Case Study Approach. Alfred Waller Publishers, Suffolk (1993)
23. Van Cauter, L., Snoeck, M., Crompvoets, J.: Removing the blinkers: what a process view learns about G2G information systems in Flanders. In: IFIP WG 8.5., Thessaloniki (2015)
24. Yang, T.M., Pardo, T., Wu, Y.J.: How is information shared across the boundaries of government agencies? An e-Government case study. Gov. Inf. Q. **31**, 637–652 (2014)
25. Yang, T.M., Zheng, L., Pardo, T.: The boundaries of information sharing and integration: case study of Taiwan e-Government. Gov. Inf. Q. **29**, S51–S60 (2012)

A Visual Uptake on the Digital Divide

Farooq Mubarak$^{(\boxtimes)}$ and Reima Suomi

Turku School of Economics, Department of Information Systems Science,
University of Turku, Turku, Finland
{farmub,reima.suomi}@utu.fi

Abstract. Factors found to influence the adoption of ICT have been explored in several studies. However, few writers have been able to produce any systematic research into the digital divide. Although, differences of opinion still exist, a growing body of literature has established that income and education are positively related to digitalization patterns. This research attempts to deepen the understanding of the present ambiguous relationship between socio-economic indicators and the ICT. This account tested the links between socio-economic variables (GDP per capita, GINI index, World Bank Education Statistics, and Transparency International's corruption perception index) and ICT diffusion across developed and developing countries. Positive correlations were found for income and ICT, education and ICT. A negative correlation was found between corruption and ICT adoption. The paper discusses implication of these findings and suggests future courses of actions for policy makers. Proceeding from the findings of this paper, this research suggests there is an urgent need to address the digital divide by initiating impactful efforts to reduce it.

Keywords: Digital divide · ICT · Digital technologies

1 Introduction

Researchers have long sought to determine how patterns of digitalization link with economic variables. To that end, a hefty volume of publications has appeared in academia that looks at such links that have a variety of well-disputed varying research methods and measurement mechanisms. Moreover, recent advancements in ICT research have laid rest to several myths concerning the nature of the digital divide. Nevertheless, much doubt still persists since there still appears to be little agreement on the leading causes of the digital divide. The past decade, in particular, has seen rapid progress in ICT research, which has reshaped the conventional concept of the digital divide from narrow to wide and made it considerably complex paradigm. This shift has resulted in mounting literature across academia, politics, and the press. Much of the existing debate on the digital divide revolves around the qualitative nature of the issue, while the quantitative uptake on the topic appears to be overlooked. Thus, Noh and Yoo [1] view the measurement of the digital divide as a controversial issue. Owing to the complexities of the subject, some researchers [2–4] state that the digital divide is a confused theme in literature. These critics justify their views by noting severe pitfalls of the data and methodologies used to quantify the digital divide.

© IFIP International Federation for Information Processing 2015
M. Janssen et al. (Eds.): I3E 2015, LNCS 9373, pp. 398–415, 2015.
DOI: 10.1007/978-3-319-25013-7_32

In the history of the digital divide, poverty has been thought of as a key factor that is responsible for the breach in the access and the utilization of the digital technologies. In their major analysis of publications from thirty leading researchers, Skok and Ryder [5] concluded that GDP per capita and education were the principal factors responsible for the digital divide. A similar result was reported by Billon et al. [6] who found GDP and infrastructure as to be the main factors for slow digital progress in developing countries. A number of researchers [e.g. 7–9] have acknowledged GDP per capita/income as a leading cause of the digital divide between and within countries. It appears safe to say that poverty alone explains a major portion of the digital divide outgrowth.

The connections between GDP per capita and patterns of digitalization have been an object of research since the evolution of the digital divide concept, in the mid-1990s. Vodoz et al. [10] noted that individuals with higher education levels are likely to adopt the digital technologies faster than people with low or no education at all. Two large-scale studies [11, 12] demonstrated the positive correlation between GDP and the Internet diffusion curve.

However, uncertainty still exists about the relation between education and ICT diffusion; and two major studies [13, 14] defy any relation between the two. Consonantly, Stanley [15] intensifies psychosocial resistances as key factor responsible for the digital divide, thus putting aside income and education. Nevertheless, considerable criticism has been levelled at quantitative research on the digital divide. Much of the research on this subject has been restricted to local and limited comparisons, while only a few studies such as those by Cruz-Jesus et al. [14] have attempted to assess the digital divide across a wide range of geographical territories.

The recent rise in public computer facilities in developed countries has allowed a large segment of the population to benefit from ICT, who otherwise could not afford computers and the Internet. Thus, it can be fairly argued that there exists a category of people in developed countries who can benefit from ICT regardless of in-come constraints. If the findings made by Vodoz et al. [10] about the high rates of ICT adoption on education being dependent on education are accurate, then what explains the exponential growth of the Internet in developing regions with low literacy rates, such as South-East Asia? What is not yet clear is the measured impact of income and education on ICT adoption in the current era. Despite the research mentioned earlier, there is still very little scientific understanding of the degree of the relationship if any, of socio-economic indicators with ICT, particularly in developed countries. In addition, the general dispute in the quantitative literature on the digital divide hints that much of the evidence so far is inconclusive at best. This indicates a need to understand the various perceptions of any possible connections of socio-economic indicators with ICT in this current age. For the purposes of this study, the chosen socio-economic variables are the World Bank GDP per capita, the GINI index, World Bank Education, and Transparency Inter-national's Corruption Perception index.

Even though, some research has studied the socio-economic links with ICT, most research has been undertaken by analyzing just a few statistical observations. Nonetheless, it is possible to further improve the research design and scope by utilizing a combination of methods. With this goal, the present research seeks to obtain visualizations of the major factors that are responsible for the digital divide. Drawing upon this stand of research into the digital divide, this paper shall attempt to verify the aforementioned claims in preceding paragraphs by examining the links between income, education, and corruption with ICT patterns across high-income Nordic countries, the low-income Indian sub-continent region, and a few middle-income countries. A secondary aim is to shed light on the implication of the findings and suggest a direction for possible future developments. This paper aims to broaden the scope of the main factors responsible for the digital divide, since the digital disparities are intermixed with social and psychological factors in addition to income and education. The hypothetical premises at this point rest on four assumptions: GDP per capita is positive related to Internet adoption, GINI index is inversely related to Internet adoption, education is positively related to Internet adoption, and Corruption is inversely related to Internet adoption.

2 Literature Review

During the past twenty years, much more information has become available on the digital divide and its rigorous threats to the world economy. However, very little was found in the literature on the question of quantitative analysis of the digital divide. The digital divide can be defined as the disparity between those who do and do not benefit from digital technologies [16]. In the context of the digital divide, the chief division is between two significant groups namely inclusion and exclusion with respect to those who benefit and those who do not benefit from digital technologies. One group consisting of developed countries is continuously reaping the benefits of ICT whereas the other group consisting of developing countries is missing out on many benefits due to lack of access to the digital technologies [17, 18]. Menou [19] and Mansell [20] already warned that if the issue of the digital divide is left unattended, world inhabitants will be living in a dual planet. Call for future research on the nature of the digital divide has been a recurring theme of many scholarly articles.

As noted in the introduction that digital divide has been viewed as a confused theme in literature: some studies [19, 21, 22] consider the digital divide to be a matter of gap in access to ICT while others [1, 23–30] consider the digital divide as a complex and broad phenomenon where several variables play their respective part. Pick and Nishida [31] found education to be the principal determinant of technology utilization. It can be inferred from the study's conclusion that the role of education is significant in increasing adoption of digital technologies. Similarly, Cooke and Greenwood [32] maintain that the educational sector has made significant progress in promoting the adoption of ICT. Pittman [33] postulates that the role of ICT is essential in fostering a globally diverse educational system.

However, recently some literature has emerged that offers contradictory findings regarding the role of education in ICT adoption. An empirical investigation led by Lee [34] suggests that demographic factors (age, gender, education) have little effect on the digital divide. Unlike Pittman [33], a survey study of 158 Small and Medium Enterprise (SME) owners by Middleton and Chambers [13] found that education has no effect on the adoption of Internet. However, this attitude would appear to be outdated even the study was published in the year 2010. There are limits to how far the idea of

Middleton and Chambers [13] can be taken because ICT is being increasingly incorporated into the education systems worldwide.

Tipton [7] and Olaniran and Agnello [8] document income disparity as the leading cause of the digital divide by noting that the digital divide reflects high income levels in developed world where as the opposite is true for the developing world. Quibria et al. [12] report a strong correlation between GDP per capita and the usage of computer. However, Tavani [27] develops the claim that there are numerous other factors responsible for the breach in access and utilization of ICT other than income alone. Brooks et al. [9] maintain that costs of Internet connectivity in developing countries is significantly higher than those in developed countries. This corroborates with the view of Norris [35] who maintain that the richer countries are better in reaping the benefits of ICT than the poorer countries. The evidence presented in this paragraph suggests that there is a strong connection between GDP per capita and the patterns of digitalization. However, a number of studies [e.g. 14, 36–39] have reported significant digital divide in developed countries, which questions the relation of GDP with the ICT diffusion. This doubt is further reinforced since some researchers [40, 41] report a regional digital divide with respect to urban and rural settlements in high-income developed countries.

Novo-Corti et al. [42] maintain that simply promoting the access to digital technologies is a simple solution to overcome the digital divide. Although, the study was targeted towards a particular region, the claim is question-able because the digital divide has been proved as a complex phenomenon and a variety of factors are responsible for the di-vide other than just access. Bach et al. [43] calls for effective policies for organizations and governments to com-bat the digital divide. This corroborates with the findings of Graham [44] who highlights the need of effective government systems with effective subsidies to minimize the digital divide. Peng [45] points out that although governments have access to household profile data such as education, income, and gender, they often lack reliable insights into psychological and cognitive profiles of individuals. This implies that efforts to fight the digital divide are required on multiple fronts since other than in-come and education; psychological factors are also responsible for adoption/non-adoption of ICT.

The insights drawn from the literature review advances us towards testing three hypothesis. First, income per capita bears a positive relation with ICT adoption. Second, education has a positive link with ICT adoption. Finally, corruption bears negative relation with ICT adoption.

3   Research Design

3.1   Methodology

To date various methods have been developed and introduced in order to determine the connection between two or more variables. Traditionally, the digital divide measurements have been studied by comparing the variables in a select geographical territory. In addition, correlations between these variables have also been calculated to measure the degree of relationship. In the present study, the visualization approach was chosen to represent the association of GDP and education with the digitalization patterns across OECD member countries. The correlation attempt to provide an estimate of the degree of association between the variables under consideration. The correlations were calculated by using the following formula.

$$r = \frac{1}{n-1} \Sigma \left(\frac{x - \bar{x}}{s_x} \right) \left(\frac{y - \bar{y}}{s_y} \right)$$

Where r is the coefficient of correlation, X and Y are variables, whereas x bar and y bar represent respective sample means. S represents the sample standard deviation of the respective variables x and y. n equals the number of items in the samples under consideration. Pearson product correlations from the above mentioned formula were calculated between the variables under consideration after each figure. The average values of the data sets were taken for the available years, wherever possible. In some cases, data was missing for some countries in specific years; thus only years with complete data for all countries were taken into consideration.

3.2   Data Selection

Although, a range of different organizations provide data sets for GDP and similar economic indicators, this research employed data from World Bank statistics database. In addition to the simplicity of the data downloads according to the customized preferences, World Bank statistics are known for providing credible information. The statistics were also found to be in close accordance with data provided by other institutions, although there were minor differences. PISA test scores were taken directly from the official PISA scores website. The corruption Perception Index was taken from the Transparency International official website. For the purposes of the present study, information from the developed Nordic countries, developing Indian subcontinent countries and a few middle-income countries was chosen. In some instances, the data for some countries was not available for a particular indicator; therefore different countries were added to the comparison. The data sources are detailed in Table 1.

Table 1. Data sources

Name	Institution	URL	Selected data source	URL
Education efficiency	PISA-test scores	www.oecd.org/pisa	Scores in mathematics	http://www.oecd.org/pisa/keyfindings/pisa-2012-results.htm
Income	World Bank	www.worldbank.org	GDP per capita	data.worldbank.org/indicator/NY.GDP.PCAP.CD
Income inequality	World Bank	www.worldbank.org	GINI index	data.worldbank.org/indicator/SI.POV.GINI
Corruption	Transparency International	www.transparency.org/country	Corruption Perception index	www.transparency.org/research/cpi/overview

3.3   Visualizations

Human brains are designed to process visual images before texts, and they need less energy to consume images than texts [46]. In particular, visualizations present clear pictures of possible trends where the data is vast. Therefore, visualizations were chosen as a means for seeing any connections between the variables. In addition to giving a clear picture of a major scenario, visualizations can also represent the predictability of a certain variable's behavior over time.

In the present study, there are a mix of developed and developing countries under consideration. After trying various visualizations in Microsoft Excel software, a few were selected because they tend to report the best trends between the variables. Charts were used to show any possible link between the two variables.

4   Results and Discussion

4.1   Results

The visualizations revealed several interesting insights, a few of which negate previous studies. In the forthcoming paragraphs, each figure shall be discussed and all possible interpretations shall be subsequently drawn in the next section of discussion (Table 2).

Figure 1 shows that a positive parallel relation was found between GDP per capita and Internet users per hundred people in developed and developing countries. The Pearson product correlation coefficient "r" was found to be 0.882. The positive correlation suggests that there is a tendency for increase in the number of Internet users with an increase in the GDP per capita of a given country. It can be seen from figure that Norway has the highest GDP per capita and the highest percentage of Internet users. Along with Norway, other Nordic countries show high rates of GDP per capita and percentage of Internet users. This is in stark contrast with the Indian subcontinent countries which show low rates of GDP per capita and Internet users (Table 3).

Figure 2 compares the GINI index with the Internet user's percentage across a range of developed, developing, and middle-income countries. A general trend that can be noticed from the Fig. 2 that higher the GINI index, lower the Internet users percentage. For instance, Iceland has the lowest value of GINI index and the highest number of Internet users percentage along with Norway. Thus, it can be inferred that higher the inequality of income in a given country, lower shall be the ICT adoption, in general. There are however exceptions to this rule. For example, GINI index values of Nepal and Bangladesh are almost same, however Nepal shows considerable high percentage of Internet users. This negative result might be due to other factors such as political and regional infrastructure differences. The correlation analysis yielded a negative value of −0,45 indicating a negative link between inequality of income and Internet usage.

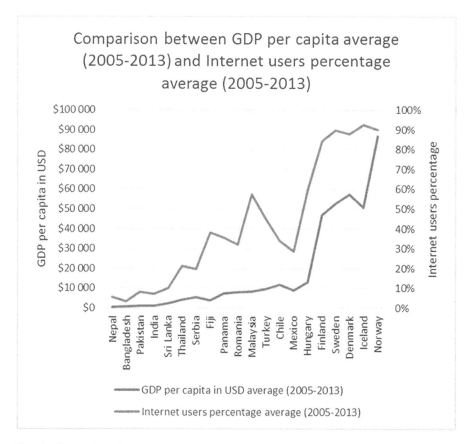

Fig. 1. Comparison between GDP per capita average (2005–2013) and Internet users' percentage average (2005–2013) (Based on World Bank 2015)

Table 2. Correlation between PISA 2012 test scores and the percentage of Internet users

Emerging correlation in Fig. 2	
Correlation between GDP per capita average (2005–2013) and Internet users percentage average (2005–2013)	0.882

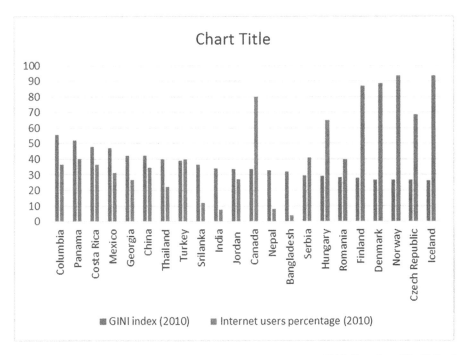

Fig. 2. Comparison between GINI index 2010 and Internet users 2010 (Based on World Bank 2015)

Table 3. Correlation between GINI index in percentage (2010) and Internet users in percentages (2010)

Emerging correlation in Fig. 2	
Correlation between the GINI index in percentages (2010) and Internet users in percentages (2010)	−0.45

Figure 3 compares percentages of secondary school enrollment and Internet users across Nordic, few Indian subcontinent countries for which data was available, and middle-income countries. Comparatively less intense than the correlation between GDP per capita and Internet users, however, still a positive correlation between secondary school enrollment and Internet diffusion was found to be 0.699 as reported in Table 4. All Nordic countries show high percentages of secondary school enrollments and

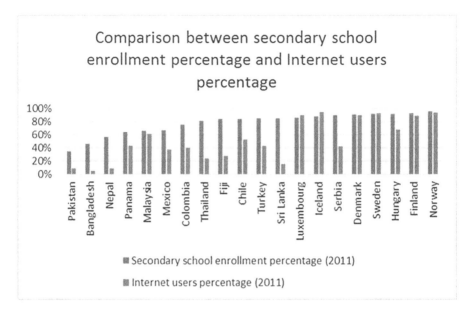

Fig. 3. Comparison between the percentage of secondary school enrollment and the percentage of Internet users for the year 2011

Table 4. Correlation between the percentages of secondary school enrollment and the percentage of Internet users

Emerging correlation in Fig. 3	
Correlation between the percentages of secondary school enrollment and Internet users	0.699

Internet users, whereas countries in Indian subcontinent region show low values of both variables under examination. There are however exceptions to this general trend of parallel growth of secondary school enrollments and Internet users. For instance, Luxembourg and Sri Lanka have almost similar values of secondary school enrollments at 86,4 % and 85,4 % respectively but they differ remarkably in percentage of Internet users. Despite the high secondary school enrollments rate, Sri Lanka has only 15 % of Internet users as compared to 90,02 % in Luxembourg.

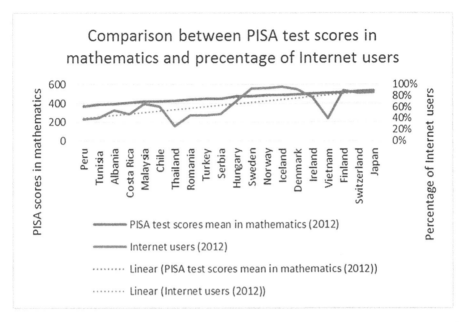

Fig. 4. Comparison between PISA 2012 test scores in mathematics and the percentage of Internet users (Based on data from OECD 2015 and World Bank 2012

Table 5. Correlation between PISA 2012 test scores and the percentage of Internet users

Emerging correlation in Fig. 4	
Correlation between PISA 2012 test scores and Internet users percentage	0.693

Advancing the comparison between education and ICT, PISA test results were included in the analysis. Since, the Indian subcontinent countries did not take part in PISA test, a different set of developing and developed countries were taken into account. It can be inferred from the Fig. 4 that countries with high PISA test scores have high percentage of Internet users, suggesting a positive relation between the two. However, there are exceptions to this assumption. For example, Vietnam and Finland show almost same values of PISA test scores, however they vary considerably in terms of Internet users percentage. Finland has far higher percentage of Internet users at 89,88 % than Vietnam at 39,50 %. The correlation between the subject variables in Fig. 4 was found to be 0,693 as reported in Table 5. The positive value of the correlation at 0,693 suggests a positive link between PISA test scores and Internet usage. The findings from Figs. 3 and 4 are close to the previous findings of Vodoz et al. [10] and Skok and Ryder [5] who determined education to be a principal factor responsible for ICT adoption.

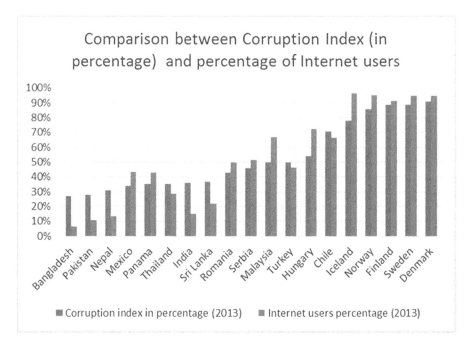

Fig. 5. Comparison between Corruption perception index and Internet users (Based on data from Transparency International Secretariat 2015 and World Bank 2015)

Table 6. Correlation between Corruption perception index and Internet users

Emerging correlation in Fig. 5	
Correlation between Corruption perception index in percentage (2013) and Internet users (2013)	0,933

Figure 5 illustrates how the corruption affects the percentage of Internet users. Corruption perception index shows the percentage of cleanliness in a given country. Figure 5 depicts that Denmark is the cleanest country in terms of corruption and has a high rate of Internet users. This is closely followed by Finland, Sweden, and Norway who are clean from corruption and show high Internet users percentage. On contrary, Bangladesh is the most corrupt country and has the lowest percentage of Internet users. It is apparent from this figure that more the cleanliness from corruption in a given country, more the percentage of Internet users. Correlation between Corruption perception index and Internet users was found to be 0,933. The high positive value of correlation at 0,933 indicates a significantly strong correlation between cleanliness from corruption in a country and Internet usage.

Taken together, these visualizations and correlations lend some support to test the hypothetical assumptions introduced in the beginning of the study. The results of the hypothetical testing is tabulated in the Table 6. In summary, the results support the assumptions presented in the introduction. The next section, therefore, moves on to discuss the findings and their implications in detail (Table 7).

Table 7. Results of hypothesis

Hypothesis	Measuring factor	Hypothetical statements	Calculated correlations	Results
H1:	Income	GDP per capita and the rate of Internet use are positively related	0.89	Supported
		Inequality of income (GINI index) is inversely related to Internet user	−0.45	Supported
H2	Education	Secondary school enrollment and the rate of Internet use are positively related	0.699	Supported
		PISA test scores and the rate of Internet use are positively related	0.693	Supported
H3	Corruption	Corruption is inversely related to the rate of Internet use	0.933	Supported

4.2 Discussion

This study set out to determine the possible links between socioeconomic indicators and ICT in developed and developing countries. It was hypothesized that income and education are positively related to ICT adoption whereas inequality of income and corruption were negatively associated with ICT adoption. As mentioned in the introduction and literature review, the abundant studies on the digital divide show no definite consensus on above mentioned links to draw any firm conclusions. Results of this study indicate that indeed income and education have a positive link with ICT whereas corruption bears a negative relation.

As shown in the literature review, there has been an inconclusive debate about whether education bears any considerable link with ICT or not. Although, there have been a few dissenters to the view that education and ICT bears any significant relation; the present research dispels this view by noticing a positive correlation between education and Internet use from two stand points. First, secondary school enrollment appears to be positively related to Internet use. Second, countries with high scores of PISA test results often show high penetration of Internet.

Correlation between GDP per capita and Internet diffusion was significantly higher than the correlation between secondary school enrollment and Internet diffusion. This leaves room for interpreting that income disparity is the ruling factor responsible for the digital divide. The result emerging from the correlation between educational factor and Internet diffusion is consistent with the views of Le (2010, 84), who maintains that education has little effect on the adoption of digital technologies. However, reader must bear in mind the ease of access to public free Internet in developed countries; opposite results are likely for developing countries where one may only/mostly tie knot with digital technologies for the purposes of higher education.

A possible explanation for a slight degree of correlation be-tween education and ICT might be that modern modes of education themselves encourage the adoption of digital technologies. ICT has been integrated with education on a global scale excluding misfortunate poverty ridden areas. It then follows that in order to continue the education; one has to utilize ICT, which may be the main motivator behind the purchase decision of digital technologies. This finding has important implications for strengthening the education systems particularly in the developing countries. Education and ICT in several developing countries are ridiculously expensive leaving millions of masses behind, for them needs of education and ICT are pushed aside by basic human needs of food and shelter. Poverty once again wins in breeding nuances of low standards of living including considerable portion of the digital divide pie. Governments particularly in the developing regions should therefore concentrate on providing ICT-enhanced education at reasonable costs for masses. Ideally, however, the solution should rest somewhere near providing free education and access to ICT wherever it is feasible.

The results of this study will now be compared to the findings of previous work. Present findings are consistent with earlier re-search (Tipton [7]; Norris [35]; Olaniran and Agnello [8]) which documents income disparity as the principal reason for the digital disparities in the world. While the current findings about positive relation between education and ICT corroborates with Lee [34], they negate with the results of Middleton and Chambers [13] who report that education has no effect on the ICT adoption. This difference urges itself as an evidence of dispute in the research concerning factors responsible for the digital divide.

Turning now to the hypothesis posed at the beginning of the study, it is now possible to state that income factor rules as the leading cause of the digital divide: however, education has slight effect on ICT adoption. Current research appears to validate the assumption that corruption and inequality of income are negatively associated with ICT use. This combination of findings provides some support for the conceptual premise that poverty is mainly responsible for breeding the nuances of the digital divide. An implication of this could be to provide subsidies on ICT related products so that masses can reach and bene-fit from the digital revolution.

Among various lasting divides on accounts of wealth, health and standards of living, the digital divide is becoming increasingly difficult to ignore due to strong ties of ICT with economic growth and wellbeing. Despite massive progress in the digital technologies over the past twenty years, the ICT diffusion re-mains regrettably uneven at a global scale. The underlying threats posed by the digital divide have been globally acknowledged across press and policy discourses, resulting in the production of various publications and action plans. Fortunately, there are good policies by OECD and International Telecommunications Union in the battle against the digital divide; unfortunately, the action plans in these policies are often poorly calibrated across different countries. Among various reasons, one obvious reason for inappropriate action plans by governments lies in the fact that the quantitative nature of the digital divide is significantly overlooked in the literature.

Researchers seem to have routinely confused technology access divide with the digital divide over the span of a decade since the evolution of the digital divide concept in the mid-1990s. The heated debate on the digital divide over the past decade left a considerable room for interpreting the digital divide as a broad and complex phenomenon. Soon it was established that breach in technology access was a small portion of the giant digital di-vide; the need for further research into the digital divide be-came even significant than earlier. Today, the digital divide breaks along multiple fronts ranging from individual to a global scale. This broadly indicates that the digital divide can be best tackled by initiating efforts from multiple ends.

As was noted in the introduction of this paper, poor quantitative understanding of the digital divide seriously impedes governments' abilities to form appropriate frameworks to minimize the digital divide; it becomes increasingly difficult to ignore the poorly understood quantitative aspect of the digital divide. Measurement of the digital divide is a classic problem often noted for deployment of unreliable data in the analysis. Once an adequate quantitative understanding of the subject is established and universally acknowledged, the mission to minimize the digital divide can be crystallized faster than ever before.

Prior research has stressed enough that governments should deploy superior strategy by focusing on weakening the roots of the digital divide rather than just providing access to ICT assuming that the market forces shall eliminate the digital divide over time. Unless governments execute impactful policies, the digital divide shall remain a dilemma for already troubled world economy.

5   Conclusion

5.1   Summary of Findings

This study was undertaken to verify the links of socio-economic indicators concerning GDP per capita, GINI index, education, and corruption with ICT. The present paper has given an account of validated positive links between income along with education and Internet usage. While there has been considerable research on the digital divide, only few studies have attempted to investigate the quantitative nature of the digital divide on a large scale. The doubt in existing quantitative accounts of the digital divide is reinforced by the distaste several researchers have in the data and methodologies used for analysis. Therefore, this study has taken a unique visualization approach with measured degree of covariance to provide a confirmatory evidence of any emerging relation between the factors under current examination.

Paired with literary clues, the visualizations paint a compelling picture in support of the hypothesis posed at the beginning of this study. This study has shown a positive association of income with levels of ICT penetration and a marginal correlation between education and ICT across an array of low-income, middle-income, and high-income countries. Present research has also shown that corruption has is inversely related to Internet usage. The present study confirms the previous findings about income and ICT relation and contributes additional evidence that the increasing GDP per capita determines increasing ICT adoption rate. Among the plausible explanation for this finding is that high purchasing power encourages investment in general. Nevertheless, the results of this study do not support the idea that education and ICT are very highly related with each other. It is, however, not an inalterable rule because in some countries, poor can benefit from free public access to computers and Internet regardless of income/education levels.

The work contributes to existing knowledge in digital divide by strengthening the views of previous research and noting that there exists a dispute in literature regarding the relation of education and ICT. However, low degree of correlation between education and ICT found in this research cannot be extrapolated to the developing world where there is not widespread free public access to ICT and one must purchase the digital technologies for education purposes. The present findings also leave significant room for blaming poverty as the leading cause of the digital divide. It is possible to state that poverty alone breeds a significant proportion of the digital divide; nevertheless other factors such as motivation to adopt ICT, education, forced adoption of ICT due to work requirements, and cultural norms are important to consider when addressing the issue of the digital divide.

The practical implication of these findings is that governments must start efforts on multiple fronts to round up the economic threats posed by the digital divide. On the basis of the evidence currently available, it is possible to suggest that the education in developing countries should be made easily affordable to the poor masses if not totally free. While the poverty breeds the digital divide, the digital inclusion would breed economic development due to enhanced workforce. Overtime, the economic and social benefits of affordable education and ICT shall outweigh the sacrifice in monetary costs by the governments.

Press and policy documents in the name of effective policies against the digital divide shall remain fruitless without pronouncing a decisive aggression against the digital divide on a global scale. The relentless objections to the existing accounts on quantification of the digital divide make it advisable to reconsider the methodology and data used for analysis in future research on the topic. It appears that the crowning success in the battle against the digital divide would require a thorough grasp of both quantitative and qualitative nature of the digital divide. It is a high time to change the course of the digital divide history by initiating impactful efforts on multiple fronts, preferably on determining the mechanism to accurately quantify the digital divide. The digital divide has caused severe havoc to socio-economic lives of millions of people, not to mention its deadly impact on economic footings. Unless there are concrete moves by the governments against the digital divide, the gap between inclusion and exclusion groups shall continue to pose rigorous threats on the world economy.

5.2   Limitations

The findings in this paper are subject to at least two main limitations. First, the chosen data was only from World Bank database in addition to official PISA website and Transparency International website; a wiser approach could have been to perform the statistical analysis with data from different databases and then compare the results. Second, few variables representing education such as PISA results and secondary school enrollments were included in analysis. The present research might have been enriched by including several representative variables of education and ICT diffusion in analysis.

5.3   Future Directions

A natural progression of this work would be to analyze the links tested in this study in the developing world. There is abundant room for future progress in determining the links between ICT and other indicators of the economy. Research questions that could be asked include link of ICT with education at different levels, effect of free public ICT access enters on the relation between income and ICT diffusion, comparing the education and income relations with ICT in OECD member states with developing countries. Future work on exploring such connections with different variables shall help to understand patterns of ICT diffusion from multiple perspectives. Further research might explore the links of cultural influences and cognitive factors with ICT adoption on a broad scale including developing and developed countries. The present slow progress in the digital divide projects a poor quantitative understanding of the subject. Therefore, the future research should concentrate on finding reasonable ways to quantify and measure the digital divide.

References

1. Noh, Y.-H., Yoo, K.: Internet, inequality and growth. J. Policy Model. **30**(6), 1005–1016 (2008)
2. Butcher, M.P.: At the foundations of information justice. Ethics Inf. Technol. **11**(1), 57–69 (2009)
3. Hilbert, M.: The end justifies the definition: the manifold outlooks on the digital divide and their practical usefulness for policy-making. Telecommun. Policy **35**(8), 715–736 (2011)
4. Van Dijk, J.A.: Digital divide research, achievements and shortcomings. Poetics **34**(4–5), 221–235 (2006)
5. Skok, W., Ryder, G.: An evaluation of conventional wisdom of the factors underlying the digital divide: a case study of the Isle of Man. Strateg. Change **13**(8), 423–428 (2004)
6. Billon, M., Lera-lopez, F., Marco, R.: Differences in digitalization levels: a multivariate analysis studying the global digital divide. Rev. World Econ. **146**(1), 39–73 (2010)
7. Tipton, F.B.: Bridging the digital divide in Southeast Asia: pilot agencies and policy implementation in Thailand, Malaysia, Vietnam, and the Philippines. ASEAN Econ. Bull. **19**(1), 83–99 (2002)
8. Olaniran, B.A., Agnello, M.F.: Globalization, educational hegemony, and higher education. Multi. Educ. Technol. J. **2**(2), 68–86 (2008)
9. Brooks, S., Donovan, P., Rumble, C.: Developing nations, the digital divide and research databases. Ser. Rev. **31**(4), 270–278 (2005)
10. Vodoz, L., Reinhard, M., Barbara, G.: Pfister, The farmer, the worker and the MP. GeoJournal **68**(1), 83 (2007)
11. Zhang, X.: Income disparity and digital divide: the internet consumption model and cross-country empirical research. Telecommun. Policy **37**(6–7), 515 (2013)

12. Quibria, M.G., et al.: Digital divide: determinants and policies with special reference to Asia. J. Asian Econ. **13**(6), 811 (2003)
13. Middleton, K.L., Chambers, V.: Approaching digital equity: is wifi the new leveler? Inf. Technol. People **23**(1), 4–22 (2010)
14. Cruz-Jesus, F., Oliveira, T., Bacao, F.: Digital divide across the European Union. Inf. Manage. **49**(6), 278–291 (2012)
15. Stanley, L.D.: Beyond access: psychosocial barriers to computer literacy. Inf. Soc. **19**(5), 407–416 (2003)
16. Cullen, R.: Addressing the digital divide. Online Inf. Rev. **25**(5), 311–320 (2001)
17. Antonelli, C.: The digital divide: understanding the economics of new information and communication technology in the global economy. Inf. Econ. Policy **15**(2), 173 (2003)
18. Qingxuan, M., Mingzhi, L.: New economy and ICT development in China. Inf. Econ. Policy **14**(2), 275 (2002)
19. Menou, M.J.: The global digital divide; beyond hICTeria. In: Aslib Proceedings. MCB UP Ltd (2001)
20. Mansell, R.: Constructing the knowledge base for knowledge-driven development. J. Knowl. Manage. **6**(4), 317–329 (2002)
21. Dewan, S., Frederick, J.R.: The digital divide: current and future research directions. J. Assoc. Inf. Syst. **6**(12), 298–336 (2005)
22. Drouard, J.: Costs or gross benefits? - What mainly drives cross-sectional variance in Internet adoption. Inf. Econ. Policy **23**(1), 127 (2011)
23. Edwards, Y.D.: Looking beyond the digital divide. Fed. Commun. Law J. **57**(3), 585–592 (2005)
24. Moss, J.: Power and the digital divide. Ethics Inf. Technol. **4**(2), 159 (2002)
25. Mordini, E., et al.: Senior citizens and the ethics of e-inclusion. Ethics Inf. Technol. **11**(3), 203–220 (2009)
26. Couldry, N.: New media for global citizens? The future of the digital divide debate. Brown J. World Aff. **14**(1), 249–261 (2007)
27. Tavani, H.T.: Ethical reflections on the digital divide. J. Inf. Commun. Ethics Soc. **1**(2), 99–108 (2003)
28. DiMaggio, P., et al.: Social implications of the internet. Annu. Rev. Sociol. **27**, 307–336 (2001)
29. Katz, J.E., Rice, R.E., Aspden, P.: The Internet, 1995-2000: access, civic involvement, and social interaction. Am. Behav. Sci. **45**(3), 405–419 (2001)
30. Fuchs, C., Horak, E.: Africa and the digital divide. Telematics Inform. **25**(2), 99–116 (2008)
31. Pick, J.B., Nishida, T.: Digital divides in the world and its regions: a spatial and multivariate analysis of technological utilization. Technol. Forecast. Soc. Chang. **91**, 1–17 (2015)
32. Cooke, L., Greenwood, H.: "Cleaners don't need computers": bridging the digital divide in the workplace. Aslib Pro. **60**(2), 143–157 (2008)
33. Pittman, J.: Converging instructional technology and critical intercultural pedagogy in teacher education. Multicultural Educ. Technol. J. **1**(4), 200–221 (2007)
34. Lee, J.-W.: The roles of demographics on the perceptions of electronic commerce adoption. Acad. Mark. Stud. J. **14**(1), 71–89 (2010)
35. Norris, P.: Digital Divide: Civic Engagement, Information Poverty, and the Internet Worldwide. Cambridge University Press, Cambridge (2001)
36. Van Dijk, J., Hacker, K.: The digital divide as a complex and dynamic phenomenon. Inf. Soc. **19**(4), 315–326 (2003)
37. Kyriakidou, V., Michalakelis, C., Sphicopoulos, T.: Digital divide gap convergence in Europe. Technol. Soc. **33**(3–4), 265–270 (2011)

38. Vicente, M.R., López, A.J.: Assessing the regional digital divide across the European Union-27. Telecommun. Policy **35**(3), 220–237 (2011)
39. Rao, S.S.: Bridging digital divide: efforts in India. Telematics Inform. **22**(4), 361–375 (2005)
40. Schleife, K.: What really matters: regional versus individual determinants of the digital divide in Germany. Res. Policy **39**(1), 173 (2010)
41. Douglas, H.: The rural-urban digital divide. J. Mass Commun. Q. **77**(3), 549–560 (2000)
42. Novo-Corti, I., Varela-Candamio, L., García-Álvarez, M.T.: Breaking the walls of social exclusion of women rural by means of ICTs: The case of 'digital divides' in Galician. Comput. Hum. Behav. **30**, 497–507 (2014)
43. Bach, M.P., Zoroja, J., Vukšić, V.B.: Determinants of firms' digital divide: a review of recent research. Procedia Technol. **9**, 120–128 (2013)
44. Graham, S.: Bridging Urban digital divides? Urban polarisation and information and communications technologies (ICTs). Urban Stud. (Routledge) **39**(1), 33–56 (2002)
45. Peng, G.: Critical Mass, diffusion channels, and digital divide. J. Comput. Inf. Syst. **50**(3), 63–71 (2010)
46. Burkhard, R.A.: Learning from architects: complementary concept mapping approaches. Inf. Visual. **5**(3), 225 (2006)

Sentiment Analysis of Products' Reviews Containing English and Hindi Texts

Jyoti Prakash Singh[1]([✉]), Nripendra P. Rana[2], and Wassan Alkhowaiter[3]

[1] National Institute of Technology Patna, Patna, Bihar, India
jyotip.singh@gmail.com
[2] School of Management, Swansea University, Swansea SA2 8PP, UK
n.p.rana@swansea.ac.uk
[3] Qassim University, Buraidah, Saudi Arabia
w.alkhowaiter@gmail.com

Abstract. The online shopping is increasing rapidly because of its convenience to buy from home and comparing products from their reviews written by other purchasers. When people buy a product, they express their emotions about that product in the form of review. In Indian context, it is found that the reviews contain Hindi text along with English. It is also found that most of the Hindi text contains opinionated words like *bahut achha, bakbas, pesa wasool* etc. We have tried to find out different Hindi texts appearing in product reviews written on Indian E-commerce portals. We have also developed a system which takes all those reviews containing Hindi as well as English texts and find out the sentiment expressed in that review for each attribute of the product as well as a final review of the product.

Keywords: Sentiment analysis · POS-Tagging · Review analysis · Product summarization

1 Introduction

The life style of society is changing with the penetration of Internet, and E-commerce in every corner of the world. Earlier, the advertisement and friends recommendations were a major source of information while buying a product. The number of recommendations was a limited one to compare similar products of different brands. Nowadays, as the e-commerce business has grown up, they are offering more products. The e-commerce websites also request their customers to write their experience about the product they brought in the form of a product review. These reviews offer significant information to buyers about the product they are planning to buy and also enable them to compare products of different brands. The reviews help consumers to choose the best products by comparing them based on other consumers' evaluation of the products. It also aids in the improvement of the product by informing the manufacturers about the advantages and defects of their products. The number of reviews about the

© IFIP International Federation for Information Processing 2015
M. Janssen et al. (Eds.): I3E 2015, LNCS 9373, pp. 416–422, 2015.
DOI: 10.1007/978-3-319-25013-7_33

products grows with the growth of e-commerce businesses. It becomes very difficult for buyers and sellers to manually analyze a large number of reviews and get any meaningful information. This attracts a lot of researchers to automate the analysis of reviews and get valuable information hidden in the reviews [3,5].

The reviews written by Indian buyers are mainly in English, but it contains some Hindi texts (written in English Scripts only) also as Hindi is a prevalent language in India. Some of the most widely used Hindi words like *bahut achha, bakbas, pesa wasool* are found in a number of reviews. Most of these words are opinionated and contain strong opinions in the form of *good* or *bad*. Most of the earlier work done in the area of finding polarity of opinions for product reviews neglect these texts as they are mainly developed for English texts only. As per the best of our knowledge, no work has yet been reported which consider the correction of these typos and includes the sentiment of the Hindi words along with English texts.

In this work, we have proposed a sentiment analysis system which works for reviews containing both English as well as Hindi opinionated texts. First of all we have gathered possible Hindi opinionated texts from reviews appearing on Indian popular E-commerce sites such as *amazon.in, flipkart.com, snapdeal.com, shopclues.com* and so on. These Hindi texts are preprocessed and their equivalent English words are found. The summarized review of the product is then calculated consulting sentiwordnet database.

The rest of the paper is organized as follows: The proposed system architecture and algorithm are discussed in Sect. 2. In Sect. 3, we present our results and finally in Sect. 4, we conclude the paper.

2 Proposed Work

Product reviews contents from popular Indian e-commerce sites like *amazon.in, flipkart.com, snapdeal.com, shopclues.com* are collected as our dataset. The dataset has a lot of typos in the form of joint words like *verygood* as well as abbreviations containing numerals such as *gr8* for *great*. The dataset also contains Hindi words like *"bakbas", "bekar", "achchha"* (written in English script only). Some sample typos gathered from various Indian e-commerce websites are given in Tables 2 and 3. Table 1 contains words of Hindi Texts typed in English along with their English equivalent text. Table 2 contains some popular abbreviations used online for review, chatting, etc. along with their correct form in English. Some joint words (missing space) are shown in Table 3.

One of the primary focuses of this work is to pre-process the product review available on Indian E-commerce sites so that the reviews contain only English text. Once reviews are converted to English text, Part of Speech (POS) Tagging to the text is done using wordnet [4] database. Once POS tagging is done, the adjective, noun, and adverb are extracted. Further, sentiwordnet [1,2] database is used to assign numerical values to the adjectives contained in the review. The proposed system architecture is shown in Fig. 1.

Table 1. List of wrong words in Hindi

Wrong words	Corrected words
Ye	This
Achha	Good
G8t	Great
N8t	Night
H	Is
Som1	Some one

Table 2. List of wrong words in English

Wrong words	Corrected words
Gud	Good
Gooood	Good
Exclent	Excellent
Bd	Bad
Awesm	Awesome

A pseudo code for our proposed system is given below.

Proposed Algorithm:

Step 1: Tokenize based on space.

Step 2: Consult wordnet

      If word is matched, then go to POS tagger.

      else correct word and go to POS Tagger

Step 4: Noun, Adverb, and Adjective are stored in frequent feature database.

Step 5: Generate the product summary with the help of SentiWordNet Lexical databases.

The working of our scheme is traced with the aid of an example presented here. *Yeh achha camera h. eski pictre quality bahut achhi hai. The pics resolution is enough. Zoom is bakbas. focus is verygd bt not g8t.*

The document (Complete review) is broken down into several sentences based on [.], [?], And [!] Mark. For example review, sentences are:

S1. Yeh achha camera h.

S2. eski pictre quality bahut achhi hai.

S3. The pics resolution is enough.

S4. Zoom is bakbas.

S5. focus is verygd bt not g8t.

Yeh is a Hindi word whose English equivalent is *This. Achha* is another Hindi word meaning *good* in English. The complete review is written in English after correcting and converting every word to English.

Table 3. List of joint words

Jointed words	Corrected words
Verygood	Very good
Verybad	Very bad
Bahutbura	Bahut bura

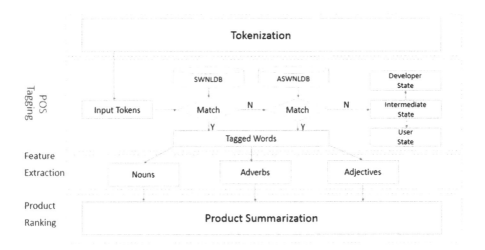

Fig. 1. Proposed system architecture

s1. This is a good camera.
s2. Its picture quality is very good.
s3. The picture resolution is enough.
s4. Zoom is bad.
s5. Focus is very good but not great.

The POS tagging is applied as given below for just one sentence. We have used Penn Treebank tagset for Part of Speech Tagging

```
This========  [ This_DT ]
Good========  [ Good_JJ ]
Camera======  [ Camera_NN ]
Is=========   [ Is_VBZ ]
. ==========  [._UH ]
```

Next step is to consult the sentiwordnet database to find the priority of the adjective to find the sentiment value of the sentence. The score of every adverb and adjective are given in the SentiWordNet lexical database. We have listed here some of them given in Table 4 which are going to be used in the above example.

Table 4. Score list of words

Word	Orientation of word	Score
Good	Positive	.75
Great	Positive	.875
Awesome	Positive	.875
Excellent	Positive	1
Well	Positive	.75
Average	Positive	.375
Enough	Neutral	.875
Bad	Negative	.65
Very	Nil	.5
Not	Negation	−1

Table 5. Sentence wise score

Sentence number	Score	Score type
S1	0.750	Positive
S2	1.250	Positive
S3	0.125	Positive
S4	0.625	Negative
S5	0.275	Positive

Where nil represents the neither positive nor negative orientation of words. And negation represents the multiplier factor which having −1 value.

For above example, sentence s1 has good as adjective whose sentiment score is +0.75. In this sentence there is no adverb or negation, so the sentiment score for sentence s1 is +0.75. For second sentence s2, adverb *(very)* is there with the adjective *(good)*, so the sentiment score of s2 is 1.25, which is a sum of scores of good (0.75) and very (0.5). The sentiment score of each sentence is shown in Table 5.

The polarity of the review of a product is determined by finding the polarity of each feature of the product across all reviews and finding a weighted sum of all features.

3   Result

We have collected 1100 reviews from *flipkart.com* and *amazon.in* of three popular mobile brands in India at the time of writing this paper. The results show that for Android based smart-phone people are talking about features like *camera, battery, memory, processor, RAM, display, price, weight and phone*. Out of these features, battery, camera and display are found to be more prominent ones across

Table 6. Summary of the review of product **Samsung Galaxy S3 Neo, Asus Zenfone 2** and **Honor 4X**

Phone feature		Galaxy S3 Neo	Zenfone 2	Honor 4X
Feature Name	*Scores Type*	*Percentage(%)*	*Percentage(%)*	*Percentage(%)*
Camera	Positive	100	66.39	60.17
	Negative	0.0	33.64	39.82
Battery	Positive	61.76	90.82	66.09
	Negative	38.23	9.18	33.9
Memory	Positive	100	100	49.24
	Negative	0.0	0.0	50.76
Processor	Positive	100	100	49.25
	Negative	0.0	0.0	50.74
RAM	Positive	100	100	48.24
	Negative	0.0	0.0	51.75
Display	Positive	90.5	0.0	100
	Negative	9.5	100	0.0
Price	Positive	0.0	No opinion	No opinion
	Negative	100	No opinion	No opinion
Weight	Positive	100	0.0	No opinion
	Negative	0.0	100	No opinion
Phone	Positive	55.07	89.09	62.02
	Negative	44.93	10.91	37.98
Overall	**Positive**	**68.86**	**75.14**	**62.60**
	Negative	**31.134**	**21.15**	**32.26**

all phones. The results are shown in Table 6. *no opinion* in both positive and negative rows shows that no one has given any opinion about that feature of that product.

4   Conclusion

We have designed a sentiment analysis system which can take reviews written in Hindi as well as English texts and find the sentiment of customers for that product. We have taken a dictionary based approach to correct the wrong words and replace Hindi text with their English equivalent. We further want to extend this system with a machine learning algorithm to correct the wrong words and Hindi words. The final opinion score is a weighted average of all the features of the product under consideration. We are also working to identify the most prominent features of a product to calculate the final opinion score.

References

1. Baccianella, S., Esuli, A., Sebastiani, F.: Sentiwordnet 3.0: An enhanced lexical resource for sentiment analysis and opinion mining. In: LREC, vol. 10, pp. 2200–2204 (2010)
2. Esuli, A., Sebastiani, F.: Sentiwordnet: a publicly available lexical resource for opinion mining. In: Proceedings of the 5th Conference on Language Resources and Evaluation (LREC 2006), pp. 417–422 (2006)
3. Hu, M., Liu, B.: Mining and summarizing customer reviews. Proceedings of the Tenth ACM SIGKDD International Conference on Knowledge Discovery and Data Mining. KDD 2004, pp. 168–177. ACM, New York (2004)
4. Miller, George A.: Wordnet: a lexical database for english. Commun. ACM **38**(11), 39–41 (1995)
5. Popescu, A.-M., Etzioni, O.: Extracting product features and opinions from reviews. Proceedings of the Conference on Human Language Technology and Empirical Methods in Natural Language Processing. HLT 2005, pp. 339–346. ACL, Stroudsburg (2005)

Adaptive Normative Modelling: A Case Study in the Public-Transport Domain

Rob Christiaanse[1,2], Paul Griffioen[1], and Joris Hulstijn[1(✉)]

[1] Delft University of Technology, Delft, Netherlands
{p.r.griffioen,j.hulstijn}@tudelft.nl
[2] EFCO Solutions, Amsterdam, Netherlands
r.christiaanse@efco-solutions.nl

Abstract. Data analytics promises to detect behavioral patterns, which may be used to improve decision making. However, decisions need to motivated, and they are often motivated by models. In this paper we explore the interplay between data analytics and process modeling, specifically in normative settings. We look specifically at value nets, mathematical models of the flow of money and goods, as used in accounting. Such models can be used to analyze the proportions of various flows, such as resources consumed and products produced. Such analyses can be used in the planning and control cycle, for forecasting, setting a budget, testing and possibly adjusting the budget. In other words, it can be used for adaptive normative modeling. We look in particular at a case study of a provider of public transport services for school children. The case shows that the use of value nets for analysis of proportions is (i) feasible, and (ii) useful, in the sense that it provides valuable new insights about the revenue model.

Keywords: Data analytics · Business process modeling · Normative systems

1 Introduction

Data analytics is heralded as the new big thing [5]. People, services, goods or processes leave traces, producing vast amounts of data, which can be analyzed. Some have claimed about 'Big Data' that the sheer volume of data and the power of the analytical tools would remove the need for scientific method [2]. Patterns in behavior emerge which can be made actionable, i.e., they can used to make better decisions. We disagree. Decisions that matter must be motivated. In many application domains, professional decisions are based on models. In particular, this is true in normative settings, where situations are compared to norms or standards, which can be either based on practices and conventions, or on formal rules and regulations. In such normative settings, the model has become a reference model or norm. For instance, in compliance checking, a 'de jure' model, based on the law, is compared with a 'de facto' model, based on data about the real situation [10].

© IFIP International Federation for Information Processing 2015
M. Janssen et al. (Eds.): I3E 2015, LNCS 9373, pp. 423–434, 2015.
DOI: 10.1007/978-3-319-25013-7_34

In this paper we explore the interplay between computational models, on the one hand, and data analytics tools on the other hand for decision making in a very specific task and domain. In particular, we focus on the planning and control cycle, that involves forecasting, setting a target, and comparing results to the target [12]. Clearly, this setting is normative and the decisions are guarded by a set of expectations and norms. In this setting there is also a challenge: data analytics is essentially empirical; it is exploratory and starts from the available data. Analyzing models in a computational setting, on the other hand, is typically based on theory, generalizations and abstractions. The challenge is to combine the power of data analytics, with the justification of models for decision making. Moreover, as the world changes, it is likely that models need to be adjusted. How can models be adapted? In other words: we need *adaptive normative modeling*. In Sect. 2 we argue that value nets model the essential process flow ratios that underly the numerical relation between data analysis and process modeling.

Value nets can be viewed as instantiations of business process models useful for process diagnosis and norm analysis. Norm analysis is a foundational notion buttressing the planning and control cycles within organizations. Just as data analytics, business process modeling is interested in finding and describing structural patterns in data. In this sense structural means that the decision structure is explicitly captured. Value nets provides in a computational framework for the modeling and analysis of business processes that focuses on the normative ratios in enterprise behavior for the purpose of quality management, quality assurance, and measurement. Here do data-analytics and value nets modeling meet.

In Sect. 3 value nets will be modeled as extended Petri Nets that capture the flow of values in an enterprise. From this model an enterprise's normative cyclic behavior can be computed as so called tours. These tours contain the ratios of the process events can be factored into what Ijiri calls causal chains of events. Section 4 shows how the causal chains can be computed by factoring the value net's incidence matrices. In the case study in Sect. 5 it is shown how this is used to interpret the data for planning and control.

The case study focuses on the planning an control cycle of a public transport provider, providing schoolbus services to schoolchildren. The governance setting is similar to that described in [6]. This is a highly regulated domain; compliance plays an important role. As you can imagine, this involves a lot of data about transport movements, which may contain many hidden pattens of behavior. Some of these patterns are relevant to the auditing task: are the invoices legitimate? Other patterns are more relevant to management, when they want to improve efficiency. We have access to this data.

The research can be positioned as part of Business Process Management (BPM) [4,7]. Here, we observe the same contrast in attitude between modeling (forward looking; engineering) and data analytics (backward looking; empirical). Related to this, we made three observations about the way BPM has developed, which make it less suitable for the management accounting domain. First, much of the research effort in BPM has traditionally been focused on process models: specification and verification of formal properties, such as termination or

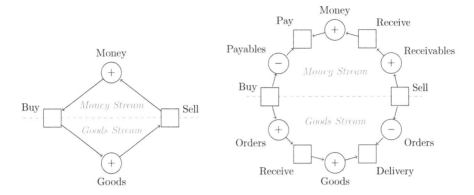

Fig. 1. Value Nets for trading companies. The horizontal dashed line splits the model into the *money flow* in the upper part and the *goods flow* in the lower part.

conformance to rules. This is done at design time. What has sometimes been lacking, is a look at real data, after execution, to see whether these generalizations apply in practice. Second, BPM models, in particular those based on work-flow, typically look at individual cases. However, in accounting and management, we are more interested in aggregated data: how many cases of a certain type per month? What was the use of resources? This aspect is also taken on by process mining. Third, BPM models tend to focus on the existence of dependencies between process steps or activities; we are just as much interested in the amount of value or the number of objects flowing along these links. In other words, we are not only interested in the control flow, but rather in the size of the flows.

The contribution of this paper is a computational framework for the modeling and analysis of business processes that focuses on the normative ratios in enterprise behavior for the purpose of quality management, quality assurance, and measurement. These normative ratios provide in the de jure models which can be compared with the de facto models extracted from the data using data analytics and vise versa.

2   Business Process Modeling and Data Analytics

The price mechanism ensures that a market price of a good and or service accurately summarizes the vast array of information held by market participants [3]. It needs no elaboration that the information aggregation characteristic of the pricing system buttresses many theories about the communicative function of prices and the decisions participants in the marketplace make to exploit business opportunities in the creation of value.

Management of enterprises decide upon a business model that depicts the transaction content, structure, and the governance designed so as to create value through the exploitation of business opportunities" [1]. The transaction content refer to goods, services or the information exchanged, where the transaction structure defines the way parties i.e. agents participate in the exchange and how they are (inter)linked. Transaction governance refers to the legal form of organization, and to the incentives for the participants in transactions. Hence data, processes and governance are intertwined notions which are strongly related to each other.

More specifically a contract is an instantiation of a business model. Consequently a contract depicts the agreed upon content, structure, the incentives and the rules of conduct among parties involved in the contract. In the case a contract is executed the buyer receives goods or services from the vendor in the agreed upon quality and the buyer pays the vendor for the agreed upon price coined as value. Actually aforementioned paying mechanism is often referred to as the revenue model of the vendor. More specifically a revenue model refers to the specific modes in which a business model enables revenue generation.

Economic transactions as described above can be formally modeled as a value cycle which is interlinked to the business model of the enterprise i.e. the revenue model. This view is inspired by accounting models in the owner ordered accounting tradition [16] and value chain theory [13]. When used in computer science, often the purpose of these models is to analyze the representations of actions and events in a business process, and study their well-formedness. Consequently an intra-organizational workflow i.e. business processes can be modeled as a value net which provides a top-level view of an enterprise that focuses on the economic events equivalent to the value cycle of an enterprise. Figure 1 shows an elementary and a more elaborate example of a value cycle for a trading company. Each event is a transfer of value.

3   Value Cycle and Value Nets

A value net is modeled with a dimensioned Petri Net, extended with a place sign and a valuation, with some special structural characteristics that make it a good representation of the intra-organizational structure. Figure 2 shows the value cycle for the case study from the next sections.

Definition 1. *A Value Net is a tuple* (P, T, F, B, s, v) *with*

- *P a finite collection of places,*
- *T a finite collection of transitions,*
- *$F, B : P \times T \to \mathbb{N}$ the net's incidence matrices*
- *$s : P \to \{asset, liability\}$ the indicator for each place's sign, and*
- *$v : P \to \mathbb{R}$ a valuation.*

Additionally a value net has the following structural properties:

- The net is cyclic in the following sense. There is a special place labeled *money* from which it is possible to reach every other node in the model and that can be reached from every node in the model. Put differently, every node in the model is on a path from *money* to *money*.

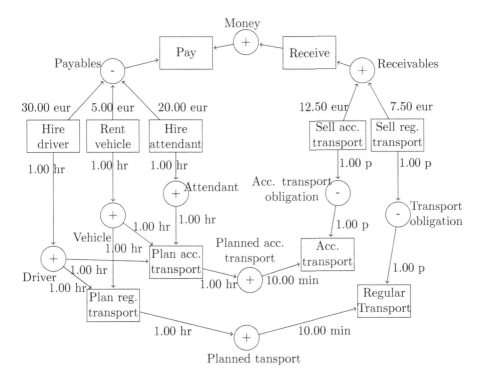

Fig. 2. Value Net for a transport provider.

– The net is divided in a money part and a production part. Place *money* is in the money part. The transitions in which the input places are in the money part and the output places in the production part are called *buy* transitions. The transitions in which the input places are in the money part and the output places in the production part are called *sell* transitions.

We use subscripts to differentiate the money and the production part. So, partitioning $P = P_{mon} \cup P_{prod}$ splits the places into the money part and the production part. Transitions in T_{mon} are between places the money part and transitions in T_{prod} between places in the production part. A T_{buy} transition consumes value from the money part and produces value in the production part. A T_{sell} transition consumes value from the production part and produces value in the money part.

A value net's semantics follows from the Petri Net semantics. A marking m is a $P \rightarrow \mathbb{N}$ vector that denotes the number of tokens in each place. A transition vector t is enabled when $F \cdot t \leq m$. When it fires it consumes $F \cdot t$ tokens and it produces $B \cdot t$ tokens. Let flow matrix G be defined by $G = B - F$. The effect of transition vector t is then the marker change computed as the matrix-vector product $G \cdot t$.

A value net extends a Petri net with a valuation, because for the computation of (normative) behavior in workflows it is essential to distinguish in monetary units and in product units. Just working with values leads to information loss, because value is the product of a valuation and a quantity.

$$\text{Value} = \text{Valuation} \times \text{Quantity}$$

A value net in monetary units has the advantage that everything is commensurable. It is even a requirement for the balancing of events in double bookkeeping [8,11,14]. However, it is unsuitable for conservation laws because the value depends on a valuation that fluctuates. As soon as money is exchanged for a good this good gets exposed to price fluctuations. Only when the good is exchanged into money again later the effect on value becomes fixed. Exact connections in the product part is only feasible in product units. Therefore it is necessary to split values into valuations and quantities and model a business model in product units. A valuation is a linear property of a marking with a monetary unit of measurement. It gives the total economic value of the tokens in a dimensioned net. A weight vector v contains the value of a token for each place. Given a marking m the inner product $\langle v, m \rangle$ is the total value.

The other extension is the distinction between assets and liabilities. Each place in a value net represents either a positive or a negative value. In a value net the positive values flow anti-clockwise and the negative values flow clockwise. In many computations it is convenient to have all value flow in the same direction, and in these case we use an unsigned value net. Let matrix G be the flow matrix of a value net and let matrix S be a diagonal matrix with $+1$ for positive and -1 for negative places, then matrix $U = S \cdot G$ is the *unsigned flow matrix*, and the corresponding value net is called the *unsigned value net*. In an unsigned net the edges connected to a negative place are switched with respect to the signed net. An unsigned value net produces the same outcome but with signs switched. Multiplying with sign matrix S again gives the identical outcome.

4   Causal Chains of Events and Reconciliation Relations

A tour relates the behavior of an enterprise to the cyclic structure of its value net i.e. a process model as depicted in Fig. 2. The cyclic structure is not directly apparent in behavior because transitions can fire at any point in the cycle. We want to group transactions that make up what Ijiri calls a causal chain of events [11]. In general terms such a chain of events is a variation on buying products and resources, producing an end product, and selling the product. The cycle starts with the consumption of money and ends with the production of money. Such cyclic behavior follows from the cyclic structure of a business process.

A formal notion of the tour concept was introduced for an audit context in [9]. It defines a tour as a constellation of events whose total effect is on money only. All other produced tokens are at some point consumed by another step in the process.

Definition 2. *A transition vector t with firing result $m = G \cdot t$ is a tour when $m[\text{money}] > 0$ and $m[x] = 0$ for any place $x \neq$ money.*

Place	$F_{\text{buy}} \cdot t$	$B_{\text{buy}} \cdot t$	$C \cdot t$	$F_{\text{sell}} \cdot t$	$B_{\text{sell}} \cdot t$
payables	3500 c	-	-	-	-
vehicle	-	3600 s	-	-	-
driver	-	3600 s	-	-	-
planned reg. transport	-	-	3600 s	-	-
reg. transport obligation	-	-	-	6 p	-
receivables	-	-	-	-	4500 c

Place	$F_{\text{buy}} \cdot t$	$B_{\text{buy}} \cdot t$	$C \cdot t$	$F_{\text{sell}} \cdot t$	$B_{\text{sell}} \cdot t$
payables	5500 c	-	-	-	-
vehicle	-	3600 s	-	-	-
driver	-	3600 s	-	-	-
attendant	-	3600 s	-	-	-
planned acc. transport	-	-	3600 s	-	-
acc. transport obligation	-	-	-	6 p	-
receivables	-	-	-	-	7500 c

Fig. 3. Factored tour for the transport case from Fig. 2. From left to right the table shows the sacrificed value, the purchased products, the intermediate products, the produced products and the received value.

Vector t is the number of times an events has to occur per tour. Every time such a tour has occurred the amount of money increases by $m[\text{money}]$. This increase is called the value jump. This formal tour concept captures the essence of the concept of value form a enterprise point of view.

With the unsigned value net we can reveal the various parts of a tour by factoring tour result $S \cdot U \cdot t$.

$$U = U_{\text{mon}} + U_{\text{buy}} + U_{\text{prod}} + U_{\text{sell}} \tag{1}$$

If we also split matrix U into $B - F$ to see the difference between token production and token consumption then we have the factoring to express connections between the value cycle's parts. Let t be a tour. Unsigned tour result $U \cdot t$ is factored in a causal chain as follows

$$(B_{\text{mon}} + B_{\text{buy}} + B_{\text{prod}} + B_{\text{sell}} - F_{\text{mon}} - F_{\text{buy}} - F_{\text{prod}} - F_{\text{sell}}) \cdot t \tag{2}$$

The various terms in this expression are the input and output of the different steps in the causal chain.

Grouping the money part and the production part immediately give the following equations:

$$(B_{\text{buy}} + B_{\text{prod}} - F_{\text{prod}} - F_{\text{sell}}) \cdot t = 0 \tag{3}$$

$$(B_{\text{sell}} + B_{\text{mon}} - F_{\text{mon}} - F_{\text{buy}}) \cdot t = G \cdot t \tag{4}$$

The matrix in $(B_{\text{sell}} + B_{\text{mon}} - F_{\text{mon}} - F_{\text{buy}}) \cdot t$ is exactly the flow matrix filtered for the production places and does not contain the money place. Because t is a tour, the product must therefore be zero according to Definition 2. The second equation then follows immediately because the tour result $G \cdot t$ must result from the other parts of Eq. 2.

The first equation is Kirchhof's law for the production process. It shows that in a tour the input and output of production cancel. The second equations shows the value jump. It is the difference between monetary value produced and consumed.

Figure 3 shows the chains of events for the transport service provider from the case study. In this case there are two tours; one for regular transport, and one for accompanied transport. The case study focuses on the tour for regular transport.

5   Case Study: Public Transport Services

In 2010 a transport provider has agreed to provide so called taxi-bus transport services for secondary school pupils, who need to attend special schools, which are at a distance. For transport of this group of passengers, many rules and regulations apply. In the request for proposals, the municipalities have laid down a number of requirements for the contract, concerning maximal waiting times, durations, and routing. Every month, the transport provider sends an invoice with a data file, detailing the number of trips, passenger details, routes, departure and arrival times, departure and arrival addresses, ordered and canceled trips etc. This data file provides evidence of the services delivered for the invoice to be paid by the municipalities.

The contract specifies the following revenue model. Hence a revenue model refers to the specific modes in which a business model enables revenue generation. Based on trip requests received from the municipality specifying travel specifications from A to B at time t, at week day w, the routing software package used by the transport provider calculates the daily routes at each time of day, entailing trip requests i.e. the number of school pupils per route. Parties agreed that the routing software package is used to predict the best route and duration, for the sake of the contract. This prediction is based on standard speeds (100 km/h for motorways, 70 km/h for main roads, 40 km/h in town etc.). For each route, this gives occupied vehicle hours, which is priced at a certain tariff: revenue per route = price per occupied hour * occupied vehicle hours. This completes the description of the revenue model.

The price per occupied hour is the result of a normative decision made by management of the transport provider offering their quote in competing for the contract in 2010. No need to say that the quotation is a best estimate of expected outcome for providing future transport services. Within the context of the planning and control cycle of the transport provider the actuals i.e. the realized data are gathered on a daily basis for monitoring purposes. Deviations to the norm are very interesting from an operational perspective because operational management learns about the quality of planning operations and more over whether the norms used in calculating the price per occupied hour is correct. From a

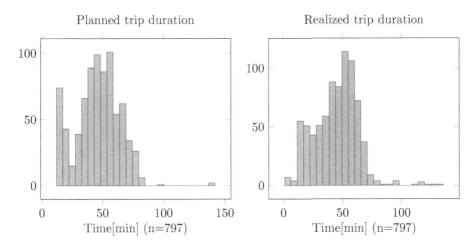

Fig. 4. Planned and realized trip durations. The average planned duration is 46.42 min, and the average realized duration is 45.10 min.

financial perspective the same information is very valuable in assessing whether the contract is profitable or not and learns the quality of the normative decision processes made by management.

In Fig. 2 the original value net buttressing the initial price bid is depicted. Factorization of the tour of the transport provider gives us the causal chain of events representing the normative representation of the inflows and outflows of business processes of the transport provider, revealing their normative ratios i.e. KPIs as input measures for analyzing deviations to the normative ratios. Hence on the right hand side we see that the sell of regular transport i.e. one trip from A to B at time t, at week day w, leads to a transport obligation. This transport obligation leads to planned transport measured in one hour. The value net shows that one trip equals 10 min drive so the occupancy rate is easily derived from these ratios. On average six trips make up one planned vehicle route. In general the occupancy rate of a vehicle is a key performance indicator for analyzing operational planning effectiveness, analyzing profitability of the contract at hand, and for most whether norms actually used in calculating pricing bids are sound. We refer to Fig. 3 for detailed exposition of the factored tours. Note that the value net is the instantiation of the revenue model of the contract and reveals the value creation within organizations.

We have access to the data detailing the number of trips, passenger details, routes, departure and arrival times, departure and arrival addresses, ordered and canceled trips etc. The period we studied comprised one month period. In Table 1 we give some preliminary figures of the data file en the computational results applying the value net logic as explained in section three and four on the planned data and realized data in that month. The occupancy rate is the reciprocal of the KIP, converted to hours.

The occupied vehicle hour planned and realized are depicted in Fig. 4. The average per route differs only slightly from each other, here 1.32 min. When we

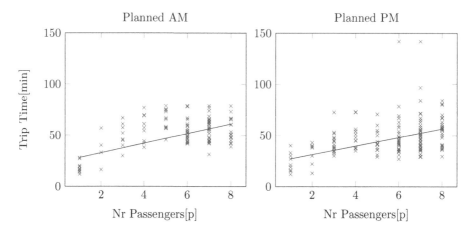

Fig. 5. Scatter plots for the planned trip durations in the morning (AM) and in the afternoon (PM).

Table 1. Comparison between planned and realized route durations, in the morning and in the afternoon.

	Planned AM	Planned PM	Realized AM	Realized PM
Mean	48.46 min	47.36 min	43.88 min	42.28 min
Std dev	17.57 min	19.06 min	18.00 min	16.83 min
KPI	9.07 min/p	8.86 min/p	8.73 min/p	8.41 min/p
Occupancy	6.62 p/hr	6.77 p/hr	6.87 p/hr	7.13 p/hr

combine this insight with the KPI and the occupancy rate then we may conclude from a financial and operational perspective that we have met our objective(s). Nothing has to be done. But if we look more closely to the data and make a more detailed analysis by splitting up the data in routes performed in the morning and in the afternoon than we must revise our opinion. The morning routes are more problematic as we thought. The overall performance in the afternoon is certainly better compared with the morning routes. The value net analysis gives us a different tour results. In this case financial management may still be content with the results but operations has now a motivation to look into the data asking why the overall performance in the afternoons is better than the morning routes. If we take a closer look at the planned data AM depicted in Fig. 5 and compare them with the realized data AM depicted in Fig. 6 than we see that the scatter plot of the KPI minutes per person per route is quite informative. In the case we compare these data with the planned and realized data PM than there are two things that are noteworthy. First we see that the morning routes show similar patters as the afternoon. After analyzing these data it turned out that traffic jams caused delays in the morning. Secondly after analyzing daily patterns it turned out that only two days in the week caused real delays. The planning

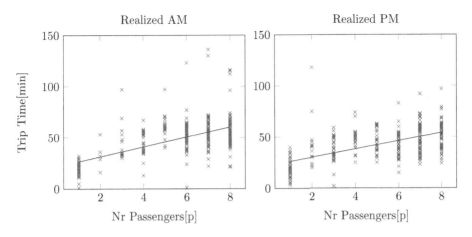

Fig. 6. Scatter plots for the realized trip durations in the morning (AM) and in the afternoon (PM).

department did not optimize routes for each route per day part. Rationally we would expect that the schedules were altered. In practice the schedules were kept on a weekly (average) basis. The value net analysis revealed that when we analyze the tours that actually one trip contains the following information: a school pupil travels from A to B at time t, at week day w. So the unit of analysis is a weekday, at time t, per school pupil which gave the correct information.

6   Conclusion

In the introduction we mentioned three aspects of BPM, which make it less suitable for the financial domain: (i) little empirical verification of applicability of models, (ii) focus on individual cases, instead of aggregate flows, (iii) focus on causal links, neglecting how much value or how many goods flow over the links.

There is an analogy between our approach and process mining [15]. Process mining also combines the exploratory and empirical aspects of data mining, with the use of linear models of processes, to analyze their formal properties. Process mining can – to a certain extend – also deal with the size of flows and use of resources. However, what it can't do is express the amount of value or number of object that flow or should flow over a link. That means that in normative settings, also process mining is left to analyze whether a process follows the procedure; it can't identify deviations in the (financial) content of transactions.

In Sects. 3 and 4 we elaborated on a computational approach for analysis of business processes models that focuses on the normative ratios in enterprise behavior for the purpose of quality management, quality assurance, and measurement. As shown in the case analysis the value net approach provides in the de jure models which can be compared with the de facto models extracted from the data using data analytics and vice versa. This approach seems very fruitful for modeling adaptive mechanisms buttressing planning and control cycles within organizations.

References

1. Amit, R., Zott, C.: Value creation in E-Business. Strat. Manag. J. **22**(6–7), 493–520 (2001)
2. Anderson, Chris: The end of theory: the data deluge makes the scientific method obsolete. Wired Mag. 16(7) (2008)
3. Atakan, A.E., Ekmekci, M.: Auctions, actions, and the failure of information aggregation. Am. Econ. Rev. **104**(7), 2014–2048 (2014)
4. Cardoso, J., van der Aalst, W.M.P.: Handbook of Research on Business Process Modeling. Information Science Publishing, Hershey (2009)
5. Chen, H., Chiang, R.H.L., Storey, V.C.: Business intelligence and analytics: from big data to big impact. MIS Q. **36**(4), 165–1188 (2012)
6. Christiaanse, R., Hulstijn, J.: Control automation to reduce costs of control. Int. J. Inf. Syst. Model. Des. **4**(4), 27–47 (2013)
7. Dumas, M., van der Aalst, W., ter Hofstede Arthur, H.M.: Process-Aware Information Systems. Wiley, New Jersey (2005)
8. Ellerman, D.P.: Economics, Accounting, and Property Theory. Lexington Books, Lexington (1982)
9. Elsas, P.: Computational auditing. Ph.D. thesis, Vrije Universiteit Amsterdam (1996)
10. Governatori, G., Sadiq, S.: The journey to business process compliance, pp. 426–445. IGI Global (2009)
11. Ijiri, Y.: The Foundations of Accounting Measurement: A Mathematical, Economic, and Behavioral Inquiry. Prentice-Hall International Series in Management. Prentice-Hall, New Jersey (1967)
12. Merchant, K.A.: Modern Management Control Systems. Text and Cases. Prentice Hall, New Jersey (1998)
13. Porter, M.E.: CompetitIve Advantage: Creating and Sustaining Superior Performance. Free Press, New york (1985)
14. Rambaud, S.C., Pérez, J.G., Nehmer, R.A.: Algebraic Models for Accounting Systems. World Scientific, Singapore (2010)
15. Rozinat, A., van der Aalst, W.M.P.: Conformance checking of processes based on monitoring real behavior. Inf. Syst. **33**(1), 64–95 (2008)
16. Starreveld, R.W., de Mare, H.B., Joëls, E.J.: Bestuurlijke informatieverzorging. 1: Algemene grondslagen, vol. deel, 2nd edn. Samson Uitgeverij, Aplhen aan den Rijn/Brussel (1988) (in Dutch)

Business Process as a Service (BPaaS)

Model Based Business and IT Cloud Alignment as a Cloud Offering

Robert Woitsch[✉] and Wilfrid Utz

BOC Asset Management GmbH, Vienna, Austria
{robert.woitsch,wilfrid.utz}@boc-eu.com

Abstract. Cloud computing proved to offer flexible IT solutions. Although large enterprises may benefit from this technology by educating their IT departments, SMEs are dramatically falling behind in cloud usage and hence lose the ability to efficiently adapt their IT to their business needs. This paper introduces the project idea of the H2020 project CloudSocket, by elaborating the idea of Business Processes as a Service, where concept models and semantics are applied to align business processes with Cloud deployed workflows. Four architectural building blocks proposed for (i) design, (ii) allocation, (iii) execution and (iv) evaluation are discussed before providing and outlook.

Keywords: Business process as a service · Business processes in the cloud · Business and IT alignment · Meta modelling and semantic

1 Introduction

Cloud Computing is undoubtedly the current mega trend that has the potential to massively influence current use of IT, especially for business applications. Estimated improvements caused by efficient, flexible and networked IT resources range up to 30 % [1]. Hence, Cloud Computing is a chance for start-ups and smart companies that enter this global IT marketplace and obviously a risk for those who do not appropriately take advantage of Cloud Computing.

The ultimate challenge is to overcome so-called business and IT alignment, and bridge the gap between the business and IT domains. With respect to cloud offerings, this means that the current application view needs a corresponding business process view. Currently, typical parameters for SaaS – which is regarded as the current and upcoming market - are of a technical nature such as pricing models considering technical parameters, computing power, availability or network capacity. Business parameters such as legal aspects, business packages, process interoperability, or avoidance of vendor lock are used for distinction between different market players. Business Domain specific parameters like customer relationship for SMEs in the health domain, or web-appearance of an IT company are potential future options.

Hence we observe the need to abstract parameters from pure technical distinctions up to business and domain specific characteristics, which describe and distinguish cloud offerings.

© IFIP International Federation for Information Processing 2015
M. Janssen et al. (Eds.): I3E 2015, LNCS 9373, pp. 435–440, 2015.
DOI: 10.1007/978-3-319-25013-7_35

In the following, the approach of the EU project CloudSocket [2] for tackling the aforementioned issues is introduced.

2   BPaaS Use Cases

Our primary targets are SMEs that are currently excluded from using the cloud due to a lack of competence and high entry barriers. There is a gap between pragmatic, legally influenced and well-defined business processes, which are understood by SMEs, and a gigantic cloud market with numerous offerings that rarely consider the business episodes of an entrepreneur but focus on technical details. Startups and SMEs typically focus on their core business. Hence, there are several business processes such as customer relations and advertising, administrative issues on registration, IT services as well as after sales support that are necessary for business success, but can only be insufficiently supported by the IT resources of those organizations.

For a complete analysis of the use cases, please refer to [3].

2.1   Business Incubator Use Case

The Business Incubator focuses on supporting the "Coaching and Finance" efforts of start-ups facilitating designing, analyzing and simulating individual business plans and processes. These aspects also demand a high degree of adaptability of Cloud Services for Start-ups, e.g. Customer Relationship Management, Order Management, Human Resources Management.

Ecological Agriculture: A 28 year old biologist has an idea to take biological waste from a restaurant and stimulates a biological decomposition process. Usually such a process takes several years but the idea of the startup is to use worms to speed up this process.

- Initial situation: The startup presented the ideas to the business incubators. After this, the consultants have discussed with her about how to transform this business idea into a solid business model.
- CloudSocket technology intervention: The startup may require a range of different BPaaS, especially for customer relationship and worm production management.

Green Energy: This startup is a small-scale virtual power plant which connects to a grid infrastructure with power generation from wind, photovoltaic, and biogas. The company serves its customers with environmentally friendly energy for households and provides smart home functions through its remote access capability for turning appliances on or off.

- Initial situation: The company is intending to expand its services to include mobile energy sources for recharging electric cars and offer them for rental as a range-extension for drivers e.g. for a long weekend trip. The startup contacted the business incubator consultants.

- CloudSocket technology intervention: The startup may require a range of different BPaaS especially for customer relationship, partner management and internal management processes.

The observation in the first use case – the Business Incubator - is that supportive business processes can be applied across several startups. So BPaaS addressing e.g. Customer relationship can be offered to a wide range of startups.

2.2 Cluster Process Broker Use Case

The Business Process Broker use case identifies typical business episodes of potential SMEs in different application domains such as eHealth, Manufacturing, Photonics, Government, Security, e-Commerce, Retail, etc. but share a common set of business processes.

Internet Research and Procurement Process: An SME sells software and integrated appliances/electronic components that make devices "internet ready" in a few seconds.

- Initial situation: The SME continuously verifies prices of the electronic and mechanical components in the market and buys only products that match specific requirements in terms of customer needs and pricing. Monitoring the prices and the quality is a costly activity, which requires an ongoing analysis and trade-off between quality and price.
- CloudSocket technology intervention: The Company needs a solution that reduces the costs for procurement activities by improving the effectiveness. Generic self-management infrastructures or specially designed research processes including crawler and result databases have the potential to run in the cloud and to raise the productivity of this SME.

Kiosk Distribution Process: A company aims at distributing newspapers and magazines to kiosks and other points of sales in an Italian town. Every day, around 250 different Italian and foreign newspaper are delivered to 600 points of sales.

- Initial situation: Current customers are small kiosks with very limited IT infrastructure. Often the orders are realized via Facebook comments. To improve the maturity of the ordering and interaction process with those points of sales, a new but still lightweight Web-application must be provided.
- CloudSocket technology intervention: A new order process can be handled in the cloud, without IT installation on either the supplier or consumer sides. Furthermore the process can reflect a better understanding of the customer needs.

The observation in the second use case – the Cluster Broker - is that most of the potential end users of the CloudSocket have the potential need for a generic business process but also for more specific business processes. Hence the flexible configuration of business processes, hiding the complexity of the cloud and providing easy to use solutions, is a promising market potential.

2.3 The SME End User Perspective

In addition to the two aforementioned use cases – that describe the targeted end users market – we identify the steps of interested end users in a process-oriented approach.

We propose three steps for a typical SME as an end user.

- Check Cloud Readiness
- Transform Business Processes to be executable in the Cloud
- Enter the marketplace to access BPaaS

The project provides a checklist for SMEs and start-ups in order to check if they are capable of entering the cloud with their business processes. This framework is provided as an online questionnaire relating to business processes, see [4].

The transformation of business processes to be executable in the Cloud is divided in two transformations, whereas the first transformation is a horizontal one that transforms from one business process to another one and the second transformation is a vertical on transforming from business process to workflows. Although both business processes are not executable, the latter one has clear anchor points, where cloud offerings make sense. Hence the horizontal transformation extracts those parts of the process where a cloud offering can actually be applied.

The next transformation is a vertical one that maps to an executable workflow in the cloud - this actually provides the cloud offerings and enables the execution in the cloud. This next step is performed by entering the market place and selecting the most appropriate workflow that runs in the cloud. This selection can be supported by smart mechanisms.

2.4 The CloudSocket Broker Perspective

The so-called CloudSocket is a market place, where BPaaS are offered in a similar way to how SaaS are offered. Hence, from a market place point of view, the same mechanisms can be applied as for SaaS, as each "executable business process" can be identified as a workflow in the cloud with a corresponding end point. The difference between a SaaS and a BPaaS is the description in the form of a business process and its semantic annotation for enabling smart selection.

For organizations aiming for becoming a CloudSocket broker we propose: Plan, Build, Run, Check.

"Plan Business Processes" denotes the use of business process management tools to acquire, design, analyse and simulate and finally release domain-specific business processes. Here we understand business processes as a know-how platform of an organisation, hence those processes have the potential for domain-specific consultancy and improvement. Traditional business process management tools such as ADONIS® [5] are used.

"Build Business Processes" denotes that each of the aforementioned business processes are made executable by a set of deployable and executable workflows. We agreed to use the term workflow for processes that are orchestrated and executable on an IT platform to strengthen the difference to human orchestrated or executed business processes. Traditional workflow design tools like yourBPM [6] may be used.

"Run Business Process" indicates the provision and operation of processes as a service within a market place that are executed and run across services offered in the cloud. Although this is technically the most challenging part, the focus of this paper and the focus of the introduced CloudSocket project is on the alignment, hence the mapping between domain specific business processes and cloud deployable and executable workflows.

"Checking Business Processes" indicates the abstraction, using conceptual models and semantic, to introduce a semantic meaning into the purely technical data and process logs from the execution environment in the cloud. The meta model platform ADOxx [7] will be used to develop conceptual and semantic models that can be analysed and mapped to business processes.

3   BPaaS Reference Architecture

CloudSocket will provide business solutions to SMEs, which can be offered in an open and interoperable form. A particular focus is on startups which do not want to invest in their own IT infrastructure but concentrate on the development of their business. IT services need to be adapted according to the changes and evolution of the organization and business.

CloudSocket comprises four phases, each phase supported by a corresponding building block: (a) the design environment to describe business processes, business requirements and workflows (b) the allocation environment linking deployable workflows with concrete services, (c) the execution environment that executes and monitors the workflow as well as (d) the evaluation environment that lifts key performance indicators back to the business level.

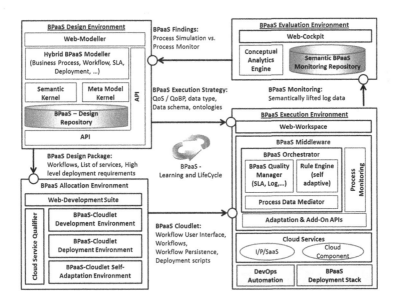

Fig. 1. CloudSocket high level architecture

Figure 1 introduces the four major building blocks, whereby each of the four building blocks supports one phase of the BPMS paradigm when applied for business process management in the cloud. The BPMS is a well-established business process management system paradigm that can be applied also on business process management in the cloud. More information on the architecture is available at [8].

4 Outlook

This paper introduces the idea of BPaaS in the context of the H2020 project CloudSocket, which started 01.01.2015. Hence this paper introduces the project idea of BPaaS and the use of a model-based approach to align domain specific business processes with cloud related executable workflows. First findings in identifying business episodes and possible business process models have been introduced and the current status of the architecture is briefly presented. By the time of the conference, the project can provide the first set of business episodes, a first reference architecture and an environment to check the cloud readiness of an end user.

All prototypes are available either as open source, open use or as provided services. A roadmap for technology provider indicates how alternative tools can be provided for a CloudSocket broker. Results on smart mapping domain specific business processes and cloud based executable workflows are expected by the end of 2015.

References

1. Liebhart, D.: Die modernen drei Musketiere, Cloud Computing, Computerwelt, März (2015)
2. CloudSocket Project. www.cloudsocket.eu. Accessed 18 May 2015
3. CloudSocket Project: D2.1 use case analysis. www.cloudsocket.eu. Accessed 18 May 2015
4. CloudSocket Project: D2.3 cloud transformation framework. https://cloudsocket.eu/transformation/. Accessed 24 July 2015
5. ADONIS Community Edition. www.adonis-community.com/. Accessed 24 July 2015
6. yourBPM. activiti.org/download. Accessed 24 July 2015
7. ADOxx. www.adoxx.org. Accessed 24 July 2015
8. CloudSocket Project: D4.1 First CloudSocket Architecture. www.cloudsocket.eu. Accessed 04 September 2015

Witness Workshop

IT-Enabled Resilient, Seamless and Secure Global Supply Chains: Introduction, Overview and Research Topics

Bram Klievink[1(✉)] and Gerwin Zomer[2]

[1] Delft University of Technology, Delft, The Netherlands
a.j.klievink@tudelft.nl
[2] TNO Netherlands Organisation for Applied Scientific Research,
Delft, The Netherlands
gerwin.zomer@tno.nl

Abstract. This paper is the introduction to the fourth Workshop on *IT-enabled Resilient, Seamless and Secure Global Supply Chains* (WITNESS 2015). In the paper, we present a synthesis of the innovation agendas of a series of international research, development and demonstration projects that seek to make supply chains more efficient and more secure. From this agenda, we highlight three main innovation areas that are central to the current transition in the domain of logistics and international trade. Against this background, we introduce the five papers that are part of the WITNESS 2015 workshop. The papers primarily cover IT related innovations and include topics such as data quality and data governance, the development and interoperability of platforms that together present an IT innovation in international trade, and serious gaming for making key players aware of the potential of the innovations. We finalise with the suggestion to put more emphasis on the non-technical parts of the innovations, as that is what is ultimately needed to ensure wide uptake of the innovations, in order to make them really count.

Keywords: Information infrastructure · International trade · Collaboration · Innovation · Supply chains · Security · Information society · WITNESS

1 Introduction

Due to globalisation and automation, international supply chains have become more dynamic and more complex over the recent years. The commercial transactions, organisation of logistics and execution of transport are performed by multiple layers of organisations, all connected by contracts of sale and carriage [1]. Furthermore, as a consequence of efficiency developments like the introduction of sea container transports and consolidation of cargo, more and more goods are concealed from supply chain visibility, negatively impacting the ability of businesses and governments to assess the security of internationally shipped cargos [2]. Concepts like outsourcing of activities, consolidating cargo from multiple shippers and transporting across multiple modes of transport (sea, air, road, rail, river) have complicated the organisation and control of the chain and demand more from the involved companies in terms of collaboration and information sharing.

© IFIP International Federation for Information Processing 2015
M. Janssen et al. (Eds.): I3E 2015, LNCS 9373, pp. 443–453, 2015.
DOI: 10.1007/978-3-319-25013-7_36

In this context, enhancing supply chain security whilst preserving trade- and logistics-related economic growth, and at the same time safeguarding societal values, is a major topic on the agendas of global and national players, both public and private. To achieve this, various projects, both international and national, have focused on studying, developing and demonstrating innovations to this end.[1] The projects aim to enhance the efficiency, speed and reliability of trade and logistics. At the same time, they seek to enhance the effectiveness and efficiency of supervising global trade, safeguarding supply chain security and integrity and safeguarding society against illegitimate trade and criminal threats. Hence, they seek to achieve both business and societal objectives. Balancing the needs for better security with safeguarding economic and public, requires two major transitions:

- Industry and the business communities aim to regain and retain control over fragmented supply chains and need to develop and apply new collaborative ways of supply chain risk management, going beyond the scope of existing enterprise risk management frameworks and building resilience into their supply chains. IT innovations need to support their supply chain visibility and control.
- Governments and border control authorities need to understand and recognise this and develop chain based control and supervision models that are both effective and result in lower compliance costs for legitimate and trusted traders. For governments, this requires a complete paradigm shift from transaction-driven control to system-based control and from a focus on national borders to understanding supply chain dynamics and impact of interventions.

A large part of the solutions currently being developed to support these transitions are IT (-related) innovations. This includes architectures and delivery infrastructures for data sharing in interoperable ways against minimum costs. Innovations are currently being undertaken to enhance data sharing and improve the timely availability of accurate data in global trade networks [3]. This requires business information infrastructures for data sharing. In addition to business-to-business information sharing, global trade networks are highly regulated and for purposes of compliance to e.g. duties and security regulation, companies are required to report a variety of information to various government organisations. This requires intensive information exchange in which a company interacts with, among others, customs authorities. Inter-organisational information systems play a key role in the simplification and harmonisation of cross-border control procedures, where paper-based procedures are replaced by electronic procedures and IT-facilitated inspection and supervision concepts.

The fourth Workshop on *IT-enabled Resilient, Seamless and Secure Global Supply Chains* (WITNESS), organized at the 14th IFIP Conference on e-Business, e-Services and e-Society (I3E 2015), focuses on innovations in the architectures and governance of smart information infrastructures, platforms and ecosystems. Amongst others, the domain deals with the challenges of interconnecting business information infrastructures

---

[1] For example, European projects in these areas include INTEGRITY, Smart-CM, e-Freight, SUPPORT, LOGSEC, IMCOSEC, CASSANDRA, COMCIS, CONTAIN, SAFEPOST, RISING, ECSIT, ITAIDE, ACXIS, and CORE.

to digital government infrastructures, inter-organizational partnerships, governance and collaboration models. This year's workshop features five papers related to innovations in the architectures of smart digital infrastructures for global supply chains. As an introduction to the workshop, this paper describes recent developments in this domain in terms of an innovation agenda, an overview of the papers, and finalises with a suggestion for additional research topics in need of attention.

2   Making Global Supply Chains Smarter: An Agenda

Among the main innovation drivers in international trade is the need to improve security of international trade supply chains. Security is hampered by a deficient information system, where supply chain actors do not or cannot always fully inform the next actor in the chain, for example about who the original shipper is, and what precisely is in a container [2]. As a consequence, the visibility that for example the consignees have on the goods being shipped to them, is limited. Also, parties further down the supply chain cannot seamlessly integrate goods flows due to a lack of accurate and timely information. What is more, as official declarations are based on transport documents that are not from the actor that really knows the specific goods being shipped, authorities such as customs have to supervise the supply chain with second-hand information that is frequently inaccurate [2].

To make supply chains more secure and to enable parties in the supply chain to enhance supply chain resilience and the effectiveness of operations, innovations that aim to smarten-up international trade are under development. ICT has a big role to play in making supply chains smarter. For instance, innovations in the ICT's used by supply chain actors enable electronic connections and information exchange between the information systems of supply chain actors, thereby enabling or improving access and re-use of these original trade data by other actors in the supply chain [4].

As indicated in the introduction, one place where these innovations are developed and tested is in research and development projects, such as a host of projects funded by the European Commission. As many of these projects are now underway or have finished already, in this section we look at these projects to see what the main innovation categories are to make supply chains more resilient, seamless and secure. We find that three main types of innovations bind these projects:

- Technology innovations;
- New supply chain risk concepts, and;
- New collaborations and supervision concepts.

In the remainder of this section we discuss the main topics that come up in the goals of a selected set of European projects and discuss the abovementioned types of innovations in more detail.

2.1   The Innovation Agenda: Main Topics in Recent EU Projects

There is a series of related EU funded projects that aim to improve the security and economy of international supply chains, by leveraging (information) technology innovations. In this paper, we focus on the following projects: CORE, CASSANDRA,

CONTAIN, SUPPORT, SAFEPOST, EUROSKY, E-FREIGHT, I-CARGO and LOGICON[2], all funded by the European Commission under the 7th Framework Programme. There are many more projects, also outside of Europe and in and by individual countries or stakeholder groups, but we focus on these as they form a 'family' (of sorts) of projects and are funded under the same programme.

A synthesis of the main goals of all aforementioned projects yields the following innovation agenda across these projects:

– Interoperability, information exchange and data sharing;
  • New IT infrastructure and ecosystems (e.g. for air in Europe, for containers globally, for ports);
  • Developing an information backbone; improving security through information-sharing;
– End-to-end supply chain visibility and visibility of supply chain risks;
– Advanced supply chain risk management;
  • Targeted screening (new solutions), improved threat handling;
  • Developing new risk models, from a supply chain perspective;
– System based supervision;
  • A systems approach to design and implement measures for a broader context (e.g. collaborating business in a chain);
  • Risk-based approach to supervision;
– Integration of effective, less-intrusive security technologies in supply chains;
  • Integrating various existing solutions;
  • Container integrated sensors, surveillance system, smart seals and other security technologies;
– Supply chain resilience;
  • Resilience capabilities of organisations and in supply chains;
– Strategic management and governance models;
  • Coordinated border management;
  • Business models for IT infrastructure, supply chain risk management and resilience;
  • Public-private governance models.

Note that optimisations in operations (which has been one of the key topics in supply chains for a long time) are often posited as a derived benefit of these innovations, and not an objective or research component itself, in these projects. Typically, these projects also aim to re-use technologies and innovations that are already present, by selecting the 'best' and integrating them.

An important part of all of these projects is that they aim to demonstrate and refine the innovations in various demonstrators, focusing on container (CASSANDRA), postal (SAFEPOST) and air (EUROSKY) supply chains, including multimodal supply chains

---

[2] http://www.coreproject.eu/ | http://www.cassandra-project.eu/ | http://www.containproject.com/ | http://www.supportproject.info/ | http://www.safepostproject.eu/ | http://www.euroskyproject.eu/ | http://www.efreightproject.eu/ | http://i-cargo.eu/ | http://www.logiconproject.eu/.

(CONTAIN). From the perspective of the workshop and similar to the CORE agenda, the innovation agenda can be grouped in three main types of innovations, which will be discussed next, based on plans and reports of the projects.

2.2 Technology Innovations

A number of technology innovations act as enablers for the innovation agenda. One of the key enabling technological developments of interest for this workshop are the advancements in ICT. In various ways, the projects mentioned above focus on capturing data from the source, i.e. from the handling actors in activities upstream the supply chain, such as container consolidation and stuffing, and ensuring reuse of high quality data through effective control and validation processes. Most of the papers in this workshop also deal with issues and developments of IT to enable many of the innovation agenda's topics primarily through data sharing. These papers are accompanied by various real demonstrations of such solutions, making the IT innovation an accelerator creating momentum for the other innovations on the agenda.

Reliable source data is a necessary ingredient for realising (amongst others) end-to-end visibility on supply chains and supply chain risk. Such visibility is required to improve company and supply chain performance (by enabling e.g. Vendor Managed Inventory), by supporting and enhancing decision-making processes, agility and resilience. Enhanced visibility promises huge benefits [5], but requires some form of seamless data interopera-bility and a supporting information infrastructure, for example in the form of an ecosystem (as in e.g. CORE) or a data pipeline (as in e.g. CASSANDRA). The nature of interna-tional trade requires a 'federated system-of-systems approach, that connects disparate systems (in supply chains, physical instruments, communities, supply chain security controls) in a global architecture providing effective access to supply chain security related information and services from anywhere within that system' [6].

Seamless data interoperability is aimed at reducing the (transaction) costs involved in searching for and exchanging information and sharing data, in both business-to-business and business-to-government relationships in international trade and logistics. Apart from yielding benefits for businesses, it is expected to contribute to enhanced effectiveness in government control. Lack of interoperability is one of the causes for poor data transforma-tion along the value chain.

Finally, technologies such as detection, scanning and cargo screening technologies, automatic identification and data capture technologies (such as smart container security devices), and tracking and tracing technologies can also contribute to supply chain security and resilience, provided the information generated by all these systems is shared timely and securely.

2.3 New Supply Chain Risk Concepts

Supply chains face various threats and vulnerabilities, ranging from external events to systemic vulnerabilities (such as the information problem discussed before), and now accompanied with concerns about cyber risks, and insurance and trade-finance costs. The visibility that can be provided by IT innovations should also be applied in the port-folio of supply chain risk management options, which is the key to advanced supply chain risk management. Control over the supply and value chain are hampered by the fragmentation of these chains in terms of actors and interests. Visibility is a key instrument for parties to regain control over the chains they are involved in. Hence, as part of a supply chain wide risk approach and supported by IT innovations, control over and insight in the chains are required for improved alignment of supply chain processes and for supply chain optimisation. However, there is also another economic justification for being in control of the chain; the intrinsic commercial and compliance value of forming a trusted and integrated supply chain. Gartner [7] identifies this as the key trend under Supply Chain Leaders in 2014. Supply chain wide visibility and control over risks across the chain form the foundations for supervision models that are based on the level of business control over the supply chain, which could lead to a lower administrative burden and higher predictability. In any case, a shared risk vocabulary is required, as is improved data and information sharing across supply chain actors (including with government), and building greater agility and flexibility into resilience strategies.

2.4 New Collaboration and Supervision Models

In the EU, DG TAXUD introduced a risk-based approach (RBA) as a customs view on supervision as an alternative to a 100 % scanning approach. Most EU customs organisations already apply an RBA (e.g. by using risk assessment systems). However, these are not always efficient, with up to 30 % of shipments targeted for further evaluation, which results in transaction-based controls disturbing trusted and secure logistics flows. The common framework for Customs Risk Management and Security of the Supply Chain of the European Commission comprises of three parts [8]:

- Identification and control of high-risk goods movements using common risk criteria;
- The contribution of Authorised Economic Operators (AEO) in a customs-trade partnership to securing and facilitating legitimate trade;
- And pre-arrival/pre-departure security risk analysis based on cargo information submitted electronically by traders prior to arrival or departure of goods in/from the EU [8].

Applying such a risk based approach to incoming trade flows highly depends on the quality, accuracy and timeliness of information that supervision agencies receive. From the CASSANDRA project we learned that the data quality of pre-arrival declarations is regularly unsatisfactory. Often, agents or freight forwarders are identified as the consignor and consignee, and customs still does not know who the true consignor and consignee really are. Also cargo related data has quality issues following from these being asked of the ocean carrier, who in many cases does not (even cannot do not not want to) know what is inside a particular container.

The RBAs as developed in the projects also cover what businesses do to identify and deal with risk in their supply chain operations [9]. Specifically, it means that businesses assess the specific risks in their supply chains at the *trade lane level*, and document the controls they have in place to address those risks. Based on that assessment, supply chain risks can be identified that are currently addressed insufficiently by the business controls, and additional controls can be identified to enhance end-to-end control over the supply chains. This will shift operational focus to higher risk business, thus creating a risk-based approach towards supply chain operations.

Again, the IT innovation plays a major role here, as a supply chain RBA is only possible when the data that circulate among the supply chain partners are accurate, timely and of sufficient quality. Furthermore, the types of risks and the specific sets of data that are required to assess and deal with risks are often product-type and trade lane specific. A supply chain wide risk approach by businesses, combined with a higher quality of crucial data elements, enables government organisations to assess the risks of the supply chain better, and apply a different mix of control mechanisms based on that assessment.

Here, other IT innovations can play a major role as well, primarily in the form of Single Windows [10], which can reduce the administrative burden for businesses, but should also act as a stepping stone for Coordinated Border Management (CBM). CBM refers to a coordinated approach by border control agencies, thus eliminating conflicts and redundancies between different policies, regulations and enforcement practices, thus enhancing their effectiveness as a whole.

An information sharing infrastructure (Sect. 2.2) and novel risk approaches (Sect. 2.3) enable government to 'piggyback' on data from better sources and on business controls; both should be part of a government risk-based approach to supervision; a system based approach (assessing the business controls) and assessing risks based on the data provided to government (for various forms of piggybacking, see [11]). This information must be reliable for government to assess the security of a goods flow, which reduces the risk of unnecessary inspection interference in safe supply chains (but which could not be assessed as being safe because the data is of insufficient quality). As part of this approach, system based control instruments can complement other instruments (such as scanning or physical inspections) in effective control mix, resulting in potentially lower trade transaction costs and lower control burden traders experience.

3   Overview of the Papers in the Workshop

These proceedings contain five papers that are part of the 4th WITNESS workshop. Most of them deal with the challenges of data sharing or with ICT solutions to address them. More specifically, two papers deal with the specifics of data in logistics and international trade, one on the effects of governance of data, the other on assuring data quality. Two other papers deal with the information infrastructure to facilitate data sharing among supply chain actors, including businesses (both traders and logistics providers) and governments (e.g. inspection agencies). Finally, one of the papers deals with a topic that is too often underestimated: whereas the technology might be capable of overcoming many hurdles, ultimately the decision to really adopt and use these technologies depends on many other factors, including the awareness that decision makers have both of problems and of potential solutions.

In the paper "Determining the Effects of Data Governance on the Performance and Compliance of Enterprises in the Logistics and Retail Sector", Nick Martijn, Joris Hulstijn, Mark de Bruijne and Yao-Hua Tan addresses the important topic of data governance. More specifically, they seek to offer a way to determine the effects that data governance has on business performance and compliance. This is important as many practitioners see or expect positive effects, but are missing a framework to actually assess the effects. This papers offers such a framework and the expected benefits for retail and logistics. The research offers interesting insights that could be added to existing models, including models that start from broader IT governance or enterprise risk management perspectives.

Another paper dealing with the specific characteristics and challenges related to data is "Data Quality Assurance in International Supply Chains: An Application of the Value Cycle Approach". In this paper, Yuxin Wang, Joris Hulstijn and Yao-Hua Tan propose a value cycle approach to assuring data quality in international supply chains. Given the numerous issues existing when it comes to data quality in international trade, the insights that the paper offers might be helpful to academics and practitioners looking into ways to ensure and verify the quality of the data that supply chain actors have to work and make decisions with.

In his paper "Towards a federated infrastructure for the global data pipeline", Wout Hofman offers set of platform services and protocols to allow interoperability of different platforms. Such interoperability is necessary given that global data exchange infrastructures (such as the data pipeline that this paper concerns) will have to be federated solutions, making use of the wide variety of solutions already in place or currently under development. The path to the vision laid out in this paper will not only require time and continued innovation, but will also be in need of new business models to ensure adoption and sustainability of the federated infrastructure.

Thomas Jensen also discusses infrastructure for improving information sharing in international trade, using the Interorganisational Systems (IOS) literature as his starting point. In "Key Design Properties for Shipping Information Pipeline", he follows a design science approach and presents a set of key design properties for a so-called Shipping Information Pipeline, an information infrastructure for international containerized trade. Not unlike Hofman's ideas, one of the principles also holds that the idea of the data pipeline is that of a 'virtual' infrastructure, whereas the actual physical infrastructure can (and will) be handled by several individual organisations.

Finally, in "Enhancing Awareness on the Benefits of Supply chain Visibility through Serious Gaming", Tijmen Joppe Muller, Rainer Müller, Katja Zedel, Gerwin Zomer and Marcus Engler discuss a serious game developed for customs-related issues and innovations in international supply chains. The gaming-background of the paper primarily concerns the learning effect, but given the background of the innovation that it concerns (data pipeline, customs innovation), we think that it can also serve other purposes as part of the innovation trajectory, notably by supporting the process of activating and including key stakeholders. The main contribution of this paper is therefore twofold: it discusses the specific game, which covers various topics of the innovation agenda discussed in this paper, but it also shows how serious gaming could (and should) be used to address non-technical aspects of the innovations.

4   Beyond the Innovation Agenda: Making Innovations Count

As demonstrated by the innovation agenda of the projects and the papers in this workshop, logistics and international are seeing a series of innovations to make logistics smarter, more competitive, improve supply chain security, enhance control and facilitate information sharing. These innovations have been and are being developed, studied and demonstrated extensively in (inter)national and European projects and are about new ways of controlling risk and collaborating in supply chains, enabled by IT innovations. Combined, these innovations should constitute a transition for the domain of logistics and government. Although the concepts that are central to these innovations are quite well known by now (as demonstrated by the many shared goals of the projects covered in this paper), there still is a big gap between the transition in theory and the actual adoption thereof in the real world. The real uptake and in-depth integration in the actual practices of organisations of what has come from these years of research, concept building, and technical development, is unfortunately limited at times. That some concepts only find limited fertile ground in the sector after the projects finish, presents a major challenge for the coming years. The limited adoption of the ideas and innovations generated by the research projects itself represents a research problem, which cannot be solved only by improving the models and technologies.

Although the European projects devote a lot of time and resources on stakeholder-driven demonstrators (which is a very important step), just how these local innovations combined can constitute a coordinated and balanced innovation for the logistics sector as a whole, is not yet well understood. A major transition through an integrated and coordinated large-scale implementation of key innovative logistics concepts is needed, but the actual adoption by the sector also requires small, local innovations, attuned to the specific stakeholders involved.

Hence, although further refinement and additional development of the topics on the innovation agenda discussed in this paper are both needed and desired, we argue that additional research is needed into the *socio* side of what ultimately are to be socio-technical artefacts. This side has challenges itself, e.g. as the parties involved in logistics and trade, which ultimately will need to adopt the innovations and solutions, have great variety in position, interests and values. To make the innovations work for them, this variety will have to be addressed and taken into account, which can only be done close to those specific stakeholders. This is difficult as the stakeholder field is very fragmented, with interests divvied up along the lines of sectors (e.g. initiatives in the horticulture industry), roles in the supply chain (e.g. transporters, freight forwarders, shippers) and the size of companies. Especially the many small and medium sized operators will have to be involved via e.g. branch organisations and collaborations as they have limited capacity focused on innovation. This complexity can explain, for example, why some of the aforementioned projects have become innovations that rely on government support, whereas they should have been (and in nature are) business innovations.

To create the right incentives for open innovation and establish a vibrant community of companies that are willing provide parts for data pipelines and to ensure that public value is realized, institutional arrangements must be developed and adopted. Typically, these institutional arrangements are developed not only by national government, but also supra-national organizations should be involved such as the European Commission or the United Nations. Furthermore, as argued by Hofman in his workshop paper, apart from the IT developments and challenges, the federation of platforms requires additional research into the (sustainability of) business models for these platforms. Making combinations of data from multiple parties (both public and private) is essential for developing new commercial services and for supporting new supervision concepts, both creating (economic) incentives for companies to contribute to realizing the innovations. However, there are also risks involved in combining data from multiple sources, especially in a competitive private environment. Some form of governance would help stakeholders (again, both public and private) to create on the one hand a level playing field (e.g. to avoid that one or a few parties can gain an unfair competitive advantage from their access to community data), and on the other hand offers enough economic incentives for businesses to make their adoption of the innovations commercially viable. As the workshop paper by Muller et al. illustrates, much of this starts with making key players aware of the issues and potential solutions. Given such awareness, the right institutional practices, incentives, process support by facilitators and viable business models for parts of the innovations, the field can develop a fertile ground for the innovations to land in, and from thereon find wide support and uptake.

Acknowledgements. This work is part of the research project "Governing public-private information infrastructures", which is financed by the Netherlands Organisation for Scientific Research (NWO) as Veni grant 451-13-020. Furthermore, for this paper, we made use of parts of CASSANDRA and CORE project reports, both funded by the European Commission as part of the 7th Framework Programme (FP7). The authors are involved in both.

References

1. Van Oosterhout, M.: Organizations and flows in the network. In: Van Baalen, P., Zuidwijk, R., Van Nunen, J. (eds.) Port Inter-Organizational Information Systems: Capabilities to Service Global Supply Chains, vol. 2, pp. 176–185. Now Publishers, Hanover (2009)
2. Hesketh, D.: Weaknesses in the supply chain: who packed the box? World Customs J. **4**, 3–20 (2010)
3. Klievink, B., Van Stijn, E., Hesketh, D., Aldewereld, H., Overbeek, S., Heijmann, F., Tan, Y.-H.: Enhancing visibility in international supply chains: the data pipeline concept. Int. J. Electron. Gov. Res. **8**, 14–33 (2012)
4. Tan, Y.-H., Bjørn-Andersen, N., Klein, S., Rukanova, B. (eds.): Accelerating Global Supply Chains with IT-Innovation. ITAIDE Tools and Methods. Springer, Berlin (2011)
5. Ngai, E.W.T., Chau, D.C.K., Chan, T.L.A.: Information technology, operational, and management competencies for supply chain agility: findings from case studies. J. Strateg. Inf. Syst. **20**, 232–249 (2011)
6. CORE Consortium: CORE (Consistently Optimised Resilient Secure Global Supply-Chains) Description of Work (DoW) (2014)
7. Gartner: The gartner supply chain top 25 for 2014 (2014)
8. European Commission: Communication from the commission to the european parliament, the council and the european economic and social committee on Customs Risk Management and Security of the Supply Chain (2013)

9. Cassandra: D2.2 – risk based approach (2012)
10. Keretho, S., Pikart, M.: Trends for collaboration in international trade: building a common Single Window Environment. United Nations (2013)
11. Klievink, B., Bharosa, N., Tan, Y.H.: Exploring barriers and stepping stones for system based monitoring: insights from global supply chains. In: Lecture Notes in Informatics (LNI), Proceedings - Series of the Gesellschaft fur Informatik (GI), pp. 35–42 (2013)

Determining the Effects of Data Governance on the Performance and Compliance of Enterprises in the Logistics and Retail Sector

Nick Martijn[✉], Joris Hulstijn, Mark de Bruijne, and Yao-Hua Tan

Faculty of Technology, Policy and Management,
Delft University of Technology, Delft, The Netherlands
nickmartijn@gmail.com,
{j.hulstijn,m.l.c.deBruijne,y.tan}@tudelft.nl

Abstract. In many of today's enterprises, data management and data quality are poor. Over the last few years, a new solution strategy has emerged, known as data governance: an overarching methodology that defines who is responsible for what data at what point in a business process. Although positive effects on the business performance and compliance of enterprises are seen in practice, a substantiated method for determining the effects of data governance has not yet been developed. This paper reports on explorative research to develop such a specification method. Through a conceptualization of data governance based on literature, case study analysis of clients of a large consultancy firm and interviews with representatives of companies that have recently implemented data governance, an effect specification framework was developed. Using the interviews, initial steps towards validation were performed.

Keywords: Data governance · Effect specification

1 Introduction

Enterprise data is becoming increasingly important. Data was initially seen as a byproduct of business processes, used for example for financial recording (Lake and Crowther 2013). Nowadays, data is considered to be a valuable asset in and of itself (Bughin et al. 2010). This value is primarily provided by two applications: measuring business performance and compliance reporting. Firstly, increasingly complex and globalizing business processes require the support of reliable data. For example, the international container shipping industry requires timely and accurate data to feed its logistical planning. Lack of data quality leads to huge losses (Steinfield et al. 2011). Secondly, enterprise data is used for financial reporting and for other kinds of compliance reporting. Companies have to comply

---

This paper summarizes (Martijn 2014). The research has been conducted with the support of Marinka Voorhout, specialist in Enterprise Data Management. We gratefully acknowledge her contribution.

© IFIP International Federation for Information Processing 2015
M. Janssen et al. (Eds.): I3E 2015, LNCS 9373, pp. 454–466, 2015.
DOI: 10.1007/978-3-319-25013-7_37

to certain laws, such as Sarbanes-Oxley for companies listed on the US stock exchange, or Solvency II for insurance companies (Eling et al. 2007). These laws demand that companies demonstrate to the regulator that they are compliant, which requires evidence. Regulatory compliance creates additional data requirements. It is not sufficient to supply evidence; an audit trail is also required (Jiang and Cao 2011). Not meeting data requirements can lead to severe financial consequences. Consider the $3.75 m fine Barclays bank received from US Financial Industry Regulatory Authority (BBC 2013). So enterprise data is used to gain insight in business performance, while also enabling compliance (Cheong Chang 2007; Golfarelli et al. 2004; Loshin 2012).

Notwithstanding its importance, the standard of data management is often poor (Haug et al. 2011). Although most companies have well-managed IT systems, the responsibilities for maintaining specific kinds of data are mostly not incorporated (Redman 2001). It is has been shown that when no clear policies, rules and controls are defined within the organization about who is responsible for what data, overall data quality will deteriorate (Batini et al. 2009). Poorly governed data may generate losses, as incomplete or erroneous information can mean a serious strategic disadvantage or lead to inefficiently organized business processes (Steinfield et al. 2011), in addition to posing the risk of being deemed non-compliant.

Over the last few years a new solution strategy has emerged: *"Data governance is a system of decision rights and accountabilities for information-related processes, executed according to agreed-upon models which describe who can take what actions with what information, and when, under what circumstances, using what methods"* (Thomas 2006). Essentially, data governance is an overarching methodology that defines who is responsible for what data at which point in the process. There is more to it though, such as internal controls, information systems architecture, standardization of data formats, corporate culture and use of technology, such as monitoring tools. Taken together, data governance measures can assure that enterprise data will be of sufficient quality. Data quality is seen as the most important aspect influencing usability of data for business processes and reporting (Friedman and Smith 2011).

It turns out to be relatively hard to specify the effects of data governance projects and interventions. How do the application of various tools and techniques affect data quality? And subsequently, how does improved data quality affect business performance and compliance reporting? These are fundamental questions that have received relatively little attention. The effectiveness of data governance projects is only known from practical experience (De Waal and De Jonge 2012). Data governance frameworks, such as DAMA DMBOK (Mosley et al. 2010), claim that they can determine these effects, but these are not fully scientifically substantiated.

In this paper, we therefore develop a framework to make it possible to specify the effects of data governance interventions. The paper is a summary of the graduation research reported in (Martijn 2014). We make use of case studies of firms, which have recently undertaken data governance projects. These cases were collected with the help of a large consultancy firm with extensive experience in helping clients improve their data quality through a data governance framework, in particular in the financial, logistics and retail sector (De Waal and De Jonge 2012).

Concerning the choice of research method, note that organizational factors may affect the effectiveness of data governance interventions, but cannot be unambiguously operationalized. Thus, the case study approach is most appropriate, as the boundaries between the phenomenon (data governance) and the context (organizational effects) is relatively unclear (Boschi 1982; Xiao et al. 2009).

The research proceeds as follows. Based on literature, we develop a conceptualization of data governance and its drivers (Sect. 2). We then develop a method to specify the expected effects of data governance interventions (Sect. 3). The conceptualization is relatively generic: it must be further specified for each case. Making use of client dossiers of companies that have recently adopted data governance measures, we show how the concepts can be further operationalized. Based on interviews with representatives of companies from the retail and logistics sector that are currently implementing data governance measures, we take initial steps towards validation of (Sect. 4). Full validation would require more cases, and would require comparison of the outcomes with other, independent, specification techniques.

2 Conceptualizing Data Governance

There is a lot of research on data quality and the effect it has on the use of information systems (Strong et al. 1997). However, not much scientific research is specifically dedicated to the reverse question. How does data governance improve data quality, and consequently increase business performance and compliance? There are several data governance frameworks, of which the DAMA Data Management Body of Knowledge is most commonly used (Mosley et al. 2010). Such frameworks provide an overview of data governance measures to increase the data quality at an organization. The DAMA approach summarizes the following best practices (Mosley et al. 2010): data architecture management, data development, database operations management, data security management, reference and master data management, data warehousing and business intelligence management, document and content management, meta data management, data quality management, all centered around the data governance. A problem with such frameworks is that there are generic and professionals have to adapt them to their own situation. The framework lists different kinds of activities, both technical and strategic. How can we structure their dependencies?

We have made a conceptualization of data governance, shown in Fig. 1. As argued in the introduction, data quality is an essential property driving business performance and compliance. This will therefore be used as the guiding notion. To structure the diagram and locate the various activities, we use an enterprise architecture, based on the layers of (Winter and Fischer 2006). From the bottom up: technology, software and integration architecture (merged here), process architecture, and business architecture. On the right we added organizational architecture, as we focus on governance aspects, involving roles, responsibilities and institutional arrangements.

Reviewing the structure from the bottom up, the first part consists of the technology, software and integration architecture. Here we find physical devices (gates; RFID readers etc.), computer systems (databases; networks etc.) and software applications

(ERP systems; workflow management etc.) to store, retrieve and process information. In addition, we also find protocols and procedures for exchanging information and for integrating different modules. In this layer the actual data is situated.

Above this infrastructure layer, the process architecture is located, including the processes that are carried out with the data. Data processes are composed of four basic operations: Create, Read, Update and Delete (Martin 1983; Polo et al. 2001). These CRUD operations determine the status of data elements at any point in the process, so we could say that this layer also contains the data architecture, which determines how data is being handled. When a certain piece of data, such as delivery address or price of a product is used, the data is mostly pulled from the infrastructure through an Enterprise Resource Planning (ERP) system.

The business architecture layer includes the value-adding processes, such as purchasing, sales, manufacturing or transport. Also internal control and risk management, compliance management, and reporting (financial statements, tax reports) are located here. To enable successful business processes, an effective data and process architecture is required. Consider for example the process of acquiring resources from a supplier, manufacturing products, and selling them on to a customer, consisting of steps like: receive, pay, manufacture, store, sell, dispatch. These steps in the primary process give rise to data operations. For example, receiving resources in a warehouse means the creation of new data objects representing the type of resources, storing the stock levels for these resources, updating inventory in the general ledger, changing the status of the corresponding purchase order, etc. The way in which basic 'CRUD' operations are implemented, largely determines data quality (Wand and Wang 1996).

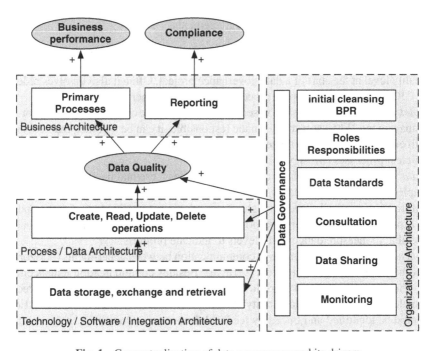

Fig. 1. Conceptualization of data governance and its drivers

Kahn et al. (2002) see data quality as the degree to which data is fit for purpose, i.e. meeting company-specific requirements, see also Juran et al. 1999 and Wang et al. (1996). Data quality in turn largely determines the effectiveness of the business processes, and also influences the reporting quality (Eppler and Helfert 2004), thereby affecting the drivers: business performance and regulatory compliance. Low data quality is pervasive, costly and can cause high inefficiencies (Eppler and Helfert 2004; Fisher and Kingma 2001; Wang et al. 2001).

The organizational architecture contains institutional bodies like the board of directors, managing staff, change advisory board (CAB) and so forth. As data governance is a top-down methodology, we also locate most of the data governance efforts here. In particular, it entails roles for making people responsible for enterprise data. Data governance can influence all other layers. There is also a direct influence link between data governance and data quality, because of initial cleansing activities that are often part of a data governance project (see Sect. 2.1).

Business performance and regulatory compliance are at the end of the causal chain. First, the data that is used in business processes, directly affects business performance (Neely et al. 2002). Errors lead to missed deliveries, dissatisfied customers, etc. Also internal reports are used for forecasting, budgeting etc. Second, the reporting function influences regulatory compliance, as evidence needs to be produced of being compliant with laws and regulations (Jiang and Cao 2011).

2.1   Elements of Data Governance

Given this conceptualization of data governance, what does it actually involve? Data governance consists of various interrelated elements:

- *Initial cleansing* and *business process redesign* interventions are required before data governance can function properly within an organization. This sets a basic data quality level at the start of a data governance project. Often it also involves redesign of CRUD processes, compare process redesign (Hammer 1990).
- *Roles and responsibilities* are essential to prevent lack of clear ownership for data management. When someone is made responsible, it can be assumed that less errors will enter the system, and that errors are detected and solved earlier, leading to more efficient processes (Mosley et al. 2010).
- *Data standards* describe how to represent, process, use and handle enterprise data. Implementing data standards, in combination with a governance structure that enforces the standards, leads to higher data and process quality. Use of data standards is a prerequisite for other data governance interventions. Standards also make it possible to measure data quality levels to indicate progress.
- *Consultation* is meant to improve communication between departments (horizontal), and between management levels (vertical). Besides communication enhancements in primary processes and workflows, so called consultation platforms are recommended to improve the adaptability of data governance measures themselves. Errors should be identified and traced by feedback from users (Orr 1988), so the company can learn from experience.

- *Data sharing* with supply chain partners is important for efficient alignment within supply chains, both internally, as well as externally (Steinfield et al. 2011). This includes monitoring of data provided by the supplier to ensure sufficient quality. Data protocols can be part of the contract provisions.
- *Monitoring* should provide continuous insight in the current quality of the data to facilitate manageability of data by responsible employees, for instance, implementing tools that produce real-time data quality overviews on a dashboard. Again, this requires the ability to measure data quality level.

This list shows that data governance intervention requires a form of governance: it cuts across all layers and departments, which requires management support. Business must be involved, as they should define the information needs. Standards must be enforced. Budget to make the required changes to the IT infrastructure must be secured. Furthermore, even if we narrowly define data governance as the implementation of roles and responsibilities over enterprise data (Thomas 2006), it cannot be abstracted from other data management aspects such as the use of standards and tools. After all, the roles and responsibilities are meaningless without technical and organizational means to support employees in executing these responsibilities. Therefore data governance is seen as a 'package deal': these elements strengthen each other.

3 Deriving a Causal Model

Using the diagram in Fig. 1, we developed a causal model to specify the effects of data governance on an organization, shown in Fig. 2. As such models are typically domain specific, the contribution lies in the method to derive the model. The model is developed on the basis of scientific literature and case study research (Sect. 4). Insights are based on dossier reviews at a large consultancy firm and interviews with clients who have recently been advised on data governance. Versions of the model were validated and adjusted on the basis of interviews with clients.

To scope the research, we decided to focus on cases from the logistics and retail sector. Data quality within supply chains is highly important (Li and Lin 2006). Business performance can be operationalized using key performance indicators from the Supply Chain Operation Reference (SCOR) model. This model provides a standard method to review the performance of a supply chain, see Lockamy and McCormack (2004), Xiao et al. (2009) and Hwang et al. (2008). The KPIs include seven elements to determine customer service level: right product, right customer, right time, right place, right condition, right quantity and right costs (Fawcett and Fawcett 2014).

The literature research, project dossiers and interviews with experts produced hypotheses for relationships between data governance measures, data quality, and ultimately business performance and compliance. The relations are shown as arrows in Fig. 2. We use the following semantics: A –[+]– > B means a positive influence: when A increases, B should also increase. Conversely, A –[–]– > B means a negative influence. When A increases, B should decrease.

Starting at the left part of Fig. 2, the various data governance elements will lead to better CRUD operations in business processes and subsequently, to better data quality. Initial cleansing will improve data quality directly. Consultation will improve communication between departments, which may lead to better IT responsiveness: the ability of the IT department to meet business demands. The enforcement of standards will reduce introduction of mistakes; in addition, it will make it easier to measure data quality. After all data quality is defined as fitness for purpose, where the purpose is reflected in company policies and data requirements, such as for instance those suggested by the SCOR model. Improved data sharing between partners in the supply chain, will improve data quality from suppliers. So also external factors play a role.

In the middle, better data quality helps to improve supply chain forecasting quality (Shankaranarayanan and Cai 2006). Inherently, if forecasts are not reliable, the primary processes will be run less efficiently (Gunasekaran et al. 2004).

Improved data quality also decreases administrative costs. According to a worldwide investigation by (GS1 2011), low data quality causes significant administrative costs referred to as shrinkage, the difference between what is shipped by the supplier and what is finally sold to the customer. Furthermore, efficiency in the primary process reduces operating expenditure (OPEX). Operating expenditure consists of all costs associated with operating a supply chain, such as transport and transaction costs. In particular, administrative costs have a large impact on OPEX.

When primary processes are more effective and efficient, for example when timeliness of deliveries is increased, the level of customer service will increase (Stevenson and Hojati 2007). Customer service crucially affects sales. In addition, customer responsiveness is defined as the manner in which the business can meet demands of customers (Friedman and Smith 2011). This property is related to the infrastructure: can it adapt. If a number of basic data elements are collected and processed reliably, new combinations can be engineered relatively easily. This improves the ability to forecast and report but also the ability to construct new customer services. These factors will help to increase sales and thus business performance (Neely et al. 2002).

In the lower part of the diagram, higher data quality leads to improved internal controls and reporting quality, which by definition increases regulatory compliance. In most cases, an audit trail of enterprise processes and cash flow is required. Reporting quality is lower, when material (i.e. crucial) errors are not detected, or when relevant aspects of behavior are not reported, or not even recorded. Data quality is closely related to the notion of reliability, which involves accuracy (data correspond to reality) and completeness (all relevant aspects of reality are recorded) (Strong et al. 1997).

In addition, reporting quality is affected by regulatory responsiveness, the ability to deal with compliance demands (Friedman and Smith 2011). Data governance also affects this responsiveness variable: when basic figures are recorded and processed reliably, with an audit trial, new combinations of reports can be constructed reliably.

The relation between compliance reporting and data quality becomes even more crucial, when we apply innovative ideas of regulatory supervision, in which data is pulled from the source. For example, a 'data pipeline' infrastructure could facilitate reliable exchange of information in a trade lane, with access for customs authorities, but also for authorized traders (Klievink et al. 2012). Also in the XBRL-GL vision, data items are recorded and 'tagged' close to the source (Cohen 2009). This makes it possible to record an audit trail with meta-data about provenance of data items. Given such basic elements with their provenance, new reports can be constructed reliably.

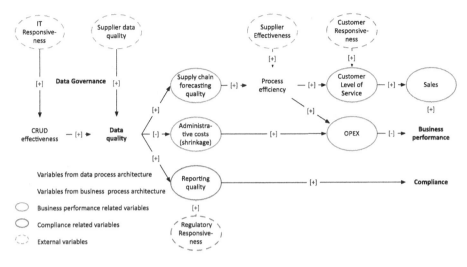

Fig. 2. Causal diagram: expected effects of data governance

The discussion above shows, that it is in fact possible to operationalize the effects of data governance. Initially this will be qualitative, but once the organization has gained some experience also quantitative measures can be used. The resulting model always depends on the specific case; what matters is the line of reasoning. We identify variables that can be used to monitor the effects of data governance interventions.

Some of these variables are well defined and are measured on a routine basis: OPEX, sales, customer service level. Other variables can be defined, once a good operationalization is found. CRUD effectiveness can be determined by a process review, as is typically done by an operational auditor. For instance, consider the property that all information needs should be covered, or that no data should be requested that is not used later. Such properties can be determined using a create-use matrix, in which all data elements are plotted against the activities and roles by which they are created, and subsequently used. Data quality can be specified as the converse of the number of deviations per volume of data from a set of agreed data requirements. So we have $DQ = volume\ of\ data/number\ of\ deviations$. In the supply chain domain, it makes sense to use the SCOR indicators as a starting point for those requirements. The point is to trace supply chain delivery errors back to the information systems that cause them. Supply chain forecasting quality is determined by comparing forecasts with actual performance. Reporting quality is typically determined as a by-product of a financial audit: the number of deviations is reported as well. The variables IT responsiveness, customer responsiveness and regulatory responsiveness should be seen as intermediate variables, to factor in the ability of the organization to adapt. Typically, these can't be measured, which is why they are indicated as dashed ellipses.

That leaves only process efficiency, business performance, and compliance. Those variables can be seen as outcomes of data governance interventions, not inputs that can be used to control and adjust. These outcomes can in fact be measured, but that topic is out of scope. Consider for instance the Business Balanced Scorecard to measure business performance. Here we focus on the effects of data governance.

Table 1. Overview of cases and main impact on the business

	Market	Main characteristic	Impact on the business
A	Food retail	Insufficient data quality in the product database, primarily caused by lack of formalized responsibilities	The company was able to improve supply chain efficiency and decrease operational expenditure. For example, 50 % of working hours on data related activities was saved
B	Food production	Strong focus on the centralization and standardization of data management	Costs of production could be lowered. Business strategy improved due to improved data quality and better insight in enterprise data
C	Logistics	Strong focus on data quality due to a supply chain in a competitive market and compliance issues	Decision making and efficiency within the supply chain improved significantly, leading to fewer compliance issues and higher customer satisfaction
D	Financial sector	Customer data stored in many different legacy systems, leading to inconsistent, incomplete, incorrect and double entries	Improved data quality increased client satisfaction and improved regulatory compliance

4   Towards Validation

To assess the adequacy and usefulness of the causal model, representatives of four companies that have recently implemented data governance were interviewed. The interviews were conducted at companies from the retail and logistics sector, which aligns with the research scope. For comparison, one case from the financial sector was included. We found no structural differences in responses from the domains.

The interviewees are mostly (IT) managers that were closely involved in the implementation of data governance measures, based on advice from the consultancy firm. Generally, their overall opinion on data governance is highly positive. They see the added value of data governance and acknowledge the positive effect on data quality and therefore on compliance and business performance. Issues identified are mostly on an operational level (Table 1). An example is erroneous product information leading to inefficiencies due to data errors (case A).

Firstly, representatives of all four companies confirmed the presence of the main causal relations in the framework. They too experienced that data governance measures have an effect on the implementation of CRUD processes, on the data quality level, and on the primary processes. This can be seen as a first step towards validation of the methodology that was developed in this research. Secondly, based on these interview outcomes, the early findings from the project dossier research and literature research, summarized in Fig. 1, can also be validated. Thirdly, the interviews led to the observation that the further 'downstream' a factor is in the framework, the more difficult it is to specify precisely. It could be that this originates from the fact that most interviewees had a strong 'data-view' on the business, in which CRUD processes and data quality play an important role. For most of the interviewees, business issues are out of their scope. Fourthly, it is confirmed that data governance is a top-down methodology, as was found in the literature research. Measures taken to implement data governance, such as standards and tooling (monitoring) are forced onto the company by management responsible for data governance, backed by general management. This result supports the organizational theory behind the conceptualization.

5 Conclusions

Business performance measurement and compliance reporting are driving an increasing demand for improved data quality in enterprise systems. Data governance has emerged as a solution concept. It is hard to specify the intended effects of a data governance project before the start of the project: what is the business case? It is even harder to specify the actual effects afterwards, both qualitatively, i.e., has data quality improved; has client market responsiveness improved, and quantitatively, i.e., how much has business performance improved; how much costs have been saved?

In this research we have studied literature about data governance, both from theory and practice. We have reviewed dossiers of recent data governance projects conducted at a consultancy firm, and have held interviews with experts and with clients of this firm. This material has led to two outcomes:

(1) a conceptualization of data governance, positioning it within the organizational part of an enterprise architecture, and indicating the effects on its drivers, namely business performance and regulatory compliance.
(2) a causal model, with hypotheses about the influence of data governance on variables representing data quality and other intermediate notions, and ultimately on business performance and compliance.

Precisely specifying the effects of data governance interventions is exceptionally complex. Data governance is an overarching and top-down methodology. Many sector-specific factors are involved. Moreover, many of the factors cannot be operationalized. For instance, every measurement of data quality – correspondence to data requirements that represent fitness for purpose – depends on the definition that is applied by a specific enterprise. The data requirements, standards or data rules are always changing, because businesses are dynamic and respond to market conditions. Furthermore, in terms of effect determination, data governance cannot be abstracted from other data management practices. Data governance involves a package deal, of measures that mutually strengthen each other. All this makes a generic specification of the effects of data governance impossible.

We can however provide a method of how data governance effects can be specified within a sector (using e.g. the SCOR model in supply chain management), or within a specific enterprise. The causal model in Fig. 2 can serve as a 'back bone' for such a method. It follows the logic of increasing positive effects (customer satisfaction; sales) and reducing negative effects (operational expenditure), by reducing data errors and improving process efficiency, as well as reporting quality.

The set-up of the research certainly has limitations. First, the research is focused on the logistics and retail sector. When another sector is considered, only the main relations in Fig. 2 can be used, not the choice of measures. Second, because data governance is an overarching methodology, in which measures strengthen each other, it is impossible to study the individual effectiveness of interventions. Such evaluations would be valuable in practice to improve efficiency of projects. Thirdly, the research is scoped towards the expected benefits of data governance. Especially for business purposes, in which benefits are often weighed against costs, it would be useful to gain insight in the costs in order to build a proper business case for data governance projects. Consider IT investments, costs of additional personnel, costs of maintenance of tooling, and increased controls. These investments should be balanced against the costs of poor data governance. This is a useful topic for future research.

References

Batini, C., Cappiello, C., Francalanci, C., Maurino, A.: Methodologies for data quality assessment and improvement. ACM Comput. Surv. (CSUR) **41**(3), 16 (2009)

Boschi, R.: Modelling exploratory research. Eur. J. Oper. Res. **10**(3), 250–259 (1982)

Bughin, J., Chui, M., Manyika, J.: Clouds, big data, and smart assets: ten tech-enabled business trends to watch. McKinsey Q. **56**(1), 75–86 (2010)

BBC: Barclays fined $3.75m after record-keeping failure, from BBC News (2013). http://www.bbc.co.uk/news/business-25525621

Cheong, L.K., Chang, V.: The need for data governance: a case study (2007)

Cohen, E.: XBRL's global ledger framework. Int. J. Discl. Gov. **6**(2), 188–206 (2009)

De Waal, A., De Jonge, A.W.A.: Data Goverance bij een grote verzekeraar. Compact **39**(2), 11–17 (2012)

Eling, M., Schmeiser, H., Schmit, J.T.: The solvency II process: overview and critical analysis. Risk Manage. Insur. Rev. **10**(1), 69–85 (2007)

Eppler, M., Helfert, M.: A classification and analysis of data quality costs. In: Paper presented at the International Conference on Information Quality (2004)

Fawcett, S.E., Fawcett, A.M.: The Definitive Guide to Order Fulfillment and Customer Service: Principles and Strategies for Planning, Organizing, and Managing Fulfillment and Service Operations. Pearson Education Inc., Upper Saddle River (2014)

Fisher, C.W., Kingma, B.R.: Criticality of data quality as exemplified in two disasters. Inf. Manage. **39**(2), 109–116 (2001)

Friedman, T., Smith, M.: Measuring the Business Value of Data Quality. Gartner, Stamford (2011)

Golfarelli, M., Rizzi, S., Cella, I.: Beyond data warehousing: what's next in business intelligence? In: Paper Presented at the Proceedings of the 7th ACM International Workshop on Data Warehousing and OLAP (2004)

GS1. Australia Data Crunch Report. GS1, Sydney, Australia (2011)

Gunasekaran, A., Patel, C., McGaughey, R.E.: A framework for supply chain performance measurement. Int. J. Prod. Econ. **87**(3), 333–347 (2004)

Hammer, M.: Reengineering work: don't automate, obliterate. Harvard Bus. Rev. 104–112 July-August 1990

Haug, A., Zachariassen, F., Van Liempd, D.: The costs of poor data quality. J. Ind. Eng. Manage. **4**(2), 168–193 (2011)

Hwang, Y.-D., Lin, Y.-C., Lyu Jr., J.: The performance evaluation of SCOR sourcing process—the case study of Taiwan's TFT-LCD industry. Int. J. Prod. Econ. **115**(2), 411–423 (2008)

Jiang, K., Cao, X.: Design and implementation of an audit trail in compliance with US regulations. Clin. Trials **8**(5), 624–633 (2011)

Juran, J.M., Godfrey, A.B., Hoogstoel, R.E., Schilling, E.G.: Juran's Quality Handbook, vol. 2. McGraw Hill, New York (1999)

Kahn, B.K., Strong, D.M., Wang, R.Y.: Information quality benchmarks: product and service performance. Commun. ACM **45**(4), 184–192 (2002)

Klievink, B., Van Stijn, E., Hesketh, D., Aldewereld, H., Overbeek, S., Heijmann, F., Tan, Y.-H.: Enhancing visibility in international supply chains: the data pipeline concept. Int. J. Electron. Gov. Res. **8**(4), 14–33 (2012)

Lake, P., Crowther, P.: Concise Guide to Databases. Springer, London (2013)

Li, S., Lin, B.: Accessing information sharing and information quality in supply chain management. Decis. Support Syst. **42**(3), 1641–1656 (2006)

Lockamy, A., McCormack, K.: Linking SCOR planning practices to supply chain performance: an exploratory study. Int. J. Oper. Prod. Manage. **24**(12), 1192–1218 (2004)

Loshin, D.: Evaluating the Business Impacts of Poor Data Quality. Knowledge Integrity Inc, Silver Spring (2012)

Martijn, N.: Exploring the Effects of Data Governance. Msc thesis, Delft University of Technology, Faculty of Technology, Policy and Management (2014)

Martin, J.: Managing the Data Base Environment. Prentice Hall, Upper Saddle River (1983)

Mosley, M., Henderson, D., Brackett, M.H., Earley, S.: DAMA guide to the data management body of knowledge (DAMA-DMBOK guide). Technics Publications (2010)

Neely, A.D., Adams, C., Kennerley, M.: The Performance Prism: the Scorecard for Measuring and Managing Business Success. Prentice Hall Financial Times, London (2002)

Orr, K.: Data quality and systems theory. Commun. ACM **41**(2), 66–71 (1988)

Polo, M., Piattini, M., Ruiz, F. Reflective persistence (Reflective CRUD: reflective create, read, update and delete). In: Paper Presented at the Sixth European Conference on Pattern Languages of Programs (EuroPLOP) (2001)

Redman, T.C.: Data Quality: the Field Guide. Butterworth-Heinemann, Wobum (2001)

Shankaranarayanan, G., Cai, Y.: Supporting data quality management in decision-making. Decis. Support Syst. **42**(1), 302–317 (2006)

Steinfield, C., Markus, M.L., Wigand, R.T.: Through a glass clearly: standards, architecture, and process transparency in global supply chains. J. Manage. Inf. Syst. **28**(2), 75–108 (2011)

Stevenson, W.J., Hojati, M.: Operations Management, vol. 8. McGraw-Hill/Irwin, Boston (2007)

Strong, D.M., Lee, Y.W., Wang, R.Y.: Data quality in context. Commun. ACM **40**(5), 103–110 (1997)

Thomas, G.: Alpha Males and Data Disasters: the Case for Data Governance. Brass Cannon Press (2006)

Wand, Y., Wang, R.Y.: Anchoring data quality dimensions in ontological foundations. Commun. ACM **39**(11), 86–95 (1996)

Wang, R.Y., Mostapha, Z., Yang, W.L.: Data Quality. Kluwer Academic Publishers, Dordrecht (2001)

Wang, R.Y., Strong, D.M., Guarascio, L.M.: Beyond accuracy: what data quality means to data consumers. J. Manage. Inf. Syst. **12**(4), 5–33 (1996)

Winter, R., Fischer, R.: Essential layers, artifacts, and dependencies of enterprise architecture. In: EDOCW 2006 (2006)

Xiao, R., Cai, Z., Zhang, X.: An optimization approach to cycle quality network chain based on improved SCOR model. Prog. Nat. Sci. **19**(7), 881–890 (2009)

Data Quality Assurance in International Supply Chains: An Application of the Value Cycle Approach

Yuxin Wang[✉], Joris Hulstijn, and Yao-Hua Tan

Section Information and Communication Technology (ICT), Faculty of Technology,
Policy and Management (TBM), Delft University of Technology,
Delft, The Netherlands
Y.Wang-12@tudelft.nl

Abstract. With increasing international trade and growing emphasis on security and efficiency, enhanced information and data sharing between different stakeholders in global supply chains is required. Currently data quality is not only problematic for traders, but also for various government agencies involved in border control, such as customs authorities and border force. We adapt principles from value cycle modelling in accounting, and show how these principles enabled by ICT can be extended to supply chain management to ensure quality of data reported to customs. We then describe a typical application scenario based on a real but anonymsed case to show that value cycle monitoring can be applied (feasibility), and if applied, what the expected benefits are (usefulness).

Keywords: Data quality · Auditing · Assurance · Information sharing · Supply chain

1 Introduction

Currently, data about shipments available in international supply chains does not provide a timely and accurate description of the goods [1]. For customs authorities, the low quality of data in reporting has proved to be a big problem, e.g. the explosion at the sea vessel MSC Flaminia, where authorities discovered that 605 of the data elements about the cargo in the containers were not accurate. Stakeholders involved have different data formats and communication channels. Also, redundancy and post processing are common problems in measures of control. For example, import declarations and bills of lading about goods are often made several days after the vessels have left the port of origin. Suppose companies in transport logistics and supply chains would be able to improve data quality, then customs can rely on business controls of enterprises, and at least for fiscal matters, additional inspecting and correcting customs related data afterwards at the port of destination would be unnecessary. Data quality issues often result from other stakeholders, further upstream in the supply chain. Under these circumstances, the so called 'push-left principle' [2] could be a solution: the consequences of deviations that are found in an audit or inspection, are 'pushed left', i.e. upstream in the supply chain to the party that caused them.

© IFIP International Federation for Information Processing 2015
M. Janssen et al. (Eds.): I3E 2015, LNCS 9373, pp. 467–478, 2015.
DOI: 10.1007/978-3-319-25013-7_38

How can we identify and develop new value adding services and accounting information systems design principles for enterprise, legislative and the audit profession community to solve these data quality problems and achieve sustainable collaboration in international supply chains? This involves enhanced cooperation between different stakeholders. Their bonds and connections are enforced by contracts. However, the principal-agent problem cannot be ignored [3]. Moral hazard and adverse selection[1] problems are sometimes inevitable. Therefore, assurance over data quality is needed, in particular for accounting information systems that are used to record such data [4].

Data Quality Management (DQM) entails the establishment and deployment of roles, responsibilities, policies, and procedures concerning the acquisition, maintenance, dissemination, and disposition of data [5]. We identify some specific challenges concerning DQM in international supply chains and provide some solution guidelines based on a case scenario afterwards in this paper.

Firstly, roles and responsibilities of different stakeholders involved needs to be analyzed, from the manufacturer, exporter and forwarder to the warehouse keeper, customs agent, cargo packers, etc. Secondly, cross-organizational boundaries are difficult to delineate. For example, customs import formalities which are formally the responsibility of the importer, are outsourced to his freight forwarder or customs broker. Thirdly, the contractual relationship is often weak and difficult to manage. For example, the importer depends for his import declaration on data about the goods provided to him by the ocean carrier, but this carrier only has a contractual relationship with the freight forwarder of the exporter of the goods in the country of origin. Business processes and data governance processes need to be well integrated. But contracts are often negotiated on price, not on service level. Other challenges, like the allocation of financial and human resources, require more cost and benefit analysis.

How can we ensure quality of data reported to customs with these challenges? Business reality can generally be modelled as a value cycle: an interrelated system of flows of money and goods [6]. The flow of money should mirror the flow of goods, but in reverse. The point of an accounting information system is to accurately and completely capture these two reverse flows using accounts information. Value-cycle models are well established in the owner-ordered audit tradition in the Netherlands that concentrates on financial reporting completeness, in addition to correctness [7]. When applied to data quality management, value-cycle models can prove to be beneficial and this will be illustrated in this paper.

Our goal here is to propose guidelines for designing and developing an information infrastructure and technology-based mechanism in international supply chains, for data quality monitoring. In this paper, data quality needs are assessed and evaluated within the context of organizational strategies, supply chain structure and existing business processes. First, we provide a brief overview of data quality and relevant definitions, as well as the general steps of data quality assurance. We then apply the value cycle approach to DQM in international supply chains. The next section is an application scenario of a real case in the Netherlands. The paper concludes with recommendations

---

[1] It refers to a market process in which undesired results occur when buyers and sellers have asymmetric information; the "bad" customers are more likely to apply for the service.

and implications on design principles for implementing value cycle (customs) controls in supply chains.

2 Data Quality Assurance and Data Quality Management

2.1 Defining Data Quality

Data quality is conformance to valid requirements. We should first [5] determine who set the requirements, then determine how the requirements are set. After that, determine the degree of conformance that is needed. In international supply chain domain, both the business and customs need to set data quality requirements. IT organizations/departments need to ensure that the business and customs can have accurate reporting data. They are aware of the existing data quality deficiencies, also the possibility and cost of overcoming them. Sometimes, changes in business processes are needed to address data quality problems. These factors must enter decision process.

Operationally, we can first define data quality in terms of data quality parameters and data quality indicators [8]: A **data quality indicator** is a data dimension that provides objective information about the data. Source, creation time, and collection method are examples. A **data quality parameter** is a qualitative or subjective dimension by which a user evaluates data quality. Source credibility and timeliness are examples. The value is directly or indirectly based on underlying quality indicator values. User-defined functions may be used to map quality indicator values to quality parameter values. For example, if the source is a RFID database, an auditor may conclude that data credibility is high. A **data quality requirement** specifies the indicators required to be documented for the data, so that at query time users can retrieve data within some acceptable range of quality indicator values.

2.2 Dimensions of Data Quality Management (DQM) Objectives

Under general accounting settings, data quality should improve from these dimensions [9]: (a) accuracy/correctness (b) completeness (c) timeliness (d) consistency, etc. For information system and IT infrastructure settings, there are more goals of DQM: (a) integrity (b) independency (c) relevance (d) confidentiality, etc.

2.3 General Processes of Data Quality Assurance (DQA)

Data quality assurance is the process of verifying the reliability of data. Protocols and methods must be employed to ensure that data are properly collected, handled, processed, used, and maintained at all stages of the scientific data lifecycle. This is commonly referred to as 'QA/QC' (Quality Assurance/Quality Control). QA focuses on building-in quality to prevent defects while QC focuses on testing for quality (e.g., detecting defects) [10]. To improve data quality, it is necessary to improve the linkage among the various uses of data throughout the system and across all business process:

1. **Data acquisition and identification**: The first step is to identify critical data areas. Normally this is manifest in two areas: (a) the basic business processes and (b) support for decision making about management of these business processes [11].

2. **Data discovery and profiling**: Data profiling is the systematic analysis of data to gather actionable and measurable information about its quality. Data discovery is achieved by executing data profiling and data monitoring tasks, analyzing data and determining the business rules used to populate the data.

3. **Data cleansing and enrichment**: Detect and correct erroneous data and data inconsistencies both within and across systems. Data enrichment involves enhancing existing data, by adding meta-data or changing data from industry standards and business insights to make it more useful downstream.

3   Data Quality Assurance with the Value Cycle Approach

3.1   Data Quality Assurance (DQA) in Supply Chain Management (SCM)

The relation between DQA and SCM is crucial, stakeholders in supply chains depend on each other, therefore information about agreements and situations must be reliable. Supply chains are generally present in enterprises across logistics, retail and other sectors. In these sectors, supply chains are crucial for business operations and SCM has a significant effect on business performance. "Supply Chain Management describes the discipline of optimizing the delivery of goods, services and related information from supplier to customer" [12]. *Enterprise Resource Planning* (ERP) systems are seen as the digital backbone for information in supply chains, especially when the supply chains are integrated over several companies or departments [13]. There are other information systems as well. Therefore, data monitoring and quality control in SCM can be continuous and automated throughout the whole DQA processes.

In SCM, data quality can have strong effects on operations. Consider for instance the bullwhip effect, which is the phenomenon of amplifying demand variability when moving up the supply chain, leading to growing inefficiencies and diminishing revenues [14]. This means that if a certain piece of data in a supply chain is erroneous or uncertain, fluctuations are increasing rapidly along the supply chain. This effect is affected by data quality, as business processes rely on data provided by others. When data cannot be relied on, it is prudential to keep extra stock. The next link in the chain will think likewise, amplifying the effect. When no specific requirements are set for a certain data element in the supply chain, this will not only cause an overall low data quality, but also amplified variances in stock levels along the supply chain.

3.2   Data Quality in Flows of Money and Goods

Supply chains have a big impact on organizations and are represented by the following flows [12]: goods flow as primary processes, information flow as CRUD (Create, Read, Update, Delete) processes, as well as financial flow. How is information flow linked with goods flow and financial flow? By CRUD operations in a database, whenever the status of the goods or money in the actual flow changes, information changes as well. CRUD processes are the four basic processes that can be performed with data in databases and describe the state of the data at a certain point in the process [15, 16], e.g.

import status changes have to do with the flow of money, as well as the flow of goods through the supply chain. These flows can often be used for cross verification. For establishing proper DQM in supply chains, it is thus required to take into account the goods flow and financial flow. The process of payments depends greatly on data quality, as errors in data can damage the relations with customers.

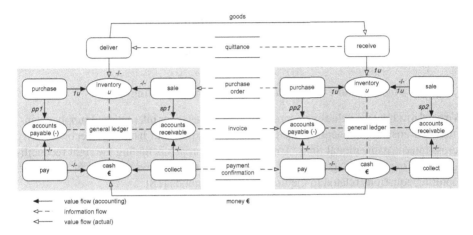

Fig. 1. Value cycle models of two trading companies linked by trade documents [18]

We look at data quality issues from an accounting perspective. Accounting has practices to deal with cross verification, using independent sources of data. One of these practices is to model businesses as a value cycle. Figure 1 shows an example of the value cycle for two trading companies, connected by trade documents (e.g. quittance, invoice, purchase order). We use the following notation. Activities are shown as a rectangle, such as a sales event. Ovals are the recordings of a state of a certain value to the company, such as inventory or accounts payable. States, i.e. accounts, are related through reconciliation relationships, indicated by dashed lines, which come together in the general ledger. The direction of the arrow indicates the influence of events. Arrows generally indicate an increment, while the sign '–/–' indicates a decrement of the corresponding account. Thus, a purchase leads to an increment of the accounts payable, while the purchased goods are added to the inventory. A sale leads to an increment of the accounts receivable and a decrement of the inventory, and so on.

The general idea of value cycle modelling is to use *Reconciliation Relations* to define a mathematically precise model of how the flow of money and goods should be (SOLL), depending on the specific manufacturing inputs and outputs for each type of business, and use it to verify actual audit business samples against (IST) [7]. The mathematical models could be instantiations of the following kinds of equations.

In Fig. 1, for all accounts S, T that are affected by an event e: $(S) \leftarrow [e] \rightarrow (T)$, e.g. $(inventory) \leftarrow [sale] \rightarrow (accounts\ receivable)$, we have the following transformation equations, where f is a constant (here *sales price*) that depends on the business model:

$$input\,(T, e) = f \cdot output\,(S, e),$$

As we record accounts in specific units of measurement (kg, 22 ft container, \$, mph), we also need conversion equations:

$$T \text{ in unit } u = f \bullet T \text{ in unit } v \text{ [17]}.$$

In addition, for all accounts S, we have the following preservation equation:

$$S(t_1) = S(t_0) + input(S, [t_0, t_1]) - output(S, [t_0, t_1]),$$

where for time interval $[t_0, t_1]$, $input(S, [t_0, t_1]) = {}_{def} Sum(input(S, e)$, for e in $[t_0, t_1])$.

3.3 Applying the Value Cycle Model to Customs Reporting

We need to adjust the value cycle model in three respects for international supply chains. First, add costs components related to goods transport and handling. Second, verify across inter-organizational links. Third, the key approach is finding the right reconciliation relationships that govern the international supply chain domain, in particular, capturing equations related to the flow of physical goods [18].

Here is a specific example of the goods flow in a bonded warehouse. Figure 2 illustrates that the data about goods entering a bonded warehouse[2] should correspond, according to many reconciliation checks, with the data about goods leaving the warehouse, either in transit or for import into free circulation in the EU. Customs have delegated controls over the warehouse to the company. To make sure the warehouse management system is reliable and no goods or documents are missing, they verify this afterwards every month, on the basis of electronic data. This is called 'electronic declaration' or 'audit file'. The so-called 'stock movement declaration' is part of the electronic declaration. The basic principle is that the total in the movement of goods must be balanced, using the following preservation equation, for any period of time:

opening balance(BV) + entries(BI) – debit entries(AF) – closing balance(EV) = 0.

From this formula, we can derive more equations for the case of boned warehouse:

opening stock at the beginning of a calendar year + entry of goods + internal changes = closing stock at the end of same calendar year + removal of goods;
opening stock at the beginning of a calendar year + entry of goods – destruction – vaporization–loss + findings + adjustment + other = closing stock at the end of same calendar year + goods released for free circulation + re-export + other.

What makes the above equations complicated is the case when several information lines together make up one mutation. For example, three different articles (A, B, C) are packed together into one article (D) (three-in-1 box) according to the rules governing usual forms of handling. In the stock records, the individual articles are registered separately from

---

[2] The bonded warehouse is under responsibility of a company, and used to store their goods under customs supervision, requiring a formal license from customs to operate. Until a customs destination is known, e.g. re-export (transit) or import (free circulation) no import duties are due. (See also https://en.wikipedia.org/wiki/Bonded_warehouse).

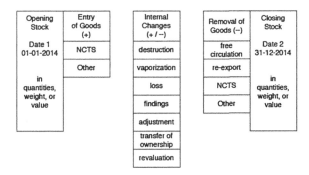

Fig. 2. Conceptual model of the inflow and outflow of a customs warehouse

the articles packed together (different article codes). When the three articles are packed together, this has an effect on the individual stock levels. The stock of the different articles (A, B, C) reduces and the stock of the article packed together (D) increases. This therefore involves 4 transactions: 3 minus-transactions and 1 plus-transaction. These information lines are linked to each other by including the same mutation number of each information line in the declaration system.

4   Application Scenario

4.1   Data Collection

Data for this application scenario description was collected from relatively open interviews with experts in trade, customs legislation, customs audit and companies. The materials from these interviews are supplemented with public sources about customs legislation and inspection policies.

4.2   Scenario Description

ABC is a Dutch company manufacturing machines for international market. Many components in manufacturing are produced abroad. Some of these parts are built into appliances and re-exported within or outside the EU. Other parts are re-exported directly. In the last case, no customs import duties are required. For these reasons, some warehouses at ABC operate as a Bonded Warehouse. Under this license it is allowed to store components from foreign origin, of which payment of import or export duties has been suspended, for an indefinite period until a customs destination is known: entering into the EU (free circulation), or re-exporting outside the EU (transit). In order to obtain and keep the license, ABC must allow regular audits and inspections from the Dutch customs. These involve IT audits of the administrative systems of the warehouse, physical security measures and occasionally inspections to find out if the inventory in the bonded warehouse corresponds to the reported goods.

We are interested in the requirements on ABC's inventory management and information systems. The most important requirement is that customs authorities should be able to rely on ABC's administration. It should be impossible to lose track of the stored goods (accurate and complete recording), to lose goods from the warehouse (physical security), or to assign the wrong customs destination (procedures). Errors that have a negative impact on accuracy or completeness of reporting data directly affect these key objectives. Data quality is crucial in two capacities:

1. Data quality is part of the requirements that follow from the Customs Warehouse license. It is therefore a key characteristic that must be audited regularly. Both the company and the auditors of the customs office therefore develop policies and procedures for assessing reliability of the company's record keeping.
2. In the audit itself, audit evidence is used and produced by the party being audited, so its reliability is not immediately guaranteed. Therefore, also to improve data quality for its own purposes, the company must build in certain additional precautions into the business processes, procedures and information systems to ensure reliability: so-called internal controls, e.g. segregation of duties; master data management; limited reading and writing access rights; access control measures; logging and monitoring; baseline security. Many controls are implemented using IT systems, hence, IT audit is necessary. These controls also need to be audited.

4.3   Issue

As part of the regular audits, both customs and the internal control department of ABC have now identified a number of weaknesses in the internal controls. Initial analysis has revealed that many errors can be attributed to the crucial process of 'reception', when goods are entered into the warehouse. At this point, ABC can still ensure that records of the goods match with the actual contents, on aspects like order numbers, container numbers, price, origin, goods description and bar-codes. Once goods have entered into the warehouse, it is much harder to trace errors and mistakes. For this reason, ABC has temporarily implemented a number of manual checks regarding the reception of goods. These controls are meant to detect and immediately correct deviations between shipping documents, such as the purchase order, invoice, customs declaration, the actual goods being received at the warehouse, etc.

What complicates the issue is that the bonded warehouse, usually consists of many physical warehouses also contains goods that are not under customs supervision.

Another complication is that at some of these premises, the reception process is not carried out by ABC itself, but has been outsourced to logistics service providers. So, ABC is dependent on logistics providers to carry out these checks adequately.

4.4   Solution Analysis

ABC has identified risks of overall processes, from general IT control, setting up purchase order and production order to sales. Also using controls in their ERP system and prescribing the right sequence of procedures in ERP, ABC sets up controls based on risks. The risk matrix ABC identified is updated on a regular basis. Experts from ABC jointly with the customs made the following steps to produce the risk matrix:

1. Identify for each individual movement type in ABC's ERP whether it is customs relevant or not. Not customs related means end of data flow, so if the goods are mistakenly categorized, it should trigger a control response in the system.
2. ABC implements for each identified customs relevant movement type a specific internal control to mitigate that risk.
3. These internal controls are built in ABC's ERP system. Financial flows rely on logistics flows and are fully automated in ERP.
4. ABC arranges regular IT audits to ensure these controls are working well in ERP.

The monthly declarations of ABC for customs are made using *Automated Periodic Reporting* (GPA), which is generated by a special information system. The EU has a special system *New Customs Transit System* (NCTS) for the reporting of so-called transit goods; i.e. goods that have entered the EU via a specific country, but have not yet been formally imported, and hence, for which no import duties have been paid yet. NCTS requires a manual step to enter the transit status of goods, and then returns a specific *Movement Reference Number* (MRN), which can be used as proof that these goods have the transit status. MRN is essential for ABC to generate an accurate GPA about goods. This manual reporting is done by the freight forwarder FF that has arranged the transport of their goods into the Netherlands.

We propose a systematic approach to improve internal controls, based on general risk management approaches (e.g. COSO ERM). First, identify remaining deficiencies in data quality; these may indicate risks related to customs compliance. Second, find the underlying root cause of these deficiencies. In many cases, the cause will be with another party, on whose data the organization depends. Try to fix deficiencies by improving information systems, processes or even conditions in the contract with other parties. Third, evaluate the remaining compliance risks. If they are unacceptable then repeat the procedure. This approach is called 'push-left' principle [2], because it aims to push any remaining control deficiencies left in the supply chain.

How can we measure the residual risks in a manageable level and indeed 'push left'? Process control in data processing is the underlying basis for data quality. If there is a gap in that process, this could mean goods are disappearing in this case. Only after being fully in control of the data processing can we go to the next level to see if the data is correct. If data at the next level is wrong, it might have financial impact but this can be fixed afterwards. Process control is on top of data quality problem, and is more about optimizing the physical goods' movement. From GPA to the risk matrix, assume which fields are mapped and covered by standard procedures, and then scope into a customs related risk matrix. After that we can delve into data quality. The suggested process controls for data quality assurance (DQA) are as follows:

1. *Acquire and record data from various sources with segregation of duties.*

Get data from different sources with adequate controls. Segregation of duties before data collection is a precondition for DQA at the source company. Despite internal controls, third parties who manage the information should also be unbiased.

Value cycle monitoring, as represented in Fig. 1, can play a crucial role in the analysis of the segregation of duties. The key auditing question from a customs point of view is how ABC can assure the accuracy of reporting data. This is an issue, because of the chain dependency of ABC on FF in providing relevant data. Also due to the manual processing of the transit status of goods transported by FF to the bonded warehouse of ABC, mistakes can happen. We will now explain how the model in Fig. 1 can be applied to analyze this auditing problem.

First, these transit status reports of FF can be viewed as an information service provided by FF to ABC. Actually this information service is just one activity in a broader portfolio of information services called customs brokerage, which are typically provided by FF and customs brokers.

The second observation is the chain perspective. The key assumption of the model is that data accuracy can be improved by using the countervailing interests between the different parties in a value network. In this case the value network consists of a simple chain of two parties: FF and ABC. FF has a different interest than ABC, because, although FF offers the transit status report as a commercial service to ABC, it does not directly affect FF's own business interest if they made a mistake. But the accuracy of these data is of direct interest for ABC, because they need to be compliant to customs, and if the report of ABC were not correct, there would be risks of being fined by customs. Therefore ABC added extra controls in ERP to double check whether the transit status reports that they receive from FF are accurate, accuracy of these data improve the accuracy of ABC's reporting to customs.

Thirdly, from a customs auditing point of view this chain can be viewed as a typical example of segregation of duties, as is depicted in Fig. 1, which enhances data accuracy. Another party, namely ABC is double-checking FF who is producing the transit status reports; the whole chain receives a positive audit assessment, because of the built-in segregation of duties between FF and ABC for this data validation.

2. *Validate data at the source against predefined data quality requirements.*

Evaluate those manual checks, set more explicit data requirements in the contract with the vendor. The 'Push left principle' requires more responsibility from the vendor.

Develop automated services for validating data records at the source. A strategic implementation enables the rules and validation mechanisms to be shared across applications and deployed at various organizations' information flow for continuous data inspection. These processes usually result in a variety of reporting schemes, e.g. flagging, documenting and subsequent checking of suspect records. Validation checks may also involve checking for compliance against applicable standards, rules, and conventions. A key stage in data validation and cleaning is to identify the root causes of the errors detected and to focus on preventing those errors from re-occurring [18].

3. *Set up unified standards, data formats and communication channels.*

All data providers need to agree on a communications protocol and the data format, to standardize data. For example, automated checks are performed during the sending of the GPA to customs. The format of data required for filing is a unified standard, and should be the same tracing back to the source manufacturers. The consequence of lack of IT and data interoperability across all stakeholders in a supply chain is that the process halts and the declarant is not informed.

4. *Build an information infrastructure to share data between stakeholders. Create a data pipeline with built-in controls, allowing more real-time collaborations.*

After negotiating with different parties involved, dedicate IT resources to build the information infrastructure and share data between all parties in the international supply chain. Transport conventions, systems, procedures and data in the Logistics Layer dominate the management of the supply chain. But the data relating to the goods to be bought, sold and moved needs to be known in the Transaction Layer to ensure the order is properly met and paid for. If that information was clarified and verified at the point of consignment completion and captured in a data system running parallel to the Logistics Layer then many of the risks associated with poor data would be reduced [11]. This means for reports about goods entry into the bonded warehouse, collect data via the data pipeline from the actual packing list of the consolidator that actually 'packed the box' with goods in the country of origin.

For automated monitoring and sufficient build-in controls, an application platform should include much more than a traditional server operating system does, e.g. a modern cloud platform could provide capabilities such as data synchronization, identity and entitlement management, and process orchestration[3]. The platform should also provide access to new technologies and ideas of enterprise computing.

5. *Check reconciliation relationships and build feedback systems to better monitor.*

With the help of normative or prescriptive equations in Sect. 3.2, deviations in the actual flows of money and goods can be identified based on actual measurements of the variables during operations. The checks could be on the net weight, number of units and money value using the equations we illustrated in Sect. 3.3. Re-valuation and transfer of ownership also need to be carefully checked with details.

Meanwhile, create automated feedback loop with human capital investment. If one data user (either internal or external) detects a data defect, he can create a flag in the system and the defect will be automated sent to the source for reviewing.

5 Conclusions

How can we get quality data with multiple standards, formats and communication channels in international supply chains? How can the value cycle approach contribute to data quality management for customs reporting? To what extent can we reduce the redundant manual checks and costs of control in data quality management?

In this paper we tried to answer these questions by introducing an approach that builds on value chain modelling from a chain perspective to application in international supply chains. This approach, specifically for data quality assurance in customs reporting, is based on segregation of duties and developing verification equations that can be used to verify data quality across the whole supply chain.

---

[3] See more on www.thesupplychaincloud.com and www.opengroup.org: Cloud Computing Open Standards, *the Supply Chain Cloud Report.*

We illustrate the approach by a case scenario of a manufacturing company. It shows that the steps of data quality assurance we proposed can be implemented and if successfully would be beneficial for different stakeholders. If the goods information generated by the vendor, at the starting point of the supply chain, is accurate and complete, those manual checks by parties at the other end of the supply chain would be unnecessary. This is only part of the research, and we leave the development of analytical detection models from reconciliation relationships for further research. Nevertheless, we believe that if information sharing could be improved this way, data quality in international supply chains could also be improved and regulatory compliance risks would be reduced, resulting in operational benefit as well.

References

1. Hesketh, D.: Weaknesses in the supply chain: who packed the box? World Customs J. **4**(2), 3–20 (2010)
2. de Swart, J., Wille, J., Majoor, B.: Het 'push left'-principe als motor van data analytics in de accountantscontrole. Maandblad voor Accountancy en Bedrijfseconomie **87**, 425–432 (2013)
3. Eisenhardt, K.M.: Agency theory: an assessment and review. Acad. Manage. Rev. **14**(1), 57–74 (1989)
4. Romney, M.B., Steinbart, P.J.: Accounting Information Systems, 10e. Prentice Hall, Upper Saddle River (2006)
5. Geiger, J.G. (ed.): Data quality management the most critical initiative you can implement. In: The Twenty-Ninth Annual SAS® Users Group International Conference (2004)
6. Starreveld, R.W., de Mare, B., Joels, E.: Bestuurlijke Informatieverzorging (in Dutch): Samsom, Alphen aan den Rijn (1994)
7. Blokdijk, J.H., Drieënhuizen, F., Wallage, P.H.: Reflections on Auditing Theory, a Contribution from the Netherlands. Limperg Instituut, Amsterdam (1995)
8. Wang, R.Y., Kon, H.B., Madnick, S.E.: Data quality requirements analysis and modeling. I: Ninth International Conference of Data Engineering; Vienna, Austria (1993)
9. Wang, R.Y., Strong, D.M.: Beyond accuracy: what data quality means to data consumers. J. Manage. Inf. Syst. **12**(4), 5–33 (1996)
10. Chapman, A.D.: Principles of Data Quality. Global Biodiversity Information Facility, Copenhagen (2005)
11. Orr, K.: Data quality and systems theory. Commun. ACM **41**(2), 66–71 (1998)
12. Cooper, M.C., Lambert, D.M., Pagh, J.D.: Supply chain management: more than a new name for logistics. Int. J. Logistics Manage. **8**(1), 1–14 (1997)
13. Gunasekaran, A., Ngai, E.W.: Information systems in supply chain integration and management. Eur. J. Oper. Res. **159**(2), 269–295 (2004)
14. Lee, H.L., Padmanabhan, V., Whang, S.: The bullwhip effect in supply chains. Sloan Manage. Rev. **38**(3), 93–102 (1997)
15. Martin, J.: Managing the Data Base Environment. Prentice Hall, Upper Saddle River (1983)
16. Polo, M., Piattini, M., Ruiz, F.: Reflective persistence (Reflective CRUD: reflective create, read, update and delete). In: Sixth European Conference on Pattern Languages of Programs (EuroPLOP) (2001)
17. Veenstra, A.W., Hulstijn, J., Christiaanse, R., Tan, Y.-H.: Information exchange in global logistics chains: an application for model-based auditing. In: PICARD2013 (2013)
18. Redman, T.C.: Data Quality: The Field Guide. Digital Press, Boston (2001)

Towards a Federated Infrastructure
for the Global Data Pipeline

Wout Hofman[(✉)]

TNO, The Hague, The Netherlands
Wout.hofman@tno.nl

Abstract. Interoperability in logistics is a prerequisite for realizing data pipelines and the Physical Internet. Forecasting data, real time data, and actual positions of shipments, containers, and transport means shared via events have to be harmonized and are expected to improve all types of processes, support synchromodal planning, and improve risk analysis from a compliance and resilience perspective. Technically, several solutions are implemented by organizations and innovations have been validated in so-called Living Labs or demonstrators in various projects. These solutions do not yet provide open systems required for a (global) data pipeline. A federation of solutions is required to construct data pipelines and to support sustainable development of applications on smart devices allowing Small and Medium sized Enterprises to collaborate. This paper proposes a set of platform services and so-called platform protocols to allow interoperability of different platforms for constructing a data pipeline. The proposed services and protocols further extend existing interoperability solutions and services for supply and logistics.

Keywords: Seamless interoperability · Data pipeline · Federated platforms · Service · Protocol · Physical internet

1 Introduction

Customs authorities require additionally data to their current declaration for risk analysis improvement and introduced the concept of data pipeline for seamless data sharing as a solution [1]. Such a data pipeline consists of a large number of stakeholders like shippers, consignees, forwarders, and carriers, exchanging value according a transaction hierarchy, called logistic chain, in an organizational network (reference). The actual implementation of a data pipeline is by interconnecting legacy systems of the stakeholders and/or support by commercial – and community solutions [2]. It is not to be expected that one global system will implement the data pipeline, but interoperability between existing systems and solutions needs to be constructed [2]. As of currently, many interoperability implementations in trade and logistics are based on the message paradigm, but also other mechanisms are explored to address for instance real time data sharing for dynamic planning or resilience [3], Service Oriented Architecture (SOA) supported by Enterprise Services Busses (ESBs) [4] or Linked Data [5]. For real time data sharing, the current generation of platforms supports an Application Programming

© IFIP International Federation for Information Processing 2015
M. Janssen et al. (Eds.): I3E 2015, LNCS 9373, pp. 479–490, 2015.
DOI: 10.1007/978-3-319-25013-7_39

Interface (API) registry [6], a particular SOA implementation based on the REST protocol. APIs are still technical specifications that require interpretation to derive semantics.

Seamless data sharing between systems and components of different stakeholders requires universal connectivity [7]. In this respect, scoping of specifications is also important like taking a bilateral or multilateral interoperability [8] or a modeling approach covering organizational chains [9]. Interconnecting internal business processes resulted in reference models either specifying both processes and data [10] or only data with a messaging choreography [11]. Implementation of these reference models still lead to closed systems, since, organizations make bilateral or community agreements based on these models [12]. Several sources [13, 14] stress the importance of unambiguous semantics as part of interoperability, but do not address the implementation of this semantics in legacy systems or other solutions. It is yet unclear how process aspects need to get addressed in interoperability. Interoperability layering [14] considers pragmatics without presenting a way to model pragmatics like taking the bilateral or multilateral, chain approach.

A complicating factor is that Small and Medium sized Enterprises (SMEs) cover some 80 % of the logistics market [15] performing some 20 % of the business. These SMEs have either simple or no IT solutions or systems, but interface manually with systems of their customers, potentially supported by web interfaces. Thus SMEs have to deal with different interfaces to become interoperable with their customers instead of having simple applications running on smart devices with cloud solutions of one or more communities and or providers, since these SMEs operate international and require interfacing with many systems and solutions.

This paper proposes a set of platform services that enables an enterprise to connect once to an infrastructure of federated platforms and compose a data pipeline. Standardization of this set of services allows development of applications on smart devices, where these applications can interconnect to any given platform thus creating a sustainable business model for app developers in logistics. Each solution provider in this infrastructure can have its particular implementation of the services, thus satisfying their customer requirements and have sufficient market share. Firstly, requirements to platform services leading to design choices are formulated and secondly the services and the protocol for platform federation are introduced. The research presented by this paper is based on an action design research approach [16] across several EU funded and Dutch projects addressing interoperability in logistics. Each project has constructed artifacts that do however not meet requirements formulated in [7, 17].

2   Design Choices

A (federation of) platform(s) has to meet particular user requirements. Since it is fairly complex to assess user requirements for all global data pipelines, those stemming from various European Union (EU) funded projects and literature will be transformed into design choices. An example of a design choice is for instance the support of the messaging paradigm, common to most interoperability implementations between organizations. By making these design choices explicit, discussion on their applicability to meet user requirements is supported. Design choices are on distinction of 'service', 'protocol', and 'interface', bilateral versus multilateral business process modeling, semantics, and data governances supported by privacy-enhanced technologies. This section presents choices based on practice inspired research [16] and briefly reflects the state of the art in research. It does not pretend to be complete, but identifies some basic research questions. The answers to these questions have to be supported by a federation of platforms; the next section shows the mechanisms to do so. Semantics is core to all choices made.

2.1 Participating in a Federated Infrastructure

Currently, organizations bilateral or multilateral develop interoperability agreements, encompassing both functional and non-functional aspects, like message implementation guidelines and process alignment, based on open standards or with proprietary formats leading to closed solutions [12]. Each time a business relation with another enterprise needs to be established, investments in agreements has to be done. Seamless interoperability [7, 17] addresses this problem, but does not provide solutions. [9, 18–19] introduce business process modeling either for bilateral or multilateral interoperability as a solution, but [20] argues to model only behavior between any two peer entities. There are different ways to specify behavior; [21] provides transaction templates for bilateral interoperability to construct chains. A generic specification of behavior will not be applicable to all resources, since they all have different goals and capabilities [22]. A generic specification can however serve as a reference framework for specifying these particular goals and capabilities.

This paper proposes to apply the concept of 'resource' offering both real time data and providing or requesting logistic services as specified by an 'Information Profile' of such a resource. In this respect, several issues need to be addressed, namely how to express the external behavior of a resource in terms of interactions and semantics. The concept of transaction templates to express external behavior for business transactions can be applied [21]; other mechanisms like events might be required to share any logistic state changes like arrival of a vessel and delivery of a container at its destination. Semantics of one's profile can be expressed as an ontology, based on a networked ontology of logistics concepts and services (see for instance ontology.tno.nl for a logistics ontology). Semantics of data and behavior need to have a technical binding to a paradigm like messaging and SOA [4] supported by a syntax like XML or JSON. In case any two communicating organizations have different technical bindings, binding negotiation and a data transformation function have to be implemented, either by a data provider, a consumer, or as service of the federation of platforms (see next section). In case Information Profiles of any two communicating organizations are based on an identical - or matched semantic models, an on-the-fly technical binding by a platform can be constructed, as long as the federation of all platforms support that technical binding.

The concept 'resource' with its 'Information Profile' needs further research and examples stemming from real use cases. These examples are currently developed in EU funded projects like EU FP7 CORE.

2.2   Data Governance and Privacy Enhanced Technologies

Organizations are hesitant in sharing information due to for instance commercial or liability reasons, e.g. the amount of free capacity of a barge might reduce prices, the location of a truck might increase vulnerability for cargo theft or providing real-time and predicted depth of a waterway might increase liability. There are currently a number of barriers that block the adoption of data sharing amongst resources, e.g. data owner-ship, privacy, commercial sensitivity, liability, and culture [23]. In this respect, data and events are classified as:

- Open data. Data is publicly available to everyone. Open data is normally considered to be available without any costs, but in some occasions like the Cadastre data, one needs to pay.
- Community data. Data is shared within a community according agreed rules. Like with open data, one might distinguish free - and paid data.
- Partner data. Data is shared with a specific partner.
- Internal data. Data is only shared within an organization, according internal data policies (Fig. 1).

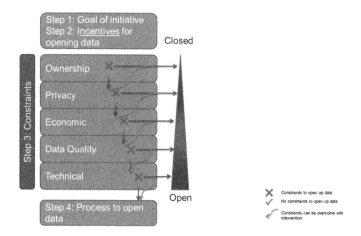

Fig. 1.   Decision support instrument for data sharing [23]

Data classification has a lifetime, e.g. it may change over time and/or may be appli-cable for one or more calls or interactions. For instance, available capacity on a trip may be shared only once in a community at the start of the trip or can be updated during the trip. Communities can also be flexible, e.g. organizations can join and leave a community over time.

The previous figure shows the decision model developed by [23]. It addresses various aspects like data ownership, privacy and commercial sensitivity, and economic aspects, resulting in a data policy supported by interventions. Many of these interventions are supported by privacy-enhanced technologies [24] like identification and authentication, access control, filtering, and homomorphic encryption [25]. One particular technology might support data sharing for one or more decisions, but we have not yet found these for logistics data sharing. For instance, Role Based Access Control expresses access control for internal data, but a more fine grained access control mechanism like Attribute Based Access Control might be required for partner – and community data. Templates for particular roles can be specified to ease the specification of data policies, e.g. a template for a role like a forwarder. These templates are a form of Role Based Access Control that can thus be refined by organizations meeting their particular requirements as they operate in one or more roles. These templates may implement formal restrictions from a liability and financial perspective, e.g. carriers should not receive any data on the content of a container, whilst they are otherwise liable for any damage or loss to the content.

Role – and Attribute Based Access Control can be expressed as an ontology of a set of rules, based on the earlier mentioned networked ontology for logistics. To support global logistics and supply, protocols are required for a federation of identities [26]. Further research is required with respect to the relation between privacy-enhanced technologies as interventions for data governance and implementing these technologies in real world cases.

2.3 Service, Protocol, and Interface

The Internet design principles of 'service', 'protocol', and 'interface' [20] are applied for specifying a federation of platforms supporting data sharing between organizations. A federated platform is said to offer a 'service' to back office systems of supply and logistic stakeholders, e.g. the ability to exchange messages, validate the message structure and content, and validate the message sequence, whereas a 'protocol' between any two platforms provides the ability to actually share data with for instance messages. The protocol is the set of agreements for sharing data between any two platforms, independent of a local implementation of the service by each of those platforms to their users, which is called 'interface'. The service is the conceptual representation of this protocol to a one or more back office systems and/or end-user. The same service can be implemented by various interfaces, e.g. a file sharing mechanism or an API can serve as an implementation of a service. In fact, 'interface' is the technical binding of a service offered to back office systems of an organization. The technical binding of an interface can differ from that of an agreed binding of the protocol, which requires transformation by a local implementation of the protocol and service by a platform (Fig. 2).

Introducing these concepts allows conceptually specification of a service and a protocol with different technical bindings, both for the service by its interface and a protocol for interoperability between two local implementations. Complexity reduction of federated platforms is achieved if all participating platforms support the same semantics and technical bindings of the protocol.

Each service – and protocol primitive has a particular structure with control information and a payload, where a semantic model specifies the semantics of the payload.

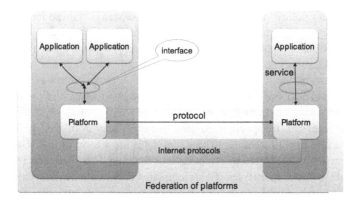

Fig. 2. The concepts service, protocol and interface

Control information of a service primitive is applied by a local implementation for processing that primitive and constructing a protocol primitive with a negotiated - or on-the-fly technical binding (see before). It allows for instance the transformation between a SOA interface of a service to a message based protocol and vice versa. Communication is already implemented by various protocols and is therefore considered out of scope. Thus, the service and protocol for logistics will be elaborated, whereas a local interface is discussed in the next section considering back office integration.

3   A Federation of Platforms

The previous section has introduced a number of research questions and concepts for federated platforms. This section further elaborates the services of a (federation of) platform(s) and its protocol. The services reflect the concept of Information Profile and the support of data governance, all based on networked ontologies.

3.1   Federated Platform Services

A (federation of) platform(s) provides generic services to resources, where these generic services are configured by semantics for a particular application area like supply and logistics. The services can be categorized in two main groups that can be further decomposed (Fig. 3): operational services consider actual sharing of logistics data utilizing various technical bindings and mechanisms; administrative services consider registration and other types of support services for operating a (federation of) platform(s). The decomposed services can be defined as:

- Registration Services. These services support an organization in its registration and connection to a platform (profile specification services) and specify its data policies (data policy specification services).

- Real time data sharing services. These support data policy negotiation, both for events and data, search for a particular (composite) service matching a goal (matching services), and sharing the state of supply and logistics chains via events (visibility services). Visibility services can be on particular objects like trucks with their location, speed, etc. in an area (geo-fencing), timeframe (time-fencing) or a combination of both [27]. Visibility services can only provide the state of an object or also evaluate the state against requirements, which is supported by (complex) event processing [28]. Matching services can be on search for a (structured set of) profile(s) meeting a customer goal, where a structured set composes a logistics chain or all retrieved profiles exactly meet a customer goal. These matching services can be applied in various ways, e.g. to support synchromodal booking [28].
- Transaction support service: validating the sequencing of interaction according an agreed transaction protocol specified by a choreography. In this particular case, the initiation and processing of transactions is in applications registered at a platform.
- Data sharing service: reliable and secure exchange of data and events according a particular technical binding, potentially supported by data transformation. Reliability services consider data resubmission in case the receiving platform or application responds and are not always required. The same is applicable for secure data sharing.
- Supporting services. These services support the operation of a federation of platforms providing particular services to platform users. These services not necessarily have a supporting protocol. The following supporting services are foreseen:
  - Semantic services providing the networked ontologies (see Sect. 2).
  - Publish/subscribe services providing the ability to subscribe to particular events, where these events provide state information. Publish/subscribe may require policy negotiation services.
  - Non-repudiation services that provide proof of actual data shared between any two users of a federation of platforms. Non-repudiation services are supported by an audit trail registering all actions in terms of data sharing between any two actors, e.g. sending or receiving particular data with a timestamp, a log containing the actual data that has been shared, and monitoring services providing both access to the audit trail and the log.
  - Accounting and billing services supporting paid data according agreed pricing structures. These services utilize the non-repudiation services.
  - Certification services providing identification and authentication of platform users.

Services are interrelated. Visibility services can for instance use publish/subscribe services and data exchange services for events, transaction support services can utilize data exchange services by messaging and reliability services to assure a reply is received in time. Secure and reliable data exchange is for instance specified by Electronic Business XML with ebMS [19] and implemented by an eFreight access point [29]. A formal service specification considers the control information and payload [20], which yet needs to be performed.

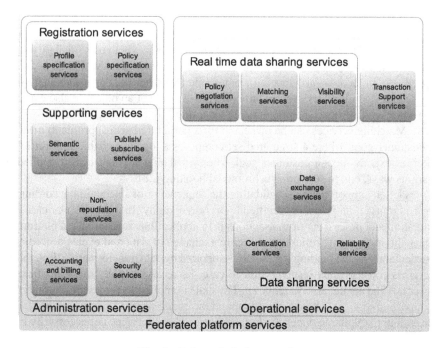

Fig. 3. Federated platform services

3.2   Federation of Platforms

Like the services, the protocol can also be decomposed. The protocol should support each of the services, but some protocols can support more than one service by a different payload of the protocol primitives [20]. The protocol is decomposed as follows:

- Data policy negotiation protocol: negotiate the data that can be shared amongst two organizations.
- Matching protocol: sharing goals and (a structured set of) service(s) to support matching services.
- Visibility protocol: how to access particular data on the supply chain status, e.g. it basically consists of events either received upon subscription or by a query.
- Transaction protocol: the agreed choreography of interactions for transaction support services.
- Data sharing protocol: sharing data between two communicating systems, where the payload is provided by the aforementioned protocols. The data sharing protocol is decomposed in:
  - Binding negotiation protocol: to establish the technical binding for sharing data.
  - Data exchange protocol: the actual sharing of data (and events) according an agreed or selected (on-the-fly) technical binding.

- Reliability protocol: resubmission and identification of resubmitted messages, potentially resulting in a receipt acknowledgement, and timers to detect timely replies to support reliability services.
- Security protocol: selecting a secure protocol with agreed certificates, e.g. https.
- Authentication protocol: a protocol for federation between certification authorities to support authentication of an identity [26].

One might consider to introduce a protocol for supporting registration services, e.g. to share complete Information Profiles and data policies. Such a protocol would provide complete transparency. In our proposed approach, the data negotiation -, matching -, and binding negotiation protocol support this type of transparency, but does not provide a generic data policy of any registered user.

3.3 Implementation and Deployment

There are several ways to implement a federation of platforms, e.g. if any two enterprises are connected to the same platform like a Port Community System, data sharing services between them might be implemented via a database and registration is based on administrative services with a proprietary format. Only in case two enterprises connected to different platforms, the protocols will be required. Currently, most of these protocols are message based (see before).

There are two dimensions to the deployment of the Connectivity Infrastructure, namely an business dimension and ICT dimension. Both will be discussed here. The ICT dimension addresses the development of a local implementation, with potentially different components providing different services and supporting particular parts of the protocol. The following options are feasible:

- Open source: the services are provided by an open source software solution that every resource can implement. Like indicated before, eFreight Access Points provide particular functionality, so do iCargo Access Points in supporting virtualization of logistics actuator objects like containers and trucks [3].
- COTS (Commercial Of The Shelve): the implementation of the protocol with a local interface to its back office systems based on COTS. The software offering the service and supporting the protocol is licensed to a resource or its owner that implements the functionality. The COTS provider is responsible for correct (and complete) implementation of the protocol.
- Proprietary: the IT department of an enterprise develops the implementation of the protocol or its implementation is outsourced to an external software developer. The implementation can be based on open source solutions and/or components like available integration brokers. The solution is owned by the enterprise (with the exception of COTS components used for implementation) and the enterprise is responsible for correctness of the protocol implementation.

From a business perspective, the service can be implemented in many ways, for instance:

- Resource Deployment: each resource implements and operates the protocol by itself. The services are internal to the resource, but utilizing an ICT solution implementing the services reduces development costs. The iCargo Access Points try to provide this functionality [3].
- Community Systems: two or more organizations might decide to implement the services themselves with different local interfaces to back office systems and/or end-user functionality. These organizations thus own the community system. Port Community Systems are examples of these types of systems.
- Cloud Platform: the service and the protocol are provided by a platform of a commercial provider. The latter ones can construct their particular solutions on top of the services. Note that this type of solution is identical to telco or other providers offering services with COTS and/or open source solutions for communication protocols.

The services can also be used for open innovation, implying the development of apps for SMEs (see also www.logicon-project.eu). To support open innovation, the services have to be more tailored to a specific target group of end-users, e.g. barge operators are expected to have other services than truck drivers.

A number of services is currently proprietary to a particular solution, e.g. registration of resources at a community system or cloud platform. These deployment solutions do not yet support the protocol, which makes it difficult to find for instance resources in an infrastructure. Furthermore, these deployment solutions have their specific data policies implemented by message implementation guides, which restricts data sharing across these platforms and require all types of transformations between them.

4   Conclusion and Further Research

The paper presents a set of services for federation of platform solutions and services supported by a protocol for logistics and supply. By standardizing supply and logistics services and their underlying protocol, the so-called data pipeline for interoperability in trade facilitation will be enabled. The services and protocol enable each individual object or actor to act as a information resource with particular capabilities for data sharing. By examples, we have mapped functionality to existing components like developed for eFreight and iCargo. Dedicated solutions and services can be developed addressing particular interoperability aspects like virtualization of actuators representing physical objects like trucks and containers, thus also contributing to the concept of the Physical Internet [7]. Standardisation of services of federated platforms also contributes to development of sustainable business models for deployment of apps on smart devices for SMEs. By separating service and protocol, each resource implementing a protocol stack known to the infrastructure will be able to participate, without additional costs and effort for development of bilateral or community guidelines. Each ICT solution – and service provider will also be able to tailor its services to optimally integrate resources in the infrastructure and provide added value like complex event processing and transformations to these resources.

The solutions provided by this paper need to be developed further, including construction of low cost connection to the federation of platforms. We have indicated that lots of existing systems and cloud solutions have a role in implementing the services. Federation of platforms requires additional research into the business models of these platforms, like a sustainable business model of community systems. Not yet, all identified services are fully supported by software solutions.

Acknowledgements. The content of this paper is based on research and development in EU FP7 SEC Cassandra, EU FP7 SEC CORE, EU FP7 INFSO iCargo, EU FP7 INFSO Logicon, several other projects for data sharing in logistics and supply chains, and discussions in various fora like ALICE on the subject of seamless interoperability.

References

1. Klievink, B., Van Stijn, E., Hesketh, D., Aldewereld, H., Overbeek, S., Heijmann, F., Tan, Y.-H.: Enhancing visibility in international supply chains: the data pipeline concept. Int. J. Electron. Gov. Res. **8**(4), 14–33 (2012)
2. Hofman, W., Bastiaansen, H.: A global IT infrastructure improving container security by data completion. In: ECITL, Zaragoza, Spain (2013)
3. Dobler, M., Schumacher, J.: A pan-european ecosystem for intelligent cargo and its impact on emission reduction. In: Transport Research Arena 2014, Paris (2014)
4. Erl, T.: Service Oriented Architecture – Concepts, Technology and Design. Prentice-Hall, Upper Saddle River (2005)
5. Heath, T., Bizer, C.: Linked data - evolving the web into a global data space. In: Synthesis lectures on the Semantic Web: Theory and Technology, Morgan & Claypool Publishers (2011)
6. Hofman, W., Rajagopal, M.: A technical framework for data sharing. J. Theoreticla Appl. Electron. Commer. **9**(3), 44–57 (2014)
7. Montreuil, B., Meller, R.D., Ballot, E.: Physical internet foundations. In: Borangiu, T., Thomas, A., Trentesaux, D. (eds.) Service Orientation in Holonic and Multi Agent Manufacturing Robots, pp. 151–166. Springer, Heidelberg (2013)
8. Schonberger, A., Wilms, C., Wirtz, G.: A requirements analysis of business-to-business integration. In: Fakultat Wirschaftsinformatik und angewandte Informatik Otto-Friedrich-Universitat, Bamberg (2009)
9. Huemer, C., Liegl, P., Motal, T., Schuster, R., Zapletal, M.: The development process of the UN/CEFACT modeling methodology. In: 10th International Conference on Electronic Commerce (IEC), Innsbruck, Austria (2008)
10. Pedersen, J.T.: One common framework for information and communication systems in transport and logistics: facilitating interoperability. In: Golinska, P., Hajdul, M. (eds.) Sustainable Transport. EcoProduction. Environmental Issues in Logistics and Manufacturing, pp. 165–196. Springer, Heidelberg (2012)
11. World Customs Organization. WCO Data model - cross border transactions on the fast track. World Customs Organization (2010)
12. Hofman, W.: Applying semantic web technology to interoperability in freight logistics. In: e-Freight 2011, Munich (2011)
13. European Commission. European Interoperability Framework for European public services. European Commission, Brussels (2010)

14. Wang, W., Tolk, A., Wang, W.: The levels of conceptual interoperability model: applying systems engineering principles to M&S. In: Spring Simulation Multiconference (2009)
15. Transport and Logistics Netherlands (TLN). Transport in figures, TLN (2015)
16. Sein, M.K., Henfridsson, O., Purao, S., Rossi, M., Lindgren, R.: Action design research. MIS Q. **35**(1), 37–56 (2011)
17. Chituc, C.-M., Azevedo, A., Toscano, C.: A framework proposal for seamless interoperability in a collaborative networked environment. Comput. Ind. **60**, 317–338 (2009)
18. Van den Heuvel, W.-J., Papazoglou, M.P.: Towards business transaction management in smart networks. IEEE Comput. Soc. (2010)
19. Kotok, A., Webber, D.: ebXML - the new global standard for doing business over the internet, New Riders (2002)
20. Tanenbaum, A.S.: Computer Networks, 3rd edn. Prentice Hall, Upper Saddle River (1996)
21. Dietz, J.: Enterprise Ontology, Theory and Methodology. Springer, Berlin (2006)
22. Spohrer, J.K.S.: Service science, management, engineering, and design (SSMED) - an emerging discipline - outline and references. In: International Journal on Information Systems in the Service Sector, May 2009
23. Eckartz, S., Hofman, W., Van Veenstra, A.F.: A decision model for data sharing. In: eGov2014, Dublin, Ireland (2014)
24. Lagendijk, I.L., Erkin, Z., Barni, M.: Encrypted signal processing for privacy protection. IEEE Signal Process. Mag. **82** (2013)
25. Gentry, C.: A fully homomorphic encryption scheme. Stanford University (2009)
26. Pruksasri, P., van den Berg, J., Hofman, W., Deskapan, S.: Multi-level access control in the data pipeline of the international supply chain system. In: iCETS, Macau (2013)
27. Merrienboer, S., Veenstra, A.W., van den Haak, W., Tavasszy, L.: Using floating truck data to optimise port logistics. In: 10th Intelligent Transport Systems European Congress (ITS2014), Helsinki (2014)
28. Hofman, W.: Control tower architecture for multi- and synchromodal logistics with real time data. In: ILS2014, Breda (2014)
29. Angholt, J., Wackerberg, M., Olovsson, T., Jonnson, E.: A First Security Analysis of a Secure Intermodal Goods Transport System. Chalmers University, Gothenburg (2013)

Key Design Properties for Shipping Information Pipeline

Jensen Thomas[1]([⊠]) and Yao-Hua Tan[2]

[1] Copenhagen Business School, Howitzvej 60, 2000 Frederiksberg, Denmark
tje.itm@cbs.dk
[2] Faculty of Technology, Policy and Management, Delft University of Technology,
Delft, The Netherlands
y.tan@tudelft.nl

Abstract. This paper reports on the use of key design properties for development of a new approach towards a solution for sharing shipping information in the supply chain for international trade. Information exchange in international supply chain is extremely inefficient, rather uncoordinated, based largely on paper, e-mail, phone and text message, and far too costly. This paper explores the design properties for a shared information infrastructure to exchange information between all parties in the supply chain, commercial parties as well as authorities, which is called a Shipping Information Pipeline. The contribution of the paper is to expand previous research with complementary key design properties. The paper starts with a review of existing literature on previous proposed solutions for increased collaboration in the supply chain for international trade, Inter-Organization Systems and Information Infrastructures. The paper argues why the previous attempts are inadequate to address the issues in the domain of international supply chains. Instead, a different set of key design properties are proposed for the Shipping Information Pipeline. The solution has been developed in collaboration with a network of representatives for major stakeholders in international trade, whom evaluate it positively and are willing to invest, develop and test the prototype of the Shipping Information Pipeline.

Keywords: International trade · Inter-organizational systems · Information infrastructure

1 Introduction

Research regarding international trade estimates that 40 % of the delays in the lead-time of supply chains for international trade in the large ports are caused by administrative burdens imposed by authorities. Typically data inaccuracy is 50 % for the information reported by businesses to authorities; since these data are used for the risk assessment, this is rather critical. The annual world-wide extra costs due to administrative burdens of crossing borders are estimated in the range 100–500 Billion US\$[1].

The organizations in international trade are characterized by utilizing a wide range of communication channels including phone, e-mail, SMS and paper based media for

---

[1] Cassandra Research Project presentation, 2014.

© IFIP International Federation for Information Processing 2015
M. Janssen et al. (Eds.): I3E 2015, LNCS 9373, pp. 491–502, 2015.
DOI: 10.1007/978-3-319-25013-7_40

information and documentation related to the shipments. This creates issues for the actors involved including lack of knowledge about status, information and documentation of shipments. Information exchange in international supply chain is considered extremely inefficient, uncoordinated and far too costly [1]. To address these challenges it's in line with previous research e.g. [2] propose to use a shared information infrastructure (II) for communication of the related shipping information whereby the information can be shared among multiple organizations[2]. The II for shipping information is named the *Shipping Information Pipeline* (SIP). It's the long term ambition to design, build, and test (ultimo 2015) plus evaluate a prototype of the SIP and this paper only covers the initial design of a prototype.

International trade plays an important role in the global economy and is a complex eco-system. The domain of international trade involves up to 40 actors/organizations in a single shipment including both private businesses and public organizations in a complex eco-system [3]. Furthermore, a serious complication in this eco-system is that the information exchange is extremely inefficient and rather uncoordinated. Complete different IT systems and message standards are used on different sides of the oceans. This has led to a practice, where information exchange is based primarily on paper and e-mail with many re-typing and copy-paste operations, phones calls and text messages. Consequently, it's relatively costly and error prone for the commercial parties as well as authorities e.g. inspection agencies in supply chain. Finally, the number of independent organizations and the huge variety of different IT systems make changes extremely difficulty even if the changes seem beneficial for the overall eco-system.

Usage of modern IT to improve the situation in international trade has been researched and attempted previously but none of the solutions have become used at an international scale. The design of an II as the SIP is different from solutions proposed in previous research which built on more traditionally centralized information systems; because the SIP has no central database, organization or control. For the previous solutions almost all communication is based on bilateral information exchanges between two organizations; whereas an II as the SIP enables sharing of information simultaneously among multiple organizations or even sharing the information by publishing it public. The SIP is comparable to the Internet and designed to be on top of the Internet and its standards. The SIP is an internet for shipping information.

In order to design the SIP one of the major players in international shipping decided to engage in research regarding the design properties for the SIP. The major player foresee that it's impossible for a single player to become successful with the SIP and search for ways to establish a collaboration with other partners including authorities in the international trade. To engage other partners the designing and testing of prototypes of the SIP are important. The research involves design theory for information infrastructure and follows the design science method to guide the research.

Accordingly the research question guiding this paper is:

What would be the key design properties of a Shipping Information Pipeline?

---

[2] Organizations can be public or private e.g. a company, a cooperation, an association, an institution, etc.- an entity on its own; an actor belongs to an organization and can be user of IS solutions. A stakeholder will have a stake e.g. in owning the SIP.

The scope of this research is limited to the containerized international trade bound for the European Union by deep sea, but this is not a severe limitation since the majority (70 to 90 %) of imported goods to the European Union is carried by container ships. Accordingly the other means of transport are not considered here. The largest volume to Europe passes through the port of Rotterdam. Accordingly the research has focused on trade lanes governed by Dutch authorities.

The rest of the paper is structured as follows: First theories behind previous proposed solutions, Inter-Organizational Systems [4] and design of Information Infrastructures (II) are presented. Secondly the research method is described. Thirdly are the proposed key design properties for a domain specific II for shipping information within the supply chain for international trade for containerized trade bound for European Union. Finally, an evaluation is provided and followed by a discussion.

2 Literature Review

The roots of the idea to share data via an electronic pipeline for international trade and subsequently attempts to develop IOS for international trade are presented below. Followed by a presentation of theories regarding IOS solutions currently used for information exchange and the Design Theory and design properties of II.

2.1 The Roots of the Idea of an Electronic Pipeline for International Trade

The idea of a sharing data via an electronic pipeline roots back to two official EU representatives from Dutch and United Kingdom customs. They came up with the idea of an electronic pipeline when visiting China, and they proposed that the European authorities should get the data about each and every shipment from the source e.g. the Chinese company packing the container. This kind of solution would dramatically reduce import authorities' serious problems with poor data quality; which is seriously affecting the customs/inspection authorities, when they are performing risk assessment of inbound cargo.. Several studies have identified that the quality of the data provided to authorities is poor, misleading and in some cases even fraudulent. An illustrative example is that when a container vessel stranded, there was a comparison of the filed information on the contents of the containers and the actual contents when the containers were opened and the authorities found that the data quality was only around 60 % [5]. Accordingly, there is a clear advantage to the EU authorities, if they via an electronic data pipeline could get the data directly from the source.

This idea caught on in the EU, and research funds were made available to fund a range of research projects addressing some of the technological challenges and demonstrating the possible solutions. Important projects addressing the issues and analyzing the potential benefits for organizations involved in international trade: ITAIDE[3], Contain[4], Integrity[5],

---

[3] www.itaide.org.

[4] www.containproject.eu.

[5] www.integrity-supplychain.eu/.

Cassandra[6], and iCargo[7]. In the ITAIDE project (2006–2010) an "I3" framework was developed for accelerated trade through networks of trusted traders utilizing an II built on IT innovations which enables four critical capabilities[8]: Real-time monitoring, Information sharing, Process control, and Partner collaboration [2]. They identified that the technology "enables designing new ways of working, i.e., new business models. The ITAIDE technologies .. facilitate a redesign of the interaction pattern between government and businesses, and second, by facilitating piggybacking as a process optimization." (ibid. p. 179). The idea of piggybacking enable actors involved to reuse electronic information instead of retyping whereby work effort is reduced and the data quality is improved. Several living labs were developed and successfully evaluated with a range of leading IT vendors illustrating the fact that by following the ITAIDE approach, it's possible to make a significant improvement in trade facilitation. In a subsequence research project Cassandra (2010–2014) the electronic data pipeline concept was elaborated for enhancing visibility in international supply chains by an event driven architecture providing more up to date/real time data to improve the logistic efficiency [6, 7]. Based on stakeholder analysis more solutions for business-to-government information have been developed e.g. a customs dashboard [8]. The above mentioned projects have successfully developed local IOS solutions primarily based on EDI (and standards) for demonstration purposes which subsequently not have been adapted by the organizations involved; one of the major reasons given is that the organizations are reluctant to share their information/documents. In summary the idea of an electronic pipeline dates years back and the solutions tend to focus on harmonizing/ standardize information. This research contribution is to provide complementary design properties which affect the architecture of the SIP.

2.2 Inter-Organizational System for International Trade

In more general terms, IOS are defined as "information systems to span boundaries between countries, organizations and the relatively separate components of large, geographically dispersed corporations" [9]. Extant literature about using IT for collaboration across organizational boundaries and borders is typically studied under the umbrella of IOS [4]. A closer look on the IOS literature reveals that there are more than 25 theories [10] and no single theory stands out as predominant. The majority of research regarding IOS is focused on EDI[9] [11], and a majority of the described IOS are successfully utilizing EDI [12]. For international trade, the benefits of facilitating IOS based on EDI are well documented [13] but the cost of change is relatively high [14]. The proposed SIP can be categorized as an IOS according the IS theory on IOS [9] since the SIP is an

---

[6] www.cassandra-project.eu.

[7] www.i-cargo.eu.

[8] "Real-time monitoring is the capability to monitor and log real time – where a shipment is and how it is handled. Process control is the capability to document and evaluate that business processes meet control standards. Information sharing means the ability to electronically exchange information regarding shipments with trading partners and authorities. Partner collaboration refers to the joint capability of trading partners and IT providers to develop end-to-end control and transparency".

[9] Based on the international EDI standard: United Nations/Electronic Data Interchange For Administration, Commerce and Transport (UN/EDIFACT) developed under United Nations.

information systems with the purpose to span boundaries between organizations with separate components/systems and borders of countries which are geographically dispersed. The examples of IOS includes "electronic data interchange .. supply chain management, electronic funds transfer, electronic forms, electronic messaging, and shared databases" [15]. Note that, though it's widely used for the collaboration, e-mails are not considered as IOS because the IOS focuses on system-to-system connection and not system-to-human. Similar the SIP focuses primarily on systems and leaves the actor/ user centric element to the individual organizations. The majority of the IOS researched are based on standardized EDI messages and in this regard the design of the SIP is different since the standardization is limited to the identifiers and the few additional data (needed for the subscribe service). So, with respect to standardization, the key focus of the SIP is on standardization of IS systems interfaces, and not on standardization of data and or messages that are exchanged between the systems.

Even IOS based on EDI is relatively successful; it's primarily used by large organizations and only covers a small part of the communication of information and documentation involved in international trade [16]. The widespread use of EDI based IOS [4] are mainly automation islands (mainly due to the standardization required primarily is successful locally), which are not properly integrated especially not internationally, accordingly there is a huge need for some type of inter-organizational reengineering to reduce mistakes, increase efficiencies and reduce time lag. The SIP can reengineer the IOS by offering communication across multiple organizations and borders complementing the existing traditional EDI based IOS. By designing the SIP to a lower cost and entrance investment than the existing IOS the SIP is expected to be used by more organizations and cover more information/documentation.

2.3 Design Theory for Information Infrastructure

The components of an IS design theory includes: (a) Requirements (b) Set of system features (c) Kernel theory and (d) Design principles/properties [17]. Hanseth and Lyytinen [18] propose a design theory for IIs based on IS design theory with a kernel theory and a set of refined properties for II. The emergent properties are: Shared, Open, Heterogeneous, and Evolving. Formulated theoretically, IIs is defined "as a shared, open (and unbounded), heterogeneous and evolving socio-technical system consisting of a set of IT capabilities and their user, operations and design communities." [19]. Additionally are identified two structural properties: Organizing principle and Control. Based on the kernel theory, a set of design properties and nineteen design rules for II has been suggested as design strategy addressing the two generic problems for IIs: bootstrapping and adaptability, bootstrapping being the initial start up and adaptability being the spreading of use.

The proposed SIP can be characterized as an II according to the IS definition of IIs since the SIP complies with both the emergent properties: Shared, Open, Heterogeneous, and Evolving; and structural properties: Organizing principle, and Control. Additionally the SIP is intended to have a global reach, being open to any organization, to be one virtual pipeline build on many pipelines, and to be realized through evolution; accordingly the SIP is regarded as an II. The IS theory to guide the design and development of a successful SIP would then be the design theory for IIs [19], which includes a set of design properties and design rules for successful bootstrapping and evolution. The organizing principle for the SIP is not settled yet but it's intended to facilitate the emergent properties where the control is distributed to the involved organizations.

Note that Hanseth and Lyytinen [19] do not distinguish between the different types of communication in the II since they have both industry specific EDI based IOS platform and the Internet as examples of II. The concept of communication makes a crucial design property difference between the EDI based IOS and the "Internet" for shipping information. Furthermore the governance of trust and protection of information is found to be a key design property where the SIP differentiates between published shared and trusted information shared only bilaterally after authentication. In above has been discussed to which extent the SIP can be categorized as IOS, a platform and/or an II according to the IS theory. The proposed SIP can be characterized as an II and an IOS, even it's development is very different from the development of standard EDI message based IOS. Accordingly the IS theory is providing guidelines for successful design and development of the SIP.

3   Research Method

The research method applied for the research reported is following an IS design science paradigm, and the interventions are described in the following. The initial focal case for the design of the SIP has been the trade lane transporting fresh cut roses from Kenya to Europe. Design science research is a particular perspective within IS research [20] which focuses on the development of artifacts related to information and communications technology. Design science research includes an evaluation of the designed artifacts. For the initial design the evaluation is artificial and not naturalistic based on real use of the SIP which follows in a later when the prototype is tested on actual shipments. Design science research places IS research between the environment (practitioners) and the knowledge base (researchers), the knowledge justifies the proposed solution to the problem in the environment. In the case of the SIP, the exchange is between a network of organizations within international shipping and the IS research field's knowledge, and the research focuses on relevance and rigor guided by a set of seven "guidelines for Design Science in Information Systems" [21]. This research has been inspired by those guidelines e.g. for the design and evaluation phases.

One of the key features of a design science project is iterations of interventions between practitioners and researchers. The interventions with both researchers and practitioners range from dedicated workshops, meetings and conference calls to workshops over 1–1½ year from spring 2014. The interventions have been documented by written material in the form of minutes of meetings and presentations, which in the subsequence interventions has been commented upon in order to validate the correctness of the documentation.

4   Design Properties for the Shipping Information Pipeline

In this section is presented the key design properties for a domain specific II for international containerized trade bound for European Union. The idea was first conceived in the IT department of one of the stakeholders (a large international container shipping line), which initiated an initial design named the SIP. Several activities have taken place towards prototyping of the SIP, which involved identifying: the conceptual idea, the business benefits and the associated business model, the major stakeholders, the issues addressed by the potential use, the potential barriers/obstacles, etc. In the following is only the design properties described.

The conceptual idea has been communicated between participants from of the potential stakeholders, who defined the SIP in the following way: "The SIP is a service based facility to allow partners in the supply chain to share accurate original data from it's source. It can connect any number of trading partners .." (minutes of meeting from a workshop 3rd September 2014 between potential stakeholders). They identified two main issues: Lacking full end-to-end supply chain visibility and lacking the ability to efficiently share common data/documents.

Another set of stakeholders focused on other but related main issues: (I) the potential for reducing the relatively high cost of the administrative barriers for international trade; (II) the security challenge for the authorities; (III) lack of visibility in the supply chain for international trade. Different stakeholders focus on different issues to be addressed by the SIP. The potential users have been identified by a mapping actors among the more than thirty organizations involved in one selected trade lane for international trade and encountered actors from both private and public organizations in more than five countries. Additionally the information primarily in the form of documents to be shared were listed and characterized. The businesses benefits for the potential organizations utilizing the SIP have been exemplified in order to identify possible business models for all relevant actors potentially affected by the new SIP. There seems to be a challenge identifying a feasible business model for using the SIP.

The potential stakeholders for the SIP have been analyzed and include e.g. IT vendors and start ups. No stakeholder stands out as being the obvious 'key-stakeholder', who is able to set the standards and enforced it. Every stakeholder holding an installed based have been positive to the idea of communicating using the SIP, but none of them have taken the lead. Instead each and every stakeholder has established collaboration with all the other relevant stakeholders, and consequently more or less everyone is in principle prepared to take part in an overall initiative, but nobody is prepared to make a commitment to lead. It seems to be a paradox that nearly all of the involved organizations are expected to benefit from the SIP but none of them sees SIP as their core business. One of the major stakeholders in the supply chain took initiative to design a first version of the SIP and the key design properties are described in the following. The design of the SIP is kept very clean and with one clear focus: sharing of information about events for shipments relevant for the actors in the supply chain for international trade. For the communication to potential stakeholder the conceptual idea behind the SIP has been formulated in <u>design properties /criteria</u>:

- **No Big Brother.** In order to avoid big brother issues where a central entity has access to central database with detailed information on global trade, no trade document are stored in or transferred via the Shipping Information Pipeline. (partly relates to the structural design properties of II: organization and control)

- **Integrate Once – Connect to Everyone.** When an actor has built the required standardized integrations, the actor will be able to exchange information seamlessly with all other actors integrated with the Shipping Information Pipeline. (relates to the design property of II: shared)
- **One Virtual Pipeline Build by Many Physical Pipelines.** For the users the Shipping Information Pipeline will look like one pipeline, but the actual physical infrastructure can be handled by several individual organizations. Standards will ensure the Shipping Information Pipelines integrates seamlessly in a way similar to how the Internet works today. (relates to the design property of II: heterogeneous)
- **No Facilitation of Commercial Agreements.** The Shipping Information Pipeline will not facilitate commercial agreements between two actors." (is not addressed by any of the design properties of II)

The above design properties/criteria from presentation at a workshop 16th October 2014 reflect the focus on creating trust for the organizations using the SIP, ease of use and global coverage. The trust is addressed by not publishing, sharing or storing detailed data about shipments in the SIP but leaving that to a direct bilateral connection among the ones wanting to share. Additionally excluding commercial agreements increase the trust since no commercial data are in the SIP to prevent any use of the SIP to gain commercial benefits on behalf of competitors are not facilitated by the SIP. When an organization has integrated to the SIP then the exchange of event information will be seamlessly and easy to use e.g. from inside the organizations' IT systems. The global coverage of the SIP is ensured behind the scene even regional set of SIPs are foreseen demanded by practicalities and by authorities.

The above description of the design considerations, including the design properties communicated by the stakeholder taking the initiative, illustrates that the design of the SIP involves many dimensions without an overall consent about the actual properties. The design of the SIP complies with all the design properties from design theory for II and is in line with the ideas from previous research: information from the source via an electronic pipeline, piggy bagging, and up to date/real time information; but to focus only the key design properties were communicated. Further the key design properties communicated changed depending on the actual utility of the particular organization and the extent to which it's possible to address their concerns. The key design properties communicated for the SIP could be characterized as not being a fixed set of properties but rather to be flexible, evolving and adapting to the audience.

5   Evaluation of the Shipping Information Pipeline

The evaluation of the SIP is an ongoing process where various potential stakeholders evaluate the SIP typically at different abstraction levels. The levels include actor level, organizational level, country/region level and society level.

The individual actor in the many organizations using the SIP will be able to get more insight into events in the supply chain for international trade for the shipments in which the individual actor is potentially interested since today none of the actors have transparency.

Additionally the SIP will provide higher quality and up to date information compared to today where information often is missing, out of date and of poor quality information and out of date information which is a major headache for the actors. Accordingly, when asked they find the service provided by the SIP useful especially when things do not go as planned.

The private organizations involved are the traders and the service providers. The traders foresee that the SIP can improve the possibilities for more efficient logistical coordination and lower the risk which will impact the international trade cost and willingness to trade. The international trade cost can be split in a physical transportation cost and an administrative cost of respectively 8 % and 20 % of retail cost [22]. The SIP addresses primarily the administrative border related part of international trade cost which is significant and amounts to approximately 20 % of the retail cost. The service providers e.g. a major shipping line (the main drivers behind the SIP is obviously interested) foresee the main benefit being that lower international trade cost will increase trade volume resulting in more business especially when being a first mover.

The public organizations are active in supporting the prototype to realize the idea (including the two officers from customs in NL and UK, who first came up with the idea of utilizing an electronic data pipeline for international trade). Further the public organizations taking part in the piloted trade lane express their willingness to collaborate regarding the pilot project accordingly they foresee potential improvements for their area of responsibility. Through the SIP the authorities will have the opportunity of getting data directly from the source whereby the quality of the information will increase compared to today, which enable the authorities to improve their risk assessments and the accuracy for the calculation of tariffs etc. but it also imply a change for the authorities' way of working since they need to follow a link to get detailed information instead of receiving it (when requested).

Another set of private organizations involved are IT vendors offering solutions that facilitates information exchange for international trade. They have been positively engaged in collaborating regarding the SIP but none of them have seen a business opportunity, which they want to pursue. Anyhow a major IT vendor has been involved in a series of workshops detailing the architectural design of the SIP and has agreed to invest and build a prototype. But the vendor is still struggling to find an attractive business model, hence it is unclear, who is prepared to fund the further development and operations of the SIP. Governing of the SIP is also a challenge for such a hugely diversified group of organizations involved in the SIP.

At country level the impact of reducing the administrative barriers are estimated to have a significant impact on trade volume which affects the economic positively. The World Economic Forum (WEF) estimates that an improvement to half-way of regional best practice and of global best practice will result in increased Gross Domestic Product (GDP) by respectively 3 % and 5 % [23]. Such improvements are important especially for developing countries e.g. in East Africa (Sub Sahara), where the similar estimate is an increase in GDP by 12 % if applying halfway global best practice. For the first piloted trade lane between East Africa and Europe several association expressed positive expectations about the SIP and committed to be actively involved in the first pilot project lead by TradeMark East Africa[10].

---

[10] Trade and Markets East Africa is an East African not-for profit Company Limited by Guarantee established in 2010 to support the growth of trade in East Africa.

European Union has a clear an interest in the SIP, since EU is actively involved and is funding the research program in which the testing of the SIP prototype is a part. The EU sponsors especially aim to improve the security for containers imported to the EU and to ease trade between US and EU.

At society level WEF estimates that by lowering barriers for international trade volume will increase and thereby fuel economical growth. "Estimates suggest that an ambitious improvement in two key components of supply chain barriers, border administration and transport and communications infrastructure, with all countries raising their performance halfway to global best practice, would lead to an increase of approximately US$ 2.6 trillion (4.7 %) in global GDP and US$ 1.6 trillion (14.5 %) in global exports. By contrast, the gains available from complete worldwide tariff elimination amount to no more than US$ 400 billion (0.7 %) in global GDP and US$ 1.1 trillion (10.1 %) in global exports" [23]. On basis of the above the SIP can potentially contribute significantly to the growth in the global economy.

In summary: (1) a venture fund of a major IT vendor has decided to fund to built a prototype of the SIP based on the design properties (2) a regional pilot implementation is planned for East Africa (3) EU sponsors the research (4) several organizations involved in international trade have committed to participate in testing with real shipments. The willingness to engage and invest in prototyping and testing of the SIP is taken as a positive evaluation of the design properties for the SIP.

6   Discussion

The overall vision of the SIP has guided the development of the key design properties. The design of previous attempted solutions for collaboration in the supply chain for international trade has been EDI based IOS; the design of the SIP is based and in line with the design properties for II plus additional ones but still in line with the initial ideas: information directly from the source, piggy bagging and real time information. One of the advantages is that an IOS based on II as the SIP is built on top of the internet and it's standards requires less standardization efforts compared to the EDI based IOS. Compared to previously attempted solutions the design knowledge is not the EDI based IOS but Design Theory for II.

Only a set of the design properties namely the key design principles have been communicated (in writing) among the stakeholders. Further focusing on the key design properties contribute to make the design clean and simple. The key design properties have evolved and have been adjusted over time primarily depending on the audience, the above described key design properties express the consensus among organizations involved in the prototype even so minor adjustment of the design properties are to be expected. As described above the key design properties for the SIP add design properties to the ones provided by previous research regarding IOS for international trade. The contribution of this paper is to expand previous research with complementary key design properties.

References

1. Jensen, T., Bjørn-Andersen, N., Vatrapu, R.: Avocados crossing borders: the missing common information infrastructure for international trade. In: Proceedings of the 5th ACM International Conference on Collaboration Across Boundaries: Culture, Distance & Technology. ACM (2014)
2. Tan, Y.H., Bjorn-Andersen, N., Klein, S., Rukanova, B. (eds.): Accelerating Global Supply Chains with IT-innovation: ITAIDE Tools and Methods. Springer, New York (2011)
3. Jensen, T., Tan, Y.-H., Bjørn-Andersen, N.: Unleashing the IT potential in the complex digital business ecosystem of international trade: the case of fresh fruit import to European Union (2014)
4. Kaniadakis, A., Constantinides, P.: Innovating financial information infrastructures: the transition of legacy assets to the securitization market. J. Assoc. Inf. Syst. 15(4) (2014)
5. Branch, M.A.I.: Report on the investigation of the structural failure of MSC Napoli, English Channel on 18 January 2007. MAIB Report (2008)
6. Klievink, A., et al.: Enhancing visibility in international supply chains: the data pipeline concept. Int. J. Electron. Gov. Res. 8(4), 2012 (2012)
7. Overbeek, S., Janssen, M., Tan, Y.-H.: An event-driven architecture for integrating information, processes and services in a plastic toys supply chain. Int. J. Coop. Inf. Syst. 21(04), 343–381 (2012)
8. Klievink, A., Janssen, M., Tan, Y.-H.: A stakeholder analysis of business-to-government information sharing: the governance of a public-private platform. Int. J. Electron. Gov. Res. 8(4), 2012 (2012)
9. Gregor, S., Johnston, R.B.: Developing an understanding of interorganizational systems: arguments for multi level analysis and structuration theory. In: ECIS 2000 Proceedings, p. 193 (2000)
10. Madlberger, M., Roztocki, N.: Cross-organizational and cross-border IS/IT collaboration: a literature review. In: Proceedings of the Fourteenth Americas Conference on Information Systems, Toronto, ON, Canada (2008)
11. Reimers, K., Johnston, R.B., Klein, S.: The shaping of inter-organisational information systems: Main design considerations of an international comparative research project. In: Tan, Y., Vogel, D., Gricar, J. Lenarts, G. (eds.) 17th Bled eCommerce Conference: "eGlobal", Faculty of Organizational Science, University of Maribor, Bled, Slovenia (2004)
12. Robey, D., Im, G., Wareham, J.D.: Theoretical foundations of empirical research on interorganizational systems: assessing past contributions and guiding future directions. J. Assoc. Inf. Syst. 9(9) (2008)
13. King, J.L.: Balance of trade in the marketplace of ideas. J. Assoc. Inf. Syst. 14(4), 3 (2013)
14. Henningsson, S., Bjørn-Andersen, N.: Exporting e-Customs to developing countries: a semiotic perspective. In: Proceedings of the Second Annual SIG GlobDev Workshop, Phoenix, USA. Idea Group Publishing, Hershey (2009)
15. Schwens, C., et al.: International entrepreneurship: a meta-analysis. In Academy of Management Proceedings. Academy of Management (2011)
16. Jensen, T., Vatrapu, R.: Ships & roses: a revelatory case study of affordances in international trade. In: 23rd European Conference on Information Systems (ECIS) 2015
17. Walls, J.G., Widmeyer, G.R., El Sawy, O.A.: Building an information system design theory for vigilant EIS. Inf. Syst. Res. 3(1), 36–59 (1992)
18. Hanseth, O., Lyytinen, K.: Theorizing about the design of Information Infrastructures: design kernel theories and principles (2004)

19. Hanseth, O., Lyytinen, K.: Design theory for dynamic complexity in information infrastructures: the case of building internet. J. Inf. Tech. **25**(1), 1–19 (2010)
20. Gregor, S., Hevner A.R.: Introduction to the special issue on design science. Inf. Syst. e-Business Manag. **9**(1) (2011)
21. Hevner, A.R., et al.: Design science in information systems research. MIS Q. **28**(1), 75–105 (2004)
22. Anderson, J.E., Van Wincoop, E.: Trade costs. National Bureau of Economic Research (2004)
23. WEF, W.E.F.i.c.w.T.B.C.G., Connected World. Transforming Travel, Transportation and Supply Chains. World Economic Forum, Insight Report, 2013

Enhancing Awareness on the Benefits of Supply Chain Visibility Through Serious Gaming

Tijmen Joppe Muller[1(✉)], Rainer Müller[2], Katja Zedel[2],
Gerwin Zomer[1], and Marcus Engler[2]

[1] TNO, PO Box 23, 3769 DE Soesterberg, The Netherlands
{tijmen.muller,gerwin.zomer}@tno.nl
[2] Institut für Seeverkehrswirtschaft und Logistik, Universitätsallee 11-13,
28359 Bremen, Germany
{rmueller,zedel,engler}@isl.org

Abstract. Improving both efficiency and security in international supply chains requires a new approach in data sharing and control measures. Instead of managing supply chain risks individually, supply chain partners need to collaborate in order to exchange cargo information and implement control measures on the level of the entire supply chain. Governmental agencies, having access to this up-to-date and complete information, can implement alternative risk assessment policies, resulting in less disruptive ways of supervising entire trade lanes. However, this paradigm shift requires awareness of these supply chain visibility concepts and increased collaboration between partners in a value chain. In order to disseminate these new concepts and initiate cooperation between key stakeholders, a serious game called 'The Chain Game' was designed, implemented and evaluated.

1 Supply Chain Visibility

One of the main challenges that international intermodal container logistics faces today is how to balance efficiency and security. Efficiency is one of the key performance indicators businesses are aiming for in order to have competitive advantages. In contrast, governmental agencies, such as customs at border control, have to maintain a high level of security to protect citizens and business partners against unlawful practices. However, increased efficiency and increased security are goals that may oppose each other. For example: in order to reach maximum efficiency, disruptive governmental interventions are undesirable from the business perspective, whereas maximum security is reached by high inspection rates and heavy enforcement measures applied to all incoming and outgoing containers. In current day practice, most organizations in a value chain manage their supply chain risks individually, by applying a combination of the 4Ts: transfer, terminate, tolerate and treat. Risk transfer practices are very mature and widely applied by the more powerful chain actors. The weaker chain actors often cannot transfer such risks and pressure on margins forces them to tolerate certain risks. If risk treatment is being considered, it is often done through internal control measures.

© IFIP International Federation for Information Processing 2015
M. Janssen et al. (Eds.): I3E 2015, LNCS 9373, pp. 503–512, 2015.
DOI: 10.1007/978-3-319-25013-7_41

Research in supply chain dependencies, vulnerabilities and resilience proposes an alternative approach, one that considers the risks from the perspective of the entire supply chain, instead of from each individual fragment of the chain. Risk based control and supervision seems the way forward, targeting high risk consignments and concentrate inspection effort on the targeted categories, whilst applying alternative less disruptive ways of supervision to the rest. This concept of *Supply Chain Visibility* [1] takes the needs of both governmental agencies and businesses in the chain into account. The basic idea is a data sharing concept that is built on integration of existing data processing systems. This so-called 'Data Pipeline' is used by business companies for end-to-end supply chain visibility, to ensure high data quality and data completeness and efficient connectivity to community systems. Customs and other authorities can connect to the Data Pipeline through a dashboard interface and use selected data for reporting purposes, such as customs declarations. It also allows them to piggy back on the validated and enriched supply chain data and advance the risk analysis and assessment they used to perform on data from declarations. An advantage for all parties is that all available information is up-to-date, since the data itself is fed by the originator of data and only retrieved by authorized users. Time-consuming search for additional information concerning a single transport could be avoided and additional data sources (from container security devices for example) could enrich the data sets without revision of company-wide or governmental data bases.

In theory this Supply Chain Visibility concept can benefit all actors within a supply chain. This approach may not only lower costs for the business partners in the supply chain because of increased data sharing efficiency, but may also have customs recognize that the corresponding trade lane partners are 'in control' of the major risks in their supply chain, resulting in less disruptive interventions from the authorities. However, some required steps need to be taken in practice to operationalize this approach. Most importantly, trade lane partners need to collaborate in implementing both the data sharing concept and additional chain control mechanisms. Examples of control mechanisms on the level of the entire chain are partner screening and coaching, applying technologies for container integrity (e.g. Container Security Devices) and data validation procedures, either through physical checks (e.g. a tallyman checking the packing list information) or by comparing data from different sources (three-way match). A second important step is new policy from the governmental agencies, that allows less disruptive interventions based on a different way of risk analysis.

Creating awareness of the aforementioned concepts and control mechanisms, having key stakeholders in a value chain recognize the benefits of such an approach and initiating collaboration to put these concepts into practice proves to be a difficult and time consuming process. In order to speed up the concept development cycle but also to disseminate the research findings we implemented an alternative approach: a serious game. We believe the characteristics of serious gaming, which allows stakeholders to have concrete experiences with the innovative concepts in interaction with each other, will disseminate the research successfully and will initiate collaboration for future implementation of supply chain visibility.

2   Serious Gaming

Serious games have proven to be an excellent medium to explain and teach complex concepts through experiential learning in a wide variety of domains. A well-known example is strategic war games that have been used in military training for centuries. At least since the mid-19th century the use in the military tactical training of officers has been documented [2, 3]. Business simulations, which can be simple board games or massive computer games, are commonly used in management education [4, 5]. An almost classic example for this is the Beer Game, which was created in early 1960s at the MIT's Sloan School of Management [6] and is still used in management education to have players experience the bullwhip effect. Technological progress has allowed for more complex and realistic simulators, such as flight simulators, nautical simulators, crane simulators or driving simulators.

Simulation games and simulators are commonly used to allow students to experience situations which are too costly, hazardous or unethical to experience in the real world. Simulation games are very useful to understand complex problems through experiential learning: experiencing the effects and dependencies of own behavior is a very strong learning method. Simulation games allow making mistakes in a safe environment and thus learning by trial and error [7]. This invites players to try new strategies and challenge long-standing patterns of behavior. In "The Chain Game" we use this to guide the players to try more collaboration.

The use of (computer) games in education - either especially designed for education or commercial off-the-shelf (COTS) games - is growing [8–10]. The learning theories which can be identified in games include well known theories of learning such as Gagne's "Conditions of Learning theory", Gardner's "Theory of Multiple Intelligences", Skinner's "Operant Conditioning theory", Thorndike's "Laws of Effect", Maslow's "Hierarchy of Needs theory" and Kolb's "Experiential Learning Model". Also incorporated are learning approaches like active learning, experiential learning and situated learning [11].

Probably the most compelling point about games is their ability to motivate and engage, far better than other teaching methods. Prensky believes that games "are the most engaging intellectual past-time that we have invented" [8]. In studies, participants report to have more interest in simulations or games than in classroom instructions. This correlates with the participants investing significantly more time in learning with games [7, 9, 11]. For our dissemination we take advantage of the engaging and interest raising nature of games. Buckley and Anderson identified three common characteristics which make games so compelling [13]:

1. Being in control: Players work at their own ability level and speed and repeat material as needed.
2. Feedback and rewards: Games give immediate feedback. Additional most games reward behaviors and actions by the player which are "positive" according to the in-game logic particularly. This increases the frequency of this behavior and teaches a positive attitude towards the content of the game.
3. Challenge and mastery: Games challenge players but remain doable. They keep the balance between too hard and tedium, giving the player feelings of self-efficacy.

Since games are an engaging medium for explaining complex systems, they can be very useful to create awareness for new concepts, such as the data sharing concept of Cassandra, with key stakeholders. Serious gaming allows meaningful interaction with both a relevant representation of the world and between key stakeholders, allowing them to analyze the situation from different perspectives and create a shared understanding of the challenges [14]. As such, gaming can play an important role in the transformation process needed to eventually implement these concepts.

3   The Chain Game

Given the goals we hope to achieve, explaining the complex and abstract concepts of supply chain visibility and initiating collaboration between key stakeholders in supply chains, we believe serious gaming is a very suitable medium – more so than reports or presentation material. In modern supply chain logistics stakeholders face multidimensional problems with complex side effects and non-linear dependencies. A game typically is a simplified world, reducing this complex multidimensional problem to its essentials in order to showcase only the relevant parts. This section describes *The Chain Game*, the serious game that was designed to transfer the abstract chain visibility and chain control concepts into practical examples and to allow players to experience the effects of the researched innovations, even though the exact effects (in reality) are not yet known.

The starting point for designing The Chain Game is the goals that should be achieved after playing the game. These learning goals were defined as follows:

1. Players understand that collaboration between supply chain partners is needed in order to operationalize the concept of supply chain visibility. Additionally, collaboration will have benefits on the performance of the supply chain, but the investing partners will not necessary receive the gains, so discussion on return-on-investments is necessary. In general, the performances of businesses within one chain are interdependable.
2. The concept of supply chain visibility and the underlying innovations need to be disseminated. The players (typically key stakeholders in a supply chain) need to understand the broader picture of supply chain visibility, but also understand typical control measures they could implement.
3. Implementation of the supply chain visibility concepts will increase robustness of the entire supply chain as risks are more easily mitigated and governmental agencies will apply less disruptive supervision measures.

These requirements are used to design components such as the world model (cause and effect), physical setting, narrative (i.e. scenario and storyline), aesthetics, but also the game session (i.e. the process of game play). This design is an iterative process in which play-testing is key: by trying out the game, the designers can observe whether the goals are reached and interaction and user experience are what they hoped for. In the sections below, we explain how the design meets the initial requirements.

3.1 Benefits of Collaboration

In order to show the effects of collaboration and initiate discussion between key stake-holders, we needed multi-player gameplay. The Chain Game has a fixed amount of five roles, of which each plays part of an international supply chain: two sellers, two freight forwarders and one buyer. The goal within the game is to create highest value for the own company, based on three key performance indicators (KPIs). In order to do so, players can invest in innovations from their limited resources. Innovations are either individual (conform the internal control measures) or collaborative (in which case collaboration obviously is necessary) and will improve specific KPIs of specific companies in the chain.

Interdepence between players is designed in the game model in multiple ways. First off, investing in collaborative innovations will increase chain value more than individual innovations on the long term. However, the collaborative innovations do not necessarily benefit the players that have invested in them: they may benefit one player more than the other, may benefit a player that has not invested at all or may even decrease company value of an investing player. These effects can be reduced by the players themselves, as they are allowed to share their benefits with each other. This game mechanic stimulates investing in collaborative innovations, but also supports discussion on return-on-invest-ment. Additionally, the fact that the players are (collectively) responsible for imple-menting innovations and they almost immediately see the effects of their actions contrib-utes to the game characteristics 'Being in control' and 'Feedback and Reward' as described by [13].

Second, company value is not only based on the key performance indicators of the *own* company, but also on the performance of the chain as a whole. But the performance of the chain depends for a large part on the weakest link of that chain. As a result, players need to convince other players to improve certain KPIs, conveying the message that chain partners are highly interdependent.

3.2 Disseminating Concepts

The supply chain innovations are made available as *actions* to the players of the game. This forces the players of The Chain Game to become familiar with the concept of Supply Chain Visibility: they are required to think about the value of the innovations in the context of their working environment and will experience the necessity to cooperate in order to create greater supply chain visibility. The innovations are focused on custom-related interruptions, in line with the research goals of the Cassandra project, and described on an abstract level, in such a way that they have an effect on one or more KPIs of one or more roles. The effects of an innovation are not disclosed until the inno-vation is implemented, creating the challenge for the players to understand these inno-vations and estimate their effects if they want to master the game (as described before [13]). As they are trying to implement the best possible innovation for their company, the main goals of the game are achieved: communicating the concept of supply chain visibility, understanding of the underlying innovations and discussion within the supply chain on these innovations.

An example of a supply chain visibility concept is 'Exception Reporting', which is made available as collaborative innovation to the players of The Chain Game (see Fig. 1). For 'Exception Reporting' three investors are required: both freight forwarders are forced to invest, and a third one (either one of the sellers or the buyer) also needs to contribute to implement this innovation. As described before, the companies that invested in the innovation are not necessarily the companies that receive the (dis)advantages: 'Exception Reporting' not only improves the KPIs for both investing freight forwarders, but also the buyer, whether he invested or not. An important step in clarifying the concepts to the players is explaining the argumentation of the effects, which is done by a game facilitator. In the case of the 'Exception Reporting' innovation, the reason for the impact is described as follows: "Exception Reporting provides the Freight Forwarders with early warnings for expected errors, allowing them to take measures in an early stage. This will give them more control over their business and they will have the ability to reduce the negative consequences, which is also of interest for the Buyer."

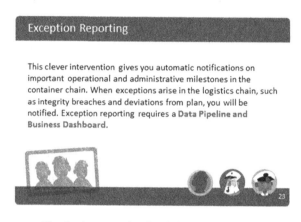

Fig. 1. An example of a chain innovation card

A big challenge of designing The Chain Game was that insufficient data is available on the exact effects of these innovative concepts, so it was not possible at this point to create a simulation model that predicts the effects on a detailed level. However, the innovations were already described on an abstract level in the CASSANDRA Project deliverables [15]. As the goal of the game is to create awareness about this knowledge, a qualitative model suffices to create this awareness and support discussion about the various concepts. As a result, the effects of each innovation were established by interviewing subject matter experts in a series of workshops to come to the content model of The Chain Game. In total, the effects of 6 individual innovations and 26 collaborative innovations were estimated and implemented this way.

3.3   Chain Robustness

One of the promises of increased supply chain visibility is that the supply chain will be less sensitive to incidents that may occur. To communicate this message, incidents occur regularly to test the current state of the supply chain. Incidents typically have a negative effect on one or more KPIs of one or more companies, but implemented innovations may (partly) mitigate these effects. The effects of an incident and the mitigating innovations for each incident have been established by the subject matter experts, similar to the effects of the innovations. An example of an incident is 'Cargo Theft': "Cargo has been stolen from trucks that were on their way from the Seller to the seaport. Trucks are totally emptied by this criminal organization." It has a negative effect on the freight forwarders, the seller and the buyer with the following argumentation: "Both of the Freight Forwarders are held responsible for their lack of security measures for their part of transportation. Their reliability is severely damaged and they have to compensate the costs of the theft and therefore their capability weakens. The Buyer doesn't get its raw product in time, which means that yet again, there will not be enough of his product available for retailers. This hurts the perceived reliability of the Buyer. The Seller is afraid to lose clients and quickly offered to deliver new cargo at cost price." The innovation 'Exception Reporting' discussed before mitigates 25 % of these effects.

A second game mechanic that gives players the feedback on the state of their supply chain is the role of Customs. Customs is strongly interested in supply chain visibility, since they profit from additional information in a way that checks on data might be executed in less time. Therefore, the partners in the supply chain may gain advantages in relation to Custom checks if they provide a certain degree of visibility. Ideally, Customs perceives the supply chain as a *trusted trade lane*, a concept strongly related to supply chain visibility. These dynamics reflect in the game by having Customs to perform a risk assessment on the state of the supply chain at the end of each game round. The risk assessment decides if a physical check of the goods is performed. Such a check has negative effects on the reliability of all stakeholders of the entire supply chain, since goods transported may be delayed for days. However, a number of innovations in the game that are of interest to Customs will have the supply chain proceed towards the certificate of trusted trade lane. Each implementation of one of these innovations decreases the probability of a physical check.

4   Evaluation

The final version of The Chain Game was disseminated by playing it with two groups of stakeholders: consultants of a large accountancy firm and managers of Dutch Customs. The goal of the play sessions was to both evaluate and disseminate the game with stakeholders. The game was evaluated by a questionnaire; additionally, observations during the reflection phase are also used for the evaluation. A play session consisted of 20 min of introduction, 10 game rounds (75 min), filling in the questionnaire and 30 min of reflection. The goal of the evaluation was to assess whether the game is valuable for creating supply chain visibility awareness and to make an inventory of possible opportunities to apply the game. The questionnaire consisted of 3 questions (see Fig. 2) that needed to be scored on a 5-point Likert scale (1 = *totally disagree* to 5 = *totally agree*), and the following open questions:

- What part of The Chain Game should stay the same?
- What part of The Chain Game should be adjusted?

- Which (business) opportunities do you see for this game?
- Any further remarks?

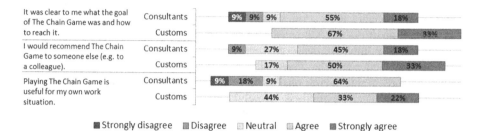

Fig. 2. Results of the evaluation questionnaire

4.1 Results

In general it can be concluded that both groups are convinced The Chain Game has value in creating awareness for supply chain visibility. Figure 2 on the next page shows the results of the three closed questions – the results of question 3 (as shown in Fig. 2) show that both groups are fairly positive about using The Chain Game in their own working environment: the 64 % of the consultants are positive (27 % negative) and 55 % of the Customs group are positive (0 % negative). We believe these score may be toned down somewhat by the fact that these groups were not the primary audience for The Chain Game. For example, Customs does not have an active playing role in the game and for consultants the focus on customs-related interruptions covers only a part of their work. On the other hand, socially desirable answers may also have influenced the results, as the questionnaire was hardly anonymous.

The answers to the open questions and the observations of the game facilitators show that both groups are positive about the way of using a game to raise awareness and understanding for complex information. This was also easily observed: players were drawn into the game from the first round on and the game could keep the attention until the final round, for more than an hour. Additionally, the interactivity and direct feedback was praised by both groups, allowing players to quickly gain insight in the effects of certain innovations. Finally, from the group of consultants several people mentioned they were drawn to the digital and visual implementation.

Both groups believed that more time should be reserved for the facilitator to explain the effects of innovations. Sometimes, the effects of innovations or incidents were challenged by a player, but this does not really stand in the way of the goal of the game: creating awareness. Discussion about the effects in real-life situations are a consequence of the fact that the model is an estimate of subject matter experts and not evidence-based. This means innovations sometimes had effects that the players did not expect. This cannot be resolved in the game, as effects need to be undisclosed to force players to think about the meaning of innovations. However, it is necessary for the game facilitator to take sufficient time for the argumentation *behind* the effects that take place. An expert opinion supports relevant discussion and counters players' frustration. Additionally, some game design choices have resulted in the omission of factors that can be of importance in the real world. For example, the number of players is fixed, in order to keep the game economy as simple as possible (i.e. a player can basically invest only in one innovation each turn). As a result, stakeholders cannot enter or leave a logistics chain, an action that can have big effects in the real world. This pragmatic choice was made to keep the game playable within a few hours.

Somewhat related is the *game process*: in the play sessions, hardly any reflection on the adopted strategy was possible during the playing of the game because of the fast pace during these sessions. As a result, players were not able to reflect upon the played innovations, the expected and the real effects and the underlying model. This prevented them in creating a broader understanding and forming and adjusting their strategy. Particularly the group of consultants requested one or two reflection phases during the game, in order to improve understanding, which can easily be adjusted by pausing the game after 3 or 4 rounds.

Interestingly, the group of consultants believed the game to be strongly focused on Customs-related innovations, while the group Customs would like a more active role for Customs in the game. The game can be improved for the consultants by adding taxes- and legislation-specific innovations; this is no problem from the perspective of the game design, as long as the effects are sufficiently realistic and in balance with each other. However, it will change the focus of the game, which was primarily aimed at Customs-related interruptions. Making Customs interactive in the game is more difficult to implement, since this would mean adding a new role that is even more asymmetrical to the roles that already exist in the game.

A number of different **opportunities** were put forward during the play sessions. The group of consultants believed the game can be played with companies to raise awareness about supply chain visibility and on the impact Customs can have on supply chains. Both groups believed the game can provide insight in the interdependence between supply chain partners and the importance of jointly investing in innovations and prevent weak links in the chain. Finally, Customs believed the game can be used as learning tool in management studies related to logistics.

5   Conclusions

Disseminating the concepts and ideas concerning supply chain visibility is a challenge, as they are abstract and hard to grasp. Additionally, future implementation of these concepts require both support and collaboration of a multitude of stakeholders, hence awareness within this group of stakeholders is essential to effectuate innovations. In order to explain the complex concepts and initiate collaboration between stakeholders, a serious game named The Chain Game was developed. The goal of The Chain Game is to have players experience the effects and advantages of increased supply chain visibility. The serious game focuses on a variety of innovative concepts, which can be implemented by the main actors in a supply chain. Feedback consists of the effects of these innovations on a supply chain level. A qualitative game model was built by a team of game designers and subject matter experts, as little evidence-based data is available as of yet.

The fully digital serious game has been tested by stakeholders in two pilots. Both groups evaluated the game as valuable in raising awareness on the subject of supply chain visibility with stakeholders and believed it can act as a supporting tool in bringing the necessary parties together. Additionally, the game can be used as learning tool in management studies related to logistics. An important observation during the tests is the crucial role of the facilitator as subject-matter expert: he needs to have the time and the knowledge to explain the argumentation behind the presented effects during the game. If this precondition is met, The Chain Game can provide an excellent starting point for further discussion on the future of supply chain visibility.

References

1. The CASSANDRA project (2014). http://www.cassandra-project.eu/
2. von Hilgers, P.: Eine Anleitung zur Anleitung. Das taktische Kriegsspiel 1812–1824. Int. J. Study Board Games cnws 200059 (2000)
3. Pias, C.: Computer Spiel Welten. (Diss) (2002)
4. Keys, B., Wolfe, J.: The role of management games and simulations in education and research. J. Manag. **16**(2), 307–336 (1990)
5. Tanner, M., Lindquist, T.: Using MONOPOLY and teams-games-tournaments in accounting education: a cooperative learning teaching resource. Acc. Educ. **7**(2), 139–162 (1998)
6. Forrester, J.W.: Industrial Dynamics. Mass, Cambridge (1961)
7. Randel, J.M., Morris, B.A., Wetzle, C.D., Whitehead, B.V.: The effectiveness of games for educational purposes: a review of recent research. Simul. Gaming **23**, 261–276 (1992)
8. Prensky, M.: Digital Game-Based Learning. McGraw-Hill Companies, New York (2001)
9. Fletcher, J.D., Tobias, S.: Using computer games and simulations for instruction: a research review. In Proceedings of the New Learning Technologies 2006 Conference, Warrenton, VA: Society for Applied Learning Technology (2006)
10. Charsky, D., Mims, C.: Integrating commercial off-the-shelf video games into school curriculums. TechTrends **52**(5), 38–44 (2008)
11. Tang, S., Hanneghan, M., El Rhalibi, A.: Introduction to games based learning. Games Based Learning Advancements for Multi-Sensory Human Computer Interfaces: Techniques and Effective Practices, pp. 1–18. IGI Global, New York (2009)
12. Foreman, Joel, et al.: Game-based learning: how to delight and instruct in the 21st century. Educause Rev. **39**, 50–67 (2004)
13. Buckley, K.E., Anderson, C.A.: A theoretical model of the effects and consequences of playing video games. Playing Video Games: Motives, Responses, and Consequences, pp. 363–378. Lawrence Erlbaum, London (2006)
14. Janssen, M., Klievink, B.: Gaming and simulation for transforming and reengineering government: towards a research agenda. Transform. Gov. People Process Policy **4**, 132–137 (2010)
15. Zomer, G., et al.: Deliverable D9.1 CASSANDRA Final Report. Delft, pp. 9–24 (2014)

Author Index

Abed, Salma S.   133
Abreu, António   48
Alalwan, Ali Abdallah   13
Albano, Cláudio Sonáglio   223
Alenljung, Beatrice   366
Alkhowaiter, Wassan   416

Baabdullah, Abdullah   95
Baur, Aaron W.   63, 183
Bick, Markus   63, 183
Bonorden, Charlotte S.   63
Brous, Paul   81
Bühler, Julian   63, 183

Cerone, Antonio   287
Christiaanse, Rob   423
Cota, Manuel Pérez   48
Crompvoets, Joep   386

da Silva Craveiro, Gisele   223
de Bruijne, Mark   454
de Reuver, Mark   354
Dohmen, Leon   338
Dwivedi, Yogesh   3, 13, 95, 133, 173

Engler, Marcus   503
Eriksson, Niklas   24

Fabricius, Susanna   24
Felden, Carsten   147

Gao, Shang   36
Göbel, Hannes   366
Gong, Yiwei   275
Griffioen, Paul   423
Guo, Hong   36

Haaker, Timber   354
Hallqvist, Carina   366
Hänel, Tom   147
Hidders, Jan   3
Hofman, Wout   479
Holgersson, Jesper   366

Homscheid, Dirk   299
Hulstijn, Joris   423, 454, 467

Janssen, Marijn   3, 81, 236, 314
Jin, Qiurui   247
Jung, Reinhard   107

Kapoor, Kawaljeet   261
Kleinveld, Bas   314
Klievink, Bram   443
Koskenvoima, Antti   326
Kosman, Ruud   354
Kuang, Jiawei   36
Kumar, Prabhat   95
Kunegis, Jérôme   299

Lal, Banita   13
Lehmkuhl, Tobias   107

Mäntymäki, Matti   326
Martijn, Nick   454
Matheus, Ricardo   236
Milić, Petar   212
Millard, Jeremy   3
Mubarak, Farooq   398
Mukala, Patrick   287
Müller, Rainer   503
Muller, Tijmen Joppe   503

Nikayin, Fatemeh   354

Palle, Madhuri S.   247
Peña-Mora, Feniosky   247

Rana, Nripendra P.   3, 13, 173, 416
Riaz, Zainab   247
Rocha, Álvaro   48
Rosenberger, Marcel   107, 372

Sabouri, Ahmad   119
Sarma, Hiren K.D.   173
Schaarschmidt, Mario   299

Shen, Charles   247
Shi, Jimin   183
Shi, Jinjing   36
Singh, Jyoti Prakash   416
Sivarajah, Uthayasankar   261
Slade, Emma L.   3, 173
Snijders, Dhoya   3
Snoeck, Monique   386
Söderström, Eva   366
Stoimenov, Leonid   212
Strang, Kenneth   200
Sun, Zhaohao   200
Suomi, Reima   398

Tan, Yao-Hua   454, 467, 491
Thomas, Jensen   491
Turini, Franco   287

Utz, Wilfrid   435

Van Cauter, Lies   386
van Helvoirt, Stefhan   160
Veljković, Nataša   212

Wang, Yuxin   467
Weerakkody, Vishanth   3, 261
Weigand, Hans   160
Williams, Michael   13, 95, 133
Woitsch, Robert   435

Xu, Yibing   36

Zedel, Katja   503
Zomer, Gerwin   443, 503
Zou, Huasheng   200

Ying Tan   Yuhui Shi
Carlos A. Coello Coello (Eds.)

Advances
in Swarm Intelligence

5th International Conference
ICSI 2014, Hefei, China, October 17-20, 2014
Proceedings, Part II

 Springer

Volume Editors

Ying Tan
Peking University
Key Laboratory of Machine Perception (MOE)
School of Electronics Engineering and Computer Science
Department of Machine Intelligence
Beijing 100871, China
E-mail: ytan@pku.edu.cn

Yuhui Shi
Xi'an Jiaotong-Liverpool University
Department of Electrical and Electronic Engineering
Suzhou 215123, China
E-mail: yuhui.shi@xjtlu.edu.cn

Carlos A. Coello Coello
CINVESTAV-IPN
Investigador Cinvestav 3F, Depto. de Computación
México, D.F. 07300, Mexico
E-mail: ccoello@cs.cinvestav.mx

ISSN 0302-9743   e-ISSN 1611-3349
ISBN 978-3-319-11896-3   e-ISBN 978-3-319-11897-0
DOI 10.1007/978-3-319-11897-0
Springer Cham Heidelberg New York Dordrecht London

Library of Congress Control Number: 2014949260

LNCS Sublibrary: SL 1 – Theoretical Computer Science and General Issues

Typesetting: Camera-ready by author, data conversion by Scientific Publishing Services, Chennai, India

Printed on acid-free paper

Springer is part of Springer Science+Business Media (www.springer.com)

Preface

This book and its companion volume, LNCS vols. 8794 and 8795, constitute the proceedings of the fifth International Conference on Swarm Intelligence (ICSI 2014) held during October 17–20, 2014, in Hefei, China. ICSI 2014 was the fifth international gathering in the world for researchers working on all aspects of swarm intelligence, following the successful and fruitful Harbin event (ICSI 2013), Shenzhen event (ICSI 2012), Chongqing event (ICSI 2011) and Beijing event (ICSI 2010), which provided a high-level academic forum for the participants to disseminate their new research findings and discuss emerging areas of research. It also created a stimulating environment for the participants to interact and exchange information on future challenges and opportunities in the field of swarm intelligence research.

ICSI 2014 received 198 submissions from about 475 authors in 32 countries and regions (Algeria, Australia, Belgium, Brazil, Chile, China, Czech Republic, Finland, Germany, Hong Kong, India, Iran, Ireland, Italy, Japan, Macao, Malaysia, Mexico, New Zealand, Pakistan, Romania, Russia, Singapore, South Africa, Spain, Sweden, Taiwan, Thailand, Tunisia, Turkey, United Kingdom, United States of America) across six continents (Asia, Europe, North America, South America, Africa, and Oceania). Each submission was reviewed by at least two reviewers, and on average 2.7 reviewers. Based on rigorous reviews by the Program Committee members and reviewers, 105 high-quality papers were selected for publication in this proceedings volume with an acceptance rate of 53.03%. The papers are organized in 18 cohesive sections, 3 special sessions and one competitive session, which cover all major topics of swarm intelligence research and development.

As organizers of ICSI 2014, we would like to express sincere thanks to University of Science and Technology of China, Peking University, and Xi'an Jiaotong-Liverpool University for their sponsorship, as well as to the IEEE Computational Intelligence Society, World Federation on Soft Computing, and International Neural Network Society for their technical co-sponsorship. We appreciate the Natural Science Foundation of China for its financial and logistic support. We would also like to thank the members of the Advisory Committee for their guidance, the members of the International Program Committee and additional reviewers for reviewing the papers, and the members of the Publications Committee for checking the accepted papers in a short period of time. Particularly, we are grateful to Springer for publishing the proceedings in the prestigious series of Lecture Notes in Computer Science. Moreover, we wish to express our heartfelt appreciation to the plenary speakers, session chairs, and student helpers. In addition, there are still many more colleagues, associates, friends, and supporters who helped us in immeasurable ways; we express our sincere gratitude to them all. Last but not the least, we would like to thank all the speakers, authors, and participants for their great contributions that made ICSI 2014 successful and all the hard work worthwhile.

July 2014

Ying Tan
Yuhui Shi
Carlos A. Coello Coello

Organization

General Chairs

Russell C. Eberhart — Indiana University-Purdue University, USA
Ying Tan — Peking University, China

Programme Committee Chairs

Yuhui Shi — Xi'an Jiaotong-Liverpool University, China
Carlos A. Coello Coello — CINVESTAV-IPN, Mexico

Advisory Committee Chairs

Gary G. Yen — Oklahoma State University, USA
Hussein Abbass — University of New South Wales, ADFA, Australia
Xingui He — Peking University, China

Technical Committee Chairs

Xiaodong Li — RMIT University, Australia
Andries Engelbrecht — University of Pretoria, South Africa
Ram Akella — University of California, USA
M. Middendorf — University of Leipzig, Germany
Kalyanmoy Deb — Indian Institute of Technology Kanpur, India
Ke Tang — University of Science and Technology of China, China

Special Sessions Chairs

Shi Cheng — The University of Nottingham, Ningbo, China
Meng-Hiot Lim — Nanyang Technological University, Singapore
Benlian Xu — Changshu Institute of Technology, China

Competition Session Chairs

Jane J. Liang — Zhengzhou University, China
Junzhi Li — Peking University, China

Publications Chairs

Radu-Emil Precup   Politehnica University of Timisoara, Romania
Haibin Duan   Beihang University, China

Publicity Chairs

Yew-Soon Ong   Nanyang Technological University, Singapore
Juan Luis Fernandez Martinez University of Oviedo, Spain
Hideyuki Takagi   Kyushu University, Japan
Qingfu Zhang   University of Essex, UK
Suicheng Gu   University of Pittsburgh, USA
Fernando Buarque   University of Pernambuco, Brazil
Ju Liu   Shandong University, China

Finance and Registration Chairs

Chao Deng   Peking University, China
Andreas Janecek   University of Vienna, Austria

Local Arrangement Chairs

Wenjian Luo   University of Science and Technology of China,
                                  China
Bin Li   University of Science and Technology of China,
                                  China

Program Committee

Kouzou Abdellah   University of Djelfa, Algeria
Ramakrishna Akella   University of California at Santa Cruz, USA
Rafael Alcala   University of Granada, Spain
Peter Andras   Newcastle University, UK
Esther Andrés   INTA, USA
Sabri Arik   Istanbul University, Turkey
Helio Barbosa   Laboratório Nacional de Computação
                                  Científica, Brazil
Carmelo J.A. Bastos Filho   University of Pernambuco, Brazil
Christian Blum   Technical University of Catalonia, Spain
Salim Bouzerdoum   University of Wollongong, Australia
Xinye Cai   Nanhang University, China
David Camacho   Universidad Autonoma de Madrid, Spain
Bin Cao   Tsinghua University, China
Kit Yan Chan   DEBII, Australia

Mu-Song Chen	Da-Yeh University, Taiwan
Walter Chen	National Taipei University of Technology, China
Shi Cheng	The University of Nottingham Ningbo, China
Leandro Coelho	Pontifícia Universidade Católica do Parana, Brazil
Chenggang Cui	Shanghai Advanced Research Institute, Chinese Academy of Sciences, China
Chaohua Dai	Southwest Jiaotong University, China
Arindam K. Das	University of Washington, USA
Prithviraj Dasgupta	University of Nebraska at Omaha, USA
Kusum Deep	Indian Institute of Technology Roorkee, India
Mingcong Deng	Tokyo University of Agriculture and Technology, Japan
Yongsheng Ding	Donghua University, China
Madalina-M. Drugan	Vrije University, The Netherlands
Mark Embrechts	RPI, USA
Andries Engelbrecht	University of Pretoria, South Africa
Fuhua Fan	Electronic Engineering Institute, China
Zhun Fan	Technical University of Denmark, Denmark
Komla Folly	University of Cape Town, South Africa
Shangce Gao	University of Toyama, Japan
Ying Gao	Guangzhou University, China
Shenshen Gu	Shanghai University, China
Suicheng Gu	University of Pittsburgh, USA
Ping Guo	Beijing Normal University, China
Haibo He	University of Rhode Island, USA
Ran He	National Laboratory of Pattern Recognition, China
Marde Helbig	CSIR: Meraka Institute, South Africa
Mo Hongwei	Harbin Engineering University, China
Jun Hu	Chinese Academy of Sciences, China
Xiaohui Hu	Indiana University Purdue University Indianapolis, USA
Guangbin Huang	Nanyang Technological University, Singapore
Amir Hussain	University of Stirling, UK
Hisao Ishibuchi	Osaka Prefecture University, Japan
Andreas Janecek	University of Vienna, Austra
Changan Jiang	RIKEN-TRI Collaboration Center for Human-Interactive Robot Research, Japan
Mingyan Jiang	Shandong University, China
Liu Jianhua	Fujian University of Technology, China
Colin Johnson	University of Kent, USA
Farrukh Khan	FAST-NUCES Islamabad, Pakistan
Arun Khosla	National Institute of Technology, Jalandhar, India

Franziska Klügl	Örebro University, Sweden
Thanatchai Kulworawanichpong	Suranaree University of Technology, Thailand
Germano Lambert-Torres	Itajuba Federal University, Brazil
Xiujuan Lei	Shaanxi Normal University, China
Bin Li	University of Science and Technology of China, China
Xiaodong Li	RMIT University, Australia
Xuelong Li	Chinese Academy of Sciences, China
Yangmin Li	University of Macau, China
Jane-J. Liang	Zhengzhou University, China
Andrei Lihu	Politehnica University of Timisoara, Romania
Fernando B. De Lima Neto	University of Pernambuco, Brazil
Ju Liu	Shandong University, China
Wenlian Lu	Fudan University, China
Wenjian Luo	University of Science and Technology of China, China
Jinwen Ma	Peking University, China
Chengying Mao	Jiangxi University of Finance and Economics, China
Michalis Mavrovouniotis	De Montfort University, UK
Bernd Meyer	Monash University, Australia
Martin Middendorf	University of Leipzig, Germany
Sanaz Mostaghim	Institute IWS, Germany
Jonathan Mwaura	University of Pretoria, South Africa
Pietro S. Oliveto	University of Sheffield, UK
Feng Pan	Beijing Institute of Technology, China
Bijaya Ketan Panigrahi	IIT Delhi, India
Sergey Polyakovskiy	Ufa State Aviation Technical University, USA
Thomas Potok	ORNL, USA
Radu-Emil Precup	Politehnica University of Timisoara, Romania
Kai Qin	RMIT University, Australia
Quande Qin	Shenzhen University, China
Boyang Qu	Zhongyuan University of Technology, China
Robert Reynolds	Wayne State University, USA
Guangchen Ruan	Indiana University Bloomington, USA
Eugene Santos	Dartmouth College, USA
Gerald Schaefer	Loughborough University, USA
Kevin Seppi	Brigham Young University, USA
Zhongzhi Shi	Institute of Computing Technology, CAS, China
Pramod Kumar Singh	ABV-IIITM Gwalior, India
Ponnuthurai Suganthan	Nanyang Technological University, Singapore
Mohammad Taherdangkoo	Shiraz University, Iran
Hideyuki Takagi	Kyushu University, Japan
Ying Tan	Peking University, China

Ke Tang	University of Science and Technology of China, China
Peter Tino	University of Birmingham, UK
Mario Ventresca	Purdue University, USA
Cong Wang	Northeasten University, China
Guoyin Wang	Chongqing University of Posts and Telecommunications, China
Jiahai Wang	Sun Yat-sen University, China
Jun Wang	Peking University, China
Lei Wang	Tongji University, China
Ling Wang	Tsinghua University, China
Lipo Wang	Nanyang Technological University, Singapore
Qi Wang	Xi'an Institute of Optics and Precision Mechanics of CAS, China
Zhenzhen Wang	Jinling Institute of Technology, China
Man Leung Wong	Lingnan University, China
Shunren Xia	Zhejiang University, China
Bo Xing	University of Johannesburg, South Africa
Ning Xiong	Mälardalen University, Sweden
Benlian Xu	Changsu Institute of Technology, China
Bing Xue	Victoria University of Wellington, New Zealand
Pei Yan	University of Aziz, Japan
Yingjie Yang	De Montfort University, UK
Wei-Chang Yeh	National Tsing Hua University, China
Gary Yen	Oklahoma State University, USA
Peng-Yeng Yin	National Chi Nan University, Taiwan
Ivan Zelinka	FEI VSB-Technical University of Ostrava, Czech Republic
Zhi-Hui Zhan	Sun Yat-sen University, China
Jie Zhang	Newcastle University, UK
Jun Zhang	Waseda University, Japan
Junqi Zhang	Tongji University, China
Lifeng Zhang	Renmin University of China, China
Mengjie Zhang	Victoria University of Wellington, New Zealand
Qieshi Zhang	Waseda University, Japan
Qingfu Zhang	University of Essex, UK
Zhenya Zhang	Anhui University of Architecture, China
Qiangfu Zhao	The University of Aizu, Japan
Wenming Zheng	Southeast University, China
Yujun Zheng	Zhejiang University of Technology, China
Cui Zhihua	Complex System and Computational Intelligence Laboratory, China
Aimin Zhou	East China Normal University, China
Zexuan Zhu	Shenzhen University, China
Xingquan Zuo	Beijing University of Posts and Telecommunications, China

Additional Reviewers

Alves, Felipe
Bi, Shuhui
Chalegre, Marlon
Cheng, Shi
Dong, Xianguang
Gonzalez-Pardo, Antonio
Haifeng, Sima
Hu, Weiwei
Jun, Bo
Lacerda, Marcelo
Lee, Jie
Li, Yexing
Ling, Haifeng
Menéndez, Héctor
Pei, Yan

Rakitianskaia, Anna
Singh, Garima
Singh, Pramod
Wang, Aihui
Wen, Shengjun
Wenbo, Wan
Wu, Peng
Wu, Zhigang
Xiao, Xiao
Xin, Cheng
Yang, Wankou
Yang, Zhixiang
Yu, Czyujian
Zheng, Zhongyang

Table of Contents – Part II

Classification Methods

Semi-supervised Ant Evolutionary Classification 1
  *Ping He, Xiaohua Xu, Lin Lu, Heng Qian, Wei Zhang, and
  Kanwen Li*

Evolutionary Ensemble Model for Breast Cancer Classification 8
  *R.R. Janghel, Anupam Shukla, Sanjeev Sharma, and
  A.V. Gnaneswar*

Empirical Analysis of Assessments Metrics for Multi-class Imbalance
Learning on the Back-Propagation Context 17
  *Juan Pablo Sánchez-Crisostomo, Roberto Alejo,
  Erika López-González, Rosa María Valdovinos, and
  J. Horacio Pacheco-Sánchez*

A Novel Rough Set Reduct Algorithm to Feature Selection Based on
Artificial Fish Swarm Algorithm 24
  Fei Wang, Jiao Xu, and Lian Li

Hand Gesture Shape Descriptor Based on Energy-Ratio and Normalized
Fourier Transform Coefficients 34
  *Wenjun Tan, Zijiang Bian, Jinzhu Yang, Huang Geng,
  Zhaoxuan Gong, and Dazhe Zhao*

A New Evolutionary Support Vector Machine with Application to
Parkinson's Disease Diagnosis 42
  *Yao-Wei Fu, Hui-Ling Chen, Su-Jie Chen, LiMing Shen, and
  QiuQuan Li*

GPU-Based Methods

Parallel Bees Swarm Optimization for Association Rules Mining Using
GPU Architecture 50
  Youcef Djenouri and Habiba Drias

A Method for Ripple Simulation Based on GPU 58
  Xianjun Chen, Yanmei Wang, and Yongsong Zhan

cuROB: A GPU-Based Test Suit for Real-Parameter Optimization 66
  Ke Ding and Ying Tan

Scheduling and Path Planning

A Particle Swarm Optimization Based Pareto Optimal Task Scheduling
in Cloud Computing   79
    A.S. Ajeena Beegom and M.S. Rajasree

Development on Harmony Search Hyper-heuristic Framework for
Examination Timetabling Problem   87
    Khairul Anwar, Ahamad Tajudin Khader,
    Mohammed Azmi Al-Betar, and Mohammed A. Awadallah

Predator-Prey Pigeon-Inspired Optimization for UAV
Three-Dimensional Path Planning.................................   96
    Bo Zhang and Haibin Duan

Research on Route Obstacle Avoidance Task Planning Based on
Differential Evolution Algorithm for AUV   106
    Jian-Jun Li, Ru-Bo Zhang, and Yu Yang

Wireless Sensor Network

An Improved Particle Swarm Optimization-Based Coverage Control
Method for Wireless Sensor Network   114
    Huimin Du, Qingjian Ni, Qianqian Pan, Yiyun Yao, and Qing Lv

An Improved Energy-Aware Cluster Heads Selection Method for
Wireless Sensor Networks Based on K-means and Binary Particle
Swarm Optimization   125
    Qianqian Pan, Qingjian Ni, Huimin Du, Yiyun Yao, and Qing Lv

Power System Optimization

Comparison of Multi-population PBIL and Adaptive Learning Rate
PBIL in Designing Power System Controller   135
    Komla A. Folly

Vibration Adaptive Anomaly Detection of Hydropower Unit in Variable
Condition Based on Moving Least Square Response Surface   146
    Xueli An and Luoping Pan

Capacity and Power Optimization for Collaborative Beamforming with
Two Relay Clusters   155
    Bingbing Lu, Ju Liu, Chao Wang, Hongji Xu, and Qing Wang

Solving Power Economic Dispatch Problem Subject to DG Uncertainty
via Bare-Bones PSO...   163
    Yue Jiang, Qi Kang, Lei Wang, and Qidi Wu

Other Applications

Extracting Mathematical Components Directly from PDF Documents
for Mathematical Expression Recognition and Retrieval 170
    Botao Yu, Xuedong Tian, and Wenjie Luo

An Efficient OLAP Query Algorithm Based on Dimension Hierarchical
Encoding Storage and Shark 180
    Shengqiang Yao and Jieyue He

The Enhancement and Application of Collaborative Filtering in
e-Learning System 188
    Bo Song and Jie Gao

A Method to Construct a Chinese-Swedish Dictionary via English
Based on Comparable Corpora 196
    Fang Li, Guangda Shi, and Yawei Lv

The Design and Implementation of the Random HTML Tags and
Attributes-Based XSS Defence System 204
    Heng Lin, Yiwen Yan, Hongfei Cai, and Wei Zhang

Special Session on Swarm Intelligence in Image and Video Processing

DWT and GA-PSO Based Novel Watermarking for Videos Using Audio
Watermark 212
    Puja Agrawal and Aleefia Khurshid

Application and Comparison of Three Intelligent Algorithms in 2D
Otsu Segmentation Algorithm 221
    Lianlian Cao, Sheng Ding, Xiaowei Fu, and Li Chen

A Shape Target Detection and Tracking Algorithm Based on the Target
Measurement Intensity Filter 228
    Weifeng Liu, Chenglin Wen, and Shuyu Ding

Multi-cell Contour Estimate Based on Ant Pheromone Intensity
Field 236
    Qinglan Chen, Benlian Xu, Yayun Ren, Mingli Lu, and Peiyi Zhu

A Novel Ant System with Multiple Tasks for Spatially Adjacent Cell
State Estimate 244
    Mingli Lu, Benlian Xu, Peiyi Zhu, and Jian Shi

A Cluster Based Method for Cell Segmentation 253
    Fei Wang, Benlian Xu, and Mingli Lu

Research on Lane Departure Decision Warning Methods Based on
Machine Vision 259
    Chuncheng Ma, Puheng Xue, and Wanping Wang

Searching Images in a Textile Image Database 267
    Yin-Fu Huang and Sheng-Min Lin

IIS: Implicit Image Steganography 275
    K. Jithesh and P. Babu Anto

Humanized Game Design Based on Augmented Reality 284
    Yanhui Su, Shuai Li, and Yongsong Zhan

Special Session on Applications of Swarm Intelligence to Management Problems

A Hybrid PSO-DE Algorithm for Smart Home Energy Management 292
    Yantai Huang, Lei Wang, and Qidi Wu

A Multiobjective Large Neighborhood Search for a Vehicle Routing
Problem 301
    Liangjun Ke and Laipeng Zhai

A Self-adaptive Interior Penalty Based Differential Evolution Algorithm
for Constrained Optimization 309
    Cui Chenggang, Yang Xiaofei, and Gao Tingyu

A Novel Hybrid Algorithm for Mean-CVaR Portfolio Selection with
Real-World Constraints 319
    Quande Qin, Li Li, and Shi Cheng

A Modified Multi-Objective Optimization Based on Brain Storm
Optimization Algorithm 328
    Lixia Xie and Yali Wu

Modified Brain Storm Optimization Algorithm for Multimodal
Optimization 340
    Xiaoping Guo, Yali Wu, and Lixia Xie

Special Session on Swarm Intelligence for Real-World Application

Classification of Electroencephalogram Signals Using Wavelet
Transform and Particle Swarm Optimization 352
    *Nasser Omer Ba-Karait, Siti Mariyam Shamsuddin, and
    Rubita Sudirman*

FOREX Rate Prediction Using Chaos, Neural Network and Particle
Swarm Optimization 363
    Dadabada Pradeepkumar and Vadlamani Ravi

Path Planning Using Neighborhood Based Crowding Differential
Evolution 376
    *Boyang Qu, Yanping Xu, Dongyun Wang, Hui Song, and
    Zhigang Shang*

Neural Network Based on Dynamic Multi-swarm Particle Swarm
Optimizer for Ultra-Short-Term Load Forecasting 384
    Jane Jing Liang, Hui Song, Boyang Qu, Wei Liu, and Alex Kai Qin

Dynamic Differential Evolution for Emergency Evacuation
Optimization 392
    Shuzhen Wan

Centralized Charging Strategies of Plug-in Electric Vehicles on Spot
Pricing Based on a Hybrid PSO 401
    Jiabao Wang, Qi Kang, Hongjun Tian, Lei Wang, and Qidi Wu

A New Multi-region Modified Wind Driven Optimization Algorithm
with Collision Avoidance for Dynamic Environments 412
    Abdennour Boulesnane and Souham Meshoul

Special Session on ICSI 2014 Competition on Single Objective Optimization

Evaluating a Hybrid DE and BBO with Self Adaptation on ICSI 2014
Benchmark Problems 422
    Yu-Jun Zheng and Xiao-Bei Wu

The Multiple Population Co-evolution PSO Algorithm 434
    Xuan Xiao and Qianqian Zhang

Fireworks Algorithm and Its Variants for Solving ICSI2014 Competition
Problems 442
    Shaoqiu Zheng, Lang Liu, Chao Yu, Junzhi Li, and Ying Tan

Performance of Migrating Birds Optimization Algorithm on Continuous
Functions 452
    *Ali Fuat Alkaya, Ramazan Algin, Yusuf Sahin,
    Mustafa Agaoglu, and Vural Aksakalli*

Author Index 461

Table of Contents – Part I

Novel Swarm-Based Search Methods

Comparison of Different Cue-Based Swarm Aggregation Strategies 1
   Farshad Arvin, Ali Emre Turgut, Nicola Bellotto, and Shigang Yue

PHuNAC Model: Emergence of Crowd's Swarm Behavior 9
   Olfa Beltaief, Sameh El Hadouaj, and Khaled Ghedira

A Unique Search Model for Optimization 19
   A.S. Xie

Improve the 3-flip Neighborhood Local Search by Random Flat Move
for the Set Covering Problem 27
   Chao Gao, Thomas Weise, and Jinlong Li

The Threat-Evading Actions of Animal Swarms without Active Defense
Abilities 36
   Qiang Sun, XiaoLong Liang, ZhongHai Yin, and YaLi Wang

Approximate Muscle Guided Beam Search for Three-Index Assignment
Problem 44
   He Jiang, Shuwei Zhang, Zhilei Ren, Xiaochen Lai, and Yong Piao

Novel Optimization Algorithm

Improving Enhanced Fireworks Algorithm with New Gaussian
Explosion and Population Selection Strategies 53
   Bei Zhang, Minxia Zhang, and Yu-Jun Zheng

A Unified Matrix-Based Stochastic Optimization Algorithm 64
   Xinchao Zhao and Junling Hao

Chaotic Fruit Fly Optimization Algorithm 74
   Xiujuan Lei, Mingyu Du, Jin Xu, and Ying Tan

A New Bio-inspired Algorithm: Chicken Swarm Optimization 86
   Xianbing Meng, Yu Liu, Xiaozhi Gao, and Hengzhen Zhang

A Population-Based Extremal Optimization Algorithm with
Knowledge-Based Mutation 95
   Junfeng Chen, Yingjuan Xie, and Hua Chen

A New Magnetotactic Bacteria Optimization Algorithm Based on
Moment Migration 103
   Hongwei Mo, Lili Liu, and Mengjiao Geng

A Magnetotactic Bacteria Algorithm Based on Power Spectrum for
Optimization 115
    Hongwei Mo, Lili Liu, and Mengjiao Geng

Particle Swarm Optimization

A Proposal of PSO Particles' Initialization for Costly Unconstrained
Optimization Problems: ORTHOinit............................... 126
    Matteo Diez, Andrea Serani, Cecilia Leotardi, Emilio F. Campana,
    Daniele Peri, Umberto Iemma, Giovanni Fasano, and Silvio Giove

An Adaptive Particle Swarm Optimization within the Conceptual
Framework of Computational Thinking 134
    Bin Li, Xiao-lei Liang, and Lin Yang

Topology Optimization of Particle Swarm Optimization............... 142
    Fenglin Li and Jian Guo

Fully Learned Multi-swarm Particle Swarm Optimization 150
    Ben Niu, Huali Huang, Bin Ye, Lijing Tan, and Jane Jing Liang

Using Swarm Intelligence to Search for Circulant Partial Hadamard
Matrices 158
    Frederick Kin Hing Phoa, Yuan-Lung Lin, and Tai-Chi Wang

Ant Colony Optimization for Travelling Salesman Problem

High Performance Ant Colony Optimizer (HPACO) for Travelling
Salesman Problem (TSP) 165
    Sudip Kumar Sahana and Aruna Jain

A Novel *Physarum*-Based Ant Colony System for Solving the
Real-World Traveling Salesman Problem 173
    Yuxiao Lu, Yuxin Liu, Chao Gao, Li Tao, and Zili Zhang

Three New Heuristic Strategies for Solving Travelling Salesman
Problem 181
    Yong Xia, Changhe Li, and Sanyou Zeng

Artificial Bee Colony Algorithms

A 2-level Approach for the Set Covering Problem: Parameter Tuning of
Artificial Bee Colony Algorithm by Using Genetic Algorithm 189
    Broderick Crawford, Ricardo Soto, Wenceslao Palma,
    Franklin Johnson, Fernando Paredes, and Eduardo Olguín

Hybrid Guided Artificial Bee Colony Algorithm for Numerical Function
Optimization   197
    Habib Shah, Tutut Herawan, Rashid Naseem, and Rozaida Ghazali

Classification of DNA Microarrays Using Artificial Bee Colony (ABC)
Algorithm.   207
    *Beatriz Aurora Garro, Roberto Antonio Vazquez, and
    Katya Rodríguez*

Crowding-Distance-Based Multiobjective Artificial Bee Colony
Algorithm for PID Parameter Optimization   215
    Xia Zhou, Jiong Shen, and Yiguo Li

Artificial Immune System

An Adaptive Concentration Selection Model for Spam Detection   223
    Yang Gao, Guyue Mi, and Ying Tan

Control of Permanent Magnet Synchronous Motor Based on Immune
Network Model   234
    Hongwei Mo and Lifang Xu

Adaptive Immune-Genetic Algorithm for Fuzzy Job Shop Scheduling
Problems   246
    Beibei Chen, Shangce Gao, Shuaiqun Wang, and Aorigele Bao

Evolutionary Algorithms

A Very Fast Convergent Evolutionary Algorithm for Satisfactory
Solutions   258
    Xinchao Zhao and Xingquan Zuo

A Novel Quantum Evolutionary Algorithm Based on Dynamic
Neighborhood Topology   267
    Feng Qi, Qianqian Feng, Xiyu Liu, and Yinghong Ma

Co-evolutionary Gene Expression Programming and Its Application in
Wheat Aphid Population Forecast Modelling   275
    *Chaoxue Wang, Chunsen Ma, Xing Zhang, Kai Zhang, and
    Wumei Zhu*

Neural Networks and Fuzzy Methods

Neural Network Intelligent Learning Algorithm for Inter-related Energy
Products Applications   284
    *Haruna Chiroma, Sameem Abdul-Kareem, Sanah Abdullahi Muaz,
    Abdullah Khan, Eka Novita Sari, and Tutut Herawan*

Data-Based State Forecast via Multivariate Grey RBF Neural Network
Model 294
  Yejun Guo, Qi Kang, Lei Wang, and Qidi Wu

Evolving Flexible Neural Tree Model for Portland Cement Hydration
Process 302
  *Zhi-feng Liang, Bo Yang, Lin Wang, Xiaoqian Zhang,
  Lei Zhang, and Nana He*

Hybrid Self-configuring Evolutionary Algorithm for Automated Design
of Fuzzy Classifier 310
  Maria Semenkina and Eugene Semenkin

The Autonomous Suspending Control Method for Underwater
Unmanned Vehicle Based on Amendment of Fuzzy Control Rules 318
  Pengfei Peng, Zhigang Chen, and Xiongwei Ren

How an Adaptive Learning Rate Benefits Neuro-Fuzzy Reinforcement
Learning Systems 324
  *Takashi Kuremoto, Masanao Obayashi, Kunikazu Kobayashi, and
  Shingo Mabu*

Hybrid Methods

Comparison of Applying Centroidal Voronoi Tessellations and
Levenberg-Marquardt on Hybrid SP-QPSO Algorithm for High
Dimensional Problems 332
  Ghazaleh Taherzadeh and Chu Kiong Loo

A Hybrid Extreme Learning Machine Approach for Early Diagnosis of
Parkinson's Disease 342
  *Yao-Wei Fu, Hui-Ling Chen, Su-Jie Chen, Li-Juan Li,
  Shan-Shan Huang, and Zhen-Nao Cai*

A Hybrid Approach for Cancer Classification Based on Particle Swarm
Optimization and Prior Information 350
  Fei Han, Ya-Qi Wu, and Yu Cui

Multi-objective Optimization

Grover Algorithm for Multi-objective Searching with Iteration
Auto-controlling 357
  Wanning Zhu, Hanwu Chen, Zhihao Liu, and Xilin Xue

Pareto Partial Dominance on Two Selected Objectives MOEA on
Many-Objective 0/1 Knapsack Problems 365
  Jinlong Li and Mingying Yan

Analysis on a Multi-objective Binary Disperse Bacterial Colony
Chemotaxis Algorithm and Its Convergence   374
    Tao Feng, Zhaozheng Liu, and Zhigang Lu

Multi-objective PSO Algorithm for Feature Selection Problems with
Unreliable Data..   386
    Yong Zhang, Changhong Xia, Dunwei Gong, and Xiaoyan Sun

Convergence Enhanced Multi-objective Particle Swarm Optimization
with Introduction of Quorum-Sensing Inspired Turbulence   394
    Shan Cheng, Min-You Chen, and Gang Hu

Multiobjective Genetic Method for Community Discovery in Complex
Networks   404
    Bingyu Liu, Cuirong Wang, and Cong Wang

A Multi-objective Jumping Particle Swarm Optimization Algorithm for
the Multicast Routing   414
    Ying Xu and Huanlai Xing

Multi-agent Systems

A *Physarum*-Inspired Multi-Agent System to Solve Maze   424
    *Yuxin Liu, Chao Gao, Yuheng Wu, Li Tao, Yuxiao Lu, and
    Zili Zhang*

Consensus of Single-Integrator Multi-Agent Systems at a Preset
Time   431
    Cong Liu, Qiang Zhou, and Yabin Liu

Representation of the Environment and Dynamic Perception in
Agent-Based Software Evolution   442
    Qingshan Li, Hua Chu, Lihang Zhang, and Liang Diao

Evolutionary Clustering Algorithms

Cooperative Parallel Multi Swarm Model for Clustering in Gene
Expression Profiling   450
    Zakaria Benmounah, Souham Meshoul, and Mohamed Batouche

Self-aggregation and Eccentricity Analysis: New Tools to Enhance
Clustering Performance via Swarm Intelligence   460
    Jiangshao Gu and Kunmei Wen

DNA Computation Based Clustering Algorithm.....................   470
    Zhenhua Kang, Xiyu Liu, and Jie Xue

Clustering Using Improved Cuckoo Search Algorithm   479
    Jie Zhao, Xiujuan Lei, Zhenqiang Wu, and Ying Tan

Sample Index Based Encoding for Clustering Using Evolutionary
Computation   489
    Xiang Yang and Ying Tan

Data Mining Tools Design with Co-operation of Biology Related
Algorithms   499
    Shakhnaz Akhmedova and Eugene Semenkin

Author Index   507

Semi-supervised Ant Evolutionary Classification

Ping He*, Xiaohua Xu*, Lin Lu, Heng Qian, Wei Zhang, and Kanwen Li

Department of Computer Science, Yangzhou University, Yangzhou 225009, China
{angeletx,arterx,linlu60}@gmail.com

Abstract. In this paper, we propose an ant evolutionary classification model, which treats different classes as ant colonies to classify the unlabeled instances. In our model, each ant colony sends its members to propagate its unique pheromone on the unlabeled instances. The unlabeled instances are treated as unlabeled ants. They are assigned to different ant colonies according to the pheromone that different colonies leave on it. Next, the natural selection is carried out to maintain the history colony information as well as the scale of swarms. Theoretical analysis and experimental results show the effectiveness of our proposed model for evolutionary data classification.

Keywords: Evolutionary classification, Ant colony, K-nearest neighbor.

1 Introduction

Evolutionary data comes from many application fields, such as topics in weblogs and locations in GPS sensors. Evolutionary data mining can be classified into the two categories, evolutionary clustering and evolutionary classification. Among them, evolutionary classification refers to the situation where some instances in the data flow are attached with known labels, and the target is to classify the unlabeled data in the real-time.

Various evolutionary classification methods [3]–[13] have been proposed from different aspects, including concept drifts, class distribution and temporal smoothness. However, the assumption of entire labeled data availability is often violated in the real-world problems, because labels may be scare or not readily available. As a result, semi-supervised evolutionary learning methods have been recently put forward. Yangging Jia et al. [14] proposed a semi-supervised classification algorithm for dynamic mail post categorization. They carried out temporal smoothness assumption using temporal regularizers defined in the Hilbert space, and then derived the online algorithm that efficiently finds the closed-form solution to the target function. Later, H. Borchani et al. [15] proposed a new semi-supervised learning approach for concept-drifting data streams. They aim to take advantage of unlabeled data to detect possible concept drifts and, if necessary, update the classifier over time even if only a few labeled data are available. However, both the previous works assumes that at any time stamp, at least one labeled instance for each class should be provided, which can be easily violated in the real-world applications.

---

* Corresponding author.

Y. Tan et al. (Eds.): ICSI 2014, Part II, LNCS 8795, pp. 1–7, 2014.

In this paper, we propose a semi-supervised ant evolutionary classification model, which only require users to specify the number of labels and provide at least one labeled sample in the beginning. In our work, we treat each data instance as an ant and each class of labeled instances as an ant colony. The whole swarm, i.e., the whole dataset, is composed of all the different colonies and the unlabeled ants. They evolve with time based on the simulation of natural selection. Therefore, our proposed algorithm is 'self-training' in nature. Compared to the previous research, our method can be applied to a more generalized scenario, where the class distribution is arbitrary and the number of labeled instances is unfixed (even down to 0) at each time step.

The rest of this paper is organized as follows: Section 2 describes our ant evolutionary classification model in detail. Section 3 presents some simulation results to demonstrate its classification performance. Finally, Section 4 concludes the paper.

2 Ant Evolutionary Classification Model

In semi-supervised evolutionary classification, each data is associated with not only a label y but also a time stamp $t \in \{1, \ldots, T\}$. Given a set of data subsets $X = \{X^1, X^2, \ldots, X^T\}$, where X^t represents the data at time step t, $X^t = X^t_m \bigcup X^t_u$, $X^t_m = (x^t_i)_{i=1\ldots|X^t_m|}$ is labeled, the corresponding label subset is $Y^t_m = (y^t_i)_{i=1\ldots|X^t_m|}$, and $X^t_u = (x^t_i)_{|X^t_m|+1\ldots|X^t|}$ is unlabeled, the goal is to predict the label of X^t_u, i.e., $Y^t_u = (y^t_i)_{|X^t_m|+1\ldots|X^t|}$, in the real time.

To solve this problem, we propose an Ant Evolutionary Classification (AEC) model. It treats each class as an ant colony, respectively denoted as $A_{l=1\ldots c}$, where c is the number of classes or labels. Particularly, we let the unlabeled dataset X_u form a special colony with unknown class, represented by $A_0 = X_u$. The lth colony at time step t is denoted as A^t_l. The ith member of A^t_l is denoted by a^t_{li}, a^t_{li} is labeled if $l > 0$ and unlabeled otherwise. Therefore, the swarm at time t is composed by $c + 1$ ant colonies, i.e., $A^t = \{A^t_0, A^t_1, \ldots, A^t_c\}$.

We assume that each ant colony possesses a unique pheromone. In order to expand the territory, each colony has to recruit new ants by spreading its pheromone onto the unlabeled ones. The new members joining the ant colony $A^t_{l>0}$ is composed of two groups: 1) the labeled data provided at time step t, 2) the unlabeled data assigned to $A^t_{l>0}$ at the time step t.

Instead of recording the pheromone left by each ant individual, we record the pheromone left each ant colony. We define the pheromone matrix at time step t as a $|X^t| \times c$ matrix τ^t. Each column of τ^t records the pheromone left by one colony on all the ants. The element $\tau^t_{(j+|X^t_m|)l}$ indicates the pheromone left by ants from colony A^t_l on the ant a_{0j} at time step t. The matrix τ^t is divided into two blocks. The first block with size $|X^t_m| \times c$ records the pheromone left on the labeled ants, the second block with size $|X^t_u| \times c$ records the pheromone left on the unlabeled ants. In this paper, we fix the first block of τ^t unchanged and only update its second block, whose element is τ_{ij}. The initial value of τ at time step 0 is set $\tau^0_{ij} = 1$ if and only if y_i belongs to the jth class, otherwise $\tau^0_{ij} = 0$.

Since labeled and unlabeled data is provided at each time, the pheromone matrix needs to be updated accordingly. We define τ^{t_s} as the pheromone matrix in the sth iteration of the pheromone update at time step t. Without loss of generality, given an ant a_{0i}^t and a nest $A_l^t (l > 0)$, the updated pheromone intensity on a_{0i}^t is

$$\tau_{(i+|X_m^t|)l}^{t_{s+1}} \leftarrow \sum_{j=1}^{|X_u^t|} \eta_{i(j+|X_m^t|)}^t \tau_{(j+|X_m^t|)l}^{t_s} + \sum_{k=1}^{|X_m^t|} \eta_{ik}^t \tau_{kl}^{t_0} \tag{1}$$

$\eta^t = (\eta_{ij}^t)_{|X_u^t| \times |X^t|}$ is the heuristic value matrix (or similarty matrix),

$$\eta_{ij}^t = e^{-d_{ij}^t} \tag{2}$$

d_{ij}^t is the distance between the ith and j^{th} ants at time step t,

$$d_{ij}^t = \begin{cases} (a_{li}^t - a_{0j}^t)^T (\Sigma_l^t)^{-1} (a_{li}^t - a_{0j}^t) & \text{if } l > 0 \\ \frac{||a_{0i}^t - a_{0j}^t||^2}{2\sigma^2} & \text{if } a_{0i}^t, a_{0j}^t \in A_0^t \\ \infty & \text{otherwise} \end{cases} \tag{3}$$

Σ_l^t is the covariance matrix of the lth colony at time step t, and σ is a spread parameter. Note that we define the distance between labeled and unlabeled ants as Mahalanobis distance so as to utilize the prior class distribution of labeled data, and define the distance among unlabeled ants as Euclidean distance due to the lack of class information. Similar to the partition of τ^t, we also divide η^t into two parts. The first block with size $|X_u^t| \times |X_m^t|$ records the similarity between unlabeled ants and labeled ants, and the second block with size $|X_u^t| \times |X_u^t|$ records the similarity among unlabeled ants.

To interprete eq. (1), we view as two parts, corresponding to the two blocks of η^t and τ^t. In the first term, $\eta_{i(j+|X_m^t|)}^t$ represents the similarity between the ith and j^{th} unlabeled ant, $\tau_{(j+|X_m^t|)l}^{t_s}$ is the pheromone on the jth unlabeled ant left by the lth colony in the sth iteration at time step t. Therefore, the first term computes the sum of pheromone indirectly propagated from the labeled ants via the unlabeled ants. In the second term, η_{ik}^t is the similarity between the ith ant and the kth labeled ant, τ_{kl}^0 is the initial pheromone on the kth labeled ant. Hence the second term computes the sum of the pheromone directly propagated from the labeled ants. The reason for using $\tau_{kl}^{t_0}$ instead of $\tau_{kl}^{t_s}$ is because we keep the pheromone on the labeled ants unchanged to avoid concept drifting.

After the convergence of the pheromone matrix $\tau^t = \tau^{t\infty}$, we predict the label of each unlabeled ant a_{0i}^t according to the amount of pheromone that different ant colonies leave on it.

$$y_i^t = \arg\max_l \tau_{il}^t \tag{4}$$

To determine whether an unlabeled data should be included in its predicted ant colony, we need to further evaluate its fitness to the colony. Given a colony A_l^t at time step t, we define the fitness of $a_{l_i}^t \in A_l^t$ as

$$fitness(a_{li}^t) = \frac{1}{|A_l^t|} \sum_{a_{lj}^t \in A_l^t, i \neq j} e^{-(a_{li}^t - a_{lj}^t)^T (\Sigma_l^t)^{-1} (a_{li}^t - a_{lj}^t)} \tag{5}$$

where $|A_l^t|$ is the size of A_l^t, Σ_l^t is the covariance matrix of A_l^t. Based on the fitness evaluation, the evolution of ant colonies are composed of two steps. 1) Member Addition. For each unlabeled ant, if its fitness to its predicted class is higher than a threshold $\beta \in (0,1]$, then it will be included in the target colony, used as the training set for the label prediction at next time step. 2) Member Deletion. To avoid class imbalance and allow member change, we set a maximum for the size of an ant colony ($MaxColonySize$). Once this maximum is reached, the members with the lowest fitness in that colony will be removed.

3   Experiments

We test our algorithm on three datasets, whose details are summarized in Table 1. Twomoons is a synthetic dataset including two classes of intertwining moons. Mushroom and Hyperplane datasets from the UCI repository are used to simulate the concept drift problem.

Table 1. Summarization of the test datasets

Dataset	#Size	#Attributes	#Classes
Twomoons	2000	2	2
Mushroom	8124	22	2
Hyperplane	10000	10	2

3.1   Synthetic Dataset

We first test our algorithm on the synthetic dataset Twomoons dataset for illustration. At first, the dataset is divided into $T = 100$ time intervals. Fig. 1 shows the the evolutionary classification process in five ascending time steps. The red crosses and blue circles respectively denote the two different classes of data, and the black dots represent the unlabeled instance. The subfigures at the left side depict the input data including the previous classification result, while the subfigures at the right side depict the predicted labels of those black dots in the left subfigures. As we can see, when $t = 1$, only two labeled instances are provided and we cannot recognize the intrinsic structure of the input data. Later, after more instances are provided, our algorithm gradually assigns labels to the unlabeled data points and then discovers the manifold structure of two moons.

3.2   Real-World Dataset

We use the 1% labeled ratio to generate the training data and randomly distribute them into $T = 100$ time blocks. Therefore, the provided labeled data may vary with different times blocks, and even maybe absent. For the evaluation of evolutionary classification

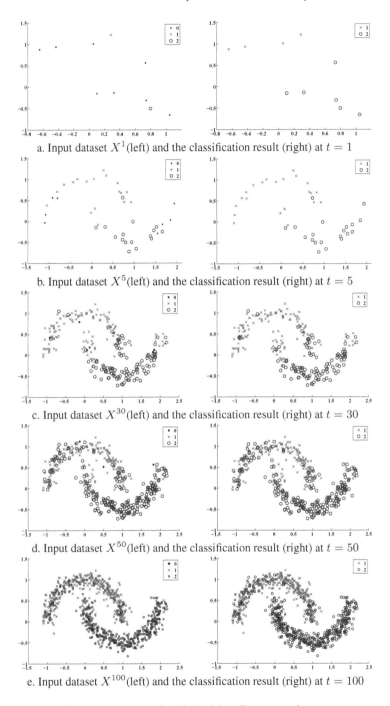

a. Input dataset X^1(left) and the classification result (right) at $t = 1$

b. Input dataset X^5(left) and the classification result (right) at $t = 5$

c. Input dataset X^{30}(left) and the classification result (right) at $t = 30$

d. Input dataset X^{50}(left) and the classification result (right) at $t = 50$

e. Input dataset X^{100}(left) and the classification result (right) at $t = 100$

Fig. 1. Illustration of AEC Model on Twomoons dataset

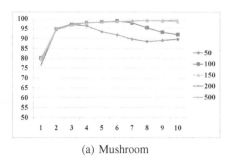
(a) Mushroom

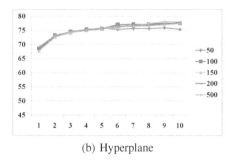
(b) Hyperplane

Fig. 2. Average block accuracy with different $MaxColonySize$ on two real-world datasets

Table 2. Average overall accuracy with different $MaxColonySize$

	50	100	150	200	500
Mushroom	92.76	95.01	96.26	**96.7**	96.6
Hyperplane	74.41	75.50	**75.65**	74.86	75.19

performance, we adopt both overall classification accuracy and local classification accuracy, which refers to the classification accuracy within each time block. To give a reliable result, 50 runs of random simulation are carried out to produce an average overall classification accuracy.

In our algorithm, we set a ceiling for the size of ant colonies, i.e., $MaxColonySize$. We adopt five values (50, 100, 150, 200, 500) as the max colony size. Fig. 2 illustrates the relationship between the block accuracy and parameter $MaxColonySize$ on the two real-world datasets. We can see that $MaxColonySize$ exerts obvious influence on Mushroom dataset.

In Table 2, each row shows the average overall accuracy on one dataset with five different $MaxColonySize$ values. We note that Twomoons and Hyperplane datasets perform best at size 150. It indicates that 150 might be a good choice for $MaxColonySize$. In addition, the setting of this parameter should also take into account of the physical memory and the runtime cost.

4   Conclusion

In this paper, we present an ant classification model for dynamic semi-supervised classification. It simulates a swarm containing varied ant colonies that will evolve with time under the rule of natural selection. Meanwhile, each generation of unlabeled instances are classified into these colonies using our proposed swarm classification method. Experimental results on a synthetic dataset demonstrate the effectiveness of our method. In the future work, we will investigate AEC and compare it with other classifiers on real-world datasets. Another interesting future line of research is to consider the scenario where labeled and unlabeled data possibly come from different distributions.

Acknowledgment. This work was supported in part by the Chinese National Natural Science Foundation under Grant nos. 61402395, 61003180, 61379066 and 61103018, Natural Science Foundation of Education Department of Jiangsu Province under contracts 13KJB520026 and 09KJB20013, Natural Science Foundation of Jiangsu Province under contracts BK2010318 and BK20140492, and the New Century Talent Project of Yangzhou University.

References

1. Chakrabarti, D., Kumar, R., Tomkins, A.: Evolutionary clustering. In: The ACM SIGKDD International Conference on Knowledge Discovery and Data Mining, pp. 554–560 (2006)
2. Chi, Y., Song, X., Zhou, D., Hino, K., et al.: Evolutionary spectral clustering by incorporating temporal smoothness. In: The ACM SIGKDD International Conference on Knowledge Discovery and Data Mining, pp. 53–62 (2007)
3. Wang, P., Wang, H.X., Wu, X.C., et al.: A low-granularity classifier for data streams with concept drifts and biased class distribution. IEEE Transactions on Knowledge and Data Engineering 19(9), 1202–1213 (2007)
4. Gao, J., Ding, B.L., Han, J.W., et al.: Classifying Data Streams with Skewed Class Distributions and Concept Drifts. IEEE Internet Computing 12(6), 37–49 (2008)
5. Anagnostopoulos, C., Tasoulis, D.K., Adams, N.M., et al.: Temporally adaptive estimation of logistic classifiers on data streams. Advances in Data Analysis and Classification 3(3), 243–261 (2009)
6. Kuncheva, L.I., Zliobaite, I.: On the window size for classification in changing environments. Intelligent Data Analysis 13(6), 861–872 (2009)
7. Peng, Z., Xingquan, Z., Jianlong, T., et al.: Classifier and Cluster Ensembles for Mining Concept Drifting Data Streams. In: IEEE 10th International Conference on Data Mining (ICDM), pp. 1175–1180 (2010)
8. Zhang, P., Li, J., Wang, P., et al.: Enabling fast prediction for ensemble models on data streams. In: The 17th ACM SIGKDD International Conference on Knowledge Discovery and Data Mining, San Diego, CA, USA, pp. 177–185 (2011)
9. Peng, Z., Gao, B.J., Xingquan, Z., et al.: Enabling Fast Lazy Learning for Data Streams. In: IEEE 11th International Conference on Data Mining (ICDM), Vancouver, Canada, pp. 932–941 (2011)
10. Zhang, P., Gao, B.J., Liu, P., et al.: A framework for application-driven classification of data streams. Neurocomput. 92, 170–182 (2012)
11. Lines, J., Davis, L.M., Hills, J., et al.: A shapelet transform for time series classification. In: The 18th ACM SIGKDD International Conference on Knowledge Discovery and Data Mining, Beijing, China, pp. 289–297 (2012)
12. Masud, M.M., Woolam, C., Gao, J., et al.: Facing the reality of data stream classification: coping with scarcity of labeled data. Knowledge and Information Systems 33(1), 213–244 (2012)
13. Li, L.J., Zou, B., Hu, Q.H., et al.: Dynamic classifier ensemble using classification confidence. Neurocomput. 99(1), 581–591 (2013)
14. Jia, Y., Yan, S., Zhang, C.: Semi-Supervised Classification on Evolutionary Data. In: The 21st International Joint Conference on Artifical Intelligence, pp. 1083–1088 (2009)
15. Borchani, H., Larranaga, P., Bielza, C.: Classifying evolving data streams with partially labeled data. Intelligent Data Analysis 15(5), 655–670 (2011)

Evolutionary Ensemble Model for Breast Cancer Classification

R.R. Janghel[1,*], Anupam Shukla[2], Sanjeev Sharma[2], and A.V. Gnaneswar[2]

[1] Sagar Institute of Research Technology and Science, Bhopal, India
rrj.iiitm@gmail.com
[2] ABV- Indian Institute of Information Technology and Management Gwalior, India
{dranupamshukla,sanjeev.sharma1868,gnani0826}@gmail.com

Abstract. A major problem in medical science is attaining the correct diagnosis of disease in precedence of its treatment. For the ultimate diagnosis, many tests are generally involved. Too many tests could complicate the main diagnosis process so that even the medical experts might have difficulty in obtaining the end results from those tests. A well-designed computerized diagnosis system could be used to directly attain the ultimate diagnosis with the aid of artificial intelligent algorithms and hybrid system which perform roles as classifiers. In this paper, we describe a Ensemble model which uses MLP, RBF, LVQ models that could be efficiently solve the above stated problem. The use of the approach has fast learning time, smaller requirement for storage space during classification and faster classification with added possibility of incremental learning. The system was comparatively evaluated using different ensemble integration methods for breast cancer diagnosis namely weighted averaging, product, minimum and maximum integration techniques which integrate the results obtained by modules of ensemble, in this case MLP, RBF and LVQ. These models run in parallel and results obtained will be integrated to give final output. The best accuracy, sensitivity and specificity measures are achieved while using minimum integration technique.

Keywords: Breast Cancer, Medical Diagnostics, Pattern Recognition, Ensemble Approach, Neural Networks, MLP, Multilayer perceptron, RBF, Radial Basis Function Network, LVQ, Learning Vector Quantization.

1 Introduction

Many real life applications are so complex that they cannot be solved by the application of a single algorithm. This necessitated the need for development of algorithms by mixing two or more of the studied algorithms. The choice of algorithms depends upon the needs and characteristics of the problem. This further helps in solving the problem to a reasonably good extent and achieving higher performances.

---

* Corresponding author.

Y. Tan et al. (Eds.): ICSI 2014, Part II, LNCS 8795, pp. 8–16, 2014.
© Springer International Publishing Switzerland 2014

In this paper, we have concentrated our efforts towards solving the problem of breast cancer diagnosis. Every year in many countries, number of woman died from breast cancer is increasing. Breast cancer is the most common cancer in women in many countries in the world. One out of eight women wills diagnosis and prognosis of breast cancer in this country. Early detection is one of the best defenses against cancer [1].

The database used in analysis of the system has been taken from Wisconsin Diagnostic Breast Cancer (WDBC) from UCI Machine Learning Repository, which comprises of data vectors from 569 patients. Then, this data is divided into training and testing data by taking 398 vectors as training data set (about 70% of the total data set) and rest as testing data set (about 30% of the total data set).

This paper is organized as follows. Section 2 reviews related work done in the concerned field. Section 3 gives the methodology used in tackling the problem. Experimental results are presented in Section 4. Conclusion and future work are given in the last section.

2   Related Work

A classification system is one that actually maps input vectors to a specific class. Hence, classification is basically the job of learning the procedure that maps the input data [2]. This has, in turn, has enthused researchers to replicate this success in the field of medical diagnostics. Their efforts have bore significant gains through the application of several standards and techniques of pattern recognition to the said problem [3]. Also it is the most widespread form of cancer among women in the world. Early detection is one of the best defenses against cancer. According to the American Cancer Society (ACS), after every thirteen minutes, four American women develop breast cancer, and one woman dies from breast cancer [1, 4-6].

Yao and Liu et.al described neural network based approaches to breast cancer diagnosis, which had displayed good generalization. The approach was based on artificial neural networks. In this approach, a feed forward neural network was evolved using BP algorithm [12]. Fogel et al. were first to derive technique to model neural networks for solving breast cancer classification [13].

Rahul et al. used multilayer perceptron neural networks (MLPNNs), radial basis function network (RBFN), competitive learning network (CL), learning vector quantization network(LVQ), combined neural networks (CNNs), probabilistic neural networks(PNNs), and recurrent neural networks (RNNs) for breast cancer diagnosis [14].

The artificial immune system with the GA in one hybrid algorithm which is the clonal selection algorithm was inspired from the clonal selection principle and affinity maturation of the human immune responses by hybridizing it with the crossover operator, which is imported from GAs to increase the exploration of the search space. [13].

Contrary to neural networks, clustering, rule induction and many other machine learning approaches, Genetic Algorithms (GAs) provide a means to encode and evolve rule antecedent aggregation operators, different rule semantics, rule base aggregation operators and defuzzification methods. Therefore, GAs remain today as one of the few and, in some sense, optimize fuzzy systems with respect to the design decisions, allowing decision makers to decide what components are fixed and which ones evolve according to the performance measures [14]. Carlos Andres Pena-Reyes et.al proposed a fuzzy-genetic approach produces systems exhibiting two prime characteristics: first they attain high classification performance and second the resulting systems involve a few simple rules and gave 97.50 % classification accuracy [15]. The goal of Fuzzy CoCo model was to evolve a fuzzy model that describes the diagnostic decision and the classification performance was 98.98%. [16].

F A good collection of methods and applications can be found in the books by Mellin and Castillo [10], and Bunke and Kendel [11, 12].

3   Methodology

Pattern Recognition and Machine Learning field have established research work on the combination of multiple classifiers (also known as ensemble of classifiers, Mixture of experts). Overall predictive accuracy can be increased by the use of multiple classifiers instead of a single classifier. The ensemble procedure constitute two steps mainly module formation and then integration of results of modules. Firstly we need to formulate the number of modules to be used, that constitute the entire ensemble architecture. Decision towards the model and architectural parameters of each of the module is made. All the networks may be initialized in this mechanism. Next the entire ensemble needs to be trained, which means the training up of the individual models making up the ensemble.

Each of the modules is trained independently and in-parallel by all the training data present in the system.

The ANNs with BPA still have some shortcomings. It is quite likely that BPA results in some local minima in place of global minima. Also we need to specify the initial parameters before the learning starts. These pose restrictions on the use of ANNs. The GA on the other hand is known for its ability of optimization. In this section we will fuse this capability of the GA along with the ANNs to train the ANN.

This solution overcomes much of the problems with the ANN training.

The block diagram of the proposed system for breast cancer diagnosis is shown in Figure 1.

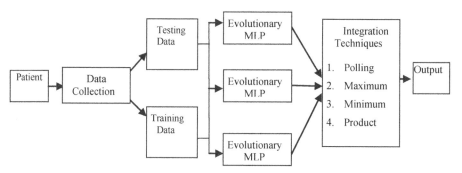

Fig. 1. Block Diagram of the Proposed System for Breast Cancer Diagnosis

In this paper, we have used the ensemble approach for classifying the inputs as malignant or benign which has 3 modules. Here, each module has evolutionary ANN and the difference between them is change in hidden neurons.

Here the GA is supposed to fix the values and the various weights as well as biases that exist in the neural networks. The GA in other words optimizes the network parameters for better performance. An ANN is a collection of various neurons. These neurons are arranged in a layered manner. Any ANN model being used in real life application normally uses a single hidden layer. The hidden layer has a specified number of neurons in it. In a fully connectionist approach, every neuron of a forward layer is connected to every neuron of the forward layer i+1 by some weight. The hidden and output neuron further have weight is adjusted during training. Besides every neuron has some bias associated with it. Now we would study the application of GA in this problem for training. Some biases that need to be optimally set.

The first task is problem encoding. The problem encoding consists of these parameters in a linear array. This is the phenotype problem representation. The population may be represented using any of double vector or a bit string representation. The Genetic Operators include Selection, Crossover, Eliticism, Mutation, etc. The Genetic Operators ensured creation of good individuals from one population to the other. Let us assume that there was a single hidden layer consisting of H neurons. The input and output layers have I an O neurons respectively. In this system, it may easily be seen that there are I x H weights between the input layer and the hidden layer and H x O weights between the hidden layer and the output layer, this makes the total number of weights as W=I x H + H x O. Further the number of biases is equal to the number of neurons. The total number of biases is H + O. This means that for a single layer ANN there would be I x H + H x O + H + O parameters to be optimized.

The fitness of any individual in the population is measured with the help of fitness function. The fitness function consists of the ANN along with its training data set. In the fitness function we initialize the ANN by the various parameters that are generated by GA. These parameters were extracted from the individual and used to set the weights and biases of the ANN. Then the training data set is passed through the ANN. The performance of the ANN against this data set is measured. This performance is the net fitness value of the GA that needs to be maximized (or the negative performance need to be minimized).Hence every time that the GA demands the measurement of fitness value of some individual, the ANN is created and the value is measured by the performance. This interfaces the GA and the ANN while training.

The neural training by GA possesses a very complex fitness landscape. Hence it is wise to use a local search strategy that places any ANN or genetic individual at the closest minima, before its fitness value is reported. This local search strategy assists the GA in the search or optimization process. In this algorithm we use Back Propagation Algorithm as the local search method. The epochs, momentum, and learning rate are kept low as per the requirements of local search.

Once the GA reaches its optimal state and terminates as per the stopping criterion, we get the final values of the weights and the biases. Then we create the ANN with these weights and bias values and this is regarded as the most optimal ANN as a result of the ANN training. We can then use this for the testing purposes. It may be seen here that validation data is not necessarily required in this type of training.

The net fitness may hence be given by equation

Fit (N) = P (N) – α C (N)

Here N is the genetic individual or ANN, α is the penalty constant, Fit() is the fitness function, P() is the performance function, C() is the number of connections.

Methods for Response Integration

Here, we use different integration schemes including polling, maximum, minimum, weighed average. These schemes are used to integrate the outputs from each of the four networks separately and the resulting detection accuracies measured. In polling scheme, each network returns the class that it considers the one to which the input belongs. After taking these classes, voting takes place between the network modules. The class with the highest votes is taken as the winner. In weighted average scheme, mean of matching scores of all the networks is taken.

The integrator receives all the probability vectors and does the task of deciding the final output of the system. For this if probability is greater than 0.5 it is marked as malignant, otherwise as benign.

4   Simulation Results

Wisconsin Diagnostic Breast Cancer (WDBC) database of UCI Machine Learning Repository is used for the experimentation of our model. The goal is to classify a tumor as either benign or malignant based on cell descriptions gathered by FNA image test. The Breast Cancer data set has vectors with a total of 30 input attributes. This database contains information about 569 patients with 212 out of 569 having malignant tumors. Attributes used here are radius mean of distances from center to points on the perimeter, texture means standard deviation of gray-scale values, smoothness means local variation in radius lengths, perimeter, area, smoothness (local variation in radius lengths), compactness (perimeter2 / area - 1.0), concavity (severity of concave portions of the contour), concave points (number of concave portions of the contour), symmetry and fractal dimension (coastline approximation - 1).They are measured for a total of 3 cells. The various integration methods are compared with respect to their ability to train, learn and generalize the data. One with the best generalizing capacity will give the best detection efficiency.

First we divide the data set into training and testing sets at 70% and 30% by taking 398 vectors as training data set and rest as testing data set. Then data set is used to train and the test the ensemble model.

The results are measured against the TP (true positive), TN(true negative), FP(false positive) and FN (false negative). The various performance measures are summarized in the table 1.

Table 1. Diagnostic performance measures Breast cancer

Cancer Test	Present	Absent	Total
Positive	True Positive [TP]	False Positive [FP]	[TP +FP]
Negative	False Negative [FN]	True negative [TN]	[FN+TN]
Total	(TP + FN)	(TN + FP)	(TP + FN+ TN + FP)
Sensitivity		TP / (TP + FN)	
Specificity		TN / (TN + FP)	
Accuracy		(TP + TN) / (TP + TN + FP + FN)	

We run the Evolutionary ANN modules to obtain optimum weights for ANN. We applied GA for the parameter optimization. The weight matrix consisted of 30*x weights between input and hidden layer, x*1 between the hidden and the output layer and a total of 18, 20, 25 hidden layer biases and 1 output layer bias. This made the total number of variables for the GA as 30*x + x*1 + x + 1. We use 18, 20, 25 hidden neurons for each module respectively.

In GA, the double vector method of population representation was used. The total number of individuals in the population was 50. A uniform creation function, rank based scaling function and stochastic uniform selection methods were used. The elite count was 2. Single point crossover was used. The program was executed till 100 generations. The crossover rate was 0.7, Best fitness is 98.78 and mean fitness is 96.61.The best performance in terms of sensitivity, specificity, accuracy, false negative and false positive are 98.70% 97.42%, 98.24%, 4.6% and 0.65% for testing respectively.

Here the results of GA are exported which are optimized weights and bias of ANN and ANN is run for 10,000 epochs. The result of this is passed through various integrators.

We then experiment ensemble model with various integration methods to find an optimized parameter which gives best performance. After getting the optimized parameter, the detection procedure is run 20 times for the same configuration. After this, the mean and standard deviation are computed. The mean is taken as the performance accuracy of the system for training and testing dataset.

The results show that the maximum accuracy was achieved when using maximum integrator with Accuracy of 99.07% along with sensitivity, specificity, FPR and FNR values as 98.79, 99.01%, 1.23%, 0.65% respectively. Figure 2 shows the spread of values of Evolutionary ANN module.

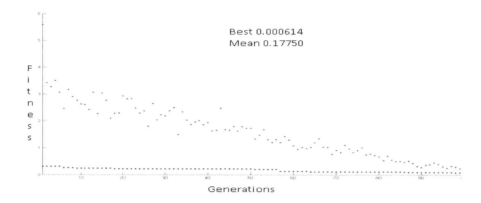

Fig. 2. Performance of Evolutionary ANN module

Table 2. Diagnostic performance of various integration techniques

Integration Techniques	S (%)	Sp (%)	A (%)	FPR (%)	FNR (%)
Maximum	**98.79**	**99.01**	**99.07**	**1.23**	**0.65**
Minimum	98.46	99.01	98.79	1.53	0.98
Polling	99.00	96.47	97.36	4.9	1.56
Sum	98.70	97.42	98.24	5.0	0.65

Now we compare our model performance with Ensemble architecture with same ANN model in all the modules. The only difference in the modules is number of hidden layers. Table 3 is given with the best results of ensemble with same modules on various integration techniques.

Here we used hidden layer of 18, 20, 25 for the 3 modules of ensemble while keeping other parameters same as that in our proposed model. Figure 3 shows the comparative analysis.

Table 3. Diagnostic performance of various integration techniques

ANN Models	Integration Techniques	Testing performances (%)		
		S (%)	Sp (%)	A (%)
MLP	Minimum	99.00	96.47	97.36
RBF	Maximum	97.01	98.05	97.17
LVQ	Polling	93.22	96.04	95.00

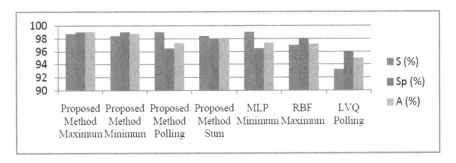

Fig. 3. Comparison of Multi Model ANN with Same ANN ensemble with different integrators

From the experimental results we can show that our proposed method performs better than that of ensemble of ANN with same ANN modules.

5   Conclusion and Future Work

In this paper we saw the working of different ensemble integration methods with three modules of Evolutionary ANN model using MLP model for detection of Breast Cancer. In all the cases we were able to solve the problem with fine accuracies using the different ensemble integration methods. The results show that, Maximum integration gives the most favorable performance when compared with other integration methods.

References

1. American Cancer Society, Cancer Facts and Figures (2011-2012)
2. Weiss, S.I., Kulikowski, C.: Computer Systems That Learn: Classification and Prediction Methods from Statistics, Neural Networks, Machine Learning and Expert Systems. Morgan Kaufmann Publishers (1991)
3. Coleman, T.F., Li, Y.: Large Scale Numerical optimization. In: Proceedings of Workshop on Large Scale Numerical Optimization, Cornell University, New York (1989)
4. Andolina, V.F., Lille, S.L., Willison, K.M.: Mammographic Imaging: A Practical Guide, New York (1992)
5. Antani, S., Lee, D.J., Long, L.R., Thoma, G.R.: Evaluation of shape similarity measurement methods for spine X-ray images. Journal of Visual Communication & Image Representation 15, 285–302 (2004)
6. Clemen, R.: Combining forecasts: A review and annotated bibliography. International Journal of Forecasting, 559–583 (1989)
7. Yao, X., Liu, Y.: Neural networks for breast cancer diagnosis. In: Proceedings of the Congress on Evolutionary Computation, vol. 3, pp. 1767–1773 (August 2002)
8. Fogel, D.B., Wasson, E.C., Boughton, E.M., Porto, V.W., Angeline, P.J.: Linear and neural models for classifying breast masses. IEEE Transactions on Medical Imaging 17(3), 485–488 (1998)

9. Rahul, K., Anupam, S., Ritu, T., Janghel, R.R.: Breast cancer diagnostic system using artificial neural networks model. In: International Conference on Information Sciences and Interaction Sciences (ICIS), pp. 89–94 (2010)

10. Melin, P., Castillo, O.: Hybrid Intelligent Systems for Pattern Recognition Using Soft Computing. STUDFUZZ, vol. 172. Springer, Heidelberg (2005)

11. Bunke, H., Kandel, A. (eds.): Hybrid Methods in Pattern Recognition. World Scientific (2002)

12. Sivanandan, S.N., Deepa, S.N.: Principle of soft computing. Wiley India Private Limited (2007)

13. Nabil, E., Badr, A., Farag, I.: An Immuno-Genetic Hybrid Algorithm. Int. Journal of Computers, Communications & Control IV(4), 374–385 (2009)

14. Alcala, R., Nojima, Y.: Special issue on genetic fuzzy systems: new advances. Evolutionary Intelligence 2, 1–3 (2009)

15. Pena-Reyes, C.A., Sipper, M.: A fuzzy-genetic approach to breast cancer diagnosis. Artificial Intelligence in Medicine, 131–155 (1999)

16. Peña Reyes, C.A.: Breast Cancer Diagnosis by Fuzzy CoCo. In: Peña Reyes, C.A. (ed.) Coevolutionary Fuzzy Modeling. LNCS, vol. 3204, pp. 71–87. Springer, Heidelberg (2004)

Empirical Analysis of Assessments Metrics for Multi-class Imbalance Learning on the Back-Propagation Context*

Juan Pablo Sánchez-Crisostomo[1], Roberto Alejo[1], Erika López-González[1],
Rosa María Valdovinos[2], and J. Horacio Pacheco-Sánchez[3]

[1] Tecnológico de Estudios Superiores de Jocotitlán, Carretera Toluca-Atlacomulco KM. 44.8,
Col. Ejido de San Juan y San Agustín, 50700 Jocotitlán, México
[2] Faculty of Engineering, Universidad Autónoma del Estado de México Cerro de Coatepec s/n,
Ciudad Universitaria C.P. 50100, Toluca, Estado de México
[3] Instituto Tecnológico de Toluca
Av. Tecnológico s/n Ex-Rancho La Virgen, 52140, Metepec, México

Abstract. In this paper we study some of the most common assessment metrics employed to measure the classifier performance on the multi-class imbalanced problems. The goal of this paper is empirically analyzing the behavior of these metrics on scenarios where the dataset contains multiple minority and multiple majority classes. The experimental results presented in this paper indicate that the studied metrics might be not appropriate in situations where multiple minority and multiple majority classes exist.

Keywords: Metrics, Multi-class Imbalance, Multiple Minority and Majority Classes.

1 Introduction

Class imbalance problems have drawn growing interest recently because of their classification difficulty caused by the imbalanced class distributions [8]. So, it has been into the 10 challenging problems identified in data mining research [9]. The class imbalance problem appears when in a training data set the number of instances in at least one class is much less than the samples in another class or classes [6]. Much work has been done in addressing the class imbalance problem [4] but, while two-class imbalance problem has been widely studied the *multi*-class imbalance problem has been relatively less investigated [8].

The multi-class imbalance problems pose new challenges, for instance, in task of assessments classifier performance it is necessary to apply different metrics than those used in traditional two-class classification problems [7]. On two-class imbalance problems, sometimes it is possible to provide more appropriate assessments metrics to measure the classifier performance, but in the multi-class imbalance problems, it is extremely difficult to provide realistic assessments of the relative severity of the classification performance [3].

---

* This work has been partially supported under grants of: PROMEP/103.5/12/4783 from the Mexican SEP and SDMAIA-010 of the TESJO.

Y. Tan et al. (Eds.): ICSI 2014, Part II, LNCS 8795, pp. 17–23, 2014.

Often in research where the multi-class imbalance is the focus of study, the authors use metrics that have been extended from two-class imbalance scenarios to compare the classifier performance, for example the geometric mean, F-measure or measures of the area under curve family [2]. However, this situation presents an interesting research question: The proposed metrics to assessment the classifier performance are appropriate on scenarios where exist multiple minority and multiple majority classes?.

We are interested in this question, so, in this paper we study the behavior of seven the most common multi-class assessment metrics on real multi-class databases with minority and majority multiple classes.

2 Assessments Metrics for Multi-class Imbalance Learning

The most studied metrics for assessment the classifier performing in class imbalance domains have been focused a two class imbalance problems and some of them have been modified to accommodate them at the multi-class imbalanced learning problems [4]. In this section we present some of the most common two-class imbalance metrics adapted at multi-class imbalance scenarios.

Macro Average Geometric (MAvG): This is defined as the geometric average of the partial accuracy of each class.

$$MAvG = (\prod_{i=1}^{J} ACC_i)^{\frac{1}{J}},\tag{1}$$

where $ACC_j = (\,correctly\ classified\ of\ class\ j)/(\,total\ of\ samples\ of\ class j)$, i.e., the accuracy on the class j. J is the number of classes.

Mean F-measure (MFM): This measure has been widely employed in information retrieval

$$F - \text{measure}(j) = \frac{2 \cdot recall(j) \cdot precision(j)}{recall(j) + precision(j)},\tag{2}$$

where $recall(j) = (correctly\ classified\ positives)/(total\ positives)$ and $precision$ $(j) = (correctly\ classified\ positives)/(total\ predicted\ as\ positives)$; j is the index of the class considered as $positive$. Finally, mean F-measure is defined for multi-class in Reference [2] as follow:

$$MFM = \sum_{j=1}^{J} \frac{FMeasure(j)}{J}.\tag{3}$$

Macro Average Arithmetic (MAvA): This is defined as the arithmetic average of the partial accuracies of each class.

$$MAvA = \frac{\sum_{i=1}^{J} ACC_i}{J}.\tag{4}$$

One the most widely used techniques for the evaluation of binary classifiers in imbalanced domains is the Receiver Operating Characteristic curve (ROC), which is a tool for visualizing, organizing and selecting classifiers based on their trade-offs between true positive rates and false positive rates. Furthermore, a quantitative representation of a ROC curve is the area under it, which is known as AUC [1]. The AUC measure has been adapted at multi-class problems [2] and can be defined as follow.

AUC of each class against each other, using the uniform class distribution (AU1U):

$$AU1U = \frac{2}{\|J\|(\|J\| - 1)} \sum_{j_i, j_k \in J} AUC_R(j_i, j_k) ,$$

(5)

where $AUC_R(j_i, j_k)$ is the AUC for each pair of classes j_i and j_k.

AUC of each class against each other, using the a priori class distribution (AU1P):

$$AU1P = \frac{2}{\|J\|(\|J\| - 1)} \sum_{j_i, j_k \in J} p(j) AUC_R(j_i, j_k) ,$$

(6)

where $p(j)$ is a priori class distribution.

AUC of each class against the rest, using the uniform class distribution (AUNU):

$$AUNU = \frac{1}{J} \sum_{j \in J} AUC_R(j, rest_j) ,$$

(7)

where $rest_j$ gathers together all classes different from class j, i.e., the area under the ROC curve is computed in the approach one against all.

AUC of each class against the rest, using the a priori class distribution (AUNP):

$$AUNP = \frac{1}{J} \sum_{j \in J} p(j) AUC_R(j, rest_j) ,$$

(8)

this measure takes into account the prior probability of each class $(p(j))$.

3   Experimental Protocols

3.1   Database Description

In this section we describe briefly the two databases (92AV3C and ALL-DATA) used in our experimentation. 92AV3C corresponds to a hyperspectral image (145 x 145 pixels) taken over Northwestern Indianas Indian Pines by the AVIRIS sensor[1]. For simplicity, in this paper we use only 38 attributes from the 220 attributes of the original dataset. The attributes were selected using a common features selection algorithm (Best-First Search [5] implemented in WEKA.[2]

ALL-DATA consists of the reflectance values of image pixels that were taken by the Compact Airborne Spectrographic Imager (CASI) and the Airborne Hyper-spectral Scanner (AHS) sensors. Corresponding chlorophyll measurements for these pixels were

---

[1] engineering.purdue.edu/biehl/MultiSpec/hyperspectral.html
[2] www.cs.waikato.ac.nz/ml/weka/

also performed. CASI set consists of the reflectance values of image pixels that were taken by the CASI sensor. Corresponding thermal measurements for these pixels were also made. The CASI sensor reflectance curves are formed by 144 bands between 370 and 1049 nm. AHS images consist of 63 bands between 455 and 2492 nm. Therefore, the input dimensionality of this dataset is 207 (the sum of the bands corresponding to the CASI and AHS sensors).

Table 1. The class distribution of the 92AV3C and ALL-DATA datasets is presented in this table. Size represents the number of samples on each class.

Database	Class	Size	Database	Class	Size
	0	95		1	8258
	1	489		2	4588
	2	834		3	11346
	3	968		4	4751
	4	54		5	1123
	5	614		6	5762
	6	497	ALL-DATA	7	3020
	7	1294		8	7013
92AV3C	8	380		9	15
	9	26		10	79
	10	234		12	82
	11	20		13	222
	12	1434		14	1733
	13	2468		15	3628
	14	747		–	–
	15	212		–	–
	16	10659		–	–

In order to study the multi-class assessment metrics behavior in domains of multiple minority classes and multiple majority classes we split the original datasets (92AV3C and ALL-DATA) in subsets (S_i, A_i and B_i).

For 92AV3C dataset the S_i subsets were integrated in the following way: the first subset (S1) contains the four more minority classes (3, 7, 12 and 13) and the most majority class (16) of the original dataset. The next subset (S2) was integrated with the union of S1 with the next minority class (not used before, class 2). This process finishes when all minority have been integrated at one subset.

The ALL-DATA was split in subsets as follow: the first subset A1 was integrated by the more minority classes (9, 10, 12 and 13) of ALL-DATA dataset, the next subset (A2) corresponds to join of A1 with the class 1, i.e., A1∪1. The rest of subsets were made of the same way.

The B_i subsets were integrated as follow: B1 contains eight majority classes (1–8) of ALL-DATA dataset, B2 = B1 ∪ with the next minority class (class 9), and the rest of B_i subsets were integrated using this process. The Table 2 shows a brief summary of the integration of S_i, A_i and B_i subsets.

In 92AV3C subsets (Si) the 10–fold cross–validation was applied. The datasets were divided into ten equal parts, using nine folds as training set and the remaining block test set. ALL-DATA subsets (Ai and Bi) were split in disjoints subsets: training (50% of the samples) and test (50% of the samples).

Table 2. Classes that ingrate the Si, Ai and Bi subsets

Subset	S1	S2	S3	S4	S5	S6	S7	S8	S9	S10	S11	S12	S13
Classes	3,7,12,13,16	S1∪2	S2∪14	S3∪5	S4∪6	S5∪1	S6∪8	S7∪10	S8∪15	S9∪0	S10∪4	S11∪9	S12∪11

Subset	A1	A2	A3	A4	A5	A6	A7	A8	A9	–	–	–	–
Classes	9,10,12,13	A1∪1	A2∪2	A3∪3	A4∪4	A5∪5	A6∪6	A7∪7	A8∪8	–	–	–	–

Subset	B1	B2	B3	B4	B5	–	–	–	–	–	–	–	–
Classes	1,2,3,4,5,6,7,8	B1∪9	B2∪10	B3∪12	B4∪13	–	–	–	–	–	–	–	–

3.2 Neural Network Configuration

In the experimental phase we use the MLP trained with the standard back-propagation in sequential mode. For each training data set, MLP was initialized ten times with different weights, i.e., the MLP was run ten times with the same training dataset. The results here included correspond to the average of those accomplished ten different initialization and of ten partitions for 92AV3C, and only the average of ten different initializations for ALL-DATA. The learning rate (η) was set at 0.1 and only one hidden layer was used. The stop criterion was established at 5000 epoch or an MSE below to 0.001. The number of neurons (n) for the hidden layer was fixed as $n =$ number of classes $+1$, because our goal it is not to find the optimal MLP configuration but to study the assessment metrics behavior.

4 Experimental Results

In order to assessment the multi-class imbalance metrics: $MAvG$, $AUNP$, $AU1P$, MFM, $AUNU$, $MAvA$ and $AU1U$ (see section 2), we have carried out an experimental comparison over twenty seven datasets with multiple minority and multiple majority classes (see Table 2).

The tables 3 and 4 present the experimental results. The first column represents the dataset used (Si, Ai or Bi), the next columns exhibit the metrics used and at least one the number of minority classes do not classified, i.e., ignored by the classifier. The rows show the values obtained from different metrics on each dataset.

Tables 3 and 4 show an interesting behavior. Observe that in some datasets all metrics (except $MAvG$) present very similar results but, in these datasets the classifier does not classify or ignored different minority classes in each dataset. For example, in Table 3 the values the $AUNP$ for the S10 and S11 datasets are 0.717557 and 0.718413 respectively, i.e., they are very similar values. However, in S10 the classifier does not classify two minority classes and S11 does not classify one class. In other words, the values the $AUNP$ for S10 and S11 are very similar but in S10 the classifier ignored more classes

Table 3. Classification performance of the subsets (Si) obtained from 92AV3C dataset (see Table 2) measured by the metrics: $MAvG$, $MAvA$, $AU1U$, $AU1P$, $AUNU$, MFM and $AUNP$

Metric	$MAvG$	$AUNP$	$AU1P$	MFM	$AUNU$	$MAvA$	$AU1U$	No. of classes ignored by the classifier
S1	0.000000	0.680849	0.680849	**0.296549**	0.606591	0.349023	0.349023	3
S2	0.000000	0.864575	0.864575	**0.298384**	0.637522	0.373081	0.373082	2
S3	0.000000	0.843382	0.843382	0.326805	0.700919	0.524812	0.524812	1
S4	0.000000	0.775310	0.775310	0.272290	**0.627219**	0.442667	0.442667	1
S5	0.000000	0.763563	0.763564	0.260260	0.624413	0.448988	0.448988	1
S6	0.000000	0.755563	0.755563	0.245079	0.645932	0.500241	0.500241	2
S7	0.000000	0.745715	0.745714	0.227627	0.628405	0.478391	0.478391	2
S8	0.000000	0.738600	**0.738599**	0.212890	**0.626571**	**0.486770**	**0.486770**	3
S9	0.000000	0.731669	0.731668	0.197799	0.628791	0.499175	0.499175	1
S10	0.000000	0.737835	**0.737835**	0.193418	0.655136	0.549927	0.549927	2
S11	0.000000	**0.717557**	0.717558	0.170630	0.615831	**0.487472**	**0.487472**	2
S12	0.000000	**0.718413**	0.718413	0.162070	0.626332	0.509208	0.509208	1

Table 4. Classification performance of ALL-DATA dataset measured by the metrics: $MAvG$, $MAvA$, $AU1U$, $AU1P$, $AUNU$ and $AUNP$

Metric	$MAvG$	$AUNP$	$AU1P$	MFM	$AUNU$	$MAvA$	$AU1U$	No. of classes ignored by the classifier
A1	0.995583	0.995543	0.995543	0.984598	0.995823	0.995625	0.995625	0
A2	0.601254	0.915319	0.915319	0.300290	0.844969	0.796271	0.796271	0
A3	0.408697	**0.775386**	**0.775386**	**0.271673**	**0.752187**	0.760755	0.760755	0
A4	0.000000	0.747106	0.747106	0.289039	0.716155	0.682684	0.682684	1
A5	0.000000	0.803038	0.803038	0.336305	**0.751786**	0.699337	0.699337	1
A6	0.000000	**0.833518**	**0.833518**	**0.350768**	0.741196	0.649088	0.649088	1
A7	0.000000	0.734872	0.734872	0.263121	0.659345	0.580251	0.580251	2
A8	0.000000	0.708736	0.708736	0.234921	0.629486	0.546548	0.546549	2
A9	0.000000	0.706589	0.706589	0.230379	0.596762	0.483704	0.483704	3
B1	0.648743	0.750903	0.750903	0.391792	0.748725	0.744061	0.744061	0
B2	0.000000	0.730440	0.730440	0.320462	0.676222	0.618496	0.618496	1
B3	0.000000	0.735042	0.735042	0.301764	0.667699	0.598593	0.598593	2
B4	0.000000	**0.708683**	**0.708683**	0.246963	0.609454	0.506425	0.506425	2
B5	0.000000	**0.706589**	**0.706589**	0.230379	0.596762	0.483704	0.483704	3

than S11. Similar situations were observed in $AU1P$ with S8 and S10, MFM with S1 and S2, $AUNU$ with S4 and S8, $MAvA$ with S8 and S11, and $AU1U$ with S8 and S11 (see Table 3). On Table 4 this behavior was observed in $AUNU$ and $AU1P$ with B4 and B5. $AUNU$ with A3 and A5.

A dramatic situation was noticed in Table 4, we observe that in some datasets the classifier presents better results when does not classify one or more classes that when it classify all classes. For example, the values for A3 and A6 with $AUNP$ are 0.775386 and 0.833518, respectively, i.e., the result the A6 is better than the A3 result, but in A6 one class is ignored for the classifier meanwhile that in A3 all classes are identified for it. This behavior was adviced too in the metrics $AU1P$ and MFM for these datasets (A3 and A6).

On the other hand, the $MAvG$ could be more appropriate in classification problems with multiple majority classes and multiple minority classes, because it notice when the classifier ignores any class (see Table 4).

5   Conclusions

In this paper we study some of the most common metrics employed to measure the classifier performance on the multi-class imbalanced problems. We focused in problems with multiple minority classes and multiple majority classes. So, some experiments have been carried out over twenty seven real data sets using a multilayer perceptron trained with the back-propagation algorithm.

From the analysis of the experimental results in this work, we might suggest that the main problem of the assessment metrics studied in this paper (except $MAvG$), is that they were designed to provide an *average* performance of the pairs of classes, so this metrics, in some cases, do not provide information when one or more classes are ignored for the classifier.

We think, therefore, that they might not be appropriate when the dataset contains multiple minority classes and multiple majority classes, in other words these metrics might not be appropriate in muti-class imbalance context as the *accuracy* was in two-class imbalance problems. However, the $MAvG$ could be more appropriate in this scenario because it notice when the classifier ignores any class.

The assessment metrics were developed with different proposes and goals, nevertheless, in the literature the researchers use they to compare the classifier performance, for this reason we consider is necessary a deeper study about of this problem than the previous one.

References

1. Fawcett, T.: An introduction to roc analysis. Pattern Recogn. Lett. 27, 861–874 (2006)
2. Ferri, C., Hernández-Orallo, J., Modroiu, R.: An experimental comparison of performance measures for classification. Pattern Recognition Letter 30(1), 27–38 (2009)
3. Hand, D.J., Till, R.J.: A simple generalisation of the area under the roc curve for multiple class classification problems. Machine Learning 45, 171–186 (2001)
4. He, H., Garcia, E.: Learning from imbalanced data. IEEE Transactions on Knowledge and Data Engineering 21(9), 1263–1284 (2009)
5. Kohavi, R., John, G.H.: Wrappers for feature subset selection. Artif. Intell. 97(1-2), 273–324 (1997)
6. Ou, G., Murphey, Y.L.: Multi-class pattern classification using neural networks. Pattern Recognition 40(1), 4–18 (2007)
7. Tsoumakas, G., Katakis, I.: Multi-label classification: An overview. Int. J. Data Warehousing and Mining, 1–13 (2007)
8. Wang, S., Yao, X.: Multi-class imbalance problems: Analysis and potential solutions. IEEE Transactions on Systems, Man and Cybernetics, Part B: Cybernetics 42(4), 1–12 (2012)
9. Yang, Q., Wu, X.: 10 challenging problems in data mining research. International Journal of Information Technology and Decision Making 5(4), 597–604 (2006)

A Novel Rough Set Reduct Algorithm to Feature Selection Based on Artificial Fish Swarm Algorithm

Fei Wang, Jiao Xu, and Lian Li

School of Information Science & Engineering, Lanzhou University,
730000 Lanzhou, China
{wangf12,xujiao12}@lzu.edu.cn

Abstract. With the purpose of finding the minimal reduct, this paper proposes a novel feature selection algorithm based on artificial fish swarm algorithm (AFSA) hybrid with rough set (AFSARS). The proposed algorithm searches the minimal reduct in an efficient way to observe the change of the significance of feature subsets and the number of selected features, which is experimentally compared with the quick reduct and other hybrid rough set methods such as genetic algorithm (GA), ant colony optimization (ACO), particle swarm optimization (PSO) and chaotic binary particle swarm optimization (CBPSO). Experiments demonstrate that the proposed algorithm could achieve the minimal reduct more efficiently than the other methods.

Keywords: feature selection, rough set, fish swarm algorithm, ant colony optimization, chaotic binary particle swarm optimization.

1   Introduction

Feature selection is the process of choosing a good subset of relevant features and eliminating redundant ones from an original feature set, which can be perceived as a principal pre-processing tool for solving the classification problem [1]. The main objective of feature selection is to find a minimal feature subset from a set of features with high performance in representing the original features [2]. In classification problems, feature selection is a necessary step due to lots of irrelevant or redundancy features. By eliminating these features, the dimensionality of feature can be reduced and the predictive performance can be improved for classification. Feature selection methods are dimensionality reduction methods often associated to data mining tasks of classification [3], which provide a reduced subset of the original features while preserving the representative power of the original features.

Rough set (RS) was proposed by Pawlak, which provides a valid tool that can be applied for both feature selection and knowledge discovery. It has been proved to be an effective feature selection approach, which can select a subset of features while preserving the meaning of the features, therefore it can predict the classification accuracy as well as the original feature set. The essence of rough set to feature selection are to find a minimal subset of the original features with the most

Y. Tan et al. (Eds.): ICSI 2014, Part II, LNCS 8795, pp. 24–33, 2014.

informative features and remove all other attributes from the feature set with minimal information loss [4]. Rough set is a powerful mathematical tool to reduce the number of features based on the degree of dependency between condition attributes and decision attributes, which has been widely applied in many fields such as machine learning and data mining. Though rough set has been used as a feature selection method with much success, it is inadequate at finding optimal reduct because of no perfect search techniques.

In order to find the optimal reduct and improve the performance, a variety of search techniques hybrid with rough set are introduced to address feature selection problems such as genetic algorithm (GA), ant colony optimization (ACO), particle swarm optimization (PSO). These swarm intelligence based algorithms such as particle swarm and ant colony optimization have been proved to be competitive in rough set attribute reduction fields. However, these algorithms have some disadvantages such as premature convergence in PSO and the performance of the reduct depending on initial parameters in ACO. In this paper, we propose a novel feature selection algorithm based on artificial fish swarm algorithm hybrid with rough set, which is not sensitive to initial parameters, has a strong robustness and has the faster convergence speed to find the minimal reduct subset.

2   Rough Set Theory

Rough set theory is an extension of traditional set theory that provides approximations in decision making, in which attribute reduction provides a valid method to extract knowledge from feature set in a concise way. In this paper, we adopt some relevant concepts of rough set theory related to our attribute reduction approach in [5] such as equivalence relation, lower approximation, positive region, and degree of dependency.

De nition of Core. The elements of feature core are those features that cannot be eliminated. In this paper, the algorithm for finding feature core is as follows: initialize $Core = \varnothing$; for every attribute $a \in C$, if $\mu_{C-\{a\}}(D) < \mu_C(D)$, then attribute a is one element of feature core, namely $Core = Core \cup \{a\}$. Where $\mu_C(D)$ represents the degree of dependency between condition attributes C and decision attribute D.

The quick reduct (QR) algorithm proposed in [6], attempts to obtain a reduct without exhaustively generating all possible subsets. It starts from an empty set, adds one attribute at a time until it generates its maximum value for the dataset.

3   Swarm Intelligence Based Rough Set Reduct Algorithm

3.1   Ant Colony Optimization Based Reduct Algorithm (ACORS)

ACO is considered a new meta-heuristic algorithm that is used successfully to solve many NP-hard combinatorial optimization problems [7]. In ACO, a swarm of artificial ants cooperate for finding good solutions to optimization problems. Every ant searches for optimal solutions in the problem space, which has a start state and one or more end conditions. The next move is determined by a probabilistic transition rule that is a function of locally available pheromone trails. Once ant has constructed a solution, then it updates the pheromone trial values which depend on the quality of solutions constructed by the ants. Finally, the ant constructs the optimal solution with the higher amount of pheromone trails. The algorithm stops iterating when an end condition is satisfied. The search for the optimal feature subset is a traversal through the graph where a minimal number of nodes are visited and the end conditions are satisfied [8]. This algorithm performs as follow: All ants start from feature core, each ant builds a solution and then the pheromone trials for every ant are updated.

Pheromone Trials and Heuristic Information. In the step, each edge is assigned a pheromone trail and heuristic information. Firstly, the initial pheromone trial on each edge is initialized to equal amount of pheromone. Secondly, each ant constructs a solution; after that, the pheromone of each edge in this solution is updated. In ACORS, the heuristic information is on the basis of the degree of dependency between the two attributes and decision attribute. The value of heuristic information η is limited in this paper, If $\eta(a,b) < \varepsilon$, then $\eta(a,b) = \varepsilon$, where ε is set to 0.001. Formally, for any two attributes $a,b \in C$, the heuristic information is defined as

$$\eta(a,b) = \frac{\left|POS_{\{a,b\}}(D)\right|}{|U|} \tag{1}$$

Where $|U|$ is the cardinality of set U and the $POS_{\{a,b\}}(D)$, called positive region, is defined in [5].

Construction of Feasible Solution. When constructing a solution, each ant should start from the feature core. Firstly, the ant selects randomly a feature, after that, it probabilistically selects the second attribute from those unselected attributes. That probability is calculated by

$$P_{ij}^k(t) = \frac{\left[\tau_{ij}(t)\right]^\alpha \cdot \left[\eta_{ij}\right]^\beta}{\sum_{l \in J}\left[\tau_{ij}(t)\right]^\alpha \cdot \left[\eta_{ij}\right]^\beta} \tag{2}$$

Where t and k represent the number of iterations and ants, respectively, J represents the set of unvisited features of ant k, η_{ij} is heuristic information of choosing feature j when at feature i, $\tau_{ij}(t)$ is the amount of pheromone between feature i and feature j at iteration t. In addition, α and β are two parameters corresponding to the importance of the pheromone trail and heuristic information. When $\mu_R(D) = \mu_C(D)$, the construction process stops, where R is the current solution constructed by an ant.

Pheromone Update. After each ant has constructed its own solution, the pheromone of only edges along the path visited by the ant is updated as

$$\tau_{ij}(t+1) = \rho\tau_{ij}(t) + q / L_{\min} \tag{3}$$

While for other edges, the pheromone trails are updated according to the following equation.

$$\tau_{ij}(t+1) = \rho\tau_{ij}(t) \tag{4}$$

Where ρ is a decay constant used to simulate the evaporation of pheromone, q is a given constant and $L_{\min}$ is the minimal feature reduct at iteration t. In ACORS, if the maximum iteration is reached, then the algorithm terminates and outputs the minimal reduct encountered. If not, then the pheromone is updated, a new colony of ants are created and the process iterates once more.

3.2   Particle Swarm Optimization Based Reduct Algorithm (PSORS)

PSO is an efficient evolutionary computation technique based on swarm intelligence and originates from the simulation of social behaviors such as birds in a flock or fishes in a school. In PSO, a particle represents a candidate solution to the problem, which has its own velocity and position in a given search space. PSO starts with the stochastic initialization of a population of particles which move in the search space to find the optimal solution by updating the position of each particle by using its own experience and its companion's experience [9]. Assume a swarm includes N particles which move around in a D-dimensional search space. The velocity of the ith particle in different space can be represented by $v_i = (v_{i1}, v_{i2}, \cdots, v_{iD})$, and the position for the ith particle in different space can be noted as $x_i = (x_{i1}, x_{i2}, \cdots, x_{iD})$. The positions and velocities of the particles are restricted to a predefined range, respectively. The personal best position recording the previous best position of the particle is called *pbest* and the best position achieved by all individual is denoted the global best position and called *gbest* . Based on *pbest* and *gbest* , the velocity and position of each particle are updated to search for the optimal solutions. When BPSO is applied to solve the feature selection problem, a binary digit is employed to stand for a feature, where the bit values 1 and 0 stand for selected and non-selected features, respectively. The velocity of each particle is updated using (5), while the position of each particle is updated using (6). The position and velocity of each particle are updated according to the following equations:

$$v_{id}^{t+1} = w \times v_{id}^{t} + c_1 \times r1 \times (pbest - x_{id}^{t}) + c_2 \times r2 \times (gbest - x_{id}^{t}) \tag{5}$$

$$x_{id} = \begin{cases} 1, & \text{if } r < S(v_{id}) \\ 0, & \text{otherwise} \end{cases} \qquad S(v_{id}) = \frac{1}{1 + e^{-v_{id}}} \tag{6}$$

Where t represents the iteration counter, $r1$ and $r2$ are random numbers between 0 and 1, c_1 and c_2 are learning factors that control how far a particle moves in a single generation, w is called the inertia weight, and the function $S(v_{id})$ is a sigmoid limiting transformation which is introduced to transform v_{id} to the range of $(0, 1)$, r is random number selected from a uniform distribution between 0 and 1.

In BPSO, the inertia weight w is the modulus that controls the influence of previous velocity on the present one, thus balancing the global exploration and local search ability. It means the appropriate control of inertia weight value is imperative to search for the optimum solution efficiently and precisely. In this paper, chaos theory and BPSO are combined into a method called CBPSO to avoid this early convergence, then CBPSO based RS reduct algorithm (CBPSORS) could be achieved to find superior reduct. Since logistic maps are the most frequently used chaotic behavior maps and chaotic sequences have been proven easy and fast to generate and store, as there is no need for storage of long sequences [10], so logistic map is used to determine the inertia weight value. The inertia weight value is substituted by sequences generated by the logistic map according to the following (7).

$$w(t+1) = \mu \times w(t) \times (1 - w(t)) \qquad w(t) \in (0,1) \qquad (7)$$

Where μ is a control parameter, which cannot be bigger than 4. When the inertia weight value is close to 0, CBPSO promotes the local search ability. For inertia weight values near 1, CBPSO strengthens the global search ability.

During the search process, each individual is evaluated using the fitness. According to the definition of RS reduct, the reduction solution must ensure that the decision ability is the same as the original decision table and the number of features in the feasible solution is kept as less as possible. Therefore, classification quality and the number of selected features are the two pivotal factors used to design a fitness function which is used to evaluate each individual. The fitness function is defined as

$$Fitness = \lambda * \mu_R(D) + \xi * \frac{|C| - |R|}{|C|} \qquad (8)$$

Where $\mu_R(D)$ represents the classification quality of selected condition attributes R relative to decision D; $|R|$ denotes the number of selected feature subset; $|C|$ denotes the number of the original feature set; λ and ξ are two parameters which determine the relative importance of classification quality and the number of features, $\lambda \in [0,1]$ and $\lambda + \xi = 1$.

4   Artificial Fish Swarm Based Reduct Algorithm (AFSARS)

AFSA is a swarm-intelligence based optimization algorithm that simulates the fish swarm behaviors such as praying, swarming and following with local search of fish individual for obtaining the global optimum, which was successfully applied to solve several combinatorial problems. It is a stochastic and parallel search algorithm. What's more, it does not need to know the concrete information of problems; instead it only needs to compare disadvantages and advantages of the solutions of the problems [11], then the final global optimum will be displayed in the population through artificial fish individual behaviors of local optimization, which has strong robustness, fast speed of convergence and being non-sensitive to initial parameters. The AFSA has been proved to be an effective global optimization algorithm using the swarm intelligence in the solution of the combinatorial problem [12].

Due to these characteristics of the AFSA, it is introduced to solve feature selection problems. Assume a fish swarm includes n particles which move around in a D-dimensional search space. The artificial fish swarm is represented as $F = \{f_1, f_2, \cdots, f_n\}$, where f_i is an artificial fish (AF). An AF can represent a subset of features, and a subset of features can be a binary vector: $X = \{x_1, x_2, \cdots, x_D\}$, $x_i \in \{0,1\}, i = 1, 2, \cdots, D$, where X is the current state of AF, D is the number of features and the bit values 1 and 0 stand for selected and non-selected features respectively. Let Y stand for the food concentration, namely the objective function value; the visual scope of AF is represented as visual distance. The Hamming distance is used to calculate the visual distance in AFSARS. The Hamming distance of two points of equal bits length is the number of positions at which the corresponding bits are different. S_m is the moving step length, trynumber is the try number and δ is the crowd factor. The representative behavior is described as follows:

Following Behavior. In the following behavior, when the AF current state is X_i, it will judge the food concentration of all its neighborhood partners. Then it will find the state X_j in the current neighborhood, which has the greatest food concentration Y_j. Let n_f represent the number of its neighbors in the current neighborhood and n represent the total number of AF. If $Y_i < Y_j$ and $\dfrac{n_f}{n} < \delta$, it denotes the state X_j has more food and is not crowded, it will moves a step toward the state X_j. Otherwise, it performs the swarming behavior.

Swarming Behavior. In the swarming behavior, when the AF current state is X_i, it will assemble in groups naturally in the moving process. Let X_c represent the center position in its visual scope. If $Y_i < Y_c$ and $\dfrac{n_f}{n} < \delta$, it denotes the center position has higher food concentration and is not crowded. It moves a step toward the center position. Otherwise, it performs the preying behavior. The center position X_c of m fishes is defined as

$$X_c(i) = \begin{cases} 1, & \sum\limits_{k=1}^{m} X_k(i) \geq \dfrac{m}{2} \\ 0, & \sum\limits_{k=1}^{m} X_k(i) \leq \dfrac{m}{2} \end{cases} \qquad i = 1, 2, 3, \cdots, D \qquad (9)$$

Preying Behavior. In the preying behavior, when the AF current state is X_i, it needs to select a state Y_j randomly in its visual scope. If $Y_i < Y_j$, it moves forward a step in this direction. Otherwise, it selects randomly a state X_j again in its visual distance, and it judges whether the forward condition is satisfied. If it can satisfy before trynumber times, it moves a step toward the state X_j, otherwise, it moves a step randomly. When the AF selects to go forward a step in this direction, the mutation operation of genetic algorithm is adopted in the proposed AFSARS. One position mutation is used to create a trial point. If the AF will go forward a step from the state X_i to the state X_j, then the number of different bits n_b is calculated. Here, if $n_b > S_m$, then $S_m = 3$, otherwise, $S_m = n_b$. Randomly generate a digit n_r which represents the number of mutations, where n_r is between 1 and S_m. Here, some indexes of the positions of mutation are selected and then the bits of selected positions are changed from 0 to 1 or vice versa.

Artificial fish swarm algorithm based rough set reduct algorithm (AFSARS)

(1) Initialize the parameters for AFSARS;
(2) Compute the feature core of feature set;
(3) Randomly initialize the population F, the length of AF equals the number of feature in C-core ;
(4) Initial the global optimum Best=0;
(5) Do
    a) Initial the local optimum besty=0;
    b) Execute the following behavior, if the following behavior succeed, return the current feature subset, otherwise go to c);
    c) Execute the swarming behavior, if the swarming behavior succeed, return the current feature subset, otherwise go to d);
    d) Execute the preying behavior, if the preying behavior succeed, return the current feature subset, otherwise go to e);
    e) Execute the random behavior, return the current feature subset;
    f) Evaluate the current feature subset ;
    g) Memorize the local and global optimal feature subset;
(6) Repeat until the total iteration number is reached.

Fig. 1. Artificial fish swarm algorithm based rough set reduct algorithm

Random Behavior. If the other fish behaviors are not executed, the AF performs the random behavior. This behavior is related with a random movement for a better position. The behavior is similar to preying behavior, but the different point is the position of mutation which can be any position of the state X_i. The pseudo-code of our proposed method is illustrated in Fig.1.

5   Experiments and Results

5.1   Datasets and Parameters Setting

To evaluate the usefulness of the proposed algorithms, we carry out experiments on six datasets of the UCI machine learning repository. In Dermatology (Der) dataset, some samples are missed in age feature, so it is removed. In the experiments, the five algorithms require additional parameter settings for their operations. The parameters of GA are set as follows: population size $P = 20$, maximum iteration $T = 500$, the default parameters of crossover and mutation are adopted in matlab 7.0. The parameters of ACORS are set as follows: $\alpha = 1$, $\beta = 0.01$, $\rho = 0.9$, $q = 0.1$ and the initial pheromone is set to 0.5, the number of ants is half the number of features and the maximum iteration equals 50. The parameters of PSORS are set as follows: the inertia weight decreases along with the iterations, varying from 1.4 to 0.4 according to the reference [9], acceleration constants $c_1 = c_2 = 2.0$, population size $P = 20$, maximum iteration $T = 500$, velocity $V_{max} = 4, V_{min} = -4$. The parameters of CBPSORS are set as follows: the inertia weight $w(0) = 0.48, \mu = 4$, acceleration constants $c_1 = c_2 = 2.0$, population size $P = 20$, maximum iteration $T = 500$, velocity $V_{max} = 4, V_{min} = -4$. These parameters are chosen based on the literature [10]. The parameters of AFSARS are set as follows: population size $P = 50$, maximum iteration $T = 50$, trynumber=20, maximum step $S_m = 3$, the visual distance of fish is half the number of features, crowd factor $\delta = 0.618$. The parameter λ of the fitness is set to 0.9 and $\xi = 0.1$ according to the reference [5]. The fitness function of GARS, PSORS, CBPSORS and AFSARS are defined as the equation (9). In CBPSORS, the core of feature set needs to compute, after that the population is initialized, and the operation is the same as AFSARS. The results achieved from 3 independent runs are employed in terms of the number of the evolved feature subsets in this paper.

5.2   Results and Analysis

Table 1 shows the reduct results of the various methods on the 6 UCI datasets. According to the experimental results, we find that AFSARS, CBPSORS and PSORS have similar efficiency and they are more effective than ACORS and GARS when dealing with datasets having less than 30 features. However, when dealing with datasets with over 30 features, PSORS is easy to fall into premature convergence, which means PSORS is not suitable to find the optimal reduct in most cases. Comparing with PSORS, AFSARS and ACORS become much effective and find successfully the global optimum in limited number of iterations on datasets with over 30 features. For those datasets having many features such as Dermatology and Lung, AFSARS and ACORS are more effective than PSORS and CBPSORS. Furthermore, PSORS hardly finds the optimal reduction until the maximum iteration is reached when it deals with datasets with many features. The performances of PSORS and CBPSORS are not improved after we change their generations to 1000. Apart from these, we find that the performance of CBPSORS is similar to the performance of PSORS when the maximum generation is 500, but when we run the two methods many times, we find that the result of CBPSO is more stable and better. On the whole, it seems to be the case that AFSARS outperforms the other methods in terms of the number of the minimal reducts. But compared to the other methods, AFSARS spends more time to find the optimum reducts.

Table 1. The experimental results of the different algorithms

Dataset	#Features	QR	GARS	ACORS	PSORS	CBPSORS	AFSARS
Momk1	6	4	3	3	3	3	3
Tic-tac	9	8	8	8	7	7	7
Zoo	16	5	6-7	5	5	5	5
Vote	16	10	8-10	9	8	8	8
Der	33	10	11-12	9-10	10	9-10	9
Lung	56	5	12-13	5	9-10	9	5

6   Conclusion

This paper starts with the concepts of rough set theory and the QR algorithm, but this technique often fails to find optimal reducts because of no perfect search strategy. Therefore, the swarm intelligence methods have been introduced to guide RS method to find the minimal reducts. Here, we have discussed four different computational intelligence based reducts: GARS, ACORS, PSORS and CBPSORS. These methods perform well on some datasets, but sometimes they cannot find the optimal solution in the limited number of iteration. In this paper, we propose a novel feature selection algorithm based on artificial fish swarm algorithm hybrid with rough set (AFSARS), which is non-sensitive to initial values, has a strong robustness and has the faster convergence speed to find the minimal reducts. Experimental results on real datasets have demonstrate our proposed method can provide competitive solutions in generating short reducts more efficiently than the other methods.

Acknowledgments. The authors would like to thank to the Natural Science Foundation of the People Republic of China (61073193, 61300230), the Key science and technology Foundation of Gansu Province (1102FKDA010), Natural Science Foundation of Gansu Province (1107RJZA188), science and technology support program of Gansu Province (1104GKCA037) for supporting this research.

References

1. Wang, X., Yang, J., Teng, X., Xia, W., Jensen, R.: Feature Selection Based on Rough Sets and Particle Swarm Optimization. Pattern Recogn. Lett. 28, 459–471 (2007)
2. Suguna, N., Thanushkodi, D.K.: A Novel Rough Set Reduct Algorithm for Medical Domain Based on Bee Colony Optimization. J. Comput. 6, 49–54 (2010)
3. Inbarani, H.H., Azar, A.T., Jothi, G.: Supervised Hybrid Feature Selection Based on PSO and Rough Sets for Medical Diagnosis. Comput. Meth. Prog. Bio. 113, 175–185 (2014)
4. Arafat, H., Elawady, R.M., Barakat, S., Elrashidy, N.M.: Using Rough Set and Ant Colony Optimization in Feature Selection. Int. J. Emerg. Trends Technol. Comput. Sci. 2, 148–155 (2013)
5. Bae, C., Yeh, W., Chung, Y.Y., Liu, S.: Feature Selection with Intelligent Dynamic Swarm and Rough Set. Expert Syst. Appl. 37, 7026–7032 (2010)

6. Velayutham, C., Thangavel, K.: Unsupervised Quick Reduct Algorithm using Rough Set Theory. J. Electron. Sci. Technol. 9, 193–201 (2011)
7. Qablan, T., Al-Radaideh, Q.A., Shuqeir, S.A.: A Reduct Computation Approach Based on Ant Colony Optimization. Basic Sci. Eng. 21, 29–40 (2012)
8. Chen, Y., Miao, D., Wang, R.: A Rough Set Approach to Feature Selection Based on Ant Colony Optimization. Pattern Recogn. Lett. 31, 226–233 (2010)
9. Xue, B., Zhang, M., Browne, W.N.: Particle swarm optimisation for feature selection in classification: Novel initialisation and updating mechanisms. Appl. Soft Comput. 18, 261–276 (2014)
10. Chuang, L., Yang, C., Li, J.: Chaotic Maps Based on Binary Particle Swarm Optimization for Feature Selection. Appl. Soft Comput. 11, 239–248 (2011)
11. Farzi, S.: Efficient Job Scheduling in Grid Computing with Modified Artificial Fish Swarm Algorithm. Int. J. Comput. Theor. Eng. 1, 13–18 (2009)
12. Liu, T., Qi, A., Hou, Y., Chang, X.: Feature Optimization Based on Artificial Fish-swarm Algorithm in Intrusion Detections. In: International Conference on Networks Security, Wireless Communications and Trusted Computing, vol. 1, pp. 542–545. IEEE Press, Wuhan (2009)

Hand Gesture Shape Descriptor Based on Energy-Ratio and Normalized Fourier Transform Coefficients

Wenjun Tan[1,2,*], Zijiang Bian[1], Jinzhu Yang[1,2],
Huang Geng[1], Zhaoxuan Gong[1], and Dazhe Zhao[1,2]

[1] Medical Image Computing Laboratory of Ministry of Education,
Northeastern University, 110819, Shenyang, China
[2] College of Information Science and Engineering,
Northeastern University, 110819, Shenyang, China
{tanwenjun,bianzijian,yangjinzhu,genghuan,
gongzhaoxuan,zhaodzh}@mail.neu.edu.cn

Abstract. The hand gesture shape is the most remarkable feature for gesture recognition system. Since hand gesture is diversity, polysemy, complex deformation and spatio-temporal difference, the hand gesture shape descriptor is a challenging problem for gesture recognition. This paper presents a hand gesture shape describing method based on energy-ratio and normalized Fourier descriptors. Firstly, the hand gesture contour of the input image is extracted by YCb'Cr' ellipse skin color model. Secondly, the Fourier coefficients of the contour are calculated to transform the point sequence of the contour to frequency domain. Then the Fourier coefficients are normalized to meet the rotation, translation, scaling and curve origin point invariance. Finally, the items of normalized Fourier coefficients are selected by calculating energy-ratio information as the hand shape descriptors. For validating the shape descriptors performance, the hand gestures 1-10 are recognized with the template matching method and the shape descriptor method, respectively. The experiment results show that the method can well describe the hand shape information and are higher recognition rate.

Keywords: Hand gesture, Shape descriptor, Fourier descriptors, Skin color.

1   Introduction

Since hand gesture has the characteristics of diversity, polysemy, complex deformation and spatio-temporal difference, hand gesture recognition is one of the current topics of new generation human-computer interaction techniques. Hand gesture recognition system based machine vision recognizes gestures by segmentation and feature extraction from 2D image sequences from camera. This hand gesture recognition system has many good performances such as simple inputs, low device requirement, freedom from interference and so on[1]. The goal of hand gesture

---

[*] Corresponding author.

Y. Tan et al. (Eds.): ICSI 2014, Part II, LNCS 8795, pp. 34–41, 2014.

segmentation is to seperate hand region from complex background and to retain gestures in foregrounds. Because the independent from data gloves or color landmarks, naturally interact and fast detect, the skin color-based model has become the most mature method at present[2-3]. Gesture feature extraction is the key recognition procedure, in which shape information is the most remarkable features and proper information of gesture recognition system[4-5].

Shape is defined as a function describing position, direction, and surrounded region of a closed curve in 2D image space. Regular descriptors originate from point coordinates on closed curve of corresponding target outline. According to the expression methods, ´shape descriptors are commonly divided into two classes: region-expression and outline-expression. Region descriptor focuses on global geometry features, including area, perimeter, axis directions, compactness, solid degrees and so on[6]. Usually several global geometry features are adopted in shape matching and recognition, which is slow in descripting calculating complex shapes and loses information seriously in feature extraction. These defects lead to low resolution to express shape inaccurately. Besides, the region descriptors use inter-target texture distribution statistics, such as the 7 invariant moment[7]. Contour descriptor includes skeleton-based morphology[8], neural network[9], fractal method[10], which usually requires complex geometry relevant function to meet the invariance limits in translation, rotation and zoom, leading to complex calculation. Fourier descriptor is proposed by Zhan C T and Roskies R Z[11], and improved by Persoon[12]. The method regards contour as a closed curve formed by a sequence of end-to-end discrete points, and transforms the point sequence to frequency domain. The Fourier coefficients are defined as Fourier descriptor. The high-frequency components of Fourier transformation corresponds to detail information and the low-frequency components corresponds to overall shape, thus the low-frequency components could be selected to descript object's shape information. Meanwhile, Fourier inverse transform is able to restore the shape information expressed by the descriptor, which is unable by other descriptors. However, the Fourier transformation coefficients are concerned with target's scale, direction curve origin. So the classical Fourier transformation method is difficult to express object's shape invariance accurately. Besides, Fourier coefficients corresponds with contour sequence point number, which is usually numerous. Thus, if the whole coefficients are calculated, the heavy burden for gesture classifiers and low recognition rate would be lead to.

On the basis of the gesture region and contour extraction method in the our previous work, this paper presents an accurate and effective gesture shape expression method based on energy-ratio and normalized Fourier descriptors, focusing on shape invariance and effective coefficients selection of the descriptors.

2   Contour of Hand Gesture Extraction

Hand gesture segmentation is the first step of the hand contour extraction. This means segmenting the region of gestures from a complex background, leaving them alone in the foreground. Skin color is so good clustering property to be able to separate 'complexion' and 'non-complexion' region.

Hsu R L proposed a way to use ellipse model to describe the skin color distribution nonlinearly transformed to YCb'Cr' in the Cb'Cr' region, and apply it to face detection, obtaining better result[14]. Hsu R L put forward that color values always have nonlinear dependence relation to the luminance value Y in YCbCr color space.

Just through the calculation that whether the pixel is in the ellipse can we detect whether it belongs to the skin color. So it has a fast computing speed and high detection accuracy. But it may not describe accurately for the particular imaging equipment. This skin model easily leads to some non-skin color point being included to cause skin point over detection. Meanwhile, the model may not include all color regions, causing the skin color detection incomplete.

Aim at this problem, we presents a YCb'Cr' space ellipse fitting under the skin modeling method based on the specific statistical properties of the color distribution to segment hand gesture[15]. So the hand regions are segmented by the method in this work. Then the contours of hand gestures are easily extracted by 8-neighbourhood tracing algorithm.

3   Shape Descriptor of Hand Gesture

3.1   Normalized Fourier Transform Coefficients

Let gesture contour be a closed curve represented by coordinate sequence $s(k) = [x(k), y(k)]$, in which $[x(k), y(k)]$ donates the coordinate pairs starting from (x_0, y_0) and going contour clockwise along the curve. The complex number is $p(l) = x(l) + jy(l)$, $(l = 0, 1, \cdots, n-1)$, $j = \sqrt{-1}$, whose discrete Fourier coefficient is expressed as:

$$z(k) = \frac{1}{n} \sum_{l=0}^{n-1} p(l) e^{-j\frac{2\pi lk}{n}} \tag{1}$$

where $(k = 0, 1 \cdots, n-1)$, $z(k)$ is the Fourier transformation of $p(l)$ and the expression of point sequence in frequency domain. $z(k)$ correlates with object's shape. After the transformation in Eq(1), the object's contour is simplified from 2D to 1D space. The high-frequency Fourier coefficients are able to describe contour details and low-frequency ones identify overall shape information. As a result, the low-frequency Fourier coefficients could be selected to express object's shape information. The inverse Fourier transformation $z(k)$ is defined as:

$$p(l) = \sum_{k=0}^{n-1} z(k) e^{j\frac{2\pi lk}{n}} \tag{2}$$

Object's shape information could be restored with the inverse transform from Eq (2). Table.1 is the properties of Fourier descriptor of contour sequence $p(l)$ for rotation, translation, zooming and origin moving procedures[16], where Δ_{xy} is defined to be $\Delta_{xy} = (\Delta x + j\Delta y) = (x_0 + jy_0)$.

The Fourier descriptors got by Eq(1) transformed were concerned with shape's scale, direction and curve origin point, thus the descriptor should be processed to meet the requirement of shape characteristic invariance. For the transformation property of translation, only the k=0 becomes impulse function and other properties are no changes. So the $z(0)$ only change the centroid position of the object and don't change the object's shape. The $z(0)$ can be set as 0 for the shape descriptors. The Eq(9) could be derived from Eq(1) expressed as[17]:

$$d(k) = \frac{\|z'(k)\|}{\|z'(1)\|} = \frac{\left\| \alpha \left\| e^{j\theta} e^{-j\frac{2\pi k l_0}{n}} z(k) \right\| \right\|}{\left\| \alpha \left\| e^{j\theta} e^{-j\frac{2\pi l_0}{n}} z(1) \right\| \right\|} = \frac{\|z(k)\|}{\|z(1)\|} \tag{3}$$

The Eq(3) shows the change of Fourier coefficients of module and phase of object's rotation, scaling and origin position. The $d(k)$ in Eq(3) is called normalized Fourier descriptor, $k = (2,\cdots,n-1)$, which conforms the rotation, scaling, translation and origin position invariance.

3.2   Fourier Coefficients Selection Based on Energy-Ratio

Usually the high-frequency Fourier coefficients could explain shape details well, but the low-frequency ones decide object's overall shape. Thus partial low-frequency Fourier coefficients could be selected to express gesture shape information. However, the coefficients are few selected, the corresponding details will be lost so that the shape information will be difficult to express accurately. So it is important to select proper descriptor numbers for the gesture shape description.

Assume the first p Fourier coefficients express gesture shape information but not all the ones, so $k > p-1$, $z(k) = 0$ in Eq(1), and the other ones remain unchanged, the curve $p(l)$ will be defined as:

$$\hat{p}(l) = \sum_{k=0}^{p-1} z(k) e^{j\frac{2\pi lk}{n}} \tag{4}$$

where, $l = (0,1\cdots,n-1)$. Though p Fourier coefficients could obtain each components $\hat{p}(l)$, the range of l is still from 0 to n-1. That is to say, similar boundaries have the same numbers of points, but these needn't to be so much Fourier coefficients.

Thus the Fourier descriptors are defined as each item coefficient, and the descriptor numbers selection corresponds with Fourier coefficient items. Fourier transformation could translate gesture shape from spatial to frequency domain, and the frequency corresponds to energy information with amplitude values. The energy information shows corresponding coefficient ratio in shape expression. Thus the Fourier coefficients are selected by calculating Fourier coefficients energy-ratio information. The Fourier coefficient energy $E(l)$ of curve $p(l)$ could be expressed as follows:

$$E(l) = \sum_{k=0}^{l} \|z(k)\| \tag{5}$$

The energy ratio of the first p Fourier coefficient is:

$$e(p) = \frac{E(p)}{E(l)} = \frac{\sum_{k=0}^{p} \|z(k)\|}{\sum_{k=0}^{l} \|z(k)\|} \tag{6}$$

From Eq(1), the Fourier transformation in translation is converted to impulse function when k=0, that is, the value of $z(0)$ varies a lot, which make significant influence of the calculation of $e(p)$. Thus, the coefficient energy of $z(0)$ is ignored in this paper and only the energy ration information is calculated of $l = (1\cdots, n-1)$. $e(p)$ is iteratively calculated by increasing the value of p, when $e(p)$ is greater than a threshold or the difference between $e(p)$ and $e(p+1)$ is relatively small, the p is regarded as the final Fourier coefficient number. The normalized Fourier descriptor $d(k-1)$, $k = (2\cdots, p)$, is the descriptor of gesture shape.

4   Experiments and Discussion

To verify the shape descriptor and the items selecting method of the Fourier coefficient, the four gesture images are adopted in this work(Fig.1). The contour of gestures is extracted by the method in section 2 and is expressed with red line in Fig.1. Fig.2(a-d) shows the amplitude scattergram of the Fourier transformation coefficient corresponding of these gesture contours. To analyze accurately, Fig.2(a-d) retained the coefficient of $z(0)$. It is shown that the amplitude of $z(0)$ had a greater difference with other coefficients, which is as same as the analysis in section 3.2. When the coefficient number was bigger than 15, the amplitude varied little. The energy ratio information of different Fourier items were calculated as Eq(6) and shown in Fig.2(e).

To verify the Fourier coefficients selection method in this paper, the results of the coefficient numbers from 9 to 15 are compared. Fig.3 is the experiment results of hand gesture 1-8, in which the green lines express the contour curves through Fourier inverse transform. The Fourier descriptor details lose seriously of gesture 3,4,5,6,8 from the figure and could not express the whole shape information with 9 coefficient numbers. And the contour boundary information could be basically expressed by 13 items; and the contour boundary information of 15 descriptors is as same as 13 descriptors. According to the energy-ratio of the Fourier descriptors and analyzing dimensions and detail lose situations, the 13 item Fourier coefficients were adopted to express shape characters in this paper. The shape descriptors $d(k-1)$ is calculated as Eq(3), where $k = 12$. Because the z(0)=0, so there are 11 items of Fourier coefficient as hand gesture shape descriptors in this work.

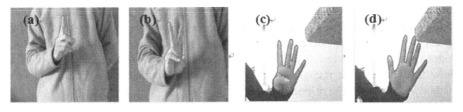

Fig. 1. The test hand gesture: (a) hand gesture 1; (b) hand gesture 2; (c) hand gesture 4; (d) hand gesture 5

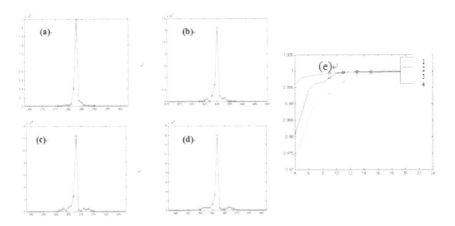

Fig. 2. Amplitude of Fourier coefficient of hand gesture contour: (a) hand gesture 1; (b) hand gesture 2; (c) hand gesture 4; (d) hand gesture 5; (e) energy ratio map of Fourier coefficient

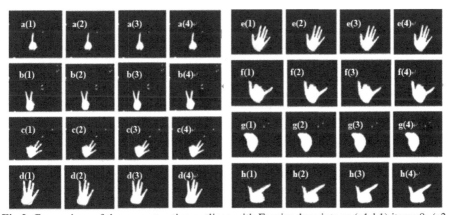

Fig.3. Comparison of the reconstruction outlines with Fourier descriptors: (a1-h1) items 9; (a2-h2) items 11; (a3-h3) items 13; (a4-h4) items 15.

The hand gestures 1-10 are recognized with the shape descriptor method in this paper to verify the descriptor validity. The template matching method is a common target recognition method, which is widely applied in pattern recognition system. For identification accuracy-ratio and time performance, the template matching method and this work are compared in this paper. In order to compare the consistency of the results of two methods, the Euclidean distance method is adopted to recognize the gesture categories. Table 1 is the recognition results of the two methods. The results show that the average recognition rate of the template matching method and our method is 79% and 90%, respectively.

Table 1. Recognition results of hand gesture with template matching method and this method

Hand gesture	Template matching method		This method	
	Correct number	Recognition rate	Correct number	Recognition rate
Gesture 1	35	87.50%	37	92.50%
Gesture 2	36	90.00%	34	85.00%
Gesture 3	32	80.00%	35	87.50%
Gesture 4	30	75.00%	36	90.00%
Gesture 5	29	72.50%	38	95.00%
Gesture 6	35	87.50%	39	97.50%
Gesture 7	27	67.50%	34	85.00%
Gesture 8	34	85.00%	38	95.00%
Gesture 9	28	70.00%	33	82.50%
Gesture 10	30	75.00%	36	90.00%

5   Conclusions

The hand gesture shape expression methods based on energy-ratio and normalized Fourier descriptor is presented in this work. The hand gesture contour is the input data for the shape descriptor, which is extracted by our previous work. Then the Fourier coefficients of the hand contour are transformed to express the point sequence of hand gesture contour in frequency domain. For meeting the rotation, translation, zooming and curve origin point invariance, the Fourier coefficients are normalized. The items of Fourier coefficients are selected by calculating energy-ratio information. Finally, to verify the shape descriptors and selection items method of the Fourier coefficients, the hand gestures 1-10 are recognized with the shape descriptor method in this paper. The experiment results show that the method can well describe the hand shape information and access higher recognition rate.

Acknowledgment. This research was partly supported by National Natural Science Foundation of China (NSFC) under Grant No. 61302012 and No. 61172002, the Fundamental Research Funds for the Central Universities under Grant N130418002 and N120518001, and Liaoning Natural Science Foundation under Grant No. 2013020021.

References

1. Elli, A.: Understanding The Color of Human Skin. In: Proceeding of the SPIE Conference on Human Vision and Electronic Imaging VI (SPIE), pp. 243–251. SPIE Press (2001)
2. Yang, J., Lu, W., Waibel, A.: Skin-Color Modeling and Adaptation. In: Chin, R., Pong, T.-C. (eds.) ACCV 1998. LNCS, vol. 1352, pp. 687–694. Springer, Heidelberg (1997)
3. Bradski, G., Yeo, B., Minerva, M.Y.: Gesture for Video Current Navigation. In: SPIE IS&T/SPIE, vol. 36, pp. 230–242 (1999)
4. Kawulok, M.: Fast Propagation-based Skin Regions Segmentation in Color Images. In: Proc. IEEE FG, pp. 1–7. IEEE Press (2013)
5. Nalepa, J., Grzejszczak, T., Kawulok, M.: Wrist Localization in Color Images for Hand Gesture Recognition. In: Gruca, A., Czachórski, T., Kozielski, S. (eds.) Man-Machine Interactions 3. AISC, vol. 242, pp. 81–90. Springer, Heidelberg (2014)
6. Ang, Y.H., Li, Z., Ong, S.H.: Image Retrieval Based on Multidimensional Feature Properties. In: SPIE, vol. 24, pp. 47–57 (1995)
7. Hu, M.K.: Visual Pattern Recognition by Moment Invariants. IEEE Trans. Informaiton Theory 8, 179–187 (1962)
8. Xu, J.: Morphological Decomposition of 2-D Binary Shapes Into Modestly Overlapped Octagonal and Disk Components. IEEE Transactions on Image Processing 16, 337–348 (2007)
9. Gupta, L., Sayeh, M.R., Tammana, R.: Neural Network Approach to Robust Shape Classification. Pattern Recognition 23, 563–568 (1990)
10. Taylor, R., Lewis, P.H.: 2D Shape Signature Based on Fractal Measurements. IEEE Proceedings -Vision, Image and Signal Processing 4, 422–430 (1994)
11. Zhan, C.T., Roskies, R.Z.: Fourier Descriptors for Plane Closed Curves. IEEE Trans. Computer 21, 269–281 (1972)
12. Persoon, E., Fu, K.S.: Shape Discrimination Using Fourier Descripters. IEEE Trans. System, Man, Cybernetics 7, 170–179 (1997)
13. Drolon, H., Druaux, F., Faure, A.: Particles Shape Analysis and Classification Using The Wavelet Transform. Pattern Recognition Letters 21, 473–482 (2000)
14. Hsu, R.L., Jain, A.K.: Face Detection in Color Images. IEEE Transactions Pattern Analysis and Machine Intelligence 24, 696–706 (2002)
15. Tan, W.J., Dai, G.Y., Su, H., Feng, Z.Y.: Gesture Segmentation Based on YCb'Cr' Color Space Ellipse Fitting Skin Color Modeling. In: The 24th Chinese Control and Decision Conference (CCDC), pp. 1905–1908. IEEE Press, Taiyuan (2012)
16. Gonzalez, R.C., Woods, R.E., Eddins, S.L.: Digital image processing using MATLAB. Publishing House of Electronics Industry, Beijing (2005)
17. Wang, T., Liu, W.Y., Sun, J.G., Zhang, H.J.: Using Fourier Descriptors to Recognize Object's Shape. Journal of Computer Research and Development 39, 1714–1719 (2002)

A New Evolutionary Support Vector Machine with Application to Parkinson's Disease Diagnosis

Yao-Wei Fu, Hui-Ling Chen[*], Su-Jie Chen, LiMing Shen, and QiuQuan Li

College of Physics and Electronic Information,
Wenzhou University, China
chenhuiling.jlu@gmail.com

Abstract. In this paper, we present a bacterial foraging optimization (BFO) based support vector machine (SVM) classifier, termed as BFO_SVM, and it is applied successfully to Parkinson's disease (PD) diagnosis. In the proposed BFO-SVM, the issue of parameter optimization in SVM is tackled using the BFO technique. The effectiveness of BFO-SVM has been rigorously evaluated against the PD Dataset. The experimental results demonstrate that the proposed approach outperforms the other two counterparts via 10-fold cross validation analysis. In addition, compared to the existing methods in previous studies, the proposed system can also be regarded as a promising success with the excellent classification accuracy of 96.89%.

Keywords: Support vector machines, Parameter optimization, Bacterial foraging optimization, Parkinson's disease diagnosis, Medical diagnosis.

1 Introduction

As a primary machine learning technique, support vector machines (SVM) [1] is rooted in the Vapnik-Chervonenkis theory and structural risk minimization principle. Thanks to its good properties, SVM has found it's applications in a wide range of classification tasks. In particular, SVM has demonstrated excellent performance on many medical diagnosis tasks. However, there is still much room for improvement of the SVM classifier. Because it has been proved that proper model parameters setting can improve the SVM classification accuracy substantially [2]. Values of parameters such as penalty parameter and the kernel parameter of the kernel function should be carefully chosen in advance when SVM is applied to the practical problems. Traditionally, these parameters were handled by the grid-search method and the gradient descent method. However, one common drawback of theses methods is that they are vulnerable to local optimum. Recently, biologically inspired metaheuristics such as genetic algorithm and particle swarm optimization (PSO) have been considered to have a better chance of finding the global optimum solution than the traditional aforementioned methods. As a relatively new member of the swarm-intelligence algorithms, BFO has been found to be a promising technique for real-world optimization problems such as optimal controller design [3], learning of artificial neural networks [4] and active power filter design [5].

---

[*] Corresponding author.

Y. Tan et al. (Eds.): ICSI 2014, Part II, LNCS 8795, pp. 42–49, 2014.

This study attempts to employ BFO to handle the parameter optimization of SVM and applied the resultant effective model BFO-SVM for effective detection of Parkinson's disease (PD). The main objective of this study is to explore the maximum generalization capability of SVM and apply it to PD diagnosis to distinguish patients with PD from the healthy ones.

The remainder of this paper is organized as follows. The related works on detection of PD is presented in Section 2. In section 3 the detailed implementation of the BFO-SVM diagnostic system is presented. Section 4 describes the experimental design. The experimental results and discussion of the proposed approach are presented in Section 5. Finally, Conclusions and recommendations for future work are summarized in Section 6.

2   Related Works on Detection of PD

PD is one kind of degenerative diseases of the nervous system, which is characterized by a group of conditions called motor system disorders because of the loss of dopamine-producing brain cells. Till now, the cause of PD is still unknown, however, it is possible to alleviate symptoms significantly at the onset of the illness in the early stage [6]. It is claimed that approximately 90% of the patients with PD show vocal impairment [7], the patients with PD typically exhibit a group of vocal impairment symptoms, which is known as dysphonia. The dysphonic indicators of PD make speech measurements an important part of diagnosis. Recently, dysphonic measures have been proposed as a reliable tool to detect and monitor PD [8].

Various researchers have studied the PD diagnosis problem. Little et al. [8] conducted a remarkable study about PD identification, they employed an SVM classifier with Gaussian radial basis kernel functions to predict PD, and also performed feature selection to select the optimal subset of features from the whole feature space, and the best accuracy rate of 91.4% was obtained by the best model. Das [9] presented a comparative study of using Neural Networks (ANN), DMneural, Regression and Decision Tree for effective diagnosis of PD, the experimental results have shown that the ANN classifier yielded the best results, the overall classification score of 92.9% was achieved. Recently, Ozcift et al. [10] combined the correlation based feature selection (CFS) algorithm with the rotation forest (RF) ensemble classifiers of 30 machine learning algorithms to identify PD, and the best classification accuracy of 87.13% was achieved by the proposed CFS-RF model. Chen et al. [11] employed the fuzzy k-nearest neighbor (FKNN) approach in combination with the principle component analysis (PCA-FKNN) to diagnose PD, and the best classification accuracy of 96.07% was obtained by the proposed diagnosis system.

3   The Proposed BFO-SVM Model

This study proposes a novel BFO-SVM model for parameter optimization problem of SVM. In the proposed model, the parameter optimization for SVM are dynamically conducted by implementing BFO algorithm, then the obtained optimal parameters are taken by the SVM model to perform the classification task. The proposed model is comprised of two main evaluation procedures:

1) Inner_Parameter_Optimization procedure: Evaluate the performance of each candidate parameters;
2) Outer_Performance_Estimation procedure: Evaluate the overall performance of the SVM classifier with the optimal parameter values obtained;

In the Inner_Parameter_Optimization procedure, the parameters C and γ are dynamically optimized by implementing BFO algorithm. The classification accuracy is taken into account in designing the fitness:

$$f = avgACC = (\Sigma_{i=1}^{K} testACC_i)/k . \tag{1}$$

where variable $avgACC$ in the function f represents the average test accuracy achieved by the SVM classifier via k-fold CV, where $k = 5$. Note that here the 5-fold CV is employed to do the model selection that is different from the outer loop of 10-fold CV, which is used to do the performance estimation. The pseudo-code of this procedure is given bellow:

Pseudo-code for the Inner_Parameter_Optimization procedure

step 1.   Initialize parameters $p, S, Nc, Ns, Nre, Ned, Ped, \theta^i$
where
p: number of dimension of the search space,
S: swarm size of the population,
Nc: number of chemotactic steps,
Ns: swimming length,
Nre: the number of reproduction steps,
Ned: the number of elimination-dispersal events,
Ped: elimination-dispersal probability, and
$C(i)$: the size of step taken in the random direction specified by the tumble.

step 2.   Elimination-dispersal loop: $l=l+1$.
step 3.   Reproduction loop: $k=k+1$.
step 4.   Chemotaxis loop: $j=j+1$.
(a) **For** $i=1,2,...,S$, take a chemotactic step for bacterium i as follows.
(b) Train SVM and compute the fitness $J(i, j, k, l)$
Let, $J(i, j, k, l)=J(i, j, k, l)+J_{ar}(\theta)$ where J_{ar} is defined in Eq. (8).
(c) Let $Jlast=J(i, j, k, l)$ to save this value since we may find a better cost via a run.
(d) Tumble: generate a random vector $\Delta(i) \in R^p$ with each element
$\Delta_m(i), m=1,2,...,p$, a uniformly distributed random number on [-1, 1].
(f) Move: let

$$\theta^i \left(j+1,k,l,di \right) = \theta^i \left(j,k,l,di \right) + C(i) \frac{\Delta(i)}{\sqrt{\Delta^T(i)\Delta(i)}}$$

(g) Train SVM and compute the fitness $J(i,j+1,k,l)$, and let

$$J(i,j+1,k,l)=J(i, j, k, l)+Jar(\theta).$$

(h) Swim.

    i) Let $n=0$;

    ii) **While** $n<Ns$

    iii) Let $n=n+1$;

    iv) **If** $J(i,j+1,k,l)<Jlast$, let $Jlast=J(i,j+1,k,l)$ and let

$$\theta^i \left(j+1,k,l,di \right) = \theta^i \left(j,k,l,di \right) + C(i)\frac{\Delta(i)}{\sqrt{\Delta^T(i)\Delta(i)}}$$

    and use this $\theta^i(j+1,k,l)$ to train SVM, and then compute the new fitness

    $J(i, j+1,k, l)$ as did in (g);

    v) Else, let $n =Ns$.

  (i) Go to next bacterium $(i+1)$ if $i \neq S$.

step 5.   If $j<Nc$, go to step 4.

step 6.   Reproduction:

    Rank all of the individuals according to the sum of the evaluation results in this period, and then removes out the last half individuals and duplicates one copy for each of the rest half.

step 7.   If $k<Nre$, go to step 3.

step 8.   Elimination-dispersal:

    For $i=1,2,...,S$ with probability Ped, eliminate and disperse each bacterium.

    If $l<Ned$, then go to step 2; otherwise end.

In the Outer_Performance_Estimation procedure, SVM model performs the classification tasks using the obtained optimal parameters via 10-fold CV analysis. The pseudo-code of this procedure is given bellow:

Pseudo-code for the Outer_Performance_Estimation procedure

/*performance estimation by using k-fold CV where $k = 10$*/

Begin

For $j = 1:k$

Training set = k-1 subsets;

Testing set = remaining subset;

    Train the SVM classifier on the training set using the parameters and feature subsets obtained from Inner_Parameter_Optimization procedure;

Test it on the testing set;

End For;

Return the average classification accuracy rates of SVM over j testing set;

End.

---

4   Experimental Setup

In this study, we have performed our conduction on the Parkinson's data set taken from UCI machine learning repository.

The BFO-SVM, PSO-SVM and Grid-SVM classification models were implemented using MATLAB platform. For SVM, LIBSVM implementation was utilized, which was originally developed by Chang and Lin [12]. We implemented the BFO, PSO and grid search algorithm from scratch. The computational analysis was conducted on Windows 7 operating system with AMD Athlon 64 X2 Dual Core Processor 5000+ (2.6 GHz) and 4GB of RAM. Normalization is firstly employed before classification, in order to avoid feature values in greater numerical ranges dominating those in smaller numerical ranges, as well as to avoid the numerical difficulties during the calculation. In order to guarantee the valid results, the k-fold CV was employed to evaluate the classification accuracy.

In this study, we designed our experiment using a two-loop scheme, which also was used in [13]. The detailed parameter setting for BFO-SVM is shown in Table 1. For PSO-SVM, the number of the iterations and particles are set to 250 and 8, respectively. v_{max} is set about 60% of the dynamic range of the variable on each dimension for the continuous type of dimensions, $c_1 = 2$, $c_2 = 2$, w_{max} and w_{min} are set to 0.9 and 0.4, respectively. The searching ranges of $C \in [2 \wedge (-5), 2 \wedge (15)]$ and $\gamma \in [2 \wedge (-5), 2 \wedge (15)]$ for BFO-SVM, PSO-SVM and Grid-SVM were set as the same.

Table 1. Common parameter setup for BFO

S	Nc	Ns	Ned	Nre	ped	datt	watt	wrepe	hrepe	C(i)
8	100	20	1	4	0.25	0.1	0.2	10	0.1	0.1

Classification accuracy, sensitivity and specificity were used to test the performance of the proposed BFO-SVM model.

5   Experimental Results and Discussions

In BFO, the parameter chemotaxis step size $C(i)$ plays an important role in controlling the search ability of BFO. Thus, we firstly present results from our investigations on the impacts of $C(i)$ and assign initial values for it. $C(i)$ can be initialized with biologically motivated values, but a biologically motivated value may not be the best for specific application [3]. In Table 2, we illustrate the relationship between the different values of $C(i)$ and the performance of BFO-SVM. The average results are presented with the standard deviation described in the parenthesis. From the table we can see that BFO-SVM reaches the best performance at $C(i) = 0.1$ in terms of accuracy, sensitivity and specificity. Therefore, we select 0.1 as the parameter value of $C(i)$ for the proposed BFO-SVM to implement the coming tasks.

Table 2. The detailed results of BFO-SVM with different C(i) on the PD data set

Chemotactic step size parameter $C(i)$	BFO-SVM		
	Accuracy (%)	Sensitivity (%)	Specificity (%)
0.05	94.84(6.43)	97.41(6.11)	88.00(17.79)
0.1	**96.90(4.34)**	**98.75(3.95)**	**90.83(16.87)**
0.15	95.39(4.53)	97.44(4.33)	90.42(10.77)
0.2	94.42(4.97)	97.57(4.19)	85.50(13.17)
0.25	93.40(6.28)	96.08(5.48)	87.71(17.36)
0.3	94.47(4.97)	96.02(4.82)	90.64(12.32)

To evaluate the effectiveness of the proposed BFO-SVM system for PD, we conducted experiments on the PD database. Table 3 shows the classification accuracy, sensitivity, specificity, and optimal pairs of (C, γ) for each fold obtained by BFO-SVM. The comparison among PSO-SVM, Grid-SVM and BFO-SVM is shown in Figure 1.

Table 3. The detailed results of BFO-SVM on the PD data set

Fold No.	BFO-SVM				
	Accuracy	Sensitivity	Specificity	C ($\times 10^4$)	γ
1	0.8947	0.8750	1.0000	1.4250	3.7726
2	1.0000	1.0000	0.8333	2.2658	3.3197
3	0.9474	1.0000	1.0000	2.2930	3.1821
4	1.0000	1.0000	0.7500	1.2372	4.6147
5	1.0000	1.0000	1.0000	2.8519	3.5251
6	0.9000	1.0000	1.0000	1.4252	3.8452
7	1.0000	1.0000	0.5000	2.0478	3.2553
8	1.0000	1.0000	1.0000	0.1331	3.7350
9	1.0000	1.0000	1.0000	2.2390	3.7168
10	0.9689	1.0000	1.0000	0.8046	4.9414
Avg.	0.0434	0.9875	0.9083	1.6723	3.7908

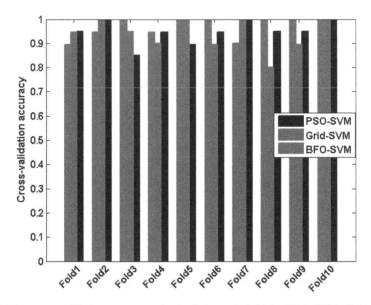

Fig. 1. The cross-validation accuracy obtained for each fold by PSO-SVM, Grid-SVM and BFO-SVM

As shown in Figure 1, we can see that BFO-SVM has dominated PSO-SVM and Grid-SVM in most folds in the process of the 10-fold CV, namely, BFO-SVM has achieved the high classification accuracy equal to or better than that of the other two models obtained for 7 folds in the whole 10 folds. The average classification accuracy of BFO-SVM is 96.89%, while the average classification accuracy of PSO-SVM and Grid-SVM are 94.89% and 93.87%, respectively.

6   Conclusions and Future Work

This work has explored a new diagnostic system, BFO-SVM, for detection of PD. The main novelty of this paper lies in the proposed BFO-based approach, which aims at maximizing the generalization capability of the SVM classifier by exploring the new swarm intelligence technique for optimal parameter tuning for PD diagnosis. The empirical experiments on the PD database have demonstrated the superiority of the proposed BFO-SVM over PSO-SVM and Grid-SVM in terms of classification accuracy. It indicates that the proposed BFO-SVM system can be used as a viable alternative solution to PD diagnosis.

Acknowledgments. This work is supported by the open project program of Wenzhou University Laboratory under Grant No. 13SK29A.

References

1. Vapnik, V.N.: The nature of statistical learning theory. Springer, New York (1995)
2. Keerthi, S., Lin, C.: Asymptotic behaviors of support vector machines with Gaussian kernel. Neural Computation 15(7), 1667–1689 (2003)
3. Passino, K.M.: Biomimicry of bacterial foraging for distributed optimization and control. IEEE Control Systems Magazine 22(3), 52–67 (2002)
4. Ulagammai, M., et al.: Application of bacterial foraging technique trained artificial and wavelet neural networks in load forecasting. Neurocomputing 70(16-18), 2659–2667 (2007)
5. Mishra, S., Bhende, C.N.: Bacterial foraging technique-based optimized active power filter for load compensation. IEEE Transactions on Power Delivery 22(1), 457–465 (2007)
6. Singh, N., Pillay, V., Choonara, Y.E.: Advances in the treatment of Parkinson's disease. Progress in Neurobiology 81(1), 29–44 (2007)
7. Ho, A.K., Iansek, R., Marigliani, C., Bradshaw, J.L., Gates, S.: Speech impairment in a large sample of patients with Parkinson's disease. Behavioural Neurology 11(3), 131–137 (1999)
8. Little, M.A., McSharry, P.E., Hunter, E.J., Spielman, J., Ramig, L.O.: Suitability of dysphonia measurements for telemonitoring of Parkinson's disease. IEEE Transactions on Biomedical Engineering 56(4), 1015–1022 (2009)
9. Das, R.: A comparison of multiple classification methods for diagnosis of Parkinson disease. Expert Systems with Applications 37(2), 1568–1572 (2010)
10. Ozcift, A., Gulten, A.: Classifier ensemble construction with rotation forest to improve medical diagnosis performance of machine learning algorithms. Computer Methods and Programs in Biomedicine 104(3), 443–451 (2011)
11. Chen, H.L., Huang, C.C., Yu, X.G., Xu, X., Sun, X., Wang, G., Wang, S.J.: An efficient diagnosis system for detection of Parkinson's disease using fuzzy< i> k</i>-nearest neighbor approach. Expert Systems with Applications 40(1), 263–271 (2013)
12. Chang, C.C., Lin, C.J.: LIBSVM: a library for support vector machines. (2001), http://www.csie.ntu.edu.tw/cjlin/libsvm
13. Chen, H.L., Yang, B., Wang, G., Liu, J., Xu, X., Wang, S.J., Liu, D.Y.: A novel bankruptcy prediction model based on an adaptive fuzzy< i> k</i>-nearest neighbor method. Knowledge-Based Systems 24(8), 1348–1359 (2011)

Parallel Bees Swarm Optimization for Association Rules Mining Using GPU Architecture

Youcef Djenouri and Habiba Drias

USTHB LRIA, Algiers, Algeria
y.djenouri@gmail.com, hdrias@hotmail.fr

Abstract. This paper addresses the problem of association rules mining with large scale data sets using bees behaviors. The bees swarm optimization method have been successfully running on small and medium data size. Nevertheless, when dealing with large benchmark, it is bluntly blocked. Additionally, graphic processor units are massively threaded providing highly intensive computing and very usable by the optimization research community. The parallelization of such method on GPU architecture can be deal large data sets as the case of WebDocs in real time. In this paper, the evaluation process of the solutions is parallelized. Experimental results reveal that the suggested method outperforms the sequential version at the order of ×70 in most data sets, furthermore, the WebDocs benchmark is handled with less than forty hours.

Keywords: bees swarm optimization, association rule mining, parallel algorithms, GPU architecture.

1 Introduction

Association Rules Mining (ARM) is one of the most important and well studied techniques of data mining tasks [1]. It aims at extracting frequent patterns, associations or causal structures among sets of items from a given transactional database. Formally, the association rule problem is as follows: let T be a set of transactions $\{t_1, t_2, \ldots, t_m\}$ representing a transactional database, and I be a set of m different items or attributes $\{i_1, i_2, \ldots, i_m\}$, an association rule is an implication of the form $X \rightarrow Y$ where $X \subset I$, $Y \subset I$, and $X \cap Y = \emptyset$. The itemset X is called antecedent while the itemset Y is called consequent and the rule means X implies Y.

Many exacts algorithms have been developed for solving ARM problem. Apriori [2] and FPgrowth [3] are the most used algorithms, Nonetheless, the exacts algorithms are high time consuming for large data sets.

Swarm intelligence algorithms have been successfully applied for association rules mining problem like: PSOARM [4], ACO_R [7], HBSO-TS [6], and BSO-ARM [5]. The experiments reveal that the bees swarm optimization outperforms

Y. Tan et al. (Eds.): ICSI 2014, Part II, LNCS 8795, pp. 50–57, 2014.
© Springer International Publishing Switzerland 2014

ACO and PSO in terms of rules quality. However, when dealing with large instances like WebDocs (the huge benchmark among the web), the complexity time still expensive. In fact, the main challenge in sequential ARM algorithms is to handle massive data-sets. ARM problem has been parallelized in different ways. On each approach, the authors took advantage of the used parallel hardware like: CD, DD algorithms for distributed memory system [8], CCPD and PCCD for shared memory system [9]. However, All these algorithms have been implemented for traditional parallel architectures (supercomputers, clusters ..) which are still expensive and not always accessible for every one.

Motivating by the forcefully of graphic processors units, in this paper, we propose a parallel GPU-based approach for association rules mining problem. It is an extended version of the bees swarm optimization algorithm. The generation of solutions and the search process are done on CPU. However, to benefit from the massive GPU threaded, the evaluation process of the solutions have been performed concurrently on GPU. The results show the effectiveness of the parallel approach compared to the sequential one.

The rest of the paper is organized as follows: Section 2 relates the state of the art of ARM algorithms followed by a brief explanation of BSO-ARM algorithm. In section 4, we present the proposed algorithm PMES. Section 5 shows the results of our algorithm using several data sets. In section 6, we conclude by some remarks and futures perspective of the present work.

2   Related Works

ARM community is many investigating on GPU architecture, for this many algorithms based on GPU have been developed. The parallel ARM on GPU was first introduced by Wenbin *et al.* in [10]. They proposed two parallel versions of Apriori called PBI (Pure Bitmap Implementation) and TBI (Trie Bitmap Implementation) respectively. In PBI, the transactions datasets and the itemsets are represented by a bitmap data structure. Indeed, the itemset structure is a bitmap $(n * m)$ where n is the number of k itemsets and m is the number of its items. In this representation, bit $(i, j) = 1$ if itemsets i contains item j otherwise 0. Similarly, a transaction structure is a bitmap $(n * m)$ where n is the number of itemsets and m is the number of transactions. Here bit $(i, j) = 1$ if transaction j contains itemsets i, 0 otherwise.

In [11], a new algorithm called GPU-FPM is developed. It is based on Apriori-Like algorithm and uses a vertical representation of the data set because of the limitation of the memory size in the GPU. The speed up reached during the reported experimentations varies from ×10 to ×15.

Syed *et al.* in [12] proposed a new Apriori algorithm for GPU architecture which performs on two steps. In the first step, the generation of itemsets is done on GPU host where each thread block computes the support of a set of itemsets. In the second step, the generated itemsets are sent back to the CPU in order to generate the rules corresponding to each itemset and to determine the confidence of each rule. The main drawback of this algorithm is the cost of the CPU/GPU communications. The speed up of this algorithm reaches ×20 with a large data set.

Another Cuda-Apriori algorithm is proposed in [13]. First, the transactional dataset is divided among different threads. Then, k-candidate itemsets are generated and allocated on global memory. Each thread handles one candidate using only the portion of the dataset assigned to its block. In each iteration, a synchronization between blocks is done in order to compute the global support of each candidate. In [14], a GPApriori algorithm is proposed using two data structures in order to accelerate the itemsets counting. For small size datasets this algorithm reaches a speed up of ($\times 100$). Nevertheless, the speed up is reduced when the number of transactions increases du to thread divergence.

In [15] another work for maximal frequent itemsets mining on GPU is reported. This work is based on the main proprieties of GPU which are data parallelism and data independence. A tree structure is used to store the frequent itemsets where each node contains an itemsets, its support and a bitmap of each itemset. A bitmap is a boolean structure that represents all transactions containing a given itemset. This algorithm works well on large datasets. Nevertheless, it requires a high space memory and many pointer links which are difficult to manage on GPU architecture.

3   BSO-ARM Algorithm

In [5], we proposed BSO-ARM algorithm. The aim of this algorithm is to find one part of association rules respecting minimum support and minimum confidence constraints with reasonable time. Each rule is considered as one solution in the search space, each of which is represented by a vector S of N bits and their positions are defined as follows:

1. $S[i] = 0$ *if* the item i is not in the solution S.
2. $S[i] = 1$ *if* the item i belongs to the antecedent part of the solution S.
3. $S[i] = 2$ *if* the item i belongs to the consequent part of the solution S.

The algorithm can be decomposed into four steps (Neighborhood Search Computation, Search Area Determination, Fitness computing and Dancing Step).

Neighborhood Search Computation. The neighborhood search is obtained by changing from a given solution S one bit in random way. Based on this simple operation, N neighborhoods are created. Notice that this operation does not generate non admissible solutions.
   Example 1
   Consider the given solution: $S = \{1, 0, 0, 1, 2\}$

1. change the first bit in S : $S1 = \{0, 0, 0, 1, 2\}$
2. change the second bit in S: $S2 = \{1, 2, 0, 1, 2\}$
3. change the third bit in S:$S3 = \{1, 0, 1, 1, 2\}$
4. change the fourth bit S: $S4 = \{1, 0, 0, 0, 2\}$
5. change the fifth bit S: $S5 = \{1, 0, 0, 1, 0\}$

Search Area Determination. It determines K search spaces, each one is associated to a bee. The jth bee builds its own search area by changing successively in the solution $Sref$ the bits $j + i \times Flip$ where $i \in [0..N - 1]$, $j \in [1..K]$ and $Flip$ is a given parameter. This strategy can be used if and only if $K \leq \frac{N}{Flip}$. If the distance between solutions is the number of different bits, then the distance between the bees and the solution reference is equal to $\frac{N}{Flip}$.

Fitness Computing. In this step for each generated solution (rule), the entire transactional database should be scanned. The solution fitness is based on the support and the confidence measures as:

$$Fitness(s) = \alpha \times conf(s) + \beta \times sup(s) \qquad (1)$$

This function should be maximized. For each invalid solution s where $Sup(s) < Minsup$ or $Conf(s) < MinConf$, the $Fitness(s)$ is set to -1 and the solution is rejected.

Dancing Step. Each bee puts in the dance table the best rule found among its search. The communication between bees is done in order to find the best dance (the best rule) which becomes the reference solution for the next pass.

The general functioning of the algorithm is as follows: First, the solution reference $(Sref)$ is initialized randomly so that each element of $Sref$ belongs to $\{0,1,2\}$.

After that, except the Fitness Computing which is applied for each generated solution, the other steps are repeated in the order until $IMAX$ is reached.

4 Parallel Single Evaluation of Solution Algorithm

The proposed algorithm parallel single evaluation of solution (PSES for short) is based on the master/slave paradigm. The master is executed on CPU and the slave is offloaded to the GPU. First, The master initializes randomly the solution reference. After that, it determines regions of the whole bees by generating the neighbors of each bee. Unlike BSO-ARM, single solution is evaluated on GPU in parallel. After, the master receives back the fitness of all rules, each bee calculates sequentially the best rule and puts it in the table dance. The best rule of the dance table become the solution reference for the next iteration. This combined CPU/GPU process is repeated until the maximum number of iterations is reached. We opted for a mapping in which all threads are mapped to one rule. Threads of the same block are launched to calculate collaboratively the fitness of a single rule with one packet of transactions. Therefore, we have as many threads as transactions. The transactions are subdivided into subsets and each subset set_i is assigned to one bloc so that each thread calculates only one transaction. After that a sum reduction is applied to aggregate the fitness value.

Such a strategy allows us to benefit from the massively parallel power of GPU by launching a large number of threads per rule. Indeed, in PSES, only the threads must synchronize after each iteration to process sum reduction technique, thus $N \times K \times IMAX$ synchronization during all the lifetime of the algorithm. The general algorithm of the GPU kernel is given in Algorithm 1.

Algorithm 1. The GPU kernel

Input: Sol: Single solution
Freq[] Array of integer
recuperate Sol from CPU
initialize Freq to zero
$idt =$ blockIdx.x $\times$ blockDim.x $+$ threadIdx.x.
for i=0 to nb transactions **do**
  if Sol$\in t_{idt}$ **then**
    freq[idt][i]=1.
  else
    freq[idt][i]=0.
  end if
end for
$Evaluation(Sol)$=Sum_Reduction(freq).
cudaMemcpy($Evaluation(Sol)$, cudaMemcpyDeviceToHost).

First, GPU recuperates the single sol containing the set of solutions generated on CPU. It initializes freq by zero. Then, each thread evaluates one solution with one transaction.

5   Performance Evaluation

To evaluate the performance of the proposed approach PSES, several data sets of different size are considered. The data sets are the well-known scientific databases that are frequently used in data mining community (Frequent and Mining Dataset Repository [17] and Bilkent University Function Approximation Repository[16].

From the smallest benchmark (*Bolts* data set) to the largest one (*WebDocs* data set), the used data sets are divided according to the number of transactions into three categories (small, average, large). The description of the different used data sets are presented in Table 1. Notice that the data sets differ according to the average size of items. On one hand, there are big data sets with a few number of items per transaction. On the other hand, there are other small data sets with a significant number of items per transaction. For instance, the number of transactions on the *Connect* data set is 100000 and the average items per transaction is only 10. Whereas, in the IBM data set the number of transactions is only 1000 and the number of items per transaction is 20.

Table 1. Data sets description

Data Set Type	Data set Name	Transactions Size	Item Size	Average Size
Small	Bolts	40	8	8
	Sleep	56	8	8
	Basket ball	96	5	5
	IBM Quest Std.	1000	40	20
	Quake	2178	4	4
	Chess	3196	75	37
	Mushroom	8124	119	23
Average	Pumbs_star	40385	7116	50
	BMS-WebView-1	59602	497	2.5
	BMS-WebView-2	77512	3340	5
	Korasak	80769	7116	50
	retail	88162	16469	10
	Connect	100000	999	10
Large	BMP POS	515597	1657	2.5
	WebDocs	1692082	526765	-

Table 2. Runtime of the proposed approach with different data sets (in Sec)

Data Set Type	Data set Name	BSO-ARM	PSES
Small	Bolts	9	2.5
	Sleep	15	5
	Pollution	22	6
	Basket ball	35	12
	IBM Q.S.	618	50
	Quake	80	20
	Chess	149956	125
	Mushroom	28815	350
Average	Pumbs_star	144120	1550
	BMS-W.V.-1	180524	6200
	BMS-W.V.-2	249985	1850
	Korasak	258451	1948
	Retail	299658	2185
	Connect	300985	4485
Large	BMP POS	Stopped After 15 Days	22565
	WebDocs	Stopped After 15 Days	45965

The suggested approach has been implemented using C-CUDA 4.0 and the experiments have been carried out using a CPU host coupled with a GPU device. The CPU host is a 64-bit quad-core Intel Xeon E5520 having a clock speed of 2.27GHz. The GPU device is an Nvidia Tesla C2075 with 448 CUDA cores (14 multiprocessors with 32 cores each), a clock speed of 1.15GHz, a 2.8GB global memory, a 49.15KB shared memory, and a warp size of 32. Both the CPU and GPU are used in single precision.

Table 2 presents the execution time of the sequential and parallel version of BSO-ARM. In order to well exploring the search space, the number of bees K, respectively the number of iterations $IMAX$ are set to 20, respectively 100.

The parallel version outperforms the sequential one in all cases. Furthermore, the GPU-based parallelization allowed us to solve two challenging large data sets (BMP POS and Web Docs) containing more than 1.5 millions of transactions and more than 0.5 million of items. To the best of our knowledge, these these two data sets have newer been solved before in the literature. Indeed, BSO-ARM blocked after 12 days whereas it takes only few hours using PSES.

6 Conclusion

In this paper, we proposed a new algorithm for association rules mining on GPU architecture. It is based on the bees behaviors, we first generate the solutions on CPU, then, the evaluation of each solution is performed in parallel using GPU threaded. The intensive multi-threaded provided on GPU conduct us to perform the single evaluation of solution at the same time. In fact, each thread is mapped with one transaction, this permits to accelerate the process of the evaluation. The experiments show that the parallel approach outperforms the sequential one in terms of the execution time. The results also reveal that using the massive threaded in GPU and the intelligent bees, the largest transactions base on the web is mined in real time.

References

1. Han, J., Kamber, M., Pei, J.: Data mining: concepts and techniques. Morgan Kaufmann (2006)
2. Agrawal, R., Imielinski, T., Swami, A.: Mining association rules between sets of items in large databases. ACM SIGMOD Record 22(2) (1993)
3. Han, J., Pei, J., Yin, Y.: Mining frequent patterns without candidate generation. ACM SIGMOD Record 29(2) (2000)
4. Kuo, R.J., Chao, C.M., Chiu, Y.T.: Application of particle swarm optimization to association rule mining. Applied Soft Computing 11(1), 326–336 (2011)
5. Djenouri, Y., Drias, H., Habbas, Z., Mosteghanemi, H.: Bees Swarm Optimization for Web Association Rule Mining. In: 2012 IEEE/WIC/ACM International Conferences on Web Intelligence and Intelligent Agent Technology (WI-IAT), vol. 3, pp. 142–146. IEEE (2012)
6. Djenouri, Y., Drias, H., Chemchem, A.: A hybrid Bees Swarm Optimization and Tabu Search algorithm for Association rule mining. In: 2013 World Congress on Nature and Biologically Inspired Computing (NaBIC). IEEE (2013)

7. Moslehi, P., et al.: Multi-objective Numeric Association Rules Mining via Ant Colony Optimization for Continuous Domains without Specifying Minimum Support and Minimum Confidence. International Journal of Computer Science (2008)
8. Agrawal, R., Shafer, J.C.: Parallel mining of association rules. IEEE Transactions on Knowledge and Data Engineering 8(6) (1996)
9. Parthasarathy, S., et al.: Parallel data mining for association rules on shared-memory systems. Knowledge and Information Systems 3(1) (2001)
10. Fang, W., et al.: Frequent itemset mining on graphics processors. In: Proceedings of the Fifth International Workshop on Data Management on New Hardware. ACM (2009)
11. Zhou, J., Yu, K.-M., Wu, B.-C.: Parallel frequent patterns mining algorithm on GPU. In: IEEE International Conference on Systems Man and Cybernetics (SMC). IEEE (2010)
12. Adil, S.H., Qamar, S.: Implementation of association rule mining using CUDA. In: International Conference on Emerging Technologies, ICET 2009. IEEE (2009)
13. Cui, Q., Guo, X.: Research on Parallel Association Rules Mining on GPU. In: Yang, Y., Ma, M. (eds.) Proceedings of the 2nd International Conference on Green Communications and Networks. LNEE, vol. 224, pp. 215–222. Springer, Heidelberg (2012)
14. Zhang, F., Zhang, Y., Bakos, J.: Gpapriori: Gpu-accelerated frequent itemset mining. In: IEEE International Conference on Cluster Computing (CLUSTER). IEEE (2011)
15. Li, H., Zhang, N.: Mining maximal frequent itemsets on graphics processors. In: Seventh International Conference on Fuzzy Systems and Knowledge Discovery (FSKD), vol. 3. IEEE (2010)
16. Guvenir, H.A., Uysal, I.: Bilkent university function approximation repository, 2012-03-12 (2000), `http://funapp.cs.bilkent.edu.tr/DataSets`
17. Goethals, B., Zaki, M.J.: Frequent itemset mining implementations repository. This site contains a wide-variety of algorithms for mining frequent, closed, and maximal itemsets (2003), `http://fimi.cs.helsinki.fi`

A Method for Ripple Simulation Based on GPU

Xianjun Chen[1,2], Yanmei Wang[3,*], and Yongsong Zhan[1]

[1] Guangxi Key Laboratory of Trusted Software, Guilin University of Electronic Technology,
Guilin, Guangxi, 541004, PR China
[2] Information Engineering School, Haikou College of Economics,
Haikou, Hainan, 571127, PR China
[3] QiongTai Teachers College, Haikou, Hainan, 571127, PR China
{Yanmei Wang,hingini}@126.com

Abstract. To improve the simulation of ripple on a personal workstation, a novel vector algebra model based on Graphic Process Unit (GPU) is proposed. First, the data structures and rules for data operation are established to meet the needs of vector algebra model. Second, the physical equation governing ripple motion is transformed discretely for vector multiplication, which will be solved by the Conjugate Gradient Method. Finally, the simulation of ripple is achieved from the height map providing normal information used by the calculation of light reflection and refraction in real time. Experiment results show that the method is robust and efficient to achieve real-time ripple simulation by making full use of the excellent computation power of programmable GPU.

Keywords: Ripple simulation, GPU, Vector algebra operation, Conjugate Gradient Method.

1 Introduction

As one of the most intriguing problems in computer graphics, the simulation of ripple has drawn the attention of a great sum of researchers. Ripples are everywhere in the nature, ranging from the streams to the rivers, from the pools to the oceans. In the applications of computer games and virtual reality, it is necessary to immerse players into plausible virtual worlds, which shall be constructed by the photorealistic simulation of natural scenes, such as ripple, smoke, and so on. Also, animators can also benefit from ripple simulation to achieve realistic effects in real time and improve the product efficiency of cartoon. With developing computer graphics, there exist many models that attempt to fake fluid-like effects. However, it is not easy to mimic the complexities and subtleties of ripple motion in a convincing manner on a graphic workstation in real time.

In this paper, a novel GPU based vector algebra operation model is proposed to improve the simulation of ripple, which is physically described by the fluid equation. First of all, the data structures and rules for data operation are established to meet the

---

* Corresponding author.

Y. Tan et al. (Eds.): ICSI 2014, Part II, LNCS 8795, pp. 58–65, 2014.

needs of vector algebra operation model. Then, the fluid dynamics equation governing ripple is transformed discretely for vector multiplication, which is to be solved by Conjugate Gradient Method. Finally, ripple rendering is achieved from the height map providing normal information used by the calculation of light reflection and refraction.

The rest of the paper is organized as follows. Related work is discussed in section 2. Section 3 introduces the GPU based vector algebra operation, including the data structures and rules for data operation. Section 4 gives a description on the Conjugate Gradient Method and the dynamics equation depicting water wave. Discretization of the fluid equation for vector multiplication is also included in this section. Section 5 presents the experimental result of a running system instance. Conclusions are drawn in section 6.

2   Related Work

In the graphics community, the early work for water phenomenon modeling placed emphasis on representations of the water surface as a parametric function, which could be animated over time to simulate wave transport[1-3]. But they are unable to easily deal with complex three-dimensional behaviors such as flow around objects and dynamically changing boundaries. To obtain water models which could be used in a dynamic animation environment, researchers turned to use two-dimensional approximations to the full 3D fluid equations[4]. Kass[5] approximated the 2D shallow water equations to get a dynamic height field surface that interacted with a static ground object. A pressure defined height field formulation was used by Chen[6] in fluid simulations with moving obstacles. O'Brien[7] simulated splashing liquids by combining a particle system and height field, while Miller[8] used viscous springs between particles to achieve dynamic flow in 3D. Terzopoulos[9] simulated melting deformable solids using a molecular dynamics approach to simulate the particles in the liquid phase. The simulation of complex water effects using the full 3D Navier-Stokes equations has been based upon the large amount of research done by the computational fluid dynamics community over the past 60 years. Foster[10] developed a 3D Navier-Stokes methodology for the realistic animation of liquids. Stam[11] replaced their finite difference scheme with a semi-Lagrangian method to achieve significant performance improvements at the cost of increased rotational damping. Foster[12] made significant contributions to the simulation and control of three dimensional fluid simulations through the introduction of a hybrid liquid volume model combining implicit surfaces and massless marker particles, the formulation of plausible boundary conditions for moving objects in a liquid, the use of an efficient iterative method to solve for the pressure, and a time step subcycling scheme for the particle and implicit surface evolution equations.

On the other hand, graphics hardware has undergone a true revolution in the past ten years. It went from being a simple memory device to a configurable unit and a fully programmable parallel processor. Although designed for fast polygon rendering, graphics hardware has been extended to various applications of general-purpose computations[13-14]. Harris[15] have implemented on the GPU the coupled map lattice, and have simulated cloud dynamics using partial differential equations. Goodnight[16] have implemented the multi-grid method on the GPU, and have applied it to heat transfer and modeling of fluid mechanics. Boltz[17] have also developed a GPU-based multi-grid solver. In addition, a conjugate-gradient solver is proposed on the GPU with a sparse matrix representation, which has been applied to the Navier-Stokes equations. At the same time, Kruger[18] has presented the GPU implementation of several linear algebra operators, which have been used to solve the Navier-Stokes equations as well. In all the above works, the computation of fluid dynamics is translated from the CPU to the GPU.

3   Vector Algebra Operation

As a main problem in the field of applied mathematics, the numerical solution for differential equations has been of prime importance in many applications of physical simulation and image procession. Transformed discretely to be linear, the differential equations are now widely used by 3D graphics applications for natural phenomena simulation. To solve the linear equations on GPU, the model of vector algebra operation is proposed to be composed of data structures and rules for data operation, both of which can be implemented by object-oriented program and extended freely.

3.1   Data Structures

The proposed model consists of three data types: float, vector and matrix. With the improvement of architecture in these years, the current GPU have been good at texture access, making it proper to define a float as a texture with the size of 1×1. A vector can be considered as either a one dimension (1D) texture or a two dimension (2D) texture. The latter is preferable in our model, for it has a better GPU support than the former. In practice, most of the current GPU set a limit to the total of 1D texture, namely 4096, which is not applicable for the complex numerical solution of differential equation. Furthermore, during the course of linear equation solving, the median is usually defined as vector, which is to be rendered as 2D texture to achieve high performance. In most computer graphics applications, matrix is usually considered as 2D texture, resulting in a high texture usage. As many matrices may be composed of a great majority of zero values and few non-zero values, it is necessary to define the matrix according to its attribute. The dense matrix can be divided into a set of vectors, which shall be considered as 2D textures. In the band matrix, the non-zero values have a distribution of band, which is considered to be a vector denoted by 2D texture as well. As to the sparse matrix, the non-zero values are distributed randomly, making it improper to be described by texture. To make full use of the parallelization operation of GPU, the sparse matrix is defined by a vertex buffer as figure 1.

Following their priority order in the sparse matrix, all the elements are inputed to the vertex buffer in turn. Each vertex will be provided with 4 non-zero values. The rows of sparse matrix are indexed by the vertex coordination, which can be adjusted by program parameter input. The texture coordination of each matrix element shall be the same as the final vertex coordination acquired by the procession of model transformation, view transformation and projection. Obviously, it can be seen from the above structure that each vertex is equipped with 6 texture coordination values, where (tu_0, tv_0), (tu_1, tv_1), (tu_2, tv_2), (tu_3, tv_4) are used as indexes for the 4 elements, (val0, val1, val2, val3) for their values, and (posX, posY) for index of the output. This definition is helpful for the multiplication between matrix and vector.

```
struct SPARSEMATRIXVERTEX
{
    FLOAT x,y,z;
    FLOAT tu_0,tv_0;
    FLOAT tu_1,tv_1;
    FLOAT tu_2,tv_2;
    FLOAT tu_3,tv_3;
    FLOAT val0,val1,val2,val3;
    FLOAT posX,posY;
    static const DWORD FVF;
};
```

Fig. 1. Data structures

3.2   Data Operation

The common data operation in the proposed model includes vector operation, vector reduction and multiplication of matrix and vector. To achieve vector addition, the following sequence of steps is performed: (1) a view frustum is set to cover a number of pixels, which is equal to the number of vector elements, and the target vector is set as Render Target; (2) a quad is rendered to cover the entire view port; (3) rasterization is implemented by the vertex program, and a mapping relationship is established between the vector elements and pixels, which can be accessed by the fragment program via texture coordination; (4) parallel processing of all the pixels by GPU is accomplished by fragment program, whose output is the target texture. The other vector operations, such as vector subtraction, multiplication between vector and scalar, and etc. can be implemented in the same way.

The operation of vector reduction is to count all the element values. Given a vector defined as a texture of $n \times n$, reduction can be implemented in $\log(n)$ rendering cycles. In the fragment program, the mean value of 4 neighboring texels is calculated and written into the Render Target, which is considered as input for the next rendering cycle. The above operation will repeat till the completion of reduction, as shown in figure 2.

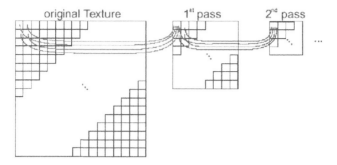

Fig. 2. Data operation

4   Ripple Rendering

The motion of ripple is controlled by 2D wave equation, which is of prime importance for the vivid simulation of wave animation. The numerical solution for this physical equation is as follows: (1) the equation is transformed to be linear by discretization; (2) the linear equation is solved by the Conjugate Gradient Method, which is completely based on the proposed vector algebra operation model accelerated by GPU.

4.1   Conjugate Gradient Method

As an applicable iteration process for solving linear equations, the Conjugate Gradient Method mainly deals with the cases with symmetric positive definite coefficient matrix. Given the linear equations as $Ax = b$, where $A \in R^{n \times n}$, the vector sequences of iteration and remainder are constructed, and the search directions are updated. These operations shall be iterated to achieve the ideal computation precision.

The procedure for Conjugate Gradient Method includes a pre-computation and a main cycle. In the course of pre-computation, the following steps shall be performed: (1) given $\forall x_0 \in R^n$; (2) $r_0 = b - Ax_0$; (3) $p_0 = r_0$; (4) a threshold constant ε is set, and $k = 0$.

4.2   Wave Equation

The inherent relationship between wave height and velocity, time and space can be described by the physical dynamics equation, which is defined as following.

$$c^2 \left(\frac{\partial^2 y}{\partial x^2} + \frac{\partial^2 y}{\partial z^2} \right) = \frac{\partial^2 y}{\partial t^2}$$

where x-z is the water surface, y is height, t is time and c is wave velocity. To achieve the numerical solution for wave equation, it shall be transformed to algebraic equation by Taylor series and Centre Differentia Method, which can be defined as follows.

$$f(x) = f(x + \Delta h) + f'(x + \Delta h) \times \Delta h + \Omega(\Delta h)$$

$$\Rightarrow (\frac{\partial y}{\partial h})_{i,j} = \begin{cases} \dfrac{y_{i+1,j} - y_{i,j}}{\Delta h} + O(\Delta h) \\[2ex] \dfrac{y_{i,j} - y_{i-1,j}}{\Delta h} + O(\Delta h) \\[2ex] \dfrac{y_{i+1,j} - y_{i-1,j}}{2\Delta h} + O(\Delta h)^2 \end{cases}$$

$$\Rightarrow (\frac{\partial^2 y}{\partial x^2})_{i,j} = \frac{y_{i+1,j} - 2y_{i,j} + y_{i-1,j}}{(\Delta h)^2} + O(\Delta h)^2$$

Finally, the 2D discrete wave equation can be denoted as follows, where

$$\beta = \frac{c^2(\Delta t)^2}{(\Delta h)^2}.$$

$$\frac{y_{i,j}^{t+1} - 2y_{i,j}^t + y_{i,j}^{t-1}}{(\Delta t)^2} = c^2 (\frac{y_{i+1,j}^t - 2y_{i,j}^t + y_{i-1,j}^t}{(\Delta x)^2} + \frac{y_{i,j+1}^t - 2y_{i,j}^t + y_{i,j-1}^t}{(\Delta z)^2})\Big|_{x=z=h}$$

$$\Leftrightarrow \frac{y_{i,j}^{t+1} - 2y_{i,j}^t + y_{i,j}^{t-1}}{(\Delta t)^2} = c^2 (\frac{y_{i+1,j}^t - 2y_{i,j}^t + y_{i-1,j}^t}{(\Delta h)^2} + \frac{y_{i,j+1}^t - 2y_{i,j}^t + y_{i,j-1}^t}{(\Delta h)^2})$$

$$\Leftrightarrow y_{i,j}^{t+1} = \beta(y_{i+1,j}^t + y_{i-1,j}^t + y_{i,j+1}^t + y_{i,j-1}^t) + (2 - 4\beta)y_{i,j}^t - y_{i,j}^{t-1}$$

5   Experiment

Based on the proposed method, an applicable implemented with C++ computer program, DirectX 9.0 and the hardware rendering program of HLSL is shown as figure 3. Furthermore, figure 4 gives a convincing representation of ripple on the workstation with NVIDIA Quadro Fx540. In this case, the 2D wave surface is presented discretely as meshes of 512X512, and simulated in real-time with the frame rate of 16fps.

In order to demonstrate the performance of the proposed model, two common-used GPU have been employed in the test, including NVIDIA Quadro FX 540 (with the memory of 256M) and NVIDIA FX 8800 (with the memory of 640M). The resultant data of table 1-2 show the efficiency and robustness of our method.

Fig. 3. Instance of system implementation   **Fig. 4.** The real-time ripple simulation

Table 1. Test result of NVIDIA Quadro FX 540.

	Image Resolution		
	512 x 512	512 x 256	256 x 256
Vector Reduction(ms)	1.00	0.71	0.62
Vector Addition(ms)	1.44	0.61	0.12
Frame Rate(fps)	17	32	64

Table 2. Test result of NVIDIA FX 8800.

	Image Resolution		
	512 x 512	512 x 256	256 x 256
Vector Reduction(ms)	0.10295	0.10076	0.09052
Vector Addition(ms)	0.01404	0.0139	0.0126
Frame Rate(fps)	295	423	440

6   Conclusion

In this paper, a novel GPU based vector algebra operation model is proposed to improve the simulation of water surface. The data structures and rules for data operation are established to meet the needs of vector algebra operation model, and the fluid dynamics equation governing ripple is transformed discretely for vector multiplication. Experimental results show the robustness and efficiency of the proposed method for the real-time simulation of water surface on GPU.

Acknowledgments. This research work is supported by the grant of Guangxi science and technology development project (No: 1355011-5), the grant of Guangxi Key Laboratory of Trusted Software of Guilin University of Electronic Technology (No:

kx201309), the grant of Guangxi Education Department (No: SK13LX139), the grant of Guangxi Undergraduate Training Programs for Innovation and Entrepreneurship (No:20121059519), the grant of Hainan Natural Science Foundation (No: 613169), and the Universities and colleges Science Research Foundation of Hainan (No: Hjkj2013-48).

References

1. Fournier, A., Reeves, W.T.: A simple model of ocean waves. In: Proc. SIGGRAPH ACM, pp. 75–84 (1986)
2. Peachy, D.: Modeling waves and surf. In: Proc. ACM SIGGRAPH, pp. 65–74 (1986)
3. Schachter, B.: Long crested wave models. Computer Graphics and Image Processing 12, 187–201 (1980)
4. Chorin, A.J., Marsden, J.E.: A mathematical introduction to fluid mechanics. Texts in Applied Mathematics, vol. 4 (1990)
5. Kass, M., Miller, G.: Rapid, stable fluid dynamics for computer graphics. In: Proc. ACM SIGGRAPH, pp. 49–57 (1990)
6. Chen, J., Lobo, N.: Toward interactive-rate simulation of fluids with moving obstacles using the Navier-Stokes Equations. Graphical Models and Image Processing 57, 107–116 (1994)
7. O'Brien, J., Hodgins, J.: Dynamic simulation of splashing fluids. In: Proc. Computer Animation, pp. 198–205 (1995)
8. Miller, G., Pearce, A.: Globular dynamics: a connected particle system for animating viscous fluids. Computers and Graphics 13, 305–309 (1989)
9. Terzopoulos, D., Platt, J., Fleischer, K.: Heating and melting deformable models (from goop to glop). In: Proc. Graphics Interface, pp. 219–226 (1989)
10. Foster, N., Metaxas, D.: Realistic animation of liquids. Graphical Models and Image Processing 58, 471–483 (1996)
11. Stam, J.: Stable fluids. In: Proc. ACM SIGGRAPH, pp. 121–128 (1999)
12. Foster, N., Fedkiw, R.: Practical animation of liquids. In: Proc. ACM SIGGRAPH, pp. 23–30 (2001)
13. Hoff, K., Culver, T.: Fast computation of generalized voronoi diagrams using graphics hardware. In: Proc. ACM SIGGRAPH, pp. 277–286 (1999)
14. Trendall, C., Stewart, A.J.: General calculations using graphics hardware with applications to interactive caustics. In: Proc. Eurographics Workshop on Rendering, pp. 287–298 (2000)
15. Harris, M., Coombe, G.: Physically-based visual simulation on graphics hardware. In: Proc. SIGGRAPH/Eurographics Workshop on Graphics Hardware, pp. 109–118 (2002)
16. Goodnight, N., Woolley, C.: A multi-grid solver for boundary value problems using programmable graphics hardware. In: Proc. SIGGRAPH/Eurographics Workshop on Graphics Hardware (2003)
17. Bolz, J., Farmer, I., Grinspun, E.: Sparse matrix solvers on the GPU: conjugate gradients and multi-grid. ACM Transactions on Graphics 22, 917–924 (2003)
18. Kruger, J., Westermann, R.: Linear algebra operators for GPU implementation of numerical algorithms. ACM Transactions on Graphics 22, 908–916 (2003)

cuROB: A GPU-Based Test Suit
for Real-Parameter Optimization

Ke Ding and Ying Tan*

Key Laboratory of Machine Perception and Intelligence (MOE), Peking University,
Department of Machine Intelligence, School of Electronics Engineering and Computer
Science, Peking University, Beijing, 100871, China
{keding,ytan}@pku.edu.cn

Abstract. Benchmarking is key for developing and comparing optimization algorithms. In this paper, a GPU-based test suit for real-parameter optimization, dubbed cuROB, is introduced. Test functions of diverse properties are included within cuROB and implemented efficiently with CUDA. Speedup of one order of magnitude can be achieved in comparison with CPU-based benchmark of CEC'14.

Keywords: Optimization Methods, Optimization Benchmark, GPU, CUDA.

1 Introduction

Proposed algorithms are usually tested on benchmark for comparing both performance and efficiency. However, as it can be a very tedious task to select and implement test functions rigorously. Thanks to GPUs' massive parallelism, a GPU-based optimization function suit will be beneficial to test and compare optimization algorithms.

Based on the well known CPU-based benchmarks presented in [1,2,3], we proposed a CUDA-based real parameter optimization test suit, called cuROB, targeting on GPUs. We think cuROB can be helpful for assessing GPU-based optimization algorithms, and hopefully, conventional CPU-based algorithms can benefit from cuROB's fast execution.

Considering the fact that research on the single objective optimization algorithms is the basis of the research on the more complex optimization algorithms such as constrained optimization algorithms, multi-objective optimizations algorithms and so forth, in this first release of cuROB a suit of single objective real-parameter optimization function are defined and implemented.

The test functions are selected according to the following criteria: 1) the functions should be scalable in dimension so that algorithms can be tested under various complexity; 2) the expressions of the functions should be with good parallelism, thus efficient implementation is possible on GPUs; 3) the functions should be comprehensible such that algorithm behaviours can be analysed in

---

* To whom the correspondence should be addressed.

Y. Tan et al. (Eds.): ICSI 2014, Part II, LNCS 8795, pp. 66–78, 2014.

the topological context; 4) last but most important, the test suit should cover functions of various properties in order to get a systematic evaluation of the optimization algorithms.

The source code and a sample can be download from `code.google.com/p/curob/`.

1.1 Symbol Conventions and Definitions

Symbols and definitions used in the report are described in the following. By default, all vectors refer to column vectors, and are depicted by lowercase letter and typeset in bold.

- $[\cdot]$ indicates the nearest integer value
- $\lfloor \cdot \rfloor$ indicates the largest integer less than or equal to
- $\mathbf{x}_i$ denotes i-th element of vector $\mathbf{x}$
- $f(\cdot)$, $g(\cdot)$ and $G(\cdot)$ multi-variable functions
- f_{opt} optimal (minimal) value of function f
- $\mathbf{x}^{opt}$ optimal solution vector, such that $f(\mathbf{x}^{opt}) = f_{opt}$
- $\mathbf{R}$ normalized orthogonal matrix for rotation
- D dimension
- $\mathbf{1} = (1, \ldots, 1)^T$ all one vector

1.2 General Setup

The general setup of the test suit is presented as follows.

- **Dimensions** The test suit is scalable in terms of dimension. Within the hardware limit, any dimension $D \geq 2$ works. However, to construct a real hybrid function, D should be at least 10.
- **Search Space** All functions are defined and can be evaluated over $\mathcal{R}^D$, while the actual search domain is given as $[-100, 100]^D$.
- f_{opt} All functions, by definition, have a minimal value of 0, a bias (f_{opt}) can be added to each function. The selection can be arbitrary, f_{opt} for each function in the test suit is listed in Tab. 1.
- $\mathbf{x}^{opt}$ The optimum point of each function is located at original. $\mathbf{x}^{opt}$ which is randomly distributed in $[-70, 70]^D$, is selected as the new optimum.
- **Rotation Matrix** To derive non-separable functions from separable ones, the search space is rotated by a normalized orthogonal matrix $\mathbf{R}$. For a given function in one dimension, a different $\mathbf{R}$ is used. Variables are divided into three (almost) equal-sized subcomponents randomly. The rotation matrix for each subcomponent is generated from standard normally distributed entries by Gram-Schmidt orthonormalization. Then, these matrices consist of the $\mathbf{R}$ actually used.

1.3 CUDA Interface and Implementation

A simple description of the interface and implementation is given in the following. For detail, see the source code and the accompanied readme file.

Interface. Only benchmark.h need to be included to access the test functions, and the CUDA file benchmark.cu need be compiled and linked. Before the compiling start, two macro, DIM and MAX_CONCURRENCY should be modified accordingly. DIM defines the dimension of the test suit to used while MAX_CONCURRENCY controls the most function evaluations be invoked concurrently. As memory needed to be pre-allocated, limited by the hardware, don't set MAX_CONCURRENCY greater than actually used.

Host interface function initialize () accomplish all initialization tasks, so must be called before any test function can be evaluated. Allocated resource is released by host interface function dispose ().

Both double precision and single precision are supported through func_evaluate () and func_evaluatef () respectively. Take note that device pointers should be passed to these two functions. For the convenience of CPU code, C interfaces are provided, with h_func_evaluate for double precision and h_func_evaluatef for single precision. (In fact, they are just wrappers of the GPU interfaces.)

Efficiency Concerns. When configuration of the suit, some should be taken care for the sake of efficiency. It is better to evaluation a batch of vectors than many smaller. Dimension is a fold of 32 (the warp size) can more efficient. For example, dimension of 96 is much more efficient than 100, even though 100 is little greater than 96.

1.4   Test Suite Summary

The test functions fall into four categories: unimodal functions, basic multimodal functions, hybrid functions and composition functions. The summary of the suit is listed in Tab. 1. Detailed information of each function will given in the following sections.

2   Speedup

Under different hardware, various speedups can be achieved. 30 functions are the same as CEC'14 benchmark. We test the cuROB's speedup with these 30 functions under the following settings: Windows 7 SP1 x64 running on Intel i5-2310 CPU with NVIDIA 560 Ti, the CUDA version is 5.5. 50 evaluations were performed concurrently and repeated 1000 runs. The evaluation data were generated randomly from uniform distribution.

The speedups with respect to different dimension are listed by Tab. 2 (single precision) and Tab. 3 (double precision). Notice that the corresponding dimensions of cuROB are 10, 32, 64 and 96 respectively and the numbers are as in Tab. 1

Fig. 1 demonstrates the overall speedup for each dimension. On average, cuROB is never slower than its CPU-base CEC'14 benchmark, and speedup of one order of magnitude can be achieved when dimension is high. Single precision is more efficient than double precision as far as execution time is concerned.

Table 1. Summary of cuROB's Test Functions

	No.	Functions	ID	Description
Unimodal Functions	0	Rotated Sphere	SPHERE	Optimum easy to track
	1	Rotated Ellipsoid	ELLIPSOID	
	2	Rotated Elliptic	ELLIPTIC	
	3	Rotated Discus	DISCUS	
	4	Rotated Bent Cigar	CIGAR	Optimum hard to track
	5	Rotated Different Powers	POWERS	
	6	Rotated Sharp Valley	SHARPV	
Basic Multi-modal Functions	7	Rotated Step	STEP	With adepuate global structure
	8	Rotated Weierstrass	WEIERSTRASS	
	9	Rotated Griewank	GRIEWANK	
	10	Rastrigin	RARSTRIGIN_U	
	11	Rotated Rastrigin	RARSTRIGIN	
	12	Rotated Schaffer's F7	SCHAFFERSF7	
	13	Rotated Expanded Griewank plus Rosenbrock	GRIE_ROSEN	
	14	Rotated Rosenbrock	ROSENBROCK	
	15	Modified Schwefel	SCHWEFEL_U	With weak global structure
	16	Rotated Modified Schwefel	SCHWEFEL	
	17	Rotated Katsuura	KATSUURA	
	18	Rotated Lunacek bi-Rastrigin	LUNACEK	
	19	Rotated Ackley	ACKLEY	
	20	Rotated HappyCat	HAPPYCAT	
	21	Rotated HGBat	HGBAT	
	22	Rotated Expanded Schaffer's F6	SCHAFFERSF6	
Hybrid Functions	23	Hybrid Function 1	HYBRID1	With different properties for different variables subcomponents
	24	Hybrid Function 2	HYBRID2	
	25	Hybrid Function 3	HYBRID3	
	26	Hybrid Function 4	HYBRID4	
	27	Hybrid Function 5	HYBRID5	
	28	Hybrid Function 6	HYBRID6	
Composition Functions	29	Composition Function 1	COMPOSITION1	Properties similar to particular sub-function when approaching the corresponding optimum
	30	Composition Function 2	COMPOSITION2	
	31	Composition Function 3	COMPOSITION3	
	32	Composition Function 4	COMPOSITION4	
	33	Composition Function 5	COMPOSITION5	
	34	Composition Function 6	COMPOSITION6	
	35	Composition Function 7	COMPOSITION7	
	36	Composition Function 8	COMPOSITION8	

Search Space: $[-100, 100]^D$, $f_{opt} = 100$

Table 2. Speedup (single Precision)

D	NO.3	NO.4	NO.5	NO.8	NO.9	NO.10	NO.11	NO.13	NO.14	NO.15
10	0.59	0.20	0.18	12.23	0.49	0.28	0.31	0.32	0.14	0.77
32	3.82	2.42	2.00	47.19	3.54	1.67	3.83	5.09	2.06	3.54
64	4.67	2.72	2.29	50.17	3.56	0.93	3.06	2.88	2.20	3.39
94	13.40	10.10	8.50	84.31	11.13	1.82	9.98	9.66	8.75	6.73

D	NO.16	NO.17	NO.19	NO.20	NO.21	NO.22	NO.23	NO.24	NO.25	NO.26
10	0.80	3.25	0.36	0.20	0.26	0.45	0.63	0.44	2.80	0.52
32	5.57	10.04	3.46	1.22	1.42	6.44	3.95	3.43	11.47	3.36
64	5.45	13.19	3.27	2.10	2.27	3.81	4.62	3.07	14.17	3.34
96	14.38	23.68	11.32	8.26	8.49	11.60	13.67	10.64	30.11	10.71

D	NO.27	NO.28	NO.29	NO.30	NO.31	NO.32	NO.33	NO.34	NO.35	NO.36
10	0.65	0.72	0.70	0.55	0.71	3.49	3.50	0.84	1.28	0.70
32	2.73	3.09	3.63	3.10	4.10	12.39	12.51	5.25	5.19	3.33
64	3.86	4.01	3.21	2.67	3.38	12.68	12.63	3.80	5.27	3.13
96	12.04	11.32	8.15	6.27	8.49	23.67	23.64	9.50	11.79	7.93

Table 3. Speedup (Double Precision)

D	NO.3	NO.4	NO.5	NO.8	NO.9	NO.10	NO.11	NO.13	NO.14	NO.15
10	0.56	0.19	0.17	9.04	0.43	0.26	0.29	0.30	0.14	0.75
32	3.78	2.43	1.80	33.37	3.09	1.59	3.52	4.81	1.97	3.53
64	4.34	2.49	1.93	30.82	3.15	0.92	2.87	2.74	2.11	3.29
96	12.27	9.24	6.95	46.01	9.72	1.78	9.62	8.74	7.87	5.92

D	NO.16	NO.17	NO.19	NO.20	NO.21	NO.22	NO.23	NO.24	NO.25	NO.26
10	0.79	2.32	0.34	0.18	0.26	0.45	0.59	0.43	1.97	0.52
32	5.10	6.79	3.28	1.13	1.29	6.10	3.63	3.14	8.15	3.23
64	4.75	8.29	3.06	1.99	2.18	3.32	4.02	2.77	9.80	2.92
96	11.91	13.81	9.75	7.37	7.78	10.24	11.55	9.57	20.81	9.40

D	NO.27	NO.28	NO.29	NO.30	NO.31	NO.32	NO.33	NO.34	NO.35	NO.36
10	0.79	2.32	0.34	0.18	0.26	0.45	0.59	0.43	1.97	0.52
32	5.10	6.79	3.28	1.13	1.29	6.10	3.63	3.14	8.15	3.23
64	4.75	8.29	3.06	1.99	2.18	3.32	4.02	2.77	9.80	2.92
96	11.91	13.81	9.75	7.37	7.78	10.24	11.55	9.57	20.81	9.40

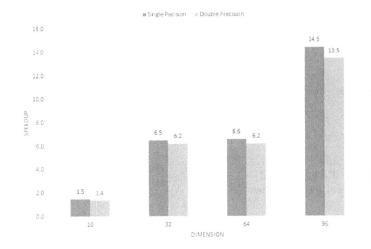

Fig. 1. Overall Speedup

3 Unimodal Functions

3.1 Shifted and Rotated Sphere Function

$$f_1(\mathbf{x}) = \sum_{i=1}^{D} \mathbf{z}_i^2 + f_{opt} \tag{1}$$

where $\mathbf{z} = \mathbf{R}(\mathbf{x} - \mathbf{x}^{opt})$.

Properties

- Unimodal
- Non-separable
- Highly symmetric, in particular rotationally invariant

3.2 Shifted and Rotated Ellipsoid Function

$$f_4(\mathbf{x}) = \sum_{i=1}^{D} i \cdot \mathbf{z}_i^2 + f_{opt} \tag{2}$$

where $\mathbf{z} = \mathbf{R}(\mathbf{x} - \mathbf{x}^{opt})$.

Properties

- Unimodal
- Non-separable

3.3 Shifted and Rotated High Conditioned Elliptic Function

$$f_2(\mathbf{x}) = \sum_{i=1}^{D}(10^6)^{\frac{i-1}{D-1}}\mathbf{z}_i^2 + f_{opt} \tag{3}$$

where $\mathbf{z} = \mathbf{R}(\mathbf{x} - \mathbf{x}^{opt})$.

Properties

− Unimodal
− Non-separable
− Quadratic ill-conditioned
− Smooth local irregularities

3.4 Shifted and Rotated Discus Function

$$f_5(\mathbf{x}) = 10^6 \cdot \mathbf{z}_1^2 + \sum_{i=2}^{D}\mathbf{z}_i^2 + f_{opt} \tag{4}$$

where $\mathbf{z} = \mathbf{R}(\mathbf{x} - \mathbf{x}^{opt})$.

Properties

− Unimodal
− Non-separable
− Smooth local irregularities
− With One sensitive direction

3.5 Shifted and Rotated Bent Cigar Function

$$f_6(\mathbf{x}) = \mathbf{z}_1^2 + 10^6 \cdot \sum_{i=2}^{D}\mathbf{z}_i^2 + f_{opt} \tag{5}$$

where $\mathbf{z} = \mathbf{R}(\mathbf{x} - \mathbf{x}^{opt})$.

Properties

− Unimodal
− Non-separable
− Optimum located in a smooth but very narrow valley

3.6 Shifted and Rotated Different Powers Function

$$f_4(\mathbf{x}) = \sqrt{\sum_{i=1}^{D}|\mathbf{z}_i|^{2+4\frac{i-1}{D-1}}} + f_{opt} \tag{6}$$

where $\mathbf{z} = \mathbf{R}(0.01(\mathbf{x} - \mathbf{x}^{opt}))$.

Properties

- Unimodal
- Non-separable
- Sensitivities of the $\mathbf{z}_i$-variables are different

3.7 Shifted and Rotated Sharp Valley Function

$$f_4(\mathbf{x}) = \mathbf{z}_i^2 + 100 \cdot \sqrt{\sum_{i=2}^{D} \mathbf{z}_i^2} + f_{opt} \tag{7}$$

where $\mathbf{z} = \mathbf{R}(\mathbf{x} - \mathbf{x}^{opt})$.

Properties

- Unimodal
- Non-separable
- Global optimum located in a sharp (non-differentiable) ridge

4 Basic Multi-modal Functions

4.1 Shifted and Rotated Step Function

$$f_3(\mathbf{x}) = \sum_{i=1}^{D} \lfloor \mathbf{z}_i + 0.5 \rfloor^2 + f_{opt} \tag{8}$$

where $\mathbf{z} = \mathbf{R}(\mathbf{x} - \mathbf{x}^{opt})$

Properties

- Many Plateaus of different sizes
- Non-separable

4.2 Shifted and Rotated Weierstrass Function

$$f_9(\mathbf{x}) = \sum_{i=1}^{D} \left(\sum_{k=0}^{k_{max}} a^k \cos\left(2\pi b^k(\mathbf{z}_i + 0.5)\right) \right) - D \cdot \sum_{k=0}^{k_{max}} a^k \cos\left(2\pi b^k \cdot 0.5\right) + f_{opt} \tag{9}$$

where $a = 0.5$, $b = 3$, $k_{max} = 20$, $\mathbf{z} = \mathbf{R}(0.005 \cdot (\mathbf{x} - \mathbf{x}^{opt}))$.

Properties

- Multi-modal
- Non-separable
- Continuous everywhere but only differentiable on a set of points

4.3   Shifted and Rotated Griewank Function

$$f_{10}(\mathbf{x}) = \sum_{i=1}^{D} \frac{\mathbf{z}_i^2}{4000} - \prod_{i=1}^{D} \cos(\frac{\mathbf{z}_i}{\sqrt{i}}) + 1 + f_{opt} \tag{10}$$

where $\mathbf{z} = \mathbf{R}(6 \cdot (\mathbf{x} - \mathbf{x}^{opt}))$.

Properties

- Multi-modal
- Non-separable
- With many regularly distributed local optima

4.4   Shifted Rastrigin Function

$$f_{11}(\mathbf{x}) = \sum_{i=1}^{D} \left(\mathbf{z}_i^2 - 10 \cos(2\pi \mathbf{z}_i) \right) + 10 \cdot D + f_{opt} \tag{11}$$

where $\mathbf{z} = 0.0512 \cdot (\mathbf{x} - \mathbf{x}^{opt})$.

Properties

- Multi-modal
- Separable
- With many regularly distributed local optima

4.5   Shifted and Rotated Rastrigin Function

$$f_{12}(\mathbf{x}) = \sum_{i=1}^{D} \left(\mathbf{z}_i^2 - 10 \cos(2\pi \mathbf{z}_i) + 10 \right) + f_{opt} \tag{12}$$

where $\mathbf{z} = \mathbf{R}(0.0512 \cdot (\mathbf{x} - \mathbf{x}^{opt}))$.

Properties

- Multi-modal
- Non-separable
- With many regularly distributed local optima

4.6   Shifted Rotated Schaffer's F7 Function

$$f_{17}(\mathbf{x}) = \left(\frac{1}{D-1} \sum_{i=1}^{D-1} \left((1 + \sin^2(50 \cdot \mathbf{w}_i^{0.2})) \cdot \sqrt{\mathbf{w}_i} \right) \right)^2 + f_{opt} \tag{13}$$

where $\mathbf{w}_i = \sqrt{\mathbf{z}_i^2 + \mathbf{z}_{i+1}^2}$, $\mathbf{z} = \mathbf{R}(\mathbf{x} - \mathbf{x}^{opt})$.

Properties

- Multi-modal
- Non-separable

4.7 Expanded Griewank plus Rosenbrock Function

$$\text{Rosenbrock Function: } g_2(x, y) = 100(x^2 - y)^2 + (x - 1)^2$$
$$\text{Griewank Function: } g_3(x) = x^2/4000 - \cos(x) + 1$$

$$f_{18}(\mathbf{x}) = \sum_{i=1}^{D-1} g_3(g_2(\mathbf{z}_i, \mathbf{z}_{i+1})) + g_3(g_2(\mathbf{z}_D, \mathbf{z}_1)) + f_{opt} \tag{14}$$

where $\mathbf{z} = \mathbf{R}(0.05 \cdot (\mathbf{x} - \mathbf{x}^{opt})) + \mathbf{1}$.

Properties

- Multi-modal
- Non-separable
-

4.8 Shifted and Rotated Rosenbrock Function

$$f_7(\mathbf{x}) = \sum_{i=1}^{D-1} \left(100 \cdot (\mathbf{z}_i^2 - \mathbf{z}_{i+1})^2 + (\mathbf{z}_i - 1)^2 \right) + f_{opt} \tag{15}$$

where $\mathbf{z} = \mathbf{R}(0.02048 \cdot (\mathbf{x} - \mathbf{x}^{opt})) + \mathbf{1}$.

Properties

- Multi-modal
- Non-separable
- With a long, narrow, parabolic shaped flat valley from local optima to global optima

4.9 Shifted Modified Schwefel Function

$$f_{13}(\mathbf{x}) = 418.9829 \times D - \sum_{i=1}^{D} g_1(\mathbf{w}_i), \qquad \mathbf{w}_i = \mathbf{z}_i + 420.9687462275036 \tag{16}$$

$$g_1(\mathbf{w}_i) = \begin{cases} \mathbf{w}_i \cdot \sin(\sqrt{|\mathbf{w}_i|}) & \text{if } |\mathbf{w}_i| \leq 500 \\ (500 - \text{mod}(\mathbf{w}_i, 500)) \cdot \sin\left(\sqrt{500 - \text{mod}(\mathbf{w}_i, 500)}\right) - \frac{(\mathbf{w}_i - 500)^2}{10000D} & \text{if } \mathbf{w}_i > 500 \\ (\text{mod}(-\mathbf{w}_i, 500) - 500) \cdot \sin\left(\sqrt{500 - \text{mod}(-\mathbf{w}_i, 500)}\right) - \frac{(\mathbf{w}_i + 500)^2}{10000D} & \text{if } \mathbf{w}_i < -500 \end{cases}$$

$$\tag{17}$$

where $\mathbf{z} = 10 \cdot (\mathbf{x} - \mathbf{x}^{opt})$.

Properties

- Multi-modal
- Separable
- Having many local optima with the second better local optima far from the global optima

4.10   Shifted Rotated Modified Schwefel Function

$$f_{14}(\mathbf{x}) = 418.9829 \times D - \sum_{i=1}^{D} g_1(\mathbf{w}_i), \qquad \mathbf{w}_i = \mathbf{z}_i + 420.9687462275036 \quad (18)$$

where $\mathbf{z} = \mathbf{R}(10 \cdot (\mathbf{x} - \mathbf{x}^{opt}))$ and $g_1(\cdot)$ is defined as Eq. 17.

Properties

- Multi-modal
- Non-separable
- Having many local optima with the second better local optima far from the global optima

4.11   Shifted Rotated Katsuura Function

$$f_{15}(\mathbf{x}) = \frac{10}{D^2} \prod_{i=1}^{D} (1 + i \sum_{j=1}^{32} \frac{|2^j \cdot \mathbf{z}_i - [2^j \cdot \mathbf{z}_i]|}{2^j})^{\frac{10}{D^{1.2}}} - \frac{10}{D^2} + f_{opt} \quad (19)$$

where $\mathbf{z} = \mathbf{R}(0.05 \cdot (\mathbf{x} - \mathbf{x}^{opt}))$.

Properties

- Multi-modal
- Non-separable
- Continuous everywhere but differentiable nowhere

4.12   Shifted and Rotated Lunacek bi-Rastrigin Function

$$f_{12}(\mathbf{x}) = \min \left(\sum_{i=1}^{D} (\mathbf{z}_i - \mu_1)^2, dD + s \sum_{i=1}^{D} (\mathbf{z}_i - \mu_2)^2 \right) + 10 \cdot (D - \sum_{i=1}^{D} \cos(2\pi(\mathbf{z}_i - \mu_1))) + f_{opt} \quad (20)$$

where $\mathbf{z} = \mathbf{R}(0.1 \cdot (\mathbf{x} - \mathbf{x}^{opt}) + 2.5 * \mathbf{1})$, $\mu_1 = 2.5$, $\mu_2 = -2.5$, $d = 1$, $s = 0.9$.

Properties

- Multi-modal
- Non-separable
- With two funnel around $\mu_1 \mathbf{1}$ and $\mu_2 \mathbf{1}$

4.13   Shifted and Rotated Ackley Function

$$f_8(\mathbf{x}) = -20 \cdot \exp\left(-0.2\sqrt{\frac{1}{D}\sum_{i=1}^{D}\mathbf{x}_i^2}\right) - \exp\left(\frac{1}{D}\sum_{i=1}^{D}\cos(2\pi\mathbf{x}_i)\right) + 20 + e + f_{opt}$$

(21)

where $\mathbf{z} = \mathbf{R}(\mathbf{x} - \mathbf{x}^{opt})$.

Properties

- Multi-modal
- Non-separable
- Having many local optima with the global optima located in a very small basin

4.14   Shifted Rotated HappyCat Function

$$f_{16}(\mathbf{x}) = |\sum_{i=1}^{D}\mathbf{z}_i^2 - D|^{0.25} + (\frac{1}{2}\sum_{j=1}^{D}\mathbf{z}_j^2 + \sum_{j=1}^{D}\mathbf{z}_j)/D + 0.5 + f_{opt}$$

(22)

where $\mathbf{z} = \mathbf{R}(0.05 \cdot (\mathbf{x} - \mathbf{x}^{opt})) - 1$.

Properties

- Multi-modal
- Non-separable
- Global optima located in curved narrow valley

4.15   Shifted Rotated HGBat Function

$$f_{17}(\mathbf{x}) = |(\sum_{i=1}^{D}\mathbf{z}_i^2)^2 - (\sum_{j=1}^{D}\mathbf{z}_j)^2|^{0.5} + (\frac{1}{2}\sum_{j=1}^{D}\mathbf{z}_j^2 + \sum_{j=1}^{D}\mathbf{z}_j)/D + 0.5 + f_{opt}$$

(23)

where $\mathbf{z} = \mathbf{R}(0.05 \cdot (\mathbf{x} - \mathbf{x}^{opt})) - 1$.

Properties

- Multi-modal
- Non-separable
- Global optima located in curved narrow valley

4.16   Expanded Schaffer's F6 Function

$$\text{Schaffer's F6 Function: } g_4(x, y) = \frac{\sin^2(\sqrt{x^2 + y^2}) - 0.5}{(1 + 0.001 \cdot (x^2 + y^2))^2} + 0.5$$

$$f_{19}(\mathbf{x}) = \sum_{i=1}^{D-1} g_4(\mathbf{z}_i, \mathbf{z}_{i+1}) + g_4(\mathbf{z}_D, \mathbf{z}_1) + f_{opt}$$

(24)

where $\mathbf{z} = \mathbf{R}(\mathbf{x} - \mathbf{x}^{opt})$.

Properties

- Multi-modal
- Non-separable

5 Hybrid and Composition Functions

Hybrid functions are constructed according to [3]. For each hybrid function, the variables are randomly divided into subcomponents and different basic functions (unimodal and multi-modal) are used for different subcomponents.

Composition functions are constructed in the same manner as in [2,3]. The constructed functions are multi-modal and non-separable and merge the properties of the sub-functions better and maintains continuity around the global/local optima. The local optimum which has the smallest bias value is the global optimum. The optimum of the third basic function is set to the origin as a trip in order to test the algorithms' tendency to converge to the search center. Note that, the landscape is not only changes along with the selection of basic function, but the optima, σ and λ can effect it greatly.

The detailed specifications of hybrid and composition functions can be found in the extended version of this paper[1], along with illustrations for all 2-D functions except hybrid functions.

Acknowledgements. This work was supported by National Natural Science Foundation of China (NSFC), Grant No. 61375119, 61170057 and 60875080.

References

1. Finck, S., Hansen, N., Ros, R., Auger, A.: Real-parameter black-box optimization benchmarking 2010: Noiseless functions definitions. Technical Report 2009/20, Research Center PPE (2010)
2. Liang, J.J., Qu, B.Y., Suganthan, P.N., Hernández-Díaz, A.G.: Problem definitions and evaluation criteria for the cec 2013 special session and competition on real-parameter optimization. Technical Report 201212, Computational Intelligence Laboratory, Zhengzhou University and Nanyang Technological University, Singapore (2013)
3. Liang, J.J., Qu, B.Y., Suganthan, P.N.: Problem definitions and evaluation criteria for the cec 2014 special session and competition on single objective real-parameter numerical optimization. Technical Report 201311, Computational Intelligence Laboratory, Zhengzhou University and Nanyang Technological University, Singapore (2013)

---

[1] Download from `http://arxiv.org/abs/1407.7737`

A Particle Swarm Optimization Based Pareto Optimal Task Scheduling in Cloud Computing

A.S. Ajeena Beegom[1] and M.S. Rajasree[2]

[1] Dept. of Computer Science and Engineering,
College of Engineering, Trivandrum, India
ajeena@cet.ac.in
[2] IIITM-K, Trivandrum, India
rajasree.ms@iiitmk.ac.in

Abstract. Task scheduling in Cloud computing is a challenging aspect due to the conflicting requirements of end users of cloud and the Cloud Service Provider (CSP). The challenge at the CSP's end is to schedule tasks submitted by the cloud users in an optimal way such that it should meet the quality of service (QoS) requirements of the user at one end and the running costs of the infrastructure to a minimum level at the other end for better profit. The focus is on two objectives, makespan and cost, to be optimized simultaneously using meta heuristic search techniques for scheduling independent tasks. A new variant of continuous Particle Swarm Optimization (PSO) algorithm, named Integer-PSO, is proposed to solve the bi-objective task scheduling problem in cloud which out performs the smallest position value (SPV) rule based PSO technique.

Keywords: Cloud Computing, Task Scheduling, Particle Swarm Optimization, Integer-PSO.

1 Introduction

Recently, cloud computing has emerged as an attractive platform for entrepreneurs as well as researchers in various domains. Cloud computing refers to leasing computing resources over the Internet. Benefits of using such a set up include reduced infrastructure cost, reduced overhead and pay only for the components he has used for the given amount of time. There are many task scheduling models available in literature for heterogeneous distributed systems but these models aim at the improvement of specific performance metrics like throughput and storage. For a cloud computing platform, apart from theses considerations, user satisfaction in terms of QoS and CSP's profit is to be considered while scheduling tasks. In this work, we consider the scheduling of large set of independent tasks of different size. The conflicting objectives of performance optimization considered are the overall execution time (makespan) of all the tasks and the cost of service. The cost include computation cost, communication cost and over all maintenance cost as well as power consumption cost.

Y. Tan et al. (Eds.): ICSI 2014, Part II, LNCS 8795, pp. 79–86, 2014.

Most of the existing task scheduling research in cloud computing addresses any one factor, namely makespan as done in [5] or cost as seen in [11] and tries to find an optimal schedule based on that factor alone. But single objective optimization solutions will try to optimize one objective making another key factor to worse. In the proposed work, we have used weighted sum approach for pareto-optimality and uses Particle swarm algorithm to solve the same.

The rest of the paper is organized as follows. In Section 2, we present Related Work in the domain and Section 3 describes the Mathematical Model and formally states the optimization problem. Section 4 details Particle Swarm Optimization Technique and the proposed Integer-PSO algorithm. Experimental Results are presented in Section 5 and Section 6 gives the Conclusion.

2   Related Work

Most of the research in computational grids and cloud systems for scheduling independent tasks to be executed in parallel tries to optimize a single objective function where the parameters are any one of makespan, profit earned, cost of service, QoS, energy consumption and average response time. Meta heuristic search techniques has been tried by many to solve the same. L.Zhu and J. Wu [15] uses PSO technique combined with Simulated annealing to solve task scheduling problem in a general scenario. M.F.Tasgetiren et al. [8] proposed Smallest Position Value (SPV) rule based PSO algorithm to solve Single Machine Total Weighted Tardiness Problem and has used a local search technique to intense the search. Lei Zhang et al. [14] used this technique to solve task scheduling problem in grid environment and has given a comparative study on the application of PSO technique with Genetic algorithm for achieving minimal completion time. PSO algorithm has also been used in solving task allocation / scheduling problems in work flows in cloud [1], [13], [7].

Multi-objective optimization for resource allocation in cloud computing has been addressed by Feng et al. [3] and uses PSO algorithm to solve the problem. They have considered total execution time, resource reservation and QoS of each task as their optimization objective and uses pareto domination mechanism to find optimal solutions. Lizheng Guo [4] addresses task assignment problem in cloud computing considering makespan and cost. They use smallest position value based PSO algorithm for finding an optimal schedule.

Our proposed method can be used to schedule tasks in pubic cloud or in private cloud for independent tasks. Our approach is unique because, to the best our knowledge, pareto optimal task scheduling using particle swarm optimization technique has not been addressed in the case of private or public cloud with independent tasks to be scheduled, but has been proposed for work flow scheduling in hybrid cloud[2]. Our work also proposes a new variant of PSO algorithm namely Integer - PSO.

3   Mathematical Model

Assume an application consists of N independent tasks, n out of N are scheduled at each time window, where the value of n is limited by the number of available VMs m and $k = \frac{N}{n}$ similar epochs are needed to complete the execution of all tasks. For each type of VM instance $I = [very\ small,\ small,\ medium,\ large,\ extra\ large,\ super]$, the associated cost of usage and the computing power are different. Let Pf_j represent the processing power of jth VM instance type where j ranges from 1 to $|I|$ and C_j represents its cost for unit time. The task length of each task $TASK_i$ is precomputed and represented as T_i, the time needed to execute each task in 'very small' type VM. The optimization objectives for N tasks are :

$$Minimize \quad Makespanfn = \sum_{p=1}^{k}\sum_{i=1}^{n} T_i * Pf_j * x_{ij} \quad for \quad some \quad j\epsilon I \quad (1)$$

$$Minimize \quad Costfn = \sum_{p=1}^{k}\sum_{i=1}^{n} C_j * T_i * Pf_j * x_{ij} \quad for \quad some \quad j\epsilon I \quad (2)$$

where x_{ij} is a decision variable, denoting $TASK_i$ is scheduled on VM_j and n tasks are there in an epoch. subject to the following constraints:

$$n \leq m \quad (3)$$

$$x_{ij} = \begin{cases} 1 & if\ TASK_i \quad scheduled to \quad VM_j \\ 0 & otherwise \end{cases} \quad (4)$$

and

$$\sum_{i=1}^{n} x_{ij} = m \quad (5)$$

3.1   Pareto Optimality

A multi objective optimization problem consists of optimizing a vector of n_{obj} where the objective function $F(x) = (f_1(x), f_2(x), \ldots, f_{n_{obj}}(x))$. The problem here is to find an optimal task schedule considering both the objectives. i.e.,cost and makespan. We have used weighted sum approach[9], [12] for solving the bi-objective optimization problem, which in essence convert a multi-objective optimization problem to a single objective one with weights representing preferences among objectives by the decision maker. This approach is easy to solve and produce a single solution to the problem. Hence our bi-objective optimization problem can now be represented using the formula:

$$Minimize \quad \theta * Costfn + (1 - \theta) * Makespanfn \quad (6)$$

where θ represents the relative weight or preference of one objective over the other, in the range $[0, 1]$. When $\theta = 0$, the optimization problem becomes that of minimizing $Makespanfn$ and when $\theta = 1$, the problem becomes minimizing $Costfn$.

4   Particle Swarm Optimization Technique

Finding an optimal schedule meeting the constraints of a bi-objective optimization problem are well-known problems in N P hard category, hence one of the heuristic techniques, Particle Swarm Optimization[6] is applied to obtain a feasible solution in reasonable time. Initially, the PSO algorithm generates a set of N solutions called *particles*, randomly in the D dimensional search space. Each *particle* is represented by a D-dimensional vector X_i where i ranges from 1 to d which stands for its location $(x_{i1}, x_{i2}, ..., x_{id})$ in space. Velocity of each particle v is constrained by v_{min} and v_{max} and its position x is updated according to the following equations:

$$v_{id}^{n+1} = w * v_{id}^n + c_1 * rand_1 * (pbest_{id}^n - x_{id}^n) + c_2 * rand_2 * (gbest_d^n - x_{id}^n) \quad (7)$$

$$x_{id}^{n+1} = x_{id}^n + v_{id}^{n+1} \quad (8)$$

where $i = 1, 2, ..., N$; $n = 1, 2, ..., iter_{max}$, the maximum iteration number, w, the inertia weight; c_1 and c_2 are two positive constants called acceleration coefficients and $rand_1$ and $rand_2$ are two uniformly distributed random numbers in the interval $[0, 1]$. Each particle maintains its position and its velocity. It also remembers the best fitness value it has achieved thus far during the search (*individual best fitness*) and the candidate solution that achieved this fitness (*individual best position (pbest)*). Also, the PSO algorithm maintains the best fitness value achieved among all particles in the swarm (*global best fitness*) and the candidate solution that achieved this fitness (*global best position (gbest)*). Equations (7) and (8) enable the particles to search around its individual best position *pbest* and update global best position *gbest*. This technique was initially proposed for solving problems in the continuous domain through the velocity updating rule. Since our problem work in the discrete domain, it has to be modified to suit the discrete domain. The Smallest Position Value rule based PSO (PSO-SPV) algorithm [8] is widely used for the same. This technique performs poor when there exist high variance in the length of the tasks submitted by end-users and when high variance exists in computational speed of resources[10]. We

---

Algorithm 1. Abstract of Particle Swarm Optimization Algorithm

---
1: $P \leftarrow$ Initial Population
2: Evaluate (P)
3: Initialize *pbest* and *gbest*
4: **while**   termination criterion not met   **do**
5:      Update Velocity(P) as indicated in equation (7)
6:      Update Position (P) as indicated in equation (8)
7:      Evaluate (P)
8:      Find *pbest* and *gbest*
9: **end while**
10: Output *gbest*

---

too observed that the same technique is not able to converge to near optimal solution with bi-objective optimization of task scheduling in cloud computing. Hence a new method for generating discrete permutations is proposed, namely integer-PSO. Here permutation encoding technique is used where every VM is assigned a number from 1 to n and a solution sequence $(5, 2, 1, 3, 4)$ means assign Task 1 to VM 5, Task 2 to VM 2 and so on. Initial populations are randomly generated. Each solution is evaluated to find its fitness based on equation (6) on different values of θ.

4.1   Integer – PSO

An update in the position of the particle based on equations (7) and (8) should result in new task assignment for a scheduling problem, but they produce floating point values in the continuous domain. Many discrete versions of PSO rounds-off the floating point position values and stores the discrete integer value for the particle's position. To preserve the stochastic nature of the continuous PSO, we have modified equation (8) in our algorithm, as shown below:

$$Y_{id}^n = ceil((x_{id}^n + v_{id}^n) * \beta) \quad where \quad \beta = 10^y \tag{9}$$

$$Pos_{id}^{n+1} = (Y_{id}^n) \quad mod \quad m \tag{10}$$

$$xc_{id}^{n+1} = \begin{cases} Pos_{id}^{n+1} & \text{if } Pos_{id}^{n+1} > 0 \\ m & \text{otherwise} \end{cases} \tag{11}$$

Table 1. Integer-PSO Example

Task J	1	2	3	4	5
x_{ij}^k	4	5	1	3	2
v_{ij}^{k+1}	-0.6015	-0.2413	0.0327	-0.0352	-0.8544
Y_{ij}^{k+1}	33985	47587	10327	29648	11456
Pos_{ij}^{k+1}	0	2	2	3	1
xc_{ij}^{k+1}	5	2	2	3	1
x_{ij}^{k+1}	5	4	2	3	1

New variables Y_{id}^n and Pos_{id}^{n+1} are introduced to store the continuous value as an integer of required accuracy and the temporary task assignment respectively. This method may create more than one task assignment to some of the VMs and may not assign any task to some other VMs. which need to be handled separately. The procedure is as shown in Table 1, assuming $m = 5$ and $k=4$.

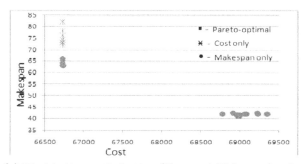

(a) Bi-objective optimization (Cost and Makespan) using Integer - PSO

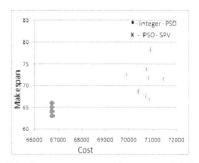

(b) Comparison on Integer - PSO and PSO-SPV algorithm

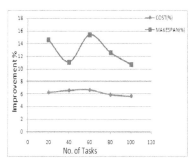

(c) Improvement with respect to number of tasks

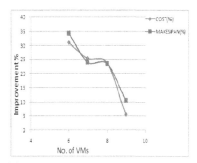

(d) Improvement with respect to number of VMs

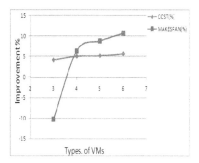

(e) Improvement with respect to type of VMs

Fig. 1. Performance Analysis

5   Results and Discussion

The algorithm has been simulated 10 times on the same data set with $\theta = 0.9$, $\theta = 0$ and $\theta = 1$ and the average value is taken to find an estimate on best convergences, on a laptop PC with PIV processor of 3 GHz clock frequency and 8GB RAM. We have tested the algorithm on a task set of 99 tasks, setting the population size ($|P|$) as 5 and other PSO parameters as $w = 0.6$, $c1 = c2 = 0.2$, $v_{max} = 4$ and $v_{min} = -4$. A suitable value of θ is found through trial and error as 0.9. Assumptions regarding the capacity of each VM in terms of speed and cost of usage per unit time and the task length are given as vectors. The algorithm runs for 500 generations per task set and is repeated 11 times. The results are shown in Figure 1(a).

For $\theta = 0.9$, the proposed algorithm (Integer-PSO) finds optimal cost in 90 percentage of time, but SPV-PSO has not converged to optimal or near optimal cost value on any value of θ, shown in Figure 1(b). When the algorithm is applied to single objective optimization scenario, for 90 percentage of time, the Integer-PSO algorithm converges to optimal value. A detailed performance analysis of both the approaches were done with different number of tasks, different number of VMs and different types of VMs and are shown in Figure 1(c) to 1(e).

6   Conclusion

Scheduling tasks in the cloud is a challenging one as the same involves many factors such as cost and profit considerations, execution time, SLAs, Quality of service parameters requested by the end user and committed by the CSP and power considerations. Also the task arrival rate is highly unpredictable and dynamic in nature. We have modelled the problem as a constraint bi-objective optimization problem, where the objectives are makespan and cost and have used Particle Swarm Optimization algorithm to solve the same, where the pareto optimality is achieved through weighted sum approach. A variant of PSO technique is proposed (Integer-PSO) whose results are promising.

References

1. Szabo, C., Kroeger, T.: Evolving multi-objective strategies for task allocation of scientific workflows on public clouds. In: Proc. of IEEE Congress on Evolutionary Computation (CEC), pp. 1–8 (2012)
2. Farahabady, R.H., Lee, Y.C., Zomaya, A.Y.: Pareto optimal cloud bursting. Accepted for publication in IEEE Transactions on Parallel and Distributed Systems (2013)
3. Feng, M., Wang, X., Zhang, Y., Li, J.: Multi-objective particle swarm optimization for reseource allocation in cloud computing. In: Proc. of 2nd International Conference on Cloud Computing and Intelligent Systems (CCIS), vol. 3, pp. 1161–1165 (2012)
4. Guo, L., Shao, G., Zhao, S.: Multi-objective task assignment in cloud computing by particle swarm optimization. In: Proc. of 8th International Conference on Wireless Communications, Networking and Mobile Computing (WiCOM), pp. 1–4 (2012)

5. Jin, J., Luo, J., Song, A., Dong, F., Xiong, R.: Bar: An efficient data locality driven task scheduling algorithm for cloud computing. In: Proc. of 11th IEEE/ACM International Symposium on Cluster, Cloud and Grid Computing, pp. 295–304 (2011)
6. Kennedy, J., Eberhart, R.C.: Particle swarm optimization. In: Proc. of IEEE International Conference on Neural Networks, pp. 1942–1948 (1995)
7. Rodrignez Sossa, M., Buyya, R.: Deadline based resource provisioning and scheduling algorithm for scientific workflows on clouds. IEEE Transactions on Cloud Computing (2014)
8. Liang, Y.-C., Tasgetiren, M.F., Sevkli, M., Gencylmaz, G.: Particle swarm optimization algorithm for single machine total weighted tardiness problem. In: IEEE Congress on Evolutionary Computation (CEC2004), vol. 2, pp. 1412–1419 (2004)
9. Marler, T., Arora, J.S.: The weighted sum method for multi-objective optimization: new insights. Springer (2009)
10. Sidhu, M.S., Thulasiraman, P., Thulasiram, R.K.: A load-rebalnce pso heuristic for task matching in heterogeneous computing systems. In: IEEE Symposium on Swarm Intelligence (SIS), pp. 180–187 (2013)
11. Sadhasivam, G.S., Selvarani, S.: Improved cost-based algorithm for task scheduling in cloud computing. In: Proc. of IEEE International Conference on Computational Intelligence and Computing Research (ICCIC), pp. 1–5 (2010)
12. Ivan, P.: Stanimirovic, Milan Lj. Zlatanovic, and Marko D Petkovic: On the linear weighted sum method for multi-objective optimization. FACTA UNIVERSITATIS (NIS), Ser. Math. Inform, 49–63 (2011)
13. Wu, Z., Ni, Z., Liu, X.: A revised discrete particle swarm optimization for cloud workflow scheduling. In: Proc. of International Conference on Computational Intelligence and Security (CIS), pp. 184–188 (2010)
14. Zhang, L., Chen, Y., Sun, R., Jing, S., Yang, B.: A task scheduling algorithm based on pso for grid computing. International Journal of Computational Intelligence Research 4(1), 37–43 (2008)
15. Zhu, L., Wu, J.: Hybrid particle swarm optimization algorithm for flexible task scheduling. In: Proc. of 3rd Internatinal Conference on Genetic and Evolutionary Computing (WGEC 2009), pp. 603–606 (2009)

Development on Harmony Search Hyper-heuristic Framework for Examination Timetabling Problem

Khairul Anwar[1], Ahamad Tajudin Khader[1],
Mohammed Azmi Al-Betar[1, 2], and Mohammed A. Awadallah[1]

[1] School of Computer Sciences, Universiti Sains Malaysia (USM),
11800 Pulau Pinang, Malaysia
[2] Department of Information Technology, Al-Huson University College,
Al Balqa Applied University, P.O. Box 50, Al-Huson, Irbid-Jordan
{ka10_com097,mama10_com018}@student.usm.my,
{tajudin,mohbetar}@cs.usm.my

Abstract. In this paper, a Harmony Search-based Hyper-heuristic (HSHH) approach is proposed for tackling examination timetabling problems. In this approach, the harmony search algorithm will operate as a high level of abstraction which intelligently evolves a sequence of low level heuristics. This sequence is a combination of improvement heuristics which consist of neighborhood structure strategies. The proposed approach is tested using the examination timetabling tracks in Second International Timetabling Competition (ITC-2007) benchmarks. Experimentally, the HSHH approach can achieve comparable results with the comparative methods in the literature.

Keywords: Examination Timetabling, Harmony Search, Hyper-heuristic.

1 Introduction

Examination timetabling is the process of scheduling a set of exams to a set of time-slots and rooms, subject to *hard* and *soft* constraints. It becomes extremely difficult when it involves a large number of events (could be hundreds or thousands) to be scheduled in limited resources in accordance with a wide variety of constraints, which need to be satisfied [1]. The hard constraints are mandatory to fulfill while the soft constraints are desired but not absolutely necessary. The solution that satisfied all the hard constraints is called *feasible*. Conventionally, the soft constraints play a major role in measuring the *quality* of the solution. The main target is to find an examination timetabling solution that satisfies all hard constraints and minimize the violation of soft constraints as much as possible.

Several optimization techniques have been introduced to solve examination time-tabling problems such as sequential techniques, constraints based techniques, local search-based techniques, population-based techniques, and many others as surveyed in [1]. But recently, the attention of the researches in the operation research and the artificial intelligence fields has shifted to be concerned with a more general approach

Y. Tan et al. (Eds.): ICSI 2014, Part II, LNCS 8795, pp. 87–95, 2014.

as hyper-heuristic. Hyper-heuristics (HH) have a higher level heuristic to choose from a set of heuristics that are applicable for the problem in hand. The main difference between the hyper-heuristics and meta-heuristics is that hyper-heuristics explore the heuristic search space while meta-heuristics explore the solution search space [2].

There are several hyper-heuristics to solve the examination timetabling problems. Sabar et al.[3] investigated a new graph coloring constructive hyper-heuristic (GCCHH) for solving examination timetabling problems using ITC-2007 dataset. The results demonstrate that GCCHH produces good results and outperforms other approaches on some of the benchmark instances. In another study, Pillay and Banzhaf [4] used genetic programming (GP) for the evolution on hyper-heuristics framework to solve uncapacitated examination timetabling problems. From the research, it shows that the genetic programming system was comparable to the other search algorithm, and in some cases it can produce better quality timetables.

In our previous work [5], Harmony search algorithm (HSA) has been employed in a hyper-heuristic framework as a high-level heuristic where some *move* heuristics (i.e. move and swap) have been employed as low-level heuristics. HSA is a population-based algorithm developed by Geem et al.[6]. HSA is a stochastic search mechanism, simple in concept, and no derivation information is required in the initial search [6]. It has been successfully applied to a wide range of optimization and scheduling problems such as course timetabling problem [7], nurse scheduling problem [8, 9], as well as examination timetabling problem [10] and many others as reported in [11].

Evidently, the initial employment produced an impressive result for the initial investigation, with a chance to improve the results by modifying the proposed method. In this research, the new low-level heuristics strategies and the pitch adjustment procedure are modified to enhance the proposed method. For purposes of evaluation, the dataset of examination timetabling tracks established in the Second International Timetabling Competition (ITC-2007) is used. The results produced by new HSHH outperformed those produced by previous versions of HSHH and comparable to the results of the ITC-2007 comparative methods.

This paper is organized as follows: Section 2 discusses the examination timetabling problem (ETP) and benchmark dataset of ITC-2007 for examination track. The description of harmony search hyper-heuristic (HSHH) algorithm is presented in Section 3. Section 4 discusses the experimental setup and the computational results. Finally, the conclusion and future works are provided in Section 5.

2   Examination Timetabling

In this research, we used ITC-2007 [12] examination version as a benchmark to evaluate the proposed method. ITC-2007 provides a capacitated examination timetabling dataset which contains eight instances and four hidden ones. For this research we experimented with the eight instances. This dataset consists of five hard constraints (i.e., H1-H5) and seven soft constraints (i.e., S1-S7) as shown in Table 1.

Table 1. ITC-2007 Hard and Soft constraints

Key	Constraints
H1	No student sits for more than one examination at the same time.
H2	The capacity of individual rooms is not exceeded at any time throughout the examination session.
H3	Period Lengths are not violated.
H4	Satisfaction of period related hard constraints (e.g., exam B must be scheduled after exam A).
H5	Satisfaction of room related hard constraints (e.g., exam A exclusively scheduled in room X).
S1	Two exams in a row.
S2	Two exams in a day.
S3	Specified spread of examinations.
S4	Mixed duration of examinations within individual periods.
S5	Larger examinations appearing later in the timetable.
S6	Period related soft constraints – some period has an associated penalty.
S7	Room related soft constraints – some room has an associated penalty.

The objective function to summarize the ITC-2007 dataset is formalized in equation (1) and the details' explanations is provided in [12].

$$min \quad f(x) = \sum_{s \in S}(w^{2R}C_s^{2R} + w^{2D}C_s^{2D} + w^{PS}C_s^{PS}) + w^{NMD}C^{NMD} + w^{FL}C^{FL} + C^P + C^R$$

(1)

where x is a complete timetabling solution; S refers to a set of students while w refers to the institutional penalty for each constraint except for period and room related soft constraint (i.e., C^P and C^R). Table 2 shows the detail notation of the variables used in equation (1).

Table 2. List of abbreviations given to the ITC-2007 soft constraint [12]

Math Symbol	Description
C_s^{2R}	"two exam in row" penalty for student s.
C_s^{2D}	"two exam in day" penalty for student s.
C_s^{PS}	"period spread" penalty for student s
C^{NMD}	"No mixed duration" penalty
C^{FL}	"Front load" penalty
C^P	"Period" penalty
C^R	"Room" penalty

3   Harmony Search Hyper-heuristic for ITC-2007

In previous work of Harmony Search Hyper-Heuristic (HSHH) [5], the pitch adjustment operator is deactivated in the improvisation step, and two neighborhood structures are utilized as low-level heuristics. In this study, the pitch adjustment operator is added during the improvisation step, and seven different neighborhood structures have been utilized as low level heuristics. They can be summarized as follows:

- **h1**: Move Exam. Select one exam at random and move to a new randomly selected feasible timeslot.
- **h2**: Swap Timeslot. Select two exams at random and swap their timeslots.
- **h3**: Swap Exam. Select two timeslots randomly and exchange all exams between them.
- **h4**: Swap Period. Select two periods and swap the exams between the periods.
- **h5**: Select two timeslots (e.g. $t1$ and $t2$) randomly and move some exams from the timeslot $t1$ to $t2$ and vice versa.
- **h6**: This heuristic similar to $h3$ but it only swaps the conflicting exams in two distinct timeslots. This heuristic is similar to kempe chain method in (Al-Betar et al.,[10]).
- **h7**: do nothing.

Basically, HSHH has five main steps as follows:

Step 1: Initialization. The HSHH begins by setting the harmony search parameters: harmony memory size (HMS), harmony memory consideration rate (HMCR), number of iterations (NI) and Harmony Memory Length (HML) which represents the length of heuristic vector. Furthermore, the Pitch Adjustment Rate (PAR) parameter also will be set. Initially, the largest degree (LD) heuristic is used to construct the initial feasible solution ($x^{feasible}$). If the solution is not feasible, then the repair procedure as used in [7] will be triggered to maintain the feasibility of the solution.

Step 2: Initializing Harmony Memory. HSHH consists of two complemented search spaces (*heuristic search space and solution search space*), each represented in a harmony memory: Heuristics Harmony Memory (HHM) and Solutions Harmony Memory (SHM). HHM contains sets of heuristic vectors determined by HMS where every vector is a heuristics sequence. The length of the vector is determined by HML. Similarly, SHM contains sets of solution vectors and the length is determined by the number of exam, N.

In initializing HHM and SHM, the HSHH, firstly, generates the new heuristics vector (h') randomly and apply this vector to the initial feasible solution ($x^{feasible}$) to produce the new solution (x'). The new solution will be evaluated using the objective function as in equation (1). If the new solution (x') is better than the initial solution ($x^{feasible}$), then the new solution (x') will be saved in the SHM and the new heuristic vector (h') will be saved in the HHM. This process will be repeated until HHM and SHM are filled (see equations (2) and (3)). After completing the process, HSHH will retain the worst solution (x^{worst}) and the best solution (x^{best}) in SHM.

Step 3: Improvise a new heuristic sequence. In this step, a new heuristics vector $h' =$ ($h'_1, h'_2, \dots h'_{HML}$) is generated from scratch, based on three HSA operators: *memory consideration, random consideration* and *pitch adjustment*.

$$HHM = \begin{bmatrix} h_1^1 & h_2^1 & \cdots & h_{HML}^1 \\ h_1^2 & h_2^2 & \cdots & h_{HML}^2 \\ \vdots & \vdots & \ddots & \vdots \\ h_1^{HMS} & h_2^{HMS} & \cdots & h_{HML}^{HMS} \end{bmatrix} \quad (2) \qquad SHM = \begin{bmatrix} x_1^1 & x_2^1 & \cdots & x_N^1 \\ x_1^2 & x_2^2 & \cdots & x_N^2 \\ \vdots & \vdots & \ddots & \vdots \\ x_1^{HMS} & x_2^{HMS} & \cdots & x_N^{HMS} \end{bmatrix}$$

(3)

Note that N refers to the number of examinations. In *memory consideration operator*, the new heuristic index of h'_i in the new heuristic vector (h') is randomly selected from the historical indexes (e.g. $h_i^1, h_i^2, \dots . h_i^{HMS}$), stored in the heuristic harmony memory with probability of HMCR, where HMCR $\in [0, 1]$.

For *Random consideration operator*, the new heuristic index is randomly assigned from the set of heuristics $h'_i \in \{h1, h2, h3, h4, h5, h6, h7\}$ with probability of (1-HMCR) as in equation (4).

$$h'_i \in \begin{cases} \{h_i^1, h_i^2, \dots . h_i^{HMS}\} & w.p. \quad HMCR \\ \{h1, h2, h3, h4, h5, h6, h7\} & w.p. \quad (1 - HMCR) \end{cases} \qquad (4)$$

In *Pitch Adjustment operator*, a simple adjustment is used. The new index of h'_i will be added/subtracted by 1 with a probability of PAR where $0 \leq PAR \leq 1$ as in equation (5).

$$Pitch\ adjust\ for\ h'_i = \begin{cases} Yes & w.p. \quad PAR \\ No & w.p. \quad (1 - PAR) \end{cases} \qquad (5)$$

In case the decision of PAR is *yes*, the index of h'_i will be recalculated as follows:

$$h'_i = h'_i \pm 1 \qquad (6)$$

Note that if the index is out of range, it will remain the same. Then the new harmony of heuristic vector h' will be applied to a solution (e.g. $x^{rand} = x_1^{rand}, x_2^{rand}, \dots . x_N^{rand}$) where x^{rand} is randomly selected from the solution search space or SHM. The HSHH used random selection to select the solution from the SHM to avoid the local optima. In this process, the heuristic in h' will be executed sequentially to the selected solution (x^{rand}). The process will continue until all the heuristics in h' have been executed, and a new solution (x') will be produced. Pseudo-code for improvisation step is shown in the Algorithm 1.

Step 4: Update HHM and SHM. In hyper-heuristic environment, this step is called a move acceptance step. HSHH will decide either to accept or neglect the new heuristic vector h'. In this process, the new solution (x') will be evaluated using the objective function. The new solution must be complete and feasible. If the new solution is better than the worst solution in solution harmony memory (SHM), the new *h'* and *x'* will be saved in the memory (*h'* in HHM and *x'* in SHM) and the worst heuristic vector and solution will be excluded from the memory (i.e., HHM and SHM).

Step 5: Check the stop criterion. Step 3 and step 4 in this approach are repeated until the stop criterion (i.e., NI) is met.

Algorithm 1: *Pseudo-code for selecting and generating heuristic vector during the improvisation process in step 3.*

$h'= 0$; //heuristic vector
for $l = 0,…,$HML do
   if $(U(0,1) \leq HMCR)$ then
      $h'_i \in \{h^1_i, h^2_i, ….. h^{HMS}_i\}$; //Memory consideration;
      if $(U(0,1) \leq PAR)$ then
         $h'_i = h'_i \pm 1$; //Pitch adjustment;
   else
      $h'_i \in \{h1, h2, h3, h4, h5, h6, h7\}$; //Random consideration;
   end if
end for
$x^{rand} \in (x^1, x^2, …, x^{HMS})$; //Select random solution from SHM;
x' = apply *h'* to x^{rand} ;

---

4   Experiments and Results

In this section, Harmony Search Hyper-heuristic is evaluated using the real world problem dataset (ITC-2007) for university examination timetabling problem. The proposed method is coded in Microsoft Visual C++ 6 under Windows 7 on Intel processor with 2G RAM. We chose to test the proposed method with each problem instances in ITC-2007.

The characteristics of the ITC-2007 dataset are provided in Table 3. This table includes information such as number of students (*Info1*), actual number of students (*Info2*), number of exams (*Info3*), number of timeslots (*info4*), and number of rooms (*Info5*). We ran each experiment 10 times for each problem due to the stochastic nature of the method [13]. The Harmony Search Hyper-Heuristic (HSHH) parameters are set as HMS=10, HML=10, PAR=0.1 HMCR=0.95, and N1=100000, where these parameter settings are used based on some experiments carried out previously.

Experimentally, the HSHH is able to find a feasible solution for seven out of eight instances in ITC-2007 dataset. Table 4 provides the comparative results of the HSHH and the other comparative methods that are working using the same dataset. The different comparative methods are provided as shown in Table 5. The numbers in table 4 referred to the penalty value of the soft constraint violations. The best results are highlighted in bold. The indicator 'x% inf' indicates that the percentage of such algorithm could not find a feasible solution.

Table 3. The Characteristics of ITC-2007 Examination Timetabling Dataset

Dataset	Info1	Info2	Info3	Info4	Info5
Exam1	7891	7833	607	54	7
Exam2	12743	12484	870	40	49
Exam3	16439	16365	934	36	48
Exam4	5045	4421	273	21	1
Exam5	9253	8719	1018	42	3
Exam6	7909	7909	242	16	8
Exam7	14676	13795	1096	80	15
Exam8	7718	7718	598	80	8

Table 4. Comparison with previous HSHH and other methods

Dataset	HSHH 2	HSHH 1	M1	M2	M3	M4	M5	M6	M7	M8	M9
Exam1	9885	11823	8559	6235	6234	4775	4370	4633	6582	**4368**	12035
Exam2	393	976	830	2974	395	**385**	**385**	405	1517	390	3074
Exam3	19931	26770	11576	15832	13002	**8996**	9378	9064	11912	9830	15917
Exam4	100% inf	100% inf	21901	35106	17940	16204	**15368**	15663	19657	17251	23582
Exam5	4065	6772	3969	4873	3900	**2929**	2988	3042	17659	3022	6860
Exam6	29935	30980	28340	31756	27000	**25740**	26365	25880	26905	25995	32250
Exam7	8801	11762	8167	11562	6214	4087	4138	**4037**	6840	4067	17666
Exam8	12145	16286	12658	20994	8552	7777	7516	**7461**	11464	7519	16184

Table 5. The comparison methods for ITC-2007 Examination Timetabling Dataset

Key	Method
HSHH 1	Harmony Search hyper-heuristic [5].
M1	Evolutionary Algorithm hyper-heuristic [14].
M2	Hybrid Approach hyper-heuristic[15].
M3	Graph Coloring Constructive hyper-heuristic[3].
M4	An improved multi-staged algorithmic[16].
M5	A Three phase constraint-based approach[17]
M6	An extended great deluge algorithm [18].
M7	Artificial Bee Colony algorithm [19] .
M8	Hybrid approach within great deluge algorithm[20].
M9	Developmental Approach [21].

As shown in Table 4, the performance of the new HSHH (i.e. HSHH 2) is much better than the performance of the previous version of HSHH (i.e. HSHH 1). Figure 1 shows the comparison between the HSHH 1 and HSHH 2 in terms of convergence behavior. Experimental results show that HSHH is able to produce good results and one of these datasets (i.e. Exam2) has achieved comparable result as shown in Table 4. Furthermore, the proposed method has also been able to obtain better results compared to the several other approaches. As compared with hybrid approach hyper-heuristic (M2), HSHH are able to produce better results in five problem instances (i.e., *Exam2, Exam5, Exam6, Exam7 and Exam8*) and six problem instances (i.e., *Exam1, Exam2, Exam5, Exam6, Exam7 and Exam8*) compared to the developmental approach (M9).

Fig. 1. Comparison of Convergence behavior between HSHH I and HSHH II

5   Conclusion and Future Work

This paper presented Harmony search-based hyper-heuristics (HSHH) for solving examination timetabling problems using the ITC-2007 dataset. The harmony search is utilized at the high-level to evolve a sequence of improvement low level heuristics. In order to evaluate HSHH, problem instances of ITC-2007 dataset have been used. The experimental result shows that HSHH is able to solve examination timetabling problems. Although, the results produced by HSHH in this study have not reached the best known results, they seem comparable or even better in some cases when compared to the previous approaches using ITC-2007. Utilizing several low-level heuristics in the HSHH framework is a very promising extension to the hyper-heuristic domain in general. This is because each low-level heuristic can deal with a region of the search and touching several regions in the search space might increase the chance of improvements.

The main objective behind proposing the HSHH is to suggest an applicable framework that is general enough to be re-implemented for other types of scheduling or combinatorial optimization problems. Therefore, we plan to apply the proposed approach to solve the nurse rostering problem using INRC2010 dataset. For future research, we plan to adapt learning mechanism within the HSHH algorithm in order to improve the heuristic selection and to enhance the speed of convergence.

References

1. Qu, R., et al.: A Survey of Search Methodologies and Automated System Development for Examination Timetabling. Journal of Scheduling 12(1), 55–89 (2009)
2. Burke, E.K., et al.: A survey of Hyper-heuristics. Computer Science Technical Report No. NOTTCS-TR-SUB-0906241418-2747, School of Computer Science and Information Technology, University of Nottingham (2009)
3. Sabar, N.R., et al.: A Graph Coloring Constructive Hyper-heuristic for Examination Timetabling Problems. Applied Intelligence, 1–11 (2011)
4. Pillay, N., Banzhaf, W.: A Genetic Programming Approach to the Generation of Hyper-Heuristics for the Uncapacitated Examination Timetabling Problem. In: Neves, J., Santos, M.F., Machado, J.M. (eds.) EPIA 2007. LNCS (LNAI), vol. 4874, pp. 223–234. Springer, Heidelberg (2007)

5. Anwar, K., et al.: Harmony Search-based Hyper-heuristic for Examination Timetabling. In: 9th International Colloquium on Signal Processing and its Applications (CSPA). IEEE (2013)
6. Geem, Z.W., Kim, J.H., Loganathan, G.: A New Heuristic Optimization Algorithm: Harmony Search. Simulation 76(2), 60–68 (2001)
7. Al-Betar, M.A., Khader, A.T.: A Harmony Search Algorithm for University Course Timetabling. Annals of Operations Research, 1–29 (2012)
8. Awadallah, M.A., et al.: Nurse Scheduling Using Harmony Search. In: Sixth International Conference on Bio-Inspired Computing: Theories and Applications (BIC-TA). IEEE (2011)
9. Awadallah, M.A., et al.: Harmony Search with Novel Selection Methods in Memory Consideration for Nurse Rostering Problem. Asia-Pacific Journal of Operational Research (2013)
10. Al-Betar, M.A., Khader, A.T., Nadi, F.: Selection Mechanisms in Memory Consideration for Examination Timetabling with Harmony Search. In: Proceedings of the 12th Annual Conference on Genetic and Evolutionary Computation (GECCO 2010). ACM, New York (2010)
11. Manjarres, D., et al.: A Survey on Applications of the Harmony Search Algorithm. Engineering Applications of Artificial Intelligence 26(8), 1818–1831 (2013)
12. McCollum, B., et al.: The Second International Timetabling Competition: Examination Timetabling Track. Technical Report QUB/IEEE/Tech/ITC2007/Exam/v4. 0/17, Queens University, Belfast, UK (2007)
13. Al-Betar, M.A., Khader, A.T., Doush, I.A.: Memetic Techniques for Examination Timetabling. Annals of Operations Research, 1–28 (2013)
14. Pillay, N.: Evolving Hyper-heuristics for a Highly Constrained Examination Timetabling Problem. In: Proceedings of the 8th International Conference on the Practice and Theory of Automated Timetabling, PATAT 2010 (2010)
15. Burke, E.K., Qu, R., Soghier, A.: Adaptive Selection of Heuristics for Improving Constructed Exam Timetables. In: Proceedings of the 8th International Conference on the Practice and Theory of Automated Timetabling, PATAT 2010 (2010)
16. Gogos, C., Alefragis, P., Housos, E.: An Improved Multi-staged Algorithmic Process for the Solution of the Examination Timetabling Problem. Annals of Operations Research 194(1), 203–221 (2012)
17. Müller, T.: ITC2007 Solver Description: a Hybrid Approach. Annals of Operations Research 72(1), 429–446 (2009)
18. McCollum, B., et al.: An Extended Great Deluge Approach to the Examination Timetabling Problem. In: Proceedings of the 4th Multidisciplinary International Scheduling: Theory and Applications (MISTA 2009), pp. 424–434 (2009)
19. Alzaqebah, M., Abdullah, S.: Artificial Bee Colony Search Algorithm for Examination Timetabling Problems. International Journal of the Physical Sciences 6(17), 4264–4272 (2011)
20. Turabieh, H., Abdullah, S.: An Integrated Hybrid Approach to the Examination Timetabling Problem. Omega 39(6), 598–607 (2011)
21. Pillay, N.: A Developement Approach to the Examination Timetabling (2007), http://www.cs.qub.ac.uk/itc2007

Predator-Prey Pigeon-Inspired Optimization for UAV Three-Dimensional Path Planning

Bo Zhang[1] and Haibin Duan[1,2,*]

[1] Science and Technology on Aircraft Control Laboratory, School of Automation Science and Electrical Engineering, Beihang University, Beijing, 100191, P. R. China
[2] Provincial Key Laboratory for Information Processing Technology, Soochow University, Suzhou, 215006, P. R. China
zhangbo0216@163.com, hbduan@buaa.edu.cn

Abstract. Pigeon-inspired optimization (PIO) is a new bio-inspired optimization algorithm. This algorithm searches for global optimum through two models: map and compass operator model is presented based on magnetic field and sun, while landmark operator model is designed based on landmarks. In this paper, a novel Predator-prey pigeon-inspired optimization (PPPIO) is proposed to solve the three-dimensional path planning problem of unmanned aerial vehicles (UAVs), which is a key aspect of UAV autonomy. To enhance the global convergence of the PIO algorithm, the concept of predator-prey is adopted to improve global best properties and enhance the convergence speed. The comparative simulation results show that our proposed PPPIO algorithm is more efficient than the basic PIO and particle swarm optimization (PSO) in solving UAV three-dimensional path planning problems.

Keywords: pigeon-inspired optimization (PIO), unmanned aerial vehicle (UAV), path planning, predator-prey.

1 Introduction

Three-dimensional path planner is an essential element of the unmanned aerial vehicle (UAV) autonomous control module [1]. It allows the UAV to compute the best path from a start point to an end point autonomously [2, 3]. Whereas commercial airlines fly constant prescribed trajectories, UAVs in operational areas have to travel constantly changing trajectories that depend on the particular terrain and conditions prevailing at the time of their flight.

Pigeon-inspired optimization (PIO), which is a new swarm intelligence optimizer based on the movement of pigeons, was firstly invented by Duan in 2014 [4]. Homing pigeons can easily find their homes by using three homing tools: magnetic field, sun and landmarks. In the optimization, map and compass model is presented based on magnetic field and sun, while landmark operator model is presented based on landmarks.

In this paper, we propose a predator-prey pigeon-inspired optimization (PPPIO) method, integrating the concept of predator-prey into PIO in order to improve its

---

* Corresponding author.

capability of finding satisfactory solutions and increasing the diversity of the population. We also solve the UAV three-dimensional path planning problem by PPPIO. Simulation results and comparisons verified the feasibility and effectiveness of our proposed algorithm.

The rest of the paper is organized as follows: Section 2 provides the representation and the cost function we developed to evaluate the quality of candidate trajectories. Section 3 describes the principle of basic PIO algorithm. Section 4 shows the implementation procedure of our proposed predator-prey PIO algorithm. Finally, we compare the quality of the trajectories produced by the PIO, particle swarm optimization (PSO) and the PPPIO in Section 5.

2 Problem Formulation

The first step of three-dimensional path planning is to discretize the world space into a representation that will be meaningful to the path planning algorithm. In this work, we use a formula to indicate the terrain environment. The mathematical function is of the form [5]:

$$z(x, y) = \sin(x/5+1) + \sin(y/5) + \cos(a \cdot \sqrt{x^2 + y^2}) + \sin(b \cdot \sqrt{x^2 + y^2}) \quad (1)$$

where z indicate the altitude of a certain point, and a, b are constants experimentally defined. Our representation of cylindrical danger zones (or no-fly zones) to be in a separate matrix where each row represents the coordinates (x_i, y_i) and the radius r_i of the ith cylinder as shown in Eq. (2). Complex no-fly zone can be built by partially juxtaposing multiple cylinders

$$\text{danger zones} = \begin{pmatrix} x_1 & y_1 & r_1 \\ x_2 & y_2 & r_2 \\ \dots & \dots & \dots \\ x_n & y_n & r_n \end{pmatrix} \quad (2)$$

The three-dimensional trajectories generated by the algorithm are composed of line segments and (x_i, y_i, z_i) represents the coordinates of the ith way point. The trajectories are flown at constant speed.

In the situation of UAV path planning, the optimal path is complex and includes many different characteristics. To take into account these desired characteristics, a cost function is used and the path planning algorithm becomes a for a path that will minimize the cost function. We define our cost function as follows [6]:

$$F_{cost} = C_{length} + C_{altitude} + C_{danger\ zones} + C_{power} + C_{collision} + C_{fuel} \quad (3)$$

In the cost function, the term associated with the length of a path is defined as follows:

$$C_{length} = 1 - (\frac{L_{p1p2}}{L_{traj}}) \quad (4)$$

$$C_{length} \in [0,1] \quad (5)$$

where L_{p1p2} is the length of the straight line connecting the starting point $P1$ and the end point $P2$ and L_{traj} is the actual length of the trajectory.

The term associated with the altitude of the path is defined as follows:

$$C_{altitude} = \frac{A_{traj} - Z_{min}}{Z_{max} - Z_{min}} \tag{6}$$

$$C_{altitude} \in [0,1] \tag{7}$$

where Z_{max} is the upper limit of the elevation in our search space, Z_{min} is the lower limit and A_{traj} is the average altitude of the actual trajectory. Z_{max} and Z_{min} are respectively set to be slightly above the highest and lowest point of the terrain.

The term associated with the violation of the danger zones is defined as follows:

$$C_{danger\ zones} = \frac{L_{inside\ d.z.}}{\sum_{i=1}^{n} d_i} \tag{8}$$

$$C_{danger\ zones} \in [0,1] \tag{9}$$

where n is the total number of danger zones, $L_{inside\ d.z.}$ is the total length of the subsections of the trajectory which go through danger zones and d_i is the diameter of the danger zone i.

The term associated with a required power higher than the available power of the UAV is defined as follows:

$$C_{power} = \begin{cases} 0, & L_{not\ feasible} = 0 \\ P + \left(\dfrac{L_{not\ feasible}}{L_{traj}} \right), & L_{not\ feasible} > 0 \end{cases} \tag{10}$$

$$C_{power} \in 0 \cup [P, P+1] \tag{11}$$

where $L_{not\ feasible}$ is the sum of the lengths of the line segments forming the trajectory which require more power than the available power of the UAV, L_{traj} is the total length of the trajectory and P is the penalty constant. This constant must be higher than the cost of the worst feasible trajectory which would have, based on our cost function, a cost of 3. By adding this penalty P, we separate nonfeasible solutions from the feasible ones.

The term associated with ground collisions is defined as follows:

$$C_{collision} = \begin{cases} 0, & L_{under\ terrain} = 0 \\ P + \left(\dfrac{L_{under\ terrain}}{L_{traj}} \right), & L_{under\ terrain} > 0 \end{cases} \tag{12}$$

$$C_{collision} \in 0 \cup [P, P+1] \qquad (13)$$

where $L_{\text{under terrain}}$ is the total length of the subsections of the trajectory which travels below the ground level and L_{traj} is the total length of the trajectory.

The term associated with an insufficient quantity of fuel available is defined as follows:

$$C_{\text{fuel}} = \begin{cases} 0, & \text{F}_{\text{traj}} \le F_{\text{init}} \\ P+1-\left(\dfrac{F_{\text{P1P2}}}{F_{\text{traj}}}\right), & \text{F}_{\text{traj}} > F_{\text{init}} \end{cases} \qquad (14)$$

$$C_{\text{fuel}} \in 0 \cup [P, P+1] \qquad (15)$$

where F_{P1P2} is the quantity of fuel required to fly the imaginary straight segment connection the starting point $P1$ to the end point $P2$, F_{traj} is the actual amount of fuel needed to fly the trajectory, F_{init} is the initial quantity of fuel on board the UAV.

The search engine will be adopted to find a solution, which can minimize the cost function during the optimization phase of our path planner algorithm. This can also be explained as to find a trajectory that best satisfies all the qualities represented by this cost function. Our cost function demonstrates a specific scenario where the optimal path minimizes the distance travelled, the average altitude (to increase the stealthiness of the UAV) and avoids danger zones, while respecting the UAV performance characteristics. This cost function is highly complex and demonstrates the power of our path planning algorithm. However, this cost function could easily be modified and applied to a different scenario.

3   Principle of Basic PIO

PIO is a novel swam intelligence optimizer for solving global optimization problems. It is based on natural pigeon behavior. Studies show that the species seem to have a system in which signals from magnetite particles are carried from the nose to the brain by the trigeminal nerve [4, 7]. Evidence that the sun is also involved in pigeon navigation has been interpreted, either partly or entirely, in terms of the pigeon's ability to distinguish differences in altitude between the Sun at the home base and at the point of release [8]. Recent researches on pigeons' behaviors also show that the pigeon can follow some landmarks, such as main roads, railways and rivers rather than head for their destination directly. The migration of pigeons is summarized as two mathematical models. One is map and compass operator, and the other is landmark operator.

3.1   Map and Compass Operator

In PIO model, virtual pigeons are used. In the map and compass operator, the rules are defined with the position X_i and the velocity V_i of pigeon i, and the positions and velocities in a D-dimension search space are updated in each iteration. The new position X_i and velocity V_i of pigeon i at the t-th iteration can be calculated with the follows [3]:

$$V_i(t) = V_i(t-1) \cdot e^{-Rt} + rand \cdot (X_g - X_i(t-1)) \tag{16}$$

$$X_i(t) = X_i(t-1) + V_i(t) \tag{17}$$

where R is the map and compass factor, $rand$ is a random number, and X_g is the current global best position, and which can be obtained by comparing all the positions among all the pigeons.

3.2   Landmark Operator

In the landmark operator, half of the number of pigeons is decreased by N_p in every generation. However, the pigeons are still far from the destination, and they are unfamiliar the landmarks. Let $X_c(t)$ be the center of some pigeons' position at the t-th iteration, and suppose every pigeon can fly straight to the destination. The position updating rule for pigeon i at t-th iteration can be given by:

$$N_P(t) = \frac{N_P(t-1)}{2} \tag{18}$$

$$X_c(t) = \frac{\sum X_i(t) \cdot fitness(X_i(t))}{N_P \sum fitness(X_i(t))} \tag{19}$$

$$X_i(t) = X_i(t-1) + rand \cdot (X_c(t) - X_i(t-1)) \tag{20}$$

where $fitness$ is the quality of the pigeon individual. For the minimum optimization problems, we can choose $fitness(X_i(t)) = \dfrac{1}{f(X_i(t)) + \varepsilon}$ for maximum optimization problems, we can choose $fitness(X_i(t)) = f(X_i(t))$.

4   PPPIO for Three-Dimensional Path Planning

4.1   Predator-Prey Concept

Predatory behavior is one of the most common phenomena in nature, and many optimization algorithms are inspired by the predator-prey strategy from ecology [9]. In nature, predators hunt prey to guarantee their own survival, while the preys need to be able to run away from predators. On the other hand, predators help to control the prey population while creating pressure in the prey population. In this model, an individual in predator population or prey population represents a solution, each prey in the population can expand or get killed by predators based on its fitness value, and a predator always tries to kill preys with least fitness in its neighborhood, which represents removing bad solutions in the population. In this paper, the concept of predator-prey is used to increase the diversity of the population, and the predators are modeled based on the worst solutions which are demonstrated as follows:

$$P_{\text{predator}} = P_{\text{worst}} + \rho(1 - t / t_{\max}) \qquad (21)$$

where P_{predator} is the predator (a possible solution), P_{worst} is the worst solution in the population, t is the current iteration, while $t_{\max}$ is the maximum number of iterations and ρ is the hunting rate. To model the interactions between predator and prey, the solutions to maintain a distance of the prey from the predator is showed as follows:

$$\begin{cases} P_{k+1} = P_k + \rho e^{-|d|}, & d > 0 \\ P_{k+1} = P_k - \rho e^{-|d|}, & d < 0 \end{cases} \qquad (22)$$

where d is the distance between the solution and the predator, and k is the current iteration.

4.2 Parallelization of the Map and Compass Operations and the Landmark Operations

In the basic model of PIO algorithm, the landmark operation is used after several iterations of map and compass operation. For example, when the number of generations N_c is larger than the maximum number of generations of the map and compass operation $N_{c\max 1}$. The map and compass operator will stop and it the landmark operation will be start. During my experiment, we found it's easy to fall into a local best solution before the number of generations got to $N_{c\max 1}$. Furthermore, half of the number of pigeons is decreased by N_p in every generation on the landmark operator. The population of pigeons is decreased too rapidly according to formula (18), which would reach to zero after a small amount of iterations. The landmark operator would make only a small impact on the pigeons' position by this way. So we make a small modification on the basic PIO algorithm. The map and compass operation and the compass operation are used parallelly at each iteration. A parameter ω is used to define the impaction of the landmark increase with a smoothly path. And a constant parameter c is used to define the number of pigeons that are in the landmark operator. Our new formula of landmark operator is as follows:

$$N_P(t) = c \cdot N_{P\max} \qquad c \in (0,1) \qquad (23)$$

$$X_c(t) = \frac{\sum X_i(t) \cdot \text{fitness}(X_i(t))}{N_P \sum \text{fitness}(X_i(t))} \qquad (24)$$

$$\omega = s + (1-s) \cdot t/ \, \mathrm{N}_{c\,\mathrm{max}} \qquad s \in (0,1) \qquad (25)$$

$$X_i(t) = X_i(t-1) + \omega \cdot rand \cdot (X_c(t) - X_i(t-1)) \qquad (26)$$

where s is a constant experimentally defined.

4.3   Proposed Predator-Prey PIO (PPPIO) Based Path Planner

In order to overcome the disadvantages of the classical PIO algorithm, such as the tendency to converge to local best solutions, PPPIO, which integrates PIO with the concept of predator-prey, was proposed in our work. After the mutation of each generation, the predator-prey behavior is been conducted in order to choose better solutions into next generation. In this way, our proposed algorithm takes the advantage of the predator-prey concept to make the individuals of sub generations distributed ergodically in the defined space and it can avoid from the premature of the individuals, as well as to increase the speed of finding the optimal solution.

The implementation procedure of our proposed PIO approach to UAV path planning can be described as follows:

Step 1: According to the environmental modeling in Section 2, initialize the detailed information about the path planning task.

Step 2: Initialize the PIO parameters, such as solution space dimension D, the population size N_p, map and compass factor R, the number of iteration N_c.

Step 3: Set each pigeon with a randomized velocity and path. Compare the fitness of each pigeons, and find the current best path.

Step 4: Operate map and compass operator. Firstly, we update the velocity and path of every pigeon by using Eqs. (16) and (17).

Step 5: Rank all pigeons according their fitness values. Some of pigeons whose fitness are low will follow those pigeons with high fitness according to Eq. (23). We then find the center of all pigeons according to Eq. (24), and this center is the desirable destination. All pigeons will fly to the destination by adjusting their flying direction according to Eq. (26). Next, store the best solution parameters and the best cost value.

Step 6: Model the predators based on the worst solution as Eq. (15) demonstrates. Then, use Eq. (16) to provide the other solutions to maintain a distance between the predator and the prey.

Step 7: If $Nc > N_{c\,\mathrm{max}}$, stop the iteration, and output the results. If not, go to step 6.

5   Comparative Experimental Results

In order to evaluate the performance of our proposed PPPIO algorithm in this work, series of experiments are conducted in Matlab2012a programing environment. Coordinates of a starting point are set as (10, 16, 0), and the target point as (55, 100, 0). The initial parameters of PIO algorithm were set as: NP =150. The comparative results of PPPIO with PIO and PSO are showed as follows:

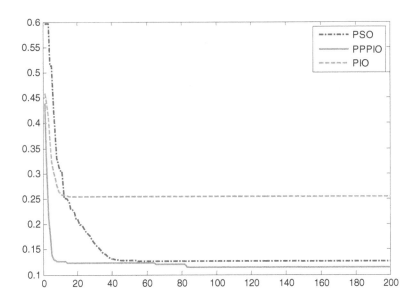

Fig. 1. Comparative evolutionary curves of PPPIO, PIO and PSO

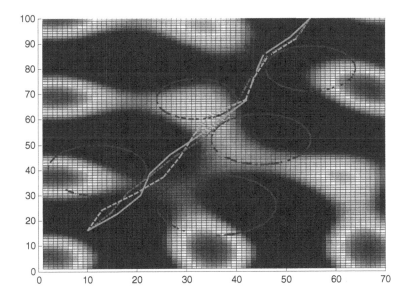

Fig. 2. Comparative path planning results of PPPIO, PIO and PSO

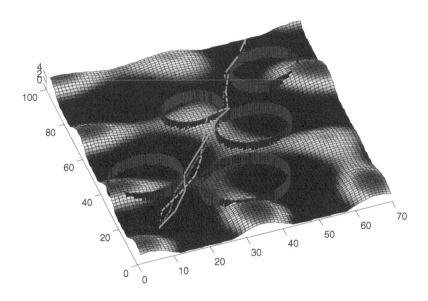

Fig. 3. Comparative path planning results of PPPIO, PIO and PSO on 3D version

6   Conclusions

This paper proposed a novel PPPIO algorithm for solving the UAV three-dimensional path planning problem in complex environments. The concept of predator-prey is adopted to improve the performance of the basic PIO algorithm. Series of comparative simulation results were given to show that our proposed PPPIO algorithm is more efficient than basic PIO and PSO in solving UAV three-dimensional path planning problems.

Acknowledgements. This work was partially supported by National Key Basic Research Program of China(973 Project) under grant #2014CB046401, Natural Science Foundation of China (NSFC) under grant # 61333004 and #61273054, National Magnetic Confinement Fusion Research Program of China under grant # 2012GB102006, and Aeronautical Foundation of China under grant #20135851042.

References

1. Chen, H., Wang, X.M., Li, Y.: A Survey of Autonomous Control for UAV. In: International Conference on Artificial Intelligence and Computational Intelligence, vol. 2, pp. 267–271 (2009)
2. Duan, H.B., Li, P.: Bio-inspired Computation in Unmanned Aerial Vehicles. Springer, Heidelberg (2014)

3. Duan, H.B., Luo, Q.N., Ma, G.J., Shi, Y.H.: Hybrid Particle Swarm Optimization and Genetic Algorithm for Multi-UAVs Formation Reconfiguration. IEEE Computational Intelligence Magazine 8(3), 16–27 (2013)
4. Duan, H.B., Qiao, P.X.: Pigeon-Inspired Optimization: A New Swarm Intelligence Optimizer for Air Robot Path Planning. International Journal of Intelligent Computing and Cybernetics 7(1), 24–37 (2014)
5. Ioannis, K.N., Athina, N.B.: Coordinated UAV Path Planning Using Differential Evolution. In: IEEE International Symposium on, Mediterrean Conference on Control and Automation, vol. 70, pp. 77–111. Springer, Heidelberg (2005)
6. Vincent, R., Mohammed, T., Gilles, L.: Comparison of Parallel Genetic Algorithm and Particle Swarm Optimization for Real-Time UAV Path Planning. IEEE Transactions on Industrial Informatics 9(1), 132–141 (2013)
7. Mora, C.V., Davison, M., Wild, J.M., Michael, M.W.: Magnetoreception and Its Trigeminal Mediation in the Homing Pigeon. Nature 432, 508–511 (2004)
8. Whiten, A.: Operant Study of Sun Altitude and Pigeon Navigation. Nature 237, 405–406 (1972)
9. Zhu, W.R., Duan, H.B.: Chaotic Predator-Prey Biogeography-Based Optimization Approach for UCAV Path Planning. Aerospace Science and Technology 32(1), 153–161 (2014)

Research on Route Obstacle Avoidance Task Planning Based on Differential Evolution Algorithm for AUV

Jian-Jun Li[1,3], Ru-Bo Zhang[1,2], and Yu Yang[3]

[1] College of Computer Science and Technology,
Harbin Engineering University, Harbin 150001, China
[2] College of Electromechanical & Information Engineering,
Dalian Nationalities University, Liaoning Dalian, 116600, China
[3] School of Computer and Information Engineering,
Harbin University of Commerce, Harbin 150028, China

Abstract. AUV mission planning route avoidance purpose is to be able to successfully avoid the threat of a number of different levels of obstacles between the start and end of the route , and plan the optimal route planning to meet certain performance indicators. Through the differential evolution algorithm analysis and description , the avoidance route mission planning problem into a multi-dimensional function optimization problems, optimization problems for AUV mission planning route avoidance functions , based on differential evolution algorithm is proposed route obstacle avoidance task planning methods and after a comprehensive analysis and simulation results validate the differential evolution algorithm in high-dimensional function optimization convergence and stability demonstrated good performance.

Keywords: Differential evolution algorithm, Autonomous Underwater Vehicle, route avoidance, mission planning.

1   Introduction

Due to the complex undersea environment, (Autonomous Underwater Vehicle, AUV), Also known as underwater robots, the need for obstructions on the route, torpedoes and other potential threats to take evasive strategy. Josep et al proposed a low cost computing underwater vehicle planning program, create a static or dynamic obstacle avoidance optimization campaign mode[1]. Zouming Cheng et al use of electronic navigation and positioning collision path planning , and application of artificial intelligence ant colony optimization algorithm to construct collision model[2]. Cruz made hydrodynamics based obstacle avoidance algorithm to determine the location of obstacles and target drones by the harmonic function, thus avoiding local optimum[3]. China Shipbuilding Industry Corporation710Research Institute, Yan Gang, who proposed an improved genetic algorithm to improve search speed and path optimization AUV levels[4]. AUV route avoidance is a key component of AUV mission planning system. Differential evolution algorithm (Differential Evolution

Y. Tan et al. (Eds.): ICSI 2014, Part II, LNCS 8795, pp. 106–113, 2014.

Algorithm, DE) by the American scholar Storn and Price propose a heuristic algorithm to solve optimization problems[5]. Differential evolution algorithm in solving global optimization problems in complex environments and continuous domain optimization problem, with outstanding advantages. Therefore, based on differential evolution algorithm AUV route avoidance, the successful completion of the task execution for AUV has important practical significance.

2   Differential Evolution Algorithm

Differential evolution algorithm remembers the evolution of individual groups and groups in the optimal solution features internal information sharing, through competition and cooperation between individuals within the group to achieve the optimal solution. Assuming a population size of NP, m to the first generation of the evolution of the population of $X(m)$,The dimension of the solution space is K .Initial population $X(0) = \left\{ x_1^0, x_2^0, \cdots, x_{NP}^0 \right\}$, Solutions of the i-th individual $x_i^0 = \left[x_{i,1}^0, x_{i,2}^0, \cdots, x_{i,k}^0 \right]$. Individual components of the formula:

$$x_{i,j}^0 = x_{j,\min} + rand(x_{j,\max} - x_{j,\min}) \tag{1}$$

$x_{j,\max}$ upper bound for the solution space, $x_{j,\min}$ lower bound for the solution space. Differential evolution algorithm including mutation , crossover and selection of three operating [6-9].

2.1   Mutation

Differential evolution algorithm mutation is the last generation of linear combinations of multiple individuals in the population , the variability of individual difference vector generation. Process variation follows the formula:

$$v_i = x_{r1} + F \cdot (x_{r2} - x_{r3}), i = 1, 2, \cdots, NP \tag{2}$$

From the previous generation arbitrary choice of three different populations of individuals $\left\{ x_{r1}, x_{r2}, x_{r3} \right\}$, and $r_1 \neq r_2 \neq r_3$,F constant factor between [0,2] between, Also known as the scaling factor, Used to control the difference vector $(x_{r2} - x_{r3}) \cdot (x_{r2} - x_{r3})$ smaller value of the difference vector, The smaller the disturbance, That is closer to the optimal value group , the disturbance value is automatically reduced.

2.2   Crossover

Differential evolution algorithm is a variation vector crossover target vectors v_i and x_i random reorganization , thereby increasing the diversity of the population of individuals. Process crossover following formula:

$$u_{ij} = \begin{cases} v_{i,j}, randk \leq CR \ or \ j = rand_j; \\ x_{i,j}, randk > CR \ or \ j \neq rand_j; \end{cases} \qquad (3)$$
$$i = 1, \cdots, NP, \ j = 1, \cdots, NP$$

randk   Is a random variable [0,1],CRIs a constant [0,1]. CR The larger the value, the greater the probability of crossover, CR=0 Cross probability 0.

2.3   Select Options

Differential evolution algorithm selection operation is to adapt to the new vector values and objectives of the individual u_i vector of individual fitness value x_i compare, When the value of u_i is better than x_i , replacing x_i . The select operation by the following formula :

$$x_i^{t+1} = \begin{cases} u_i, f(u_i) < f(x_i^t) \\ x_i^t, \text{other} \end{cases} \qquad (4)$$

3   AUV Multiple Route Avoidance Model

AUV safe navigation area, Refers to the current position of the center AUV, R_ϕ radius of the circular area , And there is no obstacle in this circular area , the AUV can be achieved without collision safe navigation. If there is an obstacle , then the AUV be single or multiple route avoidance[10-11].

AUV route avoidance of multiple models , which means that AUV underwater work space with the ability to meet multiple obstacle avoidance.

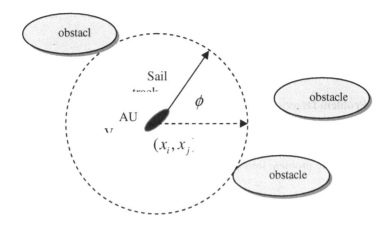

Fig. 1. AUV safe navigation area

3.1   Single Obstacle Avoidance Route Model

AUV set maximum safe radius r_a , Obstacle radius r is between $(c_x, c_y) \in R^+$, $(a_x, a_y) \in R^+$ between the starting point A and the target point B between $(b_x, b_y) \in R^+$ .If the following conditions are met.

$$\sqrt{(\Delta c_x)^2 + (\Delta c_y)^2} > \sqrt{(\Delta b_x)^2 + (\Delta b_y)^2} + r + r_a \qquad (5)$$

A starting point is the presence of AUV through navigational path and destination point B without a single obstacle collision avoidance.

3.2   Route Multiple Obstacles Avoidance Model

AUV set maximum safe radius of r_a, and set between the start and end of all obstructions are located between. A starting point is between $(a_x, a_y) \in R^+$,B target point is between $(c_x, c_y) \in R^+$ and an n the obstacle underwater space. Assuming the k the obstacle $b(k)$ is located $(b_{kx}, b_{ky}) \in R^+$, the radius of the obstacle is r_k. Then , if the following conditions.

$$\sqrt{(\Delta c_x)^2 + (\Delta c_y)^2} > \sqrt{(\Delta b_{kx})^2 + (\Delta b_{ky})^2} + r_{max} + r_a, \forall (b_{kx}, b_{ky}) \in O \qquad (6)$$

A starting point is the presence of AUV through navigational path and destination point B without collision avoidance multiple obstacles.

4   Experimental Verification

4.1   Problem Description

AUV obstacle avoidance task route planning is planning to meet the optimal route planning based on certain performance indicators mission objectives. The route mission planning problem into a k dimensional function optimization problems.

$$\alpha = \arcsin \frac{y_2 - y_1}{\left| \overrightarrow{AB} \right|} \tag{7}$$

$$\begin{pmatrix} x \\ y \end{pmatrix} = \begin{pmatrix} \cos\alpha & \sin\alpha \\ -\sin\alpha & \cos\alpha \end{pmatrix} \cdot \begin{pmatrix} x' \\ y' \end{pmatrix} + \begin{pmatrix} x_1 \\ y_1 \end{pmatrix} \tag{8}$$

According to equation (7) and Equation (8) to the original coordinate system conversion of the connection of the horizontal coordinate system for the new start and end points.

α is a coordinate rotation angle. The new coordinate system x' -axis is divided into k segments, optimizing the corresponding y' -coordinate.

After the conversion of the coordinates (x', y') connected in order to obtain a path connecting the start and end points , Thereby converting the problem into a k dimensional function route optimization problem.

4.2   Barriers Threat Level

AUV navigation path length $L_{i,j}$,The overall threat level barriers to the M_n obstacle

$$\lambda_{n, L_{ij}} = \int_0^{L_{ij}} \sum_{k=1}^{M_n} \frac{t_k}{[(x - x_k)^2 + (y - y_k)^2]^2} dl \tag{9}$$

The AUV navigation path into X segments, If the obstruction to navigation path segment from the threat within a radius of obstacles, the barrier is calculated as the threat level.

$$\lambda_{n,L_{ij}} = \frac{L_{ij}^5}{x} \sum_{k=1}^{M_n} \alpha_k \left(\frac{1}{l_{0.2,k}^4} + \frac{1}{l_{0.4,k}^4} + \frac{1}{l_{0.6,k}^4} + \frac{1}{l_{0.8,k}^4} + \frac{1}{l_{1.0,k}^4} \right) \quad (10)$$

The length of the start and end points y z edge L_{ij}, α_k obstacle to obstacle threat level, $l_{0.2,k}^4$ represents 1/5 the first pitch from the center of the k obstacle edge L_{ij}.

4.3   Program Flow

Steps are as follows :

Step1 Transformed coordinate system, the threat level obstacles to the rotating coordinate system conversion, and the horizontal axis of rotation of the coordinate system K aliquots.

Step2 Initialization K segment route , each route is calculated barriers threat level.

Step3 Iterative calculations.

Step4 For the population of feasible solutions consisting of K, perform mutation operation.

Step5 Individual against individual variation generated with the original crossover operation execution, generate new individuals.

Step6 Calculate the value of the cost function crossover operation to generate new individuals, compared to individuals with newly generated target individuals choose to perform the operation.

Step7 Iterations<maximum number of iterations, jump to Step3 iterative calculation, otherwise exit the loop.

Step8 Coordinate inverse transform, output optimal route avoidance task planning results.

4.4   Simulation

AUV starting point coordinates [10,10],End coordinates [55,100],Consideration weights0.5. Set the initial parameters , population size NP=20,Optimization dimension K=20,The maximum number of iterations $N_{c_{max}}$ =200,Variability factor F=0.5,Cross factor CR=0.9.

Table 1. AUV route barrier parameter

Obstacle radius	Barriers threat level	Disorders Center
20	5	[45,50]
15	2	[12,40]
20	6	[32,48]
18	6	[36,26]
16	4	[22,40]
22	10	[30,45]

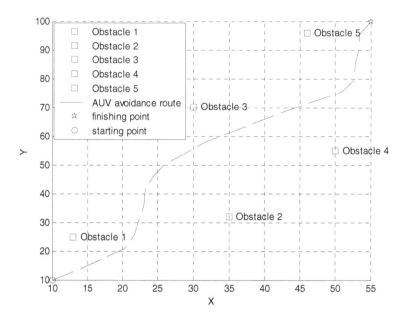

Fig. 2. AUV mission planning route avoidance

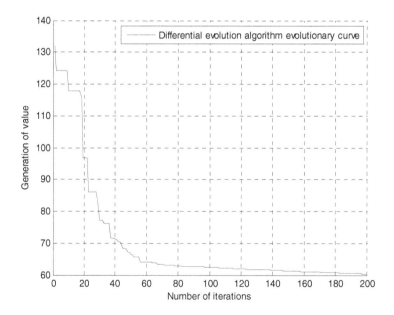

Fig. 3. Differential evolution algorithm evolutionary curve

5   Conclusion

By differential evolution algorithm route planning AUV mission planning can successfully avoid multiple obstacles, arrived in the end of the mission objectives. Simulation results show the differential evolution algorithm solves the problem of high-dimensional optimization , convergence speed and good performance.

Acknowledgments. This work was supported in part by the National Natural Science Foundation of China (60975071,61100005) , Ministry of Education, Scientific Research Project (13YJA790123).

References

1. Isern-Gonzalez, J., Hernandez-Sosa, D., et al.: Obstacle Avoidance in Underwater Glider Path Planning. Physical Agents 6(1), 11–20 (2012)
2. Tsou, M.-C., Hsueh, C.-K.: The Study of Ship Collision Avoidance Route Planningby Ant Colony Algorithm. Journal of Marine Science and Technology, 746–756 (2010)
3. Cruz, G.C.S., Encarnação, P.M.M.: Obstacle Avoidance for Unmanned Aerial Vehicles. Journal of Intelligent & Robotic Systems, 203–217 (2012)
4. Yan, G., Wang, L., Zhou, J., Zha, Z.: Path Planning Based on Improved Genetic Algorithm for AUV. Journal of Chongqing University of Technology, 115–120 (2010)
5. Storn, R., Price, K.: Differential Evolution A Simple and Efficient Heuristic for Global Optimization over Continuous Spaces. Journal of Global Optimization, 341–359 (1997)
6. Price, K.: Differential Evolution A Fast and Simple Numerical Optimizer. In: Proceedings of Biennial Conference of the North American Fuzzy Information Processing Society, pp. 524–527 (1996)
7. Islam, S.M., Das, S.: An Adaptive Differential Evolution Algorithm With Novel Mutation and Crossover Strategies for Global Numerical Optimization. IEEE Transactions on Systems, Man, and Cybernetics Part B: Cybernet ICS, 482–500 (2012)
8. Deng, Y., Beaujean, P.-P.J., An, E., Carlson, E.: Task Allocation and Path Planning for Collaborative Autonomous Underwater Vehicles Operating through an Underwater Acoustic Network. Journal of Robotics, 1–15 (2013)
9. Qu, H., Xing, K., Alexander, T.: An Improved Genetic Algorithm With Coevolution Strategy for Global Path Planning of Multiple Mobile Robots. Neuro Computing 120, 509–517 (2013)
10. Ye, W., Wang, C., Yang, M., Wang, B.: Virtual Obstacles Based Path Planning for Mobile Robots. Robot, 273–286 (2011)
11. Song, Q., Liu, L.: Mobile Robot Path Planning Based on Dynamic Fuzzy Artificial Potential Field Method. International Journal of Hybrid Information Technology 5, 85–94 (2012)

An Improved Particle Swarm Optimization-Based Coverage Control Method for Wireless Sensor Network

Huimin Du[1,2], Qingjian Ni[1,2,3,*], Qianqian Pan[4], Yiyun Yao[1], and Qing Lv[1]

[1] College of Software Engineering, Southeast University, Nanjing, China
[2] Provincial Key Laboratory for Computer Information Processing Technology,
Soochow University, Suzhou, China
[3] School of Computer Science and Engineering, Southeast University, Nanjing, China
[4] School of Information Science and Engineering, Southeast University, Nanjing, China
nqj@seu.edu.cn

Abstract. Coverage control plays a significant role in wireless sensor network (WSN) design. To meet a layout with a certain cover rate, movable nodes are maintained in deployment which accomplish self-organization through moving and changing topological structure. This paper proposes an improved discrete particle swarm optimization algorithm aimed at coverage control method of WSN, and the optimization is implemented under two processes: deployment planning and movement control. The method interpreted in this paper can be easily used solving such problems and the experiment result shows its efficiency, which will inspire new insights in this field.

Keywords: Wireless Sensor Network, Coverage Control, Discrete Particle Swarm Optimization.

1 Introduction

Wireless sensor network (WSN) in complex environment has typical characteristics like large-scale, self-organization, limited energy for nodes and inconstant topology structure, etc. Every node in the network contains a small volume, cheap, energy-saving, multifunction sensor and each sensor has the ability of signal acquiring, data handling and communicating with its neighbors. These features have made WSN topology control a challenging issue.

The quality of topology control influences directly on the lifetime and performance of networks, while a good topology scheme relies on a complete evaluation methodology. Composing those characters and system features, three following indicators are taken into major considerations[1] to evaluate the WSN topology control:

- Coverage: Coverage is a measure of WSN service quality, which is mainly focused on the coverage rate of initial nodes deployment and whether these nodes can acquire signals of the region of interest(ROI), completely and accurately.
- Connectivity: Sensor networks are usually of large scale, thus connectivity is an assurance that data information obtained by sensor can be delivered to sink nodes.

---

* Corresponding author.

Y. Tan et al. (Eds.): ICSI 2014, Part II, LNCS 8795, pp. 114–124, 2014.
© Springer International Publishing Switzerland 2014

- Network lifetime: Network lifetime is generally defined as the time duration from the start to when the percentage of dead nodes comes to a threshold.

Coverage control, or deployment design, is the cornerstone of wireless sensor networks. Node deployment can follow two trends: structured and randomized[2]. Structured method are suitable for small scope deployment where nodes positions are predefined when planning, while randomized way are more pervasive. For supervisory region with large scope which is hard to approach for humans, nodes are initialized (airdropped usually) randomly and adjusted by topology control technology to achieve monitoring. In such case, the mobility of nodes is rather crucial.

Aleksandra *et al.* stated a hexagonal repartition-based C^2 algorithm[3]. This algorithm organizes the space to hexagonal grid, and chooses Cluster Heads (CHs) in the center for each grid cell, using them to rearrange the nodes inside and adjacent to the cells to improve the coverage ratio and connectivity. Zou *et al.* proposed a Virtual Force-based deployment algorithm(VFA), dividing the nodes in the network into clusters[4], where every cluster head node collects information of nodes inside the cluster and computes their final positions and instructs the movement of nodes. Ma *et al.* put forward an Adaptive Triangular Deployment Algorithm(ATRI) to deal with large-scope situations[5]. This process adapts node deployments to regular triangles, and divide node transmission range into six sector and thus nodes can be adjusted from view of its neighbors from each sector.

The utilization of swarm intelligence has made the control processes more effective and easier to implement. Liu[6] *et al.* introduced Easidesign algorithm for WSN coverage control based on Ant Colony Optimization(ACO), which combines greedy strategy and additional pheromone evaporation methods to satisfy network connectivity of different sink positions. A Virtual force co-evolutionary PSO(VFCPSO) was proposed by Wang *et al.*[7] In this algorithm a node is moved several times by the virtual force from other nodes, and the virtual force vectors come from the distance information, their moving direction and other factors. This can also reach a higher coverage ratio. Another situation where mixing stationary and mobile nodes is solved by Li *ed al.* using a novel particle swarm genetic optimization(PSGA), combining PSO and Genetic Algorithm(GA) to repair network holes in [8]. In this method, positions of mobile nodes (or robots) are adjusted to improve the quality-of-service(QoS). This method imports mutation and selection operators to PSO and implements some extra update methods, which are proved to be well-performed.

This paper proposed a novel discrete PSO strategy and applied it to WSN coverage control, to improve QoS stated in the following part. The rest of this paper is arranged as follows: Section 2 describes the abstraction and modeling of coverage problem, Section 3 explains the basic concepts of PSO, and a new discrete strategy with redefined operators is presented in this section. Experiments are conducted and analysed in Section 4, and conclusion shows up in Section 5.

2 Problem Analyses

2.1 Problem Statement

In WSNs, every node has a certain length of sense radius R_s and communication radius R_c. Metrics of QoS include coverage rate, uniformity, time and distance[9], and we mainly consider the coverage and distance problem in this paper.

Coverage. Measuring coverage rate is to detect the ratio of scope inside sense range to the whole object range. Coverage scope is often interpreted as the amount of area. For a node v_i, its coverage scope COV_i in the object region A equals to its sense range, and the total amount of coverage range of the network is explained in formula (1):

$$COV_A = \cup_{i=1}^{k} COV_i \tag{1}$$

thus coverage rate can be represented as (2):

$$C = COV_A / M \tag{2}$$

where M is the area of object region.

Raster coverage is a meliorative strategy, where the ROI is meshing into a grid. The grid points rather than the whole area are treated as the coverage object[10]. This strategy is expanded as region-based point covering in [6]. In such a problem, some geographical points that can show the environment situation are chosen as covering points (**CP**s), and the coverage object is to cover these CPs which are not necessarily grid distributed. Hence, coverage rate becomes the ratio of covered CPs. Typical applications of such problems are Environmental Monitoring Systems and Targets Monitoring, etc. For a node v_i', its coverage scope is the number of CPs inside its sense range, coverage rate is also formulated as (1)(2), where M is the total number of CPs. Research in this paper are expended based on raster coverage.

Distance. The distance a node travels in the movement process is related to the energy limitation. Therefore, optimization strategies are taken to minimize the distance of a node and the total distance of a network. Distance a node takes is regard as the moving range from its initial position to the objective.

2.2 Problem Modeling

Nodes are deployed randomly at initial stage and after the position optimization, a higher coverage rate is obtained. The goal of this process is to maximize the coverage rate and minimize moving distances as well, achieving the minimum energy cost.

In deployment design, object region are divided into a grid, and every grid point acts as a candidate position of nodes. Deployment process contains two procedures, deployment planning and movement control, which are explained as followings:

- For a determined region and CP distribution (showing in Fig.1), design the objective deployment layout for a certain amount of nodes.
- Adjust the positions of randomly-deployed nodes to meet the objective layout.

Formula (1) and (2) are detecting indicators in covering. As for moving process, whose object sketch is shown in Fig.2, consider two sets, $P = \{p_1, p_2, ..., p_N\}$ and $Q = \{q_1, q_2, ..., q_N\}$, where P is the set of stochastic positions generated preliminarily, and Q is the objective position set, N is the number of nodes. The purpose is to pair the vertexes from different set completely in an nonredundant way, that is to make a vertex q_i of Q the moving target of vertex p_j in P, and at the same time, minimizing total moving distance. This can be measured by (3),

$$F = \sum_{i=1}^{N} Distance(q_i, p_{pair\_with\_q_i})$$ (3)

where F is the objective function to be minimized.

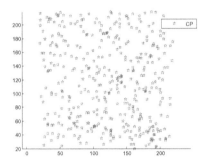

Fig. 1. Distribution of CPs

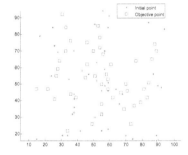

Fig. 2. Objective Sketch for Moving

3   Improved Particle Swarm Optimization for WSN Coverage Control

The section gives a brief summary of basic Particle Swarm Optimization and puts forward an improved method for discrete PSO.

3.1   Basic Principle of PSO

Particle Swarm Optimization (PSO) is a meta-heuristic algorithm that simulates the action of bird folks, proposed by J. Kennedy and R.C. Eberhart, and has been wildly used in solving NP-hard problems. Basic PSO method is:

$$v_{id}(t+1) = v_{id}(t) + c_1 \cdot r1 \cdot (p_{id}(t) - x_{id}(t)) + c_2 \cdot r2 \cdot (p_{gd}(t) - x_{id}(t))$$ (4)

$$x_{id}(t+1) = x_{id}(t) + v_{id}(t)$$ (5)

Assume that in a searching space of D-dimension, a population consists of m particles. Position of particle i in the population can be represented by a D-dimension vector: $X_i = \{x_{i1}, x_{i2}, \cdots, x_{iD}\}$, its present velocity is expressed as $V_i = \{v_{i1}, v_{i2}, \cdots, v_{iD}\}$, and its particle best position is recorded as $P_i = \{p_{i1}, p_{i2}, \cdots, p_{iD}\}$, where $i = 1, 2, \cdots, m$. During optimization, a fitness function is built to judge the quality of a particle position.

3.2 Typical Discrete Strategy

Three typical discrete particle swarm optimization(DPSO) will be introduced: Binary Strategy, Integer Strategy[11] and DPSO aimed at TSP.
Binary PSO: In Binary PSO, each particle is made of binary decisions, valid values are like 0(False) and 1(True), and the probability of those values are modeled as[12]:

$$P(X_{id} = 1) = f(X_{id}(t-1), V_{id}(t-1), p_{id}, p_{gd}) \qquad (6)$$

It's proved that the probability for a dimension to choose 1/True of a particle is a multivariate function, depending on its previous position X_{id} and velocity(trend), which are radically decided by the particle itself and the environment, i.e., p_{id} and p_{gd}. Use sigmoidal function (7)

$$s(V_{id}) = \frac{1}{1 + exp(-V_{id})} \qquad (7)$$

can educe the corresponding threshold determined by V_{id} in the probability function P. And

$$X_{id}(t) = \begin{cases} 1, & if \quad rand() < s(V_{id}) \\ 0, & otherwise \end{cases} \qquad (8)$$

gives the value conditions of X_{id}.
Integer PSO: Integer PSO comes out when solutions are constrained in only integer space but not necessarily binary. This can be seemed as a constrained problem in continuous space, where optimal values are approximated to integers. Advantage of this method is that, though solution space are truncated, algorithm performs still well[13].
DPSO for TSP: Clerc[14], Wang[15] et al., Shi[16] et al. all proposed novel DPSOs when solving traveling salesman problem(TSP). Basic thoughts of DPSO for TSP is like: particle $X_i(t) = \{x_{i1}, x_{i2}, \cdots, x_{iD}\}$ stands for a traveling routine. Start from city x_{i1}, go through $x_{i2}, x_{i3}...$,reach x_{iD}. In this kind of DPSO, velocity are defined as a swap table. That's to say, (π_{ai}, π_{bi}) means to swap city π_{ai} and π_{bi} , $v_{ij} = ((\pi_{ai1}, \pi_{bi1}), ..., (\pi_{ain}, \pi_{bin}))$, for example. For the velocity above, $|v_{ij}| = n$, vector length equals to the number of swap items. Therefore, velocity here contains no repeated item.

3.3 Proposed DPSO for WSN

Based on those existed DPSOs, this paper proposed an improved discrete strategy aimed at WSN coverage problem.

For PSO basic updating formulas (4)(5), there are elements like: position information (such as X, P), velocity information (V), weighting coefficients ($\omega, c * rand()$). In this discrete algorithm, the information are defined as follows:

Position Information: A multi-dimension vector that contains no repeated element, it stands for the spatial situation of a particle and is a sequential list.

Velocity Information: A multi-dimension vector in which elements may repeat, and is the difference of two positions.

Weighting Coefficient: A float number in [0.0, 1.0] that shows the weight of a certain element among all the elements.

Updating method for velocity and position are redefined as:

$$V_i^{(t+1)} = W(\omega) \otimes V_i^{(t)} \circ W(c_1 r_1) \otimes (P_{id} \ominus X_i^{(t)}) \circ W(c_2 r_2) \otimes (P_{gd} \ominus X_i^{(t)}) \quad (9)$$

$$X^{(t+1)} = X^{(t)} \oplus V^{(t)} = \{x_1^{(t)} + ...x_N^{(t)}\} \oplus \{v_1^{(t)}, ..., v_N^{(t)}\} \quad (10)$$

where W is a normalizing function to produce weighting coefficient. Assume there are k factors in all ($k = 3$ in (9)) for a formula, then W is built as:

$$W(factor_i) = \frac{factor_i}{\sum\limits_{i=0}^{k} factor_i} \quad (11)$$

Operators are defined as follows:

$\otimes$ **Multiplication operator for coefficient and velocity:** The result is a vector containing **null** values. Retain a certain number (round of $coefficient \times |vector|$) of dimensions using some strategy (randomly, for example) and set other dimensions to null value and composes a resultant vector. Indexes of retained items should be unique among all the vectors to be added.

$\circ$ **Addition operator for two velocities:** Merge to a new vector according to the item indexes. For example, $\{v_{i1}, v_{i2}, \textbf{null}, v_{i4}\} \circ \{\textbf{null}, \textbf{null}, v_{i3}, \textbf{null}\} = \{v_{i1}, v_{i2}, v_{i3}, v_{i4}\}$. According to (11) and multiple method above, all the sub-vectors will be merged to exact one complete vector.

$\ominus$ **Subtraction operator for two positions:** This operation results in a velocity vector. Define operate mode for each dimension as follow:

$$p_d \ominus x_d = \begin{cases} p_d, & if \quad rand() \geq \alpha \\ x_d + \gamma(p_d - x_d), & if \quad \alpha > rand() > \beta \\ x_d, & otherwise \end{cases} \quad (12)$$

where $0 < \gamma < 1, \beta \leq \alpha < 1, rand() \in [0, 1]$, and (β, α) is the probability interval of interference factor, which is optional.

$\oplus$**Addition operator for a velocity and a position:** This will produce a new position. Here, assume that velocity and position vectors are of the same dimension. Operation for each dimension are defined as:

$$x_d \oplus v_d = \begin{cases} v_d, & if \quad rand() \geq \alpha' \\ x_d + \gamma'(v_d - x_d), & if \quad \alpha' > rand() > \beta' \\ x_d, & otherwise \end{cases} \quad (13)$$

$THEN \quad do \quad x_i \leftarrow x_d \quad where \quad x_i == x_d \oplus v_d$

where $0 < \gamma' < 1, \beta' \leq \alpha' < 1, rand() \in [0, 1]$, and (β', α') is the probability interval of interference factor which is optional. This operation has two steps: calculate $x_d \oplus v_d$ and swap the result with x_d inside the vector.

4   Experiment Results and Analyses

4.1   Deployment Planning

While implementing PSO algorithm, every particle is a candidate solution maintaining the coordinates of all the nodes, i.e., a deployment layout. Thus, for N nodes in the two-dimensional space, every particle is a $2N$-dimensional vector searching those discrete vertexes in the space, i.e., the grid points. For particle i, $X_i = \{x_{i1}, x_{i2}, ..., x_{i2N}\}$.

CPs are chosen in the space as Fig.1, and N nodes are placed randomly. Appropriate decision of N can be different, but it should usually be more than needed to face unexpected conditions. Algorithm 1 and 2 deal with this situation. Use PSO process (algorithm 1) for elementary optimization. LocalBest version of PSO using ring topology [11] is adopted in this process. In particular, this is a constraint problem that the position of a node should stay in a certain scale and should have neighbor nodes near around which it can deliver message to. For a node n_i, define NC_i as the set of neighbor nodes which are inside the communication range of n_i.

$$NC_i = \{n_j | distance(n_j, n_i) \leq R_c\} \neq \emptyset \tag{14}$$

Then the position update formula (5) can be strengthened as update-method 2.

UPDATE-METHOD 2

---

1 $tempP_i^{(t+1)} = P_i^{(t)} + V_i^{(t+1)} = \{x_i^{(t+1)}, y_i^{(t+1)}, z_i^{(t+1)}, ...\}$

2 **if** $x_i^{(t+1)} >$ threshold

3   **then** $x_i^{(t+1)}$=threshold

4 judge $y_i^{(t+1)}, z_i^{(t+1)}...$

5 obtain $NC_i^{(t+1)}$

6 **if** $NC_i^{(t+1)} \neq$null

7   **then** $P_i^{(t+1)} = tempP_i^{(t+1)}$

8   **else** $P_i^{(t+1)} = P_i^{(t)}$

Result of algorithm 1 is shown in Fig.3. This algorithm is aimed at single-cover condition, and some CPs in this region are covered more than once. When multi-coverage is not required, nodes are said to be of no coverage benefit where CPs inside its coverage range are all covered by other nodes already. These nodes will be chosen as dormant

---

ALGORITHM 1: PSO PROCESS

---

1 Initialize D randomly particles, $P_1^{(0)}, \cdots, P_D^{(0)}$
2 **for** i=1 to D
3   Set $Pbests_i^{(0)} = P_i^{(0)}$
4   Judge quality of P_i and set $Lbests^{(0)}$
5 **for** i=1 to D
6   Update P_i value by (4) and Update-Method2
7   Judge quality of P_i
8   Update Pbests and Lbests
9 **if** a terminal condition is met
10   **then** go to step 5
11   **else** go back to step 12
12 Stop and output the best solution

---

nodes and adapted to sleep mode. Algorithm 2 using a traverse accomplished this job, reducing the number of working nodes to a smaller quantity.

Object deployment layout is like Fig.4 after the optimization, which reached a high coverage rate with a smaller number of nodes for single-cover case, and the connectivity of this network can then be achieved overtly. Flexibility is a significant advantage of this deployment method, since some extra nodes will be awaiting inside the region, a new layout will comes up quickly without external aid once environment changes occur or working nodes get problem. In such case the coverage rate $C \geq 98\%$.

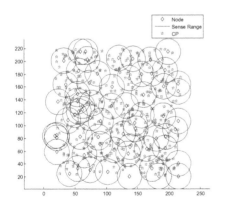

Fig. 3. Elementary Coverage Result

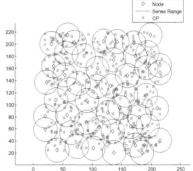

Fig. 4. Adapt Dormant Nodes

ALGORITHM 2: TRAVERSE METHOD

---

1   Nodes are represented as $\{x_1, x_2, ..., x_N\}$
2   **for** $i = 1$ to N
3       $j = i$
4   Mark all the unmarked CPs in the sense range of x_j, the amount is denoted as n
5   **if** $n = 0$
6   **then** pick x_j into dormant set
7   Let $j = (j + 1)\%N$, **if** $j \neq i$
8   **then** go back to step 4
9   **else** go to step 10
10   Get $plan_i$
11 $plan = min\{plan_1, ..., plan_N\}$

4.2   Movement Control

Geographical positions of nodes are stochastically initialized and will be changed to specific ones. Since movement consumes energy and reduces life cycle, a shortest moving plan is expected. Assume that every two vertexes in the region can be connected with one straight path (no obstacle between).

Improved DPSO proposed in this paper is the solver. P is the initial position set and Q is the object set, which are taken as sequences. Purpose of this algorithm is to realign P to $P' = \{p'_1, p'_2, ..., p'_N\}$, as to let p'_i pair with q_i, making up the moving plan. This plan are illustrated in Fig.5.

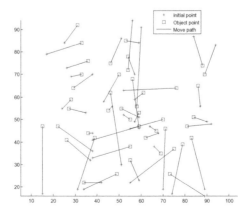

Fig. 5. Moving Plan Sketch Generated by DPSO

Parameters in formulas (9)(12)(13) are set as following: $\omega = 0.6, c_1 = 2.2, c_2 = 1.4, \alpha = 0.6, \beta = 0.4, \gamma = 0.6, \alpha' = 0.7, \beta' = 0.7, \gamma' = 0.6$.

Besides PSO, Ant Colony Optimization (ACO) also has dominant performance solving discrete problems[17]. Comparisons are made between ACO with MMAS strategy and DPSO proposed in this paper. Fig.6 shows when area scope is 200×200, performance difference with different node amount. Solid line indicates DPSO performance while dotted line is for ACO. Fig.7 represents the coverage quality for different object area scope with a certain amount of nodes ($N = 45$). Figures stand for the average result of decades runs of the algorithms.

As can be seen, discrete strategy put forward in this paper performs dominantly within a certain range. A lower total distance in the figures claims a better performance. While the node amount is within a threshold(around 100), a lower energy is obtained, and so is the result when the ROI scale is less than 450. This signifies that when implemented in a relatively small scale (of node amount and scale), DPSO proposed in this paper can provide a faster and easier way to a solution. What should be noticed is about the parameters adopted above. Adjustment to those parameters may be needed when situations changes, such as different sensing range and irregular shape of ROI. In such situations, actual concrete data of algorithm may also be different.

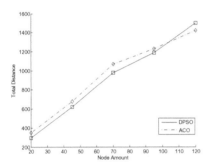
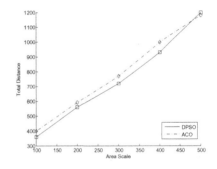

Fig. 6. Performance Comparison with Different Node Amount

Fig. 7. Performance Comparison with Different Area Scope

5   Conclusion

Coverage control for WSN includes many aspects. Besides coverage rate and connectivity, multi-dimensional environments in real world with obstacles and change still need more considerations. This paper mainly discusses application of PSO to two-dimension coverage problem and improved traditional DPSO based on characteristics of WSN coverage control. Results illustrate that in movement control, the improved DPSO which is easy to handle, is of high performance which is no less than traditional discrete problem solver. Successive research in multi-dimension covering problem is to be conducted with further applications of improved DPSO. Multi-objective problems on network uniformity and life-cycling are also crucial points following up.

Acknowledgement. This paper is supported by Provincial Key Laboratory for Computer Information Processing Technology, Soochow University (KJS1223), Suzhou, China and NSFC (Grant No.61170164).

References

1. Ghosh, A., Das, S.: Coverage and Connectivity Issues In Wireless Sensor Networks: A Survey. Pervasive and Mobile Computing 4, 303–334 (2008)
2. Younis, M., Akkaya, K.: Strategies and Techniques For Node Placement in Wireless Sensor Networks: A survey. Ad Hoc Networks 6, 621–655 (2008)
3. Mateska, A., Gavrilovska, L.: Wsn coverage and connectivity improvement utilizing sensors mobility. European Wireless, pp. 686–693 (2011)
4. Zou, Y., Chakrabarty, K.: Sensor Deployment and Target Localization Based on Virtual Forces. In: International Conference on Computer Communications (INFOCOM), NC, USA, pp. 1293–1303 (2003)
5. Ma, M., Yang, Y.: Adaptive Triangular Deployment Algorithm For Unattended Mobile Sensor Networks. IEEE Transactions on Computers 56, 946–958 (2007)
6. Wei, L., Li, C.: Ant Based Approach to The Optimal Deployment in Wireless Sensor Networks. Journal on Communications 30, 25–33 (2009)
7. Wang, X., Wang, S., Ma, J.J.: An Improved Co-Evolutionary Particle Swarm Optimization for Wireless Sensor Networks with Dynamic Deployment. Sensors 7, 354–370 (2007)
8. Li, J., Li, K., Zhu, W.: Improving Sensing Coverage of Wireless Sensor Networks by Employing Mmobile Robots. In: Proceedings of the International Conference on Robotics and Biomimetics (ROBIO), pp. 899–903 (2007)
9. Heo, N., Varshney, P.K.: A distributed self spreading algorithm for mobile wireless sensor networks. In: IEEE Conference on Wireless Communications and Networking, March 16-20, vol. 3, pp. 1597–1602 (2003)
10. Chakrabarty, K., Iyengar, S.S., Qi, H., Cho, E.: Grid coverage for Surveillance and Target Location in Distributed Sensor Networks. IEEE Transactions on Computers 51, 1148–1153 (2002)
11. del Valle, Y., Venayagamoorthy, G.K., Mohagheghi, S., Hernandez, J.-C., Harley, R.G.: Particle Swarm Optimization: Basic Concepts, Variants and Applications in Power Systems. IEEE Transactions on Evolutionary Computation 12(2), 171–196 (2008)
12. Kennedy, J., Eberhart, R.: A Discrete Binary Version of The Particle Swarm Algorithm. In: Computational Cybernetics and Simulation (ICSMC), vol. 5, pp. 4104–4108 (1997)
13. Parsopoulos, K., Vrahatis, M.: Recent Approaches to Global Optimization Problems Through Particle Swarm Optimization. Natural Computing 1, 235–306 (2002)
14. Clerc, M.: Discrete Particle Swarm Optimization Illustrated by The Traveling Salesman Problem. In: Onwubolu, G.C., Babu, B.V. (eds.) New Optimization Techniques in Engineering. STUDFUZZ, vol. 141, pp. 219–239. Springer, Heidelberg (2004)
15. Wang, K.P., Huang, L., Zhou, C.G., Pang, W.: Particle Swarm Optimization for Traveling Salesman Problem. In: International Conference on Machine Learning and Cybernetics, vol. 3, pp. 1583–1585 (2003)
16. Shi, X.H., Lianga, Y.C., Leeb, H.P., Lub, C., Wanga, Q.X.: Particle Swarm Optimization-Based Algorithms for TSP And Generalized TSP. Information Processing Letters 103, 169–176 (2007)
17. Dorigo, M., Stutzle, T.: Ant Colony Optimization: Overview and Recent Advances. International Series in Operations Research and Management Science, Handbook of Metaheuristics, IRIDIA/2009-013, pp. 227–263 (2009)

An Improved Energy-Aware Cluster Heads Selection Method for Wireless Sensor Networks Based on K-means and Binary Particle Swarm Optimization

Qianqian Pan[1,2], Qingjian Ni[2,3,4*], Huimin Du[3], Yiyun Yao[3], and Qing Lv[3]

[1] School of Information Science and Engineering,
Southeast University, Nanjing, China
[2] Laboratory of Military Network Technology, PLA University
of Science and Technology, Nanjing, China
[3] College of Software Engineering, Southeast University,
Nanjing, China
[4] School of Computer Science and Engineering,
Southeast University, Nanjing, China
nqj@seu.edu.cn

Abstract. The limited and non-replenishable energy supply is the main character of Wireless Sensor Networks (WSNs). Hence, maximizing the lifetime of WSNs becomes a critical issue in sensor networks. Clustering is one of the most effective means to extend the lifetime of the whole network. In this paper, an energy-aware cluster heads selection method, based on binary particle swarm optimization (BPSO) and K-means, is presented to prolong the network lifetime. We apply the BPSO and make it suitable for this issue. The selection criteria of the objective cost function are based on minimizing the intra-cluster distance as well as the distance between cluster heads and base station, and optimizing the energy consumption of the whole network. In addition, the sensor nodes are divided into several clusters based on K-means algorithm at the beginning, which can reduce the complexity of the whole algorithm. The performance of our technique is compared with the well-known cluster-based sensor network protocols, LEACH-C and PSO-C respectively. The simulation results demonstrate that our proposed work can achieve better network lifetime over its comparatives.

Keywords: Wireless Sensor Networks, Binary Particle Swarm Optimization, Energy-aware Cluster, K-means Algorithm.

1 Introduction

Wireless sensor network is a kind of wireless network composed of a large number of sensor nodes [1][2]. These tiny nodes are usually scattered in a sensor field

---

* Corresponding author.

Y. Tan et al. (Eds.): ICSI 2014, Part II, LNCS 8795, pp. 125–134, 2014.

and each of these scattered sensors is equipped with a capability to collect data and route data back to the sink, base station (BS). Recent advancement in wireless communication and electronics enable the development of the WSNs. The network has a wide range of applications such as health, military and home. However, sensor node, usually powered by batteries, is limited in energy supply. Therefore, energy efficiency should be considered as the critical design objective. Prolonging lifetime of the WSNs becomes a key issue.

Clustering is one of the most popular methods to prolong the lifetime of the WSNs. One of the well-known clustering protocols is called LEACH [3], which uses randomized rotation of cluster heads to evenly distribute the energy load among the sensors in the network. LEACH-C (LEACH Centralized) [4] is the extension to LEACH. In LEACH-C, BS finds the optimal cluster heads among sensor nodes whose energy are above average, using the simulated annealing algorithm [5].

Particle swarm optimization (PSO) [6][7] is a popular optimization technique, simulating the social behavior of a flock of birds flying to the food. PSO algorithm is applied to find cluster heads and produces better results [8]. A popular clustering algorithm based on PSO is PSO-C [9], which selects cluster heads considering both energy available to nodes and distances between the nodes and their cluster heads.

In this paper, we develop an energy-aware cluster heads selection method based on binary particle swarm optimization (BPSO) [10] and K-means [11] (BPSO-K). Our proposed method selects the high-energy nodes as the cluster heads and evenly distributes the energy load among nodes in the network. The main idea of our protocol is selecting cluster heads that can minimize the intra-cluster distance as well as the distance between cluster heads and BS, and optimize the energy consumption of the whole network. K-means algorithm is utilized at the beginning to divide the nodes into several initial clusters. The rest of this paper is organized as follows: the network and energy models are described in section II. The detailed description of our proposed energy-aware cluster heads selection method for WSNs based on K-means and BPSO is outlined in section III. In section IV, we discuss the simulation study of the proposed protocol. Finally, the concluding remarks appear in section V.

2   The System Model

2.1   Network Model

We assume the network model similar to those used in paper [4] and [9], with the following properties.

1. Sensor nodes are scattered in a field at random and all nodes are static.
2. All sensor nodes are energy constrained, generally powered by batteries.
3. A BS is fixed inside or outside of the sensor filed.
4. Each sensor node has capabilities of processing data and sending data to the BS.
5. Each node can compute its own location and energy level, and send them to the BS.

2.2 Energy Model

The energy model of our protocol is based on the classical model used in paper [3]. The radio hardware energy dissipation of this model is that the transmitter dissipates energy to run the radio electronics and power amplifier, and the receiver dissipates energy to run the radio electronics. Both free space and multipath fading channel model are used according to the distance between transmitter and receiver. Free space model with d^2 is used, if the distance is less than a threshold d_0. Otherwise, the multipath model with d^4 is applied. Thus, in order to transmit an l-bit message over a distance d, the radio energy extended is given in the equation (1).

$$E_{Tx}(l, d) = \begin{cases} lE_{elec} + l\epsilon_{fs}d^2, & \text{if } d < d_0 \\ lE_{elec} + l\epsilon_{mp}d^4, & \text{if } d \geq d_0 \end{cases}. \tag{1}$$

And to receive an l-bit message, the energy expended by the radio is given as equation (2).

$$E_{Rx}(l) = lE_{elec}. \tag{2}$$

where E_{elec} denotes the electronics energy, depending on the energy dissipated per bit to run the transmitter or the receiver. The amplifier energy $\epsilon_{fs}d^2$ and $\epsilon_{mp}d^4$ depend on the distance between transmitter and receiver.

3 Method Description

3.1 Binary Particle Swarm Optimization

PSO [6][12] is a simple, effective, and computationally efficient optimization algorithm for continuous optimization [14][15]. Binary PSO (BPSO) [16] is an extension of PSO based on the binary coding scheme, proposed by Kennedy and Eberhart. BPSO consists of a swarm of S particles. An individual possible solution of a problem is presented by a D-dimensional particle. A particle i has a coordinates x_{id} and a velocity v_{id} in the dth dimension, $1 \leq i \leq S$ and $1 \leq d \leq D$. The velocity is defined as changes of probabilities that decide bits of coordinate in one state or the other. Thus, each dimension of a particle moves to a state restricted to 0 or 1 depending on velocity. Each bit v_{id} of velocity represents the probability of bit x_{id} taking value 1. Velocity of a particle is determined by equation (3).

$$v_{id}(t + 1) = v_{id}(t) + c_1 r_1 (p_{id} - x_{id}) + c_2 r_2 (p_{gd} - x_{id}). \tag{3}$$

where c_1 and c_2 are positive numbers, r_1 and r_2 are two random numbers between 0 and 1 with uniform distribution, and p_{id}, p_{gd} denote particle's and global best position respectively.

Since v_{id} is a probability, it must be constrained in the interval of [0,1]. Sigmoid function is used to normalization velocity v_{id} based on equation (4).

$$s(v_{id}) = \frac{1}{1 + e^{-v_{id}}}. \tag{4}$$

where $s(v_{id})$ denotes the velocity after normalization.

The position of a particle is defined as equation (5).

$$x_{id} = \begin{cases} 1, & \text{if rand}() \leq s(v_{id}) \\ 0, & \text{otherwise} \end{cases} \quad . \tag{5}$$

3.2   Proposed Method of Initialization Using K-means

In the process of establishing clusters, the nodes of location proximity are easier to be assigned to a cluster, mainly because of lower energy for transmission among location closer nodes. Hence, we propose a method of initialization to divide sensor nodes into several initial clusters [13] according to location of nodes.

K-means [17] as a classical clustering algorithm can get different K groups which are described by their centroid of nodes. Besides, the number of clusters is determined by user. The algorithm of initialization is given as algorithm 1 in detail.

Algorithm 1. Initialization using K-means

1: **function** K-MEANS($Array, K, N$)
2:    **for** $j = 1 \to K$ **do**
3:       $C[j] \leftarrow$ random$(1, N)$
4:    **end for**
5:    **repeat**
6:       **for** $j = 0 \to K$ **do**
7:          $Q[j] = \emptyset$
8:       **end for**
9:       **for** $i = 1 \to N$ **do**
10:         MinDist $\leftarrow \infty$
11:         **for** $j = 1 \to K$ **do**
12:            **if** $Distance(Array[i], C[j]) <$ MinDist **then**
13:               MinDist $\leftarrow Distance(Array[i], C[j])$;
14:               MinNode $\leftarrow j$
15:            **end if**
16:         **end for**
17:         Q[MinNode] $\leftarrow Q$[MinNode] $+ i$
18:      **end for**
19:      $C \leftarrow$ Updata(Q)
20:   **until** $C[j]$ not changed for $j = 1 \to K$
21:   **return** Q
22: **end function**

We divide sensor nodes into K initial clusters based on K-means algorithm before the operation of the network and ensure that $K \leq M$, where M denotes the predetermined number of cluster heads. The proposed method of initialization will reduce the complexity of the whole algorithm with slight influence of lifetime.

3.3   Proposed Cluster Heads Selection Method Based on BPSO

Our proposed method is a centralized clustering algorithm to form clusters by dispersing the cluster heads throughout the network. The selection process of our method is working on the BS. The operation is divided into rounds. Each round begins with a set-up phase when the clusters are formed, followed by a steady-state phase when data are transmitted to cluster heads and to the BS [3]. During the set-up phase, each node sends information about current location and energy level to the BS. Based on these statistics, BS computes the average energy of the network. To evenly distribute energy among the whole network, only the nodes with an energy level above the average can be possible cluster heads for current round [18][19]. BS selects M cluster heads using BPSO, where M is the optimizing number of cluster heads for the network.

In the cluster heads selection method based on BPSO [14], a D-dimensional particle represents a selection of cluster heads, where D denotes the number of the possible cluster heads with sufficient energy in current round. The position x_{id} represents the state of the possible cluster head d, where $1 \leq d \leq D$. If x_{id} is restrained to 0, that means the possible cluster head d is not selected as a cluster head in this particle. Otherwise, the value of x_{id} is 1, in this particle, possible cluster head d is chosen as head node. The protocol should ensure each D-bit particle of the flock has exactly M_j bits restrained to 1 and the values of other bits are 0, where $M_j(j = 1, 2, \ldots, K)$ represents the optimizing number of cluster heads for initial cluster j and also ensure $\sum_{j=1}^{K} M_j = M$. To do this, we can adjust the value of x_i with normalized velocity $s(v_i)$. If the number of bits whose value is 1 in a particle is more than M_j, we select the nodes with larger normalized velocity as cluster heads. On the other hand, when the number is less than M_j, we obtain other cluster heads from remaining possible cluster heads according to larger $s(v_i)$.

BS determines M_j cluster heads that suit the cost function best. The main objective of cost function is to optimize the combined influences of distance and energy [20]. Thus, we define the function depending on the following factors.

(1)The longest average distance between nodes and their cluster heads, defined by equation (6).

$$dist_{j1} = \max_{i=1,2,\cdots,M_j} \left\{ \frac{\sum\limits_{k=1}^{C_{ji}} d(CM_{jik}, CH_{ji})}{C_{ji}} \right\}. \tag{6}$$

where M_j and C_{ji} denote the number of initial clusters and nodes in cluster i of the initial cluster j respectively, $d(CM_{jik}, CH_{ji})$ is distance between nodes CM_{jik} and its cluster head CH_{ji}.

(2)The longest distance between cluster heads and BS, given as equation (7).

$$dist_{j2} = \max_{i=1,2,\cdots,M_j} \{d(CH_{ji}, BS)\}. \tag{7}$$

(3)Energy consumption of whole network, represented as $E_{j\text{sum}}$. We can get it based on the system model.

All these three factors should be considered in an integrated manner. However, data for each of them has difference in dimension and magnitude between others. To eliminate such effects, we introduce a normalized function to these three factors, given as equation (8).

$$y = \frac{2}{\pi} \tanh(x). \tag{8}$$

where y is the normalized data of x.

Combining three factors mentioned above, the cost function, represented as f_j, is specified in the equation (9).

$$f_j = \alpha dist_{\text{t}j1} + \beta dist_{\text{t}j2} + \gamma E_{\text{t}j\text{sum}}. \tag{9}$$

where $dist_{\text{t}j1}$, $dist_{\text{t}j2}$ and $E_{\text{t}j\text{sum}}$ are the normalized data of $dist_{j1}$, $dist_{j2}$ and $E_{j\text{sum}}$, α, β, γ are positive factors determining the priority weighting of $dist_{\text{t}j1}$, $dist_{\text{t}j2}$ and $E_{\text{t}j\text{sum}}$, with $\alpha + \beta + \gamma = 1$. In this paper, we assume $dist_{\text{t}j1}$ and $dist_{\text{t}j2}$ have identical impact on the cost function and $E_{\text{t}j\text{sum}}$ has a bit higher influence due to energy often being considered as a main factor of WSNs. Hence, we set $\alpha = 0.3, \beta = 0.3, \gamma = 0.4$.

For a wireless sensor network with N nodes and M predetermined cluster heads, the clusters formed in each round is given as algorithm 2.

Algorithm 2. Cluster heads selection method using BPSO-K

1: $Q \leftarrow$ K-means$(Arrary, K, N)$
2: **repeat**
3:     **for** $j = 1 \rightarrow K$ **do**
4:         **repeat**
5:             Select M_j cluster heads among candidates of Q_j
6:             **for** $i = 1 \rightarrow M_j$ **do**
7:                 $P[j, i] \leftarrow \emptyset$
8:             **end for**
9:             **for** $i = 1 \rightarrow NumOfNode[j]$ **do**
10:                 ClosestHead $\leftarrow$ the closest cluster head from node i
11:                 $P[j, \text{ClosestHead}] \leftarrow P[j, \text{ClosestHead}] + i$
12:             **end for**
13:             Calculate $cost$ function
14:             Update$(LocalBest)$
15:             Update$(GlobalBest)$
16:             Update(v, x)
17:         **until** reach the set number of iteration
18:     **end for**
19: **until** *the network is dead*

After selecting cluster heads of the network and each nodes is decided which cluster it belongs to, the cluster heads act as local control centers to coordinate the data transmissions in their cluster and send the fused data to the BS.

4   Simulations and Analysis

The proposed cluster heads selection method is simulated to evaluate its performance. We define that a wireless sensor node is dead when it runs out of energy and the network is dead at the moment the first node dies. We ran the simulation for 100 nodes in a 200m × 200m network area with both equal and unequal initial energy of nodes. Paraments used in energy model are similar to paper [4], $E_{elec} = 0.5\text{nJ/bit}, \epsilon_{fs} = 10\text{pJ/bit/m}^2, \epsilon_{mp} = 0.0013\text{pJ/bit/m}^4$. The number of clusters is set to be 5 percent [9] of the nodes $M = 5$. The initial clusters divided by K-means is set as $K = 5$. For the paraments of BPSO, we use $S = 30$ particles and $c_1 = c_2 = 2$. In addition, BS is set at location $(0, 200)$ and the data message size is fixed at $l = 4000bit$. The performance of our proposed method is compared with the well-known cluster-based sensor network protocols, LEACH-C and PSO-C.

Fig.1 illustrates the system lifetime, defined by the time of the first node died. It also shows the performance of our proposed method compared with LEACH-C and PSO-C with equal initial energy of nodes. We set the total nodes have 0.5J of initial energy. Fig.2 shows the performance with unequal initial energy of nodes set from 0.3J to 0.7J randomly. The results shown in Fig.1 and Fig.2 are selected randomly from our experiences, which are simulated 20 times for each. We can find that when the first dead node occurs, the LEACH-C and PSO-C do not run exceeding 50 rounds. Whereas the BPSO-K has run about 500 rounds until the first node die.

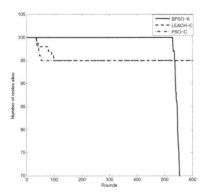   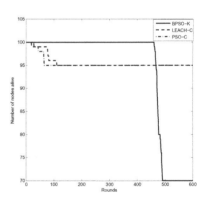

Fig. 1. Number of nodes alive with equal energy   **Fig. 2.** Number of nodes alive with unequal energy

Clearly our proposed protocol can prolong the lifetime of network significantly compared to LEACH-C and PSO-C. It fairly assigns energy consumption to each node in the field by selecting cluster heads periodically based on BPSO-K, which is helpful to avoid some sensor nodes scattered in the field dying too early. Besides, the cost function consist of both distance and energy also plays a critical role in prolonging the lifetime of the WSNs.

Fig.3 shows the clusters divided by K-means when $K = 5$. Wireless sensor nodes are assigned to different clusters based on their location. The nodes of location proximity are assigned to same cluster.

Fig.4 and 5 show the comparison of BPSO and BPSO-K with equal and unequal initial energy nodes. The results shown in Fig.4 and Fig.5 are selected randomly from our experiences, which are simulated 20 times for each. The simulation results demonstrate K-means can reduce the complexion of the method while having slight impact on the lifetime of the network.

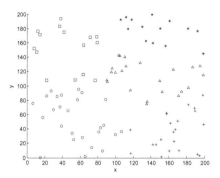

Fig. 3. Clusters divided by K-means

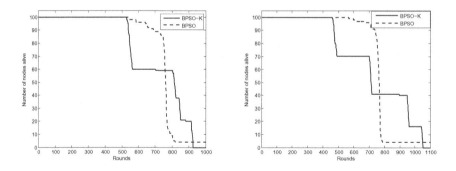

Fig. 4. Number of nodes alive with equal energy   **Fig. 5.** Number of nodes alive with unequal energy

5   Conclusion

In this paper, we presented an improved energy-aware cluster heads selection method for wireless sensor networks based on K-means and BPSO. We defined a new cost function that takes into account the maximum of the intra-cluster distance as well as the distance between cluster heads and BS, and the minimum of energy consumption. Results from the simulations indicate that the proposed protocol using BPSO and K-means algorithm gives a higher network lifetime compared to LEACH-C and PSO-C. Furthermore, the proposed protocol produces better clustering by evenly allocating the cluster heads throughout the network area. The extension of this work would be a further discussion of the parameters setting to the BPSO-K and to prolong the lifetime of networks consist of mobile nodes.

Acknowledgment. This paper is supported by Laboratory of Military Network Technology, PLA University of Science and Technology (LMNT2012-1), Nanjing, China and NSFC (Grant No.61170164).

References

1. Akyildiz, I.F., Su, W., Sankarasubramaniam, Y., Cayirci, E.: Wireless sensor networks: a survey. Computer Networks 38(4), 393–422 (2002)
2. Akkaya, K., Younis, M.: A survey on routing protocols for wireless sensor networks. Ad Hoc Networks 3(3), 325–349 (2005)
3. Heinzelman, W.R., Chandrakasan, A., Balakrishnan, H.: Energy-Efficient Communication Protocol forWireless Microsensor Networks. In: 33rd Hawaii International Conference on System Sciences. IEEE Press, Hawaii (2000)
4. Wendi, B., Heinzelman, A.P.: Chandrakasan, Hari Balakrishnan: An Application-Specific Protocol Architecture for Wireless Microsensor Networks. IEEE Transactions on Wireless Communications 1(4), 666–670 (2002)
5. Murata, T., Ishibuchi, H.: Performance Evaluation of Genetic Algorithms for Flow-shop Scheduling Problems. In: 1st IEEE Conference on Computational Intelligence, pp. 812–817. IEEE Press, Orlando (1994)
6. del Valle, Y., Venayagamoorthy, G.K., Mohagheghi, S., Hernandez, J.-C., Harley, R.G.: Particle Swarm Optimization:Basic Concepts,Variants and Applications in Power Systems. IEEE Transactions on Evolutionary Computation 12(2), 171–196 (2008)
7. Kulkarni, R.V., Venayagamoorthy, G.K.: Particle Swarm Optimization in Wireless-Sensor Networks: A Brief Survey. IEEE Transactions on Systems, Man, and Cybernetics Part C: Applications and Reviews 41(2), 262–267 (2011)
8. Bala Krishna, M., Doja, M.N.: Swarm intelligence-based topology maintenance protocol for wireless sensor networks. IET Wireless Sensor Systems 1(4), 181–190 (2011)
9. Abdul Latiff, N.M., Tsimenidis, C.C., Sharif, B.S.: Energy-aware Clustering for Wireless Sensor Networks Using Particle Swarm Optimization. In: 18th Annual IEEE International Symposium on Personal, Indoor and Mobile Radio Communications, pp. 1–5. IEEE Press, Athens (2007)
10. Luh, G.-C., Lin, C.-Y., Lin, Y.-S.: A Binary Particle Swarm Optimization for Continuum Structural Topology Optimization. Applied Soft Computing 11(2), 2833–2844 (2011)
11. Jain, A.K.: Data Clustering: 50 years beyond K-means. Pattern Recognition Letters 31(8), 651–666 (2010)
12. Shi, Y., Eberhart, R.: A Modified Particle Swarm Optimizer. In: The 1998 IEEE International Conference on Computational Intelligence, pp. 577–584. IEEE Press, Anchorage (1998)

13. Wagstaff, K., Cardie, C.: Constrained K-means Clustering with Background Knowledge. In: The Eighteenth International Conference on Machine Learning, pp. 577–584. Morgan Kaufmann Publishers Inc., San Francisco (2001)
14. Chuang, L.-Y., Tsai, S.-W., Yang, C.-H.: Improved Binary Particle Swarm Optimization Using Catfish Effect for Feature Selection. Expert Systems with Applications 38(10), 12699–12707 (2011)
15. Fernández-Martínez, J.L., García-Gonzalo, E.: Stochastic Stability Analysis of the Linear Continuous and Discrete PSO Models. IEEE Transactions on Evolutionary Computation 15(3), 405–423 (2011)
16. Kennedy, J., Eberhart, R.C.: A Discrete Binary Version of the Particle Swarm Algorithm. In: 1997 IEEE International Conference on Systems, Man, and Cybernetics, pp. 4104–4108. IEEE Press, Orlando (1997)
17. Na, S., Xumin, L., Yong, G.: Research on k-means Clustering Algorithm: An Improved k-means Clustering Algorithm. In: 3rd International Symposium on Intelligent Information Technology and Security Informatics, pp. 63–67. IEEE Press, Jinggangshan (2010)
18. Shi, S., Liu, X., Gu, X.: An Energy-Efficiency Optimized LEACH-C for Wireless Sensor Networks. In: 7th International ICST Conference on Communications and Networking, pp. 487–492. Kun Ming (2012)
19. Karaboga, D., Okdem, S., Ozturk, C.: Cluster Based Wireless Sensor Network Routing Using Artificial Bee Colony Algorithm. Wireless Networks 18(7), 847–860 (2012)
20. Singh, B., Lobiyal, D.K.: A Novel Energy-aware Cluster Head Selection Based on Particle Swarm Optimization for Wireless Sensor Networks. Human-centric Computing and Information Sciences 2(1) (2012)

Comparison of Multi-population PBIL and Adaptive Learning Rate PBIL in Designing Power System Controller

Komla A. Folly

Department of Electrical Engineering, University of Cape Town,
Private bag., Rondebosch 7701, Cape Town, South Africa
Komla.Folly@uct.ac.za

Abstract. Population-Based Incremental Learning (PBIL) is a combination of Genetic Algorithm with competitive learning derived from Artificial Neural Network. It has recently received increasing attention due to its effectiveness, easy implementation and robustness. Despite these strengths, it has been reported recently that PBIL suffers from issues of loss of diversity in the population. To deal with the issue of premature convergence, we propose in this paper a parallel PBIL based on multi-population. In parallel PBIL, two populations are used where both probability vectors (PVs) are initialized to 0.5. The approach is used to design a power system controller for damping low-frequency oscillations. To show the effectiveness of the approach, simulations results are compared with the results obtained using standard PBIL and another diversity increasing PBIL called herein as PBIL with Adapting learning rate (APBIL). It is shown that Parallel PBIL approach performs better than the standard PBIL and is as effective as APBIL.

Keywords: Adaptive learning rate, Low frequency oscillations, Population-based incremental learning, Parallel PBIL.

1 Introduction

Recently, a novel type of Evolutionary Algorithm called Population-Based Incremental Learning (PBIL) [1]-[2] has received increasing attention [3]-[6]. Population-Based Incremental Learning (PBIL) is a combination of Genetic Algorithm with competitive learning derived from Artificial Neural Network. Like other Evolutionary Algorithms such as GAs [7]-[9], Differential Evolution (DE) [10]-[11], and variants such as Particle Swarm Optimization (PSO) [12]-[13], PBIL works with a population of individuals rather than a single individual (e.g., point) [1], [2]. Over successive generations, the population "evolves" toward an optimal solution. PBIL is simpler than Genetic Algorithms GAs, and yet more effective than GAs. In PBIL, the crossover operator of GAs is abstracted away and the role of population is

Y. Tan et al. (Eds.): ICSI 2014, Part II, LNCS 8795, pp. 135–145, 2014.

redefined [1]. PBIL works with a probability vector (PV) which controls the random bit strings generated by PBIL and is used to create other individuals through learning. Learning in PBIL consists of using the current probability vector (PV) to create N individuals. The best individual is used to update the probability vector, increasing the probability of producing solutions similar to the current best individuals [2], [3]. It has been shown that PBIL outperforms standard GAs approaches on a variety of optimization problems including commonly used benchmark problems [1], [2]. PBIL has also been applied for controller design in power systems for small-signal stability improvement. In [4], PBIL based power system stabilizers (PSSs) were compared with GA based PSSs and were found to give better results GA based PSSs. In [5]-[6], it was shown that PBIL based PSS performed as effectively as BGA based PSS. However, there are still some issues related to PBIL [14]. It has been reported in [15]-[17] that PBIL suffers from diversity loss making the algorithm to converge to local optima. To cope with this problem, a PBIL with adaptive learning rate strategy was proposed in [16]-[17]. In this paper, a new approach that can improve population diversity in PBIL is presented. The idea of using parallel PBIL (PPBIL) based on multi-population to improve population diversity is explored [18]-[19]. The proposed approach is applied to a power system controller design in a multi-machine power system. The effectiveness of the proposed approach is demonstrated by comparing it to the Adaptive PBIL (APBIL) introduced in [16]-[17] and the standard PBIL (SPBIL). Simulation results show that the parallel PBIL based on multi-population performs better than the standard PBIL and is as effective as APBIL.

2   Overview of the Standard PBIL

Population–based incremental learning (PBIL) is a technique that combines aspects of Genetic Algorithms and simple competitive learning derived from Artificial Neural Networks [1], [2]. PBIL belongs to the family of Estimation of Distribution Algorithms (EDAs), which use the probability (or prototype) vector to generate sample solutions. Unlike GAs which performance depends on crossover operator, PBIL performance depends on the learning process of the probability vector. The probability vector guides the search, which produces the next sample point from which learning takes place. The learning rate determines the speed at which the probability vector is shifted to resemble the best (fittest) solution vector [3]. Initially, the values of the probability vector are set to 0.5 to ensure that the probability of generating 0 or 1 is equal. As the search progresses, these values are moved away from 0.5, towards either 0.0 or 1.0.

Like in GA, mutation is also used in PBIL presented in this paper to maintain diversity. In this paper, the mutation is performed on the probability vector; that is, a forgetting factor is used to relax the probability vector toward a neutral value of 0.5 [3], [4]. The pseudocode for the standard PBIL is shown in Fig. 1, [1]-[4].

If the learning rate is fixed during the run, it cannot provide the flexibility needed to achieve a trade-off between exploration and exploitation. To achieve a trade-off between exploration and exploitation, PBIL with adaptive learning rate strategy presented in [16]-[17] could be used. However, the approach proposed here is to use multi-population PBIL instead of a single population to achieve the same objective as discussed in section 4

```
Begin
g:= 0;
//initialize probability vector
for i:=1 to l, do PV_i^0 = 0.5;
endfor;
while not termination condition do
generate sample S(g) from (PV(g) , pop.)
Evaluate samples S(g)
Select best solution B(g)
// update probability vector PV(g) toward best
solution according to (1)
//mutate PV(g)
Generate a set of new samples using the new
probability vector
g=g+1
end while // e.g., g>G_max
```

Fig. 1. Pseudocode for standard PBIL

3 Overview of Multi-Population PBIL

For the multi-population or Parallel PBIL (PPBIL), two populations are used with two probability vectors (PV_1 and PV_2). Each probability vector is initialized to 0.5 and sampled to generate solutions independently from each other. The PVs are updated independently according to the best solution generated by each. Initially, each probability vector has equal sample solutions. That is, the total population is divided into two populations and a PV is assigned to each population. As the run progresses, the population of the probability vector (PV) that performs better is allowed to increase its share of samples. The sample sizes of the probability vectors are slightly adapted within the range [pop_{min} pop_{max}] = [0.4*pop   0.6*pop] according to their relative performances. The probability that outperforms the other is increased by a constant value $\Delta = LR$*pop, where LR is the learning rate (which was selected as 0.1 in this paper). Fig. 2 shows the pseudocode of PPBIL.

```
Begin
g:= 0;
//initialize probability vector
for i:=1 to l, do PVᵢ₁⁰ = PVᵢ₂⁰ = 0.5;
endfor;
// initialize the sizes of the probability vectors
such that: pop1= pop 2= pop/2
  while not termination condition do
  generate sample S₁(g) from (PV₁(g) , pop1.)
  generate sample S₂(g) from (PV₂(g) , pop2.)
  Evaluate samples (S₁(g), S₂(g))
  Select best solutions B₁(g)and B₂(g)
  // update probability vectors PV₁(g) and PV₂(g)
toward bests solution B₁(g)and B₂(g) according to (1)
  If f(B₁(g))> f(B₂(g) )
  then   pop₁= min [(pop₁ + Δ)     popₘₐₓ]
  If f(B₁(g))< f(B₂(g) )
  then   pop₁= max [(pop₁ -Δ)     popₘᵢₙ]
  pop₂= pop-pop₁
  //mutate PV₁(g) and PV₂(g)
  g=g+1
  end while // e.g.. g>Gₘₐₓ
```

Fig. 2. Pseudocode for parallel PBIL

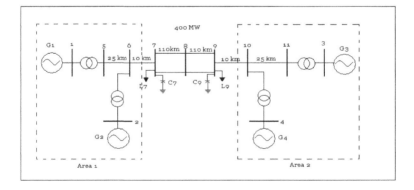

Fig. 3. Power system model

4   Problem Description and Formulation

4.1   Problem Description

The controller to be designed is also known as Power System Stabilizer (PSS) and is needed to damp low frequency oscillations ranging from 0.1 Hz to 2.5 Hz which occur in overly stressed power systems or when power is transmitted over weak transmission lines [20]-[21]. These oscillations are highly undesirable because they can lead to fatigue of machine shafts and limit the ability of the system to transfer the maximum power. It is therefore important that low-frequency oscillations are damped quickly if the security of the system is to be maintained. The power system model used in this paper is the IEEE 2-area system, 4-machine power system as shown in Fig. 3. Each machine is represented by the detailed six order differential equations. The machines are equipped with simple exciter systems. For more information on this system, the reader is referred to [18], [20]. To design the controller, several operating conditions have been considered. However, for simplicity only three operating conditions are shown in Table 1. This Table also shows the eigenvalues and damping ratios in brackets of the three operating conditions.

The system exhibits two local modes one in area 1 and the other in area 2 and one inter-area mode. For the purpose of this study, only the inter-area modes are shown in Table 1 since they are the most difficult to control.

Case 1 is the light load condition, where about 200 MW of real power is transferred from area 1 to area 2. The system is stable for this case as can be seen by the negative value of the real part of the eigenvalue. Case 2 is the nominal condition, under this operating condition, there is a transfer of 400 MW power from area 1 to area 2. The system is unstable for this case, since the real part of the eigenvalue is positive. Case three is the heavy load condition where about 500 MW of power is transferred from area 1 to area 2. This case is also unstable.

4.2   Problem Formulation and Objective Functions

The purpose of the design is to optimize the parameters of the generator excitation controls (i.e., PSSs) simultaneously and in a coordinated and decentralized manner such that adequate damping is provided to the system over a wide range of operating conditions, while keeping the structure of the PSS as simple as possible. The structure of the widely used conventional PSS was adopted here. The PBILs are applied to optimize the parameters of a fixed structure ($\Delta\omega$ input) PSS of the form:

$$K(s) = K_p \left(\frac{T_w s}{1 + T_w s} \right) \left(\frac{1 + T_1 s}{1 + T_2 s} \right) \left(\frac{1 + T_3 s}{1 + T_4 s} \right) \tag{1}$$

where, K_p is the gain, T_1-T_4 represent suitable time constants. T_w is the washout time constant needed to prevent steady-state offset of the voltage. The value of T_w is not critical for the PSS and has been set to 5sec. Therefore, five parameters are required for the optimization.

Since most of oscillation modes considered in this paper are unstable and dominate the time response of the system, it is expected that by maximizing the minimum damping ratio, a set of system models could be simultaneously stabilized over a wide range of operating conditions [10]-[12]. The following objective function was used to design the PSSs.

$$J = \max\left(\min(\zeta_{i,j}) \right) \tag{2}$$

where $i = 1, 2 \ldots n$, and $j = 1, 2, \ldots m$

and $\zeta_{i,j} = \dfrac{-\sigma_{i,j}}{\sqrt{\sigma_{i,j}^2 + \omega_{i,j}^2}}$ is the damping ratio of the i-th eigenvalue in the j-th

operating condition. σ_{ij} is the real part of the eigenvalue and the ω_j is the frequency. n denotes the total number eigenvalues and m denotes the number of operating conditions.

Table 1. Selected open-loop operating conditions including eigenvalues and damping ratios

Case	P_e [MW]	Eigenvalue (ζ)
1	200	-0.35±3.92i (0.0889)
2	400	0.0096±3.84 i (-0.0025)
3	500	0.148±3.09i (-0.0478)

4.3   Application of Standard PBIL to Controller Design

The configuration of the standard PBIL is as follows:
Length of chromosome: 15 bits
Trial solutions (population): 10
Generations: 400
Learning rate (LR): 0.1
Mutation (Forgetting factor-FF): 0.005

4.4   Application of APBIL to Controller Design

The configuration of the APBIL is as follows:
Length of chromosome: 15 bits
Trial solutions (population): 10
Generations: 400
Initial Learning rate (LR$_0$) = 0.0005

Final Learning rate (LR_{max}): 0.2
Mutation (Forgetting factor-FF): 0.005

4.5   Application of PPBIL to Controller Design

The configuration of the APBIL is as follows:
Length of chromosome: 15 bits
Trial solutions (population): 10
Initial population for PV_1: 5
Initial population for PV_2: 5
Generations: 400
Final Learning rate (LR): 0.1
Mutation (Forgetting factor-FF): 0.005
For all the controllers, the parameter domain is as follows:

$$0 \leq K_p \leq 30$$

$$0 \leq T_1, T_3 \leq 1$$

$$0.010 \leq T_2, T_4 \leq 0.3$$

5   Simulation Results

5.1   Convergence Rate

Figs. 4-6 show the convergence rate of SPBIL, APBIL and PPBIL, respectively. It can be seen that APBIL and PPBIL converge to higher fitness values of 0.514 and 0.502, respectively, compared to a value of 0.484 for SPBIL. From the simulation results, it can be seen that APBIL has more diversity in the population at the middle of the run between generation 100 and 200 than PPBIL and SPBIL. This can be attributed to the small value of learning rate, at these generations. Small learning rate increases the exploration of the algorithm and thereby introduces more diversity in the population. Unlike SPBIL which diversity is much more concentrated at the beginning of the run between generation 1 and generation 150, the diversity in PPBIL is somehow spread across all generations. At generations 300 to 400 for example, the SPBIL and APBIL have converged (i.e., almost no diversity). On the other hand, PPBIL still has some diversity. Therefore, it can still explore the search space although at a limited pace. Table 2 shows the comparison between the best, mean and worst, fitness values. It can be seen that on average SPBIL and PPBIL have practically the same fitness. The mean for PPBIL and SPBIL which is approximately 0.434 is higher than the mean of APBIL which is 0.383. The main reason for this is that APBIL has much more spread, with the worst fitness value at 0.09 compared to 0.124 for PPBIL and 0.166 for SPBIL.

In terms of the distance between the best and the worst fitness values, APBIL has the highest distance (0.424), followed by PPBIL (0.378) and then SPBIL (0.318). This suggests that both APBIL and PPBIL have more diversity in their populations than SPBIL. Table 3 shows the number of functions evaluations for each algorithm before the best fitness was found. It can be seen that that SPBIL has the lowest function evaluations (3810) and APBIL has the highest function evaluations (15950). PPBIL is somehow in the middle (10610). In terms of the speed in finding the best fitness value, SPBIL is better and APBIL is the worst. However, the best value found by SPBIL is lower than that found by APBIL and PPBIL. This suggests that although SPBIL converges faster, it converges to local optima, which may not be appropriate.

5.2   Eigenvalue Analysis

Table 4 shows the eigenvalues and damping ratios in brackets of the closed-loop systems with the three controllers. It can be seen that PPBIL and APBIL give better performances (i.e., better damping ratios) than SPBIL. However, PPBIL provides slightly a better damping than APBIL.

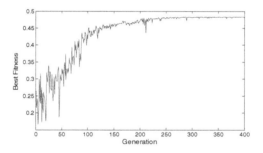

Fig. 4. SPBIL convergence rate

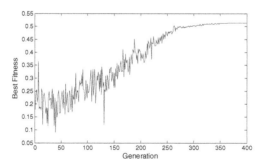

Fig. 5. APBIL convergence rate

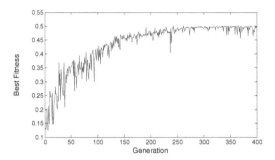

Fig. 6. PPBIL convergence rate

Table 2. Fitness Values

Fitness	SPBIL	APBIL	PPBIL
Best	0.484	0.514	0.502
Mean	0.434	0.383	0.434
Worst	0.166	0.090	0.124

Table 3. Number of Function Evaluation

Controllers	Evaluations
SPBIL	3810
PPBIL	10610
APBIL	15950

Table 4. Closed-system eigenvalues and damping ratio

Case	SPBIL	APBIL	PPBIL
1	$-1.13 \pm j2.04 \ (0.48)$	$-1.54 \pm j2.57 \ (0.51)$	$-1.53 \pm j2.53 \ (0.52)$
2	$-0.778 \pm j1.78 \ (0.44)$	$-1.26 \pm j2.11 \ (0.51)$	$-1.26 \pm j2.08 \ (0.52)$
4	$-0.582 \pm j1.25 \ (0.42)$	$-1.15 \pm j1.52 \ (0.61)$	$-1.14 \pm j1.54 \ (0.60)$

6   Conclusions

By using Parallel PBIL based on multi-population we have been able to increase the diversity in the population. This is important to prevent premature convergence that is inherent to the standard PBIL. The effectiveness of the proposed approach is demonstrated by comparing it to the Adaptive PBIL (APBIL) and the standard PBIL (SPBIL). Simulation results show that the performance of PPBIL in increasing the population diversity is as effective as that of APBIL. Both the PPBIL-PSS and the APBIL-PSS performed better than the standard SPBIL-PSS in terms of improving the damping of the system.

Acknowledgement. This work is based on the research supported in part by the National Research Foundation of South Africa, UID 83977 and UID 85503.

References

1. Baluja, S.: Population-Based Incremental Learning: A Method for Integrating Genetic Search Based Function Optimization and Competitive Learning. Technical Report, CMU-CS-94-163, Carnegie Mellon University (1994)
2. Baluja, S., Caruana, R.: Removing the Genetics from the Standard Genetic Algorithm. Technical Report CMU-CS-95-141, Carnegie Mellon University (1995)
3. Greene, J.R.: Population-Based Incremental Learning as a Simple, Versatile Tool for Engineering Optimization. In: EvCA 1996, Moscow (1996)
4. Folly, K.A.: Design of Power System Stabilizer: A Comparison Between Genetic Algorithms (GAs) and Population-Based Incremental Learning (PBIL). In: Proc. of the IEEE PES 2006 General Meeting, Montreal, Canada (2006)
5. Sheetekela, S., Folly, K.: Power System Controller Design: A Comparison Between Breeder Genetic Algorithm (BGA) and Population-Based Incremental Learning (PBIL). In: Proc. of the Int. Joint Conference on Neural Networks, IJCNN (2010)
6. Folly, K.A., Sheetekela, S.P.: Optimal design of Power System Controller using Breeder Genetic Algorithm. In: Bio-Inspired Computational Algorithms and Their Applications, Intech, pp. 303–316 (2012)
7. Goldberg, D.E.: Genetic Algorithms in Search, Optimization & Machine Learning. Addison-Wesley (1989)
8. Davis, L.: Handbook of Genetic Algorithms. International Thomson Computer Press (1996)
9. Yao, J., Kharma, N., Grogono, P.: Bi-objective Multipopulation Genetic Algorithm for Multimodal Function Optimization. IEEE Trans. On Evol. Comput. 14(1), 80–102 (2010)
10. Kennedy, J.F., Kennedy, J., Eberhart, R.C., Shi, Y.: Swarm Intelligence. Morgan Kaufmann (2001)
11. Mulumba, T., Folly, K.A.: Design and Comparison of Multi-machine Power System Stabilizer base on Evolution Algorithms. In: In Proc. of the 46th International Universities' Power Engineering Conference (UPEC), Soest – Germany, September 5-8 (2011)
12. Abido, A.A.: Particle swarm Optimization for Multimachine Power System Stabilizer Design. IEEE Trans. on Power Syst. 3(3), 1346–1351 (2001)

13. Venayagamoorthy, G.K.: Improving the Performance of Particle Swarm Optimization using Adaptive Critics Designs. In: IEEE Proceedings on Swarm Intelligence Symposium, pp. 393–396 (2005)
14. Gosling, T., Jin, N., Tsang, E.: Population-Based Incremental Learning Versus Genetic Algorithms: Iterated Prisoners Dilemma. Technical Report CSM-40, University of Essex, England (2004)
15. Rastegar, R., Hariri, A., Mazoochi, M.: The Population-Based Incremental Learning Algorithm Converges to Local Optima. Neurocomputing 69(13-15), 1772–1775 (2006)
16. Folly, K.A., Venayagamoorthy, G.K.: Effect of learning rate on the performance of the Population-Based Incremental Learning algorithm. In: Proc. of the International Joint Conf. on Neural Network (IJCNN), Atlanta Georgia, USA (2009)
17. Folly, K.A.: An Improved Population-Based Incremental Learning Algorithm. International Journal of Swarm Intelligence Research (IJSIR) 4(1), 35–61 (2013)
18. Folly, K., Venayagamoorthy, G.: Power System Stabilizer Design using Multi-Population PBIL. In: Proc. of the 2013 IEEE Symposium Series on Computational Intelligence (2013)
19. Yang, S., Yao, X.: Experimental Study on Population-Based Incremental Learning Algorithms for Dynamic Optimization Problems. Soft Computing 9(11), 815–834 (2005)
20. Kundur, P.: Power System Stability and Control, McGraw-Hill, Inc. (1994)
21. Gibbard, M.J.: Application of Power System Stabilizer for Enhancement of Overall System Stability. IEEE Trans. on Power Systems 4(2), 614–626 (1989)

Vibration Adaptive Anomaly Detection of Hydropower Unit in Variable Condition Based on Moving Least Square Response Surface

Xueli An and Luoping Pan

China Institute of Water Resources and Hydropower Research, 100038 Beijing, China
an_xueli@163.com

Abstract. It is difficult to effectively analyze and identify the conditions of hydropower unit, due to its complex operation conditions, frequent start-stop conditions, continual working status switch, less fault samples, single static alarm threshold. Lots of test research shows that active power and working head are key factors which affect the operation conditions of hydropower unit. The health standard condition of unit is determined. An adaptive real-time anomaly detection model of hydropower unit vibration parameters is proposed based on moving least square response surface. In the proposed model, active power and working head are comprehensively considered. This model can adapt variable conditions of hydropower unit. The model is used to real time detect the anomaly of hydropower unit vibration parameters. The results show that this model can effectively evaluate the performance of unit vibration, can more accurately detect the abnormal of unit vibration.

Keywords: hydropower unit, vibration parameter, adaptive anomaly detection, moving least square response surface.

1 Introduction

Hydropower plant is the most favorable power source of power system. It is used to undertake the tasks of peaking, filling valley, frequency modulation, phase modulation, and emergency reserve. Hydropower plant can improve the efficiency of thermal power plants and nuclear power plants, increase the reliability of power grid. It has a significant role in ensuring the safety of the power grid operation and improving the economy of power system [1] and [2]. Due to the complexity of hydropower units' operating conditions, frequent start-stop and working conditions conversion, making the unit easy to malfunction. To ensure units' safe and stable operation, it is needed to mine their condition monitoring data. The data mining can better get the real operating condition of units and early warn the possible abnormalities.

The research of online monitoring and fault diagnosis of hydropower units mainly focuses on the development and integration of condition monitoring system and fault diagnosis methods. The current research achievements don't meet site requirements

Y. Tan et al. (Eds.): ICSI 2014, Part II, LNCS 8795, pp. 146–154, 2014.

[1]. The studies for effectively analyzing the mass monitoring data aren't many. The studies for building the anomaly detection model based on online monitoring data are fewer [1, 3]. This makes the vast majority of hydropower plants can only use preventive maintenance strategy, namely scheduled overhaul strategy. This strategy will inevitably lead to the problem for inadequate maintenance or excess maintenance.

Anomaly detection aims to find the relationship between the abnormities of units' condition parameters and their potential failure, to reveal hidden information of abnormal parameters. The field personnel can timely take appropriate measures according to units' abnormal condition information, to restraint the further deterioration of abnormities [4], [5] and [6]. Using this method, faults can be nipped in the bud. The failure rate will be reduced. The security, stability of units operation and economy of maintenance will be increased.

In this paper, based on a long time condition monitoring data of hydropower units, the moving least square response surface is used to build adaptive anomaly detection model. This model considers the factors of active power and working head, which affect the operation condition of hydropower units. The proposed model provides a new way to online assessment of units' running condition.

2   Moving Least Square Response Surface

In a local domain Ω_x of the fitting area Ω, supposing the function values of a calculated function $f(x)$ of N sampling points x_I ($I=1, 2, \ldots, N$) in the local domain Ω_x are known, $y_I=f(x_I)$. In the local domain, the calculated function $f(x)$ can be approximated as $g(x)\approx f(x)$, the fitting function $g(x)$ [7] can be expressed as:

$$g(x) = \sum_{i=1}^{m} \beta_i(x)\rho_i(x) = \rho^T(x)\beta(x) \cdot \tag{1}$$

where $\beta(x) = \left(\beta_1(x), \ \beta_2(x), \ \cdots, \beta_m(x)\right)^T$ is calculated coefficient, $\rho(x) = \left(\rho_1(x), \ \rho_2(x), \ \cdots, \rho_m(x)\right)^T$ is m-dimensional k-order complete polynomial, m is the number of basis function.

The fitting accuracy of $g(x)$ is varietal with the different order of basis function. As a secondary basis function in two-dimensional space $\rho(x)$ is

$$\rho(x) = \left(1 \ \ x \ \ y \ \ x^2 \ \ xy \ \ y^2\right). \tag{2}$$

where $m=6$.

In moving least-squares fitting, the coefficient $\beta(x)$ is determined based on the weighted least squares. This makes the weighted sum of squares of each sampling point's errors minimized for the approximate function $g(x)$ in the neighborhood Ω_x of the point x.

$$\Psi = \sum_{I=1}^{N} w_I(x)[g(x) - f(x_I)]^2$$

$$= \sum_{I=1}^{N} w_I(x)\left[\sum_{i=1}^{m} \beta_i(x)\rho_i(x_I) - f(x_I)\right]^2 \tag{3}$$

$$\frac{\partial \Psi}{\partial \beta_i(x)} = 2\sum_{I=1}^{N} w_I(x)\left[\sum_{i=1}^{m} \beta_i(x)\rho_i(x_I) - f(x_I)\right]\rho_i(x_I) = 0 \tag{4}$$

where $i = 1, 2, ..., m$, N is number of sampling points within the neighborhood Ω_x of the point x. The $w_I(x)=w_I(x-x_I)$ is weighting function at the sampling point x_I. The $w_I(x)$ is greater than zero in a limited area Ω_I around the sample point x_I, beyond the Ω_I are zero. The Ω_I is support domain of weight function, also called impact domain of sampling points x_I. The accuracy of moving least square approximation depends largely on the weight function, which often using Gaussian weighting function in the application, detailed introduction refers to the [7].

3   Adaptive Anomaly Detection Model of Hydropower Unit Vibration

The real-time monitoring and timely warn of pumped storage unit is important for power grid's stable operation. For the moment pumped storage power units have implemented monitoring system to online collect key parts' monitoring signals. The measured values of monitoring signals are simply compared with a preset threshold to achieve the alarm, which guide the operation and maintenance of the unit. This method of a static alarm threshold is a single judge. It ignores the unit's performance differences in different working conditions. When an alarm occurs, the equipment performance of unit may have largely deviated from the design conditions. A situation may occur. That is the unit's equipment may have seriously deteriorated, but the alarm level of condition monitoring system has not yet been reached. It can be seen that static alarm threshold lacks the thorough study of monitoring signal's hidden abnormities' or faults' information. And it lacks the warning capacity for early potential failures. It is far from insufficient to fully reflect the operational condition of the unit. Meanwhile, with the constantly expanding of hydropower plant capacity and the gradual improvement of monitoring auxiliary systems, the information quantity of unit's control and monitoring data increase continuously. The operation personnel are often difficult to understand unit's situation based on such a large amount of data [8]. They can't timely find unit's abnormity and fault. This can happen that there are huge amounts of data, but lack of information to guide decision.

In the same or similar operating conditions, when the equipments of pumped storage units are normal, their monitoring parameters should be random fluctuations in the mean nearby. After a long time's running, the unit will gradually deviate normal operation condition, enter the non-normal operation condition. As time went on, unit's deterioration will gradually accelerate, from quantitative change to qualitative change, may lead to serious consequences. Therefore, unit's abnormal operating condition should be paid great attention and closely observed. The reasons for an abnormality occurs should be analyzed. The effective solutions should be searched. Unit equipments' abnormality can be early found by effectively mining the useful implicit information of online monitoring data. This can guide the operator to adjust and control unit, ensure efficient and stable operation of the unit, minimize accidents probability.

A large number of field data analysis show that the main factors affecting hydropower unit's operating condition are active power and working head. In this paper, considering the effect of active power and working head, the health standard three-dimensional surface model $c=f(P, H)$ is built, where c is the unit's condition parameter, P is active power, H is working head. Based on the three-dimensional surface model, the adaptive anomaly detection of hydropower unit is made, concrete steps are as follows:

(1) Analyzing the conditions monitoring data of pumped storage power plant units in different operating condition, determining unit's standard health condition. Selecting the characteristic parameters which can reflect unit's operating condition.

(2) Inputting the unit's characteristic parameters in the health condition into LS-SVM to train, building a three-dimensional surface model $c=f(P, H)$, and validating the model.

(3) Substituting the real-time condition monitoring data of active power and working head into the trained LS-SVM model to calculate the health standard value $c(t)$ of condition parameters in the current operating conditions. Comparing the current real value $r(t)$ and health standard value $c(t)$, calculate the vibration deviation of the unit in current condition:

$$d_v(t) = \frac{r(t) - c(t)}{c(t)} \times 100\% .$$

(5)

where t is run time of pumped storage units. If $d_v(t)$ exceeds a preset threshold, an alert of abnormal vibration can be made. This can promptly find abnormal conditions of the units.

4   Case Study

The real condition monitoring data of a pumped storage power plant unit in September 22, 2008 ~ December 15, 2011 are studied to validate the effectiveness of adaptive anomaly detection model of hydropower unit's vibration parameters in varying conditions. The model is based on moving least squares response surface. Due to the complexity of pumped storage units' operating conditions, frequent starts and stops and working conditions switch, the validity of the proposed model in changing conditions can be better reflected by using pumped storage units' monitoring data. The x-direction horizontal vibration of upper bracket is selected as the study object.

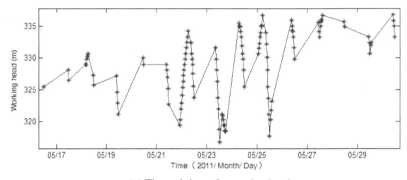

(*a*) The real data of operating head.

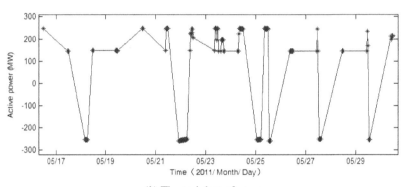

(*b*) The real data of power.

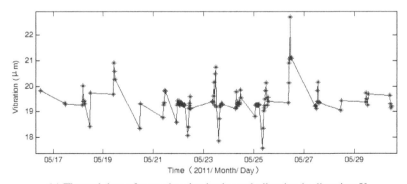

(*c*) The real data of upper bracket horizontal vibration in direction-X.

Fig. 1. The condition monitoring data of pumped storage power station unit

Figure 1 shows unit's real data of working head, active power and upper bracket horizontal vibration in direction-X in from May 16, 2011 to May 30, 2011. It can be seen from Figure 1 that unit's active power focused on 250MW for pumping conditions; power is concentrated in 150MW, 200MW and 250MW for generating operation. The working head has strong volatility. The conversion of pumping and generating conditions is frequent. So the changes of upper bracket horizontal vibration in direction-X are complex. The effective information which reflects the true condition can't be obtained only from this Figure. Research shows that active power, working head have an important impact on the unit's vibration parameters. If setting a single static alarm threshold for units, the performance difference, hidden information of abnormality and fault will be greatly neglected. And the unit's real condition can't be truly reflected. Therefore, it is need to build a three-dimensional surface model to detect the abnormality of vibration parameters. This model should be adaptive the changes of working conditions for pumped storage units.

Firstly, determining unit's standard health condition, selecting the characteristic parameter which can reflect unit's operating condition.

The online monitoring data (unit has good condition and without fault) of unit initial operation are adopted to build vibration standard model of unit in healthy condition. The 800 sets online monitoring data from September 22, 2008 to September 18, 2009 are selected. The peak-peak value of x-direction horizontal vibration of upper bracket is selected as the characteristic parameter.

Then, inputting unit's healthy condition parameter into the moving least squares response surface to train, building a three-dimensional surface model $c=f(P, H)$, and validating the model.

To real-timely get a true operating condition of hydropower units, it is need to build a health condition model. Considering the important influence of power and working head on hydropower unit's vibration characteristics, and moving least square response surface has good fitting performance for scattered data, a vibration-power-working head three-dimensional surface model $v=f(P, H)$ of hydropower unit is built. This model is based on moving least squares response surface. Through this model, the mapping relationship in health condition among power (P), working head (H) and vibration parameter (v) can be obtained. For the 800 sets data from September 22, 2008 to September 18, 2009, 600 sets data are selected to establish a health standard model, the remaining 200 sets data as the test samples to validated the model. In order to make the moving least squares response surface model has good performance, the selected 800 sets health standard data should cover possible changes in working head and active power. The 200 test samples' active power and working head are inputted this model, the results can be seen that the health standard values of upper bracket horizontal vibration in direction-X based on moving least squares response surface model is consistent with the measured values. The average relative error is 3.36 %.

Finally, substituting the unit's online monitoring data of power and working head into the trained three-dimensional surface model (moving least squares response surface), calculating the health standard value $c(t)$ of the condition parameter in current condition. Using formula (5) to online assess unit's real-time operating condition, achieving early warn to unit's vibration anomalies.

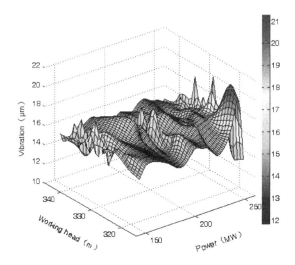

Fig. 2. The three-dimension surface of vibration-power-working head for hydropower unit

Substituting the power and working head of unit's condition monitoring data that after two years (May 12, 2011 ~ December 15, 2011) into unit's health model $v(t)=f(P(t), H(t))$, calculating the health standard values $v(t)$ of condition parameter in the current condition, and comparing the $v(t)$ and the real values $r(t)$. The comparison can be seen in Figure 3. The formula (5) is using to calculate the vibration deviation of the unit in current condition. The results are shown in Figure 4. It can be seen from Figure 4 that after two years of operation, the deterioration of pumped storage unit's component occurs, the unit is gradually deviating its healthy operation condition. If $d_v(t)$ exceeds a preset threshold, an alert of abnormal vibration is made. This can promptly detect the unit's abnormal condition.

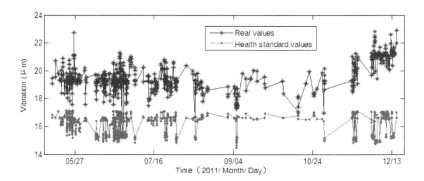

Fig. 3. The fault early warning results of pumped storage unit

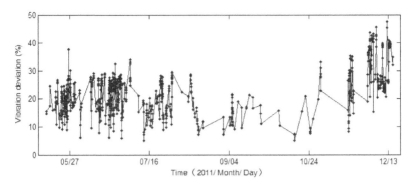

Fig. 4. The fault early warning results of pumped storage unit.

In summary, hydropower unit's vibration anomaly can be found early by using the presented three-dimensional surface anomaly detection model. When making unit's maintenance plan, the unit parts whose condition is abnormal, can be checked purposefully. This can effectively avoid the possibility of forced outages, truly realize the prevention of unit's fault. So field operator of hydropower plant can real-timely and comprehensivly obtain the health condition of unit's key components. The meaningless alarms will be reduced. The repair time will be shorten and the operation time will be increased.

5   Conclusions

Pumped storage unit has complex operating conditions. It has many condition monitoring points, less fault samples. It is difficult to effectively diagnose it's falut. Setting static alarm thresholds will ignore unit's differences of dynamic performance in varying operating conditions. So an adaptive real-time anomaly detection model of hydropower unit vibration parameters is proposed based on three-dimensional surfaces. Firstly, the unit's conditions monitoring data in different operating condition are analyzed. This reveals the key factors of active power and working head, which affect unit's performance. The unit's health standard condition is determined. Then, the characteristics parameters which can reflect of unit's operating condition are selected. The selected parameters of health condition are inputted into LS-SVM to train. Finally, unit's current condition monitoring data are inputted into the three-dimensional surface model to online assess unit's condition ment, achieve early warning of abnormal vibration. The example shows that the proposed model can effectively demonstrate the hidden information of condition monitoring data, real-timely track unit's operating condition, early warn unit's potential failures. The model has are good application prospects.

Acknowledgments. This work was supported by the National Natural Science Foundation of China (grant number 51309258) and the Special Foundation for Excellent Young Scientists of China Institute of Water Re-sources and Hydropower Research (grant number 1421).

References

1. An, X.L., Pan, L.P., Zhang, F.: Condition degradation assessment and nonlinear prediction of hydropower unit. Power System Technology 37(5), 1378–1383 (2013)
2. An, X.L.: Vibration characteristics and fault diagnosis for hydraulic generator units. Thesis for the degree of doctor of engineering, Huazhong University of Science & Technology, Wuhan (2009)
3. Lu, W.G., Dai, Y.P., Gao, F.: A hydroelectric-generator unit faults early warning method based on distribution estimation. Proceedings of the CSEE 25(4), 94–98 (2005)
4. Renders, J.M., Goosens, A., Viron, F.D.: A prototype neural network to perform early warning in nuclear power plant. Fuzzy Sets and Systems 74, 139–151 (1995)
5. Karlsson, C., Larsson, B., Dahlquist, A.: Experiences from designing early warning system to detect abnormal behaviour in medium-sized gas turbines. In: 23rd International Congress on Condition Monitoring and Diagnostic Engineering Management, pp. 117–120. Sunrise Publishing, Hikone (2010)
6. Jardine, A., Lin, D., Banjevic, D.: A review on machinery diagnostics and prognostics implementing condition-based maintenance. Mechanical Systems and Signal Processing 20(7), 1483–1510 (2006)
7. Zhang, Y., Li, G.Y., Zhong, Z.H.: Design optimization on lightweight of full vehicle based on moving least square response surface method. Chinese Journal of Mechanical Engineering 44(11), 192–196 (2008)
8. Yan, J.F., Yu, Z.H., Tian, F.: Dynamic security assessment and early warning system of power system. Proceedings of the CSEE 28(34), 87–93 (2008)

Capacity and Power Optimization for Collaborative Beamforming with Two Relay Clusters

Bingbing Lu[1,2], Ju Liu[1,2], Chao Wang[1], Hongji Xu[1,2], and Qing Wang[1]

[1] School of Information Science and Engineering,
Shandong University, Jinan, 250100, China
`juliu@sdu.edu.cn`
[2] National Mobile Communications Research Laboratory,
Southeast University, Nanjing, 210096, China
`hongjixu@sdu.edu.cn`

Abstract. In this paper, we study two approaches to optimize the problems between capacity and power in a three-hop multi-relay network. In the first approach, two beamforming weight vectors are designed to maximize the capacity under the power constraints of relay clusters. While in the second approach, we minimize the power of total relay nodes as well as meet the minimal capacity demand. In both of design schemes, we turn into two beamformer vectors to only one though a series of mathematical manipulation. Then apply genetic algorithm (GA) to obtain the optimal weight value of the nonconvex problems. Simulation results show that our proposed approaches significantly outperform the previous methods conducted.

Keywords: cooperative communication, three-hop, multi-relay, genetic algorithm.

1 Introduction

Exploiting relay nodes to improve information capacity and link reliability has attracted increasing interest recently [1–4]. Many swarm intelligence algorithms [5–8] which can achieve optimal results also quickly development. Recently, the dual-hop relay systems have attracted attentions in the research academia. As the serious signal fading and path loss problems in some specific situation, we consider a three-hop relay system which consists of a transmitter, a receiver and two clusters of relay nodes. The relays at both terminals will form like a multi-input multi-output (MIMO) beamforming system [9, 10]. Some methods are proposed to optimize this problem like in [11], the cooperative relay weight coefficients are optimized by maximizing the destination SNR under the sum-power constraints at the relay clusters.

In this paper, we develop two distributed beamforming approaches in a three-hop AF cooperative communication system. In the first approach, we aim to

Y. Tan et al. (Eds.): ICSI 2014, Part II, LNCS 8795, pp. 155–162, 2014.

maximize the information capacity subject to the separate total power constraint. In the second approach, we minimize the total transmit power which maintains the mutual information above a predefined threshold. As illustrated, we turning multi-variables into single one then solve by Genetic Algorithm (GA). Compared with the defect that be easy trapped into local optimal solution in [12], our proposed approaches can obtain the global optimal solution in statistically.

The remainder of the paper is organized as follows. In Section II, we present the system model and the optimization problem. The approaches of two relay weights optimization are presented in section III. Simulation results are provided in Section IV. Finally section V concludes the paper.

2   System Model and Optimization Problem

2.1   System Model

Consider a relay network in Fig. 1. We suppose that there is no direct link between the transmitter and receiver. Each node is equipped with a single antenna, and is subject to the half-duplex constraint. The Rayleigh flat fading channel coefficients between all the nodes are identical independent distributed.

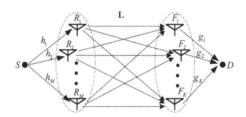

Fig. 1. System model [11]

The transmission information is divided into three parts. In the first part, the source node broadcasts the signal $\sqrt{P_0}s$ to the first relay cluster, where $E\{|s|^2\} = 1$. The signal $\mathbf{x}$ sent by $\{R_m\}_{m=1}^M$ is given as $\mathbf{x} = \sqrt{P_0}\mathbf{W}\mathbf{h}s + \mathbf{W}\mathbf{n}_R$, where $\mathbf{W} = diag([w_1, w_2, \cdots w_M])$ and $\mathbf{n}_R$ with zero mean and variance of σ^2, similar as the noise $\mathbf{n}_F$ at $\{F_k\}_{k=1}^K$ and n_D at the destination.

In the second part, $\{F_k\}_{k=1}^K$ retransmits the received signal $\mathbf{y}$ multiplied by the beamforming matrix $\mathbf{P}$, which can be expressed as $\mathbf{y} = \mathbf{P}\mathbf{L}\mathbf{x} + \mathbf{P}\mathbf{n}_F$, where $\mathbf{P} = diag([p_1, p_2, \cdots p_K])$. Finally, the signal d received at the destination can be written as

$$d = \mathbf{g}^H\mathbf{y} + n_D = \sqrt{P_0}\mathbf{p}^H\mathbf{GLHw}s + \mathbf{p}^H\mathbf{GLWn}_R + \mathbf{p}^H\mathbf{Gn}_F + n_D \qquad (1)$$

where $\mathbf{G} = \text{diag}(\mathbf{g})$, $\mathbf{H} = \text{diag}(\mathbf{h})$, $\mathbf{p} = [p_1, p_2, ..., p_K]^H$ and $\mathbf{w} = [w_1, w_2, ..., w_M]^H$.

2.2   Optimization Problem

We want to find the optimal solution from the capacity maximization and relay power minimization. By Shannon's second theorem, the mutual information between the source and destination is given by $I = \frac{1}{2}\log_2(1 + \text{SNR})$. The capacity maximization is equivalent to maximized the receiving SNR at D since I is increasing along with SNR.

The two problems will be optimized with two relay beamformers $\mathbf{w}$ and $\mathbf{p}$, which are the capacity maximization under the relay power constraints and the relay power minimization as well as meet the capacity demand. In the both problems, the transmit power P_R of relay cluster $\{\text{R}_m\}_{m=1}^{M}$ can be calculated as $P_R = \sum\limits_{m=1}^{M} E\{|x_m|^2\} = \mathbf{w}^H \mathbf{D}_R \mathbf{w}$, where $\mathbf{D}_\text{R} = P_0 diag\left(E\{|h_1|^2\}, E\{|h_2|^2\} \cdots E\{|h_M|^2\}\right) + \sigma^2\mathbf{I}$. In the same way, the total transmit power P_F can be obtained as $P_F = \sum\limits_{k=1}^{K} E\{|y_k|^2\} = \mathbf{p}^H \mathbf{D}_F \mathbf{p}$, where $[\mathbf{D}_F]_{k,k} = \sum\limits_{m=1}^{M} |l_{k,m}|^2 [\mathbf{D}_R]_{m,m} |w_m|^2 + \sigma^2, \ k = 1, 2 \cdots, K.$

3   Relay Weights Optimization Based on GA

GA is a metaheuristic optimization algorithm that use survival of the fittest evolutionary scheme to refine a set of solution candidates iteratively [13]. The computing process of GA includes mutation, crossover, selection and inheritance. The algorithm begins with creating an initial population of random individuals which represented as binary strings. Then we compute the objective values of the solution candidates to select the fittest individuals under some constraints. The selected individuals are reproduced by crossover and mutation so as to create the new individuals. The GA with the features of randomness and heuristic search nature has made the algorithm a suitable tool for finding the global optimal solution.

In the following two problems of capacity maximization and relay power minimization, the GA is a basic approach to solve the final optimization problems.

3.1   Capacity Maximization

In this subsection, our goal is to maximal the information capacity subject to the separated power constraint of the two relay clusters. Because of the relationship between I and SNR, the optimal I can be calculated by the optimal solution of SNR, then this optimization problem is equivalent to

$$\max_{\mathbf{w},\mathbf{p}} \quad \frac{P_0}{\sigma^2} \frac{\mathbf{p}^H \mathbf{GLHw}(\mathbf{GLHw})^H \mathbf{p}}{\mathbf{p}^H \left\{ \mathbf{GLw}(\mathbf{GLw})^H + \mathbf{gg}^H \right\} \mathbf{p} + 1} \tag{2}$$

$$\text{s.t.} \quad \mathbf{w}^H \mathbf{D}_R \mathbf{w} \leqslant P_R^{\max}, \mathbf{p}^H \mathbf{D}_F \mathbf{p} \leqslant P_F^{\max}$$

Our goal is to obtain the optimal vector $\mathbf{w}$ and $\mathbf{p}$ so that the SNR is maximized. We can see that, this problem is not convex, and the two beamforming vectors $\mathbf{w}$ and $\mathbf{p}$ depend on each other in problem (2), which make the problem more difficult to solve. We find that the initial problem can be regard as a question with variable $\mathbf{p}$ when $\mathbf{w}$ is a constant selected within the feasible region. And then for any available $\mathbf{w}$, the maximum achievable SNR [10] can be expressed as

$$SNR_{\max}(\mathbf{w}) = P_F^{\max} P_0 \mathbf{M}^H \mathbf{D}_F^{-1/2} \mathbf{X} \mathbf{D}_F^{-1/2} \mathbf{M} \tag{3}$$

where $\mathbf{M} = \mathbf{GLHw}$, $\mathbf{X} = (\sigma^2 \mathbf{I} + P_F^{\max} \mathbf{D}_F^{-1/2} \mathbf{Q} \mathbf{D}_F^{-1/2})^{-1}$, $\mathbf{Q} = \sigma^2 \mathbf{GLw}(\mathbf{GLw})^H$ $+ \sigma^2 \mathbf{gg}^H$. Therefore the optimization problem (2) is simplified into a question with only one variable $\mathbf{w}$ with the constraint, which can be written as

$$\max_{\mathbf{w}} \; P_F^{\max} P_0 \mathbf{w}^H (\mathbf{GLH})^H \mathbf{D}_F^{-1/2} \mathbf{X} \mathbf{D}_F^{-1/2} \mathbf{GLHw}$$
$$s.t. \quad \mathbf{w}^H \mathbf{D}_R \mathbf{w} \leqslant P_R^{\max} \tag{4}$$

Genetic Algorithm

Initial w
Calculate $\mathbf{p}$ by (3)
for g=1; g<=G; g=g+1   //G: generation limit
for i=1; i<= N; i=i+1   //N: Number of individuals
  if $\mathbf{w}_i$ don't meet constraints
  then utilize penalty function to generate new $\mathbf{w}_i$
  else
    $q(i) = \frac{\mathbf{w}_i{}^H \mathbf{A} \mathbf{w}_i}{\mathbf{w}_i{}^H \mathbf{B} \mathbf{w}_i + c}$ //q(i):fitness faction of $\mathbf{w}_i$
    if $q(i) > \varepsilon$ //ε: a predefined constant
    then reproduction $\mathbf{w}_i$
    else
      give up $\mathbf{w}_i$
    end if
  end if
  $\mathbf{w}_i$ crossover with $\mathbf{w}_{i+j}$ of probability p_c
  $\mathbf{w}_i$ mutation of probability p_m
end
end

It can be shown that, the problem (4) is not convex. GA is often applied as an approach to obtain a statistics global optimal solution for this nonconvex problems. The details process of the algorithm are expressed in **Genetic Algorithm**. The maximum I can be calculated by the obtained SNR. Because of this lengthy process, the computational complexity of GA is relatively high. Compared with using GA to solve the problem in (2) directly, the initial population of solving (4) is produced only by $\mathbf{w}$. Thus the coding length of initial population individuals will reduce to half, which leads to a decrease of computation time through abundant crossover, mutation and duplication in every generational population of GA.

3.2   Power Minimization

In this subsection, our goal is to minimize the total transmit power while keeping the capacity at the destination above a certain preconcerted threshold. Similar as the first problem, the optimization of I can be converted to the problem about SNR, which meet the threshold γ. The problem can be expressed as

$$\min_{\mathbf{w},\mathbf{p}} \quad \mathbf{w}^H \mathbf{D}_R \mathbf{w} + \mathbf{p}^H \mathbf{D}_F \mathbf{p}$$

$$\text{s.t.} \quad \frac{P_0}{\sigma^2} \frac{\mathbf{p}^H \mathbf{GLHw}(\mathbf{GLHw})^H \mathbf{p}}{\mathbf{p}^H \left\{ \mathbf{GLw}(\mathbf{GLw})^H + \mathbf{gg}^H \right\} \mathbf{p} + 1} \geq \gamma \tag{5}$$

Similar to the previous subsection, $\mathbf{w}^H \mathbf{D}_R \mathbf{w}$ in the objective function is fixed for a certain $\mathbf{w}$. Under the condition of every fixed $\mathbf{w}$, this question can be simplified as the problem about p. It can be known [10] that this optimization problem will get the global optimal solution when $\mathbf{p}$ is chosen as

$$\mathbf{p} = g(\mathbf{w}) = \left(\frac{\gamma \sigma^2}{\mathbf{u}^H \mathbf{D}_F^{-1/2}(\mathbf{R}-\gamma \mathbf{Q})\mathbf{D}_F^{-1/2}\mathbf{u}} \right)^{1/2} \mathbf{D}_F^{-1/2} \mathbf{u} \tag{6}$$

where $\mathbf{u} = \mathcal{P}\{\mathbf{D}_F^{-1/2}(\mathbf{R}-\gamma \mathbf{Q})\mathbf{D}_F^{-1/2}\}$ and the matrix $\mathbf{R} = P_0 \mathbf{GLHw}(\mathbf{GLHw})^H$. We put (6) into (5), which turns (5) into a problem with only one variable $\mathbf{w}$. Mathematically, the simplified optimization problem can be expressed as

$$\min_{\mathbf{w}} \quad \mathbf{w}^H \mathbf{D}_R \mathbf{w} + g^H(\mathbf{w}) \mathbf{D}_F g(\mathbf{w})$$

$$\text{s.t.} \quad \frac{\mathbf{w}^H \tilde{\mathbf{A}} \mathbf{w}}{\mathbf{w}^H \tilde{\mathbf{B}} \mathbf{w} + \tilde{c}} \geq \gamma \tag{7}$$

where $\tilde{\mathbf{A}} = P_0(g(\mathbf{w})^H \mathbf{GLH})^H (g(\mathbf{w})^H \mathbf{GLH})$, $\tilde{\mathbf{B}} = (g(\mathbf{w})^H \mathbf{GL})^H (g(\mathbf{w})^H \mathbf{GL})$ and the notation $\tilde{c} = \sigma^2 g(\mathbf{w})^H \mathbf{gg}^H g(\mathbf{w}) + \sigma^2$.

We solve the problem in (7) to obtain the optimal solution using GA. For the implement of GA, the initial population is produced according to the vector $\mathbf{w}$, and the beamforming weight $\mathbf{p}$ can be calculated by $\mathbf{w}$. The fitness function and the constraints in GA are the same as the objective function and the restrict function expressed in (7), respectively. We can obtain the optimal value of vector $\mathbf{w}$, and calculate the corresponding optimal $\mathbf{p}$ and the global minimum of $P_R + P_F$ through GA. By constrast, our proposed method has a lower complexity of computing than solving the problem in (5) with GA directly.

4   Simulation Result

In this section, simulations are designed to assess the performance of the proposed algorithms. Over all the simulation process, all the nodes are with the same noise power level and the transmit power P_0 of the source node is set to be

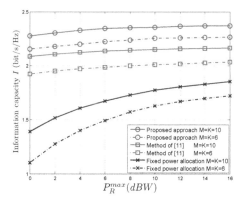

Fig. 2. Information capacity against $P_R^{\max}$ with 6 (dash line) and 10 (solid line) relay nodes, respectively

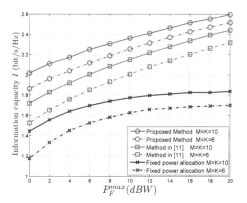

Fig. 3. Information capacity against $P_F^{\max}$ with 6 (dash line) and 10 (solid line) relay nodes, respectively

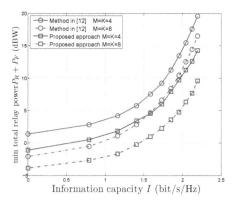

Fig. 4. Minimum total relay power $P_R + P_F$ versus capacity threshold for M=K=4, 6, 8, respectively

10dBW. In the simulation process, only beamforming vector $\mathbf{w}$ is the random variable, and $\mathbf{p}$ is calculated by $\mathbf{w}$. The objective function and constraints in GA are presented in (4) and (7).

Fig. 2 shows the maximum achievable information capacity versus the maximum allowable total transmit power $P_R{}^{\max}$. In Fig. 3, we have shown the maximum achievable mutual information versus the power $P_F{}^{\max}$. We can see that the proposed methods outperform the method presented in [11] and greatly exceed the fixed power allocation solution. As can be seen in these figures, the maximum information capacity tend to be saturated near some reaching threshold with the increasing of $P_R{}^{\max}$, while the maximum capacity almost keep linear increase with the rising of $P_F{}^{\max}$. Therefore, $P_F{}^{\max}$ has more contribution than $P_R{}^{\max}$ after they reaching a certain power.

In Fig. 4, we have plotted the minimum total power of relay nodes $P_R + P_F$ versus the capacity threshold. The total transmit power decrease with the rise of relay numbers due to the diversity gain. It can be observed that the proposed method suffers certain power reduction compared to the method in [12]. This is because the iterative computation between w and p in [12] is easy to fall into local optimal solution, while on the contrary, our proposed method based on GA can avoid this problem effectively.

5   Conclusion

In this paper, we consider the problem of distributed beamforming in a cooperative communication network which consists of a transmitter, a receiver and two relay clusters equipped at the transmitter and receiver side, respectively. The beamforming weight vectors are designed in two different approaches. In the first approach, we aim to obtain the maximum achieved information capacity subject to the power constraints of two relay clusters. In the second approach, we design the beamformer through minimizing the total transmit power of all the relay nodes subject to a constraint which guarantees the mutual information above a predefined threshold. In both of the two approaches, the two random variables can be reduced to only one. For this reason, the computational complexity will be reduced due to the halving of the initial population coding length. Simulation results show that the proposed methods can achieve great improvement compared to the existing solutions.

Acknowledgment. This work was supported by National Natural Science Foundation of China under Grant(61371188), Research Fund for the Doctoral Program of Higher Education under Grant(20130131110029), Natural Science Foundation of Shandong under Grant(ZR2011FM027), Open Research Fund of National Mobile Communications Research Laboratory under Grant (2012D10), China Postdoctoral Science Foundation funded project (2011M501092), Independent Innovation Foundation of Shandong University (2012ZD035), Special Fund for Postdoctoral Innovative Projects of Shandong Province (201103003), Scientific Research Foundation for the Excellent Young and Middle-aged Scientists of Shandong Province (BS2012DX024).

References

1. Laneman, J.N., Tse, D.N.C., Wornell, G.W.: Cooperative diversity in wireless networks: Efficient protocols and outage behavior. IEEE Transactions on Information Theory 50(12), 3062–3080 (2004)
2. Lim, G., Cimini, L.J.: Energy-efficient cooperative beamforming in clustered wireless networks. IEEE Transactions on Wireless Communications 12(3), 1376–1385 (2013)
3. Madan, R., Mehta, N.B., Molisch, A.F., Zhang, J.: Energy-efficient cooperative relaying over fading channels with simple relay selection. IEEE Transactions on Wireless Communications 7(8), 3013–3025 (2008)
4. Yang, Y., Li, Q., Ma, W.-K., Ge, J., Ching, P.C.: Cooperative secure beamforming for AF relay networks with multiple eavesdroppers. IEEE Signal Processing Letters 20(1), 35–38 (2013)
5. Ge, M., Wang, Q.-G., Chiu, M.-S., Lee, T.-H., Hang, C.-C., Teo, K.-H.: An effective technique for batch process optimization with application to crystallization. Chemical Engineering Research and Design 78(1), 99–106 (2000)
6. Precup, R.-E., David, R.-C., Petriu, E.M., Preitl, S., Radac, M.-B.: Novel adaptive gravitational search algorithm for fuzzy controlled servo systems. IEEE Transactions on Industrial Informatics 8(4), 791–800 (2012)
7. Saha, S.K., Ghoshal, S.P., Kar, R., Mandal, D.: Cat swarm optimization algorithm for optimal linear phase FIR filter design. ISA Transactions 52(6), 781–794 (2013)
8. Yazdani, D., Nasiri, B., Azizi, R., Sepas-Moghaddam, A., Meybodi, M.R.: Optimization in dynamic environments utilizing a novel method based on particle swarm optimization. International Journal of Artificial Intelligence 11(A13), 170–192 (2013)
9. Chalise, B.K., Vandendorpe, L.: MIMO relay design for multipoint-to-multipoint communications with imperfect channel state information. IEEE Transactions on Signal Processing 57(7), 2785–2796 (2009)
10. Nassab, V.H., Shahbazpanahi, S., Grami, A., Luo, Z.-Q.: Distributed beamforming for relay networks based on second-order statistics of the channel state information. IEEE Transactions on Signal Processing 56(9), 4306–4316 (2008)
11. Chen, L., Wong, K.-K., Chen, H., Liu, J., Zheng, G.: Optimizing transmitter-receiver collaborative-relay beamforming with perfect CSI. IEEE Communication Letters 15(3), 314–316 (2011)
12. Wang, C., Liu, J., Dong, Z., Xu, H., Ma, S.: Multi-hop Collaborative Relay Beamforming. In: 78th IEEE Vehicular Technology Conference, pp. 1–5. IEEE Press, Las Vegas (2013)
13. Tang, K.S., Man, K.F., Kwong, S., He, Q.: Genetic algorithms and their applications. IEEE Signal Processing Magazine 13(6), 22–37 (1996)

Solving Power Economic Dispatch Problem
Subject to DG Uncertainty via Bare-Bones PSO

Yue Jiang, Qi Kang, Lei Wang, and Qidi Wu

Department of Control Science and Engineering,
Tongji University, Shanghai 201804, China
qkang@tongji.edu.cn

Abstract. Distributed generation (DG) is becoming increasingly important in power system. However the of DG will lead risks in power system due to its failure or uncontrollable power outputs which is usually relied on renewable energy. In this work, we solve the economic dispatch (ED) problem by considering controllable and uncontrollable DG in power system. This paper applies the bare-bones particle swarm optimization (BBPSO) method to solve the ED problem. The performance of BBPSO method is evaluated via IEEE 118-bus test system, and would be compared with other methods in terms of convergence performance and solution quality. The results may verify the effectiveness and promising application of the proposed method in solving the ED problem when we are considering both controllable and uncontrollable DG in power system.

Keywords: particle swarm optimization, bare-bones PSO, economic dispatch, uncontrollable.

1   Introduction

With the increasing use of distributed generation (DG), the power system will face many risks of system disruption. Because DG mainly relies on renewable energy such as solar and wind power which is unstable, thus the real power output is changing with weather. There are many types of DG, so the changing rules are different. For instance, the real power output of wind turbines is related to wind speed, which Weibull distribution is generally considered as the optimal probability density function. Therefore, this paper adopts Weibull distribution [1] as a probability distribution of wind speed over time.

Economic dispatch (ED) is an important problem in power system. ED is used in real-time energy management power system to control the production of thermal power stations. Its objective is to minimize the total cost of operating the generators, subject to load and operational constraints [10]. This paper proposes the notion which is when some DG units have a conspicuous rand feature, we can optimize the power output of available generators in order to ensure the safe, effectiveness and efficiency operation of the power system. Its objective is to minimize the total cost of operating the generators and optimize the network voltage profile.

Y. Tan et al. (Eds.): ICSI 2014, Part II, LNCS 8795, pp. 163–169, 2014.

Traditionally, ED problem can be addressed by a series of mathematical programming methods, such as lambda-iteration method [2], gradient method [3], and dynamical programming [4]. Nevertheless, such deterministic numerical methods cannot effectively solve nonsmooth and nonconvex cost function, moreover, it has larger scale dimension which is too difficult to calculate. In order to effectively solve such problems in power systems, many researchers used a variety of computational intelligence to handle it. For example the genetic algorithm [6], differential evolution algorithm [7], neural network [8], Tabu search algorithm [9] and so on. But the methods mentioned above have many shortcomings, including premature and poor convergence performance.

Usually speaking, people consider ED problem as a static optimization problem discussing in a normal power system. In this paper, however, we want to obtain a real-time power output of controllable generators in a dynamic environment. PSO [10]-[14] has been employed to handle the ED problem for many years. We propose the BBPSO [15]-[19] to solve ED problems.

2   Mathematical Model

The objective in this paper is to minimize the total short-term cost of operating the generators and network loss. It is given by:

$$\text{Minimize } F = \sum_{j=1}^{n} F_j(P_j) + P_{ploss} \tag{1}$$

where $F_j(P_j)$ is the cost function of the jth generator, P_j is the real output of the jth generator, n is the total # of the jth generators in the power system, P_{ploss} is the network loss in the power system.

$F_j(P_j)$ is related to the real power injected into the power system, which is modeled by the function.

$$F_j(P_j) = a_j + b_j P_j + c_j P_j^2 \tag{2}$$

where a_j, b_j and c_j are cost coefficient of the jth generator.

This paper adopts the Weibull distribution as probability distribution of wind speed over time .The probability density is given by.

$$f(v) = \frac{k}{c} \left(\frac{v}{c}\right)^{k-1} \exp\left[-\left(\frac{v}{c}\right)^k\right] \tag{3}$$

where, v is wind speed, k and c are shape parameter and scale parameter of Weibull distribution respectively. k and c can be calculated by the average wind speed μ and standard deviation .

$$k = \left(\frac{\sigma}{\mu}\right)^{-1.086} \tag{4}$$

$$c = \frac{\mu}{\Gamma(1+\frac{1}{k})} \tag{5}$$

where Γ is Gamma function.

The relationship between the power output and wind is as followed[21].

$$P = k_1 v + k_2 \qquad v_{c1} < v \leq v_r \tag{6}$$

Where v_{c1} is cut-in wind speed, v_r is rated wind speed, P_r is rated power, v_{c2} is cut-out wind speed. k_1, k_2 can be calculated as follows:

$$k_1 = \frac{P_r}{v_r - v_{c1}} \tag{7}$$

$$k_2 = -k_1 v_{c1} \tag{8}$$

According to statistics, wind speeds stay between cut-in speed and the rated wind speed most of the time. In this paper DG is to simplify as PV node in the power flow caculation.

In this work, we consider the following constraints.

$$\begin{cases} \sum_{j=1}^{n} P_j = P_D + P_{ploss} \\ P_j^{min} < P_j < P_j^{man} & (j = 1, \dots, n) \\ U_j^{min} \leq U_j \leq U_j^{man} & (j = 1, \dots, n) \\ P_j - P_j^0 \leq UR_j \\ P_j^0 - P_j \leq DR_j \end{cases} \tag{9}$$

Where P_D is the load demand of the power system; The operating range of all on-line is restricted by ramp rate limits. If power generation increases, then $P_j - P_j^0 \leq UR_j$. If power generation decreases, then $P_j^0 - P_j \leq DR_j$. P_j^0 is the previous power output power, UR_j the up-ramp limit of the jth generator, and DR_j is down-ramp limit of the jth generator.

3   Bare-Bones PSO Algorithm

PSO commonly used decimal encoding. It was motivated by the social behavior of fish schooling and bird flocks. There are N particles in the PSO algorithm. The search space assumes dimension D. Each particle has three vectors. Its current position $x_i = [x_i^1, x_i^2, \dots, x_i^D]$, its velocity $v_i = [v_i^1, v_i^2, \dots, v_i^D]$ and its personal best position pBest.

However, F. van den Bergh proved the update formula which cannot guarantee convergence to the global optimal solution [22]. Therefore, Kennedy proposed the BBPSO[15].

The particles' position in BBPSO is updated by the following equations:

$$x_i^d = \begin{cases} N\left(\dfrac{pBest_i^d + gBest^d}{2}\left|pBest_i^d - gBest^d\right|\right) & \text{rand} > 0.5 \\ pBest_i^d & \text{else} \end{cases} \quad (10)$$

Where, if rand>0.5, current position $x_i^d = pBest_i^d$, otherwise the current position will be derived from a Gaussian distribution with the mean $\dfrac{pBest_i^d + gBest^d}{2}$ and standard deviation $\left|pBest_i^d - gBest^d\right|$.

By the formula (10) to update x_i^d, it avoids adjusting parameter in the PSO algorithm and easy to implement. And it has the strong global convergence.

4   Solving ED Problems via BBPSO

Step 1. Input data of the power system, and set algorithm parameters. The power system data includes load demand, minimum P_j^{min} and maximum P_j^{man} of each generator, minimum U_j^{min} and maximum U_j^{man} of each node varies.

Step 2. Set k=1. Initialize the current position of each particle. Each particle's position is represented as matrix X. Set its personal best position equal to be its current position:

Step 3. By Power flow calculation, we can obtain the fitness value for each particle.

Step 4.   Update pbest and gbest.

Step 5. Update the position according to (12). If the position cannot fulfill constraints in (9).

Step 6. Examine the termination condition. If it is not met, Set k=k+1 and return to Step 3. Otherwise, end and output results.

5   Simulations and Results

In order to verify the effectiveness of the BBPSO, the power system we used for simulation is the IEEE 118-bus system[25] in this work. This system consists of 118 buses and 54 generator nodes. 1,4,6,8,10,12,15,18,19,24 and 25 node are DGs. The rang of the real power output of GD is 0-50Mw. Node 69 is a slack one in power flow calculation. The entire load of the power system is 4242MW. It assumed that node 1 accessing a wind turbine that has a conspicuous rand feature. The wind speed will be derived from a weibull distribution with the k = 2.3466 and c = 8.0928. Rated power of wind turbine is 1Mw, cut-in wind speed is 3m / s, rated wind speed is 12m/s. The problem now is how to optimize the real power output of all controllable online generators and to satisfy the load demand.

According to (3)-(6), we can get the real power output of node 1 at interval of 10 minutes. Fig.1 shows the power output in each period.

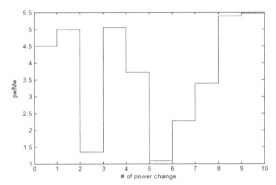

Fig. 1. Power change of node 1

The parameters of GA were set in[10], Pc=0.8 and Pm=0.1 are the crossover probability and mutation probability respectively. The parameters of RDPSO were set in[10],   and   are the thermal coefficient and the drift coefficient respectively, and the   decrease from 0.9 to 0.3 on the course of the search.   The parameters of PSO were set in[20], the range of inertia weight ω is decrease from 1.2 to 0.8 and the leaning factors were set as c1=c2=2.

Table 1 lists the total cost by each method mentioned when the power output of node 1 is 4.49Mw (in the first change). The mean cost and the standard deviation got by 100 runs of these methods. Fig.2 is convergence properties of the tested optimization methods. Both of them indicate that BBPSO has a better performance and robustness than other methods .

Table 1. The results for 100 runs

Methods	Min.Cost	Mean.Cost	Std.Dev	Mean.Time/min
GA	147856	161504	13648	10.0
PSO	150489	151815	1326	9.0
RDPSO	152352	153093	714	8.0
BBPSO	148023	148275	252	5.0

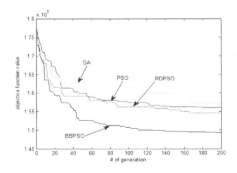

Fig. 2. Convergence properties of methods

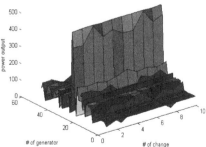

Fig. 3. The output power solution

Fig.3 is the solution of the power output of 53 generators when the power output of the node 1 keep changing in the interval of 10 minutes. It shows the power output solution of 52 generators in 10 times, when the wind speed of the node 1 is derived from a weibull distribution with the $k = 2.3466$ and $c = 8.0928$.

6   Conclusion

In this paper, we propose a BBPSO-based method to solve ED problems. The performance of the proposed method is evaluated in an IEEE 118-bus power system, and compared with other optimization methods in terms of convergence performance and solution quality. The results show that BBPSO method performs better than other compared methods in solving the ED problem. Our future work will focus on the dynamic economic dispatch problem.

Acknowledgement. This work was supported in part by the Natural Science Foundation of China (71371142, 61005090, and 61034004), the Fundamental Research Funds for the Central Universities, and the Research Fund of State Key Lab. of Management and Control for Complex systems.

References

1. Balouktsis, A., Chassapis, D., Karapantsios, T.D.: A nomogram method for estimating the energy produced by wind turbine generators. Solar Energy 72(3), 251–259 (2002)
2. Chowdhury, B.H., Rahman, S.: A review of recent advances in economic dispatch. Institute of Electrical and Electronics Engineers (1990)
3. Dhar, R., Mukherjee, P.: Reduced-gradient Method for Economic Dispatch. Proc. Inst. Elec., Eng. 120(5), 608–610 (1973)
4. Abarghooee, R., Aghaei, J.: Stochastic Dynamic Economic Emission Dispatch Considering Wind Power. In: IEEE Engineering and Automation Conference, vol. 1, pp. 158–161 (2011)
5. Loia, V., Vaccaro, A.: Decentralized Economic Dispatch in Smart Grids by Self-organizing Dynamic Agents. IEEE Trans. Systems, Man and Cybernetics: Systems 44(4), 397–408 (2014)
6. Chiang, C.: Improved Genetic Algorithm for Power Economic Dispatch of Units with Value-point Effects and Multiple Fuels. IEEE Trans. Power Syst. 20(4), 1690–1699 (2005)
7. Elsayed, S., Sarker, R., Essam, D.: An Improved Self-adaptive Differential Evolution Algorithm for Optimization Problems Industrial Informatics. IEEE Trans. on Industrial Informatics 9(1), 89–99 (2013)
8. Yang, C., Deconimck, G., Gui, W.: An Optimal Power-dispatching Control System for Electrochemical Process of Zinc Based on Back-propagation and Hopfield Neural Network. IEEE Trans. Electron 50(5), 953–961 (2003)
9. Lin, W., Cheng, F., Tsay, M.: An Improved Tabu Search for Economic Dispatch with Multiple Minima. IEEE Trans. Magn. 38, 1037–1040 (2002)

10. Sun, J., Palade, V., Wu, X.: Solving the Power Economic Dispatch Problem with Generation Constraints by Random Drift Particle Swarm Optimization. IEEE Trans. on Industrial Informatics 10(1), 222–232 (2014)
11. Faria, P., Soares, J.: Modified Particle Swarm Optimization Applied to Integrated Demand Response and DG Resources Scheduling. IEEE Transaction on Smart Grid 4(1), 606–616 (2013)
12. Chen, S., Manalu, G.: Fuzzy Forecasting Based on Two-factors Second-order Fuzzy-trend Logical Relationship Groups and Particle Swarm Optimization Techniques. IEEE Transaction on Cybernetics 43(3), 1102–1117 (2013)
13. Niknam, T., Doagou-Mojarrad, H.: Multiobjective Economic/Emission Dispatch by Multiobjective θ-particle Swarm Optimization, Generation, Transmission & Distribution, 6(5), 363–377 (2012)
14. Fu, Y., Ding, M.: Route Planning for Unmanned Aerial Vehicle (UAV) on the Sea Using Hybrid Differential Evolution and Quantum-behaved Particle Swarm Optimization. IEEE Trans. Systems, Man, and Cybernetics: Systems 43(6), 1451–1465 (2013)
15. Kennedy, J.: Bare Bone Particle Swarm. In: Proceedings of the IEEE Swarm Intelligence Symposium, pp. 80–87 (2003)
16. Blackwell, T.: A Study of Collapse in Bare Bones Particle Swarm Optimization. IEEE Transactions on Evolutionary Computation 16(3), 354–372 (2012)
17. Chen, C.: Cooperative Bare Bone Particle Swarm Optimization. Information Science and Control Engineering, 1–6 (2012)
18. Krohling, R., Mendel, E.: Bare Bones particle Swarm Optimization with Gaussian or Cauchy Jumps. In: IEEE Congress on Evolutionary Computation, pp. 3285–3291 (2009)
19. Wang, P., Shi, L.: A Hybrid Simplex Search and Modified Bare-bones Particle Swarm Optimization. Chinese Journal of Electronics 22(1), 104–108 (2013)
20. Kang, Q., Zhou, M.: Swarm Intelligence Approaches to Optimal Power Flow Problem with Distributed Generator Failures in Power Networks. IEEE Transactions on Automation Science and Engineering 10(2), 343–353 (2013)
21. Shapic, E., Balzer, G.: Power Fluctuation from a Large Wind Farm. In: International Conference on Future Power Systems, November 16-18 (2005)
22. van den Bergh, F.: An Analysis of Particle Swarm Optimizers, Pretoria, South Africa: Depatment of Computer Science, University of Pretoria (2002)
23. Pan, F., Hu, X., Eberhart, R.: An Analysis of Bare Bones Particle Swarm. In: IEEE Swarm Intelligence Symposium, pp. 1–5 (2008)
24. Kang, Q., Lan, T., Yan, Y.: Group Search Optimizer Based Optimal Location and Capacity of Distributed Generations. Neurocomputing 78(1), 55–63 (2012)
25. Power Systems Test Case Archive, http://www.ee.washington.edu/research/pstca/pf118/pg_tca118bus.htm

Extracting Mathematical Components Directly from PDF Documents for Mathematical Expression Recognition and Retrieval*

Botao Yu[1,2], Xuedong Tian[1,2,**], and Wenjie Luo[1,2]

[1] College of Mathematics and Computer, Hebei University, Baoding, Hebei, China
[2] Hebei key laboratory of Machine Learning and Computational Intelligence, Baoding, China
txdinfo@yahoo.com

Abstract. PDF document gains its popularity in information storage and exchange. With more and more documents, especially the scientific documents, available in PDF format, extracting mathematical expressions in PDF documents becomes an important issue in the field of mathematical expression recognition and retrieval. In this paper, we proposed a method of extracting mathematical components directly from PDF documents rather than cooperating indirectly with corresponding images converted from PDF files. Compared with traditional image-based method, the proposed method makes full use of the internal information of PDF documents such as font size, baseline, glyph bounding box and so on to extract the mathematical characters and their geometric information. The experimental result shows the method could meet the needs of the following processing of mathematical expressions such as formula structural analysis, reconstruction and retrieval, and has a higher efficiency than traditional image-based ways.

Keywords: PDF, Mathematical expression component, Mathematical expression recognition and retrieval, Font size, Baseline, Glyph bounding box.

1   Introduction

PDF (Portable Document Format) [1] documents present their contents and layouts in a manner independent of application software, hardware, and operating systems, which provides users with a consistent experience in sides of the displaying and printing pattern. With an increasing number of documents presented with PDF, more and more attentions are paid to this format of document for making good use of this resource.

Current researches on PDF documents involve extracting components from PDF documents or converting PDF documents into other formats such as XML and

---

* This work is supported by the National Natural Science Foundation of China (Grant No. 61375075) and the Natural Science Foundation of Hebei Province (Grant No. F2012201020).
** Corresponding author.

Y. Tan et al. (Eds.): ICSI 2014, Part II, LNCS 8795, pp. 170–179, 2014.

HTML. Chao and Fan [2] developed a method of extracting layout and content of PDF documents. The document was separated into text, image and vector graphics according to the object type. After that, words were formed to lines, then segments and images and vector graphics were saved. Marinai [3] developed a software tool to extract administrative metadata from PDF documents which could assist for building personal digital libraries. Déjean and Meunier [4] designed a system for converting PDF documents into structured XML format. In this system, streams that contain text, bitmap and vector images were extracted and converted respectively, and the extracted components were expressed in XML format. Rahman and Alam [5] proposed a method of converting PDF documents into HTML. By applying document image analysis techniques to retrieval logical and layout information of the document, the document was output in HTML format.

As an important component in PDF documents, mathematical expressions are also needed to be extracted for further recognizing and searching processes. Different from the images recognized and analyzed by mathematical formula recognition system, a PDF document that is not generated from scanned images has already contained the information of character code, baseline and font size. Therefore, the traditional operations applied to formula images for improving image quality and obtaining symbol codes such as image preprocessing (including binarization, denosing, skew detection and correction, etc.) and symbol recognition are not required. Nevertheless, the PDF documents do not provide the syntax and semantic information of symbols, it is necessary to locate the precise geometrical information of symbols and obtain their logic relationships for the following mathematical expression extraction, reconstruction and retrieval operations. Yang and Fateman [6] stated the significance of accessing mathematical expressions on line in digital documents like Postscript and PDF documents. And a method of extracting formulas from Postscript documents is proposed. First, a modified version of a program called Prescript was used to output information about strings and bounding boxes about typeset expressions. Then broken string fragments were assembled into words and items were determined as part of mathematical expressions by the characteristics of fonts and words (e.g. sin) commonly used in mathematical expressions and mathematical characters. Finally, the built-up mathematical expression in Lisp data structure was generated from stored data by applying clumping heuristics based on an existing Math/OCR program. In literature [7], Chan and Yeung summarized the existing work of mathematical expression recognition on symbol recognition and structural analysis. Lin et al. [8] combined rule-based and learning-based methods to detect both isolated and embedded mathematical expressions in PDF documents. For isolated formulas, they first used the character features to remove text lines which didn't seem to contain expressions. Then, the confidence level of classifying a line as a formula line was calculated by exploiting the geometric layout features. For embedded formulas, they focused much on the character features combined with additional layout features. Then the confidence level of a character being a math symbol was calculated. If the confidence level was higher than a threshold, the corresponding line was detected as an isolated formula or an embedded formula. In literature [9], they further discussed the identification of embedded mathematical formulas in PDF documents. First, text lines were segmented into words which are classified into formula type and ordinary text type with an SVM classifier. Then, formulas were extracted by merging formula type words as formulas. Baker et al. [10, 11] proposed a method of extracting mathematical expression by accessing the PDF document and a rasterized version of the PDF document. First, characters and their related information of font size, baseline and bounding box were extracted from the original PDF documents. In order to solve the problem of bounding box overlapping and obtain the exact character bounding boxes, they rendered PDF documents into images. After searching bounding boxes of the glyphs in the image, all the bounding boxes were registered with characters obtained from the original PDF documents. Then the expression parse tree was established with characters and related geometric information. This method paid attention to internal information in PDF documents such as font size, baseline, font name and font bounding box, which helped to locate the characters and got the minimal character bounding box from the corresponding images converted from original PDF document indirectly.

In this paper, we propose a method of extracting mathematical components directly from PDF documents for mathematical expression recognition and retrieval. Different from the method proposed in literature [10, 11], we obtain all information about glyph bounding box, baseline and font size, by directly accessing the original PDF documents, which makes full use of the internal information in the PDF documents and is also efficient. The glyph bounding boxes, together with baselines and character codes are used for the following processing.

A PDF document has complex structure. It belongs to a text and binary integrated format with compressed data, which leads to low readability of the original code of the documents. PDF documents can be generated by many tools and each tool has its own standard based on the PDF reference which has 7 editions so far. Font types used in PDF document could also frequently vary. When it comes to mathematical symbols, some symbols are generated by path construction operators (e.g. the long fraction symbol) or made of a character and a shape defined by the path construction operator (e.g. $\sqrt{}$ and a horizontal line make up the radical symbol). All these facts make it complex to extract components from a PDF document directly.

Our proposed method focuses on PDF documents with type1 font only and doesn't constrain tools that generate PDF documents.

2   Extracting Mathematical Components from PDF Documents Directly

The workflow of our method mainly contains 3 steps described as following and shown in Fig. 1.

1. Parsing of PDF documents. Parse the PDF documents to get the information of the fonts in Resource dictionary and the content of the Content stream of the Page objects.
2. Components extraction. Extract font and character information from the content parsed in step 2.
3. Expression output. Compose the mathematical expression with components in step 2 by using existing technique.

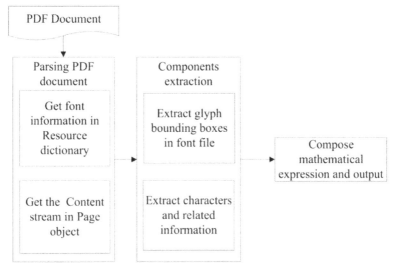

Fig. 1. Workflow of the proposed method

2.1   Parsing of PDF Documents

For a better understanding of PDF and considering our specified research purpose, we decide to develop a special parsing algorithm of PDF documents, although there are many open source SDKs or open source software such as iTextSharp, iText, PDFBox and so on that can handle PDF documents. The parsing process mainly contains the following steps:

1. Read the global objects such as Xref Table and root object to get all the Page objects.
2. Get basic font information in Resource dictionary of the Page object.
3. Decode the content stream of the Page object.

Once the PDF document is parsed and the contents of content stream and font information in resource dictionary are obtained, we move to the extraction process.

2.2   Extraction of Glyph Bounding Box in Font File

By accessing the font dictionary and the font descriptor obtained in 2.1, we get the font type, font name, font bounding box, character encoding and width of the glyph in the font. Font bounding box and width are expressed in glyph coordinate system and are measured in units in which 1000 units corresponds to 1 unit in text space. When the font bounding box is transformed by the transformation method which will be described in 2.3, we find that the font bounding box is actually larger than the area the character pixels take and cannot be treated as the parameter to calculate the precise area the character occupy, as shown in Fig. 2.

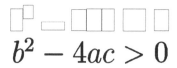

$$b^2 - 4ac > 0$$

Fig. 2. An example of transformed font bounding boxes of the corresponding characters

In Fig.2, it is obvious that character a, b and c that share the same font have the font bounding boxes with the same height which can not reflect the real areas they occupy. The bounding box for the character - is also much bigger than expected. These boxes are grosser than the exact coordinates of each character's bounding box and cannot be used as geometric information for mathematical expression recognition. To solve the problem, we can take advantage of PDF documents in consistence that different PDF customer applications can display the PDF documents with same appearance, i.e. characters at the same position with the same appearance to get more precise geometrical data from the parameters in PDF documents that control the pattern of the layouts. Now we concentrate on the font file in PDF documents.

Font file is a stream containing Type 1 font program. It's in binary format and defines logic to render the character. Positional values in font file are also expressed in glyph coordinate system. Type1 font has two types, the Compact Font Format [12] and Adobe Type 1 Font Format [13]. Each type has its own format, but uses the same method to render the character. Type1 font uses commands to draw lines and Bezier curves to describe the appearance of the characters, as shown in Fig. 3.

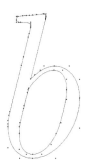

Fig. 3. An example of character b described in font file

In Fig. 3, there are many key position points on or off the outline of the character. These points control the display behaviors of the lines and Bezier curves that make up the outline, which could be used to calculate the exact bounding box, also called the glyph bounding box.

Fig. 4(a) shows one kind of binary data commands in font file that draws the character b and Fig. 4(b) shows the same commands but in decoded format which is more readable.

```
F8 34 81 9F F8 2F 9F F7 69 B3 01 B9 C2 F7 81 CC
03 F7 7B F9 31 15 8C 8F 8D 91 8B 90 08 95 81 8B
89 8A 67 88 89 79 1E 7A 8A 7C 89 79 8A 73 89 84
8A 8B 79 08 81 95 8B 95 BE 8B 82 81 84 83 6E 87
79 1E 73 2B 81 63 52 FB 77 87 79 19 86 72 8B 7A
8B 7E 08 25 C4 56 CC F7 08 F7 0C F7 29 F7 25 E7
57 CC 40 57 5C 60 77 78 1E 95 FC 04 15 6B 68 A3
D9 AC 8E 9E 9D D1 1F 8E 98 9B CB 8F 98 8D 93 C6
E4 CF 8B 08 B7 9F 5F 57 5B 6F FB 05 72 57 1F 72
55 5D 5B 5D 8B 08 0E
```

```
416 -10 20 411 20 213 40 hstem
46 55 237 65 vstem
231 699 rmoveto
1 4 2 6 0 5 rrcurveto
10 -10 0 -2 -1 -36 -3 -2 -18 vhcurveto
-17 -1 -15 -2 -18 -1 -24 -2 -7 -10 -18 rrcurveto
-10 10 0 10 51 0 -9 -10 -7 -8 -29 -4 -18 vhcurveto
-24 -96 -10 -40 -57 -227 -4 -18 rlinecurve
-5 -25 0 -17 0 -13 rrcurveto
-102 57 -53 65 116 120 149 145 92 -52 65 -75 -52 -47 -43 -20 -19 vhcurveto
10 -368 rmoveto
-32 -35 24 78 33 3 19 18 70 hvcurveto
3 13 16 64 4 13 2 8 59 89 68 0 rrcurveto
44 20 -44 -52 -48 -28 -113 -25 -52 hvcurveto
-25 -54 -46 -48 -46 0 rrcurveto
endchar
```

(a)   (b)

Fig. 4. Commands in font file rendering character b. (a) Binary format; (b) Decoded format

The process of our program that parses the decoded commands to get the glyph bounding box is composed of the following steps:

1. Read one single command and denote it as *CurrentCmd*.
2. If *CurrentCmd* is endchar, go to step 5; otherwise, go to step 3.
3. Obtain all the key points that control the boundary of the shape in *CurrentCmd* and calculate the minimal box that holds all the points. Denote the box as *CurrentBox*.
4. If *CurrentBox* is the first box got, denote it as *BOX*; otherwise calculate the minimal box that holds *CurrentBox* and *BOX*, and update *BOX* with the result. Go to step 1.
5. Output *BOX*. End.

By decoding the binary data and parsing the commands, the process of rendering characters is reproduced, from which we obtain the glyph bounding boxes. As with the font bounding box, the glyph bounding box is also expressed in glyph coordinate system. The glyph bounding box in font file is actually a unit bounding box, which means it equals to the exact area that the corresponding character whose font size is 1 takes. After the transformation in 2.3, the glyph bounding box could be used as the precise bounding box of the corresponding character with a certain font size. So far we have gotten the glyph bounding boxes in font file, the next is to extract geometric shapes and characters from content stream and transform the glyph bounding boxes.

2.3   Extraction of Mathematical Elements

Mathematical elements are classified into 4 types, single character, multi-characters, character with geometric shapes, and geometric shapes. In this process, all elements are extracted with their related attributes. We design a structure called *ElementInfo* to represent a mathematical element.

```
struct ElementInfo{
  string glyphName;/*element name, e.g. plus for + */
  double baseLine;
  TextSpaceBoundingBox actualGBB;/*transformed glyph
bounding box*/
};
```

```
character:b glyphName:b BaseLine:704.246
actualGBB left:128.95 right:133.23 bottom:704.126 top:712.543
-------------------------------------------------------------
character:2 glyphName:two BaseLine:708.584
actualGBB left:133.811 right:137.191 bottom:708.584 top:713.884
-------------------------------------------------------------
character:- glyphName:minus BaseLine:704.246
actualGBB left:141.77 right:149.075 bottom:706.996 top:707.474
-------------------------------------------------------------
character:4 glyphName:four BaseLine:704.246
actualGBB left:153.053 right:158.253 bottom:704.246 top:712.316
-------------------------------------------------------------
character:a glyphName:a BaseLine:704.246
actualGBB left:159.058 right:164.378 bottom:704.126 top:709.518
-------------------------------------------------------------
character:c glyphName:c BaseLine:704.246
actualGBB left:165.205 right:169.724 bottom:704.126 top:709.518
-------------------------------------------------------------
character:> glyphName:greater BaseLine:704.246
actualGBB left:174.068 right:181.205 bottom:703.911 top:710.546
-------------------------------------------------------------
character:0 glyphName:zero BaseLine:704.246
actualGBB left:186.007 right:190.873 bottom:703.995 top:712.196
```

Fig. 5. An example of the extracted data

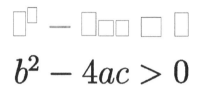

$$b^2 - 4ac > 0$$

Fig. 6. An example of transformed glyph bounding boxes

In order to get the mathematical elements, we write a stack based program to extract characters, geometric shapes and the related attributes such as font size, start point and baseline from decoded content stream obtained in 2.1. Each character's position information can be calculated with the text showing commands in content stream. During the process, the glyph bounding boxes are transformed from glyph coordinate system to text coordinate system in 3 steps. First, the value of the glyph bounding box is divided by 1000. Then, multiply the value of the bounding box by the font size. Finally, move the origin of the glyph bounding box to the corresponding character's start point. For example, the glyph bounding box of character x in font file is [0 500 0 600] which takes the form $[ll_x \; ur_x \; ll_y \; ur_y]$ specifying the lower-left x, upper-right x, lower-left y and upper-right y coordinates of the bounding box. In content stream, the start point of x is (100, 600) in text space and its font size is 6. After the first step, the value of the glyph bounding box is [0 0.5 0 0.6]. Then, it is changed to [0 3.0 0 3.6]. Finally, it becomes [100 103.0 600 603.6], which is the precise bounding box of x. All the positional information is expressed in the coordinate that sets the lower-left corner of the page as the origin. An example of the result obtained by our program for the expression is shown in Fig. 5. The transformed glyph bounding boxes of the expression are shown in Fig. 6.

From Fig. 5 and Fig. 6 we can see that all characters share the same baseline except character 2 and all bounding boxes are equal to the real sizes of the characters, which tell the exact position relationship between each character.

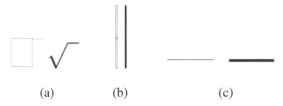

(a)   (b)   (c)

Fig. 7. Special mathematical elements. (a) Racial; (b) Long vertical bar; (c) Long fraction line.

Now the most of mathematical elements and their precise geometric information are ready for mathematical expression recognition. But some special situations still exist which would result in error results without analyzing and processing.

- Some mathematical elements are composed of characters and geometric shapes, e.g. the radical as shown in Fig. 7(a). The geometric shapes are added to the characters and the glyph bounding boxes are recalculated. For the radical in Fig. 7(a), the character $\sqrt{}$ intersects with the horizontal line, so the character and the line are treated as a whole and the glyph bounding box is recalculated to enclose the two items.
- For those who are composed of multi-characters, e.g. the vertical bar | as shown in Fig. 7(b), which may contain more than one shorter vertical bar to form a longer one, all the components are integrated as a new mathematical element with the same character name but a larger glyph bounding box and new baseline. In Fig. 7(b), the shorter vertical bars overlap with each other. The combining process is similar to the situation of the radical.
- The elements who are composed of geometric shapes, e.g. the longer fraction line as shown in Fig. 7(c) will be label as certain mathematical elements due to the context and their shapes.

2.4   Output of Expressions

After all the characters and their related attributes are extracted, the method of structural analysis of mathematical expressions proposed in [14] is used to analyze the structure to build the syntax tree for the expected expression. Once the syntax tree is build, the expression could be output in LATEX format. We will not talk about it much here.

3   Experimental Results

In this paper, MikTex and LATEX editors are used to generate PDF documents with Type1 fonts of varying font sizes and mathematical expressions in different forms such as matrix, integral, radical and so on. We use our proposed method to extract mathematical components directly from PDF documents that are similar to the layout as shown in Fig. 8(a).

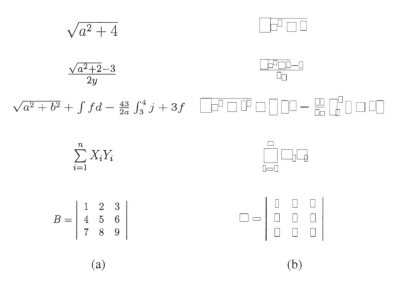

$$\sqrt{a^2 + 4}$$

$$\frac{\sqrt{a^2+2}-3}{2y}$$

$$\sqrt{a^2 + b^2} + \int fd - \frac{43}{2a} \int_3^4 j + 3f$$

$$\sum_{i=1}^{n} X_i Y_i$$

$$B = \begin{vmatrix} 1 & 2 & 3 \\ 4 & 5 & 6 \\ 7 & 8 & 9 \end{vmatrix}$$

(a)   (b)

Fig. 8. Layout of a PDF file and its experimental results. (a) Layout; (b) Results.

Fig.8 (b) is the result showing exact geometric information of mathematical elements in Fig.8 (a) by applying our method. Experimental results show that all the information of the characters in the generated mathematical expressions, including character name, glyph bounding box, position, and baseline could be extracted correctly. The mathematical components could be used for further mathematical expression recognition and retrieval.

4   Conclusions

In this paper, a method of extracting mathematical components directly from PDF documents for mathematical expression recognition and retrieval is proposed, which makes full use of the internal information of the PDF documents and doesn't draw support from any other methods like format conversion. The extracted components could be used to mathematical expression extraction, reconstruction and retrieval.

Although the proposed method could extract the components correctly, it is only for Type1 fonts. Our further work is to transferred this method to other type fonts and improve its robustness.

Acknowledgements. This work is supported by the National Natural Science Foundation of China (Grant No. 61375075) and the Natural Science Foundation of Hebei Province (Grant No. F2012201020).

References

1. Adobe Systems Incorporated, PDF Reference, 6th edn. (November 2006)
2. Chao, H., Fan, J.: Layout and Content Extraction for PDF Documents. In: Marinai, S., Dengel, A.R. (eds.) DAS 2004. LNCS, vol. 3163, pp. 213–224. Springer, Heidelberg (2004)
3. Marinai, S.: Metadata Extraction from PDF Papers for Digital Library Ingest. In: 10th International Conference on Document Analysis and Recognition, pp. 251–255. IEEE Press, New York (2009)
4. Déjean, H., Meunier, J.-L.: A System for Converting PDF Documents into Structured XML Format. In: Bunke, H., Spitz, A.L. (eds.) DAS 2006. LNCS, vol. 3872, pp. 129–140. Springer, Heidelberg (2006)
5. Rahman, F., Alam, H.: Conversion of PDF Documents into HTML: a Case Study of Document Image Analysis. In: Conference Record of the Thirty-Seventh Asilomar Conference on Signals, Systems and Computers, vol. 1, pp. 87–91. IEEE Press, New York (2004)
6. Yang, M., Fateman, R.: Extracting Mathematical Expressions from Postscript Documents. In: Proceedings of the 2004 International Symposium on Symbolic and Algebraic Computation, pp. 305–311. ACM (2004)
7. Chan, K.-F., Yeung, D.-Y.: Mathematical Expression Recognition: a Survey. J. International Journal on Document Analysis and Recognition. 3(1), 3–15 (2000)
8. Lin, X.Y., Gao, L.C., Tang, Z., Lin, X.F., Hu, X.: Mathematical Formula Identification in PDF Documents. In: 2011 International Conference on Document Analysis and Recognition, pp. 1419–1423. IEEE Press, New York (2011)
9. Lin, X.Y., Gao, L.C., Tang, Z., Hu, X., Lin, X.F.: Identification of Embedded Mathematical Formulas in PDF Documents Using SVM. In: IS&T/SPIE Electronic Imaging, pp. 82970D–82970D. International Society for Optics and Photonics (2012)
10. Baker, J.B., Sexton, A.P., Sorge, V.: Extracting Precise Data on the Mathematics Content of PDF Documents. Towards Digital Mathematics Library, Birmingham , pp. 75–79 (2008)
11. Baker, J.B., Sexton, A.P., Sorge, V.: A Linear Grammar Approach to Mathematical Formula Recognition from PDF. In: Carette, J., Dixon, L., Coen, C.S., Watt, S.M. (eds.) MKM 2009, Held as Part of CICM 2009. LNCS, vol. 5625, pp. 201–216. Springer, Heidelberg (2009)
12. Adobe Systems Incorporated, The Compact Font Format Specification, Version 1.0, 4 (December 2003)
13. Adobe Systems Incorporated, Adobe Type 1 Font Format, Version 1.1 (February 1993)
14. Tian, X.D., Li, N., Xu, L.J.: Research on Structural Analysis of Mathematical Expressions in Printed Documents. J. Computer Engineering 32(23), 202–204 (2006)

An Efficient OLAP Query Algorithm Based on Dimension Hierarchical Encoding Storage and Shark

Shengqiang Yao and Jieyue He[*]

School of Computer Science and Engineering,
MOE Key Laboratory of Computer Network and Information Integration,
Southeast University, Nanjing 210096, China
shengq_yao@163.com, jieyuehe@seu.edu.cn

Abstract. The on-line analytical processing (OLAP) queries always include multi-table joins and aggregation operations in their SQL clauses. As a result, how to reduce multi-table joins and effectively aggregate the query data with "big data" is the key issue for query processing. Therefore, the novel OLAP query algorithm is proposed in this paper based on the dimension hierarchical encoding (DHE) storage strategy with the In-Memory computing in Shark. With DHE and Shark, a star join with hierarchy level is mapped to a multidimensional range query on the fact table and the large-scale data by transformations and actions are computed on resilient distributed datasets (RDDs). The experimental results show that, compared with the data analysis operations in Hive, complex multi-table joins and I/O overhead are reduced by DHE and Shark. The query performance is greatly improved than that of the ordinary star schema.

Keywords: Big data, Data warehouse, Dimension hierarchical encoding, Shark.

1   Introduction

Data warehousing and on-line analytical processing (OLAP) [1] are essential elements of decision support system. However, the rapidly growing size of the data sets for business intelligence makes traditional warehousing solutions unsuitable. The main idea underneath this evolution is that those data sets need to be stored in the cloud and be accessed by a set of services. Following this consideration, there have been several proposals to store and process extremely large data sets.

Hadoop [2] is a popular open-source map-reduce implementation inspired by Google's MapReduce [3]. It is being widely used in web search, log analysis and other large-scale data processing filed. Hive [4] is a data warehousing solution built on top of Hadoop. It supports SQL-like declarative language (HiveQL) and is widely used. HiveQL is compiled into map-reduce jobs executed on Hadoop. However, expensive data materialization for fault tolerance and costlier execution strategies [5, 6] makes Hive slow. Shark [7, 8] is a new data warehouse system capable of deep data

---

[*] Corresponding author.

Y. Tan et al. (Eds.): ICSI 2014, Part II, LNCS 8795, pp. 180–187, 2014.

analysis using the Resilient Distributed Datasets (RDDs) memory abstraction. Though Shark shows quite better performance than hive, complex OLAP queries still take lots of system resources and affect the query efficiency, like pre-shuffle and intermediate data outputs, especially when the data sets become larger.

OLAP over big data repositories has recently received a great deal of attention from the research communities [10]. For example, there is a research focus on the performance of aggregation operations by using dimension-oriented storage in HBase [11]. A decomposed snowflake schema was proposed in [12] to get better performance in their parallel database. The work in [13] shows that HBase can't original support the complex OLAP queries. The performance of "data loading time" and "grep select time" in HBase is poorer than Hive. Other studies, like [14], they use a distributed cube model in the cloud platform. However, cube needs to be pre-computed and extra space to store. It is also quite difficult to decide which queries to pre-compute. In traditional warehouse system, a multidimensional hierarchical method [15, 16] is used to reduce table joins and improve the efficiency of queries. Therefore, an efficient OLAP query method is proposed in this paper by using dimension hierarchical encoding (DHE) storage and shark. The code for representing the hierarchies of each dimension replaces the foreign keys in the fact table. As a result, the complex table joins reduced and the efficiency of OLAP queries improved. Our experimental results on star schema benchmark (SSB) [9] show that with DHE storage and shark, the OLAP query method has better performance.

Briefly then, the outline of this paper is as follows. In Section 2, the method of OLAP query based on DHE and shark are described in detail. In Section 3, our algorithm is applied to SSB and the results are analyzed. In Section 4, the conclusions are given.

2   OLAP Query Based on DHE and Shark

2.1   Dimension Hierarchical Encoding

In data warehouse systems, a central fact table contains the measures, and the dimension tables connect to it via foreign keys. In most situations, the size of dimension table is far less than the fact table and the number of hierarchy level members is small. So we need only a little effort to tackle with the DHE when an OLAP query comes without losing any original semantics.

Assuming that L_{ij} is the j-th level attribute of dimension table D_i and $B^{L_{ij}}$ is the binary code of L_{ij}. $B^{L_{ij}}$ can be denoted as formula (1):

$$B^{L_{ij}} = \{b_{k-1} \dots b_i \dots b_0\}. \tag{1}$$

The value of b_i is 0 or 1 and k is the binary code length of L_{ij}. Its value can be calculated by formula (2):

$$k = \text{Bit}(L_{ij}) = \lceil \log_2 m \rceil \ (m = |L_{ij}|). \tag{2}$$

m is the number of   different members in L_{ij}.

Assuming that B^{D_i} is the dimension hierarchy encoding of dimension table D_i. To obtain B^{D_i}, DHE combines each hierarchy level of dimension D_i in a level-descending order (from the top level to low level). B^{D_i} can be calculated by formula (3):

$$B^{D_i} = \left(\dots \left((B^{L_{i1}} \ll Bit(L_{i2}) | B^{L_{i2}}) \ll Bit(L_{i3}) \big| B^{L_{i3}} \right) \dots \right) \ll Bit(L_{im}) | B^{L_{im}}. \tag{3}$$

"$\ll$" is the left shift operator and "$|$" is the bitwise-Or operator.

For example, as shown in table 1, the "Customer" dimension has 3 different hierarchy levels {Region, Nation, City}. "Region" is the top level attribute and has total 5 different members, which is originally stored as string type.

According to formula (1) and (2), 3 bits are used to replace the 5 different members in Region with {001,010,011,100,101}. "Nation" and "City" can be respectively represented by 4bit and 8bit. By formula (3), we can obtain the DHE of dimension table "Customer". Table 1 displays the part of the DHE of "Customer".

Table 1. Part of the DHE of Customer

Customer(Dimension)		Region(Level)		Nation(Level)		City(Level)	
CustKey	$B^{Customer}$	Region	B^{Region}	Nation	B^{Nation}	City	B^{City}
1	0010001...00001	AFRICA	001	ALGERIA	0001	ALGE 0	00000001
...	010...010100001	...	...	ARGENTINA	0010	......	......
300	...	ASIA	011	...	...	BRAZ	00010101
...	...	...	...	BRAZIL	1111	0 ...	...

In order to speed up the OLAP query with "big data" and reduce storage space, we use DHE to replace the dimension foreign keys in fact table and this phase can be done by ETL tools. Fig.1 shows the example of the storage strategy with DHE in star schema.

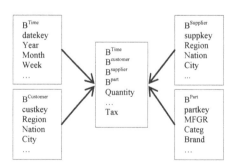

Fig. 1. Star schema with DHE

2.2   Overview of Spark and Shark

Spark [7] is a fast and general engine for large-scale data processing which outper-forms Hadoop. Spark introduces an abstraction called RDDs (resilient distributed datasets). An RDD is a read-only collection of objects partitioned across a set of ma-chines. RDDs achieve the abstraction to manipulate distributed data sets in local oper-ations. RDDs can be calculated in two ways: Transformations (e.g. map, filter) and Actions (e.g. count, collect). Transformations are lazy operations that define a new RDD, while actions launch a computation to return a value.

Spark is composed of one Master node and several Worker nodes. The program developed by users called driver program. The workers are long-lived processes that can store RDD partitions in RAM across operations. The driver defines one or more RDDs and invokes actions on them.

Shark [8] is a data warehousing implementation on Spark. Shark is compatible with Hive and uses Hive query compiler. However, it transforms the operators into the operation on RDDs rather than MapReduce jobs. The Hadoop Distributed File System (HDFS) data obtained by shark will be computed by spark.

2.3   The OLAP Query Algorithm Based on DHE and Shark

Usually OLAP users are not interested in single measures but in some form of sum-marized data. An important concept of OLAP data models is the notion of dimension hierarchies. Hierarchies provide an appropriate method of describing the level of ag-gregation for a dimension. Thus, typical OLAP queries in star schema contain restric-tions on multiple dimension tables that are then used as restrictions on the very large fact table:

```
Select F.attributes,D.attributes,Agg(Measures)
From Fact,Dimension
Where <Join Constraint> and <Dimension Restriction>
and <Fact Restriction>
Group By Gb-attributes
Order By Ob-attributes
```

Attributes in "<Dimension Restriction>", "Gb-attributes" and "Ob-attributes" mostly contain hierarchy level attributes and a few non-hierarchy-level attributes.

By DHE, a star join with hierarchy level is mapped to a multidimensional range query on the fact table. In Shark, the large-scale data is computed by transformations and actions on RDDs. Map Sets $<L_{ij}, B^{L_{ij}}>$ are preprocessed according to formula (1), (2) and B^{D_i} is got according to Formula (3). As the example shown in Fig.2, the hierarchy level restriction in dimension table "Customer" is Region="AFRICA" and Nation="ALGERIA". According to formula (3), the first 3 bits in $B^{Customer}$ represent "Region" and the middle 4 bits represent "Nation". So we create the mask code $B^{mask} = $ "111111100000000". $B^{AFRICA} = $ "001" and $B^{ALGERIA} = $ "0001" are obtained from formula (1) and (2). Thus, we create the filter code $B^{filter} = $ "001000100000000" for this restriction. Bitwise-And operator is then used between $B^{Customer}$ and B^{mask} to filter the tuples to new RDDs.

Thus, the query output can be obtained by the main steps as follows:

- Step1: Analyze the level attributes in "<Dimension Restriction>". Obtain the hierarchy level code $B^{L_{ij}}$ from Map Set $<L_{ij}, B^{L_{ij}}>$ and their corresponding offset in B^{D_i}.
- Step2: create the mask B^{mask} and the filter key B^{filter} from Step 1.
- Step3: Sequential Scan the Fact table with B^{mask} and B^{filter}. Create new RDDs for those filtered tuples in Fact table as the example shown in Fig.2.
- Step4: If there exists non-hierarchy-level attribute in <Dimension Restriction>, Group By and Order By clause, join the corresponding Dimension table with the RDDs created in Step 3 and then generate new RDDs too.
- Step5: GroupByPreShuffle (part-aggregate the data at map-side) is executed on the RDD got from Step4 according to $B^{L_{ij}}$ in B^{D_i} and the other attributes. As a result, the amount of data processed at reduce-side will decrease.
- Step6: According to Gb-attributes, value will be distributed to different reducer. Then merge the values and compute the sum at the reduce side, sort and extract data.
- Step7: Submit the final results from worker nodes to master node.

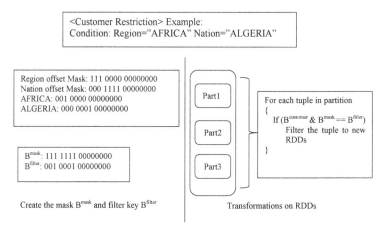

Fig. 2. Create the filter key for <Customer Restriction> and transform actions on RDDs

For example, the following complex multi-table joins query in SSB:

```
Select sum(revenue),year,brand
From datetbl join lineorder
on (datetbl.datekey=lineorder.orderdate) join part
on (part.partkey=lineorder.partkey) join supplier
on (supplier.suppkey=lineorder.suppkey)
Where category='MFGR#12' and region='AMERICA'
Group by year,brand
Order by year,brand
```

It filters data set according to region attribute in dimension table "supplier" and category attribute in "part". The summarized "revenue" is group by and order by year in dimension table "datetbl" and brand in "part". As shown in Fig.3, new RDDs reflect an operator's transformation on the RDD that resulted from the previous operator's transformation. Based on DHE and Shark, We do not need to write intermediate results to HDFS in a temporary file, and only simply write results to local disk. The complex multi-table joins and I/O overhead are also reduced by DHE.

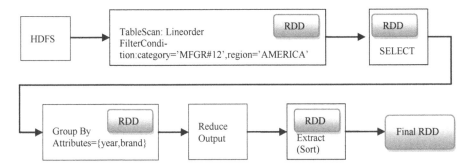

Fig. 3. Query plan based on DHE and Shark

3   Experiments and Results

SSB [9] is used in the experiment which is based on the TPC-H benchmark to measure the performance of data warehousing applications. SSB provides both functional coverage (different common types of Star Schema queries) and selectivity coverage (varying fractions of the fact table that must be accessed to answer the queries). There are four major query types in SSB. The number of dimension tables increases from query type 1 (Q1) to query type 4 (Q4). For example, Q1 is shown in the below code:

```
Select sum(lo_extendedprice*lo_discount) as revenue
From lineorder,date
Where lo_orderdate=d_datekey
and d_year=[YEAR]
and lo_discount between [DISCOUNT]-1
and [DISCOUNT]+1 and lo_quantity<[QUANTITY]
```

The above query is meant to quantify the amount of revenue increase that would have resulted from eliminating certain company-wide discounts in a given percentage range for products shipped in a given year. This is a "what if" query to find possible revenue increases.

The experiments utilized IBM high performance computing platform. The platform has total 279 computing nodes and 3500 CPU cores. We used 1 master node and 3 worker nodes. The available RAM of each worker node is set to 2GB.

We use three data sizes of 5G, 10G and 20G with DHE and pre-load them into HDFS. The records number in fact table is respectively 30 million, 60 million and 120 million. For each query type, we complete the query with different values several times (e.g. discount=0.5, discount=0.6) and use the average time as the result.

Q1 to Q4 on Hive-SSB and Shark-SSB are based on the original SSB. Q1 to Q4 on Shark-DHE are based on our OLAP query algorithm and the DHE star schema which is generated from the original SSB.

Fig.4, Fig.5 and Fig.6 illustrate that the data analysis performance built on Shark is much better than Hive. Moreover, Our OLAP algorithm based on DHE and Shark further improves the query time than the original star schema.

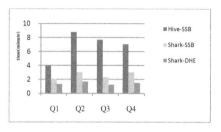

Fig. 4. Query performance on 5G data

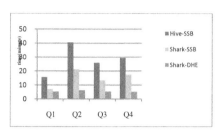

Fig. 5. Query performance on 20G data

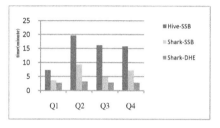

Fig. 6. Query performance on 10G data

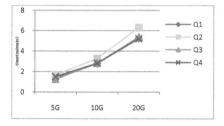

Fig. 7. Performance trends on Shark-DHE

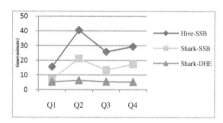

Fig. 8. Q1 to Q4 trends on 20G data

Fig.7 shows that the OLAP query time increases along with the size of the data. The trend of query time with 20G data is demonstrated in Fig.8. The trend in our method is quite smooth because most join operations can be removed by using DHE and map the query to a range query on the fact table. Q1 involves only one dimension and does not contain group by operations, so the improvement of the query time is not obvious between Shark-SSB and Shark-DHE.

4   Conclusion

With the era of "big data" and expanding amount of data size, the processing technology which computing distributed data set to satisfy the complex OLAP queries has become a research hot spot. In this paper, the novel OLAP query algorithm is proposed based on the dimension hierarchical encoding storage strategy with the In-Memory computing in Shark. This method reduces the complex multi-table joins in star schema. The results of experiment reveal that our OLAP algorithm based on DHE and Shark greatly improves the query time.

Acknowledgements. This work was supported by the Natural Science Foundation of Jiangsu Province under Grant No.BK2012742.

References

1. Chaudhuri, S., Dayal, U.: Data warehousing and OLAP for decision support. ACM Sigmod Record 26(2), 507–508 (1997)
2. Apache Hadoop, `http://wiki.apache.org/hadoop`
3. Dean, J., Ghemawat, S.: MapReduce: simplified data processing on large clusters. Communications of the ACM 51(1), 107–113 (2008)
4. Thusoo, A., Sarma, J., Jain, N.: Hive: a warehousing solution over a map-reduce framework. Proceedings of the VLDB Endowment 2(2), 1626–1629 (2009)
5. Pavlo, A., Paulson, E., Rasin, A.: A comparison of approaches to large-scale data analysis. In: Proceedings of the 2009 ACM SIGMOD International Conference on Management of Data, pp. 165–178. ACM (2009)
6. Stonebraker, M., Abadi, D., DeWitt, D.: MapReduce and parallel DBMSs: friends or foes. Communications of the AC 53(1), 64–71 (2010)
7. Zaharia, M., Chowdhury, M., Das, T.: Resilient distributed datasets: A fault-tolerant abstraction for in-memory cluster computing. In: Proceedings of the 9th USENIX conference on Networked Systems Design and Implementation, p. 2. USENIX Association (2012)
8. Pavlo, A., Paulson, E., Rasin, A.: A comparison of approaches to large-scale data analysis. In: Proceedings of the 2009 ACM SIGMOD International Conference on Management of Data, pp. 165–178. ACM (2009)
9. O'Neil, P., O'Neil, E., Chen, X.: The star schema benchmark (SSB). Pat. (2007)
10. Jing-hua, Z., Ai-mei, S., Ai-bo, S.: OLAP Aggregation Based on Dimension-oriented Storage. In: IEEE 26th International Parallel and Distributed Processing Symposium Workshops & PhD Forum (IPDPSW), pp. 1932–1936. IEEE (2012)
11. Apache HBase, `http://www.hbase.apache.org`
12. Wang, H., Qin, X., Zhang, Y., Wang, S., Wang, Z.: LinearDB: A relational approach to make data warehouse scale like mapReduce. In: Yu, J.X., Kim, M.H., Unland, R. (eds.) DASFAA 2011, Part II. LNCS, vol. 6588, pp. 306–320. Springer, Heidelberg (2011)
13. Shi, Y., Meng, X., Zhao, J.: Benchmarking cloud-based data management systems. In: Proceedings of the Second International Workshop on Cloud Data Management, pp. 47–54. ACM (2010)
14. Brezany, P., Zhang, Y., Janciak, I.: An Elastic OLAP Cloud Platform. In: IEEE Ninth International Conference on Dependable, Autonomic and Secure Computing (DASC), pp. 356–363. IEEE (2011)
15. Fa, H., Sheng, Y., Zhen, X.: A Novel Aggregation Algorithm for Online Analytical Processing Queries Evaluation Based on Dimension Hierarchical Encoding. J. Computer Research and Development. 4, 608–614 (2004)
16. Markl, V., Ramsak, F., Bayer, R.: Improving OLAP performance by multidimensional hierarchical clustering. In: International Database Engineering and Applications Symposium, pp. 165–177. IEEE (1999)

The Enhancement and Application of Collaborative Filtering in e-Learning System

Bo Song and Jie Gao

College of Software, Shenyang Normal University, Shenyang City,
Liaoning Province, China, 110034
songbo63@aliyun.com, gaojiexy@126.com

Abstract. Collaborative Filtering recommendation algorithm is one of the most popular approaches for determining recommendations at present and it can be used to solve Information Overload issue in e-Learning system. However the Cold Start problem is always one of the most critical issues that affect the performance of Collaborative Filtering recommender system. In this paper an enhanced composite recommendation algorithm based on content recommendation tags extracting and CF is proposed to make the CF recommender system work more effectively. The final experiment results show that the new enhanced recommendation algorithm has some advantages on accuracy compared with several existing solutions to the issue of Cold Start and make sure that it is a feasible and effective recommendation algorithm.

Keywords: Recommendation Algorithm, Cold Start, Collaborative Filtering, e-Learning.

1   Introduction

With the development of Internet technology, huge information resource is presented to us constantly, thus the issue of information overload formed. It becomes more difficult for users to find out what they need or they are interested in. Facing information overload issue, recommender system is one of the solutions. Collaborative Filtering is one of the most popular approaches for determining recommendations at present and the CF recommendation algorithm is proposed originally in 1992 by Goldberg, Nichols and Oki [1]. It is an intelligent and personalized information service system and describes the user's lone-term information need by user modeling, based on which it can customize the personalized information with the specific recommendation strategy [2]. The implementation of CF recommender system must rely on users' explicit rating, such as Amazon, Drugstore; these electronic commerce platforms possess a large number of users who have rated their purchased goods. The CF recommendation algorithm can be divided into two parts. One is user-based CF and the other one is item-based CF.

The e-Learning system stays on the situation that they may confront with Cold Start issue while using Collaborative Filtering. When new users register the system or new items are added into the system, the CF recommender system may recommend

Y. Tan et al. (Eds.): ICSI 2014, Part II, LNCS 8795, pp. 188–195, 2014.

no items for the new users or the new items may not be recommended for existing users because of there is no relevant rating information in the user-item rating matrix known as user modeling. This is the problem of Cold Start of CF recommendation algorithm. Aiming at above problems, in this paper an enhanced composite recommendation algorithm based on content information tags extracting and Collaborative Filtering will be proposed to solve the new users' Cold Start problem to make the recommender system works more effectively.

2   Collaborative Filtering Recommendation Algorithm

Personalized recommendation is an important part in user behavior analysis. In simple term, it is a process of searching the resource which the user might interest in. At present, Collaborative Filtering is a relatively mature and popular recommendation algorithm. In CF, an item is considered as a black box-we don't look at its content and user interactions (such as rating, saving, purchasing) with the item are used to recommend an item of interest to the user. The main idea of Collaborative Filtering is to exploit information about the past behavior or opinions of an existing user community for predicting which items the current user will most probably like or be interested in [3]. The recommender process is simply as follows: first given a user-item rating dataset and a current user as input and calculate the similarity between users; second use the similarity matrix made up of the above similarities to find similar users that had similar preferences to those of the current user, that is called nearest neighbors; third for every item that the current user has not yet seen, the prediction ratings will be calculated based on the ratings for the items made by the similar users; finally sort the prediction ratings and the top-N items will be recommended for the current user. Such a top-N list should not contain items that the current user has already rated.

The matrix of given user-item ratings is known as user modeling which is obtained by collecting and arranging users interactions. If the number of items is "n" and the number of users is "m" in the system, the user modeling is a rating matrix of "m*n". We use $U = \{User_1, User_2, ..., User_m\}$ to denote the set of users, $I = \{Item_1, Item_2, ..., Item_m\}$ for the set of items, and R as a $m*n$ matrix of ratings $r_{(i,j)}$, with $i \in \{1...m\}$ and $j \in \{1...n\}$. The possible rating values are defined on a numerical scale from 1 to 5 in this paper and the higher the rating value, the more strongly the user likes the item.

$$sim(\vec{a}, \vec{b}) = \frac{\vec{a} \cdot \vec{b}}{|\vec{a}| * |\vec{b}|} \tag{1}$$

To find nearest neighbors, the formulae applied to calculate the similarity between users or items must be used. There are several similarity calculation methods, such as the cosine similarity measure, adjusted cosine similarity measure, Pearson correlation coefficient measure. Cosine similarity is established as the standard metric in item-based Collaborative Filtering; as it has been shown that it produces the most accurate results [3]. The formula (1) shows the cosine similarity measure which calculates the rating similarity between Item a and Item b in item-based Collaborative Filtering. The range of the ratio in formula (1) represents similarity which is from 0 to 1. The larger the ratio is, the higher the similarity between two items is. However, the basic cosine measure does not take the differences in the average rating behavior of the users into account [3]. The adjusted cosine measure solves the problem and subtracts the user average. And the values for the adjusted cosine measure correspondingly range from −1 (strong negative correlation) to +1(strong positive correlation) as is shown in formula (2). Here $r_{u,a}$ is the rating given by user u to the item a, $r_{u,b}$ is the rating given by user u to the item b and $\bar{r}_u$ represents the average of user u ratings.

$$sim(a,b) = \frac{\sum_{u \in U} (r_{u,a} - \bar{r}_u)(r_{u,b} - \bar{r}_u)}{\sqrt{\sum_{u \in U} (r_{u,a} - \bar{r}_u)^2} \sqrt{\sum_{u \in U} (r_{u,b} - \bar{r}_u)^2}} \qquad (2)$$

As in the adjusted cosine measure, the Pearson correlation coefficient measure shows in the formula (3). The symbol $\bar{r}_a$ corresponds to the average rating of user a and $\bar{r}_b$ corresponds to the average rating of user b and takes the values from −1 to +1. Pearson correlation coefficient is used to calculate the similarity between users in user-based Collaborative Filtering.

$$sim(a,b) = \frac{\sum_{i \in I} (r_{a,i} - \bar{r}_a)(r_{b,i} - \bar{r}_b)}{\sqrt{\sum_{i \in I} (r_{a,i} - \bar{r}_a)^2} \sqrt{\sum_{i \in I} (r_{b,i} - \bar{r}_b)^2}} \qquad (3)$$

After the nearest neighbors were found, the recommender engine will work according to the prediction rating of non-rated items based on the current user. The prediction rating formula is shown in formula (4).

$$predictedRating = \frac{\sum (sim(a,b) * itemrating)}{\sum sim(a,b)} \qquad (4)$$

Where, the prediction rating is the ratio of similar users' rating weighted sum and similarity sum. Then add up all similar users' prediction rating and sort them according to the values of sum and the top N items will be the recommendation items for the current user. The larger the prediction rating sum of an item rated by all similar users is, the more effective recommendation results the current user will get. Correspondingly the scale of the sum will not be more than "5".

3   The Enhancement of Collaborative Filtering

The success of Collaborative Filtering relies on the availability of a sufficiently large set of quality preference ratings provided by users, but it method may suffer from the new user problem, in which there is no rating record on new users in user modeling

[4]. This is the issue of new users' Cold Start. Usually there are several solutions to Cold Start problem such as random recommendation and Mean Value recommendation. In recent years, combining content information with CF recommendation proved to be an effective solution to Cold Start problem and has become a hot research [5].

3.1 Content-Based Tags Extracting Solution to Cold Start

The recommendation process of the new enhance composite recommendation algorithm can be divided into four steps: firstly content information of new users and all existing items will be analyzed and the tags that represent their feature will also be extracted; secondly find the similar matrix between the tags matrix of new users and the tag matrix of all items with the method of linear algebra, then recommend the items for the new users, this is the first time recommendation in the new algorithm; Thirdly the new users can choose the items and rate them, then the interaction can fill in the user modeling and the new users will acquire the final more accurate recommendation results.

When new users register the e-Learning system, it is necessary for them to fill in some individual information, especially education background, preference and forte. Usually the items in the e-Learning system are added in some text depiction about them, for instance lesson name, major classification, lesson introduction. We can use the above information to build tags matrix. Tags may contain a single term or multiple terms and it is very important to convert terms into numerical value.

The typical steps involves in building tags matrix are tokenization, normalization, eliminating stop words and quantitative filling. At this stage, we will introduce the use of term frequency and document frequency to compute the weight associated with each term. Tokenization is to parse the text to generate terms and sophisticated analyzers can also extract phrases from the text. The *TokenStream* class and *Analyzer* class of Apache Lucene jar file can help to extract tags from the text [6]. Apache Lucene is a Java-based open source search engine developed by Doug Cutting [7]. And we can create a Map collection class *buildTermFrequencyMap* to build a HashMap to store the terms and their frequencies. It is with all the text information in the system as input and uses Loops and conditional statements to put the terms and frequencies into the *termFrequencyMap* Map. Then *getTopNTermFrequencies* method will find the top N terms appeared most frequently in each text information file by sorting the frequencies.

The second step is normalization to convert the terms to lowercase. To make the matrix more accurately, we will also use the *StandardAnalyzer* class of Apache Lucene to eliminate stop words such as "a, an, the" that appear in the text too often [7]. The stop words list can use a String type array to preserve. Eliminating stop words can not only enhance the importance of meaningful words, but also abate noise apparently at the same time. The Java static code block shows the above methods.

```
static {
    List<String> allStopWords = new ArrayList<String>();
    allStopWords.addAll(Arrays.asList(StandardAnalyzer.STOP_WORDS));
    allStopWords.addAll(Arrays.asList(ADDITIONAL_STOP_WORDS));
```

MERGED_STOP_WORDS = allStopWords.toArray(new String[0]);   }
public CustomAnalyzer() {
    super(CustomAnalyzer.MERGED_STOP_WORDS);   }

ADDITIONAL_STOP_WORDS is the String type array that contains stops words. It is added into the *allStopWords* ArrayList.

When the first top N terms were picked out and then they can consist of coordinate space of tags. So every new user $User_i$ can be described by term vectors: $User_i = \{U_1, U_2, ..., U_n\}$. Among them, U_j express a tag which is a column vector. For each new user' registration text information, take 80% term frequency as the cut-off point, If the term frequency is higher than 80%, we can set the term vector element value as 1, or else is 0. For each item, we can also use the above method to build the corresponding tags matrix. So, all the text information of new users and existing items can be indicated by a 0-or-1 matrix of tags by quantitative filling.

After defining the tags matrix, the tags matrixes of new users and items have the same dimensions. Then we can use the method of linear algebra to find similarity matrixes. Methods are as follows: Assume A, B as the matrixes of order n, if there exist invertible matrices P of order n, such that $P^{-1} * A * P = B$ was established, we call matrix A is similar to B, denoted by A~B. P^{-1} is the inverse matrix of P. Here also involves matrix multiplication algorithm. The steps for finding similarity matrix are as follows. Firstly calculate the Eigen values of the matrix: $|A - \lambda E| = 0$, here E is a unit matrix, from upper left to lower right corner of the main diagonal, the elements are all 1, outside are 0. The matrix A can be seen as tags matrix of random new user and matrix B can be tags matrix of existing item. Then for every Eigen value, calculate the solution space of matrix equation $(A - \lambda E)X = 0$.

We can exploit the feature vector provided by basic solutions as column vector to comprise the matrix P. The situation shows that the invertible matrix P exists and matrix A is similar to B. If the above mentioned matrix equation has no solution space, there will not be invertible matrix P. On this situation, matrix A is not similar to B; there is no similarity between the new user and the item. Using this algorithm, the tags matrix of new user will be compared with every tags matrix of existing item, if there is tags matrix of item is similar to the tags matrix of the new user, the corresponding item will be recommended to the new user. The number of matrix B that can enable the equality $P^{-1} * A * P = B$ succeeds is certainly more than one in the system. So the first recommendation list for the new user can form. Similarity matrix is an equivalence relation and has reflexive, symmetry and transitivity three properties. The items that tags matrix are similar to the tags matrix of the new user can recommend to him or her. We call it the first-time recommendation because the recommendation this time is not enough accurate and we can further make efforts to let the recommender system work more effectively. Therefore, the items of first time recommendation will be chosen by the new user. The new user then can rate the items by ones preference and education background. After this step, the new user' interactions are added into the User Modeling. After the new user' interactions are added the User Modeling, we can use Collaborative Filtering Recommendation algorithm to obtain more accurate recommendation results. A series of process, calculating similarity between the new user and existing users, predicting ratings for the new user, sorting the predicting ratings, will work as usual. The recommendation results of first-time recommendation process will be as seed candidates to create the final more accurate recommendation results. Therefore we can name the new solution to Cold Start problem as two-times recommendation. Using the new content-based tags extracting composite recommendation algorithm, Collaborative Filtering recommendation algorithm can provide better service for new users and improve the performance of recommender system.

3.2   Experimental Results and Analysis

In order to verify the effectiveness of the algorithm, this paper will use the public data set MoviLens provided by Grouplens working group to test. MoviLens is a research recommender system developed by the researchers of Grouplens working group in USA University of Minnesota based Web. It accepts user evaluation of the films and provides the corresponding movie recommendation list. At present, the system has more than 43000 users and the user ratings of the items are more than 1600 [8].

In this paper, the ML data set provided by the Movielens will be used, which is composed of 943 user evaluations of the 10000 items with 1-to-5 ratings. The data set has a total of 1682 items and each user makes evaluation on the 20 items at least [8]. In this paper we will select 10% of the data set as the experimental data randomly. Then we will choose 10% users data of the experimental data as new users' information. Their rating items are more than 100 so that enough information can be used. In this experiment, the new users' ratings will be deleted and stored in another place. Using the three different recommender algorithms of Random Recommendation, Mean Value Recommendation and the new recommendation algorithm proposed in this paper, compare the predict ratings with real ratings of the new users. This experiment process will be carried out 9 times totally. The accuracy of recommendation algorithm usually uses MAE (Mean Absolute Error) and RMSE (Root Mean Squared Error) to measure. They express that what degree the new users will like or dislike the items recommended by recommendation algorithm.

$$MAE = \frac{\sum_{i \in N} p_{i,i} - r_{i,j}}{N} \tag{5}$$

Where $p_{i,j}$ is on behalf of the predicted values. In fact, MAE is the average value of the sum of differences between predicted values and real ratings values. The lower the MAE value, the better the recommendation accuracy. Fig.1 shows the MAE curve of three recommendation algorithms that solve Cold Start problem. The recommendation results of Random Recommend are random and the MAE curve is a random fluctuating curve that the value is more than 0.6. As the ratings from new user are more, the accuracy of Random Recommend is much bad. The MAE of Mean Value Recommend is almost a fixed value. The MAE of the new recommendation algorithm is a low value firstly. When ratings from new user are added, the MAE becomes much lower and the accuracy of new recommendation is best of all.

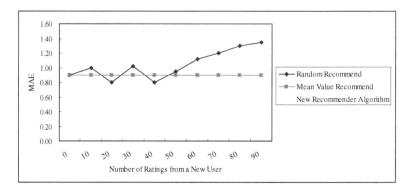

Fig. 1. The MAE Curve of Three Solutions to Cold Start

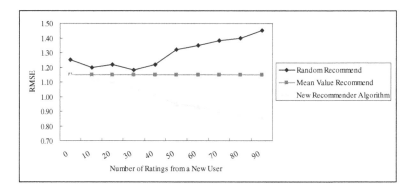

Fig. 2. The RMSE Curve of Three Solutions to Cold Start

RMES in formula (6) is another metric to measure accuracy of a recommender system. MAE is widely used because of its simple calculation, easy to understand. But MAE also has some limitations, because the low ratings that are difficult to predict accurately usually make great contribution to the MAE metric. RMSE firstly square each absolute error, so it will have heavier punishment for the larger absolute error relatively [9]. In formula (6), N is the total number of experimental items and E^p is the entire experimental data set. The lower the RMSE value, the better the recommendation accuracy, like MAE. Fig.2 shows the curve of three recommendation algorithms that solve Cold Start problem. The curve changes of RMSE are almost unanimously with MAE, just the RMSE value is more precise. The two metrics of accuracy prove that the new recommendation algorithm in this paper is a feasible and effective recommendation algorithm to solve the problem of Cold Start.

$$RMSE = \sqrt{\frac{\sum_{(u,i) \in E^p} \left(r_{u,i} - p_{u,i}\right)^2}{N}} \tag{6}$$

4   Conclusions

In this paper an enhanced composite algorithm based on content information and tags extracting is proposed to solve the issue of Cold Start to make Collaborative Filtering recommender system works more effectively. The final experiment results show that the new enhanced recommendation algorithm has some advantages on accuracy compared with several existing solutions to the problem of Cold Start and make sure that it is a feasible and effective recommendation algorithm. Therefore, we can adopt the new recommendation algorithm in e-Learning system to recommend appropriate lessons for learners. When new learners register e-Learning system, they can also acquire recommendation lessons which meet their demand. The new enhanced recommendation algorithm can be applied to the e-Learning system used in small and medium enterprises and if the scale of e-Learning system is too large, tags extracting workload will be correspondingly become too volume.

Acknowledgment. This work was supported by the Science and Technology Project of Education Department of Liaoning Province, China (Research of e-Learning System of small and medium-size Enterprises Based on SaaS, No.L2013417).

References

1. Goldberg, D., Nichols, D., Oki, B.M.: Using collaborative filtering to weave an information tapestry. Communications of the ACM 35(12), 145–147 (1992)
2. Lei, R.: The Key Technology Research of Recommender System. East China Normal University (2012)
3. Dietmar, J., Markus, Z., Alexander, F., Gerhard, F.: Recommender System an Introduction, pp.13–14, 19. Cambridge University Press (2011)
4. Liu, Q., Gao, Y., Peng, Z.: A novel collaborative filtering algorithm based on social network. In: Tan, Y., Shi, Y., Ji, Z. (eds.) ICSI 2012, Part II. LNCS, vol. 7332, pp. 164–174. Springer, Heidelberg (2012)
5. Dongting, S., Tao, H., Fuhai, Z.: Summary for Research on the Cold Start Problem in Recommender Systems. Computer and Modernization 5, 59–62 (2012)
6. Haralambos, M., Dmitry, B.: Algorithms of the Intelligent Web, pp.100–101. Publishing House of Electronics Industry (2011)
7. Satnam, A.: Collective Intelligence in Action, pp. 349–350. Manning Publications (2009)
8. Yanhong, G.: Hybrid Recommendation Algorithm of Collaborative Filtering Cold Start Problem of New Items. Computer Engineering 34(23), 11–13 (2008)
9. Yuxiao, Z.: Summary for Evaluating Indicator of Recommender System. Journal of University of Electronic Science and Technology of China 41(2), 163–172 (2012)

A Method to Construct a Chinese-Swedish Dictionary via English Based on Comparable Corpora

Fang Li, Guangda Shi, and Yawei Lv

Intellingence Engineering Lab, Beijing University of Chemical Technology,
Beijing 100029, China
lifang@mail.buct.edu.cn, shiguangdabuct@163.com,
lvyawei7@gmail.com

Abstract. Taking advantage of existing bilingual dictionaries and a third language can construct a new bilingual dictionary. However, the access of some existing bilingual dictionaries is difficult, which makes it impossible to construct a new bilingual dictionary. For solving the problem, the paper proposes a method that we can construct the bilingual dictionaries of the source language and the target language with a third language by using the corresponding comparable corpora, respectively. The proposed method is applicable to the languages which are lack of comparable corpora and the available dictionaries, such as Chinese and Swedish. This paper constructs a Chinese-Swedish bilingual dictionary by using English as a third language to prove the validity of the proposed method. The result of experiment shows that the proposed method has a good performance in the construction of the Chinese-Swedish bilingual dictionary.

Keywords: bilingual dictionary, intermediary third language, comparable corpora.

1   Introduction

With the rapid development of natural language processing such as machine translation and cross-language information retrieval, a bilingual dictionary plays a more and more important role due to the growing demand for the bilingual dictionary. In the process of constructing bilingual dictionary, especially for the languages which are not widely available, the existing bilingual dictionaries via a third language (usually English) are utilized as the usual method. But there is a problem in this process, which is difficult or even impossible to get the existing bilingual dictionaries for less common languages such as Swedish. For the problem, this paper proposes a new method of constructing a dictionary with the help of the comparable corpora.

The goal of the experiment is to construct Chinese-Swedish bilingual dictionary by using Chinese-English and English-Swedish comparable corpora to construct Chinese-English and English-Swedish dictionary, eventually we get a Chinese-Swedish bilingual dictionary.

The remainder of this paper is structured as follows: Section 2 describes the related work of the bilingual dictionary construction. Section 3 proposes a new method in detail to solve the problem of the lack of dictionary when constructing

Y. Tan et al. (Eds.): ICSI 2014, Part II, LNCS 8795, pp. 196–203, 2014.
© Springer International Publishing Switzerland 2014

Chinese-Swedish dictionary by using English as a third language. Section 4 provides our experiment and analysis. The last section is a summary of this paper and prospects.

2   Related Work of Bilingual Dictionary Construction

Tanaka and Umemura[1] built a Japanese-French bilingual dictionary by using English as third language intermediary. They made use of the existing Japanese-English bilingual dictionary and the English-French bilingual dictionary to construct Japanese-French bilingual dictionary. At the beginning, they searched the English translation candidates of Japanese words by looking up the Japanese-English bilingual dictionary, and found the French translation candidates of the words in English-French bilingual dictionary. Similarly, the English translation candidates of French translation candidates were available by looking up the French-English bilingual dictionary. After that, they got the French translation candidates of Japanese words by comparing the English translation candidates of Japanese words with the English translation candidates of French words. Bond and other scholars[2] took advantage of the Japanese-English bilingual dictionary and the English-Malay bilingual dictionary to construct Japanese-Malay bilingual dictionary. They used a method called semantic classification to sort final candidate translations.

The construction of another bilingual dictionary using comparable corpora is based on the following hypothesis: if two words have similar meaning, the contexts of the two words should also be similar. That's to say, the words in source language and its translations in target language have the similar contexts. Fung (1998)[3] extracted the contexts of the words in the comparable corpora, and then the similarity between two words is calculated by using co-occurrence vector of the words, and constructed a bilingual dictionary. Zhang Yongchen and other scholars captured the datas on the web and built a Chinese-English bilingual dictionary by using Vector Space Model. Haghighi[4] used the Matching Canonical Correlation Analysis to construct a Chinese-English bilingual dictionary. Rapp[5] considered the sequence and relationship of words based on Vector Space Model when constructing a English-German bilingual dictionary. Daille and Morin[6] used the variant methods of the Vector Space Model to construct a French-English bilingual dictionary. About the construction of bilingual dictionary, a lot of scholars did a plenty of researches, involved a lot of languages. However, the research about the dictionary construction of Chinese-Swedish is seldom seen.

3   Chinese-Swedish Dictionary Construction via English

This section detailedly describes how to extract the words of the source language and the target language to construct a bilingual dictionary when we do not have the dictionaries of the source language and the target language with a third language.

As you know, Chinese-Swedish comparable corpora are different to obtain. Our method is to construct a Chinese-English dictionary and an English-Swedish dictionary using the corresponding corpus respectively.

3.1   The Construction of Chinese-English Dictionary and English-Swedish Dictionary

In the process of constructing the Chinese-English dictionary, we make use of a method called the standard method, the technological process can be seen in Figure 1. This detailed process of the method is as follows:

1. Preprocessing. Word segmentation and filtering stop words should be done in this process. As to the preprocessing of English words, stemming and filtering stop words should be used[7].
2. Extracting the contexts. We select 3 as our window size and extract the contexts of the words in the range of window size.
3. The construction of the context space vector. We make use of the bag-of-words to create the context vectors and use the TF-IDF (term frequency-inverse document frequency) of each word to measure the importance of each word. They are calculated as follows:

$$TF_{w_i} = \frac{n_{w_i}}{N} \tag{1}$$

$$IDF_{W_i} = \log \frac{K}{k+1} \tag{2}$$

$$TF - IDF = TF_{w_i} * IDF_{W_i} \tag{3}$$

In the formulas above, w_i represents a word. n_{wi} represents the frequency of the word w_i in the documents. N represents the number of the same words in the documents. K represents the number of the documents. k represents the number of documents which contain the word w_i.

4. Calculating similarity. We select the Cosine Similarity[8] method to calculate the similarity between Chinese-English word vectors. The highest similarity words in the target language are selected as the translation candidates for words in the source language.

$$Cos(W_s, W_t) = \frac{\sum_{i=1}^{t}(w_{is} \times w_{it})}{\sqrt{\sum_{i=1}^{t} w_{is}^2 \times \sum_{i=1}^{t} w_{it}^2}} \tag{4}$$

W_S and W_t represent a source word and a target word respectively.

Analogously, the process of constructing English-Swedish bilingual dictionary is similar to the process of constructing Chinese-English bilingual dictionary.

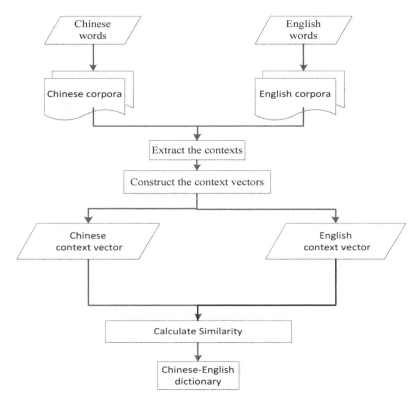

Fig. 1. The method of the Chinese-English bilingual dictionary

3.2   The Construction of Chinese-Swedish Dictionary

The process of constructing Chinese-Swedish can be seen in Figure 2.

We construct Chinese-Swedish bilingual dictionary by extracting words in the Chinese-English bilingual dictionary and the English-Swedish bilingual dictionary that we get in section 3.1.

We look up English translations for Chinese words, and then look for Swedish translations of these English translations. Then, for each Swedish word, we look up all of its English translations. After that, we count the number of shared English translations. If they have the common English translations, they are treated as the translation pairs.

We use a similarity score S for a Chinese word c and a Swedish word s is given in Equation (5), where $E(w)$ is the set of English translations of w.

$$S(c,s) = \frac{2 * |E(c) \cap E(s)|}{|E(c)| + |E(s)|} \tag{5}$$

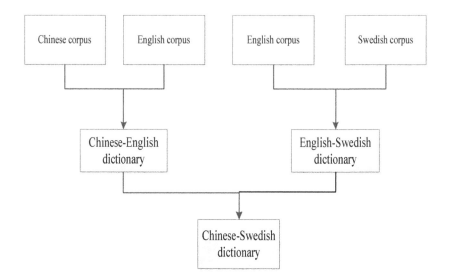

Fig. 2. The method of the Chinese-Swedish bilingual dictionary

4   Experiments

This section mainly describes the evaluation criteria of the related work and experimental results and analysis.

The paper uses the Chinese-English comparable corpora that are the XinHua news about finance in the Gigawords and the English-Swedish comparable corpora that are derived from Wikisource.

4.1   Evaluation Criterion

We use the precision and the Mean Reciprocal Rank (MRR) as the evaluation criteria. In the process of the bilingual word of construction, the precision of the evaluation criteria is often used and it is the average accuracy of the top n translation candidates. The Mean Reciprocal Rank (MRR) is that the mean reciprocal of the rank of the correct translation. It is a measure of the ranking of the correct translation in the translation candidates.

The Mean Reciprocal Rank (MRR) is a more lenient evaluation criteria than the precision. The precision and the Mean Reciprocal Rank(MRR) are calculated as follows:

$$Precision = \frac{count_{top5}}{N} \tag{6}$$

$$MRR = \frac{1}{N} \sum_{i=1}^{N} \frac{1}{rank_i} \tag{7}$$

Count$_{top5}$ represents the total numbers of the top 5 translation candidates, *rank*$_i$ represents the ranking of the correct translation, N represents the numbers of the word pairs. Different from the precision, the Mean Reciprocal Rank (MRR) is not considering the number n of the word translation candidates; therefore, it can better measure the performance of the bilingual dictionary construction.

4.2   Results and Analysis

Table 1 shows how the link is realized and the similarity scores in section 3. The similarity score shows how many English words are shared by the two dictionaries. The higher the score, the higher the probability of successful linking. As Table 1 shows, we can see that, if the number of shared English translated words is more, then we get a higher possibility of accurate matching of Chinese and Swedish. However, the accuracy reduces when the number of the shared English translated words decreases. Sometimes we have to inappropriately sort out the matched pairs such as 首都 (kapital). The reason is that a lot of intermediate words are polysemous such as capital.

Table 1. Example of linking through English translations

Score	Chinese→English	Swedish→English	Chinese-Swedish	
2	支票(check, cheque)	Check(check, cheque)	支票(check)	Yes
2	现金(cash, money)	Kassa(cash, money)	现金(kassa)	Yes
1	资本(capital)	Kapital(capital)	资本(kapital)	Yes
1	费用(cost, expense)	Kostnad (cost)	费用(kostnad)	Yes
1	首都(capital)	Kapital (capital)	首都(kapital)	No

Table 2 shows the experiment results of the method to solve the problem of the lack of dictionary when constructing Chinese-Swedish dictionary by using English as a third-party language. We evaluate the precision of the top 5 translation candidates.

As Table 2 shows, we totally extract 17368 translation pairs from the Chinese-English comparable corpora, and the precision is 54.38%, the Mean Reciprocal Rank (MRR) is 58.13%, moreover, there are 94 translation pairs that are not quite accurate, but they are accepted. We get 14273 translation pairs in total, and the precision is 51.26%, the Mean Reciprocal Rank (MRR) is 56.34%, moreover, there are 137 translation pairs that can be accepted. Finally, we have a Chinese-Swedish dictionary that contain 7369 translation pairs, its precision is 46.17%, the Mean Reciprocal Rank (MRR) is 58.26% and there are 63 translation pairs accepted.

Table 2. The Experiment Results

	Translated	Precision	MRR	Accepted
Chinese-English	17368	54.38%	58.13%	94
English-Swedish	14273	51.26%	56.34%	137
Chinese-Swedish	7369	46.17%	51.26%	63

5   Conclusion and Future Work

As a conclusion, we proposes a new method in detail to solve the problem of the lack of dictionary when constructing Chinese-Swedish dictionary by using English as a third language, and this method is applicable to other language resources. We make use of the public available resources Gigawords and Wikisource for the construction of a new language pair. The process is divided into three parts, the first part is constructing Chinese-English bilingual dictionary by using the Chinese-English comparable corpora, and the second part is constructing English-Swedish dictionary by using the English-Swedish comparable corpora, finally we can get the Chinese - Swedish dictionary based on the two above dictionary. The accuracy obtained is 46.17%, which proves the effectiveness of the proposed method when lack of the dictionary between the source language and the third language or between the target language and third language.

The current study about the Chinese-Swedish bilingual dictionary construction is relatively few, during the construction of Chinese-Swedish bilingual dictionary process, the proposed method proves its feasibility and effectiveness, and this method is also applicable to other language resources, which has an important contribution to the study of related work.

However, there is still a lot of work to do. As future work, firstly, we plan to compare different definitions of context in the process of constructing the context vector, such sentence-based context and syntax-based context. Secondly, we plan to conduct experiments on other comparable corpora such as Wikipedia and different language pairs. Finally, we plan to extend our method to deal with some compound and rare words.

Acknowledgments. This research is supported by the Chinese-English comparable corpora of Gigawords and the English-Swedish comparable corpora of Wikisource. We would also like to thank everyone who offers help.

References

1. Tanaka, K., Umemura, K.: Construction of a bilingual dictionary intermediated by a third language. In: 15th COLING International Conference on Computational Linguistics, pp. 297–303. ACL, Stroudsburg (1994)

2. Bond, F., Sulong, R., Yamazaki, T., Ogura, K.: Design and construction of a machine-tractable Japanese-Malay dictionary. In: 8th MT Summit, pp. 53–58. Santiago de Compostela (2001)
3. Fung, P.: Compiling bilingual lexicon entries from a non-parallel English-Chinese corpus. In: 3rd Annual Workshop on Very Large Corpora, Boston, pp. 173–183 (1995)
4. Haghighi, A., Liang, P., Berg-Kirkpatrick, T.: Learning Bilingual Lexicons from Monolingual Corpora. In: the Association for Computational Linguistics on Computational Linguistics, pp. 771–779. ACL, Stroudsburg (2008)
5. Rapp, R.: Automatic identification of word translations from unrelated English and German corpora. In: 37th Annual Meeting of the Association for Computational Linguistics on Computational Linguistics. Association for Computational Linguistics, pp. 519–526. ACL, Stroudsburg (1999)
6. Daille, B., Morin, E.: French-English Terminology Extraction from Comparable Corpora. In: Dale, R., Wong, K.-F., Su, J., Kwong, O.Y. (eds.) IJCNLP 2005. LNCS (LNAI), vol. 3651, pp. 707–718. Springer, Heidelberg (2005)
7. Fung, P.: A Statistical View on Bilingual Lexicon Extraction: From Parallel Corpora to Non-parallel Corpora. In: Farwell, D., Gerber, L., Hovy, E. (eds.) AMTA 1998. LNCS (LNAI), vol. 1529, pp. 1–17. Springer, Heidelberg (1998)
8. Kaji, H., Erdenebat, D.: Automatic Construction of a Japanese-Chinese Dictionary via English. In: the International Conference on Language Resources and Evaluation, Marrakech, pp. 699–706 (2008)

The Design and Implementation of the Random HTML Tags and Attributes-Based XSS Defence System

Heng Lin, Yiwen Yan, Hongfei Cai, and Wei Zhang

School of Software, Beijing Institute of Technology, Beijing 100081 China
anonymous.joker.lin@gmail.com,
278567461@qq.com

Abstract. At present, cross site scripting (XSS) is still one of the biggest threat for Internet security. But the defensive approach is still feature matching mostly; that is, to check for a matching and filter in all information submitted. However, filtering technology has many disadvantages as heavy-workload, complex-operation, high-risk and so on. For this reason, our system use the randomization techniques of HTML tags and attributes innovatively, based on the prefix of HTML tags and attributes, to determine the tags and attributes are Web designers expect to generate or other users insert in, and then we follow the results to carry out different policies, only tags and attributes that Web designers expected to generate can be rendered and implemented. By this way, we can defend against XSS attacks completely. The test results show that the system is able to solve a variety of problems in filtering technology. It uses simple and convenient operation and safe and secure effect to free developers from heavy filtering work. System has a good compatibility and portability across platforms, it also can connect with all web-based applications seamlessly. In all, system defend against XSS better and meet the need of today's XSS attacks defence.

Keywords: random, tag prefix, cross-site scripting, defence system.

1   Project Background

Cross Site Script Execution (usually abbreviated as XSS) is a kind of attack that attacker using the lack of filtering on the user's input, manufacturing malicious input which can affect other users, so as to achieve the purpose of attack, stealing user data, using user identity to make some things or using virus to attack the visitors.

In recent years, XSS vulnerability has been ranked in the top three Web security .According to the data published by OWASP(Open Web Application Security Project) in 2010 and 2013,in the top ten Web security vulnerabilities, XSS is both ranked the top three[1].

In November 2012, Anheng Institute for information security found XSS vulnerability exist in the Web applications of Baidu, Tencent, Sina and other Internet companies[2]; In April 2013,a fact was exposed by WooYun that Taobao had Cross Site Vulnerabilities [3];In June 2013, WooYun reported again that the main Web site

Y. Tan et al. (Eds.): ICSI 2014, Part II, LNCS 8795, pp. 204–211, 2014.

of Tencent and Baidu had XSS vulnerability [4-5]. XSS vulnerability has been a serious threat to the information security of majority of Internet users.

To prevent from XSS attack, someone present model-based testing evolved as one of the methodologies which offer several theoretical and practical approaches in testing the system under test (SUT) that combine several input generation strategies like mutation testing, using of concrete and symbolic execution etc. by putting the emphasis on specification of the model of an application [6]; someone detecting Cross-Site Scripting Vulnerability using Concolic Testing[7]. Existing cross-site scripting attack protection mainly contains four aspects: the input validation, escape, filtering, and character set specifies, there are Http-Only, input check, output check, three kinds of defense in total. But the reality is that each method has its own defensive shortcomings, it's also very hard to handle the text properly.

So we develop XSS-Defender, a system that allows the server to identify untrusted content and reliably convey this information to the client, and that allows the client to enforce a security policy on the untrusted content. Analogously, XSS-Defender randomizes HTML tags and attributes to identify and defeat injected malicious web content. These randomized tags and attributes serve two purposes. First, client proxy distinguish a tag or attribute is legal or not through whether the tag or attribute has a random prefix client proxy generated so that identify untrusted content. Second, they prevent the untrusted content from distorting the document tree.

We make the following contributions:
• We innovatively develop a system randomizing HTML tags and attributes to defend against XSS attacks.
• We lead a simple way for current web application to defend against popular XSS attack efficiently.

2   Systematic Realization

2.1   Processing of Client Proxy for the User Data

When a user opens a client proxy, the client proxy automatically connect to the server proxy, to get the public key that server proxy stored in the cache, after that, the public key will use as a basis of the RSA encryption algorithm to transfer randomized prefix. When a user of Web site requests a Web page or submit data to the server, first of all, the user send data to the client proxy that is deployed on the user's computer. Via the client proxy, system forwards the request to the server proxy that is deployed on the server. Meanwhile, the client proxy use Python random number generator to generate a random string as the identification of random prefix, after a non-equivalent encryption, it will be sent to server proxy with user's requests together. The random string is effective only at this time of page request and getting response to server, it will be destroyed at once after using it one time, in order to ensure the security of the system.

2.2   Processing of Server Proxy for the User Data

As shown in Fig. 1, when the server proxy starts, it generates a pair of public and private keys used for RSA encryption. When the server proxy receives requests or data, it detects the requested type first, if the request is a request of RSA public key, proxy will send the generated public key to client proxy. If the request is other networks' interactive request, proxy will judge whether it has a random prefix the client proxy generates or not. If not, the request or data will be discarded; if so, it will forward requests or submitted data to the Web server, then the one-time random prefix the client proxy generates will be stored in the cache.

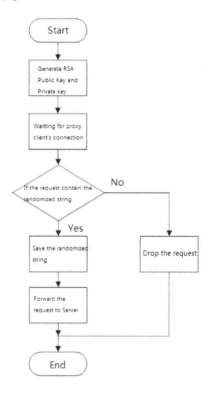

Fig. 1. Processing flow chart of server proxy for the user data

2.3   Processing of Web Server for the User Data

In the system, the Web server pre-stores a confidential 16-bit string which can be modified at any time as a tag prefix to prove that the tags or attributes come from the server. In a static page stored on the server, each HTML tag and attribute will have the confidential prefix, and in the code of Web page dynamically generated by the server, each HTML tags and attributes generated within the plan of Web site developer (legal) will have the confidential prefix, tags and attributes not in the developers' plan (illegal) does not have this prefix. Confidential prefix only transfer across the local network connection between the server proxy and server which are deployed on the same machine, in addition to the site maintainers, others all could not get the confidential prefix.

When receiving a request, the Web server dynamically generates or read the user requests' website from the cache. In this page, all legal tags and attributes have the confidential prefixes, meanwhile illegal tags and attribute does not have the confidential prefixes. In this way, the server generates a page that is able to distinguish whether each tab comes from the server or not. Then, the server sends this page to the server proxy through a local connection inside the computer.

2.4   Processing of Server Proxy for the Page Sent to Users

After receiving the Web server's response to the request, server proxy will replace all confidential prefixes in the code of Web page with confidential prefix the client proxy generates. Then, the server proxy sends modified code of Web page to the client proxy. As the prefix for the Web site's users cannot be revealed to other users, so that the replaced page still can distinguish this tag is legal or not by the tag or attribute whether the client has proxy generated random prefix. This way, the client proxy can classify and deal with them according to the different tag prefix.

2.5   Processing of Client Proxy for the Page Sent to Users

When the client proxy receives the treated page sent by server proxy, first of all, it extracts all tag and attribute information in the code of Web pages. Then, it can classify according to the different tag and attribute prefix. (Process as Fig. 2).

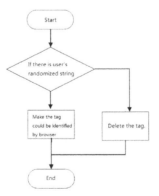

Fig. 2. Processing flow chart of client proxy for each tag in the code of Web pages

Client proxy distinguish a tag or attribute is legal or not through whether the tag or attribute has a random prefix client proxy generated. According to the result of judgments, client proxy deletes the illegal tags; but deletes the random prefix of legal tags, transforms tags into a format that can be rendered and implemented by browsers. After the sorting, the client proxy will send the last generated and transformed page to the user's browser. In this way, legal tags and attributes can be rendered and implemented on the user's browser, while illegal tags and attributes cannot be rendered and implemented on the user's browser. To avoid the browser executing malicious code submitted by the attackers efficiently, and thus system can prevent XSS occurred.

3   The Performance of Testing and Experiment

Through a large number of comprehensive testing of the system, the testing is able to detect the operations whether efficient or stable. To achieve protection against XSS attacks, we get the performance of the each module and the overall operational status of the system.

Test content is divided into system functional testing and system performance testing in two parts.

3.1   System Functional Testing

System Resources
Examine the CPU and memory usage. Turn on the system, using the Windows Task Manager to see the CPU and memory usage. See the memory usage of server and the clients which run XSS-Defender and detect the whether exist the presence of a memory leak.

Response Speed Test
Statistics the user's access speed, page load time when using XSS-Defender system and using XSS-Defender system. Then we can conclude the comparisons of running the system which affect the ability of the server's response and effect.

Server Performance Impact
Specific test operation to record the XSS-Defender server for each user request response time is calculated for each user request processing response speed; recording XSS-Defender client browser page request to the server's response time is calculated browser page request response speed.

3.2   The Results of System Testing

Functional Test for XSS Protection
We designed four different methods to attack by XSS.

Table 1. Four pieces of script code

No.	script code
1	`><script>alert(document.cookie)</script>`
2	`<img scr="javascript: alert(/XSS/)></img>`
3	`<div style={left:\0065\0078ression (alert('xss'))}></div>`
4	`<style/style" STYLE="background:expre><ssion(alert(>"\\<XSS="));">`

We established a blog site. Then we inject four script codes into the blog message board.

Based on the experimental results, the system makes the harmful input invalid, so it defends against the several of XSS attacks efficiently.

Function Test that Submission does not Affect the Contents of the User's Normal Display

Inject Code:

<script>alert("XSS");</script>

The browser of client shows:

Fig. 3. The result of injecting XSS code

The injected code is displayed as text and not executed properly in the site content: i think <script> is the best label

The browser of client shows:

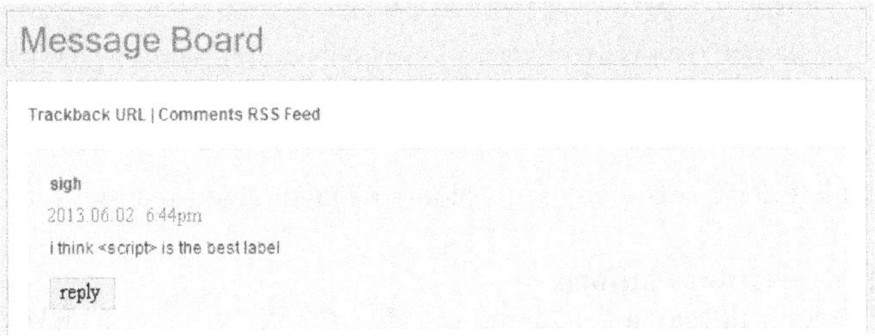

Fig. 4. Submit a normal script

Users fill out a script tag containing the 'script' contents and the browser is normally displayed.

Based on the above experimental results, the system can tell the difference between legal and illegal entry.

The Function Test of Adding the Prefix on the Service Side of the Page File
Source page on the server part:

```
|<div class="cancel-comment-reply">
|    <small>
           <a rel="nofollow" id="cancel-comment-reply-link"
        </small>
    </div>
```

Fig. 5. Source code

The client sends a request to obtain the corresponding part of the html file:

```
<sad32div class="cancel-comment-reply">
    <sad32small>
           <sad32a rel="nofollow" id="cancel-comment-reply-link"
        <sad32/small>
    <sad32/div>
```

Fig. 6. The code received

All tags add prefix 'sad 32'.

Table 2. The function test of adding the prefix on the service side of the page file

Expected test results	The actual test results	The deviation of the expected results and actual results	The conclusion of the test
Add the random prefix on the page code	As the screenshot shown	None	The function works well

Test of Log Function
Try to request the server, the log on the server side shows as follows:

```
IP:192.168.1.101  Physical Address : AC-72-89-93-4B-8A  Generated random prefix : prefix92  2013/06/06/10:52:32

IP:192.168.1.11  Physical Address : AC-71-79-23-4A-3A  Generated random prefix : asassq32  2013/06/06/10:57:27
```

Fig. 7. Test of log function

Logs completely records every request and its generated random prefix.

Function Test of Cross-Platform
The use of Python development in Linux, Mac OS X, Windows under all operating normally, the client implementation for the client agent, it can be supported by all browsers.

Table 3. Function test of adding prefix on the server side

Expected test results	The actual test results	The deviation of the expected results and actual results	The conclusion of the test
Log records	As the screenshot shown	None	No problem found

Performance Test on the Client Side

When the client program is not running, the s peed of opening a web page:

Fig. 8. The speed of opening a web page

When client program is not running.
After running the client program :

Fig. 9. The speed of opening a web page

When client program is running
The speed is almost unchanged.
When running the client, open the Task Manager, we can see that:

Name	PID	state	User name	CPU	memory(...
XSS-Defender.exe	8720	Running	admin	00	640 K

Fig. 10. The resources occupancy after running the client system

Based on the above experimental results, the system occupy very small CPU and memory.

References

1. OWASP: OWASP Top- 2013 10 rcl The Ten Most Critical Web Application Security Risks (2013)
2. eNet, http://www.enet.com.cn/article/2012/1112/A20121112190987.shtml
3. WooYun, http://www.wooyun.org/bugs/wooyun-2010-022080
4. WooYun, http://www.wooyun.org/bugs/wooyun-2010-025030
5. WooYun, http://www.wooyun.org/bugs/wooyun-2010-025002
6. Top 25 most dangerous software errors, http://cwe.mitre.org/top25/.CWE/SANS
7. Bozic, J., Wotawa, F.: XSS Pattern for Attack Modeling in Testing. In: 8th International Workshop on Automation of Software Test (AST), pp. 71–74. IEEE (2013)

DWT and GA-PSO Based Novel Watermarking for Videos Using Audio Watermark

Puja Agrawal[1] and Aleefia Khurshid[2]

[1] Department of Electronics and Communication Engineering,
Ramdeobaba College of Engineering and Mangement, Nagpur, India
[2] Department of Electronics Engineering,
Ramdeobaba College of Engineering and Mangement, Nagpur, India
{agrawalps,khurshidaa}@rknec.edu

Abstract. This paper presents a digital video watermarking scheme that can embed invisible and robust watermark information into the video streams of MPEG-1, MPEG-2, H.264/AVC, MPEG-4 standards. Watermark embedding process is in Discrete Wavelet Domain. Trade off between transparency and robustness is considered as optimization problem and is solved by Genetic Algorithm - Particle Swarm Optimization (GA-PSO) based hybrid optimization technique. An audio signal is converted into 9 bit planes by using bit plane slicing and then embedded into the frames of video signals as watermark. The performance evaluation results based on the Peak Signal to Noise Ratio (PSNR) and Normalized Correlation (NC) confirm that the proposed video processing method shows reliable improvements for various sequences compared to existing ones for geometrical attacks like rotation and cropping.

Keywords: Watermarking, Robustness, Transparency, GA-PSO, Bit Plane Slicing.

1 Introduction

Protection of multimedia data has become one of the major challenges due to the rapid growth of unauthorized access and copy of digital media objects like images, audio and video. Digital Watermarking is a process where some valuable information is embedded into the host media like images, video and audio etc. The secret message embedded as watermark can be almost anything, for example: a serial number, plain text, image, random signal, an organization's trademark, or a copyright message for copy control and authentication. Potential applications of digital watermarking includes, copy control, transaction tracking, authentication, and legacy system.

In general, digital watermarking involves two major operations: (i) Watermark embedding, and (ii) Watermark extraction. The two most important properties viz. robustness and transparency are required for preserving the security of videos from unauthorized access. The ability to detect the watermark content after application of common signal processing distortions like filtering, lossy compression, color

Y. Tan et al. (Eds.): ICSI 2014, Part II, LNCS 8795, pp. 212–220, 2014.

correction, noise, contrast distortions, geometric distortions is known as robustness. Imperceptibility/Transparency means that the presence of watermark is not noticed by the human eyes. Watermarking techniques can be classified according to the nature of host data (text, image, audio or video), or according to the working (spatial or frequency) domain.

1.1 Literature Review

Most of the proposed video watermarking scheme based on the techniques of the image watermarking and applied to raw video or the compressed video. As some issues in video watermarking are not present in image watermarking, such as video object and redundancy of the large amount video data, researchers have made use of those characteristics to develop different schemes [1].

Hui-Yu Huang et.al. have presented an approach consists of a pseudo-3-D DCT. The watermark message represents an index for selection of a particular quantizer from a set of possible quantizers. The selected quantizer is applied to the host data to encode the watermark message [1]. This is invisible, robust for Raw Videos, but complex process for implementation.

Jing Zhang et.al have described that a grayscale watermark pattern is first modified to accommodate the H.264/AVC computational constraints, and then embedded into video data in the compressed domain. With the proposed method, the video watermarking scheme can achieve high robustness and good visual quality without increasing the overall bit-rate [2].

Gwenael Doerr et.al have given the in depth overview of video watermarking and have pointed out that video watermarking is not just a simple extension of still image watermarking[3].

Ersin ELBASI [4], has proposed a novel video watermarking system based on the Hidden Markov Model (HMM). This novel watermarking scheme splits the video sequences into a Group of Pictures (GOP) with HMM. Portions of the binary watermark are embedded into each GOP with a wavelet domain watermarking algorithm.

Sanjoy Deb Roy et.al. have presented a hardware implementation of a digital watermarking system that can insert invisible, semi fragile watermark information into compressed video streams in real time [5].

Sourav Bhattacharya et.al. have presented a survey on different video watermarking techniques and comparative analysis with reference to H.264/AVC [6].

Salva A. K. Mostafa et. al. have presented a video watermarking scheme based on principal component analysis and wavelet transform [7].

Chuen-Ching Wang et. al. have presented a simple but effective digital watermarking scheme utilizing a context adaptive variable length coding method for wireless communication systems [8].

In this paper, we have presented an efficient video watermarking technique using discrete wavelet transform and GA-PSO based hybrid optimization to protect the copyright of digital videos. In the watermark, first of all the input video is segmented into shots. The frames of all the shots are decomposed into four sub-bands as HH,

HL, LH and LL. Watermark is embedded into the high middle frequency HL and LH sub-bands, where acceptable performance of imperceptibility and robustness could be achieved. An audio signal is chosen as watermark which is unusual. Before embedding into the sub-bands, the watermark audio signature is processed into a 9 bit plane slices. Then it is embedded into HL sub-band, and LH sub-band. Here, every audio bit is embedded into the chosen sub-bands with the aid of our proposed embedding process. Subsequently, the watermark audio bits are extracted with the help of our proposed extraction process. Powerful GA-PSO optimization guarantees the performance. Since, the watermarking is performed in the wavelet domain; the attained watermark image is of good quality. The efficiency of our proposed watermarking technique is proved by good PSNR and NC values obtained for the watermarked videos in the experimental results.

2   Watermark Embedding and Extraction

This process is the most important in this scheme where all the different parts of watermarks are embedded into different scenes of the video. Video Preprocessing and Embedding is done by changing position of some DWT coefficients with the following condition:

2.1   Selection of Best Embeddable Locations

Human visual system has a very strong error correction mechanism. An image contains lot of redundancies. Small changes made to an image remain undetected by the human eyes. On the other hand, it has been observed that if an effort is made to increase the invisibility of a watermark, then robustness of the scheme suffers and vice a versa. A compromise therefore has to be made in order to get an optimum system.

Wavelet based transforms gained popularity recently since the property of multi-resolution analysis that it provides [12]. The higher level sub bands are more significant than the lower level sub bands. They contain most of the energy coefficients, so embedding in higher level sub bands is providing more robustness. On the other hand lower level sub bands have minor energy coefficients so watermark in these sub bands are defenseless to attacks. The sub band LL is not suitable for embedding a watermark since it is a low frequency band that contains important information about an image and easily causes image distortions. Embedding a watermark in the diagonal sub band HH is also not suitable since the sub band can easily be eliminated, for example by lossy compression as it has minor energy coefficient. So the middle frequency sub bands LH and HL are the best choice for embedding [12].

There has been a considerable amount of research proposals on the applications of DWT in digital image and video watermarking systems by virtue of its excellent and exceptional properties mentioned above, but the scope of optimization in this area is tremendously less. An optimized DWT for digital image watermarking is capable of producing perceptual transparency and robustness among the watermarked and the extracted images [13].

Extending the above concept to videos with suitable modifications, we have proposed the DWT and GA-PSO hybrid optimization based scheme for video watermarking.

2.2   Embedding

Input: Original video sequence: V_o [a,b] , watermark audio A_w [a,b]
Output: Watermarked video V_w [a,b]

Initially, the shot segmentation technique is applied to original input video sequence V_o [a,b] is segmented into number of non-overlapping shots $D[a,b]$. For embedding purpose, we identify number of frames $E[a,b]$ in all the segmented shots $D[a,b]$. Then watermark audio signal $A_w[a,b]$ is converted into 9-bit plane $W[a,b]$, by the use of bit plane slicing. The video frames have R, G and B components. The blue channel is selected for embedding because this channel is more resistant to changes compared to red and green channels and the human eye is less sensitive to the blue channel, a perceptually invisible watermark embedded in the blue channel can contain more energy than a perceptually invisible watermark embedded in the luminance channel of a color image [12]. The blue components E_B [a,b] of all the separated frames are extracted. Then each bit of 9-bit plane sliced audio watermark W [a,b] is applied into the blue components of each frame; Discrete Wavelet Transform is applied to blue component $E_B[a,b]$. Discrete Wavelet Transform converts each frame into four sub-bands such as HH, HL, LH and LL to attain the transformed $T[a,b]$ frames. Then we select the middle frequency sub-bands (HL, LH) from the transformed frames to embed the watermark audio $A_w[a,b]$ into the appropriate sub-bands. In order to choose the embedding locations in the sub-bands, we find the similarity matrix for the video signal.   The similarity matrix for HL sub-band is denoted by $U_p(x,y)$ and the LH sub-band similarity matrix is denoted by lower part $L_p(x,y)$.Then we   calculate the mean value $T_{LH}(m)$ and the maximum   value $T_{LH}(M)$ of the chosen embedding part T_{LH}.
Watermark bits are embedded in the video frames according to following conditions:

```
Condition 1:For embedding the Watermark Pixel 1
If T_LH(a) > 1 then L_p(x,y) << [T_LH(a)]
else L_p(x,y)<< T_LH (a) + T_LH(M)
end if


Condition 2: For embedding the watermark Pixel 0
If T_LH(a)>0 then
L_p(x,y) << Abs[T_LH(a)]
else
L_p(x,y) << T_LH(a) -T_LH(M)
end if
```

Likewise the watermark bits can also be embedded into the HL band. Then the modified sub-bands are mapped into its original position and inverse wavelet transform is applied to attain the watermarked video sequence V_w $[a,b]$.

2.3   Extraction Process

Input: Watermarked video sequence $V_w[a, b]$
              & the size of the audio watermark $A_w[a, b]$.
Output: Recovered watermark audio $A_{rw}[a', b']$

The watermarked video sequence $V_w[a, b]$ is segmented into number of non-overlapping shots $D'[a, b]$ by the use of shot segmentation technique. Then we identify number of frames $E'[a, b]$ in each segmented shots. Blue components of the partitioned frames are extracted. Afterwards Discrete Wavelet Transform is applied to the each partitioned frame $E_B'[a, b]$. The DWT is splits each frame into four sub-bands such as HH, HL, LH and LL and then to attain the transformed $T'[a, b]$ frames. The middle frequency LH and HL sub-bands are selected from the transformed frames. The watermark audio bits are extracted from LH and HL sub-bands. If the embedded bit value is greater than the mean pixel value, then the extracted pixel value is one. If it is lesser, then the extracted pixel is zero.

Matrix with the size of the audio watermark is prepared and the extracted bits are placed to attain the watermark audio. The extracted audio watermark A_{rw} $[a',b']$ is obtained by the use of reverse process of vector finding operation.

3   GA-PSO Based Hybrid Optimization

In order to achieve both imperceptibility and robustness of the watermarked media, we use the Genetic algorithm (GA) and Particle swarm Optimization (PSO) based hybrid optimization. GA is applied for generating the chromosome and PSO for selecting the optimal location for embedding the watermark media into host media. GA-PSO based hybrid optimization techniques are applied in embedding as well as extraction process.

The function of the randomly generated set of genes is the generation of chromosomes. Population size plays an important role in presenting the solution to the problem at hand. The beginning population set up is done by producing a population set P that comprises of set of chromosome vectors having half size of the HL or LH sub-band. Subsequently, we initialized it with "1" according to the size of the watermark in that vector in a random manner, and the remaining places are filled down with "0" values. Then, the beginning set of chromosomes is brought forth at random with minimum number.

The watermark embedding process is iterated till the optimal locations are obtained for each chromosome in the population set. Embedding and extraction process is carried out using these procedures which were defined in the section 2. Fitness computation formula is depicted below,

$$Fitness = PSNR + NC . \tag{1}$$

PSNR: The Peak-Signal-To-Noise Ratio (PSNR) is used to measure deviation of the watermarked and attacked frames from the original video frames.

NC: The normalized coefficient (NC) gives a measure of the robustness of watermarking and its peak value is 1. For calculating the PSNR and NC we have used standard formula as mentioned in [1].

Selection of optimized chromosomes is done based on the values for fitness. Select the optimized chromosome = N_p / 2.Where, N_p is Number of Parent chromosomes. Remaining chromosomes enter the next iteration in the search of finding the optimal solutions according to fitness function.

Based on the selected optimal chromosomes, we have the value of optimal solution. This solution is fed to the crossover operation. These set of fitness value corresponding to chromosomes provide the new offspring by the use of crossover operation. Every two individuals are chosen from the better set of chromosome to produce two new offspring by single crossover point.

In mutation operation, the output of crossover operation is used as input. The process of this function is to modify one gene value that is randomly selected and then that chromosome is fed to the fitness computation operation. Here mutation operation is replaced with a velocity computation.

The velocity computation operation is the part of the Particle swarm optimization. During each iteration, each particle accelerates in the direction of its own personal best solution found so far as well as in the direction of the global best position discovered so far by any of the particle in the swarm. This means that if a particle discovers a promising new solution, all the other particles will move closer to it, exploring the region more thoroughly in the process. The velocity computation is done with the standard formula as described in [15].

Newly obtained set of chromosomes velocity computation operation can be evaluated for best fitness using fitness function. If the optimal solution for embedding the watermark media into original media is obtained then and then this process will be terminated, otherwise that solution will move to fitness computation operation again and the selection and velocity computation operators are performed iteratively. The PSO process will be iteratively performed until the desired termination is satisfied.

4 Experimental Results

We used many different videos of varying standards, framerate, framesize, payload for experimentation. Which includes Akiyo, Coastguard, Claire, Carphone, Shutttle, container, football and silent. Apart from these we used bradman.mpg, barryrichards.mpg, and chrisold.mpg and many others.

The original 30th frame and its corresponding audio watermark are shown in Fig. 1. Watermarked frame appears visually identical to the original. The performance of algorithm can be measured in terms of its imperceptibility and robustness against the possible attacks. Watermarked frame is subjected to a variety of attacks such as Salt and Pepper Noise, Median Filtering, Gaussian Noise, and geometric attacks etc. In case of geometric attacks, the scheme is tested against 90^0, 180^0, 270^0 frame rotation, and 25 % frame cropping. To evaluate the performance of any watermarking system, Peak Signal to Noise Ratio (PSNR) is used as a general measure of the visual quality. And the NC values are used as a measure of robustness. The NC values for extracted watermark for test videos by our proposed system are greater than .9. The most important aspect is the consistency of the results for different attacks. For robustness of compressions, our proposed system can effectively resist the MPEG-1, MPEG-2, MPEG-4 and H.264 compressions. Proposed system causes very slight distortion and simultaneously provides high visual quality. The embedded watermark is an audio signal which is unusual. An audio signal Example.wave, 529kb size is used and converted into suitable dimensions using wavread and signalslices functions in Matlab and transformed into a suitable watermark. Human Auditory system is more sensitive than the human visual system. Any modification in the extracted audio watermark will be noticed more easily as compared image or other watermarks. Evidently audio watermark bits are spread over LH and HL sub bands and it would be difficult to extract and reconstruct proper watermark and achieve high NC values. Our proposed system is strongly resistant to geometrical attacks like rotation and cropping in comparison to [1], [2] and [7].

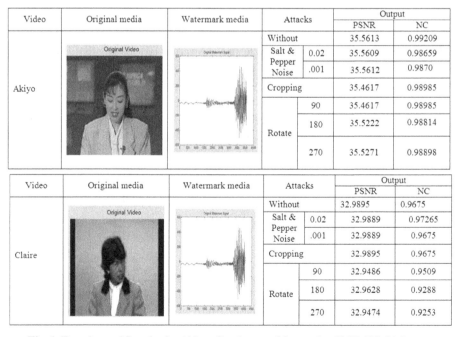

Fig. 1. Experimental Results for Akiyo, Carphone   of frame size 720X 480, 80 frames

5   Conclusion

A DWT and GA-PSO based novel video watermarking technique has been proposed using an audio signal as watermark. The performance of our purposed watermarking scheme is evaluated with common image processing attacks such as salt and pepper noises, rotation, cropping, Experimental results demonstrate this watermarking technique is robust against various attacks including the geometrical attacks. This proposed method is an extension to (HWT- Haar Wavelet Transform) HWT-GA-PSO based Image watermarking method [10].

References

1. Hui-Yu, H., Cheng-Han, Y., Wen-Hsing, H.: A Video Watermarking Technique Based on Pseudo 3-D DCT and quantization Index Modulation. IEEE Transactions on Information Forensics and Security 5(4), pp. 625–637 (2010)
2. Jing, Z., Anthony, T., Ho, S., Gang, Q., Pina, M.: Robust Video Watermarking of H.264/AVC. IEEE Transactions on Circuits and Systems-II: Express Briefs 54(2), 205–209 (2007)
3. Gwenael, D., Jean-Luc, D.: Guide Tour of Video Watermarking Signal Processing: Image Communication Elsevier. Signal Processing Image Communication 18, 263–282 (2003)
4. Elbasi, E.: Robust Multimedia Watermarking: Hidden Markov Model Approach for Video Sequences. Turk J. Elec. Eng. & Comp. Sci. 18(2), 159–170 (2010), doi:10.3906/elk-0906-85
5. Sonjoy, D.R., Xin, L., Yonatan, S., Alexander, F., Orly, Y.-P.: Hardware Implementation of a Digital Watermarking System for Video Authentication. IEEE Transactions on Circuits And Systems For Video Technology 23(2), 289–301 (2013)
6. Saurav, B., Chattopadhyay, T., Arpan, P.: A Survey on Different Video Watermarking Techniques and Comparative Analysis with Reference to H. 264/AVC. IEEE (2006)
7. Mostafa, S.A.K., Tolba, A.S., Abdelkader, F.M., Elhindy, H.M.: Video Watermarking Based on Principal Component Analysis and Wavelet Transform. International Journal of Computer Science and Network Security 9(8), 45–52 (2009)
8. Chuen-Ching, W., Yao-Tang, C., Yu-Chang, H.: Post-Compression Consideration in Video Watermarking for Wireless Communication. World Academy of Science, Engineering and Technology, 199–204 (2011)
9. Noorkami, M., Marsereau, R.M.: Digital Video Watermarking in P-Frames With Controlled Video Bit Rate Increase. IEEE Transactions on Information Forensics and Security 3(4), 441–455 (2008)
10. Puja, A., Khurshid, A.: Novel Invisible Watermarking for Various Images using HWT-GA-PSO based Hybrid Optimization. International Journal of Advanced Research in Computer Science and Software Engineering 3(8), 1093–1101 (2013) ISSN: 2277 128X
11. Martin, Z.: Master Thesis on Video Watermarking, Department of Computer Science Education, Charles University Prague (2007)
12. Shekhawat., R.S., Rao, S., Shrivastava, V.K.: A Robust Watermarking technique based on Biorthogonal Wavelet Transform. IEEE (2012) 978-1-4673-0455-9/12

13. Surekha, P., Sumathi, S.: Application of GA and PSO to the Analysis of Digital Image Watermarking Process. International Journal of Computer Science & Emerging Technologies 1(44), 350–362 (2010) (E-ISSN: 2044-6004)
14. Ramesh, S.M., Shnamugam, A.: An Efficient Robust Watermarking Algorithm in Filter Techniques for Embedding Digital Signature into Medical Images Using Discrete Wavelet Transform. European Journal of Scientific Research 60(1), 33–44 (2011) ISSN 1450-216X
15. Wang, Z., Sun, X., Zhang, D.: A Novel Watermarking Scheme Based on PSO Algorithm. In: Li, K., Fei, M., Irwin, G.W., Ma, S. (eds.) LSMS 2007. LNCS, vol. 4688, pp. 307–314. Springer, Heidelberg (2007)

Application and Comparison of Three Intelligent Algorithms in 2D Otsu Segmentation Algorithm

Lianlian Cao[1,2], Sheng Ding[1,2], Xiaowei Fu[1,2], and Li Chen[1,2]

[1] College of Computer Science and Technology,
Wuhan University of Science and Technology, China
[2] Hubei Province Key Laboratory of Intelligent Information Processing
and Real-time Industrial System, China

Abstract. 2D Otsu thresholding algorithm has been proposed based on Otsu algorithm, it is more effective in image segmentation. However, the computational burden of finding optimal threshold vector is very large for 2D Otsu method. In this paper, three kinds of intelligent algorithm are applied to improve and compare the efficiency of search. Experimental results show that these methods can not only obtain the ideal segmentation results but also greatly reduce the launch time. Moreover, it is proved that the quantum particle swarm optimization (QPSO) algorithm has the highest efficiency.

Keywords: 2D Otsu, image segmentation, intelligent algorithm, QPSO.

1 Introduction

In the image processing field image segmentation is a very important part, it is the basis of image analysis and understanding as well. The shareholding method is an effective method of image segmentation, and the most representative method is Otsu [1]. 2D Otsu [2, 3] method was presented on the basis of Otsu method, and it is based on image pixel and the two-dimensional histogram of pixel domain average. This method can get better segmentation results. However, it increases the computational complexity and limits the application of the algorithm. In order to overcome these disadvantages, the intelligent algorithms [4] are proposed to improve search efficiency. These three algorithms are respectively particle swarm optimization (PSO) algorithm [5], quantum particle swarm optimization (QPSO) algorithm [6, 7] and genetic algorithm (GA) [8].

The experiment results show that the threshold search efficiency of the three intelligent algorithms is greatly increased, and the QPSO algorithm searching efficiency is highest in these intelligent algorithms.

2 Three Intelligent Algorithms and the Main Parameter Setting

2.1 Particle Swarm Optimization Algorithm

PSO is a stochastic search method that was developed in 1995 based on the sociological behavior of bird flocking. This algorithm is easy to implement and has

Y. Tan et al. (Eds.): ICSI 2014, Part II, LNCS 8795, pp. 221–227, 2014.

been successfully applied to solve a wide range of optimization problems. Now, the PSO technique is used to solve the problem of threshold based segmentation.

Main Parameter Setting. Set the learning factor c_1 and c_2, $c_1 = c_2 = 2$; the inertia weight w, $w = w_{max} - \dfrac{w_{max} - w_{min}}{Maxiter} \times iter$, where the $w_{max} = 0.9, w_{min} = 0.4$, $Maxiter$ is the maximum number of iterations.

2.2 Quantum Particle Swarm Optimization Algorithm

QPSO is combining the classical PSO algorithm and the quantum theory, and it is based on the concept of quantum theory. The main iterative formula for particles

$$mbest(t) = \frac{1}{m} \sum_{i=1}^{m} p_i(t) = [\frac{1}{m} \sum_{i=1}^{m} p_{i1}(t), \frac{1}{m} \sum_{i=1}^{m} p_{i2}(t), ..., \frac{1}{m} \sum_{i=1}^{m} p_{iD}(t)]. \tag{1}$$

$$p_{id}(t) = (r_1 p_{id} + r_2 p_{bd})/(r_1 + r_2). \tag{2}$$

$$X_{id}(t+1) = p_{id}(t) \pm \beta \,|\, mbest(t) - X_{id}(t) \,|\, \ln(\frac{1}{u}). \tag{3}$$

where $mbest$ is the average best position in group; p_{id} as a random point between p_{id} and p_{bd}; r_1 and r_2 for the interval [0, 1] random number; t for the current iteration number; D is the dimension of particles, u is a random number in [0, 1]; β as the contraction coefficient of expansion of the algorithm.

Main parameter Setting. $\beta = m - (m-n)\dfrac{t}{Maxiter}$, where $m = 1, n = 0.5$.

2.3 Genetic Algorithm

GA is a kind of adaptive global optimization probability search algorithm, which can speed up the overall algorithm to complement real-time processing. Because the essence of 2D Otsu threshold is seeking an optimal solution process, so the available genetic algorithm has the speediness of its optimization, in order to achieve the purpose of improving the efficiency.

Main Parameter Setting. Crossover probability, $pc = 0.8$; Mutation probability, $pm = 0.02$.

3   Image Segmentation Based on Intelligent Algorithm

3.1   2D Otsu Segmentation Methods

Given an image f represented by L gray levels, size of $M \times N$, each pixel in the image of the value corresponds to a greyscale.

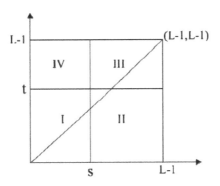

Fig. 1. Planar projection of two-dimensional gray histogram

The two-dimensional vector (s,t) in Fig.1 is the threshold, and it divides two-dimensional histogram into four parts (I, II, III and IV). The regions I and III contain the distributions of object and background classes, respectively.

Let S_0 and S_1 represent the object and the background respectively. The between-class discrete matrix is defined as

$$\sigma_B(s,t) = \sum_{k=0}^{1} P(S_k)[(\mu_k - \mu_T)(\mu_k - \mu_T)^T]. \tag{4}$$

where w_0 and w_1 are the probabilities of class occurrence; μ_0 and μ_1 are the mean value vectors of S_0 and S_1; μ_T is the total mean level vector of the 2D histogram.

$$w_0(s,t) = P(S_0) = \sum_{i=0}^{s} \sum_{j=0}^{t} p_{ij}, w_1(s,t) = P(S_1) = \sum_{i=s+1}^{L-1} \sum_{j=t+1}^{L-1} p_{ij}. \tag{5}$$

$$\mu_0(s,t) = (\sum_{i=0}^{s} \sum_{j=0}^{t} \frac{ip_{ij}}{\omega_0}, \sum_{i=0}^{s} \sum_{j=0}^{t} \frac{jp_{ij}}{\omega_0})^T, \mu_1(s,t) = (\sum_{i=s+1}^{L-1} \sum_{j=t+1}^{L-1} \frac{ip_{ij}}{\omega_1}, \sum_{i=s+1}^{L-1} \sum_{j=t+1}^{L-1} \frac{jp_{ij}}{\omega_1})^T. \tag{6}$$

$$\mu_T = (\sum_{i=0}^{L-1} \sum_{j=0}^{L-1} ip_{ij}, \sum_{i=0}^{L-1} \sum_{j=0}^{L-1} jp_{ij})^T. \tag{7}$$

The trace of discrete matrix could be expressed as

$$tr(\sigma_B) = \sum_{k=0}^{1} (\omega_k [(\mu_{ki} - \mu_{Ti})^2 + (\mu_{kj} - \mu_{Tj})^2]). \tag{8}$$

3.2   QPSO Concrete Realization of Image Segmentation

In this paper, the trace of discrete matrix function $t_r(\sigma_B)$ is the fitness function; find the maximum value, namely

$$f(s,t) = \max t_r(\sigma_B). \tag{9}$$

QPSO Algorithm Process:

1) Initialize position x_i of each particle.

2) Calculate the fitness of each particle (according to the formula (9)).

3) Update individual extremum: let the current fitness value of the i-th particle is compared with the particle individual extreme p_{id}. If the former is more optimal, then update the p_{id}, otherwise the p_{id} unchanged.

4) Update the global extremum: choose the optimal value from all p_{id} (choose the maximum value of the algorithm) as the global extremum p_{bd}.

5) Optimization process, according to the formula (1), (2) and (3) update all the particles in the QPSO algorithm.

6) Check whether meet the termination conditions, if satisfied, exit; Otherwise, $k = k + 1$ (k is the number of iterations) return to step 2, until meet the termination conditions.

4   Experimental and Result Analysis

All the algorithms are implemented on a personal computer with CPU of 2.6GHz using MATLAB R2012a programming language. In the experiments, 4 images (Lena, Cameraman, Peppers and Rice) used in this study and they are shown in Fig. 1. In order to demonstrate the effectiveness of the algorithms, below, will do some specific analysis of these algorithms and their results.

4.1   Segmentation Results

In this paper, the total number of particles is set to 10, and the largest number of iterations is set to 100. The experimental results show that the algorithm achieves the same segmentation results with the 2D Otsu algorithm. Results are as shown below:

Fig. 2. Test Images. (a) Lena; (b) Cameraman; (c) Peppers; (d) Rice

Fig. 3. Segmentation by 2D Otsu

Fig. 4. Segmentation by PSO+ 2D Otsu

Fig. 5. Segmentation by QPSO + 2D Otsu

Fig. 6. Segmentation by GA + 2D Otsu

All the figures show the experimental results of these algorithms. Because PSO, QPSO and GA algorithms can achieve the same effect as the 2D Otsu algorithm, so we can obtained that the PSO, QPSO and GA algorithms are as good as the 2D Otsu.

4.2   Calculation Results

Firstly, 2D Otsu method is used to calculate the maximum variance threshold, get the optimal threshold value of image a is (125,116), the corresponding maximum variance is 2975.5; the image b is (89,103) and corresponding maximum variance is 6209.5; the image c is (125,134) and corresponding maximum variance is 3912.3; the image d is (133,138) and corresponding maximum variance is 2433.7. The contrast of the thresholds and searching time of the various methods is shown in Table 1.

Table 1. Performance comparison

Image	2D Otsu	GA+2D Otsu	PSO+2D Otsu	QPSO+2D Otsu
Lena(125,116)	0.8452s	0.4359s	0.4164s	0.3220s
Cameraman(89,103)	0.8408s	0.4446s	0.4137s	0.3351s
Peppers(125,134)	0.8314s	0.4490s	0.4226s	0.3325s
Rice(133,138)	0.8388s	0.4638s	0.4154s	0.3644s

Table 1 lists the average elapsed time of 40 experiments. For each figure, each algorithm tested 40 times. During the experiment, the time efficiency of these algorithms is repeated comparative. The experimental results is as shown in Table 1, it can come to a conclusion: the introduction of the three intelligent algorithms has increased 2D Otsu operation efficiency and reduced the search time.

The following is the efficiency comparison among the three kinds of intelligent algorithm. It is already to know, the efficiency of these algorithms is higher than the 2D Otsu algorithm. However, in the three algorithms whose efficiency is the highest. Similarly, it can find the answer from Table 1. In respect of the consumption time of three algorithms, QPSO algorithm has the shortest search time. So it is proved that the efficiency of QPSO algorithm is optimal.

5   Conclusions

2D Otsu method is an effective method of image segmentation, and it considers the image gray level information and the space between the pixel neighborhood information. Usually, the method can get better segmentation result than one-dimensional Otsu method, but the consumption of time is greatly increased. In order to solve this problem, the GA, PSO and QPSO algorithms are used to search the optimal two-dimensional threshold vector. The experiment results show that the use of the three intelligent algorithms can reduce the search time, thereby it increase search efficiency. What is more, among these algorithms, the search time of QPSO is shortest, so we can draw a conclusion that the QPSO algorithm is optimal.

Acknowledgments. This work was supported by Open Project Program of Hubei Province Key Laboratory of Intelligent Information Processing and Real-time Industrial System (znss2013A008) and National Natural Science Foundation of China (No.61201423, 61375017)

References

1. Sthitpattanapongsa, P., Srinark, T.: A two-stage Otsu's thresholding based method on a 2D histogram. In: 2011 IEEE International Conference on Intelligent Computer Communication and Processing (ICCP), pp. 345–348. IEEE (2011)
2. Lu, C., Zhu, P.: The Segmentation Algorithm of Improvement a Two-dimensional Otsu and application research. In: 2nd International Conference on software Technology and Engineering (ICSTE) V1-76–V1-79 (2010)
3. Wang, X., Chen, S.: An improved image segmentation algorithm based on two-dimensional Otsu method. Inf. Sci. Lett 1, 77–83 (2012)
4. Kennedy, J., Eberhart, R.: Swarm Intelligence. Morgan Kaufmann Publishers, San Francisco (2001)
5. Tang, H., Wu, C., Han, L., Wang, X.: Image Segmentation Based on Improved PSO. In: The Proceedings of the International Conference on Computer and Communication Technologies in Agriculture Engineering (CCTAE 2010), pp. 191–194 (2010)
6. Yang, S., Wang, M., Jiao, L.: A quantum particle swarm optimization. In: Congress on Evolutionary Computation, CEC 2004, vol. 1, pp. 320–324. IEEE (2004)
7. Chao, Z., Jun, S.: Hybrid-Search Quantum-Behaved Particle Swarm Optimization Algorithm. In: 2011 Tenth International Symposium on Distributed Computing and Applications to Business, Engineering and Science (DCABES), pp. 319–323. IEEE (2011)
8. Sheta, A., Braik, M.S., Aljahdali, S.: Genetic Algorithms: A tool for image segmentation. In: 2012 International Conference on Multimedia Computing and Systems (ICMCS), pp. 84–90. IEEE (2012)

A Shape Target Detection and Tracking Algorithm Based on the Target Measurement Intensity Filter

Weifeng Liu, Chenglin Wen, and Shuyu Ding

Hangzhou Dianzi University, Hangzhou, Zhejiang 310018, China
{liuwf,wencl}@hdu.edu.cn

Abstract. The probability hypothesis density (PHD) is the expectation intensity in a point in state space. The intensity integral in any region of the state space is the expected number of targets contained in that region. In this paper, we propose a target measurement intensity (TMI) filter. Compared with the existing methods, the proposed approach is simpler. Since the conventional PHD filter can not directly deal with the shape target detection and tracking, we give the detection and tracking algorithm based on the TMI filter by modeling the parameter dynamics and measurement function of the shape target.

1 Introduction

The probability hypothesis density (PHD) was first proposed by Stein and Winter [1]. Literally, it is a hypothesis density and does not exist in practice. Its physical meaning is the expected number of targets in a point in state space. Therefore, its integral in certain region in state space proposes the number of targets in that region. Mahler showed that the first order moment of multitarget random finite set (RFS)[2], which is an extension of the first order moment of random point process [3], is equal to the PHD almost everywhere. He also proposed the PHD recursive filter as an alternative of RFS Bayesian equation. One can estimate target state from the PHD filter. The PHD filter is a joint decision and estimation algorithm. It can be seen as an implicit association-estimation algorithm for the association step is substituted by an estimation step. It can deal with the uncertain number of targets such as the surviving targets, the spontaneous birth of new targets, and the spawned targets. Similar to the traditional approaches, the PHD filter is used in the point target tracking.

The PHD is equal to the expected number of measurement originating from a point x in state space. This is built on the following viewpoint: under the assumption of target being a point, a target produces at most one measurement. Therefore, the number of targets statistically equals the number of measurements. The PHD thus can be seen as the target measurement intensity (TMI). In RFS framework, Mahler got the PHD filter through probability generating functionals (PGF). Erdinc et al alternatively derived the PHD filter by using the physical-space approach - a bin model [4], where the PHD is interpreted

Y. Tan et al. (Eds.): ICSI 2014, Part II, LNCS 8795, pp. 228–235, 2014.

as the bin-occupancy probability in the traditional probability framework. In another ref.[5], Streit derived a multitarget intensity filter from a Bayesian first principles approach using a Poisson point process approximation at one step. Streit's intensity filter is very similar to the PHD filter except the estimation of the target birth and measurement clutter processes. In this paper, an alternative derivation of the PHD filter is first proposed by using the TMI.

A point target produces one measurement at most in the traditional researches. In contrast, a shape target might give multiple measurements in each scan. The TMI can also be used in shape target tracking, not only a point target. we here refer it to be as the TMI filter and extend the point target tracking to the shape target tracking. Nevertheless, the detection and survival probability also have different means. For example, how to define a shape target is detected when the shape is partly covered. How to define the probability of detection in this case. Besides, the key to shape target tracking is to model dynamics and measurement function. All these implies that the original PHD filter cannot be directly used to the shape target tracking. Note that in this paper we confine the shape target to be with a parameter model, i.e., the shape target can be described by using a parameter model.

2   Background and Problem Description

Two methods were proposed to derive the PHD filter. The first is the Mahler's PGF method. The second is the physical space method given by Erdinc et al. Roy Streit proposed the intensity filter using the Poisson point process (PPP) method. In single sensor case, the intensity filter and the PHD filter have the same form. In this section, we review the PHD filter and the three methods.

Mahler's PHD filter consists of the following two recursive steps of predicted and update step [2],[6].

3   The Derivation of the Target Measurement Intensity Filter

3.1   Non-parameter State Mixture Models

The finite mixture models (FMM) are used to describe observations coming from various random sources and the models have a finite number of distributions

$$f(y|\theta) = \pi_1 f_1(y|\theta_1) + \cdots + \pi_m f_m(y|\theta_m) \tag{1}$$

Where $y \triangleq \{y_1, \cdots, y_n\}$ are observations, $\pi_1, \cdots, \pi_m$ are the mixing weights, $\theta_1, \cdots, \theta_m$ are the parameters for distributions $f_1(\cdot|\cdot), \cdots, f_m(\cdot|\cdot)$, called components here. we defined the following state mixture models:

$$f_k(z_k|\pi_k, X_k) = \pi_{k,0} f_{k,0}(z_{k,i}|x_{k,0}) + \pi_{k,1} f_{k,1}(z_{k,i}|x_{k,1}) + \cdots + \pi_{k,m_k} f_{k,m_k}(z_{k,i}|x_{k,m_k}) \tag{2}$$

For consistence, the former observations y are replaced by $z_k \triangleq \{z_{k,1}, \cdots, z_{k,n_k}\}$, k is the sampling time, , $\pi_k \triangleq \{\pi_{k,0}, \pi_{k,1}, \cdots, \pi_{k,m_k}\}$, $X_k \triangleq \{x_0, x_1, \cdots, x_{k,m_k}\}$, where $x_{k,0}$ is the clutter state, $x_{k,1}, \cdots, x_{k,m_k}$ are the states of targets, $f_{k,0}$ is the clutter density, $\pi_{k,0}$ is the clutter mixing weight, m_k can be also interpreted as the number of targets. $\{\pi_{k,j}\}_{j=1}^{m_k}$ are the mixing weights of the targets and $\{f_{k,j}(z_{k,i}|x_{k,1})\}_{j=1}^{m_k}$ are the corresponding measurement densities of targets. Streit et al introduced the state mixture models in the probabilistic multiple hypothesis tracking (PMHT) [12]-[13]. The state mixture models describes how the measurements are generated, but the derivation of the measurement likelihood function is still a difficult one. Under the independent assumption, the measurement likelihood can be given by product $L_z(x) = \prod_{i=1}^{n_k} f_k(z_{k,i}|\pi_k, X_k)$. But this expression is an implicit form for the state X_k need to be first estimated. Then, a problem is: can we obtain the states of targets while avoid the calculation of the likelihood function? This needs one to consider certain characteristic function of the target state x in the target state space. Thus the states of targets can be estimated from the characteristic function. For RFS, the characteristic function is corresponding to its first order, i.e., the PHD $D_k(x)$. In this paper we consider the target measurement intensity (TMI) $e_k(x)$ in the state space. The TMI describes the distribution of the number of the target measurements in the state space. The further research shows that the TMI are statistically equal to the PHD under certain assumptions.

Assume that the target states are all in the same state space. Consider the state models of a point x in state space

$$f_k(z_{k,i}|x) = P(D_0|x)f_k(\phi|x, D_0) + P(D_1|x)f_k[z_{k,i}|x, D_1]$$
$$= (1 - P_D(x))\pi_{k,\phi}f_k[\phi|x, e_{k,0}(x)] +$$
$$P_D(x)\{\pi_{k,c}(x)c_k[z_{k,i}|x, e_{k,i}(x) = 0] + \pi_{k,t}(x)g_k[z_{k,i}|x, e_{k,i}(x) = 1]\} \quad (3)$$

Where D_1, D_0 are the events that the target is detected and is not detected, respectively. ϕ is the event of no measurements, $c_k(\cdot|\cdot)$ is the clutter distribution, $g(\cdot|\cdot)$ is the target measurement distribution, $e_{k,i}(x)$ is the indicating variables defined in $\{0,1\}$. $e_{k,i}(x) = 1$ implies that the ith measurement produced by state x. We extend the above state mixture models to the total state space S as follows.

$$f_k(z_{k,i}|S) = f_k(z_{k,i}|S, D_0)P(D_0) + f_k(z_{k,i}|S, D_1)P(D_1)$$
$$= \int_{x \in s} [f_k(z_{k,i}|x, D_0)P(D_0|x) + f_k(z_{k,i}|x, D_1)P(D_1|x)]dx$$
$$= \int_{x \in s} \{[1 - P_D(x)]f_k[\phi|x, e_{k,0}(x)]\}dx +$$
$$\int_{x \in s} \{\pi_{k,c}(x)c_k[z_{k,i}|x, e_{k,i}(x) = 0] + P_D(x)\pi_{k,t}(x))g_k[z_{k,i}x, e_{k,i}(x) = 1]\}dx(4)$$

Where $\pi_{k,c}(x)$ and $\pi_{k,t}(x)$ are the weights of clutter measurements and target measurements. We define the mixing weight to be the probability of target existing in the point $\boldsymbol{x}$

$$\pi_{k,c}(\boldsymbol{x}) = p_k[e_{k,i}(\boldsymbol{x}) = 0|\boldsymbol{x}]$$
$$\pi_{k,t}(\boldsymbol{x}) = p_k[e_{k,i}(\boldsymbol{x}) = 1|\boldsymbol{x}]$$

Then the target measurements intensity in the state $\boldsymbol{x}$ is defined by:

$$e_k(\boldsymbol{x}) = e_{k,0}(\boldsymbol{x}) + e_{k,1}(\boldsymbol{x}) + \cdots + e_{k,n_k}(\boldsymbol{x}) \tag{5}$$

The TMI consists of two types of terms: the intensity $e_{k,0}(\boldsymbol{x})$ of no measurements and individual measurement intensities $\{e_{k,i}(\boldsymbol{x})\}_{i=1}^{n_k}$. In the following subsection, we focus on deriving the recursive equations of the TMI and the mixing weights.

3.2 Derivation of the Recursive Equation for the TMI Filter

The recursive equations of the TMI filter involve two step, i.e., the predicted step and the update step.

The Predicted TMI. The assumption in the predicted step is proposed as follows.
A.1: Each target moves and generates individually and independently of all the other targets.

Proposition 1. *Under the assumption A1. Assume that at time k the intial TMI is $e_k(\boldsymbol{x})$. Then, the predicated TMI is given by:*

$$e_{k+1|k}(\boldsymbol{x}) = e_{k+1}^{\gamma}(\boldsymbol{x}) + e_{k+1|k}^{s}(\boldsymbol{x}) + e_{k+1|k}^{\beta}(\boldsymbol{x}) \tag{6}$$

Where

$$e_{k+1|k}^{s}(\boldsymbol{x}) = \int_{\omega \in S} P_S(\boldsymbol{x}) f_{k+1|k}(\boldsymbol{x}|\omega) e_k(\omega) d\omega \tag{7}$$

$$e_{k+1|k}^{\beta}(\boldsymbol{x}) = \int_{\omega \in S} P_S(\boldsymbol{x}) \beta_{k+1|k}(\boldsymbol{x}|\omega) e_k(\omega) d\omega \tag{8}$$

Where $e_{k+1}^{\gamma}(\boldsymbol{x}), e_{k+1|k}^{s}(\boldsymbol{x}),\ e_{k+1|k}^{\beta}(\boldsymbol{x})$ are respectively the TMI for the spontaneous births, the surviving target and the spawned target. $P_S(\boldsymbol{x})$ is the survival probability. $f_{k+1|k}(\boldsymbol{x}|\omega)$ and $\beta_{k+1|k}(\boldsymbol{x}|\omega)$ are respectively the Markov models of the surviving target and the spawned target, which are the same as eq.(1). It can be derived based on the weighted sums of the TMI. This is similar as the PHD filter.

3.3 The Update TMI

In this step, we propose the following assumptions
A.2 The number of target measurements and the number of clutter measurement follow Poisson with intensities M_{k+1} and λ_{k+1}, respectively.
A.3 Clutter measurements are independent of target states.

Proposition 2. *Assume that predicted TMI is $e_{k+1|k}(\boldsymbol{x})$. Then, under the assumptions A.2 and A.3, the update TMI is given by:*

$$e_{k+1}(\boldsymbol{x}) = e_{k+1,0}(\boldsymbol{x}) + e_{k+1,1}(\boldsymbol{x}) + \cdots + e_{k+1,n_{k+1}}(\boldsymbol{x}) \qquad (9)$$

Where $e_{k+1,0}(\boldsymbol{x})$ is the TMI of no measurements, $e_{k+1,i\geq 1}(\boldsymbol{x})$ is the TMI of the ith measurement. And they can be calculated by the following equation:

$$e_{k+1,0}(\boldsymbol{x}) = (1 - P_D(\boldsymbol{x}))e_{k+1|k}(\boldsymbol{x}) \qquad (10)$$

$$e_{k+1,i\geq 1}(\boldsymbol{x}) = \frac{P_D(\boldsymbol{x})g_{k+1}(z_{k+1,i}|\boldsymbol{x})e_{k+1|k}(\boldsymbol{x})}{\lambda c_{k+1}(z_{k+1,i}) + \int P_D(\boldsymbol{x})g_{k+1}(z_{k+1,i}|\boldsymbol{x})e_{k+1|k}(\boldsymbol{x})d\boldsymbol{x}} \qquad (11)$$

4   The Shape Detection and Tracking

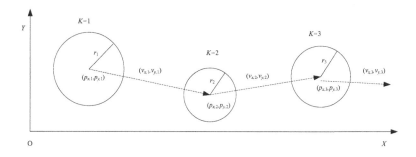

Fig. 1. A movement of a circular shape in x-y plane which is described by a five-dimensional parameter vector

4.1   The Target Dynamics and Measurement Function

A shape can be modeled in the parameter space and be described by a parameter vector. We therefore propose the dynamics and measurement function in the parameter space. The movement of the shape target can be modeled by the dynamics of the parameter vector. This is the same as the usual state function. For a nonlinear function and a linear system function, we can describe it by using the following functions:

$$\boldsymbol{X}_k = f(X_{k-1}, \boldsymbol{\omega}_k) \qquad \text{nonlinear parameter function} \qquad (12)$$

$$\boldsymbol{X}_k = A_{k-1}X_{k-1} + B_{k-1}\boldsymbol{\omega}_k \qquad \text{linear parameter function} \qquad (13)$$

We example two targets with a circular shape and a linear shape, respectively. Fig.1 shows a circular target moving in the planar. We select the center point and the radius of the circle as the parameter vector $\boldsymbol{X} = (p_{x,k}, \dot{p}_{x,k}, p_{y,k}, \dot{p}_{y,k}, r_k)$, where $(p_{x,k}, p_{y,k})$ is the center coordinate of the circular shape, $(\dot{p}_{x,k}, \dot{p}_{y,k})$ is the velocity of the center point, r_k is the circle radius.

In Figure 2, a movement of a linear target is given. It involves two types of movements which include rotation around the center of the linear shape and CV movement of the center point. A seven-dimensional parameter vector is proposed as $X_k = (p_{x,k}, \dot{p}_{x,k}, p_{y,k}, \dot{p}_{y,k}, \theta_k, \dot{\theta}_k, l_k)$, where $(p_{x,k}, p_{y,k})$ is the the center coordinate of the linear shape, $(\dot{p}_{x,k}, \dot{p}_{y,k})$ is the velocity of the center point, θ_k and $\dot{\theta}_k$ are respectively the angle and the angular velocity of the linear shape, l_k is the length of the target.

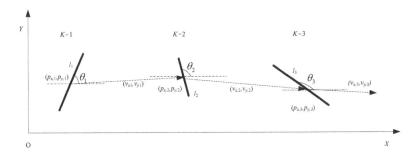

Fig. 2. A movement of a linear shape in x-y plane which is described by a seven-dimensional parameter vector

5   Simulations

In this section, a simulation of detection and tracking of three circular targets is proposed to verify the proposed TMI filter. Three circular targets with CV movement are proposed in this simulation. The parameter vector is given in

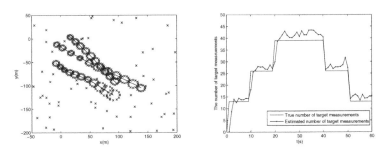

(a) The clutter measurements and target measurements

(b) The measurement intensities

Fig. 3. The measurements and intensities

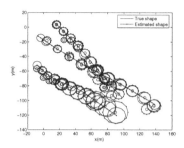 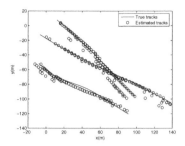

(a) The true shapes and the esti-   (b) The true tracks and the estimated shapes   mated tracks

Fig. 4. The Estimated shapes and tracks

eq.(26). The initial states for the three targets are respectively:
$X_0^1 = (-10\text{m}, 3.2\text{m/s}, -10\text{m}, -2.2\text{m/s}, 5\text{m})$, $X_0^2 = (10\text{m}, 2.2\text{m/s}, 10\text{m}, -2.2\text{m/s}, 5\text{m})$ and $X_0^3 = (-10\text{m}, 1.8\text{m/s}, -50\text{m}, -1.9\text{m/s}, 5\text{m})$. Covariance matrixes:
$P_0^i = \text{diag}([10, 2, 2.5, 10, 2.5, 1])$, $Q_k^i = \text{diag}([0.01, 0.01, 0.2])$ and
$R_k^i = \text{diag}([0.04, 0.04])$, $i = 1, 2, 3$. Clutter density $\rho(\boldsymbol{x}) = 1.0 \times 10^{-3}\text{m}^{-2}$. The surveillance region is $[-50, 200] \times [-200, 50]\text{m}^2$. The measurements are produced by eqs.(27)-(29), where parameters $\{\psi_k(i)\}$ samples in $[0, 2\pi]$ per $\pi/6$ and thus $\lambda_k = 13$.

The Gaussian mixture based TMI filter, which is analogous to the Gaussian mixture PHD filter, is proposed here. Sub figure (a) of Fig.3 shows the clutter measurements and shape measurements every 4 seconds. Obviously, these two types of measurements are mixed. Thus, our first step is to detect the target. It can be seen from (b) of Fig.3 that the measurement intensities and number of targets indicate the existing of targets and their values are approximated to the true value. In estimation of shape parameter, it can be seen from sub figure (a) of Fig.4 that the circle radiuses are close to the true values. Sub figure (b) of Fig.4 shows that the estimations of position are near the true tracks.

6   Conclusion

This paper proposed a target measurement intensity (TMI) filter based on the mixture distributions. We provided the predicted TMI and the update TMI. Under some assumptions, the TMI filter is equal to the original PHD filter. The PHD filter focus on point target tracking. A potential advantage of the TMI filter is that it can be extended the target tracking with parameter shape. Thus we can use it in the extended target tracking. Correspondingly, our next work is to extend the TMI to the target with parameter shape. The key is to model the parameter dynamics and the parameter measurement function. Based on these functions, we extend the TMI filter to the shape target tracking. Finally, we propose two experiments involving three circular targets and three linear targets to verify the proposed TMI filter.

Nevertheless, under the multiple measurements condition, estimation of the number of targets is still an intractable problem. We propose a simple formulation under the Poisson assumption of the target measurements. Besides, in this paper we confine our object to the targets with parameter shapes and they are in the same parameter space. Future works are still needed for the general shape target.

Acknowledgements. This work was supported in part by the NSFC (61175030, 61273170, 61333011, and 61271144)

References

1. Winter, C.L., Stein, M.C.: IES/BTI system overview. In: Proceedings of 8th National Symposium on Sensor Fusion, Dallas TX, vol. I, pp. 15–17, 27–46 (1995)
2. Mahler, R.: Multitarget Bayes Filtering via First-Order Multitarget Moments. IEEE Transactions on Aerospace and Electronic System 39(4), 1152–1178 (2003)
3. Daley, D.J., Vere, J.D.: An Introduction to the Theory of Point Processes, Elementary Theory and Methods, 2nd edn., vol. I, pp. 123–131. Springer, New York (2003)
4. Erdinc, O., Willet, P., Bar-Shalom, Y.: A Physical-Space Approach for the Probability Hypothesis Density and Cardinalized Probability Density Filters, Signal and Data Processing of Small Targets. In: Proc. of SPIE, vol. 6236, pp. 1–12 (2006)
5. Streit, R.L., Stone, L.D.: Bayes Derivation of Multitarget Intensity Filters. In: The 11th International Conference on Information Fusion, Colgon, Germany, pp. 1686–1693 (July 2008)
6. Mahler, R.: An Introduction to Multisource-Multitarget Statistics and Its Applications. Technical Monograph, Lockheed Martin: 1–20 (Mar 15, 2000)
7. Vo, B.N., Ma, W.K.: The Gaussian Mixture Probability Hypothesis Density Filter. IEEE Transactions on Signal Processing 54(11), 4091–4104 (2006)
8. Vo, B.T., Vo, B.N., Cantoni, A.: Bayesian Filtering with Random Finite Set Observations. IEEE Transactions on Signal Processing 56(4), 1313–1326 (2008)
9. Challa, S., Vo, B.N., Wang, X.Z.: Bayesian Approaches to Track Existence - IPDA and Random Sets. In: The 11th International Conference on Information Fusion, Annapolis, Maryland, USA, pp. 1228–1235 (2002)
10. Ulmke, M., Erdinc, O., Willett, P.: Gaussian Mixture Cardinalized PHD Filter for Ground Moving Target Tracking. In: The 10th International Conference on Information Fusion, Quebec, Canada, pp. 1–8 (July 2007)
11. Liu, W.F., Han, C.Z., Lian, F.: An alternative derivation of a Bayes tracking filter based on finite mixture models. In: Proceedings of the 12th International Conference on Information Fusion, Seattle, USA, pp. 842–849 (2009)
12. Streit, R.L., Luginbuhl, T.E.: Maximum likelihood method for probabilistic multihypothesis tracking. In: Proceedings of SPIE International Symposium, Signal and Data Proceeding of Small Targets, Bellingham, WA, USA, pp. 394–405 (1994)
13. Streit, R.L., Luginbuhl, T.E.: A Probabilistic multi-hypothesis tracking algorithm without enumeration and pruning. In: Proceedings of the Sixth Joint Service Data Fusion Symposium, Laurel, MD, pp. 1015–1024 (1993)

Multi-cell Contour Estimate
Based on Ant Pheromone Intensity Field

Qinglan Chen[1], Benlian Xu[2], Yayun Ren[2], Mingli Lu[2], and Peiyi Zhu[2]

[1] School of Mechanical Engineering,
Changshu Institute of Technology, 215500 Changshu, China
[2] School of Electrical & Automatic Engineering, Changshu Institute of Technology,
215500 Changshu, China
chenql@cslg.cn, xu_benlian@cslg.cn, zpy2000@126.com

Abstract. In this paper, we propose an ant pheromone based approach to accurately extract the contours of multiple small cells in low contrast biomedical images. With the local information of intensity variation of each pixel, the initial distribution of ant colony is generated as ants' starting positions. Following the heuristic information, such as the pixel grayscale variance, the ant inertial heading and the image intensity, ant's searching behavior is modeled appropriately to make each of ants move along the edge of interested object as possible. Due to modeling an accurate depositing mechanism of pheromone, the corresponding ring pheromone field is formed and used to extract interested cells' contours after simple morphological operations. Experiment results show that our algorithm could give an accurate contour estimate of each cell for several different image sequences.

Keywords: Image Processing, Ant Colony, Contour Estimate.

1   Introduction

As an important branch of cell motion analysis, the estimate of cell contour could directly or indirectly encompass rich contents about each individual cell, and the related research is challenging and emerging due to poor image quality, small size and discontinuities or sharp changes in intensity, etc.. In most computer applications, image contour extraction constitutes a crucial initial step before performing the task of object recognition and representation. The conventional approaches are computationally expensive because each set of operations is conducted for each pixel. Thus, many researchers resort to other promising techniques. An ACO-based approach, a nature-inspired optimization algorithm, has the potential of solving these intractable problems because of its parallelized and intelligent searching mechanisms, which makes the algorithm easily adaptable for processing multiple objects simultaneously. In terms of the combination between the ACO-based approaches and image segmentation, the related work can be divided into two main strands of research. The first strand focuses on the fusion of ACO and other edge detection and contour extraction algorithms [1, 2], mainly because of the strong and effective

Y. Tan et al. (Eds.): ICSI 2014, Part II, LNCS 8795, pp. 236–243, 2014.

optimization capabilities of ACO. Oliver *et al.*[3] formulate the shape correspondence as a Quadratic Assignment Problem (QAP), incorporating proximity information into the point matching objective function, and propose the first ACO algorithm directly aimed at solving the point and contour correspondence problems. Lai *et al.* [4] present a novel system for active contour tracking of moving objects in video sequences, incorporating the use of edge flows in the ACO algorithm to improve the efficiency. Li [5] presents a novel image contour extraction by ant colony algorithm and B-snake model. The method can enhance the flexibility of B-snake to describe complex shape. The second strand is those approaches [6-9] which construct the edge map based on the pheromone matrix. In these methods, each entry of the pheromone represents the edge information at each pixel of the image. Anna *et al.* [10] propose a ACO-based edge detection method which takes advantage of the improvements introduced in ant colony system, a extension of AS. Aminu *et al.* [11] use the discrete wavelet transform (DWT) as a preprocessing step with ACO to enhance image edge detection. Carla *et al.* [12] employ the ACO-based algorithm preceded by anisotropic diffusion to segment the optic disc in color images, and good performance is achieved as the optic disc was detected in most of all the images, even in the images with great variability.

2   Ants for Multi-cell Contour Estimate

2.1   Ants' Initial Distribution

In the original ant system, the initial ant colony is usually uniformly distributed in a searching space, however, an alternative layout is proposed to assign a given number of ants to pixels of an image where objects (cells) probably occur. For this purpose, the grayscale distribution of current image is utilized to measure the relative intensity variation of each pixel within its local region. As illustrated in Fig.1, an 8-neighbouring pixel configuration, a neighboring region of a given pixel (i, j) with intensity $I_{(i,j)}$, is defined, and its local region grayscale variance is computed as

$$\Delta\sigma_{(i,j)} = \frac{1}{|N_{(i,j)}|} \sum_{(i',j')\in N_{(i,j)}} (I_{(i',j')} - \overline{I}(N_{(i,j)}))^2 \quad , \tag{1}$$

where $N_{(i,j)}$ denotes the neighboring pixels of pixel (i, j), $\overline{I}(N_{(i,j)})$ denotes the average gray intensity of $N_{(i,j)}$, and $|N_{(i,j)}|$ is the number of neighboring pixels of pixel (i, j). As implicated in Eq. (1), the grayscale variance has a smaller value in the area of background and interior of cell, whereas a larger value is probably taken between two sides of edge, as well as in the vicinity of each edge pixel.

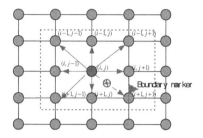

Fig. 1. 8-neighbouring pixel configuration

2.2 Ants' Working Mechanism

To emulate clumping behavior in ant colony (i.e., aggregation), ant working environment and ant decision constraint should be defined appropriately. The ant operating environment corresponds to the current cell image and is denoted by a tuple $\langle P, N \rangle$, where P is a finite set of image pixels, and $N \subseteq P \times P$ is a *closest* propagation neighboring relation among pixels. For any pixel index $g \in P$, an 8-neighbouring configuration is represented as $N_{(g)} = \{s \in P : gNs\}$ with ($|N_{(g)}| \leq 8$). With this definition, each ant can move directly towards one of its neighbors at a time, and any pixel can be visited simultaneously by several ants.

Cell contour extraction is challenging due to incomplete or discrete grayscale variance curves and irregular or partial ant colony distribution in the neighboring region of each edge pixel. Therefore, a novel ant decision is designed and modeled as

$$(i', j') = \begin{cases} \arg\max_{\substack{(i',j') \in N_{(i,j)} \\ (i',j') \notin \Omega_v}} \left((\tau_{(i',j')}^C(\hat{t}))^\lambda (\Delta\tilde{\sigma}_{(i',j')})^\varsigma (W(\Delta\theta_{(i',j')}))^\gamma \right) & \text{if } q' < \tilde{q}_0 \\ (\tilde{i}, \tilde{j}) & \text{otherwise} \end{cases}, \qquad (2)$$

where λ, ς, and γ are the adjustment parameters of contour pheromone $\tau_{(i',j')}^C(\hat{t})$, heuristic grayscale variance $\Delta\tilde{\sigma}_{(i',j')}$, and weight factor of ant heading change $W(\Delta\theta_{(i',j')})$, respectively; Ω_v is the set of pixels visited by the current ant; $\tilde{q}_0$ is a threshold which takes the value between 0 and 1.

The pixel $(\tilde{i}, \tilde{j})$ will be visited according to the probability distribution given by:

$$P_{(i,j) \to (\tilde{i},\tilde{j})}^C(\hat{t}) = \begin{cases} \dfrac{(\tau_{(\tilde{i},\tilde{j})}^C(\hat{t}))^\lambda (\Delta\tilde{\sigma}_{(\tilde{i},\tilde{j})})^\varsigma (W(\Delta\theta_{(\tilde{i},\tilde{j})}))^\gamma}{\displaystyle\sum_{\substack{(m,n) \in N_{(i,j)} \\ (m,n) \notin \Omega_v}} (\tau_{(m,n)}^C(\hat{t}))^\lambda (\Delta\tilde{\sigma}_{(m,n)})^\varsigma (W(\Delta\theta_{(m,n)}))^\gamma} & \text{if } N_{(i,j)} \not\subset \Omega_v \\ 0 & \text{otherwise} \end{cases}. \quad (3)$$

It is implicated that the above model not only propels an ant towards edge pixels of cell, but also forces an ant keeping on moving along the edge of cell instead of staying in the vicinity of its starting pixel. According to the configuration in Fig.1, the destination candidates for each ant decision are up to eight, and the angle between two neighboring candidates is 45^0. Therefore, we assume that if the current heading of an ant is known, the weight factor of ant heading change in the following decision is defined as: $W(\pm 0^0) = 1/3$, $W(\pm 45^0) = 1/3$, $W(\pm 90^0) = 1/10$, $W(\pm 135^0) = 1/16$, and $W(\pm 180^0) = 1/20$.

Once an ant has made $\bar{m}$ decisions, it will deposit an amount of pheromone on corresponding visited pixels (up to $\bar{m}$ pixels at a time) with two levels

$$r^C(\hat{t}) = \begin{cases} c_2 \cdot (1 - e^{-\hat{t}/T2}) / \sum_{m'=1}^{\bar{m}} \left(1/\max\left\{\Delta\tilde{\sigma}_{(m')}, \bar{\mu}_{\min}\right\}\right) & \text{if } std\{1/\Delta\tilde{\sigma}_{(m')}\}_{m'=1}^{\bar{m}} < \delta_0, d_{\bar{m}}^{\max} > \dfrac{\bar{m}}{6} \\ c_0 \cdot (1 - e^{-\hat{t}/T2}) & \text{otherwise} \end{cases} \quad , \quad (4)$$

where $c_2, \delta_0, T2$, and $\bar{\mu}_{\min}$ are constants, $std\{\cdot\}$ denotes the standard deviation of a given set, and $\Delta\tilde{\sigma}_{(m')}$ denotes the grayscale variance value of pixel corresponding to the m'-th decision of ant.

We observe that not all ants deposit the same amount of pheromone according to the above two constraints, but they do offer a hint of where continuous cell edges are located when an ant makes a decision. Specifically, the first constraint $std\{1/\Delta\tilde{\sigma}_{(m')}\}_{m'=1}^{\bar{m}} < \delta_0$ tries to strengthen tour pixels within the same level of grayscale invariance, while the second constraint $d_{\bar{m}}^{\max} > \bar{m}/6$ encourages ant to move as far as possible along the edges.

2.3 The Formation of Pheromone Field

In this work, we model three pheromone working mechanisms to jointly produce pheromone field. First, the pheromone deposited by different ant individuals are aggregated and merged on the corresponding pixel, which results in pheromone peaks in the interested areas. Second, pheromone performs evaporation over time, and this simulates the memory ability of ant individuals as we observe in nature. Finally, pheromone propagation is considered to build a connection between neighboring pixels, and it builds a bridge for access by neighboring agents.

Pheromone aggregation is defined as a combination of evaporation, external input, and propagation. For any pixel (i, j), the evolution of pheromone amount follows

$$\tau_{(i,j)}^C(\hat{t}+1) = E \cdot \tau_{(i,j)}^C(\hat{t}) + r_{(i,j)}(\hat{t}) + q_{(i,j)}(\hat{t}) \quad , \quad (5)$$

where $r_{(i,j)}(\hat{t})$ denotes the pheromone external input to pixel (i, j) at the $\hat{t}$-th iteration, and $q_{(i,j)}(\hat{t})$ models the propagation input to pixel (i, j). Note that the above model applies to both location and contour pheromone fields.

It is observed that, unlike the traditional ant system (AS), the pheromone propagation $q_{(i,j)}(\hat{t})$ is introduced to coincide with the pixel intensity continuity in an image, and its evolution form is defined as

$$q_{(i,j)}(\hat{t}) = \sum_{(i',j') \in N_{(i,j)}} \frac{D}{|N_{(i',j')}|} \left(r_{(i',j')}(\hat{t}-1) + q_{(i',j')}(\hat{t}-1) \right) ,\qquad (6)$$

where $|N_{(i',j')}|$ defines the cardinality of $N_{(i',j')}$, D denotes the propagation coefficient with $0 < D < 1$, and $\dfrac{D}{|N_{(i',j')}|}$ characterizes the averaged propagation proportion of total received pheromone intensity on pixel (i', j') at the $\hat{t}-1$ -th iteration to its neighboring pixels.

As defined in Eq. (6), the propagation pheromone field at next iteration is the propagating results of the field of itself and the external input pheromone field both at current iteration. Furthermore, we assume that the pheromone amount on each pixel is an un-weighted sum of pheromones of its neighbors, thus the Eq. (6) could be divided into two parts and rewritten as

$$q_{(i,j)}(\hat{t}) = \sum_{(i',j') \in N_{(i,j)}} \frac{D}{|N_{(i',j')}|} r_{(i',j')}(\hat{t}-1) + \sum_{(i',j') \in N_{(i,j)}} \frac{D}{|N_{(i',j')}|} q_{(i',j')}(\hat{t}-1)$$

$$= D \cdot r_{(i,j)}(\hat{t}-1) + D \cdot q_{(i,j)}(\hat{t}-1) \qquad (7)$$

Upon the contour pheromone field $\tau^{C}(\hat{t})$ is formed, we first treat it as an input image, and then three steps of morphological operations, including bridging unconnected pixels, filling image regions and holes, and removing interior pixels, are done to generate the contour of each cell.

3   Experiments

The performance of our proposed cell contour estimate algorithm is evaluated using two challenging low-contrast multiple cell image sequences, which considers various cases including cell dynamic difference, cell shape variation, and varying number of cells. In terms of the initial ant colony distribution, a predefined threshold is used to allocate ants to corresponding pixels. A larger value of threshold could generate fewer ants and save computational burden at the expense of the loss of more useful information, whereas a smaller value improves tracking accuracy with more ants generated and more computational cost required. To obtain the desirable tracking results, we set the threshold to be 0.1 in both image sequences.

Figs. 2 and 3 give the multi-cell contour estimates for these two sequences, and it can be observed that our method could give an accurate contour estimate of each existing cell of an image.

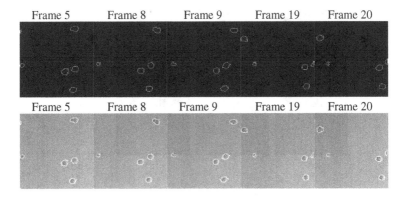

Fig. 2. Cell contour estimates of image sequence 1

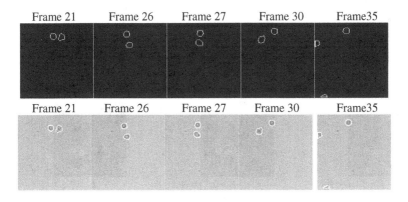

Fig. 3. Cell contour estimates of image sequence 2

Also, we note that, despite the weakness in intensity and the discontinuity of potential edges in the normalized grayscale variance field $\Delta\tilde{\sigma}_{(i,j)}$ (Fig.4(a)), our algorithm could obtain the cell contour pheromone field in the form of a series of continuous and close loops, as shown in Fig.4(b). Through three steps of morphological operations, all cell-related contours are extracted in the end, as illustrated in Fig.4 (c).

Since cell contour estimate is dependent directly on the contour pheromone field, and the working mechanism of pheromone is appropriately adjusted to form multiple close, smooth and continuous belt loops. As shown in Fig.5 (a), if the propagation coefficient D increases, it means that the effect of propagation increases as well, and the continuity of each contour is guaranteed but the size of contour is larger than the true one. For a smaller value of D, it will result in contour discontinuity and debris due to lack of link bridge between neighboring pixels in terms of pheromone. Similarly, for a smaller value of E, which means that more current pheromones evaporate and only few are used for the following iteration, an undesirable estimate of each contour is achieved as a net structure. However, with less pheromone evaporation, i.e., a larger value of E, more pheromone are utilized as a guide for ant to search for possible segment of contour, as illustrated in Fig.5 (b).

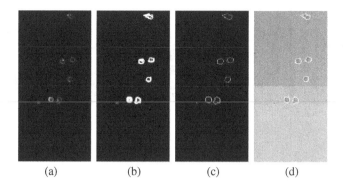

(a)   (b)   (c)   (d)

Fig. 4. Cell contour estimate

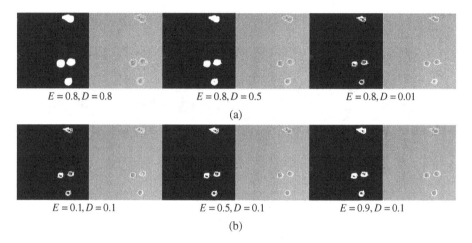

$E = 0.8, D = 0.8$ $E = 0.8, D = 0.5$ $E = 0.8, D = 0.01$

(a)

$E = 0.1, D = 0.1$ $E = 0.5, D = 0.1$ $E = 0.9, D = 0.1$

(b)

Fig.5. Sensitivity analysis of various combination of E and D on contour estimate.

4   Conclusions

Cell motion analysis has become a major research direction for understanding the full potential of time-lapse microscopy in biological research or drug discovery. In this paper, we propose an ant pheromone based approach to accurately extract the contours of multiple small cells. Experiment results show that 1) our algorithm could give an accurate estimate of contour of each cell in various scenarios; 2) the tracking accuracy depends on how the pheromone field models, which is affected mainly by parameters E and D. As part of future work, we would like to expand the scope of cell contour estimate to multi-parameter joint estimate, which could give us a broad quantitative view of cell cycle progression.

Acknowledgments. This work is supported by national natural science foundation of China (61273312) and natural science foundation of higher education colleges in Jiangsu (14KJB510001).

References

1. Li, L., Ren, Y.M., Gong, X.P.: Medical Image Segmentation Based on Modified Ant Colony Algorithm with GVF Snake Model. In: 2008th IEEE International Seminar on Future BioMedical Information Engineering, pp. 11–14. IEEE Press, Washington (2008)
2. Ruberto, C.D., Morgera, A.: ACO contour matching a dominant point approach. In: 2011 4th IEEE International Congress on Image and Signal Processing, pp. 1391–1395. IEEE Press, Shanghai (2011)
3. Kaick, O.V., Hamarneh, G., Zhang, H., Wighton, P.: Contour Correspondence via Ant Colony Optimization. In: 2007 15th Pacific Conference on Computer Graphics and Applications, pp. 271–280. IEEE Press, Washington (2007)
4. Chang, Y.H., Lai, D.X., Zhong, Z.H.: Active Contour Tracking of Moving Objects Using Edge Flows and Ant Colony Optimization in Video Sequences. In: Wada, T., Huang, F., Lin, S. (eds.) PSIVT 2009. LNCS, vol. 5414, pp. 1104–1116. Springer, Heidelberg (2009)
5. Li, J.: Image Contour Extraction Based on Ant Colony Algorithm and B-snake. In: Huang, D.-S., Zhao, Z., Bevilacqua, V., Figueroa, J.C. (eds.) ICIC 2010. LNCS, vol. 6215, pp. 197–204. Springer, Heidelberg (2010)
6. Tao, W., Jin, H., Liu, L.: Object segmentation using ant colony optimization algorithm and fuzzy entropy. J. Pattern Recognition Letters 28(7), 788–796 (2007)
7. Agrawal, P., Kaur, S., Kaur, H., Dhiman, A.: Analysis and Synthesis of an Ant Colony Optimization Technique for Image Edge Detection. In: ICCS 2012th International Conference on Computing Sciences, pp. 127–131. IEEE Press, Washington (2012)
8. Haase, R., Böhme, H.J., Perrin, R., Zöphel, K., Abolmaali, N.: Self-reproduction versus transition rules in ant colonies for medical volume segmentation. In: Dorigo, M., Birattari, M., Blum, C., Christensen, A.L., Engelbrecht, A.P., Groß, R., Stützle, T. (eds.) ANTS 2012. LNCS, vol. 7461, pp. 316–323. Springer, Heidelberg (2012)
9. Huang, P., Cao, H.Z., Luo, S.Q.: An artificial ant colonies approach to medical image segmentation. J. Comput Methods Programs Biomed 92, 267–273 (2008)
10. Baterina, A.V., Oppus, C.: Image Edge Detection Using Ant Colony Optimization. J. WSEAS Transactions on Signal Processing 6, 58–67 (2010)
11. Muhammad, A., Bala, I., Salman, M.S., Eleyan, A.: Discrete Wavelet Transform-based Ant Colony Optimization for Edge Detection. In: 2013th International Conference on Technological Advances in Electrical, Electronics and Computer Engineering (TAEECE), pp. 280–283. IEEE Press, Konya (2013)
12. Pereira, C., Goncalves, L., Ferreira, M.: Optic disc detection in color fundus images using ant colony optimization. J. Med Biol Eng Comput 51, 295–303 (2013)

A Novel Ant System with Multiple Tasks for Spatially Adjacent Cell State Estimate

Mingli Lu, Benlian Xu, Peiyi Zhu, and Jian Shi

School of Electrical & Automatic Engineering, Changshu Institute of Technology,
215500 Changshu, China
luxiaowenwp@sohu.com, xu_benlian@cslg.cn,
zpy2000@126.com, yievans2010@hotmail.com

Abstract. Multi-cell tracking is an important problem in studies of dynamic cell cycle behaviors. This paper models a novel multi-tasking ant system that jointly estimates the number of cells and their individual states in cell image sequences. Our ant system adopts an interactive mode with cooperation and competition. In simulations of real cell image sequences, the multi-tasking ant system integrated with interactive mode yielded better tracking results . Furthermore, the results suggest that our algorithm can automatically and accurately track numerous cells in various scenarios, and is competitive with state-of-the-art multi-cell tracking methods.

Keywords: Ant System, Cell Tracking, Object Motion Analysis.

1   Introduction

During cell image sequencing, two or more cells will very likely contact or present occlusions. In such cases, the image is not easily associated with spatially adjacent images because the joint observation cannot be easily segmented. Because the components of corresponding cell states are now coupled, the tracking of spatially adjacent cells or cell occlusions becomes a challenging task, rendered more complicated by low SNR in the image data. For efficiency and accuracy, the development of automated tracking methods for spatially adjacent cell is of great importance.

Many efforts have been made over the past decades. Dufour *et al.* [1] presented a fully automated technique for segmenting and tracking cells in 3-D+time microscopy data. This method uses coupled active surfaces with or without edges, together with a volume conservation constraint and several optimizations to handle touching and dividing cells, and cells entering the field of view during the sequence. In the method of Nguyen *et al.* [2], multiple cell collisions cells are automatically tracked by modeling the appearance and motion of each collision state, and testing collision hypotheses of possible state transitions. Although some of the above algorithms have resolved special challenges in spatially adjacent objects, they are problem-dependent and not applicable to generic cell tracking problems.

Y. Tan et al. (Eds.): ICSI 2014, Part II, LNCS 8795, pp. 244–252, 2014.
© Springer International Publishing Switzerland 2014

In this paper, we employ a novel ant system with multiple tasks that jointly estimates the number of cells and their individual states in sequences of cell images. Depending on the initial distribution of the ant colony, the colony is roughly divided into several groups, each assigned the task of finding a potential cell through the defined ant working mode, namely interactive mode with cooperation and competition.

2   Methods

In this section, we introduce a novel ant system with multiple tasks (AS-MT) for estimating multi-cell parameters. It is noted that our algorithm builds solutions in a parallel way on $N+1$ pheromone fields (where N is the initial number of divided ant groups),whereas the conventional ACO algorithm computes solutions in an incremental way on only one pheromone field.

2.1   Initial Distribution of Ant Colony

Since the background in most cell image sequences exhibits slowly varying background signals, such a problem can be solved by simplistic, static-background models. In our work, the approximate median method, a kind of recursive technique, is employed for the purpose of fast background subtraction [3], in which each pixel in the background model is compared to the corresponding pixel in the current frame, and finally to be incremented by one if the new pixel is larger than the background pixel or decremented by one if smaller. With the iteration evolves, a pixel in the background model converges to a value where half of the incoming pixels are greater than and half are less than its value, and this value is known as the median.

The approximate median foreground detection compares the current frame to the background model and further identifies the foreground pixels $I_1(i)$ and binary image pixels $I_2(i)$ as

$$I_1(i) = \begin{cases} I(i), & if \ |I(i) - B(i)| > Th \\ 0, & otherwise \end{cases} \tag{1}$$

$$I_2(i) = \begin{cases} 1, & if \ |I(i) - B(i)| > Th \\ 0, & otherwise \end{cases} \tag{2}$$

where $I(i)$ denote the current frame pixels, $B(i)$ are background pixels estimated by recursive technique , $B(i) = \begin{cases} B(i)+1 & if \ I(i) > B(i) \\ B(i)-1 & if \ I(i) < B(i) \end{cases}$, and Th is a predefined threshold. If the absolute difference between current frame pixel and background pixel is larger than the predefined threshold Th , the foreground pixel $I_1(i) = I(i)$, otherwise $I_1(i) = 0$. Binary image pixel $I_2(i)$ follows the same rule as Eq. (2). We further assume that an ant is distributed at the location of pixel i if $I_2(i) = 1$ in the binary image, and thus a given number of birth ants are generated in the current frame. Using the K-means clustering method[4], these ants are further divided into N subgroups.

2.2   Ant System with Multiple Tasks

In this section, working mode in ant system with multiple tasks (AS-MT) is proposed and investigated in details. Each ant can move directly towards one of it neighbors at each time, and any pixel can be visited simultaneously by several ants with the guidance of our proposed pheromone update mechanism.

Interactive Mode with Cooperation and Competition (IMCC). In our defined interactive mode with cooperation and competition (IMCC), ants with different tasks are modeled to work together with appropriate cooperation and repulsion. The repulsion term is characterized by $\dfrac{\tau_j^s(t)}{\tau_j(t)}$ indicated the ratio of pheromone level of task s to total pheromone level at the t-th iteration, and the larger the pheromone amount of task s, the more important this pheromone field will play in ant decision. while the cooperation is represented by the total pheromone $\tau_j(t)$. Therefore, the model of ant decision is a function of the pheromone amount $\tau_j(t)$, heuristic information function η_j and the pheromone amount of current corresponding task $\tau_j^s(t)$, which is formulated as

$$P_{i,j}^{s,k}(t) = \begin{cases} \dfrac{\left[\tau_j(t)\right]^\alpha \eta_j^\beta \left[\dfrac{\tau_j^s(t)}{\tau_j(t)}\right]^\gamma}{\displaystyle\sum_{j \in H(i)} \left[\tau_j(t)\right]^\alpha \eta_j^\beta \left[\dfrac{\tau_j^s(t)}{\tau_j(t)}\right]^\gamma}, & \text{if } j \in H(i) \\ \\ 0, & \text{otherwise} \end{cases} \tag{3}$$

where $H(i)$ denotes the set of neighbors of pixel i, $\tau_j(t)$ is the total sum of pheromone amount left by all ants with different tasks on pixel j, and parameters α, β and γ regulate the relative importance of corresponding terms. It is noted that, during the process of searching solutions, each ant of a given task is assumed to sense some information of its neighboring pixels such as the total pheromone amount $\tau_j(t)$, the ratio of pheromone level of a given task to total pheromone level

$\dfrac{\tau_j^s(t)}{\tau_j(t)}$, and the heuristic function η_j. In the definition of IMCC, if both the relative

proportion of pheromone s and the total pheromone amount keep in high level at pixel j, the ant of task s will select the corresponding pixel j as its next position with a lower probability than the relative proportion term is considered only, since both the ant cooperation and competition between different tasks are in effect simultaneously in a trade-off mode.

Heuristic Information. If an ant moves from pixel i to pixel j, the corresponding heuristic value can be defined as

$$\eta_j = e^{-u\left(1-\frac{1}{|T|}\sum_{t=1}^{|T|}\sum_{j=1}^{M}\min(w_i(j),\tilde{w}_i(j))\right)^{\upsilon}} \qquad (4)$$

Where μ and υ are the adjustment coefficients designed for achieving better likelihood difference comparison between the candidate blob and cell sample blobs, η_j lies in the range of 0 and 1, $\tilde{w}_i(j)$ denotes the value of the j-th element of $\tilde{w}_i$ in cell sample pool, $w_i(j)$ denotes the histogram at pixel j, M is the total number of elements in histogram w, and $|T|$ is the number of cell samples in template pool.

2.3   Merge and Prune Processes

Considering the different ant pheromone fields, if more than one ant groups tend to search for the same cell, the corresponding pheromone fields are probably partially overlapped and the absolute distance between pheromone peaks is relatively small. However, for those spatially distant ant groups, they naturally search for different objects, and the peak distances between ant pheromone fields are easily discriminated and well separated. Therefore, the overlapping ratio $O_{overlap}$ based on pheromone peak is calculated between pheromone peak blobs and treated as a criterion. In our experiments, the merging procedure is performed between two pheromone peak blobs if the overlapping ratio $O_{overlap} > \sigma$, where σ is threshold and set to $\sigma = 0.3$ in our studied cell image data.

In order to remove the false alarms caused by noise and clutters, prune procedure is employed. Suppose that we have the prior information on cell size, if the number of ants of group is less than threshold, the prune processes is carried out and the irrelevant object is removed. Finally, data association based on the easily-implementing nearest neighboring method between frames is done to establish individual trajectories of interested cells.

To visualize our proposed algorithm in a full view, we summarize the procedure in Table 1.

Table 1. Pseudo-code of our proposed algorithm (not considering data association)

Input: Image frame by frame

Generate initial distribution of ant groups by the approximate median method;

The initial ant distribution is roughly divided into N groups using K-means method;

$t = 1, q_j^s(0) = 0, \tau_j^s(0) = c$;

While $t < t_{max}$

For task $s = 1 : N$

  For ant $k = 1 : K$

    Ant k moves from pixel i to pixel j with a probability:

$$P_{i,j}^{s,k}(t) = \begin{cases} \dfrac{\left[\tau_j(t)\right]^\alpha \eta_j^\beta \left[\dfrac{\tau_j^s(t)}{\tau_j(t)}\right]^\gamma}{\sum\limits_{j \in H(i)} \left[\tau_j(t)\right]^\alpha \eta_j^\beta \left[\dfrac{\tau_j^s(t)}{\tau_j(t)}\right]^\gamma}, & if\ j \in H(i) \\ 0, & otherwise \end{cases}$$

    Deposit corresponding pheromone amount according to $\Delta r_j^{k,s}(t) = \Delta \tau_0$;

  end

  Propagated input to pixel j

$$q_{1,j}^s(t) = \sum_{j' \in H(j)} \frac{P}{|H(j')|} r_{j'}^s(t) = \frac{P}{|H(j')|} \sum_{j' \in H(j)} \sum_k \Delta r_{j'}^{s,k} ;$$

  Pheromone update on each pixel at task s $\tau_j^s(t+1) = \rho \tau_j^s(t) + r_j^s(t) + q_j^s(t)$;

  Propagated input evolution $q_j^s(t) = q_{1,j}^s(t) + \sum\limits_{j' \in H(j)} \frac{P}{|H(j')|} q_{j'}^s(t)$;

end

    Total pheromone $\tau_j(t+1) = \sum\limits_s \tau_j^s(t+1)$;

end

If the blob overlap ratio between two pheromone field peaks is greater than a given threshold *then*

  The merge process is performed.

end

If the number of ant group is less than the given threshold *then*

  the prune processes is carried out.

end

Output: Cell state

---

3   Experiments

In this section, we will test the tracking performance of our proposed algorithm in terms of cells dividing, different dynamics and varying number in cell image sequences. To evaluate the tracking accuracy between frames, we adopt three measure criterions, namely, label switching rate (LSR), lost tracks ratio (LTR) and false tracks ratio (FTR). The label switching rate is the number of label switching events normalized over total number of ground truth tracks crossing event which happen when two objects get very close each other(and they are sometimes merged into one object) and after they are separated, one object would be treated as a new object and its label is changed. The lost tracks ratio is the number of tracks lost over total number of ground truth tracks. The false tracks ratio is the number of false object that are tracked over total number of ground truth tracks. All experiments were performed in MATLAB (R2012a) on a 1.7 GHz processor computer with 4G random access memory.

Fig.1 presents an example of tracking results of selected images with our proposed algorithm.

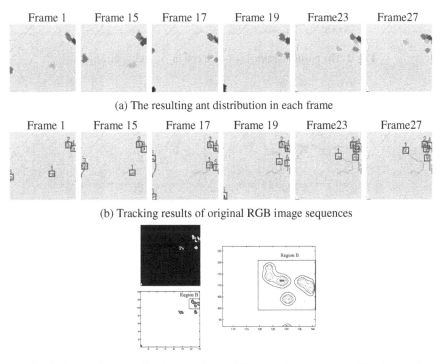

(a) The resulting ant distribution in each frame

(b) Tracking results of original RGB image sequences

(c) Initial ant colony distribution and the resulting of ant pheromone field in frame 23

Fig. 1. Tracking results of multi close moving cells($\rho = 0.8, P = 0.6, \alpha = 2.5, \beta = 1, \gamma = 1.1$)

According to the tracking results presented in Fig.1(b), our proposed algorithm could tackle the following challenging cases: cell 3 partly enter the field of view in frame 1, then moves left, partially leaves the field of view in frames 15, and fully leaves the field of view in frame 17. New cells 6 and cell 5 enter the field of view in frame 15, cells 6 leaves the field of view in frame 19. Cell 4 divides into two cells (cell 4 and cell 7) in frame 23. All cells are kept on being tracked with our algorithm in the following frames. It can be observed that initial ant distribution of three spatially adjacent cells is adhered in frame 23, with the cooperation and compete of our proposed algorithm all spatially adjacent cells are successfully separated and tracked. After 50 times of iteration, the adhesion of pheromone field is well separated. All these are illustrated in Fig. 1(c). In addition, Figs. 2 and 3 plot the position and instant velocity estimates of each cell. It can be seen that cell 1 undergoes fast dynamics, and cell 3 also moves quickly both in x and y direction.

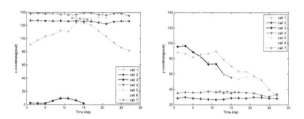

Fig. 2. The position estimate of each cell in x and y directions

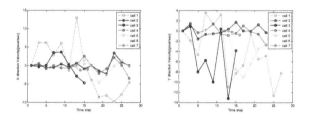

Fig. 3. The instant velocity estimate of each cell in x and y directions

To get insight into tracking performance of our proposed algorithm, we have thoroughly compared our proposed algorithm based on mode IMCC with other three recently developed multi-cell tracking algorithms, i.e., the particle filter (PF) [5] , the multi-Bernoulli filter [6] and Gaussians Mixture Probabilistic Hypothesis Density (GM-PHD) filter [7]. To perform an objective and fair comparison, both the PF and GM-PHD filter use the same detection data obtained by a hybrid cell detection algorithm [8] due to the fact that these two belong to "detect-before-track" methods; meanwhile, for the multi-Bernoulli filter, the likelihood function takes the same form as the heuristic information used in our ant system, denoted by Eq. (4), since both fall in the category of "track-before-detect" techniques. We record all label switching reports, tracks lost reports and false track reports in each frame over 50 Monte-Carlo simulations, and the averaged LSR, LTR and FTR are computed as illustrated in Table 2. According to the statistic results in Table 2, the averaged LSR, LTR and FTR are only *1.57%*, *3.17%* and *1.57%*, respectively, using our algorithm. The comparison results demonstrate that our algorithm performs better than the other methods in the case cells closely.

Table 2. Comparison results for tracking performance of various methods

Method	LSR(%)	LTR(%)	FTR(%)
PF	15.87	9.52	19.05
Multi-Bernoulli filter	12.39	11.11	17.46
GM-PHD	9.52	6.34	3.17
Our method	1.57	3.17	1.57

Without loss of generality, we present the averaged position errors using the manual tracking result as the ground truth. The comparison of cell 1 position error estimates per frame by various methods is shown in Fig.4 and the same conclusions are drawn as the above.

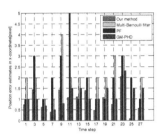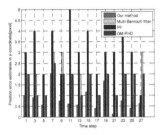

Fig. 4. The comparison of cell 1 position error estimates by various methods

4   Conclusions

The problem of properly tracking spatially adjacent objects is one of the most difficult issues in automated cell tracking. In this paper, a novel ant system with multiple tasks is modeled for jointly estimating the number of cells and individual states in cell image sequences. According to statistic results, our ant system with multiple tasks algorithm demonstrates a robust tracking performance in terms of the measures of LSR, LTR and FTR when comparing with other three recently developed multi-cell tracking algorithms.

Acknowledgments. This work is supported by national natural science foundation of China (No.61273312) and the natural science fundamental research program of higher education colleges in Jiangsu province(No. 14KJB510001).

References

1. Dufour, A., Shinin, V., Tajbakhsh, S., Guillen-Aghion, N., Olivo-Marin, J.C., Zimmer, C.: Segmenting and tracking fluorescent cells in dynamic 3-D microscopy with coupled active surfaces. IEEE Transactions on Image Processing 14, 1396–1410 (2005)

2. Nguyen, N.H., Keller, S., Norris, E., Huynh, T.T., Clemens, M.G., Shin, M.C.: Tracking Colliding Cells In Vivo Microscopy. IEEE Transactions on Biomedical Engineering 58, 2391–2400 (2011)

3. Bandi, S.R., Varadharajan, A., Masthan, M.: Performance evaluation of various foreground extraction algorithms for object detection in visual surveillance. Comput. Eng. Res. 2, 1339–1443 (2012)

4. Hartigan, J.A., Wong, M.A.: Algorithm AS 136: A K-Means Clustering Algorithm. Journal of the Royal Statistical Society. Series C (Applied Statistics) 28, 100–108 (1979)

5. Smal, I., Draegestein, K., Galjart, N., Niessen, W., Meijering, E.: Particle Filtering for Multiple Object Tracking in Dynamic Fluorescence Microscopy Images: Application to Microtubule Growth Analysis. IEEE Transactions on Medical Imaging 27, 789–804 (2008)

6. Hoseinnezhad, R., Vo, B.-N., Vo, B.-T., Suter, D.: Visual tracking of numerous targets via multi-Bernoulli filtering of image data. Pattern Recognition 45, 3625–3635 (2012)

7. Juang, R.R., Levchenko, A., Burlina, P.: Tracking cell motion using GM-PHD. In: IEEE International Symposium on Biomedical Imaging: From Nano to Macro, ISBI 2009, pp. 1154–1157 (2009)

8. Lu, M., Xu, B., Sheng, A.: Cell automatic tracking technique with particle filter. In: Tan, Y., Shi, Y., Ji, Z. (eds.) ICSI 2012, Part II. LNCS, vol. 7332, pp. 589–595. Springer, Heidelberg (2012)

A Cluster Based Method for Cell Segmentation

Fei Wang, Benlian Xu, and Mingli Lu

School of Electrical & Automatic Engineering,
Changshu Institute of Technology,215500 Changshu, China
wangleea@aliyun.com

Abstract. In the field of cell biology, cell segmentation is an essential task in biomedical application. For this purpose, a cluster based method for cell segmentation is proposed. Firstly, an ant colony clustering algorithm is used to make pre-segmentation from which cell candidates are identified, then some noise spots are filtered with area feature, after that, a novel cluster algorithm is proposed to divide adhering cells into individuals. Finally, good results of segmentation can be achieved. Experimental result show that the method remains both the advantage of image segment of ant colony cluster and the ability of further process of pre-segmentation, which improves the performance of cell segmentation.

Keywords: Ant colony algorithm, clustering algorithm, cell segmentation.

1   Introduction

Many scientific biological applications as well as experiments for drug development require the observation of cell responses to a variety of stimuli. Some of the responses that need to be quantified are cell migration, cell proliferation, and cell differentiation. The corresponding conclusions require the observation of cells over extended periods of time. An effective way to achieve this is with microscopy images. However, the resulting data sets of images are large and their manual analysis is tedious, subjective, and restrictive. Thus, an automated technology for analysis is needed urgently, of all the image based research of cells, cell segmentation is the very first step.

Many image segmentation method have been used in cell segment, like thresholding [1], gradient based methods [2], the watershed algorithm [3], level sets [4], dynamic programming [5] and various other pattern analysis and machine learning algorithms [6]. Ant colony clustering algorithm is a new swarm intelligence method that has been used in many fields, like data mining, document retrieval, image segmentation and so on. In this paper, an ant colony clustering algorithm is applied to achieve cell pre-segmentation process, then, a noise filter process is implemented, and after that a novel clustering algorithm is proposed to divide the adhering cells into individuals.

Y. Tan et al. (Eds.): ICSI 2014, Part II, LNCS 8795, pp. 253–258, 2014.

2   Framework of Cell Segmentation

In the process of cell segmentation, an ant cluster based method is applied to execute rough segmentation, then another cluster algorithm and features filter step are used to improve the performance of cell segmentation, its flow chart is shown as follows.

Fig. 1. Flow chart of cell segmentation

2.1   Ant Colony Clustering

Ant colony clustering algorithm is an improved ant colony method which was proposed by Indian researcher P.S. Shelokar. As a globally optimized heuristic method, it is used to address clustering problems and applied to integrate data according to the information around the clustering centers [7]. During application of the algorithm, every data sample is treated as an ant who has different attributes, and the process for searching food by ants is regarded as clustering procedure.

Give the initial image X, and look every pixel X_i (i = 1, 2, …, n) as an ant, every ant stands for the feature vector of pixel. Image segmentation is the process that these ants with different feature vector searching food source. The distance of any pixel X_i to X_j is d_{ij}, using Euclidean distance to calculate:

$$d_{ij} = \sqrt{\sum_{k=1}^{m} p_k (x_{ik} - x_{jk})^2} \quad . \tag{1}$$

where m is the number of feature vector, p is the weighted factor, which is set by its attribute and subjected to equation: $\sum p_k = 1, p_k \geq 0$. $\tau_{ij}(t)$ is the pheromone information value of the arc linking the data X_i to data X_j at time t, and is given as:

$$\tau_{ij}(t) = \begin{cases} 1 & d_{ij} \leq r \\ 0 & d_{ij} > r \end{cases} \quad . \tag{2}$$

r is the radius of clustering. The probability of path X_i to X_j is given by the following equation:

$$P_{ij}(t) = \frac{\tau_{ij}^{\alpha}(t)\eta_{ij}^{\beta}(t)}{\sum_{s \in S} \tau_{ij}^{\alpha}(t)\eta_{ij}^{\beta}(t)} \quad . \tag{3}$$

$$\eta_{ij} = r/d_{ij} \ .\tag{4}$$

where η_{ij} is the Heuristic Guide function, which reflects the similarity between the pixels and the clustering center, α and β are regulating factors, which respectively reflect the accumulated information during the ants moving process, and the relative importance of the heuristic information in the ants path selecting process. If $p_{ij}(t) \geq p_0$, X_i would be clustered into the cluster where its X_j belongs to. Let $C_j = \{X_k \mid d_{kj} \leq r, k = 1, 2, ..., J\}$, C_j denotes the collection of data which converge into the neighborhood of X_j . Then, the clustering center is given as:

$$C_j = \frac{1}{J}\sum_{k=1}^{J} X_k \ .\tag{5}$$

With the ants moving, the pheromone information value on every path is changing. Through one circulation, the pheromone information value on each path is adjusted according the following formula:

$$\tau_{ij}(t) = \rho\tau_{ij}(t) + \Delta\tau_{ij} \ .\tag{6}$$

where ρ represents the evaporating degree of pheromone information value with the elapse of time, $\Delta\tau_{ij}$ is the augmentation of path pheromone information value in this circulation.

$$\Delta\tau_{ij} = \sum_{k=1}^{n}\Delta\tau_{ij}^{k} \ .\tag{7}$$

$\Delta\tau_{ij}^{k}$ is the pheromone information value remained by the kth ant.

The following sections discuss a new way to solve the pre-segmentation image problems by combining a novel clustering algorithm and a noise spots filtering process.

2.2   Refinement Process

After ant colony clustering process, a rough cell segmentation image is obtained, from which we can get adhering cells and some noise spots. So a refinement process is needed in which the noise spots filtering process and adhering cells segment process are included.

Aim at filtering noise spots, we use area feature of cells which expressed as:

$$S = \sum_x s_x \qquad s_x = \begin{cases} 1 & x \in cell \\ 0 & x \notin cell \end{cases} \ .\tag{8}$$

s_x is unit pixel and subjects to constraints: $\min(S) < S < \max(S)$, where $\max(S)$ and $\min(S)$ are areas that corresponding to the maximum area and minimum area of cells. Any objects that do not satisfied cell features are considered as noise. Comparing with noise spots, cells have relatively stable feature, which facilitates removing mismatch objects and reduces or eliminates noise spots.

After noise filtering process, a clustering method is used to find individuals of cells, the clustering algorithm is shown below:

1. Initialize the set of clusters to the empty set, $S = \phi$

2. Find a cluster C in S, such that for all C_i in S $dist(C, x_i) \leq dist(C_i, x_i)$

3. If $dist(C, x_i) \leq w$, then associate x_i with the cluster C. otherwise a new cluster is created $S \leftarrow S \bigcup \{C_n\}$, where C_n is a cluster with x_i

4. Repeat step 2 and 3, until now instances are left.

where x_i is data samples, C_i is the cluster in S. After this step all cells should be identified.

3   Results and Discussion

The experiment was made under the environment of Inter I5 2410M, win7 64bit, Matlab 2011b. As for parameters $\alpha = 2, \beta = 3, r = 10, p_0 = 0.9$ were selected. Fig 2 is a simple cell image which has few cells. We use level set method, k-means method and our method to process this image, and the results are shown as follows:

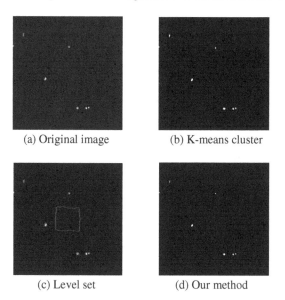

(a) Original image   (b) K-means cluster

(c) Level set   (d) Our method

Fig. 2. Simple cell image segmentation

From the results shown above we can tell that at this degree, three methods share the same detection results, but level set had the longest time consuming for its iteration. We then choose an image from a frame of the sequence images which is shown as Fig 3(a). From the image, we can see that there are many cells, some of which are adhering cells or the noises of background. Fig 3(b) is the gray scale image of Fig 3(a). The gray scale image is operated by ant colony clustering method and the pre-segmentation image is obtained as Fig 3(c).

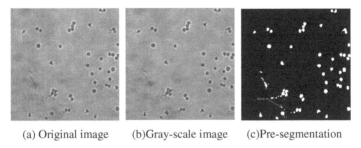

(a) Original image   (b)Gray-scale image   (c)Pre-segmentation

Fig. 3. Pre-segmentation of multi-cell image

In Fig 3(c), most of cells are classified as white spots. Some spots are cells, some are noise spots, which must be filtered. Then filtering process is carried out as mentioned in section 2, the result is shown as Fig 4(a). From Fig 4(a), we can see that many noise spots have been erased but still one big noise spot remained. After this step, the cluster method that proposed in this paper is used to make refinement. We first use this method find all possible cells in the image, as we can see from Fig 4(b), each red dot denote one cell. Then, find the connected objects, we get the conclusion that it is the adhering cell if more than one clusters are included in one connected object. The clusters were found with the method of cut off the line between two

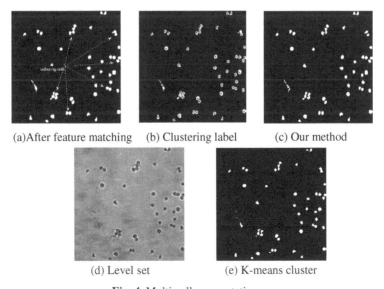

(a)After feature matching   (b) Clustering label   (c) Our method

(d) Level set   (e) K-means cluster

Fig. 4. Multi-cell segmentation

clusters to divide adhering cells, which is shown as Fig 4(c). Comparing with Fig 4(b) and Fig 4(d), we can see that some adhering cells are separated successfully.

Fig 4(d) is cell segmentation with level set, from this image we can see that some of cells are identified but some connected cells are clustered as one cell and some individuals are not detected. The last figure is k-means based segmentation method, most of cells are classified with this method, but some of them are adhering cells that can hardly be used at subsequent applications. In addition k- means based method still remains more noise spots compared with our method.

4   Conclusions

The cell segmentation is an important problem in biological study. A cluster based method is proposed in this paper, which combines ant colony cluster algorithm with image refinement process. This method remains both the advantage of image segment of ant colony cluster and the ability of further process of pre-segmentation, which improves the performance of cell segmentation. The experiment results confirmed the above point that the method proposed in this paper is effective.

Acknowledgments. This work is supported by National Natural Science Foundation of china (No.61104186) and the Natural Science Foundation of the Jiangsu Higher Education Institutions of China (No.14KJB510001).

References

1. Xiaobo, Z., Fuhai, L., Jun, Y.: A Novel Cell Segmentation Method and Cell Phase Identification Using Markov Model. J. IEEE Transaction on Information Technology in Biomedicine 13(2), 152–157 (2009)
2. Li, G., Liu, T., Nie, J., Guo, L., Wong, S.T.C.: Segmentation of touching cells using gradient flow tracking. In: 4th IEEE International Symposium on Biomedical Imaging, pp. 77–80. IEEE Press, New York (2007)
3. Muhimmah, I., Kurniawan, R., Indrayanti, I.: Automated cervical cell nuclei segmentation using morphological operation and watershed transformation. In: 2012 IEEE International Conference on Computational Intelligence and Cybernetics, pp. 163–167. IEEE Press, New York (2012)
4. Bergeest, J.-P., Rohr, K.: Efficient globally optimal segmentation of cells in fluorescence microscopy images using level sets and convex energy functionals. J. Medical Image Analysis 16(7), 1436–1444 (2012)
5. McCullough, D.P., Gudla, P.R., Meaburn, K., Kumar, A., Kuehn, M., Lockett, S.J.: 3D Segmentation of whole cells and cell nuclei in tissue using dynamic programming. In: 4th IEEE International Symposium on Biomedical Imaging, pp. 276–279 (2007)
6. Chankong, T., Theera-Umpon, N., Auephanwiriyakul, S.: Automatic cervical cell segmentation and classification in Pap smears. J. Computer Methods and Programs in Biomedicine 13(2), 539–556 (2014)
7. Shelokar, P.S., Jayaraman, V.K., Kulkarni: An ant colony approach for clustering. J. Analytica Chimica Acta 509(2), 187–195 (2004)

Research on Lane Departure Decision Warning Methods Based on Machine Vision

Chuncheng Ma, Puheng Xue, and Wanping Wang

CCCC First Highway Consultants Co., Ltd., Xi'an, Shaanxi, China
{496056424,1464925547,329833165}@qq.com

Abstract. To improve the driving safety of drivers, an effective lane detection algorithm was proposed upon the research on lane departure decision warning system based on machine vision. Firstly, the lane images were preprocessed to adapt to various lighting conditions and improve the efficiency of the lane detection. Then, by means of hough transform, actual lane line features were extracted according to the different image lane line features. Finally, after the study of different lane departure models based on lane line detection, this article put forward a lane departure decision algorithm. Experimental results demonstrate that the developed system exhibits good detection performances in recognition reliability and warning decision. It has proved that this system has high accuracy, large detection range and high practicability.

Keywords: machine vision, lane detection, hough transform, lane departure decision, warning methods.

1 Introduction

Safety Driving Assist (SDA) is the current concern in the research of intelligent transportation system, which mainly solves the traffic security problems. As an essential component in the research field of security auxiliary driving, lane departure decision warning system gets more and more attention in recent 20 years. If the vehicle deviates from the lane or there is any trend of vehicle deviation, the system will warn the tired or absent-minded drivers to alter driving directions, thus reduce lane accidents. As a result, it is of great significance for driving safety.

2 Lane Departure Decision Warning System Based on Machine Vision

Lane departure decision warning system based on machine vision mainly consists of CCD camera, video capture card, PC processor, alarm unit, display equipment and other components [1]. During the high speed running of vehicles, the system uses camera and video capture card to get lane images, and computer processes the digital images to detect real-time image of left and right lane, on which the lane departure decision is made. If the vehicle deviates from the normal lane or has the trend to

Y. Tan et al. (Eds.): ICSI 2014, Part II, LNCS 8795, pp. 259–266, 2014.

deviate, the system will send warning information through display and alarm circuit, to remind or warn the driver to alter driving direction.

3   Research on Lane Detection Algorithm

Lane detection in road image is the basis of lane departure decision warning system. The model of lane detection is shown in Figure 1. The road detection image includes some interferential information, while lane line is mainly in the below of image. For comprehensive feature analysis of road image, a threshold was set, and the region below is our research part, and only that part is processed, so that we can enhance the real-time of lane detection.

Researches show that the vehicle lane departure decision can be made by lane field image of myopia [2]. In order to simplify the processing, we can select an appropriate cut point to process the lane line locating in the below of as straight line Therefore, the lane detection becomes the detection of straight line in the region of interest, whose algorithm includes image equalization grayscale processing, edge detection, Hough transform, lane feature extraction and so on.

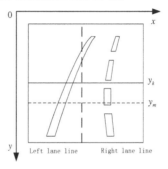

Fig. 1. The model diagram of lane detection

3.1   Image Equalization Grayscale Processing

Lamps and lighting directly affect the quality of machine vision system. As for the road images under different lights, the gray values vary greatly. In strong or low light images, the lane line presents low contrast to the background, and the range of the gray changes is narrow. To improve the adaptability and immunity of the system, the research applies the method of histogram equalization grayscale processing for lane image process.

Histogram equalization is expressed as the cumulative distribution function:

$$S_k = \sum_{i=0}^{k} \frac{n_i}{N} = \sum_{i=0}^{k} p_r(r_i) \tag{1}$$

In this function, $i = 0,1,...k$, $k = 0,1,...L$, L is the gray level range, $p_r(r_i)$ is the appearance probability of i level gray in the image, N is the total pixels of the road image.

3.2   Edge Detection and Binary Image Processing

Lane detection mainly focuses on the detection of the edge of lane line, and edge detection is realized through the use of an algorithm to extract the lines between objects and background in the image [3]. Changes in the gray image can be reflected by the gradient of image gray distribution, so that differential technique based on local image can be used to obtain edge detection operator [4].

To achieve the purpose of rapid detection, Sobel operator is applied to provide relatively accurate information about edge direction [5].

After edge detection, the road image is processed by binary to reduce the image storage capacity, thus raise lane detection speed. Figure 2 shows the image after edge detection and binary image processing.

Fig. 2. Edge detection and binary image processing

3.3   Image Hough Transform

The key of lane detection is to detect the line representing real lane. The common method for straight line extraction is Hough transform, the polar coordinates equation of straight line based on Hough transform is as follows [6]:

$$\rho = x\cos\theta + y\sin\theta \tag{2}$$

ρ is the distance from straight line to the origin in image space; θ is the angle between the line and the x-axis. There are two planes: xy is the image plane, $\rho\theta$ is the parameter plane. Linear equation is used to transform the points of the same line on the image planes to a coordinate point (ρ,θ) on the parameter plane. The image after Hough transform is shown in Figure 3.

Fig. 3. The image after Hough transform

3.4   Lane Feature Extraction

From Figure 3, it can be seen that there are still some interferential lines after Hough transform, so that it is necessary to filter out such lines. After Hough transform, parameters of the linear can be extracted, gaining all coordinate equations as well as the starting and ending coordinates of line segments [7].

Generally speaking, left lane line locates in the left half of the image and right lane line in the right half; accordingly, the interest image region is divided into two parts, left and right, to identify the coordinating lane line. In the interest image region, the middle part is the road which is relatively flat, with minimal disruption to the Hough transform. Moreover, the road lines locate in the middle of the road, so line searching algorithm from bottom to up and middle to left or right is employed to extract lane features.

Steps for feature extraction of lane lines are as follows:

1) The angles formed by the left and right lane lines with x-axes are set to α and β; the range of α is $(100°, 170°)$, and range of β is $(10°, 80°)$. At the same time the straight lines which do not meet the requirements are removed. Detection results are shown in Figure 4.

Fig. 4. Lane detection

2) Extract three factors: the length of straight line itself L_Line, the length from the middle of the image to the line L_Center, and the length from the bottom to the

line L_Bottom .Calculate the value of the three factors with each range as R_l, R_c, R_b in the left and right half of the interest region.

3) Figure out the straight lines with the biggest value of comprehensive determine coefficient in left and right half interest region respectively. Comprehensive determine coefficient of lane lines comes to:

$$A = \sum_{i=1}^{3} w_i V_i \tag{3}$$

w_i is the weight of factor values, which is set to $W = (0.4, 0.3, 0.3)$, V_i is the value of factor values.

$$V = \begin{cases} L\_Line / R_l & i = 1 \\ 1 - L\_Center / R_c & i = 2 \\ 1 - L\_Bottom / R_b & i = 3 \end{cases} \tag{4}$$

The straight lines with the biggest value of comprehensive determine coefficient in the left and right half interest region are the lane lines in left and right. Extracted lane lines are shown in Figure 5.

Fig. 5. Extraction Lane Line

4   Lane Departure Decision Warning

When vehicles are on their ways, because of environment changes, lane turning and changing, some errors may appear, resulting in lane line changes between the current frame and former frame in the video, or detection failure of lane line. For such situation, the current image frames are determined to be ineffective and current lane departure warning decision is also eliminated.

Vehicles' driving states mainly contain normal driving, slight left skew, slight right skew, vehicle departure from left lane, and vehicle departure from the right lane. The model of lane departure is given in Figure 6. Two auxiliary horizontal straight lines are set as $y = y_b$ and $y = y_t$ for the lane departure decision. In the Hough transform, the upper and lower endpoints of right and left lane segments are set as (x_{Lt}, y_{Lt}), (x_{Lb}, y_{Lb}) and (x_{Rt}, y_{Rt}), (x_{Rb}, y_{Rb}). Then the horizontal coordinates for intersections of left and right lane line with y_b are produced, X_{Lb} and X_{Rb}:

$$X_{Lb} = \begin{cases} x_{Lb} & x_{Lt} = x_{Lb} \\ x_{Lb} - \dfrac{y_b - y_{Lb}}{y_{Lb} - y_{Lt}}(x_{Lt} - x_{Lb}) & x_{Lt} \neq x_{Lb} \end{cases} \tag{5}$$

$$X_{Rb} = \begin{cases} x_{Rb} & x_{Rt} = x_{Rb} \\ x_{Rb} + \dfrac{y_b - y_{Rb}}{y_{Rb} - y_{Rt}}(x_{Rb} - x_{Rt}) & x_{Rt} \neq x_{Rb} \end{cases} \tag{6}$$

$$X_{Lt} = \begin{cases} x_{Lb} & x_{Lt} = x_{Lb} \\ x_{Lb} - \dfrac{y_t - y_{Lb}}{y_{Lb} - y_{Lt}}(x_{Lt} - x_{Lb}) & x_{Lt} \neq x_{Lb} \end{cases} \tag{7}$$

$$X_{Rt} = \begin{cases} x_{Rb} & x_{Rt} = x_{Rb} \\ x_{Rb} + \dfrac{y_t - y_{Rb}}{y_{Rb} - y_{Rt}}(x_{Rb} - x_{Rt}) & x_{Rt} \neq x_{Rb} \end{cases} \tag{8}$$

The middle point A of the intersection between Left and right lane line with $y = y_t$ is $\left(\dfrac{X_{Lt} + X_{Rt}}{2}, y_t\right)$, and the middle point B of the intersection between Left and right lane line with $y = y_b$ is $\left(\dfrac{X_{Lt} + X_{Rt}}{2}, y_t\right)$. The slope of the straight line AB is K_{AB}.

$$K_{AB} = \begin{cases} \infty & \dfrac{X_{Lt} + X_{Rt}}{2} = \dfrac{X_{Lb} + X_{Rb}}{2} \\ \dfrac{y_b - y_t}{\dfrac{X_{Lt} + X_{Rt}}{2} - \dfrac{X_{Lb} + X_{Rb}}{2}} & \dfrac{X_{Lt} + X_{Rt}}{2} \neq \dfrac{X_{Lb} + X_{Rb}}{2} \end{cases} \tag{9}$$

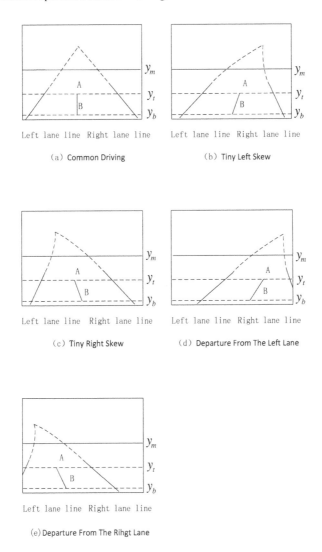

Fig. 6. The model of lane departure

After analyzing a great quantity of vehicle driving states and lane images, setting a left and a right threshold X_{Lb}' and X_{Rb}' as well as threshold K_L and K_R for the slope of straight line AB are assigned to decide whether there is lane departure.

5   Experimental Results and Analysis

The lane departure decision system based on machine vision is developed by the combination of Visual C++ and Open CV Open computer vision library. The processed video image size is 640*480. After a large number of simulation experiments on 1000 continuous frames image sequences from Video, through the lanes detection and the analysis of lane departure decision, we can come to the conclusion as follows: the rate of detection driveway is 98.23%; the success rate of lane departure decision is 97.64%; the time that deciding a road image nearly need 30ms. The final results show that the algorithm of lane detection and lane departure decision can effectively detect and track lane line and accurately make the right decision of lane departure. It is more real-time and anti-jamming. Therefore the system can improve the safety of drivers driving, playing the role of Vehicle Auxiliary Navigation.

6   Conclusion

The paper studies and implements a kind of lane departure decision warning system based on machine vision. It has shown that it can effectively and instantly detect the left, right lane lines and make right decisions on the condition of lane deviations. With favorable ability of lane detection and lane departure decision, this approach can satisfy the requirements of the lane departure decision warning system. Still this algorithm needs a further improvement so as to make sure that the system has the ability to make more accurate decision and lane detection under different atrocious weather circumstances and promotes the lane departure decision warning system to be perfect.

References

1. Jiuqiang, H.: Machine vision technology and applications. Higher Education Press, Beijing (2009)
2. Jung, C.R., Kelber, C.R.: A lane departure warning system using lateral offset with uncalibrated camera. In: Proceedings of the 2006 IEEE Intelligent Transportation Systems, pp. 102–107. IEEE (2005)
3. Huang, S.S., Chen, C.J., Hsiao, P.Y., Fu, L.C.: On-board vision system for lane recognition and front-vehicle detection to enhance driver's awareness. In: 2004 IEEE International Conference on Robotics and Automation, Proceedings. ICRA 2004, vol. 3, pp. 2456–2461. IEEE (2004)
4. Jung, C.R., Kelber, C.R.: A lane departure warning system based on a linear-parabolic lane model. In: 2004 IEEE Intelligent Vehicles Symposium, pp. 891–895. IEEE (2004)
5. Li, X., Zhang, W.: Research on lane departure warning system based on machine vision. Chinese Journal of Scientific Instrument 29, 1554–1558 (2008)
6. Yang, Y., Farrell, J.A.: Magnetometer and differential carrier phase GPS-aided INS for advanced vehicle control. IEEE Transactions on Robotics and Automation 19(2), 269–282 (2003)
7. Zhang, R., Wang, Y., Yang, R.: Researches on Road Recognition in Landsat TM Images. Journal of Remote Sensing 2, 016 (2005)

Searching Images in a Textile Image Database

Yin-Fu Huang and Sheng-Min Lin

Department of Computer Science and Information Engineering
National Yunlin University of Science and Technology
huangyf@yuntech.edu.tw, bhujmn2007@gmail.com

Abstract. In this paper, a textile image search system is proposed to query similar textile images in an image database. Five feature descriptors about the color, texture, and shape defined in the MPEG-7 specification, which are relevant to textile image characteristics, are extracted from a dataset. First, we tune the feature weights using a genetic algorithm, based on a predefined training dataset. Then, for each extracted feature descriptor, we use K-means to partition it into four clusters and combine them together to obtain an MPEG-7 signature. Finally, when users input a query image, the system finds out similar images by combining the results based on MPEG-7 signatures and the ones in three nearest classes. The experimental results show that the similar images returned from an image database to a query textile image are acceptable for humans and with good quality.

Keywords: CBIR, genetic algorithm, K-means, MPEG-7 specification, weight tuning.

1   Introduction

At present, multimedia data have played an important role in our daily life. However, querying a multimedia database by keywords is gradually insufficient to meet users' needs. Thus, facing a huge amount of images in an image database, content-based image retrieval has become a popular and required demand.

In the past years, many general-purpose image retrieval systems have been developed [5, 6, 10], and these systems rely mainly on visual features. King and Lau used MPEG-7 descriptors to retrieve fashion clothes [5]. In order to improve query results, Lai and Chen proposed a user-oriented image retrieval system by iteratively interacting with users about query results [6]. Smeulders et al. presented a review of 200 references in content-based image retrieval [10].

In this paper, we propose a textile image search system for querying similar textile images in an image database. This system consists of an offline phase and an online phase. In the offline phase, we tune the feature weights using a genetic algorithm [11], based on a predefined training dataset. Then, for each extracted feature descriptor, we use K-means to partition it into four clusters and combine them together to obtain an MPEG-7 signature [3]. In the online phase, when users input a query image, the system extracts its MPEG-7 visual features first, and then finds out similar images by combining the results based on MPEG-7 signatures and the ones in three nearest classes.

Y. Tan et al. (Eds.): ICSI 2014, Part II, LNCS 8795, pp. 267–274, 2014.

The remainder of this paper is organized as follows. In Section 2, we present the system architecture and briefly describe the procedure of searching similar textile images. In Section 3, we introduce five feature descriptors relevant to textile image characteristics. In Section 4, the methods of tuning the feature weights and generating MPEG-7 signatures are proposed to facilitate searching similar images. In Section 5, we present the experimental results to evaluate the effectiveness of three search modes provided in our system. Finally, we make conclusions in Section 6.

2   System Architecture

In this paper, we propose a textile image search system consisting of an offline phase and an online phase, as shown in Fig. 1. In the offline phase, five feature descriptors about the color, texture, and shape defined in the MPEG7 specification are extracted from training images; i.e., ColorLayout descriptor, ColorStructure descriptor, EdgeHistogram descriptor, HomogeneousTexture descriptor, and RegionShape descriptors including a total of 221 dimensions. Because each feature plays a different role in distinguishing a textile image from others, feature weights should be determined in order that the discrimination among textile images could be boosted. Here, we use a genetic algorithm to determine feature weights. Then, we build MPEG-7 signatures using k-means clustering on all textile images where the weighted Euclidean distance calculated in k-means clustering takes the feature weights determined in the genetic algorithm.

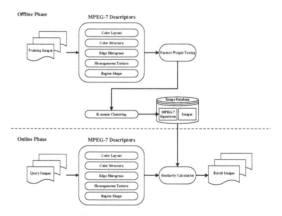

Fig. 1. System architecture

In the online phase, we also extract the same features, as mentioned, from a query image. Then, we find out 1) the images with the same MPEG-7 signature as the query image as the first candidates and 2) the images in three nearest classes to the query image as the second candidates. Finally, we can find out result images most similar to the query image, which appear in both groups of candidates.

3   Feature Extraction

In this paper, we adopt the bag-of-feature MPEG-7 [1, 2, 4, 7-9] defined by the MPEG organization, which consists of color description (i.e., two color descriptors), texture description (i.e., two texture descriptors), and shape description (i.e., one shape descriptor), as shown in Table 1. Among them, the descriptors relevant to textile image characteristics are as follows.

1) ColorLayout Descriptor describes the layout and variation of colors and this reflects the color combinations in a textile image.

2) ColorStructure Descriptor counts the contents and structure of colors in a textile image by using a sliding window.

3) EdgeHistogram Descriptor counts the number of edge occurrences in five different directions in a textile image.

4) HomogeneousTexture Descriptor calculates the energies of a textile image in the frequency space, which are the level of gray-scale uniform distribution and texture thickness, and this reflects the texture characteristics in a textile image.

5) RegionShape Descriptor relates to the spatial arrangement of points (pixels) belonging to an object or a region in a textile image.

Table 1. MPEG-7 visual descriptor features

Type	Feature description	Dim.	Overall statistics	Total number
1	ColorLayout	12	1	12
2	ColorStructure	32	1	32
3	EdgeHistogram	5	16	80
4	HomogeneousTexture	2	31	62
5	RegionShape	35	1	35

4   Method

K-means clustering is an effective approach to find similar images for a query image, but it is usually dependent on how well stored images are clustered. In reality, the Euclidean distance between two images, used in K-means clustering, plays a major role in determining good clustering. In calculating the Euclidean distance between two images, each kind of involved features (or descriptors) mentioned in Section 3 has its own semantic in measuring their similarity; thus, these features should be assigned with weights in measuring the similarity between two images.

In our system, a finest weight set is tuned to represent the weight of each feature involved in an image by using a genetic algorithm. Next, we generate MPEG-7 signatures based on K-means clustering with weighted features. Finally, in the online phase, we find out result images most similar to a query image after the similarity calculation.

4.1   Feature Weight Tuning

In this step, we use a genetic algorithm to tune the weights of features. A genetic algorithm is a search heuristic used to generate useful solutions to optimization and search problems. In general, a typical genetic algorithm requires 1) a genetic representation of a solution domain and 2) a fitness function to evaluate the solution domain. Here, we also use the weighted Euclidean distance as the fitness function to measure the similarity between two images as follows.

$$Dist(A, B) = \sqrt{\sum_{i=1}^{n}(a_i - b_i)^2 \cdot w_i}$$ (1)

where A, B are two images with feature sets $A = \{a_1, a_2, a_3, ..., a_n\}$ and $B = \{b_1, b_2, b_3, ..., b_n\}$.

Before starting the genetic algorithm, the values of all the 221 features extracted from an image have been normalized in the range [0, 1]. For an initial population of 100 individuals in the genetic algorithm, each one is a set of weight values randomly generated. Only the best 20 individuals with higher fitness values can be alive in the next generation. Then, the 20 individuals are randomly selected to generate 40 children by crossover; in crossover, each feature weight of a child is selected from the corresponding feature weight of parent A or parent B, respectively with 50% probability. Furthermore, in order to avoid being trapped into a local optimal, we also generate another 40 children by crossover plus mutation. In mutation, 10% probability of the feature weights of a child are replaced with new random values.

To measure the fitness of each individual, 24 centroids are used to represent 24 pre-defined classes of 679 training textile images, which are calculated from the images in each class. Then, the weighted Euclidean distance can be treated as a classifier; if a training image has the shortest distance with the centroid of a class using the weights of individual x and the matching class is indeed the class of the training image, 1 point is added to the score of individual x, and this score is the fitness of individual x.

By iteratively doing so, the best individual (or best feature weights) with the highest score would be found. The genetic algorithm will terminate when the finest weight set becomes stable; i.e., the finest weight set is always the best for 1000 iterations. During the iterations, if a new individual has higher fitness than the old best one, the iteration counter is reset to 0 and the new individual will be examined for the next 1000 iterations.

4.2   MPEG-7 Signatures Based on K-means Clustering

For each extracted visual descriptor, we use K-means to partition it into four clusters respectively, and number them from 0 to 3. Then, we combine the cluster numbers from the five visual descriptors together and obtain a 5-digit MPEG-7 signature. Thus, an MPEG-7 signature can represent the characteristics of an image. An MPEG-7 signature has 5 digits and each digit can be 4 different values, so that $(4)^5 = 1024$ bins could be used to distinguish the characteristics of images. Since K-Means compresses images into clusters, we would be able to build an index structure more easily using these signatures. Besides, the centroids of K-means on the five visual descriptors are also stored for the similarity measures in the online phase.

4.3   Similarity Calculation

First, we extract the MPEG-7 visual features of a query image. Then, the similarity measures between the visual features extracted from the query image and the recorded centroids of K-means on the five visual descriptors are performed to determine cluster numbers, respectively. The cluster numbers of the most similar centroids are combined together to become the query signature. Then, we can find out the images with the same MPEG-7 signature as the query image as the first candidates. This approach can be treated as the similarity measures based on local views. Next, we find out the images in three nearest classes to the query image as the second candidates where 24 centroids of pre-defined classes have been mentioned in Section 4.1. This approach can be treated as the similarity measures based on global views. Finally, we can find out the most similar images to the query image, which appear in both groups of candidates.

5   Implementation

We have implemented an "Image Search Engine" system to search similar images from an image database to a query textile image. Totally the 4069 images in the textile image database are from Globle-Tex Co., Ltd. [13], where 679 images are training images with pre-defined classes and the others are input to their nearest classes subsequently according to the weighted Euclidean distance as mentioned in Section 4.1.

5.1   User Interface

The user interface of the ISE system is shown in Fig. 2. The **"Initialization"** button is used to initialize the system (or to do data preprocessing in the offline phase). The **"Input"** button can be clicked to input a query image, and after the query image is input, Area 1 will record the path name of the query image. Furthermore, the **"Execution"** button is used to show result images. The radius buttons shown in Area 2 are page switches used to display result images of different pages in Area 3. The number of radius buttons is dynamic and dependent on the number of all result images. The check boxes shown in Area 4 are used to select a search mode; i.e., full (or default) mode, texture-concern mode, and color-concern mode. For the texture-concern mode, we use only the EdgeHistogram Descriptor, the HomogeneousTexture Descriptor, and the RegionShape Descriptor to search similar images. On the contrary, for the color-concern mode, we use only the ColorLayout Descriptor and the ColorStructure Descriptor to search similar images.

The result images displayed in Area 3 are ranked according to the similarity degree to a query image. The most similar image is put at the upper-left and the others are sequentially shown in the left-to-right and top-to-bottom way. The similarity degree to a query image is also according to the weighted Euclidean distance.

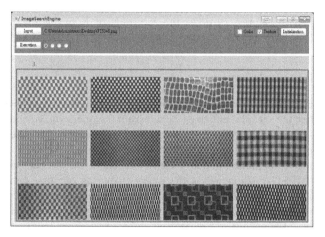

Fig. 2. User interface

5.2   Similarity Evaluation

Here, we invite ten evaluators to rate the similar images returned by the ISE system, to their query images. In order to observe the effectiveness of the ISE system, we use the acceptable percentage measure defined by Zhang et al. [12] for rating each result image as a 1-to-5 scale (1: not similar, 2: poorly similar, 3: fairly similar, 4: well similar, and 5: strongly similar). Besides, we also use the quality value measure to evaluate the quality of result images as follows.

The acceptable percentage measure:

$$m_1 = \frac{n_3 + n_4 + n_5}{\sum_{i=1}^{5} n_i} \tag{2}$$

The quality value measure:

$$m_2 = \frac{\sum_{i=1}^{5} n_i \times i}{\sum_{i=1}^{5} n_i} \tag{3}$$

where n_1, n_2, n_3, n_4, and n_5 are the number of result images with a score of 1, 2, 3, 4, and 5, respectively.

5.3   Experiments and Discussions

In Experiment 1, each evaluator tests the effectiveness of three search modes using images in the database. Since the interpretations of the evaluators on textures in the same image are diverse (i.e., someone focuses on patterns, but someone focuses on tiny formations), their measures on the full and texture-concern modes are with major differences. However, on average, the acceptable percentages on the full and texture-concern modes are 81% and 83%, and the quality values on the full and texture-concern modes are 3.6 and 3.7, respectively. Furthermore, since the interpretations of the evaluators on colors are more consistent, the average acceptable percentage and quality value on the color-concern mode are 92% and 4.1, respectively. Thus, the system works well in all three modes, when using images in the database.

In Experiment 2, each evaluator tests the effectiveness of three search modes using images out of the database. We found that their measures on these three modes are a little decreased, when compared with using images in the database. The average acceptable percentages on the three modes are 73%, 80%, and 73%, and the average quality values are 3.1, 3.6, and 3.6, respectively. The reason could be that the query images out of the database do not pertain to the pre-defined classes in the system, and the system cannot but return the similar images in three nearest classes to the query images.

6   Conclusions

In this paper, we propose and implement a textile image search system to search similar images from an image database to a query textile image. In the system, a finest weight set is tuned for the extracted features involved in an image by using a genetic algorithm. Then, we generate MPEG-7 signatures based on K-means clustering with weighted features. In the online processing, users can find out result images most similar to a query image after the similarity calculation. The experimental results show that the similar images returned from an image database to a query textile image are acceptable for humans and with good quality in all three modes.

Although our content-based ISE system can work well for searching textile images, there are still two issues to be overcome in the future. First, the descriptors we used here are still not good enough to describe all the classes of images in our system so that some of them cannot be well classified in the system. Second, for a query image without a pre-defined class, this will lead the system to return unpredictable results. For example, when users input a car image in the worst case, the system has no way to exclude this situation.

Acknowledgments. This work was supported by National Science Council of R.O.C. under grant MOST 103-2221-E-224-049.

References

1. Bober, M.: MPEG-7 visual shape descriptors. IEEE Transactions on Circuits and Systems for Video Technology 11(6), 716–719 (2001)
2. Chang, S.F., Sikora, T., Puri, A.: Overview of the MPEG-7 standard. IEEE Transactions on Circuits and Systems for Video Technology 11(6), 688–695 (2001)
3. Huang, Y.F., Chen, H.W.: A multi-type indexing CBVR system constructed with MPEG-7 visual features. In: Zhong, N., Callaghan, V., Ghorbani, A.A., Hu, B. (eds.) AMT 2011. LNCS, vol. 6890, pp. 71–82. Springer, Heidelberg (2011)
4. ISO/IEC 15938-3, Information Technology – Multimedia Content Description Interface-Part3: Visual (2002)

5. King, I., Lau, T.K.: A feature-based image retrieval database for the fashion, textile, and clothing industry in Hong Kong. In: Proc. International Symposium on Multi-technology Information Processing, pp. 233–240 (1996)

6. Lai, C.C., Chen, Y.C.: A user-oriented image retrieval system based on interactive genetic algorithm. IEEE Transactions on Instrumentation and Measurement 60(10), 3318–3325 (2011)

7. Manjunath, B.S., Ohm, J.R., Vasudevan, V.V., Yamada, A.: Color and texture descriptors. IEEE Transactions on Circuits and Systems for Video Technology 11(6), 703–715 (2001)

8. Martinez, J.M., Koenen, R., Pereira, F.: MPEG-7: the generic multimedia content description standard, part 1. IEEE Multimedia 9(2), 78–87 (2002)

9. Martinez, J.M.: MPEG-7 overview (version 10). ISO/IEC JTC1/SC29/WG11 N6828 (2004)

10. Smeulders, A.W.M., Worring, M., Santini, S., Gupta, A., Jain, R.: Content-based image retrieval at the end of the early years. IEEE Transactions on Pattern Analysis and Machine Intelligence 22(12), 1349–1380 (2000)

11. Whitley, D.: A genetic algorithm tutorial. Statistics and Computing 4(2), 65–85 (1994)

12. Zhang, Y., Milios, E., Zincir-Heywood, N.: Narrative text classification for automatic key phrase extraction in web document corpora. In: Proc. the 7th Annual ACM International Workshop on Web Information and Data Management, pp. 51–58 (2005)

13. Globle-Tex Co., Ltd., http://www.globle-tex.com/

IIS: Implicit Image Steganography

K. Jithesh[1,*] and P. Babu Anto[2]

[1] Department of Computer Science, M.G College, Iritty, Kannur Uniersity, Kerala,India
[2] Department of Information Technology, Kannur University, Kerala, India
jithukotheri@gmail.com

Abstract. In steganography secrets are imposed inside the cover medium either by replacing bits in the spatial domain or changing the frequency domain. Instead the proposed Implicit Image Steganography (IIS) scheme does not alter or replace bits in the original cover image for hiding information. As the name implies there is no explicit embedding of data inside the image. Before beginning the communication, entities should agree upon a cover image with maximum ranges of intensity values. At least, it should contain intensity values that can represent ASCII of all characters. Coordinate positions of each pixel with intensity which can be the ASCII of a letter in the secret will be the stego-key. In this scheme, transferring of cover image is not done as in the case of usual procedures. The communicating entities have to transfer only the key. The big advantage of this technique is that the cover is not required to transmit each other. Hence nobody can even know about the cover. So it is not only difficult but impossible to attack this communication. Also it is not required to worry whether distortion happens while embedding.

Keywords: Cover, Steganography, Stego-key, Stego-image, Steganalysis.

1 Introduction

The standard concept and practice of ''What You See Is What You Get (WYSIWYG)'' which we encounter sometimes while printing images or other materials, is no longer precise and would not deceive a steganographer as it does not always hold true. Images can be more than what we see with our Human Visual System (HVS); hence, an image can convey more than merely 1000 words. For decades people have been trying to develop innovative methods for secret communication. Networking and digitization have become part of the technological features in the rapid economic development of the society. The convenient and timely acquisition of on-line services through accessing the internet has assumed the proportion of a tidal current for individuals and organizations in their pursuit of work. However, the relay of sensitive information via an open Internet channel increases the risk of attacks. Thus many techniques have been proposed to deal with this problem. Data hiding, known as information hiding, plays an important role in the information security. For

---

* Corresponding author.

Y. Tan et al. (Eds.): ICSI 2014, Part II, LNCS 8795, pp. 275–283, 2014.

content authentication and perceptual transparency, the main idea of data hiding is to conceal the secret data into the cover medium, and thereby to avoid attracting the attention of attackers in the Internet channel. The growing numbers of internet-based applications nowadays have made digital communication an essential part of infrastructure. Confidentiality in some digital communication is absolutely necessary when sensitive information is being shared between two communicating parties over a public channel.

Steganography [1, 2, 3] and Cryptography [4] are two sides of the same coin for providing confidentiality and protecting sensitive information. The former is the art and science of hiding sensitive information within innocuous documents in an undetectable way. A thorough history of steganography can be found in the literatures. The latter is the art and science of writing sensitive information in such a way that no one but the intended recipient can decode it. The innocuous documents (also known as hosts/covers/carriers) can also be of any kind as long as they do not seem suspicious. However, with the advent of the digital technology, digital hosts such as image, audio and video files etc. have become nowadays the most commonly used host files. On account of their insensitivity for the human visual system, digital images [5, 6] can be regarded as an excellent choice for hiding sensitive information. One of the most commonly used data hiding approaches is the substitution technique [7, 8], [12, 13]. The embedding algorithm may require a secret key, referred to as stego-key.

Each and every method introduced in the field of information security has the sole purpose of achieving the triple pillars of information security. They are confidentiality, availability and authentication. The available algorithms of steganography have been embedding secret data inside the cover medium. A few exceptions are there like Quantum steganography protocol [9] and Multi-party covert communication with steganography and quantum secret sharing [10]. Abbas Cheddad et al. reported the current techniques of image stegnography with its pros and cones [11]. The problem associated with these methods is that they cause image distortion. Also the current techniques lose its security when the algorithm behind the communication is once revealed or an intruder has hacked the very content. Once a new method is introduced it is possible for many types of attacks to happen.

A different methodology has been introduced here for the purpose of accomplishing the goals of information security. The proposed study is not intended to go along with the usual way. Usually information is hidden inside the cover by replacing the bits or changing the frequency level. But in this scheme we do not change or make any distortion in the spatial or in the frequency domain of the cover medium. Here the cover will not be transmitted. Instead, before beginning, they should agree upon the stego-image and should keep a copy of the same. They need to send only the bit positions inside the cover which comprises the secret. They will just inform each other about the bit positions inside the image. Since they hold the same copy of the cover, they can easily extract the secret

From this perspective the authors have tried to develop a system which can give maximum security by avoiding steganalysis and distortion to the cover medium. Our primary objective here is to present a new and simple steganographic scheme that gives high security. On account of the complexities in steganography and progressive power of steganalysis methods, it has turned out to be a challenge to systematically develop techniques with much better performance. Key exchanging is also a big problem in the area of communication. This study aims to deal with these problems by proposing a scheme to increase the non-detecting power of the secret. Since the original cover image is not transmitted it can thwart all conventional attacks. The results of our experiments illustrate that since there is no stego-image, the steg-analysis is not possible. Hence the security of the system would be compromised only if communicating entities cheated each other.

The remainder of this paper is organized as follows: Section 2 contains a brief discussion of related works. In Section 3, the proposed IIS scheme is discussed. Section 4 presents the experimental results and performance of the proposed method, in terms of comparison with conventional schemes. Section 5 provides a comparative visual analysis of cover images and finally section 6 gives the conclusion of this study.

2   Related Work

Some literature proposed varieties of data hiding techniques. Most of them are irreversible. It means that, after the secret data are extracted from the stego image, the original image suffers some distortion and cannot be completely reconstructed. Nevertheless, in some fields (i.e., medical, military applications), the restoration of original image is essential after extracting the embedded secret data. Therefore, reversible data hiding schemes, also called as distortion-free data hiding or lossless data hiding, have drawn much attention of researchers. In principle, reversible data hiding schemes can be classified into three types, i.e., spatial domain, frequency domain, and compressed domain. In the spatial domain schemes, all pixel values are modified directly to embed secret data. In the frequency domain schemes, the coefficient values of image are computed by using some transformation methods (i.e., integer discrete cosine transform, integer wavelet transform). In the compressed domain schemes, the original image are first compressed based on some popular compression algorithms, such as vector quantization, block truncation coding etc. Then, according to the peculiarity of compressed codes, the compressed image is encoded to conceal secret data.

There are numerous techniques available in the literatures of steganography. The Least Significant Bit [LSB] replacement steganographic methods [7, 8], [12, 13] are the simplest one and are widely used in the fields of information security due to its high hiding capacity and quality. It can embed a secret bit stream into the LSB plane of an image. LSB replacement, LSB matching (LSBM), LSB matching revised (LSBMR) [12], and LSBMR-based edge-adaptive (LSBMR-EA) [13] image steganography are well-known LSB steganographic methods. The LSB-replacement embedding method replaces the LSB plane with embedded message bits, but the others do not. In LSB matching, if the embedded bit does not match the LSB of the cover image, then the pixel value of the corresponding pixel is randomly changed by ±1. Unlike LSB replacement and LSBM, which embed message bits pixel by pixel, LSBMR deals with two pixels at a time and allows fewer changes to the cover image. The steganalysis resistance and image distortion of LSBMR are better than those of the previous two methods. In general, the choice of embedding positions within a cover image depends on a pseudorandom sequence without consideration of the relationship.

The proposed method is a new one of its kind. From the literature survey and to the best of our knowledge there is no such steganography scheme similar to the proposed one. In the available technologies of steganography the embedding is done inside the cover image. A work which is an exception and which can be related to the proposed work in terms of not embedding secret inside an image is introduced by Guo et al. [9] in 2003 with the title Quantum Secret Sharing without Entanglement. Also, Xin Liao et al. [10] in 2010 presented a novel multi-party covert communication scheme by elegantly integrating steganography into Guo et al.'s QSS. This scheme is good in terms of security but the payload of this method is comparatively very less. That is, it communicates only one bit per transaction. Nevertheless, their idea of not embedding secret inside a digital image motivated us to introduce a method which is better in payload capacity and security. The proposed IIS scheme does nothing over the cover image. It only keeps a location map. Another important feature of the proposed stego system is that it does not require a secret key. Thus, the constructions presented demonstrate that in order to achieve perfect steganographic security no secret has to be shared between the communicating parties. The main idea behind the stegosystems we propose is to conceal the cover from outside.

3   Proposed IIS Scheme

As stated earlier this is a simple but innovative technique. Here nothing is done with the cover, but copy of the cover is shared by the communicating entities. The algorithm is as follows:

1. Select appropriate cover image carefully.
2. It can be either grey-scale or color.
3. If possible it should contain intensity values that can easily represent ASCII of each character.
4. Share the copy of the digital cover medium with the other end.
5. Exchanging of the cover should be made very confidential.
6. Hand to hand exchange is more reliable and secure, otherwise use a trusted third party.
7. After the successful exchanging of the cover the communication can be started.
8. Take the secret.
9. Digitize it.
10. Find out pixels which can fully represent characters of the secret.
11. Mark up the coordinates of the respective pixels.
12. This is treated as the stego-key.
13. If pixel [intensity] values are not enough to represent all characters then try to find consecutive bits of a pixel inside the image to represent such letters.

14. If consecutive bits of a pixel are used, the starting and ending position is required to be kept.
15. In such case two or more pixels can be used to hold a character.
16. In such cases key should be the combination of coordinates and starting as well as ending positions of the bits.
17. If color image is used coordinate position and RGB position should be the key.
18. If RGB is used a single pixel can represent 3 characters a time.
19. Stego-key can be sent to the other side with or without encryption.
20. With the key the receiving end can easily extract the secret from the copy of the image.

4   Experimental Results

Always there will be new threat in the field of communication. So no technique is good for a long time. A method that combines both logic and craft can only survive in the contest of constant changes in the field. In order to improve the security of the secret communication, here we propose a technique which gives better performance. In this section, we present the results from the experiments we conducted in order to evaluate the reliability of this method. We have used both color and grey-scale images. Care has been taken while selecting images. That is, when grey-scale image is the cover; it is assured that it contains maximum shades. Images with 0-255 ranges of intensity are the optimal case. Any character of English language can be represented with 0-255 ASCII values. So a grey scale image is enough. In a grey-scale cover, coordinate positions will be the stego-key. When the cover is a color image it is possible to represent characters of all human languages. If range of color variations of the image is more, it is very easy to represent secret.

As mentioned earlier, this method does not require anything to be done over the cover image. It requires selecting a cover image of grey-scale or color with maximum ranges. Then, hand over the copy of the same image to the other entity that belongs to the same channel of communication. The exchanging should be very confidential. It is better to exchange in a handshake mode. If it is not possible, trust a third party. The secrecy and security of the communication rely on the cover transfer. If it fails, the entire security will be compromised. As the title of this paper indicates, there is no direct or explicit hiding of information inside the cover. In fact, it is a steganogrpahy without steganography. Here follows a case in point.

Let the secret to be transmitted is "Pay".

The binary form of the information is

$$P = 80 \quad \rightarrow 1010000, \quad a = 97 \quad \rightarrow 1100001, \quad y = 121 \rightarrow 1111001$$

Fig.1 is the stego-image. Make a copy and hand it over to the other end. As mentioned the security of this approach lies in this step. Information about the stego-image should not be revealed at any cost. If it is revealed, the entire essence of the security will be easily broken.

In this experiment the first letter is P. Its binary is 101000. Select the appropriate pixel position from the cover. We have a number of techniques to select the pixel coordinate from an image with a single mouse click. For example matlab provides impixelinfo function for the same. Keep the coordinate position as the stego-key. The (x, y) coordinates of letters inside the image are (101, 63), (119, 60) and (149, 29) respectively. Here the stego-key is 101631196014929. Send this stego-key to the other end. We can use either public key or private key cryptography to transfer the key. Public key cryptosystem is the best method to exchange the key. Even if the key is lost or broken, it does not affect the security of the system. As far as the cover image is not known to anybody else other than communicating entities, the confidentiality of the information is preserved at the maximum. It is possible to send the stego-key without doing any encryption or hiding.

The proposed scheme is compared with the currently popular steganographic schemes namely, Secure Bit-Plane Based Steganography for Secret Communication [14], Adaptive Image Steganography Based on Depth-Varying Embedding [15] and A Novel Technique for Image Steganography Based on a High Payload Method and Edge Detection [16].   Table 1 shows the results of comparison. The default optimal parameter settings suggested in the respective works are adapted for experimentation.

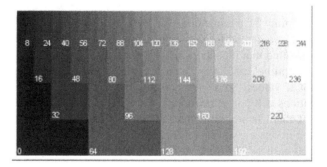

Fig. 1. Cover or Stego Image

Table 1. Comparison between conventional methods and IIS

Factors	Proposed IIS Scheme	Conventional schemes
Distortion	NIL	Caused due to hiding
Payload	Large	Comparatively less
Security	high	Comparatively less
Key exchange	It Can be revealed	Big problem
Authentication	Very easy	Difficult.

5   Visual Analysis

As indicated earlier, the available steganography leads to visual distortions to the original view of the cover image. Fig.2 shows the distortion caused to the image Pepper while using popular Chang et al.'s [17] steganography scheme. It is obvious that using any scheme except the one proposed by Xiu-Bo Chen et al. will lead to image distortion. But Xiu-Bo Chen et al.'s scheme has other limitation in that it can only send one bit of data at a single transaction.

To evaluate the visual quality of stego images by using the human eye, we have enlarged the partial area of original cover image and stego-image of Chang et al.'s scheme, as shown in Fig.2. Fig.2 (a) shows the cropped area in the original Pepper image and Fig.2 (b) shows the cropped area in the stego-image. The distortion between original cover image and stego-image is visually almost imperceptible. The table.2 shows the MSE and PSNR of popular Yang et al.'s Scheme [18], Lin-Tsai Scheme [19] and Chang et al.'s Scheme. The results reveal that even though they are good techniques, they are not free from image distortion.

Steganalysis is an interesting topic that focuses on the detection of the presence of embedded secret messages. RS attack [20], proposed by Fridrich et al., and x2 detection [21], proposed by Westfeld and Pfitzmann, are the two most effective LSB steganalytic techniques. The RS-attack technique can detect both sequentially embedded messages and randomly embedded messages. No current steganogrpahy method can claim that they are free of image distortion while embedding secret inside because PSNR never becomes infinity. Hence these techniques cannot escape from steganalysis. They are not absolutely secure against attacks. Now that the proposed method never embeds a secret directly or explicitly to an image, it can claim that it is free of any such kinds of steganalysis. Enhancing the visual quality and authentication capability of stego-image are new challenges for researchers working on the development of novel secret image sharing scheme. The results and discussion clearly indicate that the proposed scheme provide high confidentiality and authentication.

Table 2. The PSNR and MSE of stego-images with the same payload capacity (The unit of PSNR is db)

Secret image (256×256)	Stego-image (512×512)	Lin-Tsai Scheme		Yang et al's Scheme		Chang et al.'s Scheme	
		MSE	PSNR	MSE	PSNR	MSE	PSNR
	Cameraman	7.74	39.25	4.43	41.66	5.49	40.73
	Baboon	7.85	39.18	4.55	41.55	6.59	39.94
General test	Lena	7.80	39.20	4.50	41.60	5.98	40.37
pattern	Pepper	7.86	39.17	4.54	41.56	7.63	39.30
	Sailboat	7.89	39.16	4.59	41.51	8.45	38.86
	Average	7.83	39.19	4.52	41.58	6.83	39.84

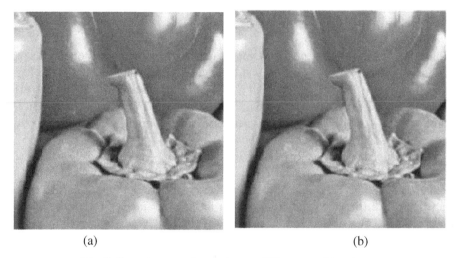

(a)   (b)

Fig. 2. Cover image and stego image of Change et al.'s scheme

6   Conclusion

Here a novel spatial steganography scheme realized with IIS paradigm is introduced. It can be also referred as steganography without steganography. Nothing is done with the cover image. Hence, as usual distortion to the medium is not caused. In fact, it is a different strategy to achieve high security to communication. Its biggest advantage is that it is impossible to break the confidentiality even if we lost the key. Since stego-key alone can do nothing, it is not necessary to encrypt or hide the key. The greatest risk of this method lies in transmitting the cover. Exchanging by hand is absolutely secure. The additional concern of this approach is to get images with ranges varying from 0 to 255. For any stego-system the next property to be considered after its security is its capacity. The capacity of a stego-system can be defined as the number of hidden bits transmitted per letter of the cover image. We show that our stego-system has the maximum possible capacity: Though it can adopt large payload, it is optimal for small secrets. In public key cryptography like RSA and others the length of the key is very big. It is common to have a key length of 1024 bits and more. In decimal notation it can easily exceed 300 digits. Here a key length with 300 digits can easily represent a secret of about 150 letters. The length of the key is directly proportional to the secret to be encoded. The common problem of recovering the original cover image from the stego-image is also resolved here. Further studies can be undertaken to increase the payload. This study has also taken advantages of human psychology in its formulation so that it become more effective and practical. Since the cover image is not transmitted all kinds of conventional threats become insignificant and ineffective thereby making it more viable for its purpose.

References

1. Johnson, N.F., Jajodia, S.: Exploring steganography: seeing the unseen. IEEE Computer 31(2), 26–34 (1998)
2. Shih, Y.F.: Digital Watermarking and Steganography Fundamentals and Techniques. CRC Press, Taylor & Francis Group, Boca Raton (2008)
3. Petitcolas, F.A.P., Stefan, K.: Information Hiding Techniques for Steganography and Digital Watermarking, Canton Street, p. 685. Artech House, Inc., Norwood (2000)
4. Stalling, W. (ed.) Cryptography and Network Security-Principles and Practices, 4th edn., Pearson Prentice Hall, India P. Ltd., India (2000)
5. Salivahanan, S.: Digital Signal Processing. Tata McGraw-Hill, India (2000)
6. Gonzalez, Woods: Digital Image Processing, 3rd edn. Prentice Hall, USA (2008)
7. Hu, C.H., Lou, D.C.: LSB Steganographic method based on reversible histogram transformation function for resisting statistical steganalysis. Information Sciences 188, 346–358 (2012)
8. Chan, C.K., Chen, L.M.: Hiding data in images by simple LSB substitution. Pattern Recognition 37, 469–474 (2004)
9. Bo, X.C., Guo, Z.Q., Jie, X.Z., Xin, X.N., Xian, Y.Y.: Novel quantum steganography with large payload. Optics Communications 283, 4782–4786 (2010)
10. Liao, X., Wena, Q.V., Suna, Y., Zhangb, J.: Multi-party covert communication with steganography and quantum secret sharing. The Journal of Systems and Software 83, 1801–1804 (2010)
11. Cheddad, A., Condell, J., Curran, K., McKevitt, P.: Digital image steganography: Survey and analysis of current methods. Signal Processing 90, 727–752 (2010)
12. Mielikainen, J.: LSB Matching Revisited. IEEE Signal Process. Lett. 13(5), 285–287 (2006)
13. Huang, J., Luo, W.: Edge adaptive image steganography based on LSB matching revisited. IEEE Trans. Inf. Forens. Security 5, 201–214 (2010)
14. Cong-Nguyen, B., Lee, H.Y., Joo, J.J.C., Lee, H.K.: Secure bit-plane based steganography for secret communication. IEICE Transactions on Information and Systems E93, 79 (2010)
15. He, J., Tang, S., Wu, T.: An adaptive image steganography based on depth-varying embedding. In: Image and Signal Processsing, CISP 2008, vol. 5, pp. 660–663 (2008)
16. Halkidis, S.T., Ioannidou, A., Stephanides, G.: A novel technique for image steganography based on high payload method and edge detection. Expert Systems with Application 39, 11517–11524 (2012)
17. Chang, C.C., Hsieh, Y.P., Lin, C.H.: Sharing secrets in stego images with authentication. Pattern Recognition 41, 3130–3137 (2008)
18. Yang, C.N., Chen, T.S., Yu, K.H., Wang, C.C.: Improvements of image sharing with steganography and authentication. Journal of Systems and Software 80, 1070–1076 (2007)
19. Lin, C.C., Tsai, W.H.: Secret image sharing with steganography and authentication. Journal of Systems and Software 73, 405–414 (2004)
20. Fridrich, J., Goljan, M., Du, R.: Reliable detection of LSB steganography in color and grayscale images. In: Proceedings ACM Workshop Multimedia and Security, pp. 27–30 (2001)
21. Westfeld, A., Pfitzmann, A.: Attacks on steganographic systems. In: Proceedings of the 3rd International Workshop on Information Hiding, Dresden, Germany, pp. 61–76 (1999)

Humanized Game Design Based on Augmented Reality

Yanhui Su[1,2], Shuai Li[1], and Yongsong Zhan[1]

[1] Guangxi Key Laboratory of Trusted Software, Guilin University of Electronic Technology,
Guilin, 541004, China
[2] Institute of Computer, Zhejiang University, Hang Zhou, 310058, China
suyanhui@gmail.com

Abstract. Currently, the primary research of AR (Augmented Reality) is focus on how to improve the accuracy of identification and reduce the dependency to markers. Successful cases are scarce on how to play the special advantage of augmented reality, combine it with mobile devices and generate practical value. We start form the actuality of AR technology characteristics, achieved the first blind game system which based on mobile platform Android. This system is not only regard game`s emotion design as starting point and pay closely attention to the vulnerable groups, especially the blind community, but also provide them opportunities that compete with average person. Besides, it offers a diversification choice of game against for blind users by set different difficulty level and game player modes. The user experience indicates that the system combines AR and the innovation of game design together, and help vulnerable group achieve flexible user experience and operation.

Keywords: Augmented reality, Game design, Humanized design, Emotional design.

1   Introduction

Augmented Reality (AR) is an emerging field and a hot spot of research in recent years as an important branch of virtual reality technology. Based on information in real scenarios, AR technology superposed virtual objects or other information that computer generated to the real scenario and fused them. By this way, a bridge would be set up between virtual and reality world, thus, we could realize the "Augment" to reality world and presented a new environment with real sensory effect to users.

AR technology has a widely application and a more obvious superiority than VR technology. Zhongwang Jiang`s article[1] introduced the development history and application field of AR technology in detail. Sui Yi`s paper[2] introduced the AR technology based on a handheld device. From 1990, Tom Caudell and David Mizell, engineer of Boeing Co, proposed the concept of "Augmented reality" firstly when they designed the auxiliary wiring system[3]. HIT laboratory at university of Washington released a develop tool of AR system which is named ARToolKit in 1999. Now, just several decades, AR technology has formed a relatively complete workflow and implementation system, its basic principle is shown as Fig.1.

Y. Tan et al. (Eds.): ICSI 2014, Part II, LNCS 8795, pp. 284–291, 2014.
© Springer International Publishing Switzerland 2014

Games developed rapidly as an emerging industry and various types of games emerged, but games' control mode and scene effect still changed little. In recent years, with the improvement of smartphone's processing capacity and PDA, AR technology applied to game gradually, so that the game scene is more realistic and users' interactive experience is more immersive. In 2000, Bruce H. Thomas developed the first outdoor mobile AR game which is named AR Quake. In 2004, Mohring[4] et al developed the first application which could completely do all the AR tasks by smart phone. In recent years, research of AR mobile game attracted more and more scholars' attention, and kinds of handheld games based on AR technology constantly emerge. Mobile Maze game[5] require user through the maze by push a car ball. In 2005, Video Processing at VIT developed a multi-user table tennis game which is named Symball[6]. HIT lab NZ developed an AR Tennis game[7]. In 2006, Siemens developed a free throw game, AR Soccer[8]. Besides, Kurt D.Squire team developed an AR scientific education game, Mad City Mystery[9]. From last year, a series of mobile AR game appeared constantly, such as Drop Defender, Zombie Room AR and so on. The application of AR technology made our visual information more intuitive and richer, the real-time interactive experience more real, it will be a development tendency for mobile application in the future.

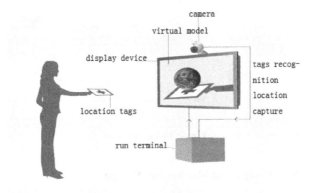

Fig. 1. Augmented reality principle

2   Humanization of Game Design

Except the play ability, game design also needs to pay attention to players' emotional needs. The book "Creating Emotion in Games"[10] told us that emotional design could help game players to realize the additional value of game as an important part of game design, especially for vulnerable groups. Such as: blind person, how to let them to fully experience game's joy as normal people, the key point is to provide them the opportunities that competitive with normal players. So Shaking World, a puzzle action game which we will introduce in this paper, used the third-person design and AR technology, built a fair competitive platform for blind players. Through the humanized touch board design and with the help of the blind touch habits, blind community could fully experience fun of this game.

3   Game System and Mechanism

3.1   Game System

Shaking World is a fantastic third-person puzzle action game developed by Unity 3d. It has two modes: online mode or console mode, where online mode set up two characters in the same scene, player A and player B. Player A is responsible for shaking the game world and player B should try best to control the character which named little fool against the shaking. They can earn coins and buy props they like by play games. Besides, Shaking World also designed custom mode what players can increase the fun by building themselves level with game props. In a word, this game easy to play, numerous levels plus thrilling scenes make it the best choice for recreation.

Fig. 2. Game system

3.2   Game Props Design and Mechanism Design

Game set in a lot of props in order to increase its playability, such as: bomb, alarm clock, lollipop and so on. The relationship of every element in the game is as shown in Fig. 2. Where, mostly props used to decorate the scene or hinder player B win the game. For example, if character falling into the swimming pool, player B needs a quick click on the character to climb back on the ground, but if player B's clicking too slowly, character will drown and the player A will win the game; The fan has a fixed direction to blow, when the character close to the source of wind, he will be blown away; If player A bought bombs and let player B touched it, player B will be killed, and player A will win the game, but if player B bought the alarm clock and touched it, player A will lose control of the game world 3 seconds, and he will win if he lets the character get the lollipop. In a word, what Player A need to do is that he should try his best to shake player B out of the game world and do the winner, as shown in the left of Fig. 3. On the contrary, player B should make a great effort to overcome player A's shaking and get the lollipop successfully, as shown in the right of Fig. 3.

Fig. 3. Game mechanism design

3.3   Technology Implementation

Vuforia SDK is based on AR application and display mobile device as a "magic camera" or show the scene as a world that coexistence of virtual and reality. We used this SDK in Shaking World, adopted the AR technology which based on mobile terminal and transplanted the style of AR to cellphone platform. The virtual scene could be stacked to the paper when used the camera of cellphone to shoot it. Besides, in order to assure the consistency between operation and braille, we used the style of virtual button, set a touch area on the paper for blind person. Game runs on Android platform, and its technical principle hierarchy Chart is showed as Fig. 4. Players take photos by camera, find out the target image and determine its coordinate; besides, identify the target images, overlay virtual image on real image and use virtual button interactive with real world. Its bottom technology is as follows: bottom is on Android system, around the operating system to build two big modules, including: camera and rendering module; besides, on the basis of the camera module, we add target tracking module, including: virtual buttons, multi-target tracking, target image recognition, space target recognition and so on, and in the form of SDK provides the chance that developers call low-level interface, can achieve all kinds of special effects effectively.

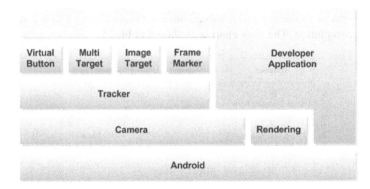

Fig. 4. Bottom framework of AR engine

Virtual Button System. Virtual buttons are developer-defined rectangular regions on image targets that when touched or occluded in the camera view, can trigger an event.

Virtual buttons can be used to implement events such as a button press or to detect if specific areas of the image target are covered by an object. Virtual buttons are evaluated only if the button area is in the camera view and the camera is steady. Evaluation of virtual buttons is disabled during fast camera movements. Define virtual buttons in the Database Configuration XML as children of an image target. To add virtual buttons, insert a section similar to the following:

```
<ImageTarget size="247 173" name="wood">
<VirtualButton name="red" rectangle="-108.68 -53.52 -
75.75 -65.87" enabled="true" />
<VirtualButton name="blue" rectangle="-45.28 -53.52 -
12.35 -65.87" enabled="true" />
</ImageTarget>
```

The Virtual Button state can be requested from active targets in the scene by iterating through the button child objects:

```
// Iterate through this targets virtual buttons:
for (int i = 0; i< target->getNumVirtualButtons(); ++i)
{ constVirtual Button* button = target-
  >getVirtualButton(i);
  if (button->isPressed())
    { textureIndex = i+1;
      break; }
}
```

User-defined the identified image. In this section we show how to use the user-defined target feature to instantiate objects of classes from TrackableSource which can be used to create new Trackables at runtime.

Two new classes, ImageTargetBuilder and ImageTargetBuilderState are introduced: where, class ImageTargetBuilder exposes an API for controlling the building progress, retrieving a TrackableSource for instantiating a new trackable upon successful completion. The flow chart is as shown in Fig. 5.

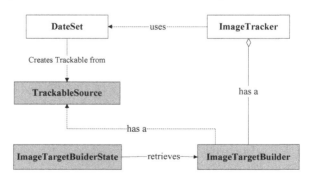

Fig. 5. The process of instantiate object

Expression form of identified images. The expression forms of identified images are as follows:

```
?xml version="1.0" encoding="UTF-8"?>
<QCARConfigxmlns:xsi="http://www.w3.org/2001/XMLSchema-
instance"
xsi:noNamespaceSchemaLocation="qcar_config.xsd">
<Tracking>
<ImageTarget size="247 173" name="stones" />
<ImageTarget size="247 173" name="chips" />
</Tracking>
</QCARConfig>
```

4   Game Results

Relative to other games, Shaking World has few bright spots: (1)Pay attention to user`s emotional experiences, provide humanized attention and care design for vulnerable groups. (2) Using augmented reality mode to increase fun of the game and expand the original dimension of game experience which based on screen to a three-dimensional game space. (3) Using the virtual button mode of AR technology, provided game operation based on braille contact for blind user.(4) Combined the play ways of virtual and reality, and provide an interaction with virtual game by manipulating physical paper. (5) With the aid of AR technology and use the levels mode of blind person book, blind man could touch the dots on graph to operate this game. Besides, different maps represent different level, so with different maps, blind person can take part in every level of game as normal people.

The game Shaking World which based on AR technology could achieve different effects. The left of Fig. 3 stands that player B was shaking out the game world, the right of Fig. 3 stands that the character eats the lollipop. Fig.6 is the whole scene of this game. Fig.7 is the game mode that designed for blind person. They could play the game by touch dots on the graph paper when run the game. Every elements of this game drawn exquisite, character designed personality, and visual effect is preferably.

Fig. 6. Humanized design of game scene

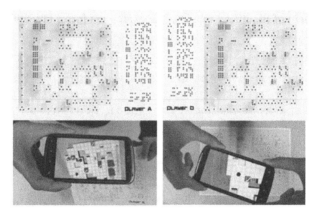

Fig. 7. Blind person mode of the AR game

5   Conclusion

Shaking world game used the AR technology, fully considered humanized design, the research result indicated that it has a certain appeal and also is a better choice for entertainment. Beyond that, it also has much commercial value: (1) Convenient development peripheral products: the "Shaking World" building blocks, gift boxes, store content boxes, alarm clocks and so on. (2) Can implant the SDK advertising on the top of houses in the game scenes and don't destroy the beautiful game. (3) Can implant some products in the art style of the game for joint operations of products. (4) Can promote to many operators, such as: App Store, CUCC, CM, CHA. (5)To cooperate with shopping website: Play the game to earn the points or gold of the shopping website. The more important is that the game also has much public value: (1) Pay attention to vulnerable groups, to provide fair and competitive opportunities for the blind, let them feel the game happiness. (2)With the help of a third party social media channels, for example the game feeds, cause people concern for the vulnerable groups and give them more help.

In the future, AR technology is an emerging and active field of research. It could bring people new visual experience, and has a broad market prospect and application scope. For the game industry, apply the AR technology to game could greatly rich the content of game frames and increase game's entertainment.

Acknowledgments. This research work is supported by the grant of Guangxi science and technology development project (No: 1355011-5), the grant of Guangxi Key Laboratory of Trusted Software of Guilin University of Electronic Technology (No: kx201309), the grant of Guangxi Education Department (No: SK13LX139), the grant of Guangxi Undergraduate Training Programs for Innovation and Entrepreneurship (No: 20121059519).

References

1. Jiang, Z.: The development of Educational Augmented Reality Game. East China Normal University, Shang Hai (April 2012)
2. Sui, Y.: Research and application of augmented reality technology based on handheld device. Qing Dao University, Qing Dao (March 6, 2009)
3. Caudell, T.P., Mizell, D.W.: Augmented reality: An application of heads-up display technology to manual manufacturing processes. In: 1992. Proceedings of the Twenty-Fifth Hawaii International Conference on System Sciences, vol. 2, pp. 659–669. IEEE (1992)
4. Mohring, M., Lessig, C., Bimber, O.: Video see-through ar on consumer cell-phones. In: Proceedings of the 3rd IEEE/ACM International Symposium on Mixed and Augmented Reality, pp. 252–253. IEEE Computer Society (2004)
5. Bucolo, S., Billinghurst, M., Sickinger, D.: Mobile maze: a comparison of camera based mobile game human interfaces. In: Proceedings of the 7th International Conference on Human Computer Interaction with Mobile Devices & Services, pp. 329–330 (2005)
6. Hakkarainen, M., Woodward, C.: SymBall: camera driven table tennis for mobile phones. In: ACM International Conference Proceeding Series, vol. 265, pp. 391–392 (2005)
7. Henrysson, A., Billinghurst, M., Ollila, M.: Face to face collaborative AR on mobile phones. In: 2005. Proceedings. Fourth IEEE and ACM International Symposium on Mixed and Augmented Reality, pp. 80–89. IEEE (2005)
8. Geiger, C., Paelke, V., Reimann, C.: Mobile entertainment computing. In: Göbel, S., Spierling, U., Hoffmann, A., Iurgel, I., Schneider, O., Dechau, J., Feix, A. (eds.) TIDSE 2004. LNCS, vol. 3105, pp. 142–147. Springer, Heidelberg (2004)
9. Squire, K.D., Jan, M.: Mad City Mystery: Developing scientific argumentation skills with a place-based augmented reality game on handheld computers. Journal of Science Education and Technology 16(1), 5–29 (2007)
10. Freeman, D.: Creating Emotion in Games. Beijing Hope Electronic Press, Beijing (2005)

A Hybrid PSO-DE Algorithm
for Smart Home Energy Management

Yantai Huang[1,*], Lei Wang[1,2], and Qidi Wu[1]

[1] Department of Electronics and Information Engineering, Tongji University, Shanghai, China
huangyantai@sina.com
[2] Shanghai Key Laboratory of Financial Information Technology, Shanghai, China
wanglei@tongji.edu.cn

Abstract. Home energy management system is an important part in smart home, smart home is assumed to be equipped with smart meter which smart control of generators, storages, and demand response programs. In this paper, we study a versatile convex optimization framework for the automatic energy management of various household loads in a smart home. The scheduling algorithm determines how energy resources available to the end-users considering a number of constraints. Hence, a hybrid PSO-DE algorithm approach is proposed. We devise a model accounting for a typical household user, and present computational results showing that it can be efficiently solved in real-life instances.

Keywords: smart home, home energy management, hybrid PSO-DE algorithm.

1   Introduction

With the increase of the electric demand, energy crisis worldwide is one of the most serious challenges in the 21st century. The residential sector is experiencing the strongest increase on its electric demand. Therefore, the study of the electric demand in the residential sector is an important task for the electric grid controllers as already shown in [1, 2].

Home energy management system provides an opportunity for improving the efficiency of energy consumption in residential sector [3, 4]. A typical Smart home for residential sector integrates the operation of electrical and thermal energy supply and demand.

Developing efficient demand response models of electrical appliances is a key problem in an energy management system in a smart home, which have received considerable attention recently [5-12]. In general, the main objective of Home energy management system is to minimize the electricity bill or maximize their users' satisfaction by allocating available resources and managing the load of appliances.

The home energy management in a smart home can be formulated as a complex mathematical optimization problem. Dynamic programming may be used if the

---

*Corresponding author.

Y. Tan et al. (Eds.): ICSI 2014, Part II, LNCS 8795, pp. 292–300, 2014.

optimal schedule is required, however, the immense computational effort involved might require a powerful computer and long computation times. On the other hand, the complex optimization problem can be solved by the commercial solver such as CPLEX and MOSEK. However, these commercial solvers are specialized optimization package whose computational requirements are not suitable for a smart meter. Thus, heuristic techniques have been applied to the original problem for obtaining a global optimal solution to the problem. Thus, this paper proposes a particle swarm optimization algorithm to solve the original problem for obtaining a global optimal solution to the problem. In order to improve the performance of the algorithm, we integrate particle swarm optimization with differential evolution (PSO-DE) to solve constrained numerical and engineering optimization problems.

The rest of this paper is organized as follows. In Section 3, the mathematical problem formulation is presented. In Section 4, the scheduling problem is resolved by a hybrid PSO-DE algorithm. Section 5 includes necessary figures, results, and discussions of the typical study case. The conclusion is drawn in Section 6. References are appended last.

2   Mathematical Formulation

It is assumed that most electric appliances are networked together and are controlled by a home energy management system. The smart home consists of micro-CHP unit, storage devices, local loads, thermal loads and must run electrical loads. The output power energy of micro-CHP unit is supplied to the kinds of appliance loads, the surplus power energy is used to charge storage batteries or sail to the main grid. In contrast, the deficit power energy is provided by batteries and main grid. In this paper, the whole problem of the smart home energy management system is defined as an optimization problem. In so doing, we have used simple models of the micro grid's components which will be described in the following subsections.

2.1   Objective Function

The objective function is to minimize the operation cost: the cost of the power purchased from the main grid during a day, and the cost of gas consumed by the micro-CHP unit (G_{CHP}). In this study, the micro-CHP's startup time and consequent cost is neglected.

$$\cos t = \sum_{h=1}^{24} [TOU(h) p_{grid}(h) + G_p G_{CHP}(h)] \tag{1}$$

where TOU(h) is the main grid's time-of-use price tariff; Pgrid(h) is the power transferred between main grid and smart home; Gp is the natural gas price; And Gchp(h) is the micro-CHP's consumed gas.

2.2   Constraints

Electrical Demand Supply Balance. The loads contain thermal appliances (hot water and air conditioner) and electrical appliances. We assume electrical appliances are must run loads that can not be scheduled. While thermal appliances consumption can be scheduled to avoid peak hours. For each household, let $P_{ms}(h)$ denote the total energy consumption of electrical(must run) appliances. while $p_{heat}(h)$ and $p_{water}(h)$ represent the air conditioner and hot water energy consumption one-hour respectively. $P_{batt}(h)$ is the battery's output power and $P_{chp}(h)$ denote the micro-CHP's electrical output power. So, electrical demand-supply balance constraint would be as follow:

$$P_{grid}(h) - p_{batt}(h) + p_{CHP}(h) = p_{ms}(h) + p_{heat}(h) + p_{water}(h) \tag{2}$$

Constraint of Battery. The battery can charge or discharge at different condition, so their can avoid peak time energy and can more efficient use of the overall energy.

$$0 \le \frac{p_{batt}^{ch}(h)}{\eta_{ch}} \le p_{ch}^{max} \tag{3}$$

$$0 \le p_{batt}^{dch}(h) \cdot \eta_{dch} \le p_{dch}^{max} \tag{4}$$

$$p_{batt}(h) = \frac{p_{batt}^{ch}(h)}{\eta_{ch}} - p_{batt}^{dch}(h) \cdot \eta_{dch} \tag{5}$$

$$SOC(h+1) = SOC(h) + \frac{p_{batt}^{ch}(h) - p_{batt}^{dch}(h)}{E_{batt}} \tag{6}$$

$$SOC^{min} \le SOC(h) \le SOC^{max} \tag{7}$$

where $SOC(h)$ is the state of charge in the battery; SOC^{min} and SOC^{max} are the minimum and the maximum SOC, respectively; $P_{batt}^{ch}(h)$ is the battery's charging power; $P_{batt}^{dch}(h)$ is the batter's discharging power; η_{ch} and η_{dch} are the battery's charging and discharging efficiency, respectively.

Desired Hot Water and Building Temperature. The building is modeled within the context of desired hot water temperature and building temperature.

In modeling the hot water storage, the energy equivalent of the hot water storage at each time step is taken in consideration and the dynamic of the water flow is not considered. So, the water storage temperature at each hour is calculated as the following equation [13], [8]:

$$T_{st}(h+1) = \frac{V_{cold}(h)*(T_{st}^{cold} - T_{st}(h)) + V_{total}*T_{st}(h)}{V_{total}} + \frac{H_{air}(h)}{V_{total}*C_{water}} \tag{8}$$

$$T_{st}^{min} \le T_{st}(h) \le T_{st}^{max} \tag{9}$$

where T_{st}^{min} and T_{st}^{max} are the minimum and the maximum desired hot water temperature respectively.

According to the thermal modeling for a building presented in [13], the building temperature at each hour is obtained by:

$$T_{in}(h+1) = T_{in}(h)*e^{\frac{-\Delta}{\tau}} + (R*H_{air}(h) + T_{out}(h))*(1 - e^{\frac{-\Delta}{\tau}}) \tag{10}$$

$$T_{in}^{min} \le T_{in}(h) \le T_{in}^{max} \tag{11}$$

where $\Delta = 1h$ and $\tau = RC$. The values used are R=18°C/kW, C=0.525kWh/°C, and the initial room temperature=20°C.

Constraint of Micro-CHP Operation. Electrical and thermal output power limits for micro-CHP:

$$V_{CHP}(h)*P_{CHP}^{min} \le P_{CHP}(h) \le V_{CHP}(h)*P_{CHP}^{max} \tag{12}$$

$$V_{CHP}(h)*H_{CHP}^{min} \le H_{CHP}(h) \le V_{CHP}(h)*H_{CHP}^{max} \tag{13}$$

Micro-CHP electrical and thermal efficiency:

$$P_{CHP}(h) = H_{CHP}(h) * \frac{\eta_e}{\eta_{th}} \tag{14}$$

Micro-CHP output power ramp rate:

$$-P_{rr} \leq P_{CHP}(h) - P_{CHP}(h-1) \leq P_{rr} \tag{15}$$

$$-\frac{\eta_{th}}{\eta_e} * P_{rr} \leq H_{CHP}(h) - H_{CHP}(h-1) \leq \frac{\eta_{th}}{\eta_e} * P_{rr} \tag{16}$$

where $H_{chp}(h)$ is the micro-CHP's thermal output power. $V_{chp}(h)$ is the micro-CHP's status. η_e and η_{th} are the micro-CHP's electrical and thermal efficiency, respectively.

3   Scheduling Using Hybrid PSO-DE Algorithm

The objective of the scheduler is to maximize the net benefits. Based on (1), the corresponding optimization problem is to find the real-value variable for battery, micro-CHP and thermal appliances. Therefore, it can be considered as a constraint real-value optimization problem. This paper mixes particle swarm optimization and differential evolution (PSO-DE) to solve the constraint optimization problem.

3.1   Overview of PSO

Particle swarm optimization is a stochastic global optimization method inspired by the choreography of a bird flock. PSO relies on the exchange of information between individuals, called particles. In PSO, each particle adjusts its trajectory stochastically towards the positions of its own previous best performance and the best previous performance of its neighbors or the whole swarm. At the each iteration, the velocity and position updating rules are given by:

$$v_{i,j}^{t+1} = wv_{i,j}^t + c_1 r_1 (pbest_{i,j}^t - x_{i,j}^t) + c_2 r_2 (gbest_j^t - x_{i,j}^t) \tag{17}$$

$$x_{i,j}^{t+1} = x_{i,j}^t + v_{i,j}^t \tag{18}$$

where w is an inertia weight factor, c_1 is a cognition weight factor, c_2 is a social weight factor, r_1 and r_2 are two random numbers uniformly distributed in the range of [0,1]. In this version, the variable $V_{i,j}$ is limited to the range $\pm V_{max}$ [17-19] analyzed and introduced the velocity adjustment as

$$v_{i,j}^{t+1} = R_1 (pbest_{i,j}^t - x_{i,j}^t) + R_2 (gbest_j^t - x_{i,j}^t) \tag{19}$$

where R1 and R2 are generated using abs(N(0,1)) According to the statistical knowledge, the mean of abs(N(0,1))is 0.798 and the variance is 0.36.

3.2   Overview of DE

Different evolution (DE) [15] has become a popular algorithm in global optimization. DE starts the search with an initial population containing NP individuals, which are randomly sampled from the search space. Then, one individual called the target vector in the population is used to generate a mutant vector by the mutation operation. So far, several mutation strategies have been proposed [14].

Subsequently, DE applies a crossover operator to generate the offspring individual, the crossover is employed and executed as follows:

$$u_{i,j} = \begin{cases} y_{i,j}^t \; if & rand \leq CR \quad OR \quad j = j_{rand} \\ x_{i,j}^t & otherwise \end{cases} \tag{20}$$

where $i \in \{1, 2, ..., NP\}$, $j \in \{1, 2, ..., n\}$, rand is a uniformly distributed rand number between $[0,1]$, j_{rand} is a randomly selected integer from $[1,n]$, CR is the crossover control parameter, $u_{i,j}^t$ is the jth component of the trial vector .

Finally, the target vector x_i is compared with the trial vector u_i in terms of the objective function value and the better one survives into the nest generation:

$$x_i^{t+1} = \begin{cases} u_i^t \, , if & f(u_i^t \leq f(x_i^t)) \\ x_i^t & , otherwise \end{cases} \tag{21}$$

3.3   PSO-DE for Real-Value Constraint Optimization

The PSO-DE's main procedure can be summarized in Fig.1 [19].

1、 Initialize:Initialize pop that contains NP particles with random position within [L,U]
    upper bound of variables U={u(1),···.u(n)}
    lower bound of variables L={l(1),···.l(n)}
2、 Evaluate:Evaluate objective function f and the degree of constrained violation G for all particles;
    pbest=previous pbest positon;
    gbest=the optimum of pbest according to feasibility-based rule
repeat for each generation
3、 updata 50% particles :sort pop in descending order according to G ;
    updata pop first half particles velocity and position

$$v_{i,j}^{t+1} = R_1(pbest_{i,j}^t - x_{i,j}^t) + R_2(gbest_j^t - x_{i,j}^t)$$
$$x_{i,j}^{t+1} = x_{i,j}^t + v_{i,j}^t$$
$$x_{i,j}^{t+1} = \begin{cases} 0.5*(l(j) + x_{i,j}^t), & x_{i,j}^{t+1} < l(j), \\ 0.5*(u(j) + x_{i,j}^t), & x_{i,j}^{t+1} > u(j) \\ x_{i,j}^{t+1}, & otherwise \end{cases}$$

replace pbest, if the new particle win the pbest according to the feasibility based rule;
gbest=optimum of pbest according to feasibility-based rule;
4、 updata pbest:generate three offspring by follow three mutation strategies for each pbest respectivily;modify variables if violates boundary,
Replace pbest if the new pbest win the previous pbest.

$$rand \, / 1 : y_{i,j}^t = x_{r(1),j}^t + F(x_{r(2),j}^t - x_{r(3),j}^t)$$
$$current-to-best/1 : y_{i,j}^t = x_{i,j}^t + F(x_{best,j}^t - x_{i,j}^t) + F(x_{r(1),j}^t - x_{r(2),j}^t)$$
$$rand/2 : y_{i,j}^t = x_{r(1),j}^t + F(x_{r(2),j}^t - x_{r(3),j}^t) + F(x_{r(4),j}^t - x_{r(5),j}^t)$$
$$z_{i,j}^t = \begin{cases} 2*l(j) - z_{i,j}^t, & if(z_{i,j}^t < l(j)) \\ 2*u(j) - z_{i,j}^t, & if(z_{i,j}^t > u(j)) \\ z_{i,j}^t, & otherwise \end{cases}$$
$$pbest_i^{t+1} = \begin{cases} z_i^t, & if(f(z_i^t < f(pbest_i^t) \cap G(z_i^t) \leq G(pbest_i^t)) \\ pbest_i^t, & otherwise \end{cases}$$

r[k](k={1···.5}) is a uniformly distributed random number in the range[1,NP]
gbest=optimum of pbest according to feasibility-based rule;
end repeat
5、 return the gbest ;

Fig. 1. PSO-DE main procedure

As described above, the paper is to solve a constrained optimization problem, the strategy for handling constraints is usually the use of penalty function methods. However, the main problem is that they require a careful fine tuning of the penalty factors .In order to overcome the drawback of choice penalty factors, this paper applies PSO-DE method to handle constraints, which does not require setting any additional parameters in comparison to the original PSO. The feasibility-based rules are applied to handle constraints of the problem.

In order to handle the constraints, we minimize the original objective function f (x) as well as the degree of constraint violation G(x). At each generation, pop is sorted according to the degree of constraint violation in a descending order. Only the first half of pop are evolved by using Krohling and dos Santos Coelho's PSO [18].

In order to compensate the convergence speed and supply more valuable information to adjust the particles' trajectories, the pBest is updated by Different evolution. DE-based search process motivates the particles to search for new regions including some lesser explored regions and enhance the particles capability to explore the vast search space [19].

4   Simulation Result

The case study is a typical residential building. A micro-CHP with 3kW capacity is considered for the building. The water storage capacity is 80L. The building's hot water demand is shown in Fig.2. The building loads include must run electrical appliances and thermal loads. Fig.3 shows the total electrical demand by the must run appliances and the price of electricity supplied to terminal loads [8, 20]. Which the must run include lights, cook, fridge, computers, washing machine, dryer, dish washer and pool pump.

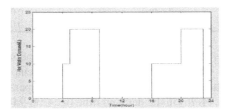

Fig. 2. Hot water demand in building

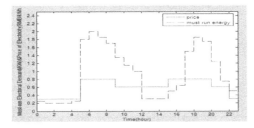

Fig. 3. Total electrical demand and the price of electricity

In this case the price of natural gas is 2.05RMB/m3, and the price of electricity fed into grid is 0.457 RMB/kWh [20]. Table I shows assumed parameters in solving(1).

Table 1. Assume values for parameters

Parameter	Value	Unit
$T_{in}^{min}, T_{in}^{max}\ T_{st}^{min}, T_{st}^{max}\ T_{st}^{cold}$	22,24,72,74,10	$^{\circ}C$
C_{water}	$11.61*10^{-4}$	$\frac{kWh}{L^{\circ}C}$
$P_{CHP}^{min}, P_{CHP}^{max}\ H_{CHP}^{min}, H_{CHP}^{max}\ P_{ch}^{max}, P_{dch}^{max}$	0.3,3,0.5,5,3,3	kW
P_{rr}	2.5	$\frac{kW}{h}$
$\eta_e, \eta_{th}\ \eta_{ch}, \eta_{dch}$	30,50,0.9,0.9	%
SOC_{min}, SOC_{max}	0.3,1	p.u.
G_{ref}	$92.59*10^{-3}$	$\frac{m^3}{h}$

The PSO-DE algorithm is programmed using the C++ programming language using a PC with an Intel dual core processor 2.6GHz on a Windows XP operation system. A visual C++ compiler was used in this work.

The proposed algorithm takes 5 trials to get the final best cost. And in each trial, the population size and maximum iteration take 300 and 5000 respectively. The best result with minimum cost is shown in Table II.

Table 2. Five trials of algorithm

Generation	Fitness	Vilolation
93	19.21	0.0
80	15.89	0.0
0	17.90	0.0
132	17.48	0.0
72	16.81	0.0

The simulation results show the algorithm can achieve optimum result under kinds of constraints. As been shown in Fig. 4, the building's temperature have been set within desired temperature (ie.,22-24), the hot water temperature also been set to comfort temperature (ie.,74-76) in the demand time (ie.,4:00-9:00,16:00-22).

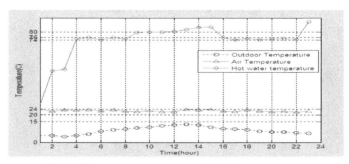

Fig. 4. Hot water temperature and building temperature

The micro-CHP output power, battery output power and its SOC is depicted in Fig. 5. The positive and negative values represent battery charging and discharging respectively.

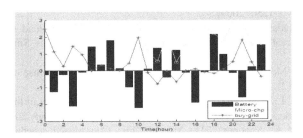

Fig. 5. Micro-CHP& Buy_grid& Battery

As expected, the battery coordinates with the micro-CHP output power to achieve economic operation of micro grid. The battery charges and discharges during low price and peak price hours, respectively. While the micro-CHP operation at its maximum capacity of peak load price hours.

Also ,as shown in the Fig.4 it achieve optimum cost operation by buying minimum electrical power during peak hours and selling its extra electrical power to the main grid at off-peak price hours.

5   Conclusion

This paper pioneers a problem of a residential smart home user equipped with kinds of appliances. We developed an optimal control algorithm for the smart home. The objective of the optimal control algorithm is to reduce the total electricity cost over a billing period (i.e., a day).This proposed PSO-DE algorithm can be used in home energy management systems and help in realization of optimum cost operation. As future study, it is suggested to look into issues such as optimization under price uncertain environment.

Acknowledgments. This work was supported in part by the Natural Science Foundation of China (61075064, 61034004, 61005090), Program for New Century Excellent Talents in University of Ministry of Education of China. Ministry of Education (NCET-10-0633), the Fundamental Research Funds for the Central Universities, and the Research Fund of State Key Lab. of Management and Control for Complex systems.

References

1. Mansouri, I., Newborough, M., Probert, D.: Energy consumption in UK households: impact of domestic electrical appliances. Appl. Energy 54(3), 211–285 (1996)
2. Hamidi, V., Li, F., Robinson, F.: Demand response in the UK's domestic sector. Electr. Power Syst. Res. 79(12), 1722–1726 (2009)

3. Katz, J.S.: Educating the smart grid. Presented at the IEEE En- ergy 2030 Conf., Atlanta, GA (November 17-18, 2008)
4. Colson, C.M., Nehrir, M.H.: A review of challenges to real-time power management of microgrids. Presented at the 2009 Power Energy Soc. Gen. Meet., Calgary, AB, Canada, PESGM2009-001250 (2009)
5. Mohsenian-Radand, A., Leon-Garcia, A.: Optimal residential load control with price prediction in real-time electricity pricing environments. IEEE Trans. Smart Grid 1(2), 120–133 (2010)
6. Conejo, A.J., Morales, J.M., Baringo, L.: Real-time demand response model. IEEE Trans. Smart Grid 1(2), 236–242 (2010)
7. Li, N., Chen, L., Low, S.H.: Optimal demand response based on utility maximization in power networks. In: Proc. IEEE Power Energy Soc. Gen. Meet. (July 2011)
8. Pedrasa, M.A.A., Spooner, T.D., MacGill, I.F.: Coordinated scheduling of residential distributed energy resources to optimize smart home energy services. IEEE Trans. Smart Grid 1(2), 134–143 (2010)
9. Mohsenian-Rad, A., Wong, V.W.S., Jatskevich, J., Schober, R., Leon-Garcia, A.: Autonomous demand-side management based on game- theoretic energy consumption scheduling for the future smart grid. IEEE Trans. Smart Grid 1(3), 320–331 (2010)
10. Sane, H., Guay, M.: Minimax dynamic optimization over a finite- time horizon for building demand control. In: Proc. Amer. Control Conf., pp. 1469–1474 (June 2008)
11. Samadi, P., Mohsenian-Rad, A., Schober, R., Wong, V.W.S., Jatskevich, J.: Optimal real-time pricing algorithm based on utility maximization for smart grid. In: Proc. IEEE Int. Conf. Smart Grid Commun., pp. 415–420 (2010)
12. Wi, Y.-M., Lee, J.-U., Joo, S.-K.: Electric Vehicle Charging Method for Smart Homes/Buildings with a Photovoltaic System. IEEE Transactions on Consumer Electronics 59(2) (May 2013)
13. Tasdighi, M., Ghasemi, H., Rahimi-Kian, A.: Residential Microgrid Scheduling Based on Smart Meters Data and Temperature Dependent Thermal Load Modeling. IEEE Trans. Smart Grid (2013)
14. Price, K., Storn, R., Lampinen, J.: Differential Evolution: A Practical Approach to Global Optimization. Springer-Verlag (2005) ISBN:3-540-20950-6
15. Neri, F., Tirronen, V.: Recent advances in differential evolution: a survey and experimental analysis. Artificial Intelligence Review 33(1-2), 61–106 (2010)
16. Mohsenian-Rad, A.-H., Wong, V.W.S., Jatskevich, J., Schober, R.: Optimal and autonomous incentive-based energy consumption scheduling algorithm for smart grid. In: Proc. IEEE Conf. Innov. Smart Grid Technol., Gaithersburg, MD, USA (2010)
17. Clerc, M., Kennedy, J.: The particle swarm-explosion, stability, and convergence in a multidimensional complex space. IEEE Transactions on Evolutionary Computation 6(1), 58–73 (2002)
18. Krohling, R.A., dos Santos Coelho, L.: Coevolutionary particle swarm optimization using Gaussian distribution for solving constrained optimization problems. IEEE Transactions on Systems, Man, Cybernetics Part B: Cybernetics 36(6), 1407–1416 (2006)
19. Liu, H., Cai, Z., Wang, Y.: Hybridizing particle swarm optimization with differential evolution for constrained numerical and engineering optimization. Applied Soft Computing 10, 629–640 (2010)
20. Guan, X., Xu, Z., Jia, Q.-S.: Energy-Efficient Buildings Facilitated by Microgrid. IEEE Trans. Smart Grid 1(3) (December 2010)

A Multiobjective Large Neighborhood Search for a Vehicle Routing Problem

Liangjun Ke and Laipeng Zhai

The State Key Laboratory for Manufacturing Systems Engineering,
Xian Jiaotong University, Xi'an, 710049, China
keljxjtu@xjtu.edu.cn

Abstract. In this paper, a multiobjective adaptive large neighborhood search is proposed for a vehicle routing problem (VRP) of which the objectives are the total travel time and the cumulative time, i.e., the total arrival time at all customers. It hybrids destroy-repair operators with local search for generating new solutions. An adaptive probabilistic rule based on Pareto dominance is proposed to select a combination of destroy-repair operator. The effectiveness of the proposed algorithm is supported by the experimental study.

Keywords: multiobjective optimization, vehicle routing problem, adaptive large neighborhood search.

1 Introduction

Vehicle routing problem (VRP) is one of the most important combinatorial optimization problems. In this problem, a set of customers are dispersed in a graph. Each customer is associated with a demand. vehicles are scheduled to serve these customers so as to achieve one or more optimal objectives whilst the route of each vehicle must satisfy specific requirements. The most common studied objectives are the total travel time, the number of vehicles, makespan, balance, and others [1].

Recently, cumulative VRP becomes a hot topic [2,3]. It aims at minimizing the cumulative time, that is, the total arrival time of all customers. This problem was extended from the delivery man problem [4,5], and can be used to model many problems such as the routing schedule during the disaster aids [2].

Although many researchers considered VRP with only one single objective, VRP is multi-objective in nature [1]. Multiobjective VRP (MVRP) has attracted great research interests [1]. A lot of approaches have been used to deal with various MVRPs. A popular approach is the scalar approach, which transforms a multiobjective problem into a single objective problem by weighted sum method or other methods, then solves it by a single objective heuristic or exact algorithm [6,7]. Based on the concept of Pareto dominance, Pareto methods are also widely used [8–10]. Other approaches, e.g., genetic algorithm [11], lexicographic method [12] and ant colony optimization [13], etc, were also adopted. Since VRP is

Y. Tan et al. (Eds.): ICSI 2014, Part II, LNCS 8795, pp. 301–308, 2014.

a NP-hard problem, metaheuristics are widely adopted for finding satisfactory solutions within acceptable computational time.

In this paper, we consider an MVRP. In the MVRP, there are n customers. Customer i, staying at node i has a demand d_i. A fleet of vehicles start from node 0 and each node $n+1$ to serve these customers. The travel time between every two nodes $i, j(i, j \in \{0, 1, \cdots, n, n+1\})$ is w_{ij}. A feasible solution consists of a set of paths such that each customer is served by one path, and the total demand of a path is not more than the vehicle capacity. The objectives are the total travel time and the cumulative time at all customers. Intuitively, a solution with smaller total travel time may have smaller cumulative time. Nevertheless, these two objectives may be not always compatible [2]. Therefore this problem can not be reduced to a single objective problem. By optimizing this problem, one can obtain a solution for minimizing the total travel time or the cumulative time in a single run. Moreover, one can obtain a set of tradeoff solutions for decision making.

To deal with this problem, we propose a new algorithm extended from adaptive large neighborhood search (ALNS). ALNS has been proven to be a powerful metaheuristic for many variants of single objective VRP. It was firstly proposed by Ropke and Pisinger [14]. It is closely related to the large neighborhood search developed by Shaw [15]. ALNS generates new solutions by using destroy and repair operators. Local search is optionally adopted for exploiting the neighborhood of those solutions. An adaptive probabilistic rule is used to schedule destroy and repair operators according to their weights, which are renewed based on the previous search experience.

The remainder of this paper is structured as follows. Section 2 describes the details of the proposed multiobjective ALNS algorithm. The experimental results are presented in section 3. Finally, the main results are concluded in section 4.

2   Multiobjective Adaptive Large Neighborhood Search

Our algorithm, deonted as MALNS, evolves a population of N solutions over time where N is population size, and employs an external archive EA to store the nondominated solutions. It works as follows. At first, a population of solutions are initialized. After that, starting from each individual in the population, a new solution is generated by a combination of destroy-repair operator. A combination is probabilistically chosen from a set of candidate combinations, depending on their weights. Subsequently, local search is adopted to improve each new solution. By using the obtained solutions, our algorithm updates the population, the external archive EA, and the weights of the combinations of destroy-repair operators. It terminates when a stopping condition is satisfied.

2.1   The Single Objective Function

The single objective function is used to evaluate the quality of a move in destroy-repair operators or local search. Two things are considered to define the single objective function: 1) it is a weighted aggregation of the objectives in MVRP; 2) to obtain a wide and evenly nondominated front, it is desirable to guide the destroy-repair operator or local search to explore different directions. Therefore, the weight vector (or search direction) should be carefully assigned.

Formally, the single objective function is given as follows.

$$\mathcal{F}(x, \lambda) = \lambda^1 \frac{f_1(x)}{f_1^*} + \lambda^2 \frac{f_2(x)}{f_2^*} \tag{1}$$

where $\lambda = (\lambda^1, \lambda^2)$ is a weight vector, $\lambda^1 + \lambda^2 = 1, \lambda^1, \lambda^2 \geq 0$. Since the difference between f_1 and f_2 is very big, scale transformation is carried out in (1). f_1^* and f_2^* are the minimal values of these two objectives respectively. In practice, they are approximated by the minimal values obtained so far. Each weight vector is randomly selected.

2.2   Destroy and Repair Operators

Many different destroy and repair operators have been proposed in the literature. An interesting survey is available in [14]. MALNS heavily depends on simpler destroy-repair operators for exploring new competitive search areas, and uses local search for exploitation. The destroy operators we used are random removal, worst removal, relatedness removal, and cluster removal [14,16]. These operators remove some customers from the routes. The set of unvisited customers, called request bank [14], is denoted as U. During the running, removed customers are saved in U. The maximal size of U is denoted as u. Parameter u significantly affects the behavior of MALNS. With a larger value, the operators are able to explore larger search area. The repair operators consist of the basic greedy insertion heuristic and regret insertion heuristic.

2.3   Selecting a Combination of Destroy-Repair Operator

There are eight combinations of destroy-repair operators. During the search, a roulette-wheel selection method is employed to select one combination every time. Each combination c_i is associated with a weight w_i which is used to measure how well combination c_i has performed in past iterations. At the beginning of the algorithm, the initial value of w_i is set to 1. The combination c_i is selected with probability $p_i = w_i / \sum_{j=1}^{8} w_j$.

2.4   Local Search

Local search plays an very important role in the design of metaheuristics for VRP [17]. A local search operator iteratively improves a solution by exploring its neighborhood in terms of the single objective function given in (1). Given a feasible solution to VRP represented by a set of routes $x = \{R_1, \ldots, R_l, \ldots, R_v\}$, where R_l is the set of customers serviced by route (or vehicle) l. Its neighborhood is denoted as $N(x)$.

The following components are critical in the implementation of a local search operator [17]. The first is the starting solution. The second is the mechanism to generate neighboring solutions of a given solution. The third is the acceptance criterion. Two popular acceptance strategies are first-accept (FA) and best-accept (BA). The FA strategy chooses the first neighboring solution that satisfies the pre-specified acceptance criterion (e.g., the objective variation after a move). The BA strategy checks all neighboring solutions which satisfy a criterion and chooses the best among them. The fourth is the condition when to stop the local search operator.

The 2-opt, exchange, cross and relocation operators [17] are adopted to generate neighborhood. FA strategy is used. These operators are invoked one by one. Local search will be ended when no more improvement can be achieved.

Unlike 2-opt, exchange, cross and relocation operators are inter-route operators. To speed up these inter-route operators, we first propose the concept of neighborhood of routes based on polar angle. Polar angle is the basic tool in the famous sweep algorithm [18]. For a route R_l, its polar angle is defined as the polar angle of its center of gravity. Two routes are said to be close if their polar angles are close. For a route, only the N closest neighboring routes are permitted to inter-change. When to select a move, a local search operator only checks its neighborhood for each route. By interchanging with the routes out of smaller neighbor, it is more likely to find better solution with smaller travel time or cumulative time.

2.5   Update of Weights

In MALNS, a combination c_i is associated with a score, denoted by φ_i. At each iteration, φ_i is initialized to 0. After an iteration, φ_i of the chosen combination c_i will be renewed based on the quality of the solutions constructed at the iteration. In detail, starting from each solution (in current population) x_s, a new solution x_n is generated by a combination c_i and improved by local search. If x_n is nondominated by solutions found so far, the score of c_i is increased by 15; If x_n dominates x_s, the score of c_i is increased by 10; If x_n is nondominated by x_s, the score of c_i is increased by 5; Otherwise, no score is obtained. Formally, it is updated as follows:

$$\varphi_i = \begin{cases} \varphi_i + 15 & \text{if the new solution } x_n \text{ is nondominated} \\ & \text{by solutions found so far} \\ \varphi_i + 10 & \text{if the new solution } x_n \text{ dominates} \\ & \text{its starting solution } x_s \\ \varphi_i + 5 & \text{if the new solution } x_n \text{ is nondominated} \\ & \text{by its starting solution } x_s \\ \varphi_i & \text{otherwise} \end{cases} \qquad (2)$$

Every iteration, each weight w_i is updated based on the scores obtained.

$$w_i = (1 - \rho)w_i + \rho \frac{\varphi_i}{\max(Freq_i, 1)} \qquad (3)$$

where $Freq_i$ denotes the times the combination c_i has been applied in the past iteration. ρ is a parameter which controls the forgotten rate of the past experience. ρ is set to 0.05. MALNS re-initializes the weights to 1 once no new nondominated solutions can be found during consecutive 50 iterations.

2.6 Population Initialization

N solutions are constructed by the initial procedure. At first, each customer is inserted in request bank, and then each customer is inserted by the regret insertion heuristic. Note that the single objective function is given by (1). To construct the the lth solution ($l \in \{1, \cdots, N\}$), the weight vector is $((l-1)/(N-1), (N-l)/(N-1))$.

2.7 Update of Population

At each iteration, we only accept a new generated solution of which request bank is empty and the number of routes is $|R|$. The population is updated by using nondominance ranking and crowding distance in [19]. From the last population and new generated solutions, a set of fronts are obtained by nondominance ranking. The first front F_1 consists of the nondominated solutions. The second front is the set of solutions which are only dominated by the first front, and so on. Crowding distance is the average side length of the cuboid formed by the objective values of these nearest neighboring solutions [19].

2.8 Update of External Archive

Once a solution x is accepted, the external archive EA is renewed as follows: If no vector in EA dominates $F(x)$, $F(x)$ will be added to EA. At the mean time, all the vectors dominated by $F(x)$ will be removed from EA.

3 Experimental Results

MALNS was coded in C++ and tested on a PC with Pentium 4, 2.4G CPU, and 4GB RAM. It was tested on 20 large-scale instances with 240 to 483 customers in [20]. The travel time between every two nodes is their Euclidean distance. All travel time is rounded to double precision [2,3].

Based on the preliminary test, the population size was set to 30. As done in [3], the maximal size u of request bank was randomly chosen from [10,60]. For each instance, MALNS was test the same times in [3] (i.e., 5) independently and stopped when a given time limit was achieved. In our experiment, the time limit was chosen as follows. At first the computational time in [3] was transformed, then the transformed time T was set. For example, the computational time of GWKC1 in [3] is 1038, then the transformed time T is 865, since our CPU is 1.2 times as fast as the one of [3].

3.1   Performance Metrics

It is widely accepted that, given two nondominated fronts, the better front has the following properties: it is closer to Pareto-optimal front; it distributes more evenly and widely in objective space [21]. Many performance metrics have been proposed to evaluate the solution quality. Among them, hypervolume is a very nice and popular metric [22], therefore we adopted it in this paper. The Hypervolume value of a nondominated front is the volume of the area in the objective space which is bounded by a reference point and dominated by the front itself. Let ζ be all the nondominated solution sets found in a test, the reference point was chosen as $(\max\{f_1(x)|x \in \zeta\}, \max\{f_2(x)|x \in \zeta\})$. Larger hypervolume value indicates better solution quality.

In order to pictorially illustrate the nondominated fronts, a statistical tool, called summary-attainment surface [23], is employed. Summary-attainment surface refers to the union of all tightest points in the objective space obtained by an algorithm. If an algorithm is tested l times, there will be l summary-attainment surface. The first, $l/2$th, lth summary-attainment surface is called the best, median, worst summary-attainment surface respectively.

3.2   Comparison with a Weighted Sum ALNS

To study MALNS, we implemented a weighted sum ALNS which works as follows: the same procedure of MALNS with only one single weight vector is performed N times ($N = 30$). In the lth time ($l = 1, \cdots, N$), only weight vector $((l-1)/(N-1), (N-l)/(N-1))$ is used and the computational time is T/N.

As seen from Fig. 1, MALNS provides better hypervolume than the weighted sum ALNS. According to the summary-attainment surfaces shown in Fig. 2, MALNS can provide wider nonodimianted front. We also note that the weighted sum ALNS performs better in some central parts. The reason may be that the computational resource in the weighted sum ALNS is biased to search some specific areas.

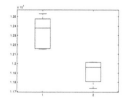

Fig. 1. The hypervolume values obtained by MALNS and weighted sum ALNS (shown from left to right) are tested for GWKC20.

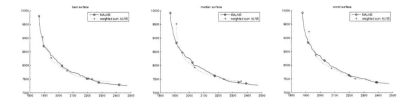

Fig. 2. The best, median, and worst summary-attainment surfaces obtained by MALNS and weighted sum ALNS for GWKC20

4 Conclusion

This paper presented a multi-objective adaptive large neighborhood search for an MVRP of which the objectives are the total travel time and the cumulative time. Although the total travel time and the cumulative time has been separately studied before, this paper investigated these objectives together. It selects a combination of destroy-repair operator for generating new solutions and improves them by local search. According to the yielded solutions, the preference of each combination is renewed based on the Pareto dominance. Compared with a weighted sum ALNS, the proposed algorithm performs better.

References

1. Jozefowiez, N., Semet, F., Talbi, E.: Multiobjective vehicle routing problems. European Journal of Operational Research 189, 293–309 (2008)
2. Ngueveu, S.U., Prins, C., Wolfler-Calvo, R.: An effective memetic algorithm for the cumulative capacitated vehicle routing problem. Computers and Operations Research 37, 1877–1885 (2010)
3. Ribeiro, G., Laporte, G.: an adaptive large variable neighborhood search heuristic for cumulative capacitated vehicle routing problem. Computers and Operations Research 39(3), 728–735 (2012)
4. Lucena, A.: Time-dependent traveling salesman problem:The deliveryman case. Networks 20, 753–763 (1990)
5. Fischetti, M., Laporte, G., Martello, S.: The delivery man problem and cumulative matroids. Operations Research 41, 1055–1064 (1993)
6. Bowerman, R., Hall, B., Calamai, P.: A multi-objective optimization approach to urban school bus routing: Formulation and solution method. Transportation Research Part A 29, 123–197 (1995)
7. Riera-Ledesma, J., Salazar-Gonzalez, J.: The biobjective travelling purchaser problem. European Journal of Operational Research 160, 599–613 (2005)
8. Geiger, M.: Genetic algorithms for multiple objective vehicle routing. In: Meta-Heuristic International Conference 2001 (MIC 2001), pp. 348–353 (2001)
9. Lacomme, P., Prins, C., Sevaux, M.: A genetic algorithm for a bi-objective capacitated arc routing problem. Computers and Operations Research 33, 3473–3493 (2006)

10. Jozefowiez, N., Semet, F., Talbi, E.G.: The bi-objective covering tour problem. Computers and Operations Research 34, 1929–1942 (2007)
11. Ombuki, B., Ross, B., Hanshar, F.: Multi-objective genetic algorithm for vehicle routing problem with time windows. Applied Intelligence 24, 17–30 (2006)
12. Current, J., Schilling, D.: The median tour and maximal covering tour problems: Formulations and heuristics. European Journal of Operational Research 73, 114–126 (1994)
13. Doerner, K., Focke, A., Gutjahr, W.: Multicriteria tourplanning for mobile health-care facilities in a developing country. European Journal of Operational Research 179, 1078–1096 (2007)
14. Ropke, S., Pisinger, D.: A unified heuristic for a large class of vehicle routing problems with backhauls. European Journal of Operational Research 171, 750–775 (2006)
15. Shaw, P.: A new local search algorithm providing high quality solutions to vehicle routing problems. In: Technical report, University of Strathclyde, Glasgow (1997)
16. Ropke, S., Pisinger, D.: An adaptive large neighborhoodsearch heuristic for the pickup and delivery problem with time windows. Transportation Science 40, 455–472 (2006)
17. Braysy, O., Gendreau, M.: Vehicle routing problem with time windows, part I: route construction and local search algorithms. Transportation Science 39(1), 104–118 (2005)
18. Gillett, B., Miller, L.: A heuristic algorithm for the vehicle dispatch problem. Operations Research 22, 340–349 (1974)
19. Deb, K., Agrawal, S., Pratap, A., Meyarivan, T.: A fast and elitist multiobjective genetic algorithm: NSGA?II. IEEE Trans. Evolutionary Computation 6(2), 182–197 (2002)
20. Golden, B.L., Wasil, E.A., Kelly, J.P., Chao, I.M.: Metaheuristics in vehicle routing. In: Fleet Management and Logistics, pp. 33–56. Springer, Heidelberg (1998)
21. Coello, C., Veldhuizen, D.V., Lamont, G.: Evolutionary Algorithms for Solving Multi-objective Problems. Kluwer (2002)
22. Fonseca, C.M., Paquete, L., Lopez-Ibanez, M.: An improved dimension-sweep algorithm for the hypervolume indicator. In: Proceedings of the 2006 Congress on Evolutionary Computation (CEC 2006), pp. 1157–1163. IEEE Press, Piscataway (2006)
23. Knowles, J.: A summary-attainment-surface plotting method for visualizing the performance of stochastic multiobjective optimizers. In: Proceedings of the Fifth International Conference on Intelligent Systems Design and Applications (ISDAV), pp. 552–557. IEEE Computer Society (2005)

A Self-adaptive Interior Penalty Based Differential Evolution Algorithm for Constrained Optimization

Cui Chenggang[*], Yang Xiaofei, and Gao Tingyu

Shanghai Advanced Research Institute, Chinese Academy of Sciences, Shanghai, China
cuicg@sari.ac.cn

Abstract. A self-adaptive interior penalty method is proposed for the constrained optimization problems by using interior penalty method to handle constraints. A set of interior penalty rules are designed to evaluate feasible solutions and infeasible solutions separately. A self-adaptive penalty factor method is proposed to prevent the interior penalty method from being sensitive to the values of penalty factor and to minimize the interior penalty function value of the optimal solution. As an instance of implementation, a different evolution algorithm is improved by means of the method proposed in this paper, based on which 10 benchmark problems are tested. The numerical solution results indicate that the performance of the method is better than four existing state-of-the-art techniques.

Keywords: Constrained optimization, Evolutionary algorithm, Interior penalty method, Differential evolution.

1 Introduction

Evolutionary algorithms (EAs) have been widely used to solve constrained optimization problems (COPs). However, EAs are normally used as "blind heuristics" in the sense of lacking an explicit mechanism to bias the search in constrained search spaces [1]. Several researchers have proposed different mechanisms to incorporate constraints into the fitness function of an EA [2]. Penalty functions are the most common approaches used to handle constraints with EAs [3]. There are two basic types of penalty functions: exterior penalty functions, which penalize infeasible solutions, and interior penalty functions, which penalize feasible solutions. Compared to interior penalty functions, exterior penalty functions are more common in EAs. The main reason is that there is no need to start with a feasible solution in exterior penalty functions. Another category of constraint handling techniques involves the preference of feasible solutions over infeasible solutions [4]. In these methods, a heuristic rule that feasible solutions are preferred over infeasible ones is used to process infeasible solutions. Multiobjective optimization techniques have also been used in the solution of constrained single objective optimization problems [5]. These techniques can be classified based on the way they transform the COP into a multiobjective optimization problem.

---

[*] Correspondig author.

Y. Tan et al. (Eds.): ICSI 2014, Part II, LNCS 8795, pp. 309–318, 2014.

In this paper, a set of interior penalty (IP) based selection rules is proposed to balance the avoidance of the constraint boundaries and the minimization of the objective function in the search process of Differential Evolution (DE) algorithms. Feasible solutions closed to the constraint boundaries are penalized to balance the conflict aims of reducing the objective function and approaching the constraint boundaries; infeasible solutions are evaluated by constraint violations to reach feasible region quickly. Three elements are employed to make these rules more effective in DEs: (1) Logarithmic penalty functions are used to make DEs yield a rapid convergence; (2) Penalty factors are updated according to the types of constraints which are determined by the Spearman's rank-order correlation coefficients; (3) Equality constraints are handled by an adaptive relaxing rule. In this paper, a self-adapt interior penalty based differential evolution algorithm is implemented as an example of this constraint handling approach. Finally, the efficiency and effectiveness of the proposed method are evaluated on 10 benchmark problems.

2   Interior Penalty Based Selection Rules

2.1   Statement of the Problem

Generally, a COP can be expressed as follows:

$$\min f(x) , \quad x = (x_1, ..., x_n) \in R^n \tag{1}$$
$$\text{s.t.} \quad g_j(x) \leq 0 , j \in \{1, ..., m\} ,$$
$$h_j(x) = 0 , j = q+1, ..., m .$$

where x is the decision vector, $f(x)$ is the objective function, q is the number of inequality constraints, and $m-q$ is the number of equality constraints. Let $S \subset R^n$ define the search space bounded by the parametric constraints $\underline{x_i} \leq x_i \leq \overline{x_i}$, $i \in \{1, 2, ..., n\}$, where $\underline{x_i}$ and $\overline{x_i}$ are the lower bound and the upper bound of x_i, respectively.

2.2   Interior Penalty Method

The interior penalty method is very popular in the traditional mathematical programming techniques, motivated by minimizing a composite function that reflects the original objective function as well as the influence of the constraints [6].

Given the constrained optimization problem (1), the interior penalty function can be formulated as follows:

$$\phi(x, r(t)) = f(x) + r(t)B(x) . \tag{2}$$

$B(x)$ is a penalty term that is nonnegative and approaches ∞ as the constraint boundaries are approached from the interior. $r(t)$ is a penalty factor.

The interior penalty method replaces a COP with a sequence of unconstrained optimization problems, defined as:

$$\lim_{t \to \infty} \min \phi(x, r(t)).$$

(3)

Solving problem (3) sequentially for a monotonously decreasing sequence $\{r(t)\}$ such that $\lim_{t \to \infty} r(t) = 0$ gives a sequence $\{x(r(t))\}$ yielding $\lim_{t \to \infty} f(x(r(t))) = f(x^*)$, where $x*$ is the optimal solution of problem (1).

2.3   Implementation of the IP Based Selection Rules

In order to balance the avoidance of the constraint boundaries and the minimization of the objective function in the search process, a set of interior penalty based selection rules is proposed to improve the efficiency of search and avoid the violations.

The IP rules can be formulated as follows:

1) Between two feasible solutions, the one with a better interior penalty function value is preferred.2) If both solutions candidates are infeasible, the one with a smaller constraint violation is preferred.3) A feasible solution is always preferred to an infeasible one.

For the IP based selection rules, there are three important properties:

1) The interior penalty term is defined only if a solution is feasible. Thus, the interior penalty function cannot handle infeasible solutions. Therefore, a preference of feasible solutions to infeasible ones is used in rule (3). In this way, feasible solutions and infeasible solutions can be evaluated by different methods.

2) In the interior penalty method, feasible solutions are penalized in order to avoid the boundaries of feasible region. Rule 1) penalizes feasible solutions approaching to the boundaries of feasible region. In this way, the search process of EAs will avoid the boundaries of feasible region.

3) The objective function is completely disregarded in rule 2). Therefore, the entire search effort is directed toward finding a feasible solution. This rule is especially suitable for highly constrained problems wherein finding a feasible solution may be extremely difficult [3]. In this way, the penalty rules can be applied to highly constrained problems without initial feasible solutions.

3   Self-adapt Interior Penalty

3.1   Form of Interior Penalty Function

The logarithmic penalty function is used as the interior penalty function in this paper since it has a superlinear convergence in the traditional mathematical programming [7]. The penalty term $B(x)$ in Eq.(2) can be formulated as follows:

$$B(x) = -\sum_{i=1}^{m} \ln\left(-v_i(x)\right),$$

(4)

where $v_i(x)$ is the normalized constraint value, defined as:

$$v_i(x) = \frac{g_i(x)}{|\min g_i(x)|} \quad . \tag{5}$$

The scaling factor $|\min g_i(x)|$ for each constraint is taken as the minimal value of constraint value $g_i(x)$ in the search process.

3.2   Self-adapt Penalty Factor

The candidate solutions are selected based on interior penalty function value in the IP based selection rules. Therefore, the optimal solution must be the one with minimum internal penalty value of all the candidates. According this rule, we proposed a self-adapt interior penalty method as follows.

Given the constrained optimization problem (1), the penalty factor r must make the penalty value of the optimal solution is less than the one of any other solution.

The Penalty factor selection rules can be expressed as the following formula:

$$f(x^*) + r(t)B(x^*) \le f(x_i) + r(t)B(x_i), \tag{6}$$

where x_i is any candidate, x^* is the optimal solution.

The Eq.(6) can be converted to:

$$(B(x^*) - B(x_i))r(t) \le f(x_i) - f(x^*) \quad . \tag{7}$$

Considering three conditions as follows:

(1) $B(x^*)\text{-}B(x_i)=0$

The left of Eq.(7) is 0 in this condition. Therefore, for any $r(t)$, the formula $f(x_i) \ge f(x^*)$ is established, i.e. Eq. (7) is established. This condition is not considered.

This condition is not considered.

(2) $B(x^*)\text{-}B(x_i)<0$

The Eq.(7) can be convert to:

$$r(t) \ge \frac{f(x^*) - f(x_i)}{B(x_i) - B(x^*)} \quad . \tag{8}$$

The right of Eq.(8) is less than or equal to 0 since $f(x_i) \ge f(x^*)$. Therefore, for any $r(t) \ge 0$, Eq.(8) is established. This condition is also not considered.

(3) $B(x^*)\text{-}B(x_i)>0$

We can get the upper bounder of $r(t)$ by Eq.(7) in this condition:

$$r(t) \le \frac{f(x^*) - f(x_i)}{B(x_i) - B(x^*)} \quad . \tag{9}$$

We can get the follows since the optimal solution x^* satisfies all the constraints, i.e. $\forall j, \; g_j(x^*) \le 0$:

$$B(x^*) = -\sum_{j=1}^{q} \ln|g_j(x^*) - \varepsilon_j| = -\sum_{j=1}^{q} \ln(\varepsilon_j - g_j(x^*)) \le -\sum_{j=1}^{q} \ln(\varepsilon_j) \text{ Let } B_d = -\sum_{j=1}^{q} \ln(\varepsilon_j) \text{, then:}$$

$$B_d - B(x_i) \ge B(x^*) - B(x_i) > 0 \quad . \tag{10}$$

Alternative $B(x^*)$ with B_d in Eq.(10). We can get an upper bounder of $r(t)$:

$$r(t) \le \min_{i \in N} \frac{f(x_i) - f(x^*)}{B_d - B(x_i)} \quad . \tag{11}$$

Further, the optimal solution can't be obtained before the original constrained optimization problem solved. However, we can use the best feasible solution in the current population instead of the optimal solution in the search process, i.e.

$$r(t) \le \min_{i \in N} \frac{f(x_i) - f(\overline{x}^*)}{B_d - B(x_i)} \quad , \tag{12}$$

where x^* is the feasible solution with minimum objective in the current population. $r(t) \le \min_{i \in N} \dfrac{f(x_i) - f(\overline{x}^*)}{B_d - B(x_i)} \le \min_{i \in N} \dfrac{f(x_i) - f(x^*)}{B_d - B(x_i)}$ is established since $f(\overline{x}^*) \le f(x^*)$.

Therefore, the penalty factor obtained by Eq. (12) satisfies Eq.(6). We can get an appropriate penalty factor without the optimal solution.

According to the above analysis, we can obtain the penalty factor by Eq.(12).

There may not be a feasible solution when the algorithm starts. We use a larger penalty factor in the early search process to ensure evolutionary algorithm can quickly find a feasible solution.

4   Self-adaptive Interior Penalty Based Differential Evolution

To illustrate validity of the interior penalty based selection rules, we introduce them to Differential Evolution algorithm called a self-adaptive interior penalty based differential evolution algorithm. The DE algorithm proposed by Storn and Price [8] is a heuristic method for real parameter optimization problems.

Let x_t^i denote an individual in the population of the DE algorithm and NP the size of the population, where i indicates the index of the individual, j the index of the variable, and t the current generation. A new mutated individual $v^i_{j,t+1}$ is generated according to the following equation:

$$v^i_{j,t+1} = x^{d_3}_{j,t} + \eta(x^{d_1}_{j,t} - x^{d_2}_{j,t}), \tag{13}$$

where the random indexes d_1, d_2, $d_3 \in [0, NP]$ are mutually different integers and also different from the running index i, and $\eta \in (0, 1]$ is called the scaling factor or the amplification factor.

According to Eq. (13), a crossover operator is used to generate the trial individual $u^i_{j,t+1}$ based on the original individual $x^{d_3}_{j,t}$ and the new individual $v^i_{j,t+1}$.

$$u^i_{j,t+1} = \begin{cases} v^i_{j,t+1}, & \text{if Rand}[0,1) \leq CR \text{ or } j = \text{randint}(1,D), \\ x^i_{j,t}, & \text{otherwise,} \end{cases} \tag{14}$$

where Rand[0, 1) is a function that returns a real number between 0 and 1, randint(min, max) is a function that returns an integer between min and max, $CR \in [0, 1]$ is a crossover factor. The probability of the mutated individuals being preserved in the next generation is determined by the crossover factor CR.

A selection operator is used to choose an individual for the next generation $(t+1)$ according to the following rule:

$$x^i_{t+1} = \begin{cases} u^i_{t+1}, & \text{if } u^i_{t+1} \text{ is better than } x^i_t, \\ x^i_t, & \text{otherwise,} \end{cases} \tag{15}$$

where u^i_{t+1} and x^i_t are compared by the IP based selection.

In this way, an individual will replace the one with a lower IP with respect to it; an individual will replace the one with the same IP depending on different conditions, where an infeasible individual will replace the one with a larger violation of the maximal non-common satisfied constraint and a feasible individual will replace the one with a worse objective with respect to it, respectively.

The capability of finding the global minimum and a fast convergence speed of DE are both highly sensitive to the control parameters CR and η [9]. Therefore, a self-adaptive approach is developed to adjust these parameters based on the success rate φ_t, where φ_t is defined by the percentage of original individuals replaced by trial individuals in the population at every generation, through the following updating law:

$$\eta_t = \begin{cases} \eta_l + \text{rand}_1 \eta_u, & \phi_t \leq 0.5, \\ \eta_{t-1}, & \text{otherwise,} \end{cases} \tag{16}$$

$$CR_t = \begin{cases} \text{rand}_2, & \phi_t \leq 0.5, \\ CR_{t-1}, & \text{otherwise,} \end{cases} \tag{17}$$

where η_t and CR_t are the scaling factor η and the crossover factor CR at generation t, respectively; rand_1 and rand_2 are uniformly distributed random numbers in [0, 1]; $\eta_l = 0.1$, $\eta_u = 0.9$. The updating of η_t and CR_t is conducted before the mutation is performed. Eqs. (14) and (15) ensure that $\eta_t \in [0.1, 1] \subset (0, 1]$, $CR_t \in [0, 1]$, $\forall t$.

The pseudo code of the DE with the IP based selection rules is shown as follows, the rules keep the operators of DE algorithms unchanged.

```
Begin
t=0;
Create NP random solutions for the initial population;
Evaluate all individuals;
For t=1 to MAX_GENERATION Do
    For i=1 to NP Do
        Select randomly d1•d2•d3;
```

```
  If (Rand[0, 1]•CR or j=randint(1, D)) Then
```
$$u^i_{j,t+1} = v^i_{j,t+1} \text{ ;}$$
```
  Else
```
$$u^i_{j,t+1} = x^i_{j,t} \text{ ;}$$
```
  End If
End For
```
Compare $u^i_{j,t+1}$ and $x^i_{j,t}$ by the IP based selected rules;
```
If
```
u^i_{t+1} is better than x^i_t Then
$$x^i_{t+1} = u^i_{t+1} \text{ ;}$$
```
Else
```
$$x^i_{t+1} = x^i_t \text{ ;}$$
```
End If
t=t+1;
update interior penalty factior;
End For
End
```

5   Experiments and Results

We performed the self-adapt interior penalty based differential evolution algorithm (SIPDE) algorithm 30 independent runs for ten benchmark problem described in Runarsson and Yao [9]. Equality constraints were transformed into inequalities using a tolerance value of 0.0001. The parameters were set the same as those of Mezura-Montes et al. [10]: NP=60, MAX_ GENERATIONS=5800. The control parameters CR and η were adjusted using a self-adaptive method. We compared our approach against four state-of-the-art approaches: the stochastic ranking (SR) algorithm [9], the simple multimembered evolution strategy (SMES) algorithm [11], the adaptive tradeoff model evolution strategy (ATMES) algorithm [12], and the constraint handling differential evolution (CHDE) algorithm [10]. The best, mean, worst results, and the standard deviations obtained by each approach are shown in Table 1.

5.1   Statement of the Problem

As described in Table 1, our approach was able to find the global optimum in ten benchmark problems. For problems g01, g03, g04, g05, g06, g07, g08, g09 and g10, the optimal solutions were consistently found in all 30 runs. For problems g02, the optimal solutions were not consistently found since the ratio of the size of the feasible region to the size of the search space was very large. Furthermore, feasible solutions were continuously found for all the benchmark problems in 30 runs. These results reveal that SIPDE has the substantial capability to deal with various kinds of COPs.

Table 1. Comparison of the best, the mean, the worst solutions, and the standard deviations found by our SIPDE against SR, SMES, ATMES, and CHDE

Prob	optimal	stat	methods				
			SR	SMES	ATMES	CHDE	SIPDE
g01		best	-15.000	-15.000	-15.000	-15.000	-15.000
	-15.000	mean	-15.000	-15.000	-15.000	-14.792	-15.000
		worst	-15.000	-15.000	-15.000	-12.743	-15.000
g02		best	-0.803515	-0.803601	-0.803388	-0.803619	-0.803619
	-0.803619	mean	-0.781975	-0.785238	-0.790148	-0.746236	-0.801758
		worst	-0.726288	-0.751322	-0.756986	-0.302179	-0.780843
g03		best	1.000	1.000	1.000	1.000	1.000
	1.000	mean	1.000	1.000	1.000	0.640326	1.000
		worst	1.000	1.000	1.000	0.029601	1.000
g04		best	-30665.539	-30665.539	-30665.539	-30665.539	-30665.539
	-30665.539	mean	-30665.539	-30665.539	-30665.539	-30592.154	-30665.539
		worst	-30665.539	-30665.539	-30665.539	-29986.214	-30665.539
g05		best	5126.497	5126.599	5126.498	5126.497	5126.498
	5126.497	mean	5128.881	5174.492	5127.648	5218.729	5126.498
		worst	5142.472	5304.167	5135.256	5502.410	5126.498
g06		best	-6961.814	-6961.814	-6961.814	-6961.814	-6961.814
	-6961.814	mean	-6875.940	-6961.284	-6961.814	-6367.575	-6961.814
		worst	-6350.262	-6952.482	-6961.814	-2236.950	-6961.814
g07		best	24.307	24.327	24.306	24.306	24.306
	24.306	mean	24.374	24.475	24.316	104.599	24.306
		worst	24.642	24.843	24.359	1120.541	24.306
g08		best	-0.095825	-0.095825	-0.095825	-0.095825	-0.095825
	-0.095825	mean	-0.095825	-0.095825	-0.095825	-0.091292	-0.095825
		worst	-0.095825	-0.095825	-0.095825	-0.027188	-0.095825
g09		best	680.630	680.632	680.630	680.630	680.630
	680.630	mean	680.656	680.643	680.639	692.472	680.630
		worst	680.763	680.719	680.673	839.783	680.630
g10		best	7054.316	7051.903	7052.253	7049.248	7049.248
	7049.248	mean	7559.192	7253.047	7250.437	8442.657	7049.255
		worst	8835.655	7638.366	7560.224	15580.370	7049.399

5.2   Comparison with Four State-of-the-Art Approaches

The performance of SIPDE was compared in detail with four state-of-the-art techniques using the selected performance metrics (Table 1). For benchmark problems g01, g03, g04, and g08, RFDDE, SR, SMES, and ATMES consistently found the optimal solutions in all 30 runs. For problem g06, the optimal solutions were consistently found by SIPDE and ATMES in all 30 runs. For problem g05, SR, SMES, and ATMES found better 'mean' and 'worst' results than SIPDE. However, SIPDE was also able to find the optimal solution in 30 runs and the 'mean' results were very close to the optimal solution. For all the other 4 problems, SIPDE found better 'best', 'mean', and 'worst' results than SR, SMES, and ATMES. As against CHDE, our approach found "similar" best results in all the problems, and furthermore located better 'mean' and 'worst' results in all the problems.

In summary, we can conclude that SIPDE outperforms or has similar performances to SR, SMES, ATMES, and CHDE in all the problems.

6   Conclusion

In order to combine constraints into the evaluation of feasible solutions, a set of interior penalty rules for handling COPs was proposed in this paper. In these rules, interior penalty functions are used to evaluate feasible solutions and constraint violations are used to evaluate infeasible solutions. Three elements are proposed to make these rules effective in an EA: (1) a logarithmic penalty function is used to make the algorithm convergence quickly; (2) the penalty factors are updated according to the type of constraints which determined by a Spearman's rank-order correlation coefficient; (3) the equalities are handled by an adaptive relax method. Furthermore, the interior penalty rules are implemented based on a DE, namely, SIPDE. Finally, the experiment results show that the proposed approach is competitive with four other state-of-the-art techniques.

References

1. Mezura-Montes, E., Coello, C.: Constraint-handling in nature-inspired numerical optimization: past, present and future. Swarm and Evolutionary Computation 1(4), 173–194 (2011)
2. Kramer, O.: A review of constraint-handling techniques for evolution strategies. Applied Computational Intelligence and Soft Computing 1, 1–11 (2010)
3. Tessema, B., Yen, G.: An adaptive penalty formulation for constrained evolutionary optimization. IEEE Transactions on Systems, Man, and Cybernetics, Part A: Systems and Humans 39(3), 565–578 (2009)
4. Deb, K.: An efficient constraint handling method for genetic algorithms. Computer Methods in Applied Mechanics and Engineering 186(2-4), 311–338 (2000)
5. Mezura-Montes, E.: Constraint-handling in evolutionary optimization. Springer, Heidelberg (2009)
6. Wright, M.: The interior-point revolution in optimization: history, recent developments, and lasting consequences. Bulletin of the American Mathematical Society 42(1), 39–56 (2005)
7. Wright, M.: The interior-point revolution in constrained optimization. High-Performance Algorithms and Software in Nonlinear Optimization, 359–381 (1998)
8. Storn, R., Price, K.: Differential Evolution–A Simple and Efficient Heuristic for global Optimization over Continuous Spaces. Journal of Global Optimization 11(4), 341–359 (1997)
9. Runarsson, T., Yao, X.: Stochastic ranking for constrained evolutionary optimization. IEEE Transactions on Evolutionary Computation 4(3), 284–294 (2000)

10. Mezura-Montes, E., Coello Coello, C.A., Tun-Morales, E.I.: Simple feasibility rules and differential evolution for constrained optimization. In: Monroy, R., Arroyo-Figueroa, G., Sucar, L.E., Sossa, H. (eds.) MICAI 2004. LNCS (LNAI), vol. 2972, pp. 707–716. Springer, Heidelberg (2004)
11. Mezura-Montes, E., Coello, C.: A simple multimembered evolution strategy to solve constrained optimization problems. IEEE Transactions on Evolutionary Computation 9(1), 1–17 (2005)
12. Wang, Y., Zixing, C.: An Adaptive Tradeoff Model for Constrained Evolutionary Optimization. IEEE Transactions on Evolutionary Computation 12(1), 80–92 (2008)

A Novel Hybrid Algorithm for Mean-CVaR Portfolio Selection with Real-World Constraints

Quande Qin[1,2], Li Li[1], and Shi Cheng[3,4]

[1] Department of Management Science, Shenzhen University, Shenzhen, China
[2] Research Institute of Business Analytics & Supply Chain Management,
Shenzhen University, Shenzhen, China
[3] Division of Computer Science, University of Nottingham Ningbo, China
[4] International Doctoral Innovation Centre, University of Nottingham Ningbo, China
qinquande@gmail.com, llii318@163.com, shi.cheng@nottingham.edu.cn

Abstract. In this paper, we employ the Conditional Value at Risk (CVaR) to measure the portfolio risk, and propose a mean-CVaR portfolio selection model. In addition, some real-world constraints are considered. The constructed model is a non-linear discrete optimization problem and difficult to solve by the classic optimization techniques. A novel hybrid algorithm based particle swarm optimization (PSO) and artificial bee colony (ABC) is designed for this problem. The hybrid algorithm introduces the ABC operator into PSO. A numerical example is given to illustrate the modeling idea of the paper and the effectiveness of the proposed hybrid algorithm.

Keywords: Conditional Value at Risk, CVaR, Hybrid algorithm, Portfolio selection.

1 Introduction

Portfolio selection is concerned with the allocation of a limited capital to a combination of securities in order to trade off the conflicting objectives of high profit and low risk [13,17]. Since the introduction of mean-variance (MV) model developed by Markowitz, variance has become the most popular risk measure in portfolio selection. Variance considers high returns as equally undesirable as low returns because high returns will also contribute to the extreme of variance. Both theory and practice indicate the variance is not a good risk measure. Some alternative risk measures have been proposed [11, 18]. Value at Risk (VaR) is widely used by financial institution. However, it has its limitations, such as it is not a coherent risk measure [1]. Rockafellar and Uryasev [15] proposed the Conditional Value at Risk (CVaR), which is the conditional expectation of losses above the VaR.

In practice, problem of portfolio selection has some real-world constraints, which exacerbates the complexity. For example, it assumes that there exists a perfect market with no tax or transaction cost. In the present study, we will consider transaction cost, and floor and ceiling constraints. In addition, the least

Y. Tan et al. (Eds.): ICSI 2014, Part II, LNCS 8795, pp. 319–327, 2014.

unit of trading is 100 shares in stock market of China, and shares must be sub-scribed a round lot. The modeling of such constraints involves the introduction of integer variables. We employ CVaR to measure the risk of portfolio, and a Mean-CVaR (MC) portfolio selection model with real-world constraints is pro-posed. In view of the difficulty to solve this model using classical optimization techniques, a hybrid meta-heuristics algorithm based Particle Swarm Optimiza-tion (PSO) and Artificial Bee Colony (ABC) is designed to handle this problem. The hybrid algorithm introduces the ABC operator into PSO. The added ABC operator is used to evolve personal experience of the particles. The hybrid ap-proach elegantly combines the exploitation ability of PSO with the exploration ability of ABC.

The rest of the paper is organized as follows. Section 2 presents the back-grounds including PSO, ABC and CVaR. Section 3 the proposed MC portfolio selection model with real-world constraints. A hybrid algorithm based on PSO and ABC is provided in Section 4. In Section 5, a numerical example is given. The conclusions are drawn in Section 6.

2   Backgrounds

2.1   Particle Swarm Optimization

PSO was originally developed to emulate the flocking behavior of birds and fish schooling [5,9]. Each individual, called a particle, in the PSO population repre-sents a potential solution of the optimization problem [2,19]. The population of PSO is referred to as a swarm, which consists of a number of particles. Particle i at iteration t is associated with a velocity vector $\boldsymbol{v}_i^t = [v_{i1}^t, v_{i2}^t, \cdots, v_{iD}^t]$ and a position vector $\boldsymbol{x}_i^t = [x_{i1}^t, x_{i2}^t, \cdots, x_{iD}^t]$ where $i \in \{1, 2, \cdots, NP\}$, NP is the population size. $x_{id} \in [l_d, u_d]$, $d \in \{1, 2, \cdots, D\}$, where D is the number of di-mensions, and l_d and u_d are the lower and upper bounds of the dth dimension of search space, respectively. Each particle flies through space with a velocity. The new velocities and the positions of the particles for the next iterations are updated using the following two equations [3–5,9]:

$$v_{id}^{t+1} = wv_{id}^t + c_1 r_1 (pbest_{id}^t - x_{id}^t) + c_2 r_2 (gbest_d^t - x_{id}^t) \tag{1}$$

$$x_{id}^{t+1} = x_{id}^t + v_{id}^{t+1} \tag{2}$$

where w is the inertia weight; $\boldsymbol{pbest}_i = [pbest_{i1}, pbest_{i2}, \cdots, pbest_{iD}]$ is the best position has been found by particle i, $\boldsymbol{gbest}_i = [gbest_{i1}, gbest_{i2}, \cdots, gbest_{iD}]$ is the historically best position has been found by the whole swarm so far; c_1 and c_1 are acceleration coefficients. The inertia weight w is used to trade off the exploration and exploitation; r_1 and r_2 represent two independently random numbers uniformly distributed on $[0, 1]$.

2.2   Artificial Bee Colony

ABC algorithm was proposed by simulating waggle dance and intelligent for-aging behaviors of honeybee colonies [7]. In the ABC algorithm, there are two components: the foraging artificial bees and the food source [8]. The position of the a food source, $\boldsymbol{x}_i = [x_{i1}, x_{i2}, \cdots, x_{iD}]$, represents a possible solution and the nectar amount of a food source corresponds to the fitness of the associated solution. The colony of artificial bees contains three groups of bees: employed bees, onlookers and scouts [14].

The ABC algorithm consists of four phases: initialization, employed bee, onlooker bee and scout bee. In the initialization phase of the ABC, SN food source positions are randomly produced with the search space. After producing food sources and assigning them to the employed bees. In the employed bee phase of ABC, each employed bee tries to find a better quality food source based on $\boldsymbol{x}_i$. The new food source, denoted as $\boldsymbol{u}_i = [u_{i1}, u_{i2}, \cdots, u_{iD}]$, is calculated from the equation below.

$$u_{ij} = x_{ij} + \phi(x_{ij} - x_{sj}) \tag{3}$$

where $i \in \{1, 2, \cdots, SN\}$, where SN denotes the number of food source; j is a randomly generated integer number in the range $[1, D]$, ϕ is a randomly number uniformly distributed in the range $[-1, 1]$, and s is the index of a randomly chosen solution. ABC changes each position in only one dimension at each iteration. The source position $\boldsymbol{x}_i$ in the employed bee's memory will be replaced by the new candidate food source position $\boldsymbol{u}_i$ if the new position has a better fitness value. Each onlooker bee chooses one of the proposed food sources depending on the probability value p_i associated with the fitness value, where

$$p_i = fit_i / \sum_{j=1}^{SN} fit_j \tag{4}$$

where fit_i is the fitness of the food source i. After the food source is selected, a new candidate food source can be expressed by Eq. (3). If a food source, $\boldsymbol{x}_i$, cannot be improved for a predetermined number of cycles, referred to as *limit*, this food source is abandoned. Then, the scout produces a new food source randomly to replace $\boldsymbol{x}_i$.

2.3 Conditional Value at Risk

Let $L(x, y)$ be the loss function with weight vector x and the return rate vector y. Let $p(r)$ be the density function of the return rate vector y. Then $L(x, y)$ is random variable dependent on x. The probability of $L(x, y)$ not exceeding a threshold α is given by

$$\psi(x, \alpha) = \int_{L(x,y) \leq \alpha} p(y) dy \tag{5}$$

The VaR of the loss associated with x and a specified probability level β in $(0, 1)$ is the value

$$VaR_\beta(x) = \min\{\alpha \in R^m : \psi(x, \alpha) \geq \beta\} \tag{6}$$

As an improved risk measure, CVaR, is the expected portfolio return, conditioned on the portfolio returns being lower than VaR. It is defined as the Eq. (7). Compared with VaR, CVaR has some superior mathematical properties.

$$CVaR_\beta(x) = E[L(x,y)|L(x,y)) \geq VaR_\beta(x)]$$

$$= (1-\beta)^{-1} \int_{L(x,y)\geq VaR_\beta(x)} L(x,y)p(y)dy \tag{7}$$

CVaR can be obtained by the following equation based on reference [15]

$$F_\beta(x,\alpha) = \alpha + (1-\beta)^{-1} \int_{y\in R^M} [L(x,y)-\alpha]^+ p(y)dy \tag{8}$$

where $(a)^+$ is defined as $\max(a,0)$.

3   The Proposed Portfolio Selection Model

In this section, we discuss the MC portfolio selection model. Assume there n risky asset and one risk-free asset in a financial market for trading. An investor hopes to allocate his/her initial wealth m_0. For notational convenience, we first introduce the following notations:

- r_i: the return of risky asset i.
- r_f: the return of risk-free asset.
- t_i: the transaction cost of risky asset i;
- $s(x)$: the total return of the portfolio.
- p_i: the price of risky asset i each round lot;
- k_i: the round lot of risky asset i invested;
- σ_i: the highest limits on risky asset i;
- ε_i: the lowest limits on risky asset i;
- λ the acceptable return of the portfolio.

The capital invested in risk assets is $\sum_{i=1}^n k_i p_i$ and the remaining capital $m_0 - \sum_{i=1}^n k_i p_i$ invested in the risk-free asset. Obviously, it holds that $\sum_{i=1}^n k_i p_i \leq m_0$. The transaction cost are consider, and it denotes as $\sum_{i=1}^n t_i k_i$. Thus, the total return $s(x)$ of the portfolio can be described as follows:

$$s(x) = \sum_{i=1}^n k_i p_i r_i + r_f(m_0 - \sum_{i=0}^n k_i p_i) - \sum_{i=1}^n t_i k_i$$

$$= r_f m_0 + \sum_{i=1}^n [k_i p_i(r_i - r_f) - t_i k_i] \tag{9}$$

The intention of the proposed model is to minimize the CVaR in the case of the return of the portfolio is equal or greater than λ.

$$\min z = CVaR \tag{10}$$

$$s.t. \begin{cases} \varepsilon_i \leq x_i \leq \sigma_i & i = 1, 2, \cdots, n \\ s(x)/m_0 \geq \lambda \\ \sum_{i=1}^n k_i p_i \leq m_0 \\ k_i \geq 0, & \text{integer, } i = 1, 2, \cdots, n \end{cases}$$

where $x = (k_1p_1/m_0, k_2p_2/m_0, \cdots, k_np_n/m_0)$ is the weight vector. In practice, asset i is chosen to be invested and the weight lies in $[\varepsilon_i, \sigma_i]$, where $0 \le \varepsilon_i \le \sigma_i \le 1$. The first constraint is called floor and ceiling constraints. The second constraint is used to ensure the return of the portfolio.

4   A Hybrid Algorithm Based on PSO and ABC

Due to the simple concept and efficiency of converging to reasonable solution fast, PSO has been successfully applied to a wide range of real-world problems. Despite the competitive performance of PSO, researchers have noted a major problem associated with the PSO is its premature convergence when solving complex problems [12]. ABC algorithm is good at exploration but poor at exploitation [20]. From the analysis of the merits and demerits of PSO and ABC, it is intuitive that hybridizing the PSO and ABC is a potential way to design an effective algorithm.

Generally, the locality of personal best position in PSO algorithm is distant from the global optimum. Once the swarm aggregates to such position, little opportunity is afforded for the swarm to explore for other solution and find the global optimum. This leads to the swarm suffer from premature convergence easily, especially when solving complicated multimodal problems. Thus, the evolution of the personal experience will promote the exploration of the personal experience space, which could potentially enhance PSO's performance. ABC has better ability to explore, which is beneficial to global search, but poor ability of exploitation. In this paper, we utilize the ABC operator to evolve the personal best position when the personal best position stagnated. It is expected that the proposed hybrid algorithm, PSOABC, combines the merits of PSO and ABC, and have capabilities of escaping from local optima and converge fast.

In PSOABC algorithm, we use PSO in the main loop. When the fitness of $pbest_i$, denoted as $fit(pbest_i)$, has not improved within a predefined number of successive iterations, denoted as k, it is considered to be stagnated and trapped into local optima. The setting of k is set to 3 in this paper. We only use the employed bee operator in ABC algorithm to evolve $pbest_i$ in this work. The pseudo-code of the PSOABC algorithm is described in Algorithm 1. When $pbest_i$ stagnated, we can use the employed bees operator to evolve $pbest_i$. The mathematical expressions of this ABC operator described as follows:

$$z_{ij} = pbest_{ij} + \phi(pbest_{ij} - pbest_{sj}) \tag{11}$$

where s are randomly selected integers from the index of all solution with $s \ne i$. j is a randomly selected dimension number. ϕ is a randomly number uniformly distributed within the interval $[-1, 1]$.

5   Numerical Example

The portfolio selection model constructed is a non-linear discrete optimization problem. The proposed hybrid algorithm based on PSO and ABC is suitable

---

Algorithm 1. The pseudo-code of PSOABC algorithm

---

1 Initialization: set up all parameters;
2 Set the maximum iteration number FEs; $t = 1$, $Stop = 0$;
3 Evaluate the fitness of the swarm and determine **pbest**$_i$ and **gbest** ;
4 **while** *the stopping criteria is not satisfied* **do**
5   **for** $i = 1 : NP$ **do**
6   **for** $d = 1 : D$ **do**
7              $v_{id}^{t+1} = wv_{id}^{t} + c_1 r_1(pbest_{id}^{t} - x_{id}^{t}) + c_2 r_2(gbest_{d}^{t} - x_{id}^{t})$;
8              $x_{id}^{t+1} = x_{id}^{t} + v_{id}^{t+1}$;
9          $i = i + 1$;
10   Evaluate the fitness of the particle i; Update **pbest**$_i$ and **gbest** ;
11   **if** $fit(\textbf{pbest}_i^t) - fit(\textbf{pbest}_i^{t-1}) = 0$ **then**
12             $Stop(i) = Stop(i) + 1$;
13   **else**
14             $Stop(i) = 0$;
15   **for** $i = 1 : NP$ **do**
16   **if** $Stop(i) \geq k$ **then**
17             $z_{ij} = pbest_{ij} + \phi(pbest_{ij} - pbest_{sj})$;
18   **if** $fit(\textbf{z}_i < fit(\textbf{pbest}_i)$ **then**
19   **pbest**$_i = \textbf{z}_i$;
20     $t = t + 1$

---

for real-valued problems. Kitayama *et al.* utilized penalty function approach handle the discrete decision variables [10]. In this approach, the discrete decision variables are handled as the continuous ones by penalizing at the intervals. The penalty function is given as the following the Eq. (12).

$$\phi(x) = \sum_{i=1}^{n} \frac{1}{2} \left[\sin \frac{2\pi\{x_{m+i}^c - 0.25(d_{i,j+1} + 3d_{i,j})\}}{d_{i,j+1} - d_{i,j}} + 1 \right] \tag{12}$$

where $d_{i,j}$ and $d_{i,j+1}$ represents the discrete decision variables. x_{m+i}^c is the continuous decision variables between $d_{i,j}$ and $d_{i,j+1}$.

We select 20 stocks from Chinese security market, as shown in Table 1. The symbol of $m(\%)$ in Table 1 denotes the expected return. The requirement of selecting the average yield is greater than 0. This paper selected raw data for the weekend's closing price.

Assuming the investor has 500 million investment funds. According to the tax and commission in Chinese securities market, the transaction cost rate is set to 0.4%. The minimum invest weigh of each stock is 0, and the maximum weight is 10%. The risk-free return rate is equal to 4.14% based on one-year deposit rate in China, and λ is 4.5%.

Experimental results among genetic algorithm (GA), PSO-w [16], basic ABC [6] and PSOABC are compared. For a fair comparison, the population size is

Table 1. Stocks selected and expected return rate

Ticker	m(%)	Ticker	m(%)
000002	0.45	600631	0.25
000039	0.46	600642	0.5
600058	0.77	600649	0.18
600098	0.63	600663	0.1
600100	0.12	600688	0.26
600115	0.35	600690	0.09
600183	0.4	600776	0.22
000541	0.26	600811	0.3
000581	0.53	600812	0.29
600600	0.37	600887	0.18

Table 2. Experimental results comparison

Algorithm	$\beta = 90\%$		$\beta = 95\%$		$\beta = 99\%$	
	Mean	SD	Mean	SD	Mean	SD
GA	0.0454	0.0019	0.0527	0.0064	0.0737	0.0042
PSO-w	0.0428	0.0014	0.0519	0.0044	0.0743	0.0057
ABC	0.0412	0.0015	0.0479	0.0027	0.0632	0.0024
PSOABC	**0.0336**	0.0009	**0.0343**	0.0016	**0.0443**	0.0013

set to 40 for all algorithms, the maximum iteration is 3500. The selection rate, crossover rate and mutation rate is set to 0.9, 0.7 and 0.03, respectively. Other parameter settings in each algorithm are used according to their original references. All algorithms run 30 times independently. The experimental results are shown in the Table 2. In Table 2, "Mean" indicate the mean values of CVaR, and "SD" stands for the standard deviation. From Table 2, it can be seen that PSOABC has a good performance and is a good alternative for the proposed portfolio selection model.

6 Conclusions

In this work, we proposed a MC portfolio selection model. In this model, the portfolio risk is measured by CVaR and some real-world constraints are added. Note that the round lot, which involves the introduction of integer variables, is considered. We have proposed a novel hybrid algorithm to solve the portfolio selection problem. The proposed algorithm introduces the ABC operator to PSO in order to balance exploration and exploitation. A penalty function is adopted to transform the discrete portfolio selection model into a continuous one. A numerical example is given to illustrate the modeling idea of the paper, and the experimental results show that the proposed hybrid algorithm outperforms is highly competitive for this portfolio problem.

Acknowledgment. This work is partially supported by Natural Science Foundation of China under grant NO.71240015, 60975080, 61273367, 51305216, Natural Science Foundation of Guangdong Province under grant No.S2011010001337, Foundation for Distinguished Young Talents in Higher Education of Guangdong, China, under grant 2012WYM_0116 and the MOE Youth Foundation Project of Humanities and Social Sciences at Universities in China under grant 13YJC630123, and Ningbo Science & Technology Bureau (Science and Technology Project No.2012B10055). This work was carried out at the International Doctoral Innovation Centre (IDIC). The authors acknowledge the financial support from Ningbo Education Bureau, Ningbo Science and Technology Bureau, China's MOST and The University of Nottingham.

References

1. Artzner, P., Delbaen, F., Eber, J.M., Heath, D.: Coherent measures of risk. Mathematical Finance 9(3), 203–228 (1999)
2. Chen, X., Li, Y.: A modified pso structure resulting in high exploration ability with convergence guaranteed. IEEE Transactions on Systems, Man, and Cybernetics, Part B: Cybernetics 37(5), 1271–1289 (2007)
3. Cheng, S.: Population Diversity in Particle Swarm Optimization: Definition, Observation, Control, and Application. Ph.D. thesis, Department of Electrical Engineering and Electronics, University of Liverpool (2013)
4. Cheng, S., Shi, Y., Qin, Q.: Population diversity of particle swarm optimizer solving single and multi-objective problems. International Journal of Swarm Intelligence Research (IJSIR) 3(4), 23–60 (2012)
5. Eberhart, R., Kennedy, J.: A new optimizer using particle swarm theory. In: Proceedings of the Sixth International Symposium on Micro Machine and Human Science, pp. 39–43 (1995)
6. Karaboga, D.: An idea based on honey bee swarm for numerical optimization. Tech. rep., Erciyes University, Engineering Faculty, Computer Engineering Department (October 2005)
7. Karaboga, D., Basturk, B.: A powerful and efficient algorithm for numerical function optimization: artificial bee colony (ABC) algorithm. Journal of Global Optimization 39(3), 459–471 (2007)
8. Karaboga, D., Basturk, B.: On the performance of artificial bee colony (ABC) algorithm. Applied Soft Computing 8(1), 687–697 (2008)
9. Kennedy, J., Eberhart, R.: Particle swarm optimization. In: Proceedings of IEEE International Conference on Neural Networks (ICNN), pp. 1942–1948 (1995)
10. Kitayama, S., Arakawa, M., Yamazaki, K.: Penalty function approach for the mixed discrete nonlinear problems by particle swarm optimization. Structural and Multidisciplinary Optimization 32(3), 191–202 (2006)
11. Konno, H., Yamazaki, H.: Mean-absolute deviation portfolio optimization model and its applications to tokyo stock market. Management Science 37(5), 519–531 (1991)
12. Liang, J.J., Qin, A.K., Suganthan, P.N., Baskar, S.: Comprehensive learning particle swarm optimizer for global optimization of multimodal functions. IEEE Transactions on Evolutionary Computation 10(3), 281–295 (2006)
13. Markowitz, H.: Portfolio Selection. The Journal of Finance 7(1), 77–91 (1952)

14. Qin, Q., Cheng, S., Li, L., Shi, Y.: Artificial bee colony algorithm: A survey. CAAI Transactions on Intelligent Systems 9(2), 127–135 (2014)
15. Rockafellar, R.T., Uryasev, S.: Optimization of conditional value-at-risk. Journal of Risk 2(3), 21–41 (2000)
16. Shi, Y., Eberhart, R.: A modified particle swarm optimizer. In: Proceedings of the 1998 Congress on Evolutionary Computation (CEC1998), pp. 69–73 (1998)
17. Yoshimoto, A.: The mean-variance approach to portfolio optimization subject to transaction costs. Journal of the Operations Research Society of Japan 39(1), 99–117 (1996)
18. Young, M.R.: A minimax portfolio selection rule with linear programming solution. Management Science 44(5), 673–683 (1998)
19. Zhang, G., Li, Y.: Orthogonal experimental design method used in particle swarm optimization for multimodal problems. In: The Sixth International Conference on Advanced Computational Intelligence (ICACI 2013), pp. 183–188 (October 2013)
20. Zhu, G., Kwong, S.: Gbest-guided artificial bee colony algorithm for numerical function optimization. Applied Mathematics and Computation 217(7), 3166–3173 (2010)

A Modified Multi-Objective Optimization Based on Brain Storm Optimization Algorithm

Lixia Xie, Yali Wu

Xi'an University of Technology, Xi'an Shaanxi
710048, China

Abstract. In recent years, many evolutionary algorithms and population-based algorithms have been developed for solving multi-objective optimization problems. In this paper, A new Multi-objective optimization algorithm-Modified Multiobjective Brain Storm Optimization (MMBSO) algorithm is proposed. The clustering strategy acts directly in the objective space instead of in the solution space and suggests potential Pareto-dominance areas in the next iteration. A Density-Based Algorithm for Discovering Clusters in Large Spatial Databases with Noise (DBSCAN) clustering and Differential Evolution (DE) mutations are used to improve the performance of MBSO. A group of multi-objective problems with different characteristics were tested to validate the usefulness and effectiveness of the proposed algorithm. Experimental results show that MMBSO is a very promising algorithm for solving these tested multi-objective problems.

Keywords: Brain Storm Algorithm, Clustering Technique, Multi-objective Optimization, Pareto-dominance.

1 Introduction

Many real world problems are commonly looked at from a variety of perspectives, and therefore are represented as multiple objectives which usually conflict with each other. These problems are called Multi-objective problems, which have gained much attention in the study of sciences, economic, engineering, etc. The optimum solution for a multi-objective optimization problem is not unique but a set of candidate solutions. In the candidate solution set, no solution is better than any other one with regards to all objectives. This set is named as Pareto-optimal set, and the associated objective vectors form the trade-off surface, also called Pareto-front, in the objective space.

During the last decades, a number of evolutionary algorithms and population-based methods have been successfully used to solve multi-objective optimization problems. For example, there are Multiple Objective Genetic Algorithm (MOGA) [1], Nondominated Sorting Genetic Algorithm (NSGA, NSGA II)[2][3] , Strength Pareto Evolutionary Algorithm (SPEA, SPEA II) [4][5], Multi-objective Particle Swarm Optimization (MOPSO) [6], to name just a few. Most of the above algorithms can improve the convergence and distribution of the Pareto-front more or less.

Y. Tan et al. (Eds.): ICSI 2014, Part II, LNCS 8795, pp. 328–339, 2014.

Human beings, as one kind of social animals, are the most intelligent in the world. When we face a difficult problem which every single person cannot solve, group person, especially with different background, get together to brain storm, the problem can usually be solved with high probability. Being inspired by this human idea generation process, Shi [6] proposed a novel optimization algorithm - Brain Storm Optimization (BSO) algorithm. The simulation results on two single-objective benchmark functions validated the effectiveness and usefulness of the BSO to solve optimization problems. In [8], two novel component designs were proposed to modify the BSO algorithm and it has significantly enhanced the performance of BSO. In [9] and [10], a multi-objective optimization algorithm based on the brainstorming process was developed. Simulation results illustrated that it can be a good optimizer for solving multi-objective optimization problems.

In this paper, a modified Multi-objective BSO (MMBSO) algorithm with clustering strategy in the objective space is proposed to solve multi-objective optimization problems. Instead of action on the population and on the obtained Pareto front, in the MMBSO, the clustering strategy acts directly on the objective vectors in the objective space. Then this operation gives a feedback to the decision space to decide which candidate solution should survive. The novel using of the clustering technique, especially for the multi-objective optimization problems with high dimensional decision vectors could reduce computational burden. Clustering and mutation, the main operators of BSO were analyzed by using a Density-Based Algorithm for Discovering Clusters in Large Spatial Databases with Noise (DBSCAN) clustering and Differential Evolution (DE) mutation which is different from the previous operator. Then the different dimensions of bench functions that named ZDT [3] were tested. The simulation results showed that MMBSO would be a promising algorithm in solving multi-objective optimization problems.

The remaining paper is organized as follows. Section 2 briefly reviews the related works about the BSO and the MOP. In Section 3, the Modified Multi-objective BSO (MMBSO) is introduced and described in detail. Section 4 contains the simulation results and discussion. Finally, Section 5 provides the conclusions and some possible paths for future research.

2   Related Work

2.1   Multi-Objective Optimization Problem (MOP)

Without loss of generality, all of the multi-objective optimization problems can be formulated as minimization optimization problems. Let us consider a multi-objective optimization problem:

$$\text{Minimize } \mathbf{F}(\mathbf{X}) = (f_1(\mathbf{X}), f_2(\mathbf{X}), \cdots, f_M(\mathbf{X})) \tag{1}$$

Where $\mathbf{X} = (x_1, \cdots, x_D) \in \mathfrak{R}^D$ is called the decision vector in the D dimensional search space and $\mathbf{F} \in \Omega^M$ is the objective vector with M objectives in the M dimensional objective space. The basic concepts of a minimization MOP can be described in [11]. Two goals of a multi-objective optimization are the convergence to the true Pareto-optimal set, and the maintenance of diversity of solutions in the Pareto front set. Many performance metrics have been suggested to measure the performance of multi-objective optimization algorithms. In this paper, we use the metric Υ and metric Δ, which were defined by Deb et al. in [3], to measure the performance of the MBSO algorithm.

2.2   Brainstorm Optimization Algorithm

The BSO algorithm is designed based on the brainstorming process [12]. In the brainstorming process, the generation of the idea obeys the Osborn's original four rules [12]. The people in the brainstorming group will need to be open-minded as much as possible and therefore generate more diverse ideas. Any judgment or criticism must be held back until at least the end of one round of the brainstorming process, which means no idea will be ignored. The algorithm is described as follows. In the initialization, N potential individuals were randomly generated. During the evolutionary process, BSO generally uses the clustering technique, mutation operator and selection operator to create new ideas based on the current ideas, so as to improve the ideas generation by generation to approach the problem solution. In the clustering technique, BSO uses a k-means clustering [11]. In the mutation operator, BSO creates N new individuals one by one based on the current ideas. To create a new individual, BSO first determines whether to create the new individual based on one selected cluster or based on two selected clusters. After the cluster(s) have been selected, BSO then determines whether create the new idea based on the cluster center(s) or random idea(s) of the cluster(s). No matter to use the cluster center or to use random idea of the cluster, we can regard the selected   based   idea   as $X_{selected}$ which   can   be   expressed as $X_{selected} = (x_{selected}^1, x_{selected}^2, \cdots, x_{selected}^d)$, then applying a mutation of the $X_{selected}$ to get the new idea X_{new} which can be expressed as $X_{new} = (x_{new}^1, x_{new}^2, \cdots, x_{new}^d)$. After the new idea X_{new} has been created, BSO evaluates X_{new} and replaces $X_{selected}$ if X_{new} has a better fitness than $X_{selected}$. The procedure of the BSO algorithm is shown in [6].

In fact, there have been several recent proposals to extend BSO to handle optimization   problems.   For   example,   A   Modified   Brain   Storm Optimization[8],Predator–Prey Brain Storm Optimization for DC Brushless Motor [15],Brain Storm Optimization [13],Solution Clustering Analysis in Brain Storm Optimization Algorithm [14], Brain Storm Optimization Algorithm for Multi-objective Optimization Problems[9], and Multi-objective Optimization Based on Brain Storm Optimization Algorithm(MBSO)[10]. In [10], The BSO is used to solve multi-objective problems in which the k-means cluster and Gaussian mutation was used. Besides, Cauchy mutation was also utilized.

3 Modified Multi-Objective Brain Storm Optimization Algorithm (MMBSO)

The paper makes improvements about clustering and mutation operations for the paper [10]. DBSCAN clustering and differential mutation was used to improve the original algorithm. Also a probability of generating a random individual is added to increase the diversity of algorithm.

3.1   Clustering Technique

In the Multi-objective Brain Storm Optimization Algorithm, the k-means cluster algorithm was used in the clustering technique, but the k-means cluster algorithm has two disadvantages: First, it must specify the cluster center in advance, thus it could have been the idea of the same class was assigned to a different class and then makes the clustering lost its original role in the algorithm. And the second is that it must determine the number of cluster, and the number of cluster is fixed which is not changed with the idea of change in each iteration. In MMBSO, the clustering technique is implemented by a clustering method based on density named DBSCAN(A Density-Based Algorithm for Discovering Clusters in Large Spatial Databases with Noise)[19] .We can image that when the similar ideas beyond a certain number, they can be put together as a class, then the ideas generated in each iteration can be fully used and thus makes different number of clusters according to the different ideas in each iteration. There are a lot of density-based clustering algorithms currently, in this article, in order to make the algorithm simple we use the simplest and most classical density-based clustering algorithm which is named DBSCAN. The procedure of it is shown in [19].

3.2   Generation Progress

The generation progress of MMBSO contains the mutation and selection operator which can be referred as follows:

- Mutation Operator:
  The mutation operator plays an important role in the generation progress. Gaussian and Cauchy mutation are typical mutation operators which were used in [10]. In solving some multimodal or multi-objectives optimization problems, Gaussian mutation as one of the main operators in the algorithm may lead to a result with a slow convergence to a good near-optimum. In MMBSO, Differential Mutation is used to improve the performance of the algorithm.

  As to BSO, a new idea X_{new} is created by adding Gaussian random noise to a based idea $X_{selected}$. It also can be known that the noise is large in early evolutionary phase and gradually become smaller during the running by the control of the logarithmic sigmoid transfer function according to [8]. Such a time varying noise strategy is consistent with the commonly intuition that large noises are needed in the early phase for global search while small noises are needed in the late phase for local fine-tuning.

In this paper, we propose to use the Differential Mutation to produce the noise value. The Differential Mutation is based on such a consideration. In the human being's brainstorming process, we can image that at the beginning of the process, everyone's idea would be much different. When they create new ideas based on the current ideas, they should take the differences of the current ideas into consideration. For example, when creating a new idea X_{new} based on a current idea $X_{selected}$, two distinct random ideas X_a which can be expressed as $X_a = (x_a^1, x_a^2, \cdots, x_a^d)$ and X_b which can be expressed as $X_b = (x_b^1, x_b^2, \cdots, x_b^d)$ from all the current ideas are taken to represent the idea difference, and the X_{new} is created as:

$$x_{new}^d = x_{selected}^d + rand(0,1)_d \times (x_a^d - x_b^d)$$

(2)

Where $rand(0,1)_d$ is a random number between (0, 1).

Using Eq. (2) to create new ideas, there are two advantages. Firstly, the computational burden of (2) is much lighter than that of the mutation of BSO that involves logarithmic sigmoid transfer function, Gaussian distribution function, random function, addition, subtraction, multiplication, and division, while (2) involves random function, multiplication, and subtraction for making up the noise value. Secondly, Eq. (2) can match the search environment of the evolutionary process. Be consistent with the brainstorming process for human being in solving problem, the ideas are much different from each other in the beginning, therefore the term ($x_a^d - x_b^d$) in (2) is larger and the new created ideas can keep the diversity in the early phase. In the late phase of the brainstorming process, the people may reach a consensus and the idea difference may be smaller. In this condition, the term ($x_a^d - x_b^d$) in (2) is also smaller to help refine the ideas. Therefore, Eq. (2) may be good at balance the global search and local search abilities according to the search information during the evolutionary process.

- Selection Operator:

It is also quite important to decide whether any newly generated solution should survive to the next generation. The selection based on Pareto dominance is utilized in this paper.

3.3   Update the Pareto Set

The Pareto Set is updated by the new non-dominated solutions. In this step, each new non-dominated solution obtained in the current iteration will be compared with all members in the Pareto Set. If the size of the Pareto Set exceeds the maximum size limit, it is truncated using the diversity consideration. In this paper, the circular crowded sorting operator [18] is adopted to guide the points toward a uniformly spread-out Pareto-optimal front.

3.4   The Procedure of the MMBSO

The MMBSO contains clustering, mutation, selection, and updating the Pareto set, which have been described above. The whole procedure of the MMBSO is shown in Fig. 1.

In the process of MMBSO algorithm, population size N imitate the number of ideas generated during the course of each round; randomly selects an elite cluster imitate the process of generating new ideas which excited by a single thought. Randomly selects two clusters was imitated by the idea of two people from different clusters inspired by the process of generating new ideas. Select an individual from the archive set was used to keep a few good ideas which imitate to pick up several good ideas in the brain storm process. The clustering technique is used to classify the idea of people with different backgrounds and ideas. And in each cluster, the cluster center acts as a facilitator of the problem to solve problems with better idea. The mutation operator is used to generate new ideas on the basis for the existing ideas. Also different from MBSO, a probability of generating a random individual is added to increase the diversity of algorithm.

4   Experiments and Discussions

In this section the MMBSO will be tested. Without loss of generality, all the multi-objective optimization problems tested in this paper are minimization problems.

4.1   Test Problems

In order to evaluate the performance of MMBSO, the ZDT test functions [3] are used in this paper. The ZDT suite is comprised of six problems, each one presenting a specific characteristic that generally cause difficulties to major evolutionary optimization strategies [20]. The bi-objective test functions used to examine the effect of the introduced MMBSO are the ZDT1, ZDT2, ZDT3, ZDT4 and ZDT6. Test problem ZDT1 and ZDT3 have convex Pareto fronts while other test problems have non-convex Pareto fronts; ZDT3 also possesses a disconnected Pareto front; the Pareto front of ZDT6 is non-uniformly spaced; and ZDT4 as a complex multimode problem is difficult to find the global Pareto front. The information of these test functions in detail can be seen in [3].

4.2   Parameter Settings

During the test, a lot of parameters are used to test the algorithm. Finally a set of parameter that is relatively good for these test functions is used. In all the simulation runs, the population size is set to be 200 and the maximum size of the Pareto set is fixed at 100. After conducting a series of experiments, the pre-determined probability values P1 is set to be 0.99, P2 and P3 are set to 0.8, P4 and P5 are set to 0.2. For the DBSCAN clustering, MinPts is set to 7 while ε is set according to [21]. All of the algorithms are implemented in MATLAB using a real-number representation for decision variables. For each experiment, 30 independent runs were conducted to collect statistical results. Each test problem will be run with different dimension, 5, 10, 20, and 30, respectively.

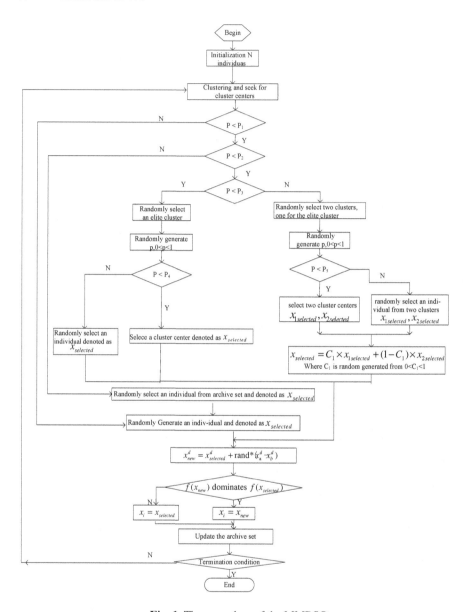

Fig. 1. The procedure of the MMBSO

4.3   Results

In all simulation runs, the metric Υ [3]and metric Δ [3]will be calculated and recorded for all the test problems. Table 1 compares the best and mean values of the convergence metric Υ obtained using MMBSO (denoted as DE), MBSO-G (MBSO with Gaussian mutation)[10] and MBSO-C (MBSO with Cauchy mutation)[10]. The diversity metric Δ about the test problems are listed in Table 2. The best results are marked with italics and bold.

Table 1. The comparisons of best and mean value of γ between MMBSO,MBSO-G and MBSO-C

Dimension	Algorithm	ZDT1		ZDT2		ZDT3		ZDT4		ZDT6	
		best	mean	best	mean	best	mean	best	mean	best	mean
5	DE	*0.0010*	*0.0011*	*0.6806 e-003*	*0.7904 e-003*	*0.0010*	*0.0012*	0.0062	0.1898	*0.0037*	*0.0040*
	G	0.0011	0.0016	0.0007	0.0008	0.0012	0.0015	*0.0009*	*0.0019*	0.0040	0.0048
	C	0.0011	0.0017	0.0009	0.0011	0.0012	0.0014	0.0010	0.0048	0.0039	0.0050
10	DE	*0.0009*	*0.0011*	*0.6803 e-003*	*0.7888 e-003*	*0.0010*	*0.0012*	0.0015	1.9661	*0.0036*	*0.0041*
	G	0.0050	0.0079	0.0014	0.0050	0.0024	0.0033	0.0029	2.4179	0.0046	0.0079
	C	0.0031	0.0062	0.0016	0.0029	0.0018	0.0027	*0.0011*	*0.1335*	0.0046	0.0072
20	DE	*0.0010*	*0.0011*	*0.7047 e-003*	*0.8018 e-003*	*0.0011*	*0.0012*	*0.0016*	6.7146	*0.0037*	*0.0040*
	G	0.0328	0.0472	0.0294	0.0498	0.0168	0.0248	2.8367	18.768	0.0323	0.0412
	C	0.0193	0.0312	0.0186	0.0311	0.0111	0.0150	1.4905	*4.4188*	0.0151	0.0241
30	DE	*0.0010*	*0.0011*	*0.0007*	*0.0008*	*0.0011*	*0.0012*	*2.9322*	*13.8379*	*0.0037*	*0.0040*
	G	0.1060	0.1347	0.1006	0.1308	0.0863	0.1136	10.3624	38.2581	0.0928	0.1385
	C	0.0695	0.0912	0.0725	0.0905	0.0443	0.0589	6.4966	15.2905	0.0580	0.0813

Table 2. The comparisons of best and mean value of Δ between MMBSO,MBSO-G and MBSO-C

Dimension	Algorithm	ZDT1		ZDT2		ZDT3		ZDT4		ZDT6	
		best	mean	best	mean	best	mean	best	mean	best	mean
5	DE	*0.1074*	*0.1266*	*0.1031*	*0.1228*	*0.4128*	*0.4180*	0.6160	1.0436	*0.5294*	*0.5422*
	G	0.3478	0.4187	0.3363	0.4016	0.5667	0.6013	0.3550	0.5058	0.6635	0.7021
	C	0.3325	0.4066	0.3422	0.4126	0.5016	0.5763	*0.3444*	*0.4786*	0.6627	0.7002
10	DE	*0.1087*	*0.1259*	*0.1031*	*0.1213*	*0.4113*	*0.4195*	*0.1413*	1.2555	*0.5325*	*0.5473*
	G	0.3765	0.4370	0.3475	0.4258	0.5420	0.6054	0.5159	0.8372	0.6751	0.7050
	C	0.3882	0.4542	0.3682	0.4115	0.5357	0.5931	0.4330	*0.7865*	0.6633	0.6989
20	DE	*0.1098*	*0.1275*	*0.0955*	*0.1208*	*0.4117*	*0.4173*	*0.1138*	1.2331	*0.5339*	*0.5443*
	G	0.4238	0.4789	0.3975	0.4869	0.5381	0.5868	0.9342	0.9748	0.6823	0.7103
	C	0.4215	0.4717	0.4371	0.4855	0.5073	0.5795	0.8660	*0.9569*	0.6724	0.7038
30	DE	*0.1008*	*0.1257*	*0.0997*	*0.1253*	*0.4126*	*0.4188*	1.2958	1.3968	*0.5322*	*0.5436*
	G	0.4823	0.5340	0.4834	0.5494	0.6026	0.6364	0.9619	0.9890	0.7080	0.7452
	C	0.5105	0.5529	0.4898	0.5588	0.5708	0.6293	*0.8362*	*0.9699*	0.6969	0.7425

In the Table 1 and 2, DE is represents MMBSO,G is MBSO-G and C is MBSO-C. The result from table 1 ~ 2 showed that MMBSO has better convergence and diversity than MBSO-G and MBSO-C except ZDT4. For the ZDT4, it is a complex multimode problem which is difficult to find the global Pareto front. DBSCAN which used in MMBSO uses the current ideas to get the clustering result, so it is limited the ZDT4 find the global optimum to some extent while k-means cluster can made a slightly good performance for ZDT4.

To further verify the performance of the algorithm, the hypervolume(HV)[22] which can evaluate convergence and diversity at the same time is added to illustrate the performance of the algorithm. The Table 3 compares the best and mean values of the ratio of HV that get by the solution set of algorithm and the corresponding true solution on the Pareto front. The ratio is between 0-1. If the ratio equals 1, then the solution obtained by the algorithm is on the true Pareto frontier. The more the ratio is close to 1, the solution obtained by algorithm is closer to the true Pareto frontier. The best results are marked with italics and bold.

Table 3. The comparisons of best and mean value of reaching the target HV value between MMBSO,MBSO-G and MBSO-C

Dime-nsion	Algo-rithm	ZDT1		ZDT2		ZDT3		ZDT4		ZDT6	
		best	mean	best	mean	best	mean	best	mean	best	mean
5	DE	0.9945	*0.9945*	0.9892	*0.9891*	*0.9975*	*0.9974*	*0.9988*	*0.9944*	*0.9945*	*0.9943*
	G	0.9908	0.9891	0.9811	0.9795	0.9958	0.9951	0.9952	*0.9915*	0.9938	0.9909
	C	*0.9976*	0.9910	*0.9986*	0.9874	1.5494	1.3569	0.9961	0.8582	0.9989	0.9939
10	DE	*0.9945*	*0.9944*	*0.9931*	*0.9890*	*0.9975*	*0.9974*	*0.9998*	*0.9962*	*0.9944*	*0.9938*
	G	0.9861	0.9831	0.9773	0.9700	0.9922	0.9904	0.9890	0.7215	0.9869	0.9825
	C	0.9939	0.9779	0.9924	0.9476	0.9939	0.9774	*0.9797*	*0.7009*	0.9942	0.9802
20	DE	*0.9950*	*0.9944*	*0.9956*	*0.9888*	*0.9975*	*0.9973*	*0.9999*	*0.9983*	*0.9942*	*0.9933*
	G	0.9729	0.9579	0.9493	0.9146	0.9706	0.9553	0.7565	0.4163	0.9786	0.9518
	C	0.9720	0.8955	0.9450	0.8078	1.1744	1.1016	*0.9775*	*0.7821*	0.9577	0.9130
30	DE	*0.9945*	*0.9943*	*0.9891*	*0.9889*	*0.9974*	*0.9973*	*0.9991*	*0.9989*	*0.9943*	*0.9930*
	G	0.9393	0.9069	0.8898	0.8203	0.9333	0.8824	0.7124	0.3065	0.9455	0.8990
	C	0.9298	0.7910	0.8709	0.6943	0.9847	0.9403	0.9732	0.7143	0.8725	0.7973

In the Table 3, DE is represents MMBSO,G is MBSO-G and C is MBSO-C.As can be seen from Table 3, the results obtained by MMBSO are better than MBSO-G and MBSO-C for the dimensions 5, 10, 20 and 30. Most of the obtained solution reached by MMBSO is more than 0.99, which illustrates the algorithm has good performance.

The Fig. 2-5 shows the simulation result. In the figure, the fine line represents the true Pareto edge and the point represents the solution get by the algorithm.

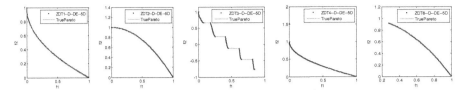

Fig. 2. Pareto-optimal front of ZDT1, 2, 3, 4 and 6 obtained by the MMBSO (5 dimension)

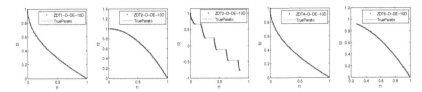

Fig. 3. Pareto-optimal front of ZDT1, 2, 3, 4 and 6 obtained by the MMBSO (10 dimension)

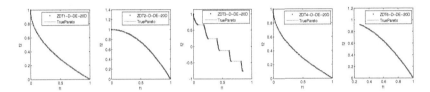

Fig. 4. Pareto-optimal front of ZDT1, 2, 3, 4 and 6 obtained by the MMBSO (20 dimension)

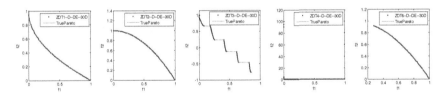

Fig. 5. Pareto-optimal front of ZDT1, 2, 3, 4 and 6 obtained by the MMBSO (30 dimension)

The result from table 1~2, also with the Fig. 2-6 show that MMBSO does better than MBSO on all functions except some result of ZDT4. This may be due to that DBSCAN clustering can make good use of the ideas generated in each iteration and get a reasonable clustering result according to the current idea, thus get a good current optimal solution. In addition, Differential Evolution mutation can match the search environment to provide suitable noise to create better ideas around the global optimal region to refine the solution for high accuracy. In MBSO, the new created idea was disturbed based on the current idea by Gaussian noise. However, this noise may be coarse. In the contrast, MMBSO uses the difference between two ideas as the disturbed noise. This way, the disturbed noise can be within a comparable order of magnitude with the current ideas. With the combination of DBSCAN clustering and Differential Evolution mutation, MMBSO get a good convergence and the diversity.

5   Conclusions and Discussion

In this paper, a modified multi-objective brain storm optimization algorithms, called MMBSO, was introduced. In the MMBSO, A clustering operator named DBSCAN has been utilized to cluster the different individuals and a different mutation operator is utilized to generate new individuals in the generation progress of the algorithm.

The results of the MMBSO have been evaluated according to three performance measures. From the simulation results, it was observed that MMBSO can be a good optimizer for solving multi-objective optimization problems. Although MMBSO performs better than the MBSO on most test problems in this study, more problems need to be tested to fully confirm this observation.

The results in Table 1 and 2 showed that MMBSO has a bad preference in solving ZDT4. In order to improve the shortcoming of the MMBSO, adaptive and mixing mutations based on niching techniques should be investigated. Another interesting area is to exploit the MMBSO for solving multi-objective optimization problems with constraints , many-objective problems and the other new test functions.

References

1. Fonseca, C.M., Fleming, P.J.: Genetic Algorithms for Multiobjective Optimization: Formulation, Discussion and Generalization. In: Proceedings of the Fifth International Conference on Genetic Algorithms. University of Illinois at Urbana-Champaign, pp. 416–423. Morgan Kaufmann Publishers (1993)
2. Srinivas, N.: Deb. K.: Multiobjective Optimization Using Nondominated Sorting in Genetic Algorithms. Evolutionary Computation 2(3), 221–248 (1994)
3. Deb, K., Pratap, A., Agarwal, S., Meyarivan, T.: A Fast and Elitist Multiobjective Genetic Algorithm: NSGA–II. IEEE Transactions on Evolutionary Computation 6(2), 182–197 (2002)
4. Zitzler, E., Thiele, L.: Multiobjective Evolutionary Algorithms: A Comparative Case Study and the Strength Pareto Approach. IEEE Transactions on Evolutionary Computation 3(4), 257–271 (1999)
5. Zitzler, E., Laumanns, M., Thiele, L.: SPEA2: Improving the Strength Pareto Evolutionary Algorithm. In: Evolutionary Methods for Design, Optimization and Control with Applications to Industrial Problems, pp. 95–100 (2001)
6. Coello, C.A.C., Pulido, G., Lechuga, M.: Handling multi-objective with particle swarm optimization. IEEE Transactions on Evolutionary Computation 8(3), 256–279 (2004)
7. Shi, Y.: Brain Storm Optimization Algorithm. In: Tan, Y., Shi, Y., Chai, Y., Wang, G. (eds.) ICSI 2011, Part I. LNCS, vol. 6728, pp. 303–309. Springer, Heidelberg (2011)
8. Zhan, Z., Zhang, J., Shi, Y., Liu, H.: A Modified Brain Storm Optiization. In: IEEE World Congress on Computational Intelligence, pp. 10–15 (2012)
9. Xue, J., Wu, Y., Shi, Y., Cheng, S.: Brain Storm Optimization Algorithm for Multi-objective Optimization Problems. In: Tan, Y., Shi, Y., Ji, Z. (eds.) ICSI 2012, Part I. LNCS, vol. 7331, pp. 513–519. Springer, Heidelberg (2012)
10. Shi, Y., Xue, J., Wu, Y.: Multi-objective Optimization Based on Brain Storm Optimization Algorithm. Journal of Swarm Intelligence Research (IJSIR) 4(3) (2013)

11. Coello, C.A.C., Becerra, R.L.: Evolutionary Multiobjective Optimization using a Cultural Algorithm. In: Proceedings of IEEE Swarm Intelligence Symposium (SIS 2003), pp. 6–13 (2003)
12. Smith, R.: The 7 Levels of Change, 2nd edn. Tapeslry Press (2002)
13. Zhan, Z., Chen, W., Lin, Y., Gong, Y., Li, Y., Zhang, J.: Parameter Investigation in Brain Storm Optimization. In: IEEE Symposium on Swarm Intelligence (SIS), pp. 103–110 (2013)
14. Cheng, S., Shi, Y., Qin, Q., Gao, S.: Solution Clustering Analysis in Brain Storm Optimization Algorithm. In: IEEE Symposium on Swarm Intelligence (SIS), pp. 111–118 (2013)
15. Duan, H., Li, S., Shi, Y.: Predator–Prey Brain Storm Optimization for DC Brushless Motor. IEEE Transactions on Magnetics 49(10), 5336–5340 (2013)
16. Xu, D., Wunsch II, D.: Survey of Clustering Algorithms. IEEE Transactions on Neural Networks 16(3), 645–677 (2005)
17. Jain, A.K.: Data clustering: 50 years beyond K-means. Journal of Pattern Recognition Letters 31, 651–666 (2010)
18. Luo, C., Chen, M., Zhang, C.: Improved NSGA-II algorithm with circular crowded sorting. Control and Decision 25(2), 227–232 (2010)
19. Ester, M., Kriegel, H., Sander, J., Xu, X.: A Density-Based Algorithm for Discovering Clusters in Large Spatial Databases with Noise. In: Proceedings of 2nd International Conference on Knowledge Discovery and Data Mining, KDD 1996 (1996)
20. Adra, S.F., Dodd, T.J., Griffin, I.A., Fleming, P.J.: Convergence Acceleration Operator for Multiobjective Optimization. IEEE Transactions on Evolutionary Computation 13(4), 825–847 (2009)
21. Daszykowski, M., Walczak, B., Massart, D.L.: Looking for Natural Patterns in Data. Part 1: Density Based Approach, Chemmon Intell. Lab. Syst. 56, 83–92 (2001)
22. Zitzler, E., Thiele, L.: Multiobjective evolutionary algorithms: a comparative case study and the strength Pareto approach. IEEE Transactions on Evolutionary Computation 3(4), 257–271 (1999)

Modified Brain Storm Optimization Algorithm for Multimodal Optimization

Xiaoping Guo, Yali Wu, and Lixia Xie

Xi'an University of Technology, Xi'an Shaanxi, China
710048

Abstract. Multimodal optimization is one of the most challenging tasks for optimization. The difference between multimodal optimization and single objective optimization problem is that the former needs to find both multiple global and local optima at the same time. A novel swarm intelligent method, Self-adaptive Brain Storm Optimization (SBSO) algorithm, is proposed to solve multimodal optimization problems in this paper. In order to obtain potential multiple global and local optima, a max-fitness grouping cluster method is used to divide the ideas into different sub-groups. And different sub-groups can help to find the different optima during the search process. Moreover, the self-adaptive parameter control is applied to adjust the exploration and exploitation of the proposed algorithm. Several multimodal benchmark functions are used to evaluate the effectiveness and efficiency. Compared with the other competing algorithms reported in the literature, the new algorithm can provide better solutions and show good performance.

Keywords: Brain Storm Algorithm , Max-fitness Grouping Cluster, Self-adaptive Parameter Control, Multimodal Optimization.

1 Introduction

Multimodal optimization is a most challenging task in the area of optimization. Unlike the single objective optimization problems, multimodal optimization needs to provide multiple optimal solutions simultaneously. Since in the practical optimization problems, it is very common that the best solution cannot be realized at times due to the physical constraints. If multiple satisfaction solutions are known, the implementation can be quickly switched to an alternative solution while the system performance is still maintaining. As the name suggests, multimodal optimization requires optimization algorithms to find multiple optimal solutions (both local and global) and not just one single optimum as is done in a typical optimization study.

Evolutionary algorithms (EAs), due to their population-based approach, provide a population of possible solutions processed at every iteration. If multiple solutions can be preserved over all these iterations, we can have multiple good solutions at termination of the algorithm. Although evolutionary algorithm shows the potential power on multimodal optimization, two difficulties may be faced in the multimodal optimization algorithm. The first is how to locate multiple global and local optima

Y. Tan et al. (Eds.): ICSI 2014, Part II, LNCS 8795, pp. 340–351, 2014.
© Springer International Publishing Switzerland 2014

during the evolutionary process. And the second is that how to maintain the identified optima until the end of the search.

With the development of computer science and technology, numerous techniques have been developed for locating multiple optima (global or local). Multipopulation based method can be incorporated into a standard EA to promote and maintain formation of multiple stable subpopulations within a single population to locate multiple optimal or suboptimal solutions. One of the multipopulation techniques is commonly referred as "niching" methods. R. Thomsen [1] uses crowding metric to force new individuals entering a population to replace similar individuals. But the algorithm suffers from higher computational complexity. And the performance relies on prior knowledge of some niching parameters. Fitness Euclidean distance ratio PSO (FERPSO) [2] and speciation-based PSO (SPSO) [3] are two commonly effective niching PSO algorithms. It is designed only for locating all global optima, while ignoring local optima. In the literature [4], Thomsen proposed a Crowding DE (CDE) to solve multimodal problems. It is generally difficult to select suitable trial vector generation strategies and control parameters for CDE that can generate satisfactory performance over all test functions. Most of existing niching methods have difficulties to be overcome before they can be applied successfully to real-world multimodal problems [5]. In recent years, the clustering technique, as another multipopulation method, is used successfully to solve multi-modal optimization by more and more scholars. Yin and Germay [8] proposed an Adaptive Clustering algorithm (ACA) to avoid the a priori estimation of σ_{share}. ACA adopts the identified cluster instead of sharing the fitness function, but this method introduces two additional variables at the same time, which need to set a reasonable maximum and minimum value as the radius of the cluster. Hanagandi and Nikolaou [10] also use the clustering method to find the global optima in a genetic search framework. However, the main difficulty of all the methods lies in how to define the area for each sub-region in the search space and how to determine the number of sub-populations.

Brain Storm Optimization (BSO) algorithm, inspired by human idea generation process, is proposed by Shi in [11] to solve single objective optimization problem. But two features make it perform well in multimodal problems. One is the clustering operator that divides all the ideas generated in the current generation into some different groups, which is possible to maintain multi optimal solutions. And the other is the creating operator that creates new idea by learning from the self-group or other group, which can maintain the diversity of each group. While the classical clustering method such as k-means cannot solve the multimodal problem well. So in this paper, a new clustering method named Max-fitness Clustering Method (MCM) is cooperated with BSO for multimodal problem. And the self-adaptive parameter control is used for the creating operator. The clustering method enables the algorithm to assign individuals to different promising sub-regions. And the self-adaptive parameter control methods are effective in maintaining the diversity of the population.

The rest of the paper is organized as follows. Section II briefly reviews the related works about BSO. The improved algorithm is described in detail in section III. The parameter setting and results are given in Section IV. And the conclusion and further research are detailed finally in Section V.

2   Brain Strom Optimization

In any swarm intelligence algorithm, each individual cooperatively and collectively move toward the better and better areas in the solution space. When Human being face a difficult problem which every single person cannot solve, group person, especially with different background, get together to brain storm, the problem can usually be solved with high probability. Being inspired by this human idea generation process, in 2011, Shi proposed a novel algorithm named Brain Storm Optimization(BSO). In the brainstorming process, the generation of the idea obeys the Osborn's original four rules (Smith, 2002). The people in the brainstorming group will need to be open-minded as much as possible and therefore generate more diverse ideas. Any judgment or criticism must be held back until at least the end of one round of the brainstorming process, which means no idea will be ignored. The algorithm is given in Fig.1 and is described as follows:

In the initialization, N potential ideas were randomly generated. BSO uses a k-means clustering [12] as its clustering technique. In the selection operator, BSO creates N new ideas one by one based on the current ideas. To create a new idea, BSO first determines whether to create the new idea based on one selected cluster or based on two selected clusters. After the cluster(s) have been selected, BSO then determines the selected ideas whether create the new idea based on the cluster center(s) or random idea(s) of the cluster(s).

```
Algorithm BSO
01 Begin
02 Randomly generate N ideas(X,,1≤i≤N)and evaluate their fitness;
03 While (Not stop)Do
04   Cluster the N ideas into M clusters;//According to the positions
05   Record the best idea in each cluster as the cluster center,
       //Probability of replacing a randomly selected cluster center,0.2
06   If(random(0,1)< p_replace )
07     Randomly selected a cluster and replace the cluster center
       with a randomly generated idea;
08   End of If
09   For( i =1 to N )
       //Probability of generating new idea based on one cluster ,0.8
10     If(random(0,1)< p_one )
11       Randomly select a cluster j with a probability p_j;
         //Probability of using the cluster center, p_one_center =0.4
12       If (random(0,1)< p_one_center )
```

```
13          Add random values to the selected cluster center
            to generate a new idea Y₁ ;
14        Else
15          Add random values to a random idea of the selected cluster
            to generate a new idea Y₁ ;
16        End of If
17      Else//Generating new idea based on two clusters
18        Randomly select two clusters j₁ and j₂ ;
            //Probability of using the cluster center, p_two_center =0.5
19        If (random(0,1)< p_two_center )
20          Combine the two selected cluster centers and add with
            random values to generate a new idea Y₁ ;
21        Else
22          Combine two random ideas from the two selected clusters
            and add with random values to generate a new idea Y₁ ;
23        End of If
24      End of If
25      Evaluate the idea Y₁ and replace X₁ if Y₁ has better fitness than
X₁ ;
26    End of For
27 End of While
28 End
```

Fig. 1. Pseudo-code of the BSO algorithm

If the selected idea $X_{selected} = (x^1_{selected}, x^2_{selected}, ..., x^d_{selected})$ is gotten, a mutation of the $X_{selected}$ is applied to get the new idea expressed as $X_{new} = (x^1_{new}, x^2_{new}, ..., x^d_{new})$. After the new idea X_{new} has been created, BSO evaluates X_{new} and replaces $X_{selected}$ if X_{new} has a better fitness than $X_{selected}$.

In the mutation process, the Gaussian mutation will be used as random values which are added to generate new ideas; it can be represented as follows:

$$x^d_{new} = x^d_{selected} + \xi * N(\mu, \sigma) \qquad (1)$$

$$\xi = logsig((0.5 * max\_iternation - current\_iteration) / K) * rand() \qquad (2)$$

In the equation (1) and (2), $x^d_{selected}$ is the d-dimensional of the idea selected to generate new idea; x^d_{new} is the d-dimensional of the idea newly generated; $N(\mu, \sigma)$ is the Gaussian random function with mean μ and σ ; ξ is a coefficient that weights the contribution of the Gaussian mutation; $log sig()$ is a logarithmic sigmoid transfer function; $max\_iternation$ and $current\_iteration$ are the maximum iteration number and the current iteration number, K is for changing $log sig()$ function's slope, and rand() is a random value within (0,1).

3   Self-adapted Brain Strom Optimization

In the multimodal optimization problem, the challenging issue is how to find all the global optimum, local optimum while maintaining the diversity of the population. The experimental results reported in [6]-[10] show that clustering operation is an ideal technique. Although many common evolutionary algorithms (EAs) are used to solve multimodal problems, most of them introduce a variety of clustering strategies into the evolutionary process. Different from other EAs, BSO adopts the clustering operation as their converging process, which makes it very suitable to solve multimodal problem. As we all know, different clustering strategies show different advantages. The K-means method in traditional BSO is a typical clustering based on distance. The biggest drawback is that the choice of initial cluster center has a great influence on its clustering result. Once the initial cluster center is not chosen well, it may have a bad influence on its clustering result. On the other hand, the preservation mechanism of traditional BSO is just for comparison of the fitness values of simple problems. For multimodal problem, this mechanism is not able to maintain all individuals which have found extreme point. In this paper we improved the classical BSO to solve multimodal optimization in two aspects. Firstly, a new clustering operation named Max-fitness Clustering method (MCM) replaces the classical k-means clustering method to assign individuals in different promising subregions. Secondly, a self-adaptive parameter control technique is used to ensure retention of individual diversity and convergence. Detailed explanation will be showed as the followings subsections.

3.1   Max-fitness Clustering Method

The K-means clustering method in traditional BSO overly depends on the selection of the initial cluster centers, which cannot solve multimodal optimization. The maximum clustering method is proposed to solve the different clustering center selection in this paper. The operation produces are listed as follows. Firstly, the largest individual fitness value is selected as the first category center from an individual original population. Secondly, the nearest to the center of each individual is emptied. Finally, the process is repeated for the remaining individuals until all of them are classified. The difference between the clustering and other clustering methods is that each category center is the best individual of all the remaining individuals, so each clustering center has a larger probability that is distributed in the extreme point, which makes the individual learning more direction. This clustering algorithm makes full use of the information about each individual, and the solution space is effectively combined with the target space. A schematic illustration of the clustering partition is shown in Fig 2. The replacing operator of BSO is given as line 4&5 in Fig.1.

```
Algorithm  Max-fitness clustering method
```

```
Step 1 Initialize a population P of Np ideas {X_i/i=1,2,...,Np}
       in the search region randomly
Step2 Find the best(fitness value) idea as seed X
Step3 Combine M-1 ideas of the population P, which are
      nearest to X, with X to form a subpopulation. Any
      tie will be broken randomly.
Step4 Eliminate these M ideas from P
Step5 Execute Step 1- Step 3 repeatedly until the population
P is divided
      into P/M subpopulation
```

Fig. 2. Pseudo-code of the max-fitness clustering

3.2   Self-adaptive Parameter Control

According to Eq.(1), a new idea is created by adding Gaussian random noise to the selected idea. By this way, all ideas in the group will quickly converge to the direction along the previous best idea. In this paper, a self-adaptive parameter control is used to make the algorithm quickly converge to different optima by different clusters. The crossover probability Cr is updated dynamically according to the evolution process. This is similar to the adaptation strategy used in JADE [13].

In the mutation operation, the new idea $u_{i,j}$ is generated by making use of a binomial crossover operation on the last iteration idea $x_{i,j}$ and the after mutation idea $v_{i,j}$.

$$u_{i,j} = \begin{cases} v_{i,j} & if\,(rand_j(0,1) \le C_r)\,or\,(j = j_{rand}) \\ x_{i,j} & otherwise \end{cases} \qquad (3)$$

Where $i = 1,2,...,Np$, $j = 1,2,...,n\_d$, j_{rand} is a randomly chosen integer from $\{1,2,...,n\_d\}$, $rand_j(0,1)$ is a random value within (0,1). Due to the use of j_{rand}, $u_{i,j}$ is guaranteed to differ from $x_{i,j}$. At each generation, the crossover rate Cr_i of each idea is independently generated according to a normal distribution of mean C_{rm} and standard deviation 0.1

$$Cr_i = randn(C_{rm},0.1) \qquad (4)$$

C_{rm} is updated as follow:

$$C_{rm} = mean(S_{cr}) \qquad (5)$$

Where S_{cr} is the set of all successful crossover rates at previous generation. The initial value of C_{rm} is 0.5 and $S_{cr} = \phi$.

The selection operation is conducted by comparing u_i and the closest individual x_s in population. The better one will enter the next generation. For the maximization problem:

$$x_s = \begin{cases} u_i & if \quad f(u_i) > f(x_s) \\ x_s & otherwise \end{cases} \tag{6}$$

A schematic illustration of the self-adaptive parameter control is shown in algorithm 2.

Algorithm2 self-adaptive parameter control

```
For  i=1:Np
Step 1 Use the Eq.3 to produce new idea ;
Step 2 Evaluate offspring using fitness function;
Step 3 According to Eq.(6) to update idea and save the
respective
       crossover probability in Scr
End for
Step 4 Update  Crm = mean(Scr).
```

Fig. 3. Pseudo-code of the self-adaptive parameter control

4   Experimental Studies

4.1   Parameter Setting

A set of multimodal optimization benchmark functions [14] is used to evaluate the ability of the proposed algorithm in this paper. A level of accuracy (typically $0 < \varepsilon < 1$) indicates the error between the fitness value and the given extreme point. The extreme point will be considered been found if the error is less than ε. In order to compare with other algorithms fairly, the parameters will follow the guideline given in [15] and [16] and apply these parameters to all compared algorithms. The level of accuracy (ε), niching radius (γ), population size and maximal number of function evaluations (FES) allowed are listed in Table II. From the table 2, we can see that a larger number of optima require a large population size and more function evaluations.

Table 1. Parameter setting of different benchmark functions

Test function	Population size	M	ε	γ	No. of function evaluations
F1	50	5	0.05	0.5	10,000
F2	50	5	0.05	0.5	10,000
F3	50	5	0.05	0.5	10,000
F4	50	5	0.000001	0.01	10,000
F5	50	5	0.000001	0.01	10,000
F6	50	5	0.000001	0.01	10,000
F7	50	5	0.000001	0.01	10,000
F8	50	5	0.0005	0.5	10,000
F9	50	5	0.000001	0.5	10,000
F10	250	25	0.00001	0.5	10,000
F11	250	10	0.05	0.5	100,000
F12	100	10	0.0001	0.2	20,000
F13	500	50	0.001	0.2	200,000
F14	1000	100	0.001	0.2	400,000

4.2   Experimental Result

To give a clearer view, SBSO is compared with the original BSO. The distributions of population of the two algorithms at different iterations are plotted in Fig. 4. （the function of F6） and Fig. 5. （the function of F7） .

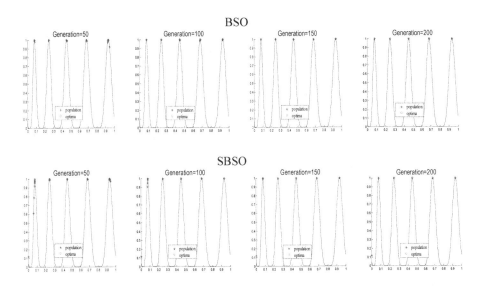

Fig. 4. Distributions of population of BSO and SBSO on different stages （F6）

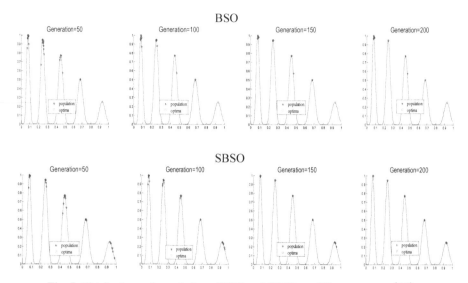

Fig. 5. Distributions of population of BSO and SBSO on different stages（F7）

From the above figure, we can clearly see the original BSO have accurately found all the extreme points of function F6 which has the same extreme points, but for the function like F7 which has different extreme points, only the SBSO can find and save all the extreme points. With the number of iterations increasing, the SBSO can converge to the extreme points. Therefore, it presents that the effectiveness of SBSO in dealing with multimodal optimization problem. In order to further demonstrate the superior performance of this algorithm, we compare the performance indicators of the proposed algorithm with the other algorithms in one of the latest multimodal works [17]. As the same as the literature [17], we adapts the following indicator as our assessment criteria.

Peak Accuracy: For each desired $peak_i,(i=1,2,...,\#\,peaks)$ to be located, the closet idea X in the population is taken and absolute difference in objective values is calculated. If the objective value of idea X is denoted by $f(X)$,the peak accuracy is calculated using

$$peak \quad accuracy = \sum_{i=1}^{\#\,peaks} \frac{\left| f\left(peak_i\right) - f\left(X\right)\right|}{\#\,peaks}$$

On all the benchmark functions, all algorithms are run until all known peaks are found or maximum number of FEs is exhausted. The proposed algorithm is compared with following standard multimodal algorithms which are generally cited in literature [17]:

1) CDE: Crowding DE;
2) SDE: Speciation-based DE
3) FER-PSO: Fitness –Euclidean distance Ratio PSO
4) SPSO: Speciation-based PSO

5) r3pso: a local best PSO with a ring topology•each member interacts with its immediate member on its left and right.

6) r2psolhc: no overlapping neighborhoods, hence acting as multiple local hill climbers, more suitable for finding global as well as local optima.

7) r3psolhc: Basically multiple PSOs search in parallel, like local hill climbers. This variant is more appropriate if the goal of optimization is to find global ad well as local optima.

The simulation results are presented in Table 3.

Table 2. Peak Accuracy comparisons of different algorithms

Fun	SBSO	CED[18]	SDE[19]	FER-PSO[14]	SPSO[14]	r3pso[14]	r2psolhc[14]	r3psolhc[14]
F1	4.33e-05(3)	9.46e-08 (2)	1.23e-08(1)	5.24e-02(5)	8.74e-02(7)	3.54e-02(4)	9.78e-02(8)	8.68e-02(6)
F2	1.78e-05(3)	8.76e-06 (2)	3.43e-07(1)	9.65e-04(4)	9.45e-02(8)	1.43e-02(5)	5.23e-02(7)	4.54e-02(6)
F3	5.49e-04(3)	9.76e-05 (2)	5.73e-05(1)	41999,.34e-02(4)	9.54e-01(8)	7.93e-02(6)	5.52e-01(7)	7.64e-02(5)
F4	2.63e-10(1)	7.43e-05 (8)	9.53e-07(7)	5.65e-07(6)	3.12e-07(5)	2.24e-07(4)	5.43e-09(2)	9.43e-08(3)
F5	8.24e-09(3)	9.43e-06 (8)	4.03e-09(2)	8.34e-09(4)	2.16e-09(1)	9.61e-07(7)	8.58e-07(6)	5.36e-07(5)
F6	7.79e-10(1)	5.39e-05 (8)	8.27e-07(7)	5.45e-09(2)	9.58e-08(4)	8.79e-08(3)	8.05e-07(6)	7.54e-08(5)
F7	1.23e-06(7)	8.97e-05 (8)	4.51e-07(3)	7.41e-07(4)	2.98e-07(1)	9.29e-07(5)	3.43e-07(2)	9.88e-07(6)
F8	4.53e-06(1)	4.27e-02 (7)	8.57e-04(2)	8.69e-04(3)	5.21e-02(8)	5.63e-03(4)	6.12e-03(5)	9.19e-03(6)
F9	2.26e-08(1)	3.42e-04 (7)	5.33e-08(2)	7.38e-08(3)	3.58e-04(8)	6.92e-05(5)	4.33e-05(4)	8.28e-05(6)
F10	5.91e-08(1)	3.96e-03 (7)	9.92e-02(8)	5.50e-06(2)	9.77e-04(6)	8.31e-05(3)	9.43e-05(4)	6.87e-04(5)
F12	1.10e-005(1)	5.24e-04 (7)	8.33e-04(8)	4.53e-04(5)	1.23e-04(2)	3.87e-04(4)	2.54e-04(3)	4.56e-04(6)
F13	7.93e-004(1)	9.87e-04 (2)	9.85e-03(8)	9.69e-03 (7)	8.45e-03 (5)	8.25e-03(4)	8.52e-03 (6)	6.32e-03 (3)
F14	1.4e-003(1)	9.23e-02 (2)	7.89e-01(5)	5.95e-01 (3)	7.86e-01 (4)	8.45e-01 (7)	9.68e-01 (8)	8.12e-01(6)
Total ranks	27	70	52	56	67	61	63	68

Please note that function F11 is a 2-D inverted Shubert function, which is not reported in[19]. Thus, the simulation results of function F11 is not included in Table III. Above the table, the numbers in brackets present the rank of Peak Accuracy, which is obtained by the different algorithms dealing with the same function. The smaller the rank, the higher the accuracy. The last line of the table is the total rank (i.e. summation of all the individual ranks). The lower the total rank, the better the performance of the algorithm. From the result, in terms of peak accuracy, we can see that the proposed algorithms show better performance. Besides, it also indicates that the proposed algorithm present a good exploitative behavior in convergence to different global and local optima.

5   Conclusion

In this paper, there are two parts being proposed to modify the BSO algorithm in multimodal optimization. The clustering strategy drives populations to search the different sub-regions to obtain the potential multiple global and local optima. At the same time, it also reduces the calculation of complexity in the proposed algorithm. Self-adaptive parameter control maintains population diversity by allowing competition to limit resources among similar idea in subpopulation. Our future works will focus on testing the performance of the algorithm on much more massive multimodal problems with high dimensionality and constraints. In some degree, the subpopulation size M affects the performance of the algorithm, so how to design an adaptive strategy to control M through the process is the future work.

References

1. Thomsen, R.: Multimodal optimization using crowding-based differential evolution. In: Proc. IEEE Congr. Evol. Comput., pp. 1382–1389 (June 2004)
2. Wang, H.F., Moon, I., Yang, S.X., Wang, D.W.: A memetic particle swarm optimization algorithm for multimodal optimization problems. IEEE Trans. Cyber. 43(2), 634–647 (2013)
3. Parrott, D., Li, X.: Locating and tracking multiple dynamic optima by a particle swarm model using speciation. IEEE Trans. Evol. Comput. 10(4), 440–458 (2006)
4. Wang, Y.J., Zhang, J.S., Zhang, G.Y.: A dynamic clustering based differential evolution algorithm for global optimization. Journal of Operational Research 183(1), 56–73 (2007)
5. Li, X.: Niching without niching parameters: Particle swarm optimization using a ring topology. IEEE Transaction on Evolutionary Computation 14(1), 150–169 (2010)
6. Rigling, B., Moore, F.: Exploitation of subpopulations in evolutionary strategies for improved numerical optimization. In: Proc. 11th Midwest Artif. Intell. Cogn. Sci. Conf., pp. 80–88 (1999)
7. Rumpler, J., Moore, F.: Automatic selection of subpopulations and minimal spanning distances for improved numerical optimization. In: Proc. Congr. Evol. Comput., pp. 38–43 (2001)
8. Zaharie, D.: A multi-population differential evolution algorithm for multimodal optimization. In: Proc. 10th Mendel Int. Conf. Soft Comput., pp. 17–22 (June 2004)
9. Hendershot, Z.: A differential evolution algorithm for automatically discovering multiple global optima in multidimensional discontinuous spaces. In: Proc. 15th Midwest Artif. Intell. Cogn. Sci. Conf., pp. 92–97 (April 2004)
10. Zaharie, D.: Extensions of differential evolution algorithms for multimodal optimization. In: Proc. SYNASC, pp. 523–534 (2004)
11. Shi, Y.: Brain storm optimization algorithm. In: Tan, Y., Shi, Y., Chai, Y., Wang, G. (eds.) ICSI 2011, Part I. LNCS, vol. 6728, pp. 303–309. Springer, Heidelberg (2011a)
12. Zhan, Z.H., Shi, Y., Zhang, J.: A modified brain storm optimization. In: IEEE World Congr. Comput. Intell. (Jule10-15, 2012)
13. Zhang, J., Sanderson, A.C.: JADE: Adaptive differential evolution with optional external archive. IEEE Trans. Evol. Comput. 13(5), 945–958 (2009)
14. Li, X.: Niching without niching parameters: Particle swarm optimization using a ring topology. IEEE Trans. Evol. Comput. 14(1), 150–169 (2010)

15. Qu, B.Y., Suganthan, P.N., Liang, J.J.: Differential evolution with neighborhood mutation for multimodal optimization. IEEE Trans. Evol. Comput. 16(5), 601–614 (2012)
16. Qu, B.Y., Suganthan, P.N., Das, S.: A distance-based locally informed particle swarm model for multi-modal optimization. IEEE Trans. Evol. Comput. 17(3), 387–402 (2013)
17. Roya, S., Islama, S.M., Dasb, S., Ghosha, S.: Multimodal optimization by artificial weed colonies enhanced with localized group search optimizers. Appl. Soft Comput. 13(1), 27–46 (2013)
18. Thomsen, R.: Multimodal optimization using crowing-based differential evolution. In: Proc. IEEE Congr. Evol. Comput., pp. 1382–1389 (June 2004)
19. Li, X.: Efficient differential evolution using speciation for multimodal function optimization. In: Proc. Conf. Genetic Evol. Comput., pp. 873–880 (2005)

Classification of Electroencephalogram Signals Using Wavelet Transform and Particle Swarm Optimization

Nasser Omer Ba-Karait[1,2], Siti Mariyam Shamsuddin[1,2], and Rubita Sudirman[3]

[1] UTM Big Data Centre, Universiti Teknologi Malaysia
[2] Faculty of Computing
Universiti Teknologi Malaysia, 81310 Skudai, Johor Bahru, Malaysia
[3] Faculty of Electrical Engineering
Universiti Teknologi Malaysia, 81310 Skudai, Johor Bahru, Malaysia
bakarait@yahoo.com, mariyam@utm.my, rubita@fke.utm.my

Abstract. The electroencephalogram (EEG) is a signal measuring activities of the brain. Therefore, it contains useful information for diagnosis of epilepsy. However, it is a very time consuming and costly task to handle these subtle details by a human observer. In this paper, particle swarm optimization (PSO) was proposed to automate the process of seizure detection in EEG signals. Initially, the EEG signals have been analysed using discrete wavelet transform (DWT) for features extraction. Then, the PSO algorithm has been trained to recognize the epileptic signals in EEG data. The results demonstrate the effectiveness of the proposed method in terms of classification accuracy and stability. A comparison with other methods in the literature confirms the superiority of the PSO.

Keywords: Particle swarm optimization, machine learning, discrete wavelet transform, EEG, epileptic seizure.

1 Introduction

The electroencephalogram (EEG) is a highly complex signal recording the brain's neural activities through changes in electrical potentials at multiple locations over the scalp. It conveys valuable clinical information about the state of the brain. Therefore in recent decades, the EEG signals (EEGs) have been used intensively to study brain function and neurological disorders. For this reason, the EEG has long been an important clinical tool in diagnosing, monitoring and managing of neurological disorders, especially those related to epilepsy [1, 2]. Epileptic seizures are caused by temporary electrical disturbance of the brain. The occurrence of a seizure seems unpredictable and its course of action is still very poorly understood. Research is therefore needed to gain a better understanding of the mechanisms causing epileptic disorders. Careful analysis of EEGs could provide valuable insight into this widespread brain disorder [3, 4].

A continuous EEG recording is required to study the epileptic seizures for presurgical evaluation. It provides essential information for locating the brain regions

Y. Tan et al. (Eds.): ICSI 2014, Part II, LNCS 8795, pp. 352–362, 2014.
© Springer International Publishing Switzerland 2014

that generate epileptic activity [5, 6]. Clearly, the analysis of the recorded EEG based on visual inspection is a very time consuming and costly task. In some cases, the seizures are uncontrollable. Recently, methods have started being developed to treat medically resistant epilepsy through delivering a local therapy to the affected regions of brain. Automatic seizures detection forms an integral part of such methods [6, 7].

It is therefore worthwhile to propose an effective algorithm for EEG changes recognition. In literature, most methods deal with this problem based on well-known classification techniques [8-12]. However, classification can be seen in multidimensional space as optimization problem where a class is identified by a centroid. In this case, particle swarm optimization (PSO) can be employed effectively to optimize coordinates of the centroids [13, 14]. To the best of our knowledge, the PSO has not been used for EEGs classification. However in many works, the classifier of EEGs is trained and/or its parameters are optimized by PSO [15, 16]. Also, it was employed to estimate the locations of sources of electrical activity, e.g. epileptic, in the brain based on the scalp EEGs [17-19]. Other EEGs issues have been addressed by PSO such as feature selection [20, 21] and optimal selection of Electrode Channels [22, 23]. In this research, the PSO algorithm is studied to evaluate its performance in detecting the epileptic seizures in EEGs using discrete wavelet transform (DWT) for feature extraction.

2   Particle Swarm Optimization

Particle swarm optimization (PSO) algorithm was originally designed by Kennedy and Eberhart [24] in 1995. The idea was inspired by the social behavior of flocking organisms, as a bird flock. PSO uses a population (called swarm) of candidate solutions (called particles) to probe promising regions of the search space. Each particle moves in the search space with a velocity that is dynamically adjusted according to its own flying experience and its companions' flying experience and it retains the best position it ever encountered in memory. The best position ever encountered by all particles of the swarm is also communicated to all particles.

The popular form of PSO algorithm is defined as:

$$v_{i,d}(t+1) \leftarrow w * v_{i,d}(t) + c_1 r_1 (p_{i,d}(t) - x_{i,d}(t)) + c_2 r_2 (p_{g,d}(t) - x_{i,d}(t)) \tag{1}$$

$$x_{i,d}(t+1) \leftarrow x_{i,d}(t) + v_{i,d}(t+1) \tag{2}$$

where $v_{i,d}$ is the velocity of particle i along dimension d, $x_{i,d}$ is the position of particle i in d, c_1 is a weight applied to the cognitive learning portion, and c_2 is a similar weight applied to the influence of the social learning portion. r_1 and r_2 are separately generated random numbers in the range between zero and one. $p_{i,d}$ is the previous best location of particle i; it is called the personal best (pbest). $P_{g,d}$ is the best location found by the entire population; it is called the global best (gbest). w is the inertia weight [25, 26].

Velocity values must be within a range defined by two parameters $-v_{max}$ and v_{max}. A PSO with an inertia weight in the range (0.9, 0.4) has the better performance on average. To get a better searching pattern between global exploration and local exploitation, researchers recommended decreasing w linearly over time from a maximal value w_{max} to a minimal value w_{min} [26-28].

$$w = w_{max} - \frac{w_{max} - w_{min}}{t_{max}} * t \qquad (3)$$

where, t_{max} is the maximum number of iterations allowed and t is the current iteration number.

3   Materials and Methods

3.1   EEG Data

The current study used the publicly available EEG data described by Andrzejak *et al.* [29]. The complete dataset contains five different sets (denoted A-E), each containing 100 single channel EEG segments of 23.6 second duration. Segments in sets A and B were recorded from five healthy volunteers. They were relaxed in an awake state with eyes open (set A) and closed (set B). The EEG archive of presurgical diagnoses was used to originate sets C, D and E by selecting EEGs from five patients. Signals in sets C and D were measured in seizure free intervals from within the epileptogenic zone and opposite the epileptogenic zone of the brain, respectively. Set E were obtained from within the epileptogenic zone during seizure activity. Fig. 1 shows typical EEG segments, one from each category.

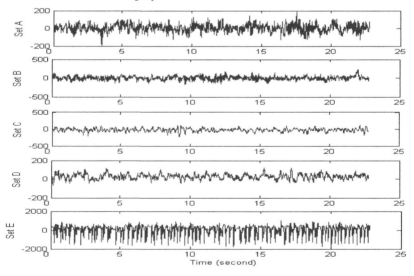

Fig. 1. Samples of five different sets of EEG data

3.2   Discrete Wavelet Transform: Feature Extraction

Discrete wavelet transform (DWT) has been particularly successful in the area of epileptic seizure detection due to its ability to capture transient features and localize them in both time and frequency domains accurately [30]. The DWT analyses a signal $s(n)$ at different frequency bands by decomposing the signal into an approximation and detail information using two sets of functions known as scaling functions and wavelet functions, which are associated with low-pass $g(n)$ and high-pass $h(n)$ filters, respectively. The DWT decomposition process is described in Fig. 2.

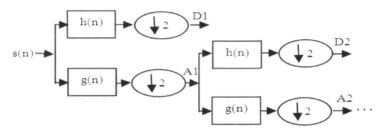

Fig. 2. Sub-band decomposition of DWT

When DWT is used to analyse the signals, two important aspects should be considered: the number of decomposition levels and the type of wavelet. The decomposition level number is selected based on the dominant frequency components of the signal. According to Subasi [31], the levels are selected such that those parts of the signal that correlate well with the frequencies required for the signal classification are retained in the wavelet coefficients. Therefore, level 4 wavelet decomposition was selected for the present study. Accordingly, the EEGs have been decomposed into the details D1-D4 and one final approximation, A4. The smoothing feature of the Daubechies wavelet of order 2 (db2) made it more suitable to detect changes in EEGs [32]. In this research, db2 has been used to compute the wavelet coefficients of the EEGs.

The computed coefficients of discrete wavelet provide a compact representation that shows the energy distribution of the signal in time and frequency. In order to decrease dimensionality of the extracted feature vectors further, statistics over the set of the wavelet coefficients are used [32]. The following statistical features were used to represent the time-frequency distribution of the EEGs: Maximum, Minimum, Mean, and Standard deviation of the wavelet coefficients in each sub-band.

3.3   PSO for EEG Classification

The PSO algorithm has been introduced in this study to classify EEGs for diagnosis purposes. In PSO for classification, each class is identified by a centroid. Therefore, for a dataset $Z = (z_1, z_2,..., z_p, ..., z_{Np})$ with N_c classes, where z_p is a pattern with N_d features, and N_P is the number of patterns in Z. The PSO is applied to optimize positions of N_c centroids in an N_d-dimensional space. Fig. 3 highlights model of

encoding PSO for EEGs classification. It includes three main steps: particle
encoding, defining the fitness function and optimization process.

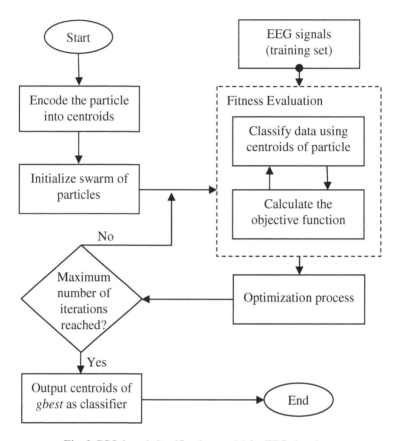

Fig. 3. PSO-based classification model for EEG signals

Each particle is represented by N_c centroids. Consequently, the position of a
particle i is encoded as $(x_i(1), ..., x_i(j)..., x_i(N_c))$ where $x_i(j)$ refers to the jth centroid of
the ith particle. The position of the jth centroid is constituted by N_d real numbers
representing its N_d coordinates in the problem space:

$$x_i(j) = \{x_{i,1}(j), x_{i,2}(j), ..., x_{i,N_d}(j)\} \tag{4}$$

As above, the velocity of each particle i is encoded as $(v_i(1), ..., v_i(j)..., v_i(N_c))$
where the velocity of the jth centroid, $v_i(j)$ is made up of N_d real numbers representing
its N_d velocity components in the problem space:

$$v_i(j) = \{v_{i,1}(j), v_{i,2}(j), ..., v_{i,N_d}(j)\} \tag{5}$$

In classification problem, the objective is to assign any pattern to its correct class. Therefore, the performance of a classification algorithm is evaluated by its accuracy, defined as the percentage of patterns correctly assigned to their classes. This study uses accuracy measure as a fitness function to evaluate the quality of solutions. The fitness of the ith particle is computed based on the dataset portion Z_D (training set) as in Eq. 6.

$$Accuracy(i, z_D) = \frac{\sum_{p=1}^{|Z_D|} Assess(z_p, i)}{|Z_D|} \tag{6}$$

$$Assess(z_p, i) = \begin{cases} 1, & if\ Classify(z_p, i) = z_p.c \\ 0, & otherwise \end{cases} \tag{7}$$

where z_p is a pattern in Z_D, $z_p.c$ is the class of z_p and $Classify(z_p,i)$ returns the class assigned to z_p by the particle i according to the nearest centroid based on Euclidean distance.

With the above premises, optimization mechanism of PSO algorithm is used to update coordinates of the centroids toward the best solution as summarized in Alg.1.

Alg.1. PSO for classification
1. Initialize each particle i to contain N_c centroids
2. For t=1 to t_{max}
   a. For each particle i
      i. Calculate fitness value using Eq. 6.
      ii. Update the personal best solution, *pbest*
   b. Update the global best solution, *gbest*
   c. For each particle i
      i. Update the centroids using Eq. 1 and Eq. 2.
   d. Update the inertia weight, *w* using Eq. 3.

4   Experimental Results

4.1   Performance Measures

In medical diagnosis tasks, the common performance measures are sensitivity, specificity and classification accuracy. Sensitivity is defined as the percentage of correctly detected epileptic EEG patterns to the total number of patterns in epileptic EEG. On the other hand, specificity is defined as the percentage of correctly detected normal EEG patterns to the total number of patterns in normal EEG. Finally, the percentage of all correctly classified patterns to the total number of patterns in both normal and seizure EEG dataset represents the accuracy. Formally, the performance of a diagnostic system is measured as

$$Sensitivity = \frac{TP}{TP + FN} \tag{8}$$

$$Specificity = \frac{TN}{TN + FP} \tag{9}$$

where TP, TN, FP and FN denote true positives, true negatives, false positives and false negatives respectively.

Accuracy: Eq.6 is calculated for testing set Z_T using the final *gbest*; *Accuracy* (*gbest*, Z_T).

4.2   Results and Discussion

The EEG dataset used consists of three categories of signals: healthy (sets A and B), seizure-free (sets C and D) and seizure (set E). Therefore, the three sets: A, D, and E of the above-described dataset are used to analyse the performance of PSO. Sets A and D are gathered to form the normal class against set E which represents the epileptic class. This is similar to real medical applications in which the EEG segments are classified into non-seizures and seizures.

In each set of EEG data, there are 100 EEGs of 4096 samples. In this research, each signal is further divided by a rectangular window composed of 256 samples. Therefore, the dataset of the considered EEG problem was formed of 4800 patterns; i.e., each set has 1600 vectors. Consequentially, the epileptic class contains 1600 patterns, while the number of patterns in the normal class is 3200. The DWT coefficients at the fourth level (D1-D4 and A4) were computed for each pattern. The statistical features that were calculated over the set of wavelet coefficients reduce the dimensionality of feature vector to 20.

It is common to partition the dataset into two separate sets: a training set and a testing set. Additionally, k-fold cross validation is often used by the researchers to evaluate the behavior of the algorithm in the bias associated with the random sampling of the training data. In this study, the EEG dataset (sets A, D and E) was randomly divided into training-testing as 50-50%, 70-30%, and with a 10-fold cross validation. The values of the PSO parameters are as follows: v_{max}=0.05, c_1=2.0, c_2=2.0, w_{max}=0.9, w_{min}=0.4. 50 particles were trained for 1000 iterations to evolve two centroids for the normal and epileptic classes. The centroids produced are then used to classify the patterns in the testing set in order to assess the effectiveness of the proposed method.

Table 1 presents the results achieved by the PSO algorithm with respect to the sensitivity, specificity and accuracy. The results are reported in terms of average, and standard deviation (SD) of ten runs for each partition of the dataset. As can be seen from Table 1, the PSO on average classified the EEGs of training-test datasets partitions: 50-50%, 70-30%, and 10-fold cross validation with accuracies of 96.91%, 97.08%, and 96.53% respectively. The results using all training-test datasets partitions are depicted in Fig. 4. These overall results illustrate that the PSO has good performance and stable behaviour for EEGs classification with accuracy of 96.84%, and standard deviation of 0.90.

Table 1. Sensitivity, specificity, and accuracy of the PSO algorithm on EEG signals

Training-testing dataset partitions (%)	Performance measures (%)		
		Sensitivity	Specificity	Accuracy
50-50	Average	96.50	97.12	96.91
	SD	1.04	0.57	0.30
70-30	Average	96.49	97.38	97.08
	SD	0.83	0.63	0.37
10-fold cross validation	Average	95.75	96.92	96.53
	SD	3.77	2.35	1.45

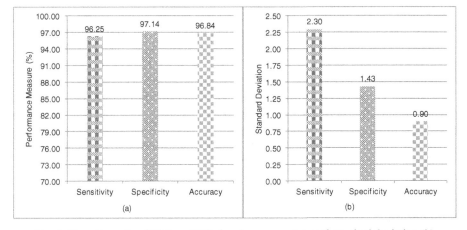

Fig. 4. Overall results of PSO on EEG signals: average (a), and standard deviation (b)

Table 2 illustrates a comparative study of the proposed algorithm with other studies in the literature. For a feasible comparison, only the studies that used the same EEG dataset with the three mentioned sets (A, D and E), and DWT for features extraction are considered. It can be concluded from this comparison that the PSO algorithm showed a promising performance compared to other methods, with a difference in accuracy varies from 0.17% to 6.46%. This proves its ability to compete with well-known classification techniques. In fact, the results reveal that combination of PSO and DWT can produce an efficient automated system for diagnosing epileptic seizures in EEGs.

Table 2. Comparison of the PSO accuracy on EEG signals with methods in the literature

Study	Method	Accuracy
Übeyli [8]	Mixture of expert	93.17
Übeyli [9]	Combined neural network	94.83
Hsu and Yu [10]	Genetic algorithm, support vector machine	90.38
Guo et al. [11]	Genetic programming, k-Nearest Neighbor	93.50
Orhan et al. [12]	K-means clustering, artificial neural network	96.67
This study	Particle swarm optimization	96.84

5   Conclusion

In the present study, discrete wavelet transform and particle swarm optimization have been hybridized to process EEGs for automatic diagnosis of epileptic seizures. The DWT is used to extract the features of signals for the PSO which separates epileptic signals from others in the EEG data. The ability of the proposed method was tested on EEG recordings with healthy, seizure-free, and seizure data. The results indicate that the PSO has very good performance in discriminating the EEGs compared to algorithms reported in the literature. Therefore, the proposed system could be a powerful tool to assist experts in facilitating the analysis of a patient's information and reducing the time and effort required to make accurate decisions on their patients.

Acknowledgment. This work is supported by Research University Grant (Vot 03H72) and Long Term Research Grant (LRGS/TD/2011/UTM/ICT/03/03). The authors would like to thanks Research Management Centre (RMC), Universiti Teknologi Malaysia (UTM) for the support in R & D, Soft Computing Research Group (SCRG) for the inspiration in making this study a success. The authors would also like to thank the anonymous reviewers who have contributed enormously to this work.

References

1. Adeli, H., Zhou, Z., Dadmehr, N.: Analysis of EEG records in an epileptic patient using wavelet transform. Journal of Neuroscience Methods 123, 69–87 (2003)
2. Übeyli, E.D.: Least squares support vector machine employing model-based methods coefficients for analysis of EEG signals. Expert Systems with Applications 37, 233–239 (2010)
3. Subasi, A.: EEG signal classification using wavelet feature extraction and a mixture of expert model. Expert Systems with Applications 32, 1084–1093 (2007)
4. Nigam, V.P., Graupe, D.: A neural-network-based detection of epilepsy. Neurological Research 26, 55–60 (2004)
5. Ocak, H.: Optimal classification of epileptic seizures in EEG using wavelet analysis and genetic algorithm. Signal Processing 88, 1858–1867 (2008)
6. Patnaik, L.M., Manyam, O.K.: Epileptic EEG detection using neural networks and post-classification. Computer Methods and Programs in Biomedicine 91, 100–109 (2008)

7. Gardner, A.B.: A novelty detection approach to seizure analysis from intracranial EEG. PhD Thesis, Georgia Institute of Technology. Georgia, United States (2004)
8. Übeyli, E.D.: Wavelet/mixture of experts network structure for EEG signals classification. Expert Systems with Applications 34, 1954–1962 (2008)
9. Übeyli, E.D.: Combined neural network model employing wavelet coefficients for EEG signals classification. *Digital Signal Processing 19, 297–308 (2009)*
10. Hsu, K.-C., Yu, S.-N.: Detection of seizures in EEG using subband nonlinear parameters and genetic algorithm. *Computers in Biology and Medicine 40, 823–830 (2010)*
11. Guo, L., Rivero, D., Dorado, J., Munteanu, C.R., Pazos, A.: Automatic feature extraction using genetic programming: An application to epileptic EEG classification. *Expert Systems with Applications 38, 10425–10436 (2011)*
12. Orhan, U., Hekim, M., Ozer, M.: EEG signals classification using the K-means clustering and a multilayer perceptron neural network model. Expert Systems with Applications 38, 13475–13481 (2011)
13. Sousa, T., Silva, A., Neves, A.: Particle Swarm based Data Mining Algorithms for classification tasks. Parallel Computing 30, 767–783 (2004)
14. De Falco, I., Cioppa, A.D., Tarantino, E.: Facing classification problems with Particle Swarm Optimization. Applied Soft Computing 7, 652–658 (2007)
15. Hema, C.R., Paulraj, M.P., Nagarajan, R., Yaacob, S., Adom, A.H.: Application of particle swarm optimization for EEG signal classification. Biomedical Soft Computing and Human Sciences 13, 79–84 (2008)
16. Chai, R., Ling, S., Hunter, G., Tran, Y., Nguyen, H.: Brain Computer Interface Classifier for Wheelchair Commands using Neural Network with Fuzzy Particle Swarm Optimization. IEEE Journal of Biomedical and Health Informatics (in Press)
17. Qiu, L., Li, Y., Yao, D.: A feasibility study of EEG dipole source localization using particle swarm optimization. In: 2005 IEEE Congress on Evolutionary Computation, Edinburgh, Scotland, UK, vol. 1, pp. 720–726 (2005)
18. Xu, P., Tian, Y., Lei, X., Yao, D.: Neuroelectric source imaging using 3SCO: A space coding algorithm based on particle swarm optimization and l_0 norm constraint. *NeuroImage 51, 183–205 (2010)*
19. Shirvany, Y., Mahmood, Q., Edelvik, F., Jakobsson, S., Hedstrom, A., Persson, M.: Particle Swarm Optimization Applied to EEG Source Localization of Somatosensory Evoked Potentials. IEEE Transactions on Neural Systems and Rehabilitation Engineering 22, 11–20 (2014)
20. Nakamura, T., Ito, S., Mitsukura, Y., Setokawa, H.: A Method for Evaluating the Degree of Human's Preference Based on EEG Analysis. In: Fifth International Conference on Intelligent Information Hiding and Multimedia Signal Processing, Kyoto, Japan, pp. 732–735 (2009)
21. Zhiping, H., Guangming, C., Cheng, C., He, X., Jiacai, Z.: A new EEG feature selection method for self-paced brain-computer interface. In: 10th International Conference on Intelligent Systems Design and Applications, pp. 845–849. IEEE, Cairo (2010)
22. Jin, J., Wang, X., Zhang, J.: Optimal selection of EEG electrodes via DPSO algorithm. In: 7th World Congress on Intelligent Control and Automation, pp. 5095–5099. IEEE, Chongqing (2008)
23. Kim, J.-Y., Park, S.-M., Ko, K.-E., Sim, K.-B.: A Binary PSO-Based Optimal EEG Channel Selection Method for a Motor Imagery Based BCI System. In: Lee, G., Howard, D., Ślęzak, D., Hong, Y.S. (eds.) ICHIT 2012. CCIS, vol. 310, pp. 245–252. Springer, Heidelberg (2012)

24. Kennedy, J., Eberhart, R.: Particle swarm optimization. In: IEEE International Conference on Neural Networks, Perth, Australia, pp. 1942–1948 (1995)
25. Ghosh, S., Das, S., Kundu, D., Suresh, K., Abraham, A.: Inter-particle communication and search-dynamics of lbest particle swarm optimizers: An analysis. Information Sciences 182, 156–168 (2012)
26. Samal, N.R., Konar, A., Das, S., Abraham, A.: A closed loop stability analysis and parameter selection of the Particle Swarm Optimization dynamics for faster convergence. In: IEEE Congress on Evolutionary Computation, Singapore, pp. 1769–1776 (2007)
27. Eberhart, R.C., Shi, Y.: Particle swarm optimization: developments, applications and resources. In: 2001 Congress on Evolutionary Computation, Seoul, Korea, pp. 81–86 (2001)
28. Lin, C.-L., Mimori, A., Chen, Y.-W.: Hybrid Particle Swarm Optimization and Its Application to Multimodal 3D Medical Image Registration. Computational Intelligence and Neuroscience 2012, 7 (2012)
29. Andrzejak, R.G., Lehnertz, K., Mormann, F., Rieke, C., David, P., Elger, C.E.: Indications of nonlinear deterministic and finite-dimensional structures in time series of brain electrical activity: Dependence on recording region and brain state. Physical Review E 64, 061907 (2001)
30. Subasi, A.: Automatic detection of epileptic seizure using dynamic fuzzy neural networks. Expert Systems with Applications 31, 320–328 (2006)
31. Subasi, A.: Epileptic seizure detection using dynamic wavelet network. Expert Systems with Applications 29, 343–355 (2005)
32. Güler, İ., Übeyli, E.D.: Adaptive neuro-fuzzy inference system for classification of EEG signals using wavelet coefficients. Journal of Neuroscience Methods 148, 113–121 (2005)

FOREX Rate Prediction Using Chaos, Neural Network and Particle Swarm Optimization

Dadabada Pradeepkumar[1,2] and Vadlamani Ravi[1,*]

[1] Center of Excellence in CRM and Analytics,
Institute for Development and Research in Banking Technology,
Hyderabad-500057, India
[2] SCIS, University of Hyderabad, Hyderabad-500046, India
dpradeepphd@gmail.com, rav_padma@yahoo.com

Abstract. This paper presents two two-stage intelligent hybrid FOREX Rate prediction models comprising chaos, Neural Network (NN) and PSO. In these models, Stage-1 obtains initial predictions and Stage-2 fine tunes them. The exchange rates data of US Dollar versus Japanese Yen (JPY), British Pound (GBP), and Euro (EUR) are used to test the effectiveness of hybrid models. We conclude that the proposed intelligent hybrid models yield better predictions compared to the baseline neural networks and PSO in terms of MSE and MAPE.

Keywords: FOREX Rate Prediction, Hybrid model, Chaos, MLP, GRNN, GMDH, PSO.

1   Introduction

The accurate prediction of Foreign Exchange (FOREX) rates helps a country to make its economy stronger [1]. The prediction process is carried out using time series analysis. A time series is said to be chaotic time series if and only if it is non-linear, deterministic and sensitive to initial conditions [2]. The exchange rates are inherently noisy, non-stationary and deterministically chaotic [34]. Chaotic time series prediction involves the prediction of chaotic system behavior in future based on information of current and past states of that system and is always a complex problem.

In order to predict today's value accurately one has to ask the following questions: (i) How many steps of relevant past values are to be considered in the process of prediction? (ii) How many such past values are needed? The answer to the first question is lag and to the second one is embedding dimension.

The following two studies form background to the current work. Ravi et al. [20] presented a forecasting model using a number of computational intelligent techniques. According to them, a variable Y_t is predicted using a vector of lagged variables $Y_{t-1}, Y_{t-2}...Y_{t-m}$ where m is count of considered lagged variables. Then, Hadvandi et al. [21] presented the PSO-based autoregression time series model

---

[*] Corresponding author.

Y. Tan et al. (Eds.): ICSI 2014, Part II, LNCS 8795, pp. 363–375, 2014.

to forecast gold price, where only last two days' gold price is considered to predict today's gold price. They employed PSO to estimate optimal coefficients of the model.

The drawbacks of these studies are: (i) Both studies considered sequential lagged variables without using any scientific method to determine the optimal lag. (ii) Both studies took arbitrary count of lagged variables which is not a scientific approach. (iii) Both studies did not check the presence of chaos in dataset. (iv) Hadvandi et al. [21] modeled only characteristic information of gold price time series but they did not model residual informaton. Hence the model needs to be extended.

The proposed two-stage prediction models have the following features:

1. The models check for presence of chaos at both stages.
2. The methodology determines both optimal lag and optimal embedding dimension scientifically as opposed to guessing them arbitraily.
3. They model both characteristic information and residual information in order to yield better predictions.

The remainder of this paper is organized as follows: A review of literature is presented in Section 2. In section 3, we presented overview of proposed hybrid models. In section 4, experimental methodology is presented. Section 5 discusses the obtained results. The paper is then concluded in Section 6.

2   Literature Survey

It is known that combining many forecasting models yields better estimates than single model [35,36,37,38] in general and in time series [6] in particular. Many previous researches had presented various hybrid FOREX rate prediction models. In this direction, Ni and Yin [7] proposed a hybrid of various regressive neural networks and trading indicators moving average convergence/divergence, relative strength index and genetic algorithm, Zhang [8] hybridized ARIMA and MLP models, Zhang and Wan [9] proposed a statistical fuzzy interval neural network, Donate et. al. [10] proposed a weighted cross-validation evolutionary artificial neural network (EANN) ensemble, Gheyas et. al. [11] proposed novel neural network ensemble, Yu et. al. [12] proposed a multistage nonlinear radial basis function (RBF) neural network ensemble, Sermpins et.al. [13] proposed hybrid neural network architecture of Particle Swarm Optimization and Adaptive Radial Basis Function (ARBF-PSO), Chang [14] proposed hybrid (PSOBPN) that is composed of particle swarm optimization and back propagation network (BPN), Huang et. al. [15] implemented a two-stage chaos and Support Vector Machines (SVMs), Aladag et. al. [16] proposed a time invariant fuzzy time series forecasting method based on PSO, Rout et. al. [17] proposed a hybrid prediction model by combining an adaptive ARMA and Differential Evolution (DE) based training of its feed-forward and feed-back parameters, Chen and Leung [18] proposed a hybrid comprising a time series model and GRNN in tandem and Ince and Trafalis [19] proposed a hybrid two-stage model consisting of ARIMA,

VAR and SVR and NN to predict foreign exchange rates. All of these researches presented that hybrid forecasting models yielded better predictions than stand-alone forecasting models. However comparing all of them is out of the scope of the paper.

3   Proposed Models

3.1   Notations

Let l_1 and m_1 be lag and embedding dimensions that are used in Stage-1; l_2 and m_2 be lag and embedding dimensions that are used in Stage-2 respectively; $e(t)$ be error at time t and $\dot{e}(t)$ be predicted error at time t; $\alpha_0, \alpha_1, \alpha_2, ...$ be coefficients to be optimized. Finally, let $\dot{y}(t)$ be the predicted value of Stage-1 at time t and $\ddot{y}(t)$ be the final predicted value at time t; $f(.)$ be a non-linear function for obtaining predictions using Multi-Layer Perceptron(MLP)/General Regression Neural Network (GRNN)/ Group Method for Data Handling(GMDH).

3.2   Hybrid Model-1

The proposed Hybrid Model-1 consists of two stages. Each stage, in turn, works in both training and test phases. The detailed description is as follows. Let $Y = \{y(1), y(2), ..., y(k), y(k+1), ..., y(n)\}$ be a dataset of n observations at times $1, 2, ..., n$ respectively. First, check for the presence of chaos in Y and rebuild Y using l_1 and m_1, if chaos is present. Later, divide this dataset as Training set $Y_1 = \{y(1), y(2), ..., y(k)\}$ and Test set $Y_2 = \{y(k + 1), y(k + 2), ..., y(n)\}$. The two-stage prediction proceeds as follows:

I. Stage-1: Using NN (MLP/GRNN/GMDH)
   A. Training Phase
      1. Input Y_1 to NN.
      2. Train NN and obtain initial training set predictions using eq.(1) and errors using eq. (2):

$$\dot{y}_1(t) = f(y(t - l_1), y(t - 2l_1), y(t - 3l_1), ..., y(t - m_1l_1))$$
$$t = l_1m_1 + 1, l_1m_1 + 2, ..., k \quad (1)$$

$$e(t) = y(t) - \dot{y}_1(t); t = l_1m_1 + 1, l_1m_1 + 2, ..., k \quad (2)$$

   B. Test Phase
      1. Input Y_2 to trained NN and obtain initial test set predictions using eq.(3) and errors using eq.(4) :

$$\dot{y}_2(t) = f(y(t - l_1), y(t - 2l_1), y(t - 3l_1), ..., y(t - m_1l_1))$$
$$t = k + 1, k + 2, ..., n \quad (3)$$

$$e(t) = y(t) - \dot{y}_2(t); t = k + 1, k + 2, ..., n \quad (4)$$

After Stage-1 ends, Initial training set predictions
$\dot{Y}_1 = \{ \dot{y}_1(l_1m_1 + 1), \dot{y}_1(l_1m_1 + 2),...,\dot{y}_1(k)\}$ and Initial test set predictions
$\dot{Y}_2 = \{\dot{y}_2(k + 1), \dot{y}_2(k + 2),...,\dot{y}_2(n)\}$ are obtained.

II. Stage-2: Using PSO-Based Autoregression model
   A. Training Phase
      1. Check $e(t)$ for the presence of chaos. Rebuild $e(t)$ using l_2 and m_2. Obtain $\dot{e}(t)$ using PSO-based Autoregression model as in eq. (5), if chaos present; otherwise, apply Polynomial Regression to obtain $\dot{e}(t)$.

$$\dot{e}(t) = \alpha_0 + \alpha_1 e(t - l_2) + \alpha_2 e(t - 2l_2) + ... + \alpha_{m_2} e(t - m_2l_2)$$
$$t = l_1m_1 + l_2m_2 + 1, l_1m_1 + l_2m_2 + 2, ..., kl_2 \quad (5)$$

      2. Compute final training set predictions using eq. (6):

$$\ddot{y}_1(t) = \dot{y}_1(t) + \dot{e}(t); t = l_1m_1 + l_2m_2 + 1, l_1m_1 + l_2m_2 + 2, ..., k$$
$$(6)$$

   B. Test Phase
      1. Obtain $\dot{e}(t)$ using PSO-based Autoregression model as in eq. (7), if chaos present; otherwise, use Polynomial Regression to obtain $\dot{e}(t)$.

$$\dot{e}(t) = \alpha_0 + \alpha_1 e(t - l_2) + \alpha_2 e(t - 2l_2) + ... + \alpha_{m_2} e(t - m_2l_2)$$
$$t = k + 1, k + 2, ..., n \quad (7)$$

      2. Compute final test set predictions using eq. (8):

$$\ddot{y}_2(t) = \dot{y}_2(t) + \dot{e}(t); t = k + 1, k + 2, ..., n \quad (8)$$

After Stage-2 ends, Final training set predictions
$\ddot{Y}_1 = \{ \ddot{y}_1(l_1m_1 + l_2m_2 + 1), \ddot{y}_1(l_1m_1 + l_2m_2 + 2),...,\ddot{y}_1(k)\}$ and Final test set predictions $\ddot{Y}_2 = \{ \ddot{y}_2(k + 1), \ddot{y}_2(k + 2),...,\ddot{y}_2(n)\}$ are obtained.

3.3 Hybrid Model-2

The proposed Hybrid Model-2 also consists of two stages. In this hybrid, PSO is invoked in Stage-1 and NN/PR is invoked in Stage-2 (if chaos is present/absent). Accordingly, in this hybrid, in place of eq. (1) and eq. (3), eq. (9) and eq. (10) are used to obtain initial predictions and in place of eq. (5) and eq. (7), eq. (11) and eq. (12) are used to obtain final predictions.

$$\dot{y}_1(t) = \alpha_0 + \alpha_1 y(t - l_1) + \alpha_2 y(t - 2l_1) + ... + \alpha_{m_1} y(t - m_1l_1)$$
$$t = l_1m_1 + 1, l_1m_1 + 2, ..., k \quad (9)$$

$$\dot{y}_2(t) = \alpha_0 + \alpha_1 y(t - l_1) + \alpha_2 y(t - 2l_1) + ... + \alpha_{m_1} y(t - m_1l_1)$$
$$t = k + 1, k + 2, ..., n \quad (10)$$

$$\dot{e}(t) = f(e(t - l_2), e(t - 2l_2), ..., e(t - m_2 l_2))$$
$$t = l_1 m_1 + l_2 m_2 + 1, l_1 m_1 + l_2 m_2 + 2, ..., k \tag{11}$$

$$\dot{e}(t) = f(e(t - l_2), e(t - 2l_2), ..., e(t - m_2 l_2))$$
$$t = k + 1, k + 2, ..., n \tag{12}$$

4   Experimental Design

The foreign exchange data used in our study are obtained from US Federal Reserve System(http://www.federalreserve.gov/releases/h10/hist/). The sets of data collected are of daily US dollar exchange rates with respect to three currencies- JPY, GBP and EUR. The daily data of USD-JPY and USD-GBP from 1st January 1993 to 31st December 2013 (6036 observations each) and USD-EUR from 3rd January 2000 to 31st December 2013 (3772 observations), are used as datasets. From both USD-JPY and USD-GBP datasets,80% of dataset is used as training set (4829 observations) and 20% of dataset is used as test set (1207 observations) and from USD-EUR dataset,80% of dataset is used as training set (3018 observations) and 20% of dataset is used as test set (754 observations).

In the proposed hybrid models, Saida's Method [22,23] implemented in MAT-LAB is used for checking the presence of chaos. Akaike Information Criterion (AIC) [24,31] available in Gretl tool is used to obtain optimal lag. Cao's Method [25,32] implemented in MATLAB is used to obtain minimum embedding dimension. Various Neural Networks (MLP/GRNN/GMDH) available in NeuroShell tool [26,27,28,29,33] are used to obtain predictions. PSO [30] implemented in Java is used to obtain coefficients of the autoregression model. Finally, Polynomial Regression available in Microsoft Excel is used to obtain predictions in the abscence of chaos.

While conducting experiments over the datasets, different user-defined parameters are tweaked in order to obtain the best performance from the techniques. While training on USD-JPY data set using MLP, the learning rate is 0.6,the momentum rate 0.9 and number of hidden nodes is 10, and using GRNN, the smoothing factor is 0.1144531 and using PSO, number of particles is 50, dimensions are 11, inertia is 0.8, iterations are 40000 and $c_1 = c_2 = 2$ are tweaked. Similarly, while training on USD-GBP data set using MLP, the learning rate is 0.5,the momentum rate 0.7 and number of hidden nodes is 30, and using GRNN, the smoothing factor is 0.0915294 and using PSO, number of particles is 60, dimensions are 16, inertia is 0.6, iterations are 40000 and $c_1 = c_2 = 2$ are tweaked.Similarly, while training on USD-EUR data set using MLP, the learning rate is 0.6,the momentum rate 0.8 and number of hidden nodes is 20, and using GRNN, the smoothing factor is 0.0179688 and using PSO, number of particles is 60, dimensions are 16, inertia is 0.8, iterations are 40000 and $c_1 = c_2 = 2$ are tweaked. Since the performance of machine learning techniques in general and Neural Networks techniques, in particular, is, by and large dataset dependent, we need to tweak parameter values in order to get best results for that Neural Network architecture in a given dataset.Mean Squared Error (MSE) and Mean Absolute Percentage Error (MAPE) are used as performance measures as in (13) and (14) :

$$MSE = \frac{\sum_{t=1}^{k}(y(t) - \dot{y}(t))^2}{k} \tag{13}$$

$$MAPE = \frac{100}{k}\sum_{t=1}^{k}\left|\frac{y(t) - \dot{y}(t)}{y(t)}\right| \tag{14}$$

where k is number of forecasting observations, $y(t)$ is actual observation at time t, $\dot{y}(t)$ is predicted value at time t.

5   Results and Discussion

In tables 1, 2 and 3, the MSE and MAPE values for both training and test sets after modeling chaos using stand-alone models are presented. In Tables 4, 5 and 6, the MSE and MAPE values for both training and test sets after modeling chaos using hybrid models are presented along with the optimal lag and the optimal embedding dimension. The experiments #1,#2 and #3 are variants of Hybrid Model-1 and #4 is Hybrid Model-2. From Tables 4, 5 and 6, it is observed that the GRNN, GMDH and PSO were extremely effective in modeling the chaos present in the datasets in Stage-1 whereas MLP was ineffective in doing so. The observation is corroborated by the figures 1-6 where the winning hybrid variant and corresponding neural network in stand-alone mode (Stage-1) behaved pretty closely. Further, it is observed that Stage-2 fine tunes the predictions only when chaos is completely modeled in Stage-1.

Moreover, GMDH turned out to be the best technique in Stage-1, for obtaining initial predictions of USD-JPY and USD-GBP datasets, while PSO-based auto-regression model turned out to be the best in the case of USD-EUR dataset. The best hybrid models in tables 4,5 and 6 are highlighted in boldface. The variant of Hybrid Model-1, involving GMDH, turned out to be the best hybrid model in the case of both USD-JPY and USD-GBP. whereas the Hybrid Model-2 yielded the best result in the case of USD-EUR. Figures 1-6 depict the predictions of proposed hybrid models for both training and test sets of all datasets along with actual values, MLP, GRNN, GMDH and PSO predictions. It can be easily observed that the proposed hybrid models outperformed the standalone MLP, GRNN, GMDH and PSO.

Table 1. Results of Stand-alone models for USD-JPY data

Model	Training set MSE (MAPE)	Test set MSE (MAPE)
MLP	486.5760 (15.6530)	10.7855 (3.0728)
GRNN	5.7430 (1.6291)	33.4999 (5.7749)
GMDH	0.6581 (0.5064)	0.3560 (0.4793)
PSO	2.4686 (1.0316)	2.0915 (1.3491)

Table 2. Results of Stand-alone models for USD-GBP data

Model	Training set MSE (MAPE)	Test set MSE (MAPE)
MLP	0.0417 (8.7604)	0.0027 (2.6912)
GRNN	0.0001765 (0.6146)	0.000540 (1.1062)
GMDH	0.0001024(0.4324)	0.0000856 (0.4530)
PSO	2.5353 (0.7056)	2.1005 (0.7044)

Table 3. Results of Stand-alone models for USD-EUR data

Model	Training set MSE (MAPE)	Test set MSE (MAPE)
MLP	0.0527 (17.4361)	0.0024 (2.9137)
GRNN	0.00004707 (4.4992)	0.00010758 (0.5918)
GMDH	0.00006252 (0.4880)	0.000064089 (0.4473)
PSO	0.00006741 (0.5115)	0.000073068 (0.4842)

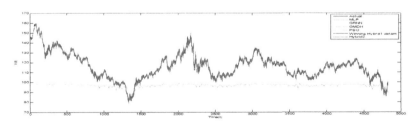

Fig. 1. Predictions of Training set of USD-JPY

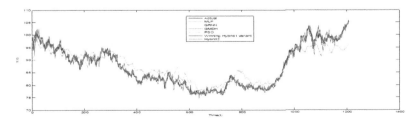

Fig. 2. Predictions of Test set of USD-JPY

Table 4. Results of proposed hybrid models for USD-JPY data

Experi-ment #	Stage #	Technique / Chaos paramters	Training set MSE (MAPE)	Test set MSE (MAPE)
#1	Stage -1 (Chaos present)	MLP ($l_1 = 4, m_1 = 20$)	0.7093 (0.5374)	0.5211 (0.6147)
	Stage-2 (Chaos present)	PSO-based Model ($l_2 = 10, m_2 = 11$)		
#2	Stage-1 (Chaos present)	GRNN ($l_1 = 4, m_1 = 20$)	5.8905 (1.6450)	5.0712 (7.9670)
	Stage-2 (Chaos absent)	2nd degree polynomial regression		
		3rd degree polynomial regression	5.9475 (1.6660)	2.9160 (4.4905)
		4th degree polynomial regression	7.0218 (1.8343)	1.8247 (3.4829)
#3	Stage-1 (Chaos present)	**GMDH** ($l_1 = 4, m_1 = 20$)	0.7314 (0.5487)	0.3552 (0.4757)
	Stage-2 (Chaos absent)	2nd degree polynomial regression		
		3rd degree polynomial regression	**0.7314 (0.5487)**	**0.3548* (0.4757)**
		4th degree polynomial regression	0.7348 (0.5501)	0.3556 (0.4759)
#4	Stage-1 (Chaos present)	PSO-based Model ($l_1 = 4, m_1 = 20$)	2.6083 (1.0825)	1.4666 (1.0676)
	Stage-2 (Chaos absent)	2nd degree polynomial regression		
		3rd degree polynomial regression	3.1846 (0.5147)	1.3127 (0.9866)
		4th degree polynomial regression	3.8139 (1.3870)	0.00007616 (0.4975)

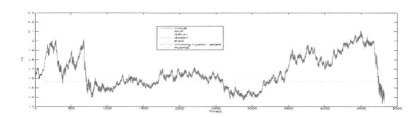

Fig. 3. Predictions of Training set of USD-GBP

Table 5. Results of proposed hybrid models for USD-GBP data

Experiment #	Stage #	Technique / Chaos paramters	Training set MSE (MAPE)	Test set MSE (MAPE)
#1	Stage -1 (Chaos present)	MLP ($l_1 = 5, m_1 = 16$)	0.0002266 (0.6717)	0.0001967 (0.6933)
	Stage-2 (Chaos present)	PSO-based Model ($l_2 = 5, m_2 = 16$)		
#2	Stage-1 (Chaos present)	GRNN ($l_1 = 5, m_1 = 16$)	0.0008580 (0.6335)	0.0008841 (1.5039)
	Stage-2 (Chaos absent)	2nd degree polynomial regression		
		3rd degree polynomial regression	0.0001883 (1.6660)	0.001000 (1.6830)
		4th degree polynomial regression	0.0001883 (0.6409)	0.001800 (2.3914)
#3	Stage-1 (Chaos present)	**GMDH ($l_1 = 5, m_1 = 16$)**	**0.0001097 (0.4595)**	**0.000094* (0.4773)**
	Stage-2 (Chaos absent)	**2nd degree polynomial regression**		
		3rd degree polynomial regression	0.0001118 (0.4623)	0.000094 (0.4773)
		4th degree polynomial regression	0.0001097 (0.4595)	0.0001029 (0.4959)
#4	Stage-1 (Chaos present)	PSO-based Model ($l_1 = 5, m_1 = 16$)	0.0002641 (0.7346)	0.0002058 (0.7013)
	Stage-2 (Chaos absent)	2nd degree polynomial regression		
		3rd degree polynomial regression	0.0002550 (0.7114)	0.0002027 (0.9866)
		4th degree polynomial regression	0.0002550 (0.7114)	1.3352 (0.9993)

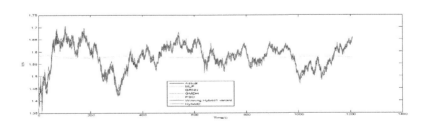

Fig. 4. Predictions of Test set of USD-GBP

Table 6. Results of proposed hybrid models for USD-EUR data

Experi-ment #	Stage #	Technique / Chaos paramters	Training set MSE (MAPE)	Test set MSE (MAPE)
#1	Stage -1 (Chaos present)	MLP ($l_1 = 1, m_1 = 10$)	0.0002328 (0.9916)	0.0001922 (0.7970)
	Stage-2 (Chaos present)	PSO-based Model ($l_2 = 10, m_2 = 16$)		
#2	Stage-1 (Chaos present)	GRNN ($l_1 = 1, m_1 = 10$)	0.0000522 (0.4807)	0.0001245 (0.6445)
	Stage-2 (Chaos absent)	2nd degree polynomial regression		
		3rd degree polynomial regression	0.0000540 (0.4888)	0.0001388 (0.6843)
		4th degree polynomial regression	0.0000574 (0.5069)	0.0001468 (0.7084)
#3	Stage-1 (Chaos present)	GMDH ($l_1 = 1, m_1 = 10$)	0.0000697 (0.5226)	0.0000744 (0.4899)
	Stage-2 (Chaos absent)	2nd degree polynomial regression		
		3rd degree polynomial regression	0.0000697 (0.5226)	0.0000749 (0.4941)
		4th degree polynomial regression	0.0000697 (0.5226)	0.0000749 (0.4941)
#4	Stage-1 (Chaos present)	**PSO-based Model** ($l_1 = 1, m_1 = 10$)	**0.0000684 (0.5147)**	**0.0000736* (0.4870)**
	Stage-2 (Chaos absent)	**2nd degree polynomial regression**		
		3rd degree polynomial regression	0.0000684 (0.5147)	0.0000736 (0.4870)
		4th degree polynomial regression	0.0000674 (0.7114)	0.00007616 (0.4975)

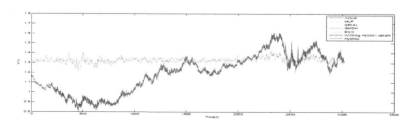

Fig. 5. Predictions of Training set of USD-EUR

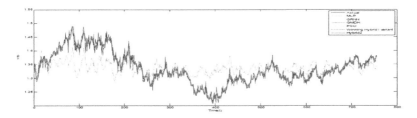

Fig. 6. Predictions of Test set of USD-EUR

6   Conclusion

For predicting FOREX rates, the paper proposes two 2-stage hybrid models comprising chaos theory, various neural network architectures viz. MLP, GRNN and GMDH and PSO or Polynomial Regression. The results of the hybrids in terms of MSE and MAPE on test datasets indicate that the proposed hybrid models outperformed the stand-alone forecasting models: MLP, GRNN, GMDH and PSO. This is the significant outcome of this study. And also, systematic modeling of chaos present in the datasets along with the application of powerful neural networks and PSO for prediction is the single most advantage of the current research. Future directions include applying Multi-objective-PSO and other competing techniques.

References

1. Hoag, A.J., Hoag, J.H.: Introductory Economics. 4th edn. World Scientific Publishing Co. Ptc. Ltd., Singapore (2006)
2. Dhanya, C.T., Nagesh Kumar, D.: Nonlinear ensemble prediction of chaotic daily rainfall. Advances in Water resources 33, 327–347 (2010)
3. Yu, L., Wang, S., Lai, K.K.: Foreign-Exchange-Rate with Artificial Neural Networks. Springer, USA (2007)
4. Zhang, G., Patuwo, B.E., Hu, M.Y.: Forecasting with artificial neural networks: The state of the art. Intl. J. of Forecasting 14, 35–62 (1998)
5. Ozkan, F.: Comparing the forecasting performance of neural network and purchasing power parity: The case of Turkey. Elsevier, Economic Modelling 31, 752–758 (2013)
6. Temizel, T.T., Casey, M.C.: A comparitive study of autoregressive neural network hybrids. Neural Networks 18, 781–789 (2005)
7. Ni, H., Yin, H.: Exchange rate prediction using hybrid neural networks trading indicators. Neurocomputing 72, 2815–2823 (2009)
8. Zhang, G.P.: Time sereis forecasting using a hybrid ARIMA and neural network model. Neurocomputing 50, 159–175 (2003)
9. Zhang, Y.Q., Wan, X.: Statistical fuzzy interval neural networks for currency exchange rate time series prediction. Applied Soft Computing 7, 1149–1156 (2007)
10. Donate, J.P., Cortez, P., Sanchez, G.G., Migue, A.S.: Time sereis forecasting using a weighted cross-validation evolutionary artificial neural network ensemble. NeuroComputing 109, 27–32 (2013)

11. Gheyas, I.A., Smith, L.S.: A novel neural network ensemble architecture for time sereis forecasting. NeuroComputing 74, 3855–3864 (2011)
12. Yu, L., Lai, K.K., Wang, S.: Multistage RBF neural network ensemble learning for exchange rates forecasting. NeuroComputing 71, 3295–3302 (2008)
13. Sermpins, K., Theofilatos, K., Karanthanpoulos, A., Georgopoulos, E.F., Dunis, C.: Forecasting foreign exchange rates with adaptive neural networks using radia-basis functions and particle swarm optimization. European Journal of Operational Research 225, 528–540 (2013)
14. Chang, J.F., Hsieh, P.Y.: Particle Swarm Optimization based on BackPropagation Network Forecasting Exchange Rates. ICIC International 7(12), 6837–6847 (2011)
15. Huang, S.C., Chuang, P.J., Wu, C.F., Lai, H.J.: Chaos-based support vector regressions for exchange rate forecasting. Expert Systems with Applications 37, 8590–8598 (2010)
16. Aladag, C.H., Yolcu, U., Egrioglu, E., Dalar, A.Z.: A new time invariant fuzzy time series forecasting method based on particle swarm optimization. Applied Soft Computing 12, 3291–3299 (2012)
17. Rout, M., Majhi, B., Majhi, R., Panda, G.: Forecasting of currency exchange rates using an adaptive ARMA model with differential based evolution. Journal of King Saud University-Computer and Information Sciences 26(1), 7–18 (2014)
18. Chen, A.S., Leung, M.T.: Regression neural network for error correction in foreign exchange forecasting and trading. Computers and Operations Research 31(7), 1049–1068 (2004)
19. Ince, H., Trafils, T.B.: A Hybrid model for exchange rate prediction. CDecision Support Systems 42, 1054–1062 (2004)
20. Ravi, V., Lal, R., Raj Kiran, N.: Foreign Exchange Rate Prediction using computational Intelligence methods. Int. J. of Computer Information Systems and Industrial Management Applications 4, 659–670 (2012)
21. Hadavandi, E., Ghanbari, A., Naghneh, S.A.: Developing a Time Series Model Based On Particle Swarm Optimization for Gold Price Forecasting. In: IEEE Third Int. Conf. on Business Intelligence and Financial Engineering (2010)
22. Saida, A.B.: Using the Lyapunov exponent as a practical test for noisy chaos (Working paper), http://ssrn.com/abstract=970074
23. http://www.mathworks.in/matlabcentral/fileexchange/22667-chaos-test
24. Akaike, H.: A new Look at the Statistical Model Identification. IEEE Transactions on Automatic Control AC-19(6) (1974)
25. Cao, L.: Practical Method for determining the minimum embedding dimension of a scalar time series. Physica D 110, 43–50 (1997)
26. Rumelhart, G.E., Hinton, G.E., Williams, R.J.: Learning internal representations by error propagation (1). MIT Press, Cambridge (1986)
27. Specht, D.F.: A General Regression Neural Network. IEEE Transactions on Neural Networks 2(6), 568–576 (1991)
28. Ivakhnenko, A.G.: The GMDH: A rival of stochastic approximation. Sov. Autom. Control 3(43) (1968)
29. Farlow, S.J.: Self-Organizing Methods in Modeling: GMDH type Algorithm, Bazel, Marcel Dekker Inc. Newyork (1984)
30. Kennedy, J., Eberhart, R.: Particle Swarm Optimization. In: Proc of IEEE International Conference on Neural Network, Perth, Australia, pp. 1942–1948 (1995)
31. http://gretl.sourceforge.net/
32. http://www.mathworks.in/matlabcentral/fileexchange/
    36935-minimum-embedding-dimension/content/cao_deneme.m

33. http://www.neuroshell.com/
34. Yao, J., Tan, C.L.: A case study on using neural networks to perform technical forecasting of forex. Neurocomputing 12(4), 79–98 (2000)
35. Bates, J.M., Granger, C.W.J.: The combination of forecasts. Oper. Res. Quart. 20, 451–468 (1969)
36. Clemen, R.: combining forecasts: a review and annotated bibliography with discussion. Int. J. Forecast 5, 559–608 (1989)
37. Makridakis, S., Anderson, A., Carbone, R., Fildes, R., Hibdon, M., Lewandowski, R., Newton, J., Parzen, E., Winkler, R.: The accuracy of extraploation (time series) methods: results of a forecasting competition. J. Forecast. 1, 111–153 (1982)
38. Pelikan, E., De Groot, C., Wurtz, D.: Power consumption in West-Bohemia: improved forecasts decorrelating connectionist networks. Neural Network World 2, 710–712 (1992)

Path Planning Using Neighborhood Based Crowding Differential Evolution

Boyang Qu[1,3], Yanping Xu[1], Dongyun Wang[1], Hui Song[2], and Zhigang Shang[2]

[1] School of Electric and Information Engineering, Zhongyuan University of Technology,
Zhengzhou, China
[2] School of Electrical Engineering, Zhengzhou University, Zhengzhou, China
[3] School of Information Engineering, Zhengzhou University, Zhengzhou, China
qby1984@hotmail.com, 120828633@qq.com, wdy1964@aliyun.com,
hsong320@163.com, zhigang_shang@zzu.edu.cn

Abstract. Path planning problems are known as one of the most important techniques used in robot navigation. The task of path planning is to find several short and collision-free paths. Various optimization algorithms have used to handle path planning problems. Neighborhood based crowding differential evolution (NCDE) is an effective multi-modal optimization algorithm. It is able to locate multiple optima in a single run. In this paper, Bezier curve concept and NCDE are used to solve path planning problems. It is compared with several other methods and the results show that NCDE is able to generate satisfactory solutions. It can provide several alternative optimal paths in one single run for all the tested problems.

Keywords: Evolutionary Computation, Path Planning, Different Evolution, Constrained Optimization.

1   Introduction

In recent years, due to the development of artificial intelligence and electrical integration technologies, robots have been used in various fields such as lunar exploration, navigation, underground exploration, rescue, etc. Mobile robot is an integrated system which combines different functions. Path planning is one of the key technologies used in mobile robot. The performance of the robot highly depends on the path planning method used. Therefore, developing efficient path planning method has attracted many researchers' attention. The aim of path planning is to generate a collision-free path between an initial location and a desired destination in an environment which is full of obstacles. The planned path refers to the optimal or suboptimal trajectory under certain specific conditions, such as the shortest path or the safest path. A good strategy of robot path planning can make a robot fulfill a desired task safely and effectively.

In an environment full of obstacles, the mobile robot must arrange its trajectory to avoid obstacles and find the shortest path when it travels form starting point to target point [1]. In literature, many algorithms have been used to solve the path planning

Y. Tan et al. (Eds.): ICSI 2014, Part II, LNCS 8795, pp. 376–383, 2014.

problems, such as self-adjusting fuzzy control algorithm [2], genetic algorithms [3] and [4], ant colony optimization [5] and particle swarm optimization [6]. However, only a few works use differential evolution (DE) to solve this problem [7] and [8]. DE is one of the most powerful stochastic real-parameter optimization algorithms in current use. It is effective in solving single global optimization problems. However, the canonical DE is not suitable to solve multimodal problems or locate multiple peaks in one run. Therefore, a newly developed niching DE algorithm called Neighborhood based Crowding Differential Evolution (NCDE) is used to handle path planning problem in this paper. This algorithm is able to generate multiple optimal paths simultaneously.

The remainder of this paper is organized as follows: Section 2 introduces the definition of Bezier curves and how to use it to solve path planning problem. Section 3 presents the novel DE algorithm which is used to optimize the path. The experimental preparation and simulation results are presented and discussed in section 4 and 5 respectively. Finally, section 6 concludes the paper.

2   Bezier Curves

For designing automobile bodies, French engineer Pierre Bezier invented the Bezier Curves in 1962, which is a new parameter curve [9]. Bezier Curves have become an essential tool in many areas, especially that it has been widely used in computer graphics and animation. Bezier curve is suitable to describe the path because of its space properties. Through controlling the anchor points, different Bezier curves can be obtained. The path planning problem can be transformed into an optimization problem with limited control points to be optimized [10].

2.1   The Definition and Properties of Bezier Curves

In Bezier Curves, $n+1$ vertices can define polynomials of degree n, and a Bezier Curve of basis polynomial of degree n can be expressed using the following equation [11]:

$$P(t) = \sum_{i}^{n} P_i B_{i,n}(t) \qquad t \in [0\ 1].\qquad (1)$$

where P_i represents the coordinates of i_{th} vertex, while $B_{i,n(t)}$ stands for a Bernstein polynomial, which is given as:

$$B_{i,n}(t) = C_n^{\ i} t^i (1-t)^{n-i} \quad (i = 0,1,...,n).\qquad (2)$$

$C_n^{\ i}$ is the binomial coefficient. The parameter equation of three times Bezier curve could be obtained by combining formula (1) and (2) as follows:

$$P(t) = P_0 (1-t)^3 + 3P_1 t (1-t)^2 + 3P_2 t^2 (1-t) + P_3 t^3.\qquad (3)$$

From formula (3), we can observe that the three times Bezier curve starts at $t=0$ and ends at $t=1$.

The properties of Bezier curves can be described as follows [12] and [13]:

1. Bezier curve is decided by four points. The curve goes through the first point and the last point. However, the shape of the curve is determined by the two other points.
2. The curve is a straight line if and only if all the control points are in a straight line.
3. The start/end of the curve is tangent to the first/last section of the Bezier polygon (The lines are used to connect the Bezier curve points, and they start from the first point and end at the last point. The lines formed the Bezier polygon.). this property can be described using the following formula:

$$P'(0) = 3 \times (P_1 - P_0), \quad P'(n) = 3 \times (P_n - P_{n-1}). \tag{4}$$

Where $P'(0)$ is the first derivative of the start point and $P'(n)$ is the first derivative of the end point.

3   DE and Neighborhood Based Differential Evolution

3.1   DE

Differential Evolution (DE) is a stochastic search technique developed by Storn and Price in 1995 [14]. It is a simple but efficient search algorithm. It starts with a randomly initialized population. The optimal solution is obtained through a cycle of three stages known as mutation, crossover and selection. These three stages are applied to every solution member x_i in each generation to create a new solution. In each generation, DE employs the mutation operation to produce a mutant vector vp associated with each parent xp at each generation.

After the mutation phase, the binary (or uniform) crossover operation is applied to each pair of the generated mutant vector and its corresponding parent vector. The process can be represented as:

$$u_{p,i} = \begin{cases} v_{p,i} & if \ rand_i \leq CR \\ x_{p,i} & otherwise. \end{cases} \tag{5}$$

Where $u_{p,i}$ is the offspring vector. The crossover rate CR is a user-specified constant within the range [0,1], which determines what parameter dimensions of $u_{p,i}$ are copied from.

F and CR are the two most important control parameters in the DE. These two parameters can significantly influence the optimization performance of the DE. Therefore, to successfully solve an optimization problem, it is generally required to perform a trial-and-error search for the most appropriate values for these parameters [14].

Different from traditional evolutionary algorithms, DE employs the difference of the parameter vectors to explore the objective function landscape. DE has been proved to be one of most powerful evolutionary algorithm for solving the real-valued optimization problems.

3.2   NCDE

The standard DE searches for a global optimum in a D-dimensional hyperspace. It is efficient when searching for single global solution. However, it is often desirable to locate multiple optimal solutions in real world problems. Taking path planning problem as an example, it is always better to find several alternative optimal paths for the decision makers to choose, as the prime objective for different decision makers can be different. Therefore, the neighborhood based crowding differential evolution (NCDE) is applied to solve the path planning problem. Different from canonical DE, vector generations are limited to a number of similar individuals as measured by Euclidean distance.   In this way, individuals are evolved towards its nearest optimal point and the possibility of between niche difference vector generation is reduced [15]. The neighborhood concept facilitates multiple convergences, because it allows a higher exploitation of the areas which pilot the moves.

4   Experiment Preparation

4.1   Experiment Setup

For the experiment, Matlab R2008a is used as the programming language and the computer configurations are Intel Pentium® 4 CPU 3.00 GHZ, 4 GB of memory. The DE parameters used are list as below:

Population size=30, F=0.5, CR=0.5

4.2   Cost Function

In order to obtain a satisfying path, an objective function for the path planning problem need to be defined. Security and length of the path are the two most important criteria for path planning problems. Therefore, these two criteria are used to form the objective function in this work.

(1) Security

Considering that the path cannot intersect with the obstacles, a punishment function f_{safe} can be designed as below:

$$f_{safe} = \begin{cases} 0 & \text{if } d_{min} > D_{safe} \\ d_{min} & \text{if } 0 \leq d_{min} \leq D_{safe.} \end{cases} \tag{6}$$

where d_{min} is the minimum distance between the path and all obstacles. D_{safe} is a predefined security distance.

(2) Length of the path

The length of the curve should be as short as possible. The length function is defined as:

$$f_{len} = L = \int_{0}^{1} \sqrt{(x'(t))^2 + (y'(t))^2} \, dt. \tag{7}$$

Where $x(t)$, $y(t)$ are the coordinates of points.

(3) The overall objective function

Considering the two objectives above weight coefficient a, the final cost function can be defined as:

$$f = f_{len} + af_{safe}. \tag{8}$$

where a is the weight factor to balance the two objectives and it is chosen as 1000 in this experiment.

5   Simulation Results

To assess the performance of the proposed algorithm, four predefined path planning problems are used. The results are plotted in Fig.1. The green circles in the figure mean the dangerous areas around the obstacles. The red path describes the best path while the black paths illustrate the possible paths for every problem. Four algorithms are tested on these problems with two Bezier curves ($n=2$, $D=4$ $n=8$):

1. PSO+: Classical Particle Swarm Optimizer with crossover operator
2. DMS-PSO+: Dynamic Multi-Swarm Optimizer with crossover operator
3. DE: Classical Differential Evolution
4. NCDE: Neighborhood Based Crowding Differential Evolution

The max fitness evaluation is 20000 for all algorithms. The population size of PSO with crossover is set to 30 and the number of sub-swarms of DMS-PSO with

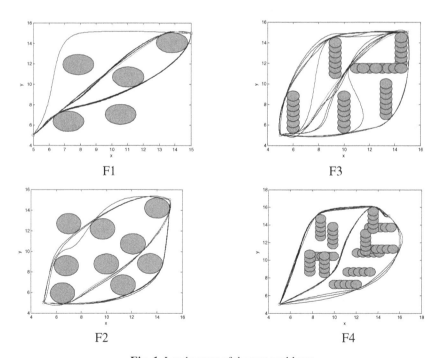

F1

F3

F2

F4

Fig. 1. Landscapes of the test problems

crossover is set to 10 (with 3 particles in each sub-swarm). All performances are calculated and averaged over 25 runs with the random initialization. The results are presented in Table 1. And to evaluate the difference of the two algorithms, Nonparametric statistical method ttest is used. $h=1$ indicates a rejection of the null hypothesis at the 5% significance level. $h=0$ indicates a failure to reject the null hypothesis at the 5% significance level.

In Fig.1, F1, F2, F3 and F4 show that the distributions of the solutions achieved by NCDE for each test problem, where the red line describes the best path. We can observer that it is easy for NCDE to find the best suitable path for every problem as well as other alternative satisfying paths which are not around the best solution at the same time.

Table 1. Results for F1-F4

Problems		PSO+	DMS-PSO+	DE	NCDE
F1	Mean	14.7698	14.7465	14.8774	**14.7237**
	Std.	0.0016	0.0006	0.1717	0.0000
	Min	14.7007	14.7290	14.6842	14.7000
	Max	14.8418	14.7797	15.9563	14.7404
	h	1	1	1	0
F2	Mean	15.9863	15.8101	16.5603	**15.6768**
	Std.	0.5050	0.1726	0.9181	0.1010
	Min	15.6221	15.0865	15.2764	15.3144
	Max	17.3518	17.2887	17.5268	16.7618
	h	1	0	1	0
F3	Mean	17.0764	16.8388	17.4209	**16.6135**
	Std.	0.2402	0.2563	0.1594	0.1398
	Min	16.4046	16.3139	16.2838	16.3900
	Max	18.1084	17.8895	17.8122	17.0463
	h	1	1	1	0
F4	Mean	15.6119	15.1971	16.5413	**14.9887**
	Std.	0.5745	0.4369	0.1743	0.2086
	Min	14.8479	14.8024	14.8028	14.7138
	Max	16.6730	16.6179	17.6099	16.6336
	h	1	1	1	0

In [16], compared with PSO and DMS-PSO, PSO+ and DMS-PSO+ perform better respectively. For every problem, every algorithm is compared with the best algorithm with ttest. From the table, some conclusions could be made as follows:

1. DMS-PSO+ performs better than PSO+.
2. NCDE outperforms DE on all the four problems.
3. NCDE performs best among all algorithms on mean value.

4. Except for problem 2, NCDE is better than the other three algorithms obviously which can be seen from the result of ttest. The result of ttest on problem 2 is no difference between NCDE and DMS-PSO+, which means that the result of NCDE accepts DMS-PSO+ at the 5% significance level. However, from the mean value we could know that NCDE has a smaller value which means NCDE has a stable ability to find good solutions during the process of search compared with DMS-PSO+.

In the process of searching, NCDE has a better local search ability, which makes multi-paths possible, so various satisfied paths could be acquired for every problem in this task. From the above four points we can know that, NCDE performs betters.

6   Conclusion

In this work, Bezier curves and neighborhood based crowding differential evolution algorithm are used to tackle path planning problem. To assess the performance of the neighborhood based crowding differential evolution algorithm in solving path planning problem, four different path problems are tested. The experiments show that neighborhood based crowding differential evolution is effective in solving all four problems. In future work, dynamic environment and constraints will be added to increase the complexity of the path planning problems. High order Bezier curves will also be used to improve the quality of the solutions.

Acknowledgments. This work was supported in part by National Natural Science Foundation of China (61305080, U1304602), Postdoctoral Science Foundation of China (Grants 20100480859, 2014M552013), Specialized Research Fund for the Doctoral Program of Higher Education (20114101110005), Scientific and Technological Project of Henan Province (132102210521, 122300410264), and Key Foundation of Henan Educational Committee (14A410001).

References

1. Chakraborty, J., Konar, A., Chakraborty, A.K., Jain, L.C.: Distributed Cooperative Multi-Robot Path Planning Using Differential Evolution. In: 2008 IEEE Congress on Evolutionary Computation, CEC 2008, Hong Kong, China, pp. 718–725 (2008)
2. Wu, C.D., Zhang, Y., Li, M.X.: A Rough Set GA-based Hybrid Method for Robot Path Planning. Journal of International Automation and Computing, 29–34 (2006)
3. Sugihara, K., Smith, J.: Genetic algorithms for adaptive motion planning of an autonomous mobile robot. In: Proc. of IEEE Intl. Symposium on Computational Intelligence in Robotics and Automation, Monterey, CA, USA, pp. 138–143 (1997)
4. ALtaharwa, S.A., Alweshah, M.: A mobile robot path planning using genetic algorithm in static environment. Journal of Computer Science 4, 341–344 (2008)
5. Bell, J.E., McMullen, P.R.: Ant colony optimization techniques for the vehicle routing problem. Advanced Engineering Informatics, 41–48 (2004)

6. Saska, M., Macas, M., Preucil, L., Lhotska, L.: Robot path planning using particle swarm optimization of ferguson splines. In: IEEE Conference on Emerging Technologies and Factory Automation, Prague, pp. 833–839 (2006)
7. Jayasree, C., Amit, K., Jain, L.C., Chakraborty, U.K.: Cooperative multi-robot path planning using differential evolution. Journal of Intelligent and Fuzzy Systems, 13–27 (2009)
8. Mo, H.W., Li, Z.Z.: Bio-geography based differential evolution for robot path planning. In: 2012 IEEE International Conference on Information and Automation, ICIA 2012, pp. 1–6. Inner Mongolia, China (2012)
9. Bashir, Z.A., Hawary, M.E.: Short-term Load Forecasting using Artificial Neural Network based on Particle Swarm Optimization Algorithm. In: 2007 Canadian Conference on Electrical and Computer Engineering, CCECD, Canadian, pp. 272–275 (2007)
10. Ho, Y.J., Liu, J.S.: Collision-free Curvature-bounded Smooth Path Planning using Composite Bezier Curve based on Voronoi Diagram. In: Proceedings of IEEE International Symposium on Computational Intelligence in Robotics and Automation, CIRA, pp. 463–468 (2009)
11. Yang, L.Q.: Path Planning Algorithm for Mobile Robot Obstacle Avoidance Adopting Bezier Curve Based on Genetic Algorithm. In: Control and Decision Conference, pp. 3286–3289 (2008)
12. Liu, H.G., Qin, G.L.: A Bezier Curve Based on The Military Arrow Mark Realized. Ordnance of Sichuan 30, 67–68 (2009)
13. Gao, S., Zhang, Z.Y., Cao, C.G.: Particle Swarm Algorithm for The Shortest Bezier Curve. In: International Workshop on Intelligent Systems and Applications, pp. 1–4 (2009)
14. Storn, R., Price, K.: Differential Evolution—a Simple and Efficient Adaptive Scheme for Global Optimization over Continuous Spaces. Journal of Global Optimization 11, 22–25 (1995)
15. Qu, B.Y., Suganthan, P.N., Liang, J.J.: Differential Evolution with Neighborhood Mutation for Multimodal Optimization. IEEE Transactions on Evolutionary Computation, 601–614 (2012)
16. Liang, J.-J., Song, H., Qu, B.-Y., Mao, X.-B.: Path Planning Based on Dynamic Multi-Swarm Particle Swarm Optimizer with Crossover. In: Huang, D.-S., Ma, J., Jo, K.-H., Gromiha, M.M. (eds.) ICIC 2012. LNCS, vol. 7390, pp. 159–166. Springer, Heidelberg (2012)

Neural Network Based on Dynamic Multi-swarm Particle Swarm Optimizer for Ultra-Short-Term Load Forecasting

Jane Jing Liang[1], Hui Song[1], Boyang Qu[1,2], Wei Liu[3], and Alex Kai Qin[4]

[1] School of Electrical Engineering, Zhengzhou Univerisity, China
[2] School of Electric and Information Engineering, Zhongyuan University of Technology, China
[3] State Grid Henan Economic Research Institute, Zhengzhou, China
[4] School of Computer Science and Information Technology RMIT University,
Melbourne 3001, Victoria, Australia
LIANGJING@zzu.edu.cn, qby1984@hotmail.com, liuwei830610@163.com,
{hsong320,kai.qin}@rmit.edu.au

Abstract. Ultra-Short-Term Load Forecasting plays an important role in Power Load Forecasting. Back Propagation Neural Network(BPNN) has become one of the most commonly used methods in Power System Ultra-Short-Term Load Forecasting for its ability of computing complex samples and training large-scale samples. However, traditional BPNN algorithm needs to set up a large amount of network training parameters, and it is easy to be trapped into local optima. A new algorithm which is Neural Network based on Dynamic Multi-Swarm Particle Swarm Optimizer (DMSPSO-NN) is proposed for Ultra-Short-Term Load Forecasting in this paper. DMSPSO-NN overcomes the shortage of traditional BPNN and has a good global search and higher accuracy which shows that it is suitable to be used for Ultra-Short-Term Load Forecasting.

Keywords: Ultra-Short-Term Load Forecasting; Back Propagation Neural Network; Dynamic Multi-Swarm Particle Swarm Optimizer

1 Introduction

Power load forecasting is an indispensable part for managing and researching power system, and it can make the full use of electricity and ease the conflict between supply and demand based on the analysis of the existing electric energy [1]. Power system load forecasting method based on electric power, economic, social and meteorological factors and so on. According to the time length of the prediction, power load forecasting can be classified as ultra-short-term load forecasting, short-term load forecasting, medium long-term load forecasting and long-term load forecasting. In terms of power system dispatching and management, ultra-short-term load forecasting which varies from an hour to a week is the most important. Accurate ultra-short-term load forecasting is very important in maintaining ultra-short-term analysis for electric power, power exchange, trading evaluation as well as the analysis of the network function, security and trend, the safety strategy of reducing load and so on[2].

Y. Tan et al. (Eds.): ICSI 2014, Part II, LNCS 8795, pp. 384–391, 2014.

In recent years, there are various methods to solve this problem, such as Expert Systems(ES)[3], Support Vector Machines(SVM)[4][5], Back Propagation Neural Network(BPNN)[6][7] and so on. The main drawback of the ES is that it learns nothing from the environment and has ambiguous relationship between rules, low efficiency and adaptability. The main disadvantage of SVM is difficult to achieve large-scale training samples and solve multi-classification problems. BPNN is widely used in power load forecasting in recent years, to calculate large-scale complex training samples as well as to slow the speed of convergence during training process.

With the development of Evolutionary Algorithm (EA), some researchers have found its advantage in handling large-scale, non-differentiable and complex multi-mode problem without any information about optimized problems for its global convergence ability and strong robustness. EA can optimize the weight, structure and learning rules of NN by searching for optimal solutions in search space with the help of evolutionary strategy, genetic algorithm or evolutionary programming. Genetic Algorithm (GA) is the most widely used in EA because it can deal with many complex problems. However, GA is easy to be trapped into local optimum and the process is difficult to control when GA is used to train NN.

Recently, the emergence of Swarm Intelligence such as Particle Swarm Optimizer (PSO) has overcome the drawbacks of EA. Compared with other algorithms, Particle Swarm Optimizer is simple, easy to realize, and need less parameter to adjust, making it an effective optimal tool. So PSO has been used to optimize NN widely such as power load forecasting [8], Fault Diagnosis of Power Transformer [9], Reservoir Parameter Dynamic Prediction [10], Modeling and Simulation of Screw Axis [11]and so on[12]. However, the traditional PSO can't maintain the diversity of particles and is difficult to reach global optimum. So Dynamic Multi-Swarm Particle Swarm Optimizer (DMSPSO) which not only overcomes the drawback of the Particle Swarm Optimizer, but also has strong global search ability, is proposed to optimize NN in this paper. The result shows DMSPSO is easy to find global optimum when it is used to optimize NN.

The rest of this paper is organized as follows. Section II gives a brief introduction about the basic Particle Swarm Optimizer and describes the search process of the Dynamic Multi-Swarm Particle Swarm Optimizer. The Back Propagation Neural Network and Neural Network Based on Dynamic Multi-Swarm Particle Swarm Optimizer model employed in this work are described in detail in Section III. Section IV introduces the experimental setup and presents the results. Conclusions and future work are given in Section V.

2   Dynamic Multi-swarm Particle Swarm Optimizer

Particle Swarm Optimizer was proposed by Kennedy and Eberhart in 1995[13][14]. The basic idea of PSO simulates the behavior of flying birds. In PSO, each bird is regarded as a potential solution in search space which is called a "particle". There exists a fitness value in each particle obtained by the fitness function. Each particle adjusts the distance and its flying direction according to its velocity. The model of PSO is shown as:

$$V_i^d = \omega * V_i^d + c_1 * rand1_i^d * (pbest_i^d - X_i^d) + c_2 * rand2_i^d * (gbest^d - x_i^d) \tag{1}$$

$$X_i^d = X_i^d + V_i^d \tag{2}$$

Where, ω is the inertia weight and the range is [0.4 0.9]; c_1 and c_2 are balance factors which are set 2.05 in general; $rand$ is a random number in [0, 1].The weakness of standard PSO is premature and easy to be trapped into local optimum.

Dynamic Multi-Swarm Particle Swarm Optimizer (DMS-PSO) is developed from local Particle Swarm Optimizer, where neighborhood structure is used in a small population [15][16]. In order to increase the distribution of population and accelerate the speed of convergence, the entire population is divided into sub-swarms equally in DMS-PSO and each sub- swarm searches in space with its own particles. The population is regrouped randomly every L generation (L is called regroup period), then the population starts the search with new topology structure. Due to this method, the information obtained from sub-swarms exchanges among them, and the diversity of the population is also increased. The updating formula is given as follows:

$$V_i^d \leftarrow \omega * V_i^d + c_1 * rand1_i^d * (pbest_i^d - X_i^d) + c_2 * rand2_i^d * (lbest_k^d - X_i^d)$$
$$V_i^d = \min(V_{max}^d, \max(-V_{max}^d, V_i^d)) \tag{3}$$
$$X_i^d \leftarrow X_i^d + V_i^d$$

where, V_i^d represents the velocity of ith particle in dimension d; X_i^d is the position of ith particle in dimension d. $lbest_k^d$ is the position of local optimum in dimension d of kth sub-swarm; $pbest_i^d$ is the best personal position in dimension d of ith particle.

3   Back Propagation Neural Network Based on Dynamic Multi-swarm Particle Swarm Optimizer

3.1   Back Propagation Neural Network

Back Propagation Neural Network can solve learning problems for connecting weights between hidden units in multi-layers network, so it has become one of the most important modal of Artificial Neural Network. The BPNN is made up of three layers: input layer, hidden layer and output layer. In order to get satisfied forecast, backward transmission error and error correction methods are used to adjust the network parameters (weights and threshold)[17].

Input layer neurons are responsible for receiving input information from the outside world, and transmitting to the middle layer neurons (which is hidden layer). The hidden layer is the internal information processing layer, which is responsible for transforming information. According to the requirements of changed information, hidden layer can be designed for single hidden layer or several hidden layers[18][19]. The structure of the BP Neural Network is shown as follows:

Where, $x_0, x_1, ..., x_j ..., x_n$ is the input value of BPNN, $o_1, o_2, ..., o_k ..., o_l$ is predictive value, v_{ij} and w_{jk} ($i = 1, 2, ..., n, j = 1, 2, ..., m, k = 1, 2, ..., l$) are the input and output weights of BPNN respectively. If the input node is n and the output node is l, BPNN expresses the mapping relationship from n independent input variables to l independent output variables. BPNN acquires associative memory and ability to predict through training the network.

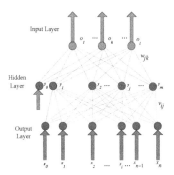

Fig. 1. Structure of BPNN

3.2   The Process of Neural Network Based on Dynamic Multi-swarm Particle Swarm Optimizer

Dynamic Multi-Swarm Particle Swarm Optimizer is used to optimize the input weights, input thresholds, output weights output thresholds of NN. The process of DMSPSO which is used to train the parameter of NN includes the following steps:

a.   Ensure the parameters of NN

According to the input and output(x, o) determine the number of input layer nodes n, hidden layer nodes m and output layer nodes l. Then the total dimension can be calculated through the following formula:

$$D=n*m+m+m*l+l \tag{4}$$

Sigmoid is used as the excitation function.

$$f(x) = \frac{1}{1+e^{-x}} \tag{5}$$

At the same time, we normalize the original data input and output samples in order to increase the learning efficiency of NN.

$$x_k^{new} = \frac{0.1+0.8\times(x_k^{old} - \min x_k^{old})}{\max x_k^{old} - \min x_k^{old}}$$

$$o_k^{new} = \frac{0.1+0.8\times(o_k^{old} - \min o_k^{old})}{\max o_k^{old} - \min o_k^{old}} \tag{6}$$

where, $K=1, 2,...m$ is the number of samples. x_k^{old} and y_k^{old}, x_k^{new} and y_k^{new} represent the input and output of the network which are unprocessed and processed respectively.

b.   Initialize the parameters

Initialize the weight between input layer and hidden layer v_{ij} and between hidden layer and output layer w_{jk}, hidden threshold a and output threshold b, regroup period L, number of Sub-warms P, population of every sub-swarm ps and every particle's velocity.

c.   Calculate fitness value

Firstly, calculate the output of hidden layer H according to the input data x, weights v_{ij} between input and hidden layer and threshold a.

$$H_j = f(\sum_{i=1}^{n} v_{ij} x_i - a_j) \quad j = 1,2,...,m \tag{7}$$

m is the nodes of hidden layer and f is the excitation function of hidden layer.

Secondly, according to the output of hidden layer H, weights w_{jk} between hidden layer and output layer, threshold b calculate o which is the predicted output of NN.

$$o_k = \sum_{j=1}^{m} H_j w_{jk} - a_k \quad k = 1,2,...,l \tag{8}$$

Then, according to predicted output o and expected output p calculate predicted error e of neural network.

$$e_t^k = P_t^k - o_t^k \quad k = 1,2,...,l \ t = 1,2,...ps*P \tag{9}$$

The fitness function is set as:

$$fit(t) = \sum_{k=1}^{l} abs(e_t^k) \tag{10}$$

Calculate every particle's fitness value by (10), search for each particle's best position achieved so far.

d.   Get sub-swarm

Divide the whole population into sub-swarms equally and get the local optimum $lbest_P$ of every sub-swarm according to the idea of DMSPSO.

e.   Update

Update every particle's position and velocity, and then enter into b.

f.   Judge the times of loop

If the iteration of regroup is satisfied, all the sub-swarms will be regrouped again.

g.   If iteration ends, stop, else return e.

4   Experimental Results

This data of experiment is got through the system of monitoring and analyzing key power industry. Two models which are BPNN and DMSPSO-NN are used to predict the power load about one day which is based on the data achieved 29 day previously.

According to the recorded load data of October, November, December, precious 29 days in December, related days (related days of everyday are 20 days) and related sampling points(related points for every sampling point are 42 points) of these 29 days are regarded as train data to predict the load of 30th, December. The prediction of each algorithm contains the whole day's prediction and every point's prediction during one day. They are defined as follows:

BPNN1: The data of the whole day is regarded as a whole prediction output (the dimension of the output is 96) and the predicted mode is BPNN.

DMSPSO-NN1: The data of the whole day is regarded as a whole prediction output and the predicted mode is DMSPSO-NN.

BPNN2: Every point in one day is regarded as a whole prediction output (the dimension of the output is 1) and the predicted mode is BPNN.

DMSPSO-BPNN2: Every point in one day is regarded as a whole prediction output and the predicted mode is DMSPSO-NN.

The parameters used in this task are set as follows:

The structure of NN1: 62 input nodes, 20 hidden nodes, 96 output nodes

The structure of NN2: 62 input nodes, 20 hidden nodes, 1output node

Dimension: 1281

Population size: 30

Number of sub-swarms:5

MaxFES (Maximum Fitness Evaluation): 40000

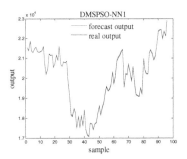
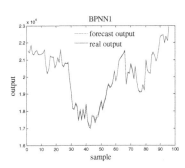

Fig. 2. The whole day as a sample to be optimized

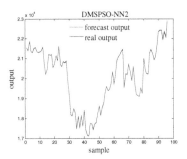
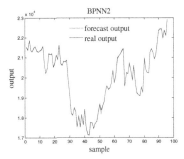

Fig. 3. Every point in one day as a sample to be optimized

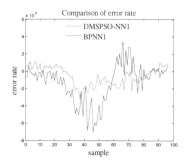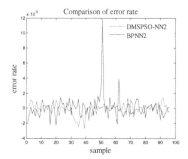

Fig. 4. Comparison of error rate under different conditions

Table 1. The error rate which is below 0.3%

	BPNN1	DMSPSO-NN1	BPNN2	DMSPSO-NN2
Error rate	0.9796	1	0.9892	1

Some conclusions could be made from the result as follows:

- DMSPSO-NN performs better than BPNN which shows that dynamic sub-swarm has improve the diversity of the solutions;
- To predict every point has a better result than regard the whole day as a sample;
- The accuracy of DMSPSO-NN is higher than PSO.

From these three points, we could know that DMSPSO has a better global search ability which helps it avoid being trapped into local optimum. After combining the NN with the DMS-PSO, the prediction is much more better which shows that DMSPSO-NN is suitable for Ultra-Short-Term Load Forecasting.

5   Conclusions

In this paper, an improved PSO(DMSPSO) is employed to optimize NN. What's more, two methods are used to test the property of DMSPSO-NN. The result shows that DMSPSO-NN has better global search ability when it is used in Ultra-Short-Term Load Forecasting problem. The result of error rate also makes us know that when every point is regared as training sample, the result is much more better. In the future, more algorithms will be used to predict load, and the more better and fast algorithm will be used for online forcast.

Acknowledgment. This research is partially supported by The Second Batch Project of Science and Technology of Henan Electric Power Company in 2013 (5217L0135029)and National Natural Science Foundation of China (61305080) and Postdoctoral Science Foundation of China (20100480859) and Specialized Research Fund for the Doctoral Program of Higher Education (20114101110005) and Postdoctoral Science Foundation Grand (2014M552013).

References

1. Gross, G., Galiana, F.D.: Short-term load forecasting. Proceedings of the IEEE 75(12), 1558–1573 (1987)
2. Abdel-Aal, R.E.: Improving electric load forecasts using network committees. Electric Power Systems Research 74(1), 83–94 (2005)
3. Hsu, Y.Y.: Fuzzy expert systems: an application to short-term load forecasting. IEE Proceedings C: Generation Transmission and Distribution 139(6), 471–477 (1992)
4. Francis, E., Tay, H.: Application of support vector machines in financial time series forecasting. Omega 29(4), 309–317 (2001)
5. Rüping, S.: Incremental learning with support vector machines. In: Proceedings-IEEE International Conference on Data Mining, ICDM, pp. 641–642 (2001)
6. Luo, X., Zhou, Y.H., Zhou, H.: Forecasting the daily load based on ANN. In: Control theory and Application, pp. 1–4 (2007)
7. Kim, C.I., Yu, I.K.: Kohonen neural network and transform based approach to short-term load forecasting. Elect. Elecr. Power Syst. Res. 63(3), 169–176 (2002)
8. Hu, J., Zeng, X.: A hybrid PSO-BP algorithm and its application. In: 2010 Sixth International Conference on Natural Computation (ICNC), vol. 5, pp. 2520–2523. IEEE (2010)
9. Li, H., Yang, D., Ren, Z.: Based on PSO-BP network algorithm for fault diagnosis of power transformer. In: 2010 International Conference on Computer, Mechatronics, Control and Electronic Engineering (CMCE), vol. 4, pp. 484–487. IEEE (2010)
10. Zhang, L., Ma, J., Wang, Y.: PSO-BP neural network in reservoir parameter dynamic prediction. In: 2011 Seventh International Conference on Computational Intelligence and Security (CIS), pp. 123–126. IEEE (2011)
11. Zhang, P.Y., Sheng, Y.L., Wan, L.L.: Modeling and simulation of screw axis based on PSO-BP neural network and orthogonal experiment. In: Second International Symposium on Computational Intelligence and Design, ISCID 2009, pp. 272–275. IEEE (2009)
12. Ren, J., Yang, S.: An Improved PSO-BP Network Model. In: 2010 International Symposium on Information Science and Engineering (ISISE), pp. 426–429. IEEE (2010)
13. Kennedy, J., Eberhart, R.C.: Particle Swarm Optimization. In: IEEE International Conference on Neural Networks, Piscataway, NJ, pp. 1942–1948 (1995)
14. Shi, Y., Eberhart, R.C.: A Modified Particle Swarm Optimizer. In: The IEEE Congress on Evolutionary Computation, pp. 69–73 (1998)
15. Liang, J.J., Suganthan, P.N.: Dynamic Multi-Swarm Particle Swarm Optimizer with Local Search. In: IEEE Congress on Evolutionary Computation, vol. 1, pp. 522–528 (2005)
16. Liang, J.J., Suganthan, P.N.: Dynamic Multi-Swarm Particle Swarm Optimizer. In: IEEE International Swarm Intelligence Symposium, pp. 124-129 (2005)
17. Gavrilas, M., Ciutea, I., Tanasa, C.: Short-term Load Forecasting with Artificial Neural Network Models. Proceedings of IEE CIRED 25(12), 28–31 (2001)
18. Wang, Q., Zhou, B., Li, Z.: Forecasting of short-term load based on fuzzy clustering and improved BP algorithm. In: International Conference on Electrical and Control Engineering (ICECE), pp. 4519–4522. IEEE (2011)
19. Yao, S.J., Song, Y.H., Zhang, L.Z., Cheng, X.Y.: Wavelet transform and neural networks for short-term electrical load forecasting. Energy Conversion and Management 41(18), 1975–1988 (2000)

Dynamic Differential Evolution for Emergency Evacuation Optimization

Shuzhen Wan

School of Computer and Information Technology,
Three gorges University of ChinaYichang, China
wanshuzhen@163.com

Abstract. Emergency evacuation in public places has become the hot area of research in recent years. Emergency evacuation route assignment is one of the complex dynamic optimization problems in emergency evaluation. This paper proposed the modified dynamic differential evolution algorithm and studied the emergency evacuation, then applied the multi-strategy dynamic differential evolution for emergency evacuation route assignment in public places. We use the Wuhan Sport Center in Wuhan China as the experiment scenario to test the performance of the proposed algorithm. The results show that the proposed algorithm can effectively solve the complex emergency evacuation route assignment problem.

Keywords: dynamic differential evolution, emergency evacuation assignment, dynamic optimization.

1 Introduction

Evolutionary Algorithms (EAs) have been applied to solve dynamic optimization problems. Many real world optimization problems are dynamic optimization problems (DOPs)[1-3]. Dynamic optimization algorithms are different from the traditional one which focuses on the static conditions, while the former one admits that both the problems and the solutions may be changed in time.

More and more scholars pay special attention to the study of applying the EAs to solve dynamic optimization problems in recent years. Several approaches have been developed in EAs to address DOPs, such as maintaining diversity during the run via random immigrants [4, 5], using memory to store and reuse useful information[3, 4, 6], multi-population approach[1, 7, 8], increasing diversity after a change[2, 9, 10].

Although all the above approaches are effectively for solving some dynamic optimization problems, there are some points of criticism, Such as: how to trace the different changing optimums in searching space. We use the prediction based multi-strategy differential evolution algorithm [11]to meet the challenges. We use a hybrid method that combines population core based multi-population strategy and prediction strategy and new local search scheme to enhance differential evolution (DE) performance for solving DOPs. The population core based multi-population strategy

Y. Tan et al. (Eds.): ICSI 2014, Part II, LNCS 8795, pp. 392–400, 2014.

is useful to maintain the diversity of the population by using the multi-population and population core concept. The prediction strategy is helpful to rapidly adapt to the dynamic environment by using the prediction area. The local search scheme is useful to improve the searching accuracy by suing the new chaotic local search method. Experimental results on the moving peaks benchmark show that the proposed schemes enhance the performance of DE in the dynamic environments. In this paper, we apply the proposed algorithm to solve the real world dynamic optimization problems-emergency evacuation route assignment optimization problem.

Emergency evacuation is an important issue for the large public spaces. There have been a lot of literatures focus on the problem of emergency evacuation[12, 13] Emergency evacuation is the study of how to evacuate the people from dangerous locations to safe places in hurry.[3] Evacuation planning is a very complex problem which needs to satisfy the consideration of many aspects. From the perspective operation research, evacuation planning is a dynamic optimization problem.

Studies on evacuation in buildings mainly focus on the simulation [13] some other researches focus on the evacuation of the traffic network [14].

This paper applied the proposed dynamic differential evolution algorithm to solve the emergency evacuation route assignment optimization problem and achieved good results comparing with other algorithms.

The remaining sections of this paper are organized as follows: Section II describes the proposed algorithm. Section III details the emergency evacuation route assignment model. Section IV presents the applying of the proposed algorithm on the emergency evacuation route assignment. Section V introduces the experimental study and discussion based on the experimental results. Finally, Section VI draws conclusions.

2   Prediction based Multi-Strategy Differential Evolution

In order to enhance the performance of differential evolution for dynamic optimization problems, we use the population core based multi-population strategy to maintain the population diversity and avoid the premature convergence. And the new local search scheme is employed in our algorithm to improve the exploitation precision. Moreover, the prediction strategy is utilized to trace the optima when the environment changes. Some of the details of the three schemes will be given as follows. This algorithm is also detailed in our previous work [11].

2.1   Multi-Population Strategy

The multi-population method can be used to maintain multiple populations on different optima when dealing with DOPs. To make this method work, the search space is divided into several local search spaces, and each local search space can contain more than one peak. Each subpopulation covers one local search space and searches the local optimum in it. However, there are some important issues to be considered such as how to divide the local search space, how to decide the number of the subpopulation, how to generate the subpopulations. In this paper, we use the hierarchical clustering method to achieve the goal of dividing subpopulations. The population core concept is used also to maintain the diversity of populations after the dividing.

2.2   Tent Map Based Local Search

Because of the ergodicity and randomicity, a chaotic system changes randomly, but eventually goes through every state if the time duration is long enough. This characteristic of chaotic systems can be utilized to build up a search operator for optimizing objective functions. To improve the local search ability, we introduce the tent map based local search method in the proposed algorithm. The tent map based local search scheme is described in [11].

2.3   Prediction Strategy

Tracing the new optima in the changed environment is the most important issue for the algorithms which are applied to deal with DOPs. The changes of environments might be very complex, and the width, the height and the location of the optima can all be changed. So, how to find the optima accurately and quickly have become the challenges for all the dynamic algorithms. We propose the prediction strategy to predict the location of the new optima before the environments are changed. The prediction strategy is based on the idea that we create the prediction areas before the changes occur, if the prediction areas just cover the areas containing the new optima, the algorithm will quickly find the optima, if not, the searching speed can also be improved due to the prediction areas are nearer to the new optima, and the probability of finding the optima will be increased. The details of the prediction scheme are shown in [11].

3   Evacuation Route Assignment Model

In this paper, we utilize the model [15] as the base of our evacuation route assignment model. In literature [15], three optimization objects are presented: minimizing the evacuation time, minimizing the total travel distance of all evacuees and minimizing the congestion during the evacuation process. The evacuation time is the most important in the three objects because if all evacuees can be evacuated within the set time, the evacuation route assignment will be effective. Thus, in this model, we select the evacuation time as the single optimization object, as the same, we take the congestion as the constraints. If the congestion degree during the evacuation process exceeds a threshold value which is defined according to the extent of the evacuation environment can afford, the evacuation route assignment proves infeasible. This object is defined as follows:

$$\min T = \max\{t_{D1}, t_{D2}, \ldots t_{Dr}, \ldots t_{Dn}\}, r \in M$$

$$s.t. f_{max} < f_{pre}$$

(1)

Where, f_{max} is the max congestion of the evacuation passageway in the system during the evacuation, f_{pre} is threshold of the congestion. The congestion in the evacuation passageway during the evacuation is changed with time, so the evacuation route assignment is the dynamic optimization problem.

4 Modified Dynamic DE for Evacuation Route Assignment

4.1 Frame of the Proposed Algorithm for Evacuation Route Assignment

In this paper, we apply the prediction based multi-strategy differential evolution algorithm [11]to solve the evacuation route assignment optimization problem. We employ the k-shortest paths algorithm [16] to initialize the evacuation routes. The frame of the proposed algorithm for evacuation route assignment is detailed as follows:

Step 1. Initialize the evacuation routes with k-shortest paths algorithm.

Step 2. Initialize the population randomly in the searching space and the archive for prediction strategy

Step 3. Use the clustering method to create the subpopulations

Step 4. Apply DE operators (mutation, crossover and selection) on each individual in each subpopulation.

Step 5. Apply the diversity control operator to maintain the population diversity.

Step 6. If optimization environments are changed, Set the prediction zone then go back to Step 3. Else, goto next step.

Step 7. If a termination condition is met then goto end, else go back to Step 4.

4.2 Frame of the Proposed Algorithm for Evacuation Route Assignment

The goal of the evacuation route assignment is to construct an effective and safe evacuation scheme to guide the evacuees evacuating from the dangerous orderly. Considering the complexity of the evacuation route assignment, we adopt the sorting based precoding method to encode each individual.

Firstly, because the evacuees often select the shortest paths while evacuating, so we apply k-shortest paths algorithm to initialize the paths from the starting point to the destination of evacuation in the evacuation zone.

Next, encoding each route with an integer and then we assign the coded paths to the individuals randomly. The individual of the population is coded as follows:

$$individual = \left\{\lambda_1 \lambda_2 \ldots \lambda_{N_1} \mid \lambda_{N_1+1} \ldots \lambda_{N_2} \mid \ldots \right.$$
$$\left. \mid \lambda_{N_{n-1}+1} \ldots \lambda_{N_n} \right\}, \lambda_1 \lambda_2 \ldots \lambda_{N_n} \in N$$

(2)

where λ_{N_i} is the evacuation passageway assigned to i individual. Each evacuee can randomly select these paths to evacuate, all the evacuation paths selected by evacuees can be constructed the evacuation routes assignment scheme.

4.3   Algorithm Design

For the evacuation route assignment, the dynamic DE[11]should be modified to adopt this optimization problem. The mutation operation and the crossover operation of DE are modified as follows.

For the mutation operation, firstly, we set a mutation factor F (value range 0~1), then mutate the individual A with the mutation factor F, thus, the evacuation paths change. The change of evacuation paths is to reselect the evacuation path whose start point is corresponding to variables of the individual A and being calculated by the k-shortest paths algorithm. After the muting, we can obtain the mutated individual A'.

For the crossover operation, we perform the crossover operation on each variable of the mutated individual A' with the crossover rate (CR). The variable will be retained if that variable is selected, otherwise, it will be replaced by the variable selected from the individual A. The process will continue till the crossover operation performs on all variables. Where, the variables are referred to the evacuation paths. The mutation operation and crossover operation are shown in Fig. 1 and Fig. 2.

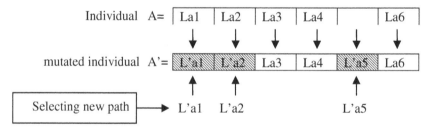

Fig. 1. Mutation operation for evacuation route assignment with $F = 0.5$

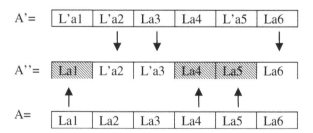

Fig. 2. Crossover operation for evacuation route assignment with $CR = 0.5$

5   Experimental Study and Discussion

5.1   Experimental Settings

The settings for the proposed dynamic evolution algorithm are shown in Table 1.

Table 1. Settings for the proposed algorithm

Parameter	Setting
Population size	100
Maximum subpopulation size	15
Mutation factor(F)	0.5
Crossover rate(CR)	0.9
Radius of the subpopulation core r_{core}	5.0
Number of optional paths for each start point	50
Threshold of congestion degree f_{pre}	0.75

The Wuhan Sport Center in Wuhan city China is taken as the experimental area to test the performance of the proposed algorithm. Wuhan Sport Center can hold about 60000 people, and there are 42 grandstands to accommodate the spectators. Suppose in a massive activity, the spectators should be evacuated to the safe area as quickly as possible for some reasons such as fire disaster and horrible attack. The number of evacuees is about 24727. The spectator is assigned to each grandstand randomly according to the number limitation of each grandstand.

The evacuation network of this stadium which contains 158 nodes and 224 arcs is converted by its structure (Fig. 3).The original locations of the spectators are the 42 grandstands, and the exits are the 5 ticket entrances, final destinations of evacuation in this scenario, as it is shown below.

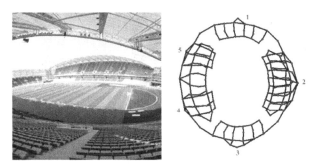

Fig. 3. Experiment area and evacuation network

In order to measure the performance of the proposed algorithm, we compare our proposed algorithm with DynDE[17]which is the classic dynamic differential evolution. In the experiments, each algorithm runs independently for 20 times with the 200 initialized paths for each start point. Though, the object is the minimum evacuation time, we also calculate the congestion degree and the total length of all evacuation paths.

5.2   Results and Analysis

The results are shown in Table 2.

Table 2. Results of the two algorithms for the evacuation route assignment

Algorithm		Evacuation time(T)	Congestion	Total length of evacuation paths (L)
Proposed algorithm	Avg _best	721.101	0.2731	34240.3
	Avg _mean	735.476	0.4353	34266.2
	Avg _worst	767.150	0.5252	34301.5
	STD	9.2346	0.0863	36.07
DynDE	Avg _best	732.119†	0.2907†	34304.8†
	Avg _mean	740.167†	0.4410	34329.6
	Avg _worst	770.234†	0.5723†	34378.3†
	STD	15.3028†	0.2057†	33.482‡

†indicates Proposed algorithm is significantly better than DynDE by the Wilcoxon signed-rank test at α = 0. 05.
‡means that DynDE is significantly better than our algorithm.

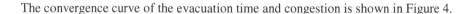

The convergence curve of the evacuation time and congestion is shown in Figure 4.

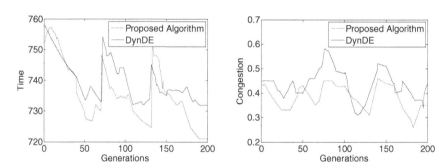

Fig. 4. The convergence curve of the evacuation time (left) and the convergence curve of the congestion (right)

From Table 2 can be seen that the proposed algorithm is superior to DynDE for the evacuation route assignment optimization. The satisfactory results are achieved by the proposed algorithm. It has the shorter evacuation time than DynDE with the better congestion. The total length of evacuation paths achieved by the proposed in the optimization process is shorter than the DynDE's.

It is clearly seen from Figure 4 that the proposed algorithm outperforms the DynDE not only on the evacuation time and but also on the evacuation congestion.

6   Conclusions

Making a feasible and effective evacuation plan in the emergency situation is an important issue. In this paper, a prediction based multi-strategy differential evolution is adopted to solve this evacuation route assignment optimization problem based on the evacuation model. The experimental results of the evacuation scenario of a large public activity in Wuhan Sport Center in Wuhan City of China show that the proposed algorithm can solve the real world complex dynamic optimization problems such as the evacuation route assignment optimization problem. The analyses of the results imply that the congestion value restricts the performance of an evacuation plan. Thus, the congestion should be given a high priority.

Acknowledgments. This work was partially supported by the Scientific Research Foundation for Introduced Excellent Scholars, China Three Gorges University (Grant No. KJ2012B055).

References

1. Blackwell, T., Branke, J.: Multiswarms, exclusion, and anti-convergence in dynamic environments. IEEE Transactions on Evolutionary Computation 10(4), 459–472 (2006)
2. Branke, J., Saliho, U. G., Lu, E., Uyar, C.S.: Towards an analysis of dynamic environments. In: Proceedings of the 2005 Conference on Genetic and Evolutionary Computation, pp. 1433–1440 (2005)
3. Branke, J.: Memory enhanced evolutionary algorithms for changing optimization problems. In: Proceedings of the 1999 Congress on Evolutionary Computation, Washington, DC, USA, pp. 1875–1882 (1999)
4. Mavrovouniotis, M., Yang, S.: Memory-based immigrants for ant colony optimization in changing environments. In: EvoCOMPLEX, EvoGAMES, EvoIASP, EvoINTELLIGENCE, EvoNUM, and EvoSTOC, EvoApplications, pp. 324–333. Springer Verlag (2011)
5. Grefenstett, J.J.: Genetic algorithms for changing environments. In: Proceedings of the 2nd Conference on Parallel Problem Solving from Nature, Brussels, Belg, pp. 137–144 (1992)
6. Yang, S.: Explicit Memory Schemes for Evolutionary Algorithms in Dynamic Environments. In: Evolutionary Computation in Dynamic and Uncertain Environments, vol. 51, pp. 3–28 (2007)
7. Novoa-Hernández, P., Corona, C., Pelta, D.: Efficient multi-swarm PSO algorithms for dynamic environments. Memetic Computing 3(3), 163–174 (2011)
8. Mendes, R., Mohais, A.S.: DynDE: a differential evolution for dynamic optimization problems. In: Proceedings of the 2005 IEEE Congress on Evolutionary Computation, pp. 2808–2815. IEEE Press (2005)
9. Blackwell, T.M., Bentley, P.: Dynamic Search with Charged Swarms. In: Proceedings of the 2002 Genetic and Evolutionary Computation Conference, pp. 19–26. Morgan Kaufman (2002)
10. Cobb, H.G.: An Investigation into the Use of Hypermutation as anAdaptive Operator in Genetic Algorithms Having Continuous, Time-Dependent Nonstationary Environments Naval Res. Lab., Washington, DC: Tech.Rep. AIC-90-001 (1990)

11. Shuzhen, W., Shengwu, X., Yi, L.: Prediction based multi-strategy differential evolution algorithm for dynamic environments. In: 2012 IEEE Congress on Evolutionary Computation (CEC), pp. 1–8. IEEE Press, Brisbane (2012)
12. Lämmel, G., Grether, D., Nagel, K.: The representation and implementation of time-dependent inundation in large-scale microscopic evacuation simulations. Transportation Research Part C: Emerging Technologies 18(1), 84–98 (2010)
13. Zheng, X., Zhong, T., Liu, M.: Modeling crowd evacuation of a building based on seven methodological approaches. Build Environ. 44(3), 437–445 (2009)
14. Yamada, T.: A network flow approach to a city emergency evacuation planning. Int. J. Syst. Sci. 27(10), 931–936 (1996)
15. Qiuping, L., Zhixiang, F., Qingquan, L., Xinlu, Z.: Multiobjective evacuation route assignment model based on genetic algorithm. In: 18th International Conference on Geoinformatics, Beijing, pp. 1–5. (2010)
16. Eppstein, D.: Finding the k shortest paths. In: Proceedings of 35th Annual Symposium on Foundations of Computer Science, Santa Fe, NM, pp. 154–165, (1994)
17. Mendes, R., Mohais, A.S.: DynDE: a differential evolution for dynamic optimization problems. In: Proceedings of the 2005 IEEE Congress on Evolutionary Computation, vol. 3, pp. 2808–2815. IEEE Press (2005)

Centralized Charging Strategies of Plug-in Electric Vehicles on Spot Pricing Based on a Hybrid PSO

Jiabao Wang[1], Qi Kang[1], Hongjun Tian[1], Lei Wang[1,2], and Qidi Wu[1]

[1] Department of Control Science and Engineering, Tongji University, Shanghai, China
wangjiabao0316@163.com, qkang@tongji.edu.cn
[2] Shanghai Key Laboratory of Financial Information Technology, Shanghai, China
wanglei@tongji.edu.cn

Abstract. This work proposes an efficient charging regulation strategy based on optimal charging priority and location of plug-in electric vehicles (PEVs). It employs a hybrid particle swarm optimization for optimal charging priority and location of PEVs in distribution networks, with the objectives of minimization of charging cost, power loss reduction and voltage profile improvement. The algorithm is executed on IEEE 30-bus test system. The results are compared with those that are gained by executing sample genetic algorithm (SGA) with diverse parameters on the same system. The results indicate the effectiveness and promising application of the proposed methodology.

Keywords: PEVs, spot pricing, centralized charging strategy, HPSO, optimal charging priority and location.

1 Introduction

Nowadays, more and more vehicles are on roads, thereby increasing the consumption of fossil fuel. Consequently, the environment is being seriously polluted. Under this circumstance, many countries have proposed their energy policies with objectives of economic effectiveness improvement, achievement of energy security, and environment pollution reduction, which promotes electrification of transportation and, especially, the rapid development of plug-in electric vehicle (PEV) industry [1].

However, PEVs, a new kind of power load, would exert a tremendous influence on the daily residential load curve of distribution network if they widely connected to the power grid for battery charging [2]. Due to the uncertainty of their charging behaviors, uncoordinated random charging of PEVs may lead to unforeseen effect on normal operation of distribution system, such as aggravating the load peak and off-peak difference in network, etc. Meanwhile, taking the spot pricing into consideration, the owners of PEVs may afford much higher cost for battery charging. Therefore, the appropriate dispatch of PEVs in a distribution system will be a challenging demand side management (DSM) [3]. Fortunately, PEVs are more flexible than traditional load, because majority of PEVs owners usually return home early in the evening and have no request for the special time that their vehicle will be charged, as long as the

Y. Tan et al. (Eds.): ICSI 2014, Part II, LNCS 8795, pp. 401–411, 2014.

batteries are full by the next morning [4]. According to the statistical data of National Household Travel Survey (NHTS), more than 90% of vehicles are parked at home between 9 P.M and 6 A.M [4]. Taking this opportunity into account, several centralized charging strategies of PEVs have been researched for utilizing less expensive electricity and shifting the PEVs load to off-peak hours. For example, under a spot pricing based on electricity market environment, a demand side response based charging strategy is proposed, and a dynamic estimation interpolation based algorithm is designed to optimize the mathematical model which is established taking into account the valley-filling effect of supply side and the users' cost [1]. Wu et al. [4] proposes a novel minimization of charging cost based heuristic approach by analyses of PEVs travel pattern and spot pricing. The results show that the strategy can lower the peak-valley difference and save users' cost effectively, but they ignore the influence of PEVs charging behavior on power qualities, such as power loss, voltage fluctuation, etc. Deilami et al. [5] proposes a real-time smart load management control strategy which is developed for the coordination of PEVs charging based on real-time minimization of total cost of generating the energy plus the associated grid energy losses. The results indicate that the approach can reduce the power losses and improve the voltage profile by considering the maximum sensitivities selection optimization based priority charging. Lan et al. [6] presents a nonlinear electric vehicle (EV) battery model, and a dynamic programming-based algorithm for optimizing an EV's charging schedule with given electricity price and driving pattern. But only one EV's charging schedule is researched here.

In this paper, taking advantage of the flexibility of the PEVs, we arrange them to charge at the relatively inexpensive electricity which occurs during off-peak hours at night. An efficient charging regulation based on optimal charging priority and location of PEVs is proposed under a spot pricing based electricity market environment. And then, PSO-GA [7], a hybrid particle swarm optimization by incorporating genetic algorithm, hereinafter to be referred as HPSO, is used for optimal charging priority and location of PEVs in distribution networks system. We take power quality and economy objectives into account to define the optimization objectives in this paper, including power losses, voltage profile and charging cost. The proposed approach is executed on the IEEE 30-bus test system.

2   Problem Formulation

Under a spot pricing based electricity market environment, uncoordinated charging of many PEVs may exert negative impacts on the security and economy of power system operation, such as power losses increment, overload, voltage fluctuation aggravation and charging cost increment. To minimize such impacts, we may proceed as follows:

1) Different locations for PEVs charging may have different influences on power quality. Appropriate locations for PEVs charging benefit power quality improvement.

2) In order to abate the vacancy of electric power, PEVs are scheduled for charging during the off-peak period of the day.

3) The charging cost should be as low as possible. Because different periods of the day correspond to different electricity price under a spot pricing based electricity market environment, if all PEVs owners had a preference for the exact time of which electricity price is lowest, a new peak demand would occur. Hence, maximum power consumption should be set for every time slot in order to avoid overload. Under this circumstance, PEVs charging priority that determined a sequence of PEV choosing charging slots impacts on daily residential load curve and total cost of electricity heavily.

2.1 Charging Rule

In this paper, the maximum demand level has been defined as the maximal value of residential load during a scheduling period. The power for PEV charging is constrained by:

$$P_i \leq P_i^{\text{limit}}, i = 1,2,\ldots, T . \tag{1}$$

$$P_i^{\text{limit}} = P_{\text{max}} - P_i^{\text{load}} . \tag{2}$$

where i and T are the time slot number and total number of slots, P_{max} is the maximum residential load demand level with PEVs being charged, P_i^{load} is the total residential power consumption at the ith time slot without PEVs plug in, P_i^{limit} is the maximum permissible power consumption for PEVs charging at ith time slot, P_i is the total power consumption for PEVs charging at ith time slot.

If the PEVs charging priority and locations are known, load scheduling is transformed into a PEVs charging rule. The basic idea is to charge each vehicle in the time slots where the lowest electricity price occur and power consumption meet the Eq. (1).

The flowchart of PEVs charging rule is shown as Fig.1. And the relevant parameters can be define as follow: $Priority_j$ is the charging priority number of jth PEV , Bus_j is the charging location number of jth PEV , $Duration_j$ is the number of time slots which jth PEV need for charging , n is the total number of PEVs, $Price_i$ is the electricity price at ith time slot, P_i^{limit} is the maximum permissible power consumption for PEVs charging at ith time slot, T is the total number of time slots, ρ_k is the serial number of PEV whose priority rank is k, τ_l is the serial number of slot whose price rank is l, P is the rated power of PEV, $Power_j^s$ is the total power consumption of PEVs at ith time slot and sth charging node, N is the number of locations, ε_j is the set of time slots where jth PEV is being charged.

2.2 Objective Outline

The objective of problem model includes minimization of charging cost, power loss reduction, voltage profile improvement. Therefore, a three-fold objective function is given by

$$OBJ : = f = \min (P_{loss} + \sigma V_{devi} + \gamma Cost) . \tag{3}$$

where P_{loss} is active power loss, σ and γ are the non-negative weighting factor used to indicate the relative important of three items, (here $\sigma=20$, $\gamma=0.01$), V_{devi} denotes load bus voltage deviations from 1.0 per unit, $Cost$ is the total electricity cost.

P_{loss} can be obtained with power flow calculation and is represented as $P_{loss} = \sum_{i=1}^{T} \sum_{j=1}^{L} |I_{ij}|^2 * R_j$, where T is the number of slots, L is the number of lines in the power system, I_{ij} is the current of jth line at ith time slot, R_j is the resistance of jth line.

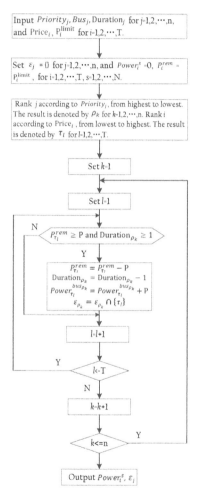

Fig. 1. The flowchart of PEVs charging rule

V_{devi} can be denoted as $V_{devi} = \sum_{i=1}^{T} \sum_{j=1}^{N} |V_{ij} - 1.0|$ (p. u.), where T plays the same role in P_{loss}, N is the number of buses in power system, and V_{ij} is node voltage (p. u.) of jth line at ith time slot [8].

Cost is denoted as $Cost = \sum_{i=1}^{T} P_i * price_i * \Delta T$, where T plays the same role in P_{loss}, P_i is the power consumption of PEVs at ith time slot, $price_i$ is the electricity price at ith time slot, ΔT is time span of a time slot.

3   Scheduling Algorithms

3.1   HPSO: An Improved Particle Swarm Optimization

Particle swarm optimization (PSO) was proposed by Kennedy and Eberhart in 1995. In [9], dynamic weights are integrated into a standard PSO to improve its global and local convergence.

HPSO [7] discards the method with which particles update their positions by tracking the individual and group optimal position in PSO. Instead, it introduces the crossover and mutation operation of a genetic algorithm (GA) into PSO. Its new particles are   refreshed by   crossover and   mutation   operators according to optimal solutions of entire population and individual.

The steps of HPSO are as follows:

Step 1. Set the optimal position of individual ($p_{best,i}$) as the initial position of ith particle (i=1, 2,..., M), where M is the population size. Select the best one among $\{p_{best,1}, ..., p_{best,M}\}$ as optimal position of population (g_{best}).

Step 2. Calculate the objective function values of all members.

Step 3. Update the $p_{best,i}$ and g_{best} according to the fitness of all members.

Step 4. Execute a crossover operation with $p_{best,i}$ and g_{best}. Update the position of the ith particle by executing a crossover operation with $p_{best,i}$ and g_{best}, respectively.

Step 5. Perform a mutation operation on each particle in the swarm.

Step 6. If the termination condition is met, stop. Otherwise, return to Step 2.

4   Simulation and Result

4.1   Simulation Data

In order to explain the problem more specifically and clearly, some assumptions are given and listed below:

1) All information of EVs and control signals generated by aggregators can be delivered immediately between EVs and aggregators [6].

2) PEV battery capacities typically range from a few kWh to over 50 kWh [5]. The capacity and rated power of PEV can be defined as 50kWh and 10kW, respectively.

3) The number of charging time slots of PEV is assumed to follow approximately Gaussian distribution whose mean value and standard deviation equal to 20 and 5 respectively.

4) 1000 PEVs are scheduled as a whole, thereby they can be treated as a PEV set, hereinafter still referred to as "PEV". And all vehicle batteries in PEV have same state of charge (SOC).

In this paper, PEVs are dispatched for charging during off-peak hours from 9 P.M. to 7 A.M., because most vehicles are vacant and the electricity price is generally low during this period. The daily residential load curve [5] shows in Fig.2. Furthermore, other relevant data are given as follow:

1) The total time for charging dispatch is segmented into 40 time slots, where each time slot has a duration of 15 minutes.

2) Data of spot pricing of electricity is released by Long Island, New York on Jan. 1, 2010 [2]. The value of electricity price ($/MW) during off-peak hours from 9 P.M. to 7 A.M is given as {71.50, 71.16, 63.46, 58.86, 62.67, 39.84 44.78, 53.03, 65.34, 57.82}.

3) The total number of PEVs which participate in scheduling is 50.

4) The distribution system used for simulation and analysis of PEVs ordered charging strategy in this paper is the IEEE 30-bus test system [10].

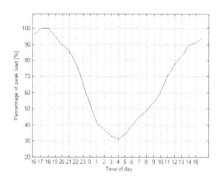

Fig. 2. Daily residential load curve

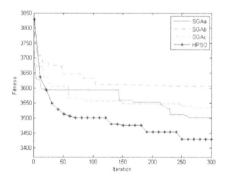

Fig. 3. Optimal fitness dynamics of system

IEEE 30-bus test system [11] is adopted for simulation and analysis in this work. In this paper, 10 buses, i.e., buses 3-12, are chosen as PEVs charging nodes. The HPSO-based method scheme can be described as follows, which is used as a solver of this optimization problem of charging priority and location of PEVs.

Step 1. Initialization

Set the time count $t=0$, dimension D (here D=50), maximum iteration number $\text{Iter}_{\max}$, and population size M. The current position of all members is $X(0) = \{X_1^0, ..., X_i^0, ..., X_M^0\}$, where

$$X_i^0 = \begin{bmatrix} x_{i,1}^0 \ x_{i,2}^0 \ \cdots \ x_{i,D}^0 \\ y_{i,1}^0 \ y_{i,2}^0 \ \cdots \ y_{i,D}^0 \end{bmatrix} \quad (i=1,2,...,\text{M}).$$

X_i^t represents the coding scheme of ith particle at tth iteration, $x_{i,j}^t (j=1,2,...,\text{D})$ is the charging priority index of jth PEV of ith particle at tth iteration, $y_{i,j}^t (j=1,2,...,\text{D})$ is the charging location (bus index) of jth PEV of ith particle at tth iteration. The first line of X_i^0 is a random permutation of integers 1~50 and $y_{i,j}^0$ is a random integer in {3,4,...,12}.

PEVs charging rule and power flow calculation are applied to all members, by the results of which the fitness of value $f(X_i^0)$ of X_i^0 is obtained. For each individual, set $p_{best,i} = X_i^0$ and $Pbest_i = f(X_i^0)$, $i=1,2,\ldots, M$. Select the best one among $\{p_{best,1}, \ldots, p_{best,M}\}$ as g_{best}, and set $Gbest = f(g_{best})$.

Step 2.   Update the time counter.

$t = t + 1$

Step 3.   Execute crossover operation.

(1) Execute a crossover operation with $p_{best,i}$

Update the position of X_i^t $(i=1,2,\ldots, M)$ by executing crossover operation with $p_{best,i}$. If $f(X_i^t) <$ Pbest$_i$, then update individual best as $Pbest_i = f(X_i^t)$ and set $p_{best,i} = X_i^t$.

(2) Execute a crossover operation with g_{best}

Update the position of X_i^t $(i=1,2,\ldots, M)$ by executing crossover operation with g_{best}. If $f(X_i^t) <$ Pbest$_i$, then update individual best as $Pbest_i = f(X_i^t)$ and set $p_{best,i} = X_i^t$.

Step 4. Execute mutation operation

Execute a mutation operation by exchanging any two elements of first line of X_i^t and change corresponding element of second line using a random integer in $\{3,4,\ldots,12\}$. If $f(X_i^t) <$ Pbest$_i$, and then update individual best as $Pbest_i = f(X_i^t)$ and set $p_{best,i} = X_i^t$.

Step 5. Carry out fitness evaluation and update individual best and population best.

For all members, PEVs charging rule and power flow calculation are applied, and then evaluates the fitness of every member X_i^t $(i=1,2,\ldots, M)$ according to Eq.(3). If $f(X_i^t) <$ Pbest$_i$, and then update individual best as $Pbest_i = f(X_i^t)$ and set $p_{best,i} = X_i^t$. Select the best one among $\{p_{best,1}, \ldots, p_{best,M}\}$ as g$_{best}$, and set $Gbest = f(g_{best})$.

Step 6. Check the stopping criteria

If $t \leq$ Iter$_{max}$, then go to **Step 2**; Else, stop the algorithm.

4.2   Parameters Setting

The parameters setting of SGA and HPSO are given in Table 1. In SGA•a population of M solutions is maintained and two probability-based operations, i.e., crossover operator and mutation operator are employed. Whether the two operators work depends on crossover rate Rate$_c$ and mutation rate Rate$_m$, respectively. Meanwhile, the SGAs with different crossover and mutation parameters are denoted as SGAa, SGAb and SGAc.

Table 1. The parameters setting of SGA and HPSO

	a	M=20; Iter$_{max}$=300; Rate$_c$=0.8; Rate$_m$=0.05.
SGA	b	M=20; Iter$_{max}$=300; Rate$_c$=0.9; Rate$_m$=0.005.
	c	M=20; Iter$_{max}$=300; Rate$_c$=0.7; Rate$_m$=0.025.
HPSO		M=20; Iter$_{max}$=300.

By executing two types of algorithms 20 times separately, the Monte Carlo simulation results are shown in Table 2, where *iteration* represents the number of iterations. The evolutionary trajectory of four algorithms is shown in Fig.3.

The simulation results show that the HPSO outperforms SGA on the metrics in P_{loss}, Cost, and fitness value. Further, HPSO have better global search ability. However, the superiority of HPSO in searching the metric V_{devi} is not outstanding.

The proposed PEVs charging rule is applied to the optimal individual sought out by HPSO, and yields the power consumption of every time slot. Fig.4 shows the impact of HPSO-based coordinated PEVs charging on total system power demand. Compared with the results of uncoordinated PEVs charging, HPSO-based coordinated PEVs charging strategy effectively shift electricity use from on-peak to off-peak period.

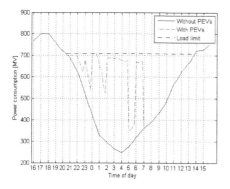 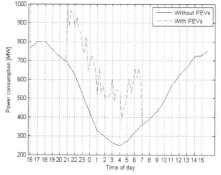

Fig. 4. The impact of HPSO-based coordinated PEVs charging on total system power demand

Fig. 5. The impact of uncoordinated PEVs charging on total system power demand

The impact of uncoordinated PEVs charging on total system power demand is shown in Fig.5, in which many peak power demands appear. With the increase of the gap between peak load and valley load in power system, it is uneconomic to install units considering the capacity of peak load. Besides, frequent start and stop are detrimental to the life of generators.

The impact of HPSO-based coordinated PEVs charging is further compared with Min-Cost Load Scheduling (MCLS) approach [4]. The basic idea of MCLS is that each vehicle is charged in the time slots where the lowest electricity price occurs. Fig.6 shows that the impact of MCLS-based PEVs charging on total system power demand. The minimum electricity cost can be achieved by carrying out the MCLS approach, but the impact of PEVs charging on system power loss and voltage deviation is ignored by MCLS approach. Besides, as we can see in Fig.6, there are still three peak loads and large gap between peak load and valley load. HPSO-based voltage (p.u.) of 30 buses at worst time slot is compared with MCSL-based voltage of 30 buses at random time slot and stochastic charging nodes, as shown in Fig.7. The HPSO-based coordinated PEVs charging improve the security and reliability of networked distribution network by minimizing voltage deviations, overloads, and power losses that would otherwise be impaired by MCSL-based PEVs charging.

Table 2. Simulation Results

Metric		SGAa	SGAb	SGAc	HPSO
$P_{loss}(MW)$	Optimum value	1374.8	1453.7	1402.4	1331.7
	Worst value	1480.5	1518.5	1461.6	1363.5
	Median value	1439.0	1474.1	1455.6	1336.8
	Mean value	1432.2	1478.1	1437.9	1344.5
	Standard deviation	41.7753	21.5690	26.3989	12.3578
$Cost(\$ \times 10^3)$	Optimum value	132.9018	133.6457	133.6895	133.9468
	Worst value	136.2426	136.6502	136.5378	135.3387
	Median value	135.4281	134.9455	135.1636	135.2335
	Mean value	134.8389	135.2212	135.1725	134.8535
	Standard deviation	1.3417	1.1264	0.9367	0.5565
V_{devi}	Optimum value	38.1942	40.3422	38.9862	37.3199
	Worst value	40.8019	41.5776	40.5185	38.0455
	Median value	39.2931	40.5403	40.1329	37.6670
	Mean value	39.6493	40.6993	39.8711	37.6834
	Standard deviation	0.9853	0.4502	0.5426	0.2360
$fitness$	Optimum value	3501.9	3606.8	3536.9	3429.5
	Worst value	3648.9	3716.5	3628.3	3477.3
	Median value	3585.0	3630.4	3606.7	3440.1
	Mean value	3573.5	3644.3	3587.0	3457.0
	Standard deviation	43.3275	38.2886	37.4149	19.320
$iteration$	Minimum value	172	82	123	252
	Maximum value	281	275	287	286
	Median value	269	245	270	274
	Mean value	249.8	218.6	244.8	269.4
	Standard deviation	39.6353	45.9125	37.3197	12.8623

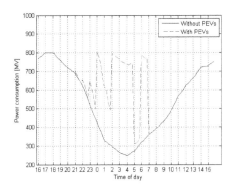

Fig. 6. The impact of MCSL-based coordinated PEVs charging on total system power demand

Fig. 7. Voltage deviation at 30 buses

5   Conclusion

This paper presents a centralized charging model of PEVs under a spot pricing-based electricity market environment. All PEVs taking part in centralized charging scheduling are equally divided into multiply, and an efficient charging rule which is based on charging priority and location of PEVs is proposed in the model. Further, in this paper, HPSO is employed to optimally determine the PEVs charging priority and location to be plugged in distribution system. Combining the HPSO with Newton method of power flow calculation can solve the problem of optimal charging priority and location of PEVs well with the objective of minimization of charging cost, power loss reduction and voltage profile improvement. The results are compared with those are obtained using SGA, and validate the superiority and effectiveness of the approach.

Under the intelligent centralized charging strategy, PEV owners are able to pay the electricity bills only in line with the number of charging time slots they use. The electricity costs resulting from the use of the proposed approach are slightly higher than those by MCLS algorithm whose objective is the maximization of energy trading profits, but it effectively reduces the gap between peak and valley load and simultaneously improves the power quality. In this case, a discount electricity price scheme can be introduced by power supplier in order to encourage PEV owners to take part in centralized charging mechanism.

The further research will be oriented to dispatching PEVs for charging under the assumption of stochastic vehicles' arrival, with the same objectives as discussed in this work.

Acknowledgments. This work was supported in part by the Natural Science Foundation of China (71371142, 61005090, 61034004), the Program for New Century Excellent Talents in University of Ministry of Education of China. Ministry of Education (NCET-10-0633), the Fundamental Research Funds for the Central Universities, and the Research Fund of State Key Lab. of Management and Control for Complex systems.

References

1. Zhao, J., Wen, F., Yang, A., Xin, J.: Impacts of electric vehicles on power systems as well as the associated dispatching and control problem. Automation of Electric Power Systems 14, 2–9 (2011)
2. Zou, W., Wu, F., Liu, Z.: Centralized charging strategies of plug-in hybrid electric vehicles under electricity markets based on spot pricing. Dianli Xitong Zidonghua (Automation of Electric Power Systems) 35(14), 62–67 (2011)
3. Masoum, A.S., Deilami, S., Moses, P.S., Masoum, M.A.S., Abu-Siada, A.: Smart load management of plug-in electric vehicles in distribution and residential networks with charging stations for peak shaving and loss minimisation considering voltage regulation. Generation, Transmission & Distribution, IET 5(8), 877–888 (2011)

4. Wu, D., Aliprantis, D.C., Ying, L.: Load scheduling and dispatch for aggregators of plug-in electric vehicles. IEEE Transactions on Smart Grid 3(1), 368–376 (2012)
5. Deilami, S., Masoum, A.S., Moses, P.S., Masoum, M.A.: Real-time coordination of plug-in electric vehicle charging in smart grids to minimize power losses and improve voltage profile. IEEE Transactions on Smart Grid 2(3), 456–467 (2011)
6. Lan, T., Hu, J., Kang, Q., Si, C., Wang, L., Wu, Q.: Optimal control of an electric vehicle's charging schedule under electricity markets. Neural Computing and Applications 23(7-8), 1865–1872 (2013)
7. Min, X.I.E.: An Improved Hybrid Particle Swarm Optimization Algorithm for TSP. Journal of Taiyuan University of Technology 4, 023 (2013)
8. Kang, Q., Lan, T., Yan, Y., Wang, L., Wu, Q.: Group search optimizer based optimal location and capacity of distributed generations. Neurocomputing 78(1), 55–63 (2012)
9. Jian-Hua, L., Rong-Hua, Y., Shui-Hua, S.: The analysis of binary particle swarm optimization. Journal of Nanjing University (Natural Sciences) 5, 003 (2011)
10. AlRashidi, M.R., El-Hawary, M.E.: Hybrid particle swarm optimization approach for solving the discrete OPF problem considering the valve loading effects. IEEE Transactions on Power Systems 22(4), 2030–2038 (2007)
11. AlRashidi, M.R., El-Hawary, M.E.: Hybrid particle swarm optimization approach for solving the discrete OPF problem considering the valve loading effects. IEEE Transactions on Power Systems 22(4), 2030–2038 (2007)

A New Multi-region Modified Wind Driven Optimization Algorithm with Collision Avoidance for Dynamic Environments

Abdennour Boulesnane[1,2] and Souham Meshoul[1]

[1] Computer Science Department, Constantine 2 University, Algeria
[2] MISC Laboratory, Constantine, Algeria
abdennour.boulesnane@gmail.com,
souham.meshoul@univ-constantine2.dz

Abstract. This paper describes a new approach to deal with dynamic optimization that uses a multi-population. Its main features include the use of a modified wind driven optimization algorithm that aims to foster impact of pressure on velocities of particles. Moreover, a concept of multi-region inspired from meteorology has been introduced along with a new collision avoidance technique to maintain good diversity while preventing collision between sub-populations. The method has been assessed using Moving Peaks Benchmark and compared to state of the art methods. Preliminary results are very encouraging and show viability of the method.

Keywords: Dynamic optimization, Swarm intelligence, Wind driven optimization, collision, multiple population methods, Moving Peaks Benchmark.

1 Introduction

In everyday life and in almost all domains, each type of optimization problem has features that make it different from the others. However, these problems usually have a common property that is their dynamic nature. Such problems are difficult to solve because the challenge is not only to locate global optima, but also to track them in environments that change over time. Therefore, a crucial requirement a dynamic optimization algorithm should fulfill is to achieve a balance between exploitation and exploration of the search space to handle optimization over time. This requires fostering diversity while ensuring very fast convergence to global optima throughout the search process because the time between two successive changes may be insufficient to converge and to follow optima at the same time. Moreover, dynamic optimization is faced to the challenge to solve both issues of outdated memory due to changes in environment and diversity loss due to traps of local optima. Outdated memory problem is usually solved by clearing the memory when a change is detected however the matter is what to do with the knowledge acquired once a change in the environment occurs: should it be reused for next changes or discarded? In [1], a study showed that the reuse of information lead to faster adaptation to changes, and thus, to

Y. Tan et al. (Eds.): ICSI 2014, Part II, LNCS 8795, pp. 412–421, 2014.
© Springer International Publishing Switzerland 2014

better solutions. Many algorithms have been proposed to address dynamic optimization problems (DOPs). A comprehensive survey can be found in [2]. More particularly, using multiple populations has been shown to be a suitable way to keep up with changes over time [7,10].

Recently, a new swarm based metaheuristic inspired from atmospheric motion has been proposed in [3] and termed as Wind Driven Optimization (WDO). WDO's model is based on the definition of trajectories of small air parcels within the earth atmosphere according to the Newton's second law of motion. WDO has been designed for static optimization and applied to electromagnetics optimization problems. In this paper, we propose investigating the potential of WDO to solve dynamic optimization. Intuitively, the motivation can be explained by the fact that the movement of air from high pressure zones to low pressure zones at velocities proportional to pressure gradient force would lead to simulation models that can be used in dynamic environments. Taking inspiration from WDO, a modified version is developed and used within a multi-population framework with the aim to adaptively detect promising regions in the search space. Therefore, we refer to the developed approach as Multi-Region Modified WDO (MR-MWDO). Furthermore, in order to maintain several sub-populations on several peaks (promising regions) and to avoid collisions between sub-populations a new strategy called Collision Avoidance Technique is introduced.

The rest of the paper is organized into four sections. Section 2 presents the WDO algorithm. The proposed Multi-Region Modified WDO is described in section 3. Section 4 is devoted to the experimental study. Finally, conclusions and perspectives are given in section 5.

2   Wind Driven Optimization

Recently, a new approach to deal with multi-dimensional and multi-modal optimization problems has been proposed by Bayraktar [3] and termed as Wind Driven Optimization. As the name suggests, WDO is inspired by the earth's atmosphere in the Troposphere layer and more specifically by the contribution of wind in the equalization of horizontal imbalances in the air pressure. In his study, Bayraktar [3] used the physical equations that govern atmospheric motion. This later is generally described by the movement of air which is a consequence of pressure gradient due to temperature differences. It is observed that wind blows from a high pressure zone to low pressure zone with a velocity proportional to the pressure gradient force. For in-depth insight into the physical model, the reader can refer to [4]. Starting from the Lagrangian model that describes atmospheric motion, Bayraktar derived WDO as an iterative metaheuristic. WDO's dynamics is similar to that of Particle Swarm Optimization (PSO). Particles in WDO refer to small air parcels that are assumed dimensionless and weightless for simplification. The trajectories of these parcels are defined according to the Newton's second law of motion. Like PSO, these air parcels are described by a position and a velocity that refer to a candidate solution and the amount of position displacement respectively. The pressure at each air parcel is used as information about the related solution quality. Updating positions and velocities of air parcels is governed by the following equations where the variable i refers to the particle and the variable t to iteration [3].

$$u_{t+1}^i = (1 - \alpha)u_t^i - gx_t^i + \left(RT\left|\frac{1}{r} - 1\right|(x_{opt} - x_t^i)\right) + \left(\frac{cu_t^{other\,dim}}{r}\right) \tag{1}$$

$$x_{t+1}^i = x_t^i + u_{t+1}^i \tag{2}$$

Where u_t^i and u_{t+1}^i are the current and the new velocity of the air parcel respectively, x_{opt} is the global best position, x_t^i and x_{t+1}^i are the current and the new positions of the air parcel, parameters α, g, R, and T are related respectively to the friction coefficient, gravity, universal gas constant and temperature in the physical model. The variable r represents the rank of the air parcel where all air parcels are ranked in descending order based on their pressure. An in-depth description WDO is available at [3].

3   The Proposed Algorithm for Dynamic Optimization

For sake of clarity, we first describe the modifications brought to WDO to properly handle optimization in dynamic environments then we present the proposed MR-MWDO for dynamic optimization.

3.1   Modified WDO Algorithm

In nature, atmospheric pressure is influenced by several factors such as temperature, humidity and altitude among others. These factors contribute in raising and lowering the air pressure according to the spatial distribution and topography on the surface of the earth. This leads to the formation of high pressure regions and low pressure regions and the gradient of the pressure force gives rise to air motion. As shown in equation (1) of WDO model, the influence of pressure on velocity of air parcels is not expressed in terms of actual values of pressure but is implicitly represented in the third and fourth terms by the rank of the particle among other particles based on their fitness values. By another side, it is known that altitude is inversely proportional to pressure. Moreover, one natural way to establish analogy between optimization and atmospheric models is to relate altitude to fitness. Therefore, the matter is how to express in a convenient manner the relationship between pressure and fitness. In [5], Chao et al. proposed a Tropical Cyclone-based Method (TCM), for solving global optimization problems with box constraints. Chao and al expressed the relationship between the pressure p^i of a particle i and its fitness $f(x_i)$ by the following equation:

$$p^i = exp\left(-n * \frac{f(x^i) - f(x^{worst})}{\sum_{k=1}^{m}(f(x^k) - f(x^{worst}))}\right) \tag{3}$$

Where $f(x^{worst})$ represents the fitness of the worst position, n is the problem dimension and m the number of particles. According to equation (3), a particle with a higher fitness possesses a lower pressure, and the pressure is scaled to be 1 at x^{worst}. Therefore, in the modified WDO we propose to use the value of pressure as given by equation (3) instead by the rank of the particle to better reflect the influence of pressure on velocities. In other words, parameter r in equation (1) is replaced by parameter p^i.

3.2   Features for Handling Dynamic Optimization

3.2.1   Concepts of Multi-region

Multi-population methods maintain multiple sub-populations concurrently. Each sub-population may handle a separate area of the search space. Each of them may also handle a specific task. For example, some sub-populations may focus on searching for the global optimum while some others may concentrate on tracking any possible changes. These two types of populations then may communicate with each other to bias the search [6].

In meteorology, a region is set as a low-pressure area (respectively a high-pressure area) if the atmospheric pressure is lower (respectively higher) than that of surrounding locations. Wind blows in an attempt to equalize horizontal imbalances in the air pressure. Meteorologists use the value of pressure at sea level as a threshold to decide the type of region: if the pressure values in a region are lower than 1013.25, then the region is low pressure else it is high pressure. The idea behind MR-MWDO is then classifying regions of a search space into low and high pressure regions could be used to guide the search process and better identify promising areas that may include optima.

In the proposed Multi-Region MWDO algorithm, this principle is implemented through the use of two different types of sub-populations with different numbers of particles. The aim is to better control the level of diversity inside and outside sub-populations knowing that after occurrence of a change in dynamic environment, the new optimum can be located near or very far compared to the previous optimum. The first type is a sub-population of observer particles (SOP), which role is exploring and calculating pressure at various search areas. The second type is a sub-population of air particles (SAP) that navigate within the search space according to MWDO equations. For example, in the case of a maximization problem, a new low pressure region is detected (a low pressure region in this case represents a promising area) if the best value of pressure in SOP is less than or equal to a threshold value and all SAP are converged. The threshold value is the median of all pressure values of all particles.

When a new low pressure region is found, a new SAP is generated within this region. Accordingly, a radius r_{region} is assigned to each SAP in order to avoid the re-exploration of already visited regions by SOP.

3.2.2   Collision Avoidance Technique

In order to ensure that each promising region (each peak) is explored by only one sub-population, researchers proposed to delimit regions using a radius. For example, Blackwell et al [7] proposed the technique of exclusion which consists in using a simple competition between the swarms that are close to each other. The swarm with the best function value is kept (best solution found so far) whereas the other will be expelled and reinitialized. Two swarms collide if and only if their attractors (best solutions found so far by each swarm) are located within an exclusion radius r_{excl} of each others [7]. Within the same spirit, we introduce a new technique called Collision Avoidance Technique to prevent collision between sub-populations and therefore maintain several sub-populations on several peaks. For each sub-population, the distance to the closest other sub-population is recorded. To determine whether this proximity between these sub-populations is desirable or not, the midpoint of the recorded distance is considered. Figure 1 shows two cases where proximity between subpopulations is desirable and one case where it is undesirable.

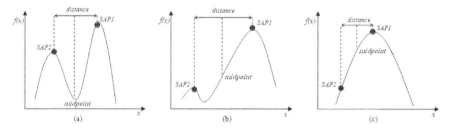

Fig. 1. Proximity between sub-populations: plots (a) and (b) show cases of desirable proximities between two SAPs related to different peaks. In plot (c), the proximity between the two SAPs is undesirable (both SAPs are related to the same peak).

From Figure 1, we can identify two major cases depending on the relative fitness of the midpoint compared to the best solutions in the two SAPs; the first case is when this value is less than the best solutions found by SAPs Figure 1.a. In this case, both SAPs will be kept. While in the second case, if this value is better than one of them as in Figure 1.b and 1.c, a new position of the midpoint must be calculated iteratively between the previous midpoint and the weakest SAP's solution. At each iteration, if the fitness of the new midpoint is worse than the weakest SAP's solution the iterative process is stopped and both SAPs will be retained Figure 1.b. Otherwise, the process will pursue until the distance between the midpoint and the lowest SAP is less than r_{region}. At this stage, if the fitness of the midpoint remains better than the weakest SAP's solution, then this SAP is removed to avoid the collision as in Figure 1.c.

3.3   Outline of the proposed Multi-region MWDO Algorithm

The algorithm starts with single sub-population of n observer particles (SOP) with the aim to explore only the search space and to find promising regions. Then the algorithm goes through an iterative process as described on Figure 2. For each generated SAP_i the convergence test is performed in the following manner:

```
For each SAP_i
  If best value of pressure is better than threshold
     & not Converged_i then
    Converged_i=true;
  Elseif best value of pressure is lower than threshold
         & Converged_i then
    Converged_i=false;
  Endif
Endfor
```

To ensure a reasonable number of sub-populations in the search space without degrading the performance of the algorithm and performing unnecessary functions evaluations knowing that the number of evaluation should be limited because the time of the next change is unknown, two successive phases are applied. The first phase is to generate a new sub-population (SAP) when a new promising region is detected as seen previously, while in the second phase, a sub-population with low pressures or being in collision with another one is removed.

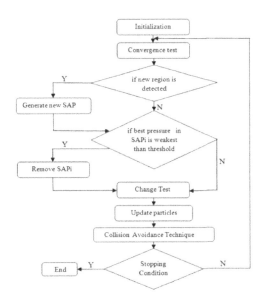

Fig. 2. Flow chart of proposed algorithm

The phase of change test is a significant step as it allows the algorithm to adapt to a new change in the environment by comparing, at each iteration, the best solutions found by the sub-populations (SAPs) to their best old achieved values. Then in the next phase, particles positions are updated using modified equations (1) and (2) for the sub-population of air particles (SAP) and random displacement for sub-population of observer particles (SOP). Once new positions are recorded, Collision Avoidance procedure is performed to keep several sub-populations on different peaks as described above. The algorithm evolves in this manner till a termination criterion is satisfied.

4   Experimental Results

In order to assess the performance of MR-MWDO, Moving Peaks Benchmark (MPB) has been used [8]. All experiments have been performed using the Scenario 2 of the MPB, proposed by Branke in [8]. This scenario has also been used by several authors and allows comparison of results obtained by different methods. The settings for this scenario are as follows: the search space has five dimensions $X^5 = [0,100]^5$ there are $p = 10$ peaks, the peak heights vary randomly in the interval $[30,70]$ and the peak width parameters vary randomly within $[1,12]$. The peaks change position every $K = 5000$ evaluations by a distance of $s = 1$ in a random direction, and their movements are uncorrelated (the MPB coefficient $\lambda = 0$). The algorithm was run for 100 consecutive changes, and each run was independently repeated 30 times, with different random seeds. The performance measure used to evaluate the algorithm is the offline error [9] given by the following equation:

$$oe = \frac{1}{Nc}\sum_{i=1}^{Nc}\left(\frac{1}{Ne(i)}\sum_{j=1}^{Ne(j)}(vbest_i - vbest_{ij})\right) \tag{4}$$

Where Nc is the total number of changes in the objective function, $Ne(i)$ is the number of evaluations performed during the ith change, $vbest_i$ is the value of the global optimum in the ith change and $vbest_{ij}$ is the value of the best solution obtained by the algorithm in the jth evaluation of ith change.

First experiment has been conducted to find suitable settings of the algorithm's parameters that consist in the size of a SOP (n), the size of a SAP (m), the radius r_{region} assigned to each SAP to avoid collisions and parameters used for the update of air particles' positions that is $[\alpha, g, c, RT]$ seen in the equation (1). These latter have been set to $(\alpha = 1, g = 0, c = 0.4, RT = 3)$. The results of this experiment are shown on tables 1 and 2.

Table 1. Offline error and Standard deviation for varying particles number. The data is for 30 runs of MPB (Scenario 2) and r_{region}=5.0.

$(m + n)$	Offline error (Std error)	$(m + n)$	Offline error (Std error)
$(5 + 5)$	6.19 ± 0.07	$(10 + 5)$	4.92 ± 0.09
$(5 + 10)$	2.26 ± 0.08	$(10 + 10)$	3.76 ± 0.10
$(5 + 20)$	1.78 ± 0.04	$(10 + 20)$	2.66 ± 0.08
$(5 + 30)$	1.92 ± 0.03	$(10 + 30)$	2.23 ± 0.06
$(5 + 50)$	2.13 ± 0.03	$(10 + 50)$	1.85 ± 0.03

Table 2. Offline error and Standard deviation depending on parameter r_{region}

r_{region}	Offline error (Std)	r_{region}	Offline error (Std)
0.5	1.91 ± 0.03	5.0	1.78 ± 0.04
1.0	1.88 ± 0.04	8.0	1.83 ± 0.04
3.0	1.79 ± 0.03	10.0	1.84 ± 0.04

As can be seen, the algorithm achieves the best performance with $(m + n) = (5 + 20)$ and $r_{region} = 5.0$. With these settings, the algorithm was found to converge quickly after each change while maintaining good diversity level within and outside sub-populations. Furthermore, it has been observed that sub-populations work within their regions delimited by r_{region} radius without collision problem.

4.1   Impact of the number of Sub-populations

The number used of sub-populations plays an important role in maintaining good diversity and occupying the promising areas that can include global optima after each change. However, using large numbers of sub-populations impacts negatively the performance of the algorithm by making unnecessary evaluations of the objective function. Figure 3 shows the total number of sub-populations during a single run of the 10 peaks MPB environment with $(m + n) = (5 + 20)$ and $r_{region} = 5.0$. As can be seen, the algorithm was able to maintain the appropriate number of sub-populations (SAP) on different peaks (10 peaks in this case), which is achieved at 35000 evaluations while improving the offline error.

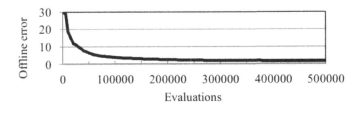

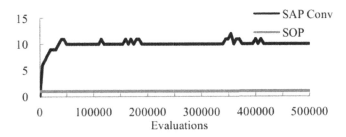

Fig. 3. Total number of sub-populations for a single instance of the 10 peaks MPB environment. Upper plot shows offline error, lower plot shows number of converged sub-populations (SAPs) and explorer sub-population (SOP).

4.2   Comparison with Other Algorithms

In order to compare MR-MWDO algorithm to other algorithms from the literature, an experiment was performed during which the offline error and its standard deviation were recorded for different shift severities (s) and different number of peaks (p) for all algorithms namely: HmSO [10], CPSO [11], mQSO [7], SPSO [12] and CPSOR [13]. Results are reported on tables 3 and 4. The shift length s is the severity of the problem dynamics. Whenever it increases, localization and tracking of peaks becomes more difficult.

Table 3. Offline error $\pm$ Standard error for different algorithms on the MPB problem with different shift severities

s	MR-MWDO	CPSOR	HmSO	CPSO	mQSO	SPSO
0	1.01	1.57	2.68	0.899	**0.601**	0.807
	±0.04	±1.23	±1.26	±1.05	±0.439	±0.972
1	1.78	**0.448**	3.49	1.35	1.69	2.28
	±0.04	±0.626	±1.65	±1.12	±0.784	±1.49
2	2.53	**0.588**	3.88	1.31	1.84	2.81
	±0.03	±0.654	±1.66	±1.15	±0.723	±1.54
3	3.57	**0.749**	4.22	1.41	2.06	3.24
	±0.03	±0.70	±1.74	±1.08	±0.775	±1.34

Table 4. Offline error $\pm$ Standard error for different algorithms on the MPB problem with different numbers of peaks

p	MR-MWDO	CPSOR	HmSO	CPSO	mQSO	SPSO
1	1.48	0.245	1.80	**$1.4e-4$**	3.15	1.86
	±0.06	±0.253	±2.09	$\pm5.9e-4$	±1.21	±1.45
5	1.46	**0.936**	3.99	0.97	1.56	1.52
	±0.06	±0.788	±1.92	±0.755	±0.691	±0.769
10	1.78	**0.448**	3.49	1.35	1.69	2.28
	±0.04	±0.626	±1.65	±1.12	±0.784	±1.49
100	2.60	2.19	3.50	**1.42**	3.39	3.50
	±0.03	±1.11	±0.953	±0.752	±1.53	±1.47

From the results shown above, we can see that MR-MWDO achieves better results than SPSO, HmSO and mQSO for all numbers of peaks except for number of peaks $p=10$ where mQSO achieves slightly better values with highest standard deviation. In general, MR-MWDO achieves intermediate results compared to the other algorithms. The best results are almost shared between CPSOR and CPSO using a hierarchical clustering method to locate and track multiple peaks. These results are very promising and show viability of the proposed approach.

5   Conclusion

In this paper, we described a Multi-Region Modified Wind Driven Optimization algorithm (MR-MWDO) to solve dynamic optimization problems. MR-MWDO employs a new technique called Multi-Region technique inspired from meteorology to detect adaptively promising regions in search space and track multiple peaks. A new Collision Avoidance Technique has been introduced as well. Sub-populations evolve according to a modified WDO algorithm. The method has been assessed using Moving Peak Benchmark and compared to state of the art methods. Obtained results are very promising and show viability of the method. As future work, we intend to foster algorithm search abilities by introducing prediction models to better track moving peaks.

References

1. Calderín, J.F., Masegosa, A.D., Suárez, A.R., Pelta, D.A.: Adaptation Schemes and Dynamic Optimization problems: A Basic Study on the Adaptive Hill Climbing Memetic Algorithm. In: Terrazas, G., Otero, F.E.B., Masegosa, A.D. (eds.) NICSO 2013. SCI, vol. 512, pp. 85–97. Springer, Heidelberg (2014)
2. Yang, S., Yao, X. (eds.): Evolutionary Computation for Dynamic Optimization Problems. SCI, vol. 490. Springer, Heidelberg (2013)
3. Bayraktar, Z., Komurcu, M., Bossard, J.A., Werner, D.H.: The Wind Driven Optimization Technique and its Application in Electromagnetics. IEEE Transactions on Antennas and Propagation 61(5), 2745–2757 (2013)
4. James, R.H.: An Introduction to Dynamic Meteorology, 4th edn., USA, vol. 88 (2004)
5. Chao, C.W., Fang, S.C., Liao, C.J.: A Tropical Cyclone-Based Method For Global Optimization. Journal of Industrial And Management Optimization 8, 103–115 (2012)
6. Nguyen, T.T.: Continuous Dynamic Optimisation Using Evolutionary Algorithms. PhD thesis, School of Computer Science, University of Birmingham (2011)
7. Blackwell, T., Branke, J.: Multiswarms, exclusion, and anti-convergence in dynamic environments. IEEE Trans. Evol. Comput. 10(4), 459–472 (2006)
8. Branke, J.: The moving peaks benchmark,
   http://www.aifb.uni-karlsruhe.de/~jbr/MovPeaks/
   (viewed November 8, 2008)
9. Branke, J., Schmeck, H.: Designing evolutionary algorithms for dynamic optimization problems. In: Advances in Evolutionary Computing: Theory and Applications, pp. 239–262 (2003)
10. Kamosi, M., Hashemi, A.B., Meybodi, M.R.: A hibernating multi-swarm optimization algorithm for dynamic environments. In: Proc. World Congr. on Nature and Biologically Inspired Computing, NaBIC 2010, pp. 363–369 (2010)
11. Yang, S., Li, C.: A clustering particle swarm optimizer for locating and tracking multi-ple optima in dynamic environments. IEEE Trans. Evol. Comput., 959–974 (2010)
12. Parrott, D., Li, X.: Locating and tracking multiple dynamic optima by a particle swarm model using speciation. IEEE Trans. Evol. Comput. 10(4), 440–458 (2006)
13. Li, C., Yang, S.: A general framework of multipopulation methods with clustering in undetectable dynamic environments. IEEE Trans. Evol. Comput. 16(4), 556–577 (2012)

Evaluating a Hybrid DE and BBO with Self Adaptation on ICSI 2014 Benchmark Problems [*]

Yu-Jun Zheng[1] and Xiao-Bei Wu[2]

[1] College of Computer Science & Technology, Zhejiang University of Technology,
Hangzhou 310023, China
[2] College of Electronics and Information Engineering,
Tongji University, Shanghai 201804, China
yujun.zheng@computer.org, xwu4@StateStreet.com

Abstract. The paper presents a new hybrid differential evolution (DE) and biogeography-based optimization (BBO) algorithm and tests its performance on the benchmark set for the ICSI 2014 Competition. The algorithm tends to perform more DE mutations in early search stage and more BBO migrations in later stage, in order to provide a good balance of exploration and exploitation. It also uses a trial-and-error method inspired by the self-adaptive DE (SaDE) to choose appropriate mutation/migration schemes during the search. Computational experiment shows that the algorithm outperforms DE, SaDE, and blended BBO on the benchmark set.

Keywords: single objective optimization, differential evolution (DE), biogeography-based optimization (BBO), self-adaptation.

1 Introduction

Evolutionary algorithms (EAs) are stochastic search methods drawing inspiration from biological evolution for optimization problems. Most EAs are initially proposed for solving single objective optimization problems, which are the basis of a wide range of real-world optimization problems.

The tradeoff between exploration and exploitation is the sticking point in search processes, having a great effect on convergence speed and accuracy of EAs [4]. Among various EAs, differential evolution (DE) [11] is a popular algorithm known for its very competitive exploration ability. Another relatively-new EA, biogeography-based optimization (BBO) [10], has shown strong exploitation ability on a variety of optimization problems.

The purpose of the paper is to present a hybrid algorithm that combines the DE's exploration ability with BBO's exploitation ability, and to evaluate the performance of the hybrid algorithm on the benchmark set for the ICSI 2014 Competition on Single Objective Optimization [12]. The hybrid method, named HSDB (Hybrid Self-adaptive DE and BBO), prefers to make more use of

---

[*] This work was supported by Natural Science Foundation (No. 61105073) of China.

Y. Tan et al. (Eds.): ICSI 2014, Part II, LNCS 8795, pp. 422–433, 2014.

DE mutations in exploration in early search stage, and is more likely to adopt BBO migrations in exploitation in later search stage. Since the performance of DE/BBO heavily depends on the mutation/migration operators, here we embed two DE mutation schemes and two BBO migration schemes in HSDB, and borrow the ideas from the SaDE algorithm [9] to dynamically choose the schemes during the search. We also compare the performance of HSDB with the DE, SaDE, and an improved BBO on the benchmark set.

The rest of this paper is as follows: Section 2 and Section 3 respectively introduce DE and BBO, Section 4 describes our HSDB algorithm, Section 5 presents the numerical experiments on the benchmark set, and Section 6 concludes.

2 Differential Evolution

DE is a population-based EA that simultaneously evolves a population P of floating-point solution vectors towards the global optimum to the given optimization problem. The most important operator in DE is mutation, which produces a mutant vector $\mathbf{v}_i$ for each individual $\mathbf{x}_i$ in the population ($1 \leq i \leq |P|$). The most-widely used mutation scheme is the DE/rand/1/bin scheme that adds the weighted difference between two randomly selected vectors to a third one:

$$\mathbf{v}_i = \mathbf{x}_{r_1} + F \cdot (\mathbf{x}_{r_2} - \mathbf{x}_{r_3}) \ . \tag{1}$$

where F is the scaling factor typically in the range [0,1], r_1, r_2 and r_3 are three mutually exclusive random indexes in $[1, |P|]$.

Afterwards, a trial vector $\mathbf{u}_i$ is generated by mixing the components of the mutant vector and the original one, where each dth component of $\mathbf{u}_i$ is determined as follows:

$$\mathbf{u}_i^d = \begin{cases} \mathbf{v}_i^d & \text{if } rand(0,1) < c_r \text{ or } d = r_i \\ \mathbf{x}_i^d & \text{else} \end{cases} \ . \tag{2}$$

where c_r is the crossover rate ranged in $(0,1)$ and r_i is a random integer within $[1, |P|]$ for each i, ensuring that the trial vector gets at least one component from the mutant vector.

In the last step of each iteration, the selection operator chooses the better one for the next generation by comparing the fitness of $\mathbf{u}_i$ with $\mathbf{x}_i$:

$$\mathbf{x}_i = \begin{cases} \mathbf{u}_i & \text{if } f(\mathbf{u}_i) > f(\mathbf{x}_i) \\ \mathbf{x}_i & \text{else} \end{cases} \ . \tag{3}$$

Nevertheless, there are other mutation schemes that can be implemented in DE. The following presents three other mutation schemes frequently used:

– DE/best/1:
$$\mathbf{v}_i = \mathbf{x}_{\text{best}} + F \cdot (\mathbf{x}_{r_1} - \mathbf{x}_{r_2}) \ . \tag{4}$$

– DE/rand-to-best/2:

$$\mathbf{v}_i = \mathbf{x}_i + F \cdot (\mathbf{x}_{\text{best}} - \mathbf{x}_i) + F \cdot (\mathbf{x}_{r_2} - \mathbf{x}_{r_3}) \ . \tag{5}$$

– DE/best/2:

$$\mathbf{v}_i = \mathbf{x}_{\text{best}} + F \cdot (\mathbf{x}_{r_1} - \mathbf{x}_{r_2}) + F \cdot (\mathbf{x}_{r_3} - \mathbf{x}_{r_4}) \ . \tag{6}$$

The performance of DE heavily depends on its mutation scheme and control parameter settings. Unfortunately, due to the variety of problems, various conflicting conclusions have been drawn with regard to parameter settings in DE literatures [9]. To avoid manual tuning, Abbass [1] proposed a self-adaptive mechanism that encodes the crossover rate c_r into each solution and simultaneously evolves the parameter and the solution. Omran et al. [8] developed a similar mechanism for adapting the scaling factor F. Qin et al. [9] proposed the SaDE algorithm, which performs a trial-and-error search for the most appropriate mutation scheme during the search. Let K be the total number of mutation schemes in the pool, at the Gth generation, the probability of choosing the kth scheme is proportional to its success rate $s_k(G)$ within the previous LP generations:

$$s_k(G) = \frac{\sum_{g=G-LP}^{G-1} ns_k(g)}{\sum_{g=G-LP}^{G-1} ns_k(g) + \sum_{g=G-LP}^{G-1} nf_k(g)} \ . \tag{7}$$

where $ns_k(g)$ is the number of trial vectors generated by the kth strategy at the gth generation that successfully enter the next generation, and $nf_k(g)$ is the number of trial vectors that fail to do so. SaDE also uses a similar strategy for adapting the c_r values, and sets F as a Gaussian random number with mean 0.5 and standard deviation 0.3, denoted by $N(0.5, 0.3)$.

3   Biogeography-Based Optimization

BBO is also a population-based EA, which was proposed by Simon [10] based on the mathematics of island biogeography. A solution is analogous to a habitat the solution components are analogous to the habitat's suitability index variables (SIVs), and the solution fitness is analogous to the species richness or habitat suitability index (HSI) of the island. High HSI habitats tend to share their features with low HSI habitats, and low HSI habitats are likely to accept many new features from high HSI habitats. For example, in a simple linear migration model, the immigration rate λ_i and the emigration rate μ_i of each habitat $\mathbf{x}_i$ are calculated as follows:

$$\lambda_i = I \frac{f_{\max} - f_i}{f_{\max} - f_{\min}} \ . \tag{8}$$

$$\mu_i = E \frac{f_i - f_{\min}}{f_{\max} - f_{\min}} \ . \tag{9}$$

where $f_{\max}$ and $f_{\min}$ are the maximum and minimum fitness value of the population, and I and E are the maximum possible immigration rate and emigration rate which are typically set to 1. However, there are also many other nonlinear migration models can be used [6].

At each generation of BBO, each SIV of a habitat $\mathbf{x}_i$ has a probability of λ_i to be immigrated, and the migrating SIV comes from an emigrating habitat $\mathbf{x}_j$, the selection probability of which is proportional to μ_j of the islands.

BBO also has a mutation operator for setting a SIV to a random value in the search range, which is mainly for increasing solution diversity and hence improving exploration.

In [7] Ma and Simon developed the blended BBO (B-BBO), which replaces the clonal migration of BBO with the following blended migration:

$$X_i^d = \alpha X_i^d + (1 - \alpha)X_j^d \ . \tag{10}$$

where α is a real number between 0 and 1.

Gong et al. [5] proposed a hybrid migration operator by combining BBO migration and DE mutation, where each habitat component has a probability of c_r to be mutated by DE mutation, and a probability of $(1 - c_r)$ to be changed by BBO migration. Experiments show that the hybrid method outperforms both DE and BBO on a set of benchmark problems. Boussaïd et al. [2] proposed another approach combining BBO and DE, which evolves the population by alternately applying BBO and DE iterations, and at each iteration always selects a fitter one from the updated solution and its parent for the next iteration. In [3] the authors extended the approach for constrained optimization, where the original mutation operator of BBO is replaced by the DE mutation operator.

By default, BBO uses a global topology where any two habitats in the population have a chance to communicate with each other. Zheng et al. [13] equipped BBO with a local topology, and any habitat can only immigrate SIVs from its neighboring habitats. In consequence, the flow of information can be moderately pass through the neighborhood, and the algorithm can achieve a better balance between exploration and exploitation. In [14] Zheng et al. enhanced BBO with two new migration operators: global migration and local migration. The former migrates features from a neighbor X_{nb} and a non-neighbor X_{far}, while the latter only considers migration from a neighbor:

$$X_i^d = X_{far}^d + \alpha(X_{nb}^d - X_i^d) \ . \tag{11}$$
$$X_i^d = X_i^d + \alpha(X_{nb}^d - X_i^d) \ . \tag{12}$$

4   A Hybrid DE and BBO with Self Adaptation

The HSDB method differentiates from the hybrid algorithms of Gong et al. [5] and Boussaïd et al. [2] in that it does not combine DE mutation and BBO migration into an integrated operator. Instead, HSDB prefers to make more use of DE mutation in exploration in early search stage, and is more likely to adopt BBO migration in exploitation in later search stage. The key to realize this is a parameter named maturity index, denoted by ν, which increases with the generation: The higher (lower) the ν, the more likely the algorithm is to conduct migration (mutation). A simple approach is to linearly increase ν from an initial value $\nu_{\min}$ to a final value $\nu_{\max}$ as follows (where g is the current generation number and $g^{\max}$ is the total generation number of the algorithm):

$$\nu = \nu^{\min} + \frac{g}{g^{\max}}(\nu^{\max} - \nu^{\min}) \ . \tag{13}$$

At each generation, each habitat has a probability of ν to be modified by BBO migration and a probability of $(1 - \nu)$ by DE mutation. Moreover, we embed a set of DE mutation schemes (denoted by S_{DE}) and a set of BBO migration schemes (denoted by S_{BBO}) in HSDB, for each scheme record its success and failure numbers within a fixed number LP of previous generations, and calculate its success rate in a similar way as SaDE [9]. However, in HSDB, we calculate the success rates for DE mutation schemes and BBO migration schemes independently. That is, at beginning every scheme is assigned with an equal selection probability; at each generation $G \geq LP$, the probability of choosing the kth DE mutation scheme and that of the k'th BBO migration scheme are respectively updated as follows:

$$p_k(G) = \frac{s_k(G)}{\sum_{k \in S_{\mathrm{DE}}} s_k(G)} \ . \tag{14}$$

$$p_{k'}(G) = \frac{s_{k'}(G)}{\sum_{k' \in S_{\mathrm{BBO}}} s_{k'}(G)} \ . \tag{15}$$

Currently, in HSDB we embed two DE mutation schemes including DE/rand /1/bin (1) and DE/rand-to-best/2/bin (5), and two BBO migration schemes including the original clonal migration and the local migration (12). This is because the two DE mutation schemes exhibit good exploration ability and the two BBO migration schemes have more exploitation ability than others.

To support local migration, HSDB also implements a local topology to establish the neighborhood structure for the population. Given an expected average neighborhood size K, the local topology, represented by an adjacent 0-1 matrix NB, is randomly set by the procedure described in Algorithm 1. As we can see, for each solution in the population p, every other solution has a probability of $K/(|P|-1)$ to be its neighbor. An advantage of this approach is that the neighborhood size K does not need to be limited to an integer. However, isolated solutions are not allowed. To increase diversity, HSDB will reset its local topology if it fails to find a new better known solution after NP generations, where NP is a control parameter typically ranges from 1 to 6.

For coherence, we replace the crossover rate c_r of DE by the immigration rate λ_i of each solution i. That is, the DE crossover operation (2) is replaced by the following equation in HSDB:

$$\mathbf{u}_i^d = \begin{cases} \mathbf{v}_i^d & \text{if } rand(0,1) < \lambda_i \text{ or } d = r_i \\ \mathbf{x}_i^d & \text{else} \end{cases} \ . \tag{16}$$

We also use the same parameter scheme for coefficient F in Eq. (1) and (5) and α in Eq. (12), all denoted by F in HSDB. We employ a Gaussian random

Algorithm 1. The procedure for setting the neighborhood topology.

1   Let NB be a $|P| \times |P|$ all-zeros matrix;
1   Let $p = K/(|P| - 1)$;
2   **for** $i = 1$ **to** $|P|$ **do**
3   **for** $j = 1$ **to** $|P|$ **do**
4   **if** $i \neq j \wedge rand(0,1) < p$ **then**
5               $NB(i,j) = NB(j,i) = 1$;
6   **if** the ith solution has no neighbor **then**
7   Randomly choose a j other than i and set $NB(i,j) = NB(j,i) = 1$;

number $N(F_\mu, F_\sigma)$ to approximate F, where F_μ and F_σ are typically set as 0.5 and 0.3 (as in SaDE [9]). Moreover, since the original BBO uses mutation mainly for exploration purpose, while here we concentrate on the exploitation ability of BBO, the HSDB method does not use the original BBO mutation scheme. Algorithm 2 describes the framework of HSDB.

Algorithm 2. The HSDB algorithm.

1   Randomly initialize a population P of solutions to the problem;
2   Initialize the local topology of the population based on Algorithm 1;
3   **while** stop criterion is not satisfied **do**
4   **for** $i = 1$ **to** $|P|$ **do**
5   Compute λ_i and μ_i for each solution $\mathbf{x}_i$;
6   **if** $rand(0,1) < \nu$ **then**
7   Select a DE mutation scheme k with probability $\propto s_k(G)$;
8   **else**
9   Select a BBO migration scheme k' with probability $\propto s_{k'}(G)$;
10   Create a mutant vector $\mathbf{v}_i$ using the selected scheme;
11   Create a trial vector $\mathbf{u}_i$ according to the crossover operation (16);
12   **if** $\mathbf{u}_i$ is fitter than $\mathbf{x}_i$ **then**
13   Replace $\mathbf{x}_i$ with $\mathbf{u}_i$ in P;
14   Update the maturity index ν;
15   **if** $G \geq LP$ **then**
16   Update the success rates of the schemes;
17   **if** no new better solution has been found for NG generations **then**
18   Reset the local topology of the population;
19 **return** the best solution found so far.

5   Computational Experiment

5.1   Experiment Setup

We test the HSDB algorithm on the ICSI 2014 benchmark set, which consists of 30 high-dimensional problems, denoted as f_1–f_{30}, all shifted and rotated. Since all the test problem are considered as black box problems, we do not tune parameters for each problem; instead we use a general parameter setting of HSDB for the whole problem set as: the range of maturity index $\nu_{\min} = 0.05$, $\nu_{\max} = 0.95$, the learning period $LP = 30$, the limit of non-improved generations $NG = 3$, and the neighborhood size $K = 3$. The population size $|P|$ is set to 50.

The experiments are conducted on a computer of Intel Core i5-2430M processor and 4GB DDR3 memory. The HSDB algorithm has been implemented using Matlab R2013a. As requested by the ICSI 2014 competition session, the algorithm has been tested on each problem with dimensions 2, 10 and 30 respectively, the search space is $[-100, 100]^D$, the maximum number of function evaluations is set as $10000D$ (where D denotes the problem dimension), and 51 simulation runs have been performed on each problem instance. Any function value smaller than $2^{-52} \approx 2.22e - 16$ (the ϵ in Matlab) is considered as zero.

In our experimental environment, the mean time for executing the benchmark program of ICSI 2014 competition over 5 runs is $T1 = 48.62s$, and the mean time of HSDB on function 9 and $D = 30$ over 5 runs is $T2 = 116.02s$, and the time complexity is evaluated by the ratio $(T2 - T1)/T1 = 1.386$.

5.2   Experimental Results

Tables 1, 2 and 3 respectively present the computational results of HSDB on 2-D, 10-D and 30-D problems (where 'Std' denotes standard deviation). As we can see, on the simple 2-D problems, HSDB achieves a function error value less than ϵ on 21 problems. With the increase of dimension, the performance of HSDB decreases on most problems. This phenomenon, known as the "curse of dimensionality", is serious on several problems such as f_1 and f_2. However, an interesting finding is that the solution accuracy of the algorithm increases on some special problems including f_{12} and f_{26}.

5.3   Comparative Results

We have compared the performance of HSDB with DE/rand/1/bin scheme, SaDE, and B-BBO on the benchmark set, and conducted nonparametric Wilcoxon rank sum tests on the results of HSDB and that of the other three methods. Due to the page limits, here we only present the comparison results on 30-D problems in Table 4 (where an h value of 1 indicates that the performances of the HSDB and DE/SaDE are statistically different with 95% confidence, h value of 0 implies there is no statistical difference, the superscript $+$ denotes HSDB has significant performance improvement over DE/SaDE and $-$ vice versa).

As we can see, HSDB significantly outperforms DE on 28 problems, DE only outperforms HSDB on f_{26}, and there is no statistical difference between them on f_{28}. In comparison with SaDE, HSDB has significant performance improvement on 15 problems; SaDE outperforms HSDB on 6 problems, and there is no statistical difference between them on 9 problems. HSDB also outperforms B-BBO on 29 problems, and there is no statistical difference only on f_{28}. The results show that the overall performance of HSDB is better than the other three methods on the benchmark set.

Table 1. The experimental results on 2-D problems

ID	Max	Min	Mean	Median	Std
f_1	0.00E+00	0.00E+00	0.00E+00	0.00E+00	0.00E+00
f_2	2.98E−27	0.00E+00	1.13E−28	0.00E+00	4.18E−28
f_3	1.67E+00	1.67E+00	1.67E+00	1.67E+00	1.12E−15
f_4	0.00E+00	0.00E+00	0.00E+00	0.00E+00	0.00E+00
f_5	1.52E−17	8.00E−20	2.45E−18	1.53E−18	2.44E−18
f_6	1.97E−31	1.97E−31	1.97E−31	1.97E−31	0.00E+00
f_7	1.21E−17	0.00E+00	6.44E−18	9.65E−18	5.26E−18
f_8	8.88E−16	8.88E−16	8.88E−16	8.88E−16	9.96E−32
f_9	-4.00E+00	-4.00E+00	-4.00E+00	-4.00E+00	2.24E−15
f_{10}	-9.80E−01	-1.00E+00	-9.96E−01	-1.00E+00	5.86E−03
f_{11}	0.00E+00	0.00E+00	0.00E+00	0.00E+00	0.00E+00
f_{12}	5.20E+01	4.86E+01	4.90E+01	4.86E+01	8.74E−01
f_{13}	1.94E−02	0.00E+00	7.00E−03	0.00E+00	9.15E−03
f_{14}	6.99E−02	1.00E−11	5.90E−03	9.41E−04	1.06E−02
f_{15}	3.60E−03	3.91E−08	7.92E−04	3.59E−04	9.40E−04
f_{16}	0.00E+00	0.00E+00	0.00E+00	0.00E+00	0.00E+00
f_{17}	0.00E+00	0.00E+00	0.00E+00	0.00E+00	0.00E+00
f_{18}	-8.38E+02	-8.38E+02	-8.38E+02	-8.38E+02	4.59E−13
f_{19}	1.35E−32	1.35E−32	1.35E−32	1.35E−32	1.11E−47
f_{20}	5.77E−23	0.00E+00	2.58E−24	1.77E−33	8.18E−24
f_{21}	0.00E+00	0.00E+00	0.00E+00	0.00E+00	0.00E+00
f_{22}	0.00E+00	0.00E+00	0.00E+00	0.00E+00	0.00E+00
f_{23}	8.97E+00	8.97E+00	8.97E+00	8.97E+00	7.64E−15
f_{24}	-4.53E+00	-4.53E+00	-4.53E+00	-4.53E+00	6.28E−15
f_{25}	-1.39E+00	-1.71E+00	-1.60E+00	-1.71E+00	1.54E−01
f_{26}	6.67E−01	5.71E−01	5.87E−01	5.71E−01	2.79E−02
f_{27}	-3.74E+07	-3.74E+07	-3.74E+07	-3.74E+07	4.01E+02
f_{28}	-4.63E+00	-5.83E+00	-5.22E+00	-5.25E+00	2.63E−01
f_{29}	2.00E+01	2.00E+01	2.00E+01	2.00E+01	8.92E−03
f_{30}	1.01E+00	2.67E−01	4.00E−01	2.70E−01	2.71E−01

Table 2. The experimental results on 10-D problems

ID	Max	Min	Mean	Median	Std
f_1	1.35E+02	5.03E−02	8.23E+00	2.28E+00	1.87E+01
f_2	2.64E+02	2.03E−01	3.26E+01	5.43E+00	5.33E+01
f_3	1.70E+02	1.70E+02	1.70E+02	1.70E+02	1.77E−02
f_4	2.96E+00	6.54E−04	3.45E−01	1.55E−01	4.67E−01
f_5	1.27E−01	1.38E−05	1.14E−02	8.59E−04	2.33E−02
f_6	9.57E+00	4.97E−01	5.02E+00	5.95E+00	2.40E+00
f_7	1.15E−03	1.00E−12	1.46E−04	1.96E−05	2.74E−04
f_8	1.16E+00	5.66E−13	1.71E−01	5.16E−03	3.71E−01
f_9	-1.95E+01	-2.00E+01	-1.98E+01	-1.99E+01	1.35E−01
f_{10}	-7.42E+00	-8.76E+00	-8.34E+00	-8.37E+00	2.85E−01
f_{11}	4.56E+00	0.00E+00	1.09E+00	1.51E−02	1.82E+00
f_{12}	5.40E−01	3.76E−01	4.52E−01	4.47E−01	4.17E−02
f_{13}	6.35E−01	1.55E−01	3.99E−01	4.02E−01	1.37E−01
f_{14}	9.91E−02	1.01E−02	5.17E−02	5.25E−02	2.19E−02
f_{15}	9.83E−02	1.74E−02	4.89E−02	4.84E−02	1.76E−02
f_{16}	4.01E−02	1.14E−08	6.05E−03	1.20E−03	9.83E−03
f_{17}	4.72E−03	3.06E−06	8.07E−04	2.72E−04	1.03E−03
f_{18}	-2.02E+03	-4.14E+03	-2.30E+03	-2.02E+03	6.89E+02
f_{19}	1.62E−03	5.20E−21	6.90E−04	9.55E−04	6.29E−04
f_{20}	3.68E−02	4.98E−06	5.61E−03	1.90E−03	8.75E−03
f_{21}	9.99E−02	9.99E−02	9.99E−02	9.99E−02	7.70E−11
f_{22}	1.08E−02	5.32E−31	8.48E−04	3.56E−05	1.90E−03
f_{23}	2.37E+01	1.55E+01	1.94E+01	1.90E+01	2.01E+00
f_{24}	-3.47E+01	-3.56E+01	-3.53E+01	-3.52E+01	2.40E−01
f_{25}	4.30E+01	4.29E+01	4.29E+01	4.29E+01	2.80E−03
f_{26}	7.59E−03	5.90E−03	6.72E−03	6.70E−03	4.87E−04
f_{27}	-6.34E+07	-1.60E+08	-1.02E+08	-1.07E+08	1.70E+07
f_{28}	-5.04E+00	-5.85E+00	-5.38E+00	-5.40E+00	1.72E−01
f_{29}	2.00E+01	2.00E+01	2.00E+01	2.00E+01	9.31E−04
f_{30}	1.01E+00	1.01E+00	1.01E+00	1.01E+00	1.86E−10

Table 3. The experimental results on 30-D problems

ID	Max	Min	Mean	Median	Std
f_1	1.69E+05	4.08E+03	7.67E+04	7.80E+04	4.15E+04
f_2	1.60E+04	1.34E+03	5.18E+03	3.71E+03	3.27E+03
f_3	4.53E+03	4.51E+03	4.52E+03	4.52E+03	5.11E+00
f_4	1.33E+01	1.59E+00	7.45E+00	7.31E+00	2.71E+00
f_5	7.87E−02	8.39E−03	3.77E−02	3.66E−02	1.72E−02
f_6	3.03E+01	2.82E+01	2.91E+01	2.90E+01	5.15E−01
f_7	1.01E−02	3.23E−03	7.37E−03	7.84E−03	1.72E−03
f_8	1.60E+00	1.86E−01	9.06E−01	9.91E−01	4.42E−01
f_9	-5.76E+01	-5.91E+01	-5.83E+01	-5.82E+01	3.35E−01
f_{10}	-1.99E+01	-2.79E+01	-2.48E+01	-2.54E+01	1.80E+00
f_{11}	8.90E−02	4.55E−04	2.98E−02	2.85E−02	1.82E−02
f_{12}	1.47E−02	1.22E−02	1.43E−02	1.44E−02	4.42E−04
f_{13}	4.02E+00	1.40E+00	3.03E+00	3.14E+00	6.88E−01
f_{14}	1.03E−01	2.53E−02	7.29E−02	7.43E−02	1.96E−02
f_{15}	1.70E−01	2.04E−02	9.48E−02	9.36E−02	3.40E−02
f_{16}	3.33E−01	8.08E−02	2.21E−01	2.24E−01	5.23E−02
f_{17}	3.93E+00	4.31E−01	2.02E+00	2.10E+00	8.17E−01
f_{18}	-3.69E+03	-6.06E+03	-5.79E+03	-6.06E+03	7.11E+02
f_{19}	8.67E−03	2.00E−03	5.05E−03	4.81E−03	1.77E−03
f_{20}	7.45E−01	9.08E−06	2.07E−01	1.11E−01	2.27E−01
f_{21}	3.00E−01	9.99E−02	2.13E−01	2.00E−01	5.30E−02
f_{22}	2.15E−01	1.01E−02	1.01E−01	9.68E−02	5.44E−02
f_{23}	5.68E+01	3.09E+01	4.43E+01	4.32E+01	7.31E+00
f_{24}	-1.13E+02	-1.16E+02	-1.15E+02	-1.15E+02	7.39E−01
f_{25}	1.13E+03	1.13E+03	1.13E+03	1.13E+03	2.06E+00
f_{26}	4.48E−04	4.12E−04	4.33E−04	4.35E−04	8.73E−06
f_{27}	-2.99E+08	-5.25E+08	-4.76E+08	-5.21E+08	8.72E+07
f_{28}	-5.24E+00	-5.81E+00	-5.48E+00	-5.48E+00	1.12E−01
f_{29}	2.00E+01	2.00E+01	2.00E+01	2.00E+01	1.53E−04
f_{30}	1.04E+00	1.01E+00	1.02E+00	1.01E+00	1.34E−02

Table 4. The Comparison of HSDB with DE and SaDE

ID	DE				SaDE				B-BBO			
	Median	Std	p-value	h	Median	Std	p-value	h	Median	Std	p-value	h
f_1	1.73E+06	1.10E+06	3.30E-18	1+	1.98E+04	2.95E+04	2.12E-08	1-	3.98E+06	9.84E+05	3.30E-18	1+
f_2	8.28E+05	1.87E+05	3.30E-18	1+	2.58E+04	8.72E+03	4.18E-18	1+	5.72E+04	2.18E+04	3.30E-18	1+
f_3	5.26E+03	7.97E+02	6.30E-17	1+	4.51E+03	5.27E+03	1.24E-05	0	4.84E+03	1.30E+02	3.30E-18	1+
f_4	3.72E+02	6.08E+01	3.30E-18	1+	1.76E+02	3.06E+01	3.30E-18	1+	1.36E+02	3.23E+01	3.30E-18	1+
f_5	1.39E+00	9.06E-01	3.30E-18	1+	2.31E-02	1.36E-02	2.41E-05	1-	1.64E+00	3.71E-01	3.30E-18	1+
f_6	3.77E+01	9.47E+00	3.72E-18	1+	2.84E+01	4.02E-01	3.77E-08	1-	6.91E+01	1.79E+01	3.30E-18	1+
f_7	3.98E-02	3.87E-03	3.30E-18	1+	8.74E-03	3.81E-03	2.06E-02	1+	3.10E-02	1.38E-02	3.30E-18	1+
f_8	2.29E+00	4.90E-01	1.24E-16	1+	9.81E-01	6.12E-01	8.94E-01	0	3.37E+00	4.41E-01	3.30E-18	1+
f_9	-5.61E+01	7.52E-01	3.30E-18	1+	-5.84E+01	3.60E-01	9.70E-02	0	-5.58E+01	5.06E-01	3.30E-18	1+
f_{10}	-1.03E+01	1.12E+00	3.30E-18	1+	-1.84E+01	1.10E+00	4.70E-18	1+	-2.42E+01	2.43E+00	2.66E-03	1+
f_{11}	2.82E-01	2.92E-01	5.29E-18	1+	1.21E-02	1.14E-02	8.30E-06	1-	2.73E+00	8.30E-01	3.30E-18	1+
f_{12}	1.53E-02	1.57E-04	3.30E-18	1+	1.48E-02	1.87E-04	3.91E-12	1+	1.46E-02	6.13E-04	1.44E-03	1+
f_{13}	7.40E+00	4.03E-01	3.30E-18	1+	5.16E+00	5.64E-01	3.50E-18	1+	4.79E+00	5.59E-01	1.32E-16	1+
f_{14}	1.40E-01	2.29E-02	1.79E-17	1+	9.16E-02	1.97E-02	1.40E-05	1+	8.37E-02	2.02E-02	4.94E-03	1+
f_{15}	1.46E-01	4.37E-02	3.24E-08	1+	1.19E-01	3.67E-02	4.42E-04	1+	4.78E-01	8.24E-02	7.51E-18	1+
f_{16}	5.63E-01	1.50E-01	3.72E-18	1+	2.07E-01	8.51E-02	5.21E-01	0	1.01E+00	2.24E-01	3.30E-18	1+
f_{17}	1.69E+02	2.91E+01	3.30E-18	1+	1.19E+01	2.83E+00	3.30E-18	1+	1.14E+01	3.15E+00	3.30E-18	1+
f_{18}	-6.05E+03	1.66E+00	2.67E-12	1+	-6.06E+03	7.60E+02	8.63E-09	1+	-3.66E+03	1.91E+03	9.65E-16	1+
f_{19}	2.85E-01	1.39E-01	3.30E-18	1+	2.35E-02	1.53E-02	1.79E-17	1+	1.90E-02	8.13E-03	3.30E-18	1+
f_{20}	6.04E+02	2.07E+02	3.30E-18	1+	7.82E+01	3.06E+01	3.30E-18	1+	1.89E+01	7.50E+00	3.30E-18	1+
f_{21}	7.78E-01	2.18E-01	5.38E-19	1+	2.00E-01	6.16E-02	2.85E-01	0	8.00E-01	2.79E-01	9.80E-19	1+
f_{22}	8.45E-01	6.23E-01	3.30E-18	1+	1.05E-02	1.94E-02	4.61E-15	1-	7.75E+00	2.32E+00	3.30E-18	1+
f_{23}	9.28E+01	5.00E+00	3.30E-18	1+	6.98E+01	4.62E+00	3.30E-18	1+	6.69E+01	4.92E+00	4.70E-18	1+
f_{24}	-1.10E+02	1.49E+00	4.18E-18	1+	-1.14E+02	8.46E-01	3.63E-01	0	-1.10E+02	1.24E+00	3.30E-18	1+
f_{25}	1.32E+03	7.89E+01	3.30E-18	1+	1.13E+03	1.69E+00	4.64E-10	1-	1.21E+03	2.57E+01	3.30E-18	1+
f_{26}	3.98E-04	1.19E-06	3.30E-18	1+	4.26E-04	1.58E-05	1.92E-01	0	6.26E-04	5.67E-05	3.30E-18	1+
f_{27}	-4.64E+08	3.13E+07	5.36E-07	1+	-5.19E+08	4.89E+07	3.29E-01	0	-2.69E+08	5.82E+07	3.30E-18	1+
f_{28}	-5.49E+00	1.26E-01	7.18E-01	0	-5.51E+00	1.44E-01	3.25E-01	0	-5.49E+00	1.14E-01	8.72E-01	0
f_{29}	2.00E+01	1.26E-04	5.95E-03	1+	2.00E+01	1.18E-04	2.77E-02	1+	2.00E+01	1.57E-04	2.72E-02	1+
f_{30}	1.04E+00	1.14E-02	8.40E-19	1+	1.01E+00	1.34E-02	4.05E-04	1+	1.04E+00	8.03E-03	6.07E-03	1+

6   Conclusion

The paper presents a new hybrid DE and BBO algorithm, named HSDB, which uses a trial-and-error method to select among two DE mutation schemes and two BBO migration schemes, and balances the exploration and exploitation based on the maturity index parameter ν. Experiments show that HSDB outperforms DE, SaDE, and B-BBO on the benchmark set for the ICSI 2014 Competition. We are currently including more DE mutation and BBO migration schemes into HSDB and testing more effective method for tuning ν.

References

1. Abbass, H.: The self-adaptive pareto differential evolution algorithm. In: Proceedings of the 2002 Congress on Evolutionary Computation, vol. 1, pp. 831–836 (2002)
2. Boussaïd, I., Chatterjee, A., Siarry, P., Ahmed-Nacer, M.: Two-stage update biogeography-based optimization using differential evolution algorithm (DBBO). Comput. Oper. Res. 38(8), 1188–1198 (2011)
3. Boussaïd, I., Chatterjee, A., Siarry, P., Ahmed-Nacer, M.: Biogeography-based optimization for constrained optimization problems. Comput. Oper. Res. 39(12), 3293–3304 (2012)
4. Chen, J., Xin, B., Peng, Z., Dou, L., Zhang, J.: Optimal contraction theorem for exploration – exploitation tradeoff in search and optimization. IEEE Trans. Syst. Man Cybern. Part A 39(3), 680–691 (2009)
5. Gong, W., Cai, Z., Ling, C.X.: DE/BBO: a hybrid differential evolution with biogeography-based optimization for global numerical optimization. Soft Comput 15(4), 645–665 (2010)
6. Ma, H.: An analysis of the equilibrium of migration models for biogeography-based optimization. Inform. Sci. 180(18), 3444–3464 (2010)
7. Ma, H., Simon, D.: Blended biogeography-based optimization for constrained optimization. Engin. Appl. Artif. Intell. 24(3), 517–525 (2011)
8. Omran, M.G.H., Salman, A., Engelbrecht, A.P.: Self-adaptive differential evolution. In: Hao, Y., Liu, J., Wang, Y.-P., Cheung, Y.-m., Yin, H., Jiao, L., Ma, J., Jiao, Y.-C. (eds.) CIS 2005. LNCS (LNAI), vol. 3801, pp. 192–199. Springer, Heidelberg (2005)
9. Qin, A.K., Suganthan, P.N.: Self-adaptive differential evolution algorithm for numerical optimization. In: 2005 IEEE Congress on Evolutionary Computation, vol. 2, pp. 1785–1791 (2005)
10. Simon, D.: Biogeography-based optimization. IEEE Trans. Evol. Comput. 12(6), 702–713 (2008)
11. Storn, R., Price, K.: Differential evolution - a simple and efficient heuristic for global optimization over continuous spaces. J. Global Optim. 11(4), 341–359 (1997)
12. Tan, Y., Li, J., Zheng, Z.: ICSI 2014 competition on single objective optimization. Tech. rep., Peking University (2014), http://www.ic-si.org/competition/ICSI.pdf
13. Zheng, Y.J., Ling, H.F., Wu, X.B., Xue, J.Y.: Localized biogeography-based optimization. Soft Comput. (2014), doi:10.1007/s00500-013-1209-1
14. Zheng, Y.J., Ling, H.F., Xue, J.Y.: Ecogeography-based optimization: Enhancing biogeography-based optimization with ecogeographic barriers and differentiations. Comput. Oper. Res. 50, 115–127 (2014)

The Multiple Population Co-evolution PSO Algorithm

Xuan Xiao and Qianqian Zhang

Beijing Institute of Technology, Beijing, China
{1521494822,289314426}@qq.com

Abstract. In order to overcome the standard particle swarm optimization algorithm which is easily trapped in local minima and optimize the shortcoming of low precision, this paper proposed a way which can make multiple information exchange between particles come true: the multiple population co-evolution PSO algorithm. This paper proposes a multiple population co-evolutionary algorithm to achieve communication among populations, and then show the feasibility and effectiveness of this algorithm through experiments.

Keywords: Particle swarm, co-evolution, PSO multiple population.

1 Introduction

Simulating the behaviors of biological populations has become a research hotspot to solve calculating problems in the field of intelligence calculation. The theory t hat the swarm intelligence is the core has formed, and it has made revolutionary progress in a number of practical applications. Particle swarm optimization(PSO) i s an intelligent optimization algorithm, which is used to handle the problem of co ntinuous variables searching the search space, and it has been applied to many ar eas, such as function optimization, constrained optimization and neural networks. This paper proposes a multiple population co-evolutionary algorithm to achieve c ommunication among populations, and then show the feasibility and effectiveness of this algorithm through experiments.

2 The Description of Basic Particle Swarm Optimization

The basic concept of PSO comes from the study of the preying of birds developed by Kennedy and Eberhart [1, 2]. Imaging a scenario like this: a flock of birds randomly search food in an area, but no one knows where the food are and how far away their current location is from the food. The strategy of flying and searching is to follow the first bird in population. PSO get inspiration from this model and is used to solve optimization problems. Every possible solution is a bird which is called "particles" in search space, and all particles have been evaluated by fitness decided by fitness function. Each particle is used to describe an alternative solution in the solution space,

Y. Tan et al. (Eds.): ICSI 2014, Part II, LNCS 8795, pp. 434–441, 2014.

and has a random velocity throughout the whole solution space. Each particle gets heuristic information from each other and guides the movement of the entire group by the exchange of information with other particles.

In the basic PSO algorithm, each particle represents a possible solution and all the particles forms swarm. Particles depend on their own historical information an d swarm information in the search space to determine the velocity and direction o f flying to find the optimal solution. Assuming that solving problems in D-dimen sion search space, swarm composed of m particles, $Swarm = \{x_1^{(k)}, x_2^{(k)}, \cdots, x_m^{(k)}\}$. At time $k+1$, the position vector is $x_i^{(k+1)} = (x_{i1}^{(k+1)}, x_{i2}^{(k+1)}, \cdots, x_{iD}^{(k+1)})$, $i = 1, 2, \ldots, m$, whic h is the location of individuals in the search space, and it is also a possible soluti on of the problem. Corresponding to the individual position vector is its velocity vector $v_i^{(k+1)} = (v_{i1}^{(k+1)}, v_{i2}^{(k+1)}, \cdots, v_{iD}^{(k+1)})$, which describes the movement of particles o f each dimension in search space. (Note: The superscript of variable represents t he iteration cycle, for example, $x_{id}^{(k+n)}$ represents the $k+n$ cycles; superscript of var iable without parentheses represents power, as ω^n represents the n-th power of ω.)

The neighborhood function of PSO generates a new location status according t o each individual's own position vector, velocity vector, individual historical infor mation, group information, and disturbance. In standard PSO algorithm, function calculating formulation of i-th particle at time $k+1$ in d-dimension is as follows:

$$\begin{cases} v_{id}^{(k+1)} = \omega\, v_{id}^{(k)} + c_1 \cdot r_1 \cdot (p_{id}^{(k)} - x_{id}^{(k)}) + c_2 \cdot r_2 \cdot (p_{ld}^{(k)} - x_{id}^{(k)}) \\ x_{id}^{(k+1)} = x_{id}^{(k)} + v_{id}^{(k+1)} \end{cases} \tag{1}$$

Standard PSO has few parameters which need to be adjusted, and we usually s elect them empirically:

Population size m is generally selected from 10 to 30, the number of particles is e nough forthe general problem with small scale, it also reduces the complexity of calc ulation. The constant limit number of particle velocity $Vmax$ determines the maximu m moving distance of particles in an iteration cycle. Achieving maximum iterations or meeting the requirement of accuracy is generally selected as terminal conditions.

3   The PSO Algorithm Consensus Analysis and the Consensus Region Boundary

In the fields of biological evolution, consensus problems are especially reflected in self-organizing aggregation of biological systems, such as flocks of birds [3], schools of fish [4], mammals [5]. Jiang and Jin [6, 7] mainly investigate the stochastic convergence of PSO system from the perspective of stochastic process. In order to solve the consensus problem of PSO, we think of a way to improve the velocity and position updating formula of single particle according to the existing standard PSO algorithm, this paper improved PSO by consensus Protocol U:

$$U = v_i^{(k+1)} = \omega\, v_i^{(k)} + \eta_{m1}^{(k)} \cdot (p_i^{(k)} - x_i^{(k)}) + \eta_{m2}^{(k)} \cdot (p_i^{(k)} - x_i^{(k)}) . \tag{2}$$

The new PSO model is further described as:

$$\begin{cases} U = \omega \cdot v_i^{(k)} + \eta \phi_{i1}^{(k)} \cdot (p_i^{(k)} - x_i^{(k)}) + \eta \phi_{i2}^{(k)} \cdot (p_l^{(k)} - x_i^{(k)}) \\ x_i^{(k+1)} = x_i^{(k)} + U \end{cases} . \tag{3}$$

and then,

$$x_i^{(k+1)} = (1 + \omega - \eta \phi_i^{(k)}) x_i^{(k)} - \omega \cdot x_i^{(k-1)} + \eta \phi_i^{(k)} \cdot p_i^{(k)} . \tag{4}$$

Particles in PSO achieve the optimization goal and consensus through mutual cooperation. The cooperative behavior is that the particle and its connected particles(neighbor) pass on information with each other, and change state according to certain strategy and the received neighbor information, resulting in a corresponding self-organizing behavior. Assuming that each particle in PSO population represents a node, then the interactions between particles form a sensing figure, which decodes relationship and interaction between particles and their neighbor.

In the improved consensus protocol U,

$$\begin{cases} 0 \langle \eta \langle 0.25 \\ \eta = 0.25 \\ \eta \rangle 0.25 \end{cases} . \tag{5}$$

The corresponding region for $0 < \eta < 0.25$, $0 \le \phi_i^{(k)} < \frac{1}{\eta}(1+\omega)$, $0 \le \phi_i^{(k)} < 4(1+\omega)$ is in complete consensus. The corresponding region for $4(1+\omega) < \phi_i^{(k)} < \frac{1}{\eta}(1+\omega)$ is inconsistent state area. When $\eta = 0.25$, $\phi_i^{(k)} = 4(1+\omega)$ is the critical value for consensus area; when $\eta > 0.25$, the corresponding region for $0 \le \phi_i^{(k)} < 4(1+\omega)$ is complete consensus area.

Based on the research of PSO consistency theory, $\phi_i^{(k)} = 4(1+\omega)$ are the critical value for consensus area. According to the boundary value, aiming at the shortcomings of the particle swarm algorithm itself, this paper puts forward an improved particle swarm optimization algorithm: particle swarm optimization algorithm of multiple population co-evolution.

When $w = 0.79$, select a few particle swarms , the values of c within these particle swarms are different from each other, and all values are in consensus area. Multiple particle swarms search solution space independently, this way can enhance global searching capability. The results of several test functions using the improved algorithm and the standard particle swarm algorithm demonstrate the effectiveness of the improved algorithm.

4   The Multiple Population Co-evolution PSO Algorithm

Experiment parameter:
Number of population: 3
Population size: 10
Particle neighborhood size: 2
Dimension: 2、10、30

The scope of the search space: [-100 100]^D

Maximum iterations: 10000*dimension

The number of goal: 1

Accuracy: 2.22e-16

Inertia weight: 0.8

Learning factor: for swarm 1, the learning factor is 1.5; for swarm 2 , the learning factor is 1.2; for swarm 3, the learning factor is 0.9.

experimental environment: MATLAB 7.11.0（R2010b）.

steps of algorithm:

Step1: Particle swarm initialization, including population number, population scale, the inertiaweight, learning factor, the initial position and velocity of the particle, etc.

The particle swarm learning factor selection strategy: the inertia weight of each particle group is 0.8, but learning factor is different, if the swarm is i, then the learning factor of each swarm is 1.8 -0.3*i respectively.

Step2: Calculate the fitness of each particle of each group

Group update strategy:

Step3: Compare the fitness of each particle of each group with its previous experience fitness at the best position, if good, then use its current fitness value as the best fitness value of particles.

Step4: Compare the fitness of each particle in the particle group (fitness) with the pbest, if good, it will be updated as the particle group best fitness value.

Step5: Compare the best fitness value of each particle group, select the smallest value as the best fitness value for the particle swarm.

Step6: According to the speed and position update formula, update each particle's velocity and position.

Step7: Algorithm set termination conditions (usually good enough to adapt to the value or reaches the maximum iterations and precision) according to specific problems, if not, then return to step 2; if do, stop the iteration, output the optimal solution.

5   Simulation Experiment and Result Analysis

In simulation experiments, we compare the multiple population co-evolution PSO algorithm procedure with the result of the given example.m, both use 30-dimension in the experiments, the test time is five, using function 9 Weierstrass Function as test function. Run example.m and record the average time of five experiments as T1, run the multiple population co-evolution PSO algorithm program and record the average time of five experiments as T2.

T1=57.7936

T2=76.2999

(T2-T1)/T1=0.320214

For dimension 2、10、30, run the multiple population co-evolution PSO algorithm program 51 times respectively, results are saved in 20140418Bit_2d、20140418Bit_10d、20140418Bit_30d. Maximum, minimum, average, median, standard variance of fitness value are recorded in analysis_2d.csv、analysis_10d.csv、analysis_30d .csv respectively.

Table 1. For 2 dimension

function number	max	min	mean	median	standard deviation
1	0.27622	7.15E-09	0.017526	0.000323	0.052023
2	381.39	0.006224	25.446	6.3034	60.777
3	1.6667	1.6667	1.6667	1.6667	2.24E-16
4	38.605	0.00205	3.9485	0.69603	7.1389
5	4.05E-05	2.18E-12	3.47E-06	2.37E-07	7.69E-06
6	1.73E-21	1.97E-31	5.66E-23	1.29E-28	2.65E-22
7	1.31E-05	1.09E-13	1.59E-06	1.01E-07	3.45E-06
8	4.44E-15	8.88E-16	1.03E-15	8.88E-16	6.96E-16
9	-3.9804	-4	-3.9992	-4	0.002984
10	-0.96952	-1	-0.99485	-0.99821	0.006531
11	2.58E-13	0	6.20E-15	0	3.64E-14
12	54.063	48.585	49.496	49.213	1.0739
13	0.019432	1.61E-08	0.010733	0.019432	0.009166
14	0.007925	5.40E-08	0.000566	0.000182	0.0012
15	0.001914	2.74E-09	0.000241	5.14E-05	0.000388
16	3.45E-16	0	6.76E-18	0	4.83E-17
17	0	0	0	0	0
18	-601.09	-837.97	-771.73	-837.68	102.84
19	1.67E-14	1.35E-32	4.44E-16	1.63E-28	2.39E-15
20	0	0	0	0	0
21	0	0	0	0	0
22	2.59E-31	0	5.08E-33	0	3.62E-32
23	8.9715	8.9715	8.9715	8.9715	0
24	-4.2842	-4.5265	-4.5063	-4.5237	0.043573
25	-1.3921	-1.7107	-1.6329	-1.7106	0.13568
26	0.67181	0.57074	0.60386	0.59685	0.02904
27	-37200000	-37400000	-37300000	-37400000	40169
28	-5.4038	-5.8966	-5.7391	-5.7558	0.11493
29	20.006	19.992	20.001	20.001	0.002386
30	1.0097	0.26706	0.31886	0.26966	0.17126

From analysis_2d.csv we can see that when dimension is 2, the maximum、 minimum 、 average、 median and standard deviation of fitness value are shown in table 1.

From analysis_10d.csv we can see that when dimension is 10, the maximum、 minimum、 average、 median and standard deviation of fitness value are shown in table 2.

Table 2. For 10 dimension

Function number	max	min	mean	median	Standard deviation
1	28464	10.482	4164.1	1115.3	6835.1
2	9820.6	206.77	3107.7	2103.8	2614.3
3	169.67	169.55	169.56	169.55	0.023359
4	255.7	10.773	78.632	74.071	51.557
5	0.10652	0.001738	0.025106	0.017069	0.024209
6	9.5293	0.090725	5.3674	6.1252	2.7106
7	0.027443	0.000625	0.008021	0.006973	0.005679
8	4.9899	0.002668	2.4819	2.5799	1.0874
9	-16.232	-19.288	-18.477	-18.623	0.6125
10	-6.3833	-8.7012	-7.9124	-7.9542	0.49787
11	14.881	9.27E-05	5.3959	4.6682	4.4417
12	0.55492	0.41675	0.4764	0.47755	0.029455
13	1.3251	0.077766	0.62293	0.57989	0.30117
14	0.12353	0.01847	0.058184	0.052282	0.025136
15	0.46897	0.015289	0.16643	0.12418	0.13413
16	0.49012	0.011708	0.21558	0.20158	0.10959
17	0.35762	0.000104	0.030355	0.015601	0.05335
18	-534.4	-3963.7	-1939.5	-2018.4	393.32
19	0.046597	0.001339	0.008284	0.005039	0.008649
20	0.020475	7.25E-17	0.001305	1.62E-06	0.004377
21	0.79987	0.099873	0.29399	0.29987	0.13916
22	0.047381	2.56E-06	0.003458	0.000649	0.007831
23	33.908	17.788	24.723	24.623	3.2122
24	-23.297	-34.026	-31.592	-31.87	1.9737
25	42.999	42.943	42.945	42.943	0.007724
26	0.010188	0.006475	0.008403	0.008328	0.000751
27	-101790000	-65600000.00	-28200000.00	-24800000.00	1.31E+07
28	-5.3948	-5.9232	-5.663	-5.642	0.13994
29	20.002	20	20	20	0.000393
30	1.0372	1.0097	1.0183	1.0097	0.012889

From analysis_30d.csv , we can see that when dimension is 30, the maximum、 minimum、 average、 median and standard deviation of fitness value are shown in table 3.

Table 3. For 30 dimension

Function number	max	min	mean	median	Standard deviation
1	1.65E+05	837.96	40120	28993	35732
2	35919	1041.7	9241.1	7925.3	6736.7
3	4554.3	4510.3	4521.8	4520.2	8.5227
4	122.78	22.076	67.24	64.817	21.807
5	0.12922	0.004461	0.040408	0.034891	0.025361
6	33.922	25.335	29.054	28.94	1.1685
7	0.034895	0.008832	0.018335	0.017374	0.00532
8	4.4911	1.5532	3.0353	3.0689	0.69854
9	-49.45	-55.663	-53.521	-53.673	1.1596
10	-20.794	-26.507	-23.939	-24.092	1.2552
11	29.852	0.001599	1.4679	0.03437	4.9221
12	0.014924	0.013084	0.013988	0.014076	0.000392
13	2.9674	0.61538	1.9113	1.8738	0.49718
14	0.036961	0.009863	0.021115	0.02106	0.005493
15	0.48666	0.044493	0.45524	0.4659	0.060491
16	1.4644	0.34813	0.76603	0.73933	0.2357
17	21.569	1.3109	7.1059	6.8476	3.7275
18	-1813.4	-6055.2	-2495	-1908.9	1145.1
19	0.13058	0.006985	0.025035	0.016779	0.023687
20	15.096	6.43E-06	0.64335	0.081668	2.2122
21	1.1999	0.29987	0.7234	0.69987	0.19554
22	0.13694	0.003823	0.048059	0.041579	0.035101
23	50.619	28.644	39.607	39.267	5.4406
24	-101.22	-110.21	-105.56	-105.43	2.1965
25	1151	1124.1	1128.2	1127.2	4.0849
26	0.000504	0.000441	0.000465	0.000459	1.67E-05
27	-47290000	-290950000	-108730000	-101790000	4.88E+07
28	-5.4465	-5.8975	-5.7366	-5.76	0.11652
29	20.005	20.004	20.005	20.005	0.00016
30	1.0782	1.0372	1.0404	1.0372	0.011123

6   Conclusion

This paper proposes a multiple population co-evolution particle swarm optimization algorithm, through which the population information exchanges a lot than before and it is more quickly and efficiently to find the optimal solution. What's more, multiple populations and changes of the learning factor can guarantee the diversity of population, which can effectively improve the inherent defects of the particle swarm algorithm. The experiment results show that the algorithm has good optimization ability, and can get satisfactory results in less time.

References

1. Kennedy, J., Eberhart, R.C.: Particle swarm optimization. In: Proceedings of IEEE International Conference on Neural Networks, vol. 4, pp. 1942–1948 (1995)
2. Eberhart, R.C., Kennedy, J.: A new optimizer using particle swarm theory. In: Proceedings of the Sixth International Symposium on Micro Machine and Human Science, pp.39–43 (1995)
3. Emlen, J.T.: Flocking behavior in birds. The Auk 69(2), 160–170 (1952)
4. Barlow: Behaviour of teleost fishes. Reviews in Fish Biology and Fisheries 4(1), 126–128 (1994)
5. Gueron, S., Levin, S.A.: R. D. I., The dynamics of herds: from individuals to aggregations. Journal of Theoretical Biology 182(1), 85–98 (1996)
6. Jiang, M., Luo, Y.P., Yang, S.Y.: Stochastic convergence analysis and parameter selection of the standard particle swarm optimization algorithm. Information Processing Letters 102(1), 8–16 (2007)
7. Jin, X.L., Ma, L.H., Wu, T.J.: The analysis of pso convergence based on stochastic process. Acta Automatica Sinica 33(12), 1263–1268 (2007)

Fireworks Algorithm and Its Variants for Solving ICSI2014 Competition Problems

Shaoqiu Zheng, Lang Liu, Chao Yu, Junzhi Li, and Ying Tan*

Department of Machine Intelligence, School of EECS, Peking University, China
Key Laboratory of Machine Perception (MOE), Peking University, China
{zhengshaoqiu,langliu,chaoyu,ljz,ytan}@pku.edu.cn

Abstract. Firework algorithm (FWA) is a newly proposed swarm intelligence based optimization technique, which presents a different search manner by simulating the explosion of fireworks to search within the potential space till the terminal criterions are met. Since its introduction, a lot of improved work have been conducted, including the enhanced fireworks algorithm (EFWA), the dynamic search in FWA (dynFWA) and adaptive fireworks algorithm (AFWA). This paper is to use the FWA and its variants to take participate in the ICSI2014 competition, the performance among them are compared, and results on 2-, 10-, 30-dimensional benchmark functions are recorded.

Keywords: ICSI2014 competition, FWA, EFWA, dynFWA, AFWA.

1 Introduction

FWA is a population based swarm intelligence algorithm proposed by Tan and Zhu [16] in 2010. It takes the inspiration from the phenomenon that the fireworks explode and illuminate the local space around the fireworks in the night sky. Its proposed explosion search manner for each firework and cooperative strategy for allocating the resources among the fireworks swarm make it a novel and promising algorithm.

Assume that objective function f is a minimization problem with the form $\min_{x \in \Omega} f(x)$, and Ω is the feasible search region. The conventional FWA works as follows: At first, a fixed number of fireworks (N) are initialized within the feasible search range, and the quality of the fireworks' positions are evaluated, based on which the explosion amplitudes and explosion sparks number are calculated. Here, the principle idea for calculating them is that: the firework with smaller fitness will have larger number of explosion sparks and smaller explosion amplitude, while the firework with larger fitness will have smaller number of explosion sparks and bigger explosion amplitude. In addition, to increase the diversity of the population of the fireworks and explosion sparks, Gaussian mutation sparks are also introduced. After these operations of generating explosion and Gaussian mutation sparks, selection strategy is performed among the candidates set which includes fireworks, explosion sparks and Gaussian mutation

---

* Corresponding author.

Y. Tan et al. (Eds.): ICSI 2014, Part II, LNCS 8795, pp. 442–451, 2014.

sparks, and a fixed number of (N) fireworks are selected for the next iteration. The algorithm continues the search until the termination criterions are reached.

Since its first presentation in [16], FWA has attracted a number of researchers to develop the conventional algorithm and apply the algorithm for optimization of real world problems. For the algorithm developments, it includes the single objective algorithm developments [13] [12] [18] [14] [11], multi-objective algorithm developments [21], hybrid version with other algorithms [20] [2] [4] and parallel implementation versions [3]. For the application, FWA has been applied for FIR and IIR digital filters design [4], the initialization of Non-negative Matrix Factorization (NMF) and iterative optimization of NMF [10], [8], [9], spam detection [5], finger-vein identification [19] and power system reconfiguration [7] [6]. Experimental results suggest that FWA is a promising swarm intelligence algorithm, which needs further research and developments.

Motivation and Synopsis: The original motivation of this paper is to let FWA and its variants to participate the competition in ICSI2014 competition, and the performance among some typical improved work are compared. The remainder of this paper is organized as follows: Section 2 briefly introduces the framework of conventional fireworks algorithm, and the FWA variants are presented in Section 3, Experiments are given in Section 4 and finally conclusions are drawn in Section 5.

2   The Conventional FWA

In FWA, it works with a population of fireworks which can generate the explosion sparks and Gaussian mutation sparks thus to maintain the fireworks swarm with global and local search abilities. After generating two kinds of sparks, the selection strategy is performed for the selection of fireworks to the next iteration. Algorithm 1 gives the framework of conventional FWA.

In FWA, to make a contract among the fireworks and balance between the exploration and exploitation capacities, the fireworks are designed to take different explosion amplitudes and explosion sparks number. Assume that the fireworks number is N, then for each firework, the explosion sparks number s_i and explosion amplitude A_i are calculated as following:

$$A_i = \hat{A} \cdot \frac{f(X_i) - y_{min} + \varepsilon}{\sum_{i=1}^{N} (f(X_i) - y_{min}) + \varepsilon}, \tag{1}$$

$$s_i = M_e \cdot \frac{y_{max} - f(X_i) + \varepsilon}{\sum_{i=1}^{N} (y_{max} - f(X_i)) + \varepsilon}, \tag{2}$$

where, $y_{max} = max(f(X_i))$, $y_{min} = min(f(X_i))$, and $\hat{A}$ and M_e are two constants to control the explosion amplitude and the number of explosion sparks, respectively, and ε is the machine epsilon. In addition, to avoid the overwhelming effects of fireworks at good/bad locations, the max/min number of sparks are

Algorithm 1. General structure of conventional FWA

1: Initialize N fireworks X_i
2: **repeat**
3:    *Explosion operator*
4:       (i) Calculate explosion amplitude A_i and explosion sparks number s_i
5:       (ii) Generate the explosion sparks
6:    **for** each firework X_i, perform s_i times **do**
7:       Initialize the location of the "explosion sparks": $\hat{X}_i = X_i$
8:       Calculate offset displacement: $\triangle X = A_i \times rand(-1, 1)$
9:       $z = round(D * rand(0, 1))$
10:      Randomly select z dimensions of $\hat{X}_i$
11:      **for** each select dimension of $\hat{X}_i^k$ **do**
12:         $\hat{X}_i^k = \hat{X}_i^k + \triangle X$
13:         **if** $\hat{X}_i^k$ out of bounds **then**
14:            $\hat{X}_i^k = X_{min}^k + |\hat{X}_i^k| \% (X_{max}^k - X_{min}^k)$
15:         **end if**
16:      **end for**
17:      (iii) Evaluate fitness of newly created explosion sparks
18:   **end for**
19:   *Gaussian mutation operator*
20:      (i) Generate the Gaussian sparks
21:   **for** perform M_g times **do**
22:      Randomly initialize the location of the "Gaussian sparks": $\tilde{X}_i = X_i$
23:      Calculate offset displacement: $e = Gaussian(1, 1)$
24:      Set $z^k = round(rand(0, 1))$, $k = 1, 2, ..., d$
25:      **for** each dimension of $\hat{X}_i^k$, where $z^k == 1$ **do**
26:         $\tilde{X}_i^k = \tilde{X}_i^k \times e$
27:         **if** $\tilde{X}_i^k$ out of bounds **then**
28:            $\tilde{X}_i^k = X_{min}^k + |\tilde{X}_i^k| \% (X_{max}^k - X_{min}^k)$
29:         **end if**
30:      **end for**
31:   **end for**
32:      (ii) Evaluate fitness of newly created Gaussian sparks
33:   *Selection strategy*
34:      (i) Select fireworks for next iteration
35: **until** termination is met.

---

bounded by

$$s_i = \begin{cases} round(aM_e) & \text{if } s_i < aM_e, \\ round(bM_e) & \text{if } s_i > bM_e, \\ round(s_i) & otherwise. \end{cases} \tag{3}$$

where, a and b are constant parameters which confine minimal/maximal sparks number (the range of the sparks number). Then for each firework, the explosion sparks are generated according to Algorithm 1.

To increase the diversity, Gaussian mutation sparks are generated based on a Gaussian mutation operator (cf. Algorithm 1).

To retain the information to the next iteration, selection strategy is performed as most of the swarm intelligence algorithms and evolutionary algorithms. In the candidates set, the individual with minimal fitness is always selected while for the rest x_i in candidates set, the selection probability p_i is calculated as

$$p(x_i) = \frac{R(x_i)}{\sum_{j \in K} R(x_j)} \tag{4}$$

$$R(x_i) = \sum_{j \in K} d(x_i, x_j) = \sum_{j \in K} ||x_i - x_j|| \tag{5}$$

where K is the set of all current locations including original fireworks and both types of sparks.

3   The Selected Typical Improvement Work

3.1   Enhanced Fireworks Algorithm

Although FWA has shown its great performance when dealing with function optimization in [16], which outperforms SPSO [1] and CPSO [15] in the selected benchmark functions, in [18], Zheng et al presented a comprehensive study of operators in conventional FWA and proposed the enhanced FWA (EFWA). Some details of the EFWA are as following.

Amplitude of Explosion: In FWA, the explosion amplitude of the best firework is usually very close to 0. In EFWA, a lower bound A_{min} is introduced to avoid this problem:

$$A_i = max(A_i, A_{min}), \tag{6}$$

and A_{min} is non-linearly decreased with the evaluation times going on:

$$A^k_{min}(t) = A_{init} - \frac{A_{init} - A_{final}}{evals_{max}} \sqrt{(2 * evals_{max} - t)t}, \tag{7}$$

where A_{init} and A_{final} are the initial and final minimum explosion amplitude, $evals_{max}$ is the maximum evaluation times and t is the current evaluation times.

Generating Sparks: In FWA, the number of to-be-mutated dimensions is calculated first and the displacement is calculated just once then used for all the selected dimensions. While in EFWA, for each dimension, an independent displacement $\Delta X^k_i = A_i \times U(-1, 1)$ is calculated with the selection probability $U(0, 1)$, ($U(a, b)$ denotes the generated random number is under the mean distribution between a and b).

$$\hat{X}^k_i = X^k_i + \Delta X^k_i. \tag{8}$$

Moreover, the way to generate Gaussian sparks makes use of the currently best location X_B to avoid the concentrated search on origin region.

$$\tilde{X}_i^k = X_i^k + e * (X_B^k - X_i^k), \tag{9}$$

where $e \sim N(0,1)$.

A new-generated spark will be mapped into a random place in the variable space with uniform distribution if the generated location exceeds the bounds.

$$\hat{X}_i^k = U(X_k^{min}, X_k^{max}). \tag{10}$$

Selection Strategy: To decrease the computational cost of selection strategy in FWA (Eq. 4), in EFWA, the best of the set will be selected first while the rest are randomly selected.

3.2   The dynFWA and AFWA

In FWA and EFWA, the explosion amplitude for fireworks is one of the most key features relevant to the performance. For each firework, its explosion amplitude is calculated by Eq. 1. In fact, the fitness of firework's position is only one kind of information to characterize the local information around X_i, good position needs further local search. However, for an optimization problem, the optimization process is dynamic, the previous static explosion amplitude calculation strategy only suggests that positions of fireworks are good or bad, not the local region within the positions of fireworks. So the explosion amplitude calculation method will lead to a bad local search ability and the experimental results in [18] are consistent with this idea.

For simplicity, the firework with minimal fitness in the fireworks swarm is defined as the core firework (CF, X_{CF}), which has the property that (i) its fitness is best among the fireworks, (ii) it is always selected to the next iteration. To overcome the limitations presented above, the dynamic search in FWA (dynFWA) and adaptive fireworks algorithm (AFWA) in [14] [11] are proposed respectively.

The dynFWA – Dynamic Explosion Amplitude Strategy: Assume that $f(\hat{X}_{best})$ is the minimal fitness among the explosion sparks and $f(X_{CF})$ denotes the fitness of CF. Here, in dynFWA, it concerns whether the generated explosion sparks can get better fitness than the CF, i.e. the $\Delta_f = f(\hat{X}_{best}) - f(X_{CF})$.

1) $\Delta_f < 0$

It means at least one of the newly generated sparks has smaller fitness than CF's fitness. If so, the $\hat{X}_{best}$ is probably created by the CF or the rest of fireworks other than CF. If $\hat{X}_{best}$ is created by CF, in order to speedup the search for the global optimum, the explosion amplitude of CF will become a bigger one compared with the current value. If $\hat{X}_{best}$ is created by one firework (X_i) other than CF, it has a high chance that the X_i is close to X_{CF}. If X_i is close to X_{CF}, the same explosion amplitude strategy for the CF in the next iteration is taken. If X_i is not close to X_{CF}, then the current explosion amplitude is in fact not effective for the newly generated CF for search any more. However, as it is hard to define the closeness and it is believed that the dynamic explosion amplitude strategy has its ability to adjust the explosion amplitude itself in the following iterations, so dynFWA just sets the explosion amplitude of newly selected CF with a increasing amplitude.

2) $\Delta_f \geq 0$

It means that none of the explosion sparks has found a position with better fitness compared to the CF. The reason for this situation is that the explosion amplitude of firework is too big for CF to search a better position. The CF needs to narrow down the search range. That is to reduce the explosion amplitude thus increasing the probability that the fireworks swarm can find a better position.

In fact, if the CF is far away from the global optimal position, increasing the explosion amplitude is one of the most effective methods to speedup the convergence. The reduction of the explosion amplitude makes it move towards the global optimal position, i.e. the CF finding a better solution.

In FWA and EFWA, to increase the diversity of the fireworks swarm, Gaussian mutation sparks are introduced. However, due to the selection method, the Gaussian mutation sparks do not work effectively as they are designed to, thus in dynFWA [14], they are eliminated.

The AFWA – Adaptive Fireworks Algorithm: The motivation of AFWA is to guarantee the progress made in current iteration is bigger than in the previous iteration [11].

AFWA tries to find the spark whose fitness is the minimal among the candidates whose fitness is worse than CF, and whose Infinite Norm is closest to the best candidate (i.e., Core firework or best spark), then the Infinite Norm between the found candidate with the firework will be taken as the explosion amplitude for the next iteration.

Under this explosion amplitude updating strategy, there are two cases. The first one is that the CF does not generate any good sparks whose fitness is smaller than the firework, then the fitness of all the sparks are larger than the firework, and the explosion amplitude will take the Infinite norm between the firework and the selected candidate. The explosion amplitude in the next iteration will be reduced. For the second case, it has two situations, and the explosion amplitude will be amplified according to the simulation with high chance. Moreover, as the Infinite Norm between the calculated candidate and the firework may change radically, in AFWA, it introduces the smoothing strategy.

4 Experiments

4.1 Experimental Setup

For the implementation of FWA, EFWA, dynFWA and AFWA in this paper, all the parameters are taken from [14] without any modifications. The experimental

Table 1. Run time results on f_9

	T1(s)	T2(s)	(T2-T1)/T1
dynFWA	28.0029	28.8269	0.0294
AFWA	28.0029	28.4186	0.0148

platform used in the experiments is MATLAB 2011b (Windows 7; Intel Core i7-2600 CPU @ 3.7 GHZ; 8 GB RAM) while ICSI-2014 competition problems are used as benchmark functions to compare the performance.

The description of the ICSI-2014 competition benchmark functions is as follows: It contains 30 functions, and for each function, the feasible range is set to $[-100, 100]$. Moreover, to make a comprehensive comparison, in the competition, three groups of experiments with dimension set as $D = 2, 10, 30$, and maximum evaluation times $D * 10000$ are designed. For each function, the max, min, mean, median value and standard deviation of 51 times results are recorded.

4.2 Experimental Results

The experimental results can be found in Table 2, Table 3 and Table 4. For the run time consuming, the experimental runs on f_9 of dynFWA and AFWA are given in Table 1 according to [17].

Table 2. Results for 2D functions

	F	1	2	3	4	5	6	7	8	9	10	11	12	13	14	15
FWA	Max	1.63E+03	1.80E+03	1.67E+00	1.13E+03	2.09E+00	1.68E-01	2.12E-03	2.58E+00	-3.74E+00	-4.14E-01	9.37E-03	6.62E+01	7.45E-02	8.05E-02	5.83E-02
	Min	8.03E-02	6.43E-01	1.67E+00	5.86E+00	1.39E-03	9.00E-05	2.00E-05	2.00E-05	-4.00E+00	-9.99E-01	1.90E-05	4.88E+01	1.90E-05	1.37E-03	3.40E-05
	Mean	1.97E+02	4.23E+00	1.67E+00	1.38E+02	3.39E-01	4.40E-02	6.91E-04	1.01E-01	-3.92E+00	-8.77E-01	1.04E-03	5.54E+01	2.10E-02	9.88E-03	1.03E-02
	Median	8.13E+01	2.10E+02	1.67E+00	7.99E+01	2.28E-01	2.37E-02	5.77E-04	4.16E-02	-3.93E+00	-9.10E-01	6.47E-04	5.48E+01	1.94E-02	4.46E-03	4.66E-03
	Std	2.96E+02	4.65E+02	2.26E-04	1.83E+02	3.94E-01	4.37E-02	4.80E-04	3.59E-01	6.26E-02	1.17E-01	1.44E-03	4.73E+00	1.14E-02	1.51E-02	1.30E-02
	F	16	17	18	19	20	21	22	23	24	25	26	27	28	29	30
	Max	4.79E-02	5.94E-01	-6.01E+02	8.60E-03	0.00E+00	2.01E-01	1.62E-01	1.27E+01	-4.01E+00	-1.37E+00	1.03E+00	-3.62E+07	-4.82E+00	2.00E-01	1.01E+00
	Min	1.13E-03	2.32E-04	-8.38E+02	9.00E-06	0.00E+00	2.60E-05	2.00E-06	8.97E+00	-4.52E+00	-1.71E+00	5.74E-01	-3.74E+07	-5.86E+00	2.00E-01	2.68E-01
	Mean	1.54E-02	1.27E-01	-7.80E+02	2.71E-03	0.00E+00	4.35E-02	1.95E-02	9.36E+00	-4.31E+00	-1.50E+00	7.15E-01	-3.72E+07	-5.45E+00	2.00E-01	4.18E-01
	Median	1.34E-02	7.02E-02	-8.38E+02	2.81E-03	0.00E+00	7.93E-03	7.86E-03	8.98E+00	-4.31E+00	-1.39E+00	6.85E-01	-3.73E+07	-5.45E+00	2.00E-01	2.96E-01
	Std	1.10E-02	1.38E-01	9.06E+01	2.19E-03	0.00E+00	5.57E-02	3.19E-02	8.58E-01	1.16E-01	1.41E-01	9.67E-02	2.22E+05	2.54E-01	6.07E-03	2.17E-01
	F	1	2	3	4	5	6	7	8	9	10	11	12	13	14	15
EFWA	Max	7.31E+02	4.06E+03	1.67E+00	1.27E+03	7.12E-04	7.14E-01	2.28E-02	2.58E+00	-3.99E+00	-6.65E-01	3.13E-06	8.33E+01	3.56E-01	3.16E-01	4.78E-01
	Min	1.16E-04	1.27E-02	1.67E+00	7.69E-01	2.32E-06	2.47E-09	7.02E-07	3.44E-04	-4.00E+00	-1.00E+00	4.02E-10	4.86E+01	6.98E-08	2.30E-02	2.11E-03
	Mean	1.32E+02	1.01E+03	1.67E+00	3.43E+02	8.86E-05	1.40E-02	5.13E-03	5.37E-02	-3.99E+00	-9.45E-01	5.37E-07	5.14E+01	5.16E-02	1.43E-01	4.00E-02
	Median	2.73E+01	6.52E+02	1.67E+00	2.26E+02	5.91E-05	1.23E-06	3.69E-04	2.19E-03	-3.99E+00	-9.70E-01	2.50E-07	4.89E+01	1.94E-02	1.45E-01	3.02E-02
	Std	1.92E+02	1.13E+03	2.93E-09	3.33E+02	1.14E-04	1.00E-01	9.23E-03	3.61E-01	2.63E-03	7.60E-02	6.92E-07	8.41E+00	6.61E-02	7.72E-02	6.60E-02
	F	16	17	18	19	20	21	22	23	24	25	26	27	28	29	30
	Max	1.88E-04	1.59E-05	-6.01E+02	4.86E-08	6.14E-32	1.08E-07	1.04E-05	1.07E-01	-2.11E+00	7.38E-01	1.31E+00	-1.46E+07	-4.51E+00	2.00E-01	1.12E+00
	Min	1.21E-05	2.68E-08	-8.38E+01	4.91E-10	1.59E-58	3.07E-11	4.61E-09	8.97E+00	-4.52E+00	-1.71E+00	5.72E-01	-3.74E+07	-5.74E+00	1.99E-01	2.67E-01
	Mean	8.94E-05	3.09E-06	-6.11E+02	5.68E-04	1.92E-33	1.46E-08	1.54E-06	9.01E+00	-4.09E+00	-1.32E+00	7.65E-01	-3.01E+07	-5.08E+00	2.00E-01	6.71E-01
	Median	8.51E-05	1.95E-06	-6.01E+02	1.16E-06	6.74E-39	4.82E-09	5.32E-07	8.97E+00	-4.34E+00	-1.39E+00	7.30E-01	-2.36E+07	-5.07E+00	2.00E-01	1.01E+00
	Std	4.64E-05	3.62E-06	4.64E+01	1.57E-03	9.39E-53	2.15E-08	2.04E-06	2.41E-01	6.85E-01	4.69E-01	1.69E-01	7.19E+06	2.68E-01	7.74E-03	3.76E-01
	F	1	2	3	4	5	6	7	8	9	10	11	12	13	14	15
dynFWA	Max	1.07E+03	5.40E+03	1.67E+00	1.45E+03	3.29E-02	1.81E-02	2.36E-02	1.60E-01	-3.87E+00	-9.13E-01	1.47E-03	5.33E+01	1.94E-02	2.85E-01	6.37E-02
	Min	3.37E-04	6.11E-02	1.67E+00	1.61E-02	5.16E-05	2.33E-07	2.39E-05	2.45E-04	-4.00E+00	-1.00E+00	1.42E-08	4.86E+01	1.90E-06	2.84E-03	7.42E-05
	Mean	1.56E+02	4.90E+02	1.67E+00	1.79E+02	3.85E-03	1.21E-03	9.70E-04	2.45E-02	-3.98E+00	-9.76E-01	7.02E-05	4.93E+01	1.55E-02	6.47E-02	1.74E-02
	Median	6.55E+01	8.70E+01	1.67E+00	3.54E+01	1.47E-03	1.21E-04	3.17E-04	1.05E-02	-3.99E+00	-9.80E-01	1.38E-05	4.91E+01	1.94E-02	4.13E-02	1.30E-02
	Std	2.44E+02	9.25E+02	1.16E-06	3.18E+02	6.32E-03	2.85E-03	3.30E-03	3.95E-02	2.45E-02	2.08E-02	2.09E-04	7.73E-01	7.75E-03	6.18E-02	1.55E-02
	F	16	17	18	19	20	21	22	23	24	25	26	27	28	29	30
	Max	1.04E-02	7.81E-02	-8.38E+02	5.10E-03	0.00E+00	9.99E-02	1.59E-02	9.13E+00	-4.22E+00	-1.39E+00	9.18E-01	-3.73E+07	-4.73E+00	2.00E+01	1.01E+00
	Min	1.19E-04	7.38E-07	-8.38E+02	4.42E-07	0.00E+00	1.97E-08	1.07E-06	8.97E+00	-4.53E+00	-1.71E+00	1.07E-14	4.86E-01	0.00E+00	5.89E+00	2.00E-01
	Mean	2.31E-03	3.63E-03	-8.38E+02	1.14E-03	0.00E+00	1.98E-03	2.27E-03	8.99E+00	-4.46E+00	-1.57E+00	6.68E-01	-3.74E+07	-5.24E+00	2.00E-01	3.39E-01
	Median	1.83E-03	1.54E-04	-8.38E+02	4.24E-04	0.00E+00	4.57E-06	5.31E-04	8.98E+00	-4.46E+00	-1.57E+00	6.45E-01	-3.74E+07	-5.23E+00	2.00E-01	2.72E-01
	Std	2.16E-03	1.24E-02	3.38E-04	1.63E-03	0.00E+00	1.40E-02	4.05E-03	3.21E-02	9.62E-02	1.50E-01	7.80E-02	9.24E-02	2.50E-01	5.86E-03	1.45E-01
	F	1	2	3	4	5	6	7	8	9	10	11	12	13	14	15
AFWA	Max	1.97E+03	4.52E+03	1.67E+00	1.31E+03	3.04E-03	1.27E-01	2.38E-03	4.44E-15	-3.86E+00	-8.82E-01	2.74E-04	5.82E+01	1.94E-02	2.84E-01	7.07E-02
	Min	5.52E-02	5.52E-02	1.67E+00	5.15E-04	5.37E-07	1.42E-10	0.00E+00	8.88E-16	-4.00E+00	-1.00E+00	1.07E-14	4.86E-01	0.00E+00	5.11E-04	1.89E-05
	Mean	2.90E+02	1.03E+03	1.67E+00	3.00E+02	4.86E-04	5.61E-03	1.35E-04	1.10E-15	-3.99E+00	-9.71E-01	1.98E-05	5.04E-01	1.83E-02	5.75E-02	2.21E-02
	Median	1.32E+02	6.46E+02	1.67E+00	1.66E+02	3.05E-04	1.93E-05	9.65E-18	8.88E-16	-4.00E+00	-9.80E-01	8.55E-07	4.89E-01	1.94E-02	3.36E-02	1.51E-02
	Std	4.12E+02	1.23E+03	4.50E-09	3.53E+02	5.87E-04	2.07E-02	4.24E-04	8.44E-16	2.08E-02	2.56E-02	4.87E-05	2.76E+00	4.62E-03	7.09E-02	2.05E-02
	F	16	17	18	19	20	21	22	23	24	25	26	27	28	29	30
	Max	0.00E+00	0.00E+00	-6.01E+02	4.85E-03	0.00E+00	0.00E+00	0.00E+00	1.07E+01	-3.12E+00	-6.50E-01	1.31E+00	-3.74E+07	-5.04E+00	2.00E-01	5.57E-01
	Min	0.00E+00	0.00E+00	-8.38E+02	0.00E+00	0.00E+00	0.00E+00	0.00E+00	8.97E+00	-4.53E+00	-1.71E+00	5.71E-01	-3.74E+07	-5.93E+00	2.00E-01	2.67E-01
	Mean	0.00E+00	0.00E+00	-8.29E+02	1.22E-03	0.00E+00	0.00E+00	0.00E+00	9.01E+00	-4.35E+00	-1.49E+00	7.05E-01	-3.74E+07	-5.53E+00	2.00E-01	3.11E-01
	Median	0.00E+00	0.00E+00	-8.38E+02	0.00E+00	0.00E+00	0.00E+00	0.00E+00	8.97E+00	-4.43E+00	-1.39E+00	6.67E-01	-3.74E+07	-5.54E+00	2.00E-01	2.70E-01
	Std	0.00E+00	0.00E+00	4.45E+01	2.03E-03	0.00E+00	0.00E+00	0.00E+00	2.41E-01	2.61E-01	1.94E-01	1.36E-01	8.99E+01	2.46E-01	8.51E-03	7.74E-02

Table 3. Results for 10D functions

	F	1	2	3	4	5	6	7	8	9	10	11	12	13	14	15
FWA	Max	7.42E+06	1.35E+05	1.82E+02	3.52E+02	4.59E+00	1.63E+02	6.03E-02	7.97E+00	-1.51E+01	-3.20E+00	2.57E+01	6.15E-01	1.76E+00	2.06E-01	2.01E-01
	Min	2.25E+05	2.49E+03	1.71E+02	2.28E+01	9.06E-02	5.90E+00	9.08E-03	2.86E+00	-1.88E+01	-8.05E+00	2.36E-01	4.35E-01	2.19E-01	1.28E-02	1.12E-02
	Mean	2.78E+06	4.15E+04	1.74E+02	1.19E+02	1.95E+00	3.83E+01	3.39E-02	5.01E+00	-1.70E+01	-5.67E+00	8.25E+00	5.35E-01	8.91E-01	9.38E-02	7.14E-02
	Median	2.32E+06	3.35E+04	1.74E+02	1.01E+02	1.57E+00	2.84E+01	3.36E-02	4.83E+00	-1.72E+01	-5.77E+00	7.47E+00	5.36E-01	8.52E-01	8.77E-02	5.92E-02
	Std	1.81E+06	2.95E+04	2.57E+00	6.01E+01	1.23E+00	3.35E+01	1.47E-02	1.30E+00	9.77E-01	1.04E+00	6.28E+00	3.99E-02	3.02E-01	4.60E-02	4.34E-02
	F	16	17	18	19	20	21	22	23	24	25	26	27	28	29	30
	Max	1.05E+00	2.54E+01	-1.66E+03	9.41E-01	4.65E+00	2.20E+00	1.71E+01	3.22E+01	-2.10E+01	4.54E+01	1.53E-02	-1.47E+07	-5.16E+00	2.00E+01	1.08E+00
	Min	9.41E-02	1.65E+00	-4.12E+03	3.36E-02	1.92E-05	9.99E-02	2.61E-01	2.09E+01	-3.27E+01	4.31E+01	8.55E-03	-9.88E+07	-5.93E+00	2.00E+01	1.01E+00
	Mean	4.22E-01	1.14E+01	-2.92E+03	2.65E-01	5.86E-01	7.65E-01	4.39E+00	2.63E+01	-2.78E+01	4.36E+01	1.12E-02	-5.30E+07	-5.53E+00	2.00E+01	1.03E+00
	Median	3.76E-01	1.16E+01	-2.95E+03	1.95E-01	5.78E-02	7.00E-01	3.30E+00	2.70E+01	-2.81E+01	4.35E+01	1.12E-02	-4.67E+07	-5.53E+00	2.00E+01	1.04E+00
	Std	2.07E-01	6.18E+00	1.04E+03	2.09E-01	1.02E+00	4.65E-01	3.29E+00	2.58E+00	2.66E+00	4.20E-01	1.77E-03	2.44E+07	1.97E-01	1.74E-03	1.50E-02
EFWA	F	1	2	3	4	5	6	7	8	9	10	11	12	13	14	15
	Max	1.37E+05	2.00E+04	1.70E+02	6.60E+01	2.53E-01	1.00E+01	3.07E+00	2.01E+01	-1.22E+01	8.07E+00	4.25E+01	5.84E-01	3.26E+00	2.02E+00	1.67E+00
	Min	1.73E+04	7.47E+02	1.70E+02	3.62E+00	1.07E-02	2.94E+00	1.98E-03	2.99E+00	-1.57E+01	-7.25E+00	3.65E+00	3.90E-01	6.40E-01	8.65E-02	8.09E-02
	Mean	5.83E+04	7.52E+03	1.70E+02	1.89E+01	1.12E-01	7.81E+00	8.51E-01	1.66E+01	-1.48E+01	-2.16E+00	2.04E+01	4.96E-01	2.11E+00	7.24E-01	3.02E-01
	Median	5.47E+04	6.49E+03	1.70E+02	1.71E+01	1.24E-01	8.46E+00	7.29E-01	2.00E+01	-1.51E+01	-2.53E+00	1.79E+01	5.03E-01	2.12E+00	3.40E-01	3.45E-01
	Std	2.81E+04	4.46E+03	4.11E-02	1.37E+01	6.73E-02	1.74E+00	9.63E-01	5.97E+00	8.49E-01	3.63E+00	1.05E+01	4.51E-02	5.64E-01	7.60E-01	2.40E-01
	F	16	17	18	19	20	21	22	23	24	25	26	27	28	29	30
	Max	1.46E+00	5.94E-02	-3.00E+03	4.82E+00	4.84E+02	6.00E-01	2.28E-01	5.51E+01	-8.86E+00	5.29E+01	1.37E-02	-3.68E+07	-5.13E+00	2.00E+01	1.08E+00
	Min	3.79E-02	7.01E-03	-3.00E+03	1.51E-02	0.00E+00	9.99E-02	2.64E-03	2.36E+01	-3.26E+01	4.29E+01	7.05E-03	-1.07E+08	-5.66E+00	2.00E+01	1.01E+00
	Mean	2.45E-01	2.44E-02	-3.00E+03	8.23E-01	9.57E+00	2.29E-01	9.15E-02	3.88E+01	-2.18E+01	4.37E+01	9.84E-03	-9.75E+07	-5.37E+00	2.00E+01	1.04E+00
	Median	2.11E-01	2.13E-02	-3.00E+03	4.68E-01	3.90E-03	2.00E-01	9.34E-02	3.95E+01	-2.21E+01	4.29E+01	9.81E-03	-1.06E+08	-5.36E+00	2.00E+01	1.04E+00
	Std	2.31E-01	1.26E-02	3.33E-02	9.42E-01	6.77E+01	1.25E-01	6.71E-02	7.04E+00	5.33E+00	2.62E+00	1.57E-03	1.87E+07	1.29E-01	4.34E-04	1.68E-02
dynFWA	F	1	2	3	4	5	6	7	8	9	10	11	12	13	14	15
	Max	1.32E+05	5.20E+04	1.70E+02	1.86E+02	2.00E-01	9.71E+00	3.32E-01	4.56E+00	-1.69E+01	-5.32E+00	1.52E+01	5.67E-01	1.66E+00	2.03E-01	1.89E-01
	Min	1.80E+03	6.97E+02	1.70E+02	4.78E+00	4.23E-03	8.88E-01	5.74E-04	2.19E-02	-1.94E+01	-8.73E+00	7.35E-04	3.89E-01	2.08E-01	2.83E-02	7.27E-03
	Mean	3.18E+04	9.49E+03	1.70E+02	6.48E+01	6.31E-02	6.56E+00	1.64E-02	2.03E+00	-1.84E+01	-7.24E+00	6.54E+00	4.70E-01	9.80E-01	8.83E-02	1.03E-01
	Median	2.49E+04	6.61E+03	1.70E+02	5.74E+01	5.46E-02	7.04E+00	4.19E-03	2.04E+00	-1.85E+01	-7.34E+00	5.86E+00	4.63E-01	9.96E-01	8.04E-02	1.04E-01
	Std	2.90E+04	9.87E+03	7.52E-02	4.78E+01	5.23E-02	2.51E+00	4.85E-02	1.23E+00	6.30E-01	7.17E-01	4.17E+00	4.42E-02	3.85E-01	4.11E-02	4.01E-02
	F	16	17	18	19	20	21	22	23	24	25	26	27	28	29	30
	Max	8.67E-01	1.68E-02	-1.68E+03	1.66E-01	8.76E+00	8.00E-01	2.33E-01	4.20E+01	-2.70E+01	5.33E+01	9.77E-03	-1.91E+07	-5.23E+00	2.00E+01	1.04E+00
	Min	3.89E-02	7.06E-05	-4.14E+03	1.32E-03	6.79E-12	9.99E-02	5.62E-05	1.96E+01	-3.51E+01	4.29E+01	6.68E-03	-1.09E+08	-5.94E+00	2.00E+01	1.01E+00
	Mean	2.67E-01	2.57E-03	-2.31E+03	2.42E-02	3.95E-01	2.88E-01	4.80E-02	2.90E+01	-3.29E+01	4.36E+01	7.92E-03	-7.83E+07	-5.77E+00	2.00E+01	1.02E+00
	Median	2.21E-01	2.00E-03	-2.02E+03	1.50E-02	1.75E-03	3.00E-01	1.78E-02	2.87E+01	-3.33E+01	4.29E+01	7.90E-03	-6.71E+07	-5.81E+00	2.00E+01	1.04E+00
	Std	1.89E-01	2.70E-03	7.54E+02	3.26E-02	1.44E+00	1.48E-01	5.98E-02	5.18E+00	1.70E+00	2.41E+00	6.37E-04	2.89E+07	1.37E-01	1.28E-04	1.39E-02
AFWA	F	1	2	3	4	5	6	7	8	9	10	11	12	13	14	15
	Max	1.06E+06	4.49E+04	1.71E+02	3.50E+02	1.20E+00	8.13E+01	9.26E-01	7.06E+00	-1.57E+01	-3.80E+00	2.51E+01	5.96E-01	2.03E+00	2.11E-01	2.32E-01
	Min	2.07E+03	8.97E+02	1.70E+02	1.04E+01	1.12E-02	1.93E+00	1.69E-03	1.18E+00	-1.95E+01	-8.54E+00	1.85E-03	3.89E-01	3.24E-01	1.16E-02	1.48E-02
	Mean	3.05E+05	1.42E+04	1.70E+02	9.24E+01	2.10E-01	1.11E+01	6.83E-02	3.60E+00	-1.77E+01	-7.00E+00	1.17E+01	5.10E-01	1.27E+00	1.01E-01	1.21E-01
	Median	2.26E+05	1.03E+04	1.70E+02	6.25E+01	8.45E-02	8.53E+00	1.52E-02	3.22E+00	-1.77E+01	-7.17E+00	1.20E+01	5.11E-01	1.29E+00	9.84E-02	1.15E-01
	Std	2.84E+05	1.16E+04	3.96E-01	7.98E+01	2.59E-01	1.24E+01	1.87E-01	1.47E+00	8.05E-01	9.47E-01	6.82E+00	4.17E-02	3.85E-01	4.52E-02	5.26E-02
	F	16	17	18	19	20	21	22	23	24	25	26	27	28	29	30
	Max	1.23E+00	1.66E+00	-1.23E+03	2.82E-01	2.09E+01	1.90E+00	1.27E+00	4.19E+01	-2.52E+01	5.34E+01	1.15E-02	-1.73E+07	-4.86E+00	2.00E+01	1.04E+00
	Min	5.55E-02	7.31E-02	-4.14E+03	1.69E-03	0.00E+00	9.99E-02	0.00E+00	2.00E+01	-3.50E+01	4.29E+01	6.74E-03	-1.06E+08	-5.92E+00	2.00E+01	1.01E+00
	Mean	3.41E-01	4.24E-01	-2.22E+03	5.90E-02	7.22E-01	5.19E-01	2.78E-01	3.16E+01	-3.10E+01	4.39E+01	8.69E-03	-5.37E+07	-5.70E+00	2.00E+01	1.03E+00
	Median	2.95E-01	3.67E-01	-2.02E+03	3.04E-02	5.00E-03	5.00E-01	1.66E-01	3.17E+01	-3.13E+01	4.29E+01	8.52E-03	-4.54E+07	-5.78E+00	2.00E+01	1.04E+00
	Std	2.47E-01	3.28E-01	7.49E+02	6.69E-02	3.20E+00	3.38E-01	3.17E-01	5.21E+00	2.02E+00	2.99E+00	9.22E-04	2.86E+07	2.15E-01	2.38E-04	1.33E-02

Table 4. Results for 30D functions

	F	1	2	3	4	5	6	7	8	9	10	11	12	13	14	15
FWA	Max	1.14E+07	5.51E+05	5.68E+03	2.25E+02	4.87E+00	2.06E+02	1.28E-01	7.67E+00	-4.62E+01	-1.26E+01	6.78E+00	1.54E-02	6.48E+00	7.64E-02	1.20E-01
	Min	2.79E+06	6.02E+04	4.79E+03	7.04E+01	1.07E+00	7.51E+01	3.25E-02	4.36E+00	-5.23E+01	-2.21E+01	1.59E+00	1.39E-02	1.48E+00	4.96E-03	8.57E-03
	Mean	6.27E+06	2.38E+05	5.16E+03	1.57E+02	2.61E+00	1.36E+02	6.65E-02	6.12E+00	-4.85E+01	-1.80E+01	4.38E+00	1.48E-02	3.68E+00	2.98E-02	3.30E-02
	Median	5.37E+06	1.98E+05	5.13E+03	1.55E+02	2.64E+00	1.31E+02	6.47E-02	6.19E+00	-4.86E+01	-1.81E+01	4.42E+00	1.48E-02	3.71E+00	2.85E-02	2.68E-02
	Std	2.36E+06	1.11E+05	2.24E+02	3.50E+01	8.88E-01	3.41E+01	2.08E-02	7.97E-01	1.37E+00	2.51E+00	1.37E+00	3.57E-04	1.19E+00	1.73E-02	2.32E-02
	F	16	17	18	19	20	21	22	23	24	25	26	27	28	29	30
	Max	2.59E+00	1.07E+02	-8.94E+03	1.34E+00	3.87E+02	3.60E+00	1.91E+01	7.52E+01	-8.74E+01	1.40E+03	9.67E-04	-3.48E+08	-5.29E+00	2.00E+01	1.18E+00
	Min	8.97E-01	1.84E+01	-1.25E+04	3.11E-01	1.29E+00	1.10E+00	4.05E+00	3.67E+01	-1.02E+02	1.18E+03	4.77E-04	-7.54E+08	-5.94E+00	2.00E+01	1.08E+00
	Mean	1.68E+00	5.81E+01	-9.31E+03	7.27E-01	4.80E+01	2.08E+00	1.09E+01	5.72E+01	-9.50E+01	1.28E+03	6.19E-04	-5.17E+08	-5.61E+00	2.00E+01	1.10E+00
	Median	1.71E+00	5.63E+01	-8.97E+03	7.16E-01	3.36E+01	2.10E+00	1.09E+01	5.59E+01	-9.48E+01	1.28E+03	6.05E-04	-4.37E+08	-5.61E+00	2.00E+01	1.08E+00
	Std	3.89E-01	1.73E+01	1.04E+03	2.59E-01	6.31E+01	5.53E-01	3.37E+00	9.05E+00	3.35E+00	4.77E+01	8.01E-05	1.38E+08	1.63E-01	2.22E-04	2.66E-02
EFWA	F	1	2	3	4	5	6	7	8	9	10	11	12	13	14	15
	Max	3.62E+05	2.71E+04	4.53E+03	2.26E+01	1.62E-01	9.03E+01	3.79E+00	2.02E+01	-4.87E+00	-3.50E-01	1.53E+02	1.51E-02	1.20E+01	2.01E+00	1.92E+00
	Min	4.11E+03	3.27E+03	4.51E+03	2.97E+00	2.26E-02	2.70E+01	9.18E-03	6.94E+00	-4.75E+01	-1.98E+01	8.66E+00	1.29E-02	1.86E+00	5.92E-02	4.26E-02
	Mean	6.63E+04	1.23E+04	4.52E+03	8.91E+00	7.12E-02	3.03E+01	2.12E+00	1.92E+01	-1.07E+01	-1.27E+01	7.82E+01	1.41E-02	7.32E+00	1.94E+00	8.91E-01
	Median	4.60E+04	1.16E+04	4.52E+03	8.13E+00	6.62E-02	2.91E+01	1.99E+00	2.01E+01	-6.34E+00	-1.40E+01	7.43E+01	1.41E-02	7.52E+00	1.98E+00	4.62E-01
	Std	7.26E+04	5.59E+03	5.27E+00	4.43E+00	2.96E-02	8.60E+00	8.40E-01	2.93E+00	1.23E+01	4.22E+00	4.07E+01	4.78E-04	2.70E+00	2.69E-01	6.66E-01
	F	16	17	18	19	20	21	22	23	24	25	26	27	28	29	30
	Max	1.35E+00	3.12E+00	-9.01E+03	4.99E+00	4.80E-05	1.20E+00	2.84E-01	1.26E+02	-7.58E+01	1.13E+03	4.73E-04	-5.08E+08	-5.31E+00	2.00E+01	1.18E+00
	Min	1.47E-01	1.39E-01	-9.01E+03	1.45E-01	3.00E-06	3.00E-01	1.63E-02	4.95E+01	-9.42E+01	1.12E+03	4.03E-04	-8.61E+08	-5.86E+00	2.00E+01	1.04E+00
	Mean	4.92E-01	1.07E+00	-9.01E+03	1.27E+00	1.20E-05	6.86E-01	6.72E-02	9.27E+01	-8.55E+01	1.13E+03	4.31E-04	-6.11E+08	-5.52E+00	2.00E+01	1.09E+00
	Median	4.11E-01	8.72E-01	-9.01E+03	7.90E-01	8.00E-06	7.00E-01	4.27E-02	9.27E+01	-8.60E+01	1.13E+03	4.28E-04	-5.21E+08	-5.50E+00	2.00E+01	1.08E+00
	Std	2.65E-01	6.63E-01	1.80E-01	1.26E+00	9.00E-06	2.10E-01	6.24E-02	2.04E+01	4.91E+00	1.30E+00	1.70E-05	1.50E+08	1.24E-01	2.31E-04	3.09E-02
dynFWA	F	1	2	3	4	5	6	7	8	9	10	11	12	13	14	15
	Max	2.43E+05	3.07E+04	4.55E+03	5.21E+01	2.43E-01	9.79E+01	3.16E-02	5.32E+00	-4.86E+01	-1.62E+01	5.21E+01	1.51E-02	5.04E+00	7.61E-02	4.73E-01
	Min	2.61E+03	4.01E+03	4.51E+03	1.04E+01	8.11E-03	2.74E+01	4.82E-03	9.59E-01	-5.63E+01	-2.61E+01	6.59E-04	1.31E-02	1.52E+00	2.39E-02	3.06E-02
	Mean	5.48E+04	1.26E+04	4.52E+03	3.03E+01	7.79E-02	3.64E+01	1.20E-02	2.75E+00	-5.33E+01	-2.22E+01	1.07E+01	1.41E-02	2.98E+00	4.61E-02	2.41E-01
	Median	3.41E+04	1.11E+04	4.52E+03	2.85E+01	6.34E-02	2.94E+01	1.15E-02	2.67E+00	-5.35E+01	-2.25E+01	2.50E+00	1.42E-02	3.00E+00	4.70E-02	1.11E-01
	Std	5.21E+04	6.72E+03	6.35E+00	1.14E+01	5.14E-02	2.01E+01	4.62E-03	8.24E-01	1.76E+00	2.02E+00	1.45E+01	5.07E-04	7.50E-01	1.31E-02	1.88E-01
	F	16	17	18	19	20	21	22	23	24	25	26	27	28	29	30
	Max	1.21E+00	5.49E+00	-3.68E+03	9.42E-01	4.44E+01	2.00E+00	3.70E-01	7.09E+01	-9.33E+01	1.14E+03	4.72E-04	-3.95E+07	-5.38E+00	2.00E+01	1.13E+00
	Min	1.39E-01	1.73E-01	-6.06E+03	8.69E-03	2.36E-03	4.00E-01	2.52E-03	3.35E+01	-1.11E+02	1.12E+03	4.01E-04	-5.18E+08	-5.96E+00	2.00E+01	1.04E+00
	Mean	6.31E-01	2.17E+00	-5.03E+03	1.95E-01	2.76E+00	8.87E-01	7.58E-02	4.98E+01	-1.04E+02	1.13E+03	4.23E-04	-2.64E+08	-5.80E+00	2.00E+01	1.06E+00
	Median	5.94E-01	1.98E+00	-6.05E+03	1.29E-01	5.70E-01	9.00E-01	3.94E-02	4.97E+01	-1.05E+02	1.13E+03	4.19E-04	-2.90E+08	-5.84E+00	2.00E+01	1.04E+00
	Std	2.65E-01	1.17E+00	1.18E+03	2.15E-01	7.05E+00	3.07E-01	8.57E-02	9.16E+00	3.57E+00	3.47E+00	1.66E-05	1.02E+08	1.40E-01	2.14E-04	2.46E-02
AFWA	F	1	2	3	4	5	6	7	8	9	10	11	12	13	14	15
	Max	2.85E+05	2.81E+04	4.55E+03	7.17E+01	2.00E-01	9.34E+01	4.94E-02	5.32E+00	-4.81E+01	-1.47E+01	4.20E+01	1.50E-02	5.82E+00	6.27E-02	4.80E-01
	Min	7.49E+02	2.83E+03	4.51E+03	1.43E+01	1.42E-02	2.69E+01	6.80E-03	2.02E+00	-5.62E+01	-2.59E+01	6.37E-03	1.31E-02	1.34E+00	1.80E-02	1.95E-02
	Mean	5.14E+04	1.19E+04	4.52E+03	3.01E+01	7.07E-02	3.26E+01	1.50E-02	3.19E+00	-5.26E+01	-2.18E+01	5.94E+00	1.42E-02	3.11E+00	4.00E-02	3.07E-01
	Median	1.97E+04	1.02E+04	4.52E+03	2.69E+01	5.79E-02	2.92E+01	1.40E-02	3.03E+00	-5.28E+01	-2.20E+01	1.35E-01	1.42E-02	2.97E+00	3.95E-02	4.22E-01
	Std	6.60E+04	5.84E+03	7.28E+00	1.18E+01	4.48E-02	1.48E+01	6.38E-03	7.93E-01	1.87E+00	2.03E+00	8.29E+00	3.87E-04	9.45E-01	1.20E-02	1.87E-01
	F	16	17	18	19	20	21	22	23	24	25	26	27	28	29	30
	Max	1.44E+00	6.34E+00	-3.69E+03	6.64E-01	5.80E+01	1.80E+00	1.87E-01	7.59E+01	-9.54E+01	1.14E+03	4.51E-04	-7.18E+07	-5.37E+00	2.00E+01	1.13E+00
	Min	2.31E-01	5.86E-01	-9.01E+03	1.06E-02	7.86E-04	3.00E-01	3.93E-03	3.32E+01	-1.10E+02	1.12E+03	4.03E-04	-8.56E+08	-5.98E+00	2.00E+01	1.04E+00
	Mean	7.35E-01	2.40E+00	-5.39E+03	1.33E-01	3.70E+00	9.78E-01	4.91E-02	5.11E+01	-1.04E+02	1.13E+03	4.23E-04	-3.07E+08	-5.81E+00	2.00E+01	1.07E+00
	Median	7.15E-01	2.21E+00	-6.05E+03	8.82E-02	3.90E-01	1.00E+00	3.38E-02	5.02E+01	-1.04E+02	1.13E+03	4.21E-04	-2.91E+08	-5.85E+00	2.00E+01	1.08E+00
	Std	2.83E-01	1.34E+00	1.45E+03	1.55E-01	9.30E+00	3.48E-01	4.71E-02	8.84E+00	3.47E+00	3.17E+00	1.20E-05	1.53E+08	1.36E-01	2.07E-04	2.76E-02

From the results in different dimension, it can be seen that with the increasing of the dimension, the results optimized by all the algorithms get worsen, which is usually called *"dimension of curse"*. From the run time results in Table 1, it can be seen that AFWA achieve smaller $(T2 - T1)/T1$ than dynFWA. Here we also need to point out that the implementation of the code is one of the core factors to influence the run time.

From the results of 2D functions in Table 2, it can be seen that AFWA achieves better results than FWA, EFWA and dynFWA. Especially on f_{16}, f_{17}, f_{20}, f_{21}, f_{22}, AFWA gets the optimum of these functions. Table 3 gives the results of 10D functions. The dynFWA and AFWA still outperform EFWA and FWA. For the comparison between dynFWA and AFWA, dynFWA achieves smaller mean fitness. Table 4 shows the results on 30D functions. None of the algorithms works well, since all the maximum and minimum are different for each function. The dynFWA and AFWA still outperform EFWA and FWA due to their great local search ability, while the performances of dynFWA and AFWA do not differ much.

5   Conclusion

In this paper, the FWA and its variants are used to take the ICSI2014 competition for solving competition problems which contains 30 functions, and the three groups of experimental results with the dimensions set to 2, 10, 30 are recorded. In the competition, the error smaller than $2^{-52} \approx 2.22e^{-16}$ is set to 0. It can be seen that for some functions, the most recent work dynFWA and AFWA still can not get the optimum, thus further research needs to be taken and it is believed that there is a long way to go for fireworks algorithm in the future.

Acknowledgements. This work was supported by National Natural Science Foundation of China (NSFC), Grant No. 61375119, No. 61170057 and No. 60875080.

References

1. Bratton, D., Kennedy, J.: Defining a standard for particle swarm optimization. In: Swarm Intelligence Symposium, SIS 2007, pp. 120–127. IEEE (2007)
2. Yu, C., Kelley, L., Zheng, S.: Fireworks algorithm with differential mutation for solving the cec 2014 competition problems. In: 2014 IEEE Congress on Evolutionary Computation (CEC). IEEE (2014)
3. Ding, K., Zheng, S., Tan, Y.: A gpu-based parallel fireworks algorithm for optimization. In: Proceeding of the Fifteenth Annual Conference on Genetic and Evolutionary Computation Conference, GECCO 2013, pp. 9–16. ACM, New York (2013), http://doi.acm.org/10.1145/2463372.2463377
4. Gao, H., Diao, M.: Cultural firework algorithm and its application for digital filters design. International Journal of Modelling, Identification and Control 14(4), 324–331 (2011)

5. He, W., Mi, G., Tan, Y.: Parameter optimization of local-concentration model for spam detection by using fireworks algorithm. In: Tan, Y., Shi, Y., Mo, H. (eds.) ICSI 2013, Part I. LNCS, vol. 7928, pp. 439–450. Springer, Heidelberg (2013)
6. Imran, A.M., Kowsalya, M.: A new power system reconfiguration scheme for power loss minimization and voltage profile enhancement using fireworks algorithm. International Journal of Electrical Power & Energy Systems 62, 312–322 (2014)
7. Imran, A.M., Kowsalya, M., Kothari, D.: A novel integration technique for optimal network reconfiguration and distributed generation placement in power distribution networks. International Journal of Electrical Power & Energy Systems 63, 461–472 (2014)
8. Janecek, A., Tan, Y.: Iterative improvement of the multiplicative update nmf algorithm using nature-inspired optimization. In: 2011 Seventh International Conference on, Natural Computation (ICNC), vol. 3, pp. 1668–1672. IEEE (2011)
9. Janecek, A., Tan, Y.: Swarm intelligence for non-negative matrix factorization. International Journal of Swarm Intelligence Research (IJSIR) 2(4), 12–34 (2011)
10. Janecek, A., Tan, Y.: Using population based algorithms for initializing nonnegative matrix factorization. In: Tan, Y., Shi, Y., Chai, Y., Wang, G. (eds.) ICSI 2011, Part II. LNCS, vol. 6729, pp. 307–316. Springer, Heidelberg (2011)
11. Junzhi Li, S.Z., Tan, Y.: Adaptive fireworks algorithm. In: 2014 IEEE Congress on Evolutionary Computation (CEC). IEEE (2014)
12. Liu, J., Zheng, S., Tan, Y.: The improvement on controlling exploration and exploitation of firework algorithm. In: Tan, Y., Shi, Y., Mo, H. (eds.) ICSI 2013, Part I. LNCS, vol. 7928, pp. 11–23. Springer, Heidelberg (2013)
13. Pei, Y., Zheng, S., Tan, Y., Hideyuki, T.: An empirical study on influence of approximation approaches on enhancing fireworks algorithm. In: Proceedings of the 2012 IEEE Congress on System, Man and Cybernetics, pp. 1322–1327. IEEE (2012)
14. Zheng, S., Andreas, J., Li, J., Tan, Y.: Dynamic search in fireworks algorithm. In: 2014 IEEE Congress on Evolutionary Computation (CEC). IEEE (2014)
15. Tan, Y., Xiao, Z.: Clonal particle swarm optimization and its applications. In: IEEE Congress on Evolutionary Computation, CEC 2007, pp. 2303–2309. IEEE (2007)
16. Tan, Y., Zhu, Y.: Fireworks algorithm for optimization. In: Tan, Y., Shi, Y., Tan, K.C. (eds.) ICSI 2010, Part I. LNCS, vol. 6145, pp. 355–364. Springer, Heidelberg (2010)
17. Tan, Y., Li, J., Zheng, Z.: Icsi 2014 competition on single objective optimization (2014)
18. Zheng, S., Andreas, J., Tan, Y.: Enhanced fireworks algorithm. In: 2013 IEEE Congress on Evolutionary Computation (CEC), pp. 2069–2077. IEEE (2013)
19. Zheng, S., Tan, Y.: A unified distance measure scheme for orientation coding in identification. In: 2013 IEEE Congress on Information Science and Technology, pp. 979–985. IEEE (2013)
20. Zheng, Y., Xu, X., Ling, H.: A hybrid fireworks optimization method with differential evolution. Neurocomputing (2012)
21. Zheng, Y.J., Song, Q., Chen, S.Y.: Multiobjective fireworks optimization for variable-rate fertilization in oil crop production. Applied Soft Computing 13(11), 4253–4263 (2013)

Performance of Migrating Birds Optimization Algorithm on Continuous Functions

Ali Fuat Alkaya[1], Ramazan Algin[1], Yusuf Sahin[2],
Mustafa Agaoglu[1], and Vural Aksakalli[3]

[1] Marmara University, Department of Computer Engineering, Istanbul, Turkey
[2] Marmara University, Department of Electrical and Electronics Engineering,
Istanbul, Turkey
[3] Istanbul Sehir University, Department of Industrial Engineering, Istanbul, Turkey
{falkaya,ysahin,agaoglu}@marmara.edu.tr,
algin.ramazan@gmail.com, aksakalli@sehir.edu.tr

Abstract. In this study, we evaluate the performance of a recently proposed metaheuristic on several well-known functions. The objective of this evaluation is to participate in a competition where several metaheuristics are compared. The metaheuristic we exploit is the recently proposed migrating birds optimization (MBO) algorithm. Our contribution in this study is to develop a novel neighbor generating function for MBO to be used in multidimensional continuous spaces. After a set of preliminary tests presenting the best performing values of the parameters, the results of computational experiments are given in 2, 10 and 30 dimensions.

Keywords: migrating birds optimization, continuous functions, single objective optimization.

1 Introduction

The MBO algorithm is a newly proposed, population-based neighborhood search technique inspired from the V formation flight of the migrating birds which is proven to be an effective formation in energy minimization. In the analogy, initial solutions correspond to a flock of birds. Likewise the leader bird in the flock, a leader solution is chosen and the rest of the solutions is divided into two parts. Each solution generates a number of neighbor solutions. This number is a determiner value on exploration and it corresponds to the speed of the flock. The higher this value, the more detailed the flock explores its surroundings.

The algorithm starts with a number of initial solutions corresponding to birds in a V formation. Starting with the first solution (corresponding to the leader bird) and progressing on the lines towards the tales, each solution is tried to be improved by its neighbor solutions. If any of the neighbor solutions is better, the current solution is replaced by that one. There is also a benefit mechanism for the solutions (birds) from the solutions in front of them. Here we define the

Y. Tan et al. (Eds.): ICSI 2014, Part II, LNCS 8795, pp. 452–459, 2014.

benefit mechanism as sharing the best unused neighbors with the solutions that follow. In other words, a solution evaluates a number of its own neighbors and a number of best neighbors of the previous solution and is replaced by the best of them. Once all solutions are improved (or tried to be improved) by neighbor solutions, this procedure is repeated a number of times (tours) after which the first solution becomes the last, and one of the second solutions becomes the first and another loop starts. The algorithm is terminated after a predetermined number of neighbors are generated. Pseudocode of our MBO is given in Figure 1.

---

1. Generate n initial solutions in a random manner and place them on an hypothetical V formation arbitrarily
2. **while** termination condition is not satisfied
3.   **for** m times
4.   Try to improve the leading solution by generating and evaluating k neighbors of it
5.   **for** each solution s_i in the flock (except leader)
6.   Try to improve s_i by evaluating $(k\text{-}x)$ neighbors of it and x unused best neighbors from the solution in the front
7.   **endfor**
8.   **endfor**
9.   Move the leader solution to the end and forward one of the solutions following it to the leader position
10. **endwhile** 11. return the best solution in the flock

Fig. 1. Pseudocode of the MBO

MBO algorithm has four parameters: number of solutions (n), number of tours (m), number of neighbor solutions to be generated from a solution (k) and number of solutions to be shared with the following solution (x). However, due to the inherent design of the algorithm n value has to be equal to or greater than $2*x+1$.

This new metaheuristic was proposed by Duman et al. [1]. They applied it to solve quadratic assignment problem instances arising from printed circuit board assembly workshops. Its performance was compared with those of metaheuristics implemented and compared in two previous studies. These metaheuristics are simulated annealing, tabu search, genetic algorithm, scatter search, particle swarm optimization, differential evolution and guided evolutionary simulated annealing. In this comparison, the MBO outperformed the best performed metaheuristic (simulated annealing) in the previous studies by approximately three percent on the average. In addition, MBO was tested with some benchmark problem instances obtained from QAPLIB and in most of the instances it obtained the best known solutions. As a result of these tests, it is concluded that the MBO is a promising metaheuristic and it is a candidate to become one of the highly competitive metaheuristics. Duman and Elikucuk [2] applied MBO to solve fraud detection problem. They also proposed a new version of MBO where a different benefit mechanism is used. They tested the original MBO algorithm and its new version on real data and compared their performance with that of genetic algorithm hybridized with scatter search (GASS). Test results showed that the MBO algorithm and its new version performed significantly better than the GASS algorithm.

In this study, we exploit MBO to solve problems in continuous environments. The set of functions used are given in [3] which are tried to be minimized on 2, 10 and 30 dimensional spaces. The search space is $[-100, 100]^D$ where D is the dimension. We believe that defining an effective neighboring function is much more important than any other modifications on the MBO. In line with this observation, our contribution in this study is to develop a novel neighbor generating function for MBO to be used in multidimensional continuous spaces.

In the next section we present an effective neighbor generating function designed for MBO. Section three presents experimental setup which includes parameter analysis of the MBO algorithm. Section four gives the results where MBO is run on 30 different functions and various dimensions. Section five gives the conclusive remarks together with some future work.

2   A Novel Neighbor Generating Function

In order to design a well performing MBO algorithm, an effective neighbor generating function is essential. To have a more effective exploration plan in the D dimensional solution space, we used D dimensional spheres (D-spheres for short throughout this paper). A neighbor of a solution can be obtained only within the D-sphere around it. A neighbor of a solution can be at most r units away from the original solution where r is the radius of the D-sphere that surrounds it. To find the radius of a D-sphere, we firstly calculate the volume allocated to it using the following formula.

$$V_D = TV/n \tag{1}$$

where V_D is the volume of a D-sphere and TV is the total volume of the solution space. In order to calculate the radii for the D-spheres in a D dimensional space, the volume of the solution space is divided by n. In this way, we try to make an effective exploration and fair distribution of volume for all birds (solutions) to fly around. When the volume of a D-sphere is calculated, we need to find the radius of the D-sphere. The following inductive formulas give the volumes of D-spheres.

$$V_1 = 2 * r \tag{2}$$

$$V_2 = \pi * r^2 \tag{3}$$

$$V_D = V_{D-2} * 2 * \pi/r^D \, for \, D > 2 \tag{4}$$

Once the volume for each sphere is calculated, the radii of each sphere can be easily calculated using Equations(2-4). After calculating the radius of D-sphere, we can develop a method to find a neighbor solution (point) within the sphere using some trigonometry. The distance that how far will the new solution be away from the original solution will be a random number in $[0, r]$ where r is the radius of the sphere.

Additionally, we also need to determine the location (coordinate in each axis) of the point in the D dimensional space. For this, we used the following set of trigonometric formula.

$x_D = l * cos(\theta_{D-1})$

$x_{D-1} = l * sin(\theta_{D-1}) * cos(\theta_{D-2})$

$x_{D-2} = l * sin(\theta_{D-1}) * sin(\theta_{D-2}) * cos(\theta_{D-3})$

...

$x_2 = l * sin(\theta_{D-1}) * sin(\theta_{D-2}) * ... * sin(\theta_2) * cos(\theta_1)$

$x_1 = l * sin(\theta_{D-1}) * sin(\theta_{D-2}) * ... * sin(\theta_2) * sin(\theta_1)$

where l is the distance that how far will the new solution be away from the original solution, x_i is the coordinate of the point in the ith axis and θ_i is the angle between i^{th} and $(i+1)^{th}$ axis. Before using this set of formula θ_i's must be obtained randomly such that $\theta_1 \in [0, 2\pi]$ and $\theta_i \in [0, \pi]$ for $i = 2, \ldots, D-1$. An example for the formulas given above is presented in Figure 2 for $D = 3$.

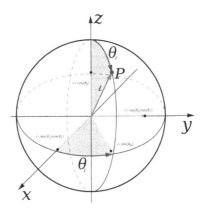

Fig. 2. Representation of a point (solution) and the vectors constituting it in three dimensions

From this setting, one can easily observe that if the number of birds (solutions) is small, then the volume that they are going to explore will be large whereas if the number of birds is large, the volume that they are going to explore will be small. Since we are limited by the number of function evaluation (neighbor generations) due to the competition rules, with a large number of solutions we will be able to explore small number of neighbors in smaller regions whereas with small number of solutions we will be able to explore large number of neighbors in larger regions. Hence, an efficient value for the n parameter must be found for the best performance of the algorithm.

3   Experimental Setup

The experiments are run on an HP Z820 workstation with Intel Xeon E5 processor at 3.0 GHz with 128 GB RAM running Windows 7. The MBO algorithm is implemented in Java language. The stopping criterion for the MBO algorithm is a given number of function evaluations which correspond to number of neighbors generated. Specifically the allowed number of function evaluations is 10000*D.

Table 1. Statistics of 51 runs on 30 different functions when D=2

Function ID	min	max	avg	med	std
1	40.82	19721.55	1951.86	1113.82	2964.70
2	2.48	4532.26	408.61	98.97	751.68
3	1.67	1.67	1.67	1.67	0.00
4	15.22	4166.54	488.89	227.25	704.77
5	0.17	1.20	0.57	0.59	0.24
6	0.03	3.58	0.95	0.82	0.80
7	0.00	0.01	0.00	0.00	0.00
8	0.91	6.44	3.99	4.17	1.17
9	0.07	0.43	0.26	0.26	0.10
10	0.02	0.39	0.20	0.21	0.10
11	0.01	1.10	0.35	0.23	0.36
12	0.08	1.28	0.55	0.55	0.30
13	0.02	0.08	0.02	0.02	0.01
14	0.23	1.21	0.73	0.75	0.21
15	0.28	2.62	0.98	0.81	0.54
16	0.01	0.15	0.07	0.06	0.04
17	0.01	0.82	0.16	0.12	0.15
18	-861.55	0.00	-840.06	-837.83	5.82
19	0.01	0.28	0.09	0.09	0.06
20	0.39	7.27	2.85	2.97	1.45
21	0.02	1.38	0.51	0.47	0.32
22	0.03	1.44	0.34	0.21	0.34
23	0.32	10.86	7.77	8.34	2.01
24	17.73	7069.62	551.74	261.12	1056.58
25	2.13	2390.53	355.57	138.30	535.51
26	1.33	12.67	2.75	2.03	1.92
27	-41721.46	0.00	-38311.70	-38587.87	2274.27
28	-1.57	0.00	-0.88	-0.87	0.26
29	8.47	17.97	12.75	12.35	2.77
30	0.05	0.84	0.33	0.33	0.16

In order to reveal the best performing parameter values of the MBO on the continuous functions, we run a set of extensive computational experiments. These preliminary tests are conducted on six functions selected out of 30 given in [3]. Best performing values for the parameters are as follows: $n = 5001, k = 3, m = 1, x = 1$.

4   Results

In this section we provide the results of the MBO algorithm on the aforementioned continuous benchmark functions. One of the results to be delivered as a rule of the competition is the $T1$ and $T2$ values. $T1$ is the average run time of five runs of the following piece of MATLAB code in our environment.

Table 2. Statistics of 51 runs on 30 different functions when D=10

Function ID	min	max	avg	med	std
1	20323203.08	109661239.94	61390076.47	63430416.62	18862528.62
2	240.72	1998.77	801.74	755.39	358.08
3	192.16	304.70	247.67	248.40	21.37
4	354.30	2204.80	1001.13	934.47	468.90
5	1.09	1.64	1.35	1.36	0.12
6	477.57	5023.73	1975.59	1786.44	1048.17
7	0.17	1.53	0.84	0.83	0.25
8	7.04	11.55	9.73	9.74	1.01
9	3.22	5.43	4.75	4.91	0.51
10	1.25	2.12	1.72	1.75	0.22
11	11.01	35.33	27.20	28.07	5.30
12	0.66	2.42	1.54	1.60	0.37
13	1.54	2.58	2.13	2.13	0.25
14	5.30	13.38	8.59	8.56	2.05
15	33.46	214.90	120.04	122.06	42.16
16	1.69	3.46	2.71	2.75	0.43
17	578.73	4028.84	1831.87	1830.40	681.23
18	-3470.64	0.00	-2717.47	-2617.14	376.96
19	1.97	8.05	4.56	4.36	1.30
20	5.90	13.83	9.96	9.91	1.73
21	5.10	19.54	13.41	13.80	3.44
22	38.72	213.23	122.00	126.23	38.49
23	29.19	43.87	38.97	39.58	2.96
24	395.74	1997.54	980.50	935.94	390.40
25	217.89	3921.61	1169.83	1048.91	596.34
26	838.83	6476.15	2421.21	2310.05	1154.72
27	-19655.46	0.00	-13727.18	-13385.63	2203.31
28	14.54	15.72	15.14	15.17	0.29
29	21.37	21.38	21.37	21.37	0.00
30	1.08	1.35	1.22	1.23	0.07

Table 3. Statistics of 51 runs on 30 different functions when D=30

Function ID	min	max	avg	med	std
1	131323505.00	371642428.18	272954269.37	278704299.47	49176717.62
2	370.11	1232.52	730.82	729.31	191.71
3	19702.97	36266.85	28845.16	29096.35	4050.11
4	397.31	2204.13	920.60	891.43	351.70
5	1.04	1.23	1.10	1.10	0.04
6	1607.06	12377.82	7458.20	7420.84	2176.52
7	1.80	3.42	2.80	2.80	0.39
8	10.20	11.89	11.05	11.11	0.46
9	13.70	17.27	15.64	15.67	0.88
10	2.97	4.82	3.96	4.06	0.42
11	81.94	140.23	114.50	117.65	13.35
12	2.77	4.47	3.74	3.74	0.39
13	8.00	10.33	9.18	9.27	0.58
14	6.30	11.09	8.99	9.18	1.14
15	194.15	371.56	298.56	311.88	42.88
16	5.71	8.75	7.47	7.60	0.72
17	35345.91	77582.40	54386.92	55674.38	10061.42
18	-5215.18	0.00	-4565.75	-4527.92	252.22
19	8.43	17.05	11.84	11.86	1.89
20	5.71	9.16	7.86	8.05	0.70
21	20.15	38.43	30.42	30.40	4.19
22	189.36	418.69	316.94	322.05	59.04
23	100.05	116.60	108.54	108.75	3.99
24	603.02	1419.04	951.32	938.95	199.66
25	11240.74	29044.32	19293.43	19268.09	4740.83
26	1980.60	11722.33	7516.65	8055.23	2296.92
27	-6756.60	0.00	-5080.52	-5009.40	753.36
28	54.70	55.72	55.18	55.18	0.28
29	21.61	21.62	21.61	21.61	0.00
30	1.38	1.47	1.43	1.44	0.02

```
for i=1:300000
    evaluate(9,rand(30,1)*200-100);
end
```

$T2$ is the average run time of five runs of the function 9 on $D=30$ in our environment.

According to our experimental work $T1$, $T2$ and $(T2\text{-}T1)/T1$ are as follows:

- $T1=29.965$
- $T2=73.369$
- $(T2\text{-}T1)/T1=1.448$

Table 1, 2 and 3 present the statistics when D=2, 10 and 30, respectively. The major observation among the tables is that in higher dimensions the performance of the MBO algorithm gets worse. This is an expected result because the search space grows much faster than the allowed number of function evaluations.

5   Conclusion

In this study we applied migrating birds optimization algorithm to 30 different functions on continuous domain. Our contribution in this study is to develop an effective novel neighbor generation function for MBO. The tests are conducted on 2, 10 and 30 dimensions. Results present that even though MBO is a recently proposed algorithm it is also promising for problems in continuous domain.

References

1. Duman, E., Uysal, M., Alkaya, A.F.: Migrating Birds Optimization: A New Meta-heuristic Approach and Its Performance on Quadratic Assignment Problem. Information Sciences 217, 65–77 (2012)
2. Duman, E., Elikucuk, I.: Solving Credit Card Fraud Detection Problem by the New Metaheuristics Migrating Birds Optimization. In: Rojas, I., Joya, G., Cabestany, J. (eds.) IWANN 2013, Part II. LNCS, vol. 7903, pp. 62–71. Springer, Heidelberg (2013)
3. Website of Fifth International Conference on Swarm Intelligence, http://www.ic-si.org/competition

Author Index

Abdul-Kareem, Sameem   I-284
Abdullahi Muaz, Sanah   I-284
Agaoglu, Mustafa   II-452
Agrawal, Puja   II-212
Akhmedova, Shakhnaz   I-499
Aksakalli, Vural   II-452
Al-Betar, Mohammed Azmi   II-87
Alejo, Roberto   II-17
Algin, Ramazan   II-452
Alkaya, Ali Fuat   II-452
An, Xueli   II-146
Anto, P. Babu   II-275
Anwar, Khairul   II-87
Arvin, Farshad   I-1
Awadallah, Mohammed A.   II-87

Ba-Karait, Nasser Omer   II-352
Bao, Aorigele   I-246
Batouche, Mohamed   I-450
Beegom, A.S. Ajeena   II-79
Bellotto, Nicola   I-1
Beltaief, Olfa   I-9
Benmounah, Zakaria   I-450
Bian, Zijiang   II-34
Boulesnane, Abdennour   II-412

Cai, Hongfei   II-204
Cai, Zhen-Nao   I-342
Campana, Emilio F.   I-126
Cao, Lianlian   II-221
Chen, Beibei   I-246
Chen, Hanwu   I-357
Chen, Hua   I-95
Chen, Hui-Ling   I-342, II-42
Chen, Junfeng   I-95
Chen, Li   II-221
Chen, Min-You   I-394
Chen, Qinglan   II-236
Chen, Su-Jie   I-342, II-42
Chen, Xianjun   II-58
Chen, Zhigang   I-318
Cheng, Shan   I-394
Cheng, Shi   II-319
Chenggang, Cui   II-309

Chiroma, Haruna   I-284
Chu, Hua   I-442
Crawford, Broderick   I-189
Cui, Yu   I-350

Diao, Liang   I-442
Diez, Matteo   I-126
Ding, Ke   II-66
Ding, Sheng   II-221
Ding, Shuyu   II-228
Djenouri, Youcef   II-50
Drias, Habiba   II-50
Du, Huimin   II-114, II-125
Du, Mingyu   I-74
Duan, Haibin   II-96

Emre Turgut, Ali   I-1

Fasano, Giovanni   I-126
Feng, Qianqian   I-267
Feng, Tao   I-374
Folly, Komla A.   II-135
Fu, Xiaowei   II-221
Fu, Yao-Wei   I-342, II-42

Gao, Chao   I-27, I-173, I-424
Gao, Jie   II-188
Gao, Shangce   I-246
Gao, Xiaozhi   I-86
Gao, Yang   I-223
Garro, Beatriz Aurora   I-207
Geng, Huang   II-34
Geng, Mengjiao   I-103, I-115
Ghazali, Rozaida   I-197
Ghedira, Khaled   I-9
Giove, Silvio   I-126
Gnaneswar, A.V.   II-8
Gong, Dunwei   I-386
Gong, Zhaoxuan   II-34
Gu, Jiangshao   I-460
Guo, Jian   I-142
Guo, Xiaoping   II-340
Guo, Yejun   I-294

Hadouaj, Sameh El   I-9
Han, Fei   I-350
Hao, Junling   I-64
He, Jieyue   II-180
He, Nana   I-302
He, Ping   II-1
Herawan, Tutut   I-197, I-284
Hu, Gang   I-394
Huang, Huali   I-150
Huang, Shan-Shan   I-342
Huang, Yantai   II-292
Huang, Yin-Fu   II-267

Iemma, Umberto   I-126

Jain, Aruna   I-165
Janghel, R.R.   II-8
Jiang, He   I-44
Jiang, Yue   II-163
Jithesh, K.   II-275
Johnson, Franklin   I-189

Kang, Qi   I-294, II-163, II-401
Kang, Zhenhua   I-470
Ke, Liangjun   II-301
Khader, Ahamad Tajudin   II-87
Khan, Abdullah   I-284
Khurshid, Aleefia   II-212
Kobayashi, Kunikazu   I-324
Kuremoto, Takashi   I-324

Lai, Xiaochen   I-44
Lei, Xiujuan   I-74, I-479
Leotardi, Cecilia   I-126
Li, Bin   I-134
Li, Changhe   I-181
Li, Fang   II-196
Li, Fenglin   I-142
Li, Jian-Jun   II-106
Li, Jinlong   I-27, I-365
Li, Junzhi   II-442
Li, Kanwen   II-1
Li, Li   II-319
Li, Lian   II-24
Li, Li-Juan   I-342
Li, Qingshan   I-442
Li, QiuQuan   II-42
Li, Shuai   II-284
Li, Yiguo   I-215
Liang, Jane Jing   I-150, II-384

Liang, Xiao-lei   I-134
Liang, XiaoLong   I-36
Liang, Zhi-feng   I-302
Lin, Heng   II-204
Lin, Sheng-Min   II-267
Lin, Yuan-Lung   I-158
Liu, Bingyu   I-404
Liu, Cong   I-431
Liu, Ju   II-155
Liu, Lang   II-442
Liu, Lili   I-103, I-115
Liu, Wei   II-384
Liu, Weifeng   II-228
Liu, Xiyu   I-267, I-470
Liu, Yabin   I-431
Liu, Yu   I-86
Liu, Yuxin   I-173, I-424
Liu, Zhaozheng   I-374
Liu, Zhihao   I-357
Loo, Chu Kiong   I-332
López-González, Erika   II-17
Lu, Bingbing   II-155
Lu, Lin   II-1
Lu, Mingli   II-236, II-244, II-253
Lu, Yuxiao   I-173, I-424
Lu, Zhigang   I-374
Luo, Wenjie   II-170
Lv, Qing   II-114, II-125
Lv, Yawei   II-196

Ma, Chuncheng   II-259
Ma, Chunsen   I-275
Ma, Yinghong   I-267
Mabu, Shingo   I-324
Meng, Xianbing   I-86
Meshoul, Souham   I-450, II-412
Mi, Guyue   I-223
Mo, Hongwei   I-103, I-115, I-234

Naseem, Rashid   I-197
Ni, Qingjian   II-114, II-125
Niu, Ben   I-150

Obayashi, Masanao   I-324
Olguín, Eduardo   I-189

Pacheco-Sánchez, J. Horacio   II-17
Palma, Wenceslao   I-189
Pan, Luoping   II-146
Pan, Qianqian   II-114, II-125